Astronomy:
From the Earth to the Universe

Gas and dust in interplanetary space, the "cometary globule" CG4, shown in true color. It is about 1 light year across and stars are forming inside. The colors indicate its composition and the effects of its surroundings. The blue is caused by dust particles scattering starlight. The red is from hydrogen gas, liberated when radiation from nearby stars breaks up molecules in the cloud. The yellow-green is the result of the absorption of red light by dust particles between the globule and us. The name "cometary" comes merely from the presence of its tail.

Astronomy:

From the Earth to the Universe

Fourth Edition, 1993 Version

Jay M. Pasachoff

Field Memorial Professor of Astronomy
Director of the Hopkins Observatory
Williams College
Williamstown, Massachusetts

Saunders Golden Sunburst Series

Saunders College Publishing
A Harcourt Brace Jovanovich College Publisher

Fort Worth Philadelphia San Diego
New York Orlando Austin San Antonio
Toronto Montreal London Sydney Tokyo

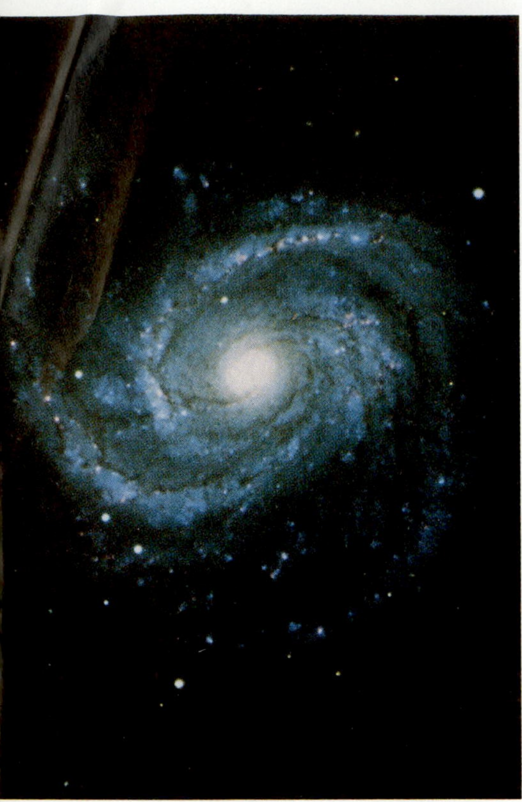

M100, a spiral galaxy in the Virgo Cluster of galaxies.

Text Typeface: Cheltenham and Helvetica
Compositor: General Graphic Services
Publisher: John J. Vondeling
Associate Editor: Lloyd W. Black
Managing Editor: Carol Field
Project Editors: Margaret Mary Anderson and Laura Maier
Copy Editor: Nanette Bendyna-Schuman
Manager of Art and Design: Carol Bleistine
Art Director: Christine Schueler
Associate Art Director: Doris Bruey
Text Designer: Bill Boehm
Cover Designer: Lawrence R. Didona
Text Artwork: George V. Kelvin, Science Graphics and J&R Technical Services
Layout Artist: Dorothy Chattin
Director of EDP: Tim Frelick
Production Manager: Joanne Cassetti
Marketing Manager: Marjorie Waldron

Cover Credit: (*Inset*) Full sky maps of variations in 3 K background radiation as measured by COBE. The top map shows variations of 3 millikelvins (0.12%) before corrections were made. The middle map shows variations of 30 microkelvins (0.01%) when the dipole anisotropy is subtracted. The red band across the center of this map results from the microwave emission of the Milky Way Galaxy. The bottom map, also showing variations of 30 microkelvins (0.01%), results when the galactic emission is then subtracted. Some of these variations—though we don't know which—are fluctuations in the background radiation rather than noise. (NASA/Goddard Space Flight Center) (*Background*) The southern sky with the Large Magellanic Cloud and its supernova. (European Southern Observatory, courtesy of Jacques M. Beckers)

Printed in the United States of America

ASTRONOMY: FROM THE EARTH TO THE UNIVERSE, Fourth Edition, 1993 Version

(ISBN) 0-03-094660-3

Library of Congress Catalog Card Number: 90-052772

2345 063 987654321

Preface

Astronomy continues to flourish. The opening of the Keck 10-meter Telescope promises important discoveries. The Hubble Space Telescope, even with its flawed mirror, is now yielding science of high value. We hope that the repair mission scheduled for late 1993 and the second-generation instruments will allow the telescope to achieve its full potential. In the meantime, the Magellan spacecraft orbiting Venus and the Compton Gamma-Ray Observatory are giving spectacular results. Further, new electronic instruments and computer capabilities, new space missions to the outer planets, and advances in computational astronomy and in theory should continue to bring forth exciting results.

In *Astronomy: From the Earth to the Universe,* I try to describe the current state of astronomy, both the fundamentals of astronomical knowledge that have been built up over decades and the exciting advances that are now taking place. I try to cover all the branches of astronomy without slighting any of them; each teacher and each student may well find special interests that may be different from my own. One of my aims in writing this book is to educate voters and prospective voters in the hope that they will endorse candidates who will support scientific research in general and astronomical research in particular.

New to This Edition

It is particularly exciting for me to be able to use color images throughout (approximately 750 photographs and 250 drawings). Over the past few years, many new color images have become available and even vital for presenting astronomy. For one thing, many new images of astronomical objects have become available in color, both through careful application of film techniques for photographs taken with telescopes on Earth and through the use of electronic techniques with sensors on both ground-based telescopes and on spacecraft. Further, the computer reduction of data, now so fundamental and widespread, has led to the ability to present data with a third dimension—color—providing additional information to an image. Of course, I have also tried to use color intelligently in the diagrams, emphasizing or bringing out whenever possible aspects on which you should spend special attention.

I am especially proud that my texts are unique in that they have separate chapters covering branches of astronomy for which we newly have extensive information: Neptune is the most prominent example in this edition, as a result of the Voyager 2 flyby. Supernovae, with the new information from the brightest supernova in nearly 400 years, and comets, with the information from the recent apparition of Halley's Comet, are other examples of new chapters since the third edition. Other topics that receive here the separate chapters they deserve are pulsars/neutron stars, black holes, quasars, and the past and future of the universe, topics that are especially interesting to students and faculty alike.

Educational Approach

I have paid special attention to making the books readable by students and the information accessible to them. The stories of the development of individual fields of study, as for the individual planets, are somewhat chronological to give students an opportunity to see how knowledge has grown. The occasional mention of the names of the contemporary astronomers performing the research discussed personalizes astronomy and shows that it is a human science as well as giving credit where credit is due. Newton was not the last scientist worthy of being mentioned by name!

In writing this book, I share the goals of a commission on the college curriculum of the Association of American Colleges, which reported that "a person who understands what science is recognizes that scientific concepts are created by acts of human intelligence and imagination; comprehends the distinction between observation and inference and between the occasional role of accidental discovery in scientific investigation and the deliberate strategy of forming and testing hypotheses; understands how theories are formed, tested, validated, and accorded provisional acceptance; and discriminates between conclusions that rest on unverified assertion and those that are developed from the application of scientific reasoning."

I have tried to keep up with recent psychological research on the teaching of science and the understanding of concepts. My books, therefore, include many concrete examples that make scientific situations more understandable to many students. I continue to consult with my psychologist colleagues so as to be able to implement correctly some of the tenets of Piaget, while taking into account the results of post-Piagetan research.

Pedagogy

I have provided aids to make this book easy to read and to study from. Many diagrams provide interpretive information not on diagrams on the same topic in other texts: professors can compare among other texts, for example, the diagrams that explain the Doppler effect, or the ones for Hubble's law. Each section of the text is numbered to allow professors to include or omit parts of chapters. Some professors will choose to include discussions, for example, of the over four dozen new worlds photographed as moons of the giant planets in our solar system; others will choose to omit this material. Further, some sections and boxes appear with asterisks to show that these are especially easy to omit.

New vocabulary is italicized in the text, listed in the key words at the end of each chapter, and defined in the glossary. The index provides further aid in finding explanations. End-of-chapter questions cover a range of material; some questions are straightforward and can be answered by merely reading the text, while others require more independent thought. Up-to-date appendixes provide some standard and much recent information on planets, star constellations, and non-stellar objects.

Ancillary Materials

Available free to all adopters of the 1993 Version are the following ancillaries:

Instructor's Resource Manual with Laboratory Experiments
This 400-page manual contains course outlines, answers to all questions in the text, lists of audiovisual aids and organizations, 11 laboratory experiments and exercises arranged for easy duplication, sample tests with an-

swers (8 quizzes and 9 exams), Test Bank objectives, and a 105-page set of notes linking each text chapter with its complementary *Project Universe* video series episode.

Saunders Astronomy Transparency Collection This comprehensive set contains 205 color overhead transparency acetates of conceptually-based artwork. Over half are enlarged reproductions of figures from the textbook. The others are supplemental illustrations that complement the text figures. A detailed guide arranged by topic accompanies the collection.

Saunders Astronomy Slide Collection The set of 205 transparency acetates is also available in 35-mm slide format.

Saunders Astronomy Video Tape This 93-minute VHS tape presents five NASA/JPL "movies," including a segment on images obtained from the Magellan Venus Orbiter. NOVA videotapes are also available.

Printed Test Bank This set of over 3200 questions in multiple choice and other formats was prepared by H. C. Snyder of St. Clair County Community College.

ExaMaster™ Computerized Test Bank for IBM and Macintosh The 3200 questions from the printed test bank are presented in a computerized format that allows instructors to edit, add questions, and print assorted versions of the same test.

RequesTest™ Instructors without access to a personal computer may contact Saunders Software Support Department at (800) 447-9457 to request tests prepared from the computerized test bank. The test will be mailed or faxed to the instructor within 48 hours.

For Information about these ancillaries contact your local Saunders sales representative or call the Harcourt Brace Jovanovich College Department at (800) 237-2665.

Acknowledgments

The publishers and I have always placed a heavy premium on accuracy in my books, and we have made certain that the manuscript and proof have been read not only by students for clarity and style and by astronomy teachers for pedagogical reasons but also by research astronomers for scientific accuracy.

I would particularly like to thank the reviewers of all or large sections of the manuscript of this edition, who include Yervant Terzian (Cornell University), Alexander G. Smith (University of Florida), Joseph Veverka (Cornell University), Bruce Margon (University of Washington), Paul Hodge (University of Washington), Hyron Spinrad (University of California, Berkeley), Gary Mechler (Pima Community College), Leonard Muldawer (Temple University), Paul B. Campbell (Western Kentucky University), Gordon E. Baird (University of Mississippi), Robert L. Mutel (University of Iowa), and Gordon B. Thompson (Rutgers University).

I also thank those astronomers who have commented on particular sections relevant to their fields of expertise, including A. W. Wolfendale (University of Durham, U.K.), cosmic rays; Bruce Partridge (Haverford College), cosmic background radiation; Ed Cheng (NASA/Goddard Space Flight Center), COBE; Roman Juskiewicz (Copernicus Astronomical Center, Warsaw, and Institute for Advanced Study, Princeton), galaxy formation; John N. Bahcall (Institute for Advanced Study, Princeton), solar neutrinos; Roger Romani (Institute for Advanced Study, Princeton), millisecond pulsars; Daniel R. Stinebring (Princeton University), pulsars; Stuart N. Vogel (University of Maryland), molecular studies; Noel Swerdlow (University of Chicago), Hipparchus; G. J. Toomer (Brown University), Hipparchus; Paul D. Steinhardt

(University of Pennsylvania and Institute for Advanced Study, Princeton), inflationary universe.

I am especially grateful to Stephen Edberg and Jurrie van der Woude of the Jet Propulsion Laboratory and Joseph Veverka of Cornell University for providing the Neptune photographs. David Malin of the Anglo-Australian Observatory has been of special help in providing his magnificent color photopgraphs, and we have even worked together to make two new Crab Nebula color images for this book. Brian Hadley of the Royal Observatory Edinburgh has provided many of his wonderful images. Stephen P. Maran of the NASA/Goddard Space Flight Center, the Press Officer of the American Astronomical Society, has been especially helpful at A.A.S. meetings. Other providers of photographs are thanked in the illustration acknowledgments section at the end.

I remain grateful to the reviewers of previous versions of my texts or sections in them, including Ronald J. Angione, Thomas T. Arny, James G. Baker, Michael Belton, Bruce E. Bohannon, Kenneth Brecher, Tom Bullock, Bernard Burke, Clark Chapman, Pamela Clark, Roy W. Clark, James W. Cristy, Martin Cohen, Leo Connolly, Peter Conti, Ernest R. Cowley, Lawrence Cram, Dale P. Cruikshank, Morris Davis, Raymond Davis, Jr., Marek Demianski, Gerard de Vaucouleurs, Dennis di Cicco, Richard B. Dunn, John J. Dykla, John A. Eddy, Farouk El-Baz, James L. Elliot, David S. Evans, J. Donald Fernie, William R. Forman, Peter V. Foukal, George D. Gatewood, John E. Gaustad, Tom Gehrels, Riccardo Giacconi, Owen Gingerich, Stephen T. Gottesman, Jonathan E. Grindlay, Alan R. Guth, Ian Halliday, U. O. Hermann, Darrel B. Hoff, James Houck, Robert F. Howard, W. N. Hubin, John Huchra, H. W. Ibser, Paul Johnson, Christine Jones, Bernard J. T. Jones, Agris Kalnajs, John Kielkopf, Robert Kirshner, David E. Koltenbah, Jerome Kristian, Edwin D. Krupp, Karl F. Kuhn, Marc L. Kutner, Karen B. Kwitter, John Lathrop, Stephen P. Lattanzio, Lawrence S. Lerner, Jeffrey L. Linsky, Sarah Lee Lippincott, Bruce Margon, Laurence A. Marschall, Brian Marsden, Janet Mattei, R. Newton Mayall, Everett Mendelsohn, George K. Miley, Freeman D. Miller, Alan T. Moffet, William R. Moomaw, David D. Morrison, R. Edward Nather, David Park, James Pierce, Carl B. Pilcher, James B. Pollack, B. E. Powell, James L. Regas, Edward L. Robertson, Thomas N. Robertson, Herbert Rood, Maarten Schmidt, David N. Schramm, Leon W. Schroeder, Richard L. Sears, P. Kenneth Seidelmann, Maurice Shapiro, Peter Shull, Jr., Joseph I. Silk, Lewis E. Snyder, Theodore Spickler, Alan Stockton, Robert G. Strom, Jean Pierre Swings, Eugene Tademaru, Laszlo Taksay, David L. Talent, Gustav Tammann, Joseph H. Taylor, Jr., Joe S. Tenn, David Theison, Laird Thompson, Aaron Todd, M. Nafi Toksöz, Juri Toomre, Kenneth D. Tucker, Brent Tully, Barry Turner, Peter van de Kamp, Gerald J. Wasserburg, Anthony Weitenbeck, Leonid Weliachew, Ray Weymann, John A. Wheeler, J. Craig Wheeler, Ewen A. Whitaker, Robert F. Wing, Reinard A. Wobus, LeRoy A. Woodward, Susan Wyckoff, Anne C. Young, and Robert N. Zitter.

I thank Nancy P. Kutner for her excellent work on the index.

I thank Haddock C. Snyder, St. Clair County Community College, for his work in preparing the Test Bank and *Project Universe* notes.

I thank many people at Saunders College Publishing for their efforts on behalf of my books. John J. Vondeling and Lloyd W. Black have been my editors since the beginning, and have contributed especially to the extensive revision of this edition to include so many color images. Joanne Cassetti has been the Production Manager. Tim Frelick has been the Director of EDP. Christine Schueler has been the Art Director, and oversaw the remake of the overall design. George Kelvin and his studio, Science Graphics, have provided many of the especially beautiful drawings, as they had previously done in black-and-white for earlier versions.

I appreciate the expert assistance in Williamstown of Susan Kaufman.

I am grateful to John N. Bahcall and the Institute for Advanced Study, Princeton, for their hospitality during the final stages of this book.

For the 1993 Version, I acknowledge the assistance of John N. Bahcall (Institute for Advanced Study), David Spergel (Princeton University), Joseph H. Taylor (Princeton University), and Ray Villard (Space Telescope Science Institute).

Various members of my family have provided vital and valuable editorial services, in addition to their general support. My wife, Naomi Pasachoff, has taken time from her own projects to give expert readings of the various stages of proof. The participation of our daughters, Eloise and Deborah, in some of the proofreading is a source of great satisfaction to me. Their continued growth continues to be monitored alongside the Ahneghito meteorite. My father, Dr. Samuel S. Pasachoff, contributed so much to my books over the years. I also appreciate the counsel of my mother, Anne T. Pasachoff.

A Final Comment

I am extremely grateful to all the individuals named for their assistance. Of course, it is I who have put this all together, and I alone am responsible for it. I appreciate hearing from readers. I invite you to write me at Williams College—Hopkins Observatory, Williamstown, MA 01267. I promise a personal response to each writer.

Jay M. Pasachoff
Williamstown, MA
June, 1992

This supernova remnant is spread over a wide area in the constellation Vela.

To the Student

Astronomy is a very varied subject, and you are sure to find some parts that interest you more than others. To get the most out of this text, you must read each chapter more than once. After you read the chapter the first time, you should go carefully through the list of key words, trying to identify or define each one. If you cannot do so, look up the word in the glossary and also find the definition that appears with the word the first time it was used in the chapter. The index will help you find these references. Next, read through the chapter again especially carefully. Read through the Summary, checking that you understand each of the major points. Finally, answer a selection of the questions at the end of the chapter.

Discovery's remote manipulator system hoists the Hubble Space Telescope (HST) over the Earth's horizon prior to deployment of HST's solar panels and antennae.

Contents Overview

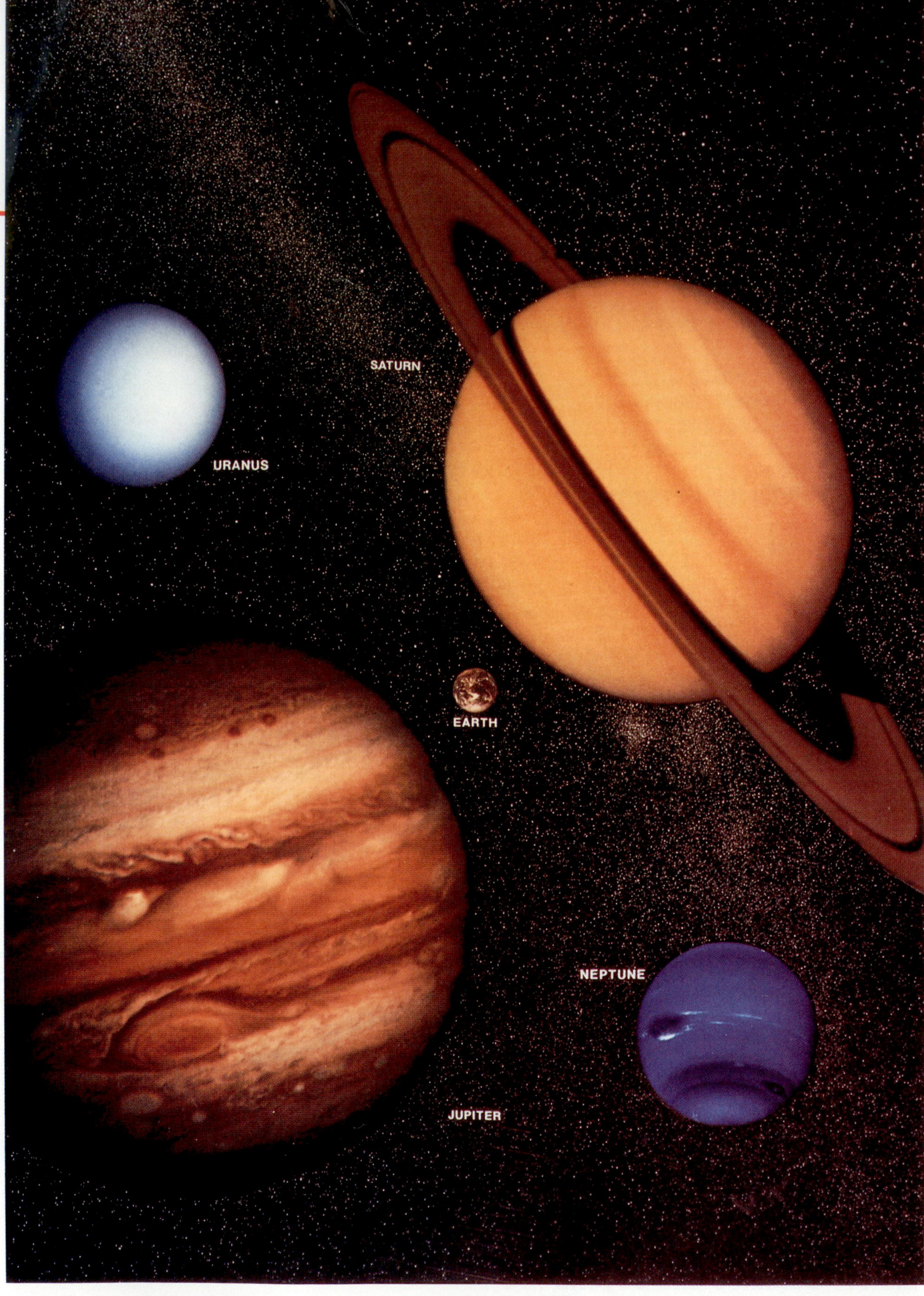

A photo montage of the giant planets and the Earth.

Contents

Part I A Sense of the Universe 1

Copernicus's original heliocentric diagram, from the first edition of his 1543 book, *De Revolutionibus*

The Hubble Space Telescope as it was launced in 1989

The annular eclipse of January 4, 1992

Part II The Solar System 108

Buzz Aldrin on the Moon

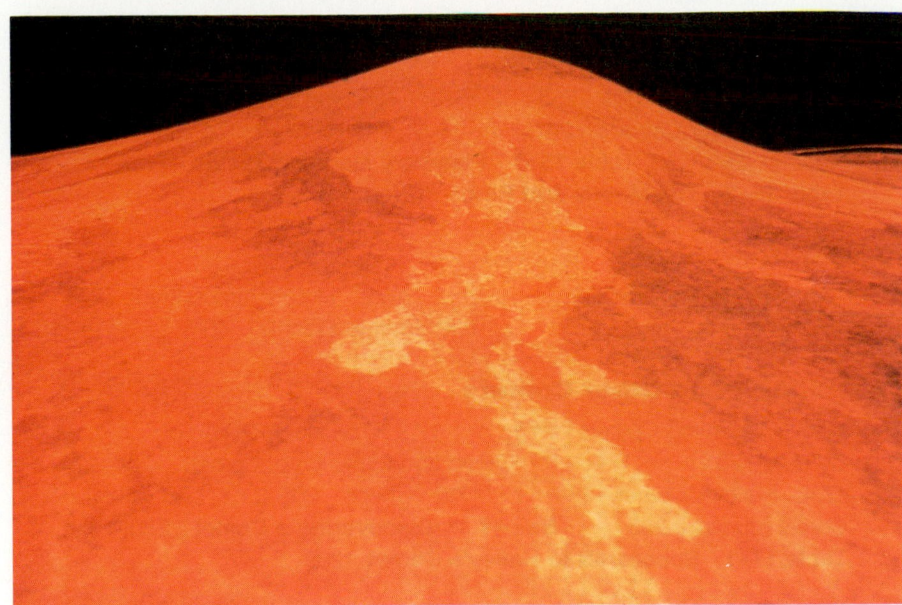

Sif Mons on Venus mapped by radar from the Magellan spacecraft

Jupiter and its Great Red Spot from Voyager 1

Neptune and its Great Dark Spot from Voyager 2

Stars and nebulosity in Sagittarius

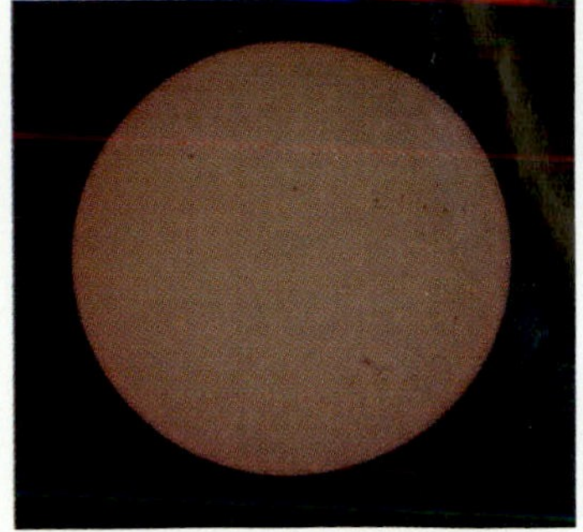

Solar photosphere with sunspots, 1989–1990 maximum

The Pleiades, an open cluster in Taurus

Galaxy NGC 253 in Sculptor

A near-infrared image of the Milky Way Galaxy obtained by COBE

The Great Nebula in Orion, printed with a technique that brings out small-scale structure.

A Sense of the Universe

1

Let us consider that the time between the origin of the universe and the year 2000 is one day. Then it wasn't until nearly 5 p.m. that the Earth formed; the first fossils date from 7 p.m. The first humans appeared only 10 seconds ago, and it is only 1/400 second since Columbus discovered America. The year 2000 will arrive in only 1/30,000 second.

Still, the Sun should shine for another 7 hours; an astronomical time scale is much greater than the time scale of our daily lives. Astronomers use a wide range of technology and theories to find out about the universe, what is in it, and what its future will be. This book surveys what we have found and how we look.

A sense of time.

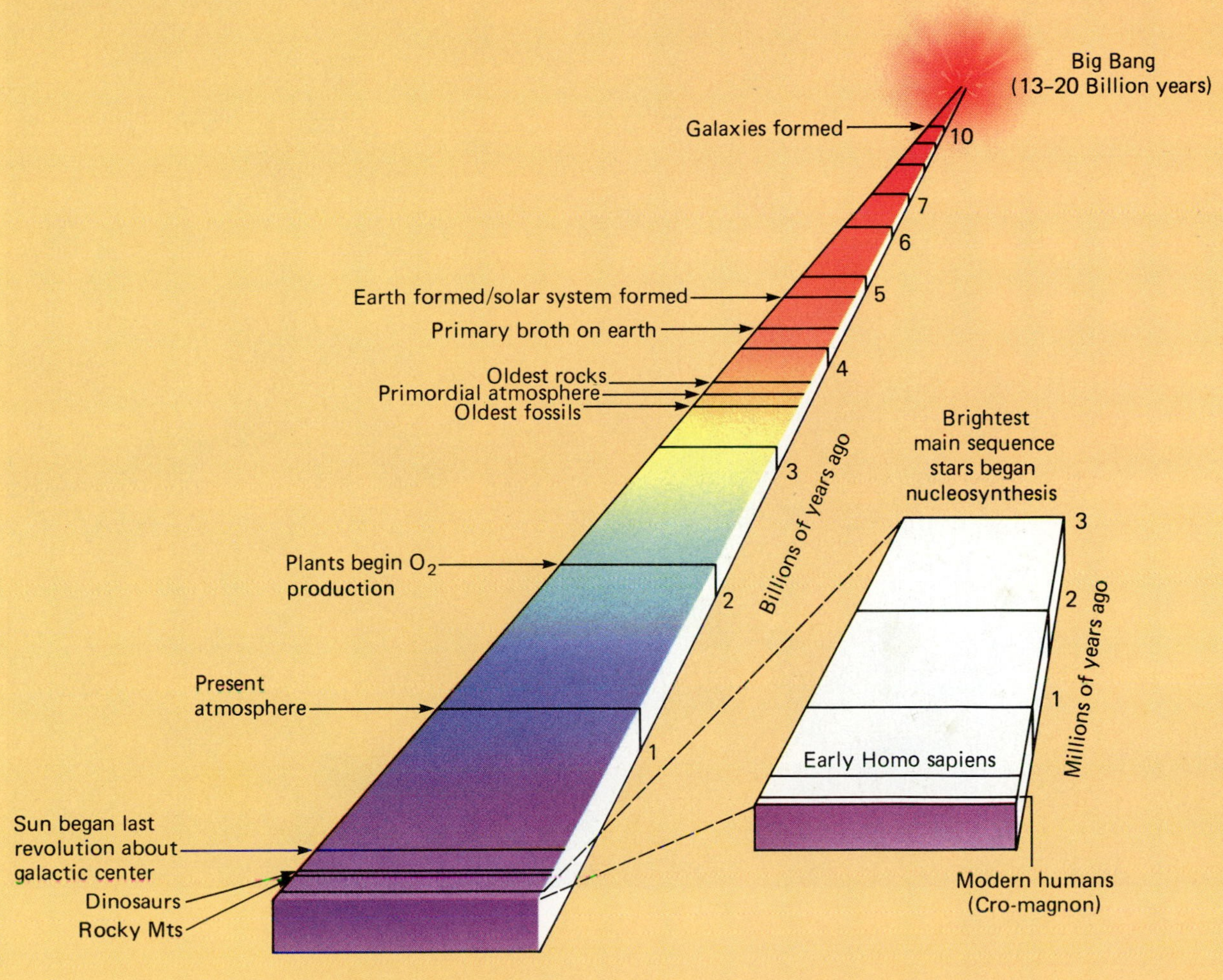

The reflection nebula NGC 6188.

The Universe: An Overview

1

Aims: To get a feeling for the variety of objects in the universe and a sense of scale

The universe is a place of great variety—after all, it has everything in it! Some of the things astronomers study are of a size and scale that we humans can easily comprehend: the planets, for instance. Most astronomical objects, however, are so large and so far away that our minds have trouble grasping their sizes and distances.

Moreover, astronomers study the very small in addition to the very large. Most everything we know comes from the study of energy travelling through space in the form of "radiation"—of which light and radio waves are examples. The radiation we receive from distant bodies is emitted by atoms, which are much too small to see with the unaided eye. Also, the properties of the large astronomical objects are often determined by changes that take place on a minuscule scale—that of atoms or their nuclear cores. Further, the evolution of the universe in its earliest stages depended on the still more fundamental particles within the nuclei. Thus the astronomer must be an expert in the study of the tiniest as well as the largest objects.

Such a variety of objects at very different distances from us or with very different properties often must be studied with widely differing techniques. Clearly, we use different methods to analyze the properties of solid particles like martian soil (using equipment in a spacecraft sitting on the martian surface) or the results of flying through Halley's Comet than we use to study the light, radio waves, or x-rays given off by a gaseous body like a quasar deep in space.

One method—*spectroscopy*—does link much of astronomy. Spectroscopy is a procedure that analyzes components of the light or other radiation that we receive from distant objects and studies these components in detail. Throughout this book, we shall return to spectroscopic methods time and again to study not only visible light but also other types of radiation. Of course, astronomers also make images—pictures—of some objects, such as planets, the Sun, and galaxies. We may also study the variation of the amount of light coming from a star or a galaxy over time, or send spacecraft to the Moon, planets, and comets.

The explosion of astronomical research in the last few decades has been fueled by our new ability to study radiation other than light—gamma rays, x-rays, ultraviolet radiation, infrared radiation, and radio waves. Astronomers' use of their new abilities to study such radiation is a major theme of this book. All the kinds of radiation together make up the *electromagnetic spectrum,* which we will discuss in Chapter 4. As we shall see there, we can think of radiation as waves, and all the types of radiation have similar properties except for the length of the waves. Still, although x-rays and visible light may be similar, our normal experiences tell us that very different techniques are necessary to study them.

The Earth's atmosphere shields us from most kinds of radiation, though light waves and radio waves do penetrate the atmosphere. Over the last 50 years, radio astronomy has become a major foundation of our astronomical knowledge. For the last 30 years, we have been able to send satellites into orbit outside the Earth's atmosphere, and we are no longer limited to the study of radio and visible (light) radiation. Many of the fascinating discoveries of recent years—the probable observations of giant black holes within our galaxy and other galaxies and quasars, for example—were made because of our newly extended senses. We will discuss how astronomers use all parts of the spectrum to help us understand the universe.

We will see, also, how we get information from direct sampling of bodies in our solar system and from cosmic rays (particles whizzing through space). These cosmic rays are atomic nuclei and subatomic particles moving through space, which are also known as "radiation," though they are different from the electromagnetic radiation we mentioned previously. We have also detected subatomic particles called neutrinos both from the Sun and, spectacularly, from the explosion of a distant star.

Further, we can now even observe the consequences of gravitational waves, whose existence is predicted by Einstein's general theory of relativity and confirmed from observations of a special pulsar. We shall discuss all these methods of astronomical research in the book; remember that the index provides page references.

1.1 A Sense of Scale

Let us try to get a sense of scale of the universe, starting with sizes that are part of our experience and then expanding toward the infinitely large. We can keep track of the size of our field of view as we expand in powers of 100: each diagram will show a square 100 times greater on a side.

In this book, we shall mostly use the International System of Units (Système International d'Unités, or SI for short), which is commonly used in almost all countries around the world and by scientists in the United States. The basic unit of length is the meter (symbol: m), which is equivalent to 39.37 inches, slightly more than a yard. The basic unit of mass is the kilogram (symbol: kg), and the basic unit of time is the second (symbol: s; we use "sec" in this book for added clarity). Prefixes (Appendix 1) are used to define larger or smaller units of these base quantities. The most frequently used prefixes are "milli-" (symbol: m), meaning $1/1000$, and "kilo-" (symbol: k), meaning 1000 times. Thus 1 millimeter (1 mm) is $1/1000$ of a meter, or about 0.04 inch, and a kilometer (1 km) is 1000 meters, or about $5/8$ mile. For mass, the prefixes are used with "gram" (symbol: g), $1/1000$ of the base unit "kilogram." We will keep track of the powers of 10 by which we multiply 1 m by writing as an exponent the number of tens we multiply together; 1000 m, for example, is 10^3 m, which is 1 km.

The meter is now defined in terms of how far light travels in a certain time. The speed of light in a vacuum (a space in which matter is absent) is, according to the "special theory of relativity" that Albert Einstein advanced in 1905 (Section 23.11), the greatest speed that is physically attainable. Light travels at about 300,000 km/sec (186,000 miles/sec), fast enough to circle the Earth 7 times in a single second. Even at that fantastic speed, we shall see that it would take years for us to reach the stars. Similarly, it has taken years for the light we see from stars to reach us, so we are really seeing the stars as they were years ago. In a sense, we are looking backward in time. The distance that light travels in a year is called a *light year* (ly); note that the light year is a unit of length rather than a unit of time even though the term "year" appears in it.

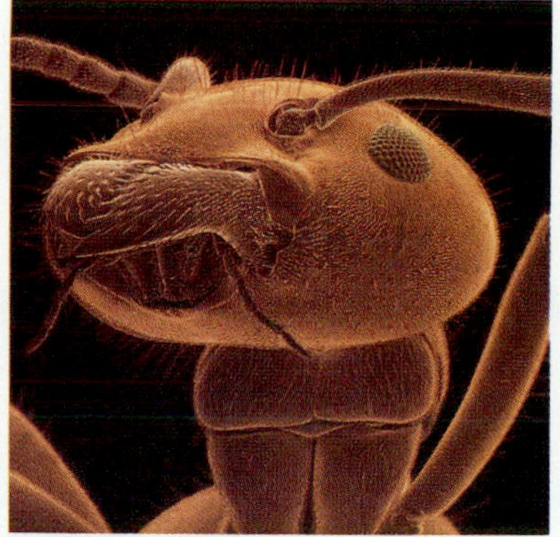

Let us begin our journey through space with a view of something 1 mm across (Fig. 1–1), an electron microscope view of an ant. Every step we take will show a region 100 times larger in diameter than the previous picture.

Figure 1–1 1 mm = 0.1 cm

A square 100 times larger on each side is 10 centimeters × 10 centimeters. (Since the area of a square is the length of a side squared, the area of a 10-cm square is 10,000 times the area of a 1-mm square.) The area encloses a flower (Fig. 1–2).

Figure 1–2 10 cm = 100 mm

Here we move far enough away to see an area 10 meters on a side (Fig. 1–3).

Figure 1–3 10 m = 1000 cm

A square 100 times larger on each side is now 1 kilometer square, about 250 acres. An aerial view of Boston shows how big an area this is (Fig. 1–4).

Figure 1–4 1 km = 10^3 m

Figure 1–5 100 km = 10^5 m

The next square, 100 km on a side, encloses the cities of Boston and Providence. Note that though we are still bound to the limited area of the Earth, the area we can see is increasing rapidly (Fig. 1–5).

A square 10,000 km on a side covers nearly the entire Earth (Fig. 1–6).

Figure 1–6 10,000 km = 10^7 m

Figure 1–7 1,000,000 km = 10^9 m = 3 lt sec

When we have receded 100 times farther, we see a square 100 times larger in diameter: 1 million kilometers across. It encloses the orbit of the Moon around the Earth (Fig. 1–7). We can measure with our wristwatches the amount of time that it takes light to travel this distance. If we were carrying on a conversation by radio with someone at this distance, there would be pauses of noticeable length after we finished speaking before we heard an answer. These pauses occur because radio waves, even at the speed of light, take that amount of time to travel. Laser pulses from Earth that are bounced off the Moon to find the distance to the Moon (Appendices 2 and 4) take a noticeable time to return. This photograph was taken by the Voyager 1 spacecraft en route to Jupiter and Saturn. Because the Moon is many times fainter than the Earth, it was artificially brightened in the computer by a factor of three so that it would show up better in this print.

When we look on from 100 times farther away still, we see an area 100 million kilometers across, ⅔ the distance from the Earth to the Sun. We can now see the Sun and the two innermost planets in our field of view (Fig. 1–8).

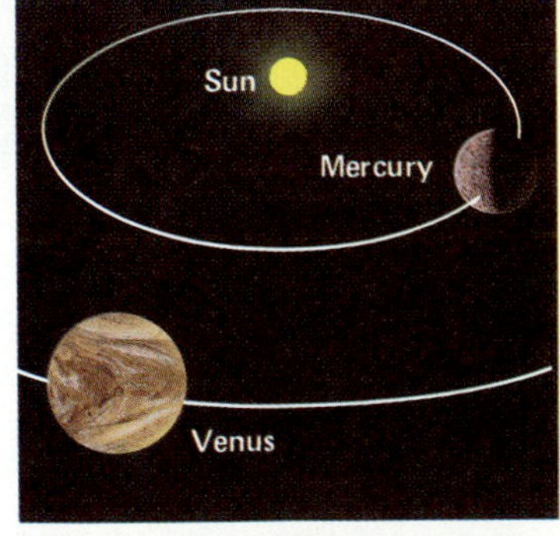

Figure 1–8 10^{11} m = 5 lt min

Figure 1–9 10^{13} m = 8 lt hrs

An area 10 billion kilometers across shows us the entire solar system in good perspective. It takes light 8 hours to travel this distance. The outer planets have become visible and are receding into the distance as our journey outward continues (Fig. 1–9). Our spacecraft have now visited or passed the Moon, Mercury, Venus, Mars, Jupiter, Saturn, Uranus, Neptune, and their moons; the results will be discussed at length in Part II. This artist's conception shows a Voyager spacecraft near Saturn.

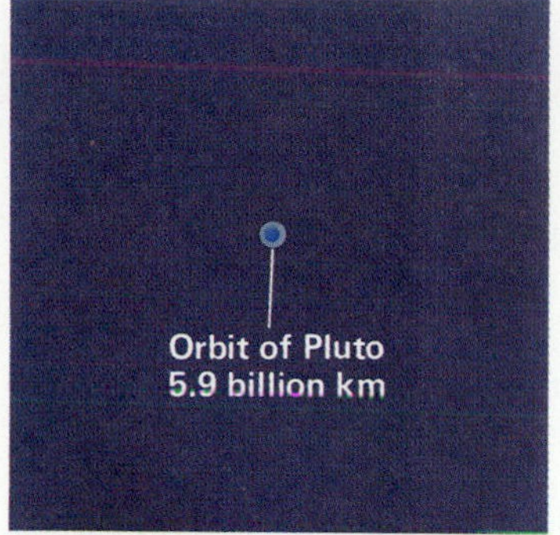

Figure 1–10 10^{15} m = 38 lt days

From 100 times farther away, we see little that is new. The solar system seems smaller, and we see the vastness of the empty space around us. We have not yet reached the scale at which another star besides the Sun is in a cube of this size (Fig. 1–10).

As we continue to recede from the solar system, the nearest stars finally come into view. We are seeing an area 10 light years across, which contains only a few stars (Fig. 1–11), most of whose names are unfamiliar (Appendix 7). Part III of this book discusses the properties of the stars.

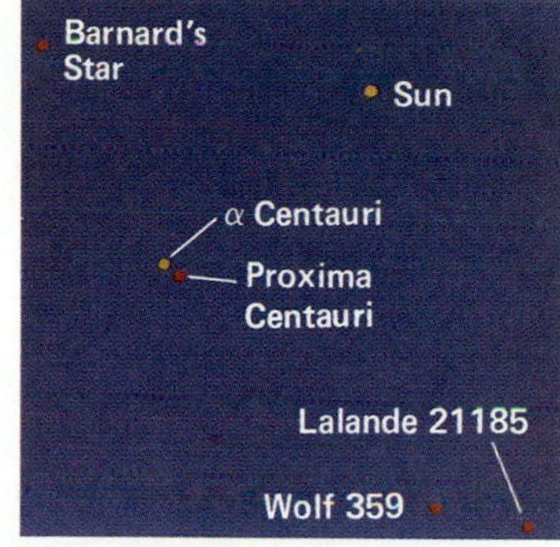

Figure 1–11 10^{17} m = 10 ly

Figure 1–12 10^{19} m = 10^3 ly

By the time we are 100 times farther away, we can see a fragment of our galaxy, the Milky Way Galaxy (Fig. 1–12). We see not only many individual stars but also many clusters of stars and many areas of glowing, reflecting, or opaque gas or dust called nebulae. Between the stars, there is a lot of material (most of which is invisible to our eyes) that can be studied with radio telescopes on Earth or in infrared, ultraviolet, or x-rays with telescopes in space. Part V of this book is devoted to the study of our galaxy and its contents. In this view we are looking past Halley's Comet to see the Milky Way.

In a field of view 100 times larger in diameter, we can now see an entire galaxy. The photograph (Fig. 1–13) shows the galaxy called M33, located in the direction of the constellation Triangulum, though it is far beyond the stars in that constellation. This galaxy shows arms wound in spiral form. The galaxy in which we live also has spiral arms, though they are wound more tightly.

Figure 1–13 10^{21} m = 10^5 ly

Figure 1–14 10^{23} m = 10^7 ly

Next we move sufficiently far away so that we can see an area 10 million light years across (Fig. 1–14). There are 10^{25} centimeters in 10 million light years, about as many centimeters as there are grains of sand in all the beaches of the Earth. Our galaxy is in a cluster of galaxies, called the Local Group, that would take up only ⅓ of our angle of vision. In this group are all types of galaxies, which we will discuss in Chapter 31. The photograph shows part of a cluster of galaxies seen in the constellation Fornax but lying far beyond the stars in it.

Focus On

Box 1.1 Scientific Notation

In astronomy we often find ourselves writing numbers that have strings of zeros attached, so we use what is called either *scientific notation* or *exponential notation* to simplify our writing chores. Scientific notation helps prevent making mistakes when copying long strings of numbers.

In scientific notation, which we used in Figures 1–4 to 1–14, we merely count the number of zeros and write the result as a superscript to the number 10. Thus the number 100,000,000, a 1 followed by 8 zeroes, is written 10^8. The superscript is called the *exponent.* We also say that "10 is raised to the eighth **power**." When a number is not a power of 10, we divide it into two parts: a number between 1 and 10 and a power of 10. Thus the number 3645 is written as 3.645×10^3. The exponent shows how many places the decimal point was moved to the left.

1.1a Survey of the Universe

If we could see a field of view 1 billion light years across, our Local Group of galaxies would appear as but one of many clusters. It is difficult to observe on such a large scale.

As we enlarge our field of view another 100 times, we might see a supercluster—a cluster of clusters of galaxies. We would be seeing almost to the distance of the quasars, which are the topic of Chapter 32. Quasars, the most distant objects known, seem to be explosive events in the cores of galaxies. Light from the most distant quasars observed may have taken 10 billion years to reach us on Earth. We are thus looking back to times billions of years ago (Fig. 1–15). Since we think that the universe began 13 to 20 billion years ago, we are looking back almost to the beginning of time.

We even think that we have detected radiation from the universe's earliest years. A combination of radio, ultraviolet, x-ray, and optical studies, together with theoretical work and experiments with giant atom smashers on Earth, is allowing us to explore the past and predict the future of the universe. All this is discussed in Chapters 33 and 34.

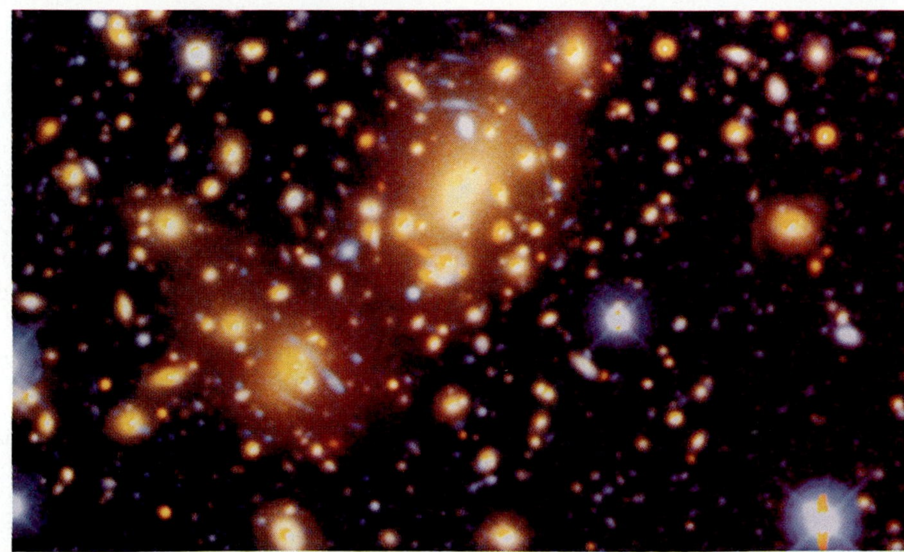

Figure 1–15 In addition to the giant cluster of galaxies (colored yellow) about 2 billion light years from us, we see dozens of bluer galaxies that are still farther away. Light from these galaxies has been distorted into arcs by the gravity of the yellower galaxies. This distortion implies that there is ten times more mass in the nearer cluster of galaxies than is visible.

1.2 The Value of Astronomy

Throughout history, observations of the heavens have led to discoveries that have had major impact on people. Even the dawn of mathematics may have followed ancient observations of the sky, made in order to keep track of seasons and seasonal floods in the fertile areas of the Earth. Observations of the motions of the Moon and the planets, which are free of such complicating terrestrial forces as friction and which are massive enough so that gravity dominates their motions, led to an understanding of gravity and of the forces that govern all motion.

Many of the discoveries of tomorrow—perhaps the discovery of new sources of energy, and perhaps something so revolutionary that it cannot now be predicted—will undoubtedly be based on discoveries made through such basic research as the study of astronomical systems. Considered in this sense, astronomy is an investment in our future.

Yet most of us study astronomy not for its technological and philosophical benefits but for its grandeur and inherent interest. We must stretch our minds to understand the strange objects and events that take place in the far reaches of space. The effort broadens us and continually fascinates us all. Ultimately, we study astronomy because of its fascination and mystery.

Key Words

spectroscopy, electromagnetic spectrum, light year, scientific notation, exponential notation, exponent

Questions

1. Why do we say that our senses have been expanded in recent years?

†2. The speed of light is 3×10^5 km/sec. Express this number in m/sec and in cm/sec.

†3. During the Apollo explorations of the Moon, we had a direct demonstration of the finite speed of light when we heard ground controllers speak to the astronauts. The sound from the astronauts' earpieces was sometimes picked up by the astronauts' microphones and retransmitted to Earth as radio signals, which travel at the speed of light. We then heard the controllers' words repeated. What is the time delay between the original and the "echo," assuming that no other delays were introduced in the signal? (The distance to the Moon and the speed of light are given in Appendix 2.)

†4. The time delay in sending commands to the Voyager 2 spacecraft when it was near Uranus was 2 hours 45 minutes. What was the distance in km from the Earth to Uranus at that time? What was the distance in Astronomical Units (A.U.), where one Astronomical Unit is the average radius of the Earth's orbit? (Appendix 2)

†5. The distance to the Andromeda Galaxy is 2×10^6 light years. If we could travel at one-tenth the speed of light, how long would a round trip take?

†This indicates a question requiring a numerical solution.

†6. How long would it take to travel to Andromeda, which is 2×10^6 light years away, at 1000 km/hr, the speed of a jet plane?

7. List the following in order of increasing size: (a) light year, (b) distance from Earth to Sun, (c) size of Local Group, (d) size of football stadium, (e) size of our galaxy, (f) distance to a quasar.

8. Of the examples of scale in this chapter, which would you characterize as part of "everyday" experience? What range of scale does this encompass? How does this range compare with the total range covered in the chapter?

9. What is the largest of the scales discussed in this chapter that could reasonably be explored in person by humans with current technology?

†10. (a) Write the following in scientific notation: 4642; 70,000; 34.7. (b) Write the following in scientific notation: 0.254; 0.0046; 0.10243. (c) Write out the following in an ordinary string of digits: 2.54×10^6; 2.004×10^2.

†11. What is (a) $(2 \times 10^5) + (4.5 \times 10^5)$; (b) $(5 \times 10^7) + (6 \times 10^8)$; (c) $(5 \times 10^3)(2.5 \times 10^7)$; (d) $(7 \times 10^6)(8 \times 10^4)$? Write the answers in scientific notation, with only one digit to the left of the decimal point.

†12. What percentage of the age of the universe has elapsed since the appearance of *Homo sapiens* on the Earth? (Hint: refer to the material opening Part I.)

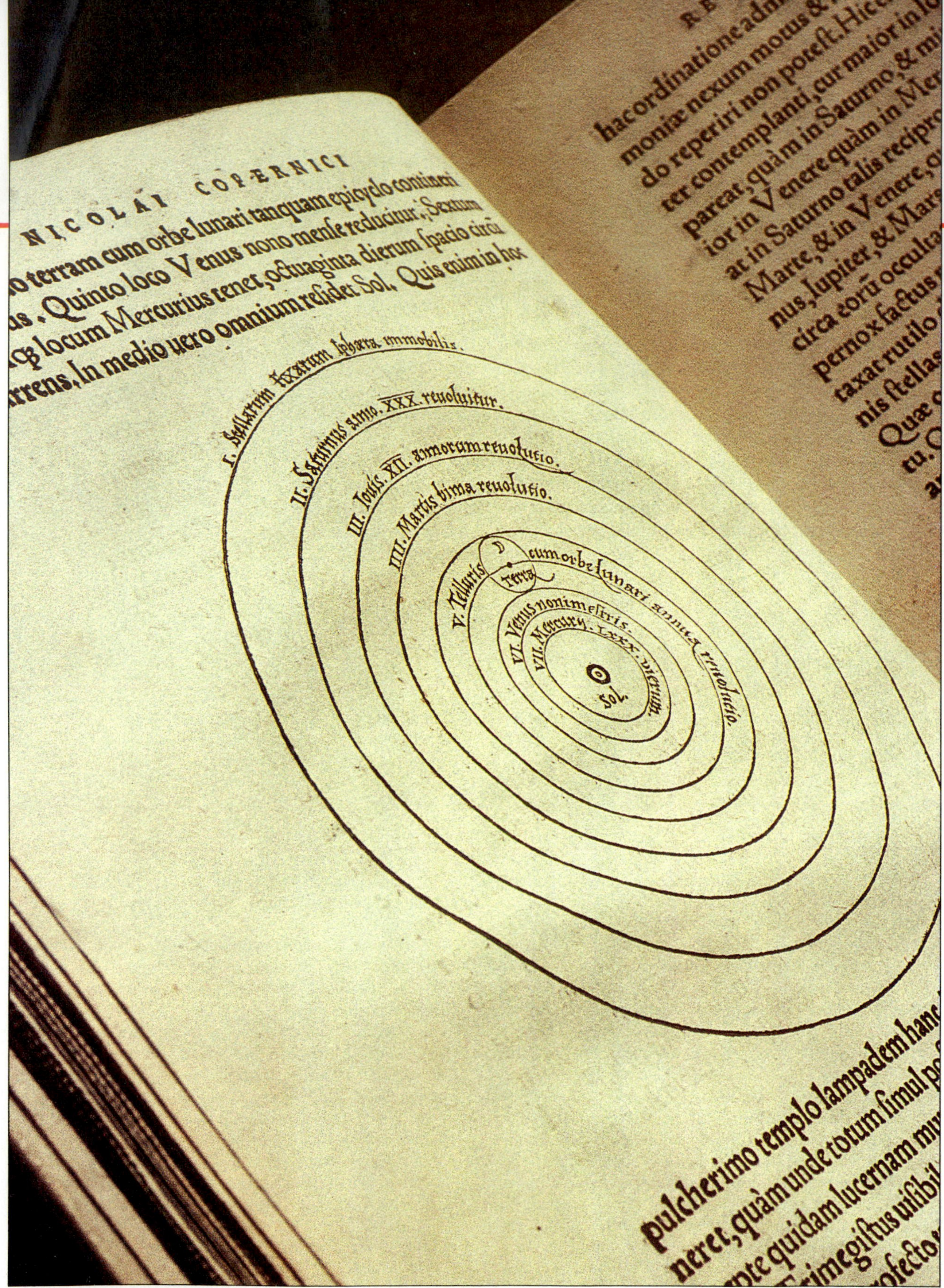

Copernicus's original heliocentric diagram, from the first edition of his 1543 book, *De Revolutionibus.*

The Early History of Astronomy

2

Aims: To follow the theories of the universe developed and held during the millennia of recorded history before the invention of the telescope

Though much of modern astronomy deals with explaining the universe, throughout history astronomy has dealt with such practical things as keeping time, marking the arrival of the seasons, and predicting eclipses of the Sun and the Moon. In this chapter we shall discuss the origins of astronomy, and trace its development, from the Earth-centered view of the solar system that dominated thought from the time of the Greeks through the Sun-centered view developed by Copernicus in the 16th century and the then unusual idea of Tycho Brahe that observations of the highest possible quality were valuable.

Figures in the sky—constellations—were recorded by the Sumerians as far back as 2000 B.C., and maybe earlier. The bull, the lion, and the scorpion are constellations that date from this time. Other constellations that we have in the present time were recorded by a Greek, Thales of Miletus, in about 600 B.C., at the dawn of Greek astronomy. Certain constellations were common to several civilizations. The Chinese also had a lion, a scorpion, a hunter, and a dipper, for example (Fig. 2–1).

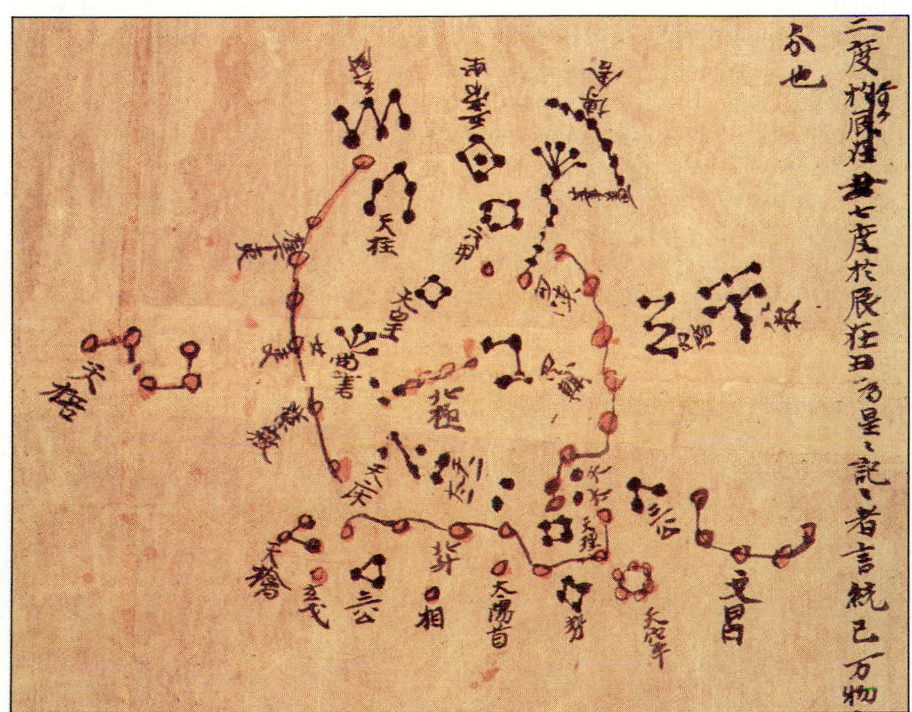

Figure 2–1 A Chinese manuscript from the 10th century, the oldest existing portable star map.

2.1 Egyptian and Other Early Astronomy

In these days of digital watches, electronic calculators, and accurate time on radio and television, it is hard to imagine what life was like in 1000 B.C. or earlier. No mechanical clocks existed, and even the idea of a calendar was vague. Yet, each spring, the Nile River flooded the neighboring farmlands. Egyptian farmers, in a land where changes in temperature do not clearly mark the seasons, wanted to know how to predict when the flooding would occur. Farmers everywhere, even those not dependent on annual flooding, wanted to know when to expect the spring and the spring rains so that they could plant their crops in ample time.

It has long been known that Egyptian pyramids and temples erected after 1500 B.C. have astronomical alignments.

Eventually, some people noticed that they could base a calendar on Sirius, the brightest star in the sky. It became visible in the eastern sky before sunrise in August, just the time of year when the crops should be planted and when the Nile was about to flood. The first visibility of an object in the pre-dawn sky, after months in which it was up only in the daytime and thus invisible, is the object's *heliacal rising* (pronounced "he-lye′e-kl"). The heliacal rising of Sirius served as a marker for Egyptian astronomers and priests, enabling them to let everybody know that the floods would soon come. We still have traces of this history in the term "the Dog Days" for the sultry period of August marked by the heliacal rising of Sirius, the Dog Star.

2.2 Greek Astronomy: The Earth at the Center

In ancient times, five planets were known—Mercury, Venus, Mars, Jupiter, and Saturn. When we observe these planets in the sky, we notice that their positions vary from night to night with respect to each other and with respect to the stars. The planets and Moon appear to move on or near the *ecliptic,* the path of the Sun across the sky. The stars, on the other hand, are so far away that their positions are relatively fixed with respect to each other. The fact that the planets appear as "wandering stars" was known to the ancients; our word "planet" comes from the Greek word for "wanderer." Of course, both stars and planets move together across our sky essentially once every 24 hours; by the wandering of the planets we mean that the planets appear to move at a slightly different rate, so that over a period of weeks or months they change position with respect to the fixed stars.

The planets do not always move in the sky in the same direction with respect to the stars. Most of the time the planets appear to drift eastward with respect to the background stars. But sometimes they drift backwards, that is, westward. We call the backward motion *retrograde motion* (Fig. 2–2).

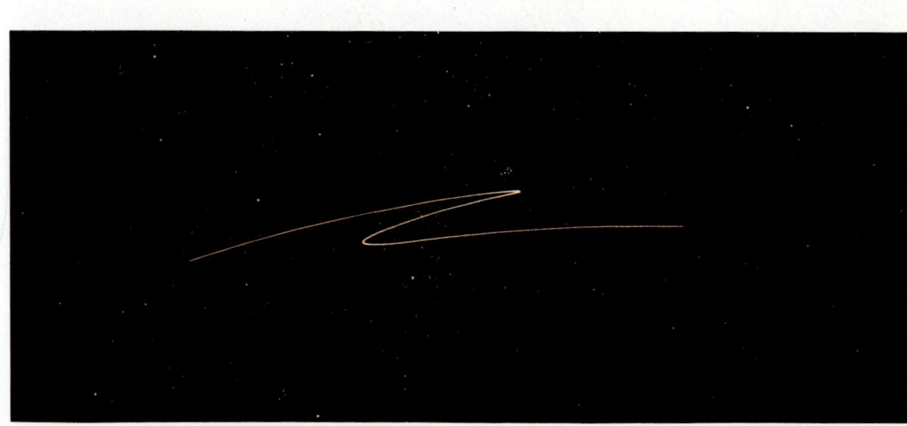

Figure 2–2 A planetarium simulation of the path of Mars from August 1, 1990, through April 1, 1991, showing the retrograde loop. Mars passes first near the V-shaped Hyades star cluster, with its reddish star Aldebaran, reverses direction, and then reverses direction again near the Pleiades star cluster. Capella is the brightest star in the pentagon of Auriga at left.

Figure 2–3 Aristotle *(right)* and Plato *(left)* in Raphael's (1483–1520) "The School of Athens."

The ancient Greeks began to explain the motions of the planets by making theoretical models of the geometry of the solar system. For example, by determining which of the known planets had the longest periods of retrograde motion, they were able to discover the order of distance of the planets.

One of the earliest and greatest philosophers, Aristotle (Fig. 2–3), lived in Greece about 350 B.C. He summarized the astronomical knowledge of his day into a qualitative cosmology that remained dominant for 1800 years. On the basis of what seemed to be very good evidence—what he saw—Aristotle thought, and actually believed that he knew, that the Earth was at the center of the universe and that the planets, the Sun, and the stars revolved around it (Fig. 2–4). The universe was made up of a set of 55 celestial spheres that fit around each other; each had rotation as its natural motion. Each of the heavenly bodies was carried around the heavens by one of the spheres. The motions of the spheres affected each other and combined to account for the various observed motions of the planets, including revolution around the Earth, retrograde motion, and motion above and below the ecliptic. The outermost sphere was that of the fixed stars, beyond which lay the prime mover, *primum mobile,* that caused the general rotation of the stars overhead.

Aristotle's theories ranged through much of science. He held that below the sphere of the Moon everything was made of four basic "elements": earth, air, fire, and water. The fifth "essence"—the quintessence—was a perfect, unchanging transparent element of which the celestial spheres were thought to be formed.

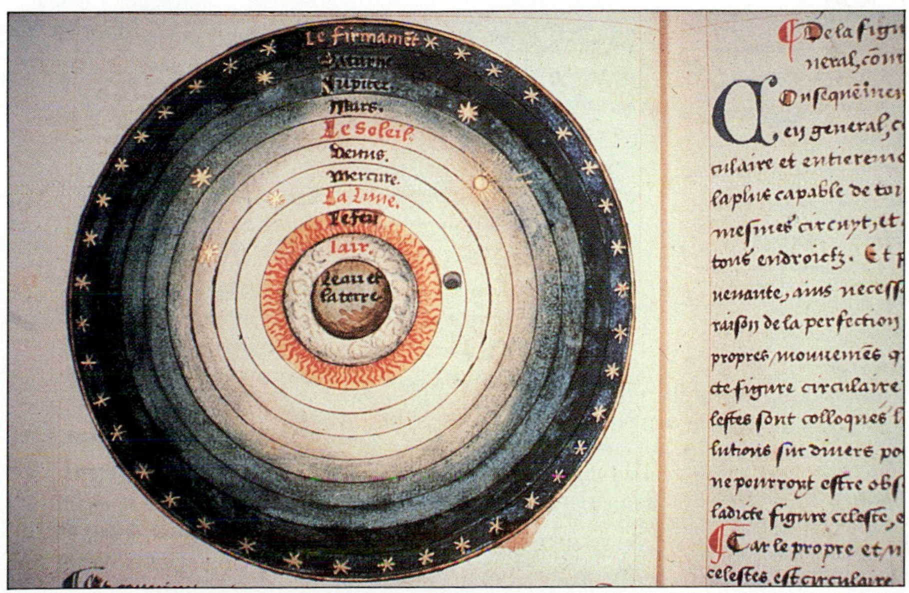

Figure 2–4 Aristotle's cosmological system, with water and Earth at the center, surrounded by air, fire, the Moon, Mercury, Venus, the Sun, Mars, Jupiter, Saturn, and the firmament of fixed stars.

Figure 2–5 Ptolemy, in a 15th-century drawing.

Focus On

Box 2.1 Ptolemaic Terms

deferent—a large circle centered approximately on the Earth
epicycle—a small circle whose center moves along a deferent. The planets move on the epicycles.
eccentric—that the deferent is not exactly centered at the Earth
equant—the point around which the epicycle moves at a uniform rate

Aristotle's theories dominated scientific thinking for almost two thousand years, until the Renaissance. Unfortunately, most of his theories were far from what we now consider to be correct, so we tend to think that the widespread acceptance of Aristotelian physics impeded the development of science.

In about A.D. 140, almost 500 years after Aristotle, the Greek astronomer Claudius Ptolemy (Fig. 2–5), in Alexandria, presented a detailed theory of the universe that explained the retrograde motion. Ptolemy's model was Earth-centered, as was Aristotle's. To account for the retrograde motion of the planets, the planets had to be moving not simply on large circles around the Earth but rather on smaller circles, called *epicycles,* whose centers moved around the Earth on larger circles, called *deferents* (Fig. 2–6). (The notion of epicycles and deferents had been advanced earlier by such astronomers as Hipparchus, whose work on star catalogues we will discuss in Section 21.1.) It seemed natural that the planets should follow circles in their motion, since circles were thought to be "perfect" figures.

Sometimes the center of the deferent was not centered at the Earth (and thus the circle was *eccentric*). The epicycles moved at a constant rate of angular motion (that is, the angle through which they moved was the same for each identical period of time). However, another complication was that the point

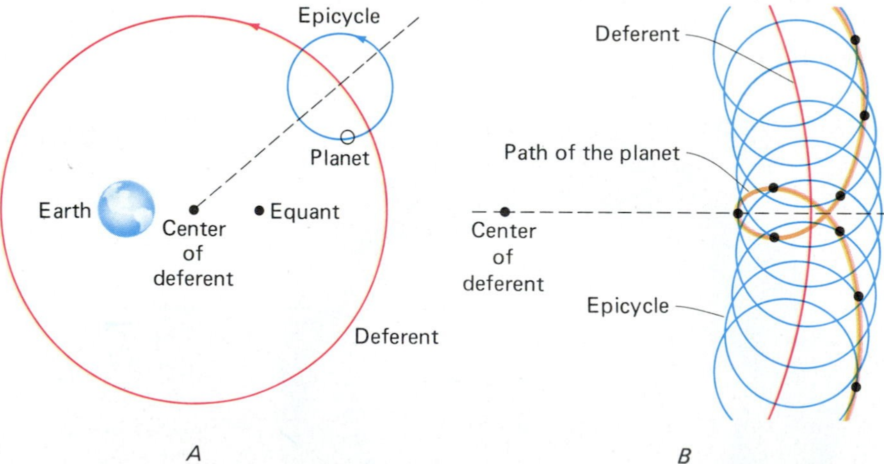

Figure 2–6 (*A*) In the Ptolemaic system, a planet would move around on an epicycle, which, in turn, moved on a deferent. Variations in the apparent speed of the planet's motion in the sky could be accounted for by having the epicycle move uniformly around a point called the equant, instead of moving uniformly around the Earth or the center of the deferent. The Earth and the equant were on opposite sides of the center of the deferent and equally spaced from it. When the planet was in the position shown, it would be in retrograde motion. (*B*) In the Ptolemaic system, the projected path in the sky of a planet in retrograde motion is shown.

Focus On

Box 2.2 Eratosthenes and the Size of the Earth

Eratosthenes (pronounced eh-ra-tos'then-ees) was born in about 273 B.C. He received his education in Athens and then spent the latter half of his life in the city of Alexandria (which is now in Egypt but was then under Greek rule).

The Greek mathematician Pythagoras and his disciples had earlier recognized that the Earth is round, but this idea was not necessarily generally accepted. Eratosthenes set out to do no less than to measure the size of the whole Earth.

He made the measurement by looking in a deep well or by using a *gnomon* (pronounced noh'mon), basically an upright stick that is allowed to cast a shadow in sunlight. (Gnomons, simple as they now seem to us when we see them on sundials, were among the prime astronomical instruments available at that time.) Eratosthenes used observations made at two cities that were on the same line of longitude; that is, one was due north of the other. The length of the shadow cast by the gnomon (or, perhaps, the shadow inside a well) at noon varied from day to day. At one of his cities (Syene, near what is now Aswan, Egypt) on one day of the year (which we now call the summer solstice), that shadow vanished. This vanishing indicated that the Sun was directly overhead in Syene at that moment. From the length of the shadow at the same date and time at his second city, Alexandria, he concluded that Alexandria was one-fiftieth of a circle around the Earth from Syene (Fig. 2–7).

Further, Eratosthenes knew the distance on the Earth's surface between Syene and Alexandria: it was 5000 stadia, where a stadium was a unit of distance equal to the size of an athletic stadium, about 160 meters. (The distance had actually been paced out by someone trained for the evenness in the size of his step.)

If one-fiftieth of the Earth's circumference was 5000 stadia, Eratosthenes concluded that the Earth must be 250,000 stadia around. The value is in reasonable agreement with the value we now know to be the actual size of the Earth. But even more important than the particular number is the idea that, with the application of logic, the size of the Earth could be measured and fathomed by mere humans.

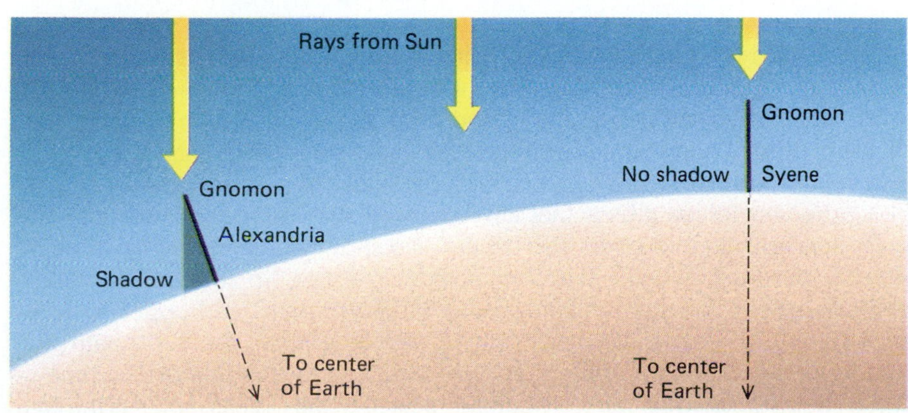

Figure 2–7 When the Sun is directly overhead at Syene, a gnomon (an upright stick) casts no shadow. At the same time, the Sun is not directly overhead at Alexandria, which is a known distance almost due north of Syene. From this angle and the distance between the cities, Eratosthenes calculated the size of the Earth. Note that the Sun is so far away from the Earth that the sunlight hitting the Earth at one city is parallel to the sunlight hitting the Earth at the other city. The angular scale is greatly exaggerated in this diagram. Tradition has it that the observation may have been made from a well at Syene; when the Sun was directly overhead at Syene, it illuminated the entire bottom of the well.

Focus On

Box 2.3 Stellar Parallaxes and the Motion of the Earth

Most Greek astronomers held that the Earth was fixed at the center of the universe. They reasoned (correctly) that if the Earth itself moved, during the course of the year the stars would be at slightly different heights in the sky because of our changing vantage point. Such an apparent movement resulting from a change in vantage point is called parallax (see Section 21.2). Parallax is similar to the apparent motion of your thumb when you view your thumb at the end of your outstretched arm, looking first with one eye and then the other. The thumb appears to jump left and right across the background. The closer you hold your thumb to your face, the greater its displacement appears.

Parallaxes are not visible to the naked eye, nor could they be observed with the instruments available to the ancient Greeks. Actually, the nearest stars do have parallaxes. The closer stars are like your thumb, and the distant stars are the background. The two points from which we observe are the positions of the Earth on different parts of its orbit around the Sun. The parallaxes of even the nearby stars are too small to have been observed before the invention of the telescope (and were not discovered until Friedrich Bessel did so in 1838).

The Greek astronomers deserve credit for the fact that, at least in this respect, their theories agreed better with observations than did the heliocentric theory. It was only much later that accurate observations changed the weight of the evidence on this point.

around which the epicycle's angular motion moved uniformly was neither at the center of the Earth nor at the center of the deferent. The epicycle moved at a uniform angular rate about still another point, the *equant.* The equant and the Earth were equally spaced on opposite sides of the center of the deferent, as the figure shows.

Ptolemy's views were very influential in the study of astronomy, because versions of his ideas and of the tables of planetary motions that he computed were accepted for nearly 15 centuries. His major work, which became known as the *Almagest* ("the Greatest"), contained both his ideas and a summary of the ideas of his predecessors (especially those of Hipparchus), and provides most of our knowledge of Greek astronomy.

The need for an equant followed from the fact that the lengths of the seasons are not equal. The present-day values for the northern hemisphere are

spring	92^d19^h
summer	93^d15^h
autumn	89^d20^h
winter	$89^d\ 0^h$.

If the Sun goes in a circular orbit at uniform speed, then the orbit cannot be centered at the Earth.

*2.3 Other Ancient Astronomy

At the same time that Greek astronomers were developing their ideas, Babylonian priests were computing positions of the Moon and planets. The major work was carried out during the period from about 700 B.C. until about A.D. 50. This work is sometimes referred to as Chaldean—pronounced something like kal-dee'an in English. Chaldea was the southern part of Babylon.

The Babylonian tablets that have survived show lists and tables of planetary positions and eclipses. The tablets also show predictions of such quantities as the times when planets would be closest to (in *conjunction* with—technically, at the same longitude along the Sun's path) and opposite to (in *opposition* to—technically, 180° away in longitude along the Sun's path) the Sun in the sky and when objects would be visible for the first or last time in a year. Babylonian methods were communicated to the Greeks, and it is through the Greeks that Babylonian astronomy influenced Western thought.

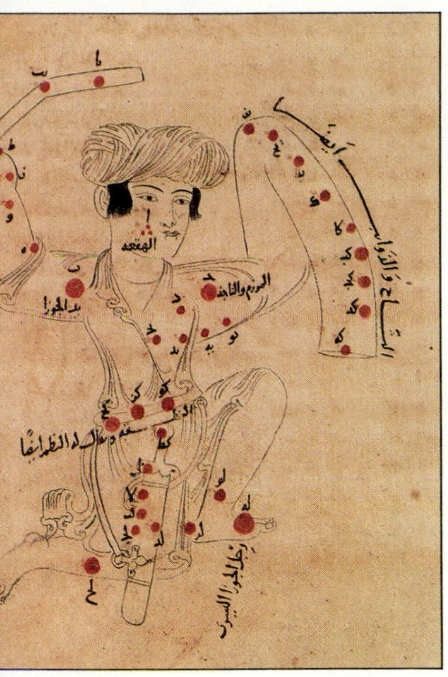
A

B

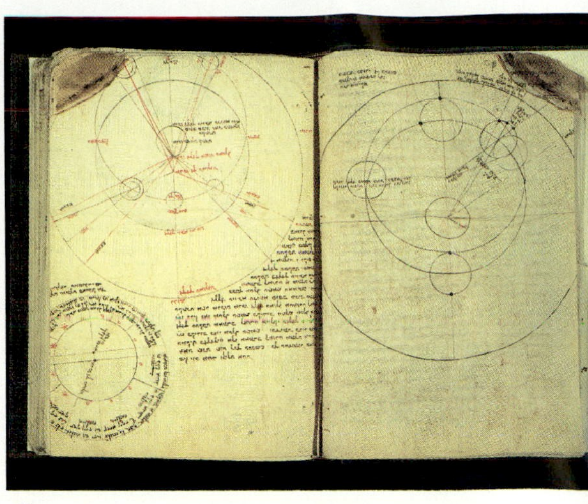
C

Figure 2–8 (*A*) The constellation Orion, from the *Book of the Stars* by Al-Sufi, a 10th-century Islamic astronomer who carried out the principal revisions of Ptolemy's work during the Middle Ages. Note the 3 stars that make Orion's belt, the star in his sword, and the bright star in his right shoulder. In this pre-telescopic age, astronomers did not know of the Orion Nebula, which is in his sword. (*B*) An astronomer is shown holding an astrolabe and discussing the laws of nature with his fellow philosophers under a sky full of stars in this page from a manuscript copy of *The Guide to the Perplexed* by the 12th-century Jewish philosopher Moses Maimonides. (*C*) Astronomical studies from a 15th-century manuscript by the Jewish astronomer Abraham bar Hiyya ha-Nasi.

*2.4 Astronomy of the Middle Ages

The Middle Ages, which we might loosely define as the thousand years up to about 1500, were not marked by epochal astronomical discoveries or significant models, such as those advanced by Ptolemy and Aristotle. Nonetheless, the period was not as dead as commonly presented.

Aristotle's ideas provided the framework for most of the medieval discussion of astronomy. Earth, air, fire, and water remained the four known "elements." And Aristotle's idea that a force has to be continuously applied to cause an object to keep moving was generally believed. (For example, the air displaced from the tip of a moving arrow was thought to come around to the back of the arrow to push it forward.) We now know the opposite—that moving objects tend to keep moving.

Arab astronomers were active (Fig. 2–8*A*), and translations of Ptolemy's major work, the *Almagest,* were made from the Greek into Arabic in the 8th and 9th centuries; only through such Arabic translations do we now know of several Greek works. (*Almagest* is, in fact, a translation of the Arabic word for "the greatest.") Major observatories were built, and careful records were kept of planetary positions, though the telescope had not yet been invented. Jewish astronomers also wrote about celestial objects (Fig. 2–8*B* and *C*).

One of the most famous medieval observatories was built in Samarkand (Fig. 2–9), Uzbekistan by Ulugh Begh, a grandson of the Tartar chief Tamerlane. Ulugh Begh's star catalogue, which we now find to be accurate to one-tenth of the diameter of the Moon, showed the high quality of the measuring instruments available in about 1420.

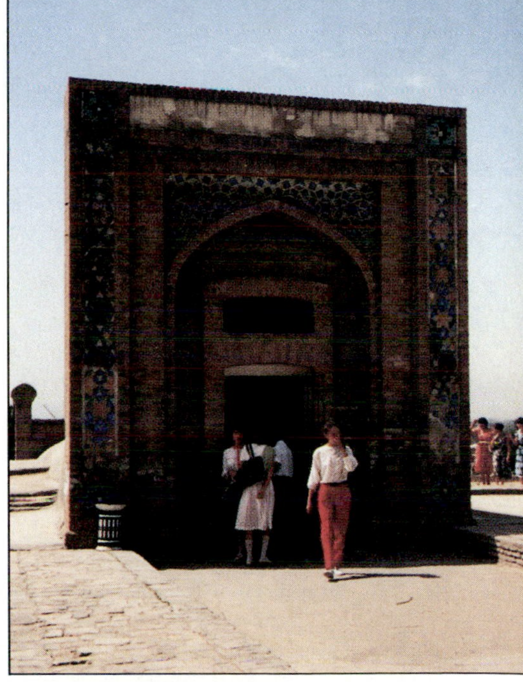

Figure 2–9 The monument building over the altitude quadrant in the center of Ulugh Begh's observatory at Samarkand.

Tabula equationũ Ueneris.

Lineę numeri communes.				Aequatio centri.		proportiõa	Lõgitudo longior.		Aequatio argumenti.		Lõgitudo propior.	
s	g	s	g	g	m	m	g	m	g	m	g	m
0	31	5	29	1	5	51	0	9	12	55	0	10
0	32	5	28	1	7	51	0	10	13	20	0	11
0	33	5	27	1	9	50	0	10	13	44	0	11
0	34	5	26	1	10	50	0	10	14	9	0	11
0	35	5	25	1	12	49	0	11	14	34	0	11
0	36	5	24	1	14	49	0	11	14	58	0	12
0	37	5	23	1	16	48	0	11	15	23	0	12
0	38	5	22	1	17	48	0	12	15	48	0	12
0	39	5	21	1	19	47	0	12	16	12	0	12
0	40	5	20	1	21	47	0	12	16	37	0	13
0	41	5	19	1	22	46	0	13	17	1	0	13
0	42	5	18	1	24	45	0	13	17	25	0	13
0	43	5	17	1	26	45	0	13	17	50	0	13
0	44	5	16	1	27	44	0	14	18	14	0	14
0	45	5	15	1	29	43	0	14	18	38	0	14
0	46	5	14	1	31	42	0	14	19	3	0	14
0	47	5	13	1	32	42	0	15	19	27	0	15
0	48	5	12	1	34	41	0	15	19	51	0	15
0	49	5	11	1	36	40	0	15	20	15	0	15
0	50	5	10	1	37	40	0	16	20	39	0	16
0	51	5	9	1	39	39	0	16	21	3	0	16
0	52	5	8	1	40	38	0	16	21	27	0	16
0	53	5	7	1	42	37	0	17	21	51	0	17
0	54	5	6	1	43	36	0	17	22	15	0	17
0	55	5	5	1	44	35	0	17	22	39	0	17
0	56	5	4	1	46	34	0	18	23	3	0	18
0	57	5	3	1	47	33	0	18	23	27	0	18
0	58	5	2	1	48	32	0	18	23	51	0	18
0	59	5	1	1	50	31	0	19	24	15	0	19
1	0	5	0	1	51	30	0	19	24	3[illegible]	0	19

Figure 2–10 Positions of Venus in the *Alphonsine Tables.*

Figure 2–11 Regiomontanus, a woodblock in the *Nuremberg Chronicle* (1493).

Figure 2–12 A comet from the *Nuremberg Chronicle* (1493), one of four woodblocks used to reproduce 13 comets.

In the late 13th century, King Alfonso X collected a group of scholars at Toledo, Spain. The astronomical tables they computed, known as the *Alphonsine Tables* (Fig. 2–10), remained in general use for centuries. Such tables provided basic information that could be used to calculate the positions of the planets at any time in the past or future.

In the mid–15th century, in Europe, Regiomontanus (as John Müller became known) and his teacher discovered how inaccurate the Alphonsine Tables could be. Two examples were the occurrence of an eclipse of the Moon an hour late and the appearance of Mars 2° (four times the diameter of the Moon) from its predicted position. Regiomontanus ultimately settled in Nuremberg (Fig. 2–11), in what is now West Germany, where he started a printing press and became responsible for much of the astronomical literature of that period. He printed theoretical works and also the first *ephemeris*—a book of the planetary positions themselves (the name is based on the Latin word for "changing," as in "ephemeral"). The positions were based on the Alphonsine Tables. Regiomontanus also completed a summary of the *Almagest* that had been begun by his teacher. Nuremberg remained a center of scientific printing; the *Nuremberg Chronicle* (1493) included reports of many comets and other astronomical events (Fig. 2–12).

2.5 Nicolaus Copernicus: The Sun at the Center

Helios was the Sun god in Greek mythology.

The major credit for the breakthrough in our understanding of the solar system belongs to Nicolaus Copernicus (Fig. 2–13), a Polish astronomer and cleric. Over 400 years ago, Copernicus advanced a *heliocentric*—Sun-centered—theory (Figs. 2–14 and 2–15). He suggested that the retrograde motion of the planets could be readily explained if the Sun, rather than the Earth, were at the center of the universe; that the Earth is a planet; and that the planets move around the Sun in circles.

Aristarchus of Samos, a Greek scientist, had suggested a heliocentric theory 18 centuries earlier, though only fragments of his work were or are known and we do not know how detailed a picture of planetary motions he presented. His heliocentric suggestion required the then apparently ridiculous

Focus On

Box 2.4 Copernicus and His Contemporaries

Nicolaus Copernicus was born in 1473, and through family connections was destined from his youth for an ecclesiastical career. At the University of Cracow, which he entered in 1490, his studies included mathematics and astronomy, the latter including the books of Regiomontanus. Though his official career was as a religious administrator, he spent much time travelling abroad, and it is clear that mathematics and astronomy were always among his greatest interests. In the few years following his return from Italy in about 1506, Copernicus probably wrote a draft of his heliocentric ideas. From 1512 on, he lived in Frauenberg (now Frombork, Poland). He wrote a lengthy manuscript about his ideas, and continued to develop his ideas and manuscript even while carrying on official duties of various kinds. Indeed, he had even studied medicine in Italy, and acted as medical advisor to his uncle, the bishop, though he probably never practiced as a physician.

Though it is not certain why he did not publish his book earlier, his reasons for withholding the manuscript from publication probably included a general desire to perfect the manuscript, and also probably involved his realization that to publish the book would involve him in controversy.

Though Copernicus had not published his work, it appears that he was well known as a mathematician and astronomer among experts in those fields. His idea that the Earth is moving and the Sun and stars are not was sufficiently unusual that it became known even to non-astronomers.

Copernicus received a copy of his book on the day of his death in 1543. Although some people, including Martin Luther, noticed a contradiction between the Copernican theory and the Bible, little controversy occurred at that time. In particular, the Pope—to whom the book had been dedicated—did not object.

The Harvard astronomer and historian of science Owen Gingerich has, by valiant searching, located about 300 copies extant today of the first edition of Copernicus's masterpiece and about an equal number of the second edition. By studying which were censored and what notes were handwritten into the book by readers of the time, he has been able to study the spread of belief in Copernicanism.

Figure 2–13 Copernicus in a painting hanging in the Torun Museum.

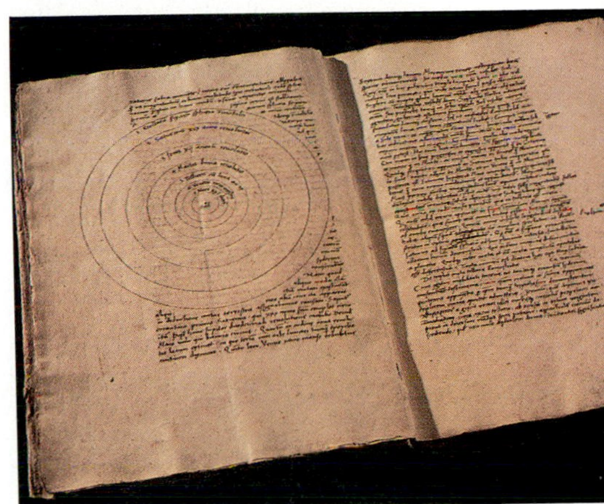

Figure 2–14 The pages of Copernicus's original manuscript in which he drew his heliocentric system. The manuscript was later set into type and published in 1543. The Sun (sol) is at the center surrounded by Mercury (Merc), Venus (Veneris), Earth (Telluris), Mars (Martis), Jupiter (Jovis), Saturn (Saturnus), and the fixed stars. Each planet is assigned the space between the circles, rather than the circles themselves as is commonly believed. The manuscript is in the University library in Cracow.

notion that the Earth itself moved, in contradiction to our senses and to the theories of Aristotle. If the Earth is rotating, for example, why aren't birds and clouds left behind the moving Earth? Only the 17th-century discovery by Isaac Newton of laws of motion substantially different from Aristotle's solved this dilemma. Aristarchus's heliocentric idea had long been overwhelmed by the *geocentric*—Earth-centered—theories of Aristotle and Ptolemy.

Copernicus's theory, although it put the Sun instead of the Earth at the center of the solar system (and, for then, the universe), still assumed that the orbits of celestial objects were circles. The notion that circles were perfect, and that celestial bodies had to follow such perfect orbits, shows that Copernicus had not been able to break away entirely from the old ideas. The so-called Copernican revolution, in the sense that our modern approach to science involves a comparison of nature and understanding, really occurred later.

Proof that the Earth rotates on its axis was not forthcoming until the 19th century, when the French physicist Foucault suspended a long pendulum from the dome of the Pantheon in Paris. The inertia of the heavy bob kept the pendulum swinging back and forth in the same plane even while the Earth rotated below it. Thus observers in the room saw the pendulum apparently rotate slowly. They were rotating with the Earth, while the plane of the pendulum was fixed with respect to the stars. Such a *Foucault pendulum* (Fig. 2–19) at the north or south pole would rotate once a day, for example. The rate of apparent rotation of a particular pendulum depends on its latitude.

currens, In medio uero omnium residet Sol. Quis enim in hoc

pulcherimo templo lampadem hanc in alio uel meliori loco poneret, quàm unde totum simul possit illuminare? Siquidem non inepte quidam lucernam mundi, alij mentem, alij rectorem uocant. Trimegistus uisibilem Deum, Sophoclis Electra intuentē

A

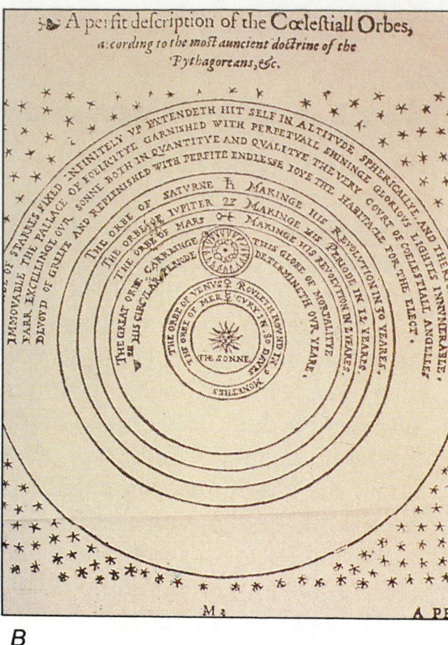

B

Figure 2–15 (*A*) Copernicus's heliocentric diagram as printed in *De Revolutionibus* (1543). (*B*) The first English diagram of the Copernican system, which appeared in 1568 as an appendix by Thomas Digges to a book by his father.

Since Copernicus's theory contained only circular orbits, he still invoked the presence of some epicycles in order to improve agreement between theory and observation. Copernicus was proud, though, that he had eliminated the equant. In any case, the detailed predictions that Copernicus himself computed on the basis of his theory were not in much better agreement with the existing observations than tables based on Ptolemy's model, because in many cases Copernicus still used Ptolemy's observations. The heliocentric theory appealed to Copernicus and to many of his contemporaries on philosophical grounds, rather than because direct comparison of observations with theory showed the new theory to be better.

Copernicus's heliocentric theory was published in 1543 (Fig. 2–16) in the book he called *De Revolutionibus* (*Concerning the Revolutions*). The theory explained the retrograde motion of the planets as follows (Fig. 2–17):

Let us consider, first, an outer planet like Mars as seen from the Earth. As the Earth approaches the part of its orbit that is closest to Mars (which is orbiting the Sun much more slowly than is the Earth), the projection of the Earth–Mars line outward to the stars (which are essentially infinitely far away compared to the planets) moves slightly against the stellar background. As the Earth comes to the point in its orbit closest to Mars, and then passes it, the projection of the line joining the two planets can actually seem to go backward, since the Earth is going at a greater speed than Mars. Then, as the Earth continues around its orbit, Mars appears to go forward again. A similar explanation can be demonstrated for the retrograde loops of the inner planets.

The idea that the Sun was at the center of the solar system led Copernicus to two additional important results. First, he was able to work out the distances to the planets, based on observation of their motion across the sky with respect to the changing position of the Sun. Second, he was able to derive the periods of the planets, the length of time they take to revolve around the Sun, based

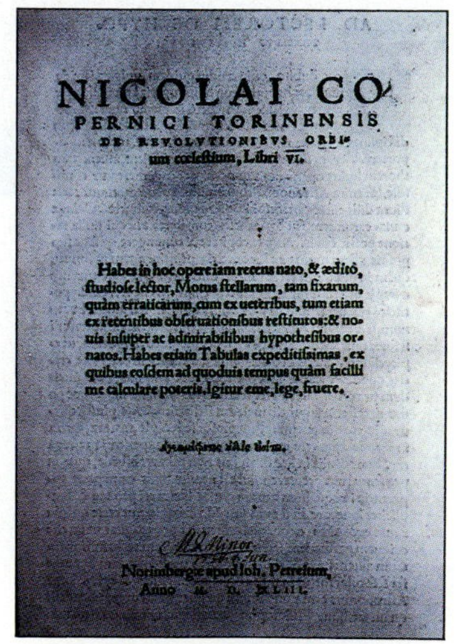

NICOLAI CO
PERNICI TORINENSIS
DE REVOLVTIONIBVS ORBIum coelestium, Libri VI.

Habes in hoc opere iam recens nato, & ædito, studiose lector, Motus stellarum, tam fixarum, quàm erraticarum, cum ex ueteribus, tum etiam ex recentibus obseruationibus restitutos: & nouis insuper ac admirabilibus hypothesibus ornatos. Habes etiam Tabulas expeditissimas, ex quibus eosdem ad quoduis tempus quàm facillime calculare poteris. Igitur eme, lege, fruere.

Norimbergæ apud Ioh. Petreium, Anno M. D. XLIII.

Figure 2–16 The title page of the first edition (1543) of *De Revolutionibus*.

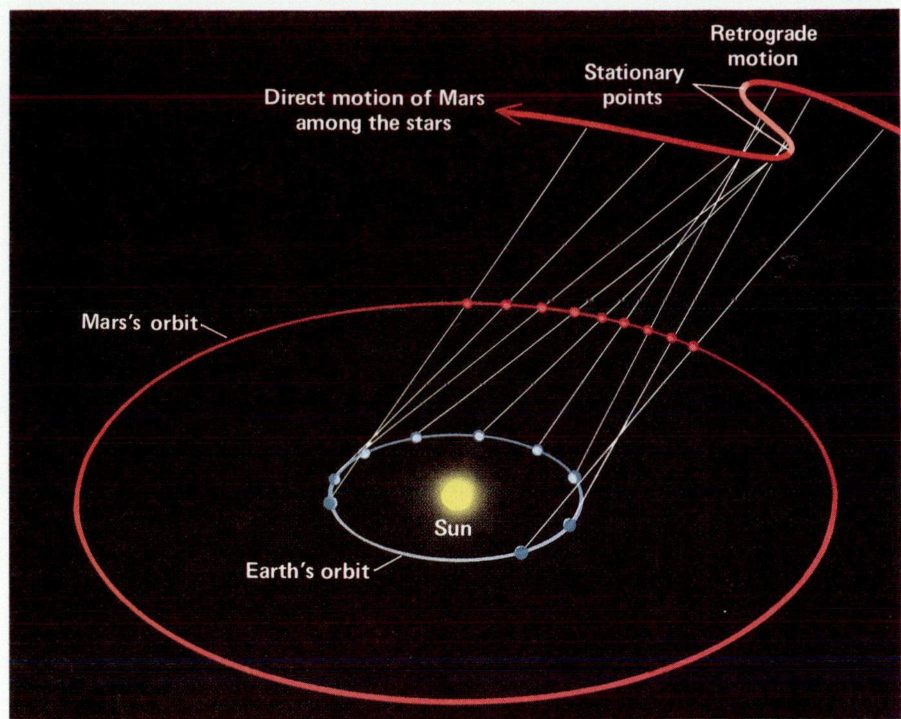

Figure 2–17 The Copernican theory explains retrograde motion as an effect of projection. For each of the nine positions of Mars shown from right to left, follow the line from Earth's position through Mars's position to see the projection of Mars against the sky. Mars's forward motion appears to slow down as the Earth overtakes it. Between the two stationary points, Mars appears in retrograde motion; that is, it appears to move backward with respect to the stars. The drawing shows the explanation of retrograde motion for Mars or for the other *superior planets,* that is, planets whose orbits lie outside that of the Earth. Similar drawings can explain retrograde motion for *inferior planets,* that is, planets whose orbits lie inside that of the Earth (namely, Mercury and Venus).

on the length of time they took to return to the position in the sky opposite to the direction of the Sun.

His ability to derive these results played a large role in persuading Copernicus of the superiority of his heliocentric system.

2.6 Tycho Brahe

In the last part of the 16th century, not long after Copernicus's death, Tycho Brahe began a series of observations of Mars and other planets. Tycho, a Danish nobleman, set up an observatory on an island off the mainland of Denmark (Fig. 2–18). The building was called Uraniborg (after Urania, the muse of astronomy). The telescope had not yet been invented, but Tycho used giant instruments to make observations of unprecedented accuracy. In 1597, Tycho lost his financial support in Denmark and moved to Prague, arriving two years later. A young assistant, Johannes Kepler, came there to work with him. At Tycho's death, in 1601, Kepler—who had worked with Tycho for only 10 months—was left to analyze all the observations that Tycho and his assistants had made (though first Kepler had to get access to the data from Tycho's family, which proved troublesome). In Section 3.1, we will see how valuable Kepler's theory, and thus Tycho's observations, turned out to be.

Figure 2–18 Tycho's observatory at Uraniborg, Denmark. Here Tycho is seen showing the mural quadrant (a marked quarter-circle on a wall) that he used to measure the altitudes at which stars and planets crossed the meridian.

Figure 2–19 The Foucault pendulum at the Griffith Observatory in Los Angeles. As the Earth rotates, the swinging ball knocks over successive sticks. A pendulum at the North Pole would rotate in 24 hours as the Earth turns beneath it; at the equator, a pendulum would not rotate at all. At the latitude of the pendulum shown, a Foucault pendulum rotates in about 42 hours.

Focus On

Box 2.5 Tycho Brahe

Tycho Brahe was born in 1546 to a Danish noble family. As a child he was taken away and raised by a wealthy uncle. In 1560, a total eclipse was visible in Portugal, and the young Tycho witnessed the partial phases in Denmark. Though the event itself was not spectacular—partial eclipses never are—Tycho, at the age of 14, was so struck by the ability of astronomers to predict the event that he devoted his life from then on to making an accurate body of observations.

Three years later, he witnessed a conjunction of Saturn and Jupiter—a time when the two planets came very close together in the sky. But the tables based on the work of Copernicus were a few days off in predicting the event (and the older Alphonsine Tables were a month off). Tycho's desire to improve such tables was renewed and strengthened. Though his family wanted him to devote himself to the law instead of wasting time on astronomy, he finally won out.

When Tycho was 20, he dueled with swords with a fellow student. During the duel some of his nose was cut off. For the rest of his life he wore a gold and silver replacement and was forever rubbing the remainder with ointment. Portraits made during his life and the relief on his tomb show a line across his nose, though just how much was actually cut off is not now definitely known.

In 1572, Tycho was astounded to discover a new star in the sky, so bright that it outshone Venus. It was what we now call a "supernova"; indeed, we now call it "Tycho's supernova." It was the explosion of a star, and remained visible in the sky for 18 months. Tycho had been building bigger instruments for measuring positions in the sky better than any of his contemporaries had, and was able to observe very precisely that the supernova was not changing in position. It was thus a real star rather than a nearby object.

He published a book about the supernova, and his fame spread. In 1576, the king of Denmark offered to set Tycho up on the island of Hveen with funds to build a major observatory, as well as various other grants. The following 20 years saw the construction on Hveen of Uraniborg. Later, he built a second observatory on Hveen. Both had huge instruments able to measure angles and positions in the sky to unprecedented accuracy for that pre-telescopic time.

Unfortunately for Tycho, a new king came into power in Denmark in 1588, and Tycho's influence waned. Tycho had always been an argumentative and egotistical fellow, and he fell out of favor in the countryside and in the court. Finally, in 1597, his financial support cut, he left Denmark. Two years later he settled in Prague, at the invitation of the Holy Roman Emperor, Rudolph II.

In 1601, Tycho attended a dinner where etiquette prevented anyone from leaving the table before the baron (or at least Tycho thought so). The guests drank freely and Tycho wound up with a urinary infection. Within two weeks he was dead of it.

2.7 Feature: Archaeoastronomy

The main line of astronomy that has led to today's conception of the universe was largely European. There is currently increasing interest in discovering the extent of astronomical understanding in other parts of the world. Astronomical development in the Orient was substantial, and many written documents were left. From still other parts of the world, documents are fewer and understanding is often demonstrated in astronomical alignments of buildings and other objects. These investigations link archaeology and astronomy, and so are called *archaeoastronomy.* We shall deal here with only three examples: Stonehenge in ancient England, Mayan sites in Central America, and North American sites. In all these cases, the recent work suggests that astronomical knowledge was greater than we might have expected.

2.7a Stonehenge

On Salisbury Plain in southern England, a set of massive stones set on end mark one of the strangest monuments known to humanity. Nowadays the stones stand next to a busy motorway and are a popular tourist site, but once they formed part of a religious site.

Stonehenge (Fig. 2–20) contains a set of giant standing stones weighing 25 tons each. They form a circle surrounding two horseshoe-shaped patterns. The circle is more than 30 meters wide and 4 meters high, and some of the pairs of stones have massive stone crosspieces (lintels) raised 4 meters above the ground. An incredible amount of work went into making this monument. Why did the site merit so much effort?

The realization that the earliest parts of Stonehenge date from 2700 B.C., and the discovery that many of the stones were brought from hundreds of miles away to this particular spot, makes the matter even more interesting. "Why?" is not the only question. How?

Actually, Stonehenge turns out to have been built over a period of many centuries, and has three distinct phases. The oldest period, Stonehenge I, was the date at which certain holes, called "post holes," were made at the site. These holes may date from as long ago as 2700 B.C. In Stonehenge II, from perhaps 2400 to 2100 B.C., and Stonehenge III, the latest period, from about 2000 to 1700 B.C., the largest stones were brought.

It has been realized since 1771 (after having been forgotten for perhaps thousands of years) that at the summer solstice—the day in the year (currently

Figure 2–20 Stonehenge, with its giant stones standing on Salisbury Plain in southern England. It may have been used for astronomical prediction.

Figure 2–21 The Sun rises over the Heelstone on the day of the summer solstice when seen from the center of the ring of stones.

June 21st) when the Sun is the farthest north and the day is the longest—the Sun rises directly over a particular stone, called the Heelstone, as seen from the center of Stonehenge (Fig. 2–21). The Heelstone is located 60 meters outside the outer circle of stones. We think that the weather was better in that region of the world 4000 years ago than it is now, so, fortunately, more days would have been clear enough for astronomical objects to be seen.

Indeed, alignments between less prominent pairs of stones point to sunrise at other significant times of the year, such as the equinoxes, which fall midway between the solstices. All of this points to the fact that the people who built Stonehenge II, and probably Stonehenge I, must have had substantial astronomical knowledge, including ideas of the year and the calendar (Fig. 2–22).

A still more elaborate theory was put forward in the 1950's by Gerald Hawkins, an American astronomer, and elaborated upon by Fred Hoyle, a British astronomer. (Having astronomers step in dismayed some professional archaeologists.) Outside the circle of the largest stones (which date from Stonehenge III) are Stonehenge I objects, a circle of white chalk-filled holes called the Aubrey Holes after John Aubrey, who discovered them in 1666. There were originally 56 of these holes, placed on the circumference of a circle 85 meters across. Hawkins and Hoyle pointed out that the Aubrey Holes could have been used as aids in counting in order to predict eclipses of the Sun and the Moon. Eclipses are awesome sights and must have been particularly fearful to people in the neolithic period. So a method of predicting eclipses would have been very valuable. The task was made all the more difficult for the Stonehenge people by the fact that many eclipses, even though correctly predicted, would not have been visible at Stonehenge, either because they occurred over different parts of the Earth or because the day was not clear.

Figure 2–22 This tiny sundial models and mocks Stonehenge as a pocket watch: "Stonehenge: 5,000 years old and still ticking! At last, you can predict an eclipse and tell the local apparent time with this beautiful pocket time piece. Delight your more erudite friends and amaze your druid neighbors," runs the advertisement.

2.7b Mayan Astronomy

About 1000 years ago, the Mayan people had a rich culture in what is now the Yucatán peninsula of Mexico and Guatemala, as can be seen from such sites as Chichen Itzá in Mexico and Tikal in Guatemala. Many features of their cities and buildings are aligned to astronomical phenomena (Fig. 2–23). Information from this period comes not only from archaeological measurement and excavation of buildings but also from the three remaining Mayan books, or codices. (All other Mayan books, incredibly, were burned by the Europeans who came later on.)

Besides setting their calendar by the Sun and the Moon, the Maya also based their calendar on the rising and setting of Venus. Though they could not observe the phases directly (after all, the phases were discovered by Galileo

Figure 2–23 The Caracol at Chichen Itzá in the Yucatán. Alignments involving windows, walls, and the horizontal shafts at the top of the building seem to point to the location of the Sun and Venus on significant days.

Figure 2–24 The Bighorn Medicine Wheel in Wyoming. Lines drawn between pairs of the major points extend toward such astronomical directions as those of the solstice sunrise and sunset and the rising positions of the stars Aldebaran, Rigel, and Sirius.

after he developed the telescope), they could tell from the positions of Venus in the sky—and by the dates of its heliacal rising—that it had a periodic cycle.

This is all set out in detail in one of the surviving Mayan books, the Dresden Codex, from at least 1000 years ago. Though we do not know how to read the text, we can read the numbers. Two large sections that contain eclipse tables and Venus tables have been deciphered. They show the Maya's high level of sophistication in mathematics and astronomy.

2.7c Native American Astronomy

In recent years, the American west has been searched for signs of Native American interest in astronomy. Evidence has been found, for example, that Native Americans may have observed the explosion of a star—a supernova—in A.D. 1054, an event that was recorded in China and the Middle East but, surprisingly, for which no European sightings are known. But the Native American evidence consists of drawings (one of which appears as Figure 26–5) and is ambiguous, so this matter is considered open.

In the Great Plains that lie on the eastern edge of the Rocky Mountains, certain Native American monuments can be found that, like Stonehenge, show important astronomical alignments. For example, in Wyoming there exists a "medicine wheel" (Fig. 2–24), the term "medicine" having connotations of magic. John A. Eddy of the High Altitude Observatory in Boulder, Colorado, has investigated the astronomical significance of the Bighorn Medicine Wheel. His analysis has shown that lines drawn between significant markings point not only to the sunrise and sunset on the day of the summer solstice but also toward the rising points of the three brightest stars that rise shortly before the Sun (have their heliacal risings) in the summer. This site probably dates from A.D. 1400 to 1700.

About 50 medicine wheels are known, ranging in size from a few meters to a hundred meters across. Some are known to be several thousand years old. And some of these show the same alignment as the Bighorn site.

As with other archaeoastronomical sites, there is always the question of whether these alignments were made intentionally or by accident. Statistical arguments have been made that indicate that it is unlikely for so many coincidences to occur. More work is going on.

Chaco Canyon, New Mexico, was the center of a complex culture developed by the prehistoric Pueblo Native Americans about A.D. 1100. On a high butte, they carved a 9 1/2-turn spiral and nearby a smaller 2 1/2-turn spiral on a cliff behind three large stone slabs. A fortuitous discovery showed that on

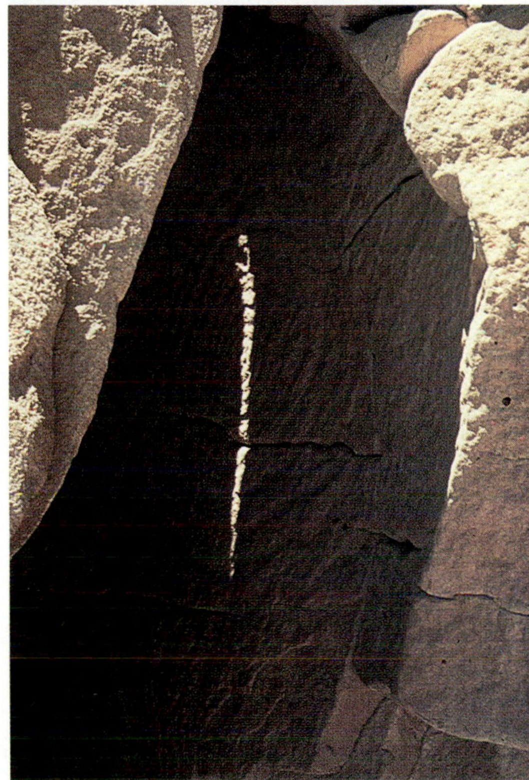

Figure 2–25 The Anasazi Sun dagger, a dagger of light that used to pass through the center of the spiral on the day of the summer solstice. The most recent observations, however, show it slightly displaced.

A

B

Figure 2–26 (*A*) A small section from the long calendar stick of the Winnebago tribe from the early 19th century. We see the opening month of the second year. Reading from right to left, we see two vertical lines with dots between them, marking two days on which the Moon was invisible, and then the new crescent Moon, facing in the orientation it has to the eye. We also see the month divided into three periods—waxing (illuminated portion growing larger), full, and waning (illuminated portion growing shorter)—and a last crescent just before the beginning of the next month. Alexander Marshack, in his analysis, shows the richness of the tribe's astronomical understanding. The way the month and the year are structured shows that it may stem from a long-previous Asian base. (*B*) Detail of the beginning of a month. (Copyright © Alexander Marshack)

days close to the summer solstice near midday, when the Sun is at its highest possible position in the sky, a dagger-shaped light pattern descends vertically through the center of the larger spiral (Fig. 2–25). At the equinoxes (three months earlier or later), the Sun's altitude is significantly lower and the vertical path of this "dagger" has shifted well to the right of the center of the spiral. A second, smaller, light pattern passes through the center of the smaller spiral at those times. At the winter solstice, when the Sun is at its lowest possible midday altitude, the two "daggers" have shifted further to the right so that they frame the larger spiral. These patterns of sunlight are formed by the openings between the stone slabs. (A recent shift of the middle slab by about 5 cm has altered these patterns significantly, showing how sensitive they were to the shape and placement of the slabs. This movement of the slab seems to have been caused by erosion around the base, accelerated by the recent increase in visitors to the site.)

Anna Sofaer, who discovered the "Sun Dagger," and her colleague Rolf Sinclair of the National Science Foundation, continue to study the site. They have shown that the full set of light patterns on the spiral reflects the 18.6-year cycle of the Moon as well as the yearly cycle of the Sun. She has also found other astronomical knowledge displayed in buildings constructed around Chaco Canyon. The idea that the prehistoric Pueblo Native Americans had such astronomical knowledge is consistent with the keen interest shown in the Sun and Moon by their descendents. Michael Zeilik, of the University of New Mexico, has argued that one must interpret such a site with caution. In agreement with Anthony Aveni of Colgate University, he points out that since there are a number of possible indications of the solar and lunar cycles, some may have been incorporated by chance.

The astronomical significance of the "Sun Dagger" site and its role in Chaco culture continues to be discussed. A number of the major ceremonial buildings are oriented to the cardinal directions or to other directions of astronomical interest. These same directions appear in the bearings between buildings and in the layout of major roadways. Also, some unusual doorways and windows that exist may have been used for observations to anticipate important astronomical events like the solstices.

Elsewhere, from the Winnebago Native Americans, the discovery of a calendar stick showing the effect of detailed astronomical record-keeping (Fig. 2–26) provides a 19th-century example of the richness of astronomical knowledge and of the accuracy in record-keeping in a North American Native American tribe.

Summary and Outline

Egyptian astronomy (Section 2.1)
- Astronomy served the practical purposes of the farmers

Greek astronomy (Section 2.2)
- The apparent motion of the planets; retrograde motion
- Geocentric theory of Aristotle and Ptolemy
- Epicycles, deferents, equants needed to explain the planetary motions

Chaldean, Chinese astronomy (Section 2.3)

Medieval astronomy (Section 2.4)
- Interpreting Aristotle
- Arab translation of Ptolemy's *Almagest*

Heliocentric theory (Section 2.5)
- Copernicus, *De Revolutionibus* published in 1543
- Earlier heliocentric idea of Aristarchus of Samos was not widely accepted
- Retrograde motion explained as a projection effect
- Copernican system still used circular orbits and epicycles
- Copernicus preferred his system on philosophical grounds rather than superiority of his predictions, because ancient observations were still used and Copernican system used circular orbits that didn't always fit the data.

Tycho Brahe (Section 2.6)
- Amassed the best set of observations that had ever been obtained during years prior to 1600

Archaeoastronomy (Section 2.7)
- Other civilizations worked out fairly complicated astronomical rules

Key Words

heliacal rising, ecliptic, retrograde motion, *primum mobile*, epicycles, deferents, eccentric, equant, gnomon, parallax, conjunction, opposition, ephemeris, heliocentric, geocentric, Foucault pendulum, archaeoastronomy

Questions

1. Discuss the velocity that a planet must have around its epicycle in the Ptolemaic theory with respect to the velocity that the epicycle has around the deferent if we are to observe retrograde motion.
2. Discuss the differences between the theories of Aristotle and Ptolemy.
3. (a) Imagining you lived on Saturn, describe the phases that the Earth would seem to go through on the Ptolemaic system. (b) Now describe the phases that the Earth would seem to go through on the Copernican system.
4. (a) Imagining you lived on Saturn, describe the phases that Pluto would seem to go through on the Ptolemaic system. (b) Now describe the phases that Pluto would seem to go through on the Copernican system.
5. What are (a) epicycle; (b) deferent; (c) eccentric?
6. Draw a diagram showing the positions of the Earth, the Sun, and Mars to show (a) Mars in conjunction and (b) Mars in opposition.
7. Could Eratosthenes have carried out his measurement using a gnomon at your latitude? Why or why not?
8. Relate measurements of Eratosthenes to the myth that it was not known at the time of Columbus that the Earth is round.
9. Discuss the following statement: "With the addition of epicycles, the geocentric theory of the solar system could be made to agree with observations. Since it was first around, and therefore better known, it should have been kept."
10. Discuss the significance of a Foucault pendulum. Describe how and why it would appear to change at the equator and at the north pole.
11. Do the astronomical alignments of Stonehenge, if established, show that the people who constructed Stonehenge III, the large stones, were astronomically sophisticated? Explain.
12. Discuss two signs of astronomy prior to A.D. 1500 in the Americas.

The frontispiece to Galileo's *Dialogue Concerning the Two World Systems* (1632), which led to his being taken before the Inquisition since he had supposedly been warned in 1615 not to teach the Copernican theory. According to the labels, Copernicus is to the right, with Aristotle and Ptolemy at the left; Copernicus was drawn with Galileo's face, however.

The Origin of Modern Astronomy

3

Aims: To follow how modern astronomy developed during the 17th century through the ideas of Kepler, Galileo, Newton, and Halley

The 17th century was a fertile time in the history of ideas. The Renaissance was under way in Italy. In what was to become the United States, Jamestown was settled in Virginia in 1609 and the Pilgrims landed on Cape Cod and Plymouth Rock in 1620.

The intellectual ferment of the 17th century set the tone for science up to our time. Aristotle's ideas of the nature of our physical world were overturned. Kepler's discovery that the planets do not orbit the Sun in circles shattered ancient ideas and led to what we now call "the Copernican revolution." Galileo peered farther into the Universe than had ever been done and started methods of inquiry we still use. And Newton (spurred on by his colleague Halley) developed the laws of physics we still treat as basic.

In this chapter, we consider the period from Kepler's starting work on the orbit of Mars in 1600, through Galileo's use of the telescope in 1609, through Newton's publication of his major work in 1687 with the aid of Halley. This period marks the origin of what we might call the modern era of astronomy; historical developments since that time are treated in this text in the context of the various separate topics.

3.1 Johannes Kepler

Johannes Kepler, as a young mathematician in the late 16th century, had a mystical view of the universe. He had studied with a professor who was one of the first to believe in the Copernican view of the universe, and came to believe in this heliocentric view. Kepler wanted to discover the cause of the regularity in the orbits of the planets. While he was teaching a high-school class in 1595, it occurred to him that the ancient Greek discovery that there are five known geometrical solid figures might be tied into explaining the orbits of the planets. Each of the five regular figures—the tetrahedron (pyramid), the cube, the octahedron (8 faces), the dodecahedron (12 faces), and the icosahedron (20 faces)—could be enclosed in a sphere. Motivated by a search for celestial harmony, Kepler sugggested that each of these regular solids came between each pair of the spheres that carried Copernicus's six planets (Fig. 3–1).

Though Tycho did not accept Kepler's idea, he was impressed with Kepler's mathematical and astronomical skill. When Tycho came to Prague, he invited Kepler to join him. Tycho's observational data showed that the tables then in use did not adequately predict planetary positions. Kepler started to study them in 1600, and carried out detailed numerical calculations. (Nowadays we could use a computer to calculate in seconds results that took Kepler weeks to work out.) Kepler was eventually able to make sense out of the observations of Mars that Tycho had made, and thus clear up the discrepancies between

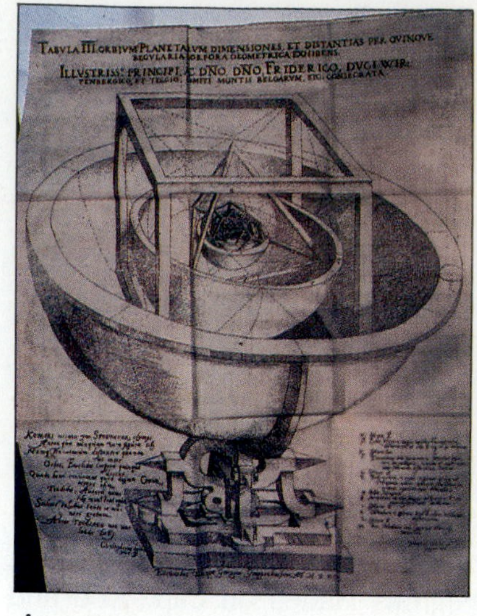

A

B

Figure 3–1 (*A*) Kepler, in his *Mysterium Cosmographicum* (1596), suggested that since there were six known planets and five regular solids known from Greek geometry, the planets could be moving on unseen spheres separated by these nested solids. (These solids are regular in that all their sides are the same: the pyramid is formed of 4 triangles, the cube is formed of 6 squares, the octahedron is formed of 8 triangles, the dodecahedron is formed of 12 pentagons, and the icosahedron is formed of 20 triangles. No other regular solids exist.) The solids were centered on the Sun; Kepler was already a Copernican. Ultimately, this stage of his search for celestial harmony proved unsatisfactory, and he proceeded to discover his laws of orbital motion. (*B*) A closeup of the central portion.

the predictions and the observations of Mars's position in the sky. In 1609, shortly before Galileo was to first turn a telescope on the sky, Kepler published in his book *Astronomia Nova* (*The New Astronomy*) the first two of his three laws based on his empirical analyses, that is, laws based on experience and observation rather than theory. His third law followed in 1618 in his book *Harmonices Mundi* (*Harmonies of the World*). Kepler and Galileo were contemporaries, and even corresponded (though Galileo did not answer all of Kepler's letters). But their influence on the development of each other's scientific thought was small.

3.1a Kepler's First Law (1609)

Kepler's first law, published in 1609, says that **the planets orbit the Sun in ellipses, with the Sun at one focus**. Even Copernicus had assumed that the planets followed "perfect" orbits, namely, circles, though he had included the admittedly imperfect Earth as a planet. The discovery by Kepler that the orbits were in fact ellipses greatly improved the accuracy of the calculations. Kepler also concluded that the planets were like the Earth in being imperfect and made of matter.

An ellipse is a curve defined in the following way: choose any two points on a plane; these points are called the *foci* (each is a *focus*). From any point on the ellipse, we can draw two lines, one to each focus. The sum of the lengths of these two lines is the same for each point on the ellipse.

It is easy to draw an ellipse (Fig. 3–2). Put two nails or thumbtacks in a piece of paper, and link them with a piece of string that has some slack in it. If you pull a pen around while the pen keeps the string taut, the pen will necessarily trace out an ellipse. The string doesn't change in length, so the sum of the lengths of the lines from the pen to the foci equals the length of the string, which is constant. The shape of the ellipse will change if you change the length of the string, or if you change the distance between the foci.

The *major axis* of an ellipse is the line within the ellipse that passes through the two foci, or the length of that line (Fig. 3–3). We often speak of the *semimajor axis*, which is just half the length of the major axis ("semi-" is a prefix from the Greek that means "half"). The *minor axis* is the part of the line lying within the ellipse that is drawn perpendicular to the major axis and

A B

C D

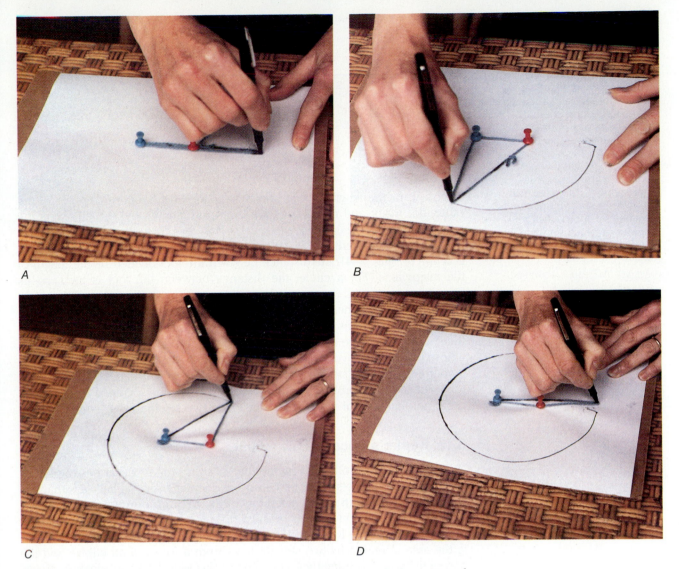

Figure 3–2 It is easy to draw an ellipse. Put two nails or thumbtacks in a piece of paper, and link them with a piece of string that has some slack in it. If you pull a pen around while the pen keeps the string taut, the pen will necessarily trace out an ellipse. It does so since the string doesn't change in length and is equal to the sum of the lengths of the lines from the point to the two foci. That this sum remains constant is the defining property of an ellipse. The shape of the ellipse will change if you change the length of the string, or if you change the distance between the foci.

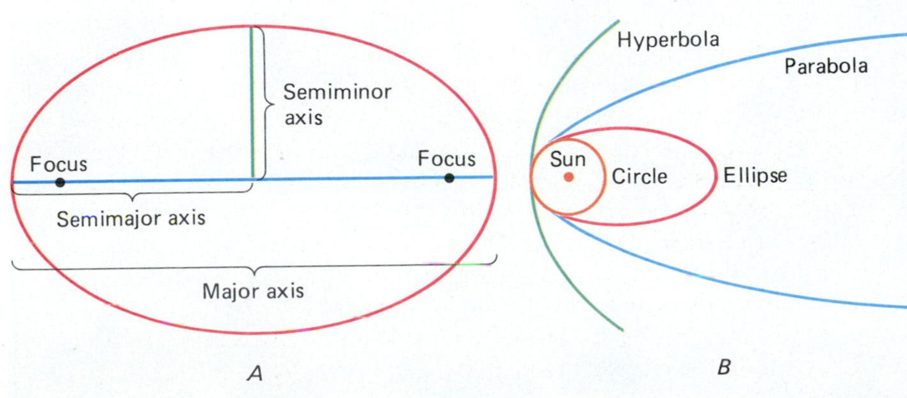

Figure 3–3 (*A*) The parts of an ellipse. (*B*) The ellipse shown has the same *perihelion* distance (closest approach to the Sun) as does the circle. Its *eccentricity*, the distance between its foci divided by its major axis, is 0.8. If the perihelion distance is kept constant but the eccentricity is allowed to reach 1, then we have a parabola. For eccentricities greater than 1, we have hyperbolas.

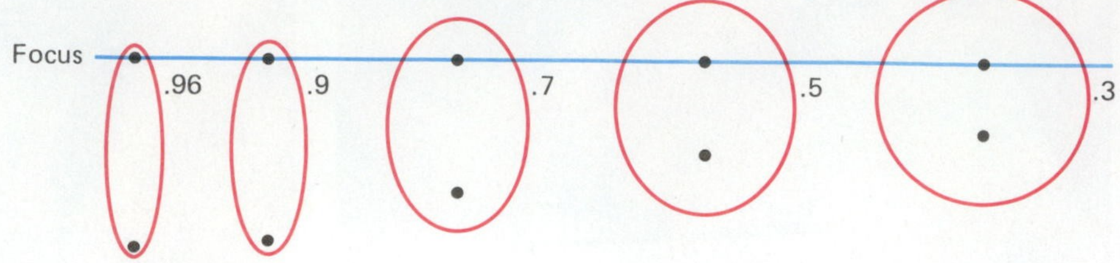

Figure 3–4 A series of ellipses of the same major axis but different eccentricities. The *foci* are marked; these are the two points inside with the property that the sum of the distances from any point on the circumference to the foci is constant. As the eccentricity—distance between the foci divided by the major axis—approaches 1, the ellipse approaches a straight line. As the eccentricity approaches zero, the foci come closer and closer together. A circle is an ellipse of zero eccentricity.

bisects it, or the length of that line. When one of the foci lies on top of the other, the major and minor axes are the same length, and the ellipse is the special case that we call a *circle.*

We often describe how far an ellipse is "out of round" by giving its *eccentricity* (Fig. 3–4), the distance between the foci divided by the length of the major axis. For a circle, the distance between the foci is zero, so its eccentricity is zero.

Ellipses (and therefore circles), parabolas, and hyperbolas are *conic sections,* that is, sections of a cone (Fig. 3–5). If you slice off the top of a cone, the shape of the slice is an ellipse. (If you cut perpendicularly to the axis, the ellipse is a circle.) If you cut parallel to the side of the cone, a parabola results. If your cut is tipped still further over, a hyperbola results.

For a planet orbiting the Sun, the Sun is at one focus of the elliptical orbit; nothing special marks the other focus (we say that it is "empty").

3.1b Kepler's Second Law (1609)

The second law, also known as the *law of equal areas,* describes the speed with which the planets travel in their orbits. Kepler's second law says that **the line joining the Sun and a planet sweeps through equal areas in equal times**. When a planet is at its greatest distance from the Sun in its elliptical orbit, the line joining it with the Sun sweeps out a long, skinny sector. (A *sector* is the area bounded by two straight lines from a focus of an ellipse and the part of the ellipse joining their outer ends.) The planet travels relatively slowly in this part of its orbit. By Kepler's second law, the long skinny sector must have the same area as the short, fat sector formed in the same period of time when the planet is closer to the Sun (Fig. 3–6). The planet thus travels faster in its orbit when it is closer to the Sun.

The uniformity of motion of a planet in its orbit had been an important philosophical concept since the time of the Greeks. Kepler's second law re-

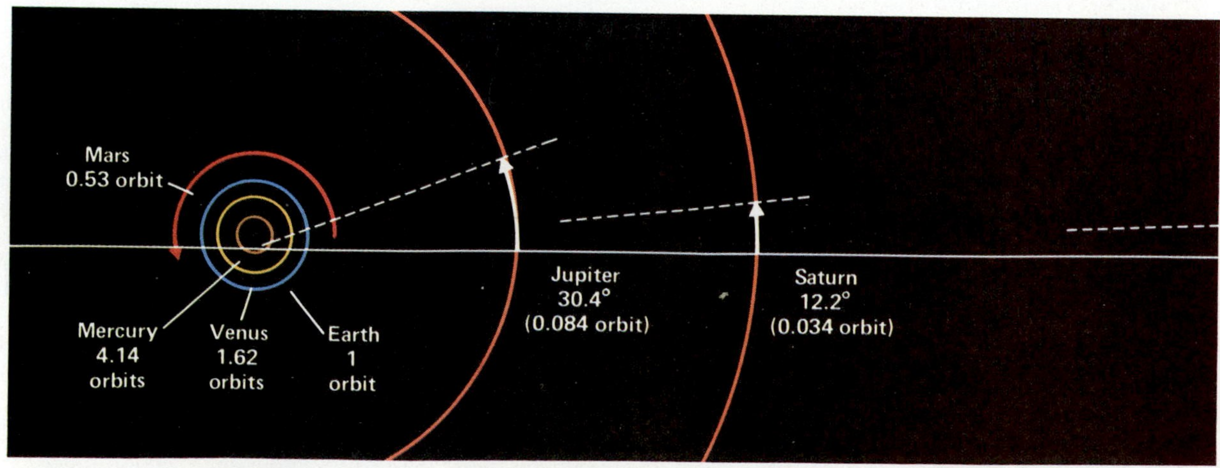

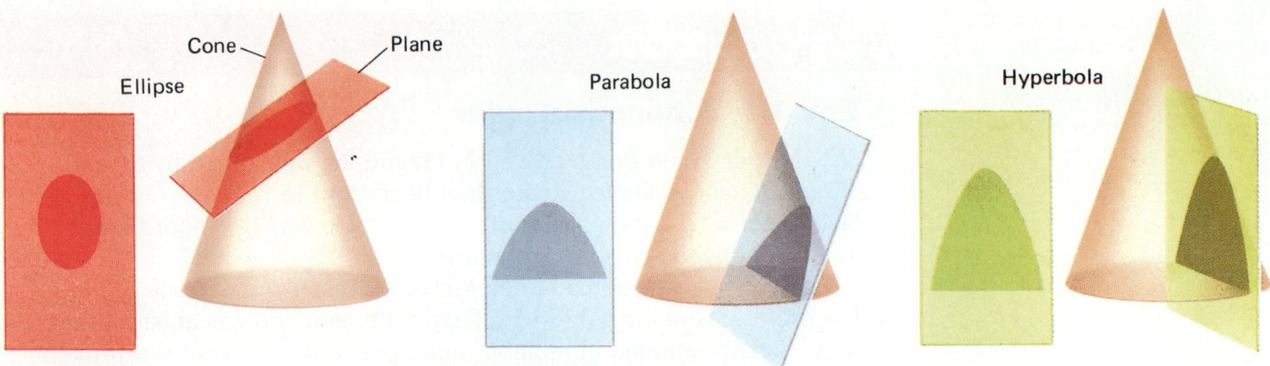

Figure 3–5 An ellipse is a conic section, in that the intersection of a cone and a plane that passes through the sides of a cone (and not the bottom) is an ellipse. If the plane is parallel to an edge of the cone, a parabola results. If the plane is tipped further over so that it is neither parallel to the edge nor intersects the side, then a hyperbola results. So parabolas and hyperbolas are conic sections as well.

placed the uniform rate of rotation with a uniform rate of sweeping out area, maintaining the concept that something was uniform.

Kepler's second law is especially noticeable for objects with very elliptical orbits such as comets. It describes why Halley's Comet sweeps so quickly, within months, through the inner part of the solar system, and takes the rest of its 76-year period moving slowly through the outer parts.

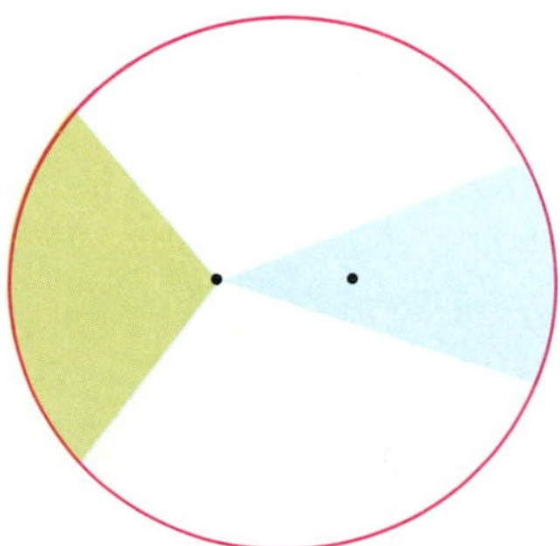

Figure 3–6 Kepler's second law states that the two shaded sectors, which represent the areas covered by a line drawn from a focus of the ellipse to an orbiting planet in a given length of time, are equal in area. The Sun is at this focus; nothing is at the other focus.

3.1c Kepler's Third Law (1618)

Kepler's third law deals with the length of time a planet takes to orbit the Sun, called its *period* of revolution. It relates the period to some measure of the planet's distance from the Sun (Fig. 3–7).

Kepler's third law says that **the square of the period of revolution is proportional to the cube of the semimajor axis of the ellipse**. That is, if the cube of the semimajor axis goes up, the square of the period goes up

Figure 3–7 Kepler's third law relates the period of an orbiting body to the size of its orbit. The outer planets orbit at much slower velocities than the inner planets and also have a longer path to follow in order to complete one orbit. The distances the planets travel in their orbits in 1 year are shown here.

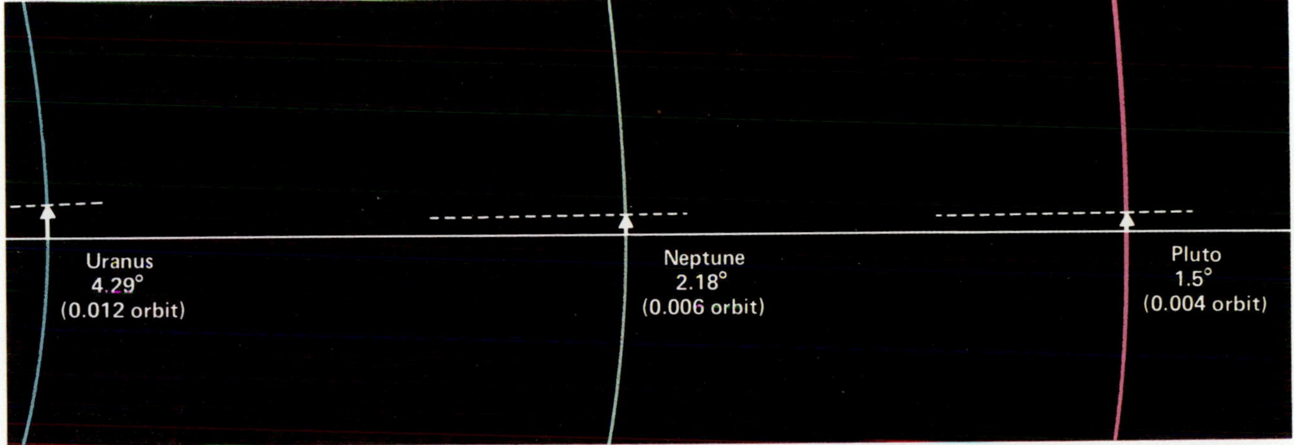

Focus On

Box 3.1 Johannes Kepler

Shortly after Tycho Brahe arrived in Prague, he was joined by an assistant, Johannes Kepler. Kepler had been born in 1571 in Weil der Stadt, near what is now Stuttgart, Germany. He was of a poor family, and had gone to school on scholarships.

At the university, he had been fortunate to study under one of the few professors in the world who taught the new Copernicanism. Kepler was officially enrolled in religion, but a position opened to teach mathematics, and Kepler thus became a teacher. Four years later all Lutheran teachers were expelled from Graz, Kepler among them, and in 1600 Kepler joined Tycho Brahe in Prague.

Kepler was eager to work with the great man's data, though he had trouble prying them out of Tycho. Tycho, on his part, recognized Kepler's promise and wanted his assistance. This common need overcame their mutual friction, though their relationship was stormy.

At Tycho's death, Kepler succeeded him in official positions. Most important, after a battle over Tycho's scientific collections, he took over Tycho's data.

In 1604, Kepler observed a supernova and wrote a book about it. It was the second supernova to appear in 32 years, and was also seen by Galileo. But Kepler's study has led us to call it "Kepler's supernova." We have not seen another supernova in our galaxy since.

By the time of the supernova, through analysis of Tycho's data, Kepler had discovered what we now call his second law. He was soon to find his first law. It took four additional years, though, to arrange and finance publication. Much later on, he found his third law.

In 1626, Kepler published a set of tables, called the Rudolphine Tables, after his patron, the Holy Roman Emperor, Rudolph II. These, of course, were based on the knowledge of his three laws and were thus much more accurate than previous tables.

The abdication of Kepler's patron, the Emperor, led Kepler to leave Prague and move to Linz. Much of his time had to be devoted to financing the printing of the tables and his other expenses—and to defending his mother against the charge that she was a witch. He died in 1630, while travelling to collect an old debt.

Figure 3–8 The frontispiece of Kepler's *Rudolphine Tables* shows Hipparchus, Ptolemy, Tycho Brahe, and Copernicus as pillars of astronomy.

by the same factor. It is often easiest to express Kepler's third law by relating values for a planet to values for the Earth. If P is the period of revolution of a planet and R is the semimajor axis,

$$\frac{P^2_{\text{Planet}}}{P^2_{\text{Earth}}} = \frac{R^3_{\text{Planet}}}{R^3_{\text{Earth}}}$$

We can choose to work in units that are convenient for us on the Earth. We call the semimajor axis of the Earth's orbit–the average distance from the Sun to the Earth–*1 Astronomical Unit* (1 A.U.). Similarly, the unit of time that the Earth takes to revolve around the Sun is defined as *one year.* Using these values, Kepler's third law appears in a simple form, since the numbers on the bottom of the equation are just 1.

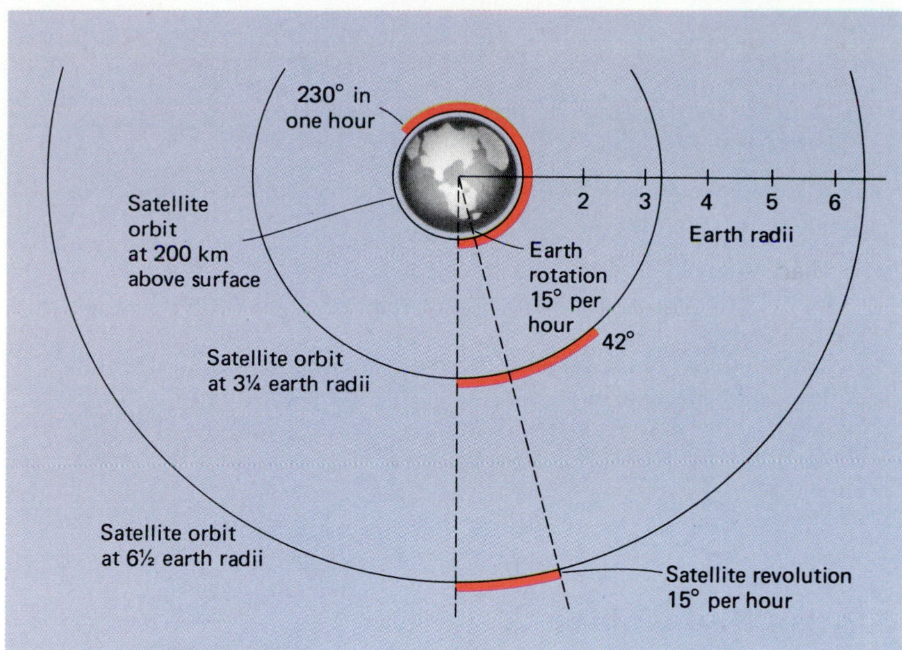

Figure 3–9 Most satellites, including the space shuttles, are only 200 km or so above the Earth's surface, and orbit the Earth in about 90 minutes. From Kepler's third law we can see that the velocity in orbit of a satellite decreases as the satellite gets higher and higher. At about 6½ Earth radii, the satellite orbits at the same velocity as that at which the Earth's surface rotates underneath. Thus the satellite is in *synchronous rotation,* and always remains over the same location on Earth. This property makes such synchronous satellites useful for monitoring weather or for relaying communications.

Example: If we know from observation that Jupiter takes 11.86 years to revolve around the Sun, we can find Jupiter's distance from the Sun in the following way:

$$\frac{(11.86)^2}{P^2_{\text{Earth}}} = \frac{R^3_{\text{Jupiter}}}{R^3_{\text{Earth}}}, \text{ and}$$

$$\frac{(11.86)^2}{1} = \frac{R^3_{\text{Jupiter}}}{1}.$$

Nowadays, we can calculate R easily with a pocket calculator. But astronomers are often content with approximate values that can be calculated in your head. For example, $(11.86)^2$ can be rounded off to $12^2 = 144$. We can then easily find the cube root of 144 by trial-and-error. It is not hard to calculate that $4^3 = 64$ (which is too small), $5^3 = 125$ (which is too small), and $6^3 = 216$ (which is too large). So the semimajor axis of Jupiter's orbit around the Sun is between 5 and 6, and is a little over 5 A.U. The actual value is 5.2 A.U.

We have used Kepler's third law to determine how the period of a planet revolving around the Sun is related to the size of its orbit. But how do we find the constant of proportionality between P^2 and R^3, that is, the number by which R^3 must be multiplied to get P^2 if we don't do the simplification of comparing with the values for the Earth? In the next section, we shall see that Isaac Newton derived mathematically that the constant depends on the mass of the central body, which is the Sun in this case.

Kepler used his laws to compute a set of tables, named the *Rudolphine Tables* (Fig. 3–8) after the Emperor Rudolph of the Holy Roman Empire, its sponsor. The *Rudolphine Tables* were printed in 1627.

The period of revolution of satellites around other bodies follows Kepler's laws as well (Fig. 3–9). The laws, with different constants of proportionality, apply also to artificial satellites in orbit around the Earth (Fig. 3–10) and to the moons of other planets.

Figure 3–10 In this time exposure, the stars trail but the series of synchronous satellites hovering over the Earth's equator appear as points.

To remember that it is the period (a measure) of time that is squared rather than cubed, think of Times Square in New York.

Focus On

Box 3.2 *Summary of Kepler's Laws of Planetary Motion*

To this day, we consider Kepler's laws to be the basic description of the motions of solar system objects.

1. The planets orbit the Sun in ellipses, with the Sun at one focus.
2. The line joining the Sun and a planet sweeps through equal areas in equal times.
3. The square of the period of a planet is proportional to the cube of the semimajor axis of its orbit—half the longest dimension of the ellipse. (Loosely: The squares of the periods are proportional to the cubes of the planets' distances from the Sun.)

3.2 Galileo Galilei

As art, music, and architecture began to flourish after the Middle Ages, astronomy also developed. Early in this Renaissance, in 1500, Leonardo da Vinci, for example, deduced that when we are able to see the dark part of the Moon faintly lighted, it is because of *earthshine,* the reflection of sunlight off the Earth and over to the Moon.

One hundred years later, also in what is now Italy, the scientist Galileo Galilei (Fig. 3–11) began his scientific work. Galileo began to believe in the Copernican system in the 1590's, and later provided important observational confirmation of the theory. In late 1609 or early 1610, simultaneously with the first settlements in the American colonies, Galileo was the first to use a telescope for astronomical observation.

In his book *Sidereus Nuncius* (*The Starry Messenger*), published in 1610, he reported that with this telescope he could see many more stars than he could with the naked eye, and could see that the Milky Way and certain other hazy-appearing regions of the sky actually contained individual stars. He described views of the Moon (Fig. 3–12), including the discovery of mountains, craters, and the relatively dark regions he called (and that we still call) *maria* (pronounced mar′ee-a), seas. And, perhaps most important, he discovered that small bodies revolved around Jupiter (Figs. 3–13 and 3–14). This discovery proved that all bodies did not revolve around the Earth, and also, by displaying something that Aristotle and Ptolemy obviously had not known about, showed that Aristotle and Ptolemy were not omniscient.

Figure 3–11 Galileo.

Subsequently, Galileo found that Saturn had a more complex shape than that of a sphere (though it took better telescopes to actually show the rings). He was one of several people who almost simultaneously discovered sunspots, and he studied their motion on the surface of the Sun. Galileo also discovered that Venus went through an entire series of phases (Fig. 3–15). This could not be explained on the basis of Ptolemaic theory (Fig. 3–16), because if Venus travelled in an epicycle located between the Earth and the Sun, Venus should always appear as a crescent. Also, more generally, Galileo's observations showed that Venus was a body similar to the Earth and the Moon in that it received light from the Sun rather than generating its own.

But although the Roman Catholic Church was not concerned about models made to explain observations, it was concerned about assertions that those

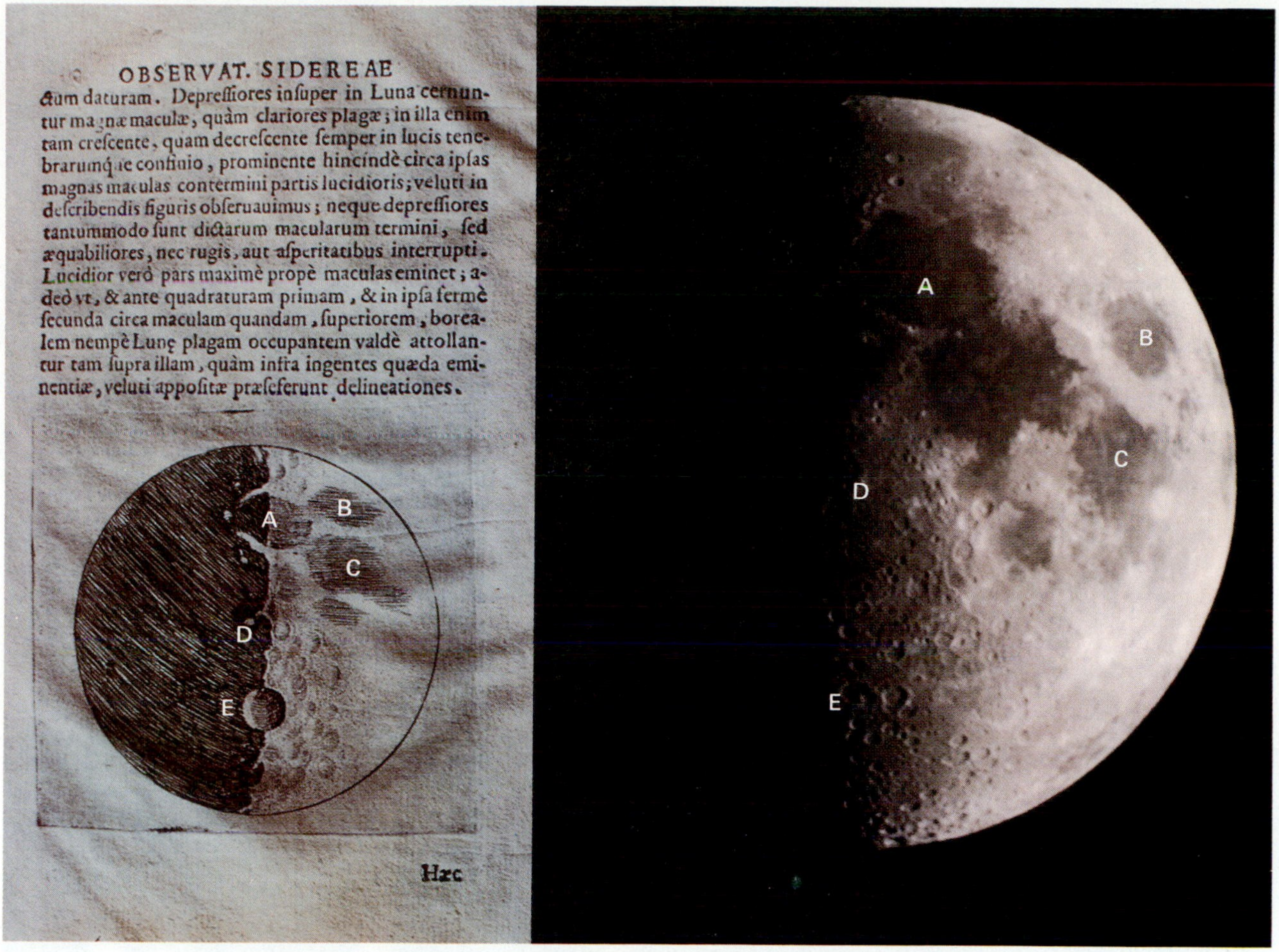

OBSERVAT. SIDEREAE

ctum daturam. Depreſſiores inſuper in Luna cernuntur magnæ maculæ, quàm clariores plagæ; in illa enim tam creſcente, quam decreſcente ſemper in lucis tenebrarumq; ae confinio, prominente hincindè circa ipſas magnas maculas contermini partis lucidioris; veluti in deſcribendis figuris obſeruauimus; neque depreſſiores tantummodo ſunt dictarum macularum termini, ſed æquabiliores, nec rugis, aut aſperitatibus interrupti. Lucidior verò pars maximè propè maculas eminet; adeò vt, & ante quadraturam primam, & in ipſa fermè ſecunda circa maculam quandam, ſuperiorem, borealem nempè Lunę plagam occupantem valdè attollantur tam ſupra illam, quàm infra ingentes quæda eminentiæ, veluti appoſitæ præſeferunt delineationes.

A B C D E

Hæc

A B C D E

Figure 3–12 One of Galileo's original drawings of the Moon, made with his small, unmounted (and therefore shaky) telescope, compared with a modern photograph. Notice how well Galileo did.

Thomas Harriot in England had observed the Moon with a telescope a few months before Galileo did, but did not report the features. It seems reasonable that Galileo, unlike Harriot, had been sensitized to interpreting surfaces and shadows by the Italian Renaissance and by his related training in drawing.

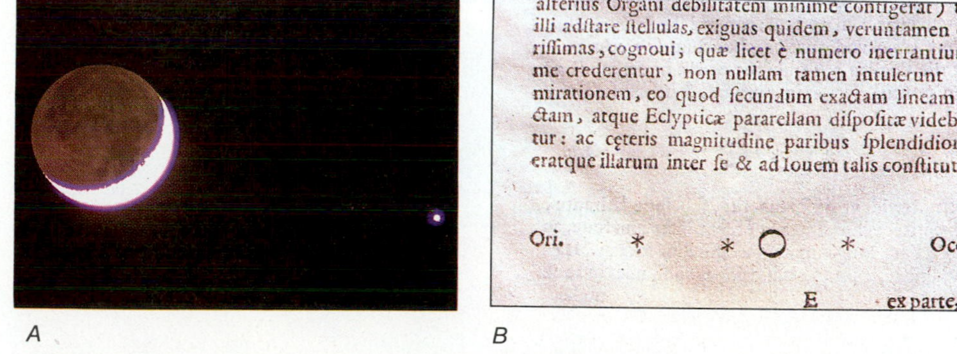

alterius Organi debilitatem minime contigerat) tres illi adſtare ſtellulas, exiguas quidem, veruntamen clariſſimas, cognoui; quæ licet è numero inerrantium à me crederentur, non nullam tamen intulerunt admirationem, eo quod ſecundum exactam lineam rectam, atque Eclypticæ pararellam diſpoſitæ videbantur: ac cęteris magnitudine paribus ſplendidiores: eratque illarum inter ſe & ad Iouem talis conſtitutio.

Ori. * * O * Occ.

E ex parte,

A B

Figure 3–13 (*A*) The crescent moon and Jupiter. Ganymede and Io are visible to the upper left of Jupiter, Europa is just below Jupiter, and Callisto is to the lower right. The crescent moon is over-exposed; earthshine illuminates the dark side of the moon. (*B*) Asterisks indicate the moons of Jupiter on this page of Galileo's *Sidereus Nuncius* (1610). Directions are east (orient) and west (occident).

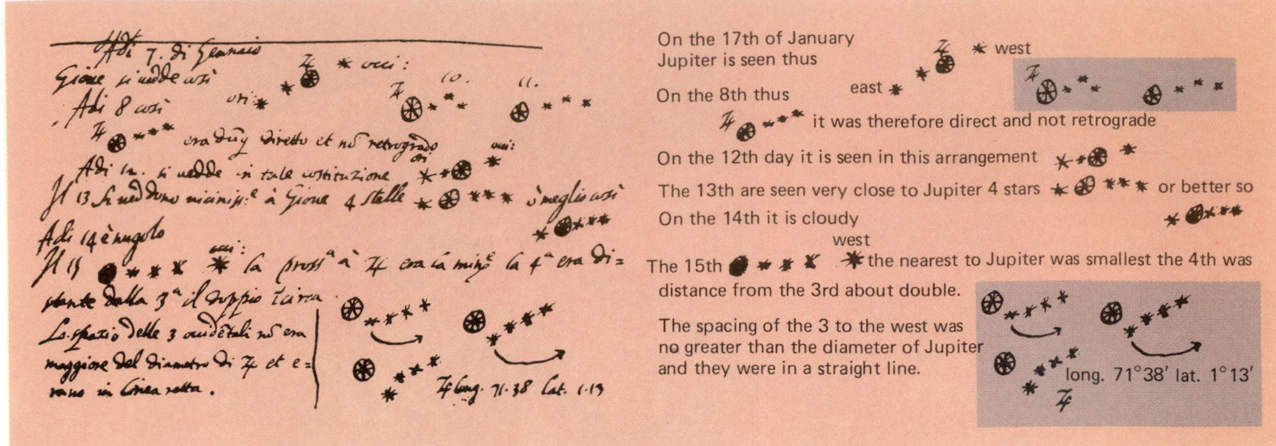

Figure 3–14 A translation (*right*) of Galileo's original notes (*left*) summarizing his first observations of Jupiter's moons in January 1610. The shaded areas were probably added later. It had not yet occurred to Galileo that the objects were moons in revolution around Jupiter.

A

B

Figure 3–15 The phases of Venus. (*A*). Venus is a crescent only when it is relatively close to the Earth, so it appears relatively large. (*B*) The phases of Venus at eight-day intervals in 1987. Note the change in size of Venus with its phase.

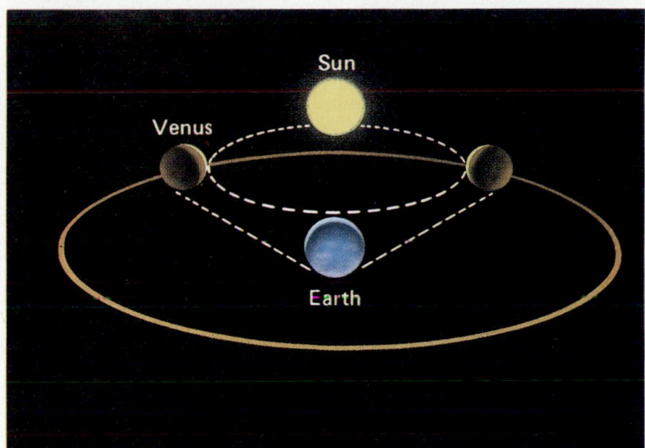

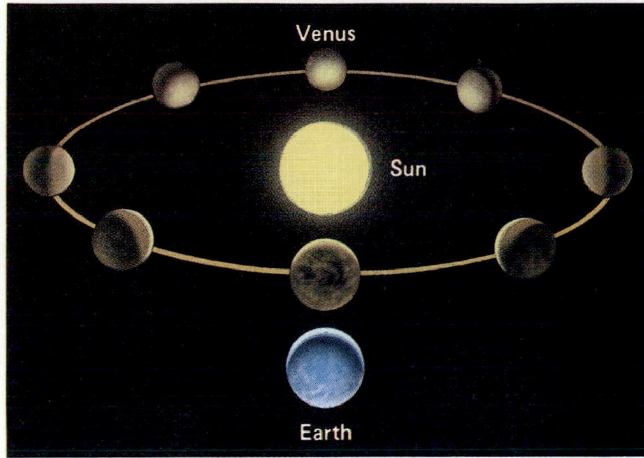

Figure 3–16 In the Ptolemaic theory (*top*), Venus and the Sun both orbited the Earth, but because it is known that Venus never gets far from the Sun in the sky, the center of Venus's epicycle was restricted to always fall on the line joining the center of the deferent and the Sun. In this diagram Venus could never get farther from the Sun than the region restricted by the dotted lines. Thus, Venus would always appear as a crescent, though before the telescope was invented, this could not be verified. In the heliocentric theory (*bottom*) Venus is sometimes on the near side of the Sun, where it appears as a crescent, and it is sometimes on the far side, where we can see half or more of Venus illuminated. This agrees with Galileo's observations, a modern version of which appears in the previous figure.

models represented physical truth. It is difficult to say how much the Church feared that Galileo would generalize his ideas to a philosophical and religious level. In his old age, Galileo was forced by the Inquisition to recant his belief in the Copernican theory. The controversy has continued to the present day. Even now, the Vatican has a group studying how Galileo was treated. Pope John Paul II set up the group in 1979 "in loyal recognition of wrongs from whatever side they come ... dispel the mistrust that still opposes, in many minds, a fruitful concord between science and faith." In 1983, he reported to a group celebrating the 350th anniversary of the publication of Galileo's *Dialogue* that it was "progressing very encouragingly" and that the Church upheld the freedom of scientific research.

Studies of Galileo and his troubles with the Church continue. An Italian historian has recently discovered a document in the Vatican files that implies that the Church's real objection to Galileo was his thinking that matter is made out of atoms, which seemed to disagree with Church doctrine. Since this objection was so serious, the historian has suggested that the trial of Galileo on the matter of teaching Copernican theory was a mere smokescreen, hiding the true problem. His conclusions are hotly debated. In any case, the suggestion reinforces the desire by many that access be granted to the Vatican archives for a search for all documents relevant to Galileo.

Focus On

Box 3.3 Galileo

Galileo Galilei was born in Pisa, in northern Italy, in 1564. Although he studied medicine, his aptitudes were greater in other areas of science and mathematics. At the age of 25 he began to teach mathematics (and astronomy) at the University of Pisa and to undertake experimental investigations. This was an unusual thing to do, because most scientific investigation of the time involved the interpretation of old texts, such as those of Aristotle.

In the 1590's, Galileo became a professor at Padua, in the Venetian Republic. At some point in this decade, he adopted Copernicus's heliocentric theory. In 1604, Galileo observed and studied a new star in the sky, a "nova" (though we now know that this event was actually a supernova). Its existence showed that, contrary to Aristotle's theories, things did change in the distant parts of the universe.

In late 1609 or early 1610, Galileo was the first to use a telescope for astronomy. We shall discuss his discoveries in the following pages. It is hard to believe that one person could have discovered so many things.

In 1613 Galileo published his acceptance of the Copernican theory; the book in which he did so was received and acknowledged with thanks by the Cardinal who, significantly, was Pope at the time of Galileo's trial. Just how Galileo got out of favor with the Church has been the subject of much study. His bitter quarrel with a Jesuit astronomer, Christopher Scheiner, over who deserved the priority for the discovery of sunspots, did not put him in good stead with the Jesuits. Still, the disagreement on Copernicanism was more fundamental than personal, and the Church's evolving views against the Copernican system were undoubtedly more important than personal feelings.

In 1615, Galileo was denounced to the Inquisition, and was warned against teaching the Copernican theory. There has been much discussion of how serious the warning was and how seriously Galileo took it. Soon thereafter, Copernicus's *De Revolutionibus* was placed on the Index of Prohibited Books, though it became permitted in 1620, when certain "corrections," making its principles appear as methods of calculation rather than as physical truths, were made in about a dozen places.

Over the following years, Galileo was certainly not outspoken in his Copernican beliefs, though he also did not cease to hold them. In 1629, he completed his major manuscript, *Dialogue on the Two Great World Systems,* these systems being the Ptolemaic and Copernican (see the photograph opening the chapter). With some difficulty, clearance from the censors was obtained, and the book was published in 1632. It was written in contemporary Italian, instead of the traditional Latin, and could thus reach a wide audience. The book was condemned and Galileo was again called before the Inquisition.

Galileo's trial provided high drama, and has indeed been transformed into drama on the stage. He was convicted, and sentenced to house arrest. He lived nine more years, until 1642, his eyesight and his health failing. Though during this time he went blind, he continued his writing (including an important book on mechanics) until the end of his life.

3.3 Isaac Newton

It was not until many years after Kepler had discovered his three laws of planetary orbits by laborious calculation that the laws were derived mathematically from basic physical principles. The credit for doing so belongs to Isaac Newton. Newton was born in England in 1642, the year of Galileo's death. He was to become the greatest scientist of his time and perhaps of all time. Later in this book we will describe his discovery that visible light can be broken down into a spectrum and his invention of the reflecting telescope. Here we shall speak only of his ideas about motion and gravitation.

Newton set modern physics on its feet by deriving laws showing how objects move on the Earth and in space, and by finding the law that describes gravity. In order to work out the law of gravity, Newton had to invent calculus! Newton long withheld publishing his results, possibly out of shyness. Edmond Halley, whose name we associate with the famous comet, persuaded him to write up and then to publish his work. The *Philosophiae Naturalis Principia Mathematica* (*Mathematical Principles of Natural Philosophy*), known as *The Principia*, appeared in 1687 (Fig. 3–17).

PHILOSOPHIÆ
NATURALIS
PRINCIPIA
MATHEMATICA.

Autore JS. NEWTON, Trin. Coll. Cantab. Soc. Matheseos Professore Lucasiano, & Societatis Regalis Sodali.

IMPRIMATUR.
S. PEPYS, Reg. Soc. PRÆSES.
Julii 5. 1686.

LONDINI,
Jussu Societatis Regiæ ac Typis Josephi Streater. Prostant Venales apud Sam. Smith ad insignia Principis Walliæ in Cœmiterio D. Pauli, aliosq; nonnullos Bibliopolas. Anno MDCLXXXVII.

Figure 3–17 The title page of the first edition of Isaac Newton's *Principia Mathematica,* published in 1687.

Newton, whether or not you believe that an apple fell on his head (Fig. 3–18), was the first to realize the universality of gravity. In other words, he was the first to realize that the same law that describes how objects fall on Earth describes how objects fall far out in space. In particular, he realized that the Moon is always falling toward the Earth. However, at the same time, the Moon moves forward in space in its orbit. Think of the Moon moving above the Earth's surface from left to right. As the Moon falls toward Earth, the Moon approaches the Earth closer on the right. But you can see that if the Earth's surface also curves downward toward the right, the Moon would never get any closer to the Earth (Fig. 3–19). For this reason, the Moon remains in orbit.

Figure 3–18 A piece of Newton's apple tree, from the Royal Astronomical Society. The donor's father was present in about 1818 when some logs were sawed from this famous apple tree, which had blown over.

Newton formulated the law describing the force of gravity (F) between two bodies. It always tends to pull the two bodies together, and its strength varies directly with the product of the masses of the two bodies (m_1 and m_2) and inversely with the square of the distance (d) between them:

$$\text{force of gravity} \propto m_1 m_2 / d^2.$$

Since "proportional to" ($\propto$) simply stands for multiplying by a constant, we can write $F = Gm_1m_2/d^2$. This is known as the *law of universal gravitation;* it is a universal law in that it works all over the universe rather than being limited to local applicability (on Earth or even in the solar system).

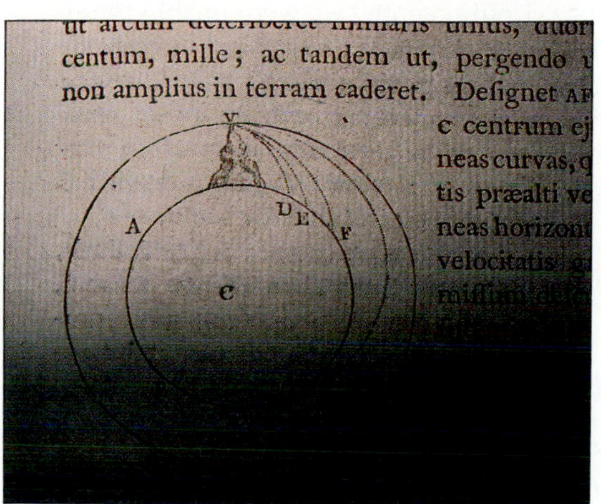
centum, mille; ac tandem ut, pergendo
non amplius in terram caderet. Designet AF
c centrum ej
neas curvas,
tis præalti ve
neas horizon
velocitatis

Figure 3–19 Newton's demonstration of the effect of increasing the velocity of a projectile, something ejected and then subject only to the force of gravity. If the projectile is moving fast enough, the Earth will curve away to match the projectile's falling, and the projectile will be in orbit.

$\propto$ means "is proportional to"; it means that the left hand side is equal to a constant times the right hand side. The constant, whose value is given in Appendix 2 (rather than here, so that you won't be tempted to memorize it), is called G or the *universal gravitational constant* because to the best of our knowledge it is constant throughout the universe.

Focus On

*Box 3.4 Kepler's Third Law, Newton, and Planetary Masses

Astronomers nowadays often make rough calculations to test whether physical processes under consideration could conceivably be valid. Astronomy has also had a long tradition of exceedingly accurate calculations. Pushing accuracy to one more decimal place sometimes leads to important results.

For example, Kepler's third law, in its original form—the period of a planet squared is proportional to its distance from the Sun cubed (P^2 = constant × R^3)—holds to a reasonably high degree of accuracy and seemed completely accurate when Kepler did his work. But now we have more accurate observations. If we consider each of the planets in turn, and if a term involving the sum of the masses of the Sun and the planet under consideration is included in the equation

$$P^2 = \frac{\text{constant}}{m_{\text{Sun}} + m_{\text{Planet}}} \times R^3,$$

the agreement with observation is improved in the fourth decimal place for the planet Jupiter and to a lesser extent for Saturn. The masses of the other planets are too small to have a detectable effect; even the effects caused by Jupiter and Saturn are tiny.

Isaac Newton derived the equation in its general form:

$$P^2 = \frac{4\pi^2}{G(m_1 + m_2)} a^3$$

for a body with mass m_1 revolving in an elliptical orbit with semimajor axis a around a body with mass m_2. The constant G is the *universal gravitational constant.*

Newton's formula shows that the planet's mass contributes to the value of the proportionality constant, but its effect is very small. (The effect can be detected even today only for the most massive planets—Jupiter and Saturn.)

Kepler's third law, and its subsequent generalization by Newton, applies not only to planets orbiting the Sun but also to any bodies orbiting other bodies under the control of gravity. Thus it also applies to satellites orbiting planets. We determine the mass of the Earth by studying the orbit of our Moon, and determine the mass of Jupiter by studying the orbits of its moons. Until recently, we were unable to reliably determine the mass of Pluto because we could not observe a moon in orbit around it. The discovery of a moon of Pluto in 1978 has finally allowed us to determine Pluto's mass, and we discovered that our previously best estimates (based on Pluto's gravitational effects on Uranus) were way off. The same formula can be applied to binary stars to find their masses.

Figure 3–20 The beginning of Newton's handwritten manuscript for the *Principia.* The manuscript is in the Royal Society in London.

3.3a Newton's Laws of Motion

In his *Principia,* Newton advanced not only the law of universal gravitation but also three laws of motion (Fig. 3–20). The laws govern the motion of objects in the normal circumstances of our everyday world.

Focus On

Box 3.5 Isaac Newton

Isaac Newton (Fig. 3–21) was born in Woolsthorpe, Lincolnshire, England, on December 25, 1642. Even now, almost three hundred fifty years later, scientists still refer regularly to "Newton's laws of motion" and to "Newton's laws of gravitation." In optics, we refer to "Newtonian telescopes" and to "Newton's rings," a series of concentric rings that appear often when two pieces of glass are placed together.

Newton's most intellectually fertile years were those right after his graduation from college when he returned home to the country (Fig. 3–22) because fear of the plague shut down many cities. Many of his basic thoughts stemmed from that period.

For many years he developed his ideas about the nature of motion and about gravitation. In order to derive them mathematically, he invented calculus. Newton long withheld publishing his results, possibly out of shyness. Finally Edmond Halley—whose name we associate with the famous comet—persuaded him to publish his work. A few years later, in 1687, the *Philosophiae Naturalis Principia Mathematica* (*Mathematical Principles of Natural Philosophy*), known as *The Principia* (prin-ki′pe-a), was published with Halley's aid. In it, Newton showed that the motions of the planets and comets could all be explained by the same law of gravitation that governed bodies on Earth. In fact, he derived Kepler's laws on theoretical grounds.

Newton was professor of mathematics at Cambridge University and later in life went into government service as Master of the Mint in London. He lived until 1727. His tomb in Westminster Abbey bears the epitaph: "Mortals, congratulate yourselves that so great a man has lived for the honor of the human race."

Figure 3–21 Sir Isaac Newton.

Newton's first law was a development of ideas that had earlier been advanced by Galileo and others. The law states that every object tends to remain either at rest or in uniform motion unless some outside force changes its state. This was in direct conflict with the Aristotelian idea that forces have to be continually applied to keep a body in motion. The property of mass that resists change in its motion is called *inertia,* and the first law is often called the *law of inertia.*

Newton's second law of motion states that the strength of a force is equal to the amount of mass involved times the acceleration it undergoes: $F = ma$. We actually define mass (technically, "inertial mass") from this formula: a larger mass requires a larger force to give it the same acceleration obtained for a smaller mass and a smaller force.

Newton's third law is usually stated as follows: "For every action there is an equal and opposite reaction." To every force corresponds another force of equal magnitude but opposite direction. It is this principle, for example, that makes a jet plane go—the force expelling gas backwards from the engine is the action, and the force moving the plane forward is the reaction. It is not significant which force is called the action and which the reaction; the important point is that the force of the plane on the gas and that of the gas on the plane are equal and opposite. The same principle causes a balloon to fly when you blow it up without tying it and let go. And it is Newton's third law that explains why the engines of space shuttles and other rockets can propel

Figure 3–22 A drawing of Newton's house at Woolsthorpe.

them even when they are in empty space. (A propeller airplane also flies forward because of Newton's third law; the force of the propeller pushing air back and that of the air pushing the propeller forward are action and reaction.)

Summary and Outline

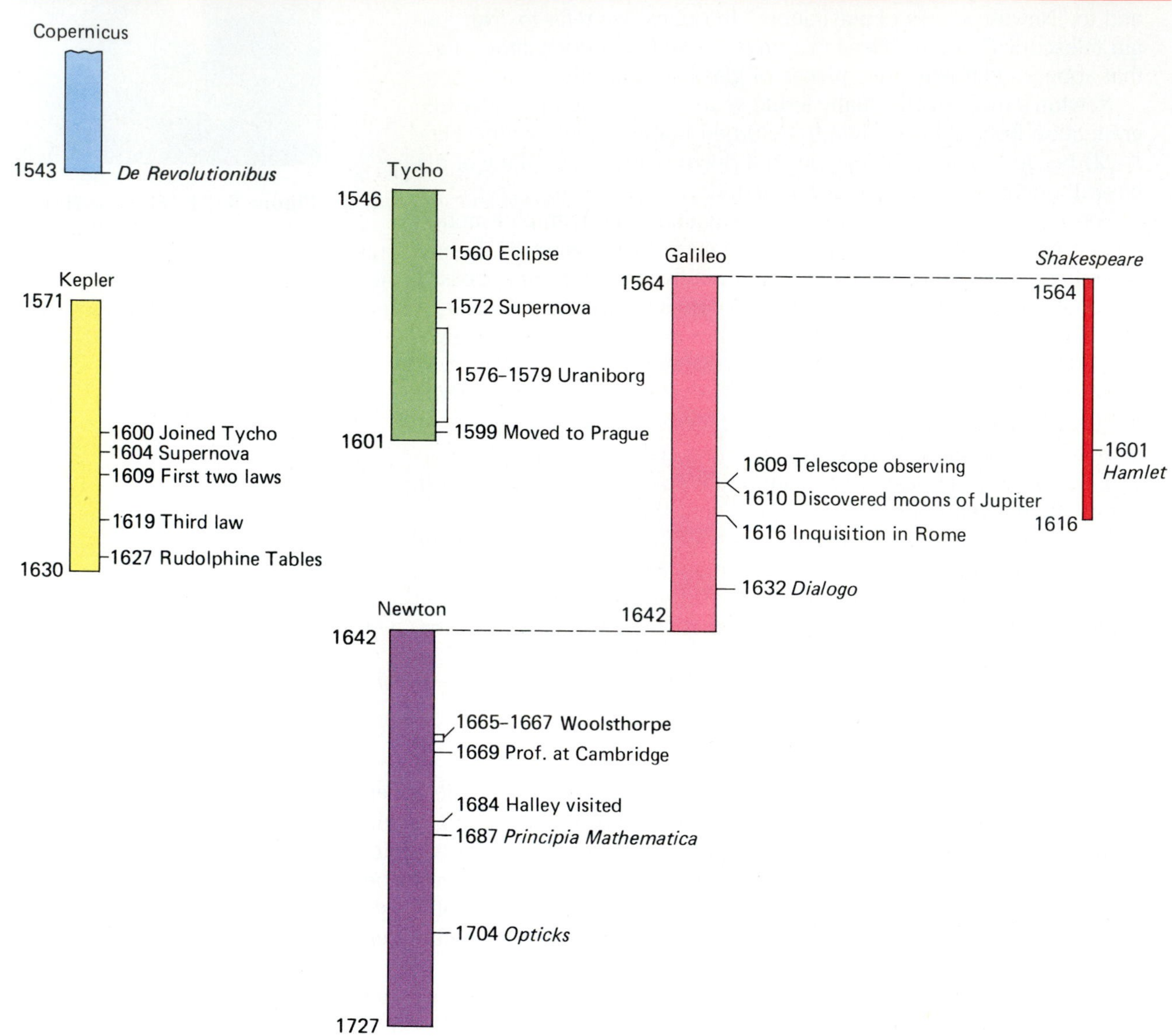

Timeline of Renaissance Astronomy

Kepler (Section 3.1)
- Studied Tycho's observations and discovered three laws of planetary motion
- Kepler's laws: (1) orbits are ellipses; (2) equal areas swept out in equal times; (3) period squared is proportional to distance cubed

Galileo (Section 3.2)
- First to use a telescope for astronomical observations
- His observations supported the heliocentric theory and further showed that Aristotle and Ptolemy had not known everything.

Newton (Section 3.3)
- Worked out laws of gravity and motion
- Was able to derive Kepler's laws and find the masses of the planets

Key Words

focus, foci, major axis, semimajor axis, minor axis, circle, eccentricity, conic sections, law of equal areas, sector, period, Astronomical Unit, one year, earthshine, maria, law of universal gravitation, universal gravitational constant, inertia, law of inertia

Questions

1. Discuss four new observations made by Galileo, and show how they tended to support, tended to oppose, or were irrelevant to the Copernican theory.
2. How do the predictions of the Ptolemaic and Copernican systems differ for the variation in **size** of Venus?
3. Do Kepler's laws permit circular orbits? Discuss.
4. At what point in its orbit is the Earth moving fastest?
5. Use Kepler's third law and the fact that Mercury's orbit has a semimajor axis of 0.4 A.U. to deduce the period of Mercury. Show your work.
6. What is the period of a planet orbiting 1 A.U. from a 9-solar-mass star?
7. The denizens of a planet far, far away around another star consider that they are 1 flibbit from their sun (a flibbit is their unit of length—a trillion times the average length of the antennae of their king and queen) and that they orbit with a period of 1 pip. They notice that a purple planet in their solar system takes 8 pips to revolve around their sun. How far is the purple planet from their sun (in flibbits)?
8. If we double the distance from the Earth's center of a satellite that has been in synchronous orbit around the Earth, what will its new period be?
9. (a) Use the data in Appendix 3 to show for which planets the square of the period does not equal the cube of the semimajor axis to the accuracy given. (b) Use the formula in Box 3.4 to show that including the effects of planetary masses removes the discrepancy.
10. What was the difference between the approaches of Kepler and Newton to the discovery of laws that controlled planetary orbits?

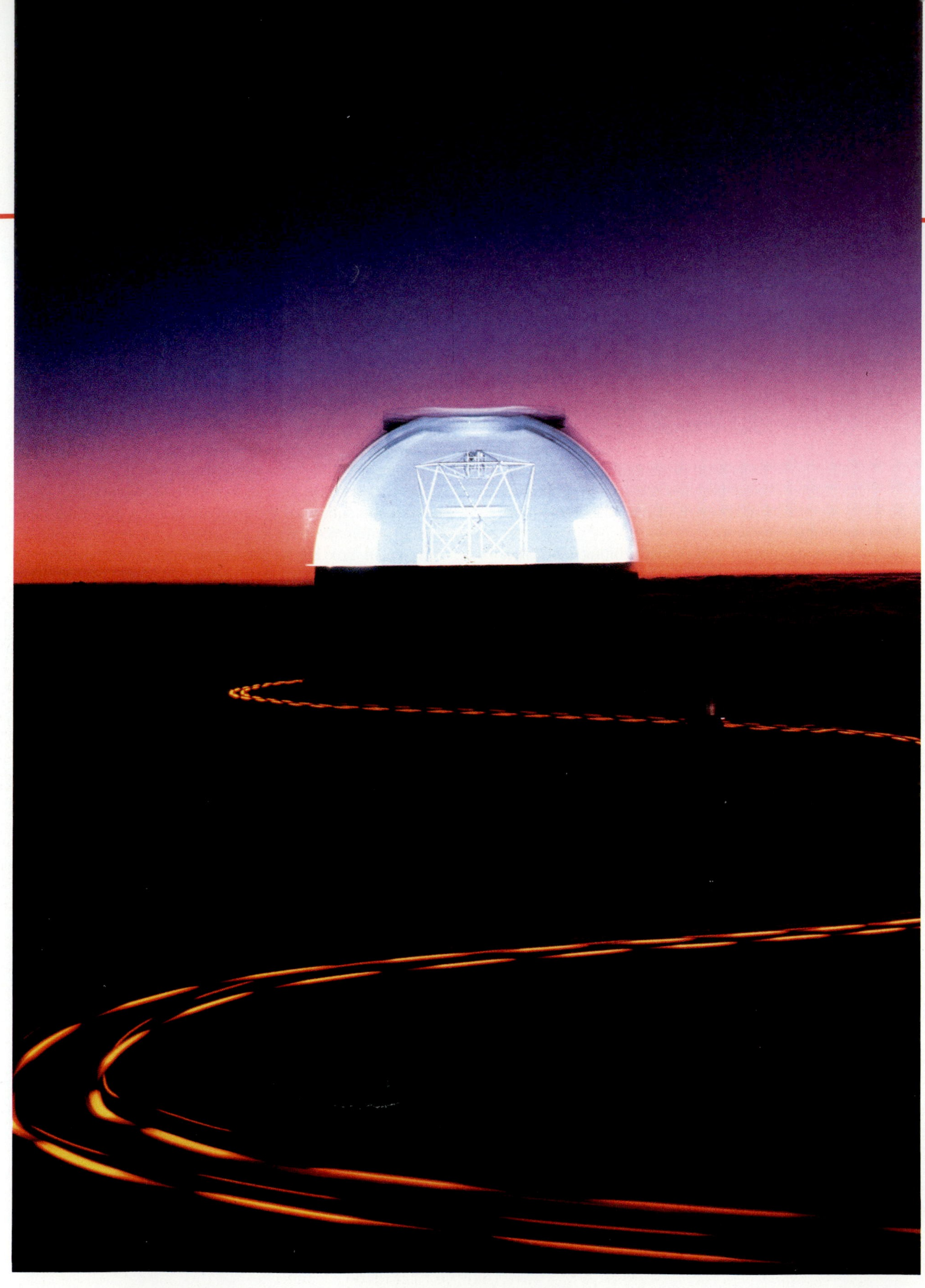

The Keck 10-m telescope atop Mauna Kea, Hawaii. The dome appears transparent because it revolved during this exposure.

Light and Telescopes

4

Aims: To understand the electromagnetic spectrum and the different types of observing techniques and instruments used in astronomy

If you forced a group of astronomers to choose one and only one instrument to be marooned with on a desert island, they would probably choose...a computer. The notion that astronomers spend most of their time at telescopes is far from the case in modern times. Still, telescopes have been and continue to be very important to the development of astronomy, and in this chapter we will discuss how we study the stars with telescopes and other observational devices.

Although the word "telescope" makes most of us think of objects with lenses and long tubes, that type of telescope is used only to observe "light." Studying ordinary "visible" light, the kind we see with our eyes, is not the only way we can study the Universe. We shall see in this chapter that light is only one type of *radiation*—a certain way in which energy moves through space. Other types of radiation are gamma rays, x-rays, ultraviolet, infrared, and radio waves. They are all fundamentally identical to ordinary light, though in practice we must usually use different methods to observe them.

In this chapter, we discuss the properties of radiation and how studying radiation enables scientists to study the Universe; after all, we cannot touch a star, and though we have brought bits of the Moon back to Earth for study, we are not able to do the same for even the nearest planets.

Figure 4–1 When white light passes through a prism, a full optical spectrum results. It ranges through ROYGBIV.

4.1 The Spectrum

It was discovered over 300 years ago that when ordinary light is passed through a prism, a band of color like the rainbow comes out the other side. Thus "white light" is composed of all the colors of the rainbow (Fig. 4–1).

These colors are always spread out in a specific order, which has traditionally been remembered by the initials of the friendly fellow ROY G. BIV, which stand for Red, Orange, Yellow, Green, Blue, Indigo, Violet. No matter what rainbow you watch, or what prism you use, the order of the colors never changes.

We can understand why light contains the different colors if we think of light as waves of radiation. These waves are all travelling at the same speed, 3×10^8 m/sec (that is, meters per second; 186,000 miles/sec; the equivalent of seven times around the Earth each second). This speed is normally called the *speed of light* (even though light normally travels at a slightly lower speed because the light is not usually in a perfect vacuum; the phrase "the speed of light" usually really means "the speed of light in a vacuum").

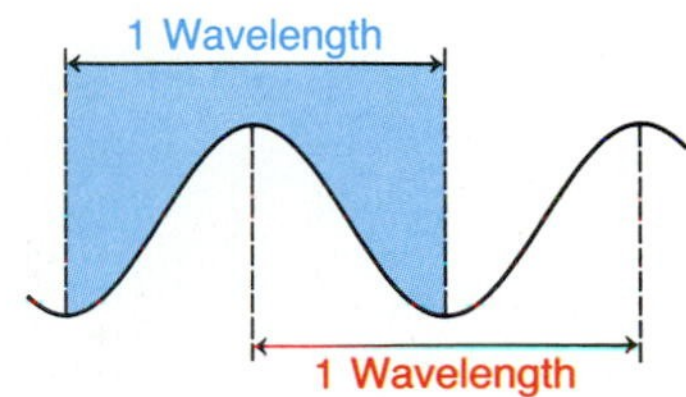

Figure 4–2 The wavelength is the length over which a wave repeats.

The distance between one crest of the wave and the next or one trough and the next, or in fact between any point on a wave and the similar point on the next wave, is called the *wavelength* (Fig. 4–2). Light of different wavelengths appears as different colors. It is as simple as that. Red light has approximately

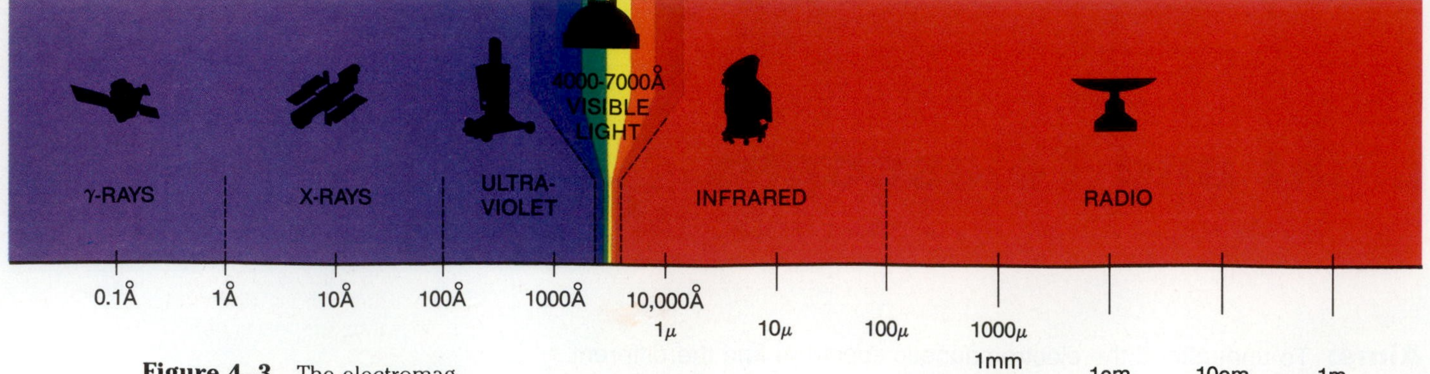

Figure 4–3 The electromagnetic spectrum. The silhouettes shown represent telescopes or spacecraft used or planned for observing that part of the spectrum: the Gamma-Ray Observatory, the Advanced X-Ray Facility, the International Ultraviolet Explorer, the dome of a ground-based telescope, the Infrared Astronomical Satellite, and a ground-based radio telescope.

1½ times the wavelength of blue light. Yellow light has a wavelength in between the two. Actually there is a continuous distribution of wavelengths, and one color blends subtly into the next.

The wavelengths of light are very short: just a few ten-millionths of a meter. Astronomers use a unit of length called an angstrom, named after the Swedish physicist A. J. Ångstrom. One angstrom (1 Å) is 10^{-10} m. (Remember that this is 1 divided by 10^{10}, which is .000 000 000 1 m.) The wavelength of violet light is approximately 4000 angstroms (4000 Å). Orange light is approximately 6000 Å, and red light is approximately 6500 Å in wavelength. (10 Å make 1 nanometer, abbreviated 1 nm, the SI unit, so violet light is 400 nm and red light is 650 nm in wavelength.)

The human eye is not sensitive to radiation of wavelengths much shorter than 4000 Å or much longer than 6600 Å, but other devices exist that can measure light at shorter and longer wavelengths. At wavelengths shorter than violet, the radiation is called *ultraviolet;* at wavelengths longer than red, the radiation is called *infrared*. (Note that this is not "infared," a common misspelling; "infra-" is a Latin prefix that means "below.")

All these types of radiation—visible or invisible—have much in common. Many of us are familiar with the fields of force produced by a magnet—a magnetic field—and the field of force produced by a charged object like a hair comb on a dry day—an electric field. It has been known for over a century that fields that vary rapidly behave in a way that no one would guess from the study of magnets and combs. In fact, light, x-rays, and radio waves are all examples of rapidly varying electric fields and magnetic fields that have become detached from their sources and move rapidly through space. For this reason, these radiations are referred to as *electromagnetic radiation*.

We can draw the entire *electromagnetic spectrum,* often simply called the "spectrum" (plural: *spectra*), ranging from radiation of wavelength shorter than 1 Å to radiation of wavelength many meters long (Fig. 4–3).

Note that from a scientific point of view there is no qualitative difference between types of radiation at different wavelengths. They can all be thought of as electromagnetic waves, waves of varying electric and magnetic fields. Light waves comprise but one limited range of wavelengths. When an electromagnetic wave has a wavelength of 1 Å, we call it an x-ray. When it has a wavelength of 5000 Å, we call it light. When it has a wavelength of 1 cm (which is 10^8 Å), we call it a radio wave. Of course, there are obvious practical differences in the methods by which we detect x-rays, light, and radio waves, but the principles that govern their existence are the same.

Note also that light occupies only a very small portion of the entire electromagnetic spectrum. It is obvious that the new ability that astronomers have to study parts of the electromagnetic spectrum other than light waves enables us to increase our knowledge of celestial objects manyfold.

Only certain parts of the electromagnetic spectrum can penetrate the Earth's atmosphere. We say that the Earth's atmosphere has "windows" for the

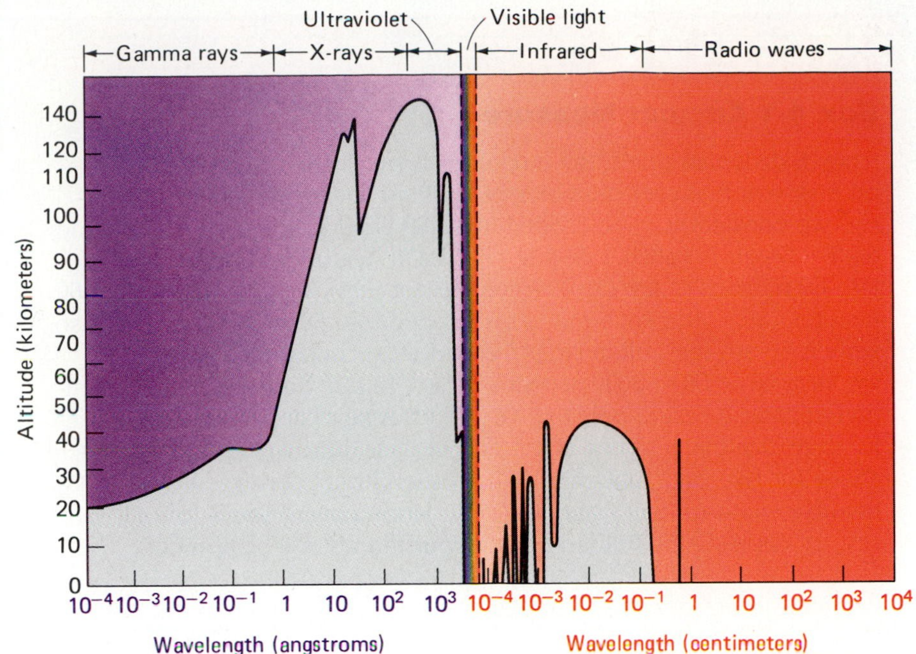

Figure 4–4 In a window of transparency in the terrestrial atmosphere, radiation can penetrate to the Earth's surface. The curve specifies the altitude at which the intensity of arriving radiation is reduced to half its original value. When this happens high in the atmosphere, little or no radiation of that wavelength reaches the ground.

parts of the spectrum that can pass through it. The atmosphere is transparent at these windows and opaque at other parts of the spectrum. One window passes what we call "light," and what astronomers technically call *visible light,* or "the visible," or "the optical part of the spectrum." Another window falls in the radio part of the spectrum, and modern astronomy uses "radio telescopes" to detect that radiation (Fig. 4–4).

But we of the Earth are no longer bound to our planet's surface; balloons, rockets, and satellites carry telescopes above the atmosphere to observe in parts of the electromagnetic spectrum that do not pass through the Earth's atmosphere. In recent years, we have been able to make observations all across the spectrum. It may seem strange, in view of the long-time identification of astronomy with visible observations, to realize that optical studies no longer dominate astronomy.

4.2 What a Telescope Is

How to define a telescope is a question with no simple answer, since a "telescope" to observe gamma rays may be a package of electronic sensors launched above the atmosphere, and a radio telescope may be a large number of small aerials strung over acres of landscape. We will begin, nevertheless, by discussing telescopes of the traditional types, which observe the radiation in the visible part of the spectrum. These optical telescopes are important because of the many things we have learned over the years by studying visible radiation. And optical telescopes are the types of telescopes most frequently used by students and by amateur astronomers (some of whom are quite professional in their approach to the subject). Further, the principles of focusing and detecting electromagnetic radiation that were originally developed through optical observations are widely used throughout the spectrum.

Contrary to popular belief, the most important purpose for which most optical telescopes are used is to gather light. (For the next few sections I shall drop the qualifying word "optical.") True, telescopes can be used to magnify as well, but for the most part astronomers are interested in observing fainter and fainter objects and so must collect more light to make these objects

Focus On

Box 4.1 Angular Measure

Since Babylonian times, the circle has been divided into parts based on a system that uses multiples of 60. If we are at the center of a circle, we consider it to be divided into 360 parts as it extends all around us. Each part is called 1 degree (1°). Each degree, in turn, is divided into 60 minutes of arc (also written 60 arc min), often simply called 60 minutes (60′). And each minute is divided into 60 seconds of arc (also written 60 arc sec), often simply called 60 seconds (60″). Thus in 1°, we have $60 \times 60 = 3600$ seconds of arc. In this book we usually mention angular measure only to give you a sense of appearance. It is useful to realize that your fist at the end of your outstretched arm covers about 10° (we say "subtends 10°") as seen from your eye, and your thumb's width subtends about 2°. The Moon covers about ½° of the sky; try covering the full Moon with your thumb one clear night.

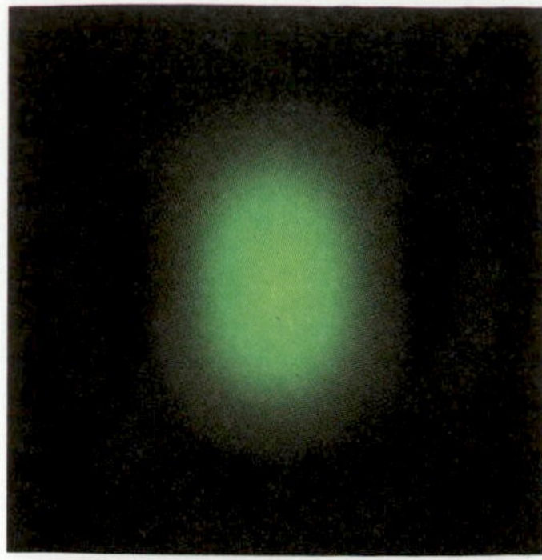

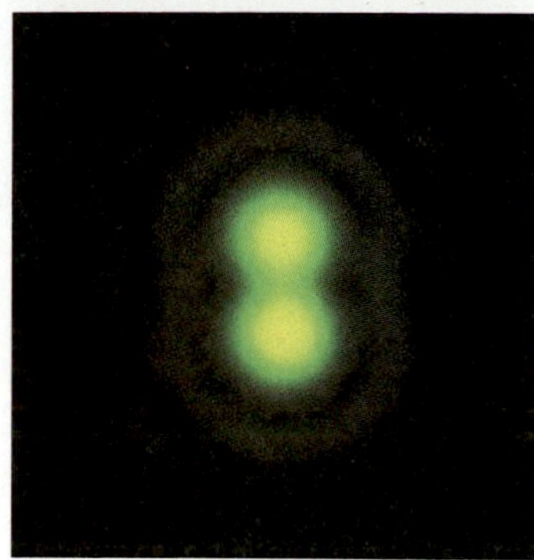

Figure 4–5 In addition to their primary use for gathering light from faint objects, telescopes are often used to increase the resolution, our ability to distinguish details in an image. We see (*top*) a pair of point sources (sources so small or far away that they appear as points) that are not quite resolved; (*middle*) sources that have passed Dawes's limit, the condition in which they can barely be resolved; and (*bottom*) sources that are completely resolved. These images were made with a mercury laser that emits green light. In this series, the images differ by different aperture size in the lab; in astronomy, as long as we are not limited by seeing, a larger image disk would come from a smaller telescope or observing at a longer wavelength.

detectable. There are certain cases in which magnification is important—such as for observations of the Sun or for observations of the planets—but stars are so far away that they appear as mere points of light no matter how much magnification is applied.

When we look at the sky with our naked eyes, several limitations come into play. First of all, our eyes see in only one part of the spectrum (the visible). Another limitation is that we can see only the light that passes through an opening (aperture) of a certain diameter—the pupils of our eyes. In the dark, our pupils dilate so that as much light as possible can enter, but the apertures are still only a few millimeters across (up to 8 millimeters for young people and decreasing to perhaps 5 millimeters for older people). Yet another limitation is that of time. Our brains distinguish a new image about 30 times a second, and so we are unable to store faint images over a long time in order to accumulate a brighter image. Astronomers overcome these additional limitations by using a telescope to gather light and a recording device, such as a photographic plate, to store the light. Other equipment may also be used, such as a spectrograph to analyze the spectral content of the light, or simple filters.

A further advantage of a telescope over the eye, or of a large telescope over a smaller telescope, is that of *resolution,* the ability to distinguish finer details in an image or to distinguish two adjacent objects (Fig. 4–5). A telescope is capable of considerably better resolution than the eye, which is limited to a resolution of roughly 1 arc min (the diameter of a dime at a distance of 60 m). (The larger the aperture accepting light, the better the resolution, and the eye's maximum opening is only a few mm across. The retina actually sets a resolution limit somewhat worse, about 3 arc min.) The aperture of each lens of binoculars is a few times larger, so binoculars can reveal details a few times finer. Thus binoculars or a telescope can distinguish the two components of a double star from each other in many cases where the unaided eye is unable to do so.

The Earth's atmosphere limits us to seeing detail on celestial objects larger than roughly ½ arc sec across (the diameter of a dime at a distance of 7 km or of a human hair two football fields away). If a telescope with a collecting area 10 centimeters across can resolve double stars that are separated by 1 arc second, a telescope 20 centimeters across, twice the diameter, can resolve stars that are separated by half that angle, ½ arc sec. In principle, for light of a given wavelength, the size of the finest details resolved (the resolution) is inversely proportional to the diameter of the telescope's primary mirror or lens. By "inversely proportional," we mean that as one quantity goes up, the other goes down by the same factor.

Even for large telescopes, the best resolution that can be achieved on the Earth's surface on an average good night is limited by turbulence in the Earth's atmosphere to about 1 arc sec. Thus increasing size no longer improves the resolution, even on the best of nights, though the advantages in terms of light-gathering power remain.

Actually, resolution depends not only on the aperture but also on the wavelength of radiation being observed; the example we just considered was green light of approximately 5000 Å.

As the wavelength of radiation doubles, resolution is halved. For example, a telescope that can resolve two sources emitting 5000 Å (green) light that are 1 arc second apart could only resolve two 10,000 Å (infrared) sources if they were 2 arc sec apart. The limit of resolution for visible light, called Dawes's limit, is approximately $2 \times 10^{-3}\ \lambda/d$ arc sec, when the wavelength, λ, is in Å and the diameter of a telescope, d, is in cm.

4.3 Lenses and Telescopes

Light (or other electromagnetic radiation) bends when it passes from one medium into another. This bending is called *refraction* (Fig. 4–6). A "medium" is simply any material (that is, a transparent object, air, interstellar space) through which the radiation travels. Electromagnetic radiation can even travel through a *vacuum*, empty space where no medium is present.

Consider parallel waves of light hitting a block of glass at an angle (Fig. 4–7). Some of the waves hit the glass before others. Since light travels 50 per cent more slowly in glass than in air, these waves travel more slowly than the waves that are still in air. As a result, a plane of radiation that had been advancing together in the air is travelling in a different direction in the glass.

A suitable curved piece of glass can be made so that all the light that travels through it is bent, bringing all the rays of each given wavelength to a focus. Such a curved piece of glass is called a *lens*. The lens and cornea of your eye do a similar thing to make images of objects of the world on your retina—each point of the object is focused to a point on the image. An eyeglass lens helps your eye's lens and cornea accomplish this task.

A lens uses the property of refraction to focus light. A telescope that has a lens as its major element is called a *refracting telescope.*

The distance of the image from the lens on the side away from the object being observed depends in part on the distance of the object from the front side of the lens. Astronomical objects are all so far away that we can treat them, for the purpose of forming images, as though they were infinitely far away. We say that they are "at infinity." The images of objects at infinity fall at a distance called the *focal length* behind the lens (Fig. 4–8).

A particular telescope forms an image of an area of the sky called its *field of view*. Objects that are at angles far from the center of the field of view may not be focused as well as objects in the center.

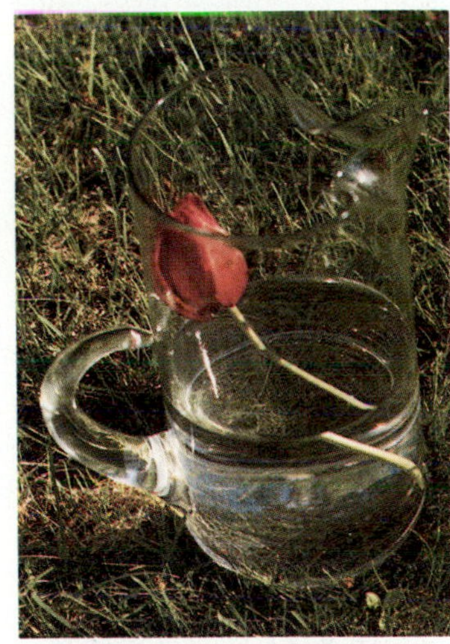

Figure 4–6 The flower stem appears both bent and displaced at the boundary between the water and the air.

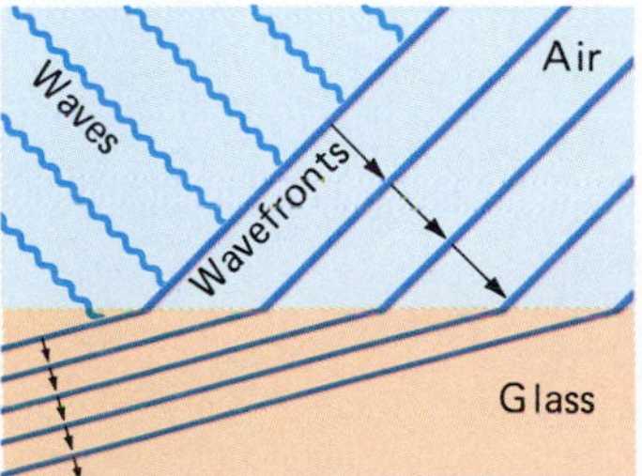

Figure 4–7 Light travels at a speed in glass that is different from the speed that it has in air. This change leads to refraction.

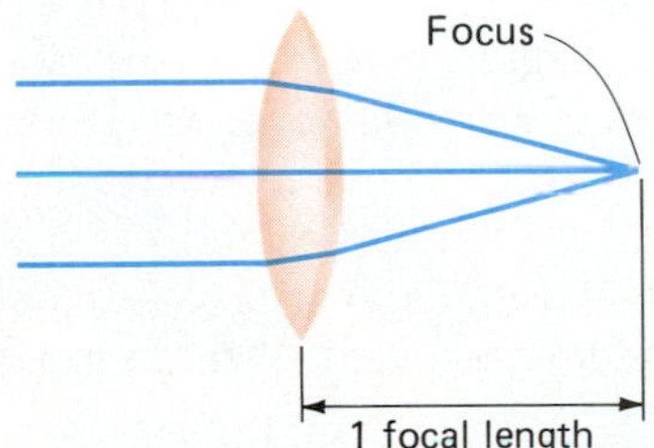

Figure 4–8 The focal length is the distance behind a lens to the point at which objects at infinity are focused. The focal length of the human eye is about 2.5 cm (1 in).

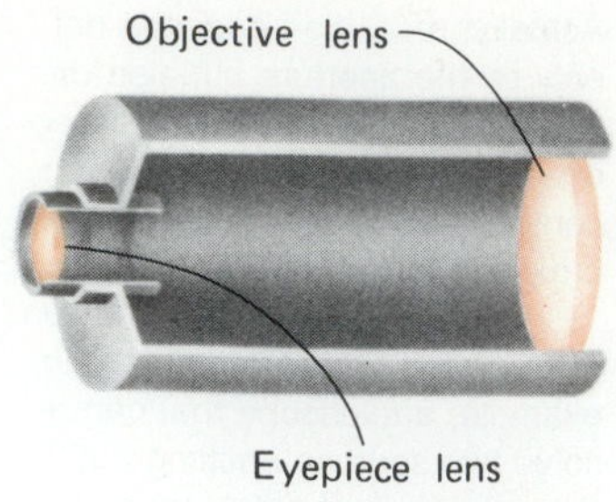

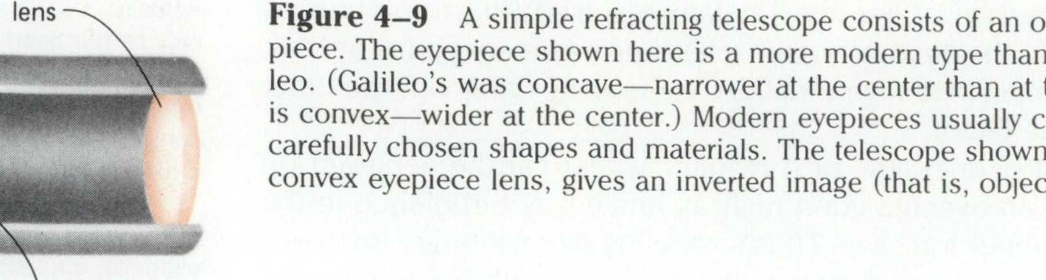
Figure 4–9 A simple refracting telescope consists of an objective lens and an eyepiece. The eyepiece shown here is a more modern type than the one used by Galileo. (Galileo's was concave—narrower at the center than at the edges—and this one is convex—wider at the center.) Modern eyepieces usually contain several pieces of carefully chosen shapes and materials. The telescope shown here, with a double convex eyepiece lens, gives an inverted image (that is, objects appear upside down).

The technique of using a lens to focus faraway objects was, we think, developed in Holland in the first decade of the 17th century. It is not clear how Galileo heard of the process, but it is known that Galileo quickly bought or ground a lens and made a simple telescope that he demonstrated in 1610 to the Senate in Venice. He put this small lens and another smaller lens at opposite ends of a tube (Fig. 4–9). The second lens, called the *eyepiece,* is used to examine and to magnify the image made by the first lens, called the *objective.* Galileo's telescope was very small, but it magnified enough to impress the nobles of Venice, who had assembled to see the new invention. The telescope had obvious commercial value, enabling them to see boats coming to Venice before the boats were visible to the naked eye.

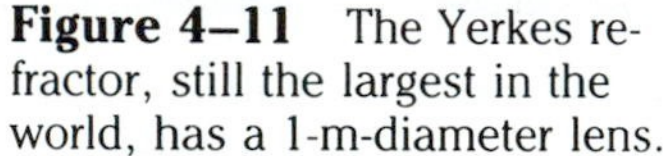
Figure 4–10 Galileo's drawing of the Milky Way from 1610.

Galileo turned his simple telescope on the heavens, and what he saw revolutionized not only astronomy but also much of seventeenth-century thought. He discovered, for example, that the Milky Way includes many stars (Fig. 4–10), that there are mountains on the Moon, that Jupiter has satellites of its own, and that Venus has phases.

The largest refracting telescope in the world has a lens 1 meter (40 inches) across. It is at the Yerkes Observatory in Williams Bay, Wisconsin (Fig. 4–11), and went into use during the 1890's.

Refracting telescopes suffer from several problems. For one thing, lenses suffer from *chromatic aberration,* the effect whereby different colors are focused at different points (Fig. 4–12). The speed of light in a substance—in glass, for example—depends on the wavelength of the light. As a result, light composed of different colors bends by different amounts, and not all wavelengths can be

Figure 4–11 The Yerkes refractor, still the largest in the world, has a 1-m-diameter lens.

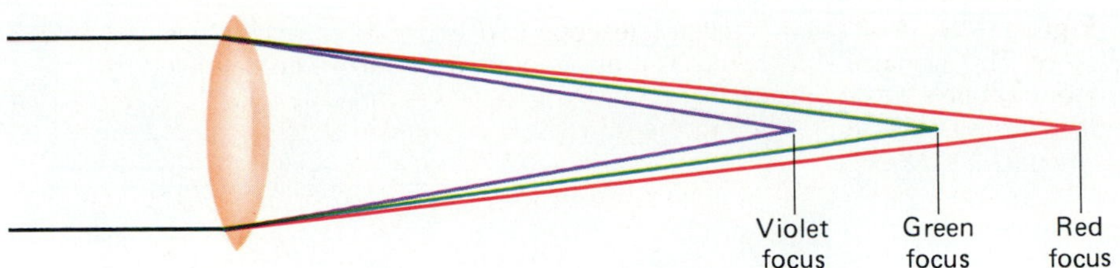

Figure 4–12 The focal length of a lens is different for different wavelengths, which leads to chromatic aberration.

brought to a focus at the same point. Ingenious methods of making "compound" lenses of different glasses that have slightly differing properties have succeeded in reducing this problem. But even after these methods of reducing chromatic aberration have been used, enough chromatic aberration remains to present a fundamental problem for the use of refracting telescopes.

4.4 Reflecting Telescopes

Reflecting telescopes are based on the principle of reflection, with which we are so familiar from ordinary household mirrors. A mirror reflects the light that hits it so that the light bounces off at the same angle at which it approached (Fig. 4–13). A flat mirror gives an image the same size as the object being reflected, but funhouse mirrors, which are not flat, cause images to be distorted.

We can construct a mirror in a shape so that it reflects incoming light to a focus. Let us consider a spherical mirror. If you were inside and at the center of a giant spherical mirror, in whatever direction you looked you would see your own image. Whatever is at a spherical mirror's center is imaged at the same point, its center (Fig. 4–14). If we were to use just a portion of a sphere, it would still image whatever was at the center of its curvature back at that same point.

But the stars are far away, and so would not be at the center of curvature if we used a spherical mirror. Thus we can use a mirror that takes advantage of the fact that a parabola focuses *parallel light* to a point (Fig. 4–15). We say that the light from the stars and planets is "parallel light" because the individual

Figure 4–13 Each individual beam of light bounces off a flat mirror at the same angle at which it hit the mirror. By arranging several flat mirrors on a curve, as shown here, incoming parallel rays of light can be bent to a single point. A telescope mirror is made so that it is curved by the exact amount necessary to reflect all incoming parallel rays of light to a single point.

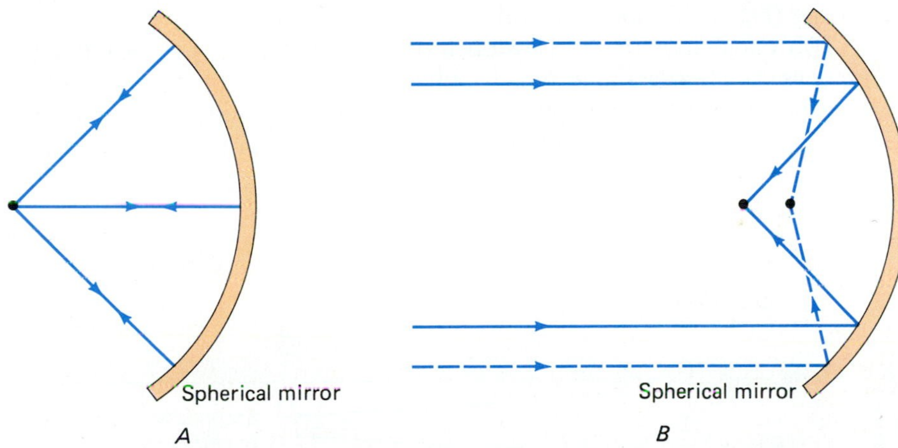

Figure 4–14 (*A*) A spherical mirror focuses light that originates at its center of curvature back on itself. (*B*) A spherical mirror, though, suffers from *spherical aberration* in that it does not perfectly focus light from very large distances, for which the incoming rays are essentially parallel.

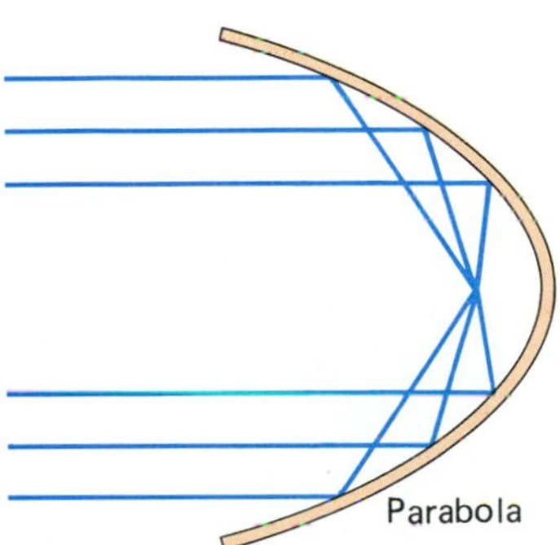

Figure 4–15 A parabola focuses parallel light to a point.

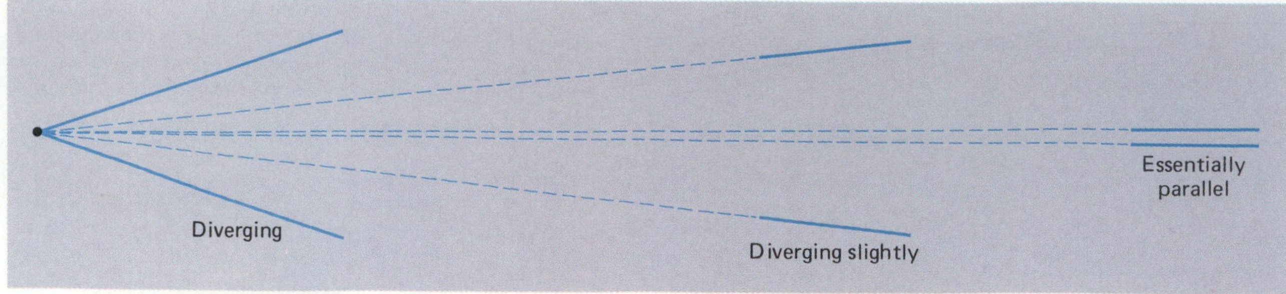

Figure 4–16 Light rays from very distant objects are diverging so slightly by the time they reach us that we speak of parallel light. Notice that if we choose two light rays the same distance apart at several points from left to right on the diagram, they are more closely parallel the farther rightward we go.

light rays are diverging by such an imperceptible amount by the time they reach us on Earth that they are practically parallel (Fig. 4–16).

A parabola is a two-dimensional curve that has the property of focusing parallel rays to a point. In many cases, a telescope mirror is actually a *paraboloid,* which is the three-dimensional curve generated when a parabola is rotated around its axis of symmetry. Over a small area, a paraboloid differs from a sphere only very slightly, and we can ordinarily make a parabolic mirror by first making a spherical mirror and then deepening the center slightly.

Reflecting telescopes have several advantages over refracting telescopes. The angle at which light is reflected does not depend on the color of the light, so chromatic aberration is not a problem for reflectors. Telescope makers normally deposit a thin coat of a highly reflecting material on the front of a suitable ground and polished surface. Since light does not penetrate the surface, what is inside the telescope mirror does not matter too much. Further, a network of supports can be placed all across the back of the telescope mirror, to prevent the mirror from sagging under the force of gravity. All these advantages allow reflecting telescopes to be made much larger than refracting telescopes. (Only reflecting telescopes can be used at wavelengths shorter than those of visible light, because most ultraviolet radiation does not pass through glass.)

Figure 4–17 A model of Newton's original telescope. It is housed in the Royal Society, London. Though often called Newton's original, it was probably made later.

A potential problem with small reflecting telescopes is the need to gather the reflected light without impeding the incoming light. If you put your head at the focal point of a small telescope mirror, the back of your head would block the incoming light and you would see nothing at all!

Isaac Newton got around this problem 300 years ago by putting a small diagonal mirror a short distance in front of the focal point to reflect the light to the side of the telescope and out (Fig. 4–17). This design is a *Newtonian* telescope (Fig. 4–18) and is in widespread use (Fig. 4–19).

Two contemporaries of Newton invented alternative types of reflecting telescopes. G. Cassegrain, a French optician, and James Gregory, a Scottish astronomer, invented telescopes in which the light was reflected by the secondary mirror through a small hole in the center of the primary mirror. These

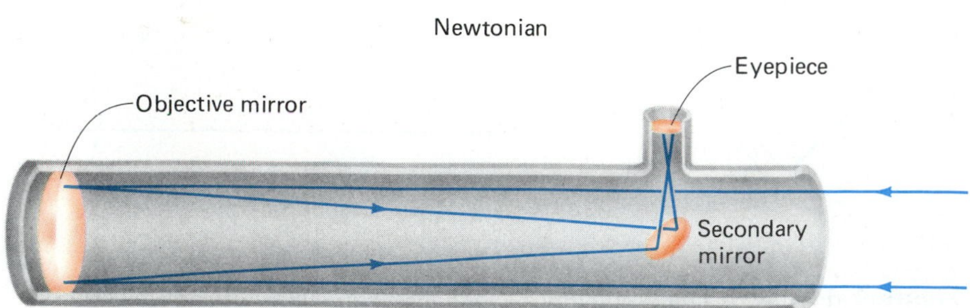

Figure 4–18 A Newtonian reflector has a flat secondary mirror tilted at a 45° angle.

designs are called *Cassegrain* and *Gregorian* telescopes (Fig. 4–20).

The largest telescopes have several interchangeable secondary mirrors built in, so that different foci can be used at different times. The very largest telescopes are so huge that the observer can even sit in a "cage" suspended

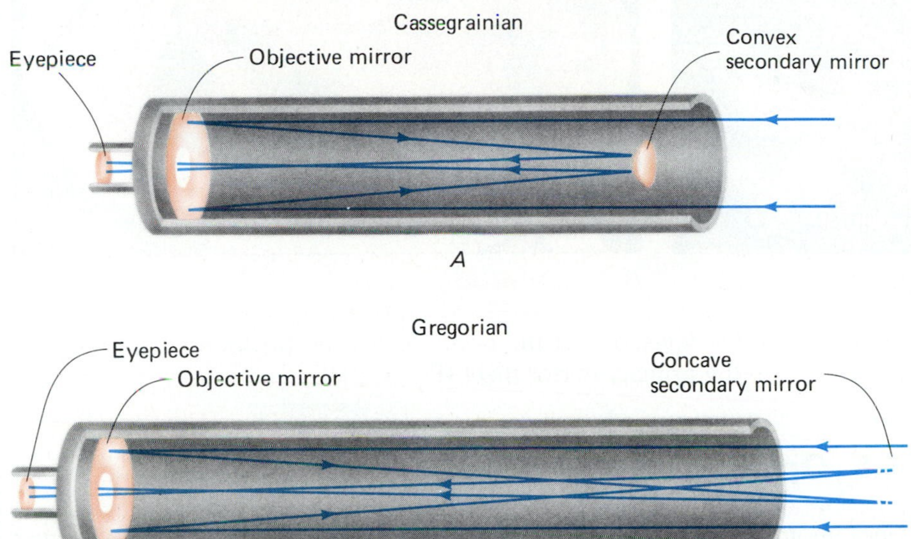

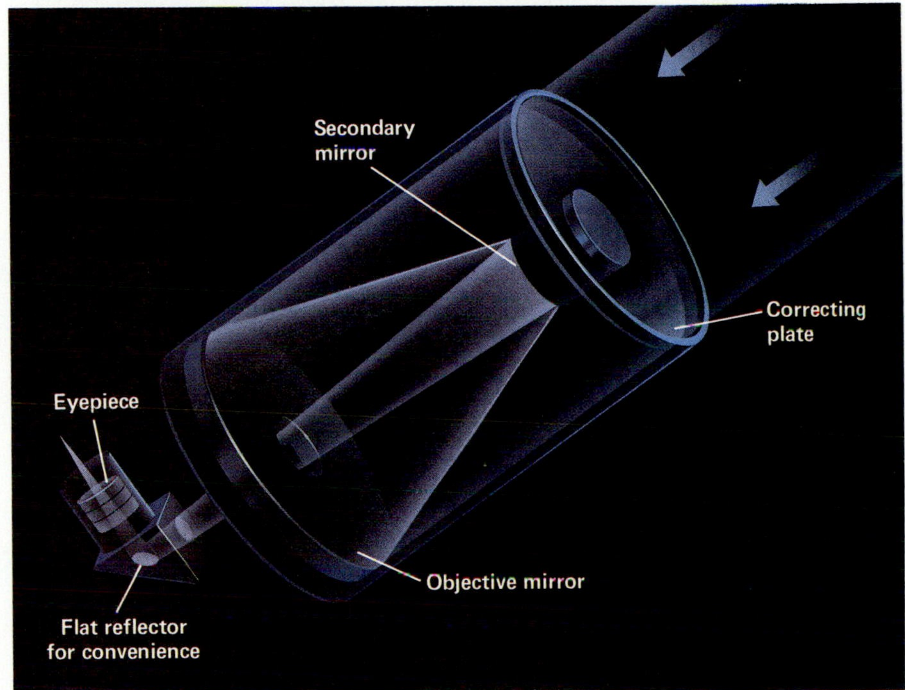

Figure 4–20 (*A*) A Cassegrain telescope has a convex secondary mirror. Cassegrains normally have short tubes relative to their mirror diameters. Most large telescopes have a Cassegrain focus. (*B*) A Gregorian telescope has a concave secondary mirror that would be located beyond the focus. The Solar Maximum Mission, a spacecraft launched in 1980, carried a Gregorian telescope. (*C*) A Schmidt-Cassegrain design, a type popular with amateur telescopes. The Schmidt-type (Section 4.6) correcting plate, a lens, provides a wide field while the Cassegrain folding of the light path shortens the telescope.

Figure 4–19 In the movie *Roxanne*, Daryl Hannah played a graduate student in astrophysics. Here we see her with Steve Martin and her telescope, which we can tell is Newtonian by the eyepiece at its top. (© 1986 Columbia Pictures Industries, Inc. All rights reserved.)

A

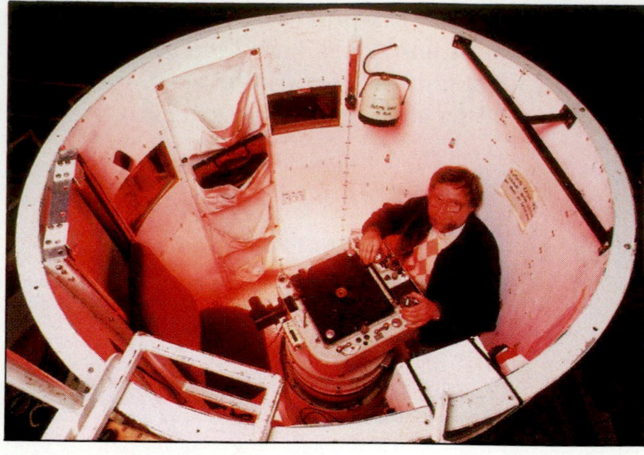

B

Figure 4–21 (*A*) The prime-focus cage of the 3.9-m Anglo-Australian Telescope is the cylinder extending upward from the top of the tube. (*B*) It is cozy inside the prime-focus cage.

at the end of the telescope at the *prime focus,* the position where the rays reflected from the primary mirror meet (Fig. 4–21).

4.5 Optical Observatories

Once upon a time, over a hundred years ago, telescopes were put up wherever astronomers happened to be located. Now all major observatories are built at remote sites chosen for the quality of their observing conditions.

4.5a Observatory Sites

On Mount Wilson, near Los Angeles, George Ellery Hale built a 2.5-m telescope, with a mirror made of plate glass. It began operation in 1917 and was the largest telescope in the world for 30 years. It was shut down in 1985, because the interests of the organization that owned it are concentrated on observations requiring a very dark sky. Though the sky from Mount Wilson, overlooking Los Angeles, is fairly bright, we could still make important stellar observations there. A new organization, the Mount Wilson Institute, is running some telescopes on Mount Wilson and educational programs for undergraduates. They hope to raise funds to reopen the 2.5-m telescope.

One of the limitations of any telescope, reflecting or refracting, is the length of time that the mirror or lens takes to reach its equilibrium shape when exposed to the temperature of the cold night air. In the 1930's, the Corning Glass Works invented Pyrex, a type of glass that is less sensitive to temperature variations than ordinary glass. Workers there cast, with great difficulty, a mirror blank a full 5.08 meters (200 inches) across (Fig. 4–22).

Figure 4–22 The Palomar 5-m mirror blank was very difficult to cast. Here we see the unsuccessful first try, which is on display at the Corning Glass Works in Corning, New York. The mirror's back is ribbed to reduce its weight.

The construction of the telescope was held up by many factors, including the Second World War. Only in 1949 did active observing begin with the telescope. It stands on Palomar Mountain in southern California, and for many years it was the largest in the world. It is named the Hale Telescope (Fig. 4–23) after George Ellery Hale, who was responsible for its construction.

After many years of being run by Caltech, the 5-m Palomar telescope is now operated 25 per cent by Cornell University. Cornell astronomers are contributing money as well as infrared and other detectors, and will receive 25 per cent of the operating time. Another 25 per cent is operated by the Carnegie Institution of Washington.

The largest reflecting telescopes today include the new 10-m Keck telescope in Hawaii; the 6-m reflector in the Soviet Union (Figs. 4–24 and 4–25),

Figure 4–23 The Palomar Observatory's 5-m Hale telescope on Palomar Mountain in California. The dome was rotated during this long time exposure, making it seem transparent. (© 1987 Roger Ressmeyer, Starlight)

the 5-m reflector on Palomar Mountain in California, the 4.75-m-equivalent Multiple Mirror Telescope in Arizona, the 4.4-m William Herschel Telescope in the Canary Islands, and the 4-m telescopes at several locations, including the Kitt Peak National Observatory in Arizona.

The Kitt Peak National Observatory's telescopes are located on Kitt Peak, a sacred mountain of the Papago Indians, about 80 km (50 miles) southwest of Tucson, Arizona (Fig. 4–26). Its headquarters are in Tucson.

Figure 4–24 The Soviet 6-m (236-inch) telescope at the Special Astrophysical Observatory on Mount Pastukhov in the Caucasus.

Figure 4–25 A replacement 6-m mirror being installed in the Soviet telescope.

Figure 4–26 Lightning flashes illuminate the Kitt Peak National Observatory in this 45-second exposure. The 4-m Mayall Telescope is the tallest.

Kitt Peak (as the Kitt Peak National Observatory is usually called) has been set up for use by all the optical astronomers in the United States. Any astronomer can apply for observing time at one of the telescopes at Kitt Peak. A 4-m (158-inch) telescope is the largest of 16 telescopes now there. The University of Arizona, a Dartmouth College–University of Michigan–MIT consortium, and others also have optical telescopes on Kitt Peak.

Astronomers of several campuses of the University of California use a 3-m (120-inch) reflector at the Lick Observatory on Mount Hamilton in California. The University of Texas has its telescopes near Fort Davis in west Texas.

Part of the sky is never visible from sites in the northern hemisphere, and many of the most interesting astronomical objects in the sky are only visible or best visible from the southern hemisphere. Therefore, in the last decades special emphasis has been placed on constructing telescopes at southern sites. Some of these sites are on coastal mountains west of the Andes range in Chile. The Cerro Tololo Inter-American Observatory there has a 4-m (158-inch) twin to the Kitt Peak 4-m telescope.

The Observatories of the Carnegie Institution of Washington boast of the Las Campanas Observatory, which includes a 2.5-m (101-inch) telescope on Cerro las Campanas, another Chilean peak. The European Southern Observatory, an intergovernmental group from several European nations, has a dozen telescopes on La Silla, still another Chilean mountaintop. It is quite common in the astronomical community to go off to Chile for a while "to observe."

Perhaps the most important new site is in Hawaii. Three of the dozen largest telescopes in the world are there, on Mauna Kea on the island of Hawaii, including the 3.8-m (150-inch) United Kingdom Infrared Telescope, the 3.6-m

Figure 4–27 The summit of Mauna Kea on the island of Hawaii (the largest island in the state of Hawaii) now boasts a variety of large telescopes. The 3.6-m Canada–France–Hawaii Telescope is in the center, on the far ridge. Then come the 0.6-m planetary patrol telescope and the 2.2-m telescope of the University of Hawaii, followed by the 3.8-m United Kingdom Infrared Telescope and another 0.6-m telescope. The 3-m NASA Infrared Telescope Facility is on a separate ridge, and appears at left center. The 10-m W. M. Keck Telescope is at the extreme left. The Maxwell submillimeter telescope and a Caltech telescope are in the foreground.

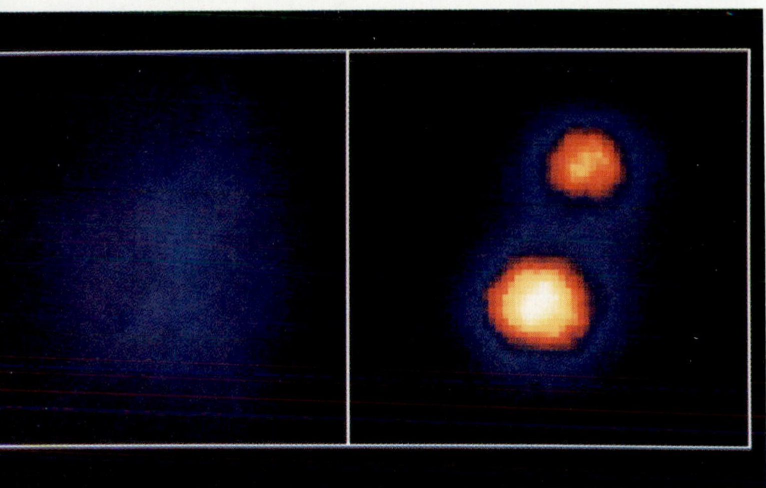

A

B

Figure 4–28 Active optics. (*A*) The double star xi Ursa Majoris with the active-optics system turned off (*left*) and on (*right*). Obviously, the active-optics system improves the image dramatically. (*B*) A laser beam is sent up to provide an artificial guide star. When the guide star's image is corrected by modifying the mirror's shape, then the images of other objects in the same field of view are corrected too. Thus, a bright object is artificially provided to allow the correction to be made.

(140-inch) Canada–France–Hawaii Telescope, and a 3.0-m (120-inch) Infrared Telescope Facility built by NASA and operated by the University of Hawaii (Fig. 4–27). This site is at such a high altitude, 4200 m (13,800 ft), that the air above is particularly dry. Since water vapor, which blocks infrared radiation, is minimal, these telescopes are especially suitable for observations at infrared and even longer wavelengths known as "submillimeter." Also, the sky is very dark, and the atmosphere is often exceptionally steady. Mauna Kea thus now includes the largest telescope of all—the 10-m W. M. Keck Telescope.

The United Kingdom, Spain, Sweden, Denmark, Holland, Ireland, Norway, and Finland are jointly constructing a new observatory, el Observatorio del Roque de los Muchachos, on the island of La Palma in the Canary Islands, Spain. The Isaac Newton telescope has been moved from the cloudy skies of the south of England to this site, with a new 2.5-m (100-inch) mirror installed, and the Nordic countries are building an additional 2.5-m telescope. A 4.2-m (165-inch) British–Irish–Dutch telescope, the world's third-largest single-mirror telescope, had "first light" in 1987. It is named after William Herschel, who discovered the planet Uranus in the 18th century.

"Active optics" is a new technology in which the shape of the mirrors can be modified from moment to moment to cancel the blurring effects of the atmosphere (Fig. 4–28).

Even though light-gathering power increases rapidly with the size of the telescope, cost used to go up even more rapidly—with the cube of the diameter. To save money, the Harvard-Smithsonian Center for Astrophysics and the University of Arizona jointly built the Multiple Mirror Telescope (MMT) (Fig. 4–29) in Arizona with six linked 1.8-m (74-inch) paraboloids. They are now replacing these several mirrors with a single 6.5-m mirror.

Tables 4–1A and B show the largest current optical telescopes.

4.5b New Technology Telescopes

Telescopes in space will solve some scientific problems, but will also suggest many new kinds of observations that can best or most efficiently be made from the ground. The need for more large ground-based telescopes continues.

Several giant telescopes planned will employ new technologies to provide large light-gathering power at relatively low cost compared to older standard designs. The first to be funded is the Keck Ten-Meter Telescope, recently completed on Mauna Kea in Hawaii. It is a joint project of the California Institute of Technology and the University of California; the University of Hawaii will

Figure 4–29 Six 1.8-m mirrors are held in a single framework to make the Multiple Mirror Telescope at the Whipple Observatory in Arizona.

Table 4–1A The Largest Optical Telescopes

Telescope or Institution*	Location	Diameter (m)	Diameter (in)	Date Completed	Mirror
Keck Telescope (Univ. Cal./Caltech)	Mauna Kea, Hawaii	10.0	400	1992	Zerodur
Soviet Special Astrophysical Obs.: BTA	Caucasus, U.S.S.R.	6.0	236	1976	Pyrex
Palomar Observatory: Hale Telescope	Palomar Mtn., California	5.0	200	1950	Pyrex
Multiple Mirror Telescope (MMT)	Mt. Hopkins, Arizona	4.5	176	1979	Fused silica
Roque de los Muchachos Obs.: Wm. Herschel Tel.	Canary Islands, Spain	4.2	165	1986	Cer-Vit
Cerro Tololo Inter-American Obs.	Cerro Tololo, Chile	4.0	156	1975	Cer-Vit
Anglo–Australian Telescope (AAT)	Siding Spring, Australia	3.9	153	1975	Cer-Vit
Kitt Peak National Obs.: Mayall Tel.	Kitt Peak, Arizona	3.8	150	1974	Quartz
United Kingdom Infrared Telescope (UKIRT)	Mauna Kea, Hawaii	3.8	150	1979	Cer-Vit
European Southern Observatory (ESO)	La Silla, Chile	3.6	142	1976	Fused silica
Canada–France–Hawaii Telescope (CFHT)	Mauna Kea, Hawaii	3.6	140	1979	Cer-Vit
New Technology Telescope (ESO)	La Silla, Chile	3.6	140	1989	Zerodur
Wash.–Princeton–Chicago–N. Mexico S.U.–Wash. S.U. (ARC)	Sacramento Peak, N.M.	3.5	138	1992	Honeycomb
German–Spanish Astronomical Center	Calar Alto, Spain	3.5	138	1983	Zerodur
Infrared Telescope Facility	Mauna Kea, Hawaii	3.0	120	1979	Cer-Vit
Lick Observatory: Shane Telescope	Mt. Hamilton, California	3.0	120	1959	Pyrex
McDonald Observatory	Mt. Locke, Texas	2.7	107	1968	Fused silica
Crimean Astrophysical Obs.: Shajn Tel.	Crimea, U.S.S.R.	2.6	102	1961	Pyrex
Byurakan Observatory	Armenia, U.S.S.R.	2.6	102	1976	Pyrex
Nordic Optical Telescope	Canary Islands, Spain	2.6	102	1988	Zerodur
Las Campanas Obs.: Irénée du Pont Tel.	Cerro Las Campanas, Chile	2.5	101	1977	Fused silica
Roque de los Muchachos Obs.: Isaac Newton Tel.	Canary Islands, Spain	2.5	101	1984	Zerodur

*Other telescopes being planned are discussed in Section 4.5.

Table 4–1B New Large Optical Telescope Projects

Project	Sponsoring Organization/Site	Area (m^2)	Equiv. Dia. (m)	Mirror
Very Large Telescope	European Southern Obs./Cerro Paranal, Chile	211	16.4	4 8.2-m glass-ceramic meniscus
Keck II	Caltech, U. California/Mauna Kea, HI	75	9.8	36 1.8-m hexagonal glass-ceramic meniscus
Spectroscopic Survey Tel.	Pennsylvania State U./Mt. Locke, TX	64	9	Pyrex
Japanese National Large Tel.	National Astron. Obs. of Japan/ Mauna Kea, HI	57	8.5	Zero-expansion meniscus
Columbus	U. Arizona, Arcetri Obs. of Florence (Italy), Research Corp. of Tucson/ Mt. Graham, AZ	55	8.4	Borosilicate honeycomb
Gemini (north and south)	NOAO, Great Britain, Canada/Hawaii and Chile	50	8 (each)	Borosilicate honeycomb
Magellan	Carnegie Inst. of Wash., Johns Hopkins U., U. AZ/Las Campanas, Chile	33	6.5	Borosilicate honeycomb
MMT Conversion	Smithsonian Astrophys. Obs., U. AZ/ Mt. Hopkins, AZ	33	6.5	Borosilicate honeycomb

share in the observing time. Its mirror will be twice the diameter and four times the surface area of Palomar's. Since we cannot construct a single 10-m mirror, U.C. scientists worked out a plan to make the mirror of many smaller segments (Fig. 4–30). A second 8-m telescope, Keck II, is under construction next to the first Keck telescope.

A

B

C

Figure 4–30 (*A*) The 10-meter W. M. Keck Telescope, a joint project of the California Institute of Technology and the University of California, under construction on Mauna Kea, with 18 segments installed. The mirror will be made of 36 contiguous hexagonal segments, which will be continuously adjustable. The telescope on Mauna Kea can be remotely controlled from California. (*B*) The Keck Telescope in its dome with all the segments installed. (*C*) One of the hexagonal segments.

Several other giant telescopes rely on a breakthrough in making relatively inexpensive, lightweight mirrors by Roger Angel of the University of Arizona (Fig. 4–31). An Astrophysical Research Consortium (ARC) of Chicago–New Mexico State–Princeton–Washington–Washington State Universities is using one of his mirrors in a 3.5-m size. A single Angel mirror 6.5 m in diameter is to replace the several mirrors of the Multiple Mirror Telescope in 1993. A Johns Hopkins University–Carnegie Institution–University of Arizona group plans a single 6.5-m telescope called Magellan in Chile. The University of Arizona, the Arcetri Observatory of Florence, Italy, and the Research Corporation of Tucson, Arizona, hope to use an 8-m Angel mirror. This "Columbus Project" is to commemorate the 500th anniversary of Columbus's first voyage to the New World.

Other huge telescopes are also being planned. The U.S. National Optical Astronomy Observatories hopes to eventually make a 15-m telescope of four 8-m mirrors, and is trying to start with a single 8-m Angel mirror in the northern hemisphere and another 8-m Angel mirror in the southern hemisphere, jointly with other countries.

Pennsylvania State University plans a segmented mirror, with limited pointing ability, for spectroscopy. The areas of the segments will add up to about 9 m.

The European Southern Observatory has received approval from its eight member nations to build the Very Large Telescope (VLT), a set of four 8-m telescopes spaced along a line in Chile (Fig. 4–32). They should be usable either separately or together, and are to be finished one at a time during the 1990's. The mirrors are to be very thin, and thus must be continuously monitored and controlled. A Japanese 8.5-m telescope, also with a "thin meniscus mirror," is to go on Mauna Kea also in the 1990's.

Figure 4–31 Roger Angel at the University of Arizona is making large lightweight mirrors by (*A*) melting raw glass and (*B*) "spin-casting" it in a rotating furnace. These views show the 6.5-m mirror made in 1992 for the former Multiple Mirror Telescope. (*C*) The melted glass takes on a paraboloidal shape because of the rotation. Here Angel (*right*) joins an engineer in inspecting a 3.5-m blank.

A

B

C

Figure 4–32 A model of the European Southern Observatory's Very Large Telescope.

4.6 Schmidt Telescopes

Bernhard Schmidt, working in Germany in about 1930, invented a type of telescope that combines some of the best features of both refractors and reflectors. In a *Schmidt camera* (Table 4–2), the main optical element is a large mirror, but it is spherical instead of being paraboloidal. Before reaching the mirror, the light passes through a thin lens, called a *correcting plate,* that distorts the incoming light in just the way that is necessary to have the spherical mirror focus it over a wide field (Fig. 4–33). The correcting plate is too thin to introduce much chromatic aberration.

One of the largest Schmidt cameras in the world is on Palomar Mountain, in a dome near that of the 5-m telescope. Renamed the Oschin Schmidt camera, it has a correcting plate 1.2 meters (49 inches) across (though it is traditionally known as the "48-inch Schmidt") and a spherical mirror 1.8 meters (72 inches) in diameter. Its field of view is about 7° across, which is much broader than the 5-m reflector's field of view (only about 2 minutes of arc across).

The field of view of the 5-m is so small that it would be hopeless to try to map the whole sky with it because it would take too long. But the 1.2-meter Schmidt has been used to map the entire sky that is visible from Palomar. The survey was carried out through both red and blue filters; 900 pairs of plates were taken over seven years, from 1949 to 1956. Plates from this Palomar Observatory–National Geographic Society Sky Survey are widely used.

Newer Schmidt cameras—one at the European Southern Observatory and the United Kingdom Schmidt Telescope at the Anglo-Australian Observatory in Siding Spring, Australia (Fig. 4–34)—have photographed the southern sky in the blue, red, and near infrared, including the one-quarter of the sky that cannot be seen from Palomar. The films they used are also considerably improved over the ones used in the earlier Palomar–National Geographic survey.

Table 4–2 Schmidt Telescopes

Telescope or Institution	Location	Diameter		Date Completed
		Corrector (m)/(in)	*Mirror (m)/(in)*	
Karl Schwarzschild Obs.	East Germany	1.3 m/53 in	2.0 m/79 in	1960
Palomar Observatory	California	1.2 m/48 in	1.8 m/72 in	1948, 1975
U.K. Schmidt	Australia	1.2 m/48 in	1.8 m/72 in	1973
Tokyo Astronomical Obs.	Japan	1.1 m/41 in	1.5 m/60 in	1975
ESO Schmidt	Chile	1.0 m/39 in	1.6 m/64 in	1972

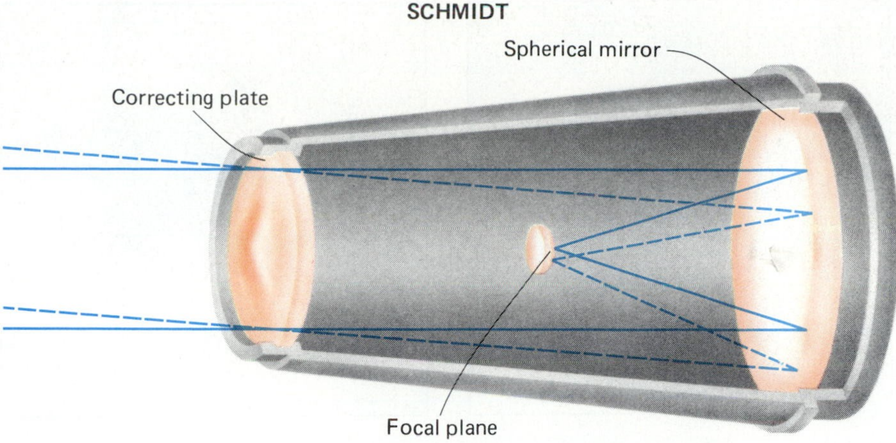

Figure 4–33 By having a non-spherical thin lens called a correcting plate, a Schmidt camera is able to focus a wide angle of sky onto a curved piece of film located at the focal plane.

A

Figure 4–34 (*A*) The U.K.–Australian Schmidt telescope in Australia. (*B*) The three individual images, each taken through a filter in one color, are printed together (*C*) to make the full-color image. This color view of the Horsehead Nebula in Orion is from the top left quarter of the three monochromatic images.

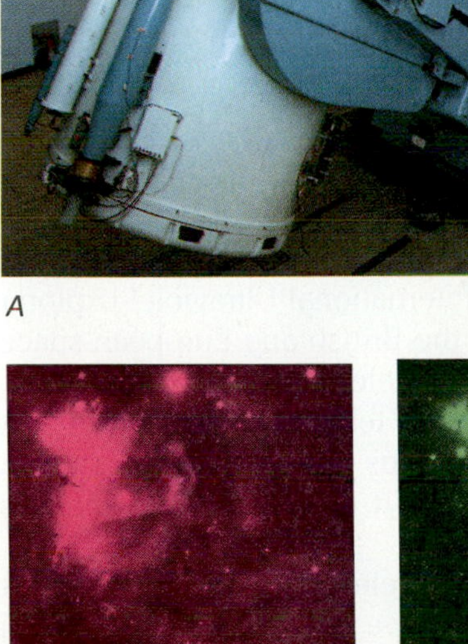
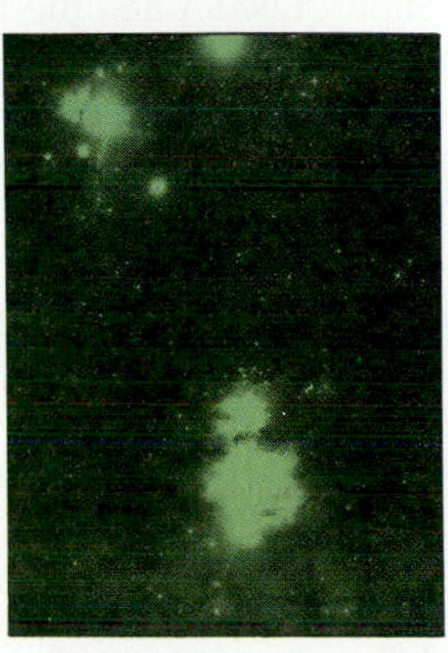

B

C

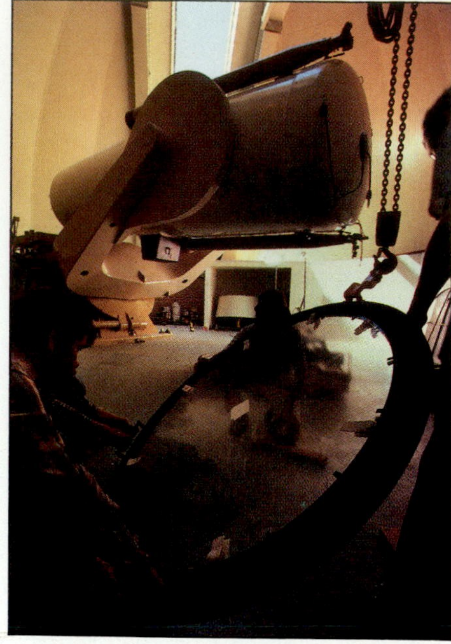

A

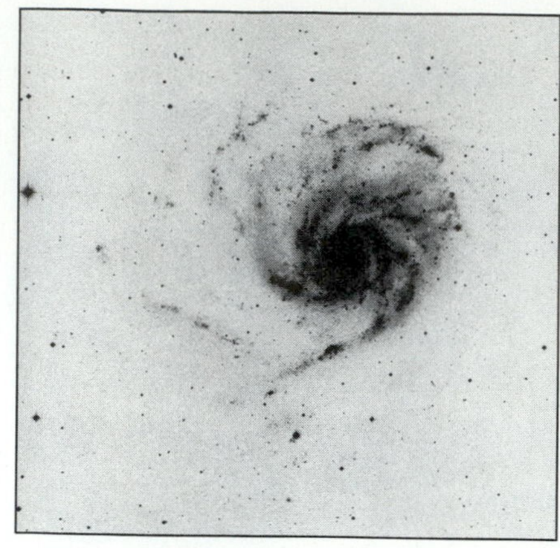

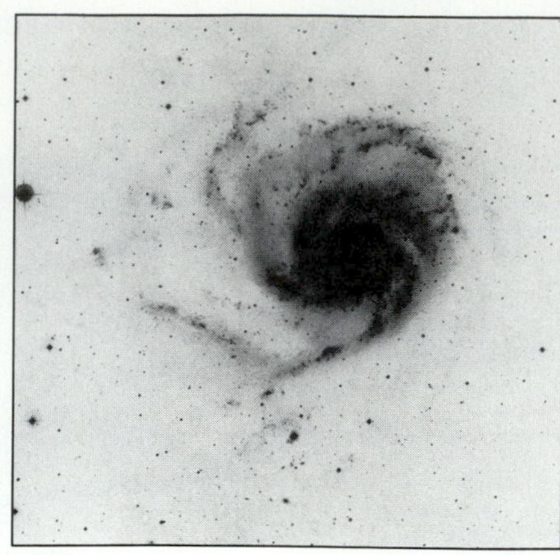

B

Figure 4–35 (*A*) The new corrector lens for the Palomar 1.2-m Schmidt telescope being used for the second-generation Palomar Observatory–National Geographic Society Sky Survey. (© 1987 National Geographic Society; photo by James Sugar) (*B*) Galaxy M101 from the old (*left*) and new (*right*) Sky Surveys. The new Survey reveals added nebulosity in the spiral arms.

A color-independent correcting plate (Fig. 4–35) has recently been made for the Palomar 1.2-m Schmidt. A second-generation sky survey is now under way, with blue, red, and infrared ("near infrared," about 8300 Å) being exposed of the whole sky north of the celestial equator. The red and blue photographs will enable astronomers to compare the skies of the 1980's and 1990's with those of 30 years earlier. The new survey will reveal objects four times fainter than those in the original survey, primarily because of Kodak's progress in providing more sensitive and finer-grain emulsions, as well as providing an infrared view. It will take until about 1992 to generate the 894 triplets of photographic plates to cover the northern sky.

The Space Telescope Science Institute, the University of Minnesota, and others are now electronically scanning not only both Palomar Observatory Sky Surveys but also special partial surveys (Fig. 4–36).

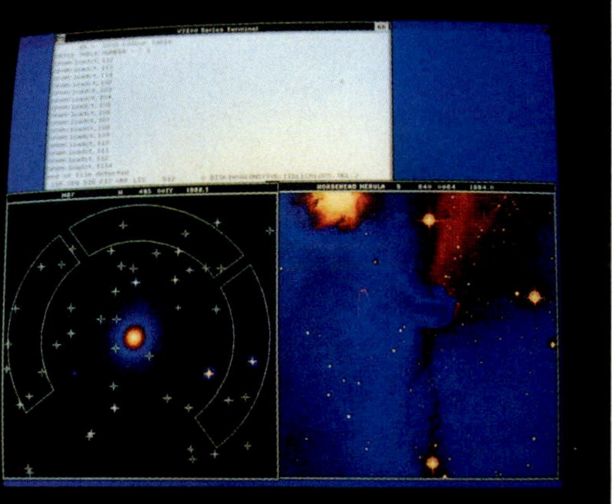

Figure 4–36 Palomar Observatory Sky Survey plates have been scanned and are being scanned and digitized at the Space Telescope Science Institute, at the University of Minnesota, and elsewhere. The digitized images can be played back, as shown in this false-color display on a computer terminal at STScI. At left, the positions of the guide stars in the field are overlaid on an image of the field surrounding the galaxy M87. The + signs in the curved boxes are usable for guiding an image of this field. At right, a false-color image is made of the Horsehead Nebula in Orion.

4.7 Telescopes in Space

Until recently, the largest astronomical telescope in space was the 0.9-m (36-inch) telescope aboard the Copernicus satellite, the third Orbiting Astronomical Observatory, launched by NASA in 1973. The International Ultraviolet Explorer (IUE), launched by a joint effort of NASA and the British and European space organizations in 1978, carries a 0.45-m (18-inch) telescope with two spectrographs for studies of the ultraviolet. It carries a television-like system that records data much more efficiently than Copernicus and hence gives spectra more quickly. In Section 25.8, we describe what it is like to work with this important spacecraft.

In 1990, NASA launched the Hubble Space Telescope (HST), the largest telescope in space to date. It is a reflecting telescope with a main mirror 2.4 m (94-inches) in diameter (Fig. 4–37). The HST was designed to have three major advantages over ground based telescopes, even though it is much smaller than the major ones. First, since it is above the Earth's atmosphere, the pinpoint structure of the images should be limited only by the mirror's size. (Telescopes on Earth, on the other hand, are limited by the atmospheric "seeing.") The HST was designed to give images only one-tenth of an arc second across, ten

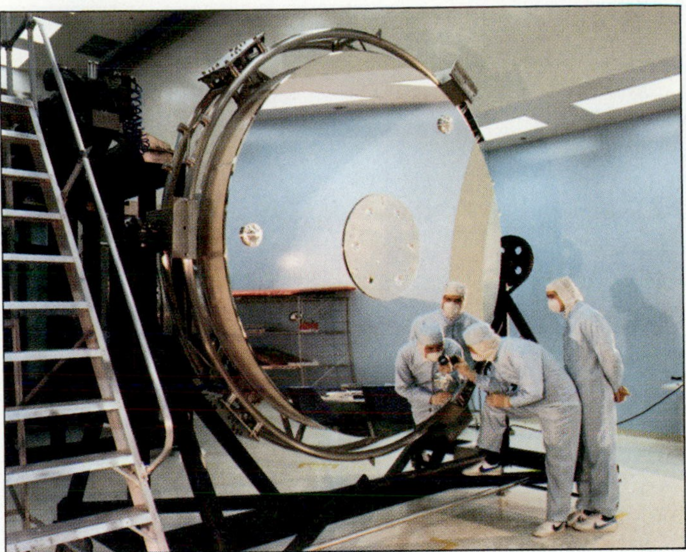

Figure 4–37 The 2.4-m mirror of the Hubble Space Telescope, after it was ground and polished and covered with a reflective coating. This single telescope should make many discoveries but will probably raise as many questions as it will answer. So we need still more ground-based telescopes as partners in the enterprise.

times finer than normal good seeing at major observatories and five times better than the best moments of seeing. Second, starlight should be concentrated into smaller images compared with images seen by ground-based telescopes, and the sky above is extremely dark, so the HST should detect fainter objects than can be seen from Earth. Third, with no atmosphere above the telescope to absorb the ultraviolet and infrared, these parts of the spectrum will be detectable.

This orbiting observatory has been named the Edwin P. Hubble Space Telescope. Hubble was the scientist who, in the 1920's, discovered that there are other galaxies comparable in scale to our own (Part VI). He also discovered that the Universe is expanding (Chapter 31).

The project is a major one. It has taken more than a decade and has cost close to 2 billion dollars. Most of the funds are from NASA. The European Space Agency provided the solar array that provides power, one of the instruments, and some of the staff. The telescope is controlled from the Space Telescope Science Institute (STScI) at Johns Hopkins in Baltimore. The HST is designed to work for at least 15 years, and to be visited by astronauts every three years or so for maintenance and for upgrading of instruments.

Figure 4–38 The Hubble Space Telescope photographed with an IMAX camera aboard the space shuttle that launched it in 1989. The telescope should one day improve our understanding of the cosmic scale of distances to galaxies, image Jupiter and Saturn with similar resolution to spacecraft that flew close to them, and maybe even discover planets around other stars.

After many delays, the Hubble Space Telescope was launched into space from the Kennedy Space Center in April 1990. It is such a complex piece of apparatus that nine months were allotted for checking out all the parts. Some problems arose, such as a vibration from the change in temperature when the telescope went from day to night every ninety minutes. These problems were well on the way to being overcome when a serious limitation was detected about two months after launch. It seemed that the telescope had no one good point of focus. All the images of stars were too large. Apparently, one of the mirrors was not made to the proper shape. The result was spherical aberration. It is as though the telescope was nearsighted.

Could eyeglasses be provided to clear the images? Scientists and engineers think so. A second-generation wide-field/planetary camera was already under construction. Its optics are being modified to put in a correcting lens that should unblur the images. A unit called COSTAR, carrying several small mirrors, should correct the images in the other instruments. The repair mission is scheduled for late 1993. It will also replace failed gyroscopes and the solar panels, which vibrate too much. An infrared instrument with correcting optics is to be launched after about six years.

In the meantime, there is plenty for the Hubble Space Telescope to do with

Figure 4–39 The Hubble Space Telescope, photographed from the space shuttle Discovery. In the background are parts of Bolivia, Chile, and Peru some 615 km away. Note the tip of the shuttle's remote-manipulator arm at bottom center. The edge of Lake Titicaca appears at upper right.

its unique ultraviolet capability. It will do ultraviolet work first and hold off visual-light studies. The HST remains orbiting the Earth (Fig. 4–39) while scientists and engineers work to bring it to full capabilities.

4.8 Light-Gathering Power and Seeing

The principal function of large telescopes is to gather light. The amount of light collected by a telescope mirror is proportional to the area of the mirror, πr^2. (It is often convenient to work with the diameter, which is twice the radius, since the diameter is the number usually mentioned. Since $r = d/2$, the formula for the area of a circle can be written as $\pi(d/2)^2 = \pi(d^2/4.)$

The ratio of the areas of two telescopes is thus the ratio of the square of their diameters $(d_1{}^2/d_2{}^2)$, which is usually more easily calculated as the square of the ratio of their diameters $(d_1/d_2)^2$. For example, the 10-meter telescope will have an area four times greater than the area of the 5-meter telescope, $(10/5)^2$ (Fig. 4–40). Thus with exposures of equal duration, the 10-m would collect four times as much light as a 5-m if all other things were equal. For the same exposure time, it would therefore record fainter stars.

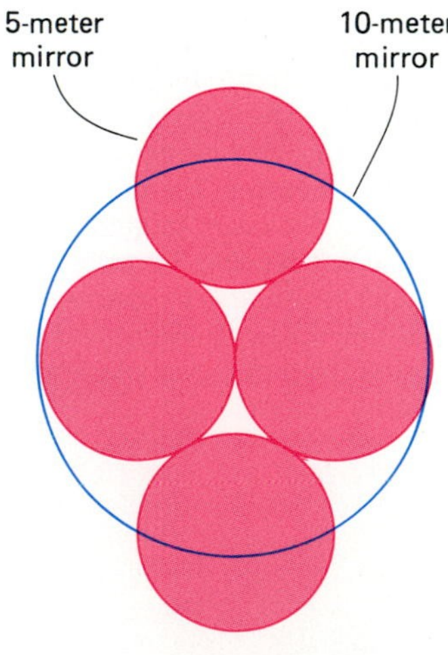

Figure 4–40 A telescope twice the diameter of another has four times the collecting area. The area enclosed by the large single circle is the same as the sum of the areas of the four shaded circles.

4.8a Seeing, Transparency, and Light Pollution

A telescope's light-gathering power (and its ability to concentrate light in as small an area as possible) is really all that is important for the study of individual stars, since the stars are so far away that no details can be discerned no matter how much magnification is used. The main limitation on the smallest image size that can be detected by ground-based telescopes is caused not in the telescope itself, but rather by turbulence in the Earth's atmosphere. The limitations are connected with the twinkling effect of stars. The steadiness of the Earth's atmosphere is called, technically, the *seeing*. We say, when the atmosphere is steady, "The seeing is good tonight." "How was the seeing?" is a polite question to ask astronomers about their most recent observing runs. Another factor of importance is the *transparency*, or how clear the sky is. It is quite possible for the transparency to be good but the seeing to be very bad,

Figure 4–41 The continental United States photographed from a satellite at night, showing how much light is escaping into the sky.

or for the transparency to be bad (a hazy sky, for example) but the seeing excellent.

A problem of modern civilization is *light pollution* (Fig. 4–41).

4.8b Magnification

For some purposes, including the study of the Sun, the planets, and the resolution of double stars, the magnification provided by the telescope does play a role. Magnification can be important for the study of *extended objects* (as distinguished from *point objects,* objects whose images are points). Nebulae and galaxies are other examples of extended objects.

Magnification is the number of times an object's angular size is enlarged; the angular size is commonly expressed in degrees, minutes, or seconds of arc, and gives the angle across the sky that an object appears to cover in any one direction. Magnification works out to be equal to the focal length of the telescope objective (the primary mirror or lens) divided by the focal length of the eyepiece. By simply substituting eyepieces of different focal lengths, you can get different magnifications. However, there comes a point at which the atmosphere limits you from getting more detail, even with a shorter eyepiece.

A telescope of 1-meter focal length will give a magnification of 25 when used with a 40-millimeter eyepiece, that is, (1 m)/(40 mm) = 25, an ordinary combination for amateur observing. With a 10-millimeter eyepiece, the same telescope has a magnification of 100. We would say that we were observing with "100 power."

The field of view of a telescope depends on the eyepiece used. The higher the magnification, the smaller the field of view. The field of view is usually indicated by degrees or minutes of arc from side to side in the sky. For example, with an eyepiece that allows the Moon to fill the image, the field of view is ½°. An eyepiece of half the focal length will have twice the magnification and a field of view of ¼°.

4.9 Spectroscopy

We have discussed the collection of quantities of light by the use of large mirrors or lenses, and the focusing of this light to a point or suitable small area. Often we want to break up the light into a spectrum instead of merely photographing the area of sky at which the telescope is pointed.

Figure 4–42 Both (*A*) prisms and (*B*) diffraction gratings make spectra. (*C*) Photography through a large prism over the telescope's objective—an "objective prism"—gives a spectrum for each star in the field of view.

A prism breaks up light into its spectrum for the same reasons that light focused by a lens has chromatic aberration. A wave front is refracted as it hits the side of the prism, but it is refracted by different amounts at different wavelengths.

Today, astronomical spectra are no longer usually made by means of prisms. Light may be broken up into a spectrum, without need for a prism, by

Figure 4–43 The light entering the instrument from the left passes through a slit that defines a line of radiation, is made parallel (*collimated;* a double-lens collimator is shown, but other alternatives are also used) and directed on the prism (or other dispersing element), and then is focused. The device is called a *spectroscope* if we observe with the eye and a *spectrograph* if we record the data with a photographic plate or an electronic sensor.

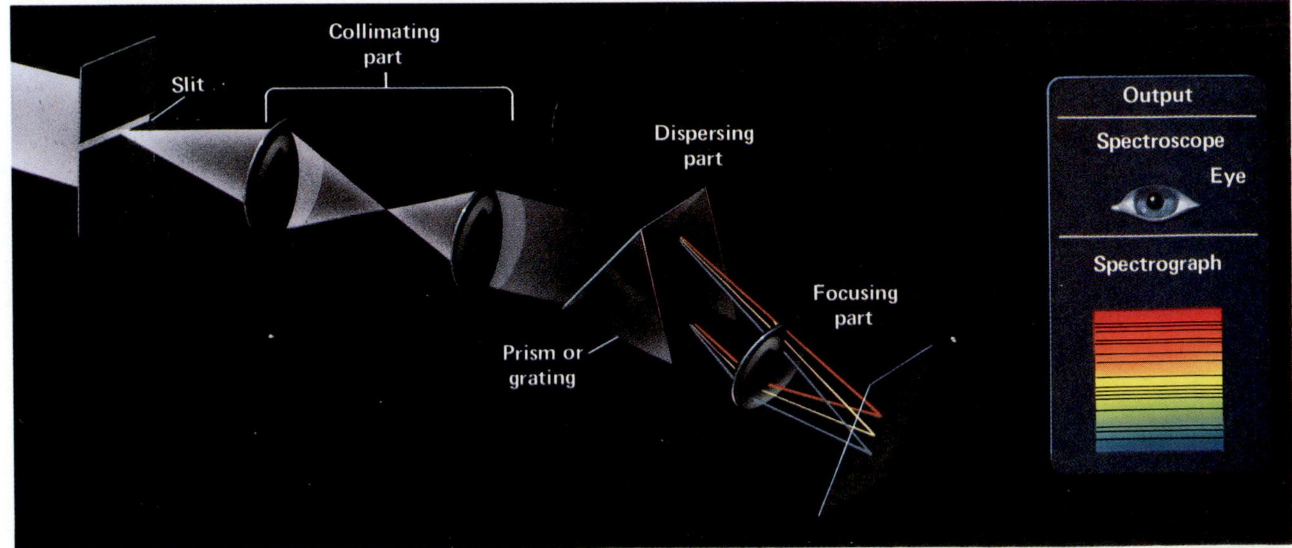

lines that are ruled on a surface very close together. And I do mean close together: often 10,000 parallel lines are ruled in each centimeter (25,000 lines in a given inch)! Such a ruled surface is called a *diffraction grating*.

When a beam of parallel light falls on a prism or grating (Fig. 4–42), the colors that result are often indistinct because the spectrum of the light in one place in the beam overlaps the spectrum of an adjacent spot in the beam. We can limit the blurring that results by allowing only the light that passes through a long, thin opening—called a *slit*—to fall on the prism. Then we have spectral lines (Section 4.10) that are sharp and relatively narrow.

Scientists use optics resembling a small telescope to examine the spectrum that results. A spectrum-making device, usually including everything between the slit at one end and the viewing eyepiece at the other, is called a *spectroscope* (Fig. 4–43). When the resultant spectrum is not viewed with the eye but is rather recorded either with a photographic plate or electronically, the device is called a *spectrograph*.

You can see the diffraction on a grating when looking at the surface of a compact disc or laser disc (Fig. 4–44).

Figure 4–44 A compact disc or laser disc contains so many lines close together that it acts like a diffraction grating and makes a spectrum.

4.10 Spectral Lines

As early as 1666, Isaac Newton showed that sunlight is composed of all the colors of the rainbow. William Wollaston in 1804 and Joseph Fraunhofer in 1811 also studied sunlight as it was dispersed (spread out) into its rainbow of component colors (Fig. 4–45). Fraunhofer was able to see that at certain colors there were gaps that looked like dark lines across the spectrum (Fig. 4–46). These gaps are thus called *spectral lines;* the continuous radiation in which the gaps appear is called the *continuum* (plural: *continua*). The dark lines in the spectrum of the Sun and in the spectra of stars are gaps that represent the lessening of electromagnetic radiation at those particular wavelengths. These dark lines are called *absorption lines;* they are also known (for the Sun, in particular) as *Fraunhofer lines.* In contrast, an ordinary light bulb (an "incandescent" bulb) gives off only a continuum; it has no spectral lines.

Note that "absor**b**" ends with a "b," but "absor**p**tion" is spelled with a "p."

It is also possible to have wavelengths at which there is somewhat more radiation than at neighboring wavelengths; these are called *emission lines.* We shall see that the nature of a spectrum, and whether we see emission or absorption lines, can provide considerable information about the nature of the body that was the source of light. We say that the lines are "in emission" or "in absorption."

Stations on the radio, places in the spectrum where there is energy, represent spectral lines. In particular, they are emission lines.

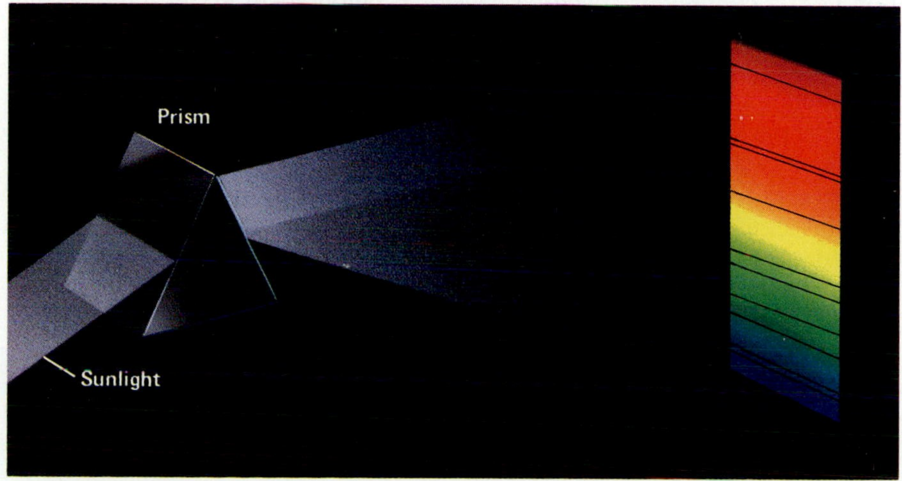

Figure 4–45 When a narrow beam of sunlight is dispersed by a prism, we see not only a continuous spectrum but also dark Fraunhofer lines.

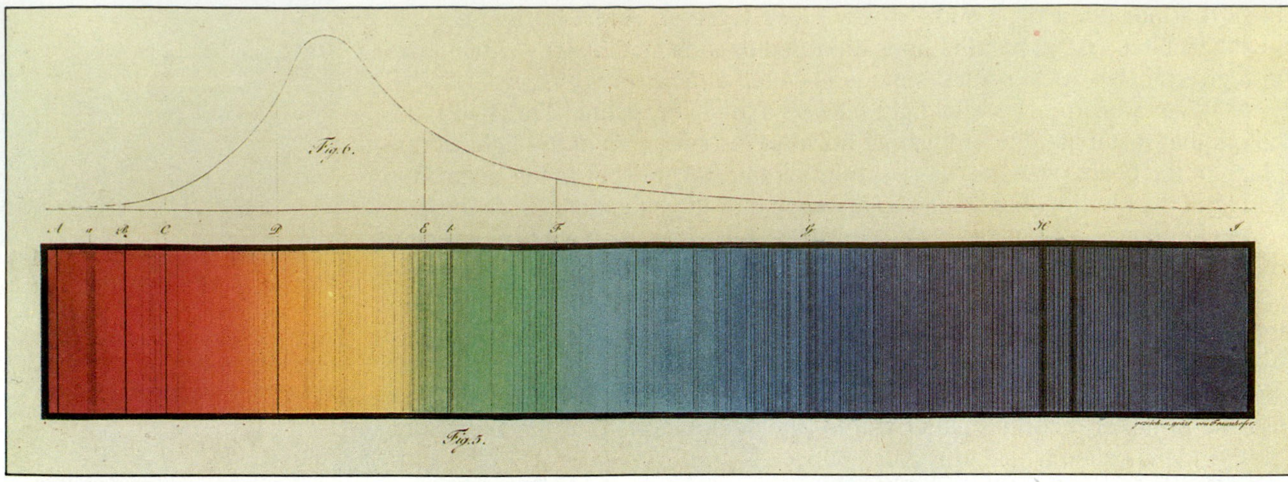

Figure 4–46 Fraunhofer's original spectrum. Only the D and H lines retain their original notation from that time. We now know that the C line is from hydrogen, the D line is from sodium, and the H line is from ionized calcium.

The explanation of the spectral lines was discovered during the nineteenth century in laboratories on Earth. Patterns of spectral lines can be explained as the absorption or emission of energy at particular wavelengths by atoms of chemical elements in gaseous form. If the gaseous form of any specific element is heated, it gives off a characteristic set of emission lines. That element, and only that element, has that specific set of spectral lines. If, on the other hand, a continuous spectrum radiated by a source of energy at a high temperature is permitted to pass through cooler gas of any specific element, a set of absorption lines (Fig. 4–47) appears in the continuous spectrum at the same characteristic wavelengths as those of the emission lines of that element. In this case, the gas has subtracted energy from the continuous spectrum passing through. It subtracted the energy at the set of wavelengths that is characteristic of that element. This rule was discovered by the German chemist Gustav Kirchhoff in 1859. If a continuous spectrum is directed first through the vapor of one absorbing element and then through the vapor of a second absorbing element, or through a mixture of the two gases, then the absorption spectrum that results will show the characteristic spectral absorption lines of both elements.

Though the exact German pronunciation is difficult for native speakers of English, the name "Kirchhoff" is normally pronounced in English with a hard "ch" (that is, like "k"): Kirk'hoff. Note the double "h."

Most solids, and gases under some conditions, can give off continuous spectra. Continuous spectra represent energy spread over a wide range of wavelengths instead of being concentrated at just a few.

The same characteristic patterns of spectral lines that we detect on Earth are observed in the spectra of stars (Fig. 4–48), so we conclude that the same chemical elements are in the outermost layers of the stars. They absorb radiation from a continuous spectrum generated below them in the star, and thus cause the formation of absorption lines. We make use of the fact that each element has its own characteristic pattern of lines when it is in a certain range of conditions of temperature and density (a measure of how closely the star's matter is packed). Thus we can study the absorption spectrum of a star (1) to identify the chemical constituents of the star's atmosphere (in other words, the types of atoms that make up the gaseous outer layers of the star), (2) to find the temperature of the surface of the star, and (3) to find the density of the radiating matter.

Figure 4–47 Almost all stars have absorption lines in their spectra. Here we see the spectrum of the star Vega.

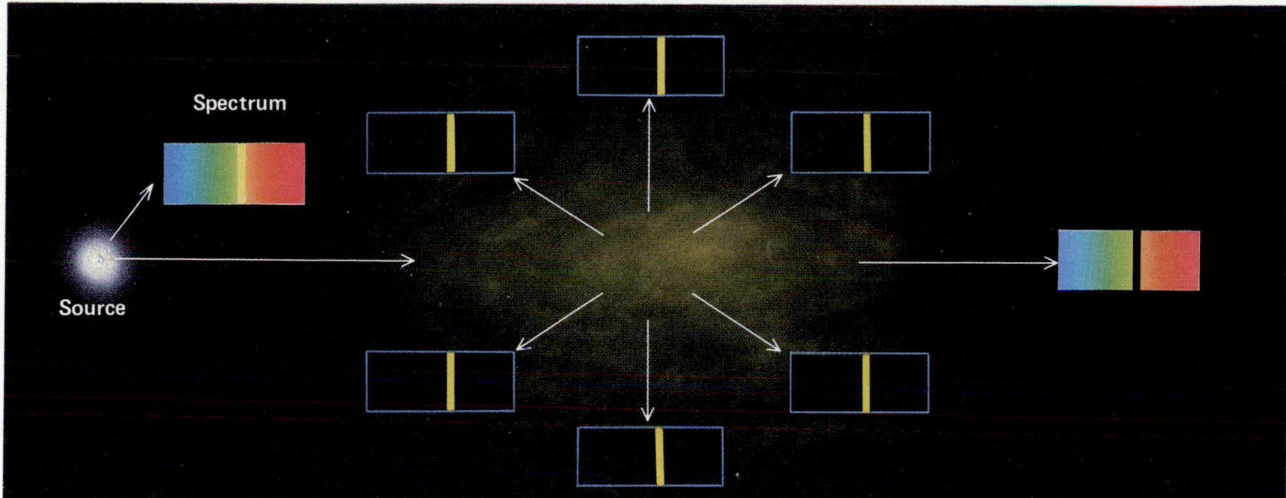

Figure 4–48 When we view a source that emits a continuum through a gas that emits emission lines, we may see absorption lines at wavelengths of the emission lines. Each of the inset boxes shows a spectrum that is the one you would measure if you were at the tip of the arrow looking back along the arrow. Note that the view from the right shows an absorption line, whereas from any other angle the gas in the center appears to be giving off an emission line. Only when you look through one source silhouetted against a hotter source do you see any absorption lines.

Not only individual atoms but also molecules—groups of linked atoms—exist in cooler bodies. (Such cooler bodies include the coolest stars, some of the gas between the stars, and the atmospheres of planets.) Molecules also have characteristic absorption—often in wide "bands"—and what we say for identifying elements goes for molecules as well.

Where does the radiation that is absorbed go? A very basic physical law called the *law of conservation of energy* says that the radiation cannot simply disappear. The energy of the radiation may be taken up in a collision of the absorbing atom with another atom. Alternatively, the radiation may be emitted again, sometimes at the same wavelength, but in random directions. Thus, fewer bits of energy proceed straight ahead at that particular wavelength than were originally heading in that direction. This leads to the appearance of an absorption line when we look from that direction.

Moreover, if one element or molecule is present in relatively great abundance, then its characteristic spectral lines will be especially strong. By observing the spectrum of a star or planet, we can tell not only which kinds of atoms or molecules are present but also their relative abundances.

The method of spectral analysis is a powerful tool that can be used to explore the Universe from our vantage point on Earth. It tells us about the planets, the stars, and the other things in the Universe. It also has many uses outside of astronomy. For example, by analyzing the spectrum, we can determine the presence of impurities in an alloy deep inside a blast furnace in a steel plant on Earth. We can measure from afar the constituents of lava erupting from a volcano; indeed, colleagues and I have done so using the same spectrometer we used to study eclipses. Sensitive methods developed by astronomers trying to advance our knowledge of the Universe often are put to practical uses in fields unrelated to astronomy.

4.11 Recording the Data

Astronomers usually want a more permanent record than is afforded by merely observing the image and more accuracy than can be guaranteed in a sketch. Traditionally, we put a photographic film or "plate" (short for photographic plate, a layer of light-sensitive material on a sheet of glass) instead of an eyepiece at the observer's end of the telescope, so as to get a permanent record of the image or spectrum.

Basically, a photographic plate (or film) consists of a glass or plastic backing covered with grains of a silver compound. When these grains are

Figure 4–49 This enlargement shows clumps of grain on a photographic plate showing an astronomical spectrum. (We shall see later on how to interpret a set of spectra like this. For each of the three sets of spectra, the spectrum of a galaxy goes from left to right across the plate. Above and below it appear, for comparison, identical sets of emission lines formed on Earth.

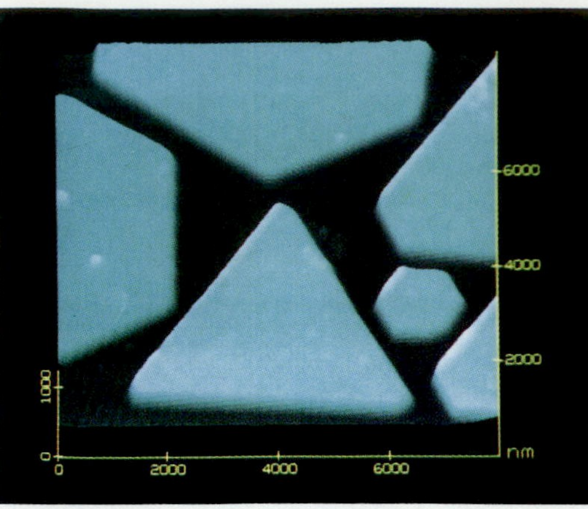

Figure 4–50 Photographic grain on an astronomical emulsion (103a-O, where the "a" stands for an astronomical emulsion that remains sensitive for long exposures and the "O" stands for blue sensitivity), enlarged 2000 times.

struck by enough light, they undergo a chemical change. The result, after development, is a "negative" image, with the darkest areas corresponding to the brightest parts of the incident image. Inspection of a photographic plate (Fig. 4–49) under high magnification shows the grainy structure of the image (Fig. 4–50).

Other methods of recording data have been developed. The day of film is past, though new emulsions now available are more sensitive and have finer grain than older emulsions. Still, the future lies in electronic devices.

The film or electronic devices significantly increase our ability to detect signals from faint sources. The eye and brain can process information for only a fraction of a second at a time and do not function cumulatively. A photographic plate or electronic sensor, however, can be left exposed to radiation for a long period. Just as a long exposure with an ordinary camera can record objects that are only dimly lit, a long exposure with a telescope can record fainter objects than would a shorter exposure.

Electronic devices are more sensitive to faint signals than is photographic film, and can also be made to be sensitive in a wider region of the spectrum. Measurements of intensity, usually made with such electronic devices, are called *photometry.* Often the intensity of objects is measured through each of several colored filters in turn, to find out how the continuous spectrum differs at different wavelengths.

Astronomers now use electronic devices built on "chips," which are related to the technology that has recently brought electronic calculators and digital watches to their present versatility and inexpensive price. These devices can be used in place of photographic plates to make either direct images of astronomical objects or spectra. They come in several types, especially charge-coupled devices, known as *CCD's* (Fig. 4–51). A *CCD* accumulates light on its surface and then scans the resulting image off that surface with internal circuitry. The name "CCD" refers to the scanning method, which involves shifting the electrical signal from each point to its neighbor one space at a time, analogous to what happens with water in a fire-fighting bucket brigade when buckets of water are passed from one individual to the next. A CCD's picture is created by scanning the picture elements line by line, as occurs in our televisions. CCD's are almost 100 times more sensitive than film. Also, adjacent objects that differ greatly in brightness can be accurately compared on a CCD image, whereas film would not permit such a comparison.

Figure 4–51 The flat square is a CCD chip with 1024 × 1024 pixels. CCD's, charge-coupled devices, are being used as detectors in many telescopes. They are normally buried in electronics and cooled by liquid nitrogen.

CCD's have become more commonplace at observatories than film. The imaging system on the Hubble Space Telescope is a 1600 × 1600 "pixel" array. The image is made up of 1600 × 1600 individually measured spots; each is known as a "picture element"—*pixel,* for short (Fig. 4–52).

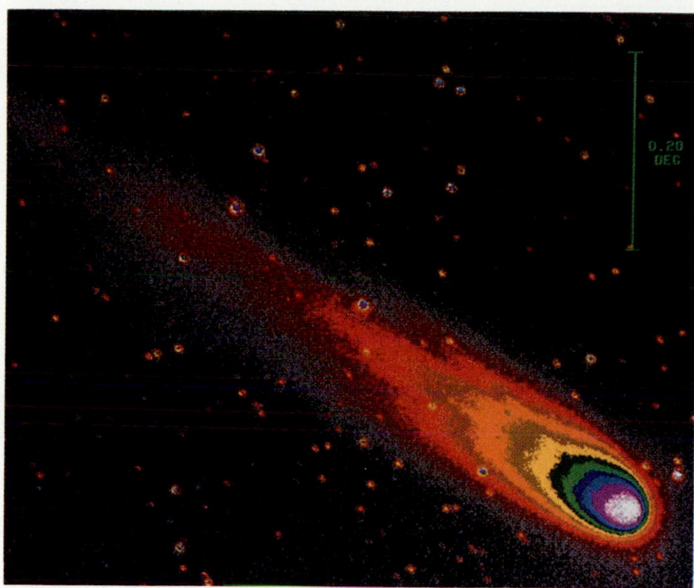

Figure 4–52 A CCD image of Halley's Comet; the individual picture elements (pixels) can be seen when this is enlarged. In this display, false color distinguishes different levels of brightness.

*4.11a Fiber Optics

Many long-distance telephone lines are now optical fibers, over which signals are sent as flashes of tiny lasers. The laser beams travel basically straight along the fibers; if they should deviate by a small angle, they are reflected off the sides so as to keep them travelling forward, even when the fiber is bent (Fig. 4–53).

In astronomy, optical fibers are beginning to be used to make telescopes more efficient. For example, several optical fibers can be arranged so that each fiber accepts the light from a different star. The other ends of the fibers can be arranged so that they form a line, making the light coming out cover the whole slit of a spectrograph. Each part of the length of the spectrograph slit thus represents light from a different star. The procedure allows a great gain in the efficiency of the spectrograph, for it now gathers several spectra at the same time instead of just one.

4.12 Observing at Short Wavelengths

From antiquity until 1930, observations in a tiny fraction of the electromagnetic spectrum, the visible part, were the only way that observational astronomers could study the Universe. Most of the beautiful images that we have in our minds of astronomical objects are based on optical studies, since most of us depend on our eyes to discover what is around us.

Thus far, we have talked of the spectrum in terms of the wavelengths of the radiation, which implies that radiation is a wave. But when light is absorbed, the energy is always transferred in discrete amounts that depend on the wavelength of the light. The radiation acts as though it were composed of particles rather than continuous waves. So it is often useful to think of light as particles of energy called *photons* instead of as waves. Photons that correspond to light of different wavelengths have different energies. We will come back to this idea in Section 20.3. But you will find it useful to have a picture in your mind

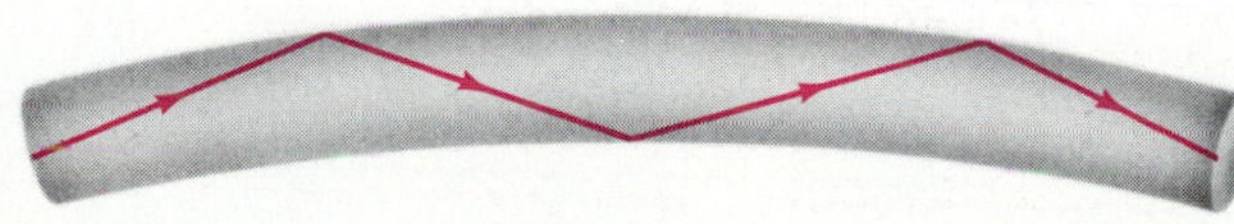

Figure 4–53 As laser light bounces off the walls of optical fibers at small angles, it keeps travelling forward along the fiber, even when the fiber is bent.

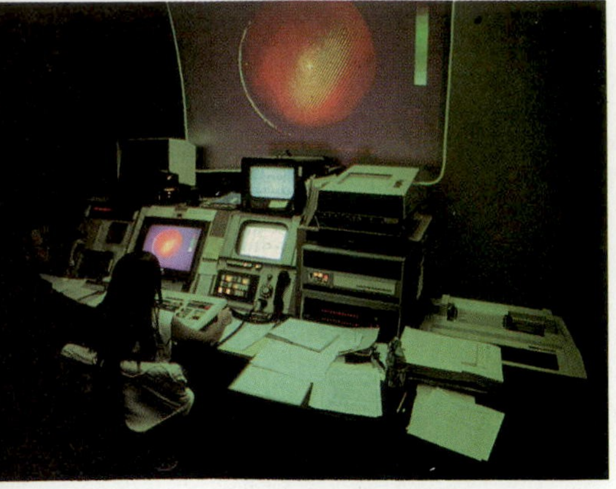

Figure 4–54 The control room of the International Explorer Spacecraft at NASA's Goddard Space Flight Center in Maryland. The image displayed on the screen shows a stellar spectrum in the ultraviolet.

of photons of high energy, and thus high penetrating power, corresponding to relatively short wavelengths (like x-rays). Similarly, photons corresponding to long wavelengths (radio waves or infrared) don't contain much energy, often not even enough to affect a photographic plate.

The wave theory and the particle theory are but different ways of understanding light. Each provides a "model"—a framework for thinking. In actuality, both the wave theory and the particle theory accurately explain how light acts at different times. The fact that we make scientific models to help us comprehend reality does not change the fact that reality may be more complicated.

4.12a Ultraviolet and X-Ray Astronomy

Ultraviolet and x-ray photons have shorter wavelengths and greater energies than photons of visible light. Thus they too have enough energy to interact with the silver grains on film. Photographic methods can be used, therefore, throughout the x-ray, ultraviolet, and visible parts of the spectrum. Electronic devices like the ones that work in the visible work in the ultraviolet too (Fig. 4–54).

At this shorter end of the spectrum, gamma rays, x-rays, and ultraviolet light do not come through the Earth's atmosphere. Ozone, a molecule of three atoms of oxygen (O_3), is located in a broad layer between about 20 and 40 kilometers in altitude. The ozone prevents all the radiation at wavelengths less than approximately 3000 Å from penetrating. To get above the ozone layer, astronomers have launched telescopes and other detectors in rockets and in orbiting satellites. The Extreme Ultraviolet Explorer (EUVE) spacecraft was launched in 1992 to map the sky in the spectral region from 10 Å to 1000 Å.

4.12b X-Ray Telescopes

In the shortest wavelength regions, we cannot merely use mirrors to image the incident radiation in ordinary fashion, since the x-rays will pass right through the mirrors! Fortunately, x-rays can still be bounced off a surface if they strike the surface at a very low angle. This is called *grazing incidence*. This principle is similar to that of skipping stones across the water. If you throw a stone straight down at a lake surface, the stone will sink immediately. But if you throw a stone out at the surface some distance in front of you, the stone could bounce up and skip along a few times.

By carefully choosing a variety of curved surfaces that suitably allow x-radiation to "skip" along, astronomers can now make telescopes that actually

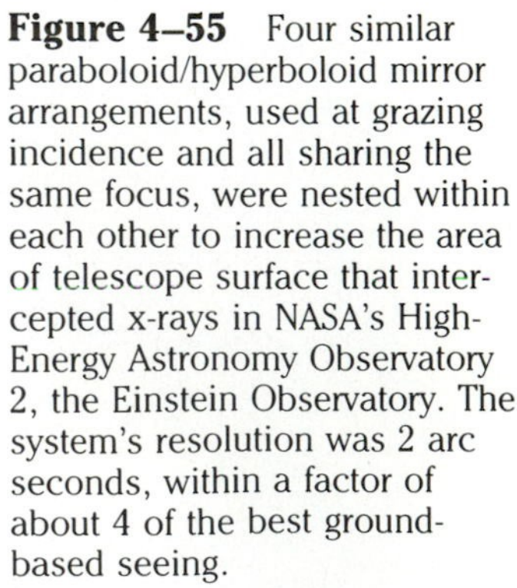

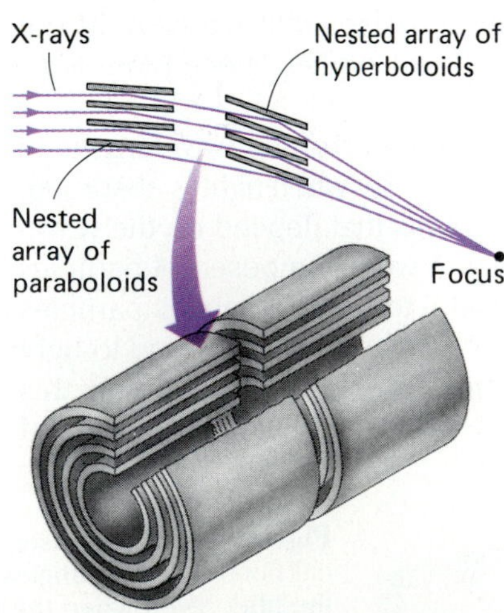

Figure 4–55 Four similar paraboloid/hyperboloid mirror arrangements, used at grazing incidence and all sharing the same focus, were nested within each other to increase the area of telescope surface that intercepted x-rays in NASA's High-Energy Astronomy Observatory 2, the Einstein Observatory. The system's resolution was 2 arc seconds, within a factor of about 4 of the best ground-based seeing.

A

B

Figure 4–56 (*A*) The nested mirrors for the Einstein Observatory. You can see a reflective surface in this oblique view. (*B*) A front view of the nested ROSAT mirrors.

make x-ray images. But the telescopes appear very different from optical telescopes (Figs. 4–55 and 4–56).

In order to form such high-energy photons, processes must be going on out in space that involve energies very much higher than most ordinary processes that go on at the surface layers of stars. The study of the processes that bring photons or particles of matter to high energies is called *high-energy astrophysics.* In the late 1970's NASA launched a series of three High-Energy Astronomy Observatories (HEAO's, pronounced "hee-ohs") to study high-energy astrophysics. HEAO-1 (Fig. 4–57) surveyed the x-ray sky with high sensitivity and excellent time resolution. HEAO-2, known as the Einstein Observatory, was able to image and study in detail the interesting objects mapped by HEAO-1. HEAO-3 studied cosmic rays (particles of matter moving with high energies) and gamma rays.

The European Space Agency's EXOSAT (European X-ray Observatory Satellite) was launched in 1983 and lasted until 1987. Though EXOSAT's images of objects were somewhat less detailed than those from the Einstein Observatory, EXOSAT had better spectral resolution than Einstein and was more sensitive at short wavelengths.

Figure 4–57 HEAO-1 (High-Energy Astronomy Observatory 1), launched in 1977 to study x-rays and gamma rays.

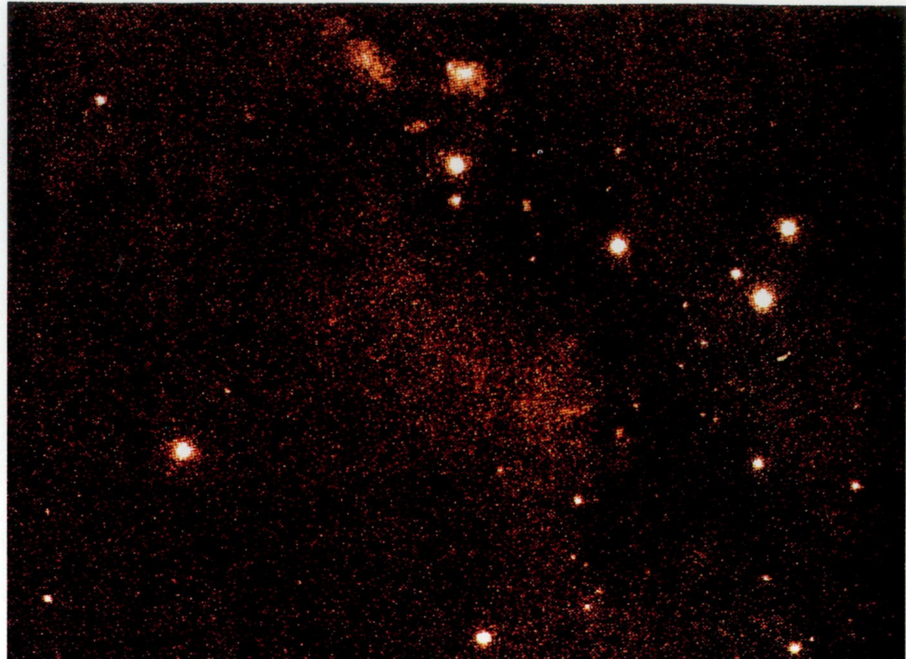

Figure 4–58 The center of our galaxy, observed from Rosat. We see a region 2° wide.

In 1987, the Japanese launched a small x-ray satellite Ginga ("Galaxy"), with the participation of American and British scientists. It reentered the atmosphere in 1991. Also in 1987, the Soviets launched the Quant ("Quantum") x-ray telescope aboard their Mir ("World") space station. The Soviet Granat satellite, launched in 1989, carries the French x-ray/gamma-ray Sigma telescope.

Larger x-ray satellites include the U.S.–West German–British Rosat (**Ro**entgen **sat**ellite, named for the discoverer of x-rays), launched in 1990. Like the Einstein Observatory, it uses a nested array of grazing-incidence mirrors and can observe in a longer-wavelength x-ray region of the spectrum than has been previously studied. Rosat provided a survey of the entire sky at a sensitivity 1000 times better than the last full-sky map (Fig. 4–58). It is slightly more sensitive than Einstein was for pointed observations in the x-ray band from 5 Å to 10 Å (Fig. 4–59); Einstein's sensitivity, though, extended to still shorter (higher-energy) wavelengths. Rosat's spatial resolution, like Einstein's, is a few seconds of arc. Observing time is distributed 50 per cent to U.S. scientists, 38 per cent to German scientists, and 12 per cent to U.K. scientists.

Figure 4–59 A Rosat image detecting x-rays from the Moon and showing that there is still a greater flux of x-rays coming from beyond.

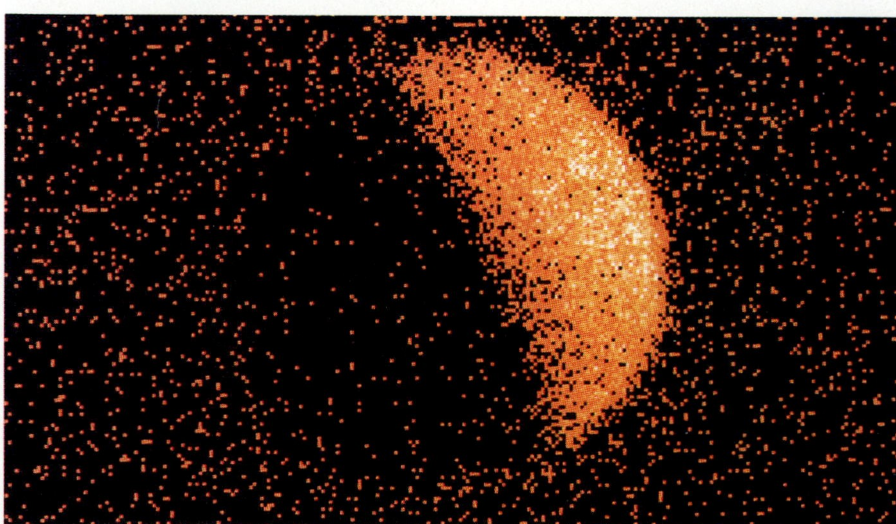

In the late-1990's we hope for completion of the American AXAF—the Advanced X-Ray Astrophysics Facility (Fig. 4–60). AXAF's collecting area and spatial resolution will be much better than those of the Einstein Observatory. AXAF received the highest priorities of the Field Report in 1982 and the Bahcall Report in 1990, both lengthy considerations of national priorities by distinguished committees. In spite of this unanimity among astronomers, and successful progress in making AXAF's mirrors, funds keep getting cut back. As of 1992, a radical plan divides AXAF into two missions, the first using the outer mirrors for imaging and some spectroscopy with the hope of launch in 1999.

Figure 4–60 An artist's conception of the single-launch version of AXAF, now divided into two launches to lower the cost-per-year over the next few years.

4.13 Observing at Long Wavelengths

4.13a Infrared Astronomy

Infrared photons have longer wavelengths and thus lower energies than visible photons. They do not have enough energy to interact with ordinary photographic plates. Some special films can be used at the very shortest infrared wavelengths, but for the most part astronomers have to employ methods of detection involving electronic devices in this region of the spectrum. Infrared wavelengths range from about 10,000 Å = 1 micrometer (μm) to about 1000 micrometers = 1 millimeter. Astronomers still mostly use the old name "micron" (μ) for the SI unit "micrometer."

Another major limitation in the infrared is that there are very few windows of transparency in the Earth's atmosphere. Most of the atmospheric absorption in the infrared is caused by water vapor, which is located at relatively low levels in our atmosphere compared to the ozone that causes the absorption in the ultraviolet. Thus we do not have to go as high to observe infrared as we do to observe ultraviolet. It is sufficient to send up instruments attached to huge balloons (Fig. 4–61).

NASA's Kuiper Airborne Observatory, an instrumented airplane, carries a 0.9-m telescope aloft (Fig. 4–62). It flies above most of the water vapor, so scientists use it especially to make infrared observations. Such observations are especially useful to study gas and dust clouds in space; they cannot be observed as well from the ground. NASA is planning a more capable successor airplane named SOFIA: Stratospheric Observatory for Infrared Astronomy.

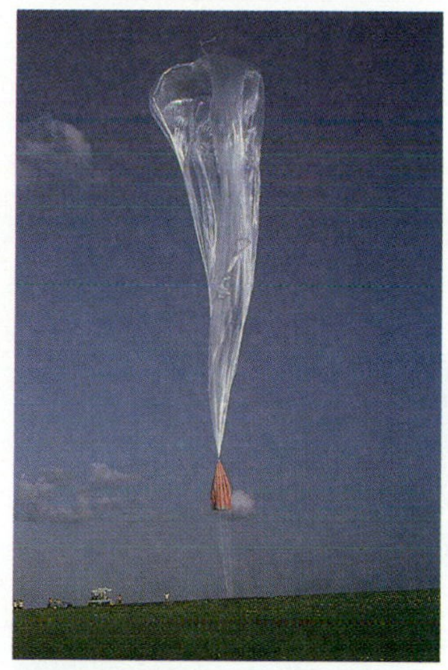

Figure 4–61 A balloon launch carries aloft a telescope used to study the infrared.

A

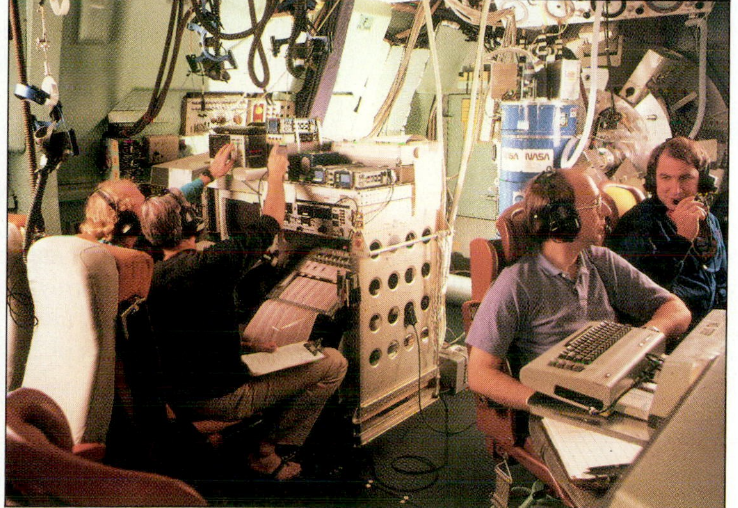

B

Figure 4–62 The Kuiper Airborne Observatory.

Figure 4–63 The American–Dutch–British Infrared Astronomical Satellite (IRAS), launched and active in 1983. It observed objects in space at the long infrared wavelengths of 12, 25, 60, and 100 microns.

Many useful observations can be made in the infrared from high mountaintops. This has led to the construction of infrared telescopes at such sites as Mauna Kea in Hawaii and to the positioning of the Multiple Mirror Telescope on one of the higher Arizona peaks. Section 4.14 describes an observing run with the Infrared Telescope Facility on Mauna Kea.

The Infrared Astronomical Satellite (IRAS) mapped the infrared sky in 1983 (Fig. 4–63). It was able to study the infrared at wavelengths longer than those observable from the ground, since its position above the atmosphere placed it above the infrared radiation from the air in our atmosphere. Its telescope was cooled to only 2 degrees above absolute zero to prevent infrared radiation from the telescope from swamping the signal from celestial objects. When the liquid helium used for cooling ran out after 10 months, the spacecraft could no longer be used.

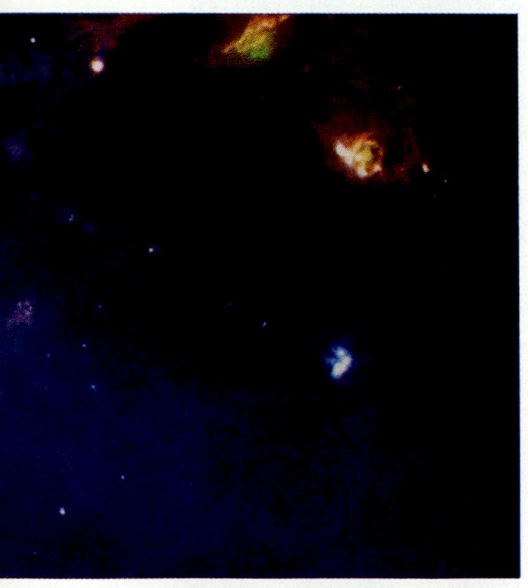

Figure 4–64 The wispy structures in the picture are "infrared cirrus," a surprise discovery of IRAS. The bright white object near the center is the Pleiades star cluster. The bright object with the ringlike feature above it is the Nebula IC348. The remaining large, bright object above and to its left is the California Nebula. The data are displayed as false-color images, each color corresponding to a different wavelength, which corresponds roughly to a different temperature.

In its brief but spectacular lifetime, IRAS made many discoveries, as we shall see throughout the book. For example, it discovered a patchy network of infrared radiation (Fig. 4–64); by analogy with cirrus clouds in the terrestrial sky, this network is called "infrared cirrus."

The constellation Orion looks quite different in the infrared than it does in the visible. Star-forming regions are of temperatures that show especially prominently in the infrared (Fig. 4–65), for they are tens or hundreds of degrees instead of the thousands of degrees typical of stars.

The European ISO (Infrared Space Observatory) is well under way for launch in 1993. It will have spectrometers for long-wavelength and short-wavelength infrared spectra, a camera, and an imaging device for photometry and polarization measurements. NASA's SIRTF (Space Infrared Telescope Facility) is also being planned.

Perhaps the most important recent instrumental development in astronomy is the "infrared array." Whereas infrared images used to be laboriously built up point by point, infrared-sensitive elements can now be put onto silicon chips in configurations of 256 × 256, making 65,536 pixels imaged at the same time. Further, the spacing between the elements is unchanging, and the atmospheric transmission is the same for all elements at a given time, both improvements on the former method of raster scanning. The center of our galaxy and star-forming regions like the Orion Nebula are among the types of objects for which infrared-array images are particularly valuable and for which examples are included later in this book. An infrared array is scheduled as a second-generation instrument on the Hubble Space Telescope in about 1995.

A

B

Figure 4–65 IRAS images show star-forming regions especially well. They also show interstellar dust. (*A*) The region near the south celestial pole, with infrared "cirrus"—interstellar dust grains that have been heated by the radiation from the stars. Part of the Large Magellanic Cloud is at the left. Stars show up as small blue dots since they are brightest at shorter infrared wavelengths, and galaxies show up as yellow dots. (*B*) The region near the diffuse nebula NGC 7822, in which massive new stars are now forming. The yellow-white regions in which the stars are forming are bright on these infrared images but are invisible in the visible part of the spectrum. In these Super-Skyflux images (a new product of the Image Processing and Analysis Center at Caltech that is continuing to study IRAS images) radiation at 12 microns is coded blue, radiation at 60 microns is coded green, and radiation at 100 microns is coded red.

Figure 4–66 Karl Jansky (*insert*) and the full-scale model of the rotating antenna with which he discovered radio astronomy.

4.13b Radio Astronomy

At even longer wavelengths, in the radio region of the spectrum, is a major window of transparency. The techniques of radio astronomy, now well established as one of the major branches of astronomy, will be discussed briefly here and further in various places in the book, especially in Parts V and VI on our galaxy and other galaxies, respectively.

We cannot always pinpoint the discovery of a whole field of research as decisively as that of radio astronomy. In 1931, Karl Jansky of the Bell Telephone Laboratories in New Jersey was experimenting with a radio antenna to track down all the sources of noise that might limit the performance of short-wave radiotelephone systems (Fig. 4–66). After a time, he noticed that a certain static appeared at approximately the same time every day. Then he made the key observation: the static was actually appearing four minutes earlier each day. This was the link between the static that he was observing and the rest of the Universe. Jansky's static kept sidereal time, just as the stars do (see Section 5.3), and thus was coming from outside our solar system! Jansky was actually receiving radiation from the center of our galaxy.

Figure 4–67 Though a radio telescope may look transparent to optical radiation, it reflects radio waves. Here we see the 64-m reflector at the Parkes Radio Observatory. Even though the holes in this radio telescope may look large to the eye, they are much smaller than the wavelength of radio radiation being observed. Thus the telescope appears smooth and shiny to the incoming radio radiation. To appear shiny, the surface of a radio telescope has to be smooth to within a small fraction of the wavelength of radiation, one-twentieth or less.

The principles of radio astronomy are exactly the same as those of optical astronomy: radio waves and light waves are exactly alike—only the wavelengths differ. Both are simply forms of electromagnetic radiation. But different technologies are necessary to detect the signals. Radio waves cause electrical changes in antennas, and these faint electrical signals can be detected with instruments that we call radio receivers.

If we want to collect and focus radio waves just as we collect and focus light waves, we must find a means to concentrate the radio waves at a point at which we can place an aerial. Lenses to focus radio waves are impractically heavy, so refracting radio telescopes are not used. However, radio waves will bounce off metal surfaces. Thus we can make reflecting radio telescopes that work on the same principle as the 5-meter optical telescope on Palomar Mountain. The reflecting surface, called a "dish" rather than a mirror because it is of less-than-optical quality, is usually made of metal. The dish needn't look shiny to our eyes, as long as it looks shiny to incoming radio waves (Fig. 4–67).

We want dishes as big as possible for two reasons: First, a larger surface area means that the telescope will be that much more sensitive. Second, a larger dish has better resolution. Since radio wavelengths are much longer than optical wavelengths, the resolution of any single radio telescope is thus far inferior to that of any single optical telescope (Section 4.2). The basic point

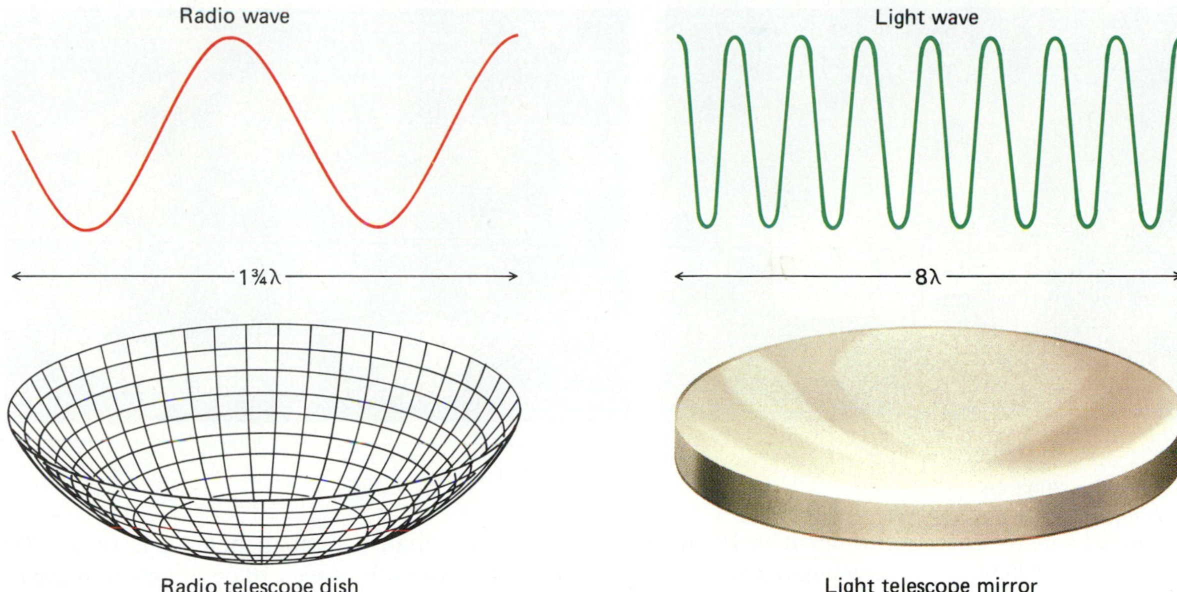

Figure 4–68 It is more meaningful to measure the diameter of telescope mirrors in terms of the wavelength of the radiation that is being observed than it is to measure it in terms of units like centimeters that have no particularly relevant significance. The radio dish at left is only 1¾ wavelengths across, whereas the mirror at right is 8 wavelengths across, making it effectively much bigger. The wavelengths are greatly exaggerated in this diagram relative to the size of any actual reflectors. For the Bonn radio telescope used to observe 1-cm waves, 100 m diameter ÷ 0.1 m per wave = 1000 wavelengths.

is that measurements of the size of a telescope are most meaningful when they are in units of the wavelength of the radiation being observed (Fig. 4–68). Thus an optical telescope one meter across used to observe optical light, which is about one two-millionth of a meter in wavelength, is two million wavelengths across. Let us consider the largest fully steerable radio telescope, at the Max Planck Institute for Radio Astronomy in Bonn, West Germany (Fig. 4–69).

Figure 4–69 The 100-m radio telescope in Bonn, West Germany, the largest fully steerable radio telescope in the world.

A

B

Figure 4–70 The 300-foot (91-m) radio telescope at the National Radio Astronomy Observatory's site at Green Bank, West Virginia (*A*) before, and (*B*) after its dramatic collapse in 1988.

Though it is 100 m in diameter, larger than a football field, it is only 1000 wavelengths across when used to observe radio waves 10 centimeters in wavelength (100 m/10 cm = 1000). The largest steerable antenna in the United States, the 300-foot (91-m) radio telescope at the National Radio Astronomy Observatory, was used for 25 years to study relatively long radio waves until it suddenly collapsed in 1988 (Fig. 4–70). Funding has been approved for a replacement at Green Bank. It will also be in the 100-m class, but will be capable of observing at much shorter wavelengths. It is to be fully steerable; the collapsed telescope had only limited tracking ability. Completion is foreseen for about 1995.

Radio telescopes used to study millimeter-length radio waves do not have to be as physically large as telescopes meant to study longer wavelengths. For example, to have the same relative size, a telescope to study waves of wavelength 2 mm need be only 2 meters across. Thus a 14-m (45-foot) radio telescope (Fig. 4–71) has 7 times better resolution at its wavelength than the physically larger 100-m telescope has for the longer-wavelength radiation.

A radio telescope dish receives radiation only from within a narrow cone called the *beam* (Fig. 4–72). The larger the telescope, measured with respect to the wavelength of radiation observed, the narrower the width of the beam. Single-dish radio telescopes have broad beams and cannot resolve any spatial details smaller than the size of their beams. Astronomers now use the technique known as interferometry to improve the resolution, using a technique to be discussed in Section 31.7. The technique has led, for example, to the giant

Figure 4–71 The radio telescope of the Five College Radio Astronomy Observatory (University of Massachusetts at Amherst, Amherst College, Smith College, Mount Holyoke College, and Hampshire College) is used to study millimeter-wavelength radiation. It is enclosed in a radome, which keeps the Sun and snow off it but is almost entirely transparent to radio waves.

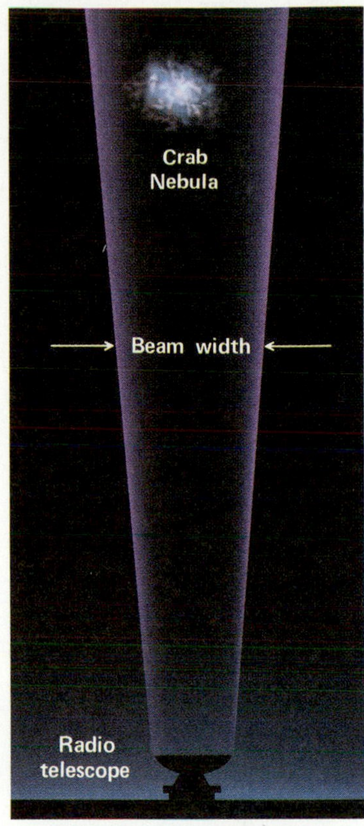

Figure 4–72 The beams of radio telescopes are ordinarily relatively large compared with radio sources. The Crab Nebula looks like a point to the radio telescope shown since it is less than a beam width across.

array of 27 radio telescopes in New Mexico called, prosaically, the VLA (Very Large Array). Arrays of millimeter-wavelength radio telescopes are now operated at the Hat Creek Radio Observatory in California (Fig. 4–73) and at IRAM's site in Spain (Fig. 4–74).

*4.14 A Night at Mauna Kea

An observing run with one of the world's largest telescopes highlights the work of many astronomers. The construction of several of the world's largest telescopes at the top of Mauna Kea in Hawaii has made that mountain the site of the world's largest observatory.

Mauna Kea was chosen because of its outstanding observing conditions. Many of these conditions stem from the fact that the summit is so high—4200 m (13,800 ft) above sea level. It is an especially good site for observations. The top of Mauna Kea is above so much of the atmosphere's water vapor, all but 10 per cent, that observations can be made in many parts of the infrared inaccessible from other terrestrial observatories.

Mauna Kea has other general advantages as an observatory site. It is so isolated that the sky above is particularly dark, allowing especially faint objects to be seen. And the flow of air across the mountaintop is often particularly smooth, leading to exceptional "seeing" for both visible and infrared observations. Moreover, the top of the mountain extends above most clouds that may cover the island below it. So Mauna Kea is perhaps the world's best site in that these advantages are linked with a large number of clear nights—about 75 per cent. Though there are taller mountains in the world, none has such favorable conditions for astronomy.

Figure 4–73 The millimeter interferometer at the Hat Creek Radio Observatory of the University of California and the University of Maryland.

Figure 4–74 Two of the four 15-m dishes in the millimeter-wavelength interferometer of IRAM (Institut de Radio Astronomie Millimetrique), a French–German–Spanish consortium. IRAM's interferometer is on the Plateau de Bure in Spain. They also have a 30-m millimeter-wavelength single dish near Grenoble in France. Most of the observing time is used to study interstellar molecules.

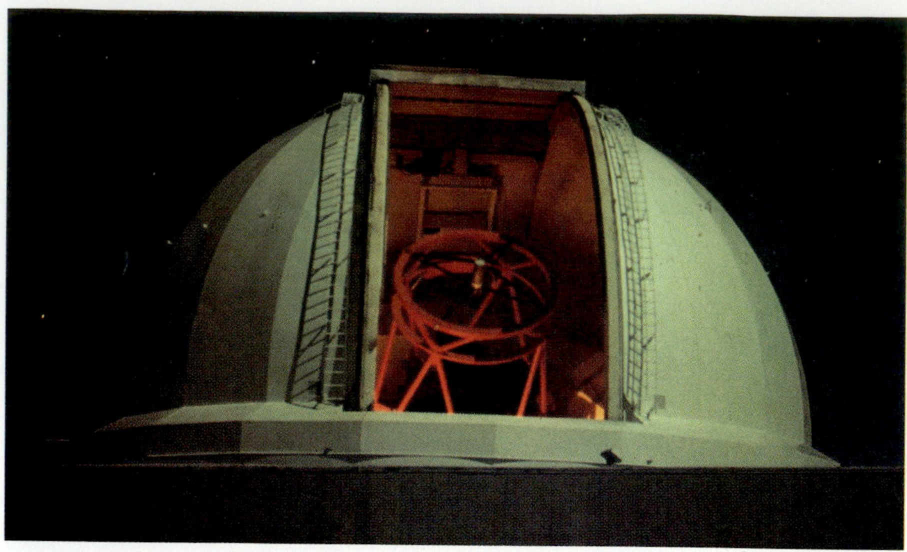

Figure 4–75 The Infrared Telescope Facility on Mauna Kea, a 3-m telescope optimized for infrared observations.

The 3-m Infrared Telescope Facility (IRTF) is sponsored by NASA and operated by the University of Hawaii (Fig. 4–75). Since it is a national facility, astronomers from all over the United States can propose observations. Every six months, a committee of infrared astronomers from all over the United States chooses the best proposals and makes out the observing schedule for the next half year.

During the months before your observing run, you make detailed lists of objects to observe, prepare charts of the objects' positions in the sky based on existing star maps and photographs, and plan the details of your observing procedure. The air is very thin at the telescope. You may not think as clearly or rapidly as you do at lower altitudes. Therefore, it is wise and necessary to plan each detail in advance.

The time for your observing run comes, and you fly off to Honolulu. Often you spend a day or two there at the Institute for Astronomy, consulting with the resident scientists, checking last-minute details, catching up with the time change, and perhaps giving a colloquium about your work. Then comes the brief plane trip to the island of Hawaii, the largest of the islands in the State of Hawaii. It is usually called simply "the Big Island."

Your plane may be met by one of the Telescope Operators, who will be assisting you with your observations. The Telescope Operators know the telescope and its systems very well, and are responsible for the telescope, its operation, and its safety. You drive together up the mountain, as the scenery changes from tropical to relatively barren, crossing dark lava flows. First stop is the Ellison Onizuka House at Hale Pohaku, the mid-level facility at an altitude of 2750 m (9000 ft), now named after the Hawaiian-born astronaut who died aboard the Space Shuttle Challenger. The astronomers and technicians sleep and eat there. Since the top of Mauna Kea is too high for people to sleep and work comfortably there for too long, everyone spends much time at the mid-level.

The rules say that you must spend a full 24 hours acclimatizing to the high altitude before you can begin your own telescope run. So you overlap with the last night of the run of the people using the telescope before you. This familiarizes you with the telescope and its operation. A combination of time and altitude adjustment sends you to bed early this first night.

The next night is yours. In the afternoon, you and the Telescope Operator may make a special trip up the mountain to install the systems you will be using to record data at the Cassegrain focus of the telescope. Since objects at

Figure 4–76 The astronomer (*right*) and telescope operator (*left*) mount their equipment, held in a container filled with liquid nitrogen. This photograph was taken at the adjacent Canada–France–Hawaii telescope.

normal outdoor or room temperatures give off enough radiation in the infrared to bother your observations, much of the instrumentation you install is cooled. The detectors used to observe wavelengths of 10 or 20 micrometers are bolometers, devices sensitive to small temperature changes that occur as a result of incoming radiation. The bolometers are cooled to a temperature of only 2 K (−271°C) by liquid helium, reducing their own infrared radiation so that it does not swamp the faint infrared celestial radiation. To keep the liquid helium from boiling away too fast, liquid nitrogen at a temperature of 77 K (−196°C) surrounds the helium. These liquids have to be carefully transferred from large holding tanks into metal flasks that contain the bolometer and its accompanying optics (Figs. 4–76 and 4–77). Then the system has to be left to cool down.

Oh, Langley invented the
bolometer,
A very good kind of thermometer.
It can measure degrees
On a polar bear's knees
At a distance of half a kilometer.

After dinner, you return to the summit to start your observing. You dress warmly, since the nighttime temperature approaches freezing at this altitude, even in the summertime. Since the sky is fairly dark at infrared wavelengths, you can start observing even before sunset. The telescope is operated by computer, and points at the coordinates you type in for the object you want to observe. A video screen (Fig. 4–78) displays an image of the object.

You drive down the mountain in the early morning Sun. The clouds you see may be a kilometer below you. The tiny pimples on the landscape that you first see turn out, as you get closer, to be giant cinder cones that tower

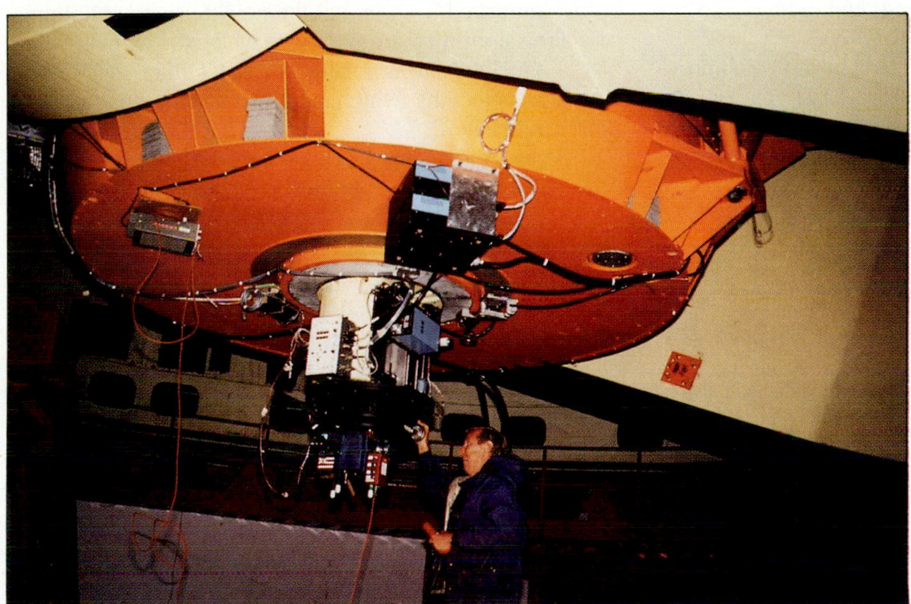

Figure 4–77 Installing a bolometer at the Cassegrain focus of the IRTF. The back plate of the telescope is at the top; the 3-m mirror is on its other side.

Figure 4–78 A telescope operator at her console. The television screen shows that the telescope is pointing at Saturn.

above you. They are from past volcanic eruptions. The mountain around you is barren, littered with huge volcanic rocks, resembling surface views from the Viking landers on Mars more than any terrestrial view you have ever seen. Then some vegetation appears, and you reach Hale Pohaku, ready for breakfast and a day's sleep. The next night you start over again.

When you leave Mauna Kea, it is a shock to leave the pristine air above the clouds. But, happily, you take home with you data about the objects you have observed. From the IRTF, the data are in the form of graphs, computer printouts, and sometimes computer tape. You may also have photographs that are the result of your work with other telescopes. It often takes you months to study the data you gathered in a few brief days at the top of the world.

Summary and Outline

The spectrum (Section 4.1)
- From short wavelengths to long wavelengths the various types of radiation are known as gamma rays, x-rays, ultraviolet, visible, infrared, and radio.
- Parts of visible spectrum are ROY G. BIV.
- All radiation travels at the "speed of light"; in a vacuum this is 3×10^8 m/sec.
- Windows of transparency of the Earth's atmosphere exist for visible light and radio waves (plus some parts of the infrared).
- Continuous radiation is called the *continuum*.

Optical telescopes (Sections 4.2 to 4.8)
- Resolution is inversely proportional to diameter (and proportional to wavelength).
- Refractors and reflectors: Ritchey-Chrétiens and Schmidt cameras have relatively wide fields.
- Optical observatories and their distribution across the world
- The Hubble Space Telescope and its increased resolution
- Light-gathering power is directly proportional to diameter squared.

Spectroscopy and spectral lines (Sections 4.9 and 4.10)
- Absorption lines are gaps in the continuum shown as narrow wavelength regions where the intensity is diminished from that of the neighboring continuum.
- Emission lines are narrow wavelength regions where the intensity is relatively greater than neighboring wavelengths (which are either continuum or zero in intensity).
- A given element has a characteristic set of spectral lines under specific conditions of temperature and pressure.

Recording the data (Section 4.11)
- Film and electronic devices (especially CCD's) for gathering photons are described.
- Fiber optics are making observing more efficient.

Ultraviolet and x-ray astronomy (Section 4.12)
- Video and grazing incidence techniques used
- Grazing incidence for HEAO-2

Infrared and radio astronomy (Section 4.13)
- Infrared: High altitude sites and electronic devices are necessary.
- Radio: Single-dish radio telescopes have lower resolution than interferometers.

Key Words

radiation, speed of light, wavelength, ultraviolet, infrared, electromagnetic radiation, electromagnetic spectrum, visible light, resolution, wave front, refraction, lens, focal length, field of view, eyepiece, objective, refracting telescope, chromatic aberration, reflecting telescope, parallel light, paraboloid, Newtonian, Cassegrain, Gregorian, prime focus, Schmidt camera, correcting plate, seeing, transparency, light pollution, extended objects, point objects, magnification, diffraction grating, slit, spectroscope, spectrograph, spectral lines, continuum (continua), absorption lines, Fraunhofer lines, emission lines, law of conservation of energy, photometry, CCD, photons, grazing incidence, high-energy astrophysics, beam

Questions

1. What advantage does a reflecting telescope have over a refracting telescope?

2. What limits the resolving power of the 5-meter telescope?

3. Why does it matter whether a telescope is in the northern or southern hemisphere?

4. List the important criteria in choosing a site for an optical observatory meant to study stars and galaxies.

†5. Two reflecting telescopes have primary mirrors 2 m and 4 m in diameter. How many times more light is gathered by the larger telescope in any given interval of time?

†6. A 4-m telescope has a Cassegrain hole 1 m in diameter. What percentage of the total area is taken up by the hole?

7. Why might some stars appear double in blue light though they could not be resolved in red light?

†8. Use Dawes's limit to calculate how far apart two double stars have to be in order to be seen as separate with a 20-cm telescope.

9. For each of the following, identify whether it is a characteristic of a reflecting telescope, a refracting telescope, both, or neither. Give any limitations on the applicability of your answer.

(a) Is free of chromatic aberration
(b) Has more severe spherical aberration
(c) Requires aluminizing
(d) Can be used for photography
(e) Has an objective supported only by its rim
(f) Can be made in larger sizes
(g) Has a prime focus at which a person can work without blocking the incoming light

10. Describe at least two methods that allow us to make large optical telescopes more cheaply than by simply scaling up designs of previous large telescopes.

11. Why can't we use ordinary photographic plates to record infrared images?

†12. The Palomar–National Geographic Sky Survey made with the Oschin Schmidt telescope covers the sky in about 900 pairs of plates. Using the field of view given in the text, and comparing with a 30-arc-min field of view for the Kitt Peak 4-m reflector, calculate how many red/blue pairs of photographs would be needed to cover the sky with the latter.

13. Why must we observe ultraviolet radiation using rockets or satellites, whereas balloons are sufficient for infrared observations?

14. Why can radio astronomers observe during the day, whereas optical astronomers are (for the most part) limited to nighttime observing?

15. What are the advantages of the Multiple Mirror Telescope? How does it compare with the 5-m telescope for studying faint objects?

16. What are the similarities and differences between making radio observations and using a reflector for optical observations? Compare the radiation path, the detection of signals, and limiting factors.

17. Why is it sometimes better to use a small telescope in orbit around the Earth than it is to use a large telescope on a mountaintop?

18. Why is it better for some purposes to use a medium-size telescope on a mountain instead of a telescope in space?

19. Compare the International Ultraviolet Explorer and the Hubble Space Telescope.

20. What are two reasons why the Hubble Space Telescope will be able to observe fainter objects than we can now study from the ground?

†This question requires a numerical solution.

This long exposure shows stars circling the celestial north pole with one of the Mauna Kea Observatory's telescopes in the foreground. (© 1987 Roger Ressmeyer—Starlight)

The Sky and the Calendar

5

Aims: To understand how astronomers locate objects in the sky, how objects appear to move in the sky, and time zones and calendars

When we look up at the sky on a dark night from a location outside a city, we see a fantastic sight. If the Moon is up, its splendor can steal the show; not only does its pearly white appearance draw our attention but also the light it gives off makes the sky so bright that we cannot see the other objects well. But when the Moon is down, we see bright jewels in the inky sky. Generally, the brightest few shine steadily, which tells us that they are planets. The rest twinkle—sometimes gently, sometimes fiercely—which reveals them to be stars.

The Milky Way arches across the sky, and if we are in a good location on a dark night, it is quite obvious to the naked eye. (City dwellers may never see the Milky Way at all.) If you know where to look, from the northern hemisphere you can see a hazy spot that is actually a galaxy, rather than individual stars. It is known as the Andromeda Galaxy from its location in the constellation Andromeda. Along with one other galaxy, it is the farthest in the Universe we can see with the naked eye.

If we are lucky, a bright comet may be in the sky, but this happens rarely. More often, meteors—"shooting stars"—dart across the sky above. And we may see the steady light of a spacecraft cross the sky in a few minutes. The lights of airplanes can be seen quite regularly.

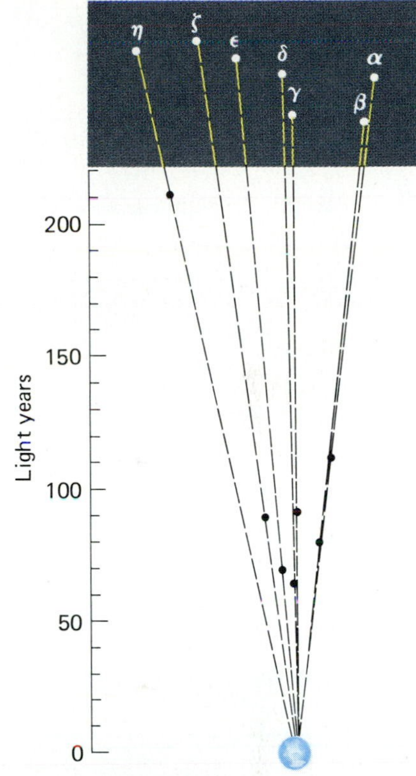

Figure 5–1 The stars we see as a constellation are actually differing distances from us. In this case we see the true distances of the stars in the Big Dipper, part of the constellation Ursa Major, in the lower part of the figure. Their appearance projected on the sky is shown in the upper part.

5.1 The Constellations

Long, long ago, when Egyptian and other ancient astronomers were beginning to study and understand the sky, they divided the sky into regions containing fairly distinct groups of stars. The groups, called *constellations,* were given names, and stories were associated with them, perhaps to make them easier to remember.

Actually, the constellations are merely areas in the sky that happen to have stars in particular directions as we see them from the Earth. There is no physical significance to the apparent groupings, nor are the stars in a given constellation necessarily associated with each other in any direct manner (Fig. 5–1).

Many of the stories we now associate with the constellations come from Greek mythology (Fig. 5–2*A*), though the names may have been associated with particular constellations more to honor Greek heroes than because the constellations actually looked like these people. Other civilizations (American Indians, for example) attached their own names, pictures, and stories to the stars and constellations (Fig. 5–2*B*). For example, the star that is nearest to stationary in the sky, with all the other stars seemingly revolving around it, we call Polaris; the Norse called it the Jeweled Nailhead, the Mongols called it the Golden Peg, the Chinese called it Emperor of Heaven, and the Skidi band

The International Astronomical Union put the scheme of constellations on a definite system in 1930. The sky was officially divided into 88 constellations (see Appendix 9) with definite boundaries, and every star is now associated with one and only one constellation.

A

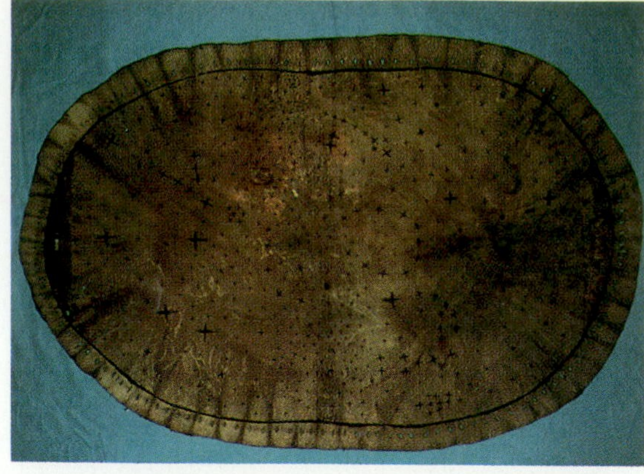

B

Figure 5–2 (*A*) Bayer, in 1603, used Greek letters to mark the brightest stars in constellations; he also used lower-case Latin letters. Here we see Cassiopeia. (*B*) A sky chart from the Pawnee Indians, who lived along the Platte River in what is now Nebraska.

of Pawnee Indians called it the Chief Star. In our Draco, the Dragon, the Egyptians saw the Hippopotamus. Our Orion, the Hunter, was the White Tiger to the Chinese of three thousand years ago.

Sometimes familiar groupings in the sky do not make up a constellation. Such groupings are called *asterisms;* the Big Dipper is an example, because it is only part of the constellation Ursa Major (the Big Bear).

In the star atlas he published in 1603, Johann Bayer assigned Greek letters (Appendix 6) in alphabetical order to the stars in each constellation, usually roughly in order of brightness. Thus α (alpha) is usually the brightest star in a constellation, β (beta) is the second brightest, and so on. (In the Big Dipper, the letters go around the bowl rather than by brightness.) The Greek letters are used with the genitival form ("of...") of the constellation name (Appendix 9), as in α Orionis, meaning "alpha of Orion," which is Betelgeuse. For fainter stars, we sometimes use the Latin letters Bayer used when he ran out of Greek letters. Or we may use the numbers assigned to stars from side to side in a constellation in order of position (right ascension, Section 5.3) by John Flamsteed in England a century later: 61 Cygni, for example.

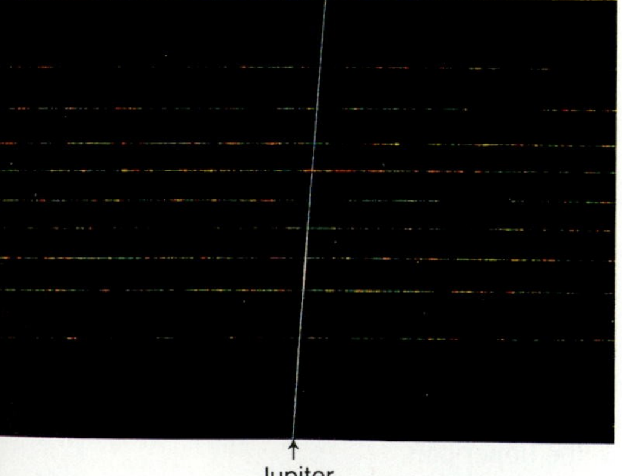

Figure 5–3 The lines represent either stars or the planet Jupiter, each of which left streaks as the camera was scanned from side to side. The steady scans are from Jupiter; the irregular scans, which show twinkling, are from the star Sirius.

5.2 Twinkling

What about the twinkling? It is not a property of the stars themselves, but merely an effect of our Earth's atmosphere. The starlight is always being bent by moving volumes of air in our atmosphere. These volumes of air bend light primarily because they are different in temperature from their surroundings and also because they are different in density and water vapor content. The effect makes the images of the stars appear to be larger than points, to dance around slightly (such image motion is called seeing, as discussed in Section 4.8a), and to change rapidly in intensity (a property called scintillation). The change in intensity—scintillation—is what we non-technically call "twinkling" (Fig. 5–3).

Planets, unlike stars, do not usually seem to twinkle. They seem steadier because they are close enough to Earth so that they appear as tiny disks large enough to be seen through telescopes. Though each point of the disk of light representing a planet may change slightly in intensity, the disk is made of so many points of light that the total intensity doesn't change. To the naked eye the planets thus appear to shine more steadily than the stars. But when the air is especially turbulent, or when a planet is so low in the sky that we see it through a long path of air, even a planet may twinkle.

*5.3 Coordinate Systems

The stars and other astronomical objects that we study are at a wide range of distances from the Earth. Though we may have to know the distance to an object to understand how much energy it is giving off, we need to know only an object's direction to observe it. It is thus often useful to think of the astronomical objects as being at a common distance, all hung on the inside of a large (imaginary) *celestial sphere.* In this section, we see how to describe the positions of objects on the celestial sphere.

Both astronomers and geographers have established systems of coordinates—*coordinate systems*—to designate the positions of places in the sky or on the Earth. The geographers' system is familiar to most of us: longitude and latitude. Lines (actually half-circles) of longitude on Earth called *meridians* run from the north pole to the south pole. The zero circle of longitude has been adopted, by international convention, to run through the former site of the Royal Greenwich Observatory in England (Fig. 5–4). We measure longitude by the number of degrees east or west an object is from the meridian that passes through Greenwich, the Prime Meridian. Latitudes are defined by parallel circles that run around the Earth, all parallel to the equator. Latitude 0° corresponds to the equator; latitude ±90° corresponds to the poles. Los Angeles, for example, is at longitude 118°W and latitude 34°N.

The astronomers' system for the sky is similar, except that astronomers use the names *right ascension* for the celestial analogue of longitude (both measure east–west) and *declination* for the celestial analogue of latitude (both measure north–south). Right ascension and declination are measured with respect to a *celestial equator,* which lies above the Earth's equator, and *celestial poles,* which are on the extensions of the Earth's axis of spin (Fig. 5–5).

Right ascension and declination form a coordinate system fixed to the stars. To observers on Earth, the stars appear to revolve every 23 hours 56 minutes. The coordinate system thus appears to revolve at the same rate. Actually, of course, the Earth is rotating while the stars and celestial coordinate system remain fixed.

Although the stars are fixed in their positions in the sky, the Sun's position varies through the whole range of right ascension each year. The path of the Sun in the sky with respect to the stars is the *ecliptic.* The ecliptic is inclined

Figure 5–4 The international zero circle of longitude in Greenwich, England.

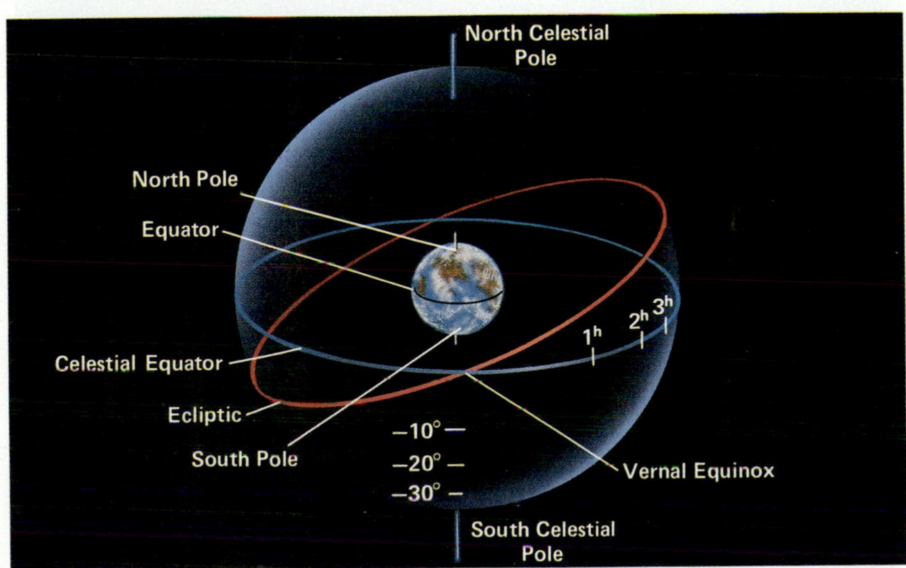

Figure 5–5 The celestial equator is the projection of the Earth's equator onto the sky, and the ecliptic is the Sun's apparent path through the stars in the course of a year. The vernal equinox is one of the intersections of the ecliptic and the celestial equator and is the zero-point of right ascension. From a given location at the latitude of the United States, the stars nearest the north celestial pole never set and the stars nearest the south celestial pole never rise above the horizon.

Right ascension is measured along the celestial equator. Each hour of right ascension equals 15°. Declination is measured perpendicularly (−10°, −20°, etc.).

by 23½° with respect to the celestial equator, since the Earth's axis is tipped by that amount. The ecliptic and the celestial equator cross at two points. The Sun crosses one of these points, the *vernal equinox,* on the first day of northern-hemisphere spring. The Sun crosses the other intersection, the *autumnal equinox,* on the first day of autumn. The vernal equinox is the zero-point of right ascension.

Technically, we measure right ascension and declination with the use of *hour circles*, great circles running through the celestial poles. (A "great circle" on a sphere is a circle that is also on a plane that goes through the center of the sphere; it is the largest possible circle that can be drawn on the sphere's surface. On a sphere, the shortest distance between two points is on a great circle.) The hour circles cross the celestial equator perpendicularly. Right ascension is the angle to a body's hour circle, measured eastward along the celestial equator from the vernal equinox. Declination is the angle of an object north (+) or south (−) of the celestial equator along an hour circle.

Astronomers have set up a timekeeping system called *sidereal time* (sidereal means "by the stars"). Each location on Earth has a *meridian,* the great circle linking the north and south poles and passing through the *zenith,* the point directly overhead. The sidereal time at any location is the length of time since the vernal equinox has crossed the local meridian. As a result, the sidereal time is equal to the right ascension of any star on that place's meridian. Also, right ascension is kept in units of time; since 360° = 24 hours, 15° = 1 hour of right ascension. One degree (°) is subdivided into 60 minutes of arc (60′), each of which in turn is subdivided into 60 seconds of arc (60″); similarly, an hour is divided into 60 minutes (^{m}), each of which in turn is subdivided into 60 seconds (^{s}). For example, the position of Sirius is right ascension $6^{h}45^{m}$ and declination −16°43′.

Each north-south on the Earth has a different sidereal time at each instant. Astronomers find sidereal time convenient because by consulting clocks that are set to run on it, they can tell whether a star is favorably placed for observing.

Sidereal time and solar time differ, though both are caused by the rotation of the Earth on its axis. A *sidereal day* (a day by the stars) is the length of time that it takes the vernal equinox to return to the celestial meridian. A *solar day*

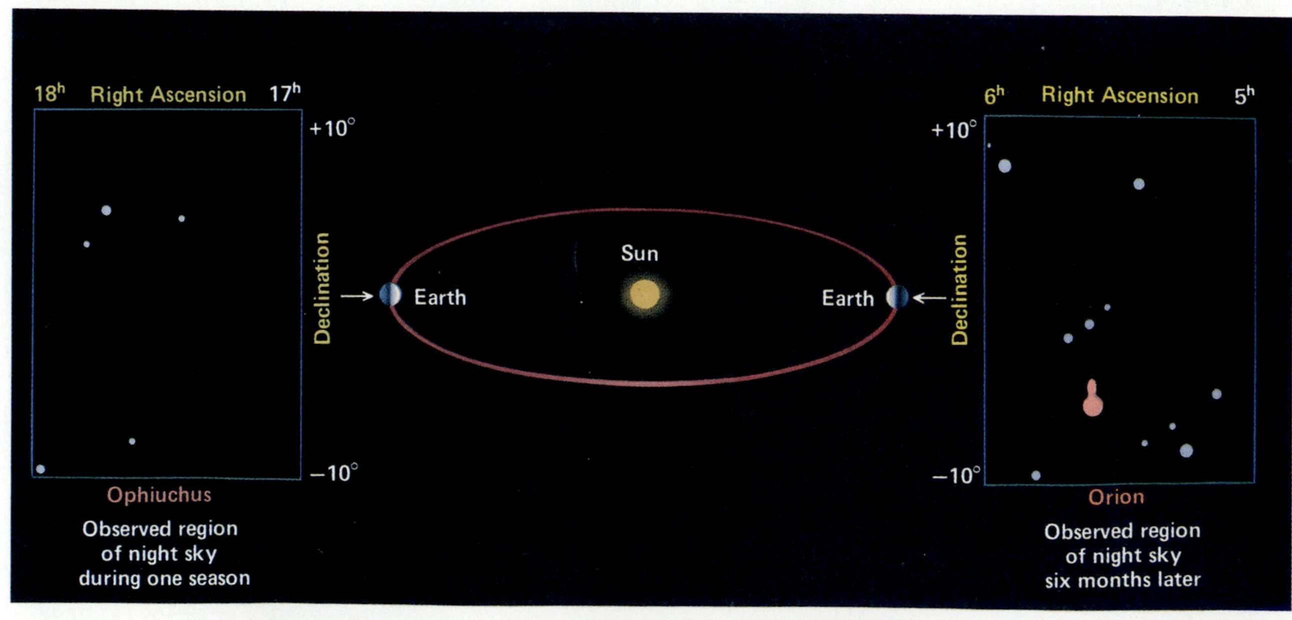

Figure 5–6 For each time of the year, we can see the constellations that are in the direction away from the Sun. Because the Earth rotates much more rapidly than it revolves around the Sun (24 hours compared with 365 days), all observers on the Earth see the same constellations in a given season, even though at any given time it is daytime for some people while it is nighttime for others.

is the length of time that the Sun takes to return to your meridian. Since the Earth revolves around the Sun once a year, by the time a day has passed, the Earth has moved 1/365 of the way around the Sun. Thus after the Earth has turned far enough for the stars to return to the same apparent positions in the sky, the Earth must still turn an additional 1/365 of 24 hours (24 hours/365 = 4 minutes) for the Sun to return to your meridian (Fig. 5–6). A solar day is thus approximately 4 minutes (actually 3 minutes 56 seconds of time) longer than a sidereal day. Thus which constellations are up at night changes from season to season (Fig. 5–7).

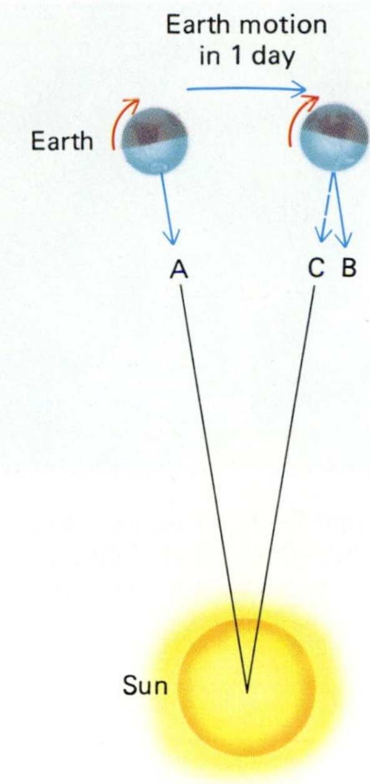

Figure 5–7 While the Earth rotates once on its own axis with respect to the stars (one sidereal day), it also moves slightly in its orbit around the Sun. Thus after one sidereal day, arrow A becomes arrow B. But one solar day has passed only when arrow B rotates a little farther and becomes arrow C. This takes an additional 4 minutes, making a solar day 4 minutes longer than a sidereal day.

Every observatory has both sidereal clocks, for the astronomers to tell when to observe their stars, and solar clocks, for the astronomers to gauge when sunrise will come and to know when to go to dinner. A solar clock and a sidereal clock show the same time (on a 24-hour system) only one instant each year, the autumnal equinox. (An equinox is both a point in the sky and the time when the Sun passes that point.) The next day the sidereal clock is 4 minutes ahead, the second day afterward it is 8 minutes ahead, and so on. Six months later the two clocks differ by 12 hours, and the stars that were formerly at their highest at midnight are then at their highest at noon, when the Sun is out. As a result, they may not be visible at all at that season.

Though the coordinates are basically fixed to the stars, a small effect called *precession* causes a slow drift of the coordinate system with respect to the stars with a 26,000-year period (Fig. 5–8). Precession takes place because the Earth's axis doesn't always point exactly at the same spot in the sky; the axis rather traces out a small circle. (The effect is like the wobbling of a spinning top.) The axis takes approximately 26,000 years to return to the same orientation. About halfway through the cycle from now—in, say, A.D. 15,000 (our A.D. 2000 plus half of 26,000 years)—the north star will be Vega. But don't worry—Polaris will be our north star again in about 26,000 years.

Because the pole moves, the celestial equator—which is always 90° from the pole—moves. Hence the equinoxes, which are the intersections of the celestial equator and the ecliptic, precess—that is, they apparently move slowly along the ecliptic. Because of this *precession of the equinoxes,* the right ascension and declination of objects in the sky change slowly. (The formulas for computing these changes are given in Appendix 2.)

As a result of precession, one has to make small corrections in any catalogue of celestial positions to update them to the present time. Precession is a small effect—the change in celestial coordinates is less than one minute of arc (one-sixtieth of a degree) per year—and is much less for some parts of the sky. Thus it need not be taken into account for casual observing, though star maps and catalogues are now being recalculated and redrawn for "epoch 2000.0."

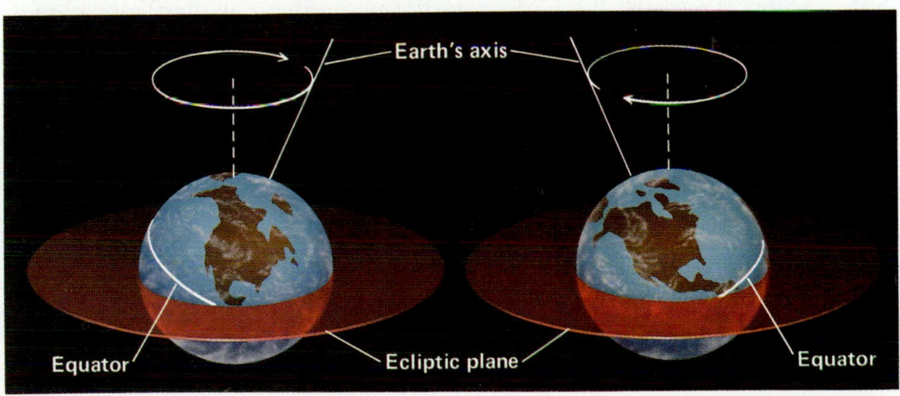

Figure 5–8 The Earth's axis precesses with a period of 26,000 years. The two positions shown are separated by 13,000 years.

As the Earth's pole precesses, the equator moves with it (since the Earth is a rigid body). The celestial equator and the ecliptic will always maintain the 23½° angle between them, but the points of intersection, the equinoxes, will change. Thus, over the 26,000-year precession cycle, the vernal equinox will move through all the signs of the zodiac. It is now in the constellation Pisces and approaching Aquarius (and thus the celebration in the musical *Hair* of the "Age of Aquarius").

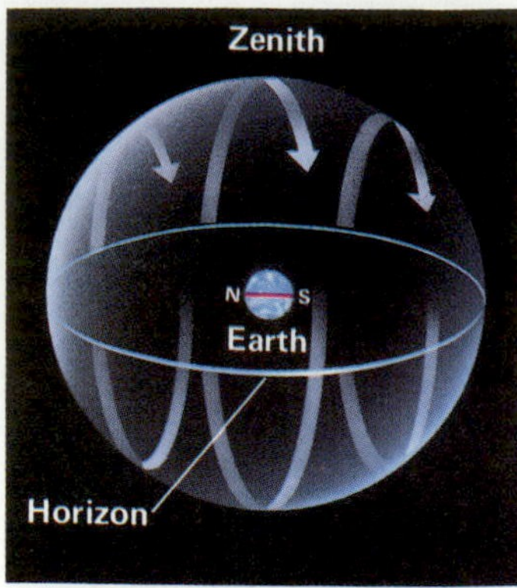

Figure 5–9 From the equator, the stars rise straight up, pass right across the sky, and set straight down.

*5.4 Motions in the Sky

At the latitudes of the United States, which range from +25° for the tip of the Florida Keys up to +49° for the Canadian border, and down to +19° in Hawaii and up to +67° in Alaska, the stars rise and set at angles to the horizon. In order to understand the situation, it is best first to visualize simpler cases. These concepts involve spherical geometry, which is difficult to visualize without practical experience with a telescope or in a planetarium.

If we were standing on the equator, the stars would rise perpendicularly to the horizon (Fig. 5–9). The north celestial pole would lie exactly on the horizon in the north, and the south celestial pole would lie exactly on the horizon in the south. Each star would rise somewhere on the eastern half of the horizon; each would remain "up" for twelve hours, and then would set. By waiting long enough, we would be able to see all stars, no matter what their declination. The Sun, no matter what its declination, would also rise, be up for twelve hours, and then set, so day and night would each last twelve hours.

If, on the other hand, we were standing on the north pole, the north celestial pole would be directly overhead, and the celestial equator would be on the horizon (Fig. 5–10). All the stars would move around the sky in circles parallel to the horizon. Since the celestial equator would be on the horizon, we could see only the stars with northern declinations. The stars with southern declinations would never be visible.

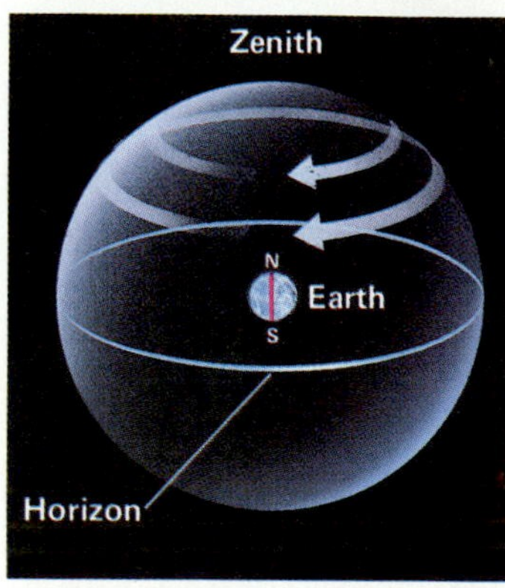

Figure 5–10 From the pole, the stars move around the sky in circles parallel to the horizon, never rising or setting.

Let us consider a latitude between the equator and the north pole, say +40°. There the stars seem to rise out of the horizon at oblique angles. The north celestial pole (with the north star nearby it) is always visible in the northern sky, and is at an *altitude* of 40° above the horizon, where "altitude" is the angle measured upward perpendicularly to the horizon. The star Polaris, a 2nd-magnitude star, happens to be located within one degree of the north celestial pole, and so is called the *pole star*. If you can see the north celestial pole, then the south celestial pole must be hidden. The celestial pole you can see is the only fixed point in the sky (Figs. 5–11 and 5–12).

If you point a camera at the sky and leave its lens open for a long time—many minutes or hours—the stars appear as trails. Those near celestial poles (Fig. 5–13 and chapter opening photograph) move in relatively small and obvious circles around the poles. The circles followed by stars farther from the celestial poles are so large that sections of them seem straight (Fig. 5–14).

When astronomers want to know if a star is favorably placed for observing, they must know both its right ascension and declination (Section 5.3). By seeing if the right ascension is reasonably close to the sidereal time, they can tell if it is the best time of year at which to observe it. But they must also know the star's declination to know how long it will be above the horizon each day.

Telescopes are often mounted at an angle such that one axis—the *polar axis*—points directly at the north celestial pole. Since all stars move across the sky in circles centered at the pole, the telescope must merely turn about

Figure 5–11 Cartoon by Charles Schulz. (© 1970 United Feature Syndicate, Inc.)

PEANUTS

IF YOU KNOW YOUR STARS, YOU'LL NEVER GET LOST IN THE WOODS..

Figure 5–12 Cartoon by Charles Schulz. (© 1970 United Feature Syndicate, Inc.)

that axis to keep up with stellar motions. The other axis of the telescope is used to point the telescope in declination.

Since motion around only one axis is necessary to track the stars, one need have only a single motor set to rotate once every 24 sidereal hours. This motor turns the polar axis in the direction opposite to the rotation of the Earth. The principle is the same for a small telescope in your backyard as for the 5-meter telescope on Palomar Mountain. Arrangements of this type are called *equatorial mounts* (Fig. 5–15), since one axis rotates perpendicularly to the celestial equator.

The alternative to this system is to mount a telescope such that one axis goes up-down (altitude) and the other goes around (azimuth). The azimuth motion is parallel to the horizon, and the altitude motion is perpendicular. Binoculars on stands at scenic overlooks are mounted this way (Fig. 5–16). To track a star, continual adjustments have to be made in both axes, which used to be very inconvenient to do smoothly enough for photography. But now, computers make the necessary calculations that allow large new telescopes to be mounted using such an *alt-azimuth* system. A flat mirror can direct the focus out to the side through the bearings on which the telescope rotates; these positions are the "Nasmyth foci."

A large alt-azimuth mount is often less expensive to construct than an equatorial mount of the same size, and can be housed in a smaller, less expensive dome. Thus the new British–Irish–Dutch 4.2-m Herschel telescope in the Canary Islands and the 10-m Keck telescope now under construction in

Figure 5–13 When we look toward the north celestial pole, stars appear to move in giant circles about the pole. Here, we are looking past the McMath Solar Telescope on Kitt Peak.

Figure 5–14 Near the celestial equator, the star circles are so large that they appear almost straight in this view past the William Herschel Telescope (*right*) and the Kapteyn Telescope (*left*) in the Canary Islands.

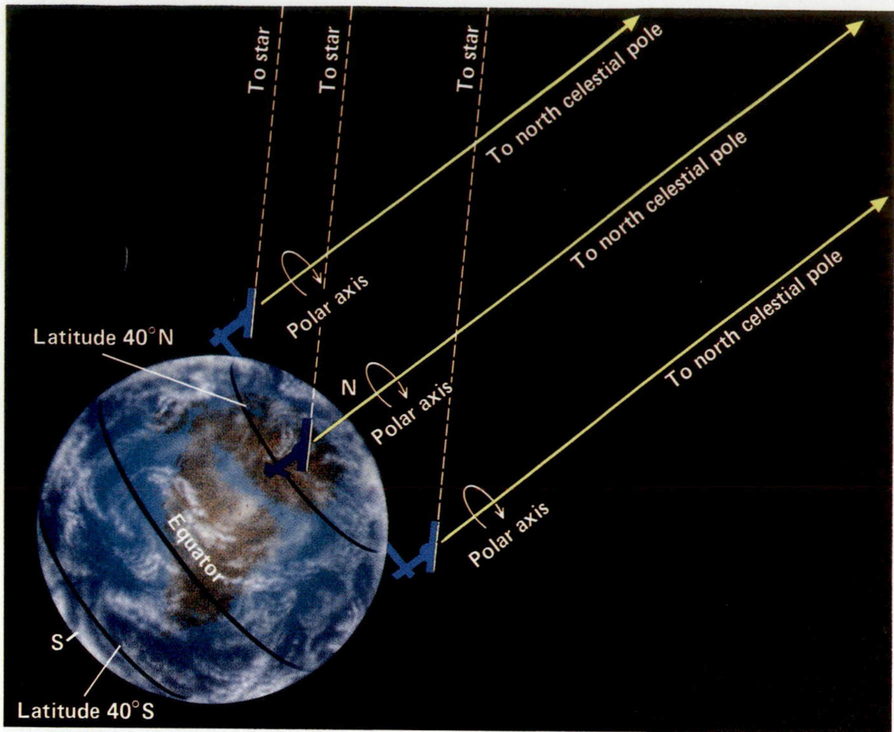

Figure 5–15 An equatorially mounted telescope need rotate on only one axis to keep pointing at a star. This axis is called the *polar axis;* it is fixed for telescopes in the northern hemisphere so that it points at the north celestial pole. Rotation of the telescope around this axis keeps up with the rotation of the Earth on its axis.

Figure 5–16 Alt-azimuth mounts, with one "up-down" motion (*altitude*) and one "around" motion (*azimuth*), are used for mounting binoculars at public viewpoints.

Hawaii have alt-azimuth mounts. Some amateur observers are using large (for amateurs, that is, up to about 30 inches in diameter), thin telescope mirrors in wooden or particle-board alt-azimuth mounts that have plastic bearings. They merely nudge the telescope along to follow the stars. These *Dobsonian* telescopes are a relatively inexpensive way to obtain large apertures for visual observing, though they cannot be used for photography because they do not track the stars.

*5.4a Positions of the Sun, Moon, and Planets

As the Sun moves along the ecliptic each year, it crosses the vernal equinox on approximately March 21st and the autumnal equinox on approximately September 23rd, respectively. On these days, the Sun's declination is 0°; the Sun's declination varies over the year from +23½° to −23½° (Fig. 5–17).

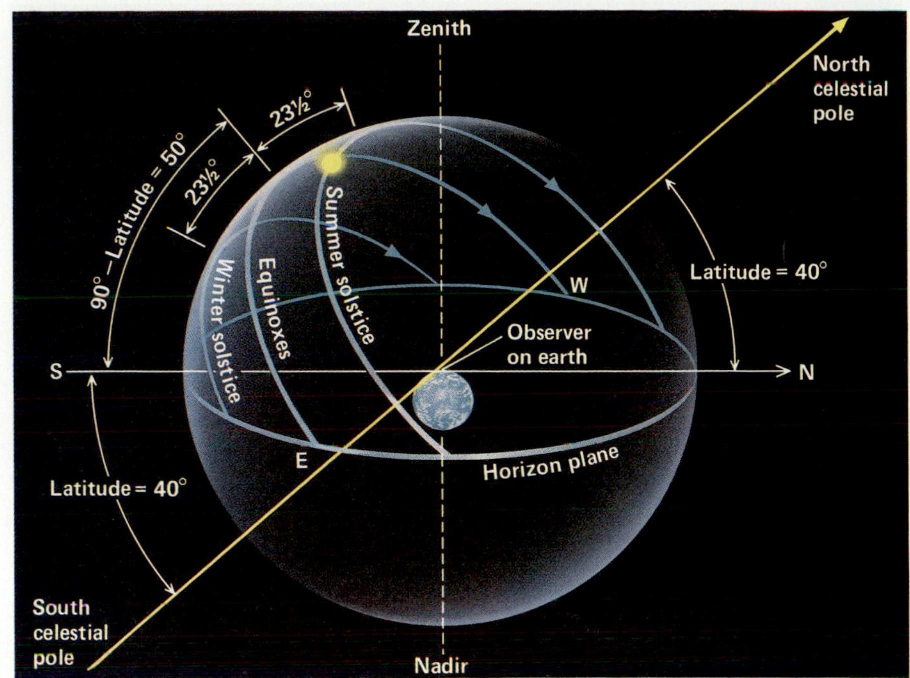

Figure 5–17 The path of the Sun at different times of the year. Around the summer solstice, June 21st, the Sun is at its highest declination, rises highest in the sky, stays up longer (because, as shown, more of its path is above the horizon), and rises and sets farthest to the north. The opposite is true near the winter solstice, December 21st. The diagram is drawn for latitude 40°.

These points are called *equinoxes* ("equal nights"; *nox* is Latin for night) because the daytime and the nighttime are supposedly equal on these days. Actually, because refraction (bending) of light by the Earth's atmosphere makes the Sun appear to rise a little early and set a little late, and the fact that the top of the Sun rises ahead of the Sun's midpoint, daytime exceeds nighttime at U.S. latitudes by about 10 minutes on the days of the equinoxes. The dates of equal daytime and nighttime precede the vernal equinox and follow the autumnal equinox by a few days.

If we were at the north pole, whenever the Sun had a northern declination, it would be above our horizon and we would have daytime. The Sun would move in a circle all around us, moving essentially parallel to the horizon. From day to day it would appear slightly higher in the sky for 3 months, and then move lower. The date when it is highest in the sky is the summer *solstice*. It occurs on approximately June 21st each year. On that date the Sun is 23½° above the horizon because its declination is +23½°. The time of the year when the Sun never sets is known as the time of the *midnight Sun* (Fig. 5–18), which

Figure 5–18 In this series taken in June from northern Norway, above the Arctic Circle, one photograph was taken each hour for an entire day. The Sun never set, a phenomenon known as the *midnight sun*. Since the site was not at the north pole, the Sun and stars move somewhat higher and lower in the sky in the course of a day.

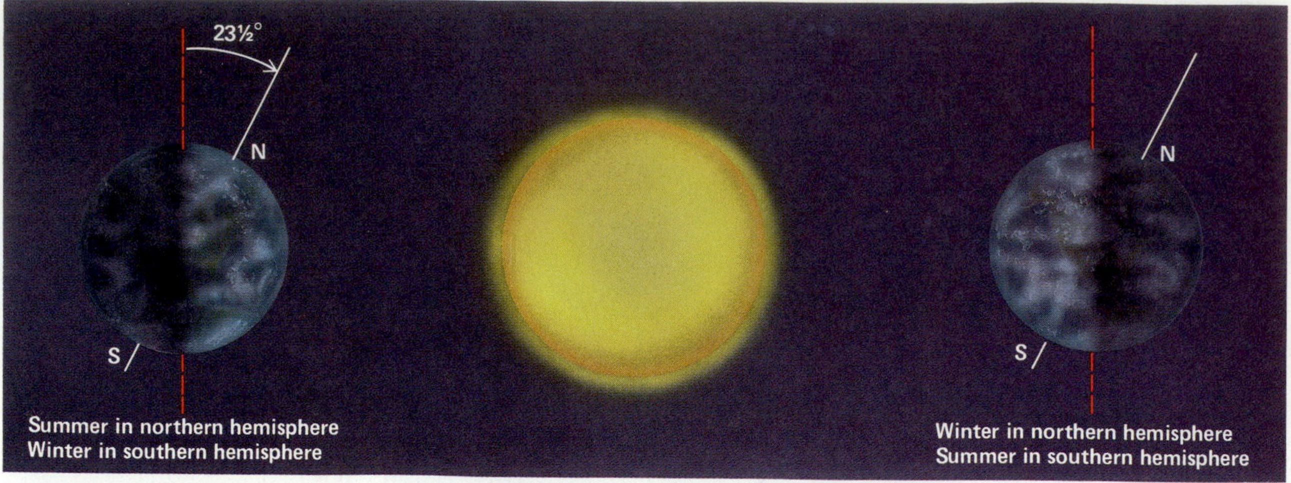

Figure 5–19 The seasons occur because the Earth's axis is tipped with respect to the plane of the orbit in which it revolves around the Sun. The dotted line is drawn perpendicularly to the plane of the Earth's orbit. When the northern hemisphere is tilted toward the Sun, it has its summertime; at the same time, the southern hemisphere is having its winter. At both locations of the Earth shown, the Earth rotates through many 24-hour day-night cycles before its motion around the Sun moves it appreciably. The diagram is not to scale.

lasts six months at the poles and shorter times at other locations within 23½° of the poles.

Since the Sun goes 23½° above the celestial equator, the midnight Sun is visible at some time anywhere within 23½° of the north pole, a boundary at 66½° latitude known as the Arctic Circle. The midnight Sun seen from within 23½° of the south pole, within the Antarctic Circle, is six months out of phase with that near the north pole.

When the Sun is at the summer solstice, it is at its greatest northern declination and is above the horizon of all northern-hemisphere observers for the longest time each day. Thus daytimes in the summer are longer than daytimes in the winter, when the Sun is at its lowest declinations. In the winter, the Sun not only is above the horizon for a shorter period each day but also never rises very high in the sky. As a result, the weather is colder. The instant of the Sun's lowest declination is the winter solstice. It is winter in the northern hemisphere when it is summer in the southern.

The seasons (Fig. 5–19), thus, are caused by the variation of declination of the Sun, which in turn is caused by the fact that the Earth's axis of spin is tipped by 23½° (Fig. 5–20). Many if not most people misunderstand the cause of the seasons. Note that the seasons are not caused by the distance between the Earth to the Sun; indeed, the Earth is closest to the Sun each year on or about January 4th, which falls in the northern-hemisphere winter.

The Moon goes around the Earth once each month, and so the Moon's right ascension changes through the entire range of ascension once each month. Since the Moon's orbit is inclined to the celestial equator, the Moon's declination also varies.

The planets' motions in the sky are less easy to categorize, but they also change their right ascension and declination from day to day. Their positions are listed or graphed in my book *A Field Guide to the Stars and Planets,* in monthly magazines and bulletins, and in *The Astronomical Almanac,* published each year under government auspices.

*5.4b The Analemma

We have seen that the inclination of the Earth's axis of rotation to the plane of the ecliptic leads to a variation over the year in the height to which the Sun rises in the sky. Thus, if we were to take a photograph of the Sun in the sky at the same hour each day, over the year, the Sun would sometimes be relatively low and sometimes relatively high.

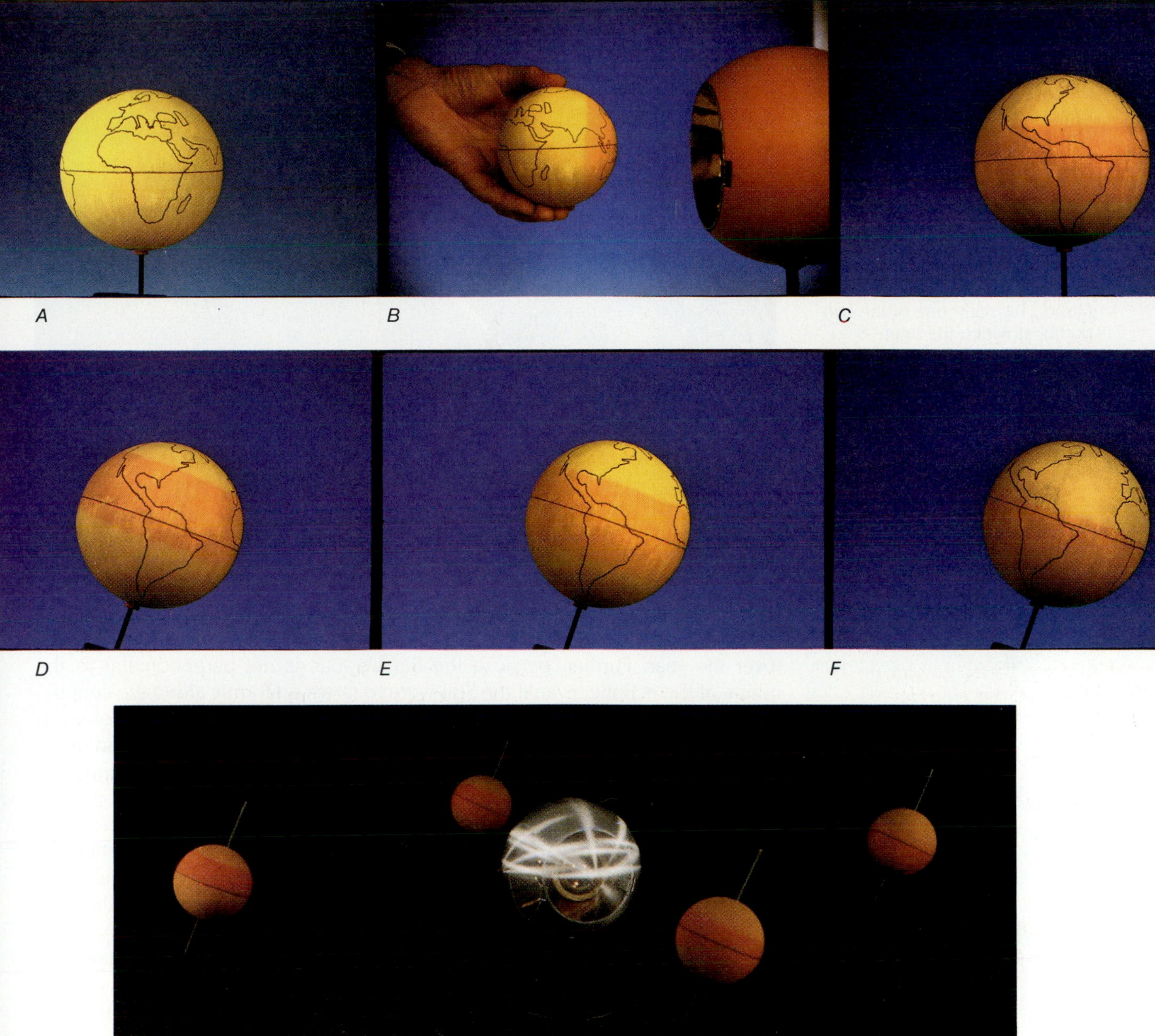

Figure 5–20 This laboratory demonstration shows the reasons for the seasons. (*A*) A ball is painted with a chemical that turns from yellow to orange when heated. (*B*) The spot nearest the heat source turns orange. (*C*) As the ball rotates on its axis, that orange region makes a band along the equator. (*D*) If the ball is inclined, as the Earth is, then the spot nearest the heat source is above the equator in the summertime. As the ball rotates on its inclined axis, a band above the equator turns orange. This reproduces summer conditions. (*E*) At the equinoxes, the band is along the equator. (*F*) In the northern hemisphere's wintertime, the band is below the equator. (*G*) The seasonal heating is reproduced in a multiexposure.

Also, if the Earth's orbit around the Sun were a circle, and if the Earth's axis of rotation were perpendicular to the circle, then we would expect the Sun to be at the same azimuth in the sky each day. But the Earth's orbit, as we have seen, is elliptical, and the speed with which the Earth travels in its orbit varies over the year. So the Sun's apparent speed across the sky varies

Figure 5–21 The analemma displayed in this unique photograph that is a multiple exposure of the Sun, photographed through a dense filter at ten-day intervals throughout a full year. Clouds have caused a few of the solar images to be relatively faint or missing. All the exposures were taken at 8:30 a.m. Eastern Standard Time. On three days of the year—close to the winter solstice (December 21st), the summer solstice (June 21st), and one of the two points at which the figure-8 crossover occurs—the shutter was left open from dawn until shortly before the exposure time for the solar image; this made the streaks. For all solar images, a very dense filter masked all but the disk of the Sun. On one day in the fall, the filter was taken off for a brief exposure to show the foreground trees and building.

over the year. Further, because the Earth's axis is not perpendicular to the plane of the ecliptic, even if the Sun were to move uniformly along the ecliptic, the part of its motion that is parallel to the celestial equator would not seem uniform. (There is an up-and-down part as well. Technically, the up-and-down part is a change in declination, and the around-the-celestial-equator part is a change in right ascension.)

Even though our clocks are based on time from the Sun's motion, we don't change from day to day the rate at which our clocks work. Rather, we tell time by a "mean Sun" (using the definition of "mean" as "average"). But the real Sun is not the mean Sun; the real Sun is almost always a bit ahead or behind the mean Sun in the sky.

These two effects—the change in height in the sky and the change in horizontal position across the sky—cause the position of the Sun at a given hour each day to map out a figure 8 in the sky (Fig. 5–21). This figure 8 is called the *analemma*. The fact that, as we plainly see, the Sun rises much higher in the sky in the summertime than in the wintertime makes the days longer and hotter in the summer.

*5.5 Time and the International Date Line

Every city and town on Earth used to have its own time system, based on the Sun, until widespread railroad travel made this inconvenient. In 1884, an international conference agreed on a series of longitudinal time zones. Now all localities in the same zone have a standard time (Fig. 5–22). Since there are twenty-four hours in a day, the 360° of longitude around the Earth are divided into 24 standard time zones, each 15 degrees wide. Each time zone is centered on a meridian of longitude exactly divisible by 15. Because the time is the same throughout each zone, the Sun is not directly overhead at noon at each point in a given time zone, but in principle is less than about a half-hour off. Standard time is based on a *mean solar day*, the average length of a solar day.

Figure 5–22 Although in principle the Earth is neatly divided into 24 time zones, in practice, political and geographic boundaries have made the system much less regular. The existence of daylight-saving time in some places and not in others further confuses the time zone system. Other countries, shown with stripes, have time zones that differ by one-half hour from a neighboring zone. India and Nepal actually differ from each other by 10 minutes! At the international date line, the date changes. In the United States, most states have daylight-saving time for seven months a year—from the first Sunday in April until the last Sunday in October. But Arizona, Hawaii, and parts of Indiana have standard time year-round. Alaska's time zones were changed in 1983, placing almost all of the state in a single zone, one hour earlier than Pacific time. Some time zones in the former Soviet Union are being realigned.

As the Sun seems to move in the sky from east to west, the time in any one place gets later. We can visualize noon, and each hour, moving around the world from east to west, minute by minute. We get a particular time back 24 hours later, but if the hours circled the world continuously, the date would not have changed. So we specify a north–south line and have the date change there. We call it the *international date line*. England won for Greenwich, then the site of the Royal Observatory, the distinction of having the basic line of longitude (the Prime Meridian), 0°. Since the international date line would disrupt the calendars of those who crossed it, that line was put as far away from the populated areas of Europe as possible—near or along the 180° longitude line. When one flies from Hawaii to Japan across the international date line, the day will change from, say, Tuesday to Wednesday. The international date line passes from north to south through the Pacific Ocean and actually bends to avoid cutting through continents or groups of islands, thus providing them with the same date as their nearest neighbor, as shown in Fig. 5–23.

In the summer, in order to make the daylight last into later hours, many countries have adopted daylight-saving time. Clocks are set ahead 1 hour on a certain date in the spring. Thus if darkness falls at 6 p.m. E.S.T. (Eastern Standard Time), that time is called 7 p.m. E.D.T. (Eastern Daylight Time), and most people have an extra hour of daylight after work. In most places, that hour is taken away in the fall, though some places have adopted daylight-

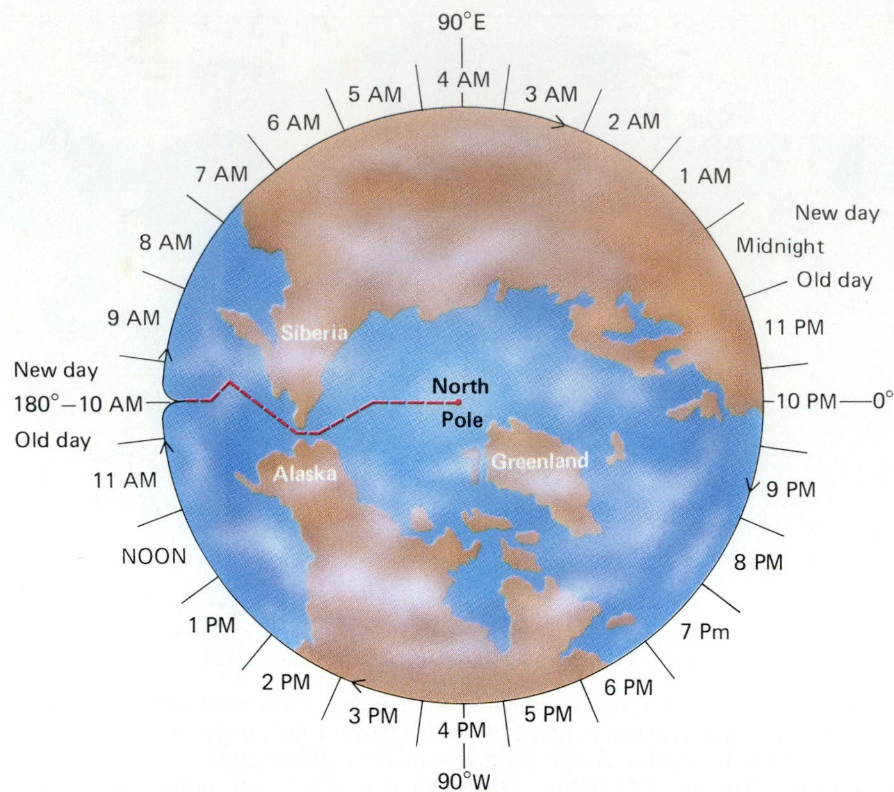

Figure 5–23 This top view of the Earth, looking down from above the north pole, shows how days are born and die at the international date line. When you cross the international date line, your calendar changes by one day.

saving time all year. The phrase to remember to help you set your clocks is "fall back, spring ahead." Of course, daylight-saving time is just a bookkeeping change in how we name the hours, and doesn't result from any astronomical changes.

*5.6 Calendars

The period of time that the Earth takes to revolve once around the Sun is called, of course, a *year*. This period is about 365¼ mean solar days. A *sidereal year* is the interval of time that it takes the Sun to return to a given position with respect to the stars. A *solar year* (in particular, a *tropical year*) is the interval between passages of the Sun through the vernal equinox, the point at which the ecliptic crosses the celestial equator travelling from south to north. Since tropical years are growing shorter by about half a second per century, we refer to a standard tropical year: 1900.

Roman calendars had, at different times, different numbers of days in a year, so the dates rapidly drifted out of synchronization with the seasons (which follow solar years). Julius Caesar decreed that 46 B.C. would be a 445-day year in order to catch up, and defined a calendar, the *Julian calendar,* that would be more accurate. This calendar had years that were normally 365 days in length, with an extra day inserted every fourth year to bring the average year to 365¼ days in length. The fourth years were, and are, called *leap years*.

The Julian calendar was much more accurate than its predecessors, but still imprecise; the actual solar year 1988 was 365 days 5 hours 48 minutes 45.3 seconds long, some 11 minutes 14.7 seconds shorter than 365¼ days (365 days 6 hours). By 1582, the calendar was about 10 days out of phase with the date on which Easter had occurred at the time of a religious council 1250 years earlier, and Pope Gregory XIII issued a bull—a proclamation—to correct the

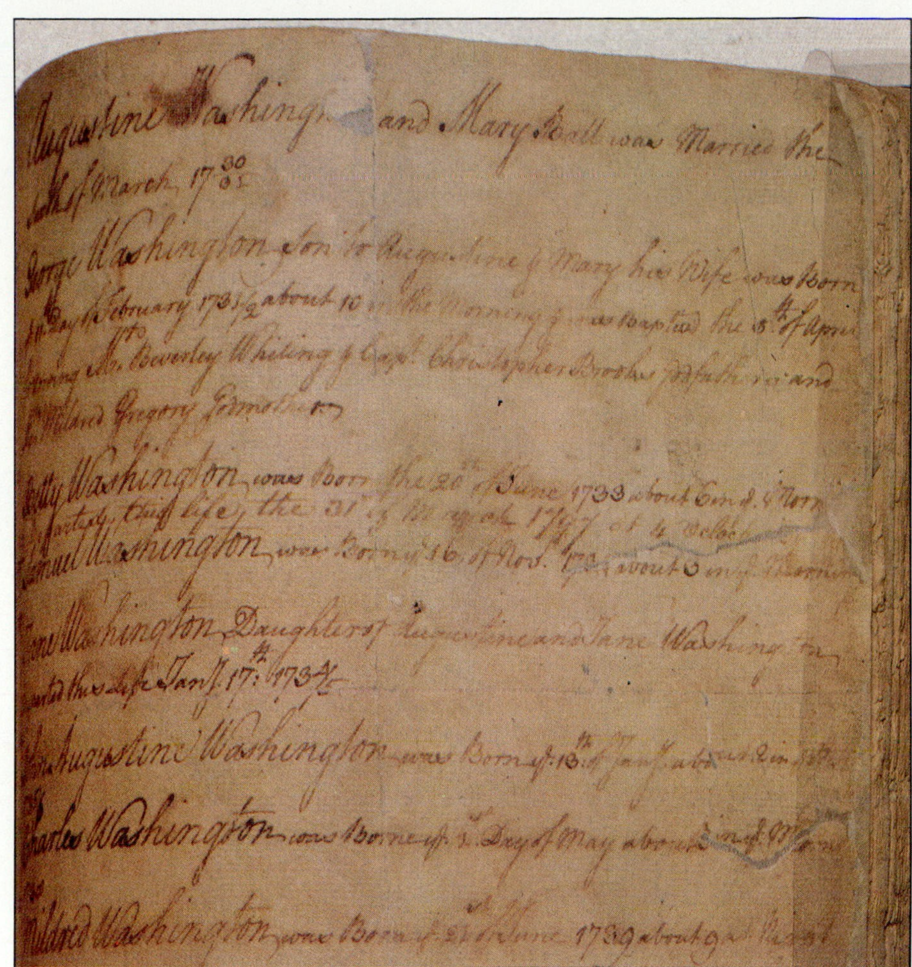

Figure 5–24 In George Washington's family bible, his date of birth is given as 1731/2. Some contemporaries would have said 1731; we now say 1732.

situation. He dropped 10 days from 1582. Many citizens of that time objected to the supposed loss of the time from their lives and to the commercial complications. Does one pay a full month's rent for the month in which the days were omitted, for example? "Give us back our fortnight," they cried.

In the *Gregorian calendar,* years that are evenly divisible by four are leap years, except that three out of every four century years, the ones not divisible evenly by 400, have only 365 days. Thus 1600 was a leap year; 1700, 1800, and 1900 were not; and 2000 will again be a leap year.

Although many countries adopted the Gregorian calendar as soon as it was promulgated, Great Britain (and its American colonies) did not adopt it until 1752, when 11 days were skipped. As a result, we celebrate George Washington's birthday on February 22nd, even though a calendar when he was born read February 11th. Also, the beginning of the year was changed from March to January. Washington was born in February 1731 (about a month before the end of 1731), often then written February 1731/2 (Fig. 5–24), but we now refer to his date of birth as February 22nd, 1732. The Gregorian calendar is the one in current use. It will be over 3000 years before this calendar is as much as one day out of step.

*5.7 Astronomy and Astrology

Astrology is not connected with astronomy, except in a historical context, so does not really deserve a place in a text on contemporary astronomy. But since so many people associate astrology with astronomy, and since astrologers claim to be using astronomical objects to make their predictions, let us use our astronomical knowledge to assess astrology's validity. Since millions of

Figure 5–25 Twelve constellations through which the Sun, Moon, and planets pass make up the zodiac. (Drawing by Handelsman; © 1978 The New Yorker Magazine, Inc.)

Americans believe in astrology—a number that shows no sign of decreasing—we cannot ignore the matter. Indeed, surveys show belief by adolescents in astrology to be increasing. Astrology, extrasensory perception (ESP), witchcraft, and several other topics of public interest are examples of *pseudoscience*, a set of topics that masquerades as having a relation to science without having real scientific content.

Science is also misrepresented to the public by "creationists," who hold that the Earth and everything in the Universe were created only a few thousand years ago. Throughout this book you will learn of the evidence found by astronomers that the Earth and Solar System are about 4.5 billion years old and that the galaxies and the Universe are even older. And you will learn about the methods scientists use to study the Universe and to validate theories. "Creationism" is simply not compatible with today's knowledge and standards.

Astrology is an attempt to predict or explain our actions and personalities on the basis of the positions of the stars and planets now and at the instants of our births. Astrology has been around for a long time, but it has never been shown to work. Believers may cite incidents that reinforce their faith in astrology, but no successful scientific tests have ever been carried out. If something happens to you that you had expected because of an astrological prediction, you would more certainly notice that this event occurred than you would notice the thousands of other unpredicted things that happened to you that day. Yet we do enough things, have sufficiently varied thoughts, and interact with enough people that if we make many predictions in the morning, some of them are likely to be at least partially fulfilled during the day. We simply forget that the rest ever existed.

In fact, even the alignments that most astrologers use are not accurately calculated, since the precession of the Earth's pole has changed the stars that are overhead at a given time of year from what they were millennia ago when astrological tables that are often still in current use were computed. At a given time of year, the Sun is usually in a different sign of the zodiac than its traditional astrological one (Figs. 5–25 and 5–26). And we know that the constellations don't even exist as physical objects. They are merely projections of the positions of stars that are usually very different distances from us.

Studies have shown that superstition actively constricts the progress of science and technology in various countries around the world and is therefore not merely an innocuous force. It is not merely that some people harmlessly believe in astrology. Their lack of understanding of scientific structure may actually impede the training of people needed to solve the problems of our age. Widespread superstitious beliefs even impeded smallpox-prevention programs. Thus many scientists are not content to ignore astrology, but actively oppose its dissemination. Further, if large numbers of citizens do not understand the scientific method and the difference between science and pseudoscience, how can they intelligently vote on or respond to scientific questions that have societal implications? The episode in which the wife of the President consulted an astrologer about his schedule was not only a serious breach of security but also shows how widespread belief in pseudoscience is.

A team of Calstate–Long Beach psychologists arranged for a magician to perform three psychic-like stunts in front of psychology classes. Even when they emphasized to the students that the performer was a magician performing tricks, 50 per cent of the class still believed the magician to be psychic.

Further, astrology just doesn't work. In 1985, Shawn Carlson, a UCLA–Lawrence Berkeley Laboratory physicist, reported on his double-blind controlled test of astrology. As part of the test, 28 astrologers, all highly respected by their peers, were asked to make a total of 116 selections, with each matching

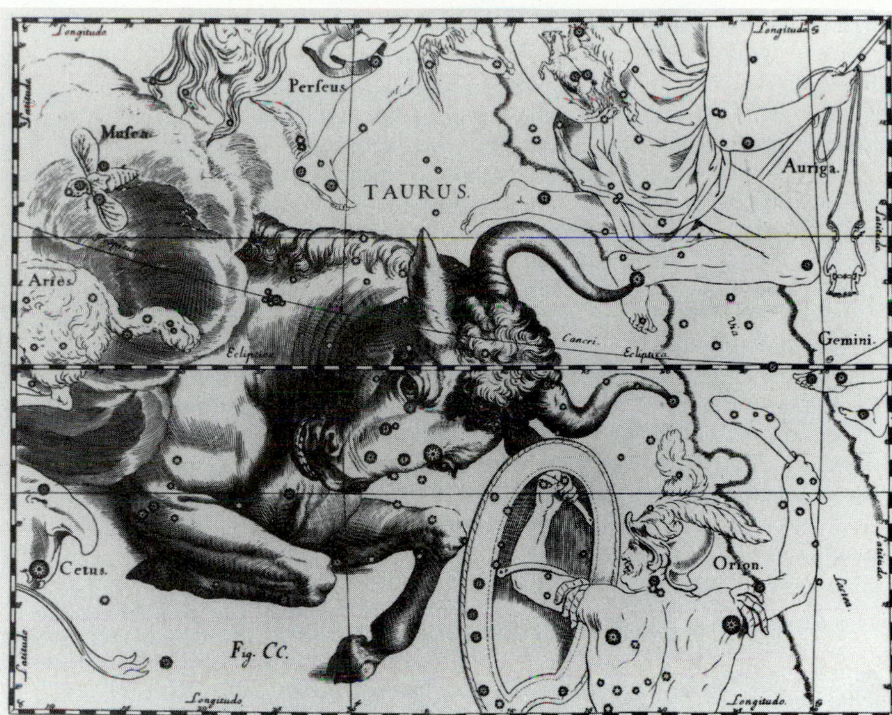

Figure 5–26 The zodiacal constellation Taurus from the star atlas of Hevelius.

one "natal chart," a horoscope based on the time and place of birth, to the results of standardized personality surveys for three different people. One of these standardized surveys came from the same person as the natal chart. Though the astrologers themselves predicted that they would get "at least" 50 per cent of these matchings correct, they scored only 34 per cent, precisely what one would expect if astrology did not work at all. Further, even when the astrologers rated a particular standardized survey as fitting the natal chart very well, they were no more likely to be correct. So even in Carlson's controlled experiment, in which he had worked with astrologers to make it a fair test to them, astrology failed.

In sum, astrology is meaningless, unnecessary, and impossible to explain if we accept the broad set of physical laws and theories we have conceived over the years to explain what happens on the Earth and in the sky. In Section 19.6b, we discuss the structure of science and how closely related it is to Occam's Razor, which states that we accept the simplest satisfactory explanation as true. Astrology snipes at the roots of all pure science. Moreover, astrology patently doesn't work. If people want to believe in astrology on an *a priori* basis, as a religion, or have a personal astrologer act as a psychologist, let them not try to cloak their beliefs in scientific astronomical gloss. The only reason people may believe that they have seen astrology work is that it is a self-fulfilling means of prophecy, conceived of long ago when we knew less about the exciting things that are going on in the Universe.

Figure 5–27 Zodiacal constellations in the book of Apianus published in 1540.

Science is more than just a set of facts, since a methodology of investigation and standards of proof are involved, but science is more than just a methodology since many facts have been well established. In this course, you are supposed to not only learn certain facts about the Universe but also to appreciate the way that theories and facts come to be accepted. (See also Section 23.11 on the verification of relativity theory at an eclipse and Section 19.6b on the definition of "truth.") Let's all learn from the stars, but let's learn the truth!

Summary and Outline

The constellations (Section 5.1)
  Stars in them are not necessarily physically grouped.
Twinkling (Section 5.2)
  Stars appear to twinkle; planets usually do not.
  Twinkling is caused in the Earth's atmosphere.
Coordinate systems (Section 5.3)
  Celestial longitude and latitude are right ascension and declination.
  The celestial equator and celestial poles are the points in the sky that are on the extensions of the Earth's equator and poles into space. The ecliptic and the celestial equator cross at the equinoxes.
  Sidereal time is time by the stars.
  Sidereal day: a given right ascension returns to your meridian.
  Solar day: the Sun returns to your meridian.
  Precession is the slow drift in the coordinate system; 26,000-year period.
Motions in the sky (Section 5.4)
  Stars rise and set; at the Earth's equator we would see them do so perpendicularly to the horizon.
  At the Earth's poles, we would see the stars move around parallel to the horizon.
  At or near the poles, the Sun and Moon rise and set only when they change sufficiently in declination; the Sun goes through this sequence once a year—thus the "midnight sun."
  The seasons are caused by the Sun's variations in declination.
  Telescopes can be mounted so that motion on one axis allows the Earth's rotation to be counteracted; such mounts are called "equatorial." Alt-azimuth mounts have separate motions in altitude and azimuth; it takes both motions to follow the stars.
Time and the International Date Line (Section 5.5)
  Standard time is based on a mean solar day.
  The date changes at the international date line.
  Daylight-saving time: "fall back, spring ahead"
Calendars (Section 5.6)
  Leap years are needed to keep up with the 365¼-day year.
  The Julian calendar, introduced by Julius Caesar, was a reasonably good calendar, but, over the centuries, the days drifted.
  We now use the Gregorian calendar, set up in 1582.
Astronomy and astrology (Section 5.7)
  No scientific basis is known for astrology.
  Belief in such pseudoscience impedes the advance of science and technology.
  Statistical tests show that astrology doesn't work.

Key Words

constellations, asterisms, celestial sphere*, coordinate systems*, meridians*, right ascension*, declination*, celestial equator*, celestial poles*, ecliptic*, vernal equinox*, autumnal equinox*, hour circles*, sidereal time*, meridian*, zenith*, sidereal day*, solar day*, precession*, precession of the equinoxes*, altitude*, pole star*, polar axis*, equatorial mounts*, alt-azimuth*, Dobsonian*, equinoxes*, solstice*, midnight sun*, analemma*, mean solar day*, international date line*, year*, sidereal year*, solar year (tropical year)*, Julian calendar*, leap years*, Gregorian calendar*, pseudoscience

*These terms are found in optional sections.

Questions

1. Explain why we cannot tell by merely looking in the sky that stars in a given constellation are at different distances, whereas in a room we can easily tell that objects are at different distances from us. What is the difference between the two situations?
2. Why is the Big Dipper only an asterism while the Big Bear is a constellation?
3. Can a star in the sky not be part of a constellation? Explain.
4. Can you reason that since all the stars in the constellation Pegasus are close together, they must have formed at about the same time? Explain.
5. What is the difference between "seeing" and "twinkling"?
6. If you look toward the horizon, are the stars you see likely to be twinkling more or less than the stars overhead? Explain.
7. Is the planet Uranus, which is in the outskirts of the solar system, likely to twinkle more or less than the nearby planet Venus?
8. Explain how it is that some stars never rise in our sky, while others never set.
9. By comparing their right ascensions and declinations (Appendix 5, the brightest stars), describe whether Sirius and Canopus, the two brightest stars, are close together or far apart in the sky. Explain.

†This question requires a numerical solution.

†10. When Arcturus (Appendix 5) is due south of you, what is the sidereal time where you are?

†11. Between the vernal equinox, March 21st, and the autumnal equinox, about 6 months later, ignoring precession,

(a) by how much does the right ascension of the Sun change?
(b) by how much does the declination of the Sun change?
(c) by how much does the right ascension of Sirius change?
(d) by how much does the declination of Sirius change?

12. (a) How does the declination of the Sun vary over the year? (b) Does its right ascension increase or decrease from day to day? Justify your answer.

13. We normally express longitude on the Earth in degrees. Why would it make sense to express longitude in units of time?

†14. By how many hours do sidereal clocks and solar clocks differ at the summer solstice?

†15. Divide the number of minutes in a day by 365 to find out by how many minutes each day the sidereal day drifts with respect to the solar day, as the Earth goes 1/365 of the way around the Sun. Show your work.

16. If a planet always keeps the same side toward the Sun, how many sidereal days are there in a year on that planet?

†17. When it is 6 p.m. on October 1st in New York City, what time of day and what date is it in Tokyo?

†18. When it is noon on April 1st in Los Angeles, describe how to use Figure 5–23 to find the date and time in China. Follow through both going westward across the international date line and going eastward across the Atlantic, and describe why you get the same answer.

19. What is the advantage of an equatorial mount?

20. What new development has led large telescopes to be placed on alt-azimuth mounts? Be explicit about the change and what made it possible.

21. Describe one of the experiments that tested astrology and discuss the results.

22. Why might people think their personality matches a horoscope even though statistical tests show no special match?

23. Describe the relation of astrology and the precession of the Earth's pole.

Topic for Discussion

1. Discuss the role of pseudoscience, like astrology, in our modern society.

Puzzle

Every row and column contains one of the official three-letter abbreviations for a constellation (Appendix 9), and no constellation appears twice in the solution. Fill in the missing letters.

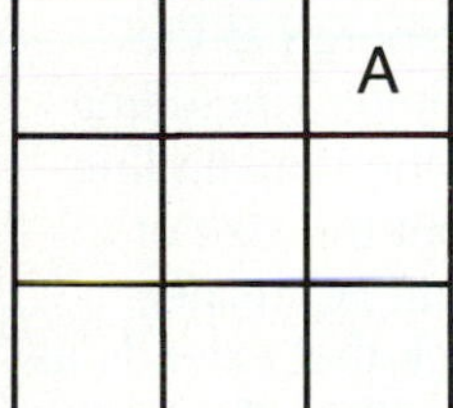

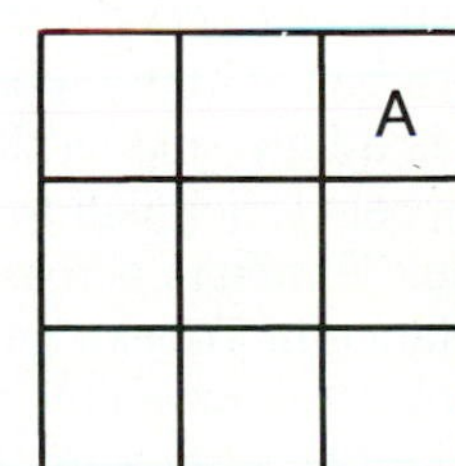

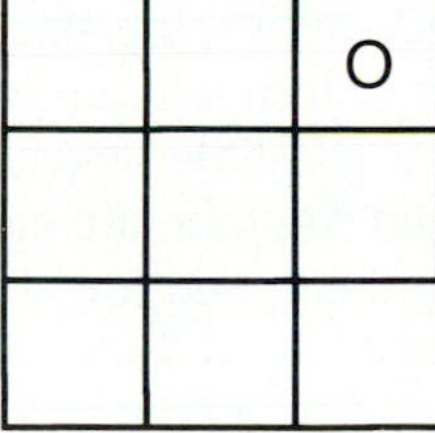

II The Solar System

The Earth and the rest of the solar system may be important to us, but they are only minor companions to the stars. In "Captain Stormfield's Visit to Heaven," by Mark Twain, the Captain races with a comet and gets off course. He comes into heaven by a wrong gate, and finds that nobody there has heard of "the world." ("**The** world, there's billions of them!" says a gatekeeper.) Finally, the gatekeepers send someone up in a balloon to try to detect "the world" on a huge map. The balloonist has to travel so far that he rises into clouds. After a day or two of searching, he comes back to report that he has found it: an unimportant planet named "the Wart."

We too must learn humility as we ponder the other objects in space. And while it is no doubt the case that the heavens are filled with a vast assortment of suns and planets more spectacular than our own, still, the solar system is our own local environment. We would like to understand it and come to terms with it as best we can. Besides, in understanding our own solar system, we may even find some keys to understanding the rest of the Universe.

To get an idea of its scale, imagine that the solar system is scaled down and placed on a map of the United States. Let us say that the Sun is a hot ball of gas taking up all of Rockefeller Center, about a kilometer across, in the center of New York City.

Mercury would then be a ball 4 meters across at the distance of mid–Long Island, and Venus would be a 10-meter ball one and a half times farther away. The Earth, only slightly bigger, is located at the distance of Trenton, New Jersey. Mars, half Earth's diameter, 5 meters across, is located past Philadelphia.

Only for the planets beyond Mars would the planets be much different in size from the Earth, and the separations become much greater. Jupiter is 100 meters across, the size of a baseball stadium, past Pittsburgh at the Ohio line. Saturn without its rings is a little smaller than Jupiter (including the rings it is a little larger), and is past Cincinnati toward the Indiana line. Uranus and Neptune are each about 30 meters across, about the size of a baseball infield, and are at the distance of Topeka and Santa Fe, respectively. And Pluto, a 2-meter ball, travels on an elliptical track that extends as far away as Los Angeles, 40 times farther away from the Sun than the Earth is. (At present, Pluto is in Montana, closer to us than Neptune.) Occasionally a comet sweeps in from Alaska, or some other random direction, passes around the Sun, and returns in the general direction from which it came.

The planets fall naturally into two groups. The first group, the *terrestrial planets,* consists of Mercury, Venus, Earth, and Mars. All are rocky in nature. The terrestrial planets are not very large, and have densities about five times that of water. (When the metric system was set up, the gram was defined so that water would have a convenient density of exactly 1 gram/cubic centimeter. Thus the terrestrial planets have densities of about 5 g/cm^3.)

The second group, the *giant planets,* consists of Jupiter, Saturn, Uranus, and Neptune. All these planets are much larger than the terrestrial planets, and also much less dense, ranging down to slightly below the density of water. Jupiter and Saturn are largely gaseous in nature, similar to the

Superimposed on a satellite photographic montage showing lights at night across the United States, the relative sizes of the orbits of the planets are visible.

Sun in composition. Uranus and Neptune also have thick gaseous atmospheres, and Voyager 2 has shown Neptune's atmosphere to be unexpectedly active. The giant planets all have fascinating moons, some the size of the inner planets. Pluto, the planet with the largest orbit, is anomalous in several of its properties, and so may have had a very different history from the other planets.

Between the orbits of the terrestrial planets and the orbits of the giant planets are the orbits of thousands of chunks of small "minor planets." These *asteroids* range up to 1000 kilometers across. Sometimes much smaller chunks of interplanetary rock penetrate the Earth's atmosphere and hit the ground. We shall discuss these *meteorites* (and where they came from) together with asteroids and comets.

Many people are interested in the planets in order to study their history—how they formed, how they have evolved since, and how they will change in the future. Others are more interested in what the planets are like today. Still others are interested in the planets mainly to consider whether they are harboring intelligent life.

Because the Moon and some other objects in the solar system have undergone less erosion on their surfaces than has the Earth, we now see them as they appeared eons ago. In this way the study of the planets and other objects in the solar system as they are now provides information about the solar system's early stages and its origin. Comets may take us even farther back. The surface appearances of other objects, like Jupiter's moon Io, change even more rapidly than does the Earth.

Now that spacecraft have explored so many planets close up—all but Pluto—we can discuss many general properties of planets or of groups of planets. This study of *comparative planetology* gives us important insights into our own planet. But studies too restricted to comparative planetology can make us lose sight of the interesting individual nature of each planet and of the stages by which our knowledge has jumped during the past decades. In the following chapters, we thus treat each planet individually and in a somewhat chronological fashion, while also stressing common features. We can then better appreciate not only the planets themselves but also the projects and individual researchers who have been uncovering their mysteries.

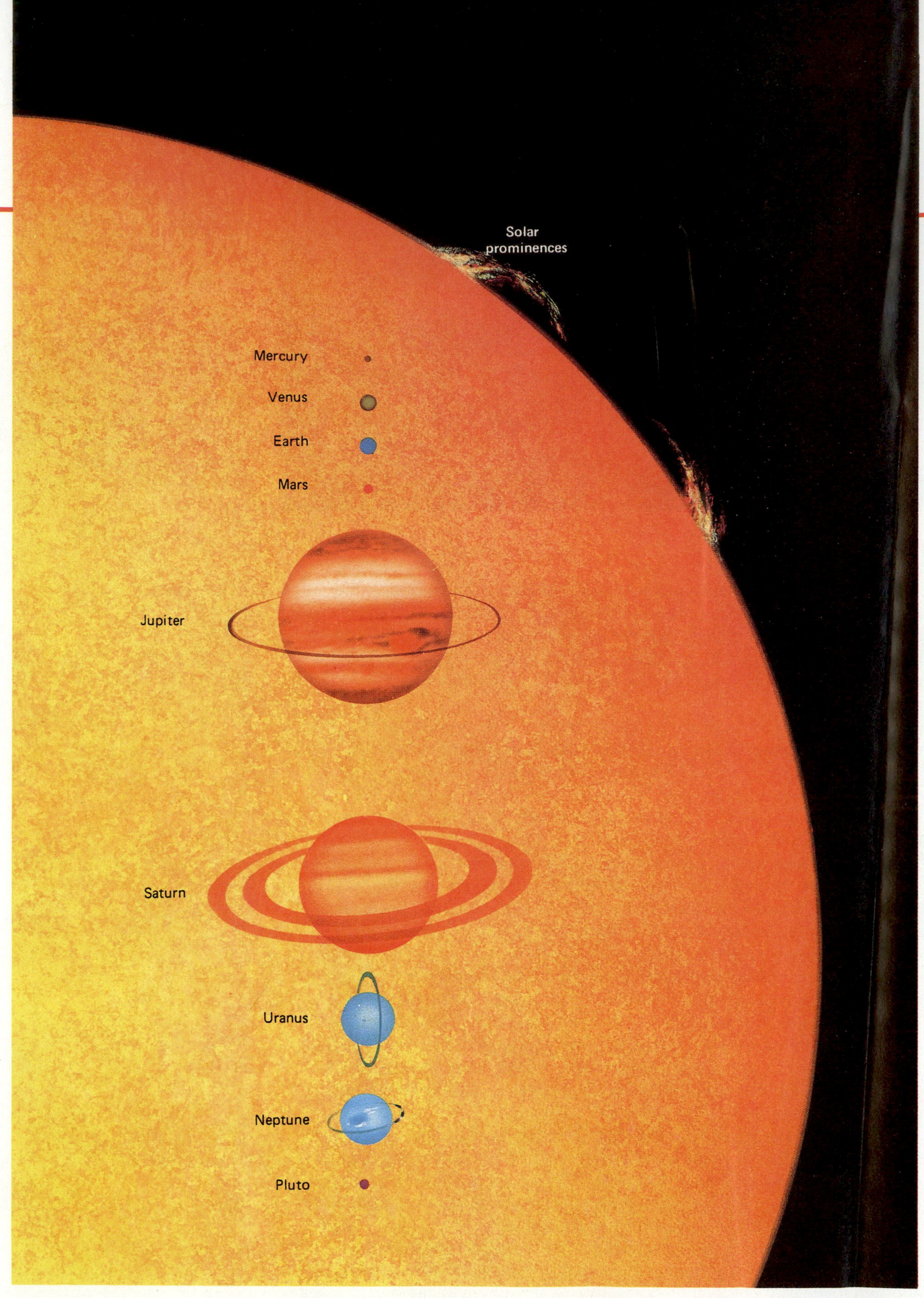

The relative sizes of the planets and the Sun.

The Structure and Origin of the Solar System

6

Aims: To discuss the Sun as the center of our solar system, the scale and structure of our solar system, and our solar system's origin

The basic motions of the objects in the solar system can be understood by the application of Kepler's laws, which we discussed in Section 3.1. Since Newton's laws of motion and gravitation provide the current framework for our analysis of planetary motions, calculations can now be carried out to great precision (and need to be in order to send spacecraft to the planets). Indeed, though Einstein's work on what we call "general relativity" is in some sense a generalization of Newton's work, for our solar system it is usually satisfactory to use the Newtonian theory.

In this chapter we describe the solar system and some aspects of its structure. First we briefly describe the central object of our solar system: the Sun. Later on, in Chapter 23, we shall treat the Sun as a star; here we merely describe its most basic structure. Next, we describe such commonly observed phenomena as phases and eclipses, and then we discuss the general arrangement of the planets and their orbits. We conclude by discussing theories that try to explain how the solar system began, although, as we shall see, this topic is not yet satisfactorily understood.

6.1 The Sun: Our Star

The Sun is the most important celestial object to us on Earth. It provides the energy that allows life to flourish. Though the Sun is just an average star, similar to thousands of pinpoints of light we perceive dimly in the nighttime sky, the Sun is so close to us that we feel its energy strongly. It is only eight light-minutes away from us, compared with over 4 light-years for the next nearest star.

The Sun is so bright that it lights up our sky. Sunlight bounces around among air molecules, making the sky blue. (Blue light bounces much more effectively than red light, since it has shorter wavelengths.) This scattering of sunlight by air molecules brightens the sky so much that the other stars are hidden from us in the daytime.

If we project an image of the Sun onto a piece of paper, we see that the Sun has dark regions on it—sunspots (Fig. 6–1). (We could observe the sunspots by looking at them through a telescope, but the telescope has to be very heavily filtered to cut out all but about one-millionth of the sunlight, or else we would burn our eyes.) The sunspots appear embedded in the everyday surface of the Sun from which we receive the solar energy. This everyday surface is called the *photosphere,* the sphere from which we get the Sun's light (from the Greek *photos,* meaning "light"). The photosphere is about 1.4 million kilometers (1 million miles) in diameter, about 1 percent of the size of Earth's orbit.

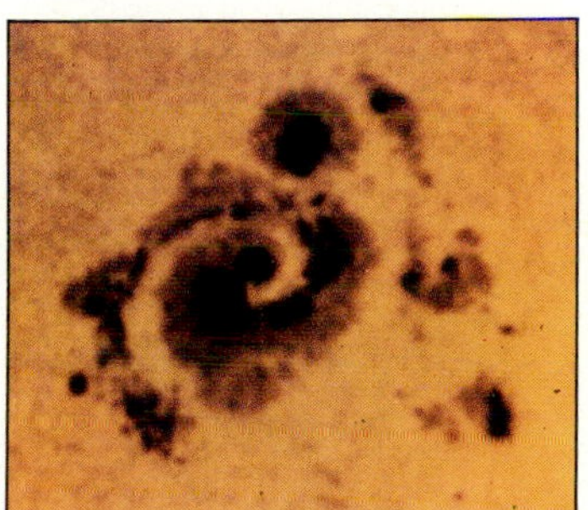

Figure 6–1 A sunspot, several times larger than the Earth.

Figure 6–2 The solar corona surrounds the dark side of the Moon in this photograph of the 1980 total solar eclipse.

The energy from the photosphere travels outward through the solar system and beyond in the form of electromagnetic radiation. Most of this solar radiation is in the visible and ultraviolet parts of the spectrum. We see the planets because they reflect the solar radiation toward us; only the Sun in the solar system forms all its own energy inside itself.

Mercury and Venus are closer to the Sun than the Earth, so receive more solar radiation on each square centimeter of their surfaces. Jupiter and its moons, for example, are farther from the Sun than the Earth, so receive less solar radiation. Thus the outer objects of the solar system are colder than inner ones.

The Sun doesn't end at the photosphere. It has layers of gas higher up, but these layers are much less dense than the photosphere and are usually invisible to us. From the ground they can be observed mainly during solar eclipses, as we shall see in Section 6.3. The uppermost level of the Sun is called the *corona* (Latin meaning "crown"), since it looks like a crown around the Sun (Fig. 6–2). The solar corona is continually expanding into space, forming a *solar wind* of particles. The solar-wind particles hit the Earth's upper atmosphere, and cause magnetic storms on Earth. Thus we must understand the Sun's atmosphere to understand the environment of the planets in space.

6.2 The Phases of the Moon and Planets

From the simple observation that the apparent shapes of the Moon and planets change, we can draw conclusions that are important for our understanding of the mechanics of the solar system. In this section, we shall see how the positions of the Sun, Earth, and other solar-system objects determine the appearance of these objects.

The *phases* of moons or planets are the shapes of the sunlighted areas as seen from a given vantage point. The fact that the Moon goes through a set of such phases approximately once every month is perhaps the most familiar everyday astronomical observation (Fig. 6–3). In fact, the name "month" comes from the word "moon." The actual period of the phases, the interval between

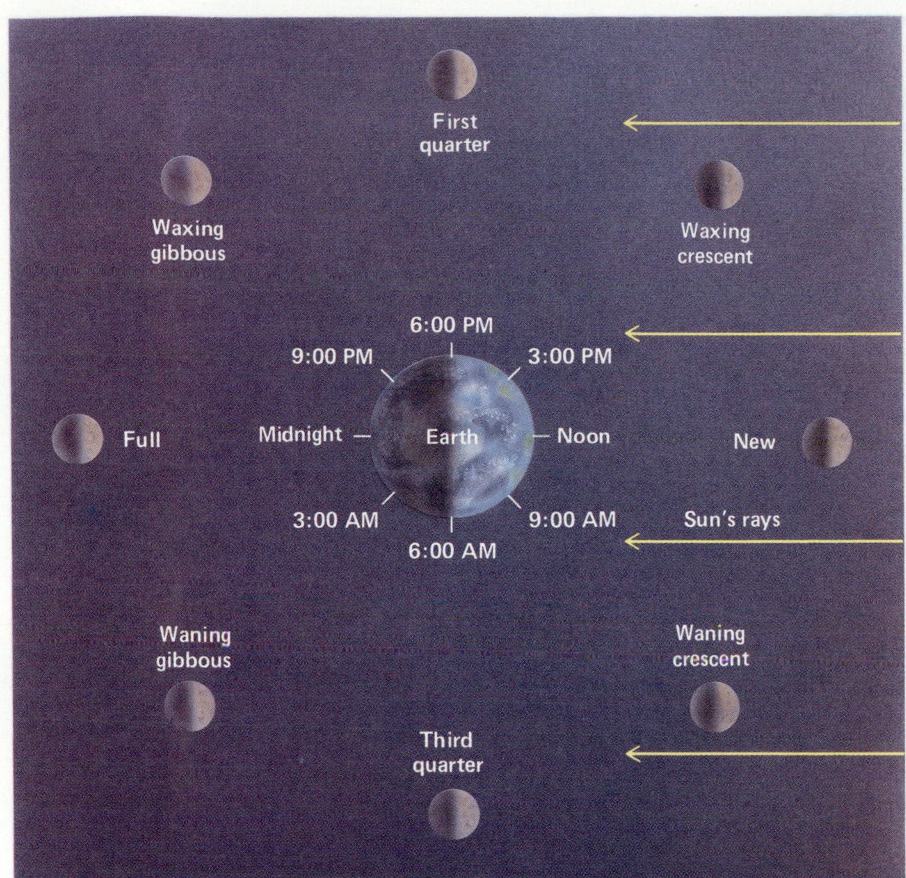

Figure 6–3 The phases of the Moon depend on the Moon's position in its orbit around the Earth. Here we visualize the situation as if we could be high above the Earth's orbit, looking down.

a particular phase of the Moon and its next repetition, is approximately 29½ Earth days (Fig. 6–4). This period can vary by as much as 13 hours.

The explanation of the phases is quite simple: the Moon is a sphere, and at all times the side that faces the Sun is lighted and the side that faces away from the Sun is dark. The phase of the Moon that we see from the Earth, as the Moon revolves around us, depends on the relative orientation of the three bodies: Sun, Moon, and Earth. The situation is simplified by the fact that the plane of the Moon's revolution around the Earth is nearly, although not quite, the plane of the Earth's revolution around the Sun.

"The Moon" is often capitalized to distinguish it from moons of other planets; in general writing, it is usually written with a small "m." In Parts I and II, we shall capitalize "Earth" to put it on a par with the other planets and the Moon. For consistency, we shall also capitalize "Sun" and "Universe."

Basically, when the Moon is almost exactly between the Earth and the Sun, the dark side of the Moon faces us. We call this a "new moon." A few days earlier or later we see a sliver of the lighted side of the Moon, and call this a "crescent." As the month wears on, the crescent gets bigger, and about 7 days after new moon, half the face of the Moon that is visible to us is lighted. We sometimes call this a "half moon." Since this occurs one-fourth of the way through the phases, the situation is also called a "first-quarter moon." (Instead of apologizing for the fact that astronomers call the same phase both "quarter" and "half," I'll just continue with a straight face and try to pretend that there is nothing strange about it.)

When over half the Moon's disk is visible, we have a "gibbous" moon. One week after the first-quarter moon, the Moon is on the opposite side of the Earth from the Sun, and the entire face visible to us is lighted. This is called a "full moon." One week later, when we see a half moon again, we have a "third-quarter moon." Then we go back to "new moon" again and repeat the cycle of phases.

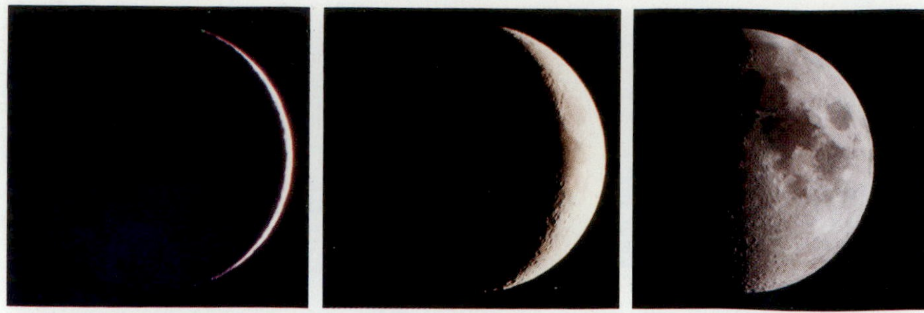

Figure 6–4 The phases of the Moon.

Note that since the phase of the Moon is related to the position of the Moon with respect to the Sun, if you know the phase, you can tell when the Moon will rise. For example, since the Moon is 180° across the sky from the Sun when it is full, a full moon is always rising just as the Sun sets (Fig. 6–5). Each day thereafter, the Moon rises about 50 minutes later (24 hours divided by the period of revolution of the Moon around the Earth), although the exact time difference also depends on the Moon's changing declination. The third-quarter moon, then, rises near midnight and is high in the sky at sunrise. The new moon rises with the Sun in the east at dawn. The first-quarter moon rises near noon and is high in the sky at sunset. It is often visible in the late afternoon.

The Moon is not the only object in the solar system that is seen to go through phases. Mercury and Venus both orbit inside the Earth's orbit, and so sometimes we see the side that faces away from the Sun and sometimes we see the side that faces toward the Sun. Thus at times Mercury and Venus are seen as crescents, though it takes a telescope to observe their shapes. Spacecraft to the outer planets have looked back and seen the Earth as a crescent and the other planets as crescents (Fig. 6–6) as well.

6.3 Eclipses

Because the Moon's orbit around the Earth and the Earth's orbit around the Sun are not precisely in the same plane (Fig. 6–7), the Moon usually passes slightly above or below the Earth's shadow at full moon, and the Earth usually passes slightly above or below the Moon's shadow at new moon. But every once in a while, up to seven times a year, the Moon is at the part of its orbit

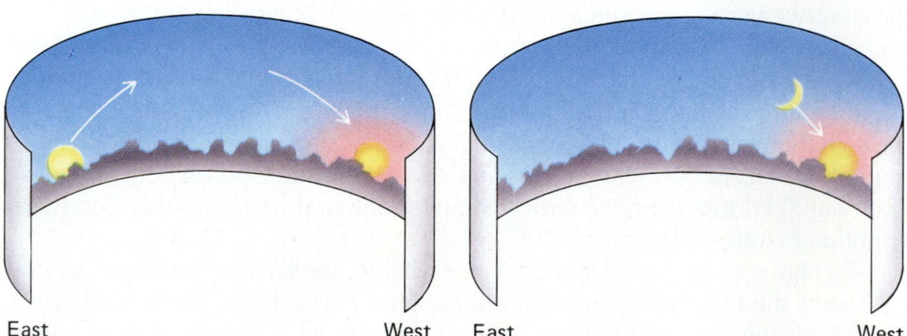

Figure 6–5 Because the phase of the Moon depends on its position in the sky with respect to the Sun, we can see why a full moon is always rising at sunset while a crescent moon is either setting at sunset, as shown here, or rising at sunrise.

that crosses the Earth's orbital plane at full moon or new moon. When that happens, we have a lunar eclipse or a solar eclipse (Fig. 6–8).

We have already discussed eclipses of the Sun, when the Moon comes directly between the Earth and the Sun. Many more people see a lunar eclipse than a solar eclipse when one occurs. At a total lunar eclipse, the Moon lies entirely in the Earth's shadow and sunlight is entirely cut off from it. So anywhere on the Earth that the Moon has risen, the eclipse is visible.

When the Moon comes directly between the Earth and the Sun, we have an eclipse of the Sun. A total solar eclipse—when the Moon covers the whole surface of the Sun that is normally visible—is a relatively rare phenomenon, occurring only every 1½ years or so. The Moon must be fairly precisely aligned between the Sun and Earth to have an eclipse; only those people in a narrow band on the surface of the Earth see the eclipse.

Figure 6–6 Crescent Uranus, photographed as the Voyager 2 spacecraft looked back on it.

6.3a Lunar Eclipses

A total lunar eclipse is a much more leisurely event to watch than a total solar eclipse. The partial phase, when the Earth's shadow gradually covers the Moon, lasts for hours. And then the total phase, when the moon is entirely within the Earth's shadow, can last for over an hour. During this time, the sunlight is not entirely shut off from the Moon. A small amount is refracted around the edge of the Earth by our atmosphere. Most of the blue light is taken out during the

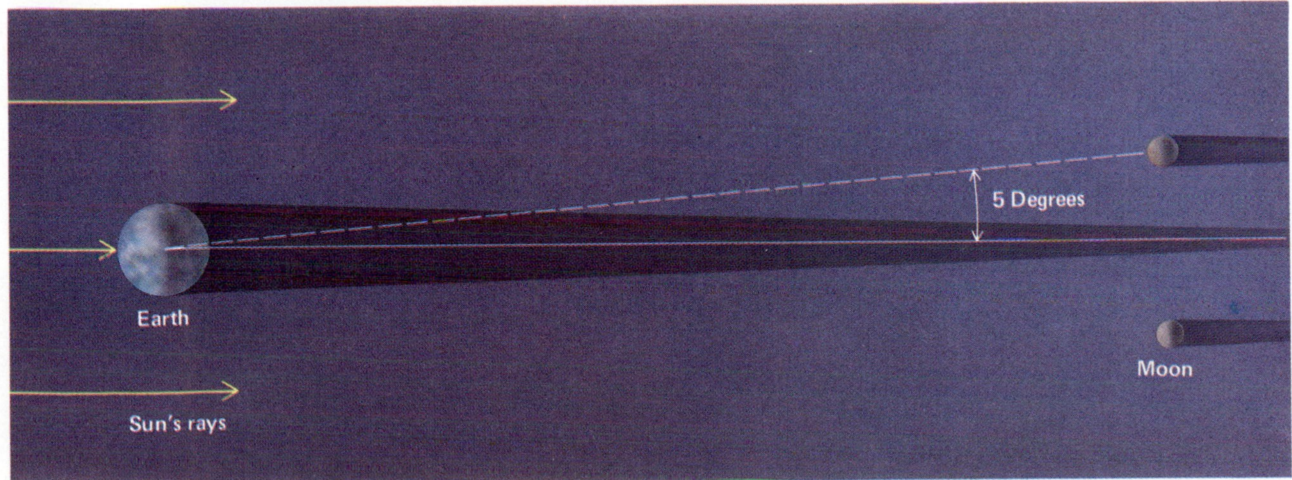

Figure 6–7 The plane of the Moon's orbit is tipped with respect to the plane of the Earth's orbit, so the Moon usually passes above or below the Earth's shadow. Therefore, we don't have lunar eclipses most months.

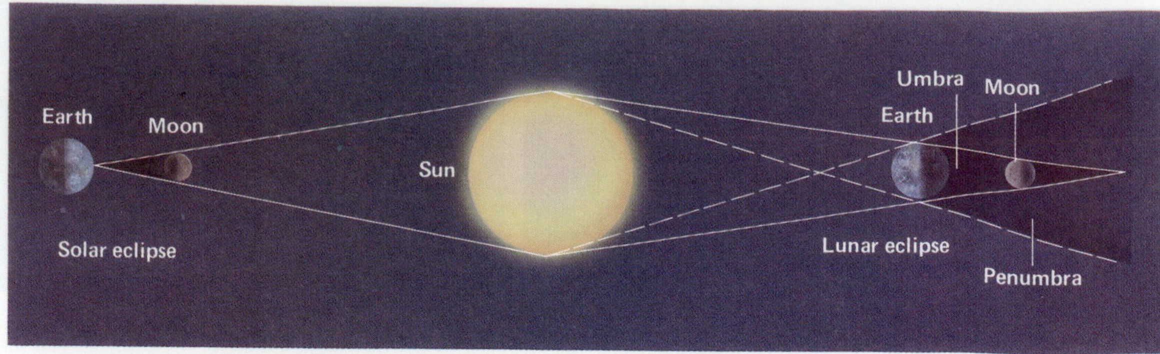

Figure 6–8 When the Moon is between the Earth and the Sun, we observe an eclipse of the Sun. When the Moon is on the far side of the Earth from the Sun, we see a lunar eclipse. The part of the Earth's shadow that is only partially shielded from the Sun's view is called the *penumbra;* the part of the Earth's shadow that is entirely shielded from the Sun is called the *umbra.* The distances in this diagram are not to scale.

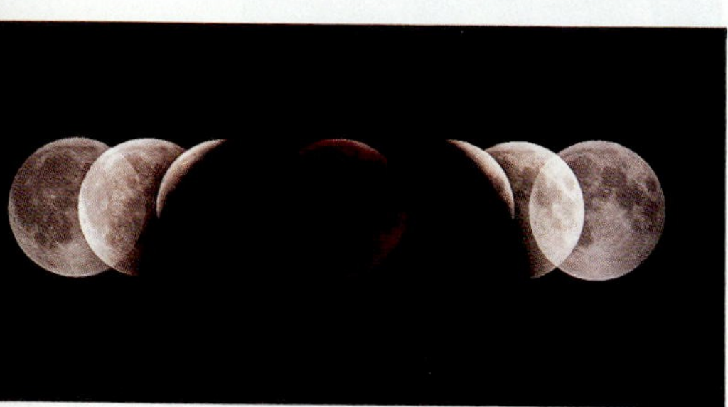

Figure 6–9 A total lunar eclipse, showing both partial and total phases. The camera was guided on the position of the Earth's umbral shadow. Note that the photograph shows that the Earth's shadow is round, and thus that the Earth is round.

sunlight's passage through our atmosphere; this explains how blue skies are made for the people part way around the globe from locations at which the Sun is setting. The remaining light is reddish, and this is the light that falls on the Moon (Fig. 6–9). Thus, the eclipsed Moon appears reddish (Fig. 6–10).

Total lunar eclipses will be visible in the United States and Canada on December 9, 1992 (eastern regions only); June 4, 1993 (western regions only); November 29, 1993; April 4, 1996 (eastern regions only); and September 27, 1996. In principle, scientists can make observations at lunar eclipses to study the properties of the lunar surface. For example, we can monitor the rate of cooling of areas on the lunar surface with infrared radiometers. But now we have landed on the Moon and have made direct measurements at a few specific places, so the scientific value of lunar eclipses has been greatly diminished. There remains little of scientific value to do during a lunar eclipse, quite differently from the situation at a solar eclipse.

6.3b Solar Eclipses

The outer layers of the Sun are fainter than the blue sky. To study them, we need a way to remove the blue sky while the Sun is up. A total solar eclipse does just that for us.

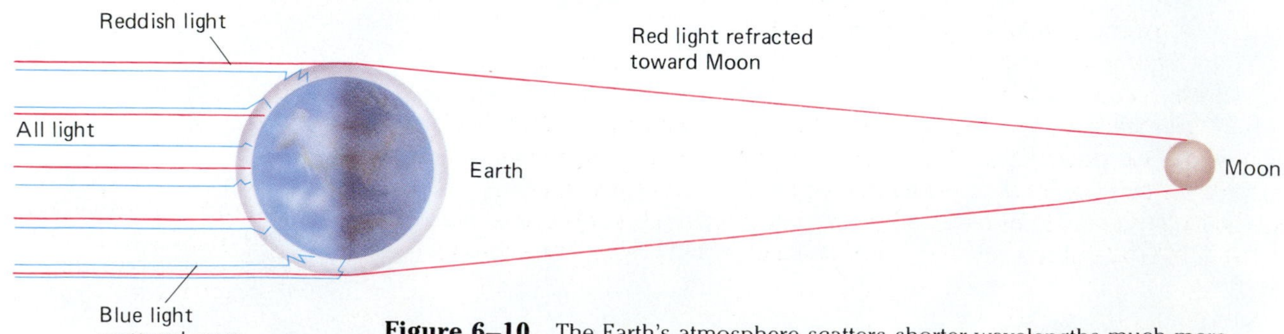

Figure 6–10 The Earth's atmosphere scatters shorter wavelengths much more than it scatters longer wavelengths, so red light is scattered much less than blue light. Some of the red light survives its passage through the Earth's atmosphere and is bent, or refracted, toward the Moon. Thus the Moon often appears reddish during the total phase of a lunar eclipse, unless there is so much volcanic dust in the Earth's atmosphere that the Moon appears gray.

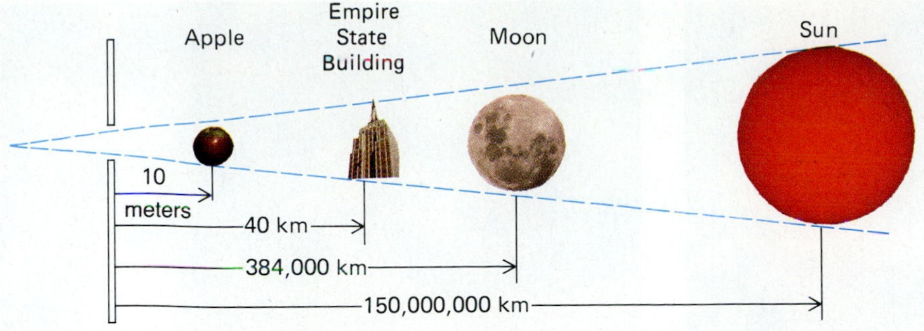

Figure 6–11 An apple, the Empire State Building, the Moon, and the Sun are very different from each other in size, but here they subtend the same angle because they are at different distances from us.

Solar eclipses arise because of a happy circumstance: though the Moon is 400 times smaller in diameter than the solar photosphere, it is also 400 times closer to the Earth. Because of this, the Sun and the Moon subtend almost exactly the same angle in the sky—about ½° (Fig. 6–11).

The Moon's position in the sky, at certain points in its orbit around the Earth, comes close to the position of the Sun. This happens approximately once a month at the time of the new Moon. Since the lunar orbit is inclined with respect to the Earth's orbit, the Moon usually passes above or below the line joining the Earth and the Sun. But occasionally the Moon passes close enough to the Earth–Sun line that the Moon's shadow falls upon the surface of the Earth.

At a total solar eclipse, the lunar shadow barely reaches the Earth's surface (Fig. 6–12). As the Moon moves through space on its orbit, and as the Earth rotates, this lunar shadow sweeps across the Earth's surface in a band up to 300 km wide. Only observers stationed within this narrow band can see the total eclipse.

From a wide area outside the band of totality, one sees only a partial eclipse. Sometimes the Moon, Sun, and Earth are not precisely aligned, and the darkest part of the shadow—called the *umbra*—never hits the Earth. We are then in the intermediate part of the shadow, which is called the *penumbra*. Only a partial eclipse is visible on Earth under these circumstances. As long as the slightest bit of photosphere is visible, even as little as 1 per cent, one cannot see the important eclipse phenomena—the faint outer layers of the Sun—that are both beautiful and the subject of scientific study. Thus partial eclipses are of little value for most scientific purposes. After all, the photosphere is 1,000,000 times brighter than the outermost layer, the corona; if 1 per cent of the photosphere is showing, then we still have 10,000 times more light from the photosphere than from the corona, which is enough to ruin our opportunity to see the corona.

If you are fortunate enough to be standing in the zone of a total eclipse, you will sense excitement all around you as the approaching eclipse is anticipated. An hour or so before totality begins, the partial phase of the eclipse starts. Nothing is visible to the naked eye immediately, but if you were to look at the Sun through a special filter, you would see that the Moon was encroaching

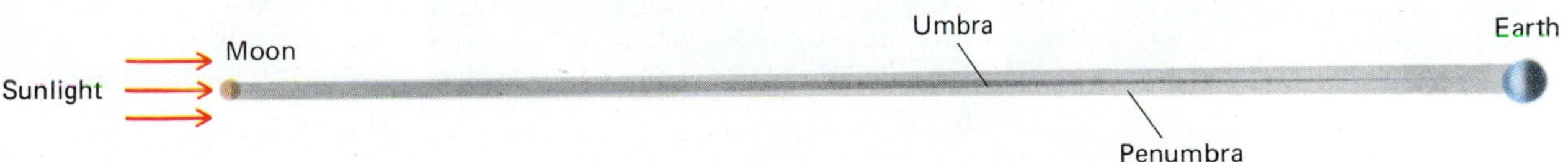

Figure 6–12 A solar eclipse with the Earth, the Moon, and the distance between them shown to actual scale.

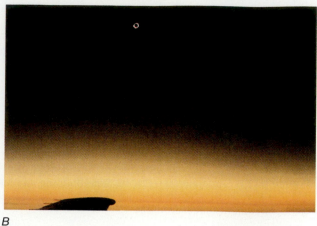

A

B

Figure 6–13 (*A*) The partially eclipsed Sun seen through clouds at our site during the eclipse of July 11, 1991. (*B*) The total solar eclipse of June 30, 1992, observed from an airplane over the south Atlantic Ocean off the coast of Africa.

upon the Sun (Fig. 6–13). At this stage of the eclipse, it is necessary to look at the Sun through a filter or to project the image of the Sun with a telescope or a pinhole camera onto a surface (Fig. 6–14). You need the filter to protect your eyes because the photosphere is visible throughout the partial phases before and after totality. Its direct image on your retina for an extended time could cause burning and blindness, though even in the few cases where eyes are harmed the damage is usually minor.

The eclipse progresses gradually, and by 15 minutes before totality the sky grows dark as if a storm were gathering. During the minute or two before totality begins, bands of light and dark race across the ground. These *shadow bands* are caused in the Earth's atmosphere, and thus tell us about the terrestrial rather than the solar atmosphere.

You still need the special filter to watch the final seconds of the partial phases. As the total phase—totality—begins, the bright light of the solar photosphere passing through valleys on the edge of the Moon glistens like a series of bright beads, which are called *Baily's beads.* The last bead seems so bright that it looks like the diamond on a ring—the *diamond-ring effect* (Fig. 6–15). For a few seconds, the chromosphere is visible as a pinkish band around the leading edge of the Moon.

Then, the corona comes into view in all its glory. You see streamers of gas near the Sun's equator and finer plumes near its poles. At this stage, the photosphere is totally hidden, so it is perfectly safe to look at the corona with the naked eye and without filters. The corona is about as bright as the full Moon and is equally safe to look at.

The total phase may last a few seconds, or it may last as long as about 7 minutes. Spectrographs are operated, photographs are taken through special

Figure 6–14 (*A*) These children, who visited our eclipse site in Loiengalani, Kenya, just prior to the 1973 total eclipse, were practicing the use of special filters to protect their eyes when observing the partial phases. (*B*) This child in Papua New Guinea used projection of the solar image by a pinhole to observe the partial phases of the 1984 eclipse. (*C*) These children are looking at crescents on the ground, pinhole images of the partially eclipsed Sun formed as the light passed through tiny spaces between the leaves of trees. The photo was taken just before the annular phase of the 1984 eclipse in the U.S.

A

B

C

filters, rockets photograph the spectrum from above the Earth's atmosphere, and tons of equipment are used to study the corona during this brief time of totality.

At the end of the eclipse, the diamond ring appears on the other side of the Sun, followed by Baily's beads and then the partial phases. All too soon, the eclipse is over.

Figure 6–15 The diamond ring effect at the June 30, 1992, total solar eclipse, observed from the south Atlantic Ocean. The sky was so clear at the plane's altitude that the corona was already well visible during the diamond ring.

6.3c Observing Solar Eclipses

On the average, a solar eclipse occurs somewhere in the world every year and a half. The band of totality, though, usually does not cross populated areas of the Earth, and astronomers often have to travel great distances to carry out their observations. Fig. 6–16 and Table 6–1 show sites and dates of future eclipses. Astronomers took advantage of the relatively long 1980 and 1983 eclipses by sending large-scale expeditions to India and Indonesia, respectively. The 1984 eclipse in Papua New Guinea and the 1988 eclipse in Indonesia and the Philippines were shorter and less observed.

The last total eclipse to cross the continental United States and Canada occurred in 1979. The last total eclipse in the United States crossed Hawaii before reaching Mexico and parts of Central and South America in 1991. The next total eclipse in the continental United States won't be until 2017. Canada won't see another total eclipse until 2024.

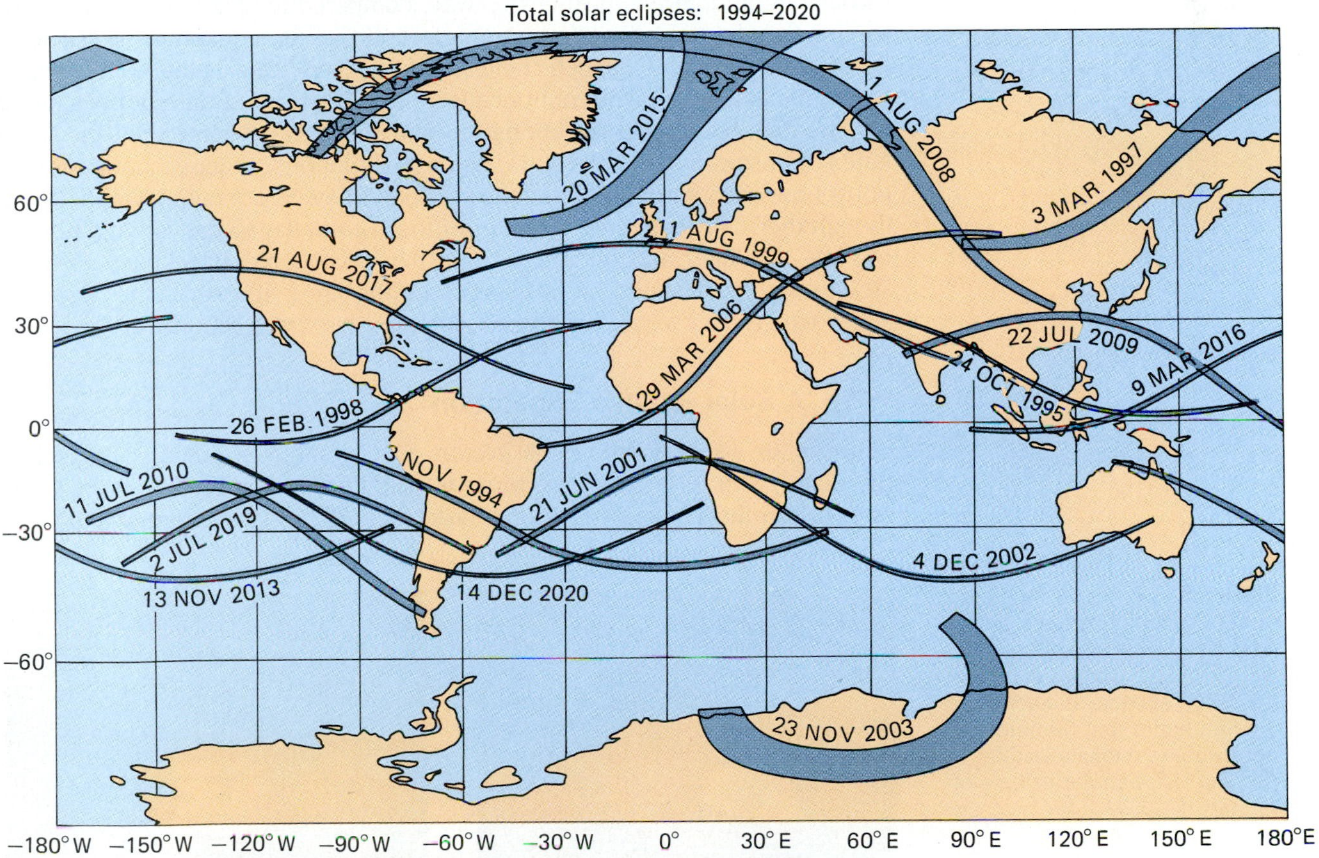

Total solar eclipses: 1994–2020

Figure 6–16 The paths of the Moon's shadow during total solar eclipses occurring between 1994 and 2020. Anyone standing along such a path on the dates indicated will see a total solar eclipse.

Table 6–1 Total Solar Eclipses

Date	Maximum Duration	Location
November 3, 1994	4^m24^s	South America
October 24, 1995	2^m10^s	India, Southeast Asia
March 9, 1997	2^m50^s	Mongolia, Russia (Siberia)
February 26, 1998	4^m09^s	Panama, Colombia, Venezuela, Guadeloupe, Montserrat, Antigua

Sometimes the Moon subtends a slightly smaller angle in the sky than the Sun, because the Moon is on the part of its orbit that is relatively far from the Earth. When a well-aligned eclipse occurs in such a circumstance, the Moon doesn't quite cover the Sun. An annulus—a ring—of photospheric light remains visible, so we call this an *annular eclipse* (Fig. 6–17). Such an annular eclipse crossed the southeastern United States on May 30, 1984. The rest of the United States saw a partial eclipse. An annular eclipse will cross the continental U.S. on May 10, 1994. People in Arizona and New Mexico through Vermont, New Hampshire, and Maine will see annularity. The rest of the continental U.S. will see partial phases.

In these days of orbiting satellites, is it worth travelling to observe eclipses? There is much to be said for the benefits of eclipse observing. Eclipse observations are a relatively inexpensive way, compared to space research, of observing the chromosphere and corona to find out such quantities as the temperature, density, and magnetic field structure. Even realizing that some eclipse experiments may not work out because of the pressure of time or because of bad weather, eclipse observations are still very cost-effective. And for some kinds of observations, those at the highest resolutions, space techniques have not yet matched ground-based eclipse capabilities. Space coronagraphs, even though they can study the outer part of the corona day to day, are not able to observe the inner and middle corona that we observe at eclipses. Coronal studies are made much more cheaply during an eclipse than the cost of a space experiment would have been, even if it had been mounted.

*6.3d A Solar Eclipse Expedition

Eclipses are like Olympic track events. Just as athletes prepare themselves thoroughly and for a long time to go to distant places every four years to be able to run the 100 meters, eclipse astronomers prepare carefully and often

A

B

Figure 6–17 (*A*) A member of my expedition using a dense filter to observe an annular eclipse. In an annular eclipse, the Moon appears smaller than the Sun and a ring (annulus) of photosphere remains visible. (*B*) The U.S. annular eclipse of January 4, 1992.

travel great distances to study total eclipses. During total eclipses, the corona and other solar phenomena outside the solar limb are best visible.

In every 18-year 11⅓-day interval, a series of a dozen total eclipses occur. One of those eclipses is especially long, approximately 7 minutes in duration. The most recent of these long eclipses occurred on July 11, 1991. Many scientific teams and many amateur astronomers went to see the eclipse because it was visible over relatively accessible areas—Hawaii and Baja California—and had excellent weather forecasts.

My own experiments dealt with the heating of the solar corona. How does the corona get to be so hot, millions of degrees? We searched for signs of magnetic waves by looking for oscillations of intensity in small regions of the corona. A second experiment was to map the coronal temperature, taking advantage of the exquisite sensitivity of the new CCD dectectors. A dozen students, some colleagues, and I were located on the Big Island of Hawaii. We were supported by the National Science Foundation and the National Geographic Society.

Figure 6–18 The corona seen through haze and clouds from Mauna Kea during the eclipse of July 11, 1991.

Our experiments did not need the high altitude or large telescopes that were available on top of Mauna Kea, though we could see some of the domes on the mountain. Because the eclipse passed over this major observatory, several kinds of observations could be made for the first time. In particular, several mountaintop experiments used the newly available detectors that could make images in the infrared. Others used the large scale of images available from a big telescope to view small details on the Sun at a rapid cadence.

We worked for years to plan our expedition and for months to carry it off. Two tons of equipment were shipped and carefully set up, aligned, and tested over the weeks before the big day. By eclipse day, all was in readiness.

A few days before the eclipse, a haze of particles from the eruption of Mount Pinatubo in the Philippines drifted overhead. Why couldn't the volcano have waited? Even more serious, the weather wasn't as good as predicted. On eclipse day, clouds blanketed the sky. An hour after sunrise, the eclipse began. Would the clouds recede as the Sun mounted? About 15 minutes before totality we glimpsed the Sun through clouds (Fig. 6-13). We were encouraged, but then the clouds mounted again, and we saw nothing during totality. The 90 per cent forecast of clear weather wasn't enough; we proved, observationally, that 90 per cent doesn't equal 100 per cent.

Even on Mauna Kea, the weather wasn't great. The observers were threatened minutes before totality by mounting fog, and cirrus clouds joined the Mount Pinatubo dust overhead. But the fog receded before totality, and the astronomers made observations during the precious 4 minutes 8 seconds of totality. After the diamond ring showed, huge prominences became visible. Then, totality showed the corona in all its glory (Fig. 6–18).

After totality, the sky cleared in time for us to see the final partial phases. The shadow of the Moon advanced to Baja California, where the diamond ring, prominences, and 6 minutes 54 seconds of totality were seen (Fig. 6–19).

Now, months later, many of the data have been studied. The infrared arrays did not find a dust ring around the Sun, even though there had been hints of such a ring from detections over a decade earlier. Perhaps the dust ring was transitory. Images with the largest telescope on the mountain showed small changes in the corona. But none of the telescopes detected the small solar flares—microflares—that had been searched for in the hope that the heating of the corona would be explained. Another telescope observed the spectrum of magnesium near the solar edge, and wound up with measurements that will require refinements in our understanding of how the temperature and density of the solar atmosphere vary with height. Other new telescopes, operating at far-infrared/radio wavelengths near one millimeter, also produced data that will affect our solar model.

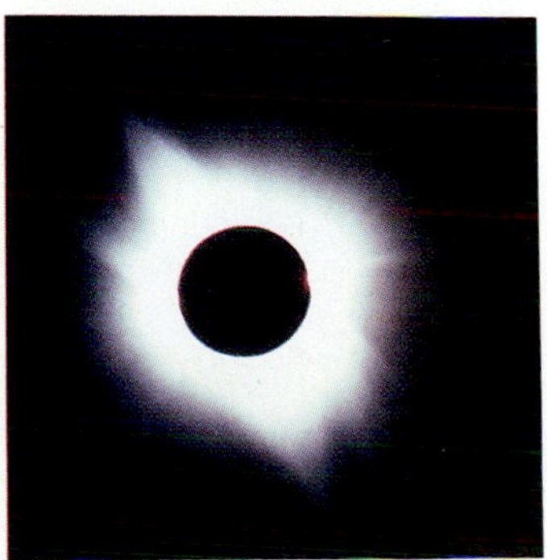

Figure 6–19 The solar corona during the July 11, 1991, total solar eclipse, surrounding the dark silhouette of the Moon.

Already, articles about eclipse results have appeared in scientific journals, and papers were given at a meeting of the American Astronomical Society. The results of the 1991 eclipse are finding their way into the thinking of those who seek to understand the Sun and the other stars.

6.4 The Revolution and Rotation of the Planets

The motion of the planets around the Sun in their orbits is called *revolution*. The spinning of a planet is called *rotation*. The Earth, for example, revolves around the Sun in 1 year and rotates on its axis in 1 day (thus defining these terms).

The orbits of all the planets lie in approximately the same plane. They thus take up only a disk whose center is at the Sun, rather than a full sphere. Little is known of the parts of the solar system away from this disk, although the comets may originate in a spherical cloud around the Sun.

The plane of the Earth's orbit around the Sun is called the *ecliptic plane* (Fig. 6–20). Of course, since we are on the Earth rather than outside the solar system looking in, the position of the Sun appears to move across the sky with respect to the stars. The path the Sun takes among the stars, as seen from the Earth, is called the ecliptic (see also Section 5.3).

The *inclination* of the orbit of a planet is the angle that the plane in which its orbit lies makes with the plane in which the Earth's orbit lies. The inclinations of the orbits of the other planets with respect to the ecliptic are small, with the exception of Pluto's 17° (Fig. 6–21). Of the other planets, Mercury's inclination is 7°; the remainder have inclinations of less than 4°.

The fact that the planets all orbit the Sun in essentially the same plane is one of the most important facts that we know about the solar system. Its explanation is at the base of most models of the formation of the solar system. Central to that explanation is a property that astronomers and physicists use in analyzing spinning or revolving objects: *angular momentum*. The amount of angular momentum of a small body revolving around a large central body is distance from the center × velocity × mass. The importance of angular momentum lies in the fact that it is "conserved"; that is, the total angular momentum of the system (the sum of the angular momenta of the different parts of the system) doesn't change even though the distribution of angular momentum among the parts may change. The total angular momentum of the solar system should thus be the same as it was in the past, unless some mechanism is carrying angular momentum away.

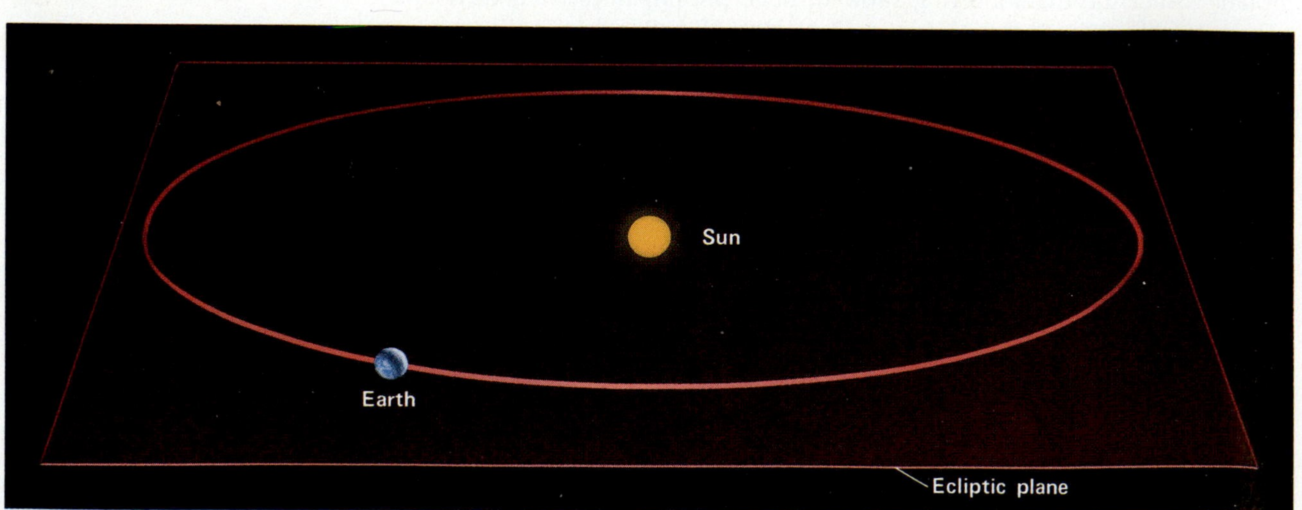

Figure 6–20 The ecliptic plane is the plane of the Earth's orbit around the Sun.

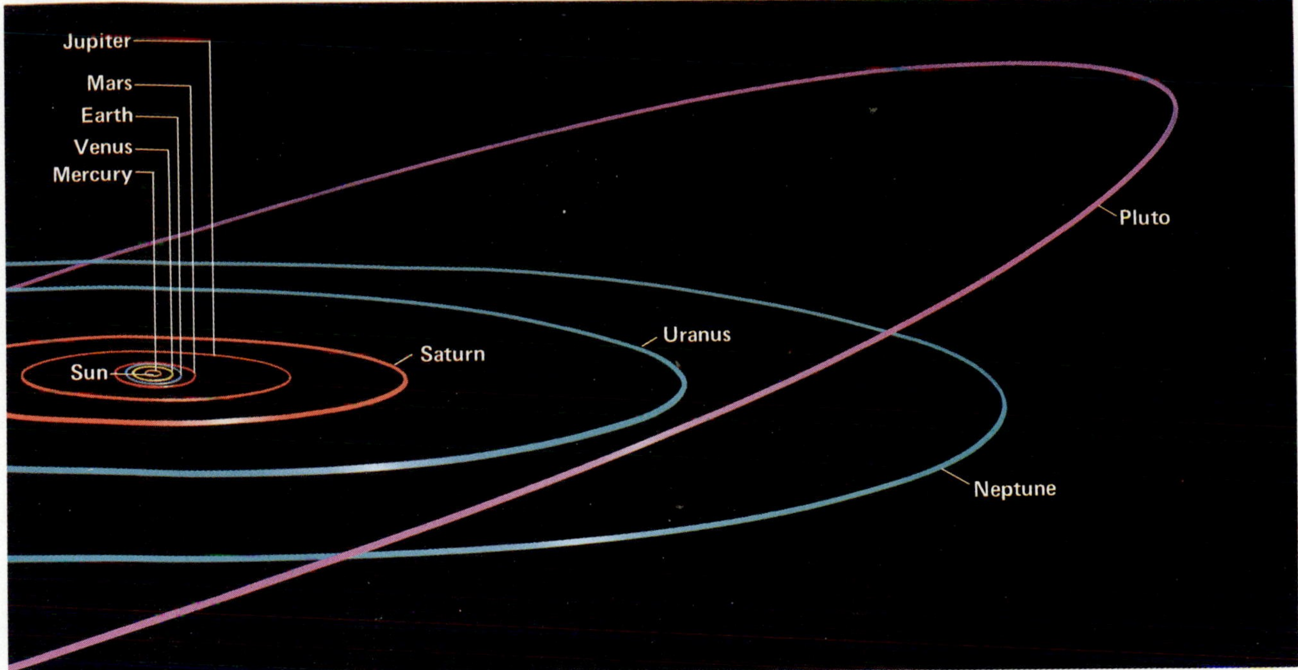

Figure 6–21 The orbits of the planets, with the exception of Pluto, have only small inclinations to the ecliptic plane.

The most familiar examples of angular momentum are ice skaters (Fig. 6–22). They may start themselves spinning by exerting force on the ice with their skates, thus giving themselves a certain angular momentum. When they want to spin faster, they draw their arms in. This changes the distribution of mass so that it is effectively closer to their centers (the axes around which they are spinning), and to compensate for this they rotate more quickly so that their angular momentum remains the same.

Thus from the fact that every planet is revolving in the same plane and in the same direction, we deduce that the solar system was formed of primordial material that was rotating in that direction. We would be very surprised to find a planet revolving around the Sun in a different direction—and we do not—or, to a lesser extent, to find a planet rotating in a direction opposite to that of its fellows. There are, however, two examples of such backward rotation—Uranus and Venus—which must each be carefully considered.

Figure 6–22 An ice skater spins faster when she draws her arms in. Since angular momentum is conserved, concentrating her mass closer to her spin axis than when her arms are outstretched leads to a compensating increase in spin. Here we see Kristi Yamaguchi, winner of the gold medal in figure skating at the 1992 winter Olympics in Albertville, France.

6.5 Theories of Cosmogony

The solar system exhibits many regularities, and theories of its formation must account for them. The orbits of the planets are almost, but not quite, circular, and all lie in essentially the same plane. All the planets revolve around the Sun in the same direction, which is the same direction in which the Sun rotates. Moreover, almost all the planets and planetary satellites rotate in that same direction. Some planets have families of satellites that revolve around them in a manner similar to the way that the planets revolve around the Sun. And cosmogonical theories must explain the spacing of the planetary orbits and the distribution of planetary sizes and compositions.

René Descartes, the French philosopher, was one of the first to consider the origin of the solar system in what we would call a scientific manner. In his theory, proposed in 1644, circular eddies—called vortices—of all sizes were formed at the beginning of the solar system in a primordial gas and eventually settled down to become the various celestial bodies.

After Newton proved that Descartes's vortex theory was invalid, it was 60 years before the next major developments. The Comte de Buffon suggested in France in 1745 that the planets were formed by material ejected from the Sun when what he called a "comet" hit it. (At that time, the size of comets was unknown, and it was thought that comets were objects as massive as the Sun itself.) Later versions of Buffon's theory, called *catastrophe theories*, followed a similar line of reasoning, although they spoke explicitly of collision with another star. A variation postulated that the material for the planets was drawn out of the Sun by the gravitational attraction of a passing star. This latter possibility is called a *tidal theory*.

Theoretical calculations show that gas drawn out of a star in collision or by a tidal force would not condense into planets, but would rather disperse. Hence, catastrophe and tidal theories are currently out of favor. Further, they predict that only very few planetary systems would exist, since only very few stellar collisions or near-collisions would have taken place in the lifetime of the galaxy. There is some observational evidence that many stars may have planets (Section 19.2), which would require a method of formation that can account for more planetary systems.

The study of the origin of the solar system is called *cosmogony* (pronounced with a hard "g").

The theories of cosmogony that astronomers now tend to accept stem from another 18th-century idea. The young Immanuel Kant, later to become noted as a philosopher, suggested in 1755 that the Sun and the planets were formed by the same type of process. In 1796, the Marquis Pierre Simon de Laplace (pronounced La Plahce), the French mathematician, independently advanced a type of theory similar to Kant's when he postulated that the Sun and the planets all formed from a spinning cloud of gas called a *nebula*. Laplace called his idea the *nebular hypothesis*, using the word "hypothesis" because he had no proof that it was correct. The spinning gas supposedly threw off rings that eventually condensed to become the planets. But though these beginnings of modern cosmogony were laid down in the time of Benjamin Franklin and George Washington, the theory still has not been completely understood or quantified, for not all the stages that the primordial gas would have had to follow are understood.

Further, some of the details of Laplace's theory were later thought for a time to be impossible. For example, it was calculated in the last century, by

Figure 6–23 The leading model for the formation of the solar system has the protosolar nebula condensing and, between stages C and D, contracting to form the protosun and a large number of small bodies called planetesimals. The planetesimals clumped together to form protoplanets (D), which in turn contracted to become planets. Some of the planetesimals may have become moons or asteroids (which are discussed in Section 18.2).

A

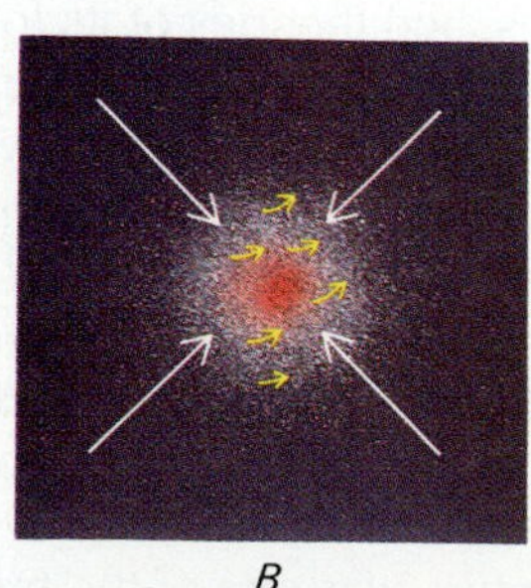

B

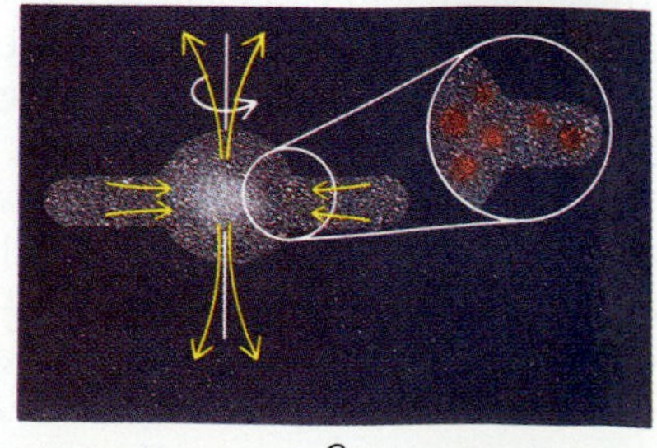

C

the great British physicist James Clerk Maxwell, that the planets would not have been able to condense out of the rings of gas and also that they could not have been set to rotate as fast as they do.

So for a time, the nebular hypothesis was not accepted. But the current theories of cosmogony again follow Kant and Laplace (Fig. 6–23). In these *nebular theories* the Sun and the planets condensed out of what is called a *primeval solar nebula.* Laplace's ring formation mechanism is no longer thought to be applicable; only the concept of joint formation of the Sun and planets from a single, rotationally flattened cloud has survived the centuries. We now think that some five billion years ago, billions of years after the galaxies began to form, smaller clouds of gas and dust began to contract out of interstellar space. Similar interstellar clouds of gas and dust can now be detected at many locations in our galaxy.

For most of the past decade, scientists increasingly believed that the collapse of gas to form the solar system was set off by shock waves (Fig. 6–24) from a nearby supernova. The evidence concerns the unusual abundances in meteorites (Section 18.1) of a certain radioactive isotope of aluminum. It had been thought that the isotopes could only have been formed in such supernova explosions. Their lifetimes are too short for them to have been formed very long before our solar system's collapse. But studies, first reported in 1984, of gamma-ray data from NASA's third High-Energy Astronomy Observatory (HEAO-3) revealed that this radioactive isotope of aluminum is widespread throughout interstellar space. The data were confirmed from the Solar Maximum Mission's gamma-ray telescope. So no supernova trigger was needed to account for the isotope in our solar system.

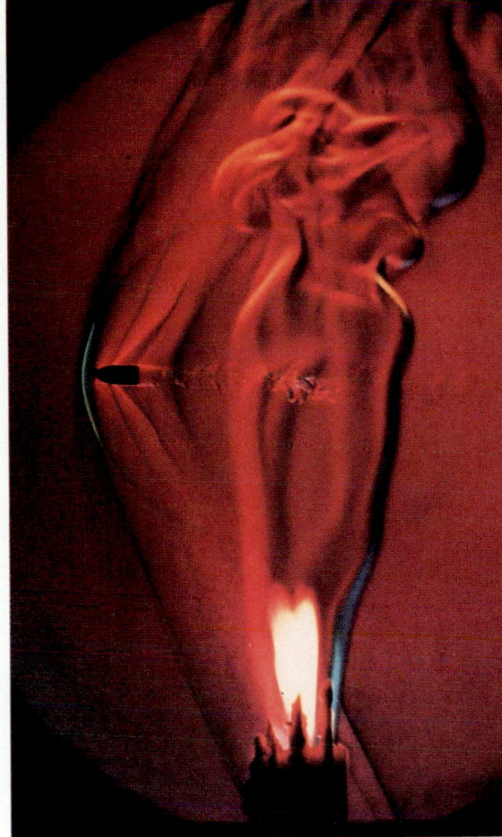

Figure 6–24 This photograph shows a shock wave in air made by a bullet travelling faster than the speed of sound. The sharp curved line at the left is the shock wave; it represents a sharp change in pressure.

However the collapse started, random fluctuations in the gas and dust would have given the primeval solar nebula a small net spin from the beginning. As the nebula contracted, it would have begun to spin faster because of the conservation of angular momentum (the same reason that the ice skater spins faster and that pulsars and black holes spin so rapidly). Gravity would have contracted the spinning nebula into a disk: In the plane of the nebula's rotation, the spin had the effect of a force that counteracts gravity. In the directions perpendicular to this plane, though, there was no force to oppose gravity's pull, and the nebula collapsed in that direction. Planetary formation in a flattened, rotating nebular naturally explains the orbital and rotational characteristics of the solar system.

Perhaps rings of material were left behind by the solar nebula as it contracted toward its center. Perhaps there were additional agglomerations beside the *protosun,* the part of the solar nebula that collapsed to become the Sun itself. These additional agglomerations, which may have been formed from interstellar dust, would have grown larger and larger. Micrometer-sized particles would have started sticking together, eventually *clumping into* large bodies centimeters or meters across; gravity would then have increased the size of

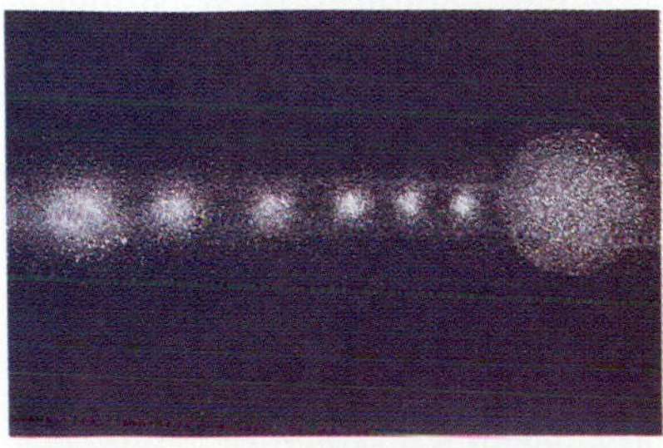

D

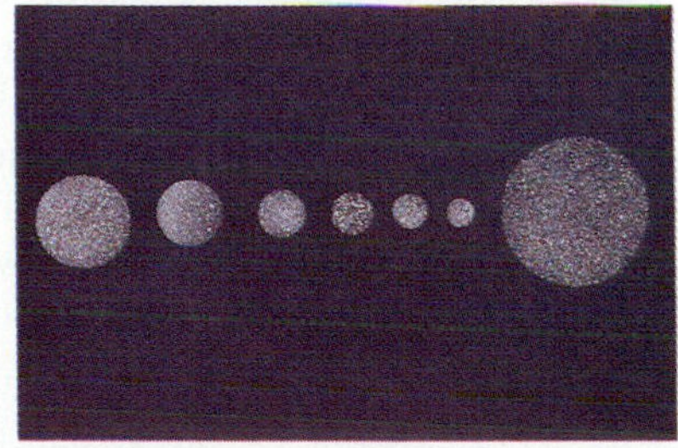

E

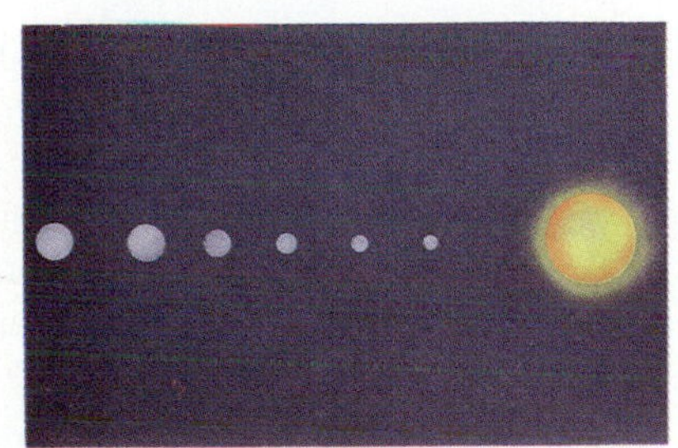

F

the clumps until there were bodies kilometers and even hundreds of kilometers across within about 10,000 years. These intermediate bodies, hundreds of kilometers across, are called *planetesimals.*

The planetesimals combined under the force of gravity over tens of millions of years to form *protoplanets.* These protoplanets may have been larger than the planets that resulted from them because they had not yet contracted, though gravity would ultimately cause them to do so. (We will discuss a possible protoplanet around T Tauri in Section 24.1.) Some of the larger planetesimals may themselves have become moons. Most of the satellites of the giant planets, though, probably formed in mini-solar nebulae around each protoplanet.

As the Sun condensed, some of the energy it gained from its contraction heated it until its center reached the temperature at which nuclear-fusion reactions began taking place. The planets, on the other hand, were simply not massive enough to heat up sufficiently to have nuclear reactions start.

Several modifications of this basic theory have been worked out to accommodate particular observational facts. For example, we have to explain why the inner planets are small, rocky, and dense, while the next group of planets out are large and primarily made of light elements. Since the protosun would have been made primarily out of hydrogen and helium, with just traces of the heavier elements formed in earlier cyclings of the material through stars and supernovae, we need a mechanism for ridding the inner regions of the solar system of this lighter material. One possible way is to say that the Sun flared fiercely and/or often in its younger days (similar to a T Tauri star, as described in Section 24.1) and that the lighter elements were blown out of the inner part of the solar system. But proto-Jupiter and the other outer protoplanets, which were much farther away from the Sun, would have retained thick atmospheres of hydrogen and helium because of their high gravity.

Billions of planetesimals may have become comets (Chapter 17), and have been far beyond the planets in the deep freeze of outer space for billions of years. The extreme interest in this most recent passage of Halley's Comet stems in part from the chance to study such pristine material.

Planetesimals may also have played additional roles in the early years of the solar system. We shall see later on how collisions with planetesimals may have fragmented Mercury and stripped off its outer layers, may have stopped Venus from spinning as fast as other planets, may have tipped Uranus onto its side, and may have stripped off material from the Earth that soon become our Moon.

A major modern method of calculation considers how the temperature decreased with distance from the center of the solar nebula, ranging from almost 2000 K close to the protosun, down to only about 20 K at the distance of Pluto. Various elements were able to condense—solidify—at different locations because of their differing temperatures. For example, in the positions occupied by the terrestrial planets, it was too hot for icy substances to form, though such ices are present in the giant planets.

Calculations by Alan P. Boss of the Carnegie Institution of Washington show that the gas falling in toward the center of the protosolar nebula can be heated to 1500 K even in the asteroid belt (2.5 A.U.), much hotter than had been expected. His model followed the collapse under its own gravity of a sphere of radius 40 A.U. (Fig. 6–25). These models can provide both a hot inner region and a cool outer region in which ice can form at Jupiter's distance and beyond. Other scientists have made different models.

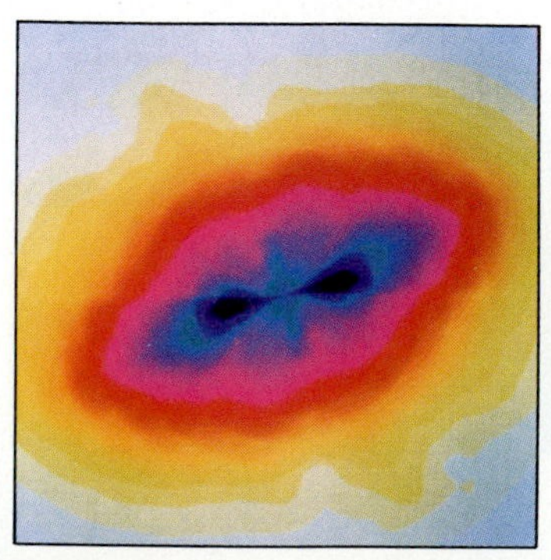

Figure 6–25 A false-color image of densities in a region 8 A.U. across in a model of the early solar nebula. The black, blue, and green regions in the center should merge to form the protosun.

Summary and Outline

The Sun, our star (Section 6.1)
The phases of the Moon and planets (Section 6.2)
Eclipse phenomena (Section 6.3)
  Partial phases, Baily's beads, diamond ring, totality
Planetary rotation and revolution (Section 6.4)
  The ecliptic plane; inclinations of the orbits of the planets; conservation of angular momentum
  All planets revolve and most rotate in the same direction
Theories of cosmogony (Section 6.5)
  Catastrophe theories
  Nebular theories of Kant and Laplace have newer versions: protosun, planetesimals, and protoplanets
  Infrared observations may show solar systems in formation

Key Words

photosphere, corona, solar wind, phases, umbra, penumbra, shadow bands, Baily's beads, diamond-ring effect, annular eclipse, revolution, rotation, ecliptic plane, inclination, angular momentum, retrograde rotation, retrograde motion, catastrophe theories, tidal theory, nebula, nebular hypothesis, nebular theories, primeval solar nebula, protosun, planetesimals, protoplanets

Questions

1. Sketch the Sun, labelling the corona and the solar wind.

2. Suppose that you live on the Moon. Sketch the phases of the Earth that you would observe for various times during the Earth's month.

3. If you lived on the Moon, would the motion of the planets appear to the eye any different than from Earth?

4. If you lived on the Moon, how would the position of the Earth change in your sky over time?

5. Discuss the circumstances under which Uranus and Neptune are seen as crescents.

6. (a) Why isn't there a solar eclipse once a month? (b) Whenever there is total solar eclipse, a lunar eclipse occurs either two weeks before or two weeks later. Explain.

7. If you lived on the Moon, what would you observe during an eclipse of the Moon? How would an eclipse of the Sun by the Earth differ from an eclipse of the Sun by the Moon that we observe from Earth?

8. Sketch what you would see if you were on Mars and the Earth passed between you and the Sun. Would you see an eclipse? Why?

9. Why can't we observe the corona every day from any location on Earth?

10. Describe why we cannot see the solar corona during the partial phases of an eclipse.

11. Where and when is the next total solar eclipse? How far will your college or university be from the 1994 annular eclipse?

12. Describe relative advantages of ground-based eclipse studies and satellite studies of the corona.

13. Explain how Pluto can have a longer period than Neptune, even though Pluto is and will be closer to the Sun until the year 2000.

14. Explain how conservation of angular momentum applies to a diver doing somersaults or twists. What can divers do to make sure they are vertical when they hit the water?

15. If two planets are of the same mass but different distances from the Sun, which will have the higher angular momentum around the Sun? (*Hint:* Use Kepler's third law.)

16. If the planets condensed out of the same primeval nebula as the Sun, why didn't they become stars?

17. Which planets are likely to have their original atmospheres? Explain.

18. Why might the ring around Vega be detectable only in the infrared and not in the visible? Explain in terms of its temperature and the radiation laws.

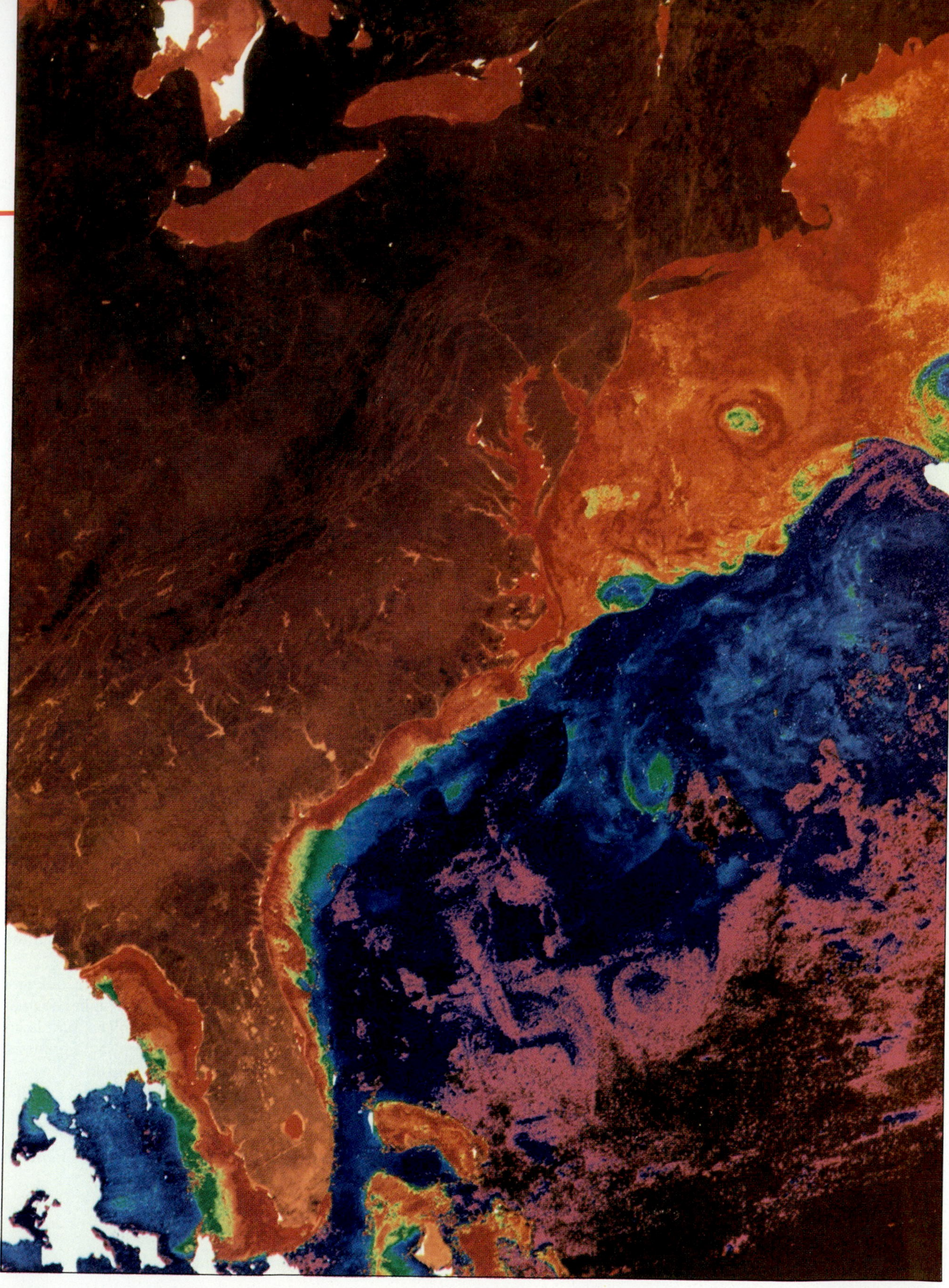

Sea surface temperatures reveal the Gulf Stream in the Atlantic. Red shows the warmest water, followed by orange, yellow, green, and then blue and purple for cold water. The results follow from measurements made in the infrared with the NOAA-7 satellite.

Our Earth

7

Aims: To describe and understand the Earth on a planetary scale

We know the Earth intimately; a mere wrinkle on its surface is a mountain to us. Weather satellites now give us a constant global view, and show us large atmospheric systems linking all parts of the Earth. More rarely, earthquakes or volcanoes bring us messages from below the Earth's surface. But no longer do we have to treat the Earth as one of a kind; by studying the other planets, we can make comparisons that allow us to understand both the other planets and our own. In this chapter, we will summarize some of our knowledge of the Earth's structure and history. In later chapters, we will compare the Earth with the other planets.

Seven of the planets in the solar system have moons—that is, smaller bodies orbiting them. Most planets have more moons than the Earth. Our Earth-Moon system and the Pluto-Charon system are the only ones, though, in which the satellite contains as much as one per cent the mass of the parent. The moons around other planets contain 800 to more than a billion times less mass than their parent planets.

The study of the Earth's interior and solid surface is called *geology.* We can study the Earth so intricately that studies of other parts of the Earth have their own names, such as oceanography and meteorology.

7.1 The Earth's Interior

Geologists study how the Earth vibrates as a result of large shocks, such as earthquakes. Much of our knowledge of the structure of the Earth's interior comes from *seismology,* the study of these vibrations. The vibrations travel through different types of material at different speeds. Also, when the *seismic waves* strike the boundary between different types of material, they are reflected and refracted, just as light is when it strikes a glass lens. From piecing together seismological and other geological evidence, geologists have been able to develop a picture of the Earth's interior (Fig. 7–1). The terms and divisions apply to many other solar-system bodies as well.

The innermost region is called the *core.* It consists primarily of iron and nickel. The central part of the core may be solid, but the outer part is probably a very dense liquid. Outside the core is the *mantle,* and on top of the mantle is the thin outer layer called the *crust.* The crust and the outer part of the mantle together are called the *lithosphere.* The lithosphere is a rigid layer, and surrounds a zone, the *aesthenosphere,* that is partially melted (Fig. 7–2).

How did such a layered structure develop? The Earth formed surrounded by a lot of debris in the form of dust and rocks. The young Earth was probably subject to constant bombardment from this debris. This bombardment heated the surface to the point where it began to melt, producing lava. However, this process could not have been responsible for heating the interior of the Earth, because the rock out of which the Earth was made conducts heat very poorly.

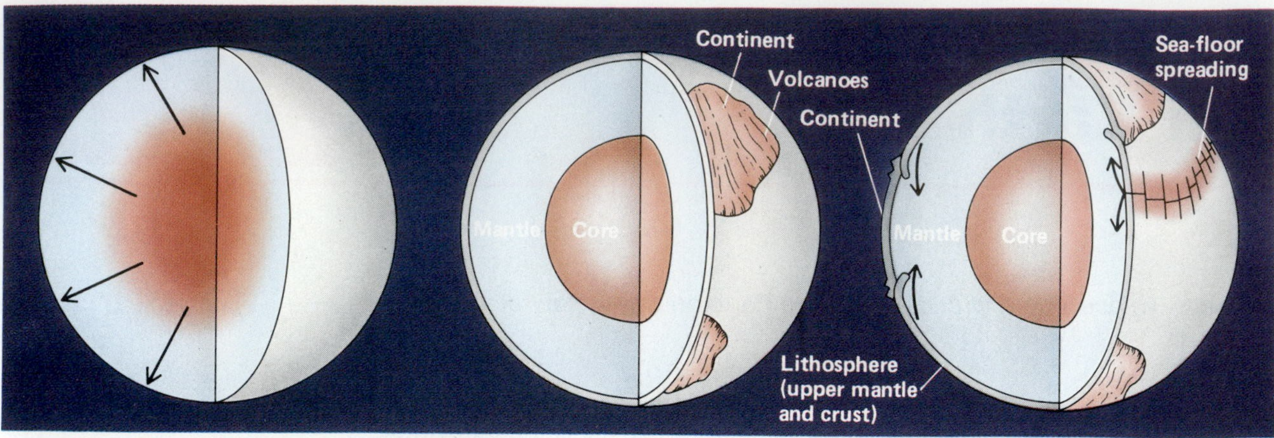

Figure 7–1 The structure of the Earth and stages in its evolution. (*A*) Tens of millions of years after its formation, radioactive elements along with gravitational compression and impact of debris produced melting and differentiation. Heavy materials sank inward and light materials floated outward. During this time, the original atmosphere (consisting mostly of hydrogen) was blown away by the solar wind. The atmosphere that replaced it contained methane, ammonia, and water. (*B*) The heaviest materials form the core, and the lightest materials form the crust. (*C*) The crust has broken into rigid plates that carry the continents and move very slowly away from areas of sea-floor spreading.

Much of the original heat for the Earth's interior came from gravitational energy released as particles accreted to form the Earth. But the major source of heat for the interior is the natural radioactivity within the Earth. The different forms of individual atoms are called *isotopes;* an isotope undergoes chemical reactions in exactly the same way as another isotope of the same element, but has a different mass (Section 24.3). Certain isotopes are unstable—they spontaneously undergo nuclear reactions. In these reactions, particles such as electrons or alpha particles (helium nuclei) are given off with high energy. These energetic particles collide with the atoms in the rock and give some of their energy to these atoms. The rock becomes hotter. When the Earth was forming, there was a sufficient amount of radioactive material in its interior to cause a great deal of heating. And since the material conducted energy poorly, the heat generated was trapped. As a result, the interior got hotter and hotter.

After about a billion years, the Earth's interior had become so hot that the iron melted and sank to the center, forming the core. Eventually other materials also melted. As the Earth cooled, various materials, because of their

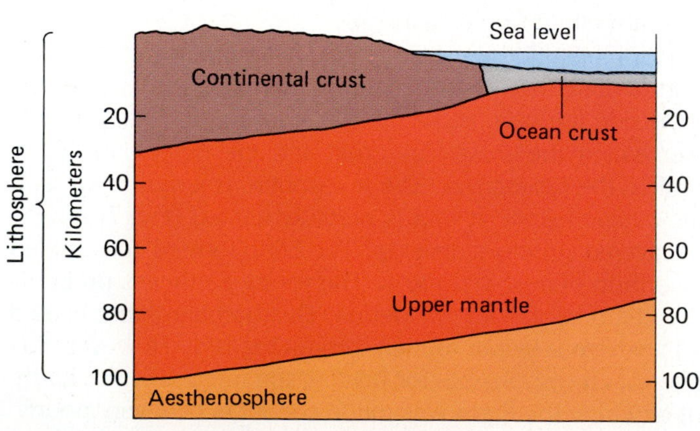

Figure 7–2 Continental crust is thicker than ocean crust and has a lower density. The lithosphere, composed of the crust and the outer upper mantle, is a rigid layer. The crust is different in composition from the mantle. Below the lithosphere is a partially melted layer known as the aesthenosphere.

different densities and freezing points (the temperature at which they change from liquid to solid), solidified at different distances from the center. This process is called *differentiation;* it is responsible for the present layered structure of the Earth.

7.2 Plate Tectonics

In the process of differentiation, most of the radioactive elements ended up in the outer layers of the Earth. Thus these elements provide a heat source not far below the ground. Their presence leads to a general *heat flow* outward through the upper mantle and crust (the lithosphere). The power flowing outward through each square centimeter of the lithosphere is small—10 million times less than that necessary to light an average light bulb and thousands of times less than that reaching us from the Sun. However, the terrestrial heat flow does have important geological consequences.

In some geologically active areas (Fig. 7–3), the heat-flow rate is much higher than average, which indicates that the source of heat is close to the surface. In a few places, the outflowing *geothermal energy* in these regions is being tapped as an energy source.

The lithosphere is a relatively cool, rigid layer; it is segmented into *plates,* thousands of kilometers in extent but only about 50 km thick. Because of the internal heating, the lithosphere sits on top of a hot layer where the rock is soft, though it is not hot enough to melt completely. The lithosphere actually floats on top of this soft layer. The hot material beneath the lithosphere's rigid plates churns very slowly in convective motions (circulation involving rising hot material; boiling is an example), carrying the plates around over the surface of the Earth. This theory, called *plate tectonics,* explains the observed *continental drift*—the slow drifting of the continents from their original positions. ("Tectonics" is from the Greek word meaning "to build.")

That continental drift took place, which once seemed unreasonable, is now generally accepted. The continents were once connected as two supercontinents, one called Gondwanaland (after a province of India of geologic interest) and a northern supercontinent called Laurasia. (These may have, in

Figure 7–3 Thermal activity beneath the Earth's surface results in geysers. The thermal area shown here is in Rotorua, New Zealand. Geothermal steam can be used to generate electricity; The Geysers, a geothermal area in California, provides much of San Francisco's electricity.

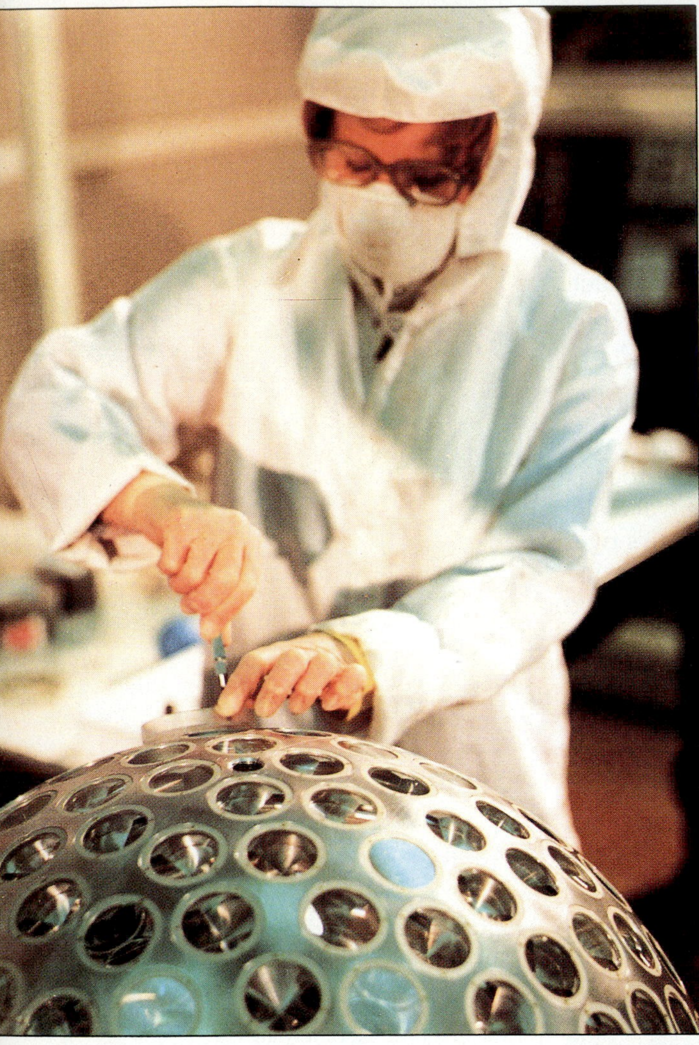

A

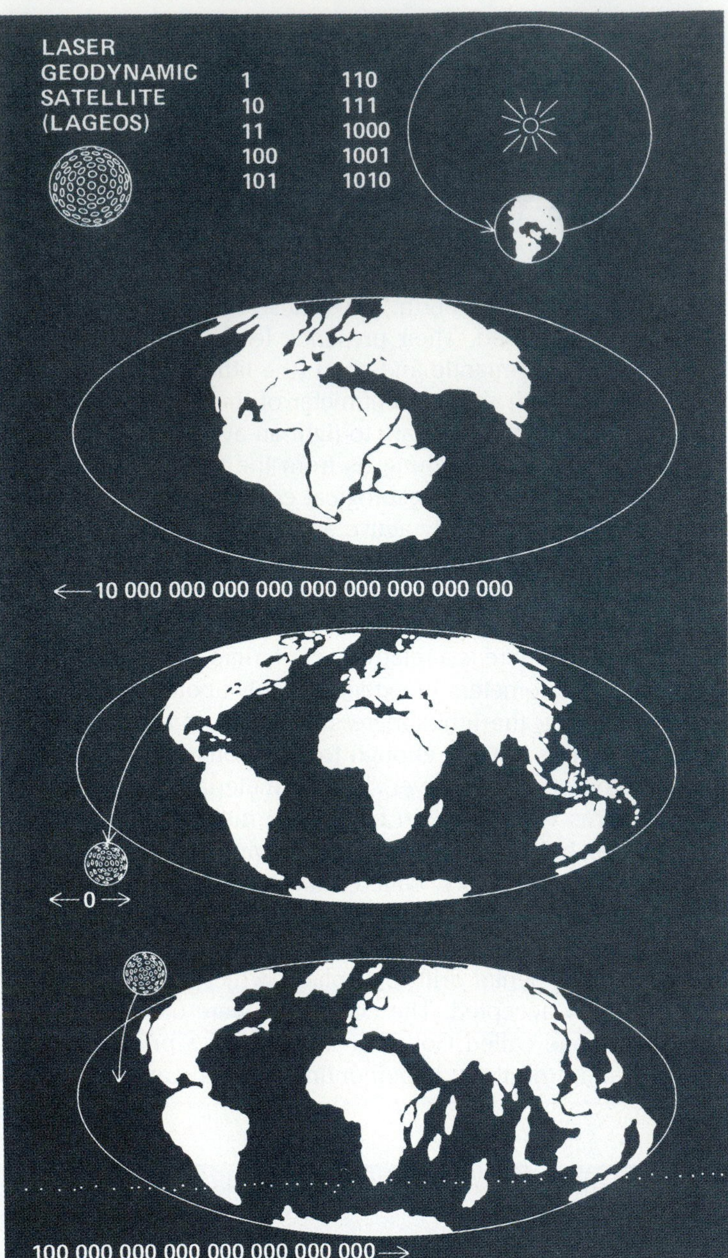

B

Figure 7–4 (*A*) The NASA satellite LAGEOS (Laser Geodynamic Satellite), launched into a circular Earth orbit in 1976, provides information about the Earth's rotation and the crust's movements. It contains 426 corner reflectors, which bounce laser beams back in exactly the direction from which they came. Scientists on Earth time the interval between their sending out a laser pulse and receiving the return signal, and use the results to figure out exactly where on Earth they are. LAGEOS is expected to survive in orbit for 8 million years. (*B*) In case the satellite is discovered in the distant future, it bears a series of views of the continents in their locations 270 million years in the past (*top*), at present (*middle*), and 270 million years in the future (*bottom*), according to our knowledge of continental drift. (The dates are given in the binary system.) Data on continental drift are gathered by reflecting laser beams from telescopes on Earth off the 426 retroreflectors that cover its surface. (Each retroreflector is a cube whose interior reflects incident light back in the direction from which it came.) Measuring the time until the beams return can be interpreted to show the accurate position of the ground station. We can now measure the position to an accuracy of about 1 cm.

turn, separated from a single supercontinent called Pangaea, which means "all lands.") Over two hundred million years or so, the continents have moved apart as plates have separated. We can see from their shapes how they originally fit together (Fig. 7–4). Finding similar fossils and rock types along two opposite coastlines, which were once adjacent but are now widely separated, provides proof. In the future, we expect California to separate from the rest of the United States, Australia to be linked to Asia, and the Italian "boot" to disappear.

The boundaries between the plates (Fig. 7–5) are geologically active areas. Therefore, these boundaries are traced out by the regions where earthquakes (Fig. 7–6) and most volcanoes (Fig. 7–7) occur. The boundaries where two plates are moving apart mark regions where molten material is being pushed up from the hotter interior to the surface, such as the *mid-Atlantic ridge* (Fig. 7–8). Molten material is being forced up through the center of the ridge and is being deposited as lava flows on either side, producing new sea floor. The motion of the plates is also responsible for the formation of the great mountain ranges. When two plates come together, one may be forced under the other, and the other may be raised. The "ring of fire" volcanoes around the Pacific Ocean (including Mount St. Helens in Washington) were formed in this way. Another example is the Himalayas, which were formed when the Indian subcontinent collided with the Asian continent.

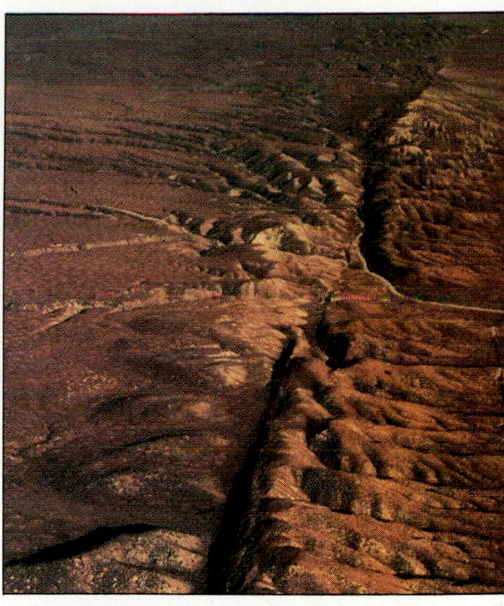

Figure 7–5 The San Andreas fault marks the boundary between the California and Pacific plates.

7.3 Tides

The tides—like the Moon's passing overhead—occur about an hour later each day. So it has long been accepted that tides are directly associated with the Moon. Tides result from the fact that the force of gravity exerted by the Moon

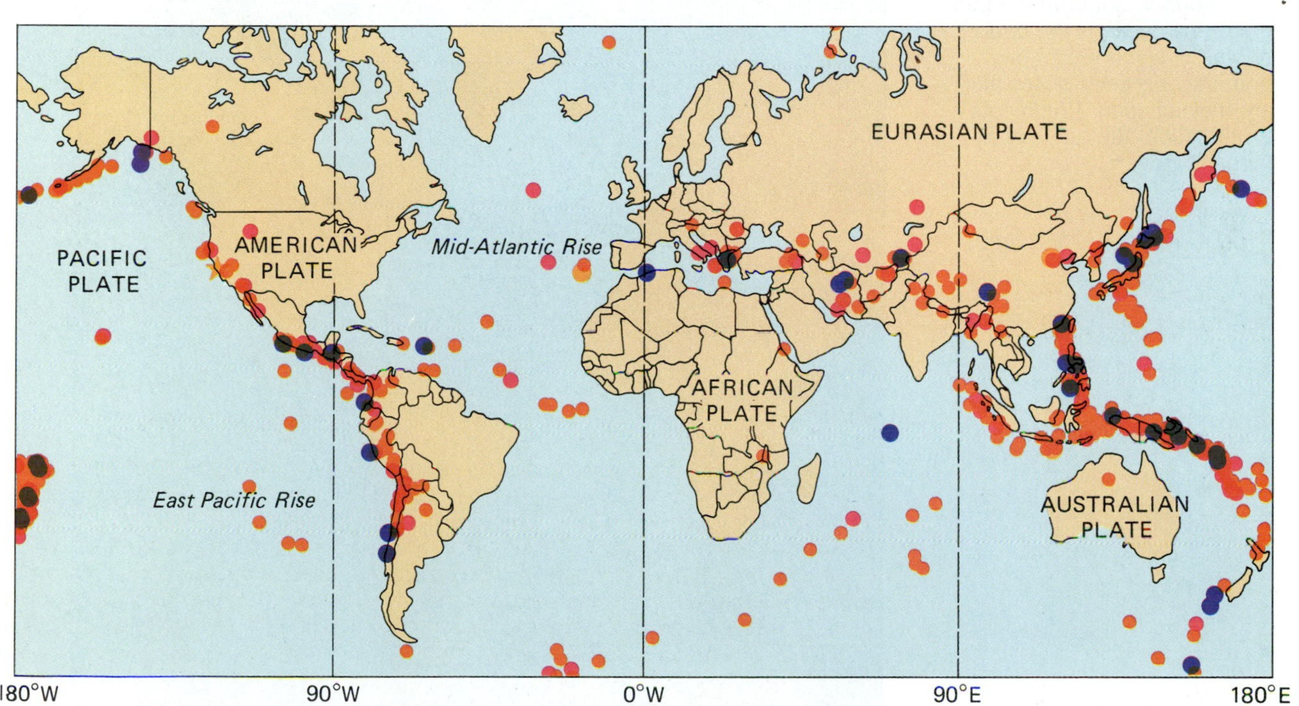

Figure 7–6 This plot of earthquakes greater than 6.7 on the Richter scale over the most recent 15 years shows that earthquakes occur preferentially at plate boundaries. The principal tectonic plates are labelled. The 1979 Loma Prieta earthquake centered in the Santa Cruz Mountains south of San Francisco is shown with a star.

Figure 7–7 (*A*) Mount St. Helens, shown here during its May 18, 1980, eruption, lies on a plate boundary. (*B*) After its devastating eruption. (*C*) Kilauea volcano on the island of Hawaii during its current series of eruptions.

Figure 7–8 This map of the Atlantic Ocean floor shows how the continents sit on the plates. North and South America are at the left; Europe and Africa are at the right. The feature running from north to south in the middle of the ocean is the *mid-Atlantic ridge.* It marks the boundary between plates that are moving apart. Continents drift at about the rate that fingernails grow.

A

B

Figure 7–9 The tidal range is so great at Mont St. Michel in France that (*A*) sheep can graze on dry land near the Mont at low tide while (*B*) boats can sail up to the Mont at high tide. The causeway that now links the Mont with the Normandy shore has been changing the tidal pattern, diminishing the great range of the tides. The shape of the ocean floor greatly influences the tidal range.

(or any other body) gets weaker as you get farther away from it. Tides caused by an object (the Moon, for example) depend on the **difference** between the gravitational attraction of that object at different points on another object, so the forces that cause tides are often called *differential forces.*

To explain the tides in Earth's oceans, consider, for simplicity, that the Earth is completely covered with water. We might first say that the water closest to the Moon is attracted toward the Moon with the greatest force and so is the location of high tide as the Earth rotates. If this were all there were to the case, high tides would occur about once a day. However, two high tides occur daily, separated by roughly 13 hours. Low tides come in between (Fig. 7–9).

To see how we get two high tides a day, consider three points, A, B, and C, where B represents the solid Earth, A is the ocean nearest the Moon, and C is the ocean farthest from the Moon (Fig. 7–10). Since the Moon's gravity

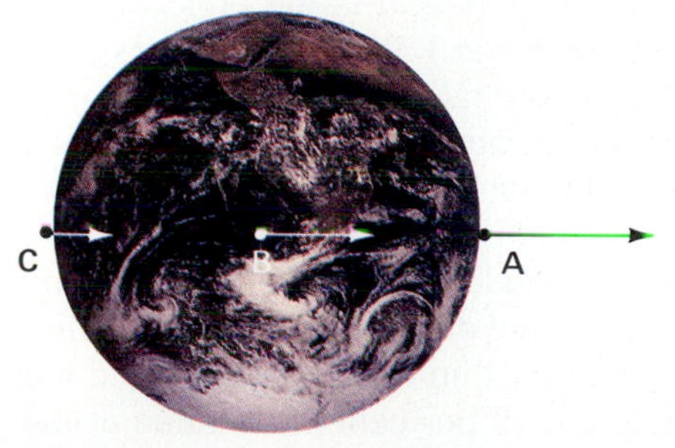

Figure 7–10 A schematic representation of the tidal effects caused by the Moon. The arrows represent the acceleration of each point that results from the gravitational pull of the Moon (exaggerated in the drawing). The water at point A has greater acceleration toward the Moon than does point B; since the Earth is solid, the whole Earth moves with point B. (Tides in the solid Earth exist, but are much smaller than tides in the oceans.) Similarly, the solid Earth is pulled away from the water at point C.

Tidal forces are important in many astrophysical situations, including accretion disks around neutron stars and black holes, and rings around planets.

weakens with distance, it is greater at point A than at B, and greater at point B than at C. If the Earth and Moon were not in orbit about each other, all these points would fall toward the Moon, moving apart as they fell. (They would actually move apart only slightly, since the pull of the Earth's gravity on the oceans is stronger than that of the Moon.)

Thus the high tide on the side of the Earth that is near the Moon is a result of the water being pulled away from the Earth. The high tide on the opposite side of the Earth results from the Earth being pulled away from the water. In between the locations of the high tides, the water has rushed elsewhere so we have low tides. As the Earth rotates once a day, a point on its surface moves through two cycles of tides: high-low-high-low. Since the Moon is moving in its orbit around the Earth, a point on the Earth's surface has to rotate longer than 12 hours to return to a spot nearest to the Moon, and so to begin a new cycle. Thus the tides repeat about every 13 hours.

The Sun's effect on the Earth's tides is only about half as much as the Moon's. Though the Sun exerts a greater gravitational force on the Earth than does the Moon, the Sun is so far away that its force does not change very much from one side of the Earth to the other. And it is only the change in force that accounts for tides.

At a time of new or full moon, the tidal effects of the Sun and Moon work in the same direction, and we have especially high tides, called *spring tides* (related to the German word "springen," meaning "to rise up"; the word has nothing to do with the season spring). At the time of the first- and last-quarter moons, the effects of the Sun and Moon do not add, and we have *neap tides,* when high and low tides differ least.

Tides vary as the distance of the Moon from the Earth changes. When the Moon is relatively close to the Earth at new moon or full moon, the tidal range is relatively great.

Many people mistakenly think that there is some special effect on the tides at the days of the equinoxes. Actually, since the equinox merely occurs when the Sun crosses the equator, there is no major effect at the equinoxes. If the Moon is relatively close to the Earth, then the new moon or full moon closest to the equinox will have a higher tidal range, but such events are not usually on the day of an equinox.

The shape of the ocean floor has a major effect on the tides. The movement of water is not immediate as the Moon passes overhead, and water may undergo reflection from distant shores. In some places, only one tide a day occurs because of such local circumstances. But all such cases can be understood as modifications of the basic tidal theory discussed previously.

7.4 The Earth's Atmosphere

The Earth's atmosphere presses down on us all the time, though we are used to it. The force pressing down per unit of area is called *pressure.* We define the average pressure at the surface of the Earth to be *one atmosphere.* (In SI units, 1 atmosphere is 101,300 pascals, where 1 pascal is the pressure of a force of 1 newton pressing on each square meter.) When we express the pressure in atmospheres, we are really taking the ratio of the pressure at any level to that at the surface. The pressure in the atmosphere (Fig. 7–11) falls off sharply with increasing height. The temperature (Fig. 7–12) varies less regularly with altitude.

It is convenient to divide the atmosphere into layers (Fig. 7–13), according to the composition and the physical processes that determine the temperature.

The Earth's weather is confined to the very thin *troposphere.* At the top of the troposphere, the pressure is only about 10 per cent of its value at the

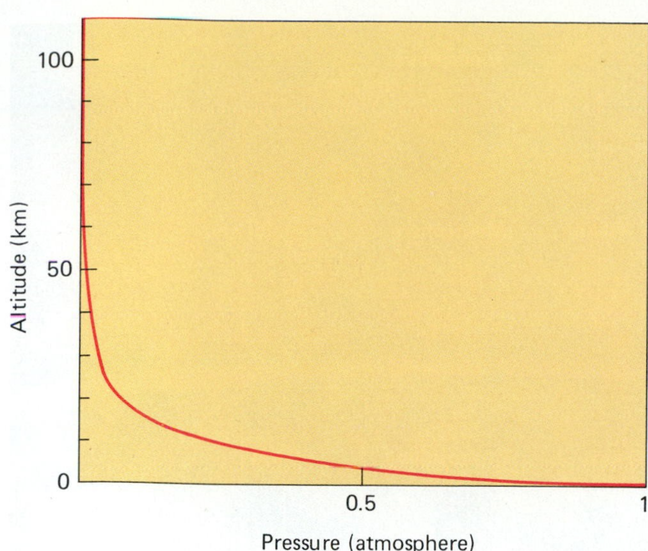

Figure 7–11 The relationship between pressure and altitude in the Earth's atmosphere. Note how rapidly the pressure decreases as the altitude increases.

ground. A major source of heat for the troposphere is infrared radiation from the ground, so the temperature of the troposphere decreases with altitude.

Above the troposphere are the *stratosphere* and the *mesosphere.* The upper stratosphere and lower mesosphere contain the *ozone layer.* (Ozone is a molecule, consisting of three oxygen atoms—O_3—that absorbs ultraviolet radiation from the Sun. The oxygen our bodies use is O_2.) The absorption of ultraviolet radiation by the ozone means that these layers get much of their energy directly from the Sun. The temperature is thus higher there than it is at the top of the troposphere.

Above the mesosphere is the *ionosphere,* where many of the atoms are ionized. The most energetic photons from the Sun (such as x-rays) are absorbed here. Thus the temperature gets quite high. Because of this rising temperature, this layer is also called the *thermosphere.* Since many of the atoms in this layer are ionized, the ionosphere contains many free electrons. These electrons reflect very long wavelength radio signals. When the conditions are right, radio waves bounce off the ionosphere, which allows us to tune in distant radio stations. When solar activity is high, as it is now, there may be a lot of this "skip."

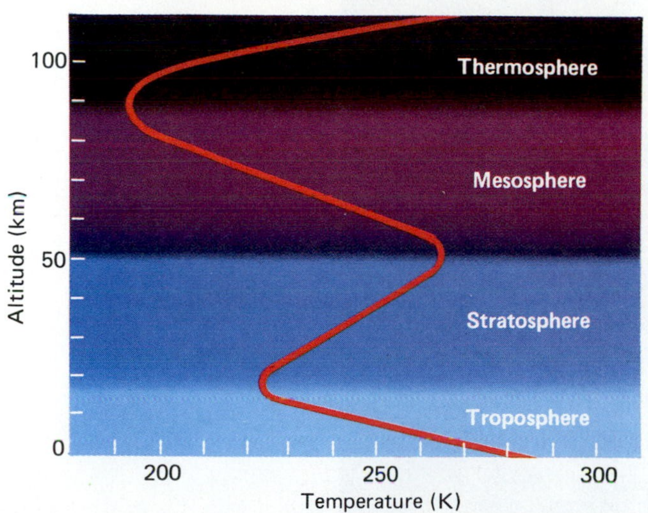

Figure 7–12 Temperature in the atmosphere as a function of altitude. In the troposphere, the energy source is the ground, so the temperature falls off with altitude. Higher temperatures in other layers result from direct absorption of solar ultraviolet and x-rays.

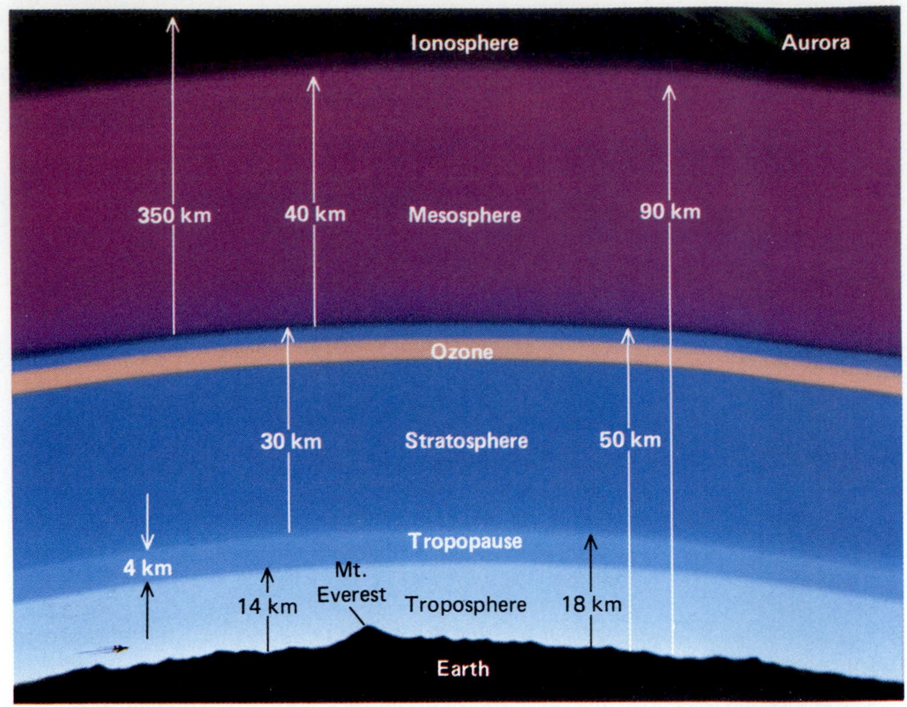

Figure 7–13 The layers of the Earth's atmosphere showing the location of the part of the stratosphere where ozone is formed. Once formed, the ozone is transported to lower stratospheric layers.

Our knowledge of our atmosphere has been greatly enhanced by observations from satellites above it and from spacecraft to the Moon (Fig. 7–14). (See also the figure opening this chapter.) Scientists carry out calculations using the most powerful supercomputers, such as the Cray supercomputers used at the National Center for Atmospheric Research, to interpret the global data and to predict how the atmosphere will behave. The equations are essentially the same as those for the internal temperature and structure of stars, except that the sources of energy are different. Our ability to predict weather (which is short term) and climate (which is long term) is improving. However, we now realize that there is a fundamental limitation in how well we will ever begin to predict the weather—the butterfly-wing effect. Even so tiny a change as a flap of a butterfly's wing can cause adjacent changes that cause adjacent

Figure 7–14 Photographs of the Earth from space, such as this one from the Geostationary Operational Environmental Satellite (GOES), clearly show weather patterns.

changes and so on until some major weather change—even a tornado—can result. Discussion of the limitation is part of the field of "chaos," in which small changes can lead to major effects.

The rotation of the Earth has a very important effect in determining how the winds blow. Comparison of the circulation of winds on the Earth (which rotates in 1 Earth day); on slowly rotating Venus (which rotates in 243 Earth days); on rapidly rotating Jupiter, Saturn and Neptune; and on Uranus (whose rotation axis points nearly at the Sun) helps us understand the weather on Earth.

7.5 Of Water, Ozone, and Life

Why is there is so much water on Earth—in our atmosphere and in our oceans? We care especially because water seems to be necessary for life. How have features that have made Earth a hospitable planet for life as we know it made the other planets so unattractive? Can we make sure we are not changing the Earth's atmosphere so that we will wind up with an atmosphere like that of, say, Venus, which has clouds of sulfuric acid drops?

The water on the Earth has spurred the incorporation of carbon dioxide into rocks. The carbon dioxide dissolves in the water and the carbon is then incorporated into the shells of sea animals; these shells eventually form rocks. On Venus the carbon dioxide is still in the atmosphere—in fact, it makes up over 95 per cent of the atmosphere. The carbon dioxide is the major cause of trapping solar energy in Venus's atmosphere, thus keeping the surface of Venus very hot, too hot for life. Apparently water and plant life have saved the Earth from suffering the same fate as Venus. Perhaps if life had arisen on Venus at the same time that it arose on Earth, it would have modified Venus's atmosphere as it did the terrestrial one.

Ozone (O_3) plays a vital role on Earth: ozone in the upper atmosphere keeps out the shortest-wavelength ultraviolet radiation from the Sun. The part of the ultraviolet that has wavelengths slightly longer than 3000 Å, where the Earth's atmosphere becomes transparent, causes suntanning at the beach. None of the wavelengths shorter than 3000 Å pass through the ozone. If they did, they would kill us. Most scientists agree that amounts of ultraviolet radiation even slightly increased over the amount we already receive, whether longer or shorter than 3000 Å, would increase the incidence of skin cancer. Plants and animals might suffer more severe consequences. (The problem of too little ozone in the upper atmosphere is quite different from the problem of too much ozone at ground level as pollution from cars and other sources.)

Scientists are very concerned about the way that our technology is threatening the ozone layer of our Earth. The most serious problem is chlorofluorocarbons (CFC's), of which the Freons (a trade name) are examples. These gases are used as refrigerants in refrigerators and air conditioners and in manufacturing silicon chips and other products. They were also often used as propellants in some aerosol cans: the United States (though not most other countries) has phased out this usage. Halons, based on bromine, are a problem too; they are used to control fires.

CFC's and halon gases are not as inert—non-interacting—as they originally had been thought to be. Once they are released into the atmosphere (Fig. 7–15), they rise to the stratosphere where they eventually break down. The individual chlorine and bromine atoms transform the ozone into other molecules.

Predictions of the global depletion of ozone have ranged from 3 to 18 per cent in the next century. But scientists have been astonished to find in the last few years that the concentration of ozone over Antarctica drops drastically

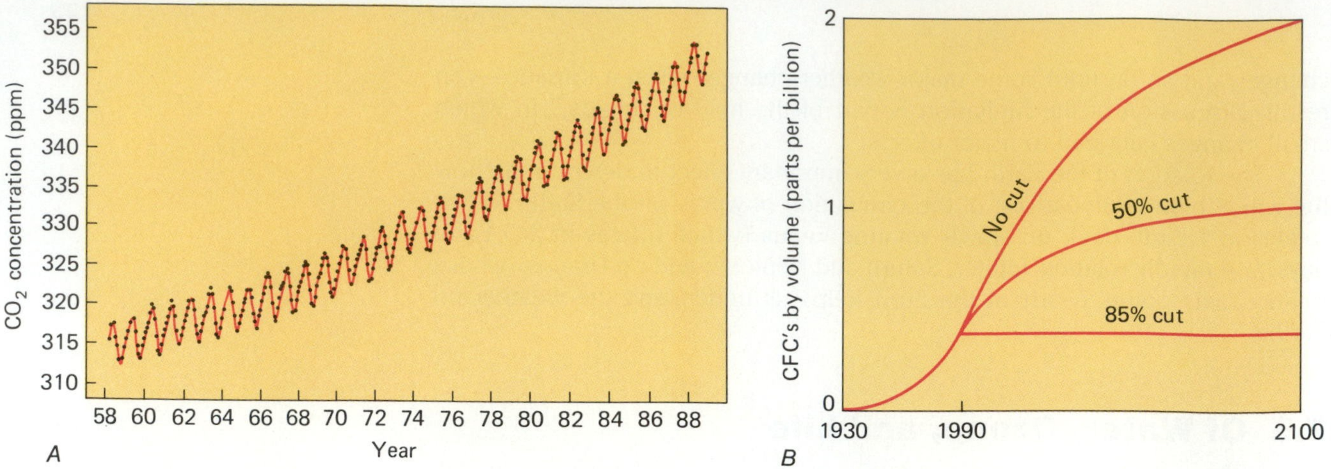

Figure 7–15 (*A*) The concentration of atmospheric carbon dioxide in parts per million (ppm) of dry air versus time, observed at the Mauna Loa Observatory in Hawaii. The dots show the monthly average concentration, and the smooth curve has been fit to them. In addition to the yearly cycle, there is a steady increase. (*B*) Because of CFC's already produced and distributed in products around the world, the concentration of CFC's would continue to rise even with a fifty per cent cut in their production. Only a very drastic cut of 85 per cent would hold the CFC concentration level, and an even more drastic cut would lead it to eventually decline.

during each Antarctic springtime (Fig. 7–16). This "ozone hole" has been getting more severe almost each year (Fig. 7–17). Is the ozone hole an isolated curiosity, with the extreme Antarctic cold magnifying the effect, or a forewarning of a significant worldwide change? Signs of an impending northern ozone hole have appeared, too, with volcanic particles from the eruption of Mt. Pinatubo taking the place of the ice crystals.

The ozone holes apparently form when stratospheric winds near the poles circle high-altitude ice crystals; this starts the chain reaction in which chlorine destroys ozone. When the CFC's break down, the products attach themselves to the ice crystals. There, the products can break down to form chlorine monox-

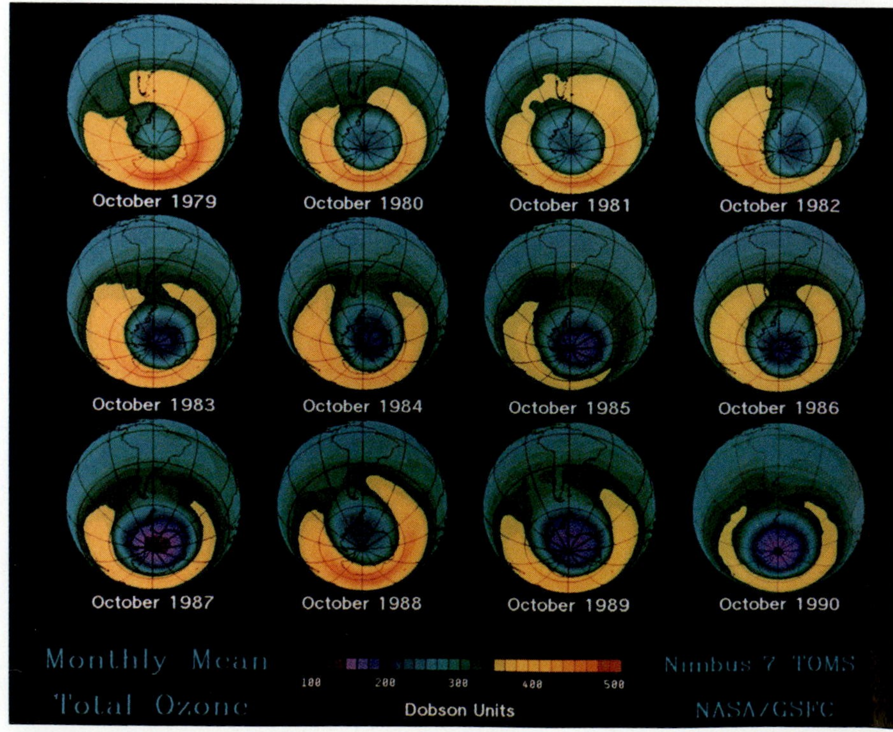

Figure 7–16 The ozone hole that appears over Antarctica each Antarctic spring; it is a result of the addition of chlorofluorocarbons to the atmosphere. Note the increased purple region over the years. Purple represents the lowest ozone values.

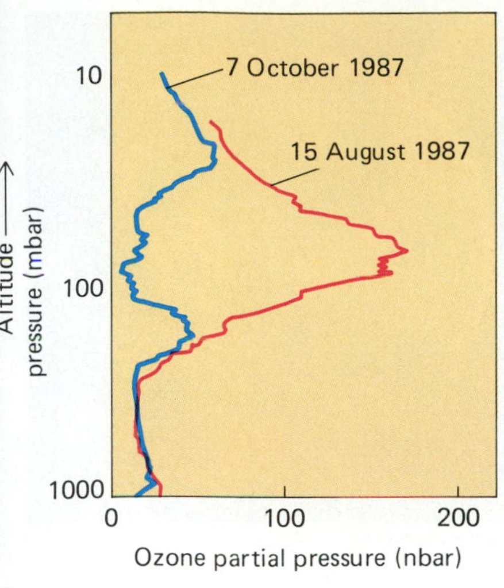

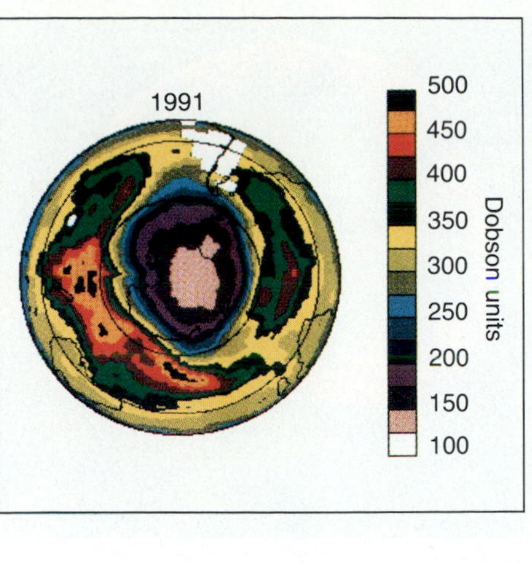

Figure 7–17 (*A*) The amount of ozone over Antarctica in the spring (October) is much lower than the amount in the winter (August). (*B*) The 1991 ozone hole.

ide, ClO. When the Sun rises at these high latitudes after months of darkness, the sunlight starts the reactions in which the chlorine monoxide breaks up the ozone. The amount of ozone is usually measured in Dobson units, the amount of atmospheric ozone in a column of cross-section 1 square centimeter and named after the British physicist G. M. B. Dobson. The ozone hole is now less than ⅓ of the normal 350 Dobson units.

We have already put enough gases into the atmosphere to make a permanent change in the ozone layer. Spacecraft are measuring the chlorine content of the upper atmosphere in order to improve our assessment of the situation, radio astronomers are measuring molecules in the atmosphere, and other astronomical methods are also being applied, as well as major Antarctic surveys carried out on the ground and from airplanes. The effect seems to be even more serious than we had first realized. The use of Freons in aerosol cans seems non-essential and has been limited. "Essential uses"—such as refrigeration, industrial processing of metals, and the making of computer chips—are still permitted, and substitutes are being developed. Worldwide meetings have led to agreements in phasing out the use of CFC's.

One lesson to learn from all these considerations is that basic scientific research exposes the possible side effects of ordinary things we otherwise take for granted. The dangers from refrigerants and aerosol propellants were understood in part because analyses were going on to understand chlorine compounds in Venus's atmosphere. Mars now provides a simpler system than the Earth to study ozone because only water vapor destroys it there. Studies of the atmospheres of other planets make us appreciate and help us understand the fundamental benefits of our own atmosphere. We must make certain that basic scientific research goes forward on all fronts so that we will be prepared in the future for whatever problems might arise. Research limited to applied problems won't be enough.

7.6 The Van Allen Belts

In January 1958, the first American space satellite carried aloft, among other things, a device to search for charged particles that might be orbiting the Earth. This device, under the direction of James A. Van Allen of the University of Iowa, detected a region filled with charged particles of high energies. We now know that there are actually two such regions—the *Van Allen belts*—surrounding the Earth, like a small and a large doughnut (Fig. 7–18).

Figure 7–18 The doughnut-shaped Van Allen belts. Though often called "radiation belts," they are actually regions of charged particles trapped by the Earth's magnetic field. The outer belt (15,000 to 25,000 km high) is mostly particles from the Sun—arriving in the solar wind and in solar flares. The inner belt (1000 to 5000 km high) contains mainly particles set free in the Earth's atmosphere by cosmic rays.

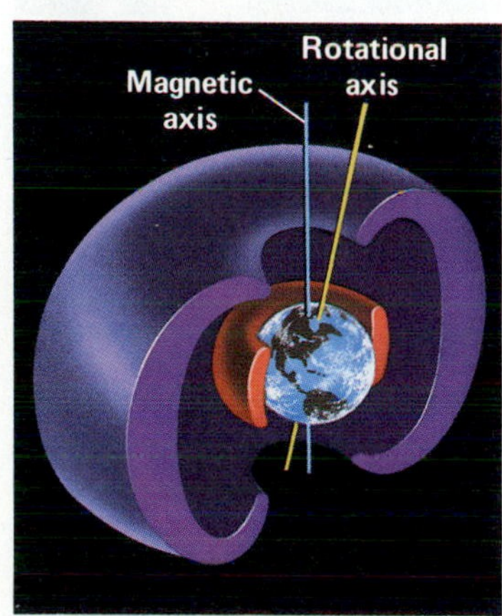

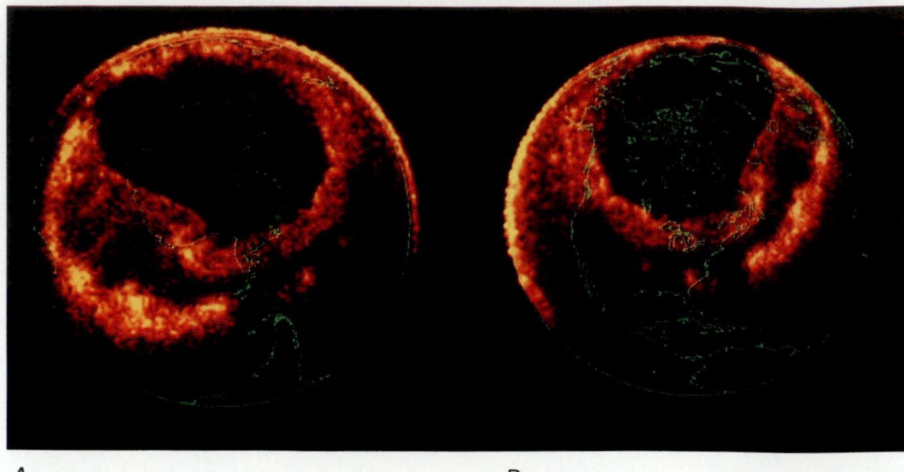

Figure 7–19 The aurora forms an oval around the south pole, which is seen in this view from the Dynamics Explorer 1 spacecraft. The observations were made in ultraviolet light mostly emitted by molecular nitrogen. The auroral oval traces the "feet" of the outer Van Allen belt and extends across the terminator between the sunlighted and night sides of the Earth. (*A*) The satellite imaged the region around the south pole. The aurora is at an altitude of about 200 km. (*B*) Though the position of the spacecraft does not allow direct imaging of the northern hemisphere, the northern-hemisphere aurora has been calculated using a model of the geomagnetic field. Both images were made for the March 13/14, 1989, aurora, which followed a huge solar flare. The regular auroral band that passes through the U.S.–Canadian border descended lower than usual, and a second, low-latitude, band also appeared.

Figure 7–20 The aurora borealis was visible over much of the United States on March 13/14, 1989. We are seeing, over the Williams College campus in Massachusetts, a bottom view of an auroral curtain hundreds of kilometers above the Earth. Atomic oxygen provides the characteristic yellow-green color and also the red color that was sometimes seen. Nitrogen molecules can also glow red.

The particles in the Van Allen belts are trapped by the Earth's magnetic field. The force on charged particles is perpendicular to the direction of magnetic-field lines, so charged particles spiral around magnetic-field lines. As a particle moves from above the equator toward one of the magnetic poles, the field lines get closer together. At some point, the particle can travel no farther. The magnetic force stops the particle's forward motion and makes it go back in the other direction as it continues its spiral. Scientists are now making such *magnetic mirrors* in terrestrial laboratories as part of one of the leading methods to keep a hot plasma (a gas of charged particles) trapped in one place long enough to start fusion. Controlled fusion may one day provide much of the energy needed to support our civilization, though major technical problems still remain to be solved.

7.7 The Aurora

Charged particles, often from solar storms, are guided by the Earth's magnetic field toward the Earth's magnetic poles. When they hit our atmosphere, molecules in our atmosphere glow. We see this glow as the beautiful northern and southern lights—the *aurora borealis* and *aurora australis,* respectively (Figs. 7–19 and 7–20). Research reported in 1988 indicates that magnetic waves from the Sun drive the aurora even when there is no solar storm.

The Earth is alone among the inner planets in having a respectable magnetic field; Mercury's magnetic field is very weak, and any magnetic fields that Venus or Mars might have are too weak to be detected. The Earth's fairly rapid spin and its internal heat source lead to the presence of our magnetic field.

Summary and Outline

Core, mantle, and crust; natural radioactivity is source of heat for interior (Section 7.1)
Continental drift (Section 7.2)
- Geothermal energy comes from heat flow at active areas
- Continents are on moving plates; the theory of plate tectonics explains the motion

Tides (Section 7.3)
- From differential forces
- Spring tides when lunar and solar tides add; neap tides when lunar and solar tides are perpendicular

Atmosphere (Section 7.4)
- All weather confined to thin troposphere
- Ozone layer between stratosphere and mesosphere
- Ionosphere (also called thermosphere) reflects radio waves

Water allows life on Earth (Section 7.5)
- We may be endangering life by modifying our atmosphere

Van Allen belts (Section 7.6)
- Discovered in data from first American satellite
- Composed of charged particles trapped by Earth's magnetic field

Aurora (Section 7.7)
- Interaction of solar wind and Earth's atmosphere

Key Words

Earth: geology, seismology, seismic waves, core, mantle, crust, lithosphere, aesthenosphere, isotopes, differentiation, heat flow, geothermal energy, plates, plate tectonics, continental drift, mid-Atlantic ridge, differential forces, spring tides, neap tides, pressure, one atmosphere, troposphere, stratosphere, mesosphere, ozone layer, ionosphere, thermosphere, Van Allen belts, magnetic mirrors, aurora borealis, aurora australis

Questions

1. Of the following in the Earth, which is the densest: (a) crust; (b) mantle; (c) lithosphere; (d) core?

2. (a) Explain the origins of tides. (b) If the Moon were twice as far away from the Earth as it actually is, how would tides be affected?

†3. Using the inverse-square law of gravity, calculate the difference between the Moon's gravitational force on a 1-gram mass located on one side of the Earth and the gravitational force on the same mass located on the other side of the Earth, given the radius of the Earth (Appendix 3) and the Earth–Moon distance (Appendix 4). Compare your result with the Moon's gravitational force on the 1-gram mass; it is sufficient to use the value of the force at the Earth's center.

4. Draw a diagram showing the positions of the Earth, Moon, and Sun at a time when there is the least difference between high and low tides.

5. What is the source of most of the radiation that heats the troposphere?

6. Look at a globe and make a list or sketches of which pieces of the various continents would fit like a jigsaw puzzle and so probably lined up with each other before the continents drifted apart.

†7. At what height above the Earth's surface is the pressure 1 per cent of its value at the ground?

8. How does the temperature vary with height in the troposphere? Does it increase, decrease, or have some different form of relation? How does it vary in the stratosphere?

9. Space-shuttle astronauts orbit 175 km above the Earth. Communications satellites are 5½ Earth radii above the surface. Compare their locations with the Van Allen belts, which can cause false readings on instruments.

†This question requires a numerical solution.

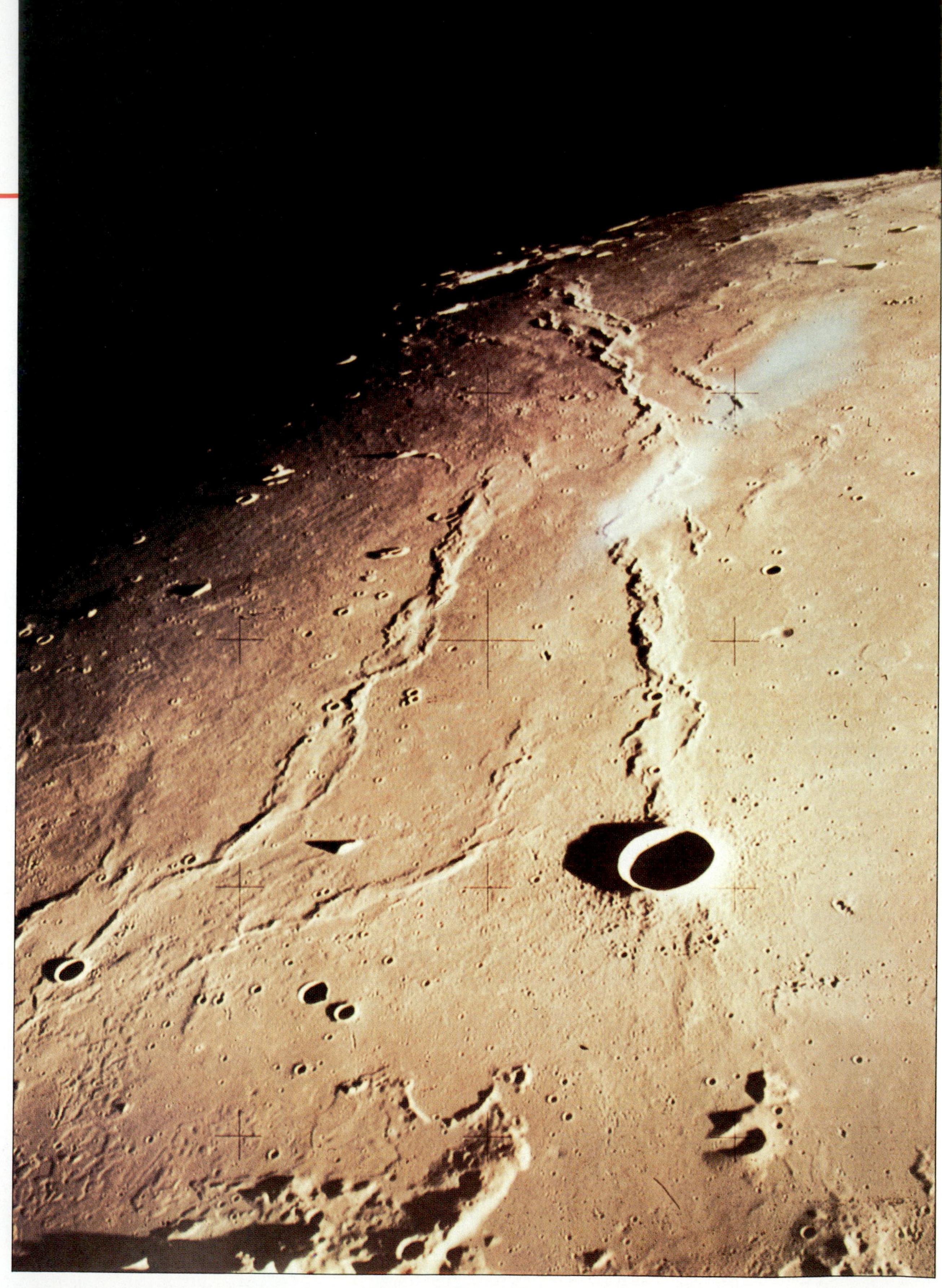

A view of the Oceanus Procellorum region of the Moon from an Apollo spacecraft in orbit around the Moon.

Our Moon

8

Aims: To see how direct exploration of the Moon has increased our knowledge manyfold, although such fundamental questions as how the Moon was formed remain the subject of current research

The Earth's nearest celestial neighbor—the Moon—is only 380,000 km (238,000 miles) away from us on the average, close enough that it appears sufficiently large and bright to dominate our nighttime sky. The Moon's stark beauty has called our attention since the beginning of history, and studies of the Moon's position and motion led to the earliest consideration of the solar system, to the prediction of tides, and to the establishment of the calendar.

8.1 The Appearance of the Moon

The fact that the Moon's surface has different kinds of areas on it is obvious to the naked eye. Even a small telescope—Galileo's, for example—reveals a surface pockmarked with craters. The *highlands* are heavily cratered. Other areas, called *maria* (pronounced mar′ee-ah; singular, *mare*, pronounced mar′ay), are relatively smooth, and indeed the name comes from the Latin word for "sea" (Fig. 8–1). But there are no ships sailing on the lunar seas and no water in them; the Moon is a dry, airless, barren place. The Moon's mass is only 1/81 that of the Earth, and the gravity at its surface is only 1/6 that of the Earth. Any atmosphere and any water that may once have been present would long since have escaped into space. The Moon is about one-fourth the diameter of the Earth, a very large fraction for a moon.

Figure 8–1 This Earth-based photograph shows a mare at top and highland with craters below.

Other types of structures visible on the Moon besides the smooth maria and the cratered highlands include *mountain ranges* and *valleys*. The mountains are formed by debris, though, unlike mountains on Earth. Lunar *rilles* are clefts that can extend for hundreds of kilometers along the surface. Some are relatively straight while others are sinuous. Raised *ridges* also occur. The craters themselves come in all sizes, ranging from as much as 295 km across for Bailly down to tiny fractions of a millimeter. Crater *rims* can be as much as several kilometers high, much higher than the Grand Canyon's rim stands above the Colorado River.

The Moon rotates on its axis at the same rate as it revolves around the Earth, always keeping the same face in our direction. The Earth's gravity has locked the Moon in this pattern, first causing a bulge in the distribution of the lunar mass and then interacting with the bulge to prevent the Moon from rotating freely. As a result of this interlock, we always see essentially the same side of the Moon from our vantage point on Earth. Because of *librations* of the Moon—the apparent slight turning of the Moon—we see, at one time or another, 5/8 of the whole surface. Librations arise from several causes, including the varying speed of the Moon in its elliptical orbit.

When the Moon is full (Fig. 8–2), it is bright enough to cast shadows or even to read by. But full moon is a bad time to try to observe lunar surface

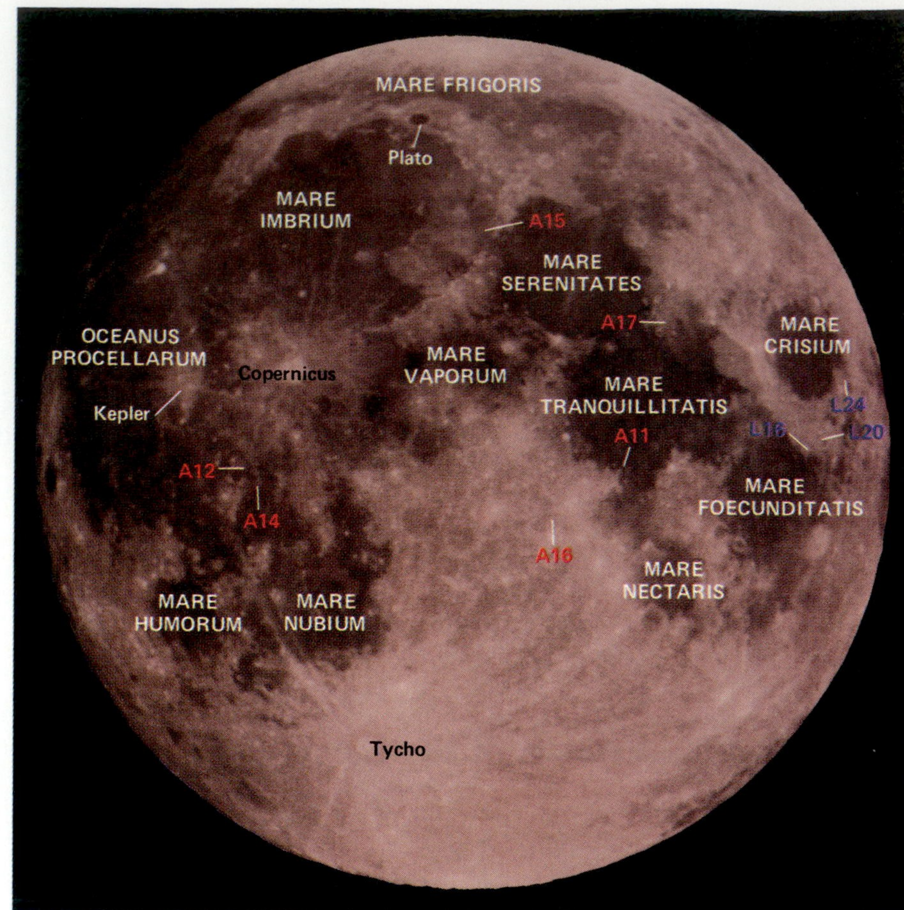

Figure 8–2 The full Moon. Note the dark maria and the lighter, heavily cratered highlands. The positions of the 6 American Apollo (A) and 3 Soviet Luna (L) missions from which material was returned to Earth for analysis (Table 8–1) are marked. This photograph is oriented with north up, as we see the Moon with our naked eyes or through binoculars; a telescope normally inverts the image.

structure, for any shadows we see are short. When the Moon is a crescent or even a half moon, however, the part of the Moon facing us is covered with long shadows. The lunar features then stand out in bold relief.

Shadows are longest near the *terminator,* the line separating day from night. The Moon makes a complete orbit of the Earth in 29½ days; the interval between successive new moons is the *synodic revolution period* of the Moon. Because the same side of the Moon always faces the Earth, different regions face the Sun as the Moon orbits. As a result, the terminator moves completely around the Moon with this 29½-day synodic period. Most locations on the Moon are thus in sunlight for about 15 days, during which time they become very hot—130°C (265°F)—and then in darkness for about 15 days, during which time their temperature drops to as low as −110°C (−170°F). The cycle of phases that we see from Earth also repeats with this 29½-day period—a *synodic month.*

In some sense, before the period of exploration by the Apollo program, we knew more about almost any star than we did about the Moon. As a solid body, the Moon reflects the solar spectrum rather than emitting one of its own, so we were hard pressed to determine even the composition or the physical properties of the Moon's surface.

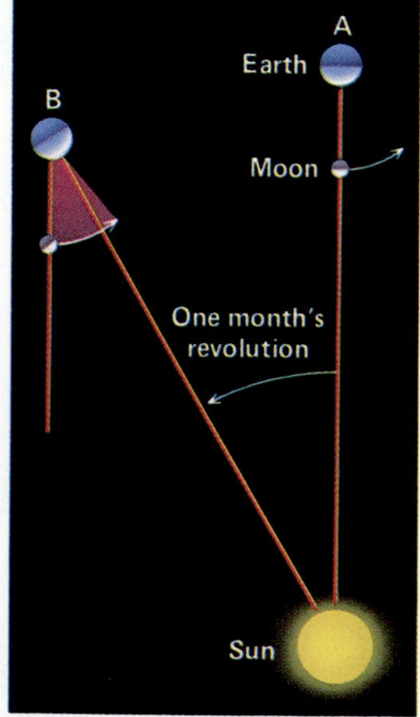

Figure 8–3 After the Moon has completed one revolution around the Earth with respect to the stars, which it does in 27⅓ days, it has moved from A to B. The Moon has still not swung far enough around to again be in the same position with respect to the Sun because the Earth has revolved one month's worth around the Sun. It takes about an extra two days for the Moon to complete its revolution with respect to the Sun, which gives a synodic rotation period of about 29½ days. The extra angle is shaded red. In the extra two days, both Earth and Moon have continued to move around the Sun (toward the left in the diagram).

Focus On

Box 8.1 Sidereal and Synodic Periods

Sidereal: with respect to the stars. For example, the Moon revolves around the Earth in about 27 1/3 days with respect to the stars. *Synodic:* with respect to some other body, usually the Earth. For example, while the Moon makes one sidereal revolution around the Earth, the Earth moves about 1/13 the way around the Sun (about 27°). Thus the Moon must travel a little farther to catch up (Fig. 8–3). The synodic period of the Moon, the synodic month, is about 29 1/2 days.

8.2 Lunar Exploration

The space age began on October 4, 1957, when the U.S.S.R. launched its first Sputnik (the Russian word for "travelling companion") into orbit. The shock of this event galvanized the American space program, and within months American spacecraft were also in Earth orbit.

In 1959, the Soviet Union sent its Luna 3 spacecraft around the Moon; Luna 3 radioed back the first murky photographs of the Moon's far side. Now that we have high-resolution maps, it is easy to forget how big an advance that was.

In 1961, President John F. Kennedy proclaimed that it would be a U.S. national goal to put a man on the Moon, and bring him safely back to Earth, by 1970. This grandiose goal led to the largest and most expensive coordinated program of any kind in world history. Only in 1989 did the Soviets admit that they too had a program to land people on the Moon.

The American lunar program, under the direction of the National Aeronautics and Space Administration (NASA), proceeded in gentle stages, starting with Alan Shepard rocketing into space for a short time and continuing with John Glenn as the first American to orbit the Earth. Three astronauts circled the Moon on Christmas Eve 1968 and returned to Earth. The next year, Apollo 11 brought humans to land on the Moon for the first time. It went into orbit around the Moon after a three-day journey from Earth, and a small spacecraft called the Lunar Module separated from the larger Command Module (Fig. 8–4). On July 20, 1969—a date that from the long-range standard of history

Figure 8–4 The Apollo 11 Lunar Module returning to the Command Module after the first landing of humans on the Moon. Smyth's Sea is in the background with the Earth rising behind it.

Figure 8–5 Neil Armstrong, the first person to set foot on the Moon, took this photograph of his fellow astronaut Buzz Aldrin climbing down from the Lunar Module of Apollo 11 on July 20, 1969. The site is called Tranquillity Base, as it is in the Sea of Tranquillity (Mare Tranquillitatis).

Table 8–1 Missions to the Lunar Surface and Back to Earth

Apollo 11	U.S.A.	1969	crewed
Apollo 12	U.S.A.	1969	crewed
Luna 16	U.S.S.R.	1970	uncrewed
*			
Apollo 14	U.S.A.	1971	crewed
Apollo 15	U.S.A.	1971	crewed
Luna 20	U.S.S.R.	1972	uncrewed
Apollo 16	U.S.A.	1972	crewed
Apollo 17	U.S.A.	1972	crewed
Luna 24	U.S.S.R.	1976	uncrewed

**An explosion on Apollo 13 while en route to the Moon led to an emergency return of the astronauts to Earth, though they had to circle the Moon to do so.*

may be the most significant of the millennium—Neil Armstrong and Buzz Aldrin left Michael Collins orbiting in the Command Module and landed on the Moon (Fig. 8–5). In the preceding days there had been much discussion of what Armstrong's historic first words should be, and millions listened as he said "One small step for man, one giant leap for mankind." (He meant to say "for a man." When, in 1984, a space shuttle astronaut became the first human to fly freely in orbit, untethered, he joked, "That may have been one small step for Neil, but it's a heck of a big leap for me." On hearing that he won the 1987 Nobel Prize in Literature, Soviet-émigré/American-citizen Joseph Brodsky said, "A big step for me, a small one for mankind.")

The Lunar Module carried many experiments, including devices to test the soil, a camera to take stereo photos of lunar soil, a sheet of aluminum with which to capture particles from the solar wind, and a seismometer. Later Lunar Modules carried additional experiments, some even including a vehicle (Fig. 8–6). Six Apollo missions in all, ending with Apollo 17 in 1972 (Fig. 8–7), carried people to the Moon. Unfortunately, the missions became confused in the popular mind with just one of the experiments—the collection of rocks. These rocks and dust, returned to Earth for detailed analysis, were indeed important, but were only a fraction of each mission.

The Soviet Union sent three unmanned spacecraft to land on the lunar surface, collect lunar soil, and return it to Earth. The first two each collected

Figure 8–6 (*A*) Eugene Cernan riding on the Lunar Rover during the Apollo 17 mission. The mountain in the background is the east end of the South Massif. (*B*) Drawing by Alan Dunn; (© 1971 The New Yorker Magazine, Inc.)

A

B

a few grams of lunar soil. The third, Luna 24, drilled to ½ meter below the lunar surface and brought a long, thin cylinder of material back to Earth. In 1970 and 1973, two other Soviet spacecraft had carried remote-controlled rovers, Lunokhods 1 and 2, that travelled across 10 km of the lunar surface over a period of months.

Figure 8–7 The Apollo 17 launch.

8.3 The Results from Apollo

The kilometers of film exposed by the astronauts, the 382 kg of rock brought back to Earth, the lunar seismograph data recorded on tape, and other data have been studied by hundreds of scientists from all over the Earth. The data have led to new views of several basic questions, and have raised many new questions about the Moon and the solar system. We had hoped, beforehand, that the data would clear up basic questions about the origin of the solar system, but that hope was not answered. In any case, the next subsections will describe what we learned.

8.3a The Composition of the Lunar Surface

The rocks astronauts found on the Moon are types that are familiar to terrestrial geologists. Almost all the rocks are *igneous,* which means that they were formed by the cooling of lava. The moon has few *sedimentary* rocks, which are formed by settling, most commonly on Earth out of water and also from deposits carried by wind. (Earth's sedimentary rocks include limestone and shale and cover ⅔ of Earth's surface. On the Moon, only matter ejected when craters were formed is available to be made into sedimentary rocks). The Moon's crust is thick—12 per cent of its volume; the Earth's continental crust is only 0.5 per cent of the Earth's volume.

In the maria, the rocks are mainly *basalts* (Fig. 8–8). The highland rocks are *anorthosites* (Fig. 8–9). (Anorthosites—defined by their particular combination of minerals—are rare on Earth, though the Adirondack Mountains are made of them.) Anorthosites, though they have also cooled from molten material, have done so under different conditions from basalts and have taken longer to cool.

In both the maria and the highlands, some of the rocks are *breccias* (Fig. 8–10), mixtures of fragments of several different types of rock that have been

Figure 8–8 A basalt brought back to Earth by the Apollo 15 astronauts. It weighs 1.5 kg. Note the many vesicles, spherical cavities that arise in volcanic rock because of gas trapped at the time of the rock's formation. A 1-cm block is present for scale.

Figure 8–9 An anorthosite. The dark part is dust covering the breccia to which the anorthosite was attached.

Figure 8–10 A breccia, dark gray and white, returned to Earth by Apollo 15. It weighs 4.5 kilograms.

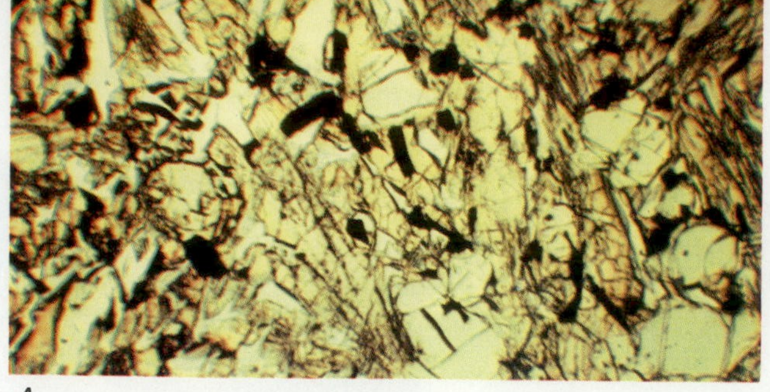

Figure 8–11 A thin section of a lunar rock, seen through a microscope (*A*) in unpolarized light, and (*B*) in polarized light. In the latter, polarized light has passed through a section of lunar rock so thin as to be transparent. Different minerals in the rock rotate the plane of polarization differently, leading to the color effect when the resulting light is passed through a second polarizer.

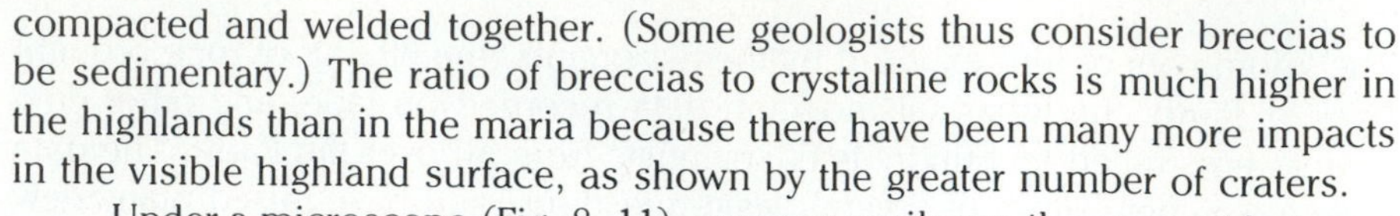

compacted and welded together. (Some geologists thus consider breccias to be sedimentary.) The ratio of breccias to crystalline rocks is much higher in the highlands than in the maria because there have been many more impacts in the visible highland surface, as shown by the greater number of craters.

Under a microscope (Fig. 8–11), one can easily see the contrast between lunar and terrestrial rocks. The lunar rocks contain no water, and so never underwent the reactions that terrestrial rocks undergo. The lunar rocks also show that no oxygen was present when they were formed.

The astronauts also collected some *lunar soils,* bits of dust and larger fragments from the Moon's surface. This *regolith* was built up by bombardment of the lunar surface by meteorites of all sizes over billions of years. Microscopes showed that some of these soils contain small glassy globules (Fig. 8–12), which are not common on Earth. Some of these glassy globules and glassy coatings undoubtedly resulted from the melting of rock during its ejection from the site of a meteorite impact and the subsequent cooling of the molten material. It was useful to have a trained geologist on the Moon, Harrison Schmitt (Fig. 8–13), though a pity that he was the only such to reach there.

Figure 8–12 An enlargement of a thin section including a glassy spherule from the lunar dust collected by the Apollo 11 mission. The shape and composition of the spherule indicate that it was created when a meteorite crashed into the Moon, melting lunar material and splashing it long distances.

Almost all of these rocks and soils have lower proportions of elements with low melting points (*volatile* elements) than do rocks and soils on the Earth. On the other hand, they contain relatively high proportions of elements with high melting points (*refractory* elements) like calcium, aluminum, and titanium. Elements that are even rarer on Earth, such as uranium, thorium, and the rare-earth elements, are also found in greater abundances. (Will we be mining on the Moon one day?)

Since none of the lunar rocks contains any trace of water bound inside their minerals, clearly water never existed on the Moon. This eliminates the possibility that life evolved there.

8.3b Lunar Chronology

One way of dating the surface of a moon or planet is to count the number of craters in a given area, a method that was used before Apollo. Even from the Earth we can count the larger lunar craters. If we assume that the events that

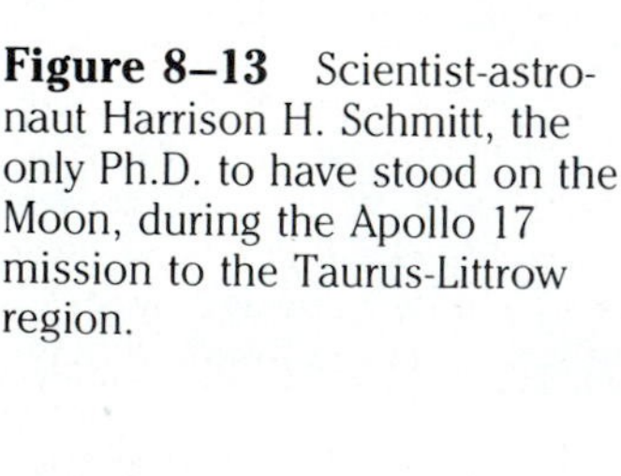

Figure 8–13 Scientist-astronaut Harrison H. Schmitt, the only Ph.D. to have stood on the Moon, during the Apollo 17 mission to the Taurus-Littrow region.

cause craters—whether they are impacts of meteors or the eruptions of volcanoes—continue over a long period of time and that no strong erosion occurs, surely those locations with the greatest number of craters must be the oldest. Relatively smooth areas—like maria—must have been covered over with volcanic material at some relatively recent time (which is still billions of years ago). When one crater is superimposed on another, we can be certain that the superimposed crater is the younger one.

A few craters on the Moon, notably Tycho and Copernicus (look back at Figure 8–2), have thrown out obvious rays of lighter-colored matter. The rays are material ejected when the crater was formed. Since these rays extend over other craters, the craters with rays must have formed later. The youngest rayed craters may be very young indeed—perhaps only a few hundred million years. The rays darken with time, so rays that may have once existed near other craters are now indistinguishable from the rest of the surface. Rays are visible for about 3 billion years.

Crater counts and the superposition of one crater on another give only relative ages. We could find the absolute ages only when rocks were physically returned to Earth. Scientists worked out the dates by comparing the current ratio of radioactive isotopes to nonradioactive isotopes present in the rocks with the ratio that they would have had when they were formed. From the known rate of radioactive decay, they can work out the ages. The oldest rocks that were found at the locations sampled on the Moon were formed 4.42 billion years ago. The youngest rocks were formed 3.1 billion years ago. (Ages are given to the accuracy to which they can be measured; the 4.42-billion-year age was especially precise because many lunar rocks apparently formed from a single catastrophic event on the Moon.)

Radioactive isotopes are those that decay spontaneously; that is, they change into other isotopes even when left alone. Stable isotopes remain unchanged over time. For certain pairs of isotopes—one radioactive and one stable—we know the proportion of the two when the rock was formed. Since we know the rate at which the radioactive one is decaying, we can calculate how long it has been decaying from a measurement of what fraction is left.

The ages of highland and maria rocks are significantly different. The highland rocks were formed between about 4.4 and 3.9 billion years ago, and the maria between 3.8 and 3.1 billion years ago. Several highland rocks are exactly the same age, all 4.42 billion years old. So 4.42 billion years ago may therefore have been the origin of the lunar surface material, that is, the time when it last cooled.

All the observations can be explained on the basis of the following general picture (Fig. 8–14): The Moon formed 4.6 billion years ago. We know that the top 100 km or so of the surface was molten after about 200 million years. The surface could have been entirely melted by the original heat or by an intense bombardment of meteorites (or debris from the period of formation of the

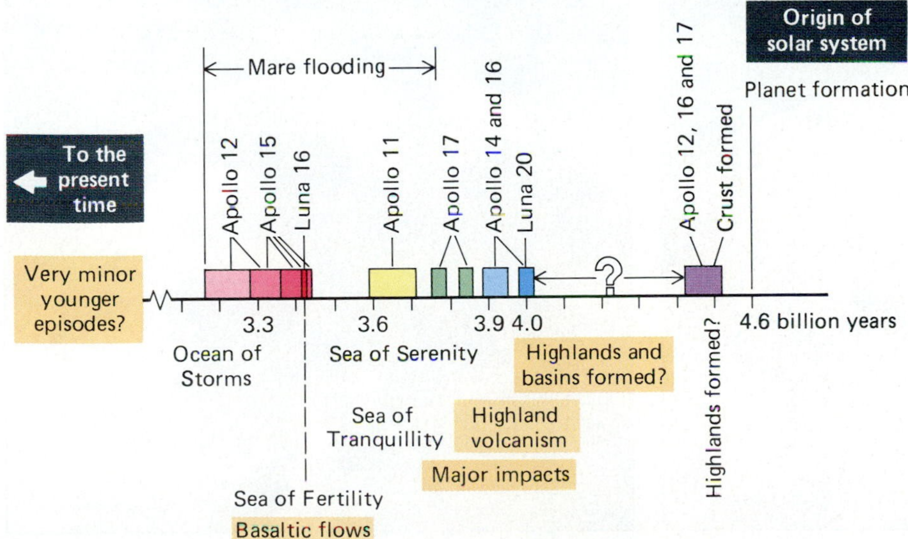

Figure 8–14 The chronology of the lunar surface, based on work carried out at the Lunatic Asylum, as the Caltech laboratory of Gerald Wasserburg is called. The ages of rocks found in eight missions are shown in boxes. The names and descriptions below the line took place at the times indicated by the positions of the words. Note how many missions it took to get a sampling of many different ages on the lunar surface.

Moon and the Earth). The decay of radioactive elements also provided heat. Then the surface cooled. From 4.2 to 3.9 billion years ago, bombardment (perhaps by planetesimals) caused most of the craters we see today. About 3.8 billion years ago, the interior of the Moon heated up sufficiently (from radioactive elements inside) that vulcanism began; lava flowed onto the lunar surface and filled the largest basins that resulted from the earlier bombardment, thus forming the maria (Fig. 8–15). By 3.1 billion years ago, the era of vulcanism

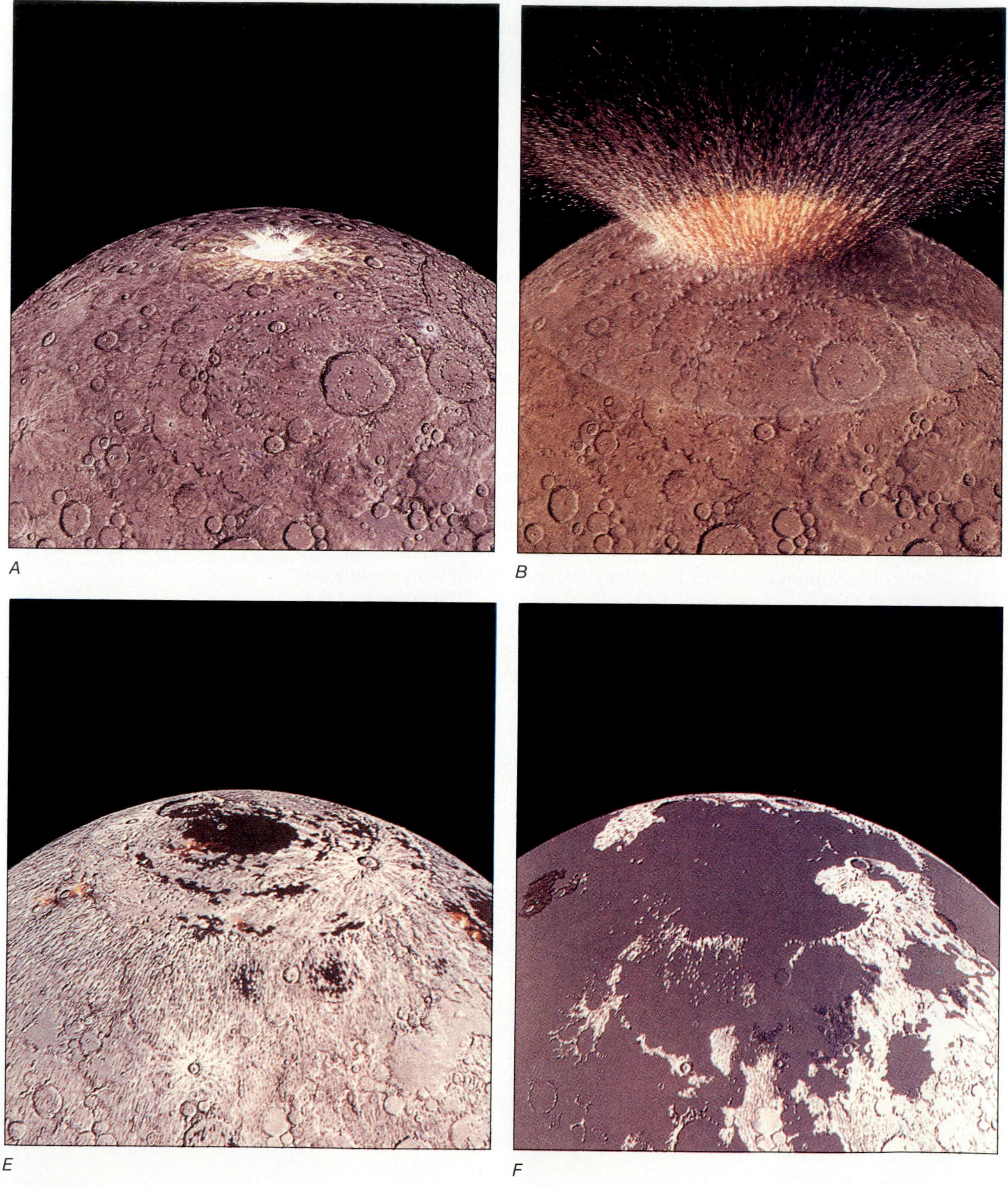

Figure 8–15 An artist's view of the formation of the Mare Imbrium region of the lunar surface. (*A*) An asteroid impact on the Moon, sometime between 3.85 and 4.0 billion years ago. (*B*) The shock of the asteroid impact began the Imbrium crater. (*C*) As the dust and heat subsided, the 1300-km Imbrium crater was left. (*D*) The lava flowed over outlying craters and cooled, leaving lunar mountains as the remainder of the rim. (*E*) 3.8 billion years ago, lava welled up from inside the Moon, filling the basin. (*F*) By 3.3 billion years ago, the lava flooding was nearly complete. (*G*) The final flow of thick lava came 2.5–3.0 billion years ago. (*H*) Subsequent cratering has left the Mare Imbrium on today's Moon. (Drawings by Donald E. Davis under the guidance of Don E. Wilhelms of the U.S. Geological Survey)

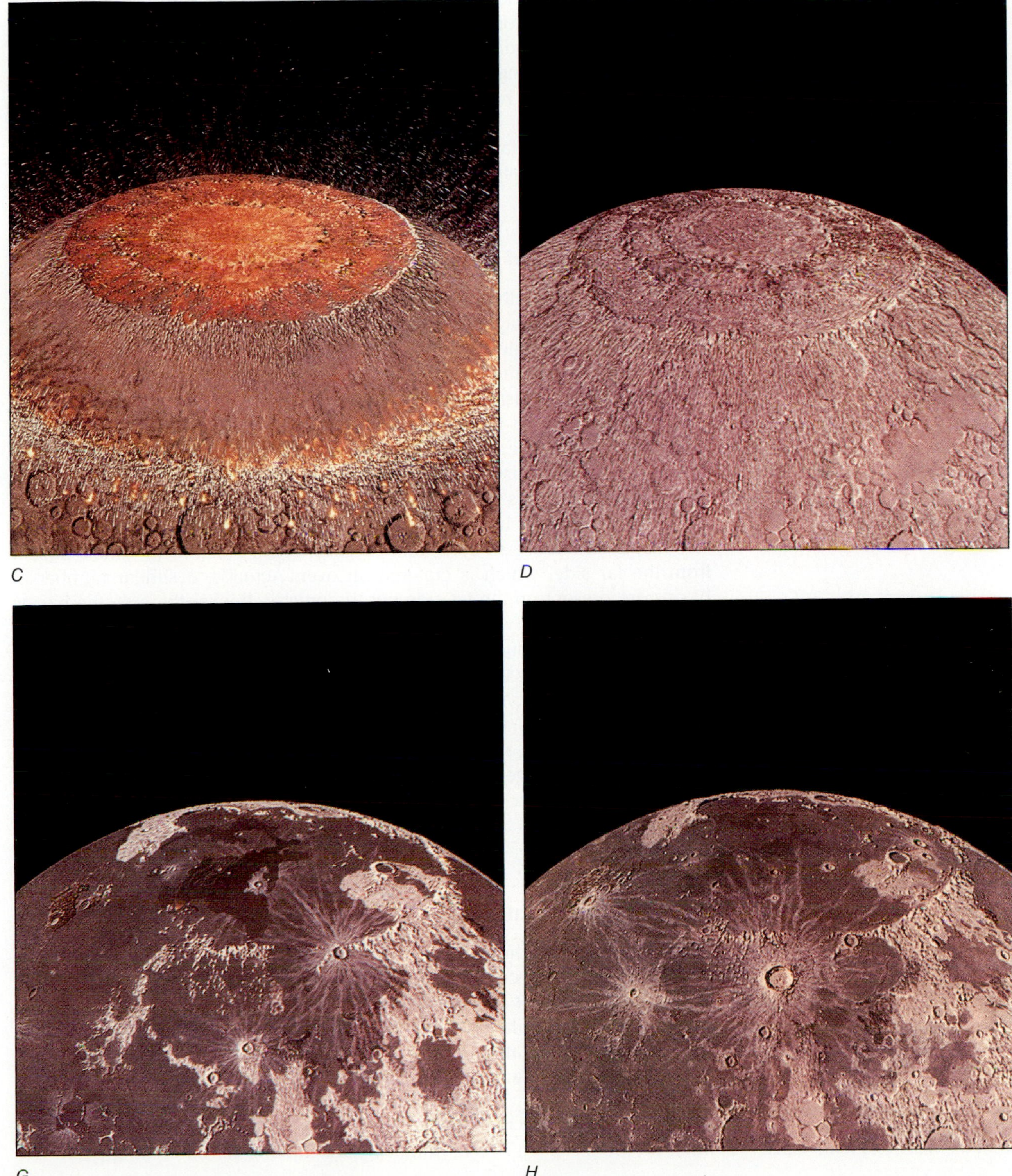

C D G H

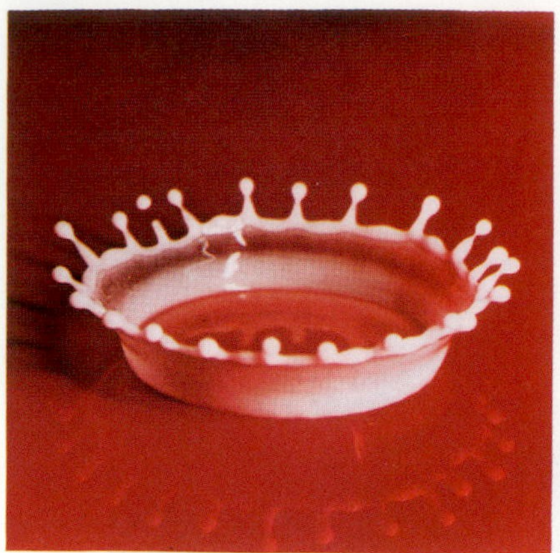

Figure 8–16 This photograph, made with a very short-exposure "strobe" light, shows the result of a falling milk drop. The formation of a lunar crater by a meteorite is similar, because the energy of the impact makes the surface material flow like a liquid.

was almost over, with only a final flow of lava between 3.0 and 2.5 billion years ago. The Moon has been geologically pretty quiet since then.

Up to this time, the Earth and the Moon shared similar histories. But active lunar history stopped about 3 billion years ago, while the Earth continued to be geologically active. Because the Earth's interior continued to send gas into the atmosphere and because the Earth's higher gravity retained that atmosphere, the Earth developed conditions in which life evolved. The Moon, because it is smaller than the Earth, presumably lost its heat more quickly and also generated a thicker crust.

Almost all the rocks on the Earth are younger than 3 billion years of age; erosion and the remolding of the continents as they move slowly over the Earth's surface, according to the theory of plate tectonics, have taken their toll. Though the oldest single rock ever discovered on Earth has an age of 4.1 or 4.2 billion years, few rocks are older than 3 billion years. So we must look to extraterrestrial bodies—the Moon, meteorites, or comets—that have not suffered the effects of plate tectonics or erosion (which occurs in the presence of water or an atmosphere) to study the first billion years of the solar system.

8.3c The Origin of the Craters

The debate over whether the craters were formed by meteoritic impact (Fig. 8–16) or by volcanic action began in pre-Apollo times. The results from Apollo indicate that most craters resulted from meteoritic impact, though some small fraction may have come from vulcanism. Though only a few craters may have resulted from vulcanism, there are many other signs of volcanic activity, including the lava flows that filled the maria. The Marius Hills in Oceanus Procellarum, for example, have many domes and rilles that apparently resulted from repeated vulcanism. Mare lavas may have flowed from such areas.

The photographs of the far side of the Moon (Fig. 8–17) have shown us that the near and far hemispheres are quite different in overall appearance. The maria, which are so conspicuous on the near side, seem almost absent from the far side, which is cratered all over. Actually, a similar number of basins are present on the far side but the crust is thicker there so the basins were less often filled with lava. Further, an asymmetry in cratering may have been a secondary result of an uneven distribution of the Moon's mass. Once any asymmetry was set up, the Earth's gravity locked one side toward us. The far side of the Moon shows many impacts and is very rough (Fig. 8–18).

In later chapters, we shall meet other cratered objects in our solar system. Careful study has apparently shown that the Moon and Mercury received their craters at the same general time. The family of small objects that made the craters on Mars came later, as did other families that made craters on the moons of Jupiter and Saturn.

Figure 8–17 The far side of the Moon looks very different from the near side in that there are few maria. This photograph was taken from Apollo 17, and shows some of the near-side maria at the upper left. Since the far-side crust is about 100-km thick, in contrast to the 65-km-thick near-side crust, less of the dark basalt can well up from the mantle to form maria.

8.3d The Lunar Interior

Before the Moon landings, it was widely thought that the Moon was a simple body, with the same composition throughout. But we now know it to be differentiated (Fig. 8–19), like the planets. It has a crust of relatively light material at its surface and a silica-rich mantle making up most or all of the interior. It may also have a metallic (iron-rich) core, though any such core would take up a much smaller fraction of the lunar interior than the Earth's core does of the Earth's interior.

The lunar crust is perhaps 65 km thick on the near side and twice as thick on the far side. This asymmetry may explain the different appearances

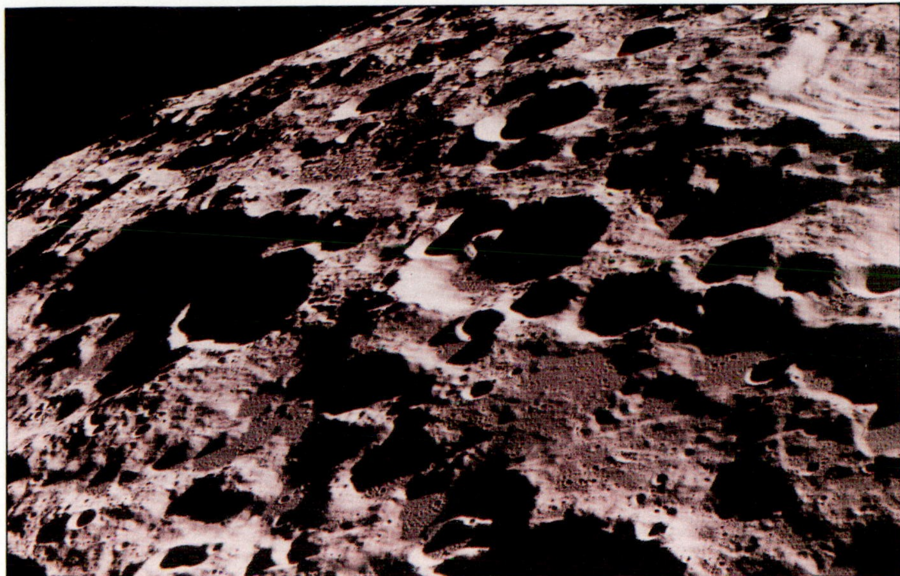

Figure 8–18 An oblique view of part of the lunar far side, from Apollo 11, shows how rough it looks.

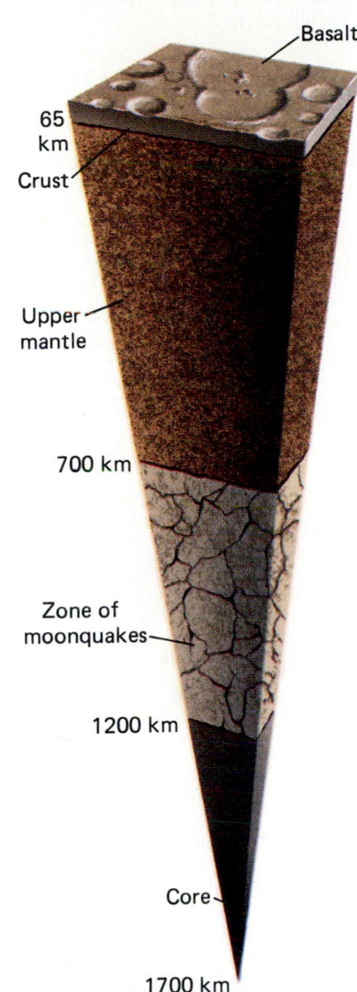

Figure 8–19 The Moon's interior. The depth of basalts is greater under maria, which are largely on the side of the Moon nearest the Earth. Almost all the 10,000 moonquakes observed originated in a zone halfway down toward the center of the Moon, a distance ten times deeper than most terrestrial earthquakes. This fact can be used to interpret conditions in the lunar interior. If too much of the interior of the Moon were molten, the zone of moonquakes probably would have sunk instead of remaining suspended there. The deep moonquakes came from about 80 locations, and were triggered at each location twice a month by tidal forces resulting from the variation in the Earth–Moon distance.

of the sides, because lava would be less likely to flow through the far side's thicker crust.

The Moon seems seismically quiet compared with the Earth. Our lunar stations worked for only 8 years, however, so we do not know if giant moonquakes occur more rarely than that. Unfortunately, the seismometers and other instruments on the Moon were shut off by NASA in 1977 as an economy measure.

If the interior of the Moon were molten, as is the Earth's interior, we would have expected to find a more intense magnetic field than the one we have detected. Yet the Moon has no general magnetic field. We do detect a weak magnetic field frozen into lunar rocks. It might be left over from a core that was molten long ago but has since cooled. We do not have a good understanding of the magnetic fields of other planets either.

Apollo astronauts made direct measurements of the rate at which heat flows upward through the top of the lunar crust. The rate is one-third that of the heat flow on Earth. The value is important for checking theories of the lunar interior.

Tracking the orbits of the Apollo Command Modules and other satellites that orbited the Moon also told us about the lunar interior. If the Moon were a perfect, uniform sphere, the spacecraft orbits would have been perfect ellipses. We interpret the deviation of the orbits from an ellipse as an effect of an asymmetric distribution of lunar mass.

One of the major surprises of the lunar missions was the discovery in this way of *mascons,* regions of **mas**s **con**centrations near and under most maria. (A few large mascons are known on Earth.) The mascons led to anomalies in the gravitational field, that is, deviations of the gravitational field from being spherical. The mascons may be lava that is denser than the surrounding matter. The existence of the mascons is evidence that the whole lunar interior is not molten, for if it were, then these mascons could not remain near the surface. However, the mascons could be supported by a crust of sufficient thickness.

In sum, the Apollo missions have given contradictory evidence as to whether the lunar interior is hot or cold, molten or solid. Still, as a result of

Figure 8–20 Buzz Aldrin and the experiments he had deployed. In the foreground, we see the seismic experiment. The laser-ranging retroreflector is behind it.

the meteorite impact detected with the seismic experiment, most scientists believe that the Moon's core is molten. If even such a small body as the Moon formed with a hot interior, then we know that the heat did not have time to be radiated away as it was forming. From this observation, we can conclude that the planetesimals (Section 6.5) would have had to coalesce very rapidly into planets. Thus the study of the Moon gives us insight into the formation of the whole solar system.

One source of knowledge about the lunar interior continues to be studied. Sets of retroreflectors were left on the Moon (Fig. 8–20) by the astronauts of Apollo 11, 14, and 15 and on the Lunakhod 2. Each corner cube in these retroreflectors has the property that a beam of light that hits it is reflected back into exactly the same direction. When illuminated by the faint pulse received from a strong laser focused by a telescope on Earth (Fig. 8–21), the retroreflectors send back enough light to telescopes on Earth to be barely detectable—only about one of the 10^{18} photons sent out in a pulse is detected when it returns about 2.5 sec later. The studies, carried out mostly by telescopes at the McDonald Observatory in Texas and on Haleakala on the island of Maui in Hawaii, allow the Earth–Moon distance to be accurately determined by measuring the time interval between sending out a brief pulse of laser light and its return. The Earth's rotation, the Moon's orbital motion, and the Moon's librations are also measured to an accuracy of about 10 cm. Only the existence of a fluid layer at the interface between a fluid lunar core of radius about 330 km and the mantle can explain the laser-ranging observations. This conclusion that a fluid core does exist is consistent with the conclusions on the basis of the far-side impact, which had otherwise been challenged as statistically uncertain.

Figure 8–21 A laser beam sent to the Moon at the McDonald Observatory of the University of Texas.

8.4 The Origin of the Moon

The leading models for the origin of the Moon that were considered at the time of the Apollo missions were the following:

1. **Fission:** The Moon was separated from the material that formed the Earth; the Earth spun up (that is, increased its rate of spin) and the Moon somehow spun off;
2. **Capture:** The Moon was formed far from the Earth in another part of the solar system, and was later captured by the Earth's gravity; and

3. **Condensation:** The Moon was formed near to and simultaneously with the Earth in the solar system.

But recent work since the mid-1980's has all but ruled out the first two of these and has made the third less likely than it seemed. The two models currently under the most study have not been adequately investigated, but research is continuing on them. They are:

4. **Interaction of Earth-orbiting and Sun-orbiting planetesimals:** Collisions of planetesimals orbiting the Earth and planetesimals orbiting the Sun led to their breakup and the eventual formation of the Moon from the debris. In the collisions, rock and iron got treated differently.

5. **Ejection of a ring:** A Mars-sized planetesimal hit the protoearth, ejecting matter. One version of such a model has the matter ejected in gaseous form. This matter would have ordinarily fallen back onto the Earth, but since it was gaseous, pressure differences could exist. These presumably caused some of the matter to start moving rapidly enough to go into orbit, and some asymmetry established an orbiting direction; the other matter fell back. Another version has clumps of matter ejected; the clumps are later broken up by tidal forces and form a disk around the Earth. In both versions, the disk of orbiting material eventually coalesced into the Moon.

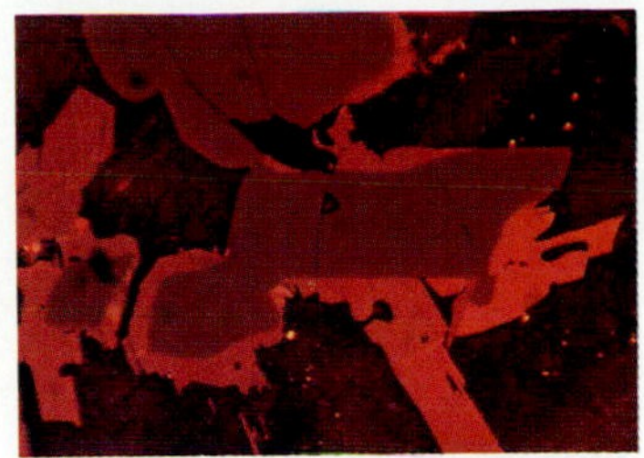

Figure 8–22 A crystal of armalcolite (**Arm**strong-**Al**drin-**Col**lins, the crew of Apollo 11), examined under a polarization microscope. This mineral has been found only on the Moon.

Comparing the chemical composition of the lunar surface with the composition of the terrestrial surface has been important in narrowing down the possibilities. The mean lunar density of 3.3 grams/cm^3 is close to the average density of the Earth's major upper region (the mantle), which had led some to believe in the fission hypothesis. However, detailed examination of the lunar rocks and soils indicates that the abundances of elements on Moon and Earth are sufficiently different to indicate that the Moon did not form directly from the Earth. (Though some minerals, such as the one shown in Fig. 8–22, that do not exist on Earth have been discovered on the Moon, this results from different conditions of formation rather than from abundance differences.)

At present, it seems that the "ejection of a ring" model must be considered the leading contender to explain the formation of the Moon. Computer simulations (Fig. 8–23) are endorsing its main points.

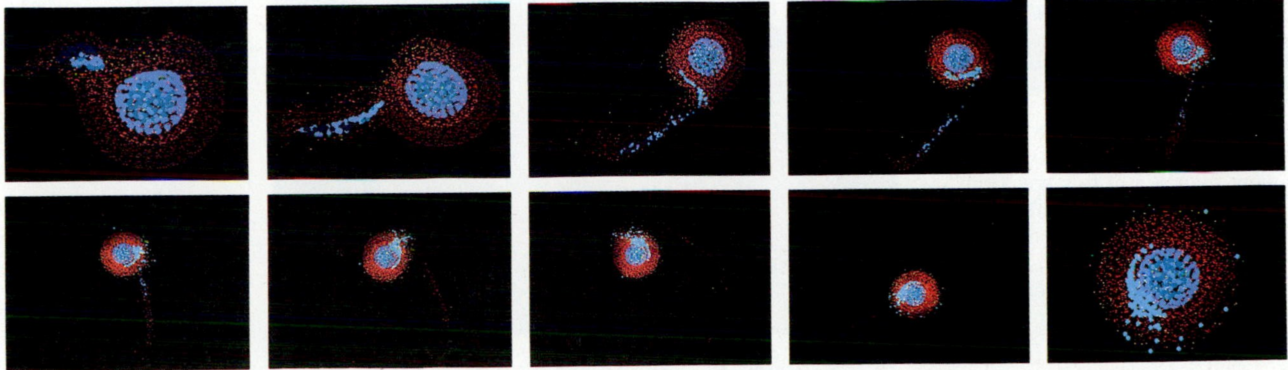

Figure 8–23 A computer simulation of a collision between the protoearth and an impactor. Each is composed of an iron core and a rock mantle. The internal energy in each increases with both temperature and pressure. Increasing internal energy is shown as dark red, light red, brown, and yellow for rock and dark blue, light blue, dark green, and light green for iron. Nearly two lunar masses of rock is left in orbit at the end of the simulation, forming a disk. In the model, the Moon would later form out of this disk.

8.5 Current and Future Lunar Studies

Figure 8–24 This rock is one of many meteorites found in the Earth's continent of Antarctica. Under a microscope and in mineralogical and isotopic analyses, it seems like a sample from the lunar highlands (including anorthosites, for example) and quite unlike any terrestrial rock or any other meteorite. Eight meteorites are thought to have come from the Moon, all breccias.

One surprise addition to lunar studies came in 1982, when it was realized that a meteorite found in Antarctica probably came from the Moon (Fig. 8–24). The rock was apparently ejected from the Moon when a crater was formed. This rock, and all the lunar samples that have not been distributed or destroyed in analysis, are kept free of contamination in a Lunar Curatorial Facility in Houston for future study. Another lunar rock, also found in Antarctica, was reported in 1984 by Japanese scientists.

Ground-based studies of the Moon still bring surprises. Following the spectroscopic discovery of sodium in Mercury's atmosphere, a small amount of sodium was similarly discovered forming a very thin atmosphere (Fig. 8–25).

The United States does not have any concrete plans for lunar exploration, either manned or unmanned, though some prospective spacecraft have been suggested. NASA's Galileo mission went around the Moon in 1991, en route to Jupiter (Fig. 8–26). In 1990, Japan sent a small spacecraft to orbit the Moon.

The orbiting manned space stations now planned for the late 1990's by the U.S. and Russia, larger than Russia's current Mir space stations, may provide the capability of sending people in space for extended periods. The U.S. space station, named Freedom, is so expensive that deciding to go ahead with the plans is a political decision. Most scientists would prefer unmanned exploration of space to the much-more-expensive manned programs; the U.S. National Academy of Sciences found that the U.S. space station is not justified on scientific grounds in the next couple of decades. A NASA committee headed by former astronaut Sally Ride concluded in 1987 that the U.S. should now explore the Moon more fully, building an outpost from lunar materials and extracting oxygen from the lunar soil. We must watch the news reports to follow NASA's current path.

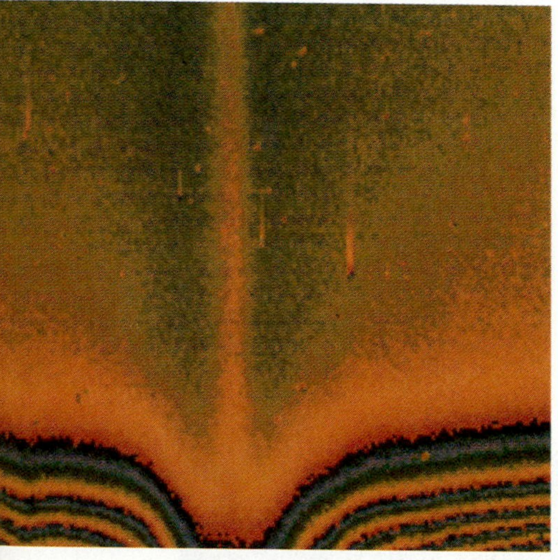

Figure 8–25 In 1988, traces of a lunar sodium atmosphere were detected in this spectrum, taken from Earth. Wavelength goes from left to right, and the false colors show intensity. The vertical yellow line reveals the presence of sodium.

Figure 8–26 The Moon, from the Galileo spacecraft. Colors, from the ratios of brightness in different filters, show variations in the Moon's composition. The Orientale basin and portions of the far side highlands (*red, at left*) that we do not see from Earth are similar to soils found at the Apollo 16 site. Highland regions shown in yellow are enhanced in iron. Some of the mare basalts have high (*blue*) titanium dioxide while others (*orange*) have lower titanium dioxide.

Summary and Outline

Lunar features (Section 8.1)
- Maria, highlands, craters, mountains, valleys, rilles, ridges
- Visibility dependent on Sun's angle; terminator

Revolution and rotation (Box 8.1)
- Sidereal: with respect to the stars
- Synodic: with respect to another body

Lunar exploration (Section 8.2)
- Manned and unmanned lines of development merge in Apollo
- Six Apollo landings: 1969 to 1972
- Three Soviet unmanned spacecraft brought samples back
- Two Soviet Lunokhods roved many kilometers

Composition of the lunar surface (Section 8.3a)
- Mare basalts and highland anorthosites from cooled lava
- More breccias—broken up and re-formed—in highlands

Chronology (Section 8.3b)
- Relative dating by crater counting
- Radioactive dating gives absolute ages

Craters (Section 8.3c)
- Almost all formed in meteoritic impact
- Signs of vulcanism also present on surface
- Far side has no maria, quite different from near side.

Interior (Section 8.3d)
- Moon is differentiated into a core, a mantle, and a crust
- Seismographs revealed that interior is molten, though lack of strong lunar magnetic field indicates the opposite
- Weak moonquakes occur regularly
- Mascons discovered from their gravitational effects.

Origin (Section 8.4)
- Condensation, fission, and capture theories were standard
- Newer theories are interaction of Earth-orbiting and Sun-orbiting planetesimals and ejection of a gaseous ring. The latter seems most viable.

Key Words

highlands, maria, mare, mountain ranges, valleys, rilles, ridges, rims, librations, terminator, synodic revolution period, synodic month, igneous, sedimentary, basalts, anorthosites, breccias, lunar soils, regolith, volatile, refractory, radioactive, stable, mascons

Questions

†1. About how many Moons would fit inside one Earth?

2. Compare the lengths of sidereal and synodic months. Which is longer? Why?

3. If the Moon's mass is 1/81 Earth's, why is gravity at the Moon's surface as great as 1/6 that at the Earth's surface?

4. To what location on Earth does the terminator on the Moon correspond?

5. Why is the heat flow rate related to the radioactive material content in the lunar surface?

6. What does cratering tell you about the age of the surface of the Moon, compared to that of the Earth's surface?

†This question requires a numerical solution.

7. Why is it not surprising that the rocks in the lunar highlands are older than those in the maria?

8. Why are we more likely to learn about the early history of the Earth by studying the rocks from the Moon rather than those on the Earth?

9. Using the lunar-chronology chart (Figure 8–14), describe the relative ages of the lava flows in the Seas of Fertility and Serenity. Which Apollo missions went there?

10. What do mascons tell us about the lunar interior?

11. Choose one of the proposed theories to describe the origin of the Moon, and discuss the evidence pro and con.

12. (a) Describe the American and Soviet lunar explorations. (b) What can you say about future plans?

Topics for Discussion

1. Discuss the scientific, political, and financial arguments for resuming (a) unmanned and (b) manned exploration of the Moon.

2. Discuss the scientific, political, and financial arguments for building the NASA Space Station Freedom.

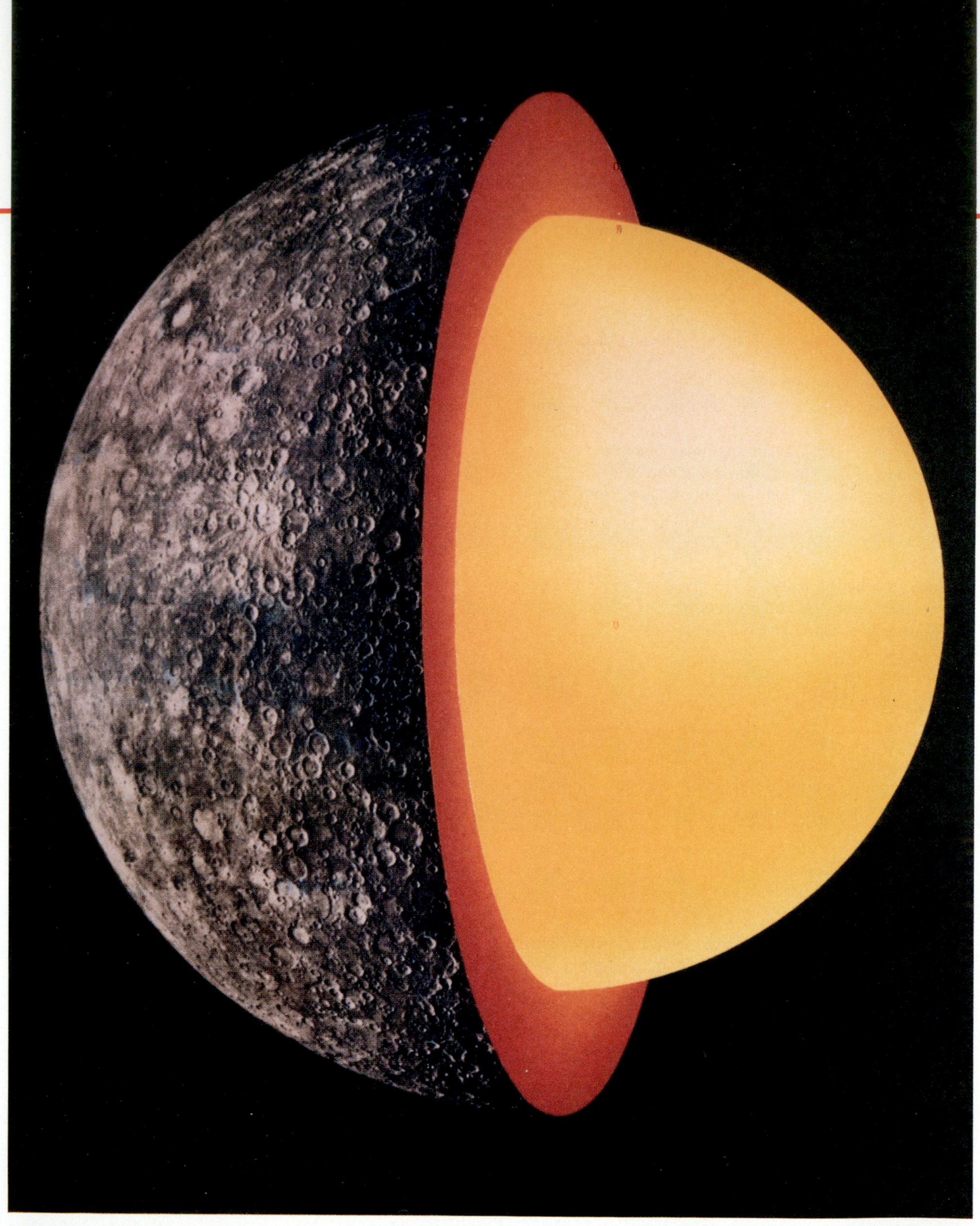

A photomosaic of Mercury from Mariner 10, showing its cratered surface, surrounding an artist's conception of Mercury's thick core (yellow) and thin lithosphere (red).

Mercury

9

Aims: To discuss the difficulties in studying Mercury from the Earth, the results of space observations, and exciting new ground-based observations

Mercury is the innermost planet, and until very recently has been one of the least understood. Except for distant Pluto, its orbit around the Sun is the most elliptical (the difference between the maximum and minimum distances of Mercury from the Sun is as much as 40 per cent of the average distance, compared with less than 4 per cent for the Earth). Its average distance from the Sun is 58 million kilometers, which is 4⁄10 of the Earth's average distance. Thus Mercury is 0.4 A.U. from the Sun. It has only 5½ per cent the mass of the Earth.

Since we on the Earth are outside Mercury's orbit looking in at it, Mercury always appears close to the Sun in the sky (Fig. 9–1). At times it rises just before sunrise, and at times it sets just after sunset, but it is never up when the sky is really dark. The maximum angle from the Sun at which we can see it is 28°, which means that the Sun always rises or sets within about two hours of Mercury's rising or setting. Of course, the difference in time is usually even less than this maximum. Consequently, whenever Mercury is visible, its light has to pass obliquely through the Earth's atmosphere. This long path through turbulent air leads to blurred images. Alternatively, Mercury is observed during daylight. Thus astronomers have never gotten a really good view of Mercury from the Earth, even with the largest telescopes. Many people have never seen it at all. (Copernicus's deathbed regret, the story has it, was that he had never seen Mercury.) Even the best photographs taken from the Earth show Mercury as only a fuzzy ball with faint, indistinct markings.

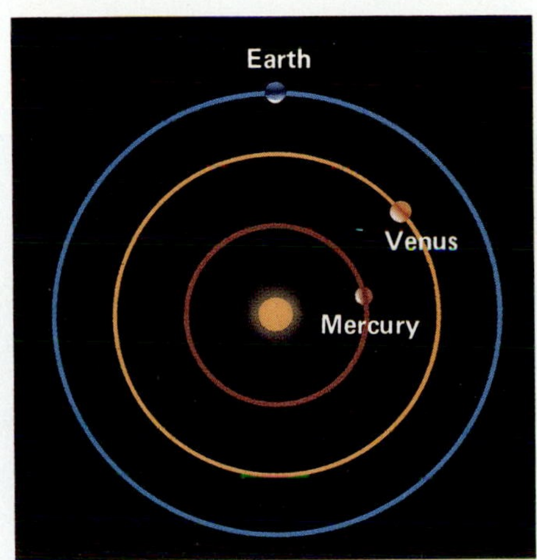

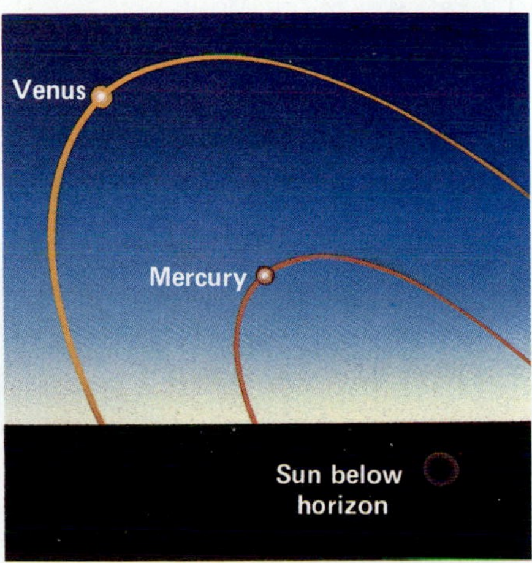

Figure 9–1 Since Mercury's orbit is inside that of the Earth (*left*), Mercury is never seen against a really dark sky. A view from the Earth appears at right, showing Mercury and Venus at their greatest respective distances from the Sun.

9.1 The Rotation of Mercury

From studies of drawings and photographs, astronomers did as well as they could to describe Mercury's surface. A few features could barely be distinguished, and the astronomers watched to see how long those features took to rotate around the planet. From these observations they decided that Mercury rotated in the same amount of time that it took to revolve around the Sun. Thus they thought that one side always faced the Sun and the other side always faced away from the Sun. This led to the fascinating conclusion that Mercury could be both the hottest planet and the coldest planet in the solar system.

It seemed reasonable that there could be a bulge in the distribution of mass of Mercury. The side that was bulging would be attracted to the Sun by gravity, locking the rotation to the revolution, just as the Moon is locked to the Earth. Such a match of periods is called *synchronous rotation.* Synchronous rotation implies that the periods of rotation and revolution are equal, and so the less massive body would always keep the same face toward the more massive body. Our moon and all of the largest moons of the outer planets are in synchronous rotation.

But early radio astronomy studies of Mercury indicated that the dark side of Mercury was too hot for a surface that was always in the shade. (Simply, the strength of the radio signal an object gives off depends on its temperature.)

Later, we became able not only to receive radio signals emitted by Mercury but also to transmit radio signals from Earth and detect the echo. This technique is called *radar.* "Radar" is an acronym for **ra**dio **d**etection **a**nd **r**anging. Since Mercury is rotating, one side of the planet is always receding relative to the other. Such motion can be measured with the Doppler effect. The results were a surprise: scientists had been wrong about the period of Mercury's rotation. It actually rotates in 58.6441 days.

Mercury's 59-day period of rotation is exactly ⅔ of the 88-day period of its revolution, so the planet rotates three times for each two times it revolves around the Sun. Although its tendency to continue rotating is too strong to be overcome by the gravitational grip of the Sun, the Sun's steadying pull is strongest every 1 1/2 rotations. At those times, Mercury is in its perihelion position, so the gravitational bulge on Mercury is in a line with Mercury's center and the Sun. Mercury's spin was probably once much faster, and was slowed down by the fact that the Sun's gravity attracted the bulge more than it attracted the rest of the planet. A *gravitational interlock* is at work. But because Mercury's orbit is elliptical, the relation of rotation to revolution is 2:3 instead of the 1:1 relation of our Moon.

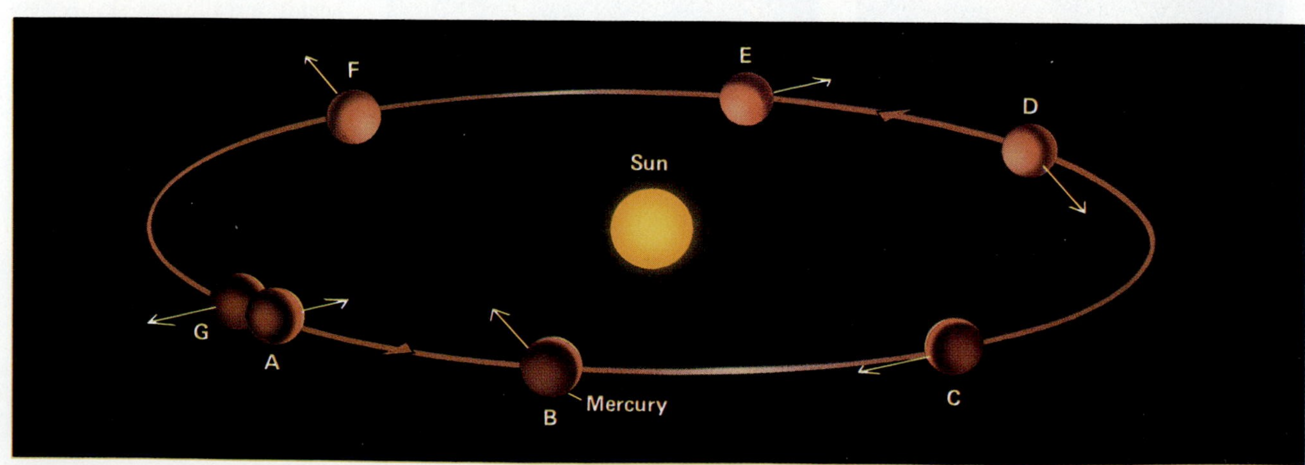

Figure 9–2 Follow the arrow that starts facing rightward toward the Sun in the image of Mercury at the left of the figure (*A*), as Mercury revolves along the line. Mercury, and thus the arrow, rotates once with respect to the stars in 59 days, when Mercury has moved only 2/3 of the way around the Sun (*E*). Note that after one full revolution of Mercury around the Sun, the arrow faces away from the Sun (*G*). It takes another full revolution, a second 88 days, for the arrow to again face the Sun. Thus the rotation period with respect to the Sun is twice 88, or 176, days.

The rotation period is measured with respect to the stars; that is, the period is one mercurian sidereal day, the interval between successive returns of the stars to the same position in the sky. Mercury's rotation and revolution combine to give a value for the rotation of Mercury relative to the Sun (that is, a mercurian solar day) that is neither the 59-day *sidereal rotation period* nor the 88-day period of revolution. As can be seen from careful analysis of the combination (Fig. 9–2), if we lived on Mercury we would measure each day and each night to be 88 Earth days long. We would alternately be fried and frozen for 88 Earth days at a time. Mercury's *solar rotation period* is thus 176 days long, twice the period of Mercury's revolution.

Since we now know that different sides of Mercury face the Sun at different times, the temperature at the point where the Sun is overhead doesn't get as hot as it would otherwise. The temperature at this point is about 700 K (800°F).

We know from Kepler's second law that Mercury travels around the Sun at different speeds at different times in its eccentric orbit. This effect, coupled with Mercury's slow rotation on its axis, would lead to an interesting effect if we could stand on its surface. From some locations we would see the Sun rise for an Earth day or two, and then retreat below the horizon from which it had just come, when the speed of Mercury's revolution around the Sun dropped below the speed of Mercury's rotation on its own axis. Later, the Sun would rise again and then continue across the sky.

On rare occasions, Mercury goes into *transit* across the Sun; that is, we see it as a black dot crossing the Sun. The transit of November 13, 1986, was visible from Europe but not from North America. The next transits will be in 1993, 2003, and 2006.

No harm was done by the scientists' original misconception of Mercury's rotational period, but the story teaches all of us a lesson: we should not be too sure of so-called facts. Don't believe everything you read here, either.

9.2 Ground-based Visual Observations

Even though the details of the surface of Mercury can't be studied very well from the Earth, there are other properties of the planet that can be better studied. For example, we can measure Mercury's *albedo,* the fraction of sunlight hitting Mercury that is reflected (Fig. 9–3). We can measure the albedo because we know how much sunlight hits Mercury (we know the brightness of the Sun and the distance of Mercury from the Sun). Then we can easily calculate at any given time how much light Mercury reflects, from both (1) how bright Mercury looks to us and (2) its distance from the Earth. Once we have a measure of the albedo, we can compare it with the albedo of materials on the Earth and on the Moon and thus learn something of what the surface of Mercury is like.

The *albedo* (from the Latin for whiteness) is the ratio of light reflected from a body to light received by it.

Let us consider some examples of albedo. An ideal mirror reflects all the light that hits it; its albedo is thus 100 per cent. (The very best real mirrors have albedoes of as much as 96 per cent.) A black cloth reflects essentially none of the light; its albedo (in the visible part of the spectrum, anyway) is almost 0 per cent. Mercury's overall albedo is low—only about 6 per cent. Its surface, therefore, must be made of a dark—that is, poorly reflecting—material. The albedo of the Moon is similar. In fact, Mercury (or the Moon) appears bright to us only because it is contrasted against a relatively dark sky; if it were

Figure 9–3 *Albedo* is the fraction of radiation reflected. A surface of low albedo looks dark.

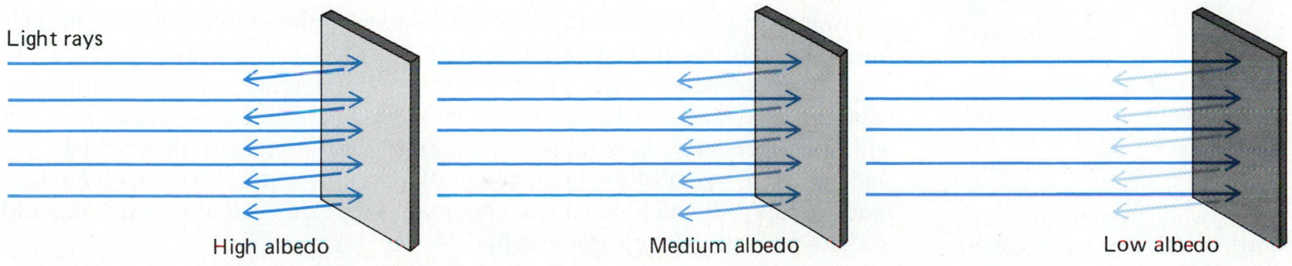

silhouetted against a bedsheet, it would look relatively dark, as if it had been washed in Brand X instead of Tide.

From Mercury's apparent angular size and its distance from the Earth—which can be determined from knowledge of its orbit—we have determined that Mercury is less than half the diameter of the Earth. Since Mercury has no moon, we can determine its mass only from its gravitational effects on bodies that pass near it, such as occasional asteroids or comets. The most accurate value we now have for Mercury's mass came from tracking the Mariner 10 spacecraft flyby. We find that Mercury's mass is five times greater than that of our Moon and 5½ per cent that of the Earth.

Mercury's density (its mass divided by its volume) can thus be calculated, and turns out to be 5.5 grams/centimeter^3, roughly the same density as that of Venus and the Earth. So Mercury's core, like Venus's and the Earth's, must be heavy; it too is made of iron. Since Mercury has less mass than the Earth and is therefore less compressed, Mercury's core must have even more iron than the Earth to give the planets the same density.

Though Venus's atmosphere, as we shall see in the following chapter, makes it hotter than Mercury, Mercury's surface undergoes the greatest variation in surface temperature. Mercury's surface, during the long solar day under the nearby Sun, reaches 427°C (800°F), hot enough to melt zinc. At night, without an atmosphere to retain energy, the surface cools to below −183°C (−300°F).

Figure 9–4 Mercury, photographed from a distance of 35,000 km as Mariner 10 approached the planet for the first time, shows a heavily cratered surface with many low hills. The valley at the bottom is 7 km wide and over 100 km long. The large, flat-floored crater is about 80 km in diameter; craters over 20 km wide have flat floors, while small craters are bowl shaped like bullet craters.

9.3 Mariner 10

So astronomers, typically, had deduced a lot from limited data. In 1974, we learned much more about Mercury in a brief time. We flew right by. The tenth in the series of Mariner (3-axis stabilized) spacecraft launched by the United States went to Mercury. First it passed by Venus and then had its orbit changed by Venus's gravity to direct it to Mercury. Tracking the orbits improved our measurements of the gravity of these planets and thus of their masses. Further, the 475-kg spacecraft had a variety of instruments on board. One was a device to measure the magnetic fields in space and near the two planets. Another measured infrared emission of the planets and thus their temperatures. Two other instruments—a pair of television cameras—provided not only the greatest popular interest but also many important data.

9.3a Photographic Results

When Mariner 10 flew by Mercury the first time (yes, it went back again), it took 1800 photographs that were transmitted to Earth. It came as close as 750 km to Mercury's surface.

The most striking overall impression is that Mercury is heavily cratered (see the photograph that opened this chapter and Fig. 9–4). At first glance, it looks like the Moon! But there are several basic differences between the features on the surface of Mercury and those on the lunar surface. We can compare how the mass and location in the solar system of these two bodies affected the evolution of their surfaces.

Mercury's craters seem flatter than those on the Moon and have thinner rims (Fig. 9–5), perhaps as a result of Mercury's higher gravity.

There are three major types of surfaces on Mercury: smooth plains (corresponding to maria on the Moon), intercrater plains (which are, in fact, covered with small craters), and rugged highlands. The intercrater plains make up perhaps 70 per cent of the hemisphere of Mercury's surface for which Mariner made images; the high density of craters on them shows that they are very old and are probably of volcanic origin.

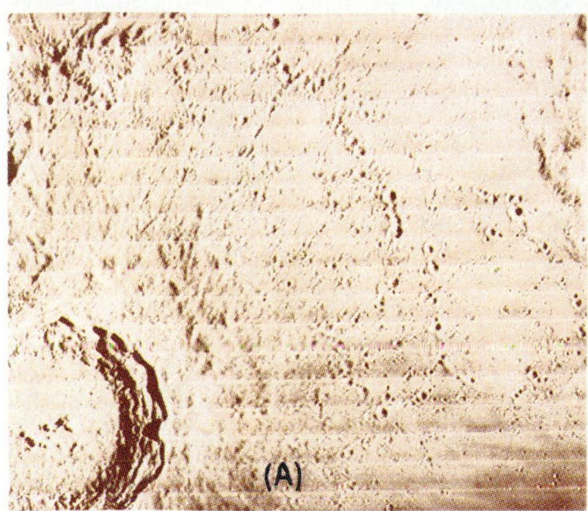

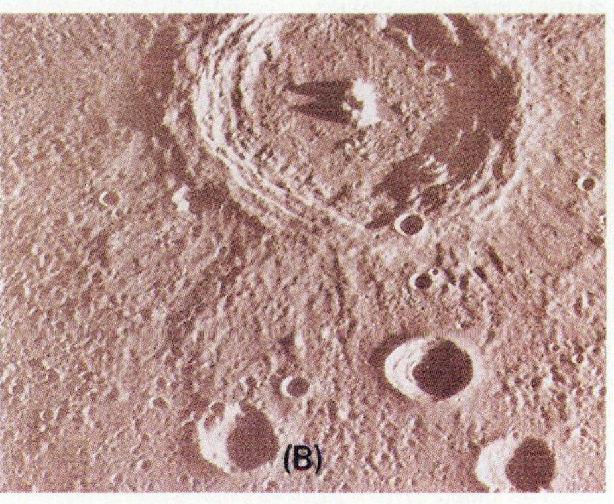

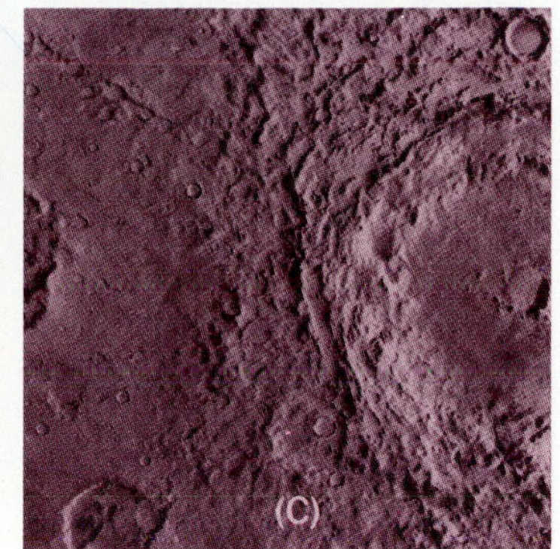

Figure 9–5 A comparison of (*A*) a lunar crater, (*B*) a crater on Mercury, and (*C*) a crater on Mars. Material has been continuously thrown out less far on Mercury than on the Moon because of Mercury's higher gravity. The martian crater shows flow across its surface, possibly resulting from the melting of a permafrost layer under the surface. The photographs are of Copernicus on the Moon, 93 km in diameter; Brahms on Mercury, 75 km in diameter; and Cerulli on Mars, 115 km in diameter.

Most of the craters themselves seem to have been formed by impacts of meteorites. The secondary craters, caused by material ejected as primary craters were formed, are closer to the primaries than on the Moon, presumably because of Mercury's higher surface gravity. The smooth plains have relatively few craters compared with the intercrater plains. The smooth plains are sufficiently extensive that they are probably volcanic. Further evidence for their vulcanism is the fact that they sometimes overlap the intercrater plains. Smaller, brighter craters are sometimes superimposed on the larger craters and thus must have been made afterward.

Some craters have rays of higher albedo emanating from them (Fig. 9–6), at least superficially resembling rays around some lunar craters. Unlike on the Moon, rays on Mercury are not always associated with the freshest craters.

Mercury has no obvious fault systems such as the San Andreas fault in California on the Earth. There appear to be some signs of small-scale geologic tensions like troughs, widely distributed around the planet. They are very old. From their orientations, they are believed to have resulted from stresses in the planet's crust as Mercury changed shape while solar tides were slowing down its rotation.

Figure 9–6 A field of rays radiating from a crater off to the top left, photographed on the second pass of Mariner 10. The crater at top is 100 km in diameter.

Focus On

*Box 9.1 Naming the Features of Mercury

The mapping of the surface of Mercury led to a need for names. The scarps were named for historical ships of discovery and exploration, such as *Endeavour* (Captain Cook's ship), *Santa Maria* (Columbus's ship), and *Victoria* (the first ship to sail around the world, which it did in 1519–22 under Magellan and his successors). Most plains were given the name of Mercury in different languages, such as Tir (in ancient Persian), Odin (an ancient Norse god), and Suisei (Japanese). Craters are being named for non-scientific authors, composers, and artists, in order to complement the lunar naming system, which honors scientists. So we find Mozart, Beethoven, Brontë, Michelangelo, and Matisse on Mercury.

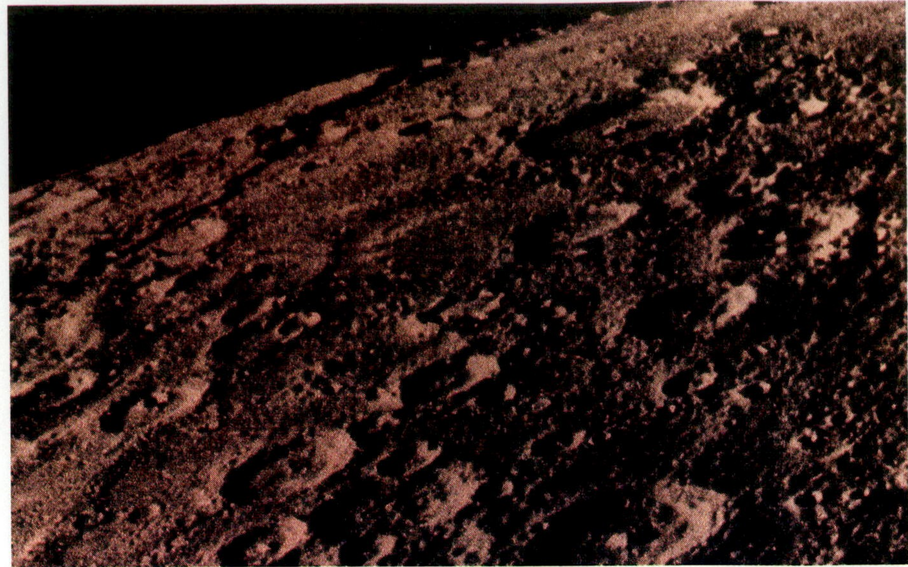

Figure 9–7 A prominent scarp near Mercury's limb.

One interesting kind of feature that is visible on Mercury is lines of cliffs hundreds of kilometers long; on Mercury as on Earth such lines of cliffs are called *scarps*. The scarps are particularly apparent in the region of Mercury's south pole (Fig. 9–7). These scarps are global in scale, not just isolated, and are relatively youthful. The scarps are wrinkles in Mercury's crust. Mercury's core, judging by the fact that Mercury's average density is about the same as the Earth's even though it does not have enough gravity to compress it as much as Earth is compressed, is probably iron and takes up perhaps 50 per cent of the volume or 70 per cent of the mass. (The Earth's core, by comparison, takes up only 16 per cent of the Earth's volume, and the Moon's core only 4 per cent of the Moon's volume.) Probably Mercury's core was once totally molten, and shrank as it solidified. This shrinking combined with shrinking of Mercury's rocky mantle as it cooled would have caused the crust to buckle, creating the scarps in at least the quantity that we now observe.

One part of the mercurian landscape seems particularly different from the rest. It seems to be grooved, with relatively smooth areas between the hills and valleys. It is called the "weird terrain," or, more technically, the hilly and lineated terrain. No other areas like this are known on Mercury and just a couple have been found on the Moon. The weird terrain is 180° around Mercury from the Caloris Basin (Fig. 9–8), the site of a major meteorite impact. Shock waves from that impact apparently were focused halfway around the planet.

The Mariner 10 mission was a navigational coup not only because it used the gravity of Venus to get the spacecraft to Mercury, but also because scientists and engineers were able to find an orbit around the Sun that brought the spacecraft back to Mercury several times over. Every six months Mariner 10 and Mercury returned to the same place at the same time. As long as the gas jets for adjusting and positioning Mariner functioned, it was able to make additional measurements and to send back additional pictures in order to increase the photographic coverage. On its second visit, for example, in September 1974, Mariner 10 was able to study the south pole and the region around it for the first time. This pass was devoted to photographic studies. The spacecraft came within 48,000 kilometers of Mercury, farther away than the 750-kilometer minimum of the first pass, but the data were still very valuable. On its third visit, in March 1975, it had the closest encounter ever—only 300 km above the surface. Thus it was able to photograph part of the surface with a high resolution of only 50 meters. Then the spacecraft ran out of gas for the small jets that control its pointing, so even though it still passes close to Mercury every few months, it can no longer take clear photographs or send them back to Earth.

In total, Mariner 10 sent back images of 45 per cent of Mercury's surface, with an average resolution of 1 km. We have little idea what the rest of the surface is like.

Figure 9–8 The fractured and ridged planes of the Caloris Basin. It is 1300 km in diameter and is bounded by mountains 2 km high. It is similar in size and appearance to Mare Imbrium on the Earth's Moon and so resulted from the impact of a body tens of kilometers in diameter. It is named Caloris from the Greek word for "hot" because it is almost directly under the Sun when Mercury is at perihelion.

9.3b Infrared Results

Mariner's infrared radiometer gave data that indicate that the surface of Mercury is covered with fine dust, as is the surface of the Moon, to a depth of at least several centimeters. Astronauts sent to Mercury, whenever they go, will leave footprints behind them.

The Mariner 10 mission gave more accurate measurements of the temperature changes across Mercury than had been determined from the Earth. Within a few hundred kilometers of the terminator, the line between Mercury's day and night, the temperature falls from about 775 K to about 425 K, and then drops even lower farther across into the dark side of the planet.

9.3c Results from Other Types of Observations

Closeup spectral measurements showed that Mercury even has an atmosphere, although an all-but-negligible one. It is only a few billionths as dense as the Earth's, so slight that even someone standing on Mercury would need special instruments to detect it. Traces of hydrogen, helium, oxygen, and argon were detected with a spectrometer that operated in the ultraviolet. The presence of helium was a particular surprise, because helium is a light element and had been expected to escape from Mercury's weak gravity within a few hours. So it must be constantly replaced. All these gases probably come from the solar wind.

One more surprise—perhaps the biggest of the mission—was that a magnetic field was detected in space near Mercury. It was discovered on Mariner 10's first pass and then confirmed on the third pass. The field is weak; extrapolated down to the surface it is about 1 per cent of the Earth's. It is believed likely that the field is due to an active dynamo in a still-molten shell of sulfur-rich iron surrounding the now frozen inner core. The magnetic field can persist only with a partially molten core, and the core would have solidified if it contained less than 0.2 per cent sulfur or more than 7 per cent sulfur. The presence of sulfur in this range, more than would be expected for an object at Mercury's current orbit, means that Mercury is partly composed of volatile materials that condensed farther from the Sun than Mercury's present location.

Mariner 10 detected lots of electrons near Mercury. Perhaps they are trapped in some sort of belt by the magnetic field, similar to the Van Allen belts around the Earth. But perhaps they are bound to Mercury for shorter times than electrons trapped by the Earth's magnetic field.

9.4 Mercury Research Rejuvenated

While studies of Mercury since Mariner 10 have continued, until recently progress was evolutionary rather than revolutionary. Mercury's geologic processes continued to be interpreted, and new thinking arose about Mercury's origin, composition, and thermal evolution. Then planetary astronomers were shocked in 1985 by the astonishing discovery of sodium in Mercury's thin atmosphere.

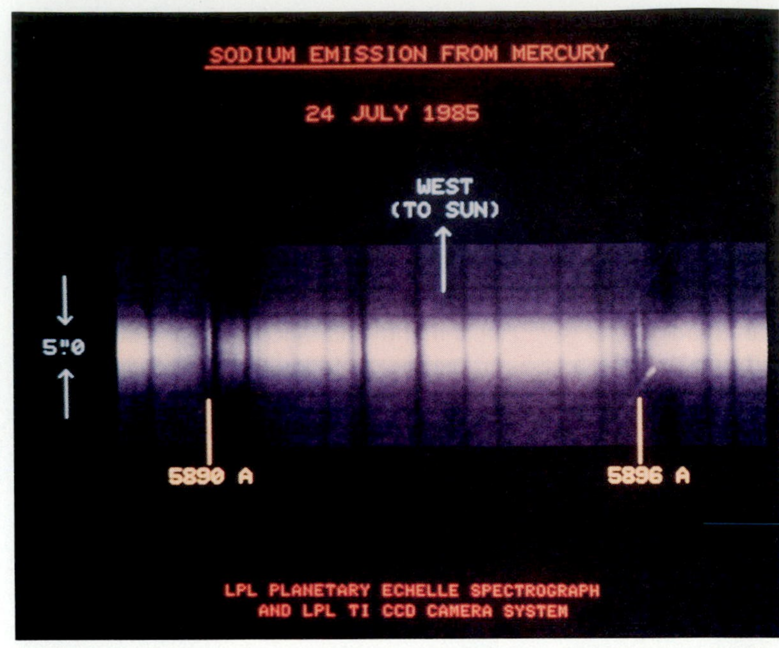

Figure 9–9 A photographic spectrum of the planet Mercury showing an enlargement of the small region of spectrum including the pair of sodium spectral lines known as the D lines. Overall, Mercury's spectrum is that of the Sun, since we see reflected sunlight; thus we see a continuous spectrum with several dark absorption lines. The surprise was that just to the side of each dark absorption line of sodium is a bright emission line marked with the D-line wavelengths 5890 Å and 5896 Å, respectively. The emission lines show the presence of Mercury's atmosphere.

9.4a Mercury's Atmosphere

The sodium (Fig. 9–9) was discovered with the ground-based telescope of the University of Texas, used for daytime observations of the planet at times when it was farthest from the Sun. The spectra of planetary surfaces show mainly the continuum and absorption lines of the solar spectrum reflected to us. Surprisingly, spectra of Mercury showed not only the sodium absorption lines (the yellow "D-lines" that you see when you throw salt into a flame) but also bright, narrow emission lines. The emission lines were shown to be from Mercury itself since they match the changing Doppler shift of Mercury as it orbits the Sun (Fig. 9–10). Some potassium has been discovered in a similar manner.

Mercury's atmosphere is very tenuous, but now we know that there is more sodium in it than any other element—150,000 atoms per cubic centimeter compared with 4,500 of helium and smaller amounts of oxygen, potassium, and hydrogen. The spectrographs aboard Mariner 10 had worked only in the ultraviolet, while the sodium lines are in the visible. It had been thought that the sodium was presumably ejected into Mercury's atmosphere as the result of impact on Mercury's surface of either solar-wind particles (a process called

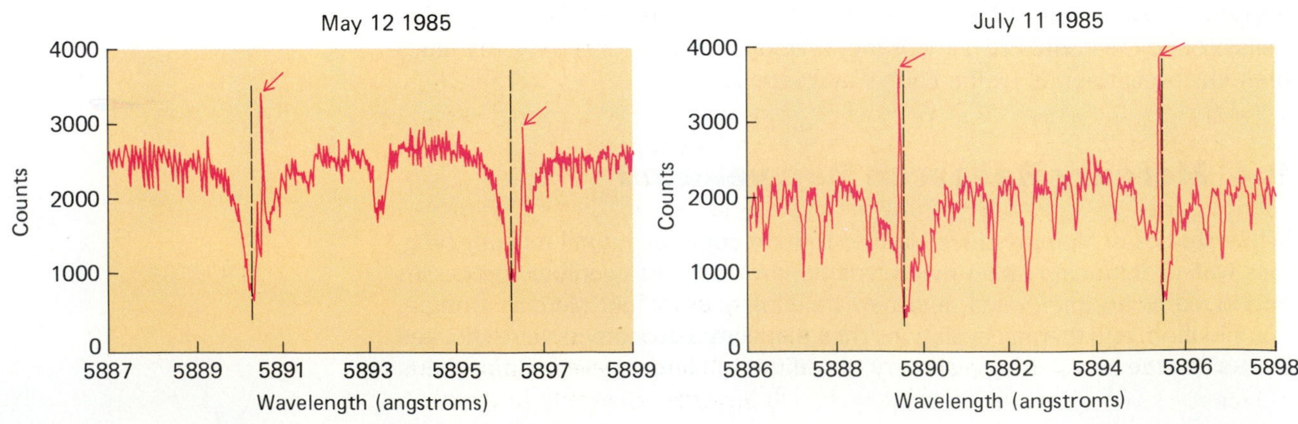

Figure 9–10 Two spectra of the planet Mercury are displayed here as graphs of intensity (the number of counts registered) vs. wavelength. Note that on May 12, 1985, the emission lines (*arrows*) are at longer wavelengths (redward) than the absorption lines, while 2 months later on July 11, 1985, the emission lines are at shorter wavelengths (blueward). This wavelength shift results from the motion of the planet Mercury and its atmosphere with respect to the Sun. For ease in comparison, dotted lines mark the central wavelengths of the absorption lines of sodium, which are relatively broad compared with the emission lines.

"sputtering") or, on a larger scale, meteorites—between 6 and 60 tons per day. We shall see in Chapter 12 that Jupiter's moon Io has a cloud of sodium particles near it; these particles have presumably been ejected from Io by a process similar to the one in force at Mercury, though the particles at Io causing the sputtering would come from Jupiter rather than directly from the solar wind. Calculations show that the sodium particles at Mercury would be continually ejected and so would have to be continually replenished. The situation apparently resembles more the "coma" of a comet (Chapter 17) than a planetary atmosphere like Earth's. New evidence in 1989 that the potassium and sodium are enhanced when Caloris is in view indicate, instead, that Mercury's atmosphere may have diffused up through Mercury's crust.

Starting in 1986, scientists used the Very Large Array set of radio telescopes to map Mercury's temperature structure slightly under its surface (Fig. 9–11). The insulating properties of Mercury's surface make the temperature constant, as the observations verified.

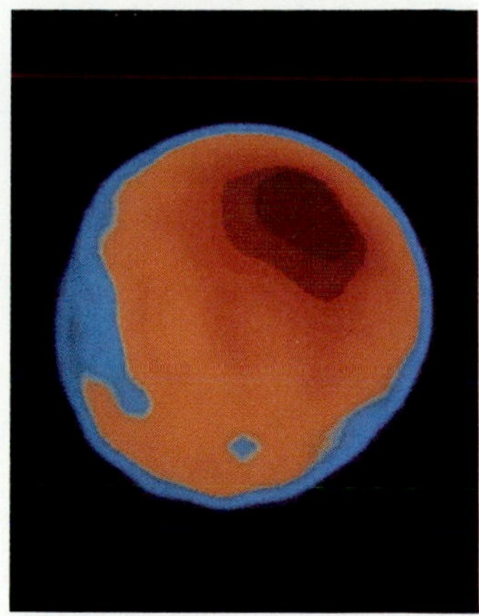

Figure 9–11 An image of Mercury in radio waves, made in 1986 with the Very Large Array, provided resolution of Mercury's subsurface for the first time. At the 5-GHz frequency (six-cm wavelength) of observation, we see to about 70 cm below the surface. The map shows the temperature distribution, which results from the unusual rotation period of Mercury's surface with respect to the Sun. The out-of-round appearance of this image is an artifact of the computer processing.

9.4b Ground-based Radar Observations

During the 1970's and 1980's, extensive radar observations were made from the Goldstone and Arecibo antennas on Earth of the band extending 15° north and south of Mercury's equator. Surface resolution was 12 to 20 km around the equator and 50 to 100 km perpendicularly. The radar observations show especially the height structure across the surface, with a resolution of 150 m, and the roughness of the surface. The radar features have been compared with the features observed during the Mariner 10 flyby, which imaged only about half the planet. The craters, with their floors flatter than the Moon's craters, and the scarps, show clearly on the radar maps. The radar data that show the side of Mercury not imaged by Mariner 10 reveal that it too is dominated by intercrater plains, though it shows that Mercury is basically asymmetric.

9.4c Mercury's History

A generally accepted scenario for Mercury's early years holds that a core like that of the Earth developed and the planet's crust expanded. After the expansion opened paths for magma to flow outward from the interior, the volcanic flooding caused the intercrater plains. As the core cooled, the crust contracted, and the scarps resulted. At about the same time, less extensive volcanic flooding formed the smooth plains. Solar tides on Mercury slowed its original rotation and led to the formation of the large-scale linear features that crisscross perpendicularly.

The sequence is quite different from that of the Moon's early years. The tentative conclusion reached soon after Mariner 10 flew by that Mercury resembled the Moon has now been replaced in the minds of researchers by the idea that Mercury is unique. (Similarly, the first images returned from Mars led researchers to think it looked like the Moon, but later observations and study revealed otherwise.) The distribution of major types of surface features (Fig. 9–12) and most of the tectonic features is very different between the Moon and Mercury. For example, most of the imaged hemisphere of Mercury is a terrain—the intercrater plains—intermediate in albedo and roughness, with lesser but equal amounts of terrain lighter and darker. On the Moon, the bright, rugged lunar highlands dominate. Further, bright lunar rays are associated with well-formed craters, and the younger craters have the most prominent rays. On Mercury, the bright streaks are often not clearly associated with prominent features, and may be remnants of the global system of linear structures, though they may be reactivated by impacts. Many tectonic features have been detected in Mercury's smooth plains; many fewer such features are visible in the anal-

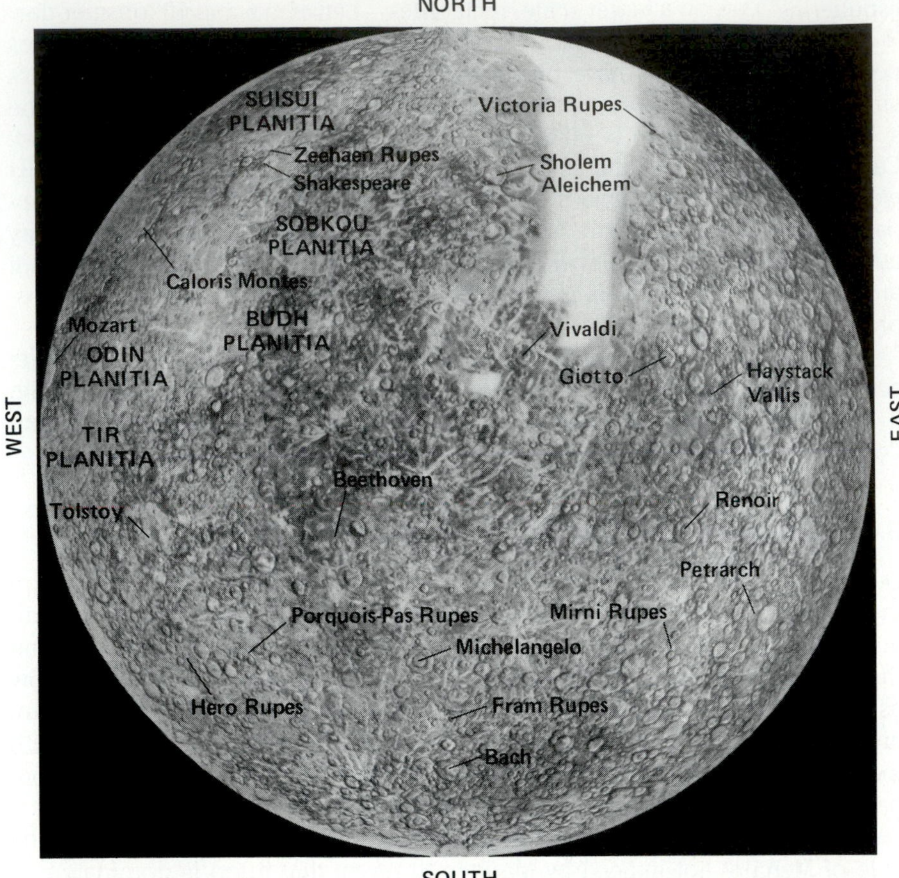

Figure 9–12 An airbrushed map of Mercury made by the U.S. Geological Survey based on the Mariner 10 observations.

ogous lunar maria. (The Moon has wrinkle ridges and rilles, but they seem more local in origin and extent. There is no comparable scarp system.)

The dominant view is that the oldest features on Mercury are 4.2 billion years old. The Caloris basin formed from a giant impact and the scarps formed as Mercury cooled and shrank. The smooth plains then formed by 3.8 billion years ago as lava that erupted from Caloris and other large impact basins. Only

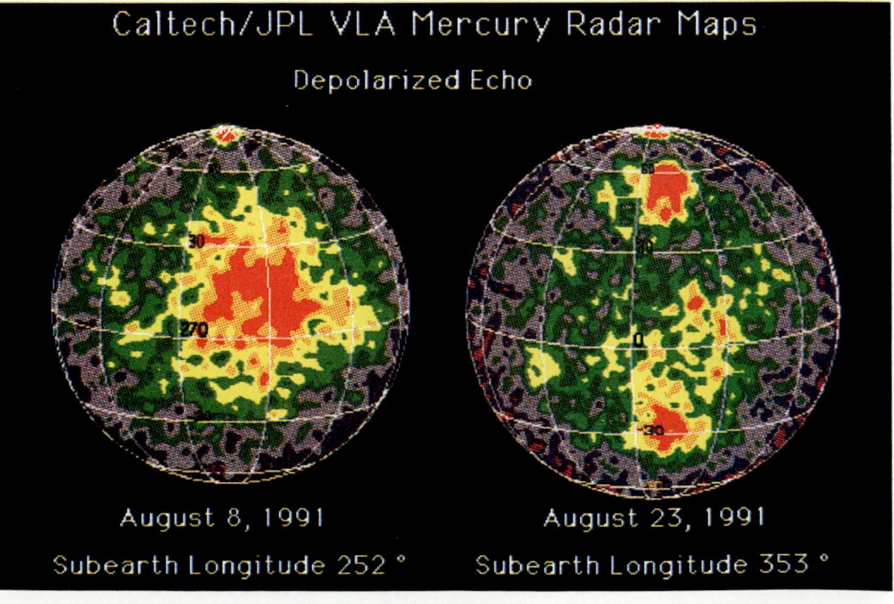

Figure 9–13 On each of these two radar images of Mercury, the brightest echo is precisely at the north pole. Further, the polarization of the radio radiation being returned is unique. Scientists have concluded that the reflecting medium is ice, and probably water ice. It would lie just under the surface. The two newly discovered "basins" in the right-hand image have different polarization properties, not to mention that they lie at warm latitudes.

On these radar maps, made at a wavelength of 3.5 cm with the VLA and Goldstone antennas, the left image shows the entire hemisphere not seen by Mariner 10 and the right image is 90° of longitude later.

if Mercury's craters were produced by an as-yet-undiscovered population of "vulcanoids" (named after the never-found and not-existing planet Vulcan that was long searched for) could the time scale be extended to still more recent dates.

A Mercury conference in 1986 brought researchers together to discuss Mercury's nature. A major new insight was that Mercury may be the fragment of a giant early collision that nearly stripped it to its core, thus accounting for its large proportion of iron.

Radar observations from Earth in 1991 revealed strange echoes that may come from an ice cap (Fig. 9–13). The radar penetrates perhaps 1 or 2 m; perhaps 10 cm of dirt could shield the ice. Since Mercury's axis of rotation is not inclined to its orbit, some regions at the poles could be perpetually shielded from sunlight and therefore be very cold. We can see them because the tilt of the Earth's axis took us high enough to look on at an 11° angle.

Summary and Outline

- Mercury is difficult to observe from the Earth; it is never far from the Sun in the sky
- Radio astronomy (Section 9.1)
    - Surface temperatures
    - Rotation period linked to orbit: 1 day to 2 years
- Albedo (Section 9.2)
    - Low albedo means a poor reflector
    - Mercury has a low albedo, only 6 per cent
- Mass and density (Section 9.2)
    - Mass is 5 1/2 per cent that of Earth
    - Density is similar to Earth's
- Mariner 10 observations (Section 9.3)
    - Photographic results
        - Types of objects: craters, maria, scarps, "weird terrain"
        - Mechanisms: impacts, vulcanism, shrinkage of the crust
    - Results from infrared observations
        - Dust on the surface
        - Temperature measurements
    - Results from other types of observations
        - A small atmosphere
            - Where does the helium come from?
        - A big surprise: a magnetic field, which tells us about the histories of Mercury and of the Earth
- New ground-based observations (Section 9.4)
    - Sodium is the major constituent of Mercury's atmosphere
    - Radar studies show that Mercury is very different from the Moon

Key Words

synchronous rotation, radar, gravitational interlock, sidereal rotation period, solar rotation period, albedo, scarps

Questions

1. Assume that on a given day, Mercury sets after the Sun. Draw a diagram, or a few diagrams, to show that the height of Mercury above the horizon depends on the angle that the Sun's path in the sky makes with the horizon as the Sun sets. Discuss how this depends on the latitude or longitude of the observer.
2. If Mercury did always keep the same side toward the Sun, does that mean that the night side would always face the same stars? Draw a diagram to illustrate your answer.
3. Explain why a day on Mercury is 176 Earth days long.
4. What did radar tell us about Mercury? How did it do so?
5. Given Mercury's measured albedo, if about 1 erg hits a square meter of Mercury's surface per second, how much energy is reflected from that area? (An erg is a unit of energy.)
6. If ice has an albedo of 70–80 per cent and basalt has an albedo of 5–20 per cent, what can you say about the surface of Mercury based on its measured albedo?
7. If you increased the albedo of Mercury, would its temperature increase or decrease? Explain.
8. List those properties of Mercury that could best be measured by spacecraft observations.
9. Think of how you would design a system that transmits still photographs taken by a spacecraft back to Earth and then produces the picture. The system can be simpler than your TV, since still pictures are involved.
10. How would you distinguish an old crater from a new one?
11. What evidence is there for erosion on Mercury? Does this mean there must have been water on the surface?
12. List three major findings of Mariner 10.

A global view of the surface of Venus as imaged with radar from NASA's Magellan spacecraft. The view is centered at 270° east longitude.

Venus ♀

10

Aims: To discuss Venus's atmosphere and surface, and to use this knowledge to improve our understanding of the Earth's structure and atmosphere

Venus and the Earth are sister planets: their sizes, masses, and densities are about the same. But they are as different from each other as the wicked sisters were from Cinderella. The Earth is lush; it has oceans and rainstorms of water, an atmosphere containing oxygen, and creatures swimming in the sea, flying in the air, and walking on the ground. On the other hand, Venus is a hot, foreboding planet with temperatures constantly over 750 K (900°F), a planet on which life seems unlikely to develop. Why is Venus like that? How did these harsh conditions come about? Can it happen to us here on Earth?

Venus orbits the Sun at a distance of 0.7 A.U. Although it comes closer to us than any other planet—as close as 45 million kilometers—we still do not know much about it because it is always shrouded in heavy clouds (Fig. 10–1). Observers in the past saw faint hints of structure in the clouds, which seemed to indicate that these clouds might circle the planet in about 4 days, rotating in the opposite sense from Venus's orbital revolution. The clouds, though, never part to allow us to see the surface.

10.1 The Atmosphere of Venus

Studies from the Earth show that the clouds on Venus are primarily composed of droplets of sulfuric acid, H_2SO_4, with water droplets mixed in. Sulfuric acid may sound strange as a cloud constituent, but the Earth too has a significant layer of sulfuric acid droplets in its stratosphere. However, the water in the lower layers of the Earth's atmosphere, circulating because of weather, washes

Figure 10–1 (*A*) Venus is the bright object over the Earth in this photograph from a space shuttle. (*B*) A crescent Venus, observed with the 2.5-m Mt. Wilson telescope. We see only a layer of clouds.

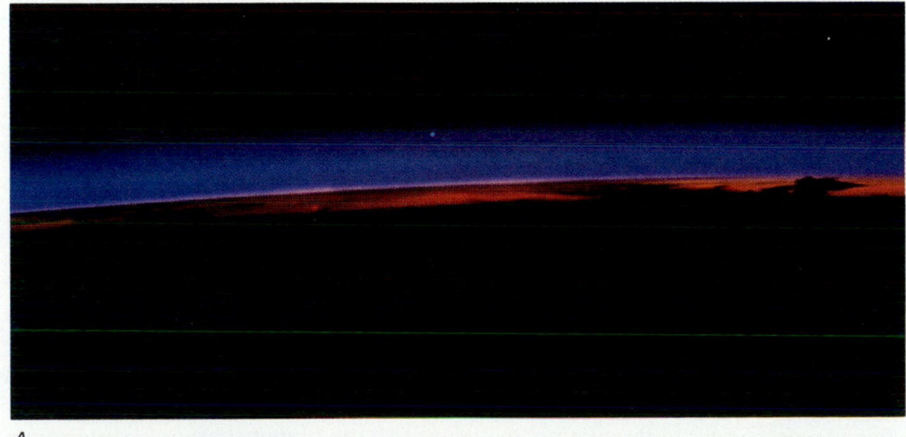

A

B

the sulfur compounds out of these layers, whereas Venus has sulfur compounds in the lower layers of its atmosphere in addition to those in its clouds.

Sulfuric acid takes up water very efficiently, so there is little water vapor above Venus's clouds. Painstaking work conducted at high-altitude sites on the Earth revealed the presence of the small amount of cytherean water vapor above Venus's clouds. This observation was difficult, because the spectral lines of water vapor from Venus were masked by the spectral water vapor lines that arise from the Earth's own atmosphere.

Astronomers often use the adjective *cytherean* to describe Venus. Cythera was the island home of Aphrodite, the Greek goddess who corresponded to the Roman goddess Venus. The adjectives "venusian" and "venerean" are now in increasing use.

Spectra taken on Earth show a high concentration of carbon dioxide in the atmosphere of Venus. In fact, carbon dioxide makes up 96 per cent of Venus's atmosphere (Fig. 10–2); nitrogen makes up almost all the rest. The Earth's atmosphere, by comparison, is mainly nitrogen, with a fair amount of oxygen as well. Carbon dioxide makes up less than 0.1 per cent of the terrestrial atmosphere.

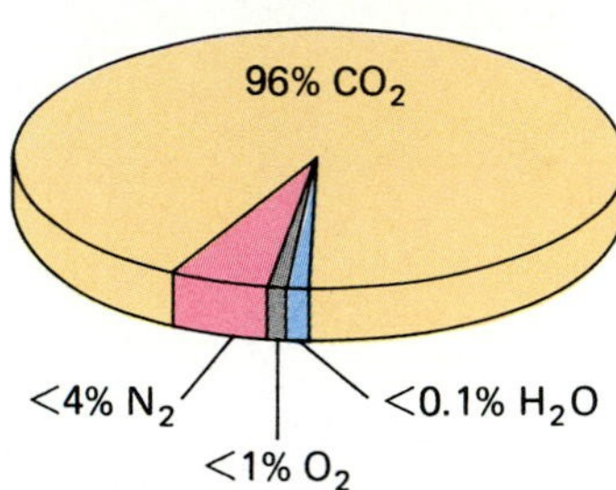

Figure 10–2 The composition of Venus's atmosphere.

The surface pressure of Venus's atmosphere is 90 times higher than the pressure of Earth's atmosphere, as a result of the large amount of carbon dioxide in the former. Carbon dioxide on Earth mixed with rain to dissolve rocks; the dissolved rock and carbon dioxide eventually flowed into the oceans, where they precipitated to form our terrestrial rocks, often with the help of life forms. (Many limestones, for example, have formed from marine life under the Earth's oceans.) If this carbon dioxide were released from the Earth's rocks, along with other carbon dioxide trapped in sea water, our atmosphere would become as dense and have as high a pressure as that of Venus. Venus, slightly closer to the Sun than Earth and thus hotter, had no oceans in which the carbon dioxide could dissolve or life to help take up the carbon. Thus the carbon dioxide remains in Venus's atmosphere.

Also, Venus has probably lost almost all the water it ever had. Since Venus is closer to the Sun than the Earth is, its lower atmosphere was hotter even early on. The result was that more water vapor went into its upper atmosphere, where solar ultraviolet broke it up into hydrogen and oxygen. The hydrogen, a light gas, escaped easily; the oxygen has combined with other gases or with iron on Venus's surface.

Could life have arisen on Venus in an early, hot ocean? Even if it had, it would have been destroyed when the oceans disappeared and Venus heated up so.

10.2 The Rotation of Venus

In 1961, radar astronomy penetrated Venus's clouds, allowing us to determine an accurate rotation period for the planet's surface. Venus, because it comes closer to Earth, is an easier target for radar than Mercury. Venus rotates in 243 days with respect to the stars in the direction opposite from the other planets; such backward motion is called *retrograde rotation,* to distinguish it from forward (direct) rotation. Venus revolves around the Sun in 105 Earth days. Venus's periods of rotation and revolution combine, in a way similar to that in which Mercury's sidereal day and year combine, so that a solar day on Venus corresponds to 127 Earth days; that is, the planet's rotation brings the Sun back to the same position in the sky every 127 days.

The notion that Venus is in retrograde rotation seems very strange to astronomers, since the other known planets revolve around the Sun in the same direction, and all the planets (except Uranus) and most satellites also rotate in that same direction. Because of the conservation of angular momentum (Section 6.4) and because the original material from which the planets coalesced was undoubtedly rotating, we expect all the planets to revolve and rotate in the same sense.

Nobody knows definitely why Venus rotates "the wrong way." One possibility is that when Venus was forming, the planetesimals formed clumps of different sizes. Perhaps the second largest clump struck the largest clump at an angle that caused the result to rotate backwards. Scientists do not like *ad hoc* ("for this special purpose") explanations like this one, since ad hoc explanations often do not apply in general.

10.3 The Temperature of Venus

Though we can't see through Venus's clouds with our eyes, radio waves emitted by the surface penetrate the clouds. The intensity of radiation of a surface depends on its temperature, so long before we landed spacecraft on Venus scientists deduced the temperature of the cytherean surface by studying Venus's radio emission. The surface is very hot, about 750 K (900°F). Infrared penetrates the clouds somewhat, though perhaps not all the way to the surface (Fig. 10–3).

In addition to measuring the temperature on Venus, scientists theoretically calculate approximately what it would be if the cytherean atmosphere allowed all radiation hitting it to pass through it. This value—less than 375 K (215°F)—is much lower than the measured values. The high temperatures derived from radio measurements indicate that Venus traps much of the solar energy that hits it.

The process by which this happens on Venus is similar to the process that is generally—though incorrectly—thought to occur in greenhouses here on the Earth. The process is thus called the *greenhouse effect* (Fig. 10–4); it was suggested for Venus by Carl Sagan of Cornell and James B. Pollack of NASA/Ames.

Sunlight passes through the cytherean atmosphere in the form of radiation in the visible part of the spectrum. The sunlight is absorbed by, and so heats up, the surface of Venus. At the temperatures that result, the radiation that the surface gives off is mostly in the infrared. But the carbon dioxide and other constituents of Venus's atmosphere are together opaque to infrared radiation, so the energy is trapped. Thus Venus heats up above the temperature it would

Figure 10–3 A false-color, infrared image of Venus, taken with an infrared array camera. Blue represents the 1.54-μ image, green is the 1.74-μ image, and red is the 2.3-μ image. The latter two wavelengths penetrate the upper visible cloud level of Venus's atmosphere, while the first wavelength does not. We see the crescent of reflected sunlight that results from Venus's position only 18° from the Sun. The light on the remainder of Venus's disk is thermal emission from deep in the atmosphere or from the surface. The dark features probably arise within the middle cloud level of Venus, 50 km above the surface and 20 km beneath the visible clouds. They rotate in the retrograde direction in about six days, which implies that the wind is blowing at 70 m/sec. These infrared features are lower and rotate more slowly than do the ultraviolet clouds.

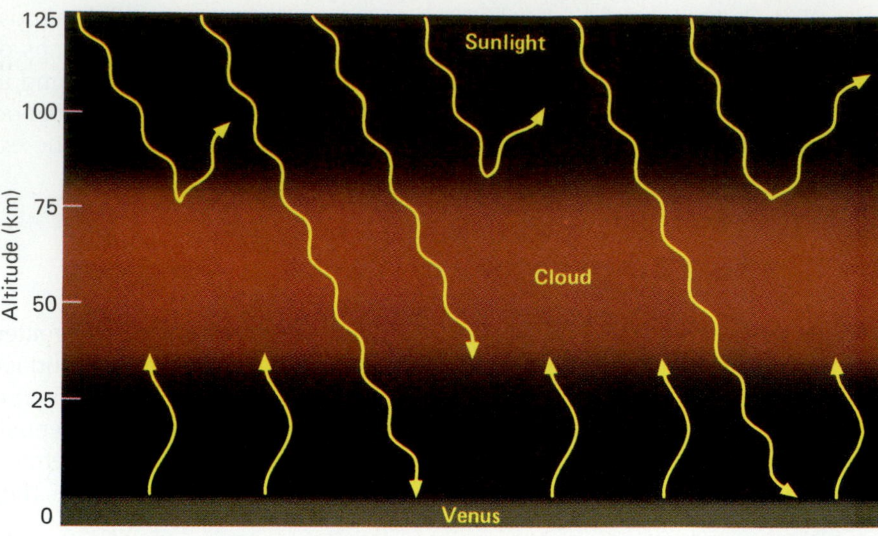

Figure 10–4 Sunlight can penetrate Venus's clouds, so the surface is illuminated with radiation in the visible part of the spectrum. Venus's own radiation is mostly in the infrared. This infrared radiation is trapped, a phenomenon called the greenhouse effect.

have if the atmosphere were transparent; the surface radiates more and more energy as the planet heats up until a balance is struck between the rate at which energy flows in from the Sun and the rate at which it trickles out (as infrared) through the atmosphere (Fig. 10–5). The situation is so extreme on Venus that we say a "runaway greenhouse effect" is taking place there. Understanding such processes involving the transfer of energy is but one of the practical results of the study of astronomy.

Greenhouses on Earth don't work quite this way. The closed glass of greenhouses on Earth prevents convection and the mixing in of cold outside air. The trapping of solar energy by the "greenhouse effect" is a less important process in an actual greenhouse. Try not to be bothered by the fact that the greenhouse in your backyard is not heated by the "greenhouse effect."

If the Earth's atmosphere were to gain enough more carbon dioxide, the oceans could boil, and the carbon in rocks could be released and enter the

Our Earth's atmosphere currently provides about 20°C of greenhouse effect, warming our air enough to make the Earth livable. But we are now adding so much carbon dioxide to our air that we risk additional warming, with dire consequences.

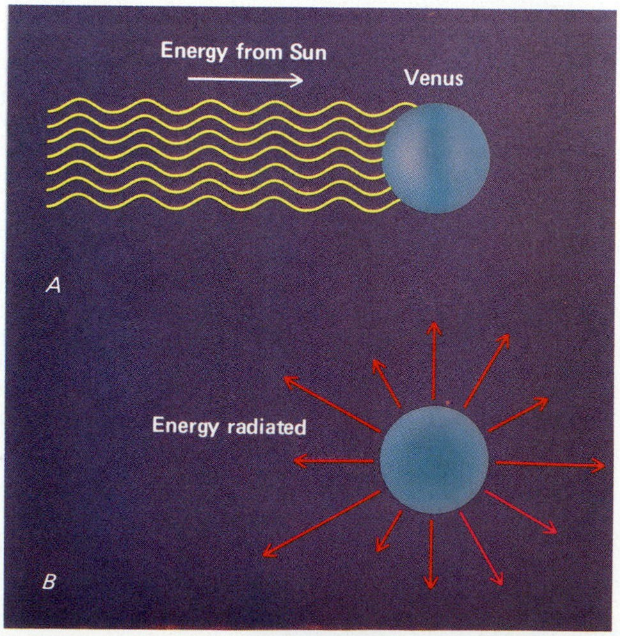

Figure 10–5 Venus receives energy from only one direction, and heats up and radiates energy (mostly in the infrared) in all directions. From balancing the energy input and output, astronomers can calculate what Venus's temperature would be if Venus had no atmosphere. This type of calculation—balancing quantities theoretically to make a prediction to be compared with observations—is typical of those made by astronomers.

atmosphere as carbon dioxide. This would increase the greenhouse effect, which could also become a runaway. We are currently increasing the amount of carbon dioxide in the Earth's atmosphere by burning fossil fuels; Venus has shown us how important it is to be careful about the consequences of our energy use. Even a rise in Earth's temperature of 1–2°C, which is foreseen during the next decades, could have catastrophic consequences on agriculture, sea level (leading to coastal flooding), and climate. We must stop using coal, oil, and gas for fuel very soon.

10.4 Radar Mapping of the Surface of Venus

We can study the surface of Venus by using radar to penetrate Venus's clouds. Radars using huge Earth-based radio telescopes, such as the giant 1000-ft (304-m) dish at Arecibo, Puerto Rico, have mapped a small amount of Venus's surface with a resolution of up to about 2 km. Regions that reflect a large percentage of the radar beam back at Earth show up as bright, and other regions as dark.

10.4a How Radar Mapping Works

Radar involves sending radio waves and studying the echoes. The point of Venus nearest Earth reflects the signal back a fraction of a second before points farther away (Fig. 10–6). In the instants following, the incoming waves spread out in concentric circles. Thus at any given time, we receive the echo from a circle around Venus.

We make use of the planet's rotation to give further information. Because of the rotation, one side of the planet is approaching relative to the other. Thus the Doppler shift gives a different frequency to the signal reflected from each

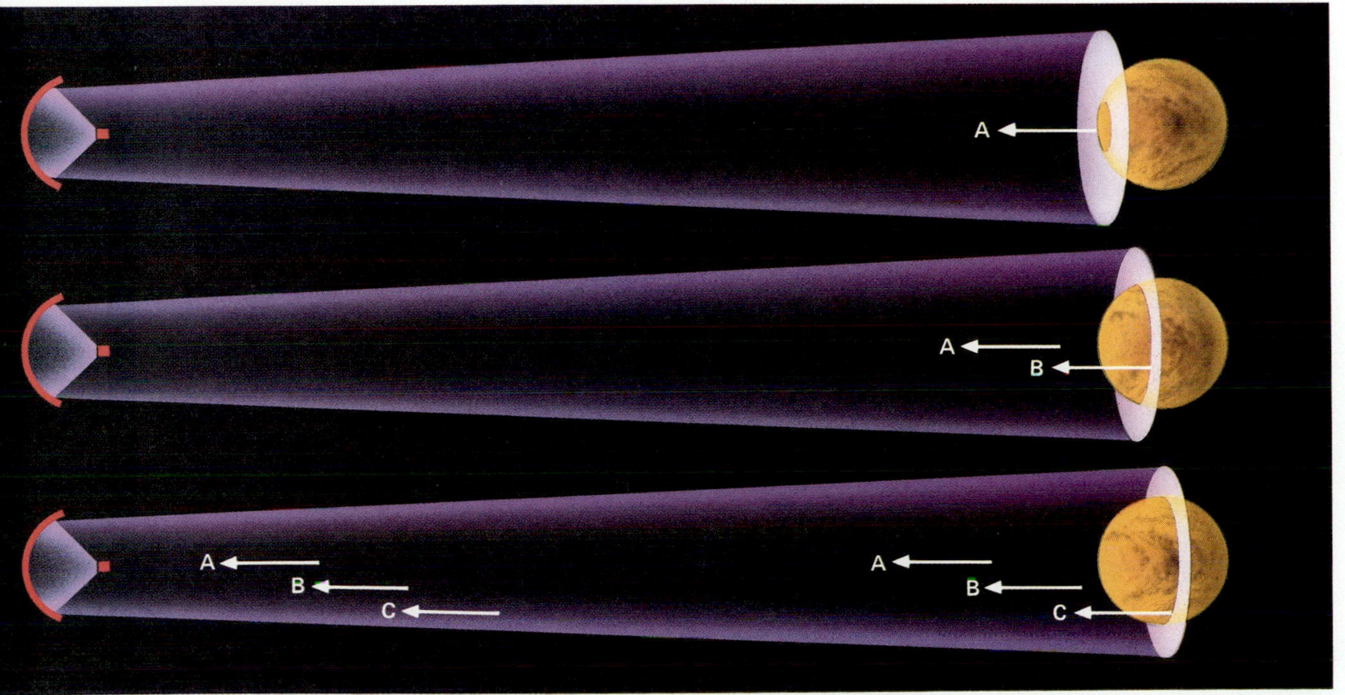

Figure 10–6 A pulse of radio waves sent from the telescope on Earth, shown at left, is reflected from Venus, shown at right. The radar pulse first hits the point on Venus nearest to the Earth (*A*), and then spreads as time goes on in concentric circles (*B* and *C*). The reflected pulse is thus spread out over a longer time interval than was the transmitted pulse. By observing the time of arrival of each part of the reflected pulse back at the Earth, we can tell from which concentric circle that part of the returned pulse was reflected.

vertical line (parallel to its rotation axis) on Venus (Fig. 10–7). The circle from the time information and the line from the frequency information intersect at two points. So even though the beam of outgoing radio waves is larger than all of Venus, we can study small regions of its surface. (We must use an interferometer to distinguish between the northern and southern of the two points whose echoes are blended.) From the strength of the echo and how its strength varies with time we can tell the roughness and altitude of the region of Venus's surface reflecting the signal.

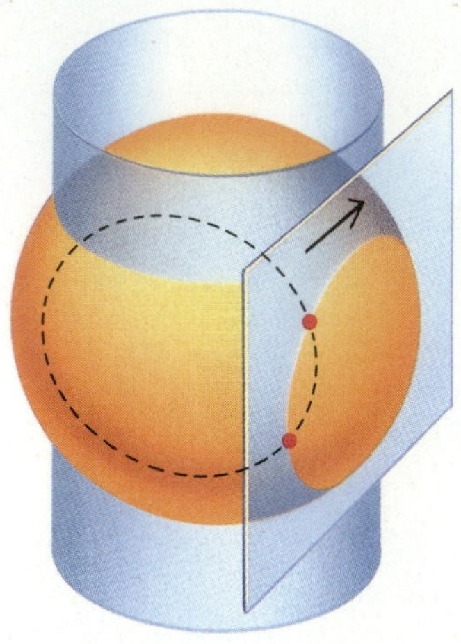

Figure 10–7 Everything the same distance from the axis of rotation of a planet—i.e., the surface of the cylinder—rotates at the same velocity. Our radar signal from Earth travels parallel to the plane at right in the direction shown by the arrow. Any signal reflected by the part of Venus that intersects the plane has the same Doppler shift. We see only the side of the circle shown that is facing us; from our vantage point it looks like a straight line. Thus by observing the echo at a given frequency (the original frequency transformed by a Doppler shift), we define a line from top to bottom across the surface of the planet. This line intersects the circle shown here and in Fig. 10–6 at the two marked points.

10.4b Earth-Based Radar Maps

The Earth-based radar maps of Venus (Fig. 10–8) show large-scale surface features, some very rough and others relatively smooth, similar to the variation we find of surfaces on the Moon. Many round areas that are probably craters have been found, ranging up to 1000 km across. Most have probably been formed by meteoritic impact.

From Venus's size and from the fact that its mean density is similar to that of the Earth, we concluded that its interior is also probably similar to that of the Earth. This meant that we expected to find volcanoes and mountains on Venus, and that venusquakes probably occur too. A bright area near a large plain has been called Maxwell; we will see later on that Maxwell turns out to be a huge mountain.

The region Alpha, long known for its very high radar reflectivity, is circular and 1100 km across. It appears to have no counterpart on Earth. A central dark region may be a volcano.

A huge rise, Beta Regio, has a base over 750 km in radius and a depression 40 km in radius at its summit. Radar observations show that it too is volcanic. Mountains on it called Rhea Mons and Theia Mons have long tongues of rough material—apparently lava flowing from the summit—extending 500 km in an irregular fashion. Theia Mons is at the junction of at least three major rift zones. A cluster of some 10 smaller peaks resembles the configuration of clusters of volcanoes on Earth.

A long trough at Venus's equator, 1500 km in length, seems to resemble the Rift Valley in East Africa, the Earth's largest canyon.

The quality of the Earth-based radar studies continues to be improved as shorter wavelengths and more powerful transmitters are used. Resolutions better than 2 km are now available for about 7 per cent of Venus's surface, and give more and more evidence for volcanic features.

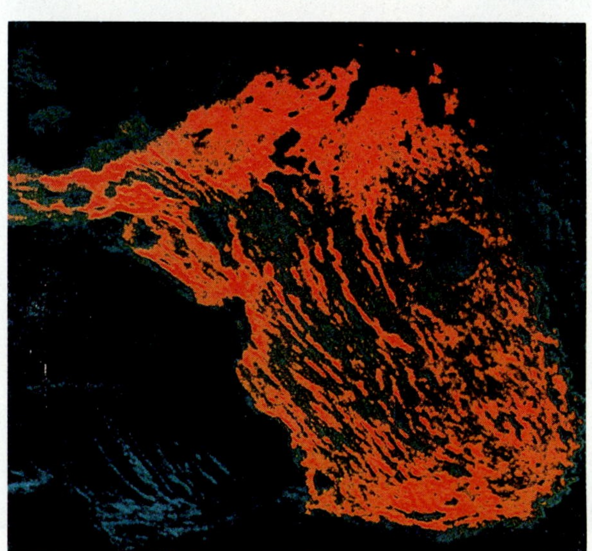

Figure 10–8 The Arecibo Observatory's high-resolution radar image of Maxwell Montes, with a resolution of about 2 km. The summit caldera is named Cleopatra. Brighter areas show regions where more power bounces back to us from Venus, which usually corresponds to rougher terrain than darker areas.

10.5 Early Spacecraft Observations

Venus was an early and has been a steady target of both American and Soviet space missions since the early 1960's. In 1970, the Soviet Venera 7 spacecraft radioed 23 minutes of data back from the surface of Venus before it succumbed to the high temperature and pressure. Two years later, the lander from Venera 8 survived on the surface of Venus for 50 minutes. Both missions confirmed the Earth-based results of high temperatures, high pressures, and high carbon dioxide content.

The United States spacecraft Mariner 10 took thousands of photographs of Venus in 1974 as it passed by en route to Mercury. It went sufficiently close and its imaging optics were good enough that it was able to study the structure in the clouds with resolution of up to 100 meters (Fig. 10–9). The structure shows only when viewed in ultraviolet light. We could observe much finer details from the spacecraft than from Earth.

Figure 10–9 The circulation of the clouds of Venus, photographed from Mariner 10 in the ultraviolet near 3650 Å. The contrast has been electronically enhanced so that small actual differences in contrast are made apparent. The general character of the features has remained constant since imaging began in 1978. The clouds on Venus are produced by reflections from small particles of unknown composition that are embedded within a thick layer of sulfuric acid particles. They are thus different from terrestrial clouds, in spite of their similar appearance.

The clouds appear as long, delicate streaks, like terrestrial cirrus clouds. In the cytherean tropics the clouds also show a mottling, which suggests that convection, the boiling phenomenon, is going on. A big "eye" of convection is visible just downwind from the point at which the Sun is overhead and, therefore, at which its heating actions are at the maximum. Having an "eye" downwind from the subsolar point is a perfect example of what atmospheric scientists would expect, since hot air rises and starts the process of convection. Strong winds blow the clouds at these upper levels around the planet at 300 km/hr, as rapidly as the jet stream blows on Earth but over a much wider area.

No magnetic field was detected. This may indicate either that Venus does not have a liquid core or that the core does not have a high conductivity.

Such studies of Venus have great practical value. The better we understand the interaction of solar heating, planetary rotation, and chemical composition in setting up an atmospheric circulation, the better we will understand our Earth's atmosphere. We then may be better able to predict the weather and discover jet routes that would aid air travel, for example. The potential financial return from this knowledge is enormous: it would be many times the investment we have made in planetary exploration.

Soviet spacecraft that landed on Venus in 1975 survived long enough to send back photographs. The single photograph that the Venera 9 lander took in the 53 minutes before it succumbed to the tremendous temperature and pressure was the first photograph ever taken on the surface of another planet. It showed a clear image of sharp-edged, angular rocks. This came as a surprise to some scientists, who had thought that erosion would be rapid in the dense cytherean atmosphere and that the rocks should therefore have become smooth or disintegrated into sand. Three days later, the Venera 10 lander sent data for 65 minutes. The rocks at its site were not sharp; they resembled huge pancakes, and between them were sections of cooled lava or debris of weathered rock.

The Veneras measured the surface wind velocity to be only 1 to 4 km/hr. The low wind speed makes it seem likely that erosion is caused not by sand blasting but by melting, temperature changes, chemical changes (which can be very efficient at the high temperature of Venus), and other mechanisms.

10.6 Venus's Surface and Atmosphere from Pioneer

The United States and the Soviet Union both sent spacecraft to Venus in 1978. NASA's Pioneer Venus 1, the Orbiter, was the first spacecraft to go into orbit around Venus, and it has been making observations ever since. Its elongated elliptical orbit sometimes has brought the spacecraft as low as 150 km in

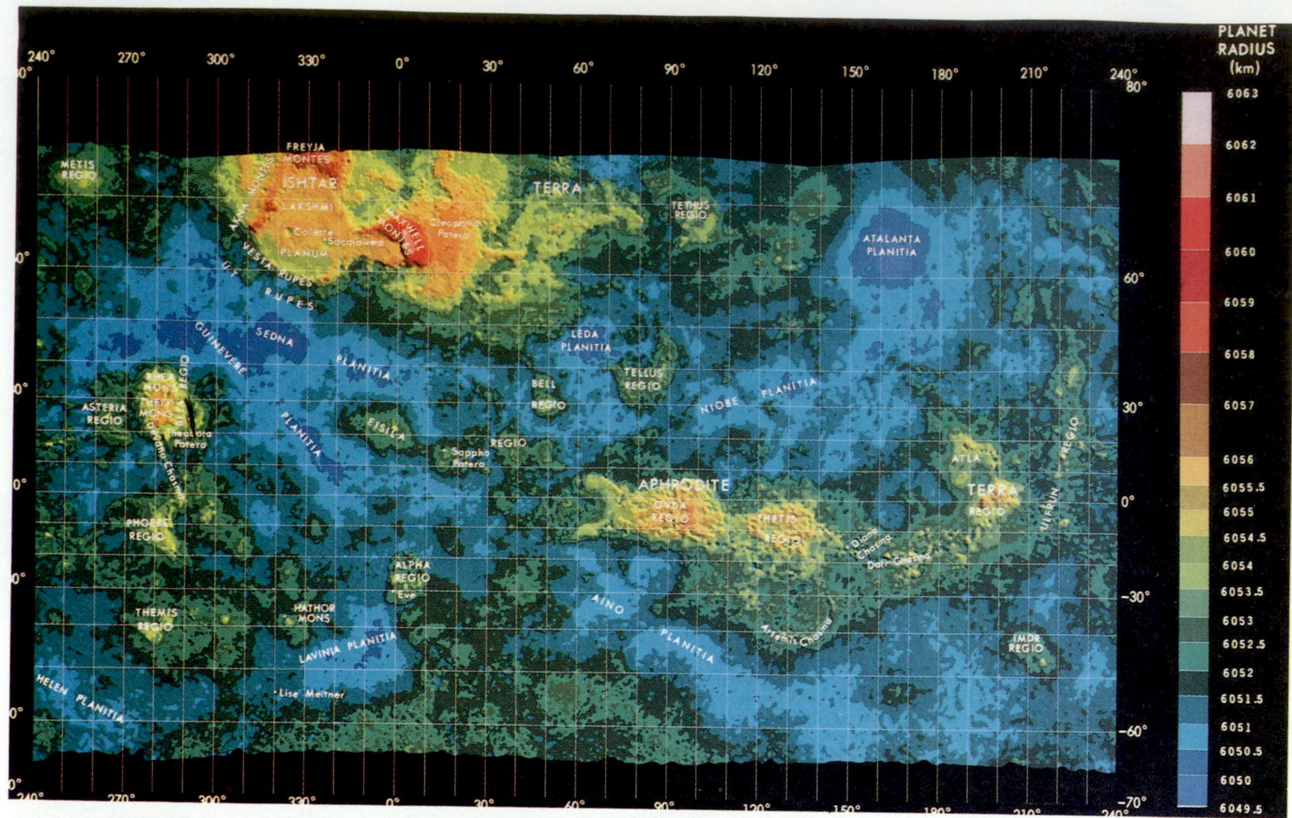

Figure 10–10 A radar map of Venus from the Pioneer Venus Orbiter. Most of Venus is covered by a rolling plain, shown in green and blue. The highlands (yellow and brown contours) sit atop the plain, like continents. Aphrodite Terra is half as big as Africa. Ishtar Terra is the size of Australia, but looks relatively large in this Mercator projection. The lowlands, although resembling Earth's ocean basins, cover only 16 per cent of the planet. (Experiment and data: MIT; maps: U.S. Geological Survey; NASA/Ames spacecraft)

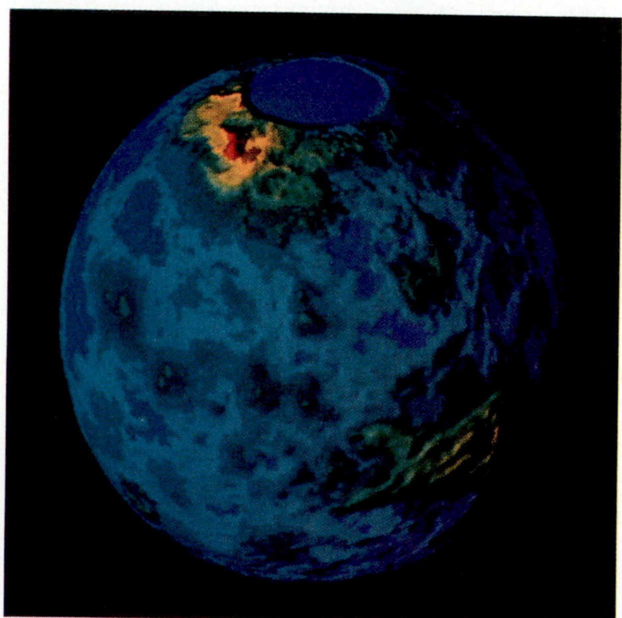

Figure 10–11 A radar map of Venus from the Pioneer Venus Orbiter, in a global projection. Artificial illumination has been added. Ishtar is at top left, and Aphrodite is at right.

Figure 10–12 A radar map of Venus from the Pioneer Venus Orbiter, in a global projection. Artificial illumination has been added. Aphrodite is at left.

altitude; it extends higher than 65,000 km. Its dozen experiments included cameras to study Venus's weather by photographing the planet regularly in ultraviolet light.

10.6a Venus's Topography

The spacecraft also carried a small radar to study the topography of Venus's surface. The resolution was not as good as that of the best Earth-based radar studies of Venus, but the orbiting radar mapped a much wider area. It mapped over 90 per cent of the surface with a resolution typically better than 100 km.

From the radar map (Figs. 10–10 to 10–12), we now know that 60 per cent of Venus's surface is covered by a rolling plain, flat to within plus or minus 1 km. Only about 16 per cent of Venus's surface lies below this plain, a much smaller fraction than the two-thirds of the Earth covered by ocean floor (Fig. 10–13).

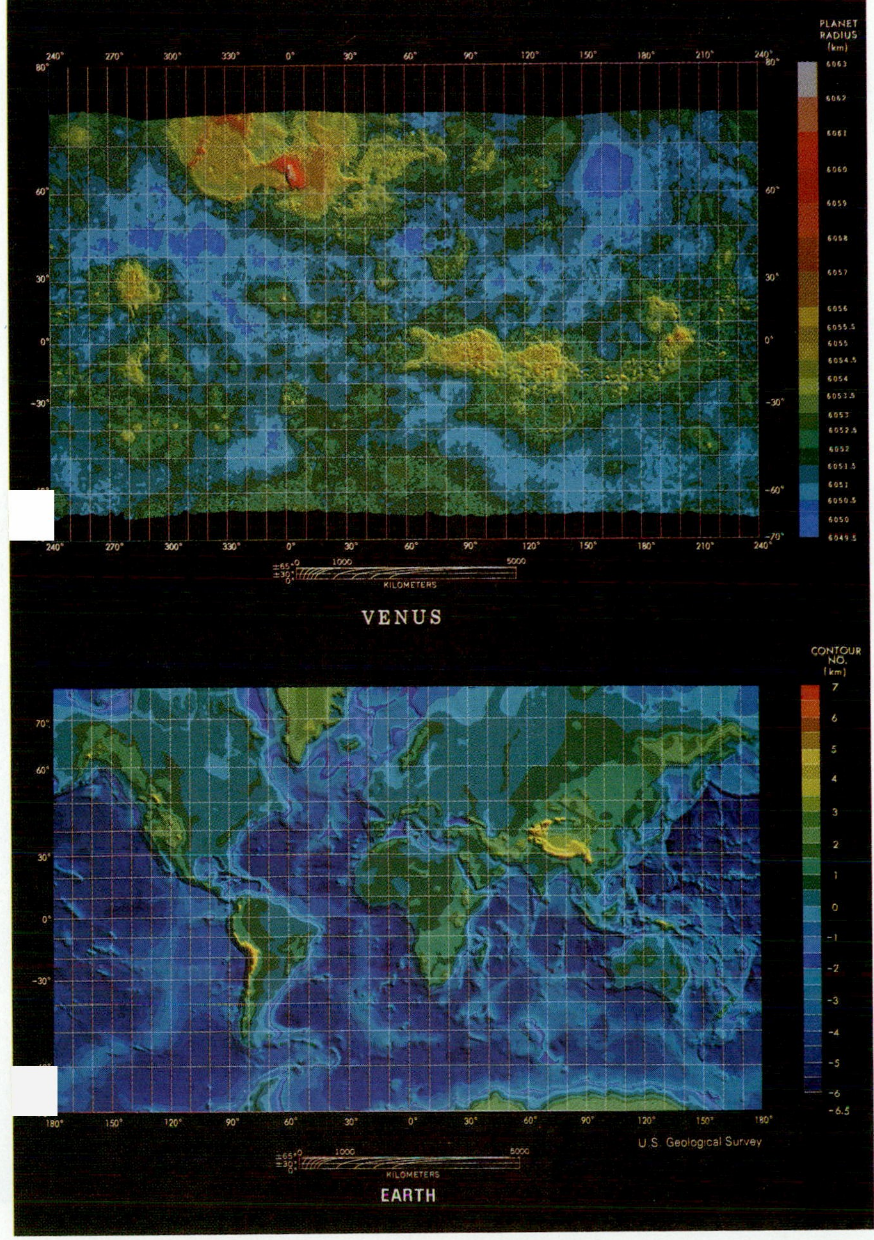

Figure 10–13 Venus's surface, based on the Pioneer Venus radar map compared with Earth's surface at the same resolution (50–100 km). Two continents exist on Venus: Aphrodite Terra, which is as large as half the size of Africa, and Ishtar Terra, which is comparable in scale with the continental United States or Australia.

Sixty per cent of Venus's surface is covered with a huge, rolling plain. Only about 16 per cent of Venus's surface is covered with lowlands; in comparison, over two-thirds of Earth's surface is covered with oceans. Mid-ocean ridges, a sign of spreading plates, are not visible on Venus, though this comparison shows that they would be detectable on Earth at this resolution.

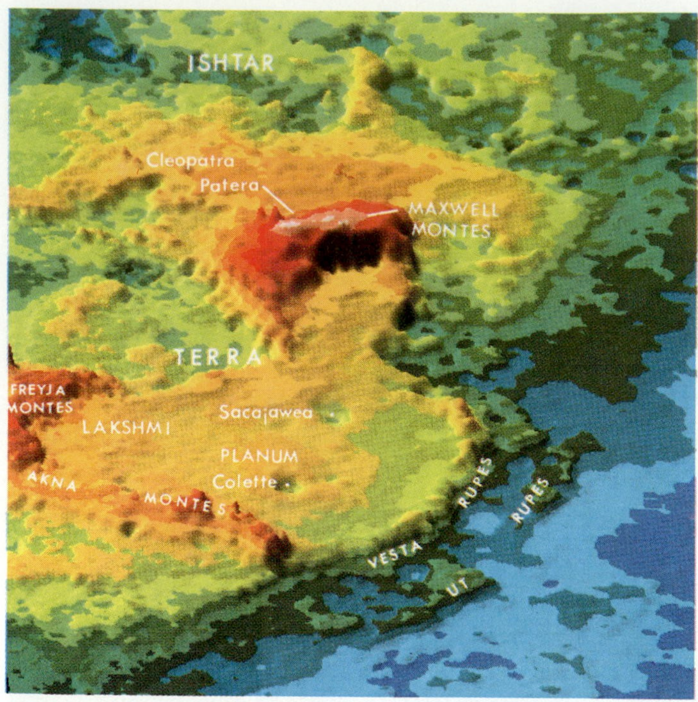

Figure 10–14 Part of Ishtar Terra, with the giant Maxwell Montes, from the Pioneer Venus Orbiter radar.

Two large features, the size of small Earth continents, extend several kilometers above the mean elevation. A northern continent, Ishtar Terra, is about the size of the continental United States. A giant mountain on it known as Maxwell Montes is 11 km high (Fig. 10–14), 2 km taller than Earth's Mt. Everest stands above terrestrial sea level. Maxwell had formerly been known only as a bright spot on Earth-based radar images. Ishtar's western part is a broad plateau, about as high as the highest plateau on Earth (the Tibetan plateau) but twice as large. A southern continent, Aphrodite Terra (Fig. 10–15), is about twice as large as the northern continent and is much rougher.

A smaller elevated feature, Beta Regio, on the basis of Pioneer Venus and terrestrial radar results and on the results of the Soviet landers, appears to be a pair of volcanoes. The fourth highland feature, Alpha Regio, has rough

Figure 10–15 Aphrodite Terra from the Pioneer Venus Orbiter radar.

Focus On

Box 10.1 Naming the Features of Venus

The two main continents on Venus are named Aphrodite Terra and Ishtar Terra. Aphrodite, the Greek goddess of love, was the equivalent of the Roman goddess Venus. Ishtar was the Babylonian goddess of love and war, daughter of the Moon and sister of the Sun.

Major features of Venus have been named after mythical goddesses, minor features after other mythical female figures, and still smaller circular features after famous women, like the physicist Lise Meitner.

terrain that may resemble the western United States. Elsewhere on the surface, a giant canyon is 1500 km long, 5 km deep, and 400 km wide. It is the largest in the solar system and may be left over from an earlier stage of Venus's geological evolution.

Several circular features on the plains may be craters. It is important to find which are craters and which aren't as part of the effort to find how old the surface of Venus is. Identifying many of the circular features with craters could indicate that the plains are geologically old, though there are many fewer craters than there are on the Moon, Mercury, or Mars. As we have seen in Chapter 7, the Earth's current surface is geologically young, formed by moving "plates" as part of continental drift or "plate tectonics." No reliable estimates of the ages of the surface of Venus exist.

10.6b Venus's Atmosphere

The Pioneer Orbiter showed changes in the clouds of Venus over time (Fig. 10–16). Venus's wind patterns undergo long-term patterns of change. The jet-stream pattern seen at mid-latitudes when Mariner 10 flew by in 1974 had given way to a pattern of cloud and wind acting like a solid body by the time the Pioneer Venus Orbiter arrived. The upper atmosphere moves westward around the planet at tremendous speed, circling the planet every 4 days. The high-altitude haze layer (over the main cloud layer) that appears and disappears over a period of years contains tiny droplets of sulfuric acid.

The Orbiter has also provided evidence that volcanoes may be active on Venus. The abundance of sulfur dioxide it found on arrival was far higher than the upper limits of previous observations, and then dropped by a factor of 10

Figure 10–16 Four photographs of Venus taken over a two-month span. They show relatively dark equatorial bands with many small features, probably from convection, superimposed. Polar bands of clouds are relatively bright. (*A*) January 14, 1979. The bright polar rings are obvious. (*B*) February 11, 1979. A Y-shaped figure in the clouds covers much of the disk. (*C*) February 14, 1979. Belts near the equator are prominent. (*D*) March 3, 1979.

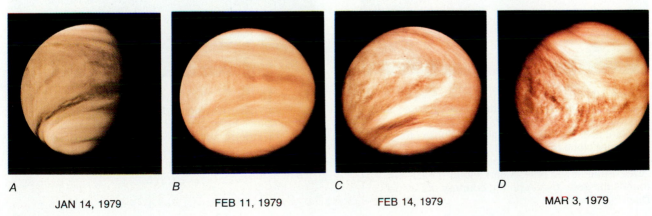

A JAN 14, 1979

B FEB 11, 1979

C FEB 14, 1979

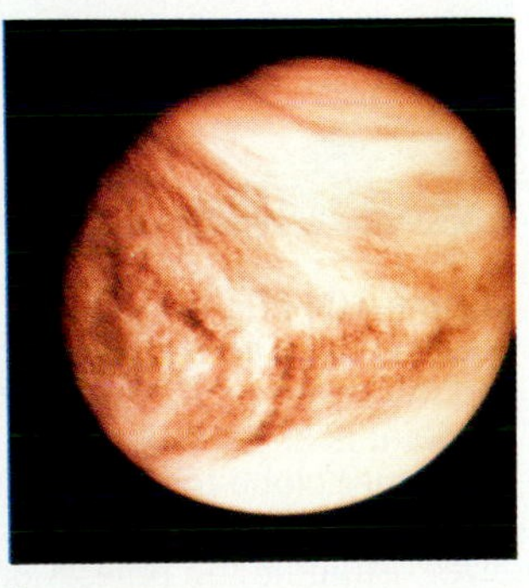

D MAR 3, 1979

in the next five years. Over the same five-year period, the haze dropped drastically, as it had about 20 years earlier. The effect could come from eruptions at least ten times greater than Mexico's El Chichón in 1982. In 1989, a report of radio-astronomy studies from Earth confirmed the diminution of sulfur dioxide and added an apparent decline of sulfuric acid vapor.

Pioneer Venus 2 arrived at Venus at about the same time as Pioneer Venus 1, and carried several spacecraft together on a basic "bus." The bus and its four probes all penetrated Venus's atmosphere and descended to the surface. The Pioneer probes found that high-speed winds at the upper levels are coupled to other high-speed winds at lower altitudes. The lowest part of Venus's atmosphere, however, is relatively stagnant. Winds in this lower half of the atmosphere do not exceed about 18 km/hr. This lower half is much denser than the upper half, however, so the amount of momentum in the lower- and upper-level winds may be the same.

Now that the Pioneer Venus results have enabled us to understand the greenhouse effect on Venus, we can much better understand the effect of adding carbon dioxide from burning fossil fuels to Earth's climate. We have already increased the amount of carbon dioxide in Earth's atmosphere by 15 per cent; some scientists have predicted that it may even double in the next 50 years, which could result in a worldwide temperature rise of a few °C. The effects on people, climate, land use, agriculture, and so on, could be major.

The probes found that carbon dioxide makes up 96 per cent of the atmosphere. They found less water vapor but more forms of sulfur than had been expected in Venus's lower atmosphere. This is particularly important since the presence of carbon dioxide alone is not sufficient to cause the greenhouse effect. Carbon dioxide's spectrum has gaps in its blockage of infrared that would let out too much energy at some infrared wavelengths. Thus Venus's carbon dioxide absorbs only 55 per cent of the solar heat entering Venus's atmosphere. But comparison of computer models with the Pioneer Venus results tells us that the gaps are plugged by water vapor, sulfuric acid droplets with other types of cloud and haze particles, and sulfur dioxide. An infrared camera studied the atmosphere's temperature (Figs. 10–17 and 10–18).

Venus's clouds, largely made of sulfuric acid droplets, start 48 km above its surface and extend upward about 30 km in three distinct layers. They extend much higher than terrestrial clouds, which are made of water droplets and which rarely go above 10 km. (The upper layers of the Earth's atmosphere,

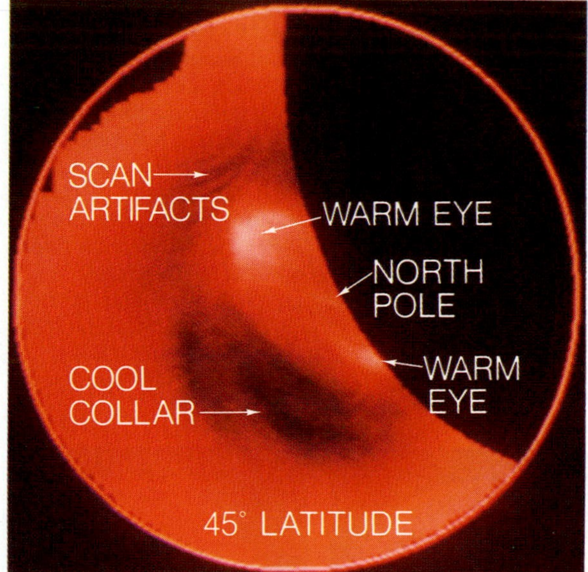

Figure 10–17 An infrared polar view from the Pioneer Venus Orbiter. The north pole is at the center, and the outer circle represents 45° latitude. The subsolar point is at the bottom. Brighter regions correspond to warmer cloud-top temperatures in these measurements at a wavelength of 11 microns. Two bright "eyes" straddle the pole. A dark cloud "collar" surrounds the pole. (NASA photo, courtesy of David J. Diner)

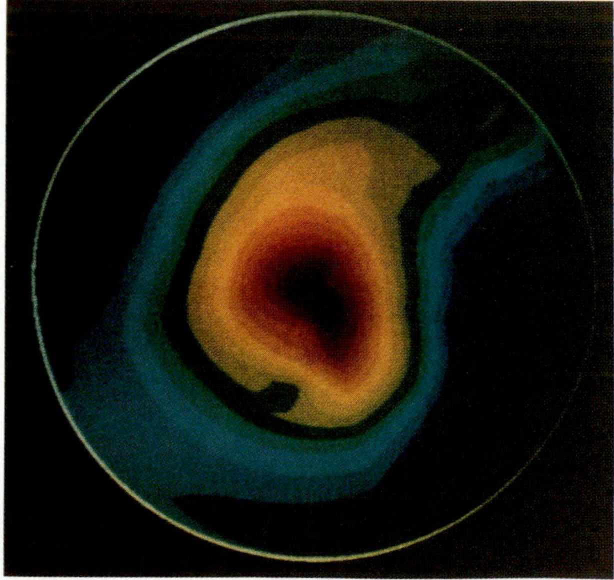

Figure 10–18 A 72-day average of 11-μ infrared radiation from Venus. The north pole is at the center. The radiation comes from about 70 km above the surface, just above the clouds, and clearly shows a circumpolar collar. The coldest temperature (blue) is about −60°C, and the warmest temperature (brown) is about −15°C.

above the clouds, also contain sulfuric acid droplets but are thin enough to be transparent, unlike Venus's layers of sulfuric acid droplets.) Below the clouds, the atmosphere is fairly free of dust and cloud particles.

Pioneer Venus observed hydrogen and oxygen around Venus (Fig. 10–19). These clouds of gas extend far into space.

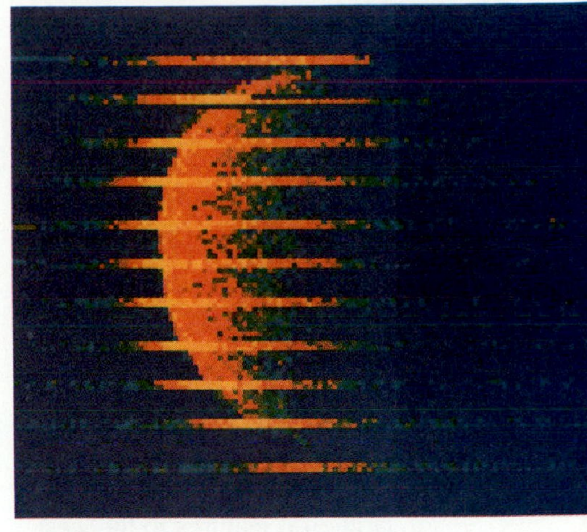

Figure 10–19 A false-color view of Venus in far ultraviolet light from atomic oxygen. The oxygen is in Venus's upper atmosphere. The stripes were caused by the periodic measurements of hydrogen, which extends thousands of kilometers into space around Venus.

10.7 Venera and Vega Studies

The two 1978 Soviet missions, Veneras 11 and 12, each consisted of a flyby, which acted as a communications relay, and a lander. They reached Venus shortly after the American missions, and the landers parachuted softly to the surface. On the way, they measured the chemical composition of the atmosphere. They did not transmit any pictures.

Both American and Soviet missions, from the radio signals they detected, reported lightning at a frequent rate. The Soviet landers detected up to 25 pulses of energy per second between 5 and 11 km above the surface, later confirmed by the Pioneer Venus Orbiter. The lightning is concentrated over Beta Regio and the eastern part of Aphrodite Terra. This is strong evidence that those regions are sites of active volcanoes. On Earth, it is known that the rubbing together of dust and ash ejected by volcanoes sometimes generates static electricity that produces lightning. But other models are possible for producing the lightning.

Two Soviet spacecraft, Veneras 13 and 14, arrived at Venus in 1982. They carried cameras that showed a variety of sizes and shapes of rocks (Figs. 10–20 and 10–21). Venus's sky is orange. The landing sites were chosen in consultation with American scientists on the basis of Pioneer Venus data, the first example of such Soviet-American cooperation on planetary exploration.

The landers carried devices that measured the composition of the surface of the planet. Samples were drilled and brought into a low-pressure container at an Earth-like temperature. X-rays emitted when the samples were irradiated revealed which elements are present. (A "fluorescent" process occurs when radiation at one wavelength is given off in response to radiation at another, so the devices are known as x-ray fluorescence spectrometers.) The gamma-ray

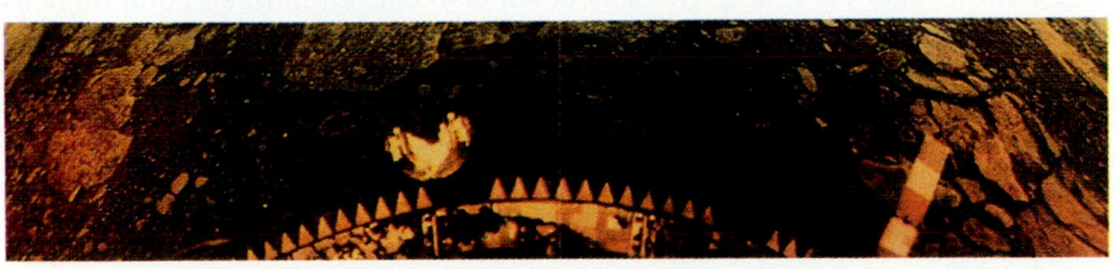

◀ **Figure 10–20** The view from Venera 13, showing a variety of sizes and textures of rocks. The spacecraft landed in the foothills of a mountainous region south of Venus's equator and below the region Beta. The spacecraft survived for 127 minutes.

The camera first looked off to the side, then scanned downward as though looking at its feet, and then scanned up to the other side. As a result, opposite horizons are visible as slanted boundaries at upper left and right.

▲ **Figure 10–21** The view from Venera 14, showing flat structures without the smaller rocks of the other site. The flat structures may be large flat rocks or may be fine material held together. The site is 1000 km east of the Venera 13 site and at a lower elevation. The chemical composition resembles that of basalts found on the Earth's ocean floor and the Moon's maria. Venera 13, on the other hand, found basalts more typical of the Earth's continental crust and the Moon's highlands.

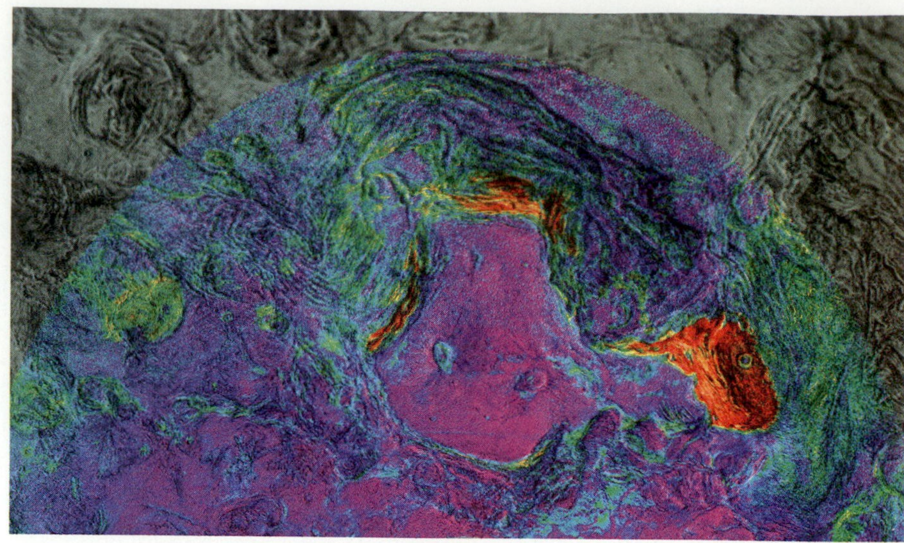

Figure 10–22 The Venera 15 and 16 radar images have been used together with Pioneer Venus altimetry (height measurements) by the U.S. Geological Survey to make shaded relief maps. This Mercator projection of the Lakshmi Planus region of Venus has been overlaid with color-coded radar brightness based on Arecibo observations. The purples and blues are relatively smooth areas, while the greens are rougher areas. The brightest areas, reddish in color here, show surface materials with high radar reflectivity. The large oval features to the west of Lakshmi may be related to large-scale vulcanism.

spectrometers on earlier Venera spacecraft had been able to measure only naturally radioactive isotopes of uranium, thorium, and potassium. The chemical makeup of the regions of Venus where the new spacecraft landed is similar to that of basalt—volcanic rock. Since they had landed in regions near Beta Regio that had the overall topography of a volcanic region, the result indicates that the same chemical processes are at work on both Venus and the Earth.

The Soviet Veneras 15 and 16 reached Venus in 1983. They carried radars that used a relatively new method of gaining resolution, called *synthetic-aperture radar.* In this method, the radar images are recorded over a period of time while the spacecraft moves. The views from these different aspects are assembled during the data reduction, giving resolution equivalent to a radar of much larger aperture; the larger aperture has been "synthesized."

The spacecraft's high resolution over a wide region has shown, for example, the geological folds around Maxwell Montes, the highest point on Venus (Fig. 10–22). Over 100 impact craters with diameters greater than 15 km were found. The number indicates that impact craters remain visible for 150 million to 1 billion years on Venus. The spacecraft also carried infrared equipment to study the atmosphere. Veneras 15 and 16 are no longer sending back data.

In 1985, a pair of joint Soviet-French probes were carried to Venus by Soviet rockets en route to Halley's Comet.

10.8 Magellan at Venus

Our space results, coupled with our ground-based knowledge, show us that Venus is even more different from the Earth than had previously been imagined. Among the differences are Venus's slow rotation, its one-plate surface, the absence of a satellite, the extreme weakness or absence of a magnetic field, the lack of water in its atmosphere, the high abundance of primordial argon, and its high surface temperature.

NASA's Magellan spacecraft (Fig. 10–23) was launched by a space shuttle on May 4, 1989. After five months in space, it reached Venus's orbit, but Venus wasn't there! It had to go around the Sun one whole time before it and Venus were in the same place at the same time, in August 1990. Then it went into a 3-hour 9-minute elliptical orbit that extends up to near Venus's poles. Its distance from Venus ranges from as close as 250 km to as far as 8000 km. During the part

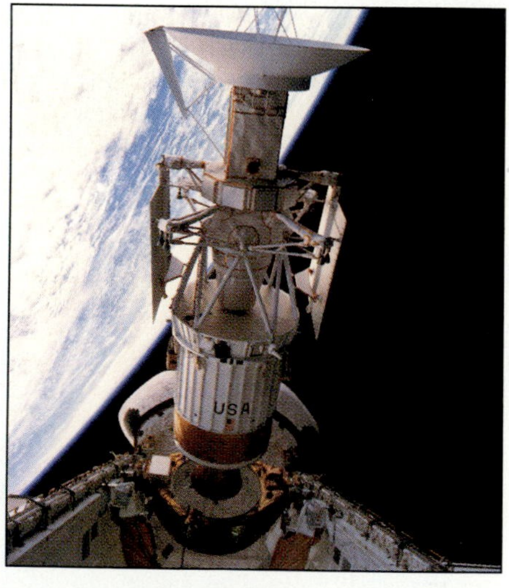

Figure 10–23 The Magellan spacecraft, as it was launched from a space shuttle in 1989.

Figure 10–24 Maat Mons, in a simulated perspective view based on radar data from Magellan. Our viewpoint is 560 km north of Maat Mons at an elevation of 1.7 km. We see lava flows extending hundreds of kilometers across the foreground fractured plains. Maat Mons is 8 km high.

of each orbit in which it is closest to the planet, it studies Venus's surface; during the rest of the orbit it transmits data to Earth.

Magellan carries a *synthetic-aperture radar,* that is, a radar that allows scientists to combine images taken from two slightly different positions and make it seem that a single radar with a bigger antenna—and thus better resolution—is present, rather than the radar antenna that is actually there. The radar uses radio waves of 12.6-cm wavelength—about the width of the column of type you are reading. The resolution that results is about 100 m—about the size of a football field—and about 10 times better than the resolution of Venera 15 and 16. The images are thus over a hundred times more detailed than those from Pioneer Venus that went into making up that satellite's global radar map (which appeared as Figure 10–10). A Magellan mosaic of half the planet appeared as the photograph opening this chapter (p. 172).

From the Magellan spacecraft, we learned that Venus's crust is built up vertically while Earth's crust is built up horizontally. On Venus, lava from the erupting volcanoes builds up. On Earth, lava erupts from mid-ocean ridges, adding to the plates.

The Magellan images show few craters, so lava flows must have resurfaced the planet. There are almost no very small craters, one-thousandth the crater density of our Moon. This observation can be explained if the smallest meteoroids burned up in Venus's atmosphere before they could hit the surface. On the whole, the surface of Venus is about half a billion years old. It isn't known, though, whether there was a widespread episode of vulcanism then, or whether the resurfacing took place more gradually, at different times in different places. Evidence for the first possibility is the few craters are seen partially covered with lava.

Computer processing can transform the radar data into perspective images, making it seem as though we were looking down at a slant at the surface. Apparent lava flows can be seen on many of the mountains and plains (Figs. 10–24 and 10–25). We must remember, though, that these are radar images and not visual ones, so what appears bright means only that it is a strong reflector of radio waves of a certain wavelength. Bright regions, thus, may have rocky, rough surfaces with typical rock size of about the wavelength of the radio waves—12.6 cm.

Figure 10–25 A view of the volcano Sif Mons made by superimposing the radar data on altimetry data. The volcano is 2 km high and 300 km in diameter. The brighter, rougher lava flows extend 120 km down the mountain and are superimposed on older flows.

The argument over whether Venus has plate tectonics continues. Perhaps Venus is too hot for most mountains to survive the planet's gravity. The crust

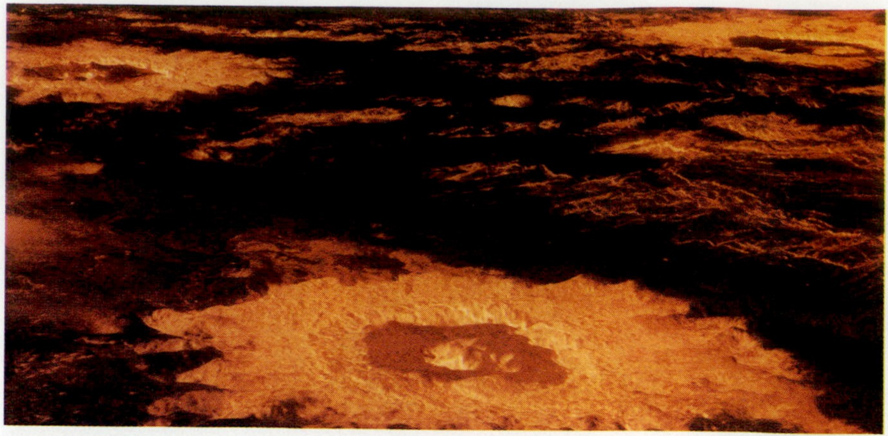

Figure 10–26 Three large impact craters in a simulated perspective view. Howe Crater, at center, is 37 km across. Danilova, at upper left, is 48 km across, and Aglaonice, 63 km across, is at upper right.

may be too weak for plate tectonics, or may be in a pre-tectonic stage. Some signs of horizontal motion appear, but the evidence for tectonics is weak. The existence of the tall mountain Maxwell Mons indicates that, at least in this case, lava flowing from a hot spot can keep up with the effect of gravity. In any case, it seems most likely that at present, Venus does not have plate tectonics.

Craters (Fig. 10–26) range in size from 2 km to 275 km. Sometimes craters have ejecta asymmetrically around them, showing the effect of Venus's atmosphere (Fig. 10–27). Other types of surface features have been found on Venus that aren't known elsewhere, such as giant circular features. There were a couple of false alarms in which scientists thought they had detected a change on Venus from one observation to another of the same location. The culprit, however, turned out to be an effect of radar imaging from different angles.

Before being shut off, Magellan will orbit Venus at an altitude from which scientists on Earth will monitor Venus's gravity.

The Magellan spacecraft can go down in the list of NASA's big successes.

Figure 10–27 Radar images from the Magellan spacecraft reveal three large impact craters, with diameters from 37 km to 50 km. Their ejecta are bright to the radar and are therefore rough. As is common with meteorite impact craters, we see terraced inner walls and central peaks.

Summary and Outline

Venus's atmosphere (Section 10.1)
- Mainly composed of carbon dioxide
- Clouds primarily composed of sulfuric acid droplets

A slow rotation period, in retrograde (Section 10.2)

The temperature of Venus (Section 10.3)
- A high temperature, 750 K, caused by the greenhouse effect
- Greenhouse effect: visible light in, turns to infrared, infrared can't escape

The surface of Venus (Section 10.4)
- Radar maps show volcanoes and mountains

Early spacecraft observations (Section 10.5)
- U.S. Mariner 10 observations of clouds and atmosphere
- Soviet Venera 9 and 10 landers: temperature measurements and photographs of the surface

U.S. Pioneer Venus 1 and 2 studies (Section 10.6)
- Orbiter studied Venus for many months
  - Followed evolution of clouds
  - Mapped surface with radar
  - Found further evidence of volcanoes
  - Temperature maps show atmospheric circulation
- Multiprobe contained bus and four probes
  - Measured composition of atmosphere and temperatures
  - Observed lightning
  - Confirmed that greenhouse effect causes the high temperature

U.S.S.R. Venera and Vega studies (Section 10.7)
- Veneras 11 and 12: radioed data from surface; observed temperature, lightning, composition of atmosphere. Veneras 13 and 14: landers, photographs showed sky is orange; soil is basaltic in this region, which has volcanic topography. Veneras 15 and 16: signs of vulcanism.

NASA Magellan, in orbit August 1990 (Section 10.8)
- Carries high-resolution radar
- Mapped 95 percent of Venus
- No plate tectonics

Key Words

cytherean, retrograde rotation, greenhouse effect, synthetic-aperture radar

Questions

1. Make a table displaying the major similarities and differences between the Earth and Venus.
2. Why does Venus have more carbon dioxide in its atmosphere than does the Earth?
3. Why do we think that there have been significant external effects on the rotation of Venus?
4. If observers from another planet tried to gauge the rotation of the Earth by watching the clouds, what would they find?
5. Suppose a planet had an atmosphere that was opaque in the visible but transparent in the infrared. Describe how the effect of this type of atmosphere on the planet's temperature differs from the greenhouse effect.
6. Why do radar observations of Venus provide more data about the surface structure than a flyby with close-up cameras?
7. If one removed all the CO_2 from the atmosphere of Venus, the pressure of the remaining constituents would be how many times the pressure of the Earth's atmosphere?
8. The Earth's magnetic field protects us from the solar wind. What does this tell you about whether or not the solar-wind particles are charged?
9. Some scientists have argued that if the Earth had been slightly closer to the Sun, it would have turned out like Venus; outline the logic behind this conclusion.
10. Outline what you think Venus would be like if we sent an expedition to Venus that resulted in the loss of almost all of the CO_2 from its atmosphere.
11. Why do we say that Venus is the Earth's "sister planet"?
12. Compare and contrast the observations of the cytherean surface made from the Venera landers.
13. Compare the ground-based and spacecraft-based radar observations of Venus.
14. Do radar observations of Venus study the surface or the clouds? Explain.
15. Describe the major radar results from the Venus Pioneer Orbiter.
16. What is the advantage of synthetic-aperture radar over ordinary radar?
17. What signs of vulcanism are there on Venus?
18. Compare and contrast the circulation of the atmospheres of Venus and the Earth.
19. Describe the path the Magellan spacecraft took between the Earth and Venus.
20. Look up current results from the Magellan spacecraft and describe them.

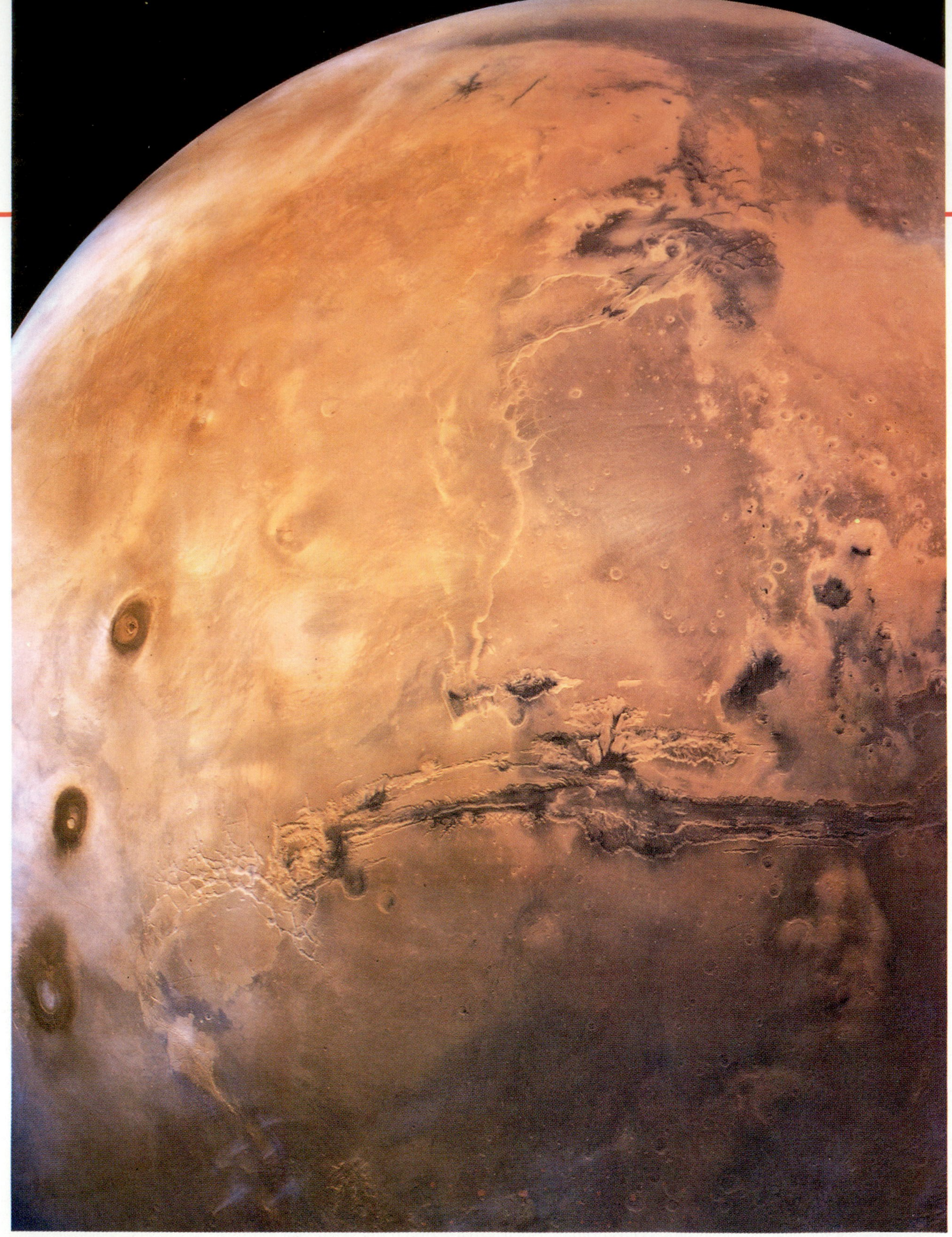

Mars, a mosaic of 102 images from Viking Orbiter. The valleys Valles Marineris appears horizontally across the center; it is as long as the U.S. is wide. At left we see the three giant volcanoes Arsia Mons (*below*), Pavonis Mons (*center*), and Ascraeus Mons (*above*). The faint sharp line curving downward between Arsia Mons and Pavonis Mons and just to their right is a shock wave in the martian atmosphere. The symbol for Mars appears at the top of the opposite page.

Mars ♂ 11

Aims: To discuss the atmosphere and surface features of Mars, to see the results of spacecraft in orbit and on the surface, and to assess the chances of finding life there

Mars has long been the planet of greatest interest to scientists and non-scientists alike. Its interesting appearance as a reddish object in the night sky and some past scientific studies have made Mars the prime object of speculation as to whether or not extraterrestrial life exists.

In 1877, the Italian astronomer Giovanni Schiaparelli published the results of a long series of telescopic observations he had made of Mars. He reported that he had seen *canali* on the surface. When this Italian word for "channels" was improperly translated into "canals," which seemed to connote that they were dug by intelligent life, public interest in Mars increased.

In this country, Percival Lowell grew interested in the problem and in 1894 established an observatory in Flagstaff, Arizona, to study Mars. Over the next decades, there were endless debates over just what had been seen. We now know that the channels or canals Schiaparelli and other observers reported (Fig. 11–1) are not present on Mars—the positions of the *canali* do not even always overlap the spots and markings that are actually on the martian surface. But hope of finding life in the solar system springs eternal, and the latest studies have indicated the presence of considerable quantities of liquid water in Mars's past, a fact that leads many astronomers to hope that life could have formed during those periods. Still, as we shall describe, the Viking spacecraft that landed on Mars found no signs of life.

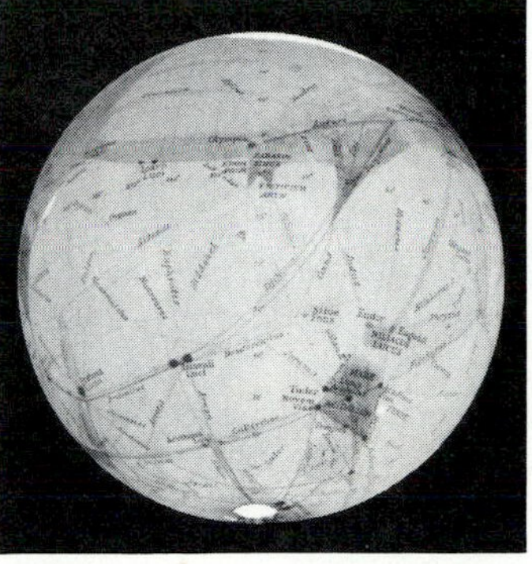

Figure 11–1 An old drawing of Mars that showed "canals."

11.1 Characteristics of Mars

Mars is a small planet, 6800 km across, which is only about half the diameter and one-eighth the volume of Earth or Venus, although somewhat larger than Mercury. Mars's atmosphere is thin—at the surface its pressure is only 1 per cent of the surface pressure of Earth's atmosphere—but it might be sufficient for certain kinds of life.

From Earth, we have relatively little trouble in seeing through the martian atmosphere to inspect that planet's surface (Fig. 11–2), except when a martian dust storm is raging. At other times, we are limited mainly by the turbulence in our own thicker atmosphere, which causes unsteadiness in the images and limits our resolution to 60 km at best on Mars. We can, however, follow features on the surface of Mars and measure the rotation period quite accurately. It turns out to be 24 hours 37 minutes 22.6 seconds from one martian noontime to the next, so a day on Mars lasts nearly the same length of time as a day on Earth.

Unlike the orbits of Mercury or Venus, the orbit of Mars is outside the Earth's, so it is much easier to observe in the night sky (Fig. 11–3). Because of Mars's elliptical orbit around the Sun, at some oppositions it is closer to

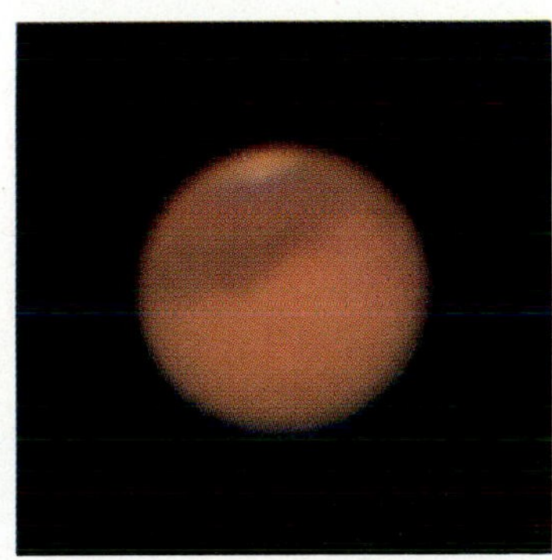

Figure 11–2 A photograph of Mars taken with a large telescope on Earth. The darker regions spread in the martian springtime. One of the polar caps is clearly visible.

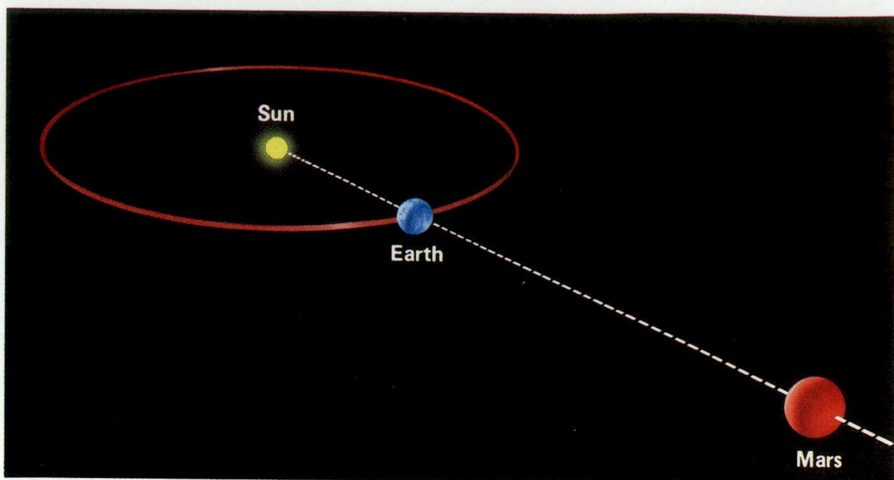

Figure 11–3 When Mars is at opposition, it is high in our nighttime sky and thus easy to observe. We can then simply observe in the direction opposite from that of the Sun. We are then looking at the lighted face.

When three celestial bodies are in a line, the alignment is called a *syzygy,* and the particular case when the planet is on the opposite side of the Earth from the Sun is called an *opposition.* Oppositions of Mars occur at intervals of 26 Earth months on the average.

Earth than at others. It was particularly close at the opposition of 1988, and electronic techniques of photographing it with CCD's allowed such short exposures ($\frac{1}{20}$ sec) compared with film that high-resolution views could be captured in the brief moments of the best seeing (Fig. 11–4).

Mars revolves around the Sun in 23 Earth months. The axis of its rotation is tipped at a 25° angle from the plane of its orbit, nearly the same as the Earth's 23½° tilt. Because the tilt of the axis causes the seasons, we know that Mars goes through a year with four seasons just as the Earth does.

We have watched the effect of the seasons on Mars over the last century. In the martian winter, in a given hemisphere, there is a polar cap. As the martian spring comes to the northern hemisphere, the north polar cap shrinks and material at more temperate zones darkens. The surface of Mars is always mainly reddish, with darker gray areas that appear blue-green for physiological reasons of color contrast. In the spring, the darker regions spread. Half a martian year later, the same thing happens in the southern hemisphere.

One possible explanation that long ago seemed obvious for these changes is biological: martian vegetation could be blooming or spreading in the spring. But there are other explanations, too. The theory that now seems most reasonable is that each year at the end of southern-hemisphere springtime, a global dust storm starts, covering Mars's entire surface with light dust. Then the winds reach velocities as high as hundreds of kilometers per hour and blow fine, light-colored dust off some of the slopes. This exposes the dark areas underneath. Gradually, as Mars passes through its seasons over the next year, the location where the dust is stripped away changes, mimicking the color change we would expect from vegetation. Finally, a global dust storm starts up again to renew the cycle. Astronomers still debate why the dust is reddish; the presence of iron oxide (rust) would explain the color in general, but might not satisfy the detailed measurements.

Mars has $\frac{1}{10}$ the mass of the Earth; from the mass and radius one can easily calculate that its average density is slightly less than 4 grams/cm^3, not much higher than that of the Moon. This is substantially less than the density of 5½ grams/cm^3 shared by the three innermost terrestrial planets, and indicates that Mars's overall composition must be fundamentally different from that of these other planets. Mars probably has a smaller core and a thicker crust than the Earth.

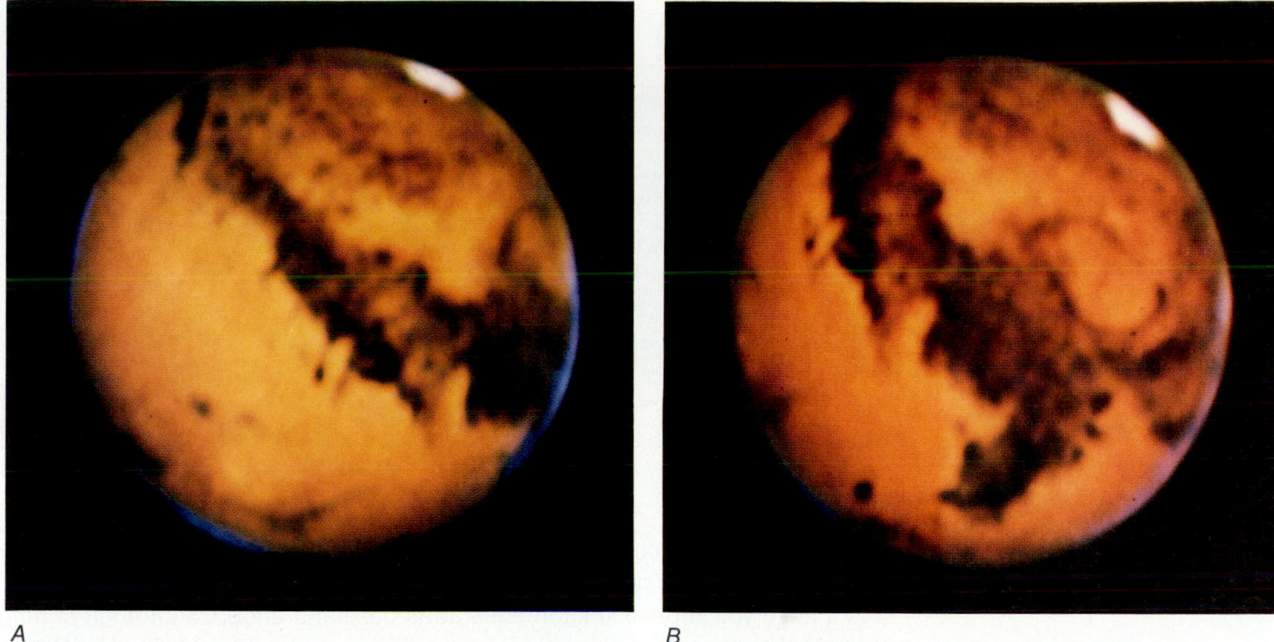

Figure 11–4 Two views of Mars taken at its extremely close recent opposition on September 26, 1988. The composite is made from blue, green, and red/near-infrared images and so is not quite true color. (*A*) 1:15 UT. (*B*) 22:30 UT, which is 21 hours 15 minutes later out of the $24^h\ 37^m$ rotation period of Mars.

11.2 Early Space Observations

Mars has been the target of series of spacecraft launched by both the United States and the Soviet Union. The earliest of these spacecraft were flybys, which were in good position for only a few hours as they passed by. The first flyby, NASA's Mariner 4, sent back photos in 1965 of a section of Mars showing only a cratered surface. This seemed to indicate that Mars resembles the Moon more than it does the Earth. Mariners 6 and 7 confirmed the cratering four years later, and also showed some signs of erosion.

In 1971, the United States sent out a spacecraft not just to fly by Mars for only a few hours, but actually to orbit the planet and send back data for a year or more. This spacecraft, Mariner 9, went into orbit around Mars after a five-month voyage from Earth. But when it reached Mars, our first reaction was as much disappointment as elation. While Mariner 9 was en route, a tremendous dust storm had come up, almost completely obscuring the entire surface of the planet. Only the south polar cap and four dark spots were visible. The storm began to settle after a few weeks, with the polar caps best visible through the thinning dust. Finally, three months after the spacecraft arrived, the surface of Mars was completely visible, and Mariner 9 could proceed with its mapping mission.

The data taken after the dust storm had ended showed four major "geological" areas on Mars: volcanic regions, canyon areas, expanses of craters, and terraced areas near the poles. These features were subsequently imaged with higher resolution from the Viking missions, NASA missions that reached Mars in 1976 and carried on studies for several years. The photographs we show here are from those latter missions.

A chief surprise of the Mariner 9 mission was the discovery of extensive areas of vulcanism. The four dark spots on the martian surface proved to be volcanoes. They are on a 2000-km-long ridge known as Tharsis (Fig. 11–5).

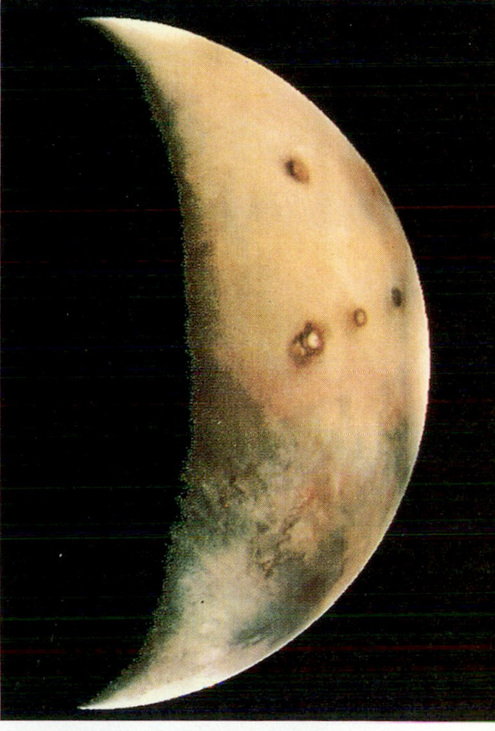

Figure 11–5 A Viking image showing several volcanoes on the Tharsis Ridge of Mars. Olympus Mons is nearest the top.

Figure 11–6 Olympus Mons, from Mariner 9. The image shows a region 800 km across.

The largest, which corresponds to the surface marking long known as Nix Olympica, "the snow of Olympus," is named Olympus Mons, "Mount Olympus." It is a huge volcano—600 kilometers at its base and about 25 kilometers high (Fig. 11–6). It is crowned with a crater 65 km wide (Fig. 11–7); Manhattan island could be easily dropped inside. (The tallest volcano on Earth is Mauna Kea, in the Hawaiian islands, if we measure its height from its base deep below

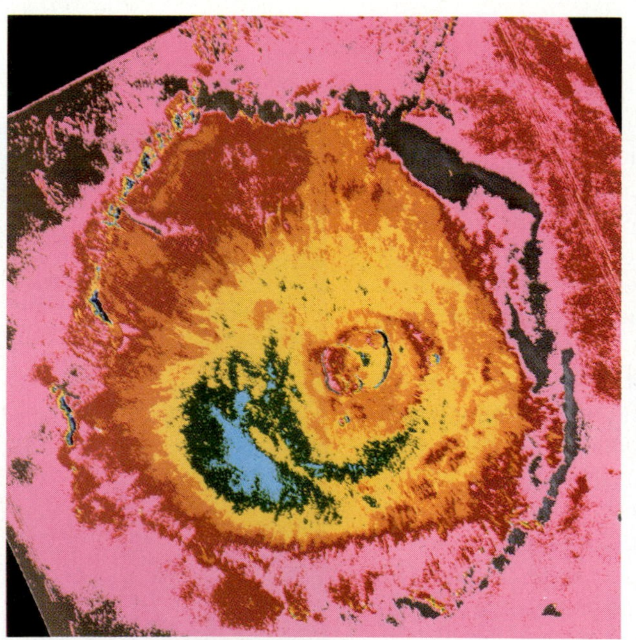

A

B

C

Figure 11–7 (*A*) A false-color view of Olympus Mons. The false color exaggerates the slight differences in the volcano's lava flows. (*B*) A high-resolution view of the caldera on top of Olympus Mons, photographed by the Viking 1 orbiter. We see a close-up of the upper half of the central crater. (*C*) The gradual rise of Mauna Kea on Earth is typical of shield volcanoes.

A

B

Figure 11–8 Stream channels with tributaries indicate to most scientists that water flowed on Mars in the past. We see the boundary scarp between the ancient cratered highlands and the northern plains. (*A*) Amazonis Planitia. (*B*) East Mangala Valles. The ancient river channels run north through the ancient cratered highlands, but north of the scarp are covered by young lava flows. The flow-like lobes of the large impact crater may reveal the presence of subsurface water or ice.

the ocean. Mauna Kea is taller than Everest, though still only 9 km high. The average level of the Tharsis ridge itself is 10 km. Volcanoes with gentle slopes like the Hawaiian volcanoes on Earth and the Tharsis volcanoes on Mars are known as "shield volcanoes.")

Another surprise on Mars was the discovery of systems of canyons. One tremendous set of canyons—about 5000 kilometers long—is as long as the United States is wide and comparable in size to the Rift Valley in Africa, the longest geological fault on Earth. The main canyon is 200 km wide and up to 7 km deep. Here again the size of the canyon with respect to Mars is proportionately larger than any geologic formations on the Earth. Venus, the Earth, and Mars each seems to have a large rift. The canyons, named after the Mariner 9 spacecraft from which they were discovered, are shown in the photograph opening this chapter.

Perhaps the most amazing discovery on Mars was the presence of sinuous channels. These are on a smaller scale than the *canali* that Schiaparelli had seen and are entirely different phenomena. Some of the channels show tributaries (Fig. 11–8), and the bottoms of some of the channels show the same characteristic features as stream beds on Earth. Even though water cannot exist on the surface of Mars under today's conditions, it is difficult to think of ways to explain the channels satisfactorily other than to say that they were cut by running water in the past.

This indication that water most likely flowed on Mars is particularly interesting because biologists feel that water is necessary for the formation and evolution of life. The presence of water on Mars, therefore, even in the past, may indicate that life could have formed and may even have survived.

How long ago was the water present? New detailed studies on Viking images of martian topography by a U.S. Geological Survey scientist indicate complicated overlays of channels cut by water and volcanic flows, all of which may have occurred within the last 200 million years. This time period is recent rather than ancient, on a planetary scale.

If there had been water on Mars in the past, and in quantities great enough to cut river beds, where has it all gone? Most of the water may be in a permafrost layer beneath middle latitudes and polar regions. A small amount of the water is bound in the polar caps (Fig. 11–9). Up to the time of Mariner 9, we thought that the polar caps were mostly frozen carbon dioxide—"dry ice"; the presence of water was controversial. We now know that the large polar caps that extend to latitude 50° during the winter are carbon dioxide. But when a cap shrinks during its hemisphere's summer, a residual polar cap of water ice remains, especially in the northern hemisphere.

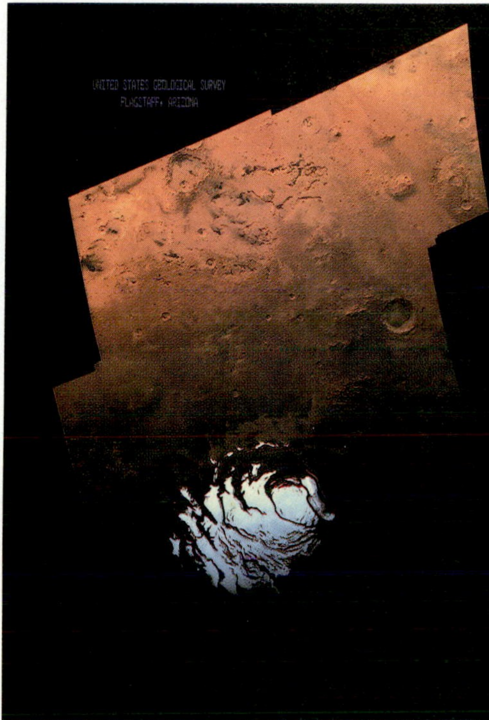

Figure 11–9 A mosaic of Viking images of the south polar cap during the martian southern summer when the cap has retreated the most. The south residual cap is about 400 km across, and at least the exposed surface is thought to be mainly frozen carbon dioxide. Water ice may be present in layers underneath. By contrast, the north polar residual cap is made of water ice with no frozen carbon dioxide.

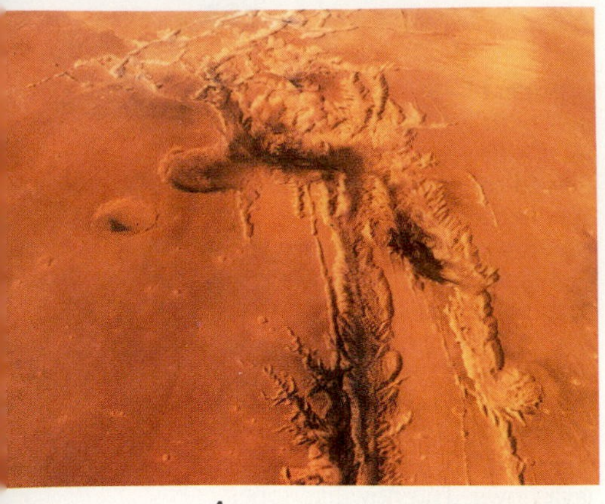

A

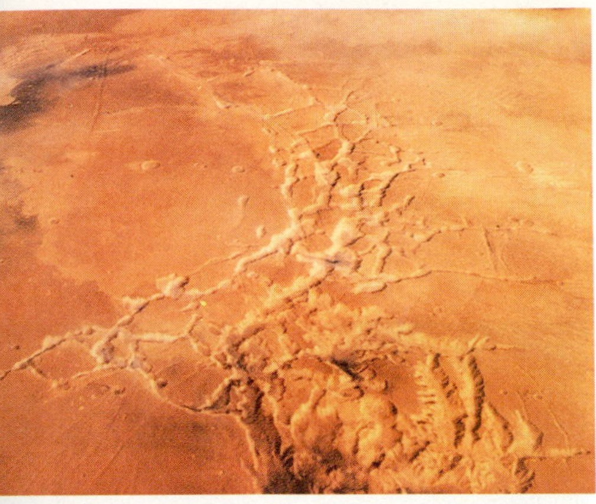

B

C

D

We found that the martian atmosphere is composed of 90 per cent carbon dioxide with small amounts of carbon monoxide, oxygen, and water. We measured the density of the atmosphere by monitoring the changes of radio signals from spacecraft as they went behind Mars. As the spacecraft were occulted by the atmosphere of the planet, we could tell the rate at which the density drops. The surface pressure is less than 1 per cent of that near Earth's surface and decreases with altitude slightly more than it does on Earth. The atmosphere is too thin to affect the surface temperature, in contrast to the huge effects that the atmospheres of Venus and the Earth have on climate.

11.3 Viking Orbiters

In the summer of 1976, two elaborate U.S. spacecraft named Viking reached Mars after 10-month flights. Each contained two parts: an orbiter and a lander. The orbiter served two roles: it mapped and analyzed the martian surface using its cameras and other instruments and relayed the lander's radio signals to Earth. The lander served two purposes as well: it studied the rocks and weather near the surface of Mars and sampled the surface in order to decide whether there was life on Mars!

Viking 1 reached Mars in June of 1976 after a flight that led to spectacular large-scale views of the martian surface. The first month of its orbit around Mars was devoted to radioing back pictures to scientists on Earth who were trying to determine a relatively safe spot for its lander to touch down.

The orbiter observed the volcanoes and the huge canyon (Figs. 11–10 and 11–11) at higher resolution than did Mariner 9. The landslides that can be seen in the walls indicate that tectonic activity has been present.

These detailed views of Mars allow us to interpret better the similarities and differences that this planet—with its huge canyons and gigantic volcanoes—has with respect to the Earth. For example, Mars has exceedingly large, gently sloping volcanoes but no signs of the long mountain ranges or deep mid-ocean ridges that on Earth tell us that plate tectonics has been and is taking place. Many of the large volcanoes on Mars are "shield volcanoes"—a type that has gently sloping sides formed by the rapid spread of lava. On Earth, we also have steep-sided volcanoes, which occur where the continental plates are overlapping, as for Mt. St. Helens, the Aleutian Islands, or Mount Fujiyama. No chains of such volcanoes exist on Mars.

Perhaps the volcanic features on Mars can get so huge because continental drift is absent there. If molten rock flowing upward causes volcanoes to form, then on Mars the features just get bigger and bigger for hundreds of millions of years, since the volcanoes stay over the sources and do not drift away.

The presence of two orbiters circling Mars allowed a division of duties. One continued to relay data from the two Viking landers to Earth, while the orbit of the other was changed so that it passed over Mars's north pole. Summertime heat in the northern hemisphere caused the main polar cap (thought to be made of dry ice—frozen carbon dioxide) to retreat, leaving only a residual

Figure 11–10 Computer-enhanced views of the giant canyon Valles Marineris. (*A*) A view looking west from 200-km altitude. (*B*) Labyrinthus Noctis, a system of interwoven canyons that continues westward of Valles Marineris. (*C*) Looking across the canyon from the edge of Ophir Chasma, in the northern part of Valles Marineris. The volcanoes Pavonis Mons (*left*) and Ascraeus Mons (*right*) are in the distance about 3000 km away. (*D*) A northern view of Tharsis Montes from 30 km above the surface. Arsia Mons is in the foreground, Pavonis Mons is in the center, Ascraeus Mons is beyond it, and Tharsis Tholus is in the distance.

Figure 11–11 Candor Chasm in Valles Marineris. The landscape has been shaped by tectonics and wind, and perhaps also by water and vulcanism.

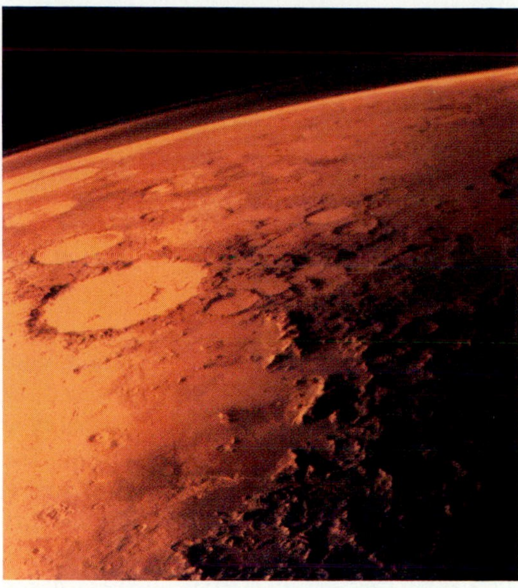

Figure 11–12 This oblique view shows layers in the atmosphere of Mars.

ice cap. From the amount of water vapor detected with the orbiter's spectrograph, and from the fact that the temperature was too high for this residual ice cap to be frozen carbon dioxide, the conclusion has been reached that it is made of water ice. This important result indicates the probable presence of lots of water on the martian surface in the past—in accordance with the existence of the stream beds—and seems to raise the possibility that life has existed or does exist on Mars. The layered look of the polar caps indicates that the general atmospheric conditions of Mars fluctuated over eons, as do Earth's.

Dating of the stream channels became possible late in the Viking mission when exceptionally high-resolution photos showed craters in the stream beds. The higher the density of craters, the older the underlying flow. So the earlier idea that all stream beds were formed at one stage early in Mars's history when the atmosphere was denser was not quite correct; volcanic flows and flows of water must have alternated over a long period of time. Note that these dates are all relative; finding absolute dates will have to await the return of samples from Mars to Earth.

Observations from the orbiters showed Mars's atmosphere (Fig. 11–12). The lengthy period of observation led to the discovery of weather patterns on Mars. With its rotation period similar to that of Earth, many features of Mars's weather are similar to our own. Surface temperatures measured from the landers ranged from a low of 150 K (−190°F) at the northern site of Lander 1 to over 300 K (80°F) at Lander 2. Temperature varied each day (Fig. 11–13) by 35–50°C (60–90°F).

Both Viking landers and orbiters lasted long enough to show seasonal changes on Mars. The yearly temperature range spans the −111°C (−190°F) freezing point of carbon dioxide, which thus sometimes forms frost.

The orbiters showed dust storms (Fig. 11–14), which generally form at a time that is in the early summer in the southern hemisphere and in winter in the northern hemisphere. The storms are seasonal partly because Mars is closest to the Sun at this time and the increased solar heating causes the atmosphere to circulate. The overall climates of the northern and southern

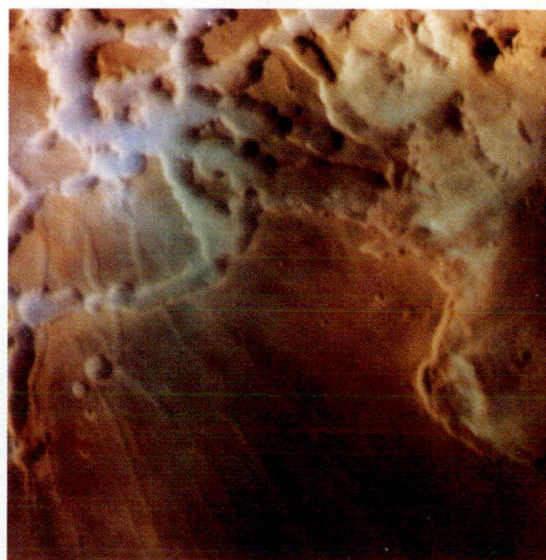

Figure 11–13 Clouds of water ice form over Labyrinthus Noctis at sunrise.

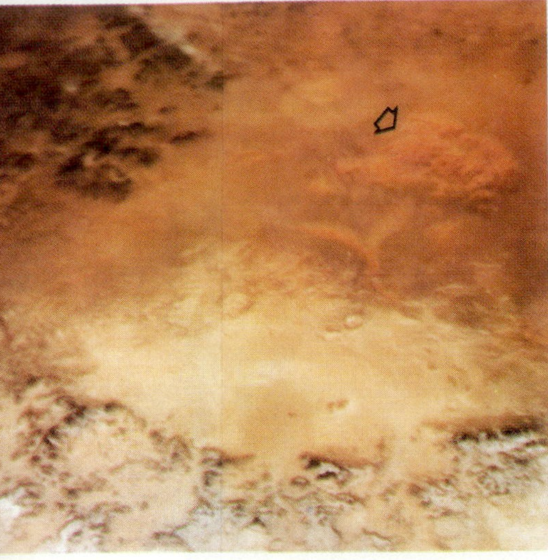

Figure 11–14 A martian dust storm over 300 km across inside the Argyre Basin. This martian storm resembles satellite pictures of storms on Earth. The temperature is too warm for the clouds to be carbon dioxide; they therefore must be made of water ice. In this view, taken at the end of southern winter, the polar cap is retreating and covers only the southern half of Argyre.

hemispheres are different, probably because Mars's orbit is more elliptical than Earth's.

As the result of careful study of the martian images, we now can categorize the surface into geological regions (Fig. 11–15).

As for magnetic field, Viking didn't find any around Mars. Just as for Venus, if Mars has any magnetic field, it is too weak to measure.

11.4 Viking Landers

On July 20, 1976, exactly seven years after the first manned landing on the Moon, Viking 1's lander descended safely onto a plain called Chryse. The views showed rocks of several kinds, covered with yellowish-brown material that is probably an iron oxide (rust) compound (Fig. 11–16). Sand dunes were also visible. The sky on Mars turns out to be yellowish-brown, almost pinkish, from dust suspended in the air as a result of one of Mars's frequent dust storms. Craters up to 600 m across were visible near the horizon, and may have supplied the rocks in the view.

A series of experiments aboard the lander was designed to search for signs of life. A long arm was deployed, and a shovel at its end dug up a bit of the martian surface. The soil was dumped into three experiments that searched for such signs of life as respiration and metabolism. The results were astonishing at first: the experiments sent back signals that seemed similar to those that would be caused on Earth by biological rather than by mere chemical processes. But later results were less spectacular, and non-biological explanations seem more likely. It is probable that some strange chemical process mimicked life in these experiments.

One experiment gave much more negative results for the probability that there is life on Mars. It analyzed the soil and looked for traces of organic compounds. On Earth, many organic compounds left over from dead forms of life remain in the soil; the life forms themselves are only a tiny fraction of the organic material. Yet these experiments found no trace of organic material. On Mars, who knows? Perhaps life forms evolved that efficiently used up their predecessors. Still, the absence of organic material from martian soil is a strong argument against the presence of life on Mars.

The Viking 2 lander descended on September 3, 1976, on a site that was thought to be much more favorable to possible life forms than the Viking 1 site because it was closer to the north pole with its prospective water supply. It too found yellowish-brown dust and a pinkish sky (Fig. 11–17). Compared with Viking 1, Viking 2 found a smaller variety of types of rocks, but more large rocks and more pitted rocks. Such a distribution would result if the surface there had been ejected and flowed from a large crater 200 km away. The

Figure 11–15 This computer-generated mosaic of Viking 2 images uses false color to show variations in the surface chemistry of Mars. Red shows rocks that contain a high proportion of iron oxides; dark blue shows fresher, less oxidized materials, probably volcanic basalts; bright turquoise shows surface frosts and fogs made from carbon dioxide or water ice; and brown, orange, and yellow show deposits of sand and dust.

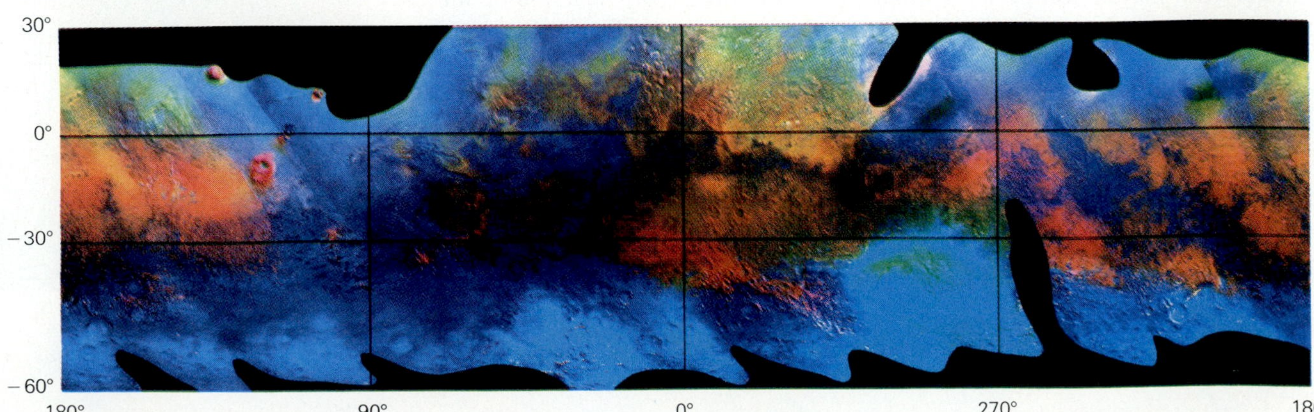

Figure 11–16 A Viking 1 view shows the red rocks and pink sky of Mars. The sampler arm at right has dug a trench at far left.

Viking 1
Orbiter:
June 19, 1976–Aug. 7, 1980
Lander:
July 20, 1976–Nov. 13, 1982
Viking 2
Orbiter:
Aug. 7, 1976–July 25, 1978
Lander:
Sept. 3, 1976–Apr. 12, 1980

atmosphere at the second site, Utopia Planitia, contains three or four times more water vapor than had been observed near Chryse (the first landing site), 7500 km away. Viking 2's experiments in the search for life sent back data similar to Viking 1's.

Both Viking landers survived on the surface of Mars to send back meteorological information over several seasons and to observe seasonal changes on Mars's surface (Fig. 11–18).

Even if the life signs detected by Viking come from chemical rather than biological processes, as seems likely, we have still learned of fascinating new chemistry going on. When life arose on Earth, it probably took up chemical processes that had previously existed. Similarly, if life began on Mars in the past or will begin there in the future (assuming our visits didn't contaminate Mars and ruin the chances for indigenous life), we might expect the life forms to use chemical processes that already existed. So even if we haven't detected life itself, we may well have learned important things about its origin.

Figure 11–17 A Viking 2 view of Mars. The spacecraft landed on a rock and so was at an angle. The boom that supports Viking's weather station cuts through the center of the picture. ("Chance of precipitation," the local newscaster would say, "is 0 per cent.")

Figure 11–18 Frost forms around some of the rocks in this view from the Viking 2 lander. The antenna that sends signals to Earth is in the foreground.

11.6 Satellites of Mars

In Greek mythology, Phobos (Fear) and Deimos (Terror) were companions of Ares, the equivalent of the Roman war god, Mars.

Mars has two moons, Phobos and Deimos. In a sense, these are the first moons we have met in our study of the solar system, for Mercury and Venus have no moons at all, and the Earth's Moon is so large relative to its parent that we may consider the Earth and Moon as a double-planet system. The moons of Mars are mere chunks of rock, only 27 and 15 kilometers across, respectively.

Phobos and Deimos are very minor satellites. They revolve very rapidly; Phobos completes an orbit of Mars in only 7 hours 40 minutes, less than a martian day. Deimos orbits in about 30 hours, slower than Mars's rotation period. Because Phobos orbits more rapidly than Mars rotates, an observer on the surface of Mars would see Phobos moving conspicuously backwards compared to Deimos, the other planets, and the stars. Because Mars's gravity continually pulls back on Phobos, it is slowly spiralling downward. It should hit Mars in only 30 million years!

It is amusing to note that in 1727, long before the moons were discovered, Jonathan Swift invented two moons of Mars for *Gulliver's Travels.* This is widely considered to be a lucky guess, but it seemed reasonable at that time for numerological purposes. It may even have dated back to Kepler, who had considered in 1610 that the planets interior to the Earth—Mercury and Venus—had no known moons, the Earth had one, while Jupiter, the next planet out from Mars, had 4 known moons. He reasoned that Mars should have the intermediate value of 2. Both Swift's moons and the real ones are very small and revolve very quickly. The two non-fictional moons were discovered in 1877 at a favorable opposition of Mars to Earth.

Mariner 9 and the Vikings made close-up photographs and studies of Phobos and Deimos. Each turns out to be not at all like our Moon, which is

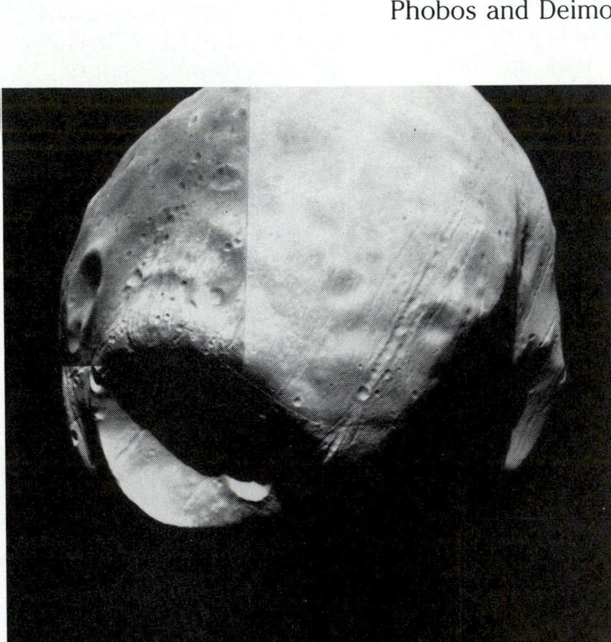

Figure 11–19 Phobos, from the Viking 1 Orbiter, is about 20 km across. Its two largest craters have been named after Asaph Hall, who discovered Phobos and Deimos in 1877 at the U.S. Naval Observatory in Washington, and after Angelina Stickney, his wife, who encouraged the search. The large crater Stickney, 10 km across, faces us, and the crater Hall is at top. The grooves were probably associated with the formation of the crater Stickney. Similar chains of small craters on Earth's Moon, on Mars, and on Mercury were formed by secondary cratering from a larger impact.

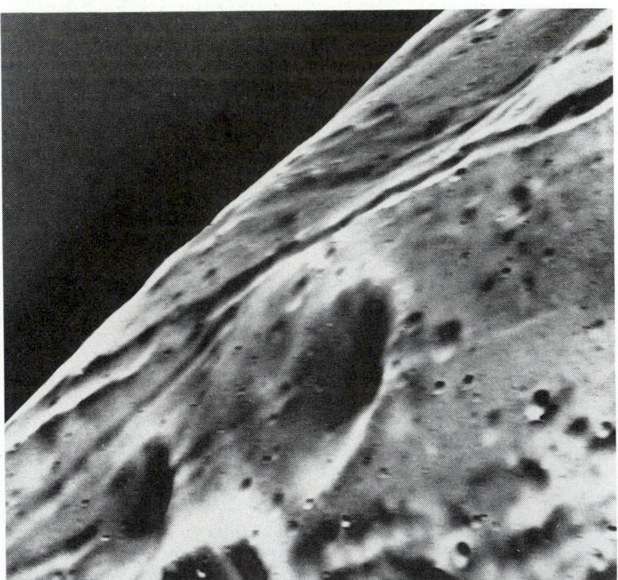

Figure 11–20 A high-resolution picture of Phobos, covering an area 3 km × 3.5 km. The craters in this view range up to 1.2 km in diameter. The striations appear to be grooves rather than crater chains.

a spherical, planet-like body. Phobos and Deimos are just cratered chunks of rock; they may be fragments of a former single martian moon that collided with a large meteoroid. Phobos and Deimos are too small to have enough gravity to have made them round.

Phobos (Figs. 11–19 and 11–20), which is only about 27 km in its longest dimension and 19 km in its shortest, has a crater on it that is 8 kilometers across, a large fraction of Phobos's circumference. Deimos (Figs. 11–21 and 11–22) is even smaller than Phobos, only around 15 km in its longest dimension and 11 km in its shortest. From their sizes, we can calculate their albedoes; Phobos and Deimos turn out to be about as dark as the darkest maria on the Moon, with albedoes of only 6 per cent. From more detailed but similar comparisons of how the albedoes vary with wavelength, we deduce that they may be made of dark carbon-rich ("carbonaceous") rock, similar to some of the asteroids.

We can conclude that the moons are fairly old because they have been around long enough for their surfaces to be extensively cratered. The rotation of both moons is gravitationally linked to Mars, with the same side always facing the planet, just as our Moon is linked to Earth.

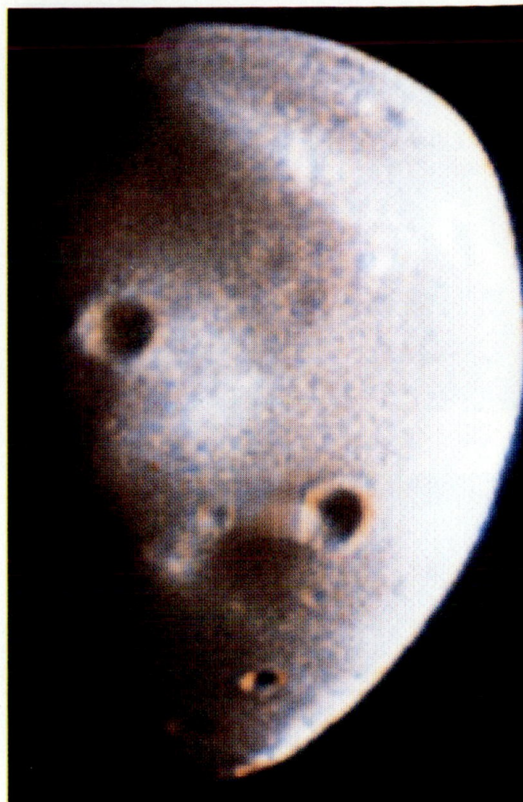

Figure 11–21 Deimos measures about 14 km × 12 km × 11 km. The surface is heavily cratered and thus presumably very old. The surface seems smoother than that of Phobos, though, because a regolith 10 m thick covers Deimos.

The image was constructed from two black-and-white images—one in violet and the other in orange. Actually, Deimos is very dark grey. The spectra of Phobos and Deimos are similar to those of one type of dark asteroid.

11.6 The Phobos Missions

A Soviet pair of spacecraft launched in 1988 was supposed to rendezvous with Phobos while orbiting Mars, and so were named Phobos 1 and Phobos 2. The orbits were to be so similar in size and shape to that of Phobos that they would essentially hover only 50 meters above Phobos's surface! They were even to shoot powerful laser beams and ion beams at Phobos's surface in order to analyze the spectrum of the debris. Part of each spacecraft was to land on Phobos. Unfortunately, an incorrect signal was sent to the first spacecraft while en route, and contact with the spacecraft was lost. The second spacecraft survived until it was in orbit around Mars, but it too failed just as it was about to arrive near Phobos. Still, it sent back 40 images of Phobos (Fig. 11–23) and mapped over 80 per cent of its surface. It measured a spectrum from the ultraviolet through the infrared that shows that Phobos's albedo is only 4 per cent, similar to the carbonaceous chondritic meteorites (Chapter 18), as are

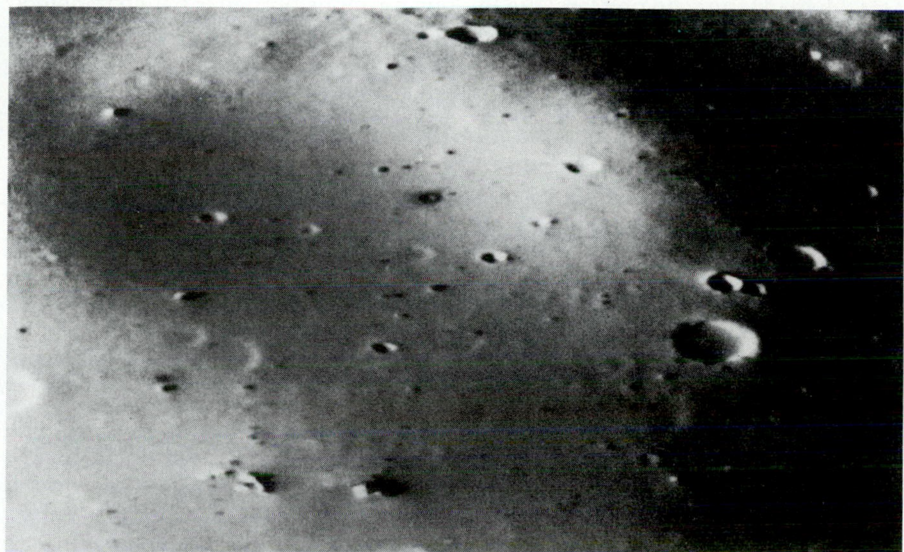

Figure 11–22 This Viking 2 close-up of Deimos covers a region only 1.2 km × 1.5 km and shows features as small as 3 m across. Boulders as big as houses litter the smooth surface.

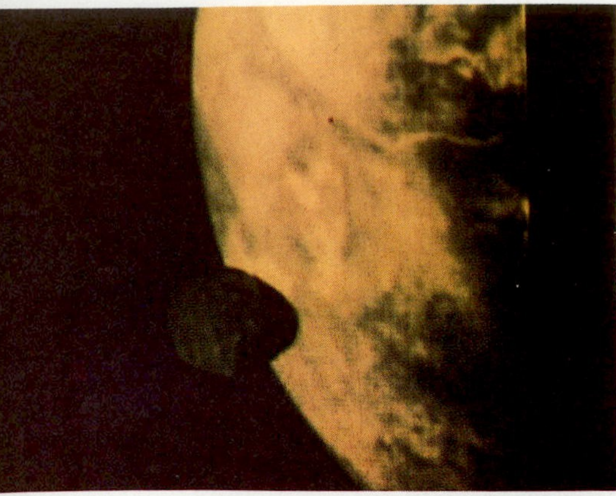

A

B

Figure 11–23 Two images from the Soviet Phobos 2 mission shortly before contact with it was lost. (*A*) Phobos against the edge of Mars. (*B*) Phobos with some of its craters.

some types of asteroids. The irregular nature of Phobos supports the idea that it is an asteroid captured by Mars. Since many asteroids are thought to have dry surfaces covering icy interiors, so may Phobos and Deimos. The interiors could provide water, oxygen, and hydrogen for expeditions around Mars. Accurate tracking of the spacecraft allowed scientists to refine their measurement of Phobos's density to 1.95 g/cm^3, a reduction of 10 per cent from the Viking measurements that makes it closer to the density of carbonaceous chondrites.

Though the Phobos 2 mission was truncated, some of the non-photographic instruments, including an infrared scanner for study of Mars itself, sent back most of the data for which they were programmed. The temperature maps were sharp, and showed the physical characteristics of different regions of martian soil (Fig. 11–24). Further, a French infrared spectrometer and a gamma-ray spectrometer aboard made a mineralogical map of Mars's surface. They showed that the water content of minerals on the slopes of the Tharsis volcanoes is as much as 20 per cent higher than the surrounding plains. Other instruments mapped water vapor and other gases in the martian atmosphere.

Before it failed, Phobos 2 had looped many times around Mars and mapped the martian magnetosphere in much greater detail than had previously been carried out. The magnetosphere is particularly interesting because Mars's relatively weak magnetic field allows the solar wind to penetrate to the top of the martian atmosphere, a different situation from earth's magnetosphere. The link between magnetic lines of force from Mars and from the solar wind even make a channel by which some of Mars's atmosphere escapes. The effect may have led to the loss of much of Mars's original water supply.

11.7 What's Next?

What's the next step? The Viking missions were so spectacularly successful that further exploration of Mars would surely be valuable.

NASA is preparing to launch Mars Observer in 1992 to map surface chemistry, study atmospheric content and seasonal weather patterns, and make a radar map. One device should be able to directly detect hydrogen as much as a meter below the surface, and so detect the subsurface ice that may exist at high latitudes. Mars Observer is to carry an antenna to pick up signals from the Russian 1994 lander, in a cooperative effort. It may be in a low enough orbit to detect a magnetic field on Mars; the Phobos missions orbited too high to detect whether or not the magnetic field is zero.

Before the failure of Phobos 1 and partial success of Phobos 2, the then Soviets were to send a spacecraft to explore the surface of Mars in 1994, followed by a sample-return mission in 1996 or 1998. Since the breakup of the Soviet Union plans are now uncertain, though they are supposed to be proceeding.

The Japanese have announced their plans to send the Planet-B spacecraft to Mars in 1996.

It would be nice to be able to implant seismometers into the surface of Mars. Many scientists would like to send a roving vehicle to travel over the martian surface. A study of layers of martian rock at a stream bed or at a canyon would reveal the ages of Mars, as the Grand Canyon does of Earth.

Many people have called for NASA to declare that a manned Mars landing would be the major goal. The declaration would focus NASA's interest, and perhaps save it from its current diffuse outlook. The hope is that people would be so interested in a manned mission to Mars that funding would be forthcoming. The project would be so expensive that it has been suggested that it be carried out jointly by the United States and the Soviet Union. The instrumental failures of the Phobos missions have raised questions about spacecraft reliability and thus about the value of such cooperation.

Figure 11–24 An infrared image of Mars made by the Phobos 2 spacecraft. We see chaotic terrain, where underground water may have been suddenly released.

A committee chaired by former astronaut Sally Ride, in a report to NASA submitted in 1987, did not recommend such a major goal. Instead they wrote of a more gradual approach, with workstations on the Moon to come first. Further, many scientists are afraid that scientific aims will again be lost to cost-overruns in manned space programs. A U.S. commitment to unmanned along-side of manned space research drew the loudest applause of an audience of astronomers at a 1990 meeting of the American Astronomical Society.

Summary and Outline

Observations from the Earth (Section 11.1)
- Rotation period; surface markings; dust storms; seasonal changes

Early space observations (Section 11.2)
- Many space probes sent from the U.S. and the U.S.S.R.
- Mariner 9 orbiter mapped all of Mars
- Volcanic regions indicating geological activity; canyons larger than Earth counterparts; craters and terraced areas near the poles; signs of water, including sinuous channels and the underlying layer of polar caps

Viking Orbiters (Section 11.3)

Viking Landers (Section 11.4)
- Close-up photographs of surface rocks; yellowish- brown rocks, dust, and sky; analysis of soil
- Search for life: biological explanation seemed possible at first but chemical explanation now seems most likely

Satellites of Mars (Section 11.5)
- Phobos and Deimos are both irregular chunks of rock

Phobos missions (Section 11.6)
- Mars was mapped but Phobos 2 failed before reaching Phobos.

Future missions (Section 11.7)

Key Words

canali, syzygy, opposition

Questions

1. Outline the features of Mars that made scientists think that it was a good place to search for life.

2. If Mars were closer to the Sun, would you expect its atmosphere to be more or less dense than it is now? Explain.

†3. From the relative masses and radii (see Apendix 3), verify that the density of Mars is about 70 per cent that of Earth.

†4. Approximately how much more solar radiation strikes a square meter of the Earth than a square meter of Mars?

5. Compare the tallest volcanoes on Earth and Mars relative to the diameters of the planets.

6. What evidence is there that there is, or has been, water on Mars?

7. Why is Mars's sky pinkish?

†This question requires a numerical solution.

8. Describe the composition of Mars's polar caps. Explain the evidence.

9. List the various techniques for determining the composition of the atmosphere of Mars.

10. Compare the temperature ranges on Mercury, Venus, Earth, and Mars.

11. List the evidence from Viking for and against the existence of life on Mars.

12. (a) Aside from the biology experiments, list three types of observations made from the Viking landers. (b) What are two types of observations made from the Viking orbiters?

13. Why aren't Phobos and Deimos regular, round objects like the Earth's Moon?

14. Describe our success and one failure of the second Soviet Phobos spacecraft.

Topics for Discussion

1. Discuss the problems of sterilizing spacecraft, and whether we should risk contaminating Mars by landing spacecraft.

2. Discuss the additional problems of contamination that manned exploration would bring. Analyze whether, in this context, we should proceed with manned exploration in the next decade or two.

3. Should a manned mission to Mars be a major goal for the U.S.?

4. Should missions to Mars be international? Comment in view of the 1988 Soviet Phobos missions.

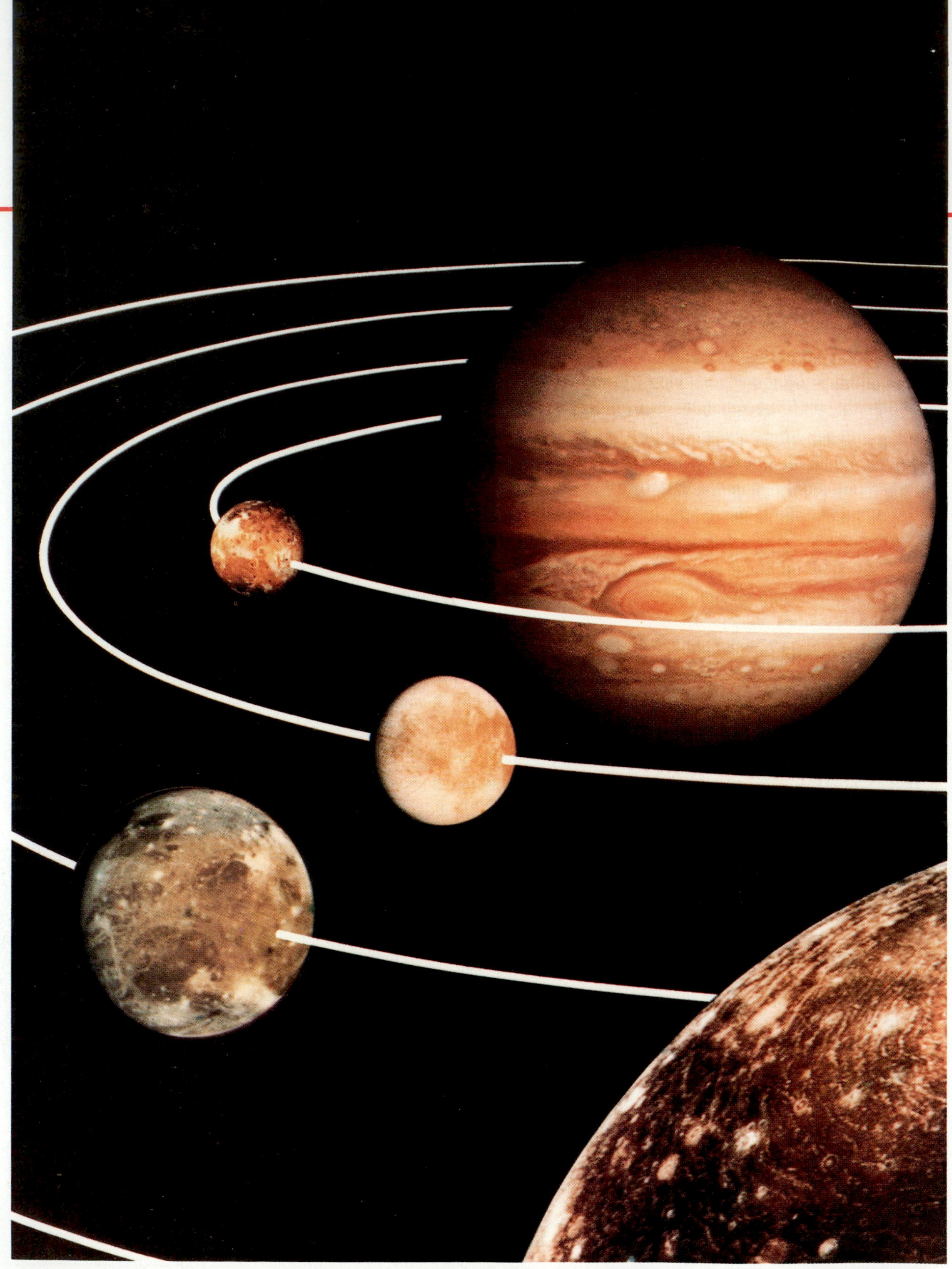

A mosaic of Jupiter and its four Galilean satellites made from the Voyager spacecraft. The satellites' orbits are drawn in. From inner to outer, the Galilean satellites are: Io, Europa, Ganymede, and Callisto. The symbol for Jupiter appears at the top of the opposite page.

Jupiter 12

Aims: To describe Jupiter, the dominant planet of the solar system; to discuss it as an example of a class of planets; and to meet a major new set of bodies—Jupiter's moons

The planets beyond the asteroid belt—Jupiter, Saturn, Uranus, and Neptune—are very different from the four "terrestrial" planets. These *giant planets,* or *jovian planets,* not only are much bigger and more massive but also are less dense. These facts suggest that the internal structure of these giant planets is entirely different from that of the four terrestrial planets.

Jupiter, also called Jove, was the chief Roman deity. "Jovian" is the adjectival form of "Jupiter."

12.1 Fundamental Properties

The largest planet, Jupiter, dominates the Sun's planetary system. Jupiter is 5 A.U. from the Sun and revolves around it once every 12 years. It alone contains two-thirds of the mass in the solar system outside of the Sun, 318 times as much mass as the Earth. Jupiter has at least 16 moons of its own and so is a miniature planetary system in itself. It is often seen as a bright object in our night sky, and observations with even a small telescope reveal bands of clouds across its surface and show four of its moons.

Jupiter is more than 11 times greater in diameter than the Earth. From its volume and mass, we calculate its density to be 1.3 grams/cm^3, not much greater than the 1 gram/cm^3 density of water. This tells us that any core of heavy elements (such as iron) that Jupiter may have does not make up as substantial a fraction of Jupiter's mass as the cores of the inner planets make up of their planets' masses. Jupiter, rather, is composed mainly of the lighter elements hydrogen and helium. Jupiter's chemical composition is closer to that of the Sun and stars than it is to that of the Earth (Fig. 12–1). It is mostly hydrogen, which occurs in Jupiter's outer parts as hydrogen molecules (H_2). In fact, 86.1 per cent of Jupiter's atmosphere, which is the only part we can measure directly, is hydrogen and 13.8 per cent is helium, making 99.9 per cent of those two gases. Because of the presence of so much hydrogen and the high pressures and temperatures in the planet's deeper layers, molecules involving hydrogen are not uncommon: we find ammonia (NH_3) instead of the nitrogen molecules (N_2) of our atmosphere and methane (CH_4) instead of the carbon dioxide of the atmospheres of the inner planets. Still, though they are more visible than the hydrogen and helium for technical reasons having to do with their spectra, the ammonia and methane total only about 0.1 per cent of the atmosphere.

Jupiter isn't solid; it has no crustal surface at all. At deeper and deeper levels, its gas just gets denser and denser, eventually liquefying. We know from Jupiter's average density that it must have heavier elements at its core, but we know little about them.

Jupiter rotates very rapidly, once every 10 hours. Undoubtedly, this rapid spin rate is a major reason for the colorful bands, which are clouds spread

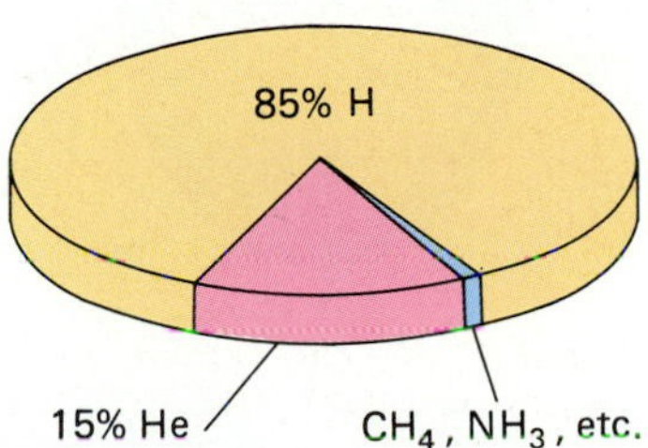

Figure 12–1 The composition of Jupiter.

Figure 12–2 Jupiter, photographed from the Earth, shows belts and zones of different shades and colors. By convention (that is, the set of terms we arbitrarily choose to use), the bright horizontal bands are called *zones* and the dark horizontal bands are called *belts*. Adjacent belts and zones rotate at different speeds.

out parallel to the equator. The bands appear in subtle shades of orange, brown, gray, yellow, cream, and light blue, and are beautiful to see (Fig. 12–2). They are in constant turmoil; the shapes and distribution of bands change in a matter of days.

The most prominent feature of the visible cloud surface of Jupiter is a large reddish oval known as the *Great Red Spot* (see also Figures 12–8 and 12–9). It is about 14,000 km × 30,000 km, larger than the Earth, and drifts about slowly with respect to the clouds as the planet rotates. The Great Red Spot is a relatively stable feature, for it has been visible for over 350 years. Sometimes it is relatively prominent, and at other times the color may even disappear for a few years. Other, smaller spots are also present. Three white ovals still present on Jupiter, for example, began to form in 1938.

Jupiter's rapid rotation makes the planet bulge at the equator. The distance from pole to pole is 7 per cent less than the equatorial diameter; we say that Jupiter is *oblate* (Fig. 12–3). Jupiter's axis of rotation is only 3° from the axis of Jupiter's orbit around the Sun, and so, unlike the Earth and Mars, Jupiter has no seasons.

In 1955, intense bursts of radio radiation were discovered to be coming from Jupiter. Though it was quite a surprise to detect any radio signals at all, several kinds of signals were detected. Some bursts are correlated with the positions of Jupiter's innermost moon, Io, with respect to Jupiter and to Earth. At shorter radio wavelengths, Jupiter emits continuous radiation.

The fact that Jupiter emits radio waves indicated that Jupiter, even more so than the Earth, has a strong magnetic field and strong *radiation belts* (actually, belts filled with magnetic fields in which particles are trapped, large-scale versions of the Van Allen belts of Earth). We can account for the presence of this radio radiation only with mechanisms that involve such magnetic fields and belts. High-energy particles passing through space interact with Jupiter's magnetic field to produce the radio emission (Fig. 12–4).

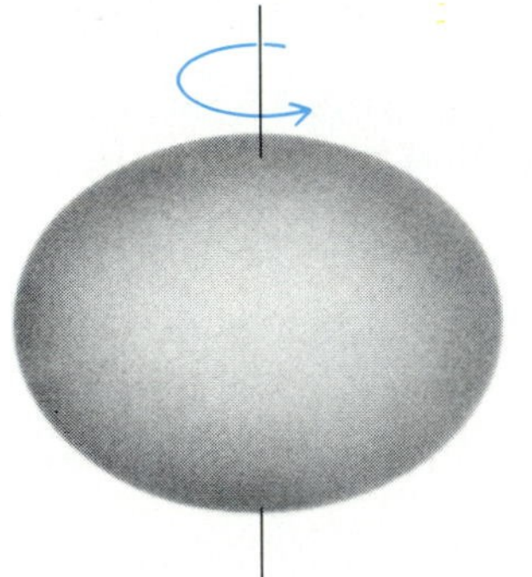

▲
Figure 12–3 The top ellipsoid is *oblate*; the bottom ellipsoid is *prolate*.

12.2 Jupiter's Moons: Early Views

Jupiter has at least 16 satellites. Four of the innermost satellites were discovered by Galileo in 1610 when he first looked at Jupiter with his telescope. These four moons are called the *Galilean satellites* (Fig. 12–5). One of these moons, Ganymede, at 5276 km in diameter, is the largest satellite in the solar system

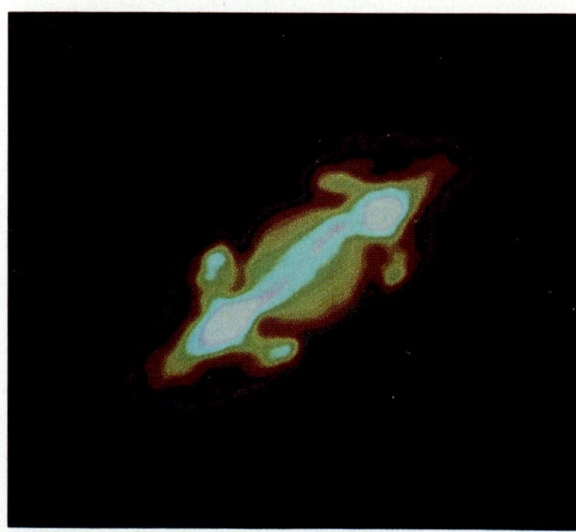

Figure 12–4 The radio emission from Jupiter, imaged with the VLA, shows its radiation belts. Observations were taken at a wavelength of 21 cm. The emission comes from electrons circling around the magnetic field in the Van Allen belts, and extends beyond 4 Jupiter radii. White shows the strongest emission.

Figure 12–5 Jupiter with the four Galilean satellites. These satellites were named Io, Europa, Ganymede, and Callisto by Simon Marius, a German astronomer who independently discovered them in 1611.

and is larger than the planet Mercury. The moons have been observed from a variety of telescopes on the ground and in space (Fig. 12–6).

The Galilean satellites have played a very important role in the history of astronomy. The fact that these particular satellites were noticed to be going around another planet, like a solar system in miniature, supported Copernicus's heliocentric model of our solar system. Not everything revolved around the Earth! A half-century later, Ole Rømer's study of the times that the moons eclipse each other led him to deduce—correctly—that light travels at a finite speed. When Jupiter and its moons are moving away from Earth, the times between eclipses of the moons by Jupiter are longer than when Jupiter and its moons are approaching.

The dozen moons of Jupiter that were discovered before the last decade fall into three groups. The first group includes the five innermost satellites then known (Amalthea and the Galilean satellites), which are also the largest. Their orbits are all less than 2 million km in radius. The orbits of the four moons in the second group are all about 12 million km in radius. The orbits of the four outermost moons in the third group are 21 to 24 million km in radius. These outermost moons revolve in the direction opposite from that of Jupiter's rotation (that is, retrograde), while the other moons revolve prograde. Moons in the outer two groups may be bodies captured by Jupiter's gravity after their formation elsewhere, for they have high inclinations and/or eccentricities.

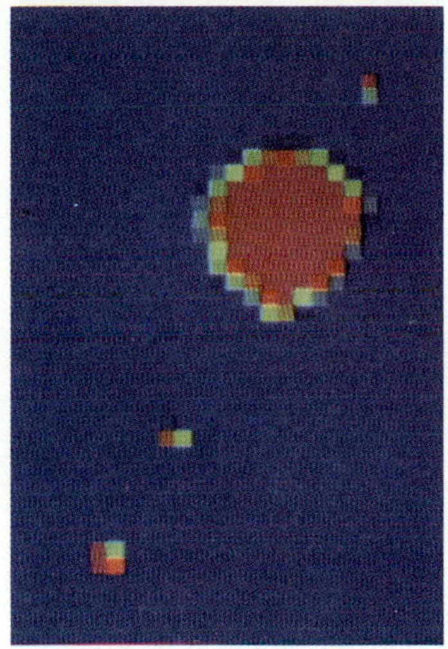

Figure 12–6 Io, Europa, and Ganymede as observed with the International Ultraviolet Explorer.

Focus On

Box 12.1 Jupiter's Satellites in Mythology

All the moons except Amalthea are named after lovers of Zeus, the Greek equivalent of Jupiter. Amalthea, a goat-nymph, was Zeus's nurse, and out of gratitude he made her into the constellation Capricorn. Zeus changed Io into a heifer to hide her from Hera's jealousy; in honor of Io, the crescent Moon has horns.

Ganymede was a Trojan youth carried off by an eagle to be Jupiter's cup bearer (the constellation Aquarius). Callisto was punished for her affair with Zeus by being changed into a bear. She was then slain by mistake, and rescued by Zeus by being transformed into the Great Bear in the sky. Jealous Hera persuaded the sea god to forbid Callisto to ever bathe in the sea, which is why Ursa Major never sinks below the horizon.

Europa was carried off to Crete on the back of Zeus, who took the form of a white bull. She became Minos's mother. Pasiphae (also the name of one of Jupiter's moons) was the wife of Minos and the mother of the Minotaur.

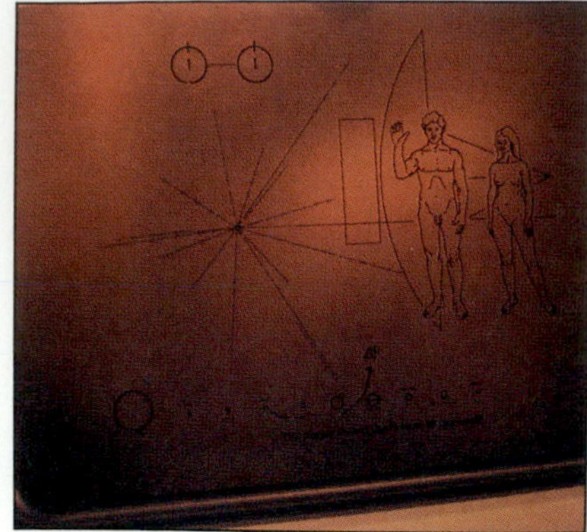

Figure 12–7 The plaques borne by Pioneers 10 and 11. A man and a woman are shown standing in front of an outline of the spacecraft, for scale. The spin-flip of the hydrogen atom (Section 30.4) is shown at top left. It also provides a scale, because even travellers from another solar system would know that the wavelength that results from the spin-flip is a certain length, which is 21 cm in our units. The Sun and planets (including distinctively ringed Saturn) and the spacecraft's trajectory from Earth are shown at bottom. (Will the visitors realize that we had not yet discovered the rings around Jupiter, Uranus, and Neptune?) The directions and periods of several pulsars are shown in the rays extending from the midpoint at left. Numbers are given in the binary system.

The 1.2-m Palomar Schmidt telescope has been used to search for faint moons. Two were found. Spacecraft to Jupiter have discovered three satellites.

12.3 Spacecraft Observations

The Pioneer 10 and Pioneer 11 spacecraft gave us our first close-up views of Jupiter in 1973 and 1974. A second revolution in our understanding of Jupiter occurred in 1979, when Voyager 1 and Voyager 2 also flew by Jupiter.

Each spacecraft carried many types of instruments to measure properties of Jupiter, its satellites, and the space around them. The observations made with the imaging equipment were of the most popular interest. The resolution of the Voyager images was five times better than that of the best images we can obtain from Earth. The Voyagers provided a quantum leap in image quality over even the Pioneers. In as little as 48 seconds, the Voyagers could send a full digitally coded picture back to Earth, where detailed computer work improved the images. This was remarkable for a vehicle so far away that its signals travelled 52 minutes to reach us. The energy in the signal would have to be collected for billions of years to light a Christmas tree bulb for just one second!

The Pioneer and Voyager spacecraft are travelling fast enough to escape the solar system. As they depart, the spacecraft are radioing back valuable information about other planets and about interplanetary conditions. In about 80,000 years the spacecraft will be about one parsec (about three light years) away from the solar system. The Pioneers bear plaques that give information about their origin and about life on Earth, in case some interstellar traveller from another solar system happens to pick it up (Fig. 12–7). The Voyagers bear more elaborate mementos, as we shall describe in Section 19.4.

12.3a The Great Red Spot

The Great Red Spot shows very clearly in many of the images (Fig. 12–8). It is a gaseous island (or isolated storm) many times larger than the Earth. Its top lies about 8 km above the neighboring cloud tops. Storms on Earth and other rotating planets usually rotate. Storms are called cyclones when their centers have low pressure and anticyclones when their centers have high pressure. The Great Red Spot is an anticyclonic storm lying in the southern hemisphere; it has counterclockwise winds (Fig. 12–9). We still have many things to learn about Jupiter's Great Red Spot. No upwelling has been detected in its center,

A

B

Figure 12–8 Jupiter from (*A*) Voyager 1, and, (*B*) 4½ months later, from Voyager 2. Note how the white ovals have drifted around the Great Red Spot. Winds blow at different speeds at different latitudes, so clouds pass each other.

Figure 12–9 The Great Red Spot. The white oval south of the Great Red Spot is similar in structure. Both rotate in the anticyclonic (counterclockwise) sense.

though some would be expected in such a huge storm. We don't even know why it is red, though traces of phosphorus have been suggested as the coloring agent.

Time-lapse photographs made from Voyager data have revealed the flow within and around the Spot. Clouds within the Spot rotate counterclockwise with a period of about 6 days. The photographs also show how the Great Red Spot interacts with surrounding clouds and smaller spots (Fig. 12–9). Similar spots, the so-called white ovals, occur at other latitudes and have been shown by Voyager to have similar internal flows, but none is as large or as colorful as the Great Red Spot.

In new ideas about Jupiter's storms and wind currents (Fig. 12–10), the so-called *Coriolis force* plays a key role. The Coriolis force is not a true force,

Figure 12–10 The surface of Jupiter as seen from Voyager 1 (*top*) and Voyager 2 (*bottom*) is unrolled. Note the relative motions from side to side of the Great Red Spot and other features.

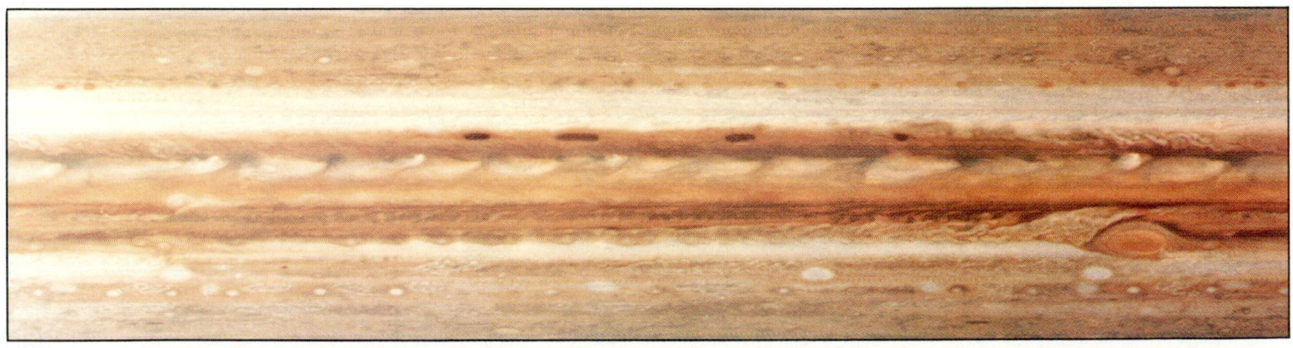

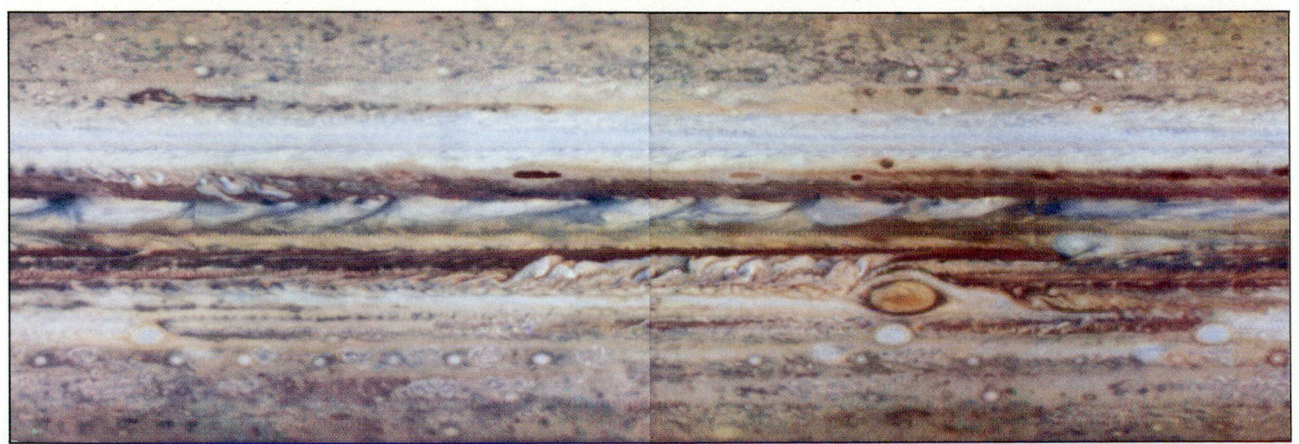

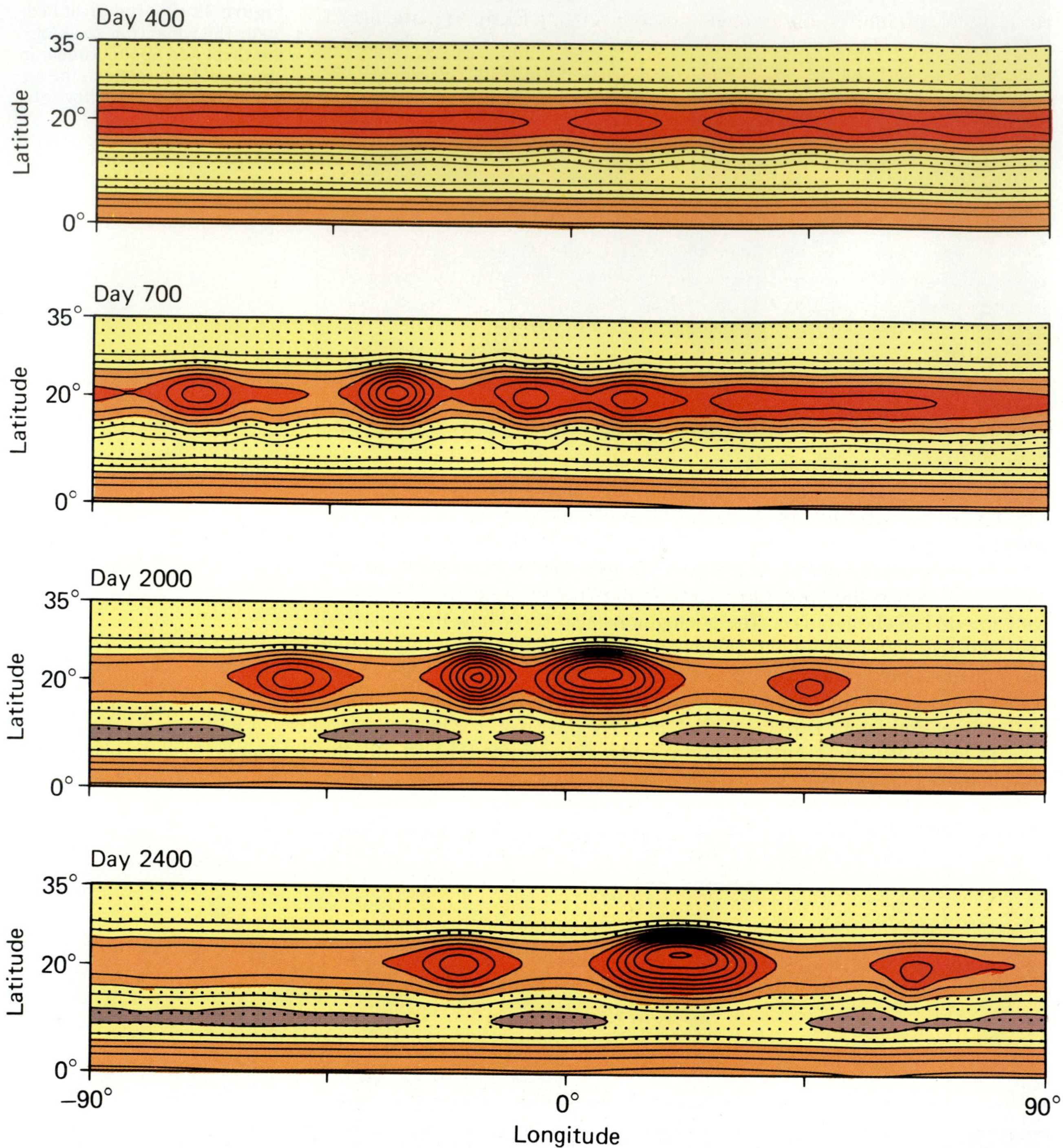

Figure 12–11 A computer simulation shows how disturbances to unstable currents in a gas on a rapidly rotating planet can develop. They can lead, as here, to multiple, unequal vortices that merge during encounters to produce a single permanent vortex. The Great Red Spot may well have formed in this way.

In this simulation, we begin with 3 Jupiter-like horizontal currents. A weak perturbation is introduced, and grows by taking excess energy from the strong westward jet (days 500–600). Because the perturbation is not uniform in longitude, it develops into a series of unequal vortices (by day 2000). The stronger vortices travel faster than the weaker ones and so collide (days 2000–2200). In the collisions, the stronger vortex absorbs most of the weaker ones. Eventually, after multiple encounters, a single dominant vortex emerges. (Computer model by Gareth P. Williams and R. John Wilson, 1985) but is merely an illusion of our motion on our rotating Earth, in particular of the system of rotating latitude and longitude coordinates we fix to the Earth. Consider the following to explain the Coriolis force: Imagine you are looking down on the north pole of the Earth as it rotates from west to east. A point on the Earth's equator rotates with a greater velocity than a point at a latitude above (or below) the equator. If you were to shoot an object northward from the equator, the object has a relatively large eastward velocity. Since the higher latitude is not moving eastward as fast, the object will land eastward of its original longitude. If we were at the higher latitude watching the projectile, we would think it were curving eastward. Similarly, an object shot southward from the equator would also appear to curve eastward. Though only the Earth's rotation and no strange force is involved, it looks cursorily as though a force has made the object swerve eastward; this fictional force is known as the Coriolis force. Using the idea of a Coriolis force often simplifies calculations.

Jupiter rotates very rapidly, so the effect that we call the Coriolis force is particularly strong. Because of it, disturbances in the atmosphere tend to be wave-like, with low- and high-pressure centers forming and growing into storms. Computer models show that small eddies generate such waves, which, in turn, generate the eastward and westward currents. Such currents can produce large vortices like the Great Red Spot. When many unequal spots are produced, the stronger ones absorb the weaker ones until only a single vortex is left (Fig. 12–11). Such vortices last a long time. A terrestrial laboratory experiment reported in 1988 matched computer calculations and developed such a devouring and persisting vortex in a water tank (Fig. 12–12).

Alternative models, also using supercomputer calculations, indicate that the condition with one giant spot is a natural state, since it minimizes the energy present in the large-scale flow on Jupiter's surface. Often, in physics, the natural state is one of minimum energy, just as a ball runs down a hill and remains in equilibrium in a valley.

Though giant vortices aren't as obvious on Earth as they are on Jupiter, some large-scale (continental) weather systems on Earth—with high-pressure regions sitting in the same place for weeks—may correspond. Circulating rings that break off from the Gulf Stream in the Atlantic Ocean are other possible parallels. So our studies of Jupiter may have applications on Earth in understanding weather and ocean currents.

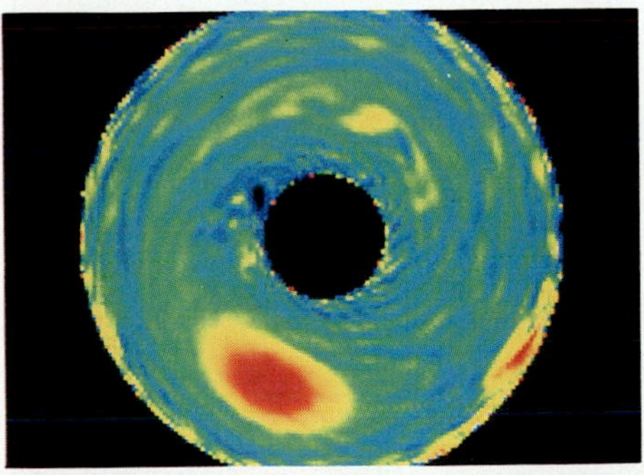

A

B

Figure 12–12 (*A*) A computer model of the Great Red Spot, the result of merging many weaker spots. The model is by Philip Marcus of the University of California at Berkeley. (*B*) A laboratory simulation of the Great Red Spot in a water-filled vat at the University of Texas designed to match the Marcus computer model. Researchers spin the vat to simulate Jupiter's rotation while pumping water in through the six valves nearest the center and out through the valves in the middle ring (the outer ring was not used). The Coriolis effect makes a current in the opposite direction from the rotating vat, and vortices form in the shear zone. Small vortices merge into larger ones, and a large vortex like Jupiter's Great Red Spot results.

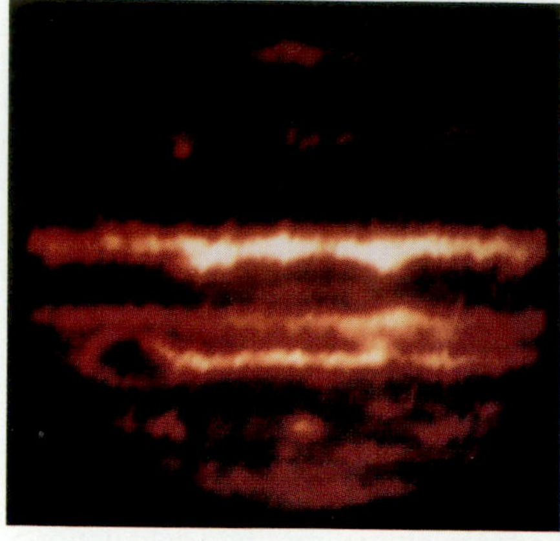

Figure 12–13 (*left*) Ground-based infrared observations with the 5-m Palomar telescope show bright regions where the temperature is the highest, presumably where we can see deepest into the clouds. The Great Red Spot appears on the left limb as a dark area encircled by a bright ring, indicating that the Spot is cooler than the region that surrounds it. (*right*) This view in the visible was taken only one hour later from Voyager.

12.3b Jupiter's Atmosphere

Jupiter has bright bands called *zones* and dark bands called *belts*. The zones are rising gas, while the belts are falling gas. The tops of these dark belts are somewhat lower (about 20 km) than the tops of the zones and so are about 10 K warmer (Fig. 12–13). The colors we see in the belts may have to do with sulfur compounds in the middle clouds, or perhaps with organic molecules.

Voyager measurements of wind velocities showed that each hemisphere of Jupiter has a half-dozen currents blowing eastward or westward. The Earth, in contrast, has only one westward current at low latitudes (the trade winds) and one eastward current at middle latitudes (the prevailing westerlies).

Extensive lightning storms, including giant-sized lightning strikes called "superbolts," were discovered from the Voyagers, as were giant aurorae.

Pioneer 11 gave scientists their first look at Jupiter's poles, which never show well from Earth. At high latitudes the band pattern we see at the equator is destroyed. The bands break up into eddies. The Voyagers allowed detailed mapping of both poles (Fig. 12–14). With information like this, we can now study "comparative atmospheres": those of the Earth, Venus, Mars, Jupiter, Saturn, Uranus, and Neptune.

Figure 12–14 Computer reconstructions of Voyager images of Jupiter's poles show the views that would be seen from directly above the poles. Irregularly shaped regions directly at the poles, for which no photographs were available at this 600-km resolution, appear black. (*A*) The south polar region. The photograph shows the Great Red Spot and a disturbance trailing from it that extends halfway around the planet. (*B*) The north polar region.

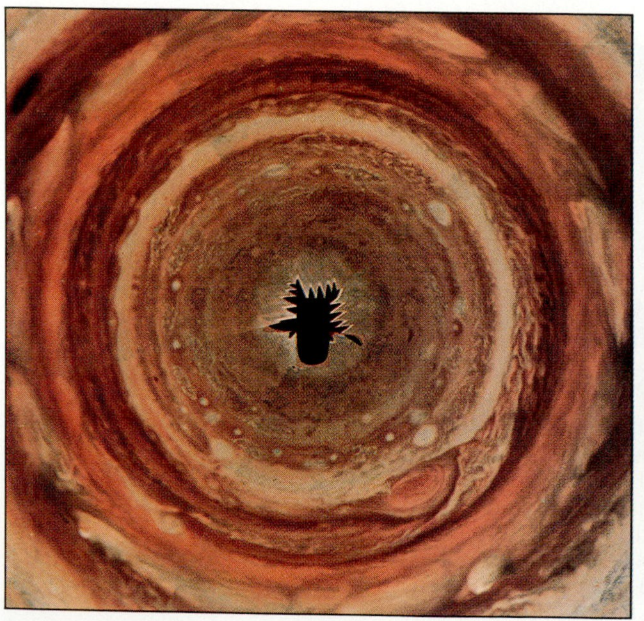

A

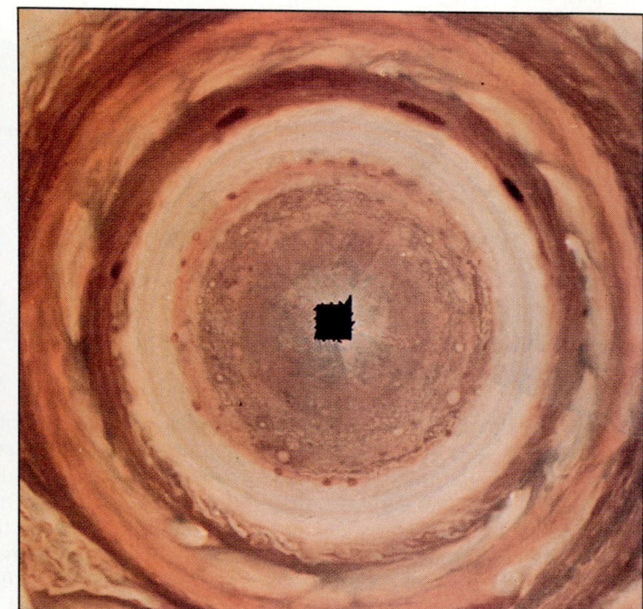

B

12.3c Jupiter's Interior

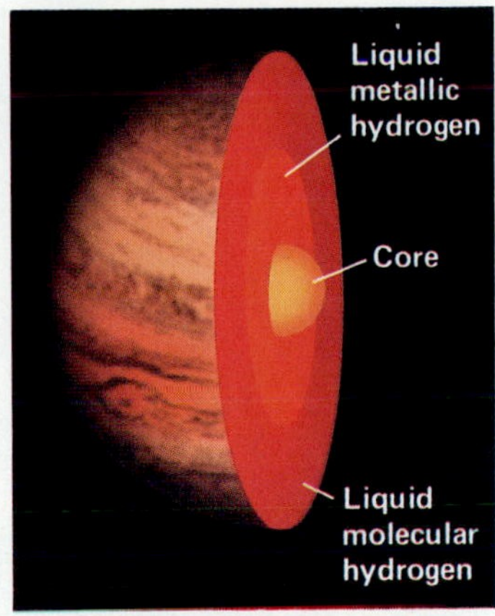

Figure 12–15 The current model of Jupiter's interior.

Data from the Pioneers and Voyagers increased our understanding not only of the atmosphere but also of the interior of Jupiter (Fig. 12–15). Most of the interior is in liquid form. Jupiter's central temperature may be between 13,000 and 35,000 K. The central pressure is 100 million times the pressure of the Earth's atmosphere measured at our sea level because of Jupiter's great mass pressing in. Because of this high pressure, Jupiter's interior is composed of ultracompressed hydrogen surrounding a rocky core consisting of 20 Earth masses of iron and silicates. The heavy elements should have sunk to the center, but we have no direct evidence that they are there.

The lower levels of the liquid hydrogen are in the form of molecules that have lost their outer electrons. In this state, the hydrogen conducts heat and electricity. Since these properties are basic to our normal terrestrial definition of "metal," we say that the hydrogen is in a "metallic" state. This liquid metallic region makes up 75 per cent of Jupiter's mass. It is probably where Jupiter's magnetic field is generated by the interaction of mass, motion, and an electric field. (This process also takes place in "dynamos" on Earth, which convert mechanical energy into electricity.)

A major limitation on our ability to model Jupiter's interior is our lack of understanding of hydrogen under extreme pressure. Scientists on Earth have been experimenting with ultrahigh pressures by using diamond anvils—bits of diamond with small ends that are forced together. Since pressure is force divided by area, the ability to have very small areas leads to very high pressures. In 1989, it was discovered that at 2.5 million times the pressure of our atmosphere, hydrogen solidifies. Experiments continue. Among future steps is the measurement of the electrical conductivity of the solid hydrogen.

Jupiter radiates 1.6 times as much heat as it receives from the Sun, leading to Jupiter's complex and beautiful cloud circulation pattern. There must be some internal energy source—perhaps the energy remaining from Jupiter's collapse from a primordial gas cloud 20 million km across to a protoplanet 700,000 km across, 5 times the present size of Jupiter. Jupiter is undoubtedly still contracting. It lacks the mass necessary by a factor of about 75, however, to have heated up enough for nuclear reactions to begin.

12.3d Jupiter's Magnetic Field

The Pioneer missions showed that Jupiter's tremendous magnetic field is even more intense than many scientists had expected, a result confirmed by the Voyagers (Fig. 12–16). At the height of Jupiter's clouds the magnetic field is 10 times that of the Earth, which itself has a strong field.

The inner field is toroidal, shaped like a doughnut, containing several shells like giant versions of the Earth's Van Allen belts. High-energy protons and electrons are trapped there. The satellites Amalthea, Io, Europa, and Ganymede travel through this region. The middle region, with charged particles being whirled around rapidly by the rotation of Jupiter's magnetic field, does not have a terrestrial counterpart. The outer region of Jupiter's magnetic field interacts with the solar wind as far as 7 million km from the planet and forms a shock wave, as does the bow of a ship plowing through the ocean. When the solar wind is strong, Jupiter's outer magnetic field (shaped like a flattened pancake) is pushed in. Jupiter's magnetic field and radiation belts not only show up in the particle sensors but also are revealed by the x-rays discovered by Voyager.

Pioneer 11 observed a 10-hour fluctuation in the magnetic field. This fluctuation occurs because the magnetic axis is tilted 10 degrees from the axis of rotation, and the center of the field is not exactly at the center of the planet. The magnetic field thus resembles a wobbly disk. Also, Jupiter's magnetic

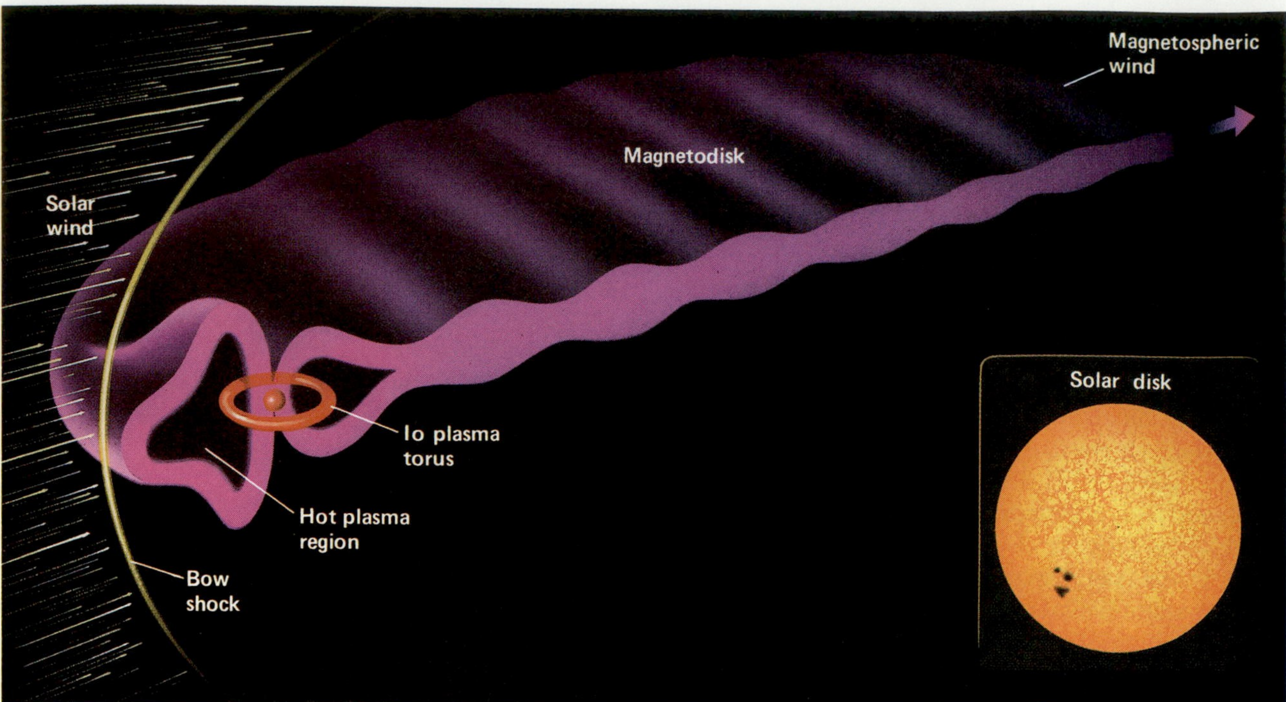

Figure 12–16 An artist's view of Jupiter's magnetosphere, the region of space occupied by the planet's magnetic field. It rotates at hundreds of thousands of kilometers per hour around the planet, like a big wheel with Jupiter at the hub. The inner magnetosphere is shaped like a doughnut with the planet in the hole. The highly unstable outer magnetosphere is shaped as though the outer part of the doughnut has been squashed. The outer magnetosphere is "spongy" in that it pulses in the solar wind like a huge jellyfish. It often shrinks to one-third of its largest size. Volcanic emissions from Io form a torus of plasma; the plasma then escapes into the rest of the magnetosphere.

Jupiter's rapid rotation heats some of the plasma to a few hundred million degrees. Eventually this hot plasma flows away from the planet at high speeds to form a magnetospheric wind.

The size of the solar disk is shown for scale.

poles are reversed compared with the Earth's: our north pole is on the same side of the plane of planetary orbits as is Jupiter's south pole.

12.4 Jupiter's Moons Close Up

Through Voyager close-ups, the satellites of Jupiter have become known to us as worlds with personalities of their own. The four Galilean satellites, in particular, were formerly known only as dots of light. Since these satellites range between 0.9 and 1.5 times the size of our own Moon, they are substantial enough (Fig. 12–17) to have interesting surfaces and histories (Fig. 12–18).

Io provided the biggest surprises. Scientists knew that Io gave off particles as it went around Jupiter, but Voyager 1 discovered that these particles resulted from active volcanoes on the satellite (Fig. 12–19). Eight volcanoes were seen actually erupting (Fig. 12–20), many more than erupt on the Earth at any one time. When Voyager 2 went by a few months later, most of the same volcanoes were still erupting. (Incidentally, astronomers do not agree whether to say "eye'oh" or "ee'oh.")

Io's surface (Fig. 12–21) has been transformed by the volcanoes, and overall is the youngest surface we have observed in our solar system. Gravitational forces from the other Galilean satellites pull Io slightly inward and

Figure 12–17 Jupiter's four moons to scale. We see Io (*top left*), Europa (*top right*), Ganymede (*lower left*), and Callisto (*bottom right*). Each is an individual, very different from the other moons.

The International Astronomical Union names features on the moons of the outer planets after the people who discovered them or other relevant associations. For example, names on Europa have to do with southern Europe.

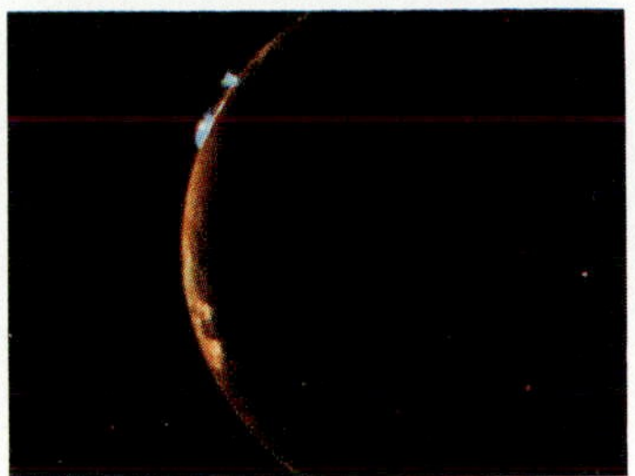

Figure 12–19 The volcanoes of Io can be seen erupting on this photograph. Linda Morabito, a JPL engineer, discovered the first volcano. At the time, she was studying images that had been purposely overexposed to bring out the stars so that they could be used for navigation. She saw the large volcanic cloud barely visible on the right limb. The plume extends upward for about 250 km and is scattering sunlight toward us.

Names on Io are those from mythology that have to do with fire, including the Greek Prometheus and the Hawaiian volcano god Pele.

Figure 12–18 Io against the Great Red Spot and Europa against a white oval.

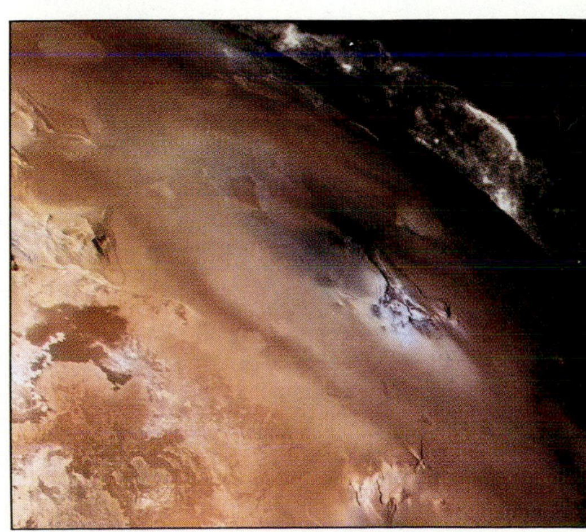

Figure 12–20 Io's surface has been transformed by the sulfur erupted from volcanoes like the one on the limb.

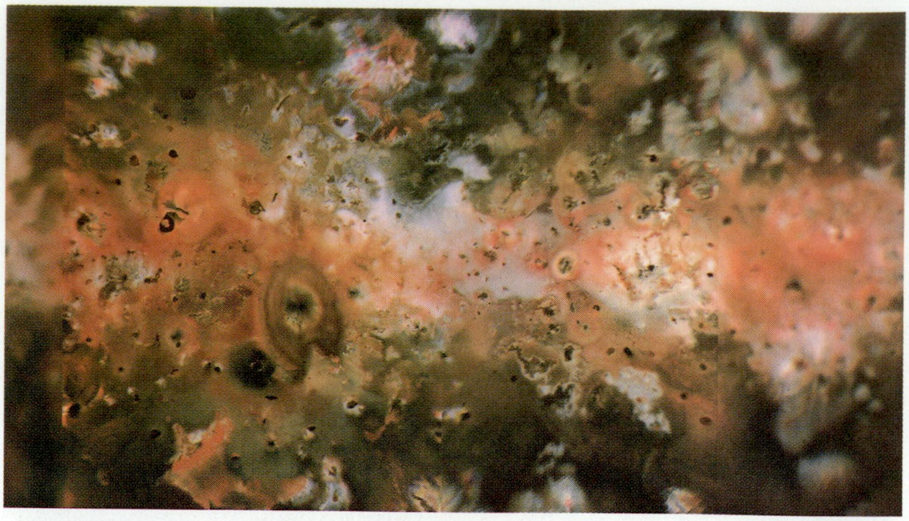

Figure 12–21 A Mercator projection of Io's surface, made in 1987. The volcano Pele is the most prominent feature.

outward as it orbits Jupiter, flexing the satellite. Its surface moves in and out by as much as 100 meters. This squeezing and unsqueezing creates heat from friction, which presumably heats the interior and leads to the vulcanism. Io may have an underground ocean of liquid sulfur. The surface of Io is covered with sulfur and sulfur compounds, including frozen sulfur dioxide, and the atmosphere is full of sulfur dioxide. Sulfur can take on many colors, depending on how it cools. Io certainly wouldn't be a pleasant place to visit.

Io has mountains up to 10 km tall. Sulfur mountains that high couldn't support themselves, so perhaps these are silicate mountains with merely a coating of sulfur.

Once we knew about them from the Voyagers, we could follow the volcanoes and hot spots with infrared detectors on telescopes on Earth. The Loki lava lake remains the major radiator, even after many years.

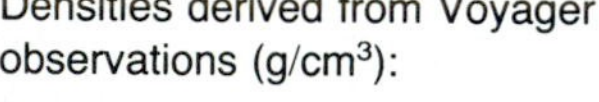

Densities derived from Voyager observations (g/cm³):

Io	3.5
Europa	3.0
Ganymede	1.9
Callisto	1.8

The low densities indicate that Ganymede and Callisto contain large amounts of water ice. The high heat of Jupiter's first 100 million years drove off lighter elements from these two moons.

What causes Io's high internal temperature, which has led to the vulcanism? Jupiter is hundreds of times more massive than Earth yet is at the same distance from Io that the Earth is from the Moon. Thus Io has a huge tidal bulge. Europa and Ganymede pull Io back and forth a bit as they all orbit Jupiter, and Io and its tidal bulge flex. The flexing generates a lot of energy.

Gases—including sulfur, sodium, potassium, and oxygen—given off by Io fill a doughnut-shaped region that surrounds Io's orbit (Fig. 12–22). Further, as Io moves in its orbit through the jovian magnetic field, a huge electric current is set up between Io and Jupiter itself in a tube of concentrated magnetic field called a *flux tube* (Fig. 12–23). A set of mutual eclipses of Jupiter's satellites in 1985 allowed the amount of sodium in Io's atmosphere to be measured. Europa was used as a mirror to look with an Earth-based telescope at the Sun through Io's atmosphere.

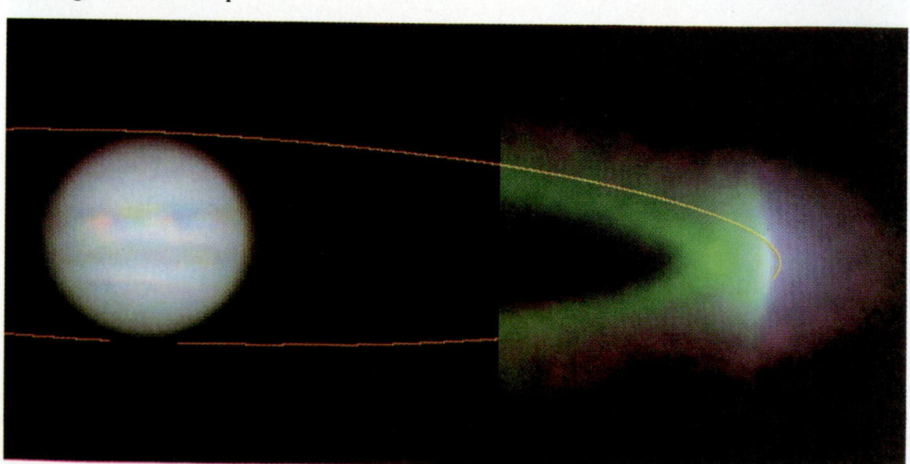

Figure 12–22 The Io Torus, sulfur and oxygen ions in Io's orbit. This false-color view shows sulfur gas with once-ionized sulfur in green and twice-ionized sulfur in red. These images were obtained with a telescope on Earth at the Las Campanas Observatory in Chile.

Figure 12–23 An artist's conception of the tube of magnetic field, a *flux tube*, that extends out of the plane of Io's orbit and carries a huge current.

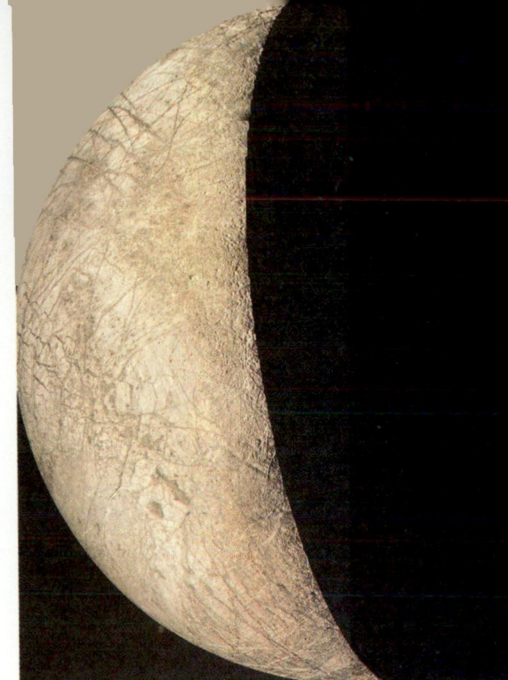

Figure 12–24 The surface of Europa, which is very smooth, is covered by this complex array of streaks. Few impact craters are visible.

Europa, the brightest of Jupiter's Galilean satellites, has a very flat uncratered surface and is covered with narrow dark stripes (Fig. 12–24). This suggests that the surface we see is ice. And spectra in the infrared prove that the surface is ice. The markings may be fracture systems in the ice, like fractures in the large fields of sea ice near the Earth's north pole. Few craters are visible, suggesting that the ice was soft enough below the crust to close in the craters. Either internal radioactivity or a gravitational heating like that inside Io may have provided the heat to soften the ice. Because it possibly has liquid water and some extra heating, some scientists consider it a worthy location to check for signs of life.

The largest satellite, Ganymede (Fig. 12–25), shows many craters alongside weird, grooved terrain (Fig. 12–26). Ganymede is larger than Mercury but less dense. It contains large amounts of water-ice surrounding a rocky core. Ganymede's rotation is linked to its orbital period. But an icy surface is hard

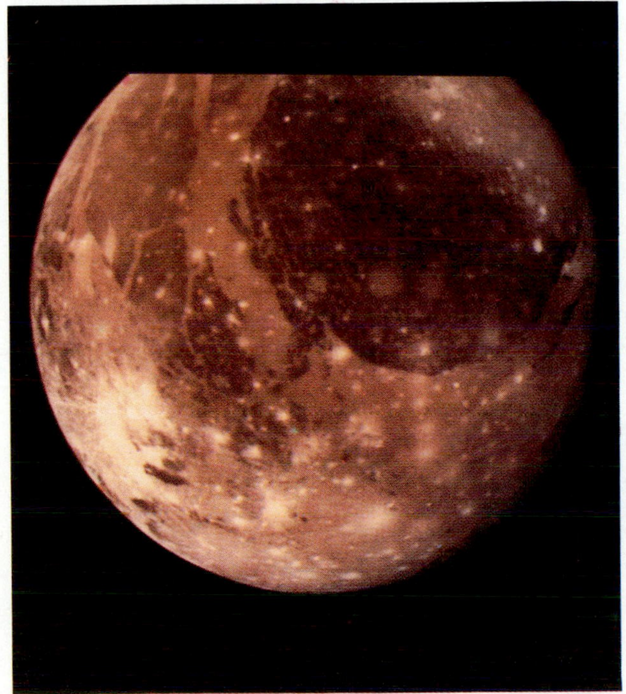

Figure 12–25 This photograph of Ganymede, Jupiter's largest satellite, shows numerous impact craters, many with bright ray systems. The large dark region has been named Galileo Regio. Low-albedo features on satellites are named for astronomers.

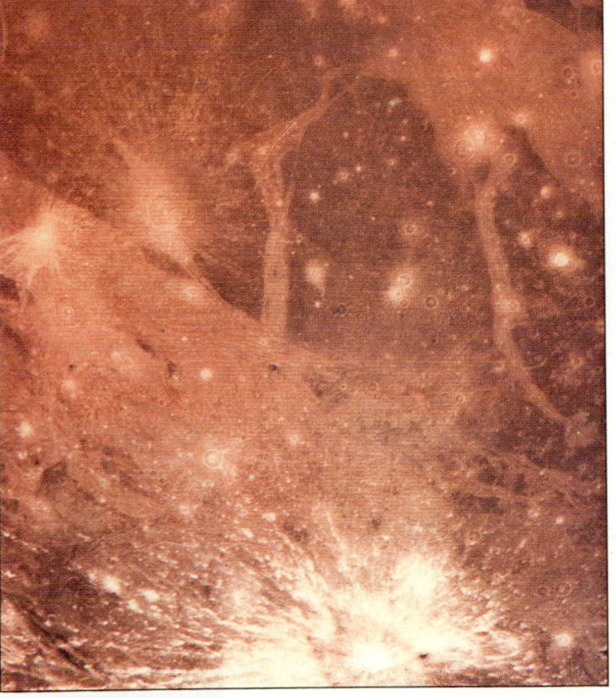

Figure 12–26 The terrain over large areas of Ganymede is covered with many grooves tens of kilometers wide, a few hundred meters high, and up to thousands of kilometers long. Younger grooves cover older grooves, and offsets sometimes occur along series of grooves.

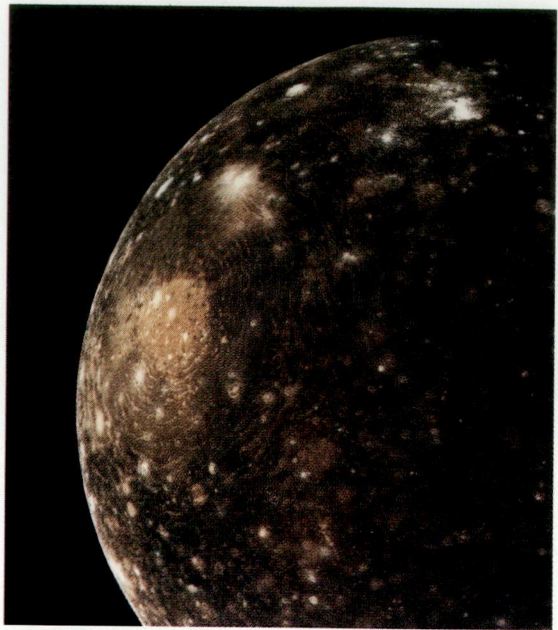

Figure 12–27 Callisto is cratered all over, very uniformly. Several craters have rays or concentric rings. No craters larger than 50 km are visible, from which we deduce that Callisto's crust cannot be very firm. The fact that the limb is so smooth indicates that there is no high relief. Because there are so many craters, the surface must be very old, probably over 4 billion years. The large basin, named Valhalla, is lighter and younger than the rest of Callisto's surface.

in the cold conditions that far from the Sun if it is not otherwise heated, so the cratered surface remains from perhaps 4 billion years ago. The grooved terrain is younger. The appearance of the side facing away from Jupiter is dominated by a dark, cratered region. Few craters are visible near its center. The formation of the younger grooved terrain may have erased the evidence of the ancient impact there.

Ganymede also shows lateral displacements, where grooves have slid sideways, like those that occur in some places on Earth (for example, the San Andreas fault in California). Ganymede is the only place besides the Earth where such faults have been found. Thus further studies of Ganymede may help our understanding of terrestrial earthquakes. Other signs of modification of Ganymede's surface by tectonic processes in Ganymede's early years have also been seen.

Callisto has so many craters (Fig. 12–27) that its surface must be the oldest of Jupiter's Galilean satellites. Spectra indicate that Callisto is probably also covered with ice. No large craters are found; perhaps the ice flows slightly at Callisto's distance from the Sun, erasing the large craters. A huge bull's-eye formation, Valhalla, contains about 10 concentric rings, no doubt resulting from a huge impact. Perhaps ripples spreading from the impact froze into the

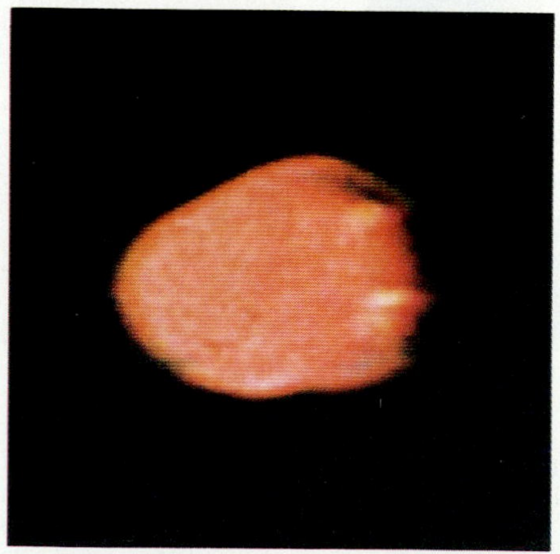

Figure 12–28 Amalthea, which orbits Jupiter every 12 hours, only 1.55 Jupiter radii above Jupiter's clouds. Amalthea's albedo is only 10%, so it is much darker than the Galilean satellites.

ice to make Valhalla. Despite its general appearance, Valhalla differs greatly in detail from ringed features on the Moon and Mercury; its rings are asymmetric, for example.

Voyager also observed Amalthea (Fig. 12–28), formerly thought to be Jupiter's innermost satellite. This small chunk of rock is irregular and oblong in shape, looking like a dark red potato complete with potato "eyes." It is comparable in size to many asteroids, ten times larger than the moons of Mars and ten times smaller than the Galilean satellites. Its reddish color may be caused by material from Io.

The Voyagers discovered three previously unknown satellites of Jupiter, making the total of known moons 16.

Studies of Jupiter's moons tell us about the formation of the Jupiter system, and help us better understand the early stages of the entire solar system.

12.5 Jupiter's Ring

Though Jupiter wasn't expected to have a ring, Voyager 1 was programmed to look for one just in case; Saturn's ring, of course, was well known, and Uranus's ring had been discovered only a few years earlier. The Voyager 1 photograph indeed showed a wispy ring of material around Jupiter at about 1.8 times Jupiter's radius, inside its innermost moon (Fig. 12–29). As a result, Voyager 2 was targeted to take a series of photographs of the ring. From the far side looking back, the ring appeared unexpectedly bright. This brightness results from very small—micron-size—particles in the ring that scatter the light toward us. Within the main ring, fainter material appears to extend down to Jupiter's surface (Fig. 12–30). A diffuse "gossamer ring" is outside the main ring.

The ring particles may come from Io, or else they may come from comet and meteor debris or from material knocked off the innermost moons by meteorites. The individual particles probably remain in the ring only temporarily. Perhaps Jupiter's newly discovered innermost moon causes the rings to end where they do.

In the next chapter, we shall discuss the fundamental ideas about how gravity leads to the formation of planetary rings, in the context of the most famous ring of all: the ring of Saturn.

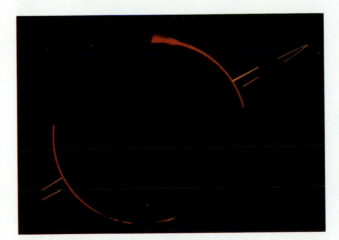

Figure 12–29 Jupiter's ring, in a backlighted view, seen from Voyager 2.

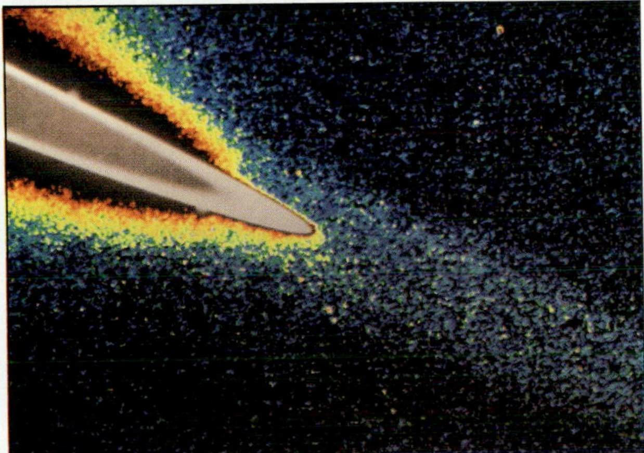

Figure 12–30 Jupiter's ring, with a fainter outer ring revealed by subsequent computer processing. The existence of this "gossamer ring" has forced a retargeting of the Galileo mission, which had been supposed to pass through this region.

Focus On

Box 12.2 Project Galileo

NASA's Project Galileo (Fig. 12–31) is carrying an orbiter and an atmospheric probe to Jupiter. It was delayed by the Challenger explosion and was finally launched in 1989 (Fig. 12–32). It must take a slower route to Jupiter because post-explosion safety concerns prevented it from using as powerful a rocket to leave Earth orbit as originally planned. After a lengthy flight from Earth, Galileo received a gravity assist as it swung around Venus (!) and then will receive two gravity assists in two passes by the Earth separated by two years. It will arrive at Jupiter on December 7, 1995. As of early 1992, its main antenna would not open fully. Most of the data will be lost if it cannot be opened.

The Probe will transmit data for 75 minutes as it falls 700 km through the jovian atmosphere (Fig. 12–33). It should give us accurate measurements of Jupiter's composition, the heights of the cloud layers, and the variations of temperature, density, and pressure (Fig. 12–34).

The Orbiter will orbit Jupiter ten times in a 22-month period (Fig. 12–35), coming so close to several of Jupiter's moons that pictures will have resolutions 100 times greater than even those from the Voyagers.

Project Galileo will travel by a VEEGA orbit: Venus-Earth-Earth Gravity Assist.

Venus flyby	Feb. 9, 1990
Earth flyby	Dec. 8, 1990
Asteroid (Gaspra)	Oct. 29, 1991
Earth flyby	Dec. 8, 1992
Asteroid (Ida)	Aug. 28, 1993
Jupiter arrival	Dec. 7, 1995

12.6 Future Exploration of Jupiter

Voyager 1 and Voyager 2 were spectacular successes. The data they provided about the planet Jupiter, its satellites, and its ring dazzled the eye and allowed us to make new theories that revolutionized our understanding.

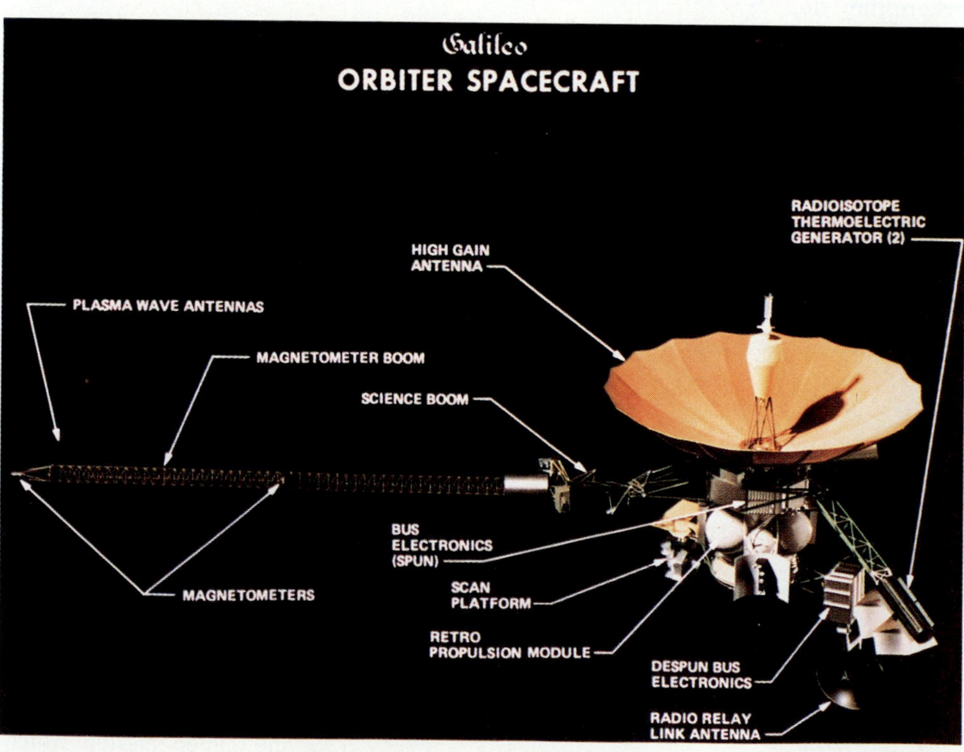

Figure 12–31 The Galileo Orbiter.

Figure 12–32 The launch of Galileo from a space shuttle on October 18, 1989.

Figure 12–33 The probe portion of Project Galileo. Six instruments inside the probe will measure composition of Jupiter's atmosphere, locate clouds, study lightning and radio emission, find where energy is absorbed, and measure precisely the ratio of hydrogen to helium.

Jupiter's large mass makes it a handy source of energy to use to send probes to more distant planets. If you bounce a ball off a wall, it comes back to you at about the same velocity at which you threw it. But if you bounce the ball off a train that is coming right at you, the train adds energy to the ball and the ball comes back at you rapidly. Similarly, some of Jupiter's energy can be transferred to a spacecraft through a gravitational interaction. This *gravity-assist*

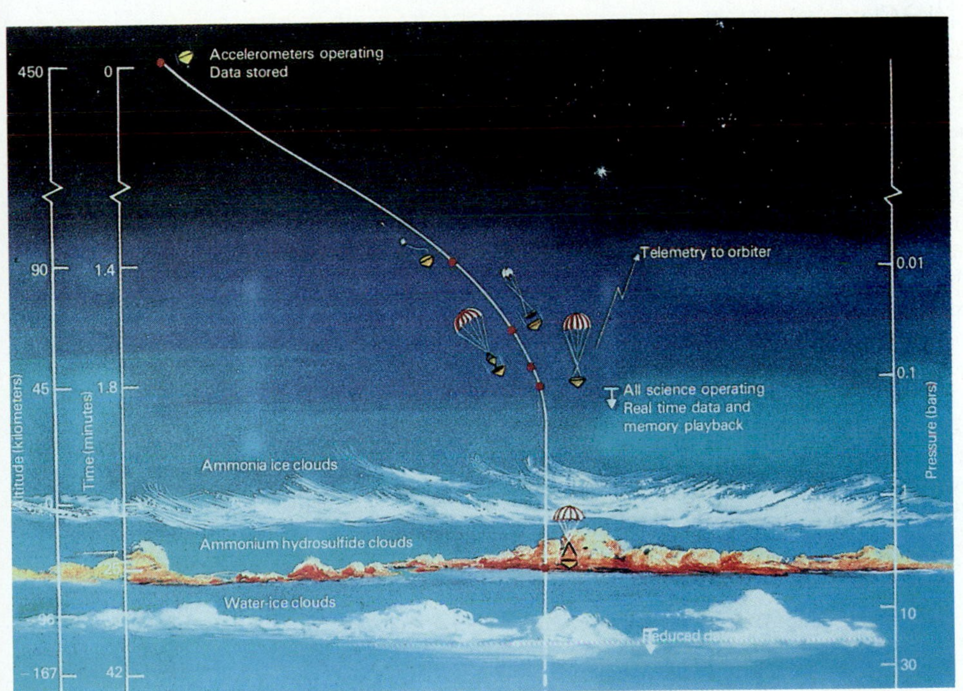

Figure 12–34 The mission of the Galileo probe.

Figure 12–35 Jupiter, photographed with the Wide-Field/Planetary Camera on the Hubble Space Telescope. True-color images like this one will be taken periodically.

method has sent Pioneer 11 and the Voyagers to Saturn, Voyager 2 beyond to Uranus and Neptune, and Galileo to Jupiter. It has been used in other planetary missions as well.

The resolution of images of Jupiter taken with the Hubble Space Telescope in Earth orbit is approximately equal to that from Voyager (Fig. 12–35).

Summary and Outline

Fundamental properties (Section 12.1)
- Highest mass and largest diameter of any planet
- Low density: 1.3 grams/cm^3
- Composition primarily of hydrogen and helium
- Rapid rotation, causing oblateness of disk
- Colored bands on surface; Great Red Spot
- Intense bursts of radio radiation
- Strong magnetic field and radiation belts

Jupiter's moons (Section 12.2)
- Historic role in Copernican theory

Spacecraft observations (Section 12.3)
- Pioneer 10 and 11 flybys in 1973 and 1974
- Voyager 1 and 2 flybys in 1979; increased resolution
- Great Red Spot is a giant, anticyclonic storm
- Gaseous atmosphere and liquid interior
- Jupiter radiates 1.6 times the heat it receives
- Tremendous magnetic field detected

Spacecraft observations of moons (Section 12.4)
- Io is transformed by volcanoes
- Europa has a flat surface, crisscrossed with dark markings
- Ganymede has many craters and grooved terrain
- Callisto is covered with craters, including a bull's-eye
- Amalthea is irregular and oblong

Jupiter's ring (Section 12.5)
- A narrow ring, composed of small particles

Future exploration of Jupiter (Section 12.6)
- Gravity-assist method: Pioneer 11, Voyagers to Saturn; Voyager 2 to Uranus and Neptune, Galileo (VEEGA)
- Project Galileo: Orbiter and jovian atmosphere Probe

Key Words

giant planets, jovian planets, Great Red Spot, oblate, radiation belts, Galilean satellites, Coriolis force, zones, belts, flux tube, gravity assist

Questions

1. Why does Jupiter appear brighter than Mars despite its greater distance from the Earth and Sun?

2. Even though Jupiter's atmosphere is very active, the Great Red Spot has persisted for a long time. How is this possible?

3. (a) How did we first know that Jupiter has a magnetic field? (b) What did spacecraft studies show about the magnetic field?

†4. From Jupiter's distance from the Earth (Appendix 3), the size of the orbits of the Galilean satellites (Appendix 4), and the speed of light (Appendix 2), calculate how "late" the moons of Jupiter appear to arrive in their predicted positions when Jupiter is at its farthest point from the Earth compared to when Jupiter is closest to the Earth. Rømer used the opposite of this calculation in 1675 to test whether light travels at finite or infinite speed.

5. It has been said that Jupiter is more like a star than a planet. What facts support this statement?

6. What advantages over the 5-m Palomar telescope on Earth did Voyagers 1 and 2 have for making images of Jupiter?

†This question requires a numerical solution.

7. What were two types of observations other than photography made from the Voyagers at Jupiter?

8. Describe applications of computer simulations to analyzing structure on Jupiter.

9. How does the interior of Jupiter differ from the interior of the Earth?

10. Which moons of Jupiter are icy? How do we know?

11. Contrast the volcanoes of Io with those of Earth.

12. Compare the surfaces of Callisto, Io, and the Earth's Moon. Show what this comparison tells us about the ages of features on the surfaces.

13. Using the information given in the text and in Appendices 3 and 4, sketch the Jupiter system, including all the moons and the ring. Mark the groups of moons.

14. (a) Describe the gravity-assist method. (b) What does it help us do?

15. Explain how the Hubble Space Telescope might be able to equal Voyager's resolution even though it will be farther from Jupiter.

†16. Given the resolution of the Hubble Space Telescope, calculate the size of features visible on Jupiter. Provide sample distances between cities on Earth for comparison.

†17. Repeat the calculation of Question 16 for Io, and comment on what kinds of structure might be seen.

Projects

1. Describe in some detail the origins in Greek mythology of the names of each of Jupiter's moons.

2. From your own observations made over a four-hour interval one night, plot the positions of the Galilean satellites with respect to Jupiter itself at half-hour intervals.

3. Over a one-week interval, plot the positions of the Galilean satellites from night to night with respect to Jupiter itself. Using the observations in Projects 2 and 3 together, deduce the projection on the sky of the orbits of the individual satellites.

Topics for Discussion

1. Although many of the most recent results about Jupiter came from spacecraft, prior ground-based studies had told us many things. Discuss the status of our pre-1973 knowledge of Jupiter, and specify both some things about which space research did not add appreciably to our knowledge and some things about which space research led to a major revision of our knowledge.

2. If we could somehow suspend a space station in Jupiter's clouds, the astronauts would find a huge gravitational force on them. What are some of the effects that this would have? (The novel *Slapstick* by Kurt Vonnegut deals with some of the problems that would arise if gravity were different in strength from what we are used to.)

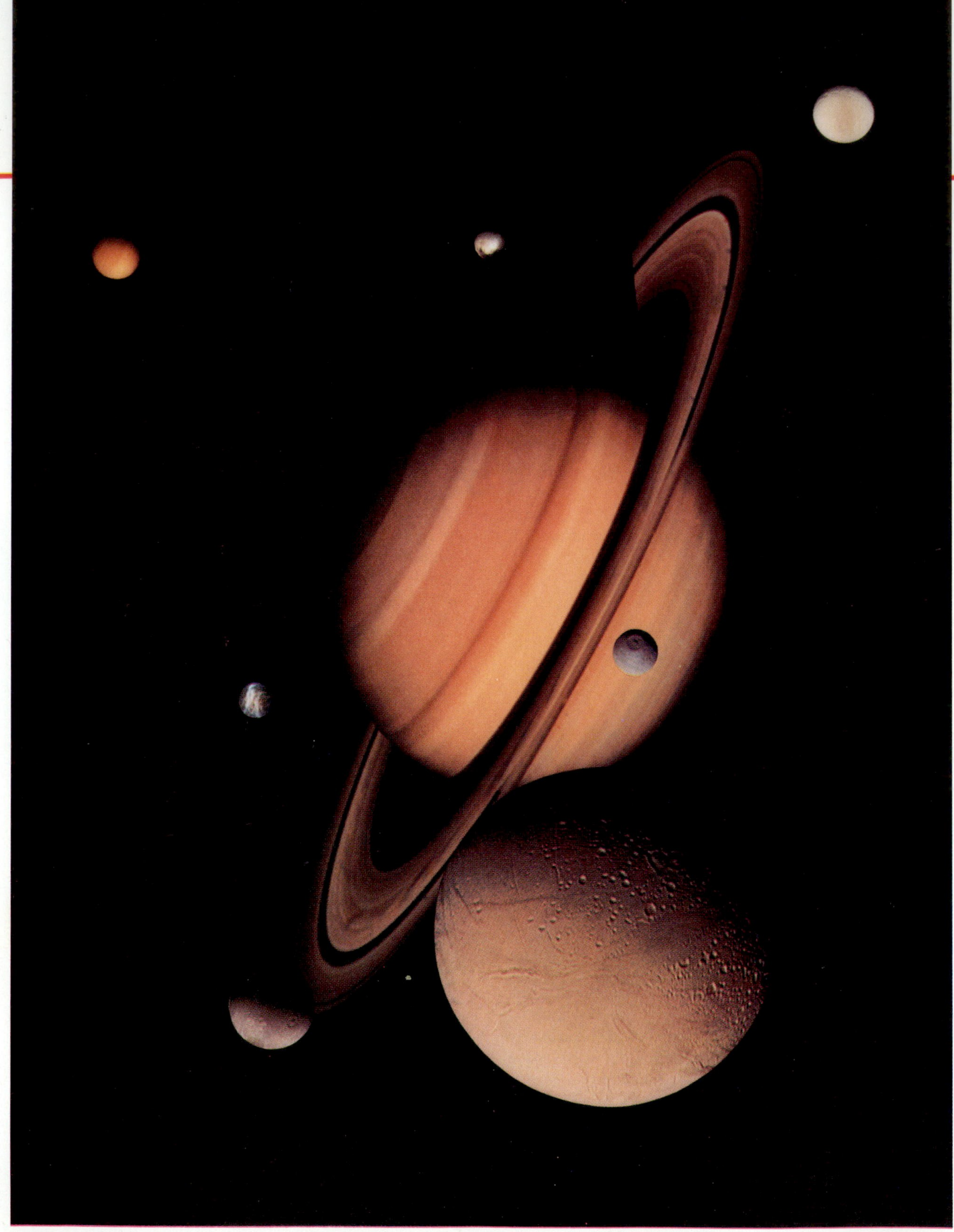

Saturn and some of its moons, photographed from the Voyager spacecraft. Enceladus is in the foreground, Mimas is above it in front of Saturn's disk, Dione is at lower left, Rhea is at middle left, Titan is at top left, Iapetus is at top center, and Tethys is at upper right. The symbol for Saturn appears at the top of the opposite page.

Saturn

13

Aims: To describe Saturn, its rings, and its moons, and to see how the Voyager spacecraft multiplied our knowledge of them

Saturn is the most beautiful object in our solar system, and possibly even the most beautiful object we can see in the sky. The glory of its system of rings makes it stand out even in small telescopes.

Saturn, like Jupiter, Uranus, and Neptune, is a giant planet. Its diameter, without its rings, is 9 times greater than Earth's; its mass is 95 times greater. Saturn is 9.5 A.U. from the Sun, and has a lengthy year, equivalent to 30 Earth years.

Figure 13–1 Since Saturn's density is lower than that of water, it would float, like Ivory Soap, if we could find a big enough bathtub. But it would leave a ring.

The giant planets have low densities. Saturn's is only 0.7 g/cm^3, 70 per cent the density of water (Fig. 13–1). The bulk of Saturn is hydrogen molecules and helium. Deep in the interior where the pressure is sufficiently high, the hydrogen molecules are converted into metallic hydrogen in liquid form. Saturn could have a core of heavy elements, including rocky material, making up 20 per cent of its interior.

Saturn's atmosphere (Fig. 13–2) has relatively less helium compared with hydrogen than Jupiter—92.4 per cent hydrogen molecules and 7.4 per cent helium for Saturn compared with 86.1 and 13.8 per cent, respectively, for Jupiter. (The remaining 0.2 per cent is almost all methane, plus some ammonia and traces of other molecules.) We think that Saturn's internal temperature is sufficiently lower than Jupiter's to prevent helium from dissolving in the liquid hydrogen. Helium droplets thus fall under the force of gravity toward Saturn's center. Energy is made available that makes the ratio of energy radiated to energy received higher for Saturn than for Jupiter.

The rings extend far out in Saturn's equatorial plane, and are inclined to the planet's orbit. Over a 30-year period, we sometimes see them from above their northern side, sometimes from below their southern side, and at intermediate angles in between. When seen edge on, they are all but invisible.

13.1 Saturn from the Earth

The rings of Saturn are either material that was torn apart by Saturn's gravity or material that failed to accrete into a moon at the time when the planet and its moons were forming. These bits of matter spread out in concentric rings around Saturn. We know they are not solid, for we can see stars through them on the rare occasions when Saturn passes in front of one.

Every massive object has a sphere, called the *Roche limit,* inside of which blobs of matter cannot be held together by their mutual gravity. The forces that tend to tear the blobs apart from each other are tidal forces (Section 7.3); they arise because some blobs are closer to the planet than others and are thus subject to higher gravity.

The radius of the Roche limit varies with the amount of mass in the parent body and the nature of the bodies but is usually about 2½ times the radius of

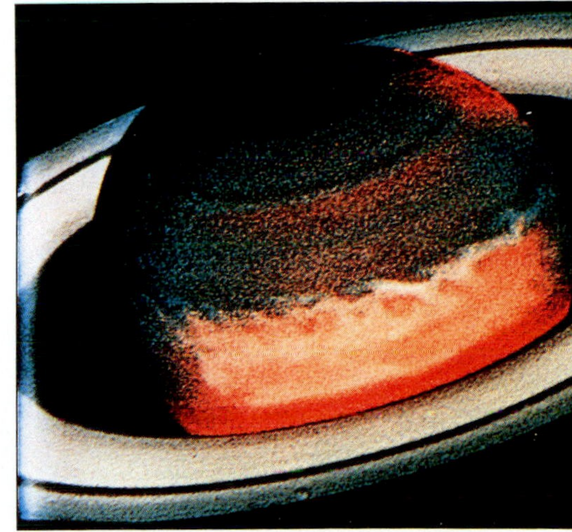

Figure 13–2 A storm that developed on Saturn in 1990 was photographed with the Hubble Space Telescope. The view here is in exaggerated color.

Focus On

Box 13.1 The Roche Limit

E. A. Roche defined the limit in 1849 as the shortest distance within which a liquid body does not break up because the body's own gravitation pulling itself together is stronger than the tidal forces pulling it apart. Similar limits can be defined for solid bodies and non-spherical bodies, and for the case where the orbiting body is taking on individual molecules. The limits depend on both the mass of the planet and the density of the ring particles. Thus the existence of rings at a given radius can help distinguish whether the rings are made of ice or of stony material. Saturn's rings, to be within the Roche limit, must be made primarily of ice.

the larger body. The Sun also has a Roche limit, but all the planets lie outside it. The natural moons of the various planets lie outside the planets' respective Roche limits. Saturn's rings lie inside Saturn's Roche limit, so it is not surprising that the material in the rings is spread out rather than collected into a single orbiting satellite.

Artificial satellites that we send up to orbit around the Earth are constructed of sufficiently rigid materials that they do not break up even though they are within the Earth's Roche limit.

Saturn has several concentric major rings visible from Earth (Fig. 13–3). The brightest ring (the "B-ring") is separated from a fainter broad outer ring (the "A-ring") by an apparent gap called *Cassini's division.* Another ring (the "C-ring," or "crepe ring") is inside the brightest ring. It has long been thought that the gaps between the major rings result from gravitational effects from Saturn's satellites, though this is uncertain. Finer structure in the rings has been studied from spacecraft, as we shall see.

The rotation of Saturn's rings can be measured directly from Earth with a spectrograph, for different parts of the rings have different Doppler shifts. We know that the rings are not solid objects, because they rotate at different angular speeds at different radii.

Figure 13–3 Saturn, from the Hubble Space Telescope.

Focus On

Box 13.2 Cassini's Division

In 1610, Galileo discovered with his new invention, the telescope, that Saturn was not round; it seemed to have "ears." In 1656, the Dutch astronomer Christian Huygens published an anagram—a coded message made by scrambling letters (a procedure then often followed to establish priority for a discovery)—explaining that Saturn has a ring. Huygens's open publication of his result in a book followed three years later. Giovanni Cassini, who worked at the Paris Observatory some years later, observed the largest gap in the ring in 1675, and this gap is named after him.

The rings are relatively much flatter than a phonograph record. Radar waves were first bounced off the rings in 1973. The result of the radar experiments showed that the particles in the ring are probably rough chunks of ice at least a few centimeters and possibly a meter across. Infrared spectra show that they are covered with ice.

Like Jupiter, Saturn rotates quickly on its axis, also in about 10 hours. As a result, it is oblate—bulges at the equator—by 10 per cent. Saturn's delicately colored bands of clouds rotate 10 per cent more slowly at high latitudes than at the equator. Methane, ammonia, and molecular hydrogen have been detected spectrographically.

Saturn gives off radio signals, as does Jupiter, an indication to earthbound astronomers that Saturn also has a magnetic field. And, like Jupiter, Saturn has a source of internal heating.

13.2 Saturn from Space

Pioneer 11, which had passed Jupiter in 1974, reached Saturn in 1979. But Pioneer 11 had the handicap of having to travel across the diameter of the solar system from Jupiter to reach Saturn, while the Voyagers travelled a much shorter route. So Voyager 1 reached Saturn in 1980, only a year after Pioneer 11. Voyager 2 arrived in 1981. Between the time of the launch of Pioneer 11 and the launches of the Voyagers, experimental and electronic capabilities had improved so much that the Voyagers could transmit data at 100 times the rate of Pioneer 11.

13.2a Saturn's Rings

From Pioneer 11, the rings were visible for the first time from a vantage point different from the one we have on Earth. The backlit view obtained by Pioneer 11 showed that Cassini's division, visible as a dark (and thus apparently empty) band from Earth, appeared bright, with its own dark line of material. The outermost major ring, the A-ring, showed structure as well. The brightest ring observed from Earth, the B-ring, appeared dark on the Pioneer 11 view, showing that it is too opaque to allow light to pass through it. The rings appear to have low mass and therefore low density, endorsing the idea that they are made up largely of ice. An icy composition would also explain why atomic hydrogen was found around the rings; it could result from dissociation of water ice.

Figure 13–4 Saturn and its rings taken by Voyager 2. Saturn's moons Tethys, Dione, and Rhea appear below it. The shadow of Tethys appears on Saturn's disk. (All Voyager photos courtesy of Jet Propulsion Laboratory/NASA)

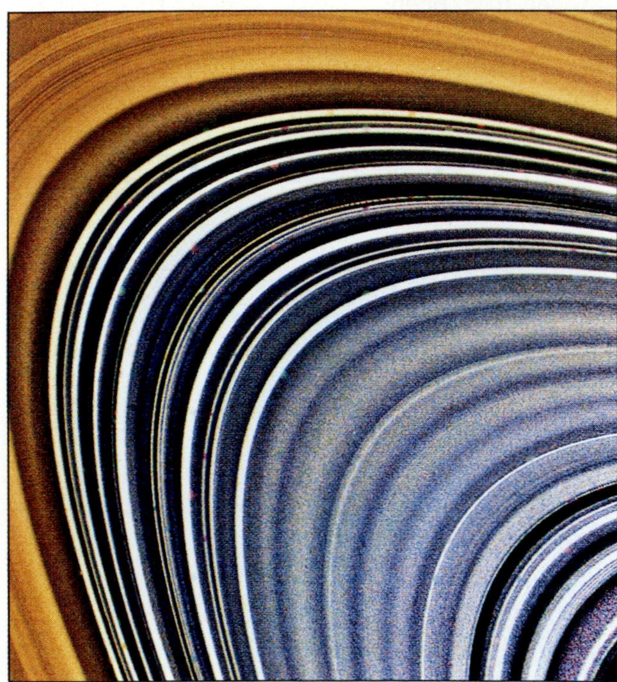

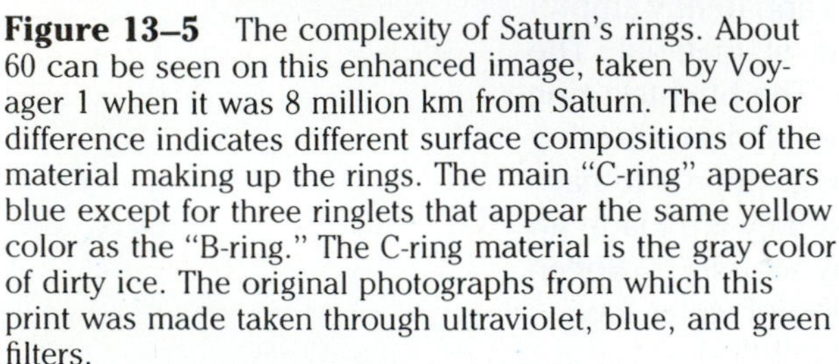

Figure 13–5 The complexity of Saturn's rings. About 60 can be seen on this enhanced image, taken by Voyager 1 when it was 8 million km from Saturn. The color difference indicates different surface compositions of the material making up the rings. The main "C-ring" appears blue except for three ringlets that appear the same yellow color as the "B-ring." The C-ring material is the gray color of dirty ice. The original photographs from which this print was made taken through ultraviolet, blue, and green filters.

Figure 13–6 Saturn's rings and their shadow on the top of Saturn's clouds in enhanced color. Sunlight passing through the Cassini Division makes the white band that divides the shadow. The planet can be seen as an orange band through the Cassini Division at lower left. Saturn's orange disk at top right shows atmospheric features. The two dark doughnuts are artifacts caused by dust on the camera lens.

Figure 13–7 Saturn's rings in their shadow in natural color.

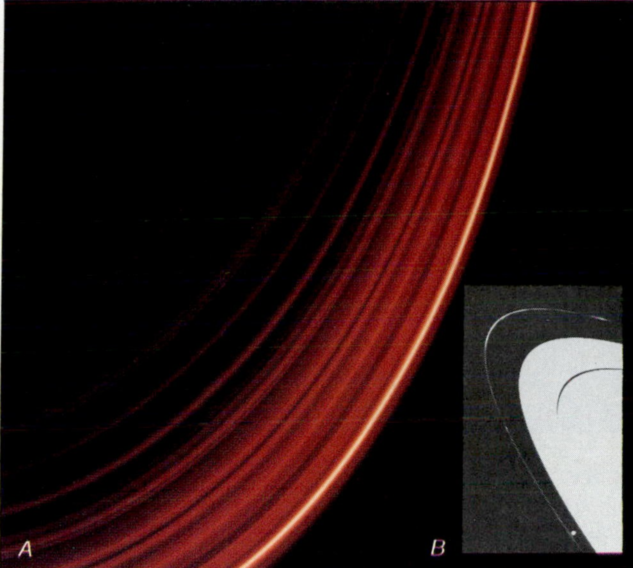

Figure 13–8 (*A*) A computer playback showing detail in Saturn's outermost ring, the "F-ring," based on the occultation of the star delta Scorpii by Saturn as observed from Voyager 2. The technique gave a resolution of about 0.2 km, 50 times better than the best resolution of the cameras. The picture is based on a single path of the star, so no variations in the F-ring appear. (*B*) The true variations of Saturn's F-ring, and the satellite discovered just inside the F-ring.

The passage of Voyager 1 by Saturn was one of the most glorious events of the space program. The resolutions of the Voyager's cameras were much higher than the resolution of Pioneer 11's instrument, which had been designed for measuring polarization rather than for photography. Voyager 2's cameras turned out to be twice as sensitive as Voyager 1's and had higher resolution. Though the subtlety of colors in Saturn's clouds made the Voyagers' photographs of the surface less spectacular than those of Jupiter, the structure and beauty of the rings dazzled everyone (Fig. 13–4).

The closer the spacecraft got to the rings, the more rings became apparent. Before Voyager, scientists discussed whether there were three, four, five, or six rings. But as Voyager 1 approached, the photographs increasingly showed that each of the known rings was actually divided into many thinner rings. By the time Voyager 1 had passed Saturn, we knew of hundreds or a thousand rings (Figs. 13–5 to 13–7). Voyager 2 saw still more. Further, a device on Voyager 2 was able to track the change in brightness of a star as it was seen through the rings, and found even finer divisions (Fig. 13–8). The number of these rings (sometimes called "ringlets") is in the hundreds of thousands. The words "rings" or "ringlets" for these features, though, perhaps imply that they persist in their current form, which may not be the case. The total amount of mass in the rings is about the mass of a middle-sized moon, such as Mimas.

Studies of the changes in the radio signals from the spacecraft when it went behind the rings showed that the rings are only about 20 m thick, equivalent to the thinness of a phonograph record 30 km across—super long play.

Everyone had expected that collisions between particles in Saturn's rings would make the rings perfectly uniform. But there was a big surprise. As Voyager 1 approached Saturn, we saw that there was changing structure in the rings aligned in the radial direction. "Spokes" can be seen in the rings when they are seen at the proper angle (look back at Figure 13–4). Differential rotation must make the spokes dissipate soon after they form. The spokes look dark from the side illuminated by the Sun (Fig. 13–9), but look bright from behind (Fig. 13–10). This information showed that the particles in the spokes were very small, about 1 micron in size, since only small particles—like terrestrial dust in a sunbeam—reflect light in this way. And it seems that the spoke material is elevated above the plane of Saturn's rings. Since gravity wouldn't cause this, electrostatic forces may be repelling the spoke particles.

Figure 13–9 Dark, radial spokes became visible in the rings as the Voyagers approached. They formed and dispersed within hours.

Figure 13–10 When sunlight scatters forward, the spokes appear bright.

It also came as a surprise that some of the rings are not circular. At least some of the rings have different brightness or are displaced from one location to another.

Before the flybys, it had been thought that the structure of the rings and the gaps between them were determined by gravitational effects of Saturn's several known moons on the orbiting ring particles. If a particle in the rings has a period that is a simple fraction of a moon's period, then the particle and the moon would come back in the same configuration regularly and be relatively close together in their elliptical orbits. This pattern reinforces the gravitational pull of the moon and sweeps away particles at those locations. But this explanation had been worked out with fewer than a half-dozen rings known. The ring structure the Voyagers discovered is too complex to be explained in this way alone. Consideration of the sixteen or so satellites cannot explain thousands of rings. And the gaps turned out to be in the wrong place for this model to work.

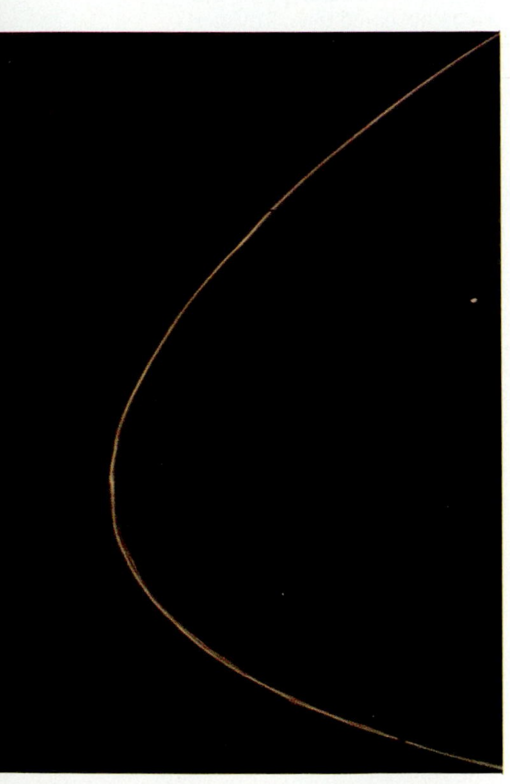

Figure 13–11 Two narrow, braided, 10-km-wide rings are visible, as is a broader diffuse component about 35 km wide, on this Voyager 1 image of the F-ring. The Voyager 2 images of the F-ring showed no signs of braiding.

The outer major ring turns out to be kept in place by a tiny satellite orbiting just outside it. And at least some of the rings are kept narrow by "shepherding" satellites. By Kepler's laws, the outer satellite is moving slightly slower than the ring particles, and the inner one slightly faster. If a ring particle should move outward, it would be pulled back by the outer satellite, diminishing its angular momentum. This would make it sink back toward the ring. Conversely, if a ring particle should move inward, it would be pulled ahead by the inner satellite, increasing its angular momentum. This would push it back toward the ring. Thus material tends to stay in the narrow ring.

Scientists were astonished to find on the Voyager 1 images that the outer ring, the narrow F-ring discovered by Pioneer 11, seems to be made of three braids (Fig. 13–11). But soon after scientists succeeded in finding explanations for the braided strands, involving the gravity of the pair of newly discovered moons shown in the preceding photograph, Voyager 2 images showed that the rings were no longer intertwined.

A post–Voyager-1 theory said that many of the narrowest gaps may be swept clean by a variety of small moons—"ringmoons"—embedded in them. These objects would be present in addition to the smaller icy snowballs that make up the bulk of the ring material. Unfortunately for the theoreticians, Voyager 2 did not find these larger objects, but no better reason for the narrow rings and gaps has been found.

Figure 13–12 Comparison of the Voyager 1 (*left*) and Voyager 2 (*right*) images show changes in Saturn's rings and atmosphere over a nine-month interval. The false-color view is based on photographs taken through ultraviolet, violet, and green filters. The blue band in the northern hemisphere in the Voyager 1 image is believed to be deeper atmosphere unobscured by high-level haze. By the time Voyager 2 arrived, high clouds or haze had formed, so the band is whitish. The rings are brighter in the Voyager 2 image because the angle of the Sun on the rings had changed.

13.2b Saturn below the Rings

The structure in Saturn's clouds is of much lower contrast than that in Jupiter's clouds. After all, Saturn is colder, so it has different chemical reactions. Even so, cloud structure was revealed to the Voyagers at their closest approaches. A haze layer present for Voyager 1 cleared up by the time Voyager 2 arrived, so more details could then be seen (Fig. 13–12). Turbulence in the belts and zones showed up clearly. A few circulating ovals similar to Jupiter's Great Red Spot and ovals were detected (Fig. 13–13). Saturn's ovals are also storms in high-pressure regions, circulating oppositely to the low-pressure storms on Earth. Study of cloud features gave the first direct view of the rotation of Saturn's clouds and allowed detailed measurements of the velocities of rotation to be made.

Extremely high winds, 1800 km/hr and 4 times higher than the winds on Jupiter (Fig. 13–14), were measured. On Saturn, the variations in wind speed do not seem to correlate with the positions of belts and zones (Fig. 13–15), unlike the case with Jupiter. Also as on Jupiter but unlike the case for Earth, the winds seem to be driven by rotating eddies, which in turn get their energy from the planet's interior. Infrared scans showed that the temperature decreases 10 K from equator to pole.

Figure 13–13 (*A*) A false-color image brings out the divisions between belts and zones. The red oval is about the diameter of Earth. The shadow of Dione appears. (*B*) The red oval shows more clearly. (*C*) This curled cloud lasted for many months.

A

B

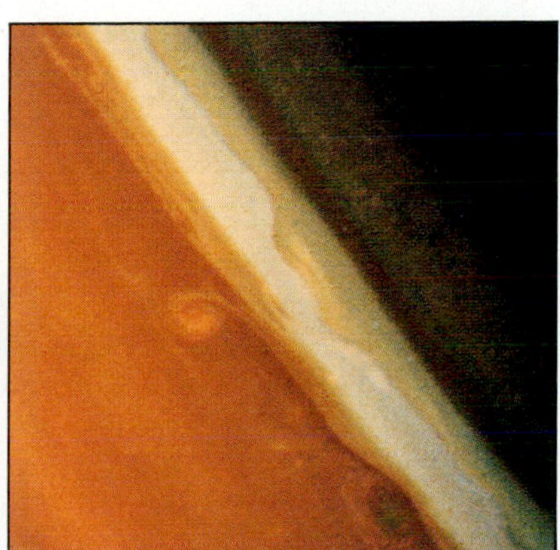

C

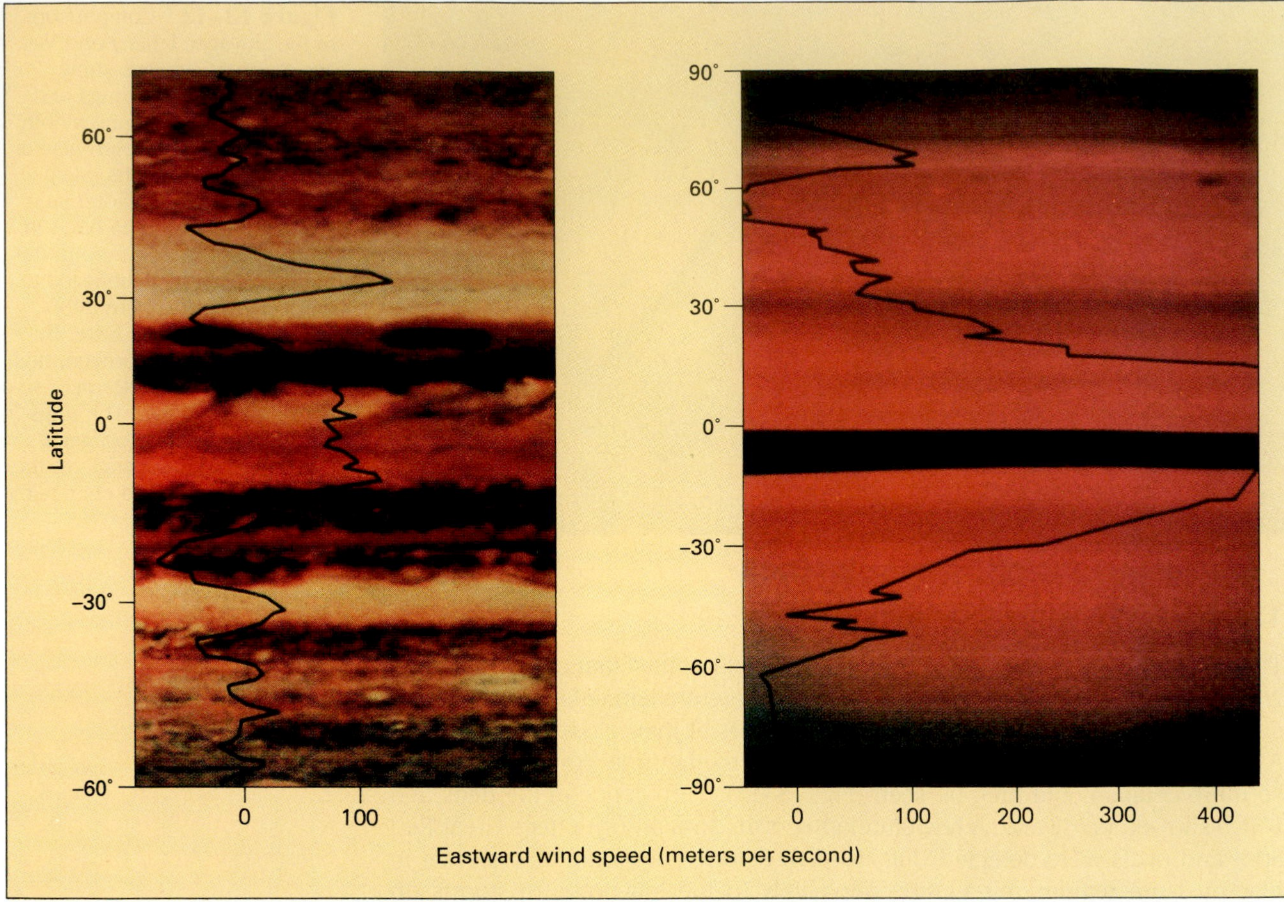

Figure 13–14 Jupiter's winds, to compare with Saturn's winds in the adjacent photo. A graph of the winds, defined with respect to a fixed rotation rate, is superimposed on a Voyager photo of the corresponding regions of Jupiter. Negative velocities correspond to currents travelling from east to west. The east–west currents, especially an eastward equatorial jet of 100 m/sec (360 km/hr), are apparent. Latitudes range from +65° at top to −70° at bottom, and velocities from −50 m/sec at left to 150 m/sec at the highest peak toward the right.

Figure 13–15 Saturn's winds, to compare with Jupiter's winds in the adjacent photo. We see a graph of the wind velocities at each latitutde, superimposed on a Voyager photo. The winds on Saturn are much higher at the equator and do not correlate with the cloud belts, in contrast with Jupiter. Thus the circulations of the atmospheres of Jupiter and Saturn are very different. Latitudes range from +70° at top to −60° at bottom, and velocities from 0 at left to 400 m/sec at right.

Saturn radiates about 2.5 times more energy than it absorbs from the Sun. One interpretation is that only ⅓ of Saturn's heat is energy remaining from its formation and from continuing gravitational contraction. The rest would be generated by the gravitational energy released by helium sinking through the liquid hydrogen in Saturn's interior. The helium that sinks has condensed because Saturn, unlike Jupiter, is cold enough.

13.2c Saturn's Magnetic Field

Pioneer 11 had revealed the magnetic field at Saturn's equator to be somewhat weaker than had been anticipated, only 2/3 of the field present at the Earth's equator. Remember, though, that Saturn is much larger than the Earth, and so its equator is much farther from its center. Saturn's magnetic axis appears to be aligned with its rotational axis, unlike Jupiter's, the Earth's, and Mercury's. This observation forces us to formulate new theories of how magnetic fields form inside planets, since it had been thought that the relative tilt of magnetic and rotational axes led to the generation of the magnetic field. The total strength of Saturn's magnetic field (which depends on both the field's value and the planet's volume, as described in Section 15.2) is 1000 times stronger than Earth's and 20 times weaker than Jupiter's. The magnetic field leads to radio emission (Fig. 13–16).

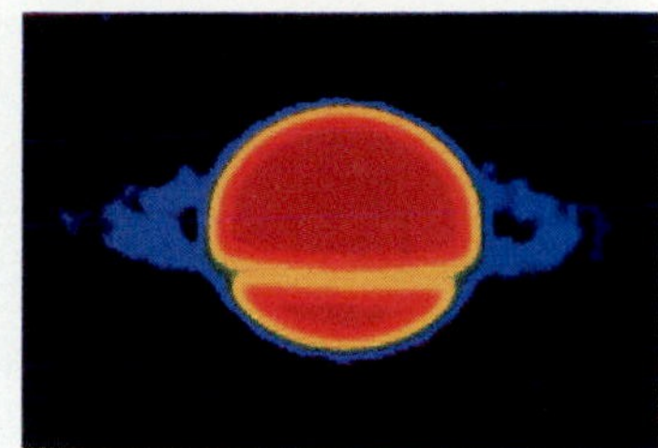

Figure 13–16 A radio image of Saturn taken with the VLA. The rings are relatively faint compared with the disk.

Saturn has belts of charged particles (Van Allen belts), which are larger than Earth's but smaller than Jupiter's. Though Saturn's magnetic field passes through the rings, the charged particles can't, so the Van Allen belts begin only beyond the rings. Even there, the number of charged particles is depleted by the satellites. Thus the belts are weaker than Jupiter's. The size of the belts fluctuates as the pressure of the solar wind varies, allowing them to expand or forcing them to contract.

Since the rotational and magnetic axes are aligned, scientists were hard pressed to explain why a radio pulse occurred every rotation. By comparing Voyager 1 and 2 results, scientists pinpointed the area from which the pulse comes. It may be the result of an internal anomaly. The existence of the pulse allowed Saturn's rotation period to be accurately measured.

Titan is sometimes within the belts and sometimes outside. Many of the particles in Saturn's magnetosphere probably come from Titan's atmosphere. A huge torus—doughnut—of neutral hydrogen from this source fills the region from Titan inward to Rhea.

A variety of instruments aboard the Voyagers studied the ionized gas around Saturn and the magnetic field. The region between the moons Dione and Rhea is filled with a "plasma torus." (A "plasma" is a gas containing both ions and electrons. It is electrically neutral, but is affected by electric and magnetic fields.) A cloud of this plasma is at the tremendously high temperature of 600 million K, making it the hottest known place in the solar system. The high temperature means that the velocities of individual particles are high. Since the density is extremely low, though, the total energy in the cloud is not large.

The three passes through the magnetic field and plasma around Saturn, by Pioneer 11 and the two Voyagers, will surely not allow us to understand this complex object as well as we would like.

13.2d Saturn's Family of Moons

Farther out than the rings, we find the satellites of Saturn. Like those of Jupiter, they now have the personalities of independent worlds and are no longer merely dots of light in a telescope.

The largest of Saturn's satellites, Titan, is larger than the planet Mercury and has an atmosphere (Fig. 13–17) with several layers of haze (Fig. 13–18). We had known something about Titan's atmosphere from ground-based studies. Spectra had shown signs of methane, polarization measurements had found signs of haze, and infrared measurements had indicated that hydrocarbons might be present. But a better understanding of Titan's atmosphere had to wait for the Voyagers. Studies of how the radio signals faded when Voyager 1 went behind Titan showed that Titan's atmosphere is denser than Earth's. Surface pressure on Titan is 1½ times that on Earth. Though we had thought that Titan was the largest moon in our solar system, Titan's atmosphere is so thick that its surface is actually slightly smaller than the surface of Jupiter's Ganymede.

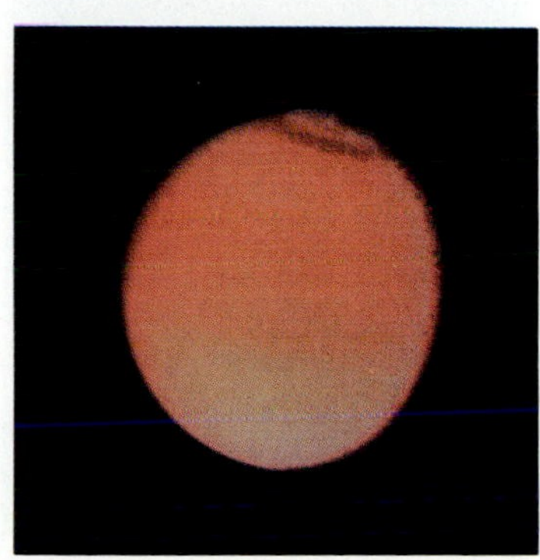

Figure 13–17 Titan was disappointingly featureless even to Voyager 1's cameras because of its thick, smoggy atmosphere. Its northern polar region was relatively dark. The southern hemisphere was lighter than the northern.

The Voyagers' ultraviolet spectrometers detected nitrogen (N_2), which makes up the bulk of Titan's atmosphere. The methane is only a minor constituent, perhaps 1 per cent. Titan's huge doughnut-shaped hydrogen cloud presumably results from the breakdown of its methane (CH_4).

The temperature near the surface, deduced from measurements made with the Voyagers' infrared radiometers, is only about −180°C, somewhat warmed by the greenhouse effect but still extremely cold. This temperature is near that of methane's "triple point," at which it can be in any of the physical states—

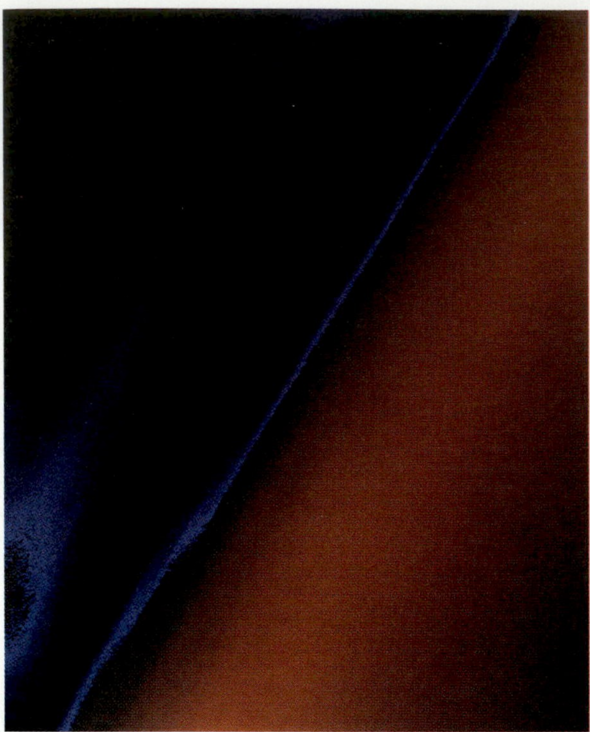

Figure 13–18 Haze layers can be seen at Titan's limb, with divisions at average altitudes of 200, 375, and 500 km. False color exaggerates the layers.

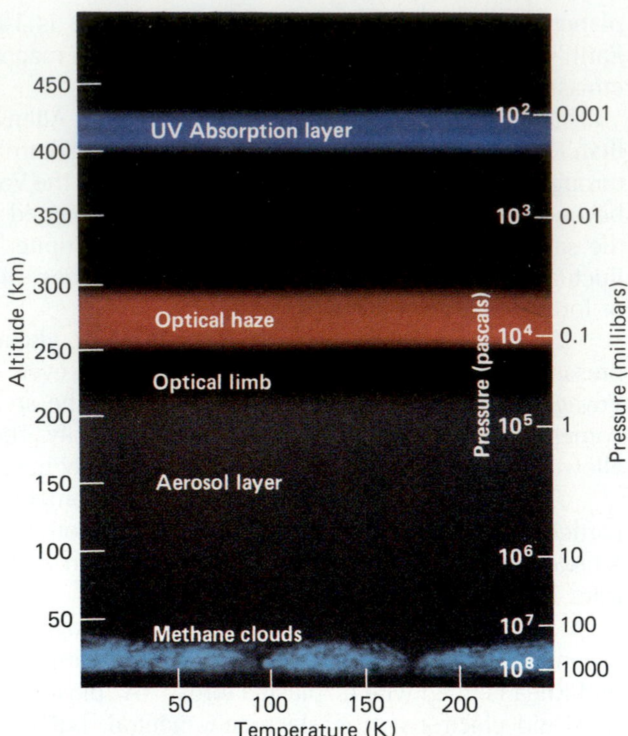

Figure 13–19 A model of Titan's atmosphere.

Titan also has an "antigreenhouse effect." A haze layer keeps solar energy out while letting heat escape. The antigreenhouse effect balances some of Titan's greenhouse effect.

solid, liquid, or gas. So methane may play the role on Titan that water does on Earth, though some evidence disagrees with this conclusion. Parts of Titan may be covered with methane lakes or oceans, and other parts may be covered with methane ice or snow. An ethane (C_2H_6) ocean, or a mixture of ethane and methane, has more recently been suggested to overcome some of the objections that the pure methane ocean did not satisfy certain predictions.

Titan's clouds are opaque, and probably contain both liquid nitrogen and liquid methane (Fig. 13–19). It seems that Titan's atmosphere is opaque because of the action of sunlight on chemicals in it, forming a sort of smog and giving it its reddish tint. Smog on Earth forms in a similar way. Some of the color may result from the bombardment of Titan's atmosphere by protons and electrons trapped in Saturn's magnetic field. Other energy to run the chemical reactions may come from the solar wind, since Titan's orbit is large enough to go in part outside Saturn's protective magnetic field. A view looking back from a Voyager showed the extent of the atmosphere (Fig. 13–20).

Figure 13–20 Looking back at Titan's dark side, Voyager 2 recorded how extensive Titan's atmosphere is by photographing sunlight scattered by small particles hundreds of kilometers high.

Some of the organic molecules formed in Titan's atmosphere may rain down on its surface. Thus the surface, hidden from our view, may be covered with an organic crust about a kilometer thick, perhaps partly dissolved in liquid methane. These chemicals are similar to those from which we think life evolved on the primitive Earth. The 1983 discovery (with the 4-m Kitt Peak telescope) of traces of carbon monoxide provides further evidence for an atmosphere resembling the atmosphere that some scientists think existed on the early Earth. But it is probably too cold on Titan for life to begin. Indeed, this cold may explain why Titan has an atmosphere while Jupiter's Ganymede and Callisto do not. Titan was sufficiently cold that its ice could trap large quantities of nitrogen and methane, from which other gases formed. Ganymede and Callisto were slightly too warm to do so. At their distance from the Sun, and possibly with additional warming from Jupiter itself, their gases escaped.

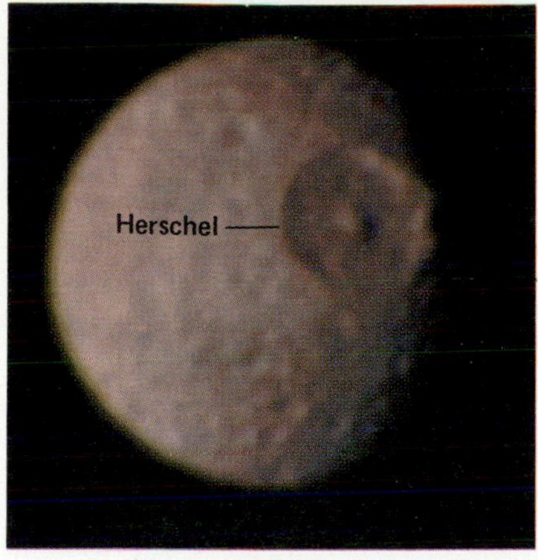

Figure 13–21 The impact feature on Mimas, named Herschel after the discoverer of Uranus, is about 130 km in diameter.

Figure 13–22 The other side of Mimas, showing the trough crossing the center that may be the result of the impact.

The surfaces of Saturn's other moons, all icy, are so cold that the ice acts as rigidly as rock and can retain craters. The Voyagers found that these moons' mean densities are all 1.0 to 1.5, which means that they are probably mostly water ice throughout with some rocky material included. Unexpectedly, the densities do not decrease with distance from the planet, as do the densities of the planets with distance from the Sun and of Jupiter's moons with distance from that planet.

In addition to Titan, four of Saturn's moons are over 1000 km across, so we are dealing with major bodies (see Appendix 4), about ⅓ the size of Earth's Moon. For Saturn as for Jupiter, most satellites perpetually have the same side facing their planet. Let us consider the major ones in order from the inside out.

Mimas boasts a huge impact structure, named Herschel, that is ¼ the diameter of the entire moon (Fig. 13–21). The crater has a raised rim and a central peak, typical of large impact craters on the Earth's Moon and on the terrestrial planets. The whole satellite is saturated with craters. The canyon visible halfway around Mimas (Fig. 13–22) may be a result of the impact. The energy of the impact may have shattered the satellite and been focused halfway around. Many craters a few kilometers across are visible.

Enceladus's surface (Fig. 13–23) has both smooth regions and regions covered with impact craters. But the spectrum of the light reflected off the surface is uniform over the whole moon, indicating that the surface layer is uniform and probably relatively young. Also, its albedo is very high. Further, the existence of smooth regions suggests that Enceladus's surface was melted comparatively recently and so may be active today. The internal heating may be the result of a gravitational tug of war with Saturn and the other moons, as for Jupiter's Io, though the situation is not as clear and no explanation is widely accepted. Linear sets of grooves on Enceladus, tens of kilometers long, are probably geologic faults in the crust, resembling those on Jupiter's Ganymede.

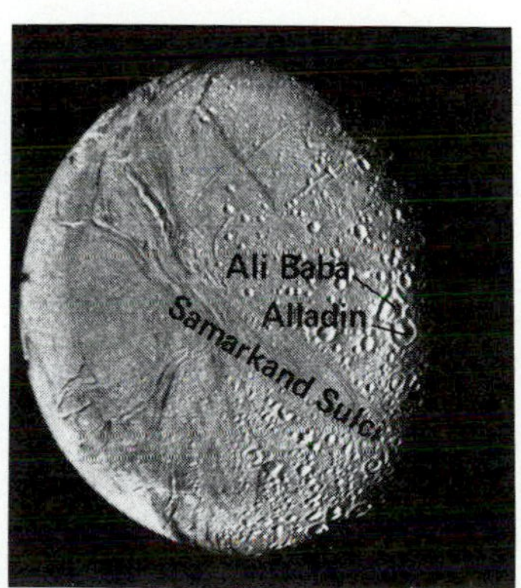

Figure 13–23 A computer-enhanced image of Enceladus from Voyager 2. Enceladus resembles Jupiter's Ganymede in spite of being 10 times smaller. Resolution is 2 km on this Voyager 2 mosaic.

Tethys also has a large circular feature (Fig. 13–24), called Odysseus, that is ½ the diameter of the entire moon. The difference between heavily and lightly cratered regions may indicate that internal activity took place early in Tethys's history. The side of Tethys that faces Saturn (Fig. 13–25) shows not only many craters (Fig. 13–26) but also a large canyon, Ithaca Chasma, that goes ⅔ the way around the satellite. The canyon, or trench, which is several

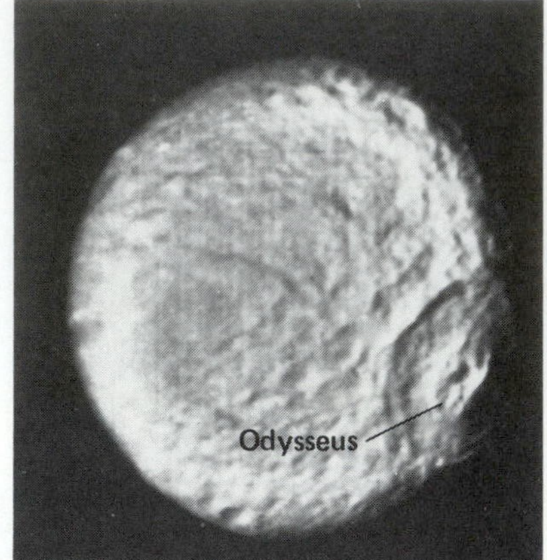

Tethys

Figure 13–24 A computer-enhanced view of Tethys, showing a large, bright, circular feature. The crater has been flattened by the flow of softer ice, and probably formed early in Tethys's history when its interior was still relatively warm and soft.

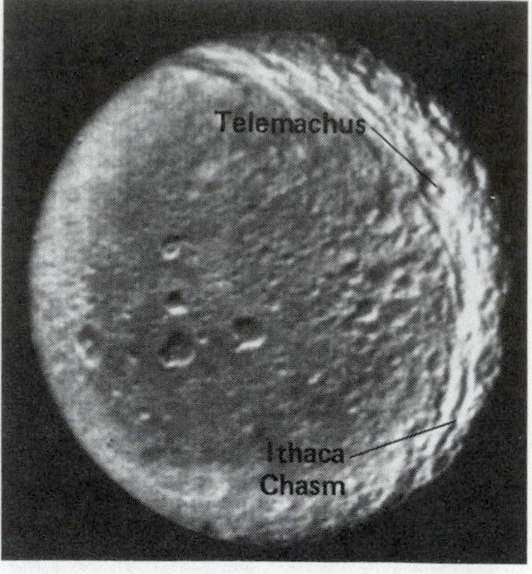

Tethys

Figure 13–25 Craters and a 750-km-long canyon or trench on Tethys. The large crater at the upper right sits on the canyon or trench. Both images are from Voyager 2, with resolutions of 9 and 5 km, respectively.

Tethys

Figure 13–26 The best color view of Tethys shows a cratered terminator.

kilometers deep, could have resulted from the expansion of Tethys as its warm interior froze. Or it could be a fault through the entire planet, a sign of the brittleness of ice under these extremely cold conditions.

Dione, too, shows many impact craters, some of them with rays of debris (Fig. 13–27). Many valleys are visible in Dione's icy crust (Fig. 13–28). Scientists wonder why Dione's two sides are so different (Fig. 13–29). The wisps of material on the side of Dione that trails as Dione orbits Saturn may have erupted from inside.

Two associated satellites share Tethys's orbit, always slightly ahead of Tethys itself. A small moon also shares Dione's orbit. No moons were known to share orbits before we studied the Saturn system.

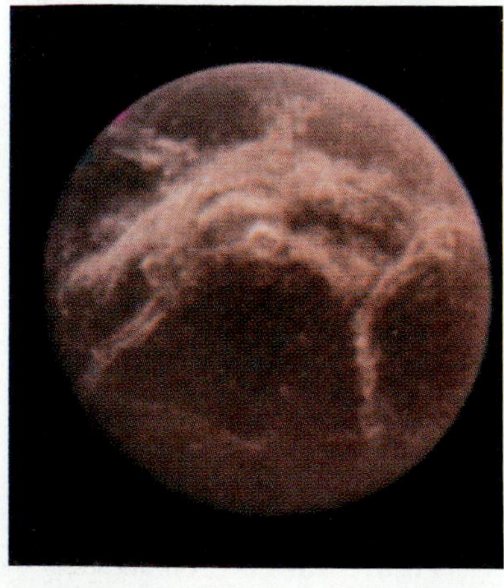

Dione

Figure 13–27 Dione, showing impact craters, debris, ridges or valleys, and wispy rays.

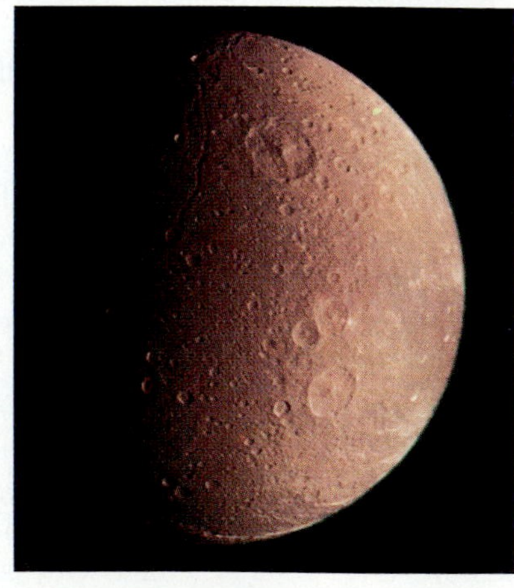

Dione

Figure 13–28 The largest crater on Dione is about 100 km in diameter. The sinuous valleys were probably formed by geologic faults.

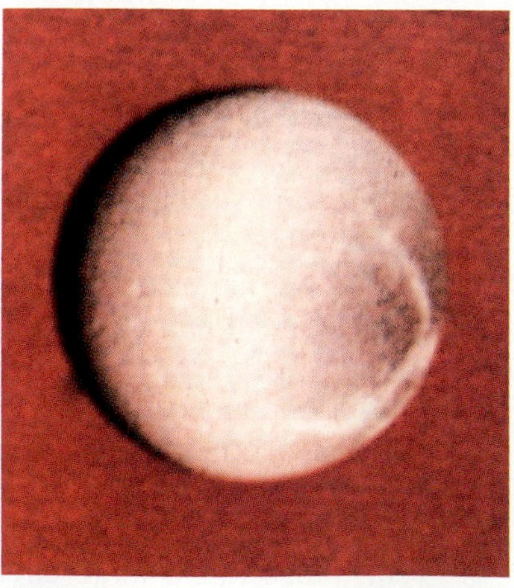

Dione

Figure 13–29 Dione, in front of Saturn's disk.

Rhea

Figure 13–30 Rhea has craters as large as 300 km across, many with central peaks. This Voyager 1 close-up of Rhea has a resolution of 2.5 km. The area shows an ancient, heavily cratered region. White areas on the edges of several craters in the upper right are probably fresh ice either visible on steep slopes or deposited by leaks from inside.

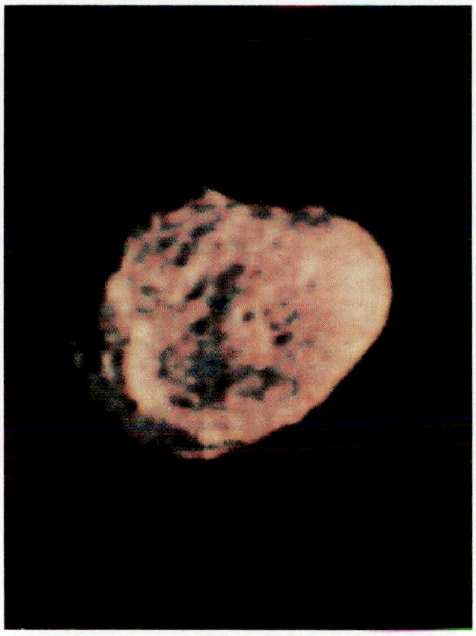

Hyperion

Figure 13–31 Hyperion, an irregularly shaped satellite roughly 410 × 220 km.

Rhea has impact craters (Fig. 13–30). Some craters, with sharp rims, must be fresh; others, with subdued rims, must be ancient. Rhea's other side, the side that is trailing as Rhea moves in its orbit and so collides with fewer objects, has wispy light markings. Slush and mud could have flowed long ago out of Rhea's interior to cover all the large craters in some areas. The debris in the Saturn system continued to make small craters.

Next out is Titan, which we have already discussed.

Hyperion, when photographed by Voyager 2, turned out to look like a hamburger or a hockey puck (Fig. 13–31). The rotation of Hyperion is chaotic, with abrupt changes in orientation occurring. The effects of Hyperion's eccentric orbit, the strong gravity from Titan, and other gravitational forces on it result in this chaotic rotation. The field of "chaos" is a new one for science. Hyperion and Mimas are essentially the same size but are vastly different in appearance. Surely they had different histories.

Strangely, the side of Iapetus (Fig. 13–32) that precedes in its orbit is over 10 times darker than the side that trails. The large circular feature, Cassini Regio, is probably an impact feature outlined by dark material. Iapetus's density indicates that it may be the sole moon to contain an interior partly made of methane ice in addition to water ice. Water ice covers the bright part of the surface, giving it its albedo of about 50 per cent. The ice is covered with craters. Perhaps the dark material, whose albedo is less than 5 per cent, is a hydrocarbon formed by sunlight reacting on some of the methane that wells up. The spectrum of the dark material resembles that of some asteroids and meteorites.

Saturn's outermost satellite, Phoebe, is probably a captured asteroid and will be discussed in that section, Section 18.2.

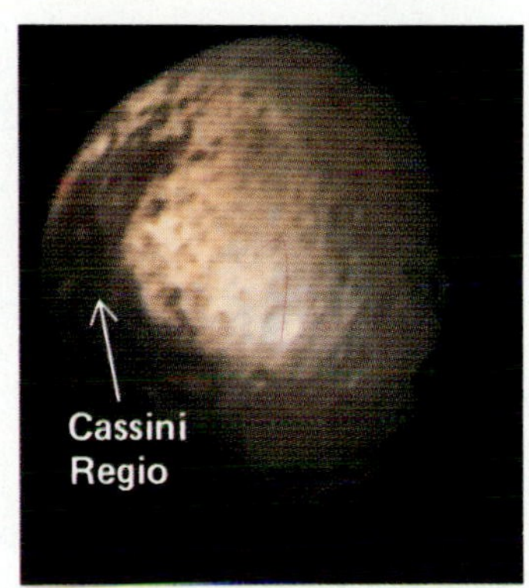

Iapetus

Figure 13–32 Iapetus, whose trailing side is 5 times brighter than its leading one. The dark side may have accumulated dust spiralling in toward Saturn or may be covered with hydrocarbons. It could be as black as pitch because it is pitch! A large circular feature, Cassini Regio, is also visible.

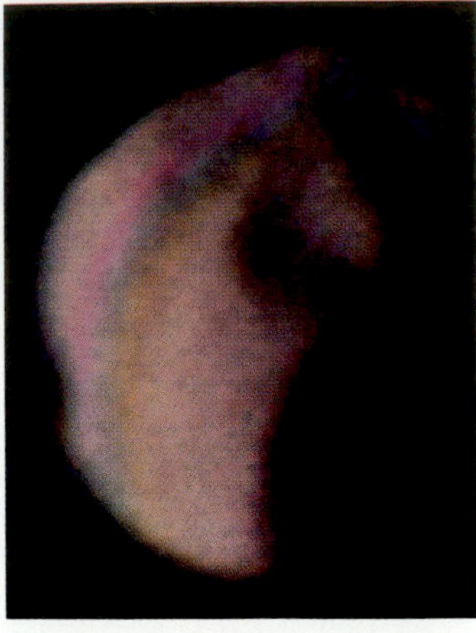

Janus

Figure 13–33 Saturn's tooth-shaped moon Janus shares an orbit with Epimetheus. It is only about 220 × 200 × 160 km. The shadow of Saturn's F-ring crossed the satellite, and was photographed in a series of six images superimposed here. The shadow appears as colored stripes, since the images were taken through different colored filters.

In addition to these 9 moons long known, several others have been detected from the ground in the last decades, and several have been confirmed or discovered by the Pioneer and Voyager spacecraft. Voyager even provided close-ups of one of the recent ground-based discoveries (Fig. 13–33). This satellite is too small for gravity to have pulled it into a spherical shape. Oddly, it apparently all but shares an orbit with another newly discovered satellite. The orbits are not absolutely identical, so the two approach each other every few years. Their gravitational fields are strong enough to affect the other only when they are very close to each other. When this happens, they twirl around each other, interchange orbits, and are off on their merry ways again, separating from each other. No other case of such a gravitational interaction is known in the solar system. They may be halves of a moon that broke apart.

It is difficult to say how many moons Saturn has. Indeed, there may be many that are 50 or 100 km across orbiting in the rings (Fig. 13–34). They are probably fragments of larger bodies that broke up in the early years of the Saturn system. They may be mostly icy with a mixture of meteoritic rock.

13.3 Beyond Saturn

Yet another superlative for Voyager 1 was the view it gave us when it looked back on the crescent Saturn (Fig. 13–35). The spacecraft is now travelling up and out of the solar system.

The Voyagers' passages through the Saturn system exchanged a lot of unknowns for a lot of knowledge plus many specific new mysteries. Monitoring Saturn with the Hubble Space Telescope should provide some new information. A joint NASA/European Space Agency mission called Cassini (after the discoverer of the gap in Saturn's rings) is being planned. It would be launched

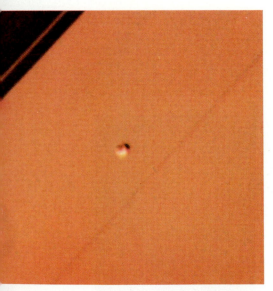

Prometheus

Figure 13–34 Prometheus, the inner shepherd of the F-ring, and the F-ring itself, cross Saturn's disk. Prometheus is only 140 × 100 × 80 km. Its albedo is clearly higher than that of Saturn's clouds, so the moon must be covered with ice like the larger satellites. The A-ring and the second largest gap in Saturn's rings, the Encke Division, appear in the corner.

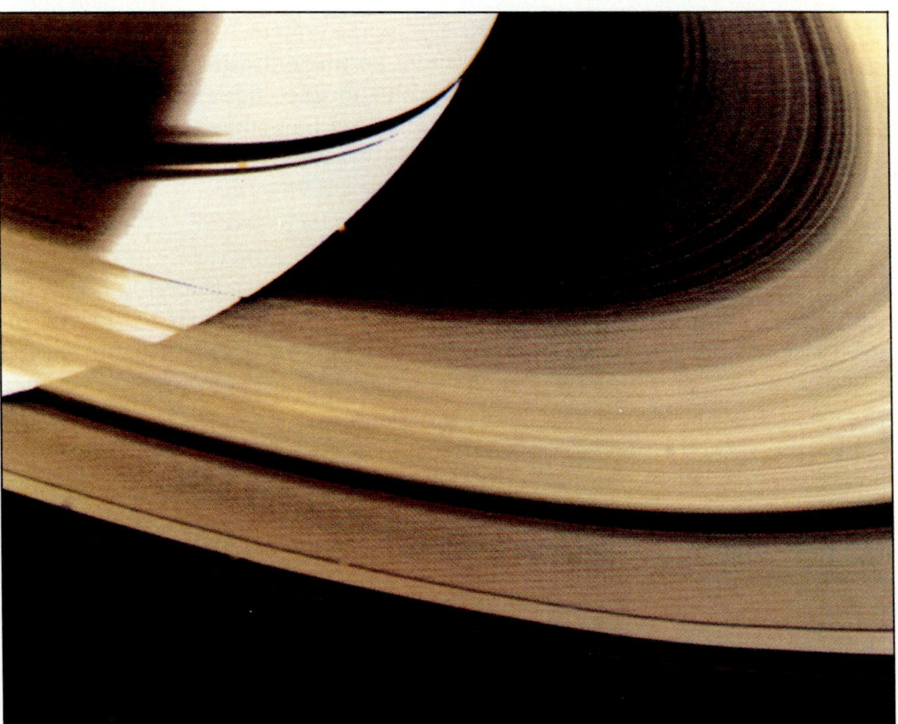

Figure 13–35 The crescent of Saturn and the planet's rings, taken from 1,500,000 km from the far side of Saturn as Voyager 1 departed. The rings' shadows are visible cutting across the overexposed crescent.

in 1998, take gravity assists from the Earth and from Jupiter, possibly pass near an asteroid, and arrive at Saturn in 2004. Cassini would tour Saturn and its moons for four years, and drop a probe named Huygens into Titan's atmosphere. Christian Huygens discovered Titan in March 1655, and the probe named after him is to be the contribution of the European Space Agency.

Summary and Outline

Giant planet with system of rings; low density: 0.7 g/cm^3
Ring system (Section 13.1)
- Inclined 27° to orbit; chunks of rock and ice orbiting inside Roche limit; Cassini's division apparently dark; radar shows the rings are very thin; rapid rotation makes planet oblate; magnetic field

Pioneer 11 (1979); Voyagers 1 and 2 (1980 and 1981)
Rings (Section 13.2a)
- Divided into thousands of ringlets; radial spikes made out of small particles held up by electrostatic force; ring structure affected by shepherding satellites

Saturn's surface and interior (Section 13.2b)
- High-pressure storms found; winds 4 times higher than those on Jupiter and do not match belts and zones

Magnetic field (Section 13.2c)
- Surface field only ⅔ that of Earth but total field is 1000 times stronger; rotational and magnetic axes aligned; plasma torus of extremely high temperature

Moons (Section 13.2d)
- Titan: atmosphere mostly nitrogen; surface pressure is 1½ Earth's; methane near its triple point; methane lakes or oceans may be present; reddish smog
- Other moons are icy
  - Mimas: huge crater and giant canyon
  - Enceladus: recent activity and internal heating
  - Tethys: large crater and giant canyon
  - Dione: craters, rays, wisps
  - Rhea: craters and wisps; no large craters
  - Hyperion: hamburger-shaped; same size as Mimas
  - Iapetus: one side very dark, the other bright
  - Phoebe: captured asteroid

Cassini Mission (Section 13.3)
- Probe, 4-year orbiter

Key Words

Roche limit, Cassini's division

Questions

1. What are the similarities between Jupiter and Saturn?

†2. How many times smaller is the angular size of the Sun as viewed from Saturn than the apparent angular size of the Sun from Earth? Compare with the 3-arc-min resolution of the naked eye.

†3. When Jupiter and Saturn are closest to each other, what is the angular size of Jupiter as viewed from Saturn?

4. What is the Roche limit, and how does it apply to Saturn's rings?

†5. Calculate the relative strength of the tidal force from Saturn itself on a moon the size of Titan (a) at Saturn's B-ring and (b) at Titan's orbit.

6. Sketch Saturn's major rings and the orbits of the largest moons, to scale.

†This question requires a numerical solution.

7. Describe the major developments in our understanding of Saturn, from ground-based observations to Pioneer 11 to the Voyagers.

8. Why are the moons of the giant planets more appealing for direct exploration by humans than the planets themselves?

9. Why does Saturn have so many rings? What holds the material in a narrow ring?

10. Explain how part of the rings can look dark from one side but bright from the other.

11. What did the Voyagers reveal about Cassini's division?

12. What are "spokes" in Saturn's rings, and how might they be caused?

13. What have we learned about Titan?

14. Explain how methane on Titan may act similarly to water on Earth.

15. Describe two of Saturn's moons other than Titan.

16. What is the Cassini mission?

A montage of Uranus and its moons. Ariel is at lower right, Umbriel is small to its left, Titania is at top left, Oberon is to the left of Uranus, and Miranda is at the top of Uranus. The symbol for Uranus appears at the top of the facing page.

Uranus

14

Aims: To study the third giant planet—Uranus—and its rings, magnetic field, and moons

The two other giant planets beyond Saturn, Uranus (U'ranus) and Neptune, are each about 50,000 km across and about 15 times more massive than the Earth. Though Uranus and Neptune are like Jupiter and Saturn in having no solid surfaces, they have higher proportions of heavier elements mixed in with the hydrogen and helium of which they are almost entirely formed.

Uranus, in Greek mythology, was the personification of Heaven and ruler of the world, the son and husband of Gaea, the Earth.

The densities of Uranus and Neptune are low (1.2 and 1.6 grams/cm^3, respectively). Thus we know that they are made out of hydrogen and helium. Their albedoes are high, so we have long known that that are covered with clouds.

Until Voyager 2's flyby of Uranus in 1986, little was known about this distant planet, though its rings had been discovered and thoroughly analyzed from the ground during the preceding decade. In this chapter, we shall describe our new knowledge.

14.1 Uranus Seen from Earth

Uranus was the first planet to be discovered that had not been known to the ancients. The English astronomer and musician William Herschel reported the discovery in 1781. Actually, Uranus had been plotted as a star on several sky maps during the hundred years prior to Herschel's discovery, but had not been singled out.

Uranus revolves around the Sun in 84 years, at an average distance of more than 19 A.U. from the Sun. Uranus never appears larger than 4.1 seconds of arc, so studying its surface from the Earth is difficult (Fig. 14–1).

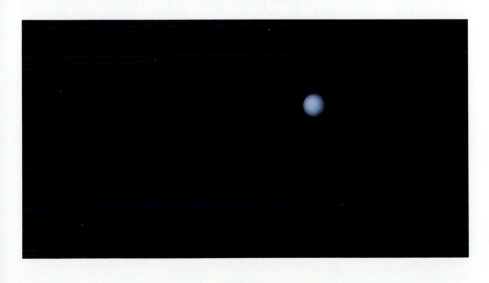

Figure 14–1 Uranus, in a Voyager photo from sufficiently far away, matches the resolution of a ground-based photograph. Neither shows any features.

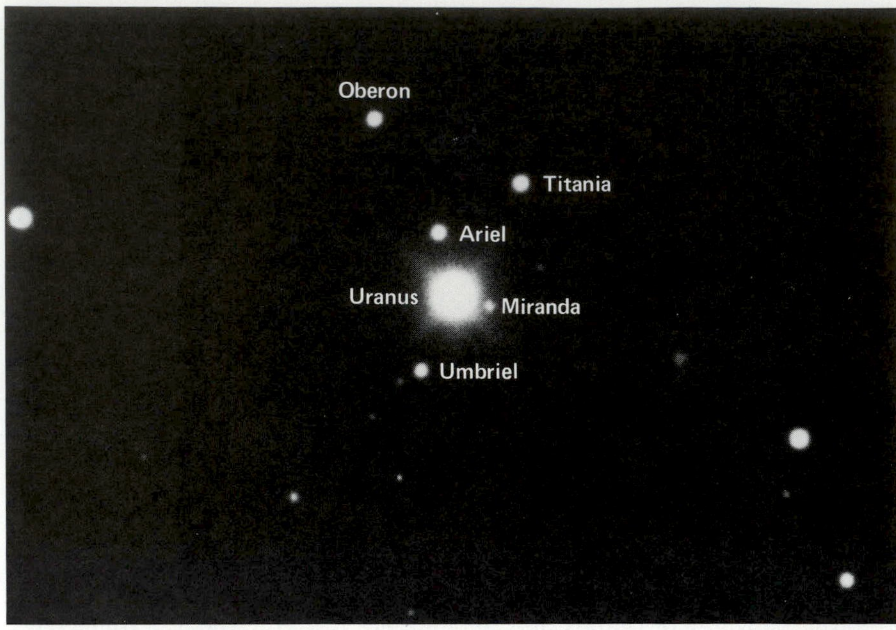

Figure 14–2 Uranus and its moons, observed from the Earth. The moons were only points of light.

Uranus is so far from the Sun that its outer layers are very cold. Studies of its infrared radiation give a temperature of 58 K. There is no evidence for an internal heat source, unlike the case for Jupiter, Saturn, and Neptune. The discovery of an aurora using the International Ultraviolet Explorer (IUE) spacecraft was our first indication that Uranus has a magnetic field.

Uranus has five moons, ranging from 320 to 1690 km across. They orbit in the plane of Uranus's equator. Little was known about them (Fig. 14–2), but spectra revealed water ice.

14.2 The Rotation of Uranus

The other planets rotate such that their axes of rotation are very roughly parallel to their axes of revolution around the Sun. Uranus is different, for its axis of rotation is roughly perpendicular to the other planetary axes, lying only 8° from the plane of its orbit (Fig. 14–3). Its rotation is considered retrograde because its axis is tilted more than 90°. Sometimes one of Uranus's poles faces the Earth, 21 years later its equator crosses our field of view, and then another 21 years later the other pole faces the Earth. Polar regions remain alternately in sunlight and in darkness for decades. Thus there could be strange seasonal effects on Uranus. When we understand just how the seasonal changes in heating affect the clouds, we will be closer to understanding our own Earth's weather systems.

It is important to know about the period of rotation because models for the structure of Uranus's interior depend on the rotation period. We must know Uranus's rotation period before we can settle on a model for the interior. Values measured from the Earth differed wildly—ranging from 12 to 24 hours.

14.3 Uranus's Rings

In 1977, Uranus occulted (passed in front of) a faint star. Predictions showed that the occultation would be visible only from the Indian Ocean southwest of Australia. James Elliot, then of Cornell and now of MIT, led a team that observed

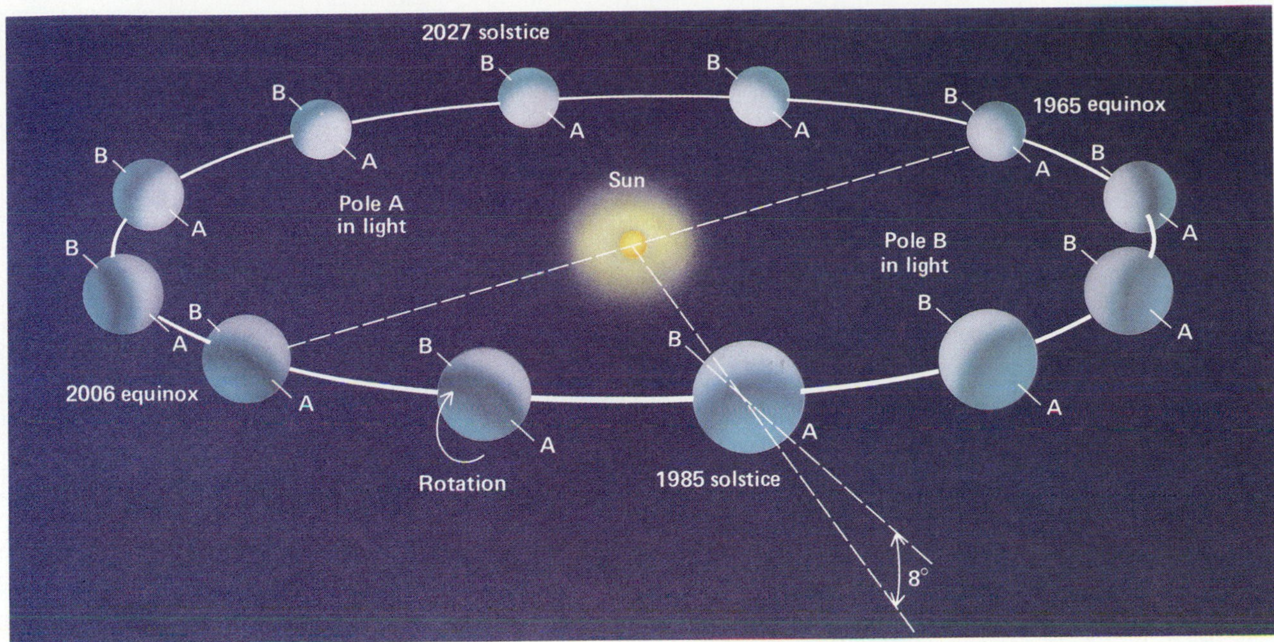

Figure 14–3 Uranus's axis of rotation lies in the plane of its orbit. Notice how the planet's poles come within 8° of pointing toward the Sun, while ¼ of an orbit before or afterward, the Sun is almost over the equator.

the occultation from NASA's Kuiper Airborne Observatory, an instrumented airplane (Section 4.13a). Surprisingly, about half an hour before the predicted time of occultation, they detected a few slight dips in the star's brightness (Fig. 14–4). (The equipment had been turned on early because the exact time of the occultation was uncertain.) They recorded similar dips, in the reverse order, about half an hour after the occultation (Fig. 14–5). Ground-based observers detected some of the dips, but none observed complete before-and-after sequences of all the rings, as was possible from the plane. The dips indicated

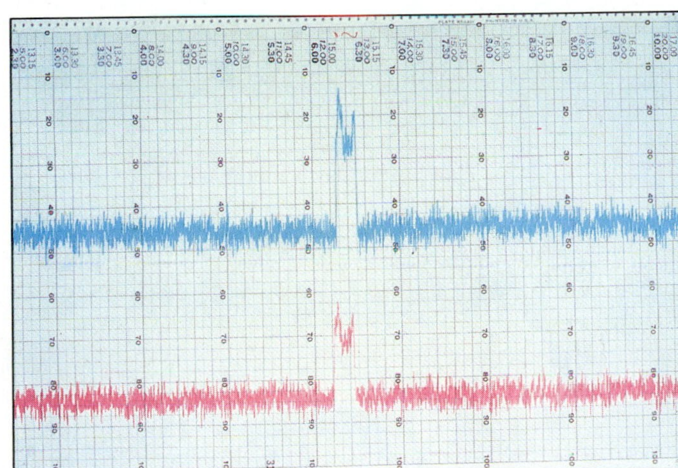

Figure 14–4 The Uranus occultation by the E-ring; different channels are recorded in different colors. This occultation started minutes after the scientists started recording data.

Figure 14–5 James Elliot studying the occultation data, lining up the immersion and emersion scans. From this procedure, he realized that Uranus has rings.

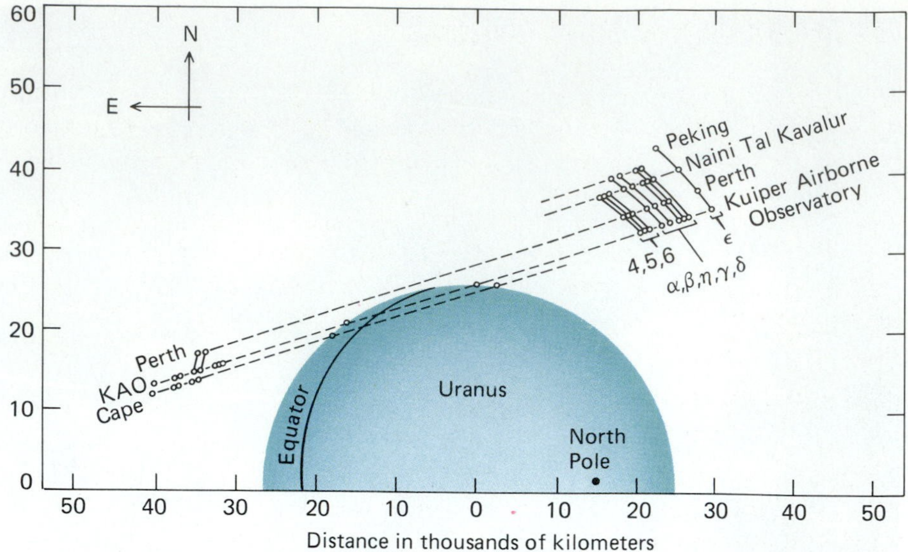

Figure 14–6 The positions of the dips in the observations by which the rings of Uranus were discovered are marked. Some of the rings have been assigned Greek letters, and others have been assigned numbers. The epsilon ring is particularly out of round, and the orientation of the orbit's axis is precessing about 2° per day.

that Uranus is surrounded by at least five rings (Fig. 14–6). More occultations confirmed the discovery of these rings and suggested the possibility of additional ones, to a total of 9. In 1987, Uranus was about to go into retrograde motion when it occulted a star, so the occultation lasted 4 days instead of a half hour! As a result, the rings passed very slowly over the star, giving especially accurate data (Fig. 14–7).

The rings have radii between 42,000 and 52,000 km, 1.7 to 2.1 times the radius of the planet. (Uranus's innermost moon is 130,000 km out.) Reexamination of the old photographs from the Stratoscope balloon revealed that the shadow of the rings could be detected on Uranus's disk. The rings can even be imaged using new infrared technology on Earth (Fig. 14–8).

The uranian rings are very narrow from side to side; some are only a few kilometers wide. No material has been detected between them. How can narrow rings exist, when the tendency of colliding particles would be to spread out? The discovery of Uranus's narrow rings led to the suggestion that a small satellite—a "ringmoon"—in each ring keeps the particles together. This model, set up for Uranus, joins the model of shepherding satellites in possibly being applicable to the narrow ringlets of Saturn discovered by the Voyagers.

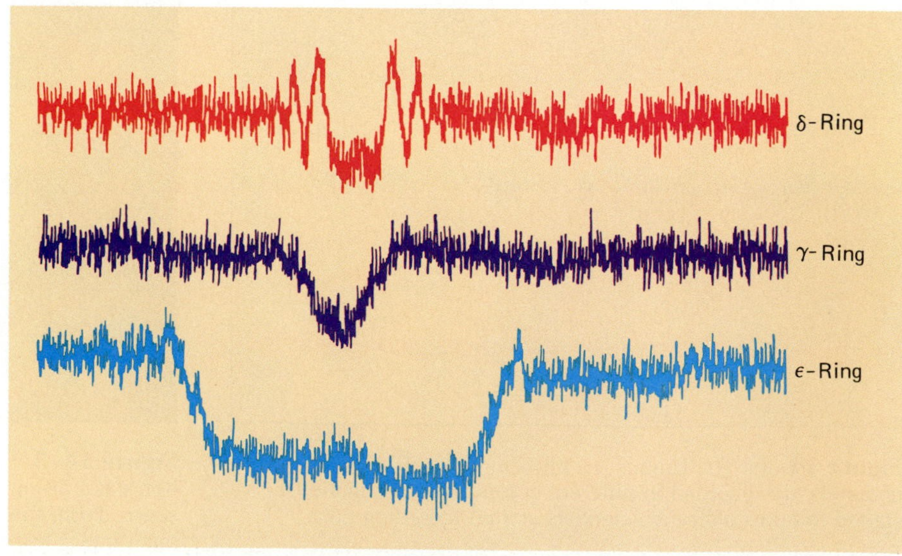

Figure 14–7 The occultation of a star by Uranus took 4 days on one occasion in 1987, since Uranus was about to enter retrograde motion. As a result, the rings occulted the star relatively slowly, giving unprecedented accuracy. Here we see 1 minute of intensity data for the δ, γ, and ϵ rings, respectively.

14.4 Voyager 2 at Uranus

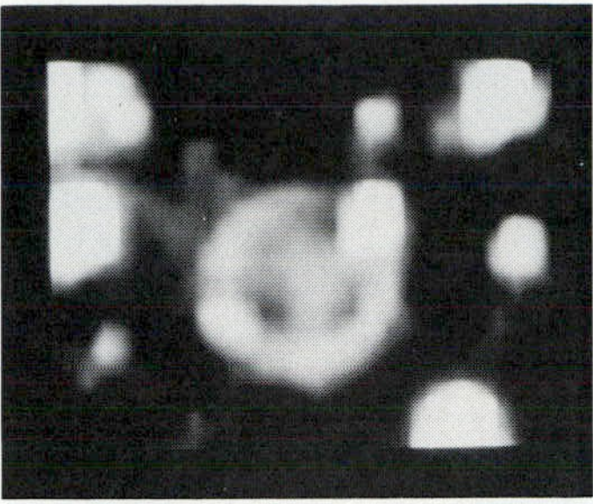

Figure 14–8 At the infrared wavelength of methane, Uranus absorbs sunlight very well and so reflects relatively little. Thus Uranus appears relatively dark compared with the rings, and the rings can be observed with an infrared array.

Voyager 2 visited the Uranus system on January 24, 1986. The spacecraft rushed through quickly, since Uranus and its moons are oriented like a bull's-eye in the sky. At its passage, it was so far from Earth that the radio signals took 2 hours 45 minutes to reach us.

Sunlight was only 1/370 as bright at Uranus as it is at Earth, so exposures had to be long. Fortunately, engineers at the Jet Propulsion Laboratory had developed a technique of turning the spacecraft during the exposure to minimize blurring.

14.4a Uranus's Surface

Voyager 2 came as close as 80,000 km—a fifth of the distance from the Earth to the Moon—from Uranus's surface. Even from that close up, the surface of Uranus appeared bland (Fig. 14–9). This observation verified scientists' expectations; after all, Uranus is farther from the Sun than Jupiter or Saturn, and thus colder, which limits chemical reactions. The clouds form only relatively deep in the atmosphere.

The Voyager observations revealed a dark polar cap. The observations are best interpreted in terms of high-level photochemical haze added to the effect of sunlight scattered by hydrogen molecules and helium atoms. At lower levels, the abundance of methane gas (CH_4) increases. It is this methane gas that absorbs the orange and red wavelengths from the sunlight that hits Uranus, leaving mostly blue-green in the light scattered back toward us. The relative amounts of hydrogen and helium are somewhat larger than for Jupiter and Saturn, but still similar to the Sun's atmosphere. The density of Uranus, though, is too high for Uranus to be entirely of solar composition, and CH_4 clouds are present, so the percentage of carbon may be 20 times higher in Uranus than in the Sun. The theory that Uranus formed from a core of rocky and icy planetesimals explains this abundance.

The polar cap is apparently caused by an increased haze. Since it is centered on Uranus's pole of rotation as defined by the orbits of the satellites and by the rings, and is offset from the subsolar point, the circulation of the atmosphere must be affecting the haze distribution.

The two elongated bright features (Fig. 14–10) may be places where convection carries the clouds, believed to be of methane ice, higher than usual. They are above more of the haze, so appear brighter. It may be that material is carried upward to form these tails at higher altitudes.

The feature at latitude 35° was observed for 15 rotations, at 16.3 hours per rotation. The feature at latitude 27° was observed for 14 rotations, at 16.9

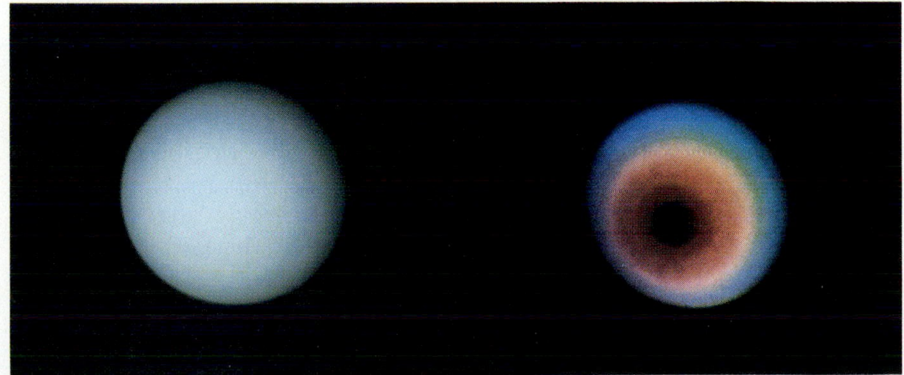

Figure 14–9 Uranus, as the Voyager 2 spacecraft approached. The picture on the left shows Uranus as the human eye would see it. On the right, false colors bring out slight contrast differences.

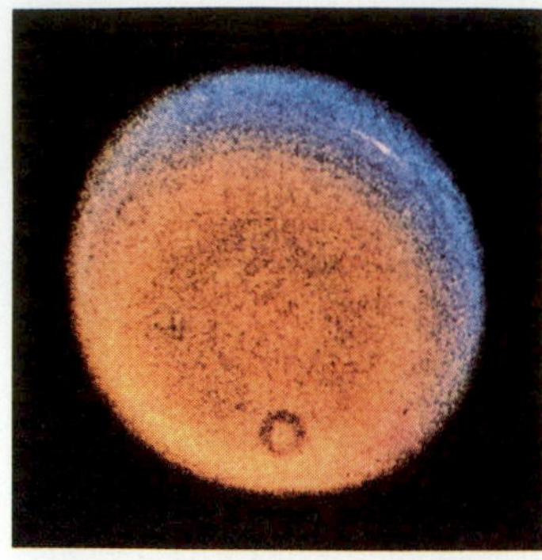

Figure 14–10 An exaggerated false-color view. (The doughnut shapes are out-of-focus dust spots.) The rotation period of the larger cloud, at 35° latitude, is 16.3 hours, and the rotation period of the smaller, fainter cloud, at 27°, is 16.9 hours. Faint latitudinal bands are also visible. Calculations based on a comparison of the orange, green, and methane (0.619-micron) filter images indicate that the feature at a latitude of 27° is about 1.3 km higher than its surroundings and the feature at latitude 35° is about 2.3 km higher, to an accuracy of a factor of 2.

hours per rotation. Several smaller features lasted for over a rotation period, allowing the rotation to be measured at several latitudes.

A few low-contrast bands were seen on Uranus, especially in mid-latitudes. The patterns in the clouds and the motions of the clouds show that the winds circulate east–west rather than north–south. Winds on Jupiter and Saturn act like these, and winds on other planets do so to a lesser extent. The fact that clouds are drawn out in latitude shows how important rotation is for weather on a planet, more important than whether the Sun is overhead, heating the pole more than the equator. It was a surprise to find that both poles, even the one out of sunlight, are about the same temperature. The equator is nearly as warm. Comparing such a strange weather system with Earth's will help us understand our own weather better.

Several experiments gathered information about Uranus's upper atmosphere. For example, an infrared radiometer measured the temperature of gases and how it varies with depth. The variation of a star's ultraviolet light as the planet apparently passed between it and the spacecraft also tells us about Uranus's atmosphere, as does the variation of Voyager's own radio beacon as detected on Earth.

A strange ultraviolet emission was discovered at Uranus. Called the *electroglow,* it occurs on the sunlighted side 1500 km above the top of the clouds. Apparently sunlight splits hydrogen molecules (H_2) into protons and electrons. The electrons somehow acquire extra energy and collide with remaining hydrogen molecules, and the electroglow results. The extra energy may involve Uranus's magnetic field. The electroglow is too strong to allow any auroras to be seen on the daylighted side. On the night side, where there is no electroglow, a faint ultraviolet aurora was detected. We now realize that a faint electroglow also exists on Jupiter, Saturn, and Titan.

Figure 14–11 All 9 previously known rings. A newly discovered 10th ring is not quite visible about midway between the wide epsilon ring at upper left and the next ring. This photograph was taken by Voyager 2 in reflected sunlight at a distance of 1.1 million km.

14.4b Uranus's Rings

Images from the Voyager showed the 9 rings known from ground-based occultation observations, plus a faint 10th ring (Fig. 14–11) and a broad, faint 11th ring. All but one are not quite circular, and most are inclined relative to the plane of Uranus's equator. They differ slightly in color and therefore composition (Fig. 14–12).

The detailed Voyager observations showed that all the rings vary in width and apparent darkness with position around the ring. The epsilon ring has the widest variation, from 20 km closest to Uranus up to 96 km farthest away. Other rings are closer to 10 km across, as had been known.

The ground-based results that the rings were only about 2 per cent reflective enabled Voyager scientists to set proper exposures. The epsilon ring was resolved into two bright inner and outer regions with a lower brightness region between them. The epsilon ring is gray; the other rings can't be measured as accurately, but no color was seen for them, either. The low albedoes and absence of color make a sharp contrast with the brighter, reddish surfaces of the ring particles and moons around Jupiter and Saturn.

Following the unexpected brightness of Jupiter's ring when seen backlighted—looking through it toward the Sun—Voyager scientists arranged many

Figure 14–12 A false-color view of the rings of Uranus. The fainter, pastel lines between the rings are computer artifacts. The epsilon ring, at top, is neutral in color, but the fainter eight other rings show color differences. Interpreting the color differences is important for understanding the composition of the ring material.

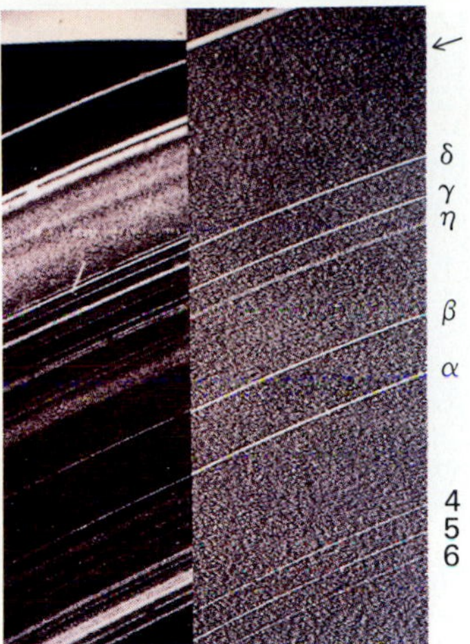

Figure 14–13 When the Voyager spacecraft viewed through the rings back toward the Sun, the backlighted view revealed new dust lanes between the known rings, which also show in this view. The streaks are stars. The forward-lighted view is provided for comparison. One of the new rings discovered from Voyager is marked with an arrow.

exposures looking toward the Sun through Uranus's rings. But, as they watched in the control room, these exposures were all turning out blank. Apparently, there is less small dust in Uranus's rings to cause such forward scattering of sunlight. Less than 0.001 of the particles in Uranus's rings are micron-sized, in comparison with a few per cent in the rings of Saturn and perhaps 50 per cent in the rings of Jupiter. Some process, perhaps an extended upper atmosphere of Uranus, must be removing these micron-sized particles from Uranus's rings.

After many blank exposures, when a photograph showing ring structure appeared on the monitoring screen, scientists at first thought that it might be a joke. But it turned out to be the single long exposure that was taken (Fig. 14–13). At the time, Uranus was hiding both Earth and Sun from the spacecraft, allowing the 96-second exposure of the backlighted rings to be taken. All nine main rings are identifiable. The bright feature next to the wide band corresponds to the newly discovered 10th ring. The dark lane just inside this ring corresponds to the orbit of a newly discovered moon, perhaps a shepherding satellite. About 50 dust bands also appeared; Uranus's upper atmosphere extends into the rings and should draw the dust out of the rings in less than 1000 years.

The rings were also studied by observing how their apparent brightness changed as they occulted a star (Fig. 14–14). As for Saturn, this method gives the most precise information on the distribution of material in the ring system. Ring particles beyond the detected rings showed up when the plasma-wave experiment registered direct hits as the spacecraft passed through the plane of the ring system. The fact that all the previously known rings affected the reception of the Voyager's radio signal on Earth when the rings occulted the spacecraft as seen from Earth means that they must contain particles about the same size as the radio wavelengths used. Since they were detected at both 3.6-cm and 13-cm wavelengths, ring particles must be at least many centimeters in size.

Two of the other new satellites discovered—Cordelia and Ophelia—are apparently the shepherds for the epsilon ring (Fig. 14–15). Shepherds for the

Figure 14–14 The epsilon ring, 100 km wide here, shows a wide bright band, a wide fainter band, and a narrow bright band. This reconstruction was made with data gathered when Voyager viewed the star sigma Sagittarii through the rings. In this false color, reddish areas represent portions of the ring containing less material, and yellow areas are portions containing more material. The ring is different widths at different positions. We see here two slices, 31 km and 22 km wide, respectively.

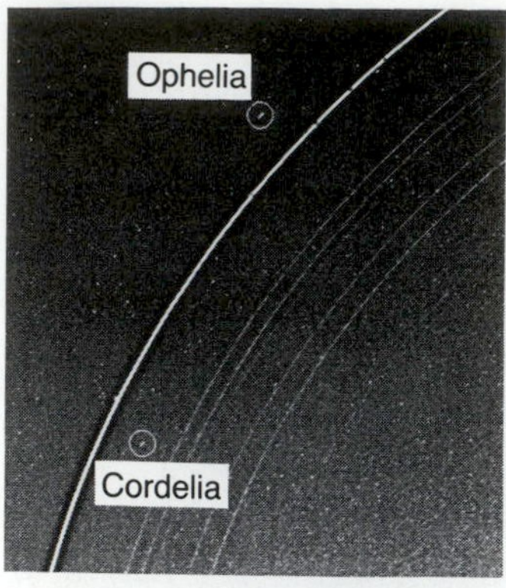

Figure 14–15 Three days before closest approach, Voyager 2 discovered the two "shepherd satellites" for the epsilon ring. Based on their brightness and their likely albedo, they are 20 and 30 km in diameter, respectively. All 9 of the previously known rings are also visible. The image was processed to enhance narrow features, giving the epsilon ring an artificial dark halo. The dark blips on the ring are also artifacts.

other rings have not been found, though. Perhaps the shepherds are too small to have been seen by Voyager.

Since the rings of Uranus are apparently younger than 100 million years of age (for one thing, the shepherds are too small to hold them longer), there must have been a source of dust to form them. That source may well have been a ringmoon destroyed by a giant impact on it by a meteoroid or comet. Particles in the dust rings seen only in the backlighted image may come from a different source—perhaps from the surfaces of the current moons.

14.4c Uranus's Interior and Magnetic Field

Though the presence of a magnetic field had been suspected, its direct detection by instruments on Voyager 2 was a major result of the mission. Voyager detected radio waves from Uranus only about two days before arrival; electrons in Uranus's ionosphere apparently block radio noise. Less than 8 hours before arrival at Uranus, about 500,000 km away, the spacecraft reached Uranus's magnetosphere. It is intrinsically about 50 times stronger than Earth's. Surprisingly, it is tipped 60° with respect to Uranus's axis of rotation. And even more surprising, it is centered around a point offset from Uranus's center by 8000 km (Fig. 14–16). Since the field is so tilted, it winds up like a corkscrew as Uranus rotates. Uranus's magnetosphere contains belts of protons and electrons, similar to the Earth's Van Allen belts.

Two possibilities for why the axis of Uranus's magnetic field is so tipped are that (a) the magnetic field is in the midst of changing direction, as the Earth's does every 200,000 years, or (b) an Earth-sized object rammed into Uranus early on and tipped it on its side. Less *ad hoc*, and therefore perhaps more plausible, is (c) the idea that the dynamo is in a thin electrically conducting shell outside the core of the planet, rather than in a deeply seated core as for the Earth and Jupiter. The discovery, as we shall see in the following chapter, that Neptune's magnetic field is also very inclined and offset makes the last idea more appealing, because it can be applied elsewhere.

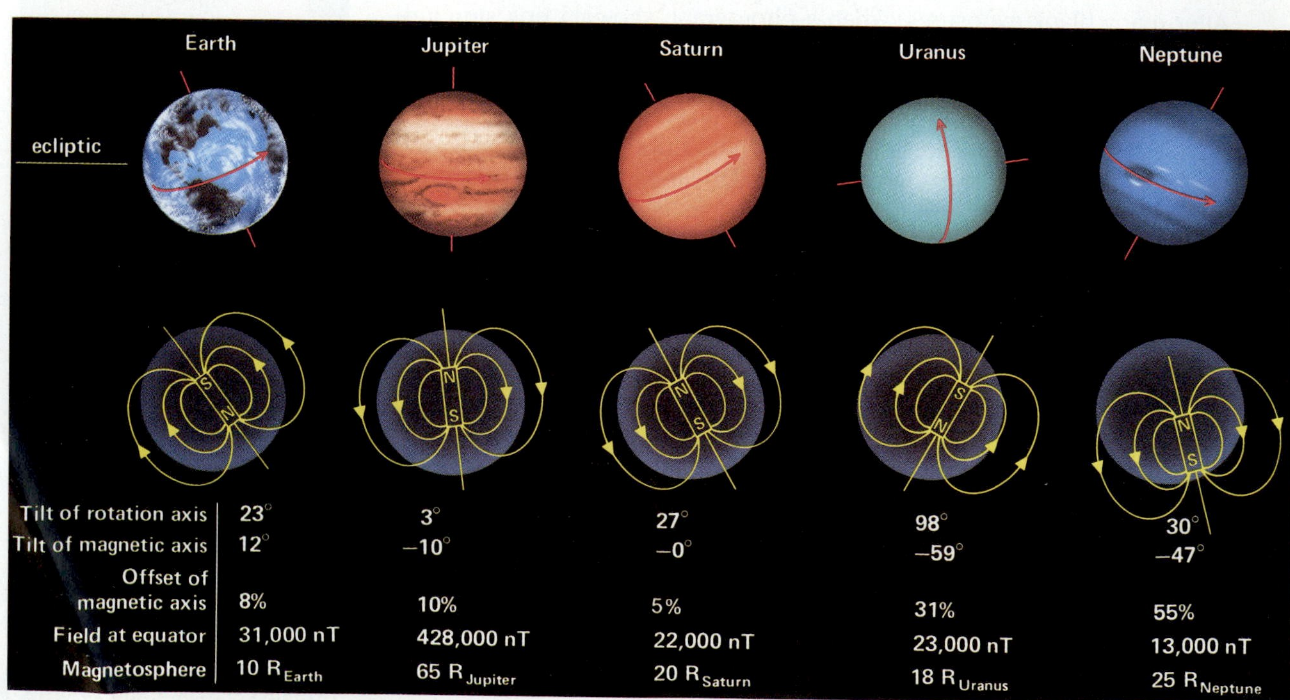

Earth
Jupiter
Saturn
Uranus
Neptune
ecliptic
Tilt of rotation axis | 23° | 3° | 27° | 98° | 30°
Tilt of magnetic axis | 12° | −10° | −0° | −59° | −47°
Offset of magnetic axis | 8% | 10% | 5% | 31% | 55%
Field at equator | 31,000 nT | 428,000 nT | 22,000 nT | 23,000 nT | 13,000 nT
Magnetosphere | 10 R_{Earth} | 65 $R_{Jupiter}$ | 20 R_{Saturn} | 18 R_{Uranus} | 25 $R_{Neptune}$

Figure 14–16 Uranus's magnetic field is tipped and offset from Uranus's center.

Studies of radio bursts, whose emission is linked to magnetic field lines anchored deep inside the planet, repeat with a period of 17.24 hours. Thus Uranus's interior rotates more slowly than its atmosphere.

All the rings and most of the moons of Uranus lie within the magnetosphere. They are thus protected from the solar wind.

14.4d Uranus's Moons

Voyager took fantastic photographs of Uranus's 5 previously known moons (Fig. 14–17). Since they were oriented in their orbits around Uranus like a bull's-eye, the spacecraft sped by them all in a few hours. Nonetheless, the images we have show us their distinctive individual characters. We can remember Miranda, Ariel, Umbriel, Titania, and Oberon by their initials: MAUTO. Titania and Oberon are the largest, about the same size as each other and Saturn's moons Iapetus and Rhea. Uranus's moons all rotate synchronously, once per orbit, leaving the same side always facing Uranus. Since the Sun shone almost directly down on their south poles, we have imaged only their southern hemispheres; the northern hemispheres remain unknown to us.

Figure 14–17 The 5 largest moons of Uranus, reproduced in their relative sizes and albedoes.

In general, as had been discovered from ground-based measurements, the moons of Uranus are darker (have lower albedoes) than the moons of Saturn, except for Phoebe and the dark regions of Iapetus. There is no obvious trend in albedo with distance from Uranus. The dark material seems evenly dark across the spectrum, that is, gray rather than colored. Carbon in the form of soot or graphite would explain this grayness; iron-bearing silicates and sulfur are ruled out because of their reddish color. The dark material in the rings is similarly gray.

Only for Miranda did the spacecraft fly sufficiently close to allow us to measure the moon's mass from its gravitational effect. The masses of the other major satellites have been found from detailed calculations of their gravitational effects on each other, measured on Voyager data as small variations of their positions from their average orbit. The measured masses plus the observed volumes of the moons allow the densities to be calculated. They turn out to be about 1.5 times the density of water, higher than those of Saturn's icy satellites, and do not vary in any consistent way with distance from Uranus. The moons, like those of Jupiter and Saturn, are apparently mixtures with varying proportions of rock and ices of water, ammonia, methane, and other chemicals. Perhaps the conditions under which the satellites are formed were affected by the event that threw Uranus over on its side. Since the moons' orbits are in the plane of Uranus's current equator, they were presumably formed after the catastrophic event, whatever it was. The moons could have formed from a disk of material thrown out around Uranus.

All the satellites have craters on them. The largest craters probably formed in collisions with planetesimals or other debris of the formation of the planets during the early years of the solar system, more than 4 billion years ago. Many of the smaller craters were probably caused by secondary debris of collisions between satellites and of impacts with satellites. We think that whatever craters are still forming result from impact with short-period comets captured into orbits near Uranus's.

Ariel and Umbriel are similar in size and density, but they differ greatly in the way they look. Their geological histories must thus be very different. Ariel is about 4 times farther out than the rings and Umbriel is about 5 times farther out than the rings.

Ariel (Fig. 14–18) has the youngest surface of Uranus's moons and is the brightest. It shows signs of geological activity on a large scale: Many of the fractures of Ariel's crust are global, extending over the entire surface we can

Figure 14–18 Ariel, 1,200 km across, is covered with small craters and shows many valleys and scarps. At this resolution, the highest obtained, we see linear grooves that presumably resulted from tectonic activity and smooth patches that show where material was deposited.

Figure 14–19 Umbriel, also 1,200 km across, is the darkest of Uranus's moons and shows the least geological activity. The nature of the bright ring is unknown, though it might be a frost deposit around an impact crater.

see. Ariel has fault valleys though few large craters, the presence of the former and the absence of the latter both being signs of geological activity. Apparently material has flowed from under Ariel's surface; in one location, a half-buried crater is seen. Sinuous valleys similar to those on the Moon are found, some with parallel ridges alongside. These features appear similar to lava channels and ridges known on the Earth and other inner planets. What material is flowing? It is too cold for water to melt. A mixture of ammonia and water ice has been suggested. And perhaps tidal effects from the other moons could have heated Ariel in the past.

Umbriel (Fig. 14–19), though similar physically, is much darker. Its ground-based spectrum shows weaker signs of water ice than do spectra of Uranus's other major moons. Umbriel's surface is strikingly uniform, also in contrast to Uranus's other major moons. It is covered with craters; Umbriel has the oldest surfaces, similar to the ancient cratered highlands of the inner planets. Its surface may date back 4.5 billion years to the formation of the solar system. The bright ring is a high-albedo deposit that covers the floor of a 40-km-diameter crater, Wunda. It must be a relatively young feature. Why is Umbriel so uniform? Perhaps a uniform blanket of dark material has coated the surface, covering earlier markings. Since only Umbriel is coated, the material would have had to come from Umbriel itself, perhaps from a large impact that sent up lots of material. The bright ring would have been deposited more recently or formed in a region where the blanket was thin. A similar model has been advanced for Saturn's Enceladus. The major objection to this model is that such an impact has a low probability. An alternate model is that something in Umbriel erupted explosively; methane has been suggested. In this model, the bright ring might be connected with the eruption. The major objection to this

Figure 14–20 Titania, 1,600 km across, shows so many impact craters that it must have an old surface. The walls of some of the fault valleys may appear bright because of frost deposits.

Figure 14–21 Oberon, also 1,600 km across, was not imaged as well as the other moons. The bright patches may be patterns caused by the formation of an impact crater in an icy surface.

model is that we see no signs on the rest of the surface that vulcanism occurred. Remember, though, we cannot see Umbriel's northern hemisphere—perhaps huge volcanoes are there. There is a lot we don't know about Umbriel, including why no craters with bright rays are seen.

Just as Ariel and Umbriel are similar in size, so are Titania and Oberon, each about 1600 km across and twice as far out. The albedoes of the latter pair are similar to each other, 20 to 30 per cent. Both are grey. The visible surface of Titania shows many craters 10 to 50 km across, but few large ones (Fig. 14–20). The large ones would have been formed earlier, at the same time as others on other moons, since only in the early years of the solar system when planetesimals were still around did a full range of size of impactors exist. (After the planets and their moons formed, the other bodies left over were all fairly small.) The large craters that would have once been on Titania were perhaps erased by material that flowed out of Titania's interior. Among the few large craters still present are a large impact basin 300 km across. If the impact occurred after the large craters were covered, the debris from the impact would have caused many of the smaller craters we now see. Titania's surface shows many scarps from 2 to 5 km high. Halves of craters split by the scarps are displaced, showing that the scarps are relatively young. A few bright deposits are visible on some of the scarps, probably showing fresh material from underground.

Oberon is heavily cratered with large craters, many greater than 100 km across (Fig. 14–21). As with Umbriel, parts resemble the ancient cratered highlands of the inner planets. Astronomers in this field believe that such craters resulted from impacts of the debris left over from the formation of the planets, and are over 4 billion years old. A mountain was seen sticking out 11 km beyond Oberon's limb; since its base is around on the other side, we do not know how much taller than 20 km it is. The presence of large circular regions shows that Oberon has been geologically active. Some very dark material occurs in the floors of a few large craters. The dark patches are similar in some

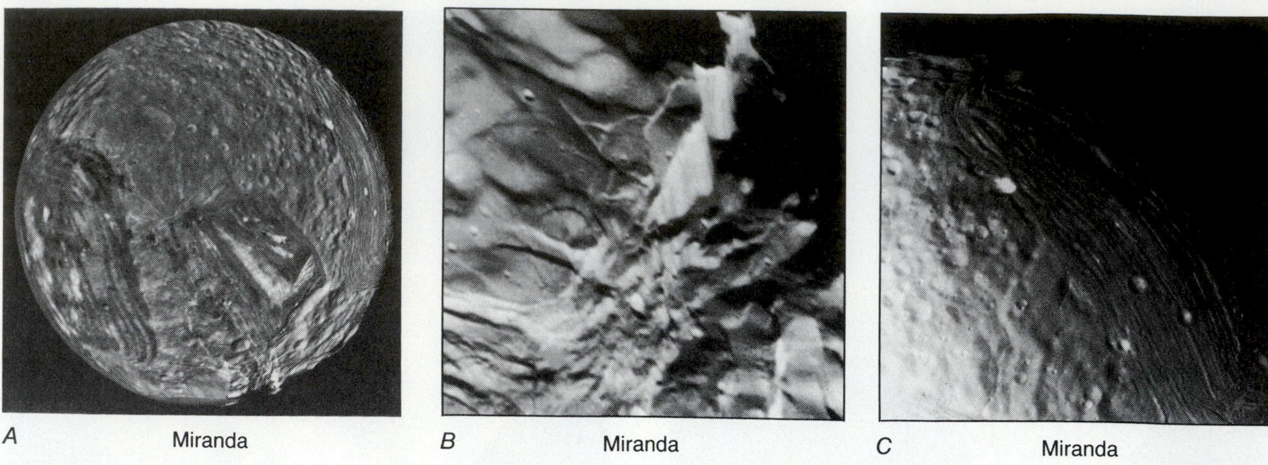

Figure 14–22 (*A*) Miranda, only 500 km in diameter, shows several different types of geological regions on its surface, in spite of its small size. The south pole is almost exactly in the center of this image. (*B*) The giant canyon on the edge of Miranda, at a resolution of 0.6 km. The grooves and troughs can be a few kilometers deep. (*C*) At left, an ancient, cratered terrain of rolling hills and degraded craters; at center, a younger, grooved terrain; and near the terminator, a still younger, complex terrain that ends abruptly at the grooves.

ways to dark patches in craters on Iapetus's trailing hemisphere and to volcanic deposits in several lunar craters. It seems that the dark patches were a fluid from inside Oberon that either was originally dark or turned dark after it reached the surface.

Inside the orbits of these four moons lies the orbit of Miranda, half the size of Ariel and Umbriel and a third the size of Titania and Oberon. Many people thought it is too small to have much activity. But they were wrong—Miranda turns out to be as varied a body as exists in the solar system. As Voyager 2 approached Miranda, a bright region shaped like a chevron was visible (Fig. 14–22). Only an internal process could have caused such a sharp angle to appear. Closer views revealed that part of Miranda's surface is covered with undulating cratered plains and the other part is covered with regions showing almost parallel dark bands with a few bright bands. The trapezoidal region that contains the chevron is 200 km on a side, with sharp corners. Another region with parallel bands is about 300 km across and has rounded corners; we can see only part of this feature, as the rest is over the limb. Both these regions have outer belts about 100 km wide that wrap, like a racetrack, around an inner core that shows no concentric pattern. The ridges in the racetrack pattern resemble the grooved terrain found on Jupiter's Ganymede. In the highest-resolution images, the dark bands proved to be scarps facing outward. In some locations, erupted material has apparently flowed over and partly buried the grooves and ridges. Miranda shows fault systems on a global scale. A huge valley showed on Miranda's edge (Fig. 14–23). Computer simulations have shown what Miranda might seem like if we were travelling around it (Fig. 14–24).

How did Miranda get so hot inside? Perhaps a collision tore it apart long ago. When the pieces came together, they gave Miranda its jumbled surface. Energy might have been released from gravity as denser blocks sank. Tidal heating, as for Io, is another source of energy that may have contributed. Though the satellites of Uranus do not interact in that manner now, they might have long ago.

Voyager 2 discovered 10 new satellites (Fig. 14–25), all in nearly circular orbits and all but one between the epsilon ring and the orbit of Miranda. One of

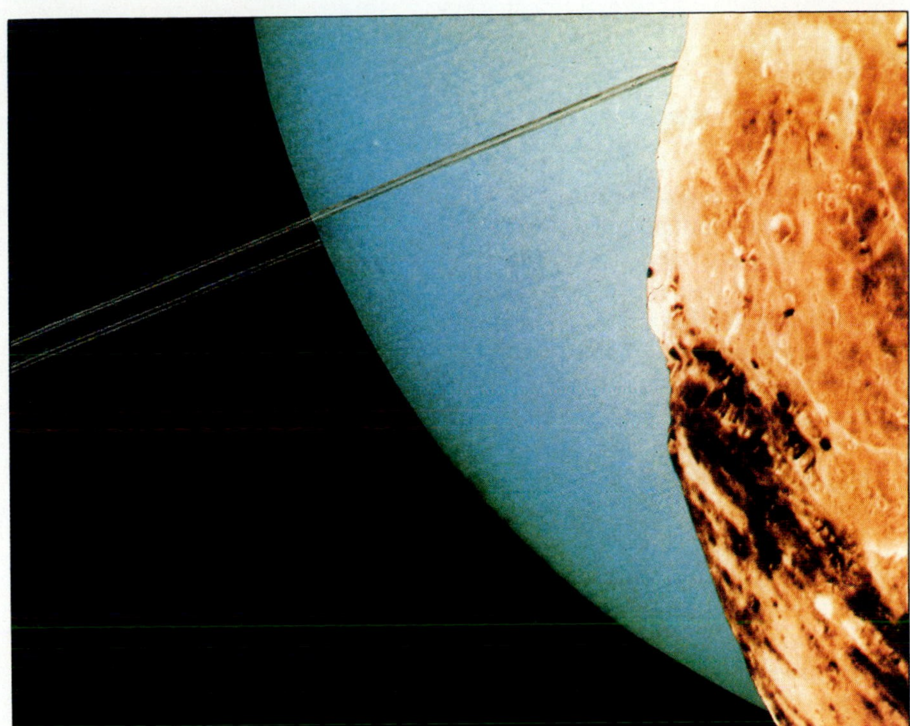

Figure 14–23 A montage of photographs of the giant valley on the edge of Miranda and Uranus, with a ring drawn around Uranus for scale.

Figure 14–24 A computer-calculated three-dimensional simulated flight over the mountains and through the canyons of Miranda. The vertical scale is expanded by a factor of three.

Focus On

Box 14.1 Naming the Moons of Uranus

Six years after he discovered Uranus, William Herschel reported the discovery of several moons around it. In 1851, William Lassell discovered two additional moons of Uranus and showed that only two of Herschel's discoveries were real. Lassell assigned names from the fairy kingdom. At his request, William Herschel's astronomer son John Herschel selected the names Oberon and Titania for the moons his father had discovered. Oberon is King of the Fairies and Titania is Queen of the Fairies in Shakespeare's *A Midsummer Night's Dream.* Umbriel and Ariel are the names of guardian spirits in Alexander Pope's *The Rape of the Lock* and Ariel is also a character in Shakespeare's *The Tempest.* Miranda wasn't discovered until 1948, when it was named for the heroine of *The Tempest.*

The first rings to be discovered, by James Elliot and colleagues in 1977, were assigned Greek letters. The rings first noted by others were assigned numbers.

Permanent names for the rings and moons were assigned by the International Astronomical Union at its General Assembly in Baltimore in 1988. The first moon to be discovered by Voyager 2 was named Puck, after the spirit in Shakespeare's *A Midsummer Night's Dream.* Most other moons were named after Shakespearean heroines: Cordelia and Ophelia are the epsilon-ring shepherds and Bianca, Cressida, Desdemona, Juliet, Portia, and Rosalind follow in increasing distance from Uranus. Then comes Belinda, whose name comes from Pope's *The Rape of the Lock.* This moon's name fulfills the poet's prophecy:

> This lock, the muse shall consecrate to fame,
> And midst the stars inscribe Belinda's name.

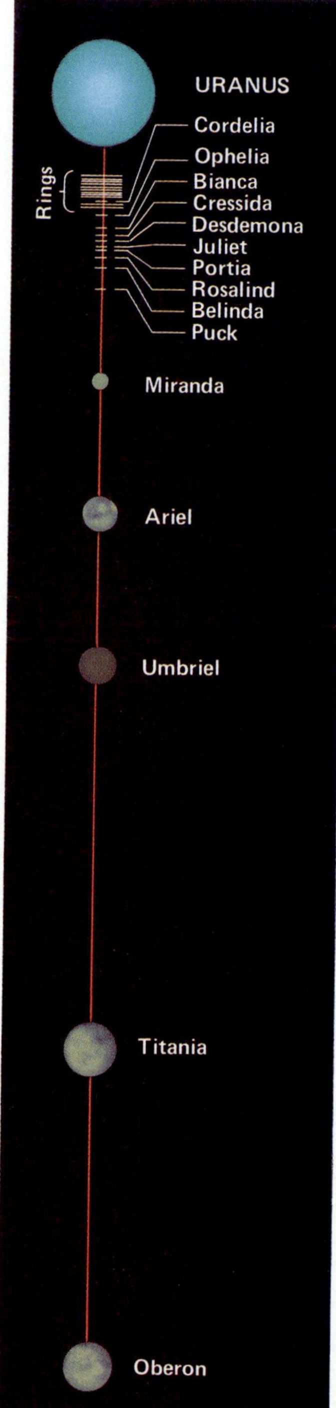

the new satellites, Puck, was discovered so long before the closest approach of the spacecraft to the Uranus system that the spacecraft could be reprogrammed to get an image. The lack of surface contrast may indicate that any fresh material exposed by impacts on its surface is quickly erased, or that the moon has dark material mixed uniformly through its interior. In the latter case, it would be a primitive object. But the image is underexposed, and Puck may not really be so bland.

The albedoes of all Uranus's smaller satellites are low, 5 to 7 per cent, in contrast to the high albedoes of Saturn's smaller satellites and even to the higher albedoes of Uranus's larger satellites. Thus the composition of these small moons or perhaps the conditions under which these small moons have formed and evolved must have been different for the Uranus system than for the Saturn system.

Figure 14–25 The rings and moons of Uranus. The moons are in correct scale with respect to each other but are at a larger scale than Uranus and the positions of the rings and moon.

Summary and Outline

Uranus and Neptune are each 50,000 km in diameter and 15 times more massive than Earth

Uranus seen from Earth (Section 14.1)

- First planet not known to ancients
- No detail visible on surface, perhaps surrounded by thick methane clouds
- No internal heat source
- Aurora indicates magnetic field

The rotation of Uranus (Section 14.2)

- Axis lies in plane of solar system

Uranus's rings (Section 14.3)

- Nine narrow rings discovered in 1977 at a stellar occultation

Voyager 2 at Uranus (Section 14.4)

- Passed through system in 1986
- Dark polar cap; mid-latitude brightening
- A few methane clouds probably above the haze; rotation period 16–17 hours
- Elongated clouds and banks indicate east–west circulation; rotation is more important than position of Sun
- Both poles the same temperature
- Electroglow—ultraviolet emission
- Little small dust in rings; particles at least several cm across
- Magnetic field intrinsically 50 times stronger than Earth's, tipped 60° and offset from center
- Five largest moons mixtures of rock and ices, all have craters
  - Ariel: youngest surface; brightest. Global fractures, fault valleys, few large craters
  - Umbriel: much darker, uniform surface covered with craters, perhaps dark coating of material; unexplained bright ring
  - Titania: few large craters, many young scarps, some fresh material
  - Oberon: many large craters, high mountain, scarps, dark patches in crater floors
  - Miranda: part of surface with undulating cratered plains and part with regions of scarps that show as dark bands, global faults

Key Words

electroglow

Questions

1. What is strange about the direction in which Uranus rotates?
2. Using Appendix 4, compare the sizes of the moons of Uranus with other objects in the solar system.
3. Explain how the occultation of a star can help us learn the diameter of a planet.
4. How were the rings of Uranus discovered?
5. What did the Voyager 2 encounter tell us about the rings that we didn't already know?
6. How do we learn whether a planet's rotation or the orientation of its axis determines its weather?
7. Why does Uranus appear blue-green?
8. What causes the electroglow on Uranus?
9. Compare frontlighted and backlighted views of Uranus's rings. What does the comparison tell us about their composition?
10. Compare Uranus's magnetosphere with Earth's.
11. Why didn't we observe the northern hemispheres of Uranus's moons from Voyager 2?
12. Describe the physical properties of two of Uranus's 5 major moons.
13. Which of Uranus's moons has the youngest surface? Why do we think so?
14. Which of Uranus's moons have the oldest surfaces? Why do we think so?
15. Compare the albedoes of Uranus's major and minor satellites.

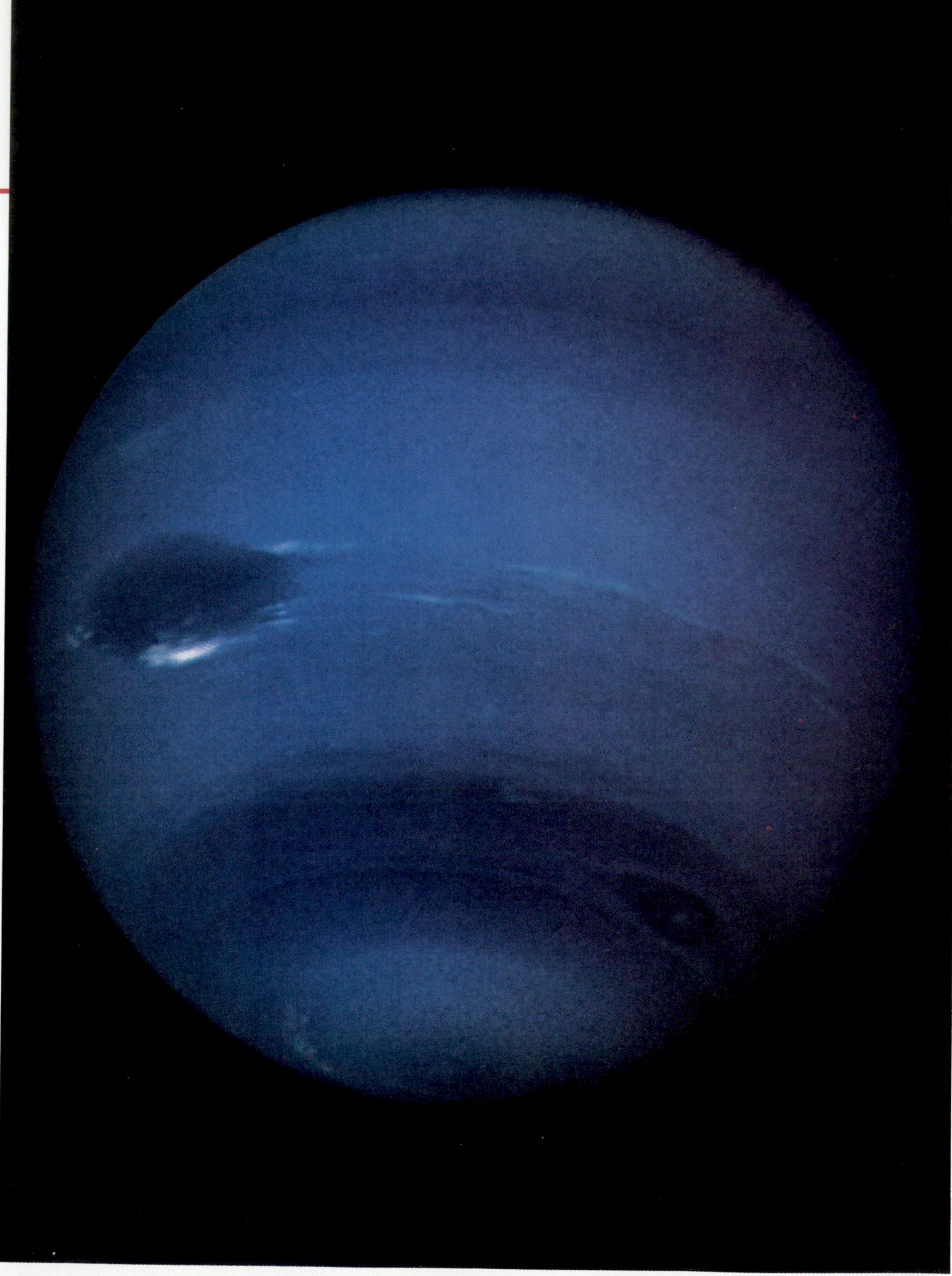

Neptune, from Voyager 2, with its Great Dark Spot. Its blue-green color comes from absorption of sunlight of the red part of the spectrum by traces of methane mixed into Neptune's atmosphere.

Neptune

15

Aims: To study the farthest giant planet—Neptune—seeing, especially, what the Voyager 2 spacecraft revealed about it and its moons and rings

Neptune is like Uranus in many ways. Neptune is also about 50,000 km across and is about 17 times more massive than the Earth. Like Uranus, Neptune has a high albedo and a low density. Both probably have rocky cores of 10 to 15 times the Earth's mass. Both are probably 10 to 20 per cent hydrogen and helium. But Neptune and Uranus are more like cousins than like siblings. Uranus's surface is bland, while Neptune has many clouds. Uranus has no internal heat source, while Neptune has a strong one.

Though Uranus and Neptune are giant planets, they are fairly different from Jupiter and Saturn. Uranus and Neptune's hydrogen/helium content is lower than those of Jupiter or Saturn. A higher proportion of Uranus and Neptune probably formed from icy and rocky planetesimals. The rocky cores of Jupiter and Saturn, though probably the same absolute size as those of Uranus and Neptune, are much smaller fractions of their masses.

Neptune, in Roman mythology, was the god of the sea, and the planet Neptune's trident symbol reflects that origin.

Another similarity that Neptune has with Uranus is that its secrets were bared to us by NASA's Voyager 2 spacecraft. Voyager 2 swept by Neptune and its rings and moons in August 1989. The visit was the last "first" of solar system exploration, aside from a visit to tiny Pluto. Over the last three decades, each of the other planets has been visited close up by spacecraft for the first time.

Scientists are also nostalgic about the Neptune visit for another reason—the knowledge we gained will have to last us for a long time. It is difficult to foresee that any further visit to Neptune will take place for many decades.

15.1 Neptune from Earth

Neptune is even farther from the Sun than Uranus, 30 A.U. compared to about 19 A.U. Neptune takes 165 years to orbit the Sun. Its discovery was a triumph of the modern era of Newtonian astronomy. Mathematicians analyzed the deviations of Uranus (then the outermost known planet) from the orbit it would follow if gravity from only the Sun and the other known planets were acting on it. The small deviations could have been caused by gravitational interaction with another planet.

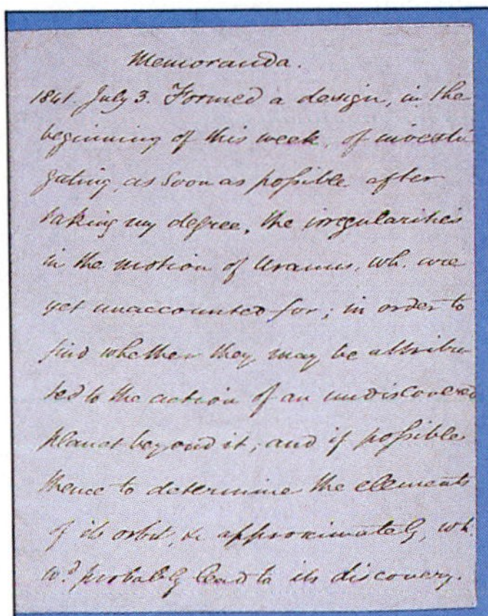

Memoranda.
1841. July 3. Formed a design, in the beginning of this week, of investigating, as soon as possible after taking my degree, the irregularities in the motion of Uranus, wh. are yet unaccounted for; in order to find whether they may be attributed to the action of an undiscovered planet beyond it; and if possible thence to determine the elements of its orbit, &c. approximately, wh. wd. probably lead to its discovery.

Figure 15–1 From Adams's diary, kept while he was in college: "1841. July 3. Formed a design in the beginning of this week, of investigating, as soon as possible after taking my degree, the irregularities in the motion of Uranus which are yet unaccounted for."

John C. Adams, in England, predicted positions for the new planet in 1845 (Fig. 15–1), but the astronomy professor at Cambridge did not bother to try to check out his prediction with a telescope; after all, Adams was to him merely a recent college graduate. Adams went to London to meet the Astronomer Royal (the chief British astronomer), but the A.R.'s butler did not permit dinner to be interrupted, and Adams went away. Though Adams left a copy of his calculations, when the Astronomer Royal requested further information, partly to test Adams's abilities, Adams did not take the request seriously and did not respond at first. The very "proper" Astronomer Royal took offense and

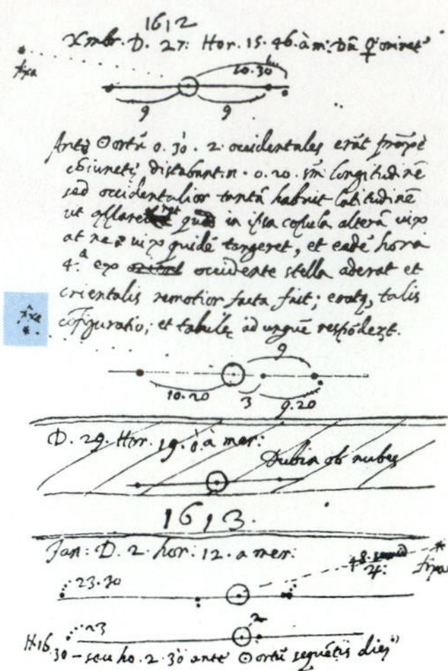

Figure 15–2 Galileo's notebook from late December 1612, showing a * marking an apparently fixed star to the side of Jupiter and its moons. The "star," at extreme left, was apparently Neptune. The horizontal line through Jupiter extends 24 Jupiter radii to each side. Jupiter's moons appear as dots on the line.

The episode shows the importance of keeping good lab notebooks, a lesson we should all take to heart.

did not choose to have further dealings with Adams. The story then continues in France, where a year later Urbain Leverrier was independently working on predicting the position of the undetected planet.

When the Astronomer Royal saw in the scientific journals that Leverrier's work was progressing well (Adams's work had not been made public), for nationalistic reasons he began to be more responsive to Adams's calculations. But the search for the new planet, though begun in Cambridge, was carried out halfheartedly. Neither did French observers take up the search. Leverrier sent his predictions to an acquaintance at Berlin, where a star atlas had recently been completed. The Berlin observer, Johann Galle, enthusiastically began observing and discovered Neptune within hours by comparing the sky against the new atlas.

Years of acrimonious, nationalistic debate followed over who (and which country) should receive the credit for the prediction. (Adams and Leverrier themselves became friends.) We now credit both Adams and Leverrier. With hindsight, we see that they each assumed a radius for the new planet's orbit based on a numerological scheme for the distances of planets. (The scheme is known as Bode's law; no justification for it has ever been found and, in any case, it fails for Pluto.) The value Adams and Leverrier assumed was incorrect, but luckily gave the right position anyway at that time.

Neptune has not yet made a full orbit since it was located in 1846. But it now seems that Galileo actually observed Neptune in 1612, which would more than double the period of time over which it has been observed. Charles Kowal, a Palomar astronomer, and Stillman Drake, a historian of science, tracked down Galileo's observing records from 1612 and 1613, when calculations showed that Neptune had passed close to Jupiter. Galileo, in fact, twice recorded in his notebooks stars that were very close to Jupiter (Fig. 15–2), stars that modern catalogues do not show. Galileo even once noted that one of the "stars" actually seemed to have moved from night to night, as a planet would. The objects that Galileo saw were very close to but not quite exactly where our calculations of Neptune's orbit show that Neptune would have been at that time. Presumably, Galileo saw Neptune, and we can use positions he measured to improve our knowledge of Neptune's orbit.

Neptune's angular size in our sky is so small that it is always very difficult to study. Even measuring its diameter accurately from Earth is hard, and is best done when it occults stars. From observations of the rate at which the stars dim, astronomers deduce information about Neptune's upper atmosphere, including its temperature and pressure structure. From the length of time that the star is hidden by the disk of Neptune, they measure Neptune's diameter. An accurate value for the diameter is important for calculating the planet's density, thus leading to deductions about its composition.

Neptune's density is 1.6 grams/cm^3, higher than Uranus's and Jupiter's 1.3 grams/cm^3 and much higher than Saturn's 0.7 grams/cm^3. Thus Neptune must have a higher proportion of heavy elements. Models predict the presence of cores of about the same mass—10 to 15 Earth masses—at the centers of all the giant planets, which makes these cores take up a larger fraction of Uranus and Neptune than they do of Jupiter and Saturn.

Neptune, like Uranus, appears bluish in a telescope because of its atmospheric methane (Fig. 15–3). In both cases, the atmosphere is mostly hydrogen, with helium next in abundance, but the trace of methane absorbs most of the red light instead of reflecting it. Like Uranus, Neptune's rotational period is difficult to determine from the Earth. Pre-Voyager assessments ranged from 17 to 22 hours.

Neptune radiates substantially more heat (2.7 times) than it receives from the Sun, similarly to Jupiter and Saturn. Uranus, on the other hand, seems not

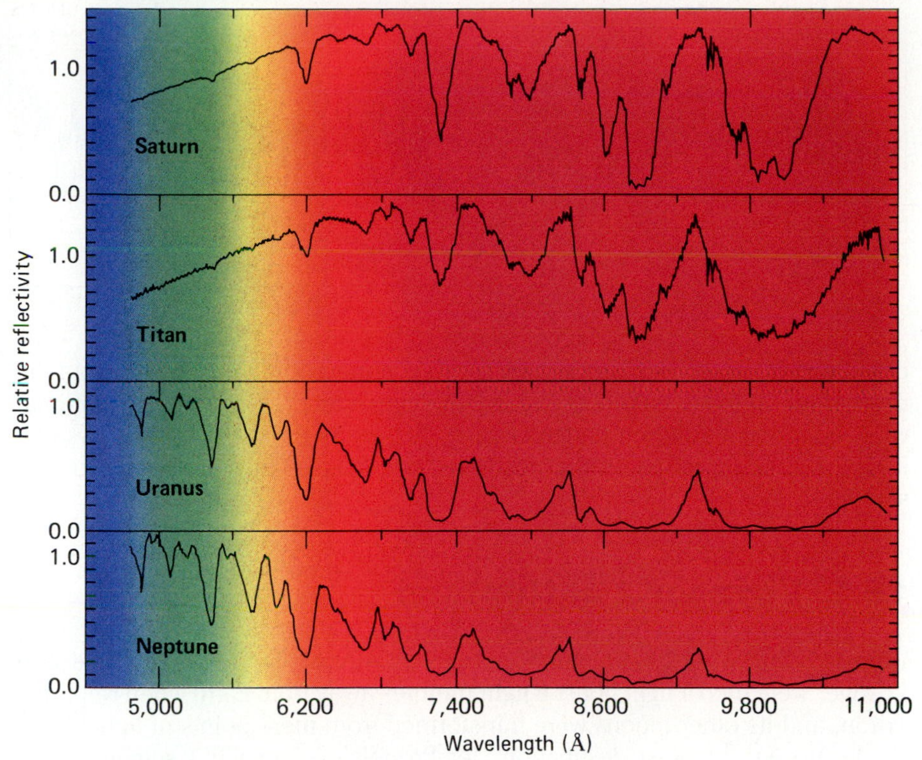

Figure 15–3 The ground-based spectra of the giant planets and Titan, divided by the solar spectrum to show the effect of the atmospheres of the planets and of Titan. The dark bands are from methane. Note how strong the methane absorption is in the spectra of Uranus and Neptune. A trace of ammonia is also detectable in the spectrum of Saturn. The graphs extend from blue into infrared.

to have an internal heat source, even after the more precise measurements made from Voyager 2.

Markings on Neptune have been detected from the ground (Fig. 15–4) on images taken electronically. The images show bright regions in the northern and southern hemispheres that are probably discrete clouds separated by a dark equatorial band. Motion caused by Neptune's rotation can also be seen.

Two of Neptune's moons were known from Earth-based studies. Triton (named after a sea god, son of Poseidon) seemed to be large, apparently a little larger than our Moon, and was discovered only a month after Neptune. It is massive enough to have an atmosphere and a melted interior. Triton orbits Neptune in the retrograde direction, the only reasonably large moon in the solar system to do so, and is thus probably a captured object. Spectroscopy seemed to show that Triton has methane ice, or methane frost, and possibly water ice and liquid or gaseous nitrogen as well. A second moon, Nereid (the Greek word meaning "sea nymph"), is much smaller, 340 kilometers across. It revolves in a very elliptical orbit with an average radius 15 times greater than Triton's. Its orbit is so elliptical that it is sometimes as close as 1.3 million kilometers and sometimes as far as 9.6 million kilometers from Neptune.

Does Neptune have rings, like the other giant planets? There was no obvious reason why it shouldn't, but no rings had been found at several occultations. Then, starting in 1983, astronomers' luck changed. During the 1983 occultation of a star by Neptune, good observations were obtained. (My own site, in Indonesia, at which students and I planned to observe the occultation, was clouded out.) Two observatories reported the same dip at the occultation, so Neptune apparently had a ring or rings after all (Fig. 15–5). Since that time, other occultations have sometimes shown signs of rings and sometimes not, so scientists concluded that any rings present would not completely surround Neptune but would rather be partial arcs. Indeed, in a 1985 occultation, only one of the two stars in a double-star system was occulted, so the arcs apparently

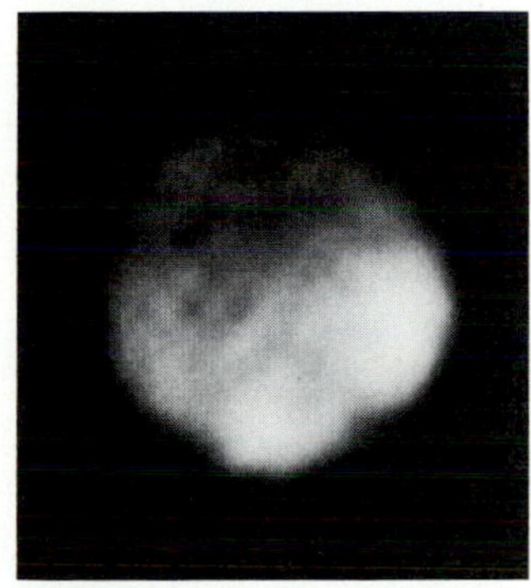

Figure 15–4 Neptune, photographed in 1986 with a CCD (Section 4.11) in the infrared light (8900 Å) that is strongly absorbed by methane gas in Neptune's atmosphere, making most of the planet appear dark. The bright regions are clouds of methane ice crystals above the methane gas, reflecting sunlight. The brightest feature is a mid-southern latitude cloud (40°S) rotated around Neptune with a 17-hour period. The CCD provided the infrared sensitivity necessary to take the picture with a short exposure. Neptune's disk was 2.3 arc sec across; the seeing was 0.5 arc sec.

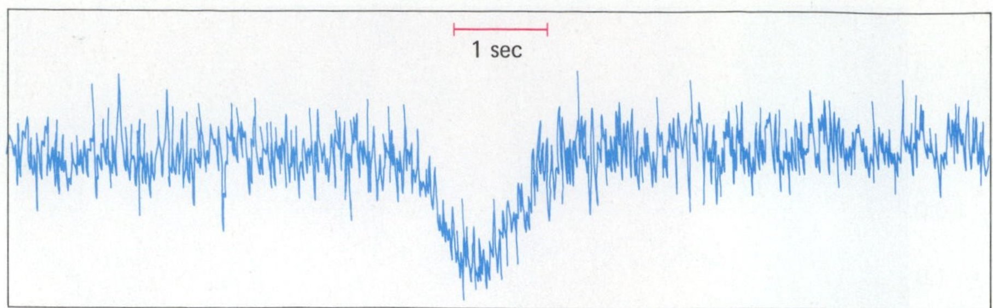

Figure 15–5 This photoelectric observation of July 22, 1984, of Neptune occulting a star revealed the presence of material orbiting Neptune at a distance of 67,000 km. This signal dropped by 32% for 1.2 sec. Since other observing sites did not detect a dip during the same occultation, the material seemed to be in the form of an incomplete arc rather than a smoothly varying ring.

could end abruptly. Neptune's rings are thus apparently very different from the rings around Jupiter, Saturn, or Uranus, and much theoretical and observational work is concentrating on understanding this fourth solar-system ring system. It may be telling us about the positions and masses of otherwise unknown satellites.

15.2 Voyager 2 at Neptune

Voyager 2 reached the Neptune system on August 24, 1989. Neptune and Voyager were then 4 light hours 6 light minutes away from Earth when Neptune, Triton, and its other moons were transformed from mere points of light. Since it was Voyager 2's last destination, Jet Propulsion Lab (JPL) scientists and engineers were able to direct the spacecraft as close to Triton as they pleased. They were thus able to get images of Triton of extremely high resolution. Ed Stone of Caltech was the Voyager project scientist. The imaging team, headed by Brad Smith of the University of Arizona, provided the views that dazzled both the eye and the mind.

As Voyager approached, even from far away it was obvious that Neptune had very active weather systems (Fig. 15–6). Why should Neptune be more active than Uranus, when Neptune is farther out from the Sun than Uranus and thus colder? It apparently has to do with the existence of a substantial extra internal heat source on Neptune.

As Voyager neared Neptune, an Earth-sized region soon called the Great Dark Spot (Fig. 15–7) became noticeable in Neptune's southern hemisphere. The Great Dark Spot, though colorless, seems analogous to Jupiter's Great Red Spot in several ways, including relative scale and position in the planet's southern hemisphere. It was very difficult for scientists to tell which way the Great Dark Spot was rotating, though, and thus whether it is cyclonic or anticyclonic, because successive exposures would show a new set of cloud tracers rather than movement of the clouds that had been previously seen. It was finally decided that the Great Dark Spot rotates counterclockwise, which is anticyclonic in its hemisphere, and has a period of 17 days. The Great Dark Spot is therefore a high-pressure region. The cirrus-like clouds form at the edge of the Great Dark Spot as the spot's high pressure forces upward methane-rich gas. Absorption by methane would darken the region if the spot lies deeper in the atmosphere than the surrounding clouds, so we can conclude that the planet-wide methane haze is diminished over the spot.

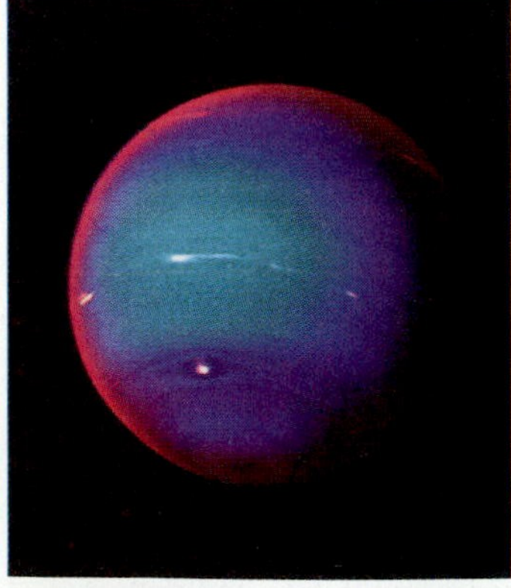

Figure 15–6 A false-color image of Neptune from Voyager 2. The areas that appear red are a semitransparent haze surrounding the planet.

Several other cloud systems were also seen on Neptune's disk (Fig. 15–8). The "Scooter" moved quite rapidly around, to the south of the Great Dark Spot. A second major dark spot had a bright core. Since each of the features moves around Neptune at a different speed, they are not usually as close to each other as they were for the photograph.

Neptune's high clouds are made of methane ice. Uniquely in the solar system, the shadows of high clouds were seen (Fig. 15–9). Thus there must

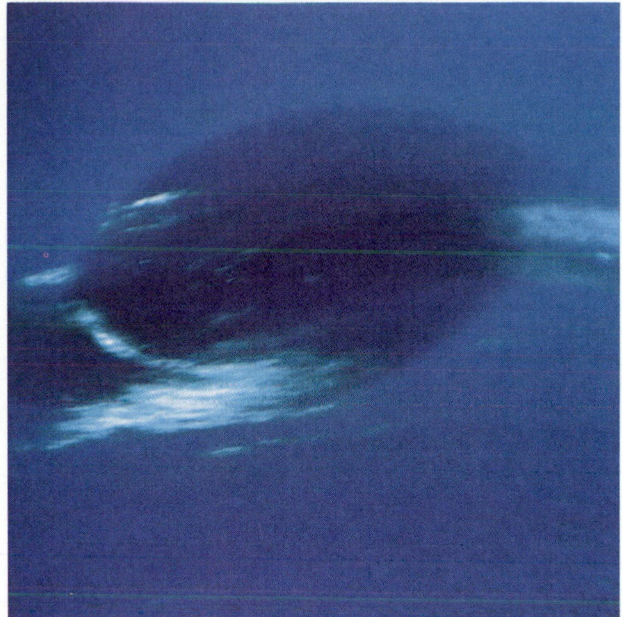

Figure 15–7 Neptune's Great Dark Spot, a giant storm in Neptune's atmosphere. It takes up about the same span of latitude and longitude as the Great Red Spot does on Jupiter. Neptune's Great Dark Spot is about the size of the Earth.

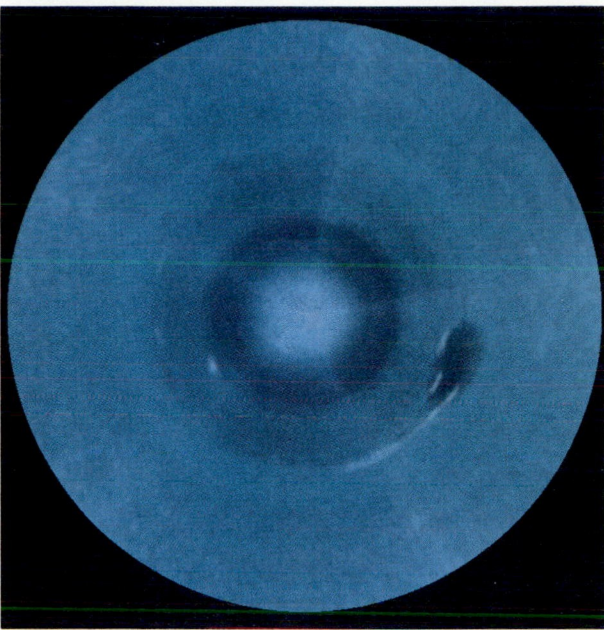

Figure 15–8 Cloud systems in Neptune's southern hemisphere. Neptune's south pole is at the center of this polar projection. The outer circle corresponds to 15° north latitude. The oval storm drifts in latitude at up to 7000 km/hr.

be two cloud decks separated by a clear region about 50 km in depth. The lower, more opaque cloud deck may be made of hydrogen sulfide ice particles. The Scooter seems to be an upwelling from this lower cloud deck, and perhaps lies over a long-lived hot spot. The cloud shadows appeared less distinct at short wavelengths (through the violet filter) than at longer wavelengths (through the orange filter). The effect apparently results from scattering by molecules in the extensive thickness of atmosphere between, which diffuses light into the shadow. The molecules scatter blue light much more efficiently than red light, so the shadows are darkest at the longest wavelengths and appear blue when illuminated with white light.

Neptune's equatorial radius was measured to be 24,760 km. The amount of radiation Neptune gives off corresponds to an average temperature of 59.3 K. Thus Neptune radiates about 2.7 times as much energy as it absorbs from the Sun. Neptune receives the most solar energy at its equator, yet both its equator and pole are about the same temperature. Mid-latitudes are several degrees cooler. Uranus, though primarily heated at one of its poles, has the same atmospheric temperature structure.

Radio bursts from Neptune were detected from Voyager at intervals of 16.11 hours, which must be the rotation period of Neptune's interior. Upper atmosphere wind speeds varied greatly, from 560 m/sec westward to 20 m/sec eastward with respect to this underlying rotation rate. On Neptune, the Earth, and Uranus, the equatorial winds blow more slowly than the interior rotates. By contrast, equatorial winds on Venus, Jupiter, Saturn, and the Sun blow more rapidly than the interior rotates. We now have quite a variety of planetary atmospheres to help us understand the basic causes of circulation.

The magnetometer on board detected an extensive magnetic field around Neptune, starting with its crossing a bow shock 35 Neptune radii out. The field, as for Uranus, is both greatly tipped and offset from Neptune's center (Table 15–1). Neptune's field is inclined 47° and its center is offset 55 per cent of

Figure 15–9 High-altitude cloud systems casting shadows, through a clear region, 50 km down to a lower cloud deck. The clouds are elongated at constant latitude, near 29° north, and the Sun is at lower left. The clouds are 50 to 200 km long.

Table 15–1 Planetary Magnetic Fields

	Tilted field character			Field strength		
Planet	*Tilt*	*Moment (tesla-meter^3)*	*Offset (R_{planet})*	*Equator*	*Minimum (nanoteslas)**	*Maximum*
Mercury	14°	0.005×10^{15}	?	300	300	600
Earth	11.7°	8.0×10^{15}	0.08	31,000	24,000	62,000
Jupiter	−9.6°	$160{,}000 \times 10^{15}$	0.1	428,000	300,000	1,440,000
Saturn	−0.0°	$4{,}700 \times 10^{15}$	0.05	22,000	18,000	84,000
Uranus	−58.6°	380×10^{15}	0.31	23,000	10,000	100,000
Neptune	−46.8°	280×10^{15}	0.55	13,000	10,000	100,000

*1 nanotesla = 10^{-5} gauss
Courtesy of Norman F. Ness, Bartol Research Institute

Neptune's radius from the center of the planet. Neptune's magnetic field was the first to be penetrated at the pole by Voyager, and was the first case observed in which the solar wind hits near the magnetic pole rather than near the magnetic equator. Faint auroras were detected on both Neptune and Triton.

Table 15–1 compares the magnetic fields on the planets out to Neptune. We give first the value of the field on the planet's surface at the magnetic equator, which is related both to the total field in the planet and to the planet's size. We also give the total field (technically, the "magnetic moment," which is the product of the field at the planet's magnetic equator and the planet's volume). The tilt given is positive if the north magnetic pole is above the ecliptic (that is, for Mercury and the Earth) and negative if it is below the ecliptic (that is, for the jovian planets).

The polarity of Neptune's magnetic field is the same as that of Jupiter, Saturn, and Uranus, all of which are opposite to the direction of the Earth's field with respect to the direction that the planets rotate. If not for the offset, the strength of magnetic field at Neptune's equator would be 13,000 nanoteslas, about one-third the Earth's equatorial field and about two-thirds the equatorial fields of Saturn and Uranus. The fact that Neptune's field is so tipped and offset indicates that there must be some fundamental reason for Neptune's and Uranus's fields being so, rather than mere chance collisions with planetesimals or chance observation during a period of magnetic reversal. The presence of an electromagnetic dynamo in an electrically conducting shell outside the planets' cores rather than deep in the core as for Earth and Jupiter would account for the phenomenon.

"I used to think that the idea of space probes was to answer questions. That's hopeless. You invariably raise more questions than you answer."
Andrew Ingersoll, Caltech, member of the imaging team

15.3 Neptune's Rings

Because of the ground-based studies that implied the existence of ring arcs, which are hard to account for theoretically, there was great interest in the Voyager observations of Neptune's rings. At a great distance, it was thought that the ring arcs had actually been seen. But as Voyager grew closer to Neptune, it became clear that the rings went all the way around Neptune (Fig. 15–10), at its equator. Two distinct rings were detected at 53,000 and 63,000 km from Neptune's center, as well as a faint inner ring at about 42,000 km and an unprecedentedly broad "plateau" ring that extends from the 53,000-km bright ring halfway out to the other bright ring. The rings showed up in backlighted images taken after Voyager had passed Neptune. The material rotates in the same direction as Neptune itself, and the rings are very circular and lie very close to Neptune's equator.

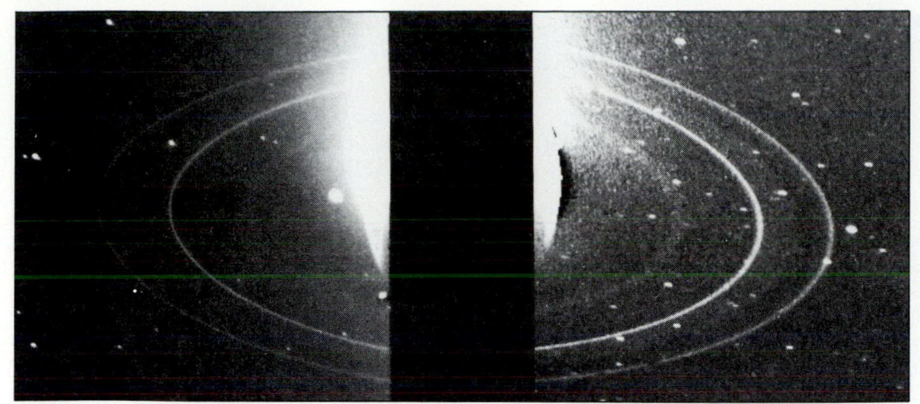

Figure 15–10 Neptune's rings, a composite of views of the two sides of the planet. Each exposure lasted nearly 10 minutes, with 1 hour 27 minutes between exposures, and was taken looking back toward the Sun from a distance of 280,000 km. Unfortunately, the bright ring clumps that had been thought to be arcs were on the opposite side of the planet during each exposure and do not show; two small gaps in the upper part of the outer ring on the left image are blemishes removed in data processing. The faint, broad inner ring at about 42,000 km from Neptune's center and the "plateau" that extends smoothly from the 53,000-km ring to about halfway between the two bright rings are also visible. Numerous stars are visible in the background.

The material in one of Neptune's rings is very clumpy (Fig. 15–11), which had led to the incorrect pre-Voyager deduction that ring arcs existed. The term "ring arcs" is expected to fade until it is a mere historical curiosity.

The fact that Neptune's rings are so much brighter when backlighted tells us about the sizes of particles in them. Two of the rings and the clumps in the third have a hundred times more dust-size grains than most of the rings of Uranus and Saturn. Since dust particles settle out of the rings, new sources—such as collisions of moonlets—must continually be active. Though much less dusty, Saturn's F ring and Encke Gap ringlet and one of Uranus's rings are similarly narrow.

15.4 Neptune's Moons

As Voyager 2 approached Neptune, a new moon was discovered fairly early on, and then five more closer to the encounter. This number was smaller than some had expected, given the number of moons discovered by Voyager around Jupiter, Saturn, and Uranus.

Everyone was waiting for the encounter with Triton, and nobody was disappointed with the results (Fig. 15–12). For one thing, as had been hoped, Triton's atmosphere is sufficiently thin that we could see through to the surface. Nitrogen gas is its dominant component. Triton turned out to have an incredibly varied surface, the weirdest in the entire solar system. Its density is 2.07 grams per cubic centimeter, so it is probably about 70 per cent rock and 30 per cent water ice. It is more dense than any jovian-planet satellite except Io and Europa. Its density is a major datum used to construct models of Triton's interior.

Much of the region imaged was near the south polar cap. It is apparently covered with seasonal ice, probably nitrogen. The temperature during this late-spring season was 37 K (−236°C = −393°F). The ice appeared slightly reddish, probably organic material formed by the action of solar ultraviolet and magnetospheric particles upon methane in the atmosphere and surface. Some bluish nitrogen frost appeared at equatorial latitudes. Methane and nitrogen are the only materials directly detected on its surface, but craters and cliffs would slump if they were made of methane; thus water ice must be the major component either by itself or with ammonia ice mixed in. Most of the craters are thought to have formed from impacts by comets over the past 3.5 billion years; Neptune's gravity captures comets into orbits that cross its own orbit with fairly short periods.

Figure 15–11 Two of Neptune's rings, showing the clumpy structure in one of them that led ground-based scientists to conclude from occultation data that Neptune had ring arcs. The rings are 53,000 and 63,000 km, respectively, from Neptune's center. These photographs were taken as Voyager 2 left Neptune, 1.1 million km later, and so are backlighted. The main clumps are 6° to 8° long.

Giant faults cross Triton's surface. About 50 dark streaks were seen to be aligned parallel to each other. They are apparently dark material vented from

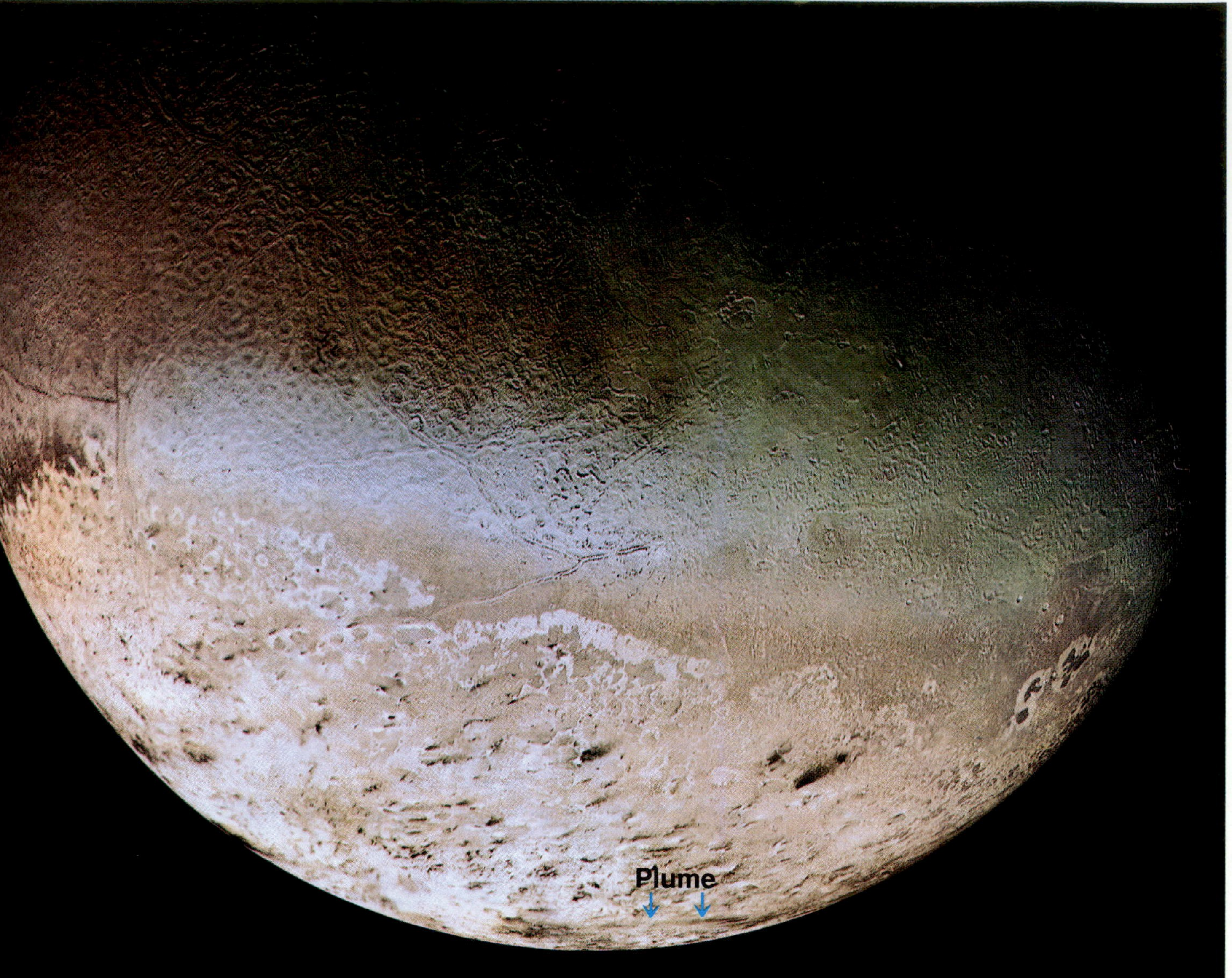

Figure 15–12 Triton's southern region, including its south polar ice cap. Most of the cap may be nitrogen ice. The pinkish color may come from the action of ultraviolet light and magnetospheric radiation upon methane in the atmosphere and surface. The dark streaks represent material spread downwind from eruptions. An erupting plume is the long, thin dark streak marked with arrows. Though the streaks are dark relative to other features on Triton, they are still ten times more reflective than the surface of our own Moon. The most detailed information on the photograph was taken in black and white through a clear filter; color information was added from lower-resolution images.

All the material—rings and satellites—within Neptune's Roche limit would form a single body 260 km across; the corresponding diameter would be 390 km for Saturn and 150 km for Uranus.

below as ice volcanoes and spread out across Triton's surface by winds in the thin atmosphere. Careful post-encounter appraisal of the images revealed several volcanic plumes. One of the plumes is seen to rise about 8 km before blowing downwind, presumably showing a change in Triton's atmosphere at that altitude. A leading model is that the Sun heats darkened methane ice on Triton's surface, vaporizing underlying nitrogen ice, which escapes through vents in the surface. This method involves a different type of energy source than the internal energy that powers the Earth's volcanoes or the flexing that

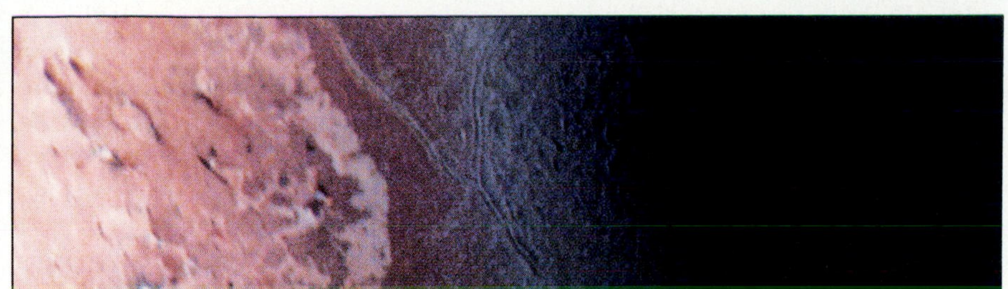

Figure 15–13 A false-color map across Triton's surface, with ultraviolet, violet, and green images projected into cylindrical coordinates and combined. Several distinct terrain and geological features appear. At center, the blue-gray region is the cantaloupe terrain, which seems older than the terrains to its left. Next to its left, appearing reddish, is a smoother region cut by a prominent fault system. This region apparently overlies material of much higher albedo, which is visible still farther left. The prominent change in albedo separates the relatively undisturbed smooth terrain from irregular patches that represent a breakup of the same material. The parallel streaks at far left show material apparently vented from ice volcanoes and blown by Triton's winds.

powers Io's. In any case, on Triton, dark material is carried by the plume as much as 150 km downwind. Since the streaks are on top of seasonal ice, they are probably less than 100 years old.

Much of the trailing (western) hemisphere of Triton is so puckered that it is called the cantaloupe terrain (Fig. 15–13). The cantaloupe terrain contains 30-km-diameter depressions crisscrossed by ridges. Other terrains seem to have been more recently covered with material, presumably material that flowed from under Triton's surface. Triton has few impact craters. Some regions of Triton are broad, flat basins (Fig. 15–14), possibly old impact basins. The ones shown have been extensively modified by flooding, melting, faulting, and collapse. The basins may have been filled and partially emptied several times. The most recent eruption is apparently in the middle of the central depression.

Triton has obviously been geologically active (Fig. 15–15). Some of the faults are quite long. We have not seen such surfaces sculpted in ice in this way on other solar-system bodies. These depressions look as though the icy surface has melted and collapsed (Fig. 15–16). These frozen lakes of water ice appear mainly in the eastern hemisphere, which leads as Triton orbits. A thin coating of methane and nitrogen ices and resulting materials apparently covers most or all of Triton.

Since Triton is in a retrograde orbit around Neptune, it was probably born elsewhere in the solar system and was later captured by Neptune. Tidal forces from Neptune would then have kept Triton molten until its orbit became cir-

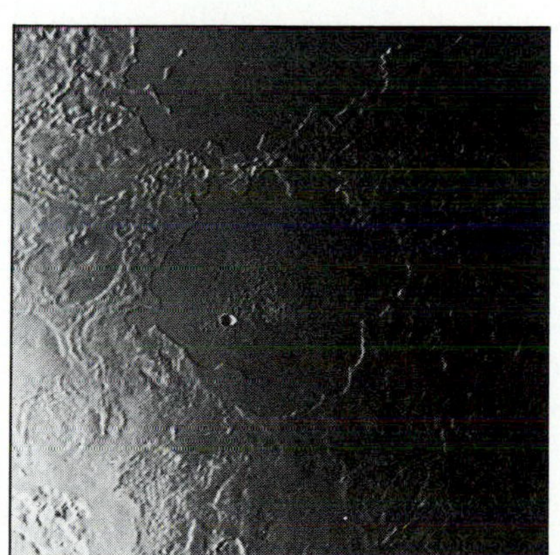

Figure 15–14 A flat "lake" region on Triton, about 300 km across. Since water on Triton acts like rock on Earth, liquid water (perhaps with ammonia mixed in) acts like lava, and we see multiple layers as in terrestrial calderas. The impact crater is one of the largest seen anywhere on Triton.

Figure 15–15 Triton from 120,000 km. The long feature is probably a narrow down-dropped fault block.

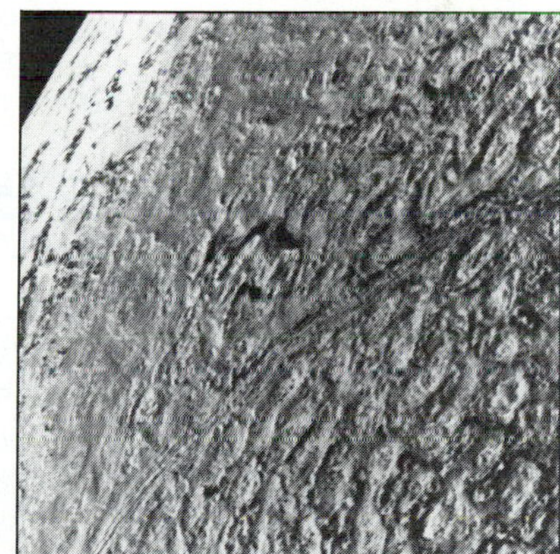

Figure 15–16 Triton from 40,000 km. The depressions may have been caused by melting and collapsing of the icy surface.

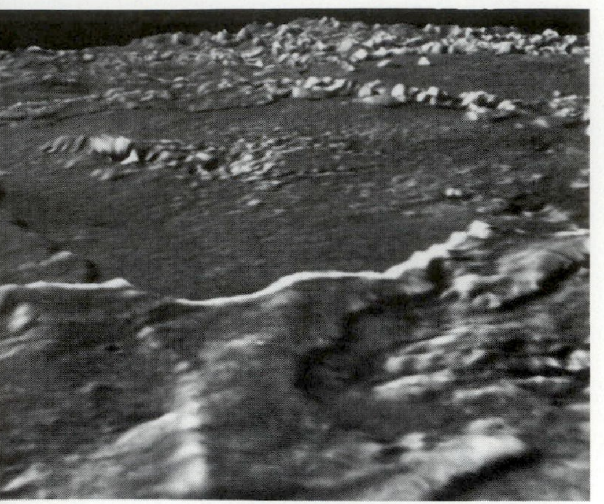

Figure 15–17 A computer-generated perspective view of one of Triton's caldera-like depressions.

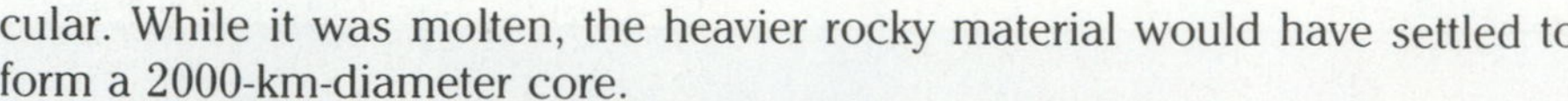

cular. While it was molten, the heavier rocky material would have settled to form a 2000-km-diameter core.

The advance of computer technology is such that the JPL scientists were able to produce calculated and corrected images in hours, instead of the months that similar images had taken for the Uranus mission only 3 years earlier. One computer-generated view shows a calculated perspective of one of the depressions on Triton (Fig. 15–17).

After Voyager 2's flyby at an altitude of only 3000 km above Triton's surface, an altitude actually too close to take unblurred pictures, the spacecraft could look back to see Triton as a crescent (Fig. 15–18). A diffuse haze was seen, probably made of photochemical smog. An occultation of the Sun by Triton was used to study details of Triton's atmosphere.

Triton orbits with the same side always facing Neptune. Its orbit is inclined 157°. Nereid (Fig. 15–19), whose orbit is inclined 29°, was not well imaged by Voyager. Its orbit is so elliptical, with its distance from Neptune ranging from 1.4 to 9.6 million kilometers, that it seems unlikely that its rotation is locked to Neptune. Voyager did not detect variations, though, that would give its rotation period.

All six of the newly discovered moons orbit Neptune in circular orbits in the same direction as Neptune's orbital motion. The orbits of five are within 1° of Neptune's equatorial plane and the other's orbit is inclined 5°. These moons probably share with Triton the property of having their rotation periods locked to their orbits so that the same surface always faces Neptune. Surprisingly, one of these newly discovered moons (Fig. 15–20) turned out to be larger than Nereid (Table 15–2). It had not been discovered from Earth, though, because of its closeness to Neptune, making it lost in Neptune's glare. The number of craters viewed on the larger moons indicates that the smaller ones could not have survived over the last 3.5 billion years so probably resulted from breakup of a larger object more recently.

Most of the newly discovered moons are closer to Neptune than Neptune's rings. Since objects cannot grow by accretion within the Roche limit, these moons must have formed elsewhere and moved into their current orbits. A few months after the encounter, names were assigned to the moons by a committee of the International Astronomical Union. Two of the moons orbit just inside two of the rings, and may serve as inner shepherds that keep ring material from spiralling inward. No outer shepherds have been detected, though they would not have been detectable if they were smaller than 12 kilometers across.

Table 15–2 The Moons and Rings of Neptune

Name	Radius of object (km)	Radius of orbit (km)	Period
Neptune's atmosphere		24,760	
Inner ring		41,900	
1989N6	27	48,000	7.1 hours
1989N5	40	50,000	7.5 hours
1989N3	90	52,500	8.0 hours
Bright ring		53,000	
Plateau ring		53,200–59,000	
1989N4	75	62,000	10.3 hours
Outer ring		62,900	
1989N2	95	73,600	13.3 hours
1989N1	200	117,600	26.9 hours
Triton	1303	354,800	141.0 hours
Nereid	170	5,513,400	8643.1 hours

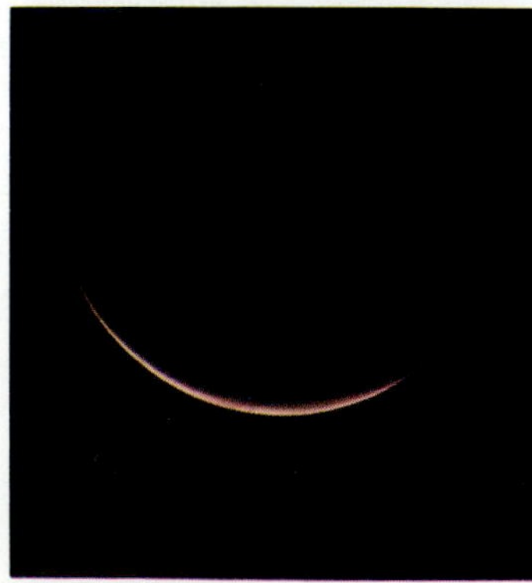

Figure 15–18 The crescent Triton, just after Voyager 2's closest approach.

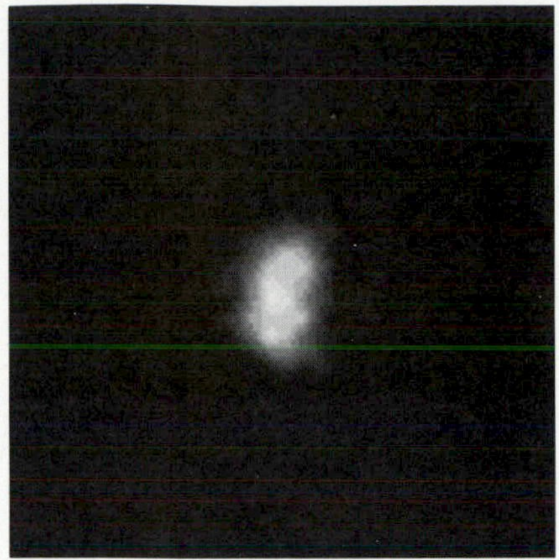

Figure 15–19 Nereid. No brightness variations were detected, and not even the position of its pole is known. We see a poorly resolved crescent, concave at the right.

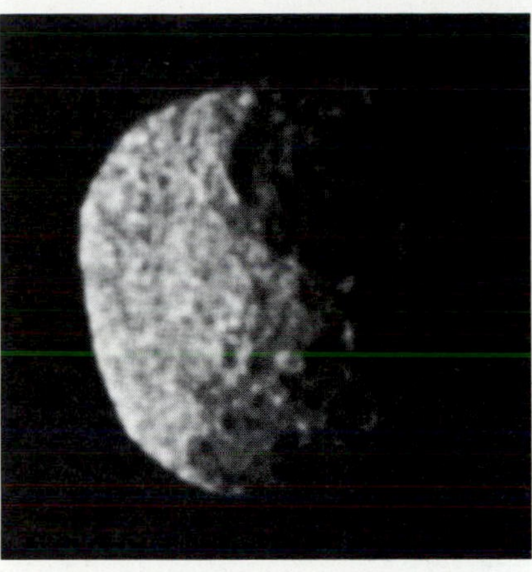

Figure 15–20 The second-largest moon in the neptunian system, approximately 210 km in radius. Its large crater shows that it was nearly broken apart.

And no moons were found in the positions that would have explained the observed clumpiness in one of the rings. No current theory satisfactorily explains the clumpiness.

15.5 Voyager Departs

All too soon, Voyager completed its tour of the solar system and departed (Fig. 15–21). Its look back at Neptune gave us a new perspective on that planet and its atmosphere. Voyager 2 was an unqualified success, sending back fantastic data over a much longer period than had been thought possible. It made the so-called "grand tour" of the outer planets a reality.

Figure 15–21 A post-encounter view of Neptune's south pole from a distance of 900,000 km. Near the bright limb, clouds located at 71° and 42° south latitude rotate onto Neptune's night side. The bright cloud at bottom center is within 1.5° of Neptune's south pole, determined from the orbits of the rings and satellites. The feature is believed to show an organized circulation, an "eye," around the south pole.

Figure 15–22 Neptune and Triton, the smaller crescent in the foreground, three days after the flyby.

We hope to stay in touch by radio with Voyager 2 for another 30 years or so. During that time, the spacecraft may reach the edge of the solar system, where the solar wind's influence ceases to dominate. We hope to receive signals about the location of that important boundary.

In the meantime, we have fond memories of Voyager 2's explorations, with its studies of Neptune and Triton (Fig. 15–22) as a superlative on top of other superlatives.

15.6 The History of the Giant Planets

Current knowledge indicates that all four giant planets have rocky cores containing 10 to 20 times the mass of the Earth. Jupiter and Saturn are much larger than Uranus and Neptune, so must have much thicker atmospheres. Perhaps the protosolar nebula from which the planets formed contained less gas at the positions of Uranus and Neptune. Since the gas was largely the light elements hydrogen and helium, Uranus and Neptune wound up with less of these light elements and thus have higher average densities than Jupiter and Saturn.

Summary

Neptune from Earth (Section 15.1)
- Large planet with low density and high albedo
- Predictions by Adams and Leverrier—part science and part luck
- Atmosphere of methane and molecular hydrogen
- 2 known moons: Triton and Nereid
- Internal heating
- A ring may have been discovered at an occultation

Neptune from Voyager 2 (Section 15.2)
- Great Dark Spot discovered
- Cloud shadows indicate a 50-km clear region in atmosphere

Neptune's rings (Section 15.3)
- They are complete, after all, though clumpy
- Two thin rings plus two broad rings discovered
- They show up best in forward scattering, indicating small particle size

Neptune's moons (Section 15.4)
- Triton has the most varied surface in the solar system
- Triton may have ice volcanoes that emitted streaks
- Giant faults, cantaloupe terrain, pinkish ice visible
- One of six newly discovered moons is bigger than Nereid

Questions

1. Which planets are known to have significant internal heat sources?
2. What fraction of its orbit has Neptune traversed since it was discovered? Since it was first seen?
3. Why do we have clearer images of Triton than we have of the moons of Jupiter?
4. Who deserves credit for the discovery of Neptune? Compare the relative claims of Galileo, Adams, Leverrier, and Galle.
5. Describe the relative roles of methane and of hydrogen gas in the atmosphere of Neptune.
6. Compare the Great Dark Spot on Neptune with the Great Red Spot on Jupiter.
7. Describe the history of ring discoveries and of "ring arcs" on Neptune.
8. Compare Triton in size and surface with other major moons of the solar system, like Ganymede and Titan.
9. Describe some terrains in Triton's surface.
10. Discuss the location of Neptune's moons and their relation with the Roche limit.

An Open Letter to Voyager 2

By John Updike

Dear Voyager:
This is to thank you for
The last twelve years, and wishing you, what's more,
Well in your new career in vacant space.
When you next brush a star, the human race
May be a layer of old sediment,
A wrinkle of the primates, a misspent
Youth of some zoömorphs. But you, your frail
Insectoid form, will skim the sparkling vale
Of the void practically forever. As
The frictionless light-years and aeons pass,
The frozen points that from Earth's vantage held
Their mythic patterns firm will shift and melt;
No wide-dish radios will strain to hear
Your whispered news, nor poets call you dear.

Ere then, let me assure you, you've been grand—
A little shaky at the outset, and
Arthritic in the swivel-joints, antique
In circuitry, virtually deaf, and weak
As a refrigerator bulb, you kept
Those picture postcards coming. Signals crept
To Pasadena, where they were enhanced
Until those planets clear as daylight danced.
The stripes and swirls of Jupiter's slow boil,
Its crazy moons, one cracked, one fried in oil,
One glazed with ice, and one too raw to eat,
Still cooking in the juice of inner heat,
Arrived on our astonished monitors.
Then, next, after a station break of years,
Fat Saturn rode your feeble beam, and lo!—
Not corny as we feared, but art deco—
The hard-edge, Technicolor rings, as thin
As cardboard, broader than Lake Michigan,
And casting flashlit shadows. Planet three
Was Uranus (accented solemnly
By anchormen on the first syllable,
Lest viewers think the "your" too personal):
A glassy globe of gas upon its side,
Its nine dark, close-knit rings at last descried,
Its corkscrew-shaped magnetic passions bared,
Its pocked attendants digitized and aired.
Last loomed, against the Oort cloud, blue Neptune,
Its counterrevolutionary moon,
Its wispy arcs of rings and whitish streaks
Of unpredicted tempests—thermal freaks,
As if an unused backyard swimming pool,
Remote from stirring sunlight, dark and cool
(Sub-sub-sub-freezing), by itself would splash.
Displays of splendid waste, of rounded trash!
Your looping miles of guided drift brought home
How barren cosmic space would be to roam.
One awful ball succeeds another, none
Fit for a shred or breath of life. Our one
Delightful, verdant orb was primed to cede
The H_2O and O and N we need.
Your survey, in its scrupulous depiction,
Purged from the solar system science fiction—
No more Uranians or Io-ites,
Just Earthlings dreaming through their dewy nights.

You saw where we could not, and dared to go
Where we could scarcely dream; you showed
A kind of metal courage, and faithfulness.
Your cryptic, ciphered, graven messages
Are for ourselves, designed to boomerang
Back like a prayer from where the angels sang,
That shining ancient blank encirclement.
Your voyage now outsoars mundane intent
And joins matter's blind motion. *Au revoir,*
You rickety free-falling man-made star!
Machines, like songs, belong to all. A man
Aloft is Russian or American,
But you aloft were simply sent by Man
At large.
Sincerely yours,
A fan.

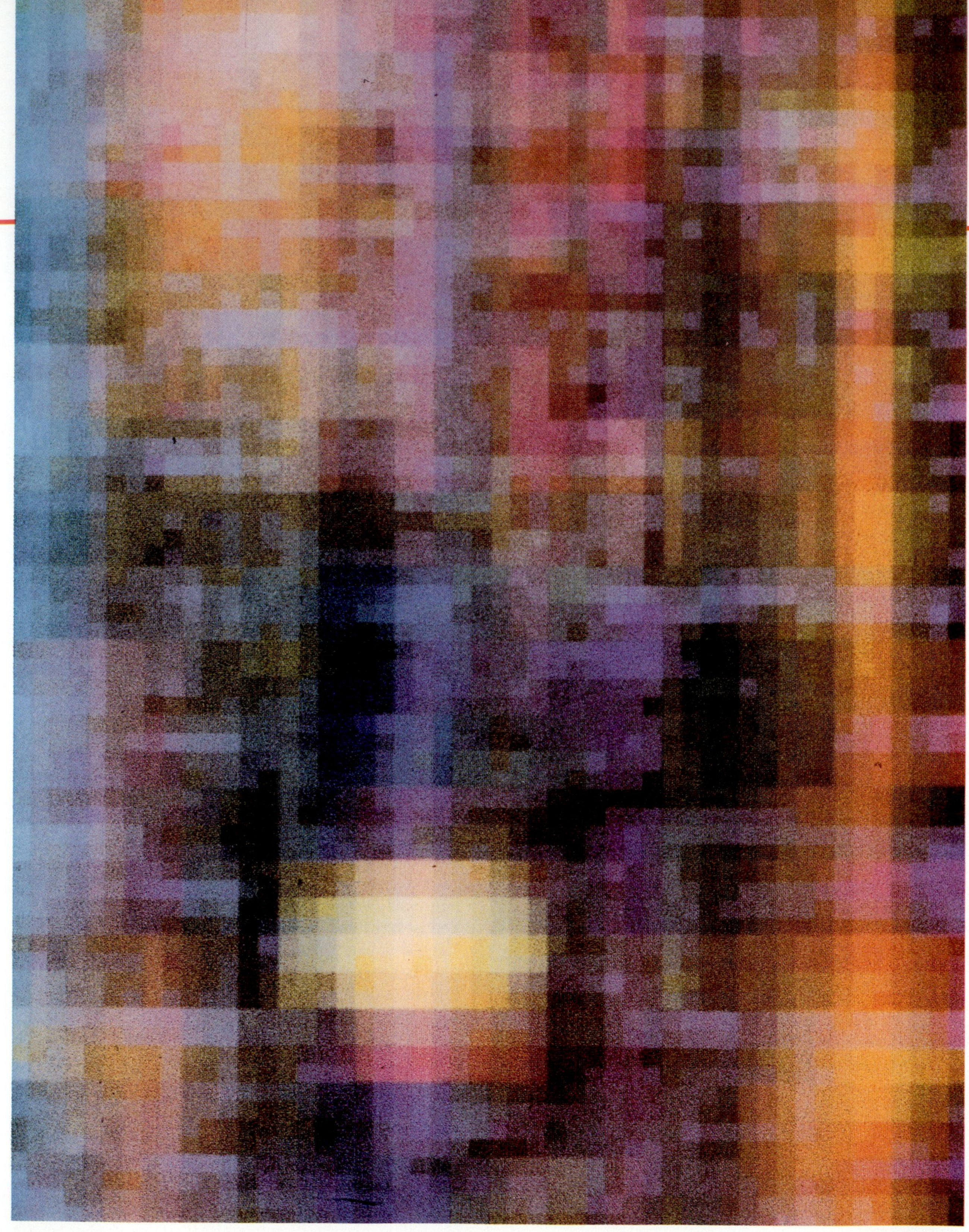

The Pluto/Charon system is the unresolved whitish region toward the bottom; Pluto and Charon actually fit within a pinhead at the center of the blur. The view is from the Infrared Astronomical Satellite at wavelengths of 12 μ (reproduced as blue), 60 μ (reproduced as green), and 100 μ (reproduced as red). In the rest of the image, blue background sources are warm and red ones are cold (just as the peak amount of radiation is toward the blue in hotter objects).

Pluto ♇ 16

Aims: To show how studies of tiny Pluto and its moon Charon have told us about the masses, sizes, and compositions of these distant solar-system objects

Pluto, the outermost known planet, is a deviant. Its orbit is the most eccentric and has the greatest inclination with respect to the ecliptic plane, near which the other planets revolve.

Pluto reached perihelion, its closest distance from the Sun, in 1989. Its 249-year orbit is so eccentric that part lies inside the orbit of Neptune. It is now on that part of its orbit, and will remain there until 1999. Thus in a sense, Pluto is the eighth planet for a while, though the significant point is really that the semi-major axis of Pluto's orbit, 40 A.U., is greater than that of Neptune. Pluto is so far away and so small that from Earth we cannot see any detail on its surface (Fig. 16–1).

Since Pluto has just passed its perihelion on September 5, 1989, of its 249-year period, it is about as bright at opposition as it ever gets from Earth. It hasn't been as bright—about magnitude 13.7—for over 200 years. It should be barely visible through even a small telescope under dark-sky conditions.

The discovery of Pluto was a result of a long search for an additional planet that, together with Neptune, was causing perturbations in the orbit of Uranus. Finally, in 1930, Clyde Tombaugh found the dot of light that is Pluto (Fig. 16–2) after a year of diligent study of photographic plates at the Lowell Observatory. From its slow motion with respect to the stars from night to night

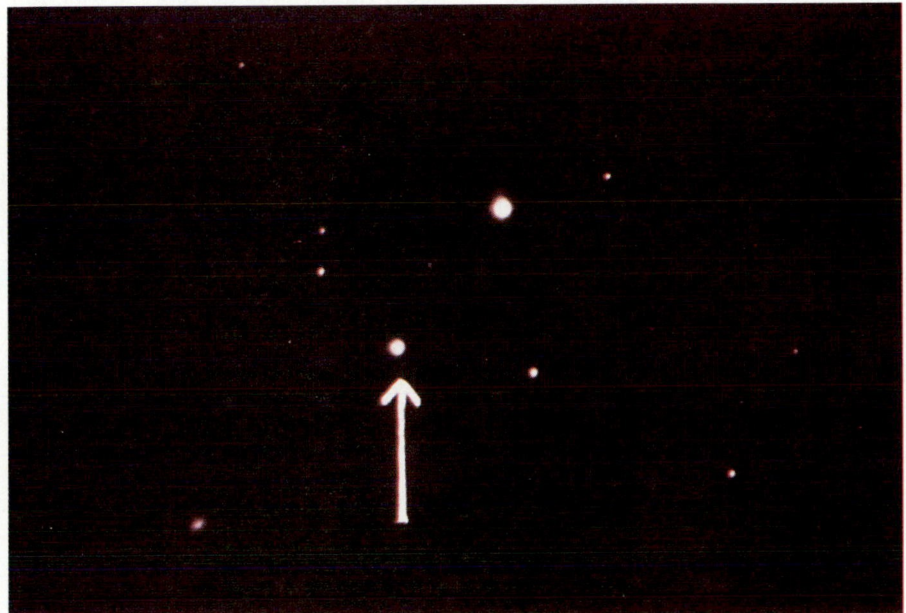

Figure 16–1 Pluto, with the 5-m Hale Telescope on Palomar. Even this huge telescope detects no surface details.

Figure 16–2 Clyde Tombaugh.

Figure 16–3 Small sections of the plates on which Tombaugh discovered Pluto. On February 18, 1930, Tombaugh noticed that one dot among many had moved between January 23, 1930 (*left*), and January 29, 1930 (*right*).

(Fig. 16–3), Pluto was identified as a new planet. As we shall see later on, though, Pluto does not have enough mass to have caused noticeable effects on Uranus's orbit. Tombaugh's discovery resulted, rather, from his hard work instead of from the predictions.

16.1 Pluto's Mass and Size

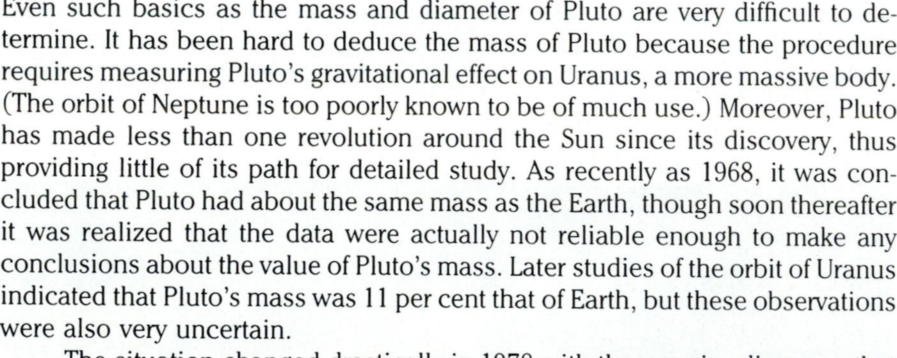

Even such basics as the mass and diameter of Pluto are very difficult to determine. It has been hard to deduce the mass of Pluto because the procedure requires measuring Pluto's gravitational effect on Uranus, a more massive body. (The orbit of Neptune is too poorly known to be of much use.) Moreover, Pluto has made less than one revolution around the Sun since its discovery, thus providing little of its path for detailed study. As recently as 1968, it was concluded that Pluto had about the same mass as the Earth, though soon thereafter it was realized that the data were actually not reliable enough to make any conclusions about the value of Pluto's mass. Later studies of the orbit of Uranus indicated that Pluto's mass was 11 per cent that of Earth, but these observations were also very uncertain.

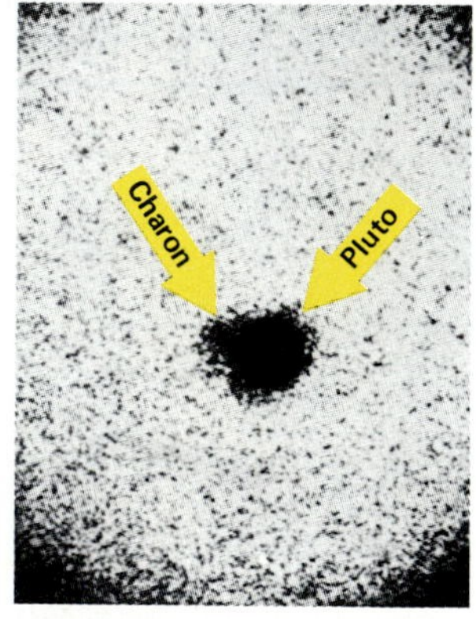

Figure 16–4 The discovery image of Charon and Pluto, taken on July 2, 1978. The bump at upper left was interpreted as the satellite. Charon is $\frac{1}{2}$ the size of Pluto and is separated from Pluto by only about 8 Pluto diameters (compared with the 30 Earth diameters that separate the Earth and the Moon). So Pluto/Charon are almost a double-planet system. The fact that almost no photographs taken since Charon's discovery show it any more clearly than this picture illustrates why it had not been previously discovered.

The situation changed drastically in 1978 with the surprise discovery that Pluto has a satellite. The presence of a satellite allows us to deduce the mass of the planet by applying Newton's form of Kepler's third law. In this age of space exploration, it is refreshing to see that important discoveries can be made with ground-based telescopes.

A U.S. Naval Observatory scientist, James W. Christy, was studying plates taken in a new series of observations to refine our knowledge of Pluto's orbit. He noticed that some of the photographs seemed to show a bump on the side of Pluto's image (Fig. 16–4). The bump was much too large to be a mountain, and turned out to be a moon orbiting with the period that we had previously measured for the variation of Pluto's brightness—6 days 9 hours 17 minutes. The elongated image of Pluto has since been seen both on older photographs of Pluto and on more recent ones in the positions that have been predicted on the basis of past observations (Fig. 16–5). The moon has been named Charon, after the boatman who rowed passengers across the River Styx to Pluto's realm in Greek mythology (and pronounced "Shar'on," similar to the name of the discoverer's wife, Charlene). We will have to wait for the Hubble Space Telescope to see Pluto and Charon resolved from each other.

From even the discovery photograph we can get an approximate idea of the distance between Pluto and Charon. This measurement allows us to calculate the sum of the masses of Pluto and Charon. If we assume that Pluto and its moon have the same albedoes and densities, we can compute the masses of each. Charon is 5 or 10 per cent of Pluto's mass, and Pluto is only 1⁄500 the mass of the Earth, ten times less than had been suspected even recently.

Only in 1979 was the diameter of Pluto measured directly. The technique of speckle interferometry, which involves studying a series of images taken rapidly enough to freeze out the effect of the Earth's turbulent atmosphere, was used with the 5-m Hale telescope (Fig. 16–6). The method gave 3000 to 3600 km for Pluto's diameter. Pluto is thus much smaller than Mars (6800 km) or Mercury (4800 km), and is approximately the same size as our Moon (3500 km).

Starting in 1985, Pluto and Charon have passed in front of each other as they orbit each other every 6.4 days. These mutual occultations last a few months each year, and continued until 1990. When we measure the apparent brightness of Pluto, we are really receiving light from both Pluto and Charon together. Their blocking each other (and an additional contribution from the shadow of one object on the other) leads to dips in the total brightness we measure (Fig. 16–7). Pluto is 2300 ± 14 km in diameter, smaller than expected, and Charon is 1186 ± 20 km. Modelling the light curve can even show what albedoes different parts of Pluto's surfaces are likely to have (Fig. 16–8). For example, the depth of the drop of the light curve is twice as great when Charon passes in front of Pluto compared with Pluto passing in front of Charon, so Pluto's pole must be twice as bright as Charon's. Pluto's average albedo is about 50 per cent, and Charon's is about 37 per cent. The mean density of Pluto and Charon together is 2 g/cm^3. Since Pluto contributes 7⁄8 of the mass, Pluto's density cannot be far from 2 g/cm^3, which is about 4 times too great for Pluto to be made entirely of methane ice. It must be largely rock, probably with about 25 per cent water ice and a bit of methane ice forming a mantle around a rocky core.

The complete hiding of Charon behind Pluto as seen from Earth in 1987 allowed the spectrum of Pluto alone to be recorded. This has then been subtracted from the Pluto+Charon spectrum to give Charon's spectrum. Charon's infrared spectrum shows that water ice but not methane or ammonia frost is

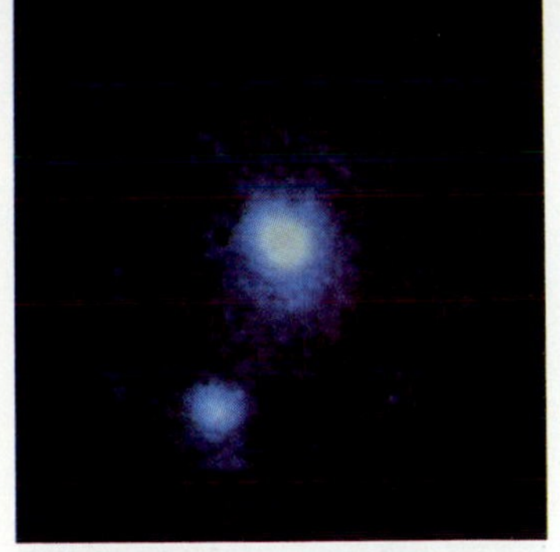

Figure 16–5 The Faint Object Camera on the Hubble Space Telescope (HST/FOC) has been able to show Pluto and Charon clearly distinguished.

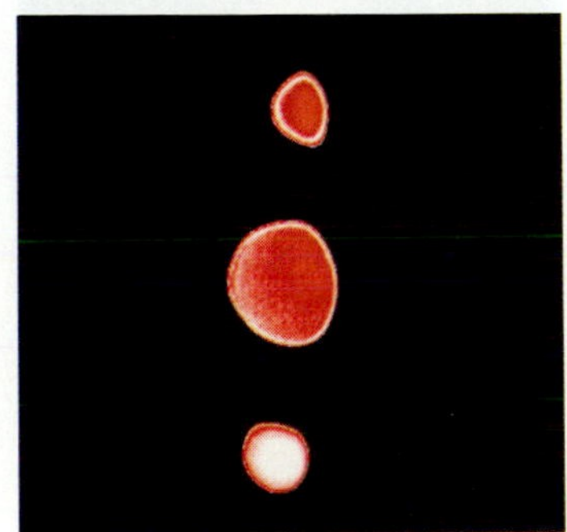

Figure 16–6 A speckle-interferometry image shows Charon separated from Pluto. Because of the method, Charon appears both above and below Pluto; the method does not reveal which position is real, but the separation from Pluto shown is accurate.

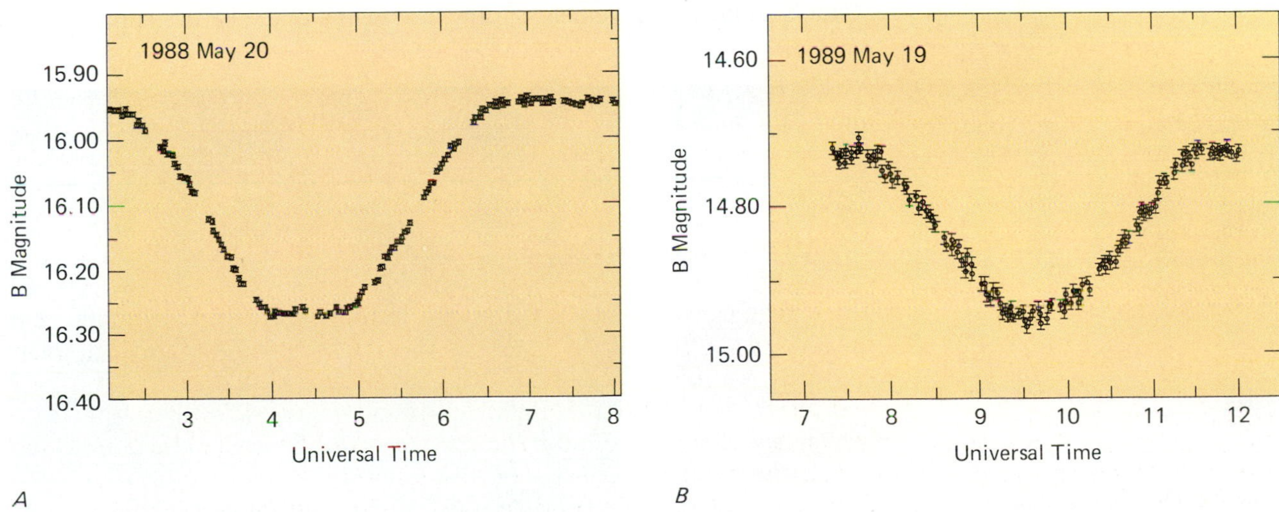

Figure 16–7 (*A*) A 1988 occultation of Pluto by Charon. Note the flat bottom of the light curve, as Charon spent some time entirely in front of Pluto. (*B*) By 1989, the light curve no longer had a flat bottom, as Charon only partly occulted Pluto. (Photometry by David J. Tholen, Mauna Kea Observatory)

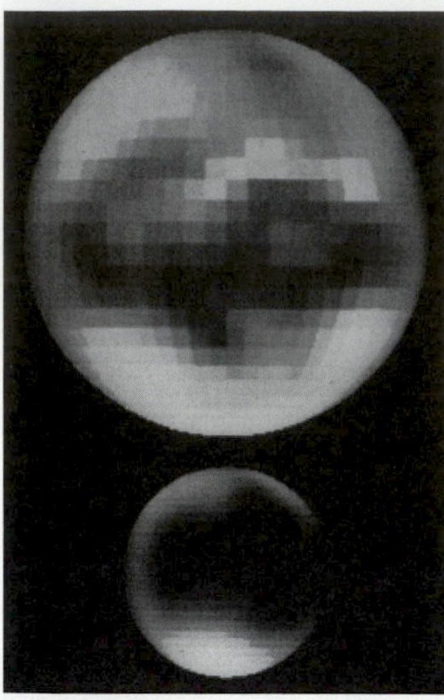

Figure 16–8 Computer-generated renderings of the surface brightness of Pluto (*top*) and Charon (*bottom*) based on their mutual occultations. Here we see an equatorial view.

present. Charon's lower mass apparently did not allow it to retain the methane that Pluto did.

IRAS observations of Pluto+Charon (see the photograph opening this chapter) have recently been interpreted to indicate that Pluto has a dark equatorial band and bright polar caps of methane ice that vary over a Pluto year. Ice caps had also been predicted from study of the way Pluto's overall brightness changed over the years as it gradually moved closer to the Sun and was seen from Earth at different angles. The IRAS observations indicated that Pluto's temperature is 54–59 K.

Pluto's equatorial region is not only darker but also redder than the polar regions, which matches how methane frost behaves when irradiated by solar ultraviolet.

16.2 Pluto's Atmosphere

Pluto is not massive enough to retain much of an atmosphere. But a tenuous atmosphere of methane has been detected from infrared spectral observations. The atmosphere could come from methane frost on its surface continually changing to methane gas. The methane is probably only a trace mixed with heavier gases.

When predictions showed that Pluto would occult a 12th-magnitude star—a relatively bright star for such an occultation—on June 9, 1988, several observers travelled to Australia and New Zealand to observe the event from the ground and from the Kuiper Airborne Observatory. The star and Pluto were observed approaching (Fig. 16–9), and then the total amount of light dimmed (Fig. 16–10). The light curves showed the light dimming gradually rather than abruptly, so Pluto has a substantial atmosphere. The curves indicate that either a layer of smog or haze absorbs all the light from the star or a temperature inversion in Pluto's atmosphere bends the light away from us. A smog or haze layer would have to extend at least 46 km from the surface. It would be so opaque that the starlight would be absorbed before its path to us passed next to Pluto's surface. Calculations show that a temperature inversion would be closer to the surface. In either case, we are no longer sure that the value that we have for Pluto's diameter from the mutual events is the true surface diameter; it may include some of Pluto's atmosphere. The atmospheric pressure deduced for Pluto's surface is, nonetheless, only 1/100,000 that of the Earth's surface pres-

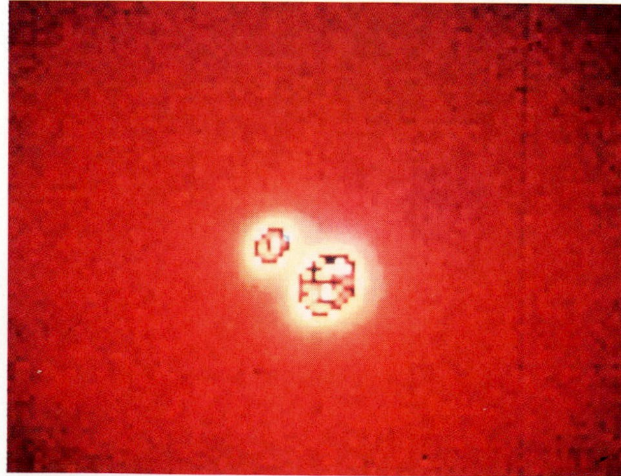

Figure 16–9 A view of Pluto and the star it occulted a few hours prior to the occultation on June 9, 1988.

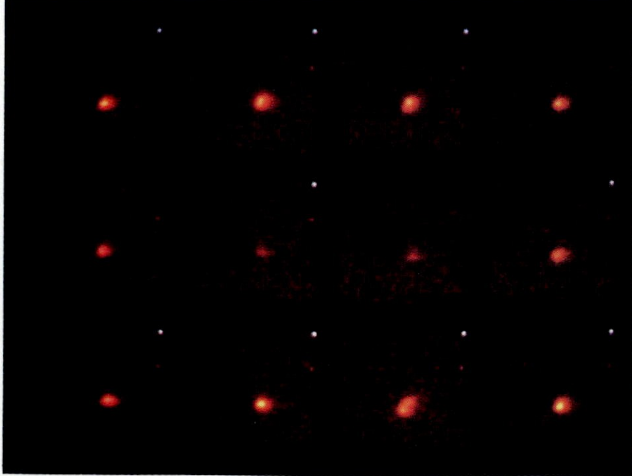

Figure 16–10 The merged image of Pluto and the star before, during, and after the occultation.

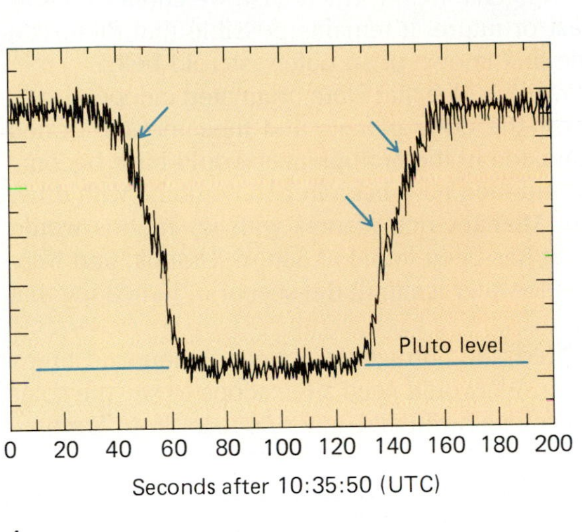

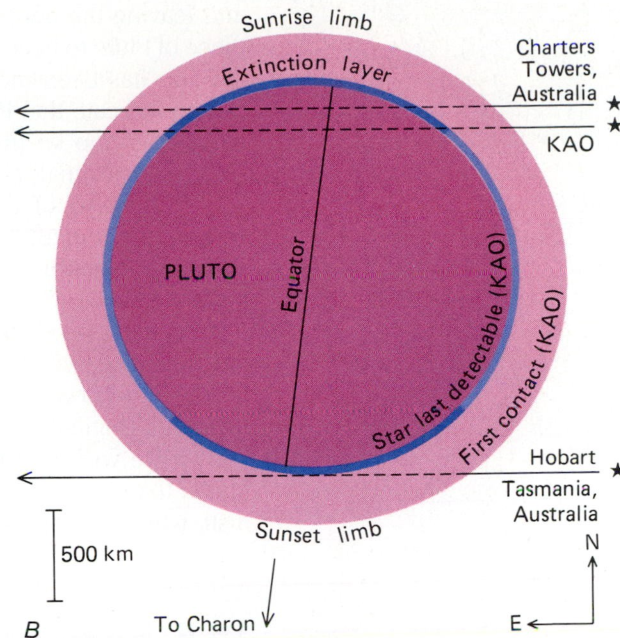

Figure 16–11 (*A*) The light curve measured by MIT scientists from the Kuiper Airborne Observatory when Pluto occulted a star in 1988. The slope is steepest near the base, showing a change in the atmosphere. Arrows mark spikes resulting from the starlight passing through structure in Pluto's atmosphere where the temperature changes by a few per cent. (*B*) The path of the star behind Pluto as visible from an airplane and two ground-based observatories.

sure. And it may yet prove, as some calculations indicate, that sharp changes in the temperature variation with altitude can explain the occultation observations without need for the haze layer. Structure in Pluto's atmosphere made "spikes" in the light curve (Fig. 16–11). It has also been concluded from the light curve that a gas heavier than methane must be present in Pluto's atmosphere, perhaps carbon monoxide or nitrogen.

16.3 What Is Pluto?

The newer values of Pluto's mass and radius can be used to derive Pluto's density, which turns out to be too low to be solid rock. Since only ices have such low density, Pluto must be made of frozen materials. Its composition is thus more similar to that of the satellites of the giant planets than to that of the terrestrial planets. Spectral evidence indicates that non-icy materials (possibly silicates) are probably present in addition to methane ice.

Ironically, now that we know Pluto's mass, we calculate that it is far too small to cause the perturbations in Uranus's orbit that originally led to Pluto's discovery. Thus the prediscovery prediction was actually wrong, and the discovery of Pluto was purely the reward of hard work in conducting a thorough search in a zone of the sky near the ecliptic.

No longer does Pluto, with its moon and its atmosphere, seem so different from the other outer planets. Pluto remains strange in that it is so small next to the giants, and its orbit is so eccentric and so highly inclined to the ecliptic. The new values of Pluto's mass and density revive the thinking that Pluto may be a former moon of one of the giant planets, probably Neptune, that escaped because of a gravitational encounter with another planet. Charon would not have been able to escape together with Pluto, however, so had to be captured later (unless both formed in their current places).

The orbits of Neptune and Pluto are affected at present by their mutual gravities in such a way that their orbital positions relative to each other repeat in a cycle every 20,000 years. The two planets can now never come closer to each other than 18 A.U., but Pluto may still have broken away from Neptune when another object passed nearby or even hit it, breaking off the outer layers and leaving the denser rocky core. The same event could have broken off a piece of Pluto to become Charon. In this model, their original joint orbit around the Sun has been modified by gravity over the years. However, some calculations indicate that Pluto's orbit is chaotic, in which case we cannot reliably calculate it too far into the past or future. It remains possible that Pluto is a large planetesimal, perhaps the first known of an outer asteroid belt.

Other lines of argument favor the idea that Pluto originated independently of Neptune. Calculations of how the outer planets and their moons formed show that plentiful carbon monoxide in the protoplanets would have become methane (CH_4). The oxygen would then have been free to combine with other hydrogen to form water (H_2O). Methane-rich planets with icy moons would have resulted, which is just what has been found at Saturn, Uranus, and Neptune. Pluto, however, has too little water ice to fit the scenario, indicating that it did not originate as a moon.

If we were standing on Pluto, the Sun would appear a thousand times fainter than it does to us on Earth. We would need a telescope to see the solar disk, which would be about the same size that Jupiter appears from Earth.

16.4 Pluto from the Hubble Space Telescope

Careful attention paid to improving the quality of images with the telescopes on Mauna Kea have allowed Pluto and Charon to be seen from the Earth slightly more clearly than they had been previously. In 1990, the Hubble Space Telescope surpassed the ground-based limit significantly. Its image showed Pluto and Charon as distinct objects for the first time (Fig. 16–12).

16.5 Further Planets?

Are there still further planets beyond Pluto? Tombaugh continued his search (Fig. 16–13) and found none. Other optical searches have been made since, also without success. But IRAS recorded thousands of objects in the infrared. A 10th planet that was sufficiently large or close or had an internal energy source might have been recorded. It may one day turn up in the analysis of the massive amounts of data IRAS collected. However, it is interesting to note that Pluto was observed by IRAS only with a special effort.

Edward Bowell at the Lowell Observatory and Eugene and Carolyn Shoemaker using a small Schmidt telescope at the Palomar Observatory are also carrying out optical searches for distant objects.

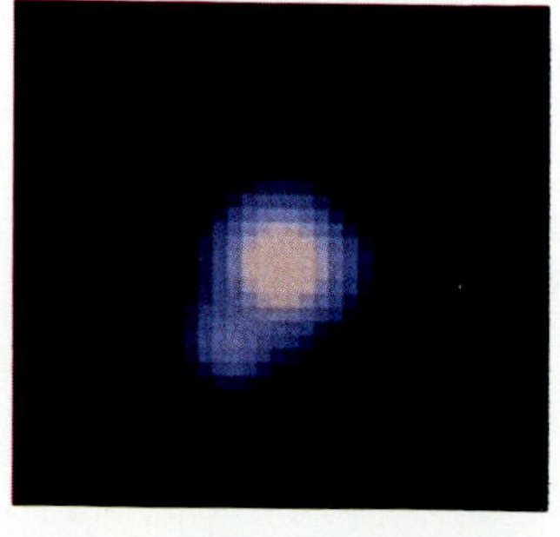

Ground based

Hubble Space Telescope

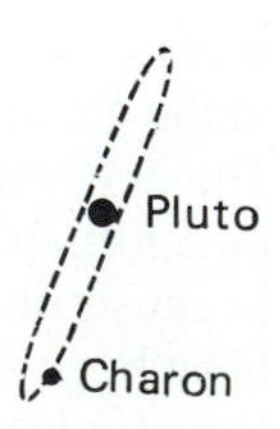

Figure 16–12 Pluto and Charon can be seen as partly separated on a ground-based image from the Canada-France-Hawaii Telescope on Mauna Kea (*left*), and have been completely resolved in an image taken with the Hubble Space Telescope (*right*).

Figure 16–13 One of the original plates on which Tombaugh discovered Pluto. This plate is still at the Lowell Observatory; the other half of the pair is in the Smithsonian Institution.

Analysis using only the twentieth-century measurements for the positions of Uranus does not find the deviation in Uranus's actual orbit from its expected orbit that is found when older data are also used. By that standard, there is no indication that a 10th planet exists.

Tracking data on the Pioneer 10 spacecraft, now out of the solar system past the orbits of Neptune and Pluto, does not show any gravitational effect from a 10th planet. The limit set for the mass that any 10th planet could have is now very low.

Summary and Outline

Most eccentric orbit; greatest inclination to ecliptic
Rotation accurately determined as a result of varying albedo
Discovered in 1930 by diligent search near ecliptic
Determining radius, mass, and thus density (Section 16.1)
- Discovery of a moon (Charon) allows mass to be deduced; it is only 0.2 per cent that of Earth; density thus is similar to that of satellites of outer planets
- Radius measured from mutual occultations of Pluto and Charon

Atmosphere measured in an occultation (Section 16.2)
Pluto's origin unknown (Section 16.3)
Search for still further planets with IRAS (Section 16.4)

Questions

†1. What fraction of its orbit has Pluto traversed since it was discovered?

2. What evidence suggests that Pluto is not a "normal" planet?

†3. At what wavelength would a cold body like a 10th planet give off the most radiation? What part of the spectrum is this, and how does it affect the chance of discovering such a planet? Use a reasonable estimate for the temperature and Wien's law (Section 20.1).

†This question requires a numerical solution.

4. Summarize the evidence that suggests that Pluto is not a giant planet.

5. From the separation of Pluto and Charon, show how to calculate the mass of Pluto.

6. Describe what mutual occultations of Pluto and Charon are and what they have told us.

†7. During what years have there been mutual occultations of Pluto and Charon? During this period, how many days and hours separated the dips in the light curve? Explain.

8. Describe the occultation observations of Pluto's atmosphere.

Halley's Comet in 1986. Note the bluish gas tail and the broad, white dust tail.

Halley's and Other Comets

17

Aims: To discuss how comets bring us information about the origin of the solar system, and what recent studies of Halley's Comet and other comets have taught us

Besides the planets and their moons, many other objects are in the family of the Sun. The most spectacular, as seen from Earth, are comets. Bright comets have been noted throughout history, instilling in observers great awe of the heavens.

It has been realized since the time of Tycho Brahe, who studied the comet of 1577, that comets are not in the Earth's atmosphere. From the fact that the comet did not show a parallax when observed from different locations on Earth, Tycho deduced that the comet was at least three times farther away from the Earth than the Moon.

Comets have long been seen as omens:
"When beggars die, there are no comets seen;
The heavens themselves blaze forth the death of princes."
Shakespeare, *Julius Caesar*

17.1 Observing Comets

Every few years, a bright comet is visible in our sky. From a small, bright area called the *head,* a *tail* may extend gracefully over one-sixth (30°) or more of the sky (Fig. 17–1). The tail of a comet is always directed away from the Sun.

The "long hair" that is the tail led to the name *comet,* which comes from the Greek for "long-haired star," *aster kometes.*

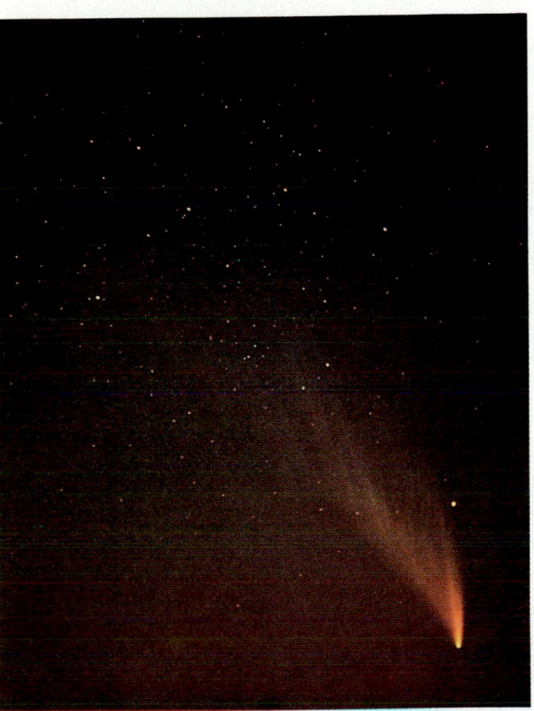

Figure 17–1 Comet West in the dawn sky in 1976. Note the delicate structure in its dust tail, which extended over 20° across the sky. (This five-minute exposure was made with a 55-mm f/1.8 and ASA 200 film.)

Focus On

Box 17.1 Discovering a Comet

Though Edmond Halley did not discover the comet named after him, nowadays a comet is generally named after its discoverer—the first person (or spacecraft) to detect it (or the first two or three if they independently find it within an interval of days). Comets are also temporarily assigned letters in their order of discovery in a given year—for example, Halley's Comet was 1982i. Then, a year or two later when all the comets that passed near the Sun in that given year are likely to be known, Roman numerals are assigned in order of their perihelion passage—for example, 1986 III.

Many discoverers of comets are amateur astronomers, including some who examine the sky each night with large binoculars in hope of finding a comet. To do this, one must know the sky very well, so that one can tell if a faint, fuzzy object is a new comet or a well-known nebula. It is for that reason that Messier made his famous eighteenth-century list of nebulae (Appendix 8).

If you find a comet, telegraph or telex the International Astronomical Union Central Bureau for Astronomical Telegrams, at the Smithsonian Astrophysical Observatory in Cambridge, Massachusetts (Telex II=TWX 710-320-6842 ASTROGRAM CAM; telegraph address: CENTRAL BUREAU FOR ASTRON, CAMBRIDGE MASS). Or send an electronic-mail message to Marsden@CFA. Telephoning (617) 495-7244 is a much less desirable method. In any case, you should identify the direction of the comet's motion, its position, and its brightness. Don't forget to identify yourself by giving your name, address, and telephone number to the person or answering machine. If you are the first (or maybe even the second or third) to find the comet, it will be named after you.

The most comets have been found by Eugene and Carolyn Shoemaker, who are still finding more. They have seen nearly thirty, surpassing the old record of Jean Louis Pons, the former caretaker of the Marseilles Observatory, who discovered 26 between 1801 and 1827. Active observers who have also discovered more than a dozen comets include David Levy in the United States, William Bradfield in Australia, and Antonin Mrkos in Czechoslovakia.

Many comets, particularly the ones discovered by professional astronomers, are discovered not by eye but only by examination of photographs taken with telescopes. The photographs were usually taken for other purposes, so the discovery of a comet exemplifies serendipity—a fortuitous extra discovery. The orbiting infrared telescope IRAS found half a dozen comets and Solar Maximum Mission discovered 11.

Figure 17–2 Comet Kohoutek in 1974 did not live up to its advance billing. The camera tracked the comet, so the foreground is trailed. Comet Austin in 1990 similarly disappointed us.

Although the tail may give an impression of motion because it extends out only to one side, the comet does not move visibly across the sky as we watch. With binoculars or a telescope, however, an observer can accurately note the position of the head with respect to nearby stars, and detect that the comet is moving at a slightly different rate from the stars as comet and stars rise and set together. Within days or weeks a bright comet will have faded to below naked-eye brightness, though it can be followed for additional weeks with binoculars and then for additional months with telescopes.

Most comets are much fainter than the one we have just described. About a dozen new comets are discovered each year, and most become known only to astronomers. Many of these are making their first passage near the Earth and Sun in recorded history (Fig. 17–2). An additional few are "rediscovered" each year—that is, from the orbits derived from past occurrences, it can be predicted when and approximately where in the sky a comet will again become visible. Halley's Comet (Section 17.4) belongs to the latter group. Up to the present time, about 1000 comets have been discovered.

Figure 17–3 Fred Whipple, originator of the dirty-snowball model.

17.2 The Composition of Comets

At the center of a comet's head is its *nucleus*, which is a few kilometers across. The most widely accepted theory of the composition of comets, advanced in 1950 by Fred L. Whipple (Fig. 17–3) of the Harvard and Smithsonian Observatories, is that the nucleus is like a *dirty snowball* a few kilometers across. The nucleus is apparently made of ices of such molecules as water (H_2O), carbon dioxide (CO_2), ammonia (NH_3), and methane (CH_4), with dust mixed in.

This dirty-snowball model explains many observed features of comets, including why the orbits of comets do not appear to accurately follow the laws of gravity. When sunlight evaporates the ices, molecules are expelled from the nucleus. This action generates an equal and opposite reaction, the same force that runs jet planes. Since the comet nucleus is rotating, the force is not always directly away from the Sun, even though the evaporation is triggered on the sunny side. So comets show the effects of non-gravitational forces in addition to the effect of solar gravity.

The nucleus itself is so small that we cannot observe it directly from Earth. Radar observations have verified that it is a few kilometers across. The rest of the head is the *coma* (pronounced cō'ma), which may grow to be over a million kilometers across. The coma shines partly because its gas and dust are reflecting sunlight toward us and partly because gases liberated from the nucleus are excited enough by sunlight that they radiate.

With rockets, we became able to observe in comets the Lyman-alpha line of hydrogen in the ultraviolet. In 1970, we discovered that a huge hydrogen cloud a million km in diameter (Fig. 17–4) surrounds a comet's head. The

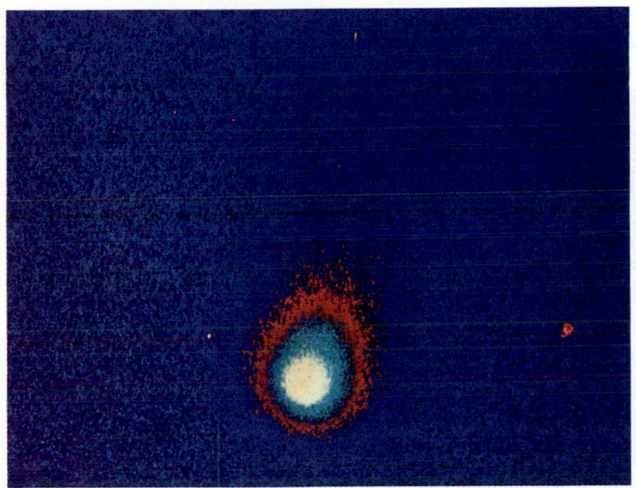

A

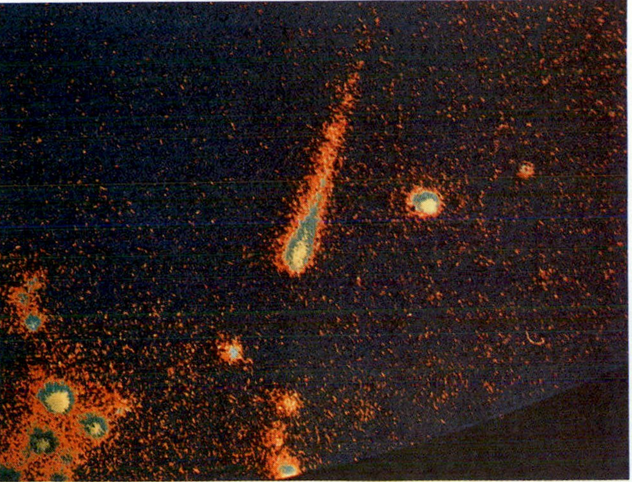

B

Figure 17–4 Contours of intensity from photographs of Comet Kohoutek taken by the Skylab astronauts outside the Earth's atmosphere. (*A*) The photograph was taken in the Lyman-alpha line of hydrogen at 1216 Å in the ultraviolet. (*B*) These contours result from a photograph that was taken immediately afterward at exactly the same scale but through a filter that did not pass the Lyman-alpha line.

The comparison shows the presence of a huge hydrogen halo, 1° across or 2,500,000 km in diameter. In *B*, we see the tail, 2° or 5 million km in length. Hot stars from the background constellation, Sagittarius, also show.

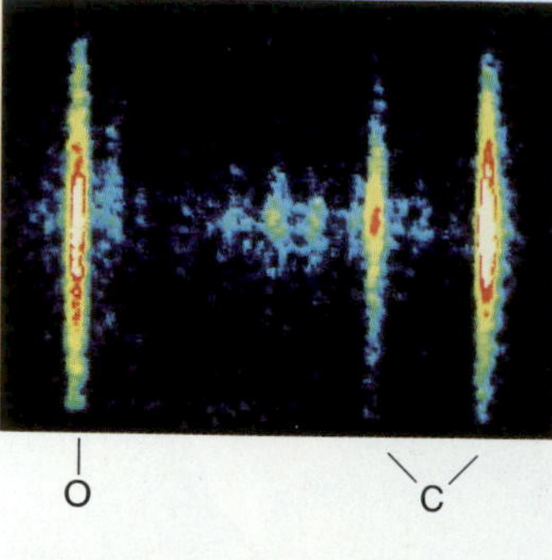

Figure 17–5 The ultraviolet spectrum of Halley's Comet showing strong spectral lines of oxygen at 1304 Å and carbon at 1651 Å and 1657 Å. The oxygen comes from the dissociation of water. These observations have shown that the carbon most likely comes from the dissociation of carbon monoxide.

hydrogen cloud probably results from the breakup of water molecules by ultraviolet light from the Sun. In the visible, we have long been able to detect spectral lines from simple molecules, and rockets and spacecraft enable us to take spectra in the ultraviolet (Fig. 17–5).

A comet's tail can extend 1 A.U. (150 million km), so comets can be the largest objects in the solar system. But the amount of matter in the tail is very small—the tail is a much better vacuum than we can make in laboratories on Earth.

Many comets actually have two tails. Both extend generally in the direction opposite to that of the Sun, but are different in appearance. The *dust tail* is caused by dust particles released from the ices of the nucleus when they are vaporized. The dust particles are left behind in the comet's orbit, blown slightly away from the Sun by the pressure caused by photons of sunlight hitting the particles. As a result of the comet's orbital motion, the dust tail usually curves smoothly behind the comet.

The *gas tail* (also called the *ion tail*) is composed of ions blown out more or less straight behind the comet by the solar wind. As puffs of ionized gas are blown out and as the solar wind varies, the ion tail takes on a structured appearance. Each puff of matter can be seen. Magnetic fields in the solar wind carry only ionized matter along with them; the neutral atoms are left behind in the coma.

Dust that may be from comets is captured by sending up sticky plates on a NASA U-2 aircraft sent high above terrestrial pollution. The abundances of the elements in the dust are similar to those of some meteorites rather than to terrestrial material. The particles may date back to the origin of the solar system.

A comet—head and tail together—contains less than a billionth of the mass of the Earth. It has been said that a comet is as close as something can come to being nothing.

17.3 The Origin and Evolution of Comets

It is now generally accepted that trillions of incipient comets surround the solar system in a sphere perhaps 50,000 A.U. (almost 1 light year) in radius, far outside Pluto's orbit. This sphere is known as the *Oort Comet Cloud* after Jan H. Oort, the Dutch astronomer who advanced the theory in 1950. The total mass of matter in the cloud is less than that of the Earth. Occasionally some of the incipient comets leave the Oort Cloud, probably because gravity of a nearby star has tugged them out of place or because the Sun has passed near a giant molecular cloud or other mass near the plane of the galaxy. (On the average, stars approximately as massive as the Sun pass through the comet cloud every few million years.) Some of these comets are ejected from the solar system, while others come closer to the Sun, approaching it in long ellipses. The comets' orbits may be altered if they pass near a jovian planet. Because the Oort Cloud is presumably spherical, comets are not limited to the plane of the ecliptic and come in randomly from all angles. Though long-period comets come from the Oort Cloud, new calculations indicate that short-period comets come directly from a reservoir just beyond the orbit of Neptune.

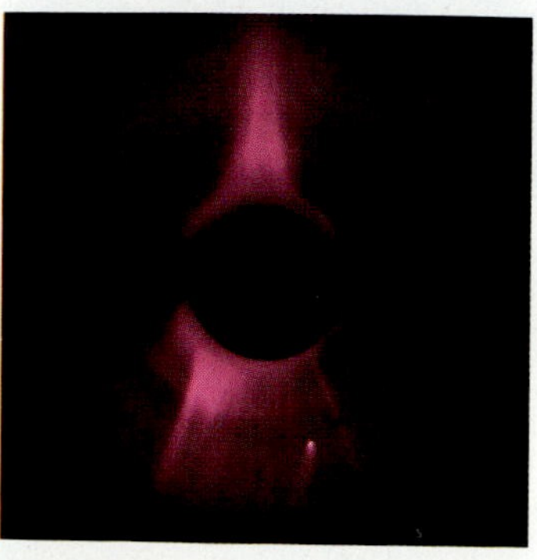

Figure 17–6 Comet Kohoutek, passing near the Sun, in a view from Skylab with the Sun blocked out in the center.

Objects in the Oort Cloud may have formed in place or perhaps from a reservoir of objects closer to the orbits of Uranus and Neptune. If so, some of these objects would be thrown by the planets into highly eccentric orbits that extend out 50,000 A.U. By Kepler's second law, these objects would spend much more of their time out in the Oort Cloud than in the inner parts of their orbits.

The Oort Cloud is huge; an Oort Cloud the size of ours around α Centauri, the nearest star to the Sun, would span 20° of sky when viewed from Earth.

As a comet gets closer to the Sun (Fig. 17–6), the solar radiation begins to vaporize the icy nucleus. From the ground, we have not generally been able to detect the molecules in the nucleus directly—the *parent molecules*—though we would clearly love to do so. However, we have mostly detected in the head the simpler molecules into which the parent molecules break down—the *daughter molecules.* Examples of daughter molecules are H, OH, and O, which might be results of the breakdown of the parent H_2O. Similarly, daughters NH and NH_2 can result from the parent NH_3. Some molecules can be observed with radio telescopes; radio radiation has been detected thus far from only three comets: Kohoutek, West, and Halley. Such parent molecules as HCN (hydrogen cyanide) and H_2CO (formaldehyde) were observed.

The tail forms, and it grows longer as more of the nucleus is vaporized. Even though the tail can be millions of kilometers long, it is still so tenuous that only 1/500 of the mass of the nucleus may be lost. Thus a comet may last for many passages around the Sun. But some comets may hit the Sun and be destroyed (Fig. 17–7).

The comet is brightest and its tail is generally longest at perihelion. However, because of the angle at which we view the tail from the Earth, it may not appear the longest at this stage. Following perihelion, as the comet recedes from the Sun, its tail fades; the head and nucleus receive less solar energy and fade as well. The comet may be lost until its next return, which could be as short as 3.3 years (Encke's Comet) or as long as 80,000 years (Comet Kohoutek) or more. With each reappearance, a comet loses a little mass and eventually disappears. We shall see in Section 18.1 that some of the meteoroids are left in its orbit. Some of the asteroids, particularly those that cross the Earth's orbit, may be dead comet nuclei.

In 1983, the IRAS satellite and then two amateur astronomers independently discovered a comet that passed exceptionally close to Earth—4,600,000 km—making it the closest comet since 1770 (Fig. 17–8). On only a few days' notice, Comet IRAS-Araki-Alcock became visible to the naked eye, appearing for a couple of days like a hazy cloud (Fig. 17–9). Scientists bounced radar off its nucleus, confirming the prediction of Whipple's model that solid material is present in cometary nuclei. IRAS scientists also picked up other comets from the million-kilometer-long narrow trails they show on the infrared image.

Because new comets come from the places in the solar system that are farthest from the Sun and thus coldest, they probably contain matter that is unchanged since the formation of the solar system. So the study of comets is important for understanding the birth of the solar system.

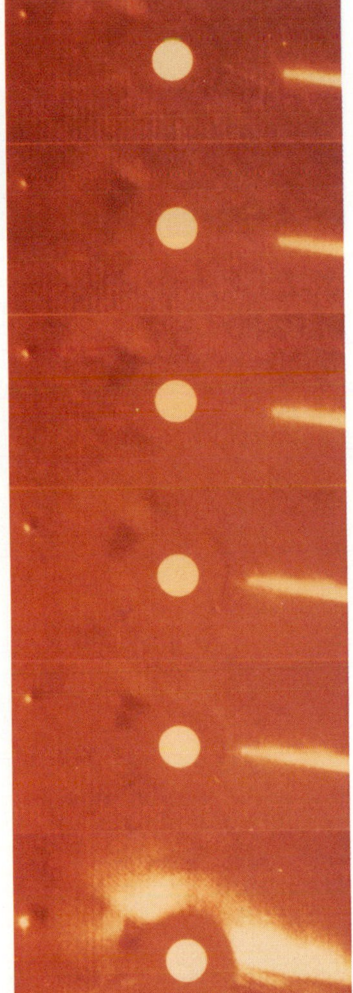

A

B

Figure 17–7 (*A*) A coronagraph in orbit photographed this comet heading for the Sun. A dark disk made an artificial eclipse of the Sun; a white spot the size of the Sun has been added. The comet disappeared behind this disk at 1 million km/hr and did not reemerge. Eleven hours after the impact, cometary material appeared. Although it looks like a splash, it may simply be a part of the comet's tail that had been blown into view by the pressure of the Sun's radiation. There may also be some contribution from a "coronal transient," a solar eruption. The white dot at upper left is Venus. (*B*) The orbiting coronagraph that took these photos was destroyed in a test of part of the Strategic Defense Initiative program ("Star Wars") in 1985, upsetting many in the astronomical community. Though it had exceeded its design lifetime, it was sending back data the day it was destroyed. It had discovered six comets in all. The coronagraph aboard the Solar Maximum Mission also discovered an additional 10 comets that hit or grazed the Sun. All these comets belong to a single family.

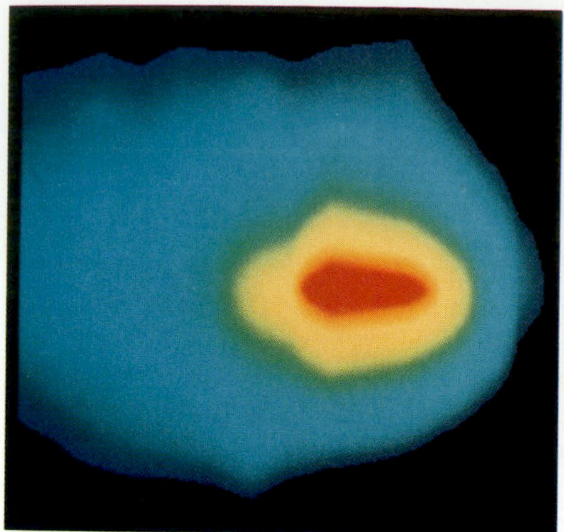

Figure 17–8 This IRAS image of Comet IRAS-Araki-Alcock shows the comet's coma at the infrared wavelength of 25 μ. The false color represents emission strength, with red the strongest. The blue-green material on the side away from the Sun is from dust in the comet's tail.

Figure 17–9 Comet IRAS-Araki-Alcock came so close to Earth in 1983 that it moved 60° across the sky in a single day. It looked like a cloud in the sky three times the diameter of the Moon.

17.4 Halley's Comet

In 1705, the English astronomer Edmond Halley (Fig. 17–10) applied the new theory of gravity developed by his friend Isaac Newton to determine the orbits of comets from observations of their positions in the sky. He reported that the orbits of the bright comets that had appeared in 1531, 1607, and 1682 were about the same. He was troubled, though, that the intervals between appearances were not quite equal. When he resolved this difficulty by analyzing the effect on the comet's orbit by the gravity of Jupiter and Saturn, Halley suggested

Figure 17–10 Edmond Halley.

Figure 17–11 The Bayeux Tapestry (Queen Matilda's Tapestry that hangs in Bayeux, France) showed people looking up at the comet before King Harold was defeated by William the Conqueror.

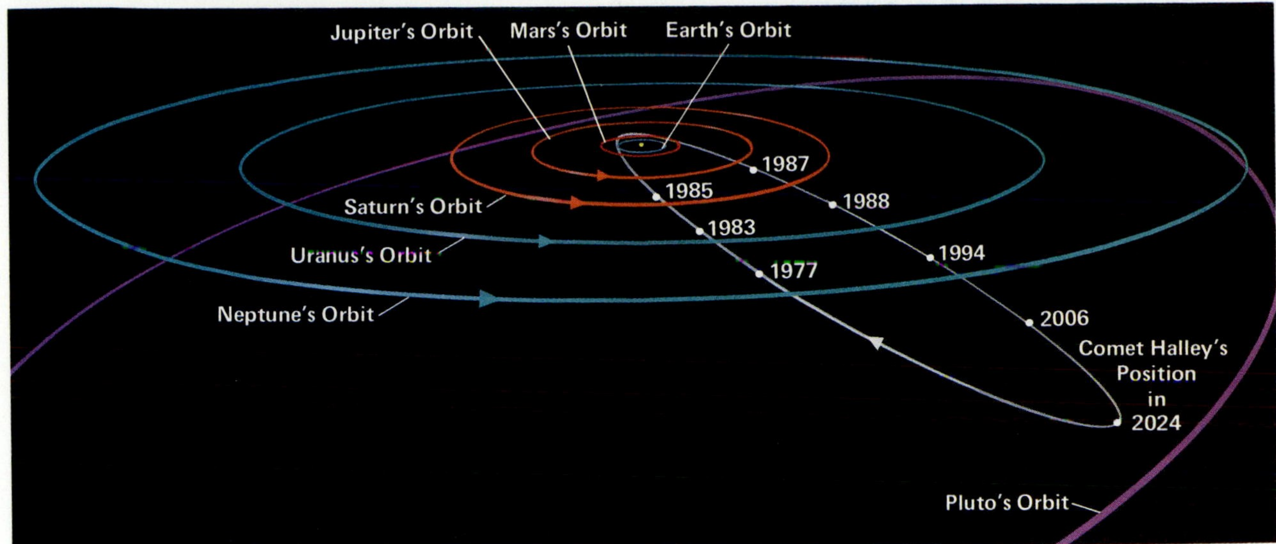

Figure 17–12 Halley's Comet moves on an elliptical orbit in the opposite direction from the revolution of the planets.

that we were observing a single comet orbiting the sun. He predicted that it would return in 1758. The reappearance of this bright comet on Christmas night of that year, 16 years after Halley's death, was the proof of Halley's hypothesis; the comet has since been known as Halley's Comet. It seems probable that the bright comets reported every 74 to 79 years since 240 B.C. were earlier appearances. The fact that Halley's Comet has been observed dozens of times (Fig. 17–11) endorses the calculations that show that less than 1 per cent of a cometary nucleus's mass is lost at each perihelion passage. Halley's Comet has a long, elliptical orbit that extends far out of the ecliptic plane (Fig. 17–12).

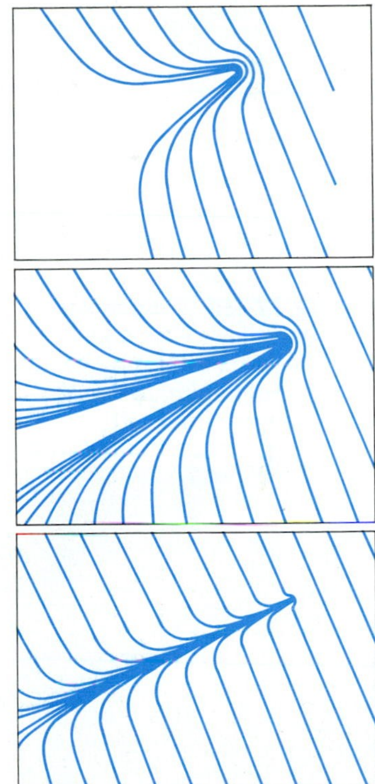

Figure 17–13 The draping of the interplanetary magnetic field, frozen in place in the solar wind, over the ionosphere of a comet. The plasma trapped between the oppositely directed magnetic field lines forms the plasma tail. The model was verified by the ICE spacecraft, which passed through the adjacent regions of oppositely directed magnetic field.

17.4a Halley's Comet Comes

The world waited for Halley's Comet to return. It was picked up in 1982 with a sensitive CCD designed as a test system for the Hubble Space Telescope. At that time, it appeared only as a fuzzy spot in the sky. For years it was followed with large telescopes to see how it changed. In particular, discovering how far from the Sun it was when it became active helps tell what its surface is made from, since the different ices that were thought to be on the surface sublime (change from solid to gas form) at different temperatures.

The United States, embarrassed that it had no space mission to Halley's Comet, succeeded in sending an old spacecraft to another comet, Comet Giacobini-Zinner, in September 1986. This first spacecraft visit to a comet was a great success, with the renamed International Cometary Explorer (ICE, a fitting name for a spacecraft to visit an icy celestial body) passing through the coma and inner tail. Among other discoveries, it measured Giacobini-Zinner's magnetic field, and showed the correctness of theoretical models of how the solar wind's magnetic field draped around the comet's magnetic field (Figs. 17–13 and 17–14). ICE discovered the existence 4 million kilometers from the nucleus of heavy ions that had been given off by the comet and then picked up by the solar wind and carried outward.

Astronomers know the orbit of Halley's Comet very well, but predicting its brightness is more difficult, since its brightness depends not only on how far the comet is from the Sun and Earth but also on how fast it gives off gas

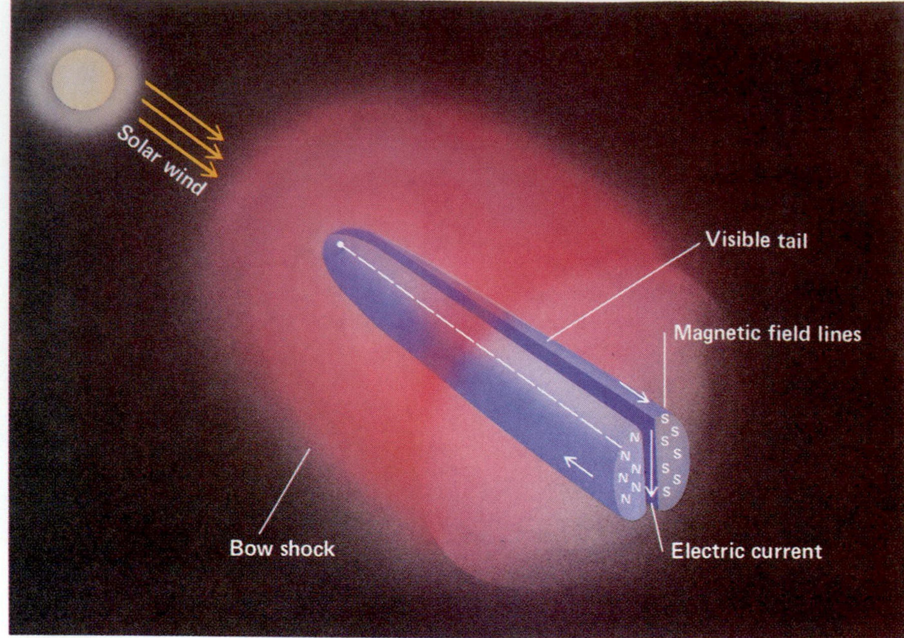

Figure 17–14 The draping of the magnetic field leads to the formation of magnetic lobes in the tails of comets. The magnetic field lines are organized into two lobes of opposite polarity. An electric current flows in the plain separating the lobes.

and dust. Still, it was certainly known that Halley's Comet would never become very bright on this apparition because of the Earth's position when the comet passed closest to the Sun—perihelion—on February 9, 1986. The Earth would then be on the opposite side of the Sun. The best chance to see the comet from Earth seemed to be in early April 1986, when the comet passed closest to Earth and we could see it broadside. Nevertheless, the predictions showed that it would be the faintest as seen from Earth in any appearance in the last 2000 years.

Further, during the post-perihelion period the comet would be far south of the celestial equator. The curve of the Earth would prevent people at temperate northern latitudes from seeing the comet well. From the northern part

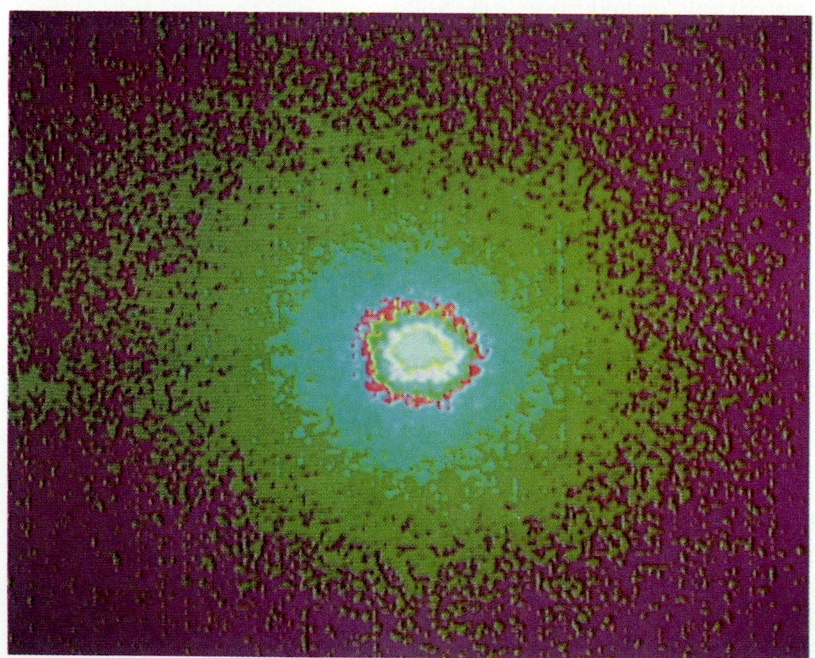

Figure 17–15 False-color view of Halley's Comet in October 1985, after the formation of a coma but before the formation of an extensive tail.

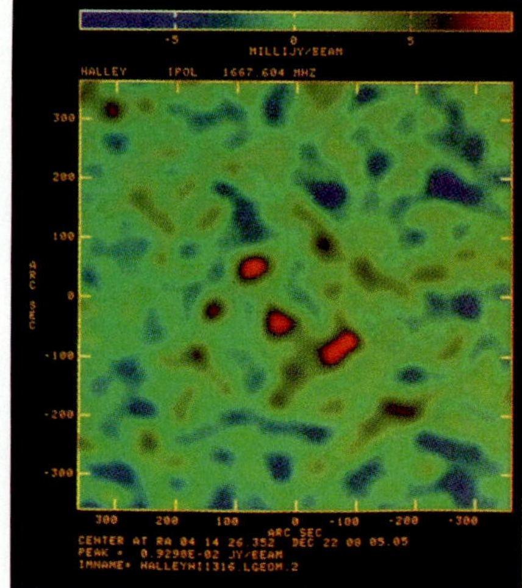

Figure 17–16 A radio image of Halley's Comet taken at the 18-cm wavelength of OH. The clumps are 10,000 km across and are 50,000 to 100,000 km from the nucleus. The image was taken with the VLA.

A

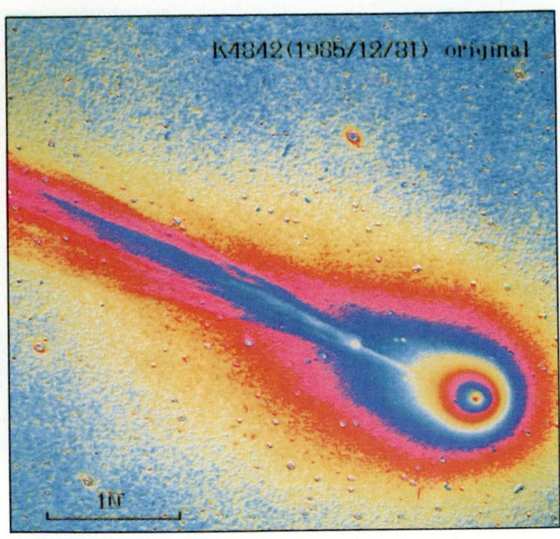

B

Figure 17–17 (*A*) A CCD image of Halley's Comet on December 3, 1985, reproduced without color coding. (*B*) A CCD image of Halley's Comet on December 31, 1985, with false-color indicating brightness.

of the continental United States, it was visible only briefly in the minutes before dawn.

There is a lore about Halley's Comet, and, even with the warnings, astronomers and the public alike wanted to see it. Observations started early, long before the tail had grown (Fig. 17–15). The comet came sufficiently close that radio techniques could be used to observe it (Fig. 17–16). The International Halley Watch was set up to coordinate and collect amateur and professional observations. Many people planned trips to southern latitudes to observe the comet during the peak periods of January and March/April.

Most people didn't try too hard to see Halley's Comet in December 1985 and January 1986 (Fig. 17–17), since it was predicted to have its longest tail in the spring. Observatories were in full swing on the project, though (Fig. 17–18). In late January, about two weeks before perihelion, the comet moved so close to the Sun in the sky that it was lost in the solar glare. Only the Solar Maximum Mission in orbit around the Earth (Fig. 17–19*A*) and the Pioneer Venus spacecraft in orbit around Venus (Fig. 17–19*B*) could see it in the weeks around perihelion.

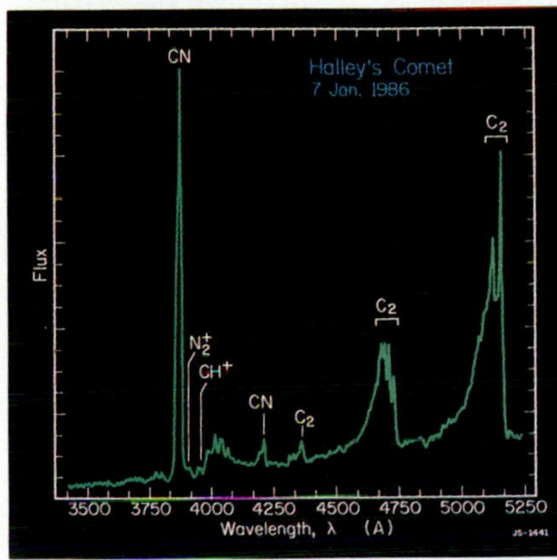

A

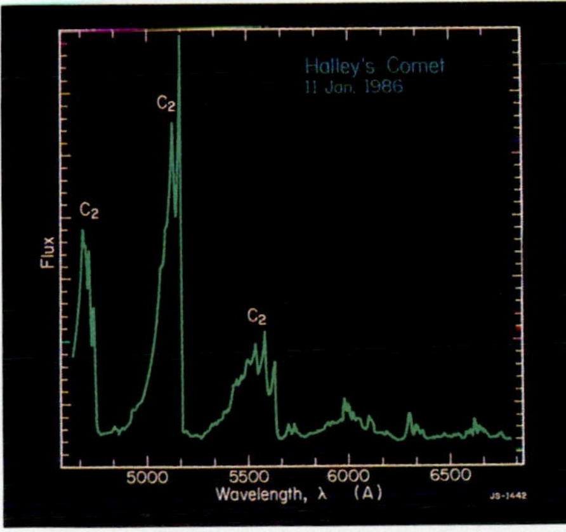

B

Figure 17–18 Spectra of Halley's Comet in January 1986.

A

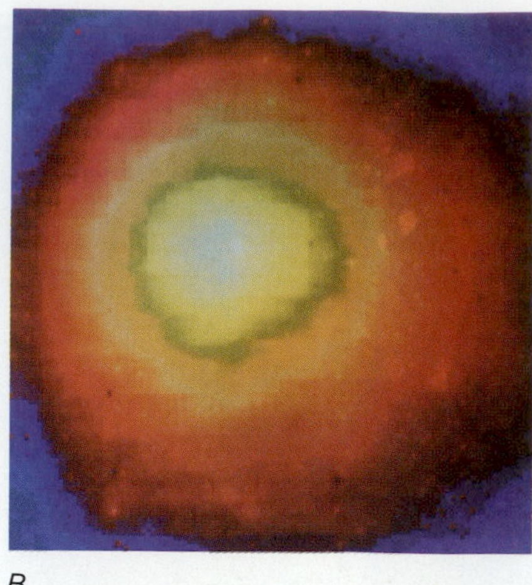

B

Figure 17–19 (*A*) Halley's Comet near perihelion, observed on January 28, 1986, twelve days before perihelion, with the coronagraph on the Solar Maximum Mission in Earth orbit. It was only 15° from the Sun. (*B*) The coma, 20 million km across, observed in the ultraviolet from Pioneer Venus in orbit around Venus. The 20,000 scans that make up the image were collected February 2–5, 1986, within a week before perihelion. The Sun is at top; the pressure of its radiation is pushing the coma in the opposite direction.

When Halley's Comet emerged from perihelion, it had indeed grown a beautiful tail (Fig. 17–20). Alas, the tail was not very bright. It could be seen, barely, with binoculars; at its peak in March it could also be seen with the naked eye, though was never spectacular (Fig. 17–21). From a professional point of view, the observations were a great success, since it was a relatively bright object for large telescopes, and high-quality images and spectra could be taken. The comet was so well watched for such a long time that a series of "tail-disconnection events" were discovered to happen very regularly (Fig. 17–22). The tail would come loose, undoubtedly when the solar wind changed its magnetic polarity (that is, the direction toward north and south magnetic poles switched). It would then drift behind the comet, and the comet would grow a new tail.

17.4b Halley Close Up

In early March, a month after perihelion, an international armada of spacecraft reached Halley's vicinity (Table 17–1 and Figure 17–23). They were timed for then since it is expensive in terms of fuel to go out of the Earth's orbital plane,

Figure 17–20 Halley's Comet on March 12, 1986, showing the dust and gas tails. The color image was composed from blue, green, and red plates, so the motion of the comet makes each star appear three different colors.

Table 17–1 Spacecraft to Halley's Comet

Spacecraft	Sponsor	Approach Date	Closest Approach
Vega 1	U.S.S.R.	March 6, 1986	8,890 km
Suisei (Comet)	Japan	March 8, 1986	151,000 km
Vega 2	U.S.S.R.	March 9, 1986	8,030 km
Sakigake (Forerunner)	Japan	March 11, 1986	7,000,000 km
Giotto	E.S.A.	March 13–14, 1986	605 km

Figure 17–21 Halley's Comet in the sky seen from Georgia in March 1986. (15-sec guided exposure with 135-mm lens at f/3.5 on ASA 400 film)

so the spacecraft intersected the comet when Halley passed through the Earth's orbital plane.

The Soviet Vega 1 (Venus-Galley, where Galley is the Russian word for Halley) spacecraft arrived first, on March 6th. Scientists from all over the world gathered in Moscow at the invitation of the Soviets to see the data come in. One American instrument was even on board. A million miles from the nucleus, the spacecraft already passed into the bow shock formed by the interaction of the magnetic fields of the comet and the solar wind. Vega 1 worked very well as it passed, eventually, within 9000 km from Halley's nucleus. It had no high-resolution imaging device, so the images were difficult to interpret, but the nucleus was surely pinpointed. The data from the other devices on board, including dust detectors, charged particle detectors, a spectrometer, a magnetometer, and several others, were collected for subsequent analysis. It was immediately obvious that there were more small dust particles than expected, down to the smallest detectable mass of 10^{-16} gram; the largest particle was 10^{-6} gram.

Two days later, the Japanese Suisei (meaning: Comet) spacecraft flew 151,000 km from the comet. Even at that relatively great distance, it encountered dust from Halley. It carried an ultraviolet imager for observing the hydrogen corona that extends far into space around the coma and a device to observe the energy of particles in the solar wind and in the comet's plasma. It found the hydrogen corona changing in brightness from day to day. From its measurements, we can estimate the total amount of water evaporated from Halley's nucleus. Even at its distance from the comet, Suisei encountered cometary ions.

On March 9th, Vega 2 flew 8000 km from Halley's nucleus with its battery of instruments (Fig. 17–24). It was apparently facing a less active side of Halley,

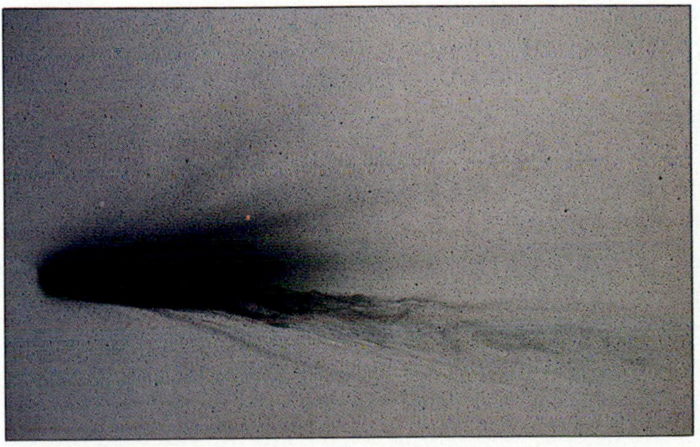

A

B

Figure 17–22 Halley's Comet in March 1986, showing a tail disconnection event. (*A*) A direct photograph. (*B*) False color shows intensity.

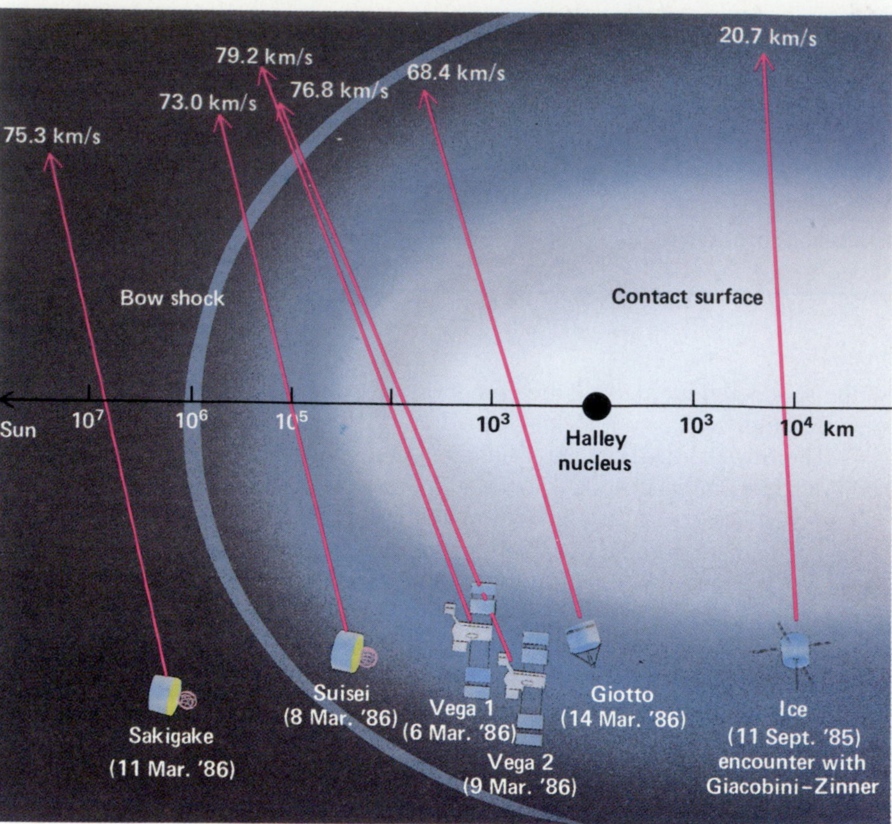

Figure 17–23 Spacecraft encounters with Halley's Comet. The encounter for ICE was with Comet Giacobini-Zinner rather than with Halley's Comet.

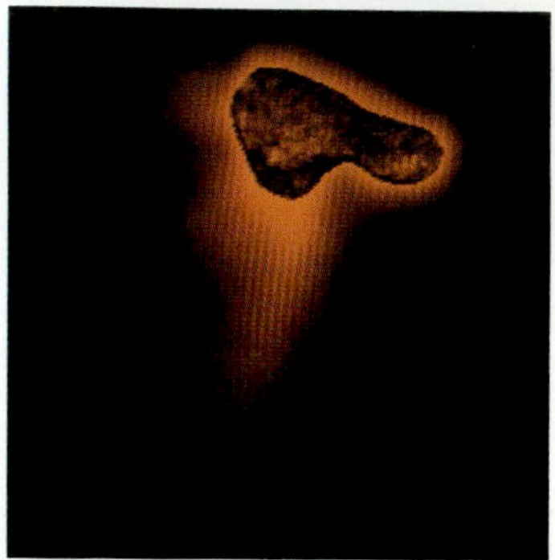

Figure 17–24 A processed Vega-2 image of the nucleus of Halley's Comet, taken March 9, 1986.

since it encountered only one-third the dust of Vega 1. Between Vega 1 and Vega 2, at least one of each type of instrument worked. A world network of radio telescopes had tracked the spacecraft carefully to locate their positions, and the spacecraft located the position of Halley's nucleus relative to themselves. This information was then hurriedly used to redirect the European Space Agency's Giotto mission so that it could pass about 600 km from the nucleus. Formerly, the position of the nucleus had not been known to that accuracy.

On March 11th, the Japanese Sakigake (meaning: Forerunner) spacecraft flew 7 million km from Halley's nucleus, sampling the solar wind there. It also measured the magnetic field and recorded plasma waves that were probably formed by shock waves travelling through Halley's coma.

Two days later, the European Space Agency's Giotto spacecraft (Fig. 17–25) flew close to Halley, reaching the nearest point 3 minutes after the day changed from March 13th to 14th. Giotto got its name from the Italian painter who had included an image of Halley's Comet, which had been a bright object in the sky in A.D. 1301, as the star of Bethlehem in a fresco he did a few years later. (Prof. Roberta J. M. Olson of Wheaton College in Norton, Massachusetts, had made the identification while the spacecraft was being planned.) It had always been known that the spacecraft would be travelling so fast relative to the comet that an impact by a dust particle could destroy the spacecraft or knock it askew. Indeed, because of that possibility, the data were sent back to Earth in "real time" rather than recorded. Excitement mounted as Giotto came closer and closer to the nucleus, sending back photos and other data all the while. The spacecraft was 8 light minutes from Earth at the time, so had some automatic systems to follow the brightest part of the image. This brightest part turned out to be the jets rather than the nucleus, so within 14,400 km the nucleus moved out of the field of view. The best images came from processing several images together, including some from a longer view and some from closer up (Fig. 17–26). At least seven jets seemed to come together to form a fan-shaped coma.

Figure 17–25 A half-sized model of the Giotto spacecraft.

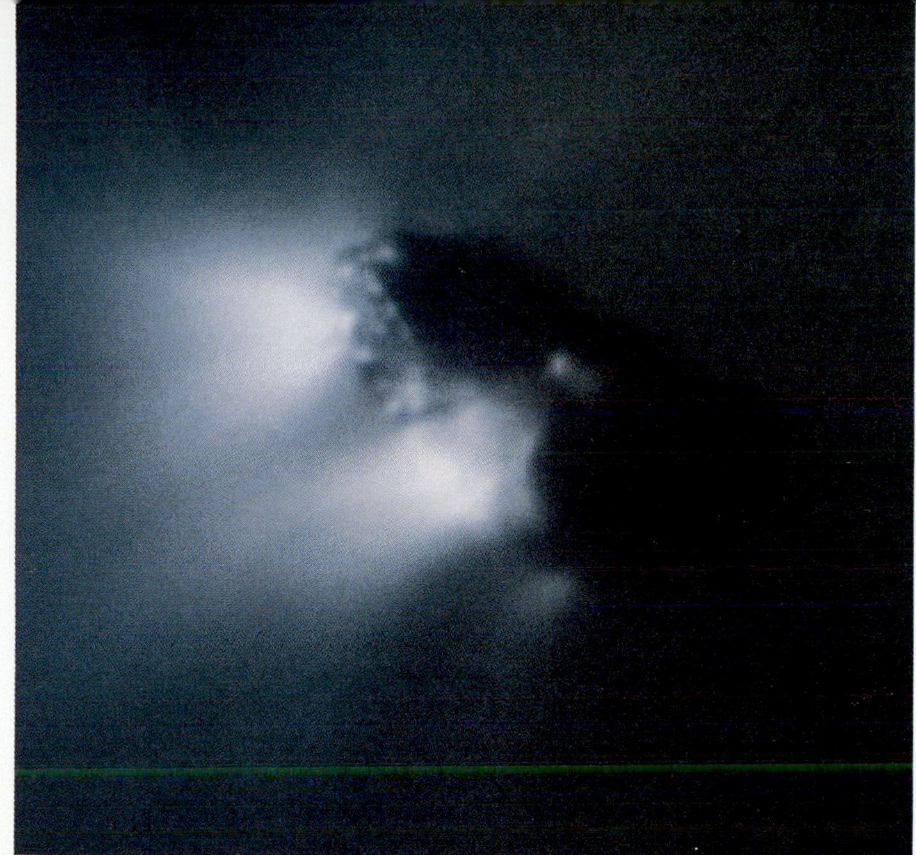

Figure 17–26 Halley's Comet photographed by Giotto; 7 frames are overlaid to make this high-resolution image. The potato-shaped nucleus is about 16 km by 8 km; the frame is 30 km across. Two bright jets are directed toward the Sun, which is at the left, 27° above the horizontal and 15° behind the plane of the image. The spacecraft was oriented so as to observe the night side of the nucleus. The 1.5-km-across bright region in the center of the nucleus is probably a raised region of its surface that is struck by sunlight. All the jets come from the sunlight side of the nucleus.

The nucleus became visible, though partly masked by the dust jets. These jets are regions of the coma in which there are three to ten times as many dust particles (1 to 10 micrometers in size) as usual. The nucleus turned out to be potato-shaped (Fig. 17–27), with a surface area of 100 square km. The surface has depressions and hills (Fig. 17–28). The long diameter is 16 km and the smaller diameters of the cigar-shaped ellipse are 8 km. These dimensions are larger than expected, and mean that the nucleus must be darker than expected to give the observed brightness. The albedo of the nucleus is only 2 per cent to 4 per cent, making the nucleus as dark as velvet, equally dark as the other dark objects in the solar system such as the rings of Uranus and the dark side of Iapetus. The jets do not seem to be active on the night side of the nucleus.

Dramatically, fourteen seconds before closest approach, Giotto was hit by a relatively large dust particle. The impact tilted the spacecraft's axis and made it wobble by up to 1.8°, enough to cause Giotto to lose steady communication with Earth for half an hour. Observations from the other side of the comet, including images of the sunlit side that would have allowed an exact determination of the comet's size and a mapping of the locations of the bases of the jets on its surface, were thus lost. Still, Giotto was certainly a spectacular success.

The reduction of the Soviet Vega data agreed with the Giotto data in size and albedo; seeing the Giotto images, which had higher resolution, was useful in helping scientists realize that the bright areas seen on the Vega data were jets rather than surface features on the nucleus.

Giotto carried 10 instruments in addition to the camera. Among them were mass spectrometers to measure the types of particles present, detectors

Figure 17–27 This potato looked remarkably like the nucleus of Halley's Comet, except 100,000 times smaller.

Figure 17–28 At a banquet of an international meeting about the results from Halley's Comet, the centerpiece was a model of the comet's nucleus based on the Giotto results.

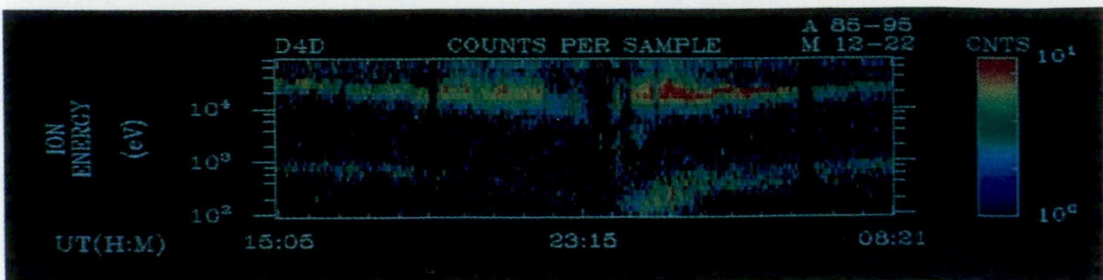

Figure 17–29 The dust sampled by one of the instruments on the Giotto spacecraft, with the closest approach at the center.

for dust (Fig. 17–29), equipment to listen for radio signals that revealed the densities of gas and dust in the coma, detectors for ions, and a magnetometer to measure the magnetic field. As expected, water turned out to be the dominant parent molecule in Halley's coma—80 per cent by volume. Many ions—such as C^+, H_2O^+, CO^+, and S^+—were observed. So many C^+ ions were detected that carbon may be released on the nucleus's surface. The distributions in the coma of some molecules—such as CH, C_2, CO^+, and OH—were studied with the "optical probe" experiment. Rockets set up from Earth at about the same time made ultraviolet images (Fig. 17–30).

Some of Halley's dust particles are made only of hydrogen, carbon, nitrogen, and oxygen (Fig. 17–31*A*); this simple composition resembles that of the oldest type of meteorite (Section 18.1b). It thus indicates that these particles may be from the earliest years of the solar system. Other particles are rich in these same elements, though with other constituents present as well (Fig. 17–31*B*); the distribution is similar but not quite the same as the composition of carbonaceous chondrites, the type of meteorite we had thought is close to being cometary material. The many tiny dust grains detected, each a million times smaller than a particle in cigarette smoke, are smaller than dust particles known in meteorites. But interstellar particles are thought to be of such sizes, endorsing the idea that the material in comets comes from the earliest period of the solar system.

Another instrument found a series of peaks of molecular mass that indicates the presence of a long polymer. It is a chain of formaldehyde—H_2CO—units, so is called polymerized formaldehyde. This polymer may explain why Halley's nucleus is so dark. Since the polymer is still being released, it may be primordial material that is frozen into the ice.

Giotto detected 12,000 impacts of dust particles, which ranged from 10^{-17} (the lower limit of detectability) to 10^{-4} gram. Similarly to the Vega results, there were more dust particles of relatively low mass than had been expected. Perhaps 0.1 to 1 gram of matter was encountered in the form of dust.

As a result of all the flybys, Whipple's model of a comet as a dirty snowball was upheld. The remaining models of the comet as self-gravitating chunks were ruled out. But though the existence of gas jets had been known from ground-based observations, including recent reduction of old photographs of Halley from its 1910 passage, nobody anticipated that the emission of gas and dust would be so localized. The rest of the surface is apparently coated with a black crust. Since no bright regions were discovered at all, the ice is "dirty" even in the active centers. About 30 tons per hour of water and 5 tons per hour of dust were lost at the time the spacecraft flew by; the comet will survive for hundreds of passes but will eventually sublime away.

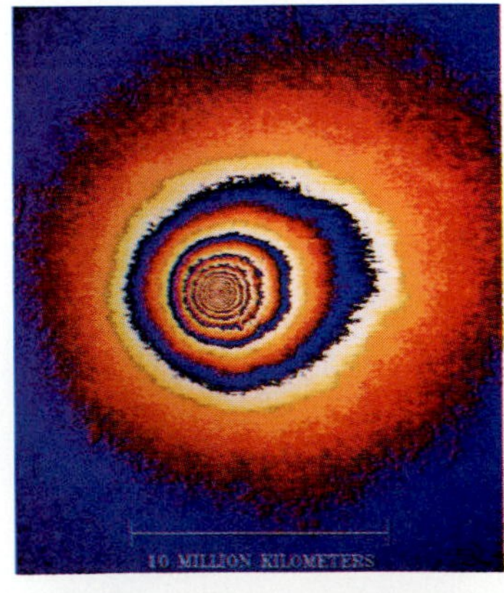

Figure 17–30 An image of Halley's Comet in the Lyman α ultraviolet spectral line of hydrogen.

One conclusion expressed at the International Astronomical Union's 1989 Colloquium on Comets in the Post-Halley Era was based on the observation of the mountains and valleys. For them to persist, they are probably not made of the volatile materials (ices). Thus there may be a dusty substructure to

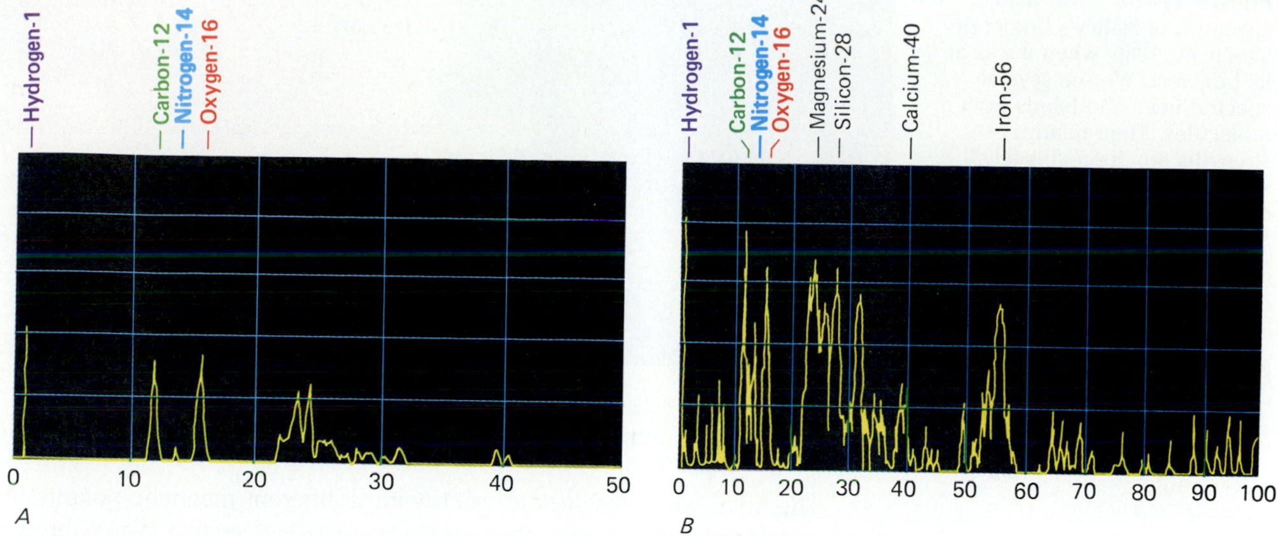

Figure 17–31 The Vega mass spectrometer showed the atomic mass number of each element in dust particles. (*A*) A CHON dust particle, composed only of carbon, hydrogen, oxygen, and nitrogen. (*B*) Other dust particles contain other elements as well.

Halley's Comet; it might be called an icy dustball rather than a dusty (dirty) snowball. Indeed, the nucleus may be made up of a number of 1-km cometesimals.

17.4c Halley from the Ground

Following the Halley space encounters, the comet was seen at its best from the ground (Fig. 17–32). Excellent spectra were obtained in the optical (Fig. 17–33) and in the infrared. Radio telescopes studied molecules. Water vapor is the most prevalent gas, but carbon monoxide and carbon dioxide were also detected.

Figure 17–32 Halley's Comet near the Milky Way on March 13, 1986.

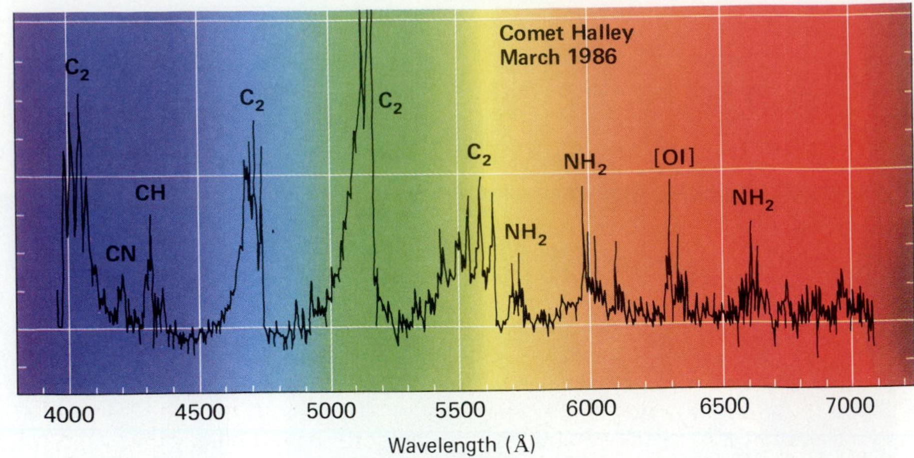

Figure 17–33 The visible spectrum of Halley's Comet on March 14, 1986, when it was at its brightest. We see several spectral lines and bands from molecules. Their relative strengths are approximately the same as they are in other comets, except for NH_2, which was slightly stronger in Halley.

The solar wind has sectors, each with a different magnetic polarity. As the Sun and solar wind rotate, the boundaries between sectors sweep through the solar system in a spiral, like the spiral of a garden sprinkler. When such a boundary passes a comet, the plasma tail can disconnect. Apparently such an event took place in late March, for when the full moon waned and the sky became dark enough to see the comet again around April 1st, the comet had little tail. Those individuals who had gone south to see the comet in the "wrong time" in March wound up with a better view than those who took April comet-watching trips. During April, the comet's head was barely visible to the naked eye, and both head and a short tail were faintly visible in binoculars (Fig. 17–34). The hoped-for tail stretching across a good fraction of the sky never materialized.

Still, the data from ground-based telescopes were valuable, and it will be years before they are all interpreted. Halley now continues to recede from the Sun (Fig. 17–35), and will do so until 2024. The Hubble Space Telescope may be able to observe Halley continually; Halley may never be "lost" again. Halley

Figure 17–34 By April 14, 1986, Halley's Comet had lost its long tail and had only a short, fan-shaped tail. Here we see it near in the sky to the globular cluster omega Centaurii.

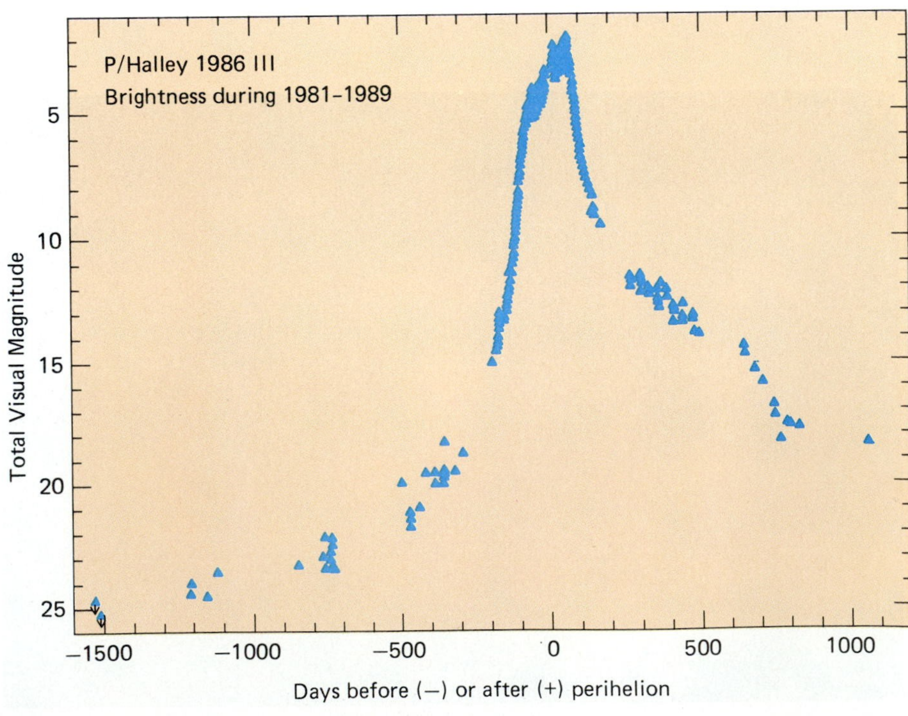

Figure 17–35 The brightness of Halley's Comet.

will be back in the inner solar system in 2061, but this next apparition won't be a favorable one either for Earth-based observers. Only in May 2134 will we have a spectacular view from the Earth's surface—though we may not be limited to the Earth's surface by that time. Actually, a more spectacular comet comes along every 5 to 10 years—though not with such advance notice. When you hear that a bright, spectacular comet is almost here, don't miss it.

Measuring the rate at which Halley loses mass has shown that it will survive for only another 30 or so returns before its activity ceases.

17.5 Chiron

Chiron (kī′ron), a small object in our outer solar system, was discovered in 1977. (In Greek mythology, Chiron was the wisest of the centaurs and the teacher of Achilles.) Calculations of its orbit permitted observations to be found as far back as 1895, so its orbit is now well determined. It has a 50-year period and an eccentricity of 0.38. Its distance from the Sun ranges from 8.5 A.U., within the orbit of Saturn, to almost as far out as the orbit of Uranus. Chiron is so faint that it must be small—100 to 300 km across. Chiron's orbit is so elliptical that it will ultimately either collide with a planet or be ejected from the solar system. A similar object was discovered in 1991.

Chiron was thought to probably be an asteroid (perhaps the first discovered of a trans-Saturnian belt of asteroids; it is catalogued as number 2060) or an exceptionally large comet far from the Sun. Chiron is approaching the Sun, though, en route to its 1995 perihelion, and has brightened and started showing signs of a coma (Fig. 17–36). It is thus a comet rather than an asteroid. It will certainly be the subject of scrutiny for the next few years.

Chiron may really be telling us that there is no sharp distinction between asteroids and comets. Traditionally, comets give off gases because they contain ices, and asteroids are rocky; but other evidence already indicates that asteroids can contain ices and comets can contain rock. Most scientists think that comets originate far beyond Pluto, while asteroids were born close to their present location in the middle solar system. But other distinctions between comets and asteroids may be slight.

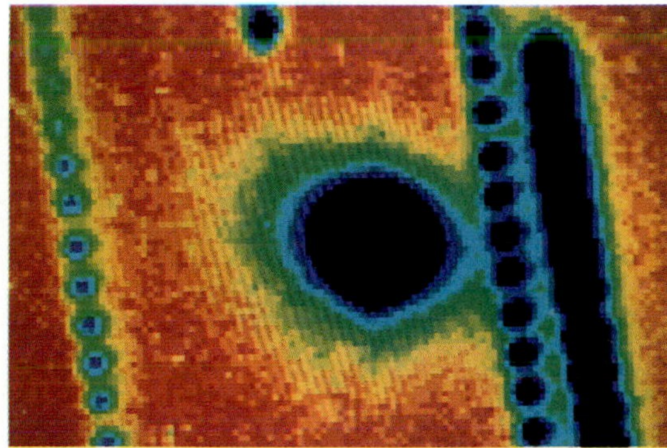

Figure 17–36 Chiron (*center*), with star trails.

17.6 Future Comet Studies

New comets appear all the time and sometimes become reasonably bright (Fig. 17–37). We wait for the very bright naked-eye comet that comes every decade or so; Comet West in 1976 was the last up to this writing.

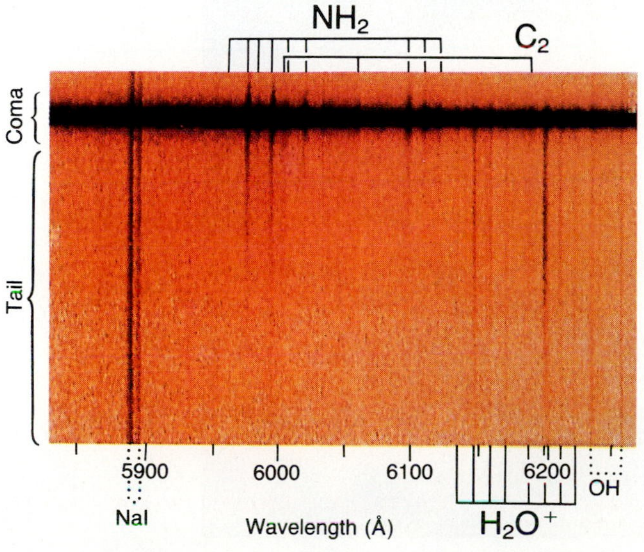

Figure 17–37 The spectrum of Comet Wilson, observed about one year after Halley's Comet was at its brightest. It is useful to compare similar results for a wide variety of comets, because comets are so different from each other.

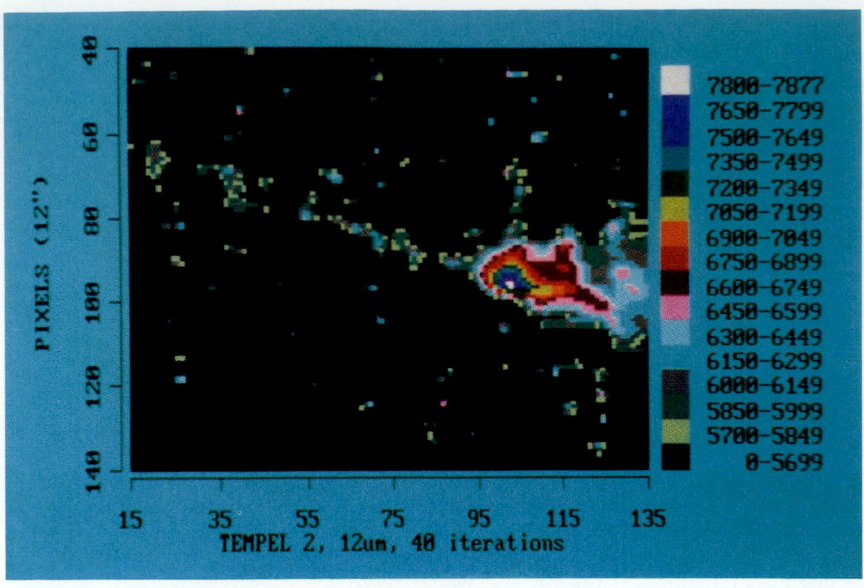

Figure 17–38 Comet Tempel 2 from IRAS.

Figure 17–39 A ground-based image of Comet Tempel 2 in the infrared.

NASA's **C**omet **R**endezvous/**A**steroid **F**lyby (CRAF) mission has been left out of the budget several times; it had been supposed to travel to periodic comet Tempel 2 (Figs. 17–38 and 17–39). As of 1991, it was definitely cancelled for budgetary reasons. After a flyby of the asteroid 449 Hamburga, it was to rendezvous with periodic comet Kopff in July 2000, when the comet will be near aphelion (the farthest point in its orbit from the Sun). CRAF would then have stayed near the comet for two years, until P/Kopff reached perihelion, its near point to the Sun.

Two comet probes that visited Halley's Comet are still functioning. Japan's Sakigake is to fly by Comet Honta-Mrkos-Padjušáková in February 1996. The European Space Agency's Giotto flew by Comet Grigg-Skjellerup on July 10–12, 1992 and sent back measurements (but no photos).

Continued processing of IRAS infrared data has revealed trails of dust left by comets (Fig. 17–40).

Figure 17–40 Dust trails in our solar system left by comets in their wakes were revealed by IRAS.

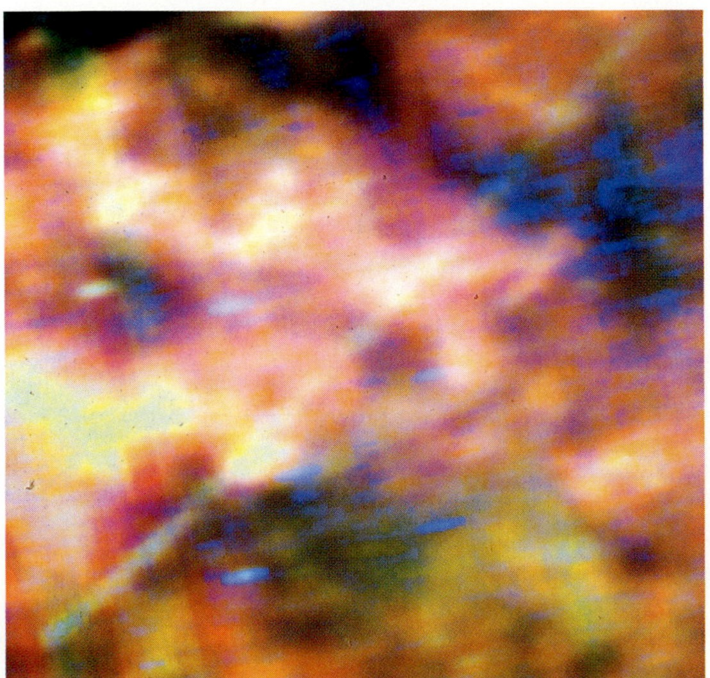

Summary and Outline

Observing comets (Section 17.1)
- The head (nucleus and coma together); the tail

The composition of comets (Section 17.2)
- Dirty-snowball theory: nucleus of ice with dust mixed in
- Coma is gases vaporized from nucleus; molecules present
- Ultraviolet space observations revealed huge hydrogen cloud
- Dust tail: sunlight reflected off particles; ion tail (gas tail): sunlight re-emitted by ions blown back by solar wind

The origin and evolution of comets (Section 17.3)
- Origin of comets: Oort comet cloud; comets detached by gravity

Halley's Comet comes (Section 17.4a)
- Followed since 1982 to February 9, 1986, perihelion
- Earth badly placed for us to see the 1986 apparition
- During perihelion observed from spacecraft

Halley close up (Section 17.4b)
- An armada of spacecraft in March 1986
- Soviet Vega 1 and Vega 2 passed within 9000 km of the nucleus; imaged gas jets; took many kinds of measurements
- Japanese spacecraft flew farther from nucleus; closer one still encountered dust; farther one measured magnetic field
- European Space Agency Giotto imaged nucleus from 605 km
    - Brightest features are dust jets
    - Nucleus is dark as velvet; 15 km × 10 km; covered with crust except for jet regions
    - Water is dominant molecule; many other molecules detected
    - Bow wave, small magnetic field detected
- Whipple's dirty-snowball model upheld

Halley goes (Section 17.4c)
- Rotation of solar wind sectors causes tail disconnection events
- Comet was fainter with shorter tail than hoped during prime observing time from Earth

Chiron (Section 17.5)
- Thought to be a distant asteroid; now known to be a comet

Future plans
- Comet Rendezvous/Asteroid Flyby (CRAF) cancelled
- Two Halley probes still alive

Key Words

comet, head, tail, nucleus, dirty snowball, coma, dust tail, gas tail, ion tail, Oort comet cloud, parent molecules, daughter molecules

Questions

1. In what part of its orbit does a comet travel head first?

2. Why is the Messier catalogue important for comet hunters?

3. Would you expect comets to follow the ecliptic? Explain.

†4. How far is the Oort Comet Cloud from the Sun, relative to the distance from the Sun to Pluto?

5. The energy that we see as light from a comet tail comes from what source or sources?

6. Many comets are brighter after they pass close to the Sun than they are on their approach. Why?

7. What part of a comet has the most mass?

8. Why was the large cloud of hydrogen that surrounds the head of a comet not detected until 1970?

9. Explain why Halley's Comet showed delicate structure in its tail. Which of its tails shows that structure?

†10. Use Kepler's third law and the known period of Halley's Comet to deduce its semimajor axis. Sketch the orbit's shape on the solar system to scale, using Kepler's first law to position the Sun.

†11. Use Kepler's third law and the 80,000-year period of Comet Kohoutek to deduce its semimajor axis. Relate the result to the size of planetary orbits.

12. What did the International Cometary Explorer discover?

13. Why wasn't the 1986 passage of Halley's Comet a good one for earthbound viewers?

14. Which spacecraft imaged Halley's jets?

15. Describe the nucleus of Halley's Comet.

16. Describe the dust discovered near the nucleus of Halley's Comet.

17. Describe the magnetic field of Halley's Comet.

18. Compare Whipple's model for comets with the discoveries of the spacecraft to Halley's Comet.

19. How steady is Halley's output of gas? How do we know?

20. Why do photographs of Halley's Comet look more spectacular than the comet looked to the eye?

21. How could Chiron, a comet, be confused for so long with an asteroid?

22. What will CRAF study?

†This question requires a numerical solution.

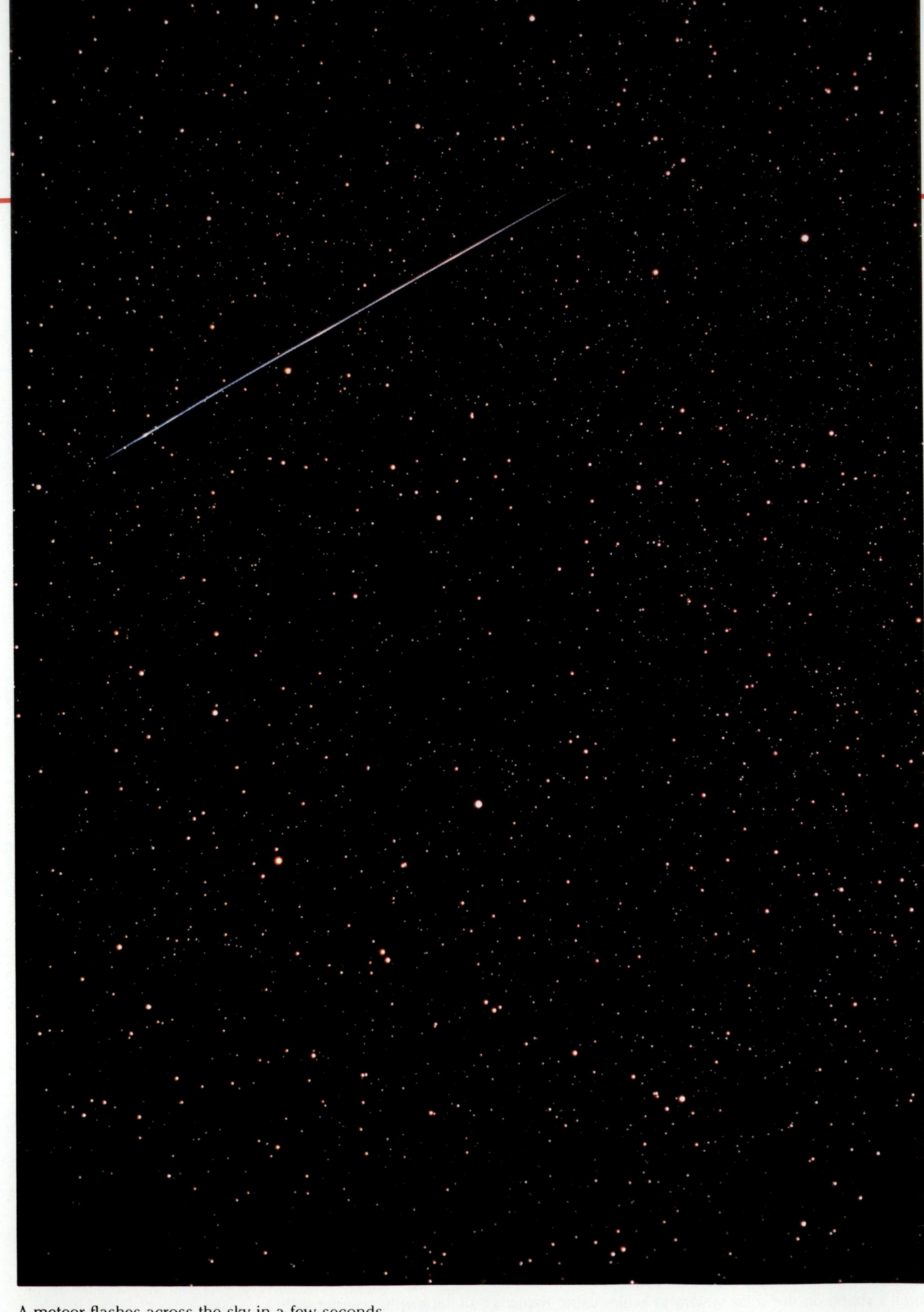

A meteor flashes across the sky in a few seconds.

Meteorites and Asteroids

18

Aims: To describe meteorites and asteroids, and to see how their history may provide us with information about the origin of the solar system

Asteroids and meteoroids are members of our solar system that are solid, like planets, but are much smaller. We shall see how they, like the comets, are storehouses of information about the solar system's origin. The meteoroids, which we discuss first, are probably chunks broken off asteroids.

18.1 Meteoroids and Meteorites

There are many small chunks of matter in interplanetary space, ranging up to tens of meters across. When these chunks are in space, they are called *meteoroids.* When one hits the Earth's atmosphere, friction slows it down and heats it up—usually at a height of about 100 km—until all or most of it is vaporized. Such events result in streaks of light in the sky, which we call *meteors* (popularly known as *shooting stars*). Most meteors, though, are the burning up of dust specks, probably given off by comets.

The brightest meteors can reach magnitude −15 or −20, brighter than the full moon. We call such bright objects *fireballs* (Fig. 18–1). We can sometimes even hear the sounds of their passage and of their breaking up into smaller bits. A fireball that breaks up in this way is a *bolide.* When a fragment of a meteoroid survives its passage through the Earth's atmosphere, it is called a *meteorite.* We now also refer to objects that hit the surface of the Moon or of other planets as meteorites.

18.1a Types and Sizes of Meteorites

Tiny meteorites less than a millimeter across, *micrometeorites,* are the major cause of small-scale erosion on the Moon. Micrometeorites also hit the Earth's upper atmosphere all the time, and remnants can be collected for analysis from balloons or airplanes. The micrometeorites are thought to be debris from comet tails and asteroids. They may have been only the size of a grain of sand, and are often sufficiently slowed down that they are not vaporized before they reach the ground. The resulting dust—100 tons a day of it—can be sampled by collecting ice from the Arctic or Antarctic, or from mountaintops.

Space is full of meteoroids of all sizes, with the smallest being most abundant. Most of the small particles, less than 1 mm across, probably come from comets. Most of the large particles, more than 1 cm across, may come from collisions of asteroids in the asteroid belt between Mars and Jupiter. It is generally these larger meteoroids that become meteorites. From matching the spectrum of reflected sunlight, some meteorite types can even be assigned to certain asteroids.

There are several kinds of meteorites. Most of the meteorites that are found (as opposed to most of those that exist) have a very high iron content—

Figure 18–1 A fireball observed on August 10, 1972, from Grand Teton National Park in Wyoming. The fireball was visible for 101 seconds and might have reached magnitude −19. Thus it was about 80 m across and weighed about one million tons. It came 58 km from hitting the Earth's surface.

Figure 18–2 A 3-metric-ton mass of the Cape York meteorite, known as the "Woman." Note how dense it has to be to have this high mass, given its small size. It is in the American Museum of Natural History in New York City. The largest meteorite found in North America is the 14-metric-ton Willamette, in the adjacent American Museum—Hayden Planetarium. The surfaces of these charred chunks of iron and nickel are dark, pitted, and rough.

Table 18–1 Meteorites

Composition	Seen Falling	Finds
Irons	6%	66%
Stony-irons	2%	8%
Stones	92%	26%

about 90 per cent; the rest is nickel. These *iron meteorites* (*irons,* for short) are thus very dense—that is, they weigh quite a lot for a given volume (Figs. 18–2 and 18–3.).

Most meteorites that hit the Earth are stony in nature, and are often referred to simply as *stones*. Because stony meteorites resemble ordinary rocks and disintegrate with weathering, they are not usually discovered unless their fall is observed. That explains why most meteorites discovered at random are irons. But when a fall is observed, most meteorites recovered are stones. The stony meteorites have a high content of silicates; only about 10 per cent of their mass is nickel and iron. Most stony meteorites are of a type called *chondrites,* because they contain rounded particles called "chondrules." Astronomers think the chondrules condensed from the original solar nebula. The stony meteorites without chondrules are called *achondrites;* they may have once been chondrites, but have melted and cooled.

A third kind of meteorite is a *stony-iron,* which contains roughly equal quantities of silicates and iron-nickel alloy. Most stony-irons are of a type called "pallasites"; they contain beautiful green crystals of the mineral olivine.

A large terrestrial center that is obviously meteoritic in origin is the Barringer Meteor Crater in Arizona (Fig. 18–4). It resulted from what was perhaps the most recent large meteorite to hit the Earth, for it was formed only 40,000 years ago (Fig. 18–5). Dozens of other impact craters are known on Earth (Fig. 18–6).

Every few years a meteorite is discovered on Earth immediately after its fall, as in the stony meteorite found in Innisfree, Canada, in 1977 (Fig. 18–7). The chance of a meteorite's landing on someone's house is very small, but it

A

B

Figure 18–3 (*A*) The largest mass of the Cape York meteorite, a 31-metric-ton piece known as "Ahnighito" or "the Tent." It was discovered by Eskimos in Greenland in the early 1800's and was brought to New York by Robert Peary and Matthew Henson in 1897. It is the largest on display in a museum (second in size only to a meteorite still in the ground in Namibia) and is in the American Museum of Natural History in New York City. (*B*) Its 1979 move into the museum.

Figure 18–4 The Barringer "meteor crater" (actually a meteorite crater) in Arizona. It is 1.2 km in diameter. Dozens of other terrestrial craters are now known, many from aerial or space photographs. The largest may be a depression over 400 km across under the Antarctic ice pack, comparable with lunar craters. Another very large crater, in Hudson Bay, Canada, is filled with water. Most are either disguised in such ways or have eroded away.

Figure 18–5 A planetary geology field trip to the Barringer Meteor Crater led by Dr. Eugene Shoemaker.

happens every three years or so (Fig. 18–8)! It has recently been calculated on the basis of the number of detectable falls, a person's size, the population and size of North America, etc., that on the average a meteorite should hit a person every 180 years in North America. The only such documented case occurred on November 30, 1954, when a nine-pound stony meteorite crashed into a house in Sylacauga, Alabama, and hit Mrs. Hewlett Hodges on a bounce. She was bruised, but not seriously injured.

Often the positions of fireballs in the sky are tracked in the hope of finding fresh meteorite falls. The newly discovered meteorites are rushed to laboratories, since studies of the isotopes in them can reveal how long they have been in space. Many meteorites have recently been found in the Antarctic, where they have been well preserved as they accumulated over the years (Fig. 18–9). They are being kept free of human contamination. It has been discovered that the overall composition of these meteorites is sufficiently different from the composition of meteorites found elsewhere that meteorites of long ago may have preferentially come from different regions of the solar system than

Tektites, small, rounded glassy objects that are found at several locations on Earth, may have splashed out from meteorite impacts. The origin of tektites, including whether they came from the Earth or from the Moon, has long been controversial, though a terrestrial origin seems most probable.

Figure 18–6 A meteorite crater in Quebec.

Figure 18–7 A bright fireball was observed and photographed on February 5, 1977. Analysis indicated that it probably landed near Innisfree, Alberta, Canada. Scientists searched the snow-covered wheat fields on snowmobiles and found this 2-kg meteorite. It is only the third meteorite recovered whose previous orbit around the Sun is known.

A

B

C

Figure 18–8 A meteorite landed in 1982 in Wethersfield, Connecticut, coming through the roof and ceiling and bouncing off a bench. (*A*) The hole in the roof. (*B*) The hole in the living room ceiling. (*C*) The meteorite.

most meteorites now. Even stranger, a few odd Antarctic meteorites seem to have come from the Moon or even from Mars (Fig. 18–10). Such lunar meteorites apparently take from hundreds of thousands to hundreds of millions of years in their transfer to the Earth. Studies of how often they were hit by particles in interplanetary space give us this information, as well as indicating that they arrived on Earth tens of thousands of years ago.

In recent years, scientists have realized that some of the meteorites that have never been heated contain grains of interstellar dust from before the formation of the solar system. These interstellar grains can be recognized by their unusual ratios of abundances of carbon, nitrogen, and other elements. So dust from the stars lands at our feet for us to study.

*18.1b Carbonaceous Chondrites

Objects that contain large quantities of carbon, often in organic compounds, are *carbonaceous. Carbonaceous chondrites*—stony meteorites with chondrules and a high carbon content—are rare. One such carbonaceous chondrite—the Murchison meteorite (Fig. 18–11)—contains simple amino acids, building blocks of life. It fell near Murchison, Victoria, Australia, in 1969.

Figure 18–9 Scientists picking up meteorites in the Antarctic. The meteorites have probably been long buried in ice, and are made visible when wind erodes the ice. Thousands of meteorites were found in this expedition.

Figure 18–10 This meteorite found in Antarctica is made of material that makes us think it is a rock from Mars. In appearance and composition, it resembles certain basaltic rocks, which is unlikely on asteroids. Yet the relative abundances of oxygen isotopes in this meteorite are unlike those from Earth or from any other meteorite. Further, this meteorite is 1.3 billion years old, one of 8 similar objects known; Mars was still forming crust from molten rock at that time. Most other meteorites are 4.5 billion years old. The separate discovery of a meteorite that probably came from the Moon (Fig. 8–23) makes it likely that other meteorites hitting the Moon or Mars can blast rocks off them without pulverizing the rocks, though we do not yet understand how. Analysis has indicated that the martian meteorites were ejected from Mars about 200 million years ago. It certainly is cheaper to pick up rocks from Mars in this way instead of sending a spacecraft there. One of the supposed Mars meteorites has been found to carry organic (carbon-containing) compounds, again raising the possibility of organic compounds existing on Mars.

Analysis of these amino acids showed that they were truly extraterrestrial and had not simply contaminated the sample on Earth. The formation of such complicated organic chemicals in cold, isolated places like meteoroids is one of several indications that the precursors of life develop naturally. We shall be developing more of this evidence in the following chapter. Some of the carbonaceous chondrites may be cometary debris.

The largest carbonaceous chondrite available for scientific study is the Allende meteorite (Fig. 18–12), 5 tons of which fell in Mexico in 1969. The 1987 discovery of deuterium in the Murchison meteorite in an abundance too high to be terrestrial contamination is new evidence that the amino acids, by analogy, are extraterrestrial.

We measure the ages of meterorites by studying the ratios of radioactive and non-radioactive isotopes in them, just as we date lunar or terrestrial rocks. The measurements show that the meteoroids were formed up to 4.6 billion years ago, the beginning of the solar system. The chondrites may date from the formation of the solar system itself; some may be pieces of planetesimals. The abundances of the elements in meteorites thus tell us about the solar nebula from which the solar system formed, or even about conditions before that (Fig. 18–13). In fact, up to the time of the Moon landings, meteorites were the only extraterrestrial material we could get our hands on.

Most of our knowledge of the abundances of the elements in the solar system has come either from analysis of the solar spectrum or from analysis of meteorites. Until recently it was thought that all the analyses showed that abundances of each element were uniform throughout the solar nebula. But evidence is growing, including, in particular, the high-quality data from the

Figure 18–11 Close-up of a part of the Murchison meteorite.

Figure 18–12 A view through a polarizing microscope of a thin section of the Allende meteorite.

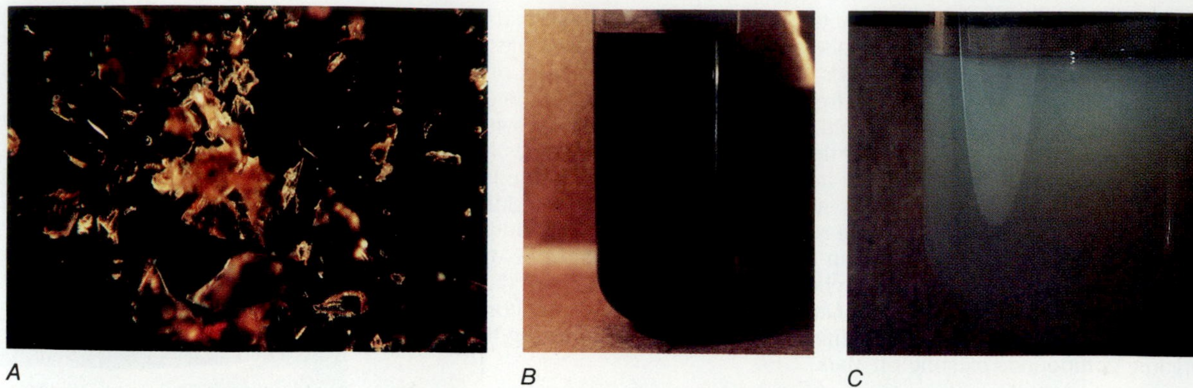

A B C

Figure 18–13 (*A*) Miniature diamonds found in the Allende meteorite. The ratios of isotopes in these diamonds resemble those of matter formed before our solar system formed. (*B*) The tiny diamonds were found in a sample like this partially dissolved one from the Allende meteorite. The sample also contains most of the noble gases (helium, neon, argon, krypton, and xenon) of the meteorite. (*C*) The sample has been treated here with acid, which dissolved away much of the organic material. Some of the grains left behind are the diamonds, apparently formed without high pressure. The isotopic composition found indicates the presence of material formed in stars during the late stages of stellar evolution. This "stardust" may have come from several stars.

Allende meteorite, that abundances varied from place to place in the solar nebula. The abundances of isotopes of common elements like oxygen may vary by 5 per cent; the abundances of rare earths (elements 57 to 71) can vary by greater amounts.

18.1c Meteor Showers

On any clear night, a naked-eye observer may see a few *sporadic* meteors an hour, that is, meteors that are not part of a meteor shower. (Just try going out to a field in the country on a night when the Moon is not up and watching the sky for an hour.) Meteors often occur in *showers,* when meteors are seen at a rate far above average. During a shower several meteors may be visible to the naked eye each minute (Fig. 18–14), though this is rare. Meteor showers occur at the same time each year (Table 18–2) and represent the Earth's passing through the orbits of defunct comets and hitting the meteoroids left behind. The duration of a shower depends on how spread out the meteoroids are in

Focus On

Box 18.1 The Tunguska Explosion

Many theories have been advanced to explain an explosion near the Tunguska River in Siberia in 1908. Most scientists think it was a fragment of a meteoroid or comet entering our atmosphere. Since no remnant has been found, even though trees were blown outward for kilometers around, it may rather have been a small fragment of a comet, and vaporized completely and/or exploded high above the Earth's surface. Signs of dust from the explosion have been found at widely separated regions of the Earth.

Table 18–2 Meteor Showers*

Name	Date of Maximum	Duration Above 25% of Maximum	Approximate Limits	Number per Hour at Maximum	Parent
Quadrantids	Jan. 4	1 day	Jan. 1–6	110	—
Lyrids	April 22	2 days	April 19–24	12	Comet 1861 I
η Aquarids	May 5	3 days	May 1–8	20	Comet Halley
δ Aquarids	July 27–28	7 days	July 15–Aug. 15	35	—
Perseids	Aug. 12	5 days	July 25–Aug. 18	68	Comet 1862 III
Orionids	Oct. 21	2 days	Oct. 16–26	30	Comet Halley
Taurids	Nov. 8	Spread out	Oct. 20–Nov. 30	12	Comet Encke
Leonids	Nov. 17	Spread out	Nov. 15–19	10	Comet 1866 I
Geminids	Dec. 14	3 days	Dec. 7–15	58	3200 Phaethon

*The number of sporadic meteors per hour is 7 under perfect conditions. The visibility of showers depends mostly on how bright the Moon is on the date of the shower, which depends on its phase. Meteors are best seen with the naked eye; using a telescope or binoculars merely restricts your field of view.

the former comet's orbit. Meteorites do not usually result from showers, so presumably the meteoroids that cause a shower are very small.

The rate at which meteors are usually seen increases after midnight on the night of a shower, because that side of the Earth is then turned so that it plows through the oncoming interplanetary debris. If the Moon is gibbous or full, then the sky is too bright to see the shower well. The meteors in a shower are seen in all parts of the sky, but their trajectories all seem to emanate from a single point in the sky, the *radiant.* A shower is usually named after the constellation that contains its radiant. The existence of a radiant is just an optical illusion; perspective makes the set of parallel paths approaching us appear to emanate from a point.

Figure 18–14 Several meteor trails cross star trails during the great 1966 Leonid meteor shower.

Some meteor showers, like the Perseids we can expect each year on August 12, are fairly steady year after year. The Perseids are the most widely seen, in large part because of the warm weather in the northern hemisphere that makes outdoor observing pleasant. Other showers, like November's Leonids, are occasionally spectacular.

The average meteor we see is about 100 km high and is moving about 30 km/sec. The small amount of sodium, calcium, silicon, iron, and other heavy elements that is left when the meteoroid vaporizes results in a radio-reflecting variable layer of the Earth's upper atmosphere. The electrons in the trails formed by the meteoroids also reflect radio waves, and can be used to reflect radio messages for up to 15 seconds each. Some trucking companies are using these free reflections to keep in touch with their fleets.

18.2 Asteroids

The nine known planets were not the only bodies to result from the agglomeration of planetesimals 4.6 billion years ago. Thousands of *minor planets,* called *asteroids,* also resulted (Fig. 18–15).

Most of the asteroids have elliptical orbits between the orbits of Mars and Jupiter, in a zone called the *asteroid belt.* A million or so larger than 1 km apparently exist there. Indeed, it was not a surprise when the first asteroid was discovered, on January 1, 1801, because Bode's law, a numerological formula for the sizes of planetary orbits (still without known justification), predicted the existence of a new planet in approximately that orbit.

A

Figure 18–15 (*A*) The position of the asteroid 1 Ceres over a two-day interval. The brightest star in the background is Aldebaran. The Hyades, an open cluster in the constellation Taurus, is the V-shaped set of stars. (*B*) When the telescope tracks the stars, the asteroid appears as a trail. Here we see 1987 HR.

B

The first asteroid was discovered by a Sicilian clergyman/astronomer, Giuseppe Piazzi, and was named Ceres after the Roman goddess of harvests. Though Piazzi followed Ceres's motion in the sky for over a month, he became ill, and before he could observe again, the object moved into evening twilight where it could no longer be seen. Ceres was lost. But the problem led to a great advance in mathematics: the young mathematician Carl Friedrich Gauss developed an important way of plotting the orbits of objects given only a few observations. The orbit Gauss worked out led to the rediscovery of Ceres. Modern versions of Gauss's methods are still in use today, though calculations are now done by computer rather than by hand.

Though the discovery of Ceres was a welcome surprise, even more surprising was the subsequent discovery in the next few years of three more

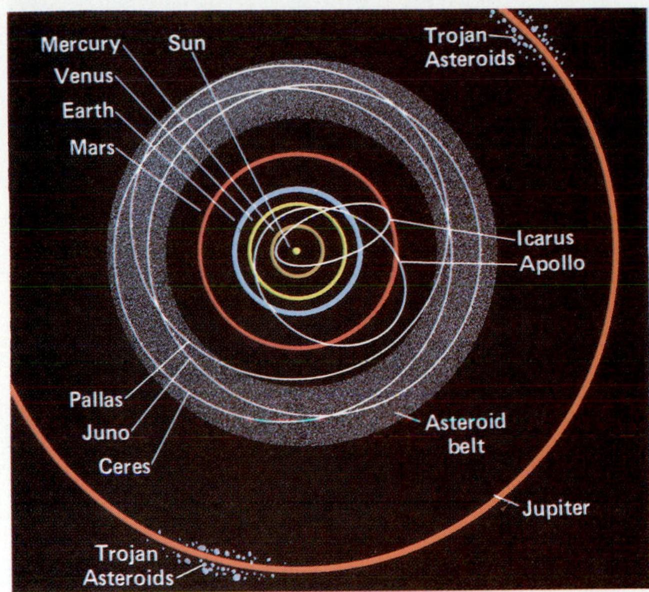

Figure 18–16 Some asteroid orbits. The asteroid belt, in which most asteroids lie, is shaded.

asteroids. The new asteroids were named Pallas, Juno, and Vesta, also after goddesses, which began the generally observed tradition of assigning female names to asteroids. The next asteroid wasn't discovered until 1845, and about 2500 are now named and numbered. They are assigned numbers when their orbits become well determined. The number and name of an asteroid are often listed together: 1 Ceres, 16 Psyche, and 433 Eros, for example. Asteroids rarely come within a million km of each other, though occasionally collisions do occur, producing meteoroids. Gravitational effects of the major planets cause the Kirkwood gaps in the asteroid belt (Figs. 18–16 and 18–17).

Only half a dozen known asteroids are larger than 300 km across, and over 200 more are larger than 100 km across. The largest asteroids known (Fig.

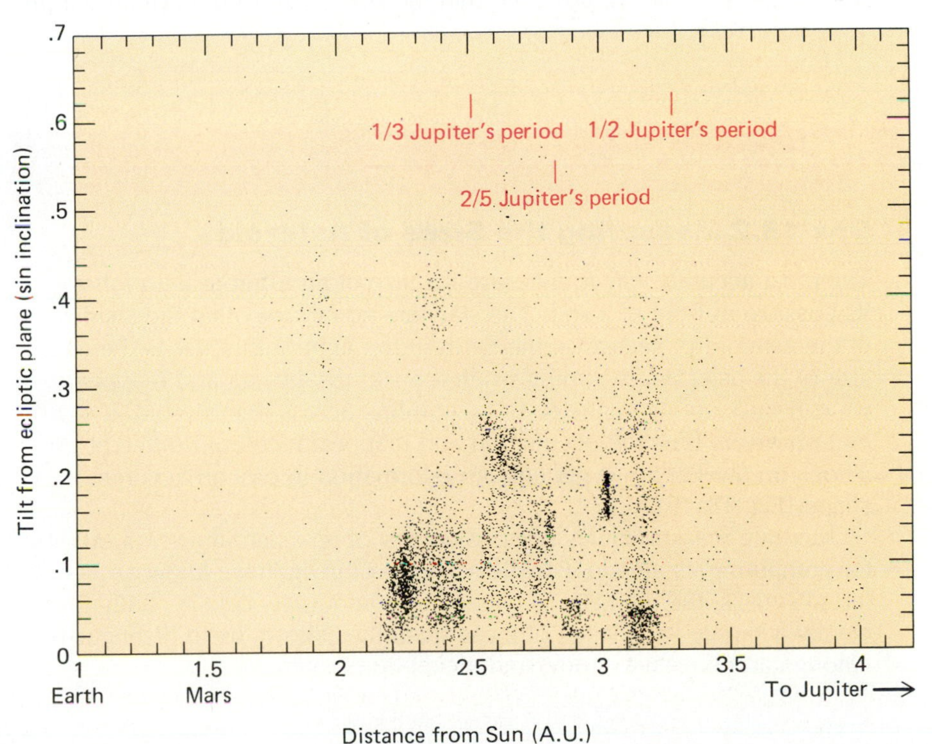

Figure 18–17 A graph of the positions of many asteroids shows gaps at certain values of semimajor axis. These Kirkwood gaps result from the gravitational effects of the planets and chaotic motion. The vertical axis shows the sine of the angle of inclination of the orbits to the ecliptic plane.

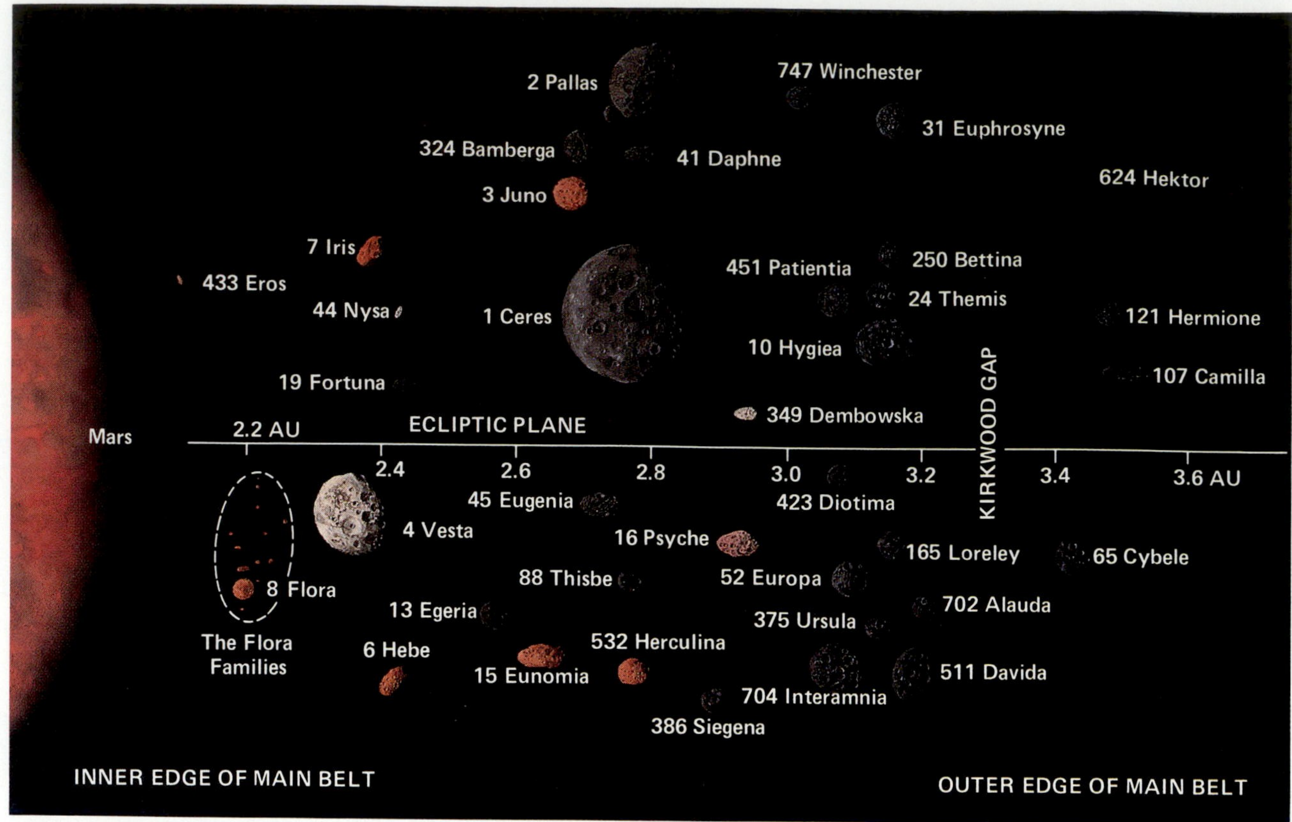

Figure 18–18 The sizes of the larger asteroids and their relative albedoes.

18–18) are the size of some of the moons of the planets. Small asteroids may be only 1 km or less across. All the asteroids together contain less than 0.1 per cent of the mass of the Earth. Perhaps 100,000 asteroids could be detected with Earth-based telescopes. The spins of almost 500 asteroids are now known; all but a dozen rotate with periods between 3 and 30 hours, like most major planets. It does not seem, however, that the spins result from primordial pro-

Focus On

Box 18.2 Measuring the Sizes of Asteroids

The most accurate way to measure the size of an asteroid is to follow its passage in front of a star. The stars are so far away that the shadow of the asteroid projected on the Earth in the light of this star is the full size of the asteroid. By timing when the star disappears and reappears, we can measure its diameter along one line across the asteroid. If several observers time the occultation and reappearance from different positions on the Earth, we get enough information to plot an accurate shape (Fig. 18–19).

Accurate measurements of the diameter of an asteroid are important for computing its albedo. For the few cases in which we know an asteroid's mass, the diameter is needed to compute its density. Sufficiently accurate knowledge of these quantities allows us to distinguish among carbonaceous, stony, and nickel-iron materials.

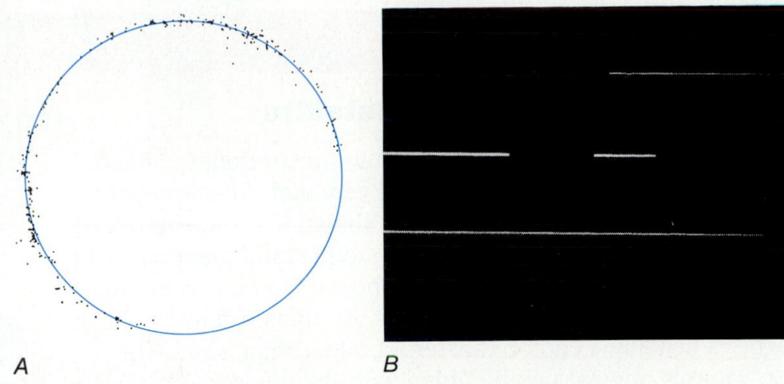

Figure 18–19 (*A*) The points show the individual observations of dozens of amateur observers who timed the moments at which a star disappeared and reappeared when it was occulted by the asteroid Pallas in 1983. The result is best fit by an ellipse 520 by 516 km. The star is so far away that the shadow of the asteroid it casts on the Earth is essentially the same size as the asteroid itself. (*B*) Star trails at the occultation of a star by Pallas. Pallas blocked the star for 43 seconds at the observing site; other stars were not blocked.

cesses; the spins are apparently telling us something about angular-momentum transfer.

The Pioneers and Voyagers en route to Jupiter and beyond travelled through the asteroid belt for many months and showed that the amount of dust among the asteroids is not increased over the amount of interplanetary dust in the vicinity of the Earth. Particles the size of dust grains are only about three times more plentiful in the asteroid belt, and smaller particles are less plentiful. So the asteroid belt will not be a hazard for space travel to the outer parts of the solar system.

The University of Arizona's Spacewatch Camera, now under way, is using CCD detectors and repetitive scanning to search for new asteroids. Scientists at the Lowell Observatory in Flagstaff, Arizona, who have discovered a number of asteroids, have named seven of them after the seven astronauts (Fig. 18–20) who died in the explosion of the space-shuttle Challenger.

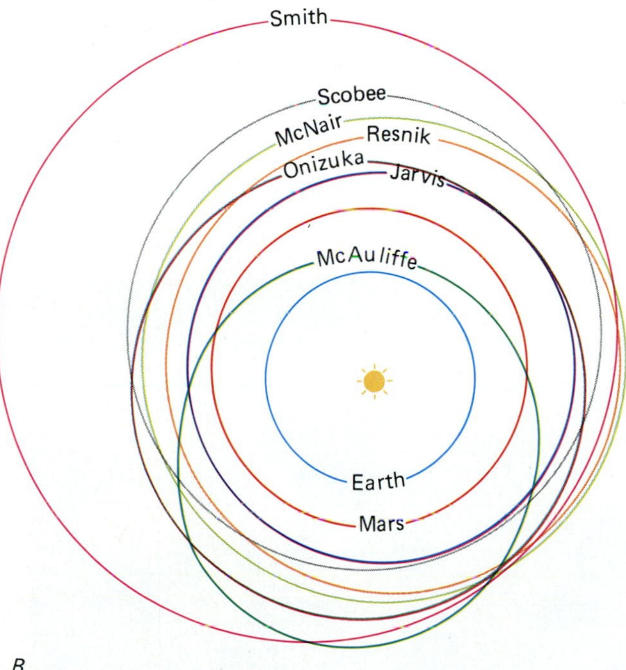

Figure 18–20 (*A*) The trail of the asteroid 3352 McAuliffe. (*B*) Asteroids have been named for each of the seven astronauts who died in the explosion of the space-shuttle Challenger.

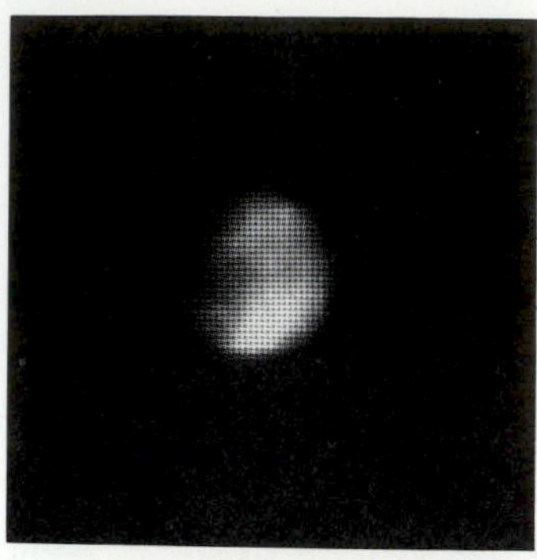

Figure 18–21 A Voyager 2 view of Saturn's moon Phoebe, taken on September 4, 1981, from a distance of 2.2 million km. Phoebe is 200 km in diameter, twice the size previously believed. Its albedo is only 5 per cent.

Focus On

Box 18.3 Phoebe, a Captured Satellite

Voyager 2 revealed much about Saturn's outermost satellite, Phoebe (Fig. 18–21). Phoebe is 200 km across and very dark (albedo about 5 per cent). Ground-based observations had shown that Phoebe orbits Saturn in the ecliptic rather than in Saturn's equatorial plane, in which the other satellites orbit Saturn, and that Phoebe's orbit is retrograde, unlike the orbits of Saturn's other satellites. All this evidence indicates that Phoebe is actually a captured asteroid rather than a satellite formed in Saturn's original nebula along with the other satellites.

*18.2a Special Groups of Asteroids

Aten, Apollo, and about half of the *Amor asteroids* have orbits that cross the orbit of the Earth (Fig. 18–22). The Aten asteroids, further, have orbits whose major axes are smaller than Earth's. (The perihelions of Amors are slightly farther from the Sun than those of Apollos.) We have observed a few dozen asteroids of these types. We see them only when they come close to Earth, and there may be 1000 in all. Two asteroids even go inside the orbit of Mercury. The first to be found was named Icarus, after the inventor in Greek mythology who flew too near the Sun. On June 14, 1968, Icarus came within 6 million km of the Earth. We may be able to send a spacecraft to such a closely approaching asteroid during the next decade. One day in 1989, we were surprised to learn that an asteroid the previous week had passed only 725,000 km from Earth, only about twice the distance to the Moon. The case demonstrates that we do not usually know when such objects are coming close to us. This latest object, now called 1989 FC, is about 1 km across. If it had hit the Earth, it would have made a 10-km crater on land or raised tidal waves hundreds of meters high if it had hit the ocean.

The asteroids at the inner edge of the asteroid belt are mostly stony in nature, and the ones at the outer edge are darker (as a result of being more

Figure 18–22 The orbits of (*A*) Apollo and (*B*) Amor asteroids, whose orbits approach or cross that of the Earth.

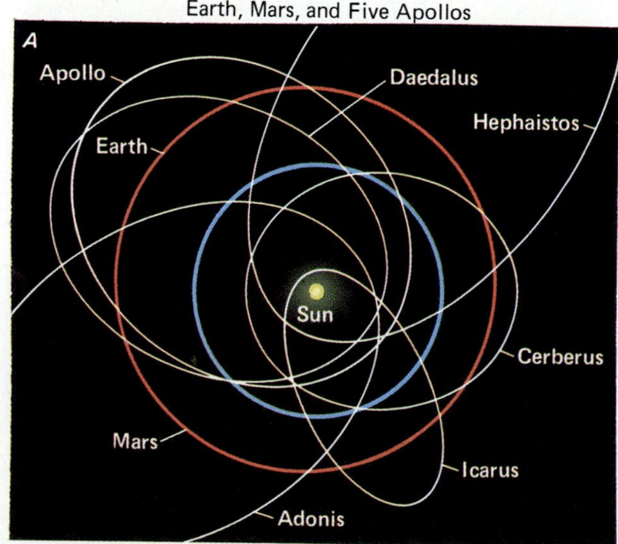

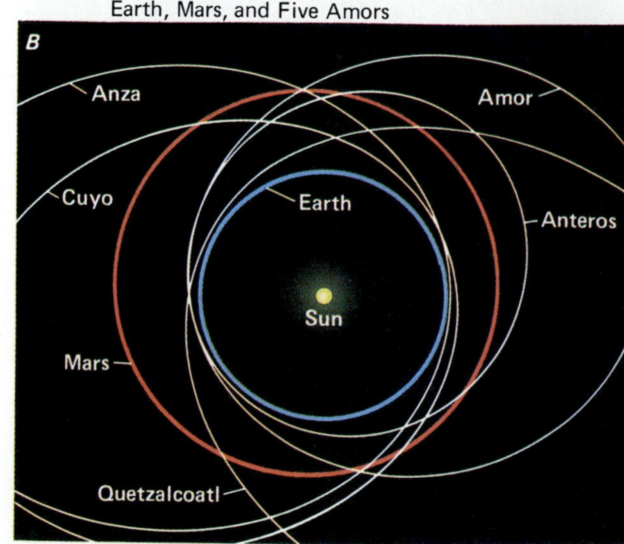

Focus On

Box 18.4 The Extinction of the Dinosaurs

On Earth 65 million years ago, dinosaurs and many other species were extinguished. Evidence has recently been accumulating that these extinctions were sudden and were caused by a 10-km-diameter asteroid hitting the Earth. Among the signs is the element iridium released from the asteroid in the impact (Fig. 18–23). The impact would have raised so much dust into the atmosphere (Fig. 18–24) that sunlight could have been shut out for months; many species of plants and animals would not have been able to survive. In particular, large animals like dinosaurs could not have taken refuge in caves the way the smaller ones like mammals may have. Further evidence in favor of this theory is a worldwide enhancement of soot in a layer of rock 65 million years old. The idea is very relevant to the theory that thermonuclear war could lead to a devastating "nuclear winter"—when sunlight would be prevented from hitting the Earth's surface by dust and smoke in the atmosphere following a nuclear exchange—or at least, as the latest computer models seem to show, a significant but less drastic "nuclear fall." Such calculations are also tied up with studies of the greenhouse effect. An impact could have released a lot of carbon dioxide into the atmosphere, raising the temperature by 20°C for years. Calculations show that chemical interactions are also disastrous. They lead to nitrogen-oxide smog, water poisoned with trace metals leached from soil and rock, and heavily corrosive acid rain.

As for the relation of an asteroid impact to the disappearance of the dinosaurs, evidence contrary to the above argument also exists. In particular, dinosaurs may have disappeared more gradually, which would mean that a catastrophic event was, therefore, not the cause.

A still newer idea is based on a possible periodicity of 28 million years between mass extinctions on Earth. It has been suggested that a faint, undiscovered companion to the Sun comes to the inner part of its orbit with that period, and its gravity then sends a number of comets in toward the Earth, where they crash. (Fortunately, the star—for which the name "Nemesis" is unofficially waiting—isn't due back for 15 million years even if it exists.) Obviously, a lot of verification both for the existence of periodic extinctions and for this theory would have to be found before it is generally accepted. Some scientists feel that it has already been disproved, especially since the data indicating that extinctions were periodic are very weak. Still, others think the Nemesis theory remains reasonable.

Some scientists continue to search for terrestrial explanations for the mass extinctions. One such theory involves instabilities that would arise deep inside the Earth, just above the core. Blobs of material would float upward, carrying the iridium that is common within the Earth as well as other materials. In the model, the material would then be ejected in volcanic eruptions. Research, and controversy, continues.

Figure 18–23 The iridium layer in rock, taken by Walter and Luis Alvarez as a sign of impact of an asteroid 65 million years ago.

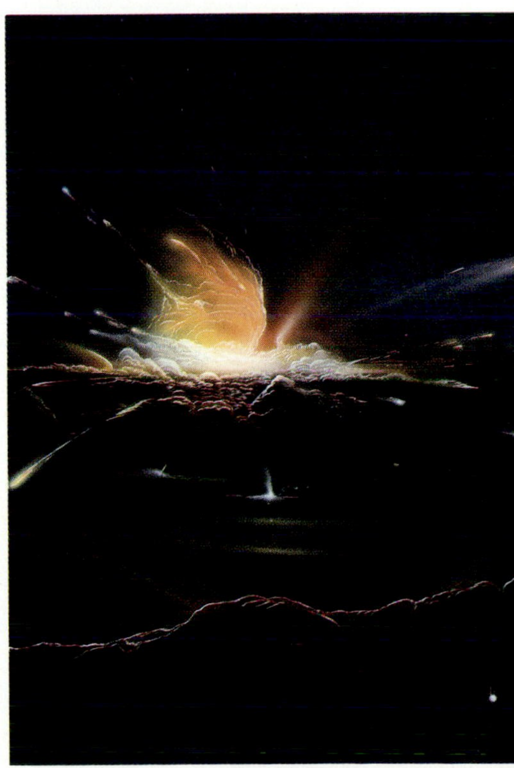

Figure 18–24 An artist's conception of the devastation resulting from the asteroid impact 65 million years ago that may have led to the extinction of many species, including the dinosaurs.

carbonaceous). Most of the small asteroids that pass near the Earth—Icarus, Eros, Toro, Geographos, and Alinda—are among the stony group. Three of the largest asteroids—Ceres, Hygiea, and Davida—are among the carbonaceous group. A third group may be mostly composed of iron and nickel. The differences may be a direct result of the variation of the solar nebula with distance from the protosun. Differences in chemical composition provide another reason not to believe the old theory that the asteroids represent the breakup of a planet that once existed between Mars and Jupiter. (There isn't enough mass present in asteroids anyway.)

Earth-crossing asteroids may well be the source of most meteorites, which could be the debris of collisions occurring when these asteroids visit the asteroid belt as they traverse their elliptical orbits. Eventually, most Earth-crossing asteroids will probably collide with the Earth. Luckily, we think that only a few dozen of them are greater than 1 km in diameter; statistics show that collisions of this tremendous magnitude should take place every few hundred million years and could have drastic consequences for life on Earth.

The new Palomar sky survey has already discovered hundreds of asteroids, including several near-Earth asteroids. A dozen such may be discovered each year as the survey continues.

The first Trojan asteroid to be discovered was named Achilles. Later it was decided to name the asteroids in the Lagrangian point ahead of Jupiter after Greek heroes and those in the point trailing Jupiter after Trojan heroes. But 617 Patroclus, a Greek, is in the Trojan camp. Similarly, there is one Trojan, 624 Hector, in the Greek camp.

Another asteroid type is the Trojans, which always oscillate about the points 60° ahead of or behind Jupiter in their solar orbits (the "Lagrangian points"). At the Lagrangian points, the gravitational pulls of the Sun and of Jupiter balance, so material does not move toward one or the other. Edward Bowell at the Lowell Observatory, Eugene and Carolyn Shoemaker using a small Schmidt telescope at the Palomar Observatory, and Linda French have more than tripled the number of Trojan asteroids known to 200.

18.2b Radar Observations

As ground-based radars become stronger, scientists have been able to reflect radar off a few asteroids. Some have proved to be irregular in shape, perhaps even in the form of a dumbbell (Fig. 18–25).

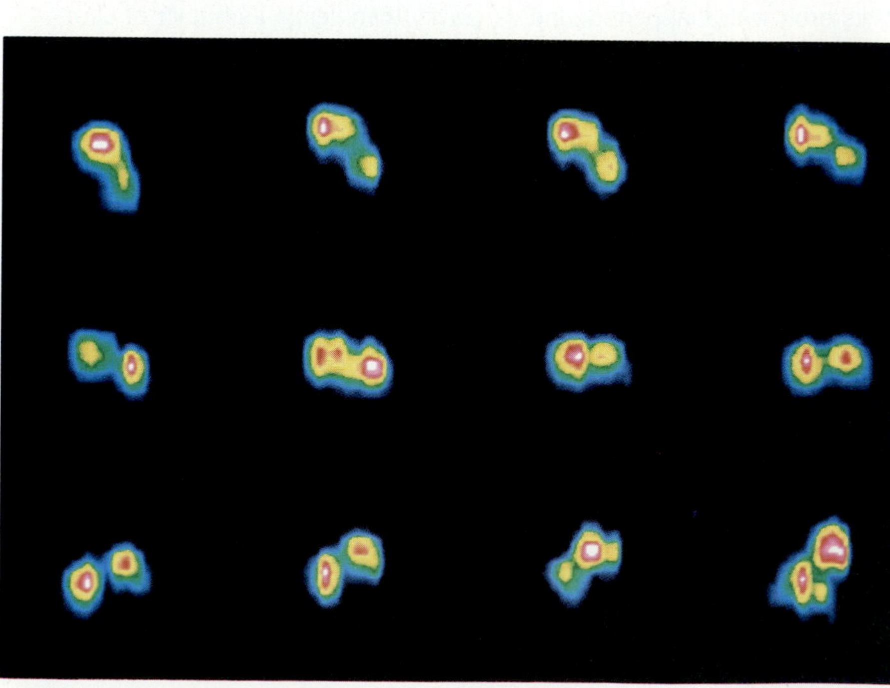

Figure 18–25 A series of radar images of asteroid 4769 Castalia, showing that it appears to be in the shape of a dumbbell.

18.3 Space Projects

The Galileo spacecraft en route to Jupiter passed close to asteroid Gaspra on October 29, 1991 (Fig. 18–26). The illuminated part is 16 × 12 km. It rotates every 7 hours. A Venera spacecraft may go close to 4 Vesta, and the Cassini spacecraft to Saturn is to fly by Maja in March 1997.

Figure 18–26 Gaspra, in a high-resolution image taken by the Galileo spacecraft, looks like a fragment of a larger body.

Summary and Outline

Meteoroids (Section 18.1)
- *Meteoroids:* in space; *meteors:* in the air; *meteorites:* on the ground
- Most meteorites that hit the Earth are chondrites, which resemble stones and are therefore often undetected; iron meteorites are more often found
- Meteorites may be chips of asteroids and thus give us material with which to investigate the origin of the solar system
- Some meteors arrive in showers, others are sporadic

Asteroids, also called minor planets (Section 18.2)
- Most in asteroid belt, 2.2 to 3.2 solar radii (between orbits of Mars and Jupiter)
- Apollo and Aten asteroids come within Earth's orbit
- Trojan asteroids in the Lagrangian points (Jupiter ± 60°)

Space Projects (Section 18.3)
- Galileo viewed Gaspra
- Venera to go to 4 Vesta
- Cassini to go to Maja

Key Words

meteoroids, meteors, shooting stars, fireballs, bolide, meteorite, micrometeorites, iron meteorites, irons, stones, chondrites, achondrites, stony-iron, tektites, carbonaceous*, carbonaceous chondrites*, sporadic, showers, radiant, minor planets, asteroids, asteroid belt, zodiacal light, Aten asteroids*, Apollo asteroids*, Amor asteroids*

*This term is found in an optional section.

Questions

1. What is the relation of meteorites and asteroids?
2. Why don't most meteoroids reach the Earth's surface to become meteorites?
3. Why do some meteor showers last only a day while others can last several weeks?
4. Why are meteorites important in our study of the solar system?
5. Compare asteroids with the moons of the planets.
6. How might the occultation of a star by an asteroid tell us whether the asteroid has an atmosphere?
7. Using Appendix 4, compare the sizes of the largest asteroids with the smallest planetary satellites.
8. What are three locations in the solar system from which meteorites come?

Observing Project

Observe the next meteor shower. No instruments will be necessary; the naked eye is best.

Yoda. © Lucasfilm, Ltd. (LFL) 1980. All rights reserved. From the motion picture *The Empire Strikes Back*, courtesy of Lucasfilm, Ltd.

19

Life in the Universe

Aims: To discuss the possibility that intelligent life exists elsewhere in the Universe besides the Earth and to assess our chances of communicating with it

We have discussed the nine planets and the dozens of moons in the solar system, and have found most of them to be places that seem hostile to terrestrial life forms. Yet some locations besides the Earth—Mars, with its signs of ancient running water, and perhaps Titan or Europa—have characteristics that allow us to convince ourselves that life may have existed there in the past or might even be present now or develop in the future. *Exobiology* is the study of life elsewhere than on Earth.

In our first real attempt to search for life on another planet, the Viking landers carried out biological and chemical experiments with martian soil. The results showed that there is probably no life on Mars.

Since it seems reasonable that life as we know it would be on planetary bodies, let us first discuss the chances of life arising elsewhere in our solar system. Then we will consider the chances that life has arisen in more distant parts of our galaxy or elsewhere in the Universe.

19.1 The Origin of Life

It would be very helpful if we could state a clear, concise definition of life, but unfortunately that is not possible. Biologists state several criteria that are ordinarily satisfied by life forms—reproduction, for example. Still, there exist forms on the fringes of life—viruses, for example, which need a host organism in order to reproduce—and scientists cannot always agree whether some of these things are "alive" or not.

In science fiction, authors sometimes conceive of beings that show such signs of life as the capability for intelligent thought (Fig. 19–1), even though the being may share few of the other criteria that we ordinarily recognize. In Fred Hoyle's novel *The Black Cloud,* for example, an interstellar cloud of gas and dust is as alive as—and smarter than—you or I. But we can make no concrete deductions if we allow such wild possibilities, and exobiologists prefer to limit the definition of life to forms that are more like "life as we know it."

Figure 19–1 "I'll tell you something else I think. I think there are other bowls somewhere out there with intelligent life just like ours." Drawing by Frank Modell; © 1987 The New Yorker Magazine, Inc.

This rationale implies, for example, that extraterrestrial life is based on complicated chains of molecules that involve carbon atoms. Life on Earth is governed by deoxyribonucleic acid (DNA) and ribonucleic acid (RNA), two carbon-containing molecules that control the mechanisms of heredity. Chemically, carbon is able to form "bonds" with several other atoms simultaneously, which makes these long carbon-bearing chains possible. In fact, we speak of compounds that contain carbon atoms as *organic.* In addition, a few scientists consider that silicon-based life could exist, since silicon is almost as abundant as carbon and can also form several simultaneous bonds (though its chemistry is less rich).

Figure 19–2 The simple solution of organic material in the oceans, from which life may have arisen, is informally known as "primordial soup."

How hard is it to build up long organic chains? To the surprise of many, an experiment performed in the 1950's showed that it was much easier than had been supposed to make organic molecules. At the University of Chicago, at the suggestion of Harold Urey, Stanley Miller put quantities of several simple molecules in a glass jar. Water vapor (H_2O), methane (CH_4), and ammonia (NH_3) were included along with hydrogen gas. Miller exposed the mixture to electric sparks, simulating the lightning that may exist in the early stages of the formation of a planetary atmosphere. After a few days, long chains of atoms had formed in the jar; these organic molecules were even complex enough to include simple amino acids, the building blocks of life. Modern versions of these experiments have created even more complex organic molecules from a wide variety of simple actions on simple molecules.

Since the original experiment of Urey and Miller, most scientists have come to think that the Earth's primitive atmosphere was not made of methane and ammonia, which would have disappeared soon after the Earth's formation. Rather, the gases in today's oceans and atmosphere were released from the Earth's interior by volcanoes. When life formed, the atmosphere might have consisted of carbon dioxide, water, and carbon monoxide, but no methane and ammonia (Fig. 19–2). The atmosphere, in short, lacked the oxygen of our present-day atmosphere. Cyril Ponnamperuma and colleagues at the University of Maryland and Carl Sagan and Bishun Khare at Cornell University are working in their laboratories on the problem of the formation of complicated molecules (Fig. 19–3). Ponnamperuma and colleagues have synthesized all five of the chemical bases of human genes in a single experiment that simulated the Earth's primitive atmosphere. They have also detected all five in the Murchison meteorite. Scientists have also found extraterrestrial amino acids in two meteorites that had been long frozen in Antarctic ice.

However, mere amino acids or even DNA molecules are not life itself. A jar containing a mixture of all the atoms that are in a human being is not the same as a human being. This is a vital gap in the chain; astronomers certainly are not qualified to say what supplies the "spark" of life.

Sidney Altman of Yale and Thomas Cech of the University of Colorado won the 1989 Nobel Prize in Chemistry for studies that indicate that RNA was a key molecule at the origin of life. They discovered that RNA can act as an enzyme, a class of molecules that accelerate chemical reactions. (For example, an enzyme in human saliva helps change starch into the glucose sugar molecule.) RNA accelerates reactions by a million times and makes it possible for life to exist. Altman and Cech discovered that not only could RNA rather than only proteins be enzymes but also that RNA can copy itself. Thus though we depend on DNA to carry the genetic code, early life may have used RNA.

In any case, many astronomers think that since it is not difficult to form complex molecules, life may well have arisen not only on the Earth but also in other locations. Even if life is not found in our solar system, there are so

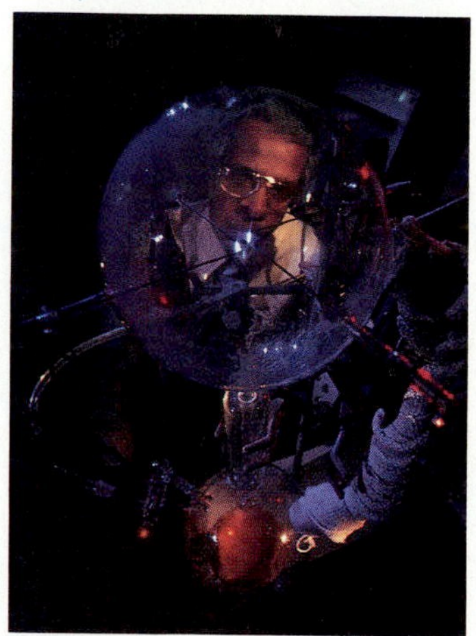

A

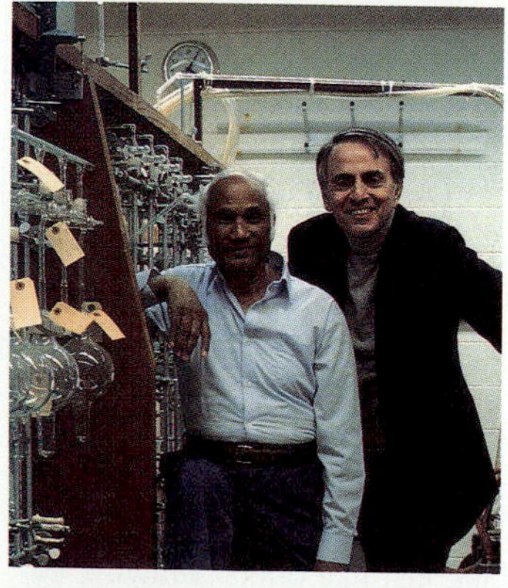

B

Figure 19–3 (*A*) Cyril Ponnamperuma is examining a "spark discharge" apparatus, used to simulate conditions on the primitive Earth, or, at times, the outer planets of our solar system. The lower chamber is charged with water, and gases of the desired composition (methane, ammonia, nitrogen, etc.) are introduced into the upper chamber. Electric coils produce a spark discharge in the upper chamber, simulating lightning. A convection current simulates atmospheric flow. After several days, the materials that have been synthesized are recovered. These materials are commonly amino acids, sometimes sugars and sometimes other complicated materials. Most recently, the laboratory has found evidence of nucleic acid bases in the products. (*B*) Carl Sagan and Bishun Khare of Cornell University in their laboratory. (*A*: © 1976 National Geographic Society/Bruce Dale; *B*: Andrew Clegg/Cornell U.)

many other stars in space that it would seem that some of them could have planets around them and that life could have arisen there independently of life on Earth.

Figure 19–4 The Thaw refractor at the Allegheny Observatory. The Hyades is the V-shaped open cluster overhead. The Thaw refractor is an astrometric telescope, used especially to take series of plates over many years. The plates are used to measure proper motions and parallaxes (Chapter 21).

19.2 Other Solar Systems?

We have seen how difficult it was to detect even the outermost planets in our own solar system. At the four-light-year distance of Proxima Centauri, the nearest star to the Sun, any planets around stars would be too faint for us to observe directly. There is hope, though, that the Hubble Space Telescope will be able to see very large ones.

We now have a better chance of detecting such a planet by observing its gravitational effect on the star itself. One such method is the same that has been used to discover white dwarfs. We study the proper motion of a star—its apparent motion across the sky with respect to the other stars (as we will discuss in Section 21.7)—and see if the star follows a straight line or appears to wobble. Any wobble would have to result from an object too faint to see that is orbiting the visible star.

The easiest stars to observe in a search for planets are the stars with the largest proper motion, for these are likely to be the closest to us and thus any wobble would cover the greatest angular distance that we could hope for. The star with the largest proper motion, Barnard's star, is only 6 light years away from the Sun. It is the fourth nearest star to the Sun; only the members of the alpha Centauri triple system are nearer. Because of its closeness, we can readily detect its apparent shift against the sky background when we view it from different parts of the Earth's orbit over the year (Fig. 19–4).

Barnard's star and a few other nearby stars appear to show slight wobbles in their *proper* motions. Even if the wobbles are confirmed as real, it is unclear whether they should be explained by large planets or mini-suns.

Spectrographic surveys have endorsed the idea that many stars have planets. Varying radial velocity corresponds to a star's going back and forth as a planet swings around it, in order to keep the gravitational center of the system moving straight. Some of the invisible companions have masses that may be too small to allow them to be stars. Some of these may be planets rather than brown dwarfs—objects just a little too small to be undergoing nuclear burning; just how many, nobody can say. Methods of measuring the radial velocities of stars with greatly improved accuracy have been developed and the first few years of observations made; only time will tell.

Figure 19–5 This picture of what may be another solar system in formation was taken with a ground-based telescope. A circular "occulting disk" blocked out the star β (beta) Pictoris; shielding its brightness revealed the material that surrounds it. About 200 times as much mass as the Earth is present in the form of solid material extending 60 billion km from the star. The image is taken through an infrared filter. β Pictoris is too far south in the sky to be seen from mid-northern latitudes.

Evidence for the existence of brown dwarfs has come and gone over the last few years. Perhaps the most persuasive evidence for this class of object is the detection by Ben Zuckerman and Eric Becklin of UCLA of an excess of infrared radiation in the spectrum of a white dwarf. The presence of a Jupiter-like brown dwarf, with a temperature of 1200 K and a radius of 0.15 the Sun's radius ($1\frac{1}{2}$ times Jupiter's radius), would explain the observations. There is no obvious source of dust to provide matter for the ring of dust that could be an alternative explanation of the radiation. And the presence of a true planet seems unlikely because we might detect a variation in its infrared brightness between its day and night sides.

The indications from IRAS that solar systems may be forming around stars, as deduced from the excess infrared measured, are not as satisfactory as a direct sighting of a planet or a detection of its gravitational effect. The ground-based observations of a cloud of material orbiting a star are more persuasive (Fig. 19–5). The material present apparently consists of ices, carbon-rich substances, and silicates. Our Earth formed from such materials, so planets may be forming there. The density of the material indicates that planets may

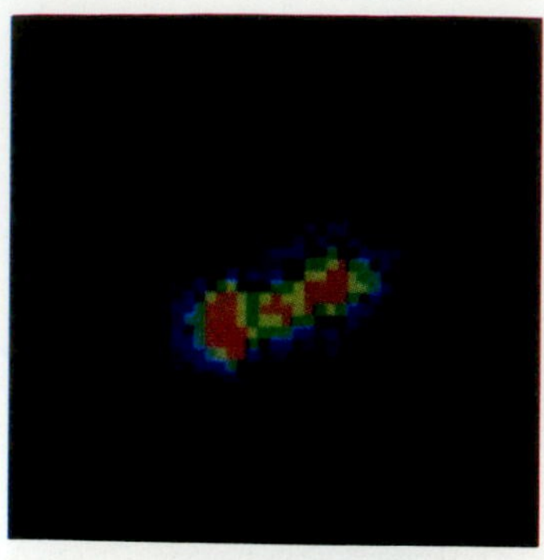

Figure 19–6 A cloud of warm dust and gas that may be the remains of a dusty, disk-shaped cloud from which a massive star has recently formed. The dust and gas are heated by the star that has been born inside it. The cloud, imaged with an infrared array, is in the region of our galaxy known as W51. Our own Sun and solar system may have originated in a similar way.

Michael Jura of UCLA, investigating another bright infrared source, found a thick cloud of dust around a star 250 light years from us in the constellation Centaurus. Since the star is on the main sequence, enough time has passed since its formation for planets to form.

indeed be present. Little material is present in the inside portion of the disk; it may have been swept away by the planets. A U. Mass.–Smith–U. Heidelberg group has found disks around 20 young stars in the Taurus molecular cloud. The disks block our view of the part of their stars' solar wind that is heading away from us. These results surely encourage the point of view that solar systems are common. The infrared observations of the stars Fomalhaut, Vega, and beta Pictoris by Kitt Peak and NASA scientists are best fit by a model in which the dust clouds around them have central regions in which dust is depleted. Large, perhaps Earth-sized, objects could be sweeping the dust away. Infrared images reveal warm clouds of dust and gas that may be surrounding newly formed stars (Fig. 19–6).

19.3 The Statistics of Extraterrestrial Life

Instead of phrasing one all-or-nothing question about life in the Universe, we can break down the problem into a chain of simpler questions. This procedure was developed by Frank Drake, a Cornell astronomer now at the University of California at Santa Cruz, and extended by, among others, Carl Sagan of Cornell and Joseph Shklovskii, a Soviet astronomer.

19.3a The Probability of Finding Life

First we consider the probability that stars at the centers of solar systems are suitable to allow intelligent life to evolve. For example, the most massive stars evolve relatively quickly, and probably stay in the stable state that characterizes the bulk of their lifetimes for too short a time to allow intelligent life to evolve.

Second, we ask what the chances are that a suitable star has planets. With the proper-motion studies (Barnard's star, etc.) no longer seeming to give clearly positive results, we have no accurate estimate of the probability, but most scientists think that the chances are probably pretty high.

Third, we need planets with suitable conditions for the origin of life. A planet like Jupiter might be ruled out, for example, because it lacks a solid surface and because its surface gravity is high. (Alternatively, though, one could consider a liquid region, if it were at a suitable temperature, to be as advantageous as were the oceans on Earth to the development of life here.) Also, planets probably must be in orbits in which the temperature does not fluctuate too much. If the planets are in multiple-star systems, as most stars seem to be, such orbits exist. The orbits are either far from a central pair of stars or, if the stars are far apart, close to one of the stars. For single stars, new calculations by NASA/Ames scientists have broadened the range of distances from the Sun over which planets are sufficiently warm for life to begin. Their more sophisticated treatment of the carbon-dioxide balance and the greenhouse effect indicates that the habitable zone would extend from 0.95 A.U. to 1.5 A.U. Their model may explain the "Goldilocks effect": why Venus is too hot, Mars is too cold, but the Earth is just right.

Fourth, we have to consider the fraction of the suitable planets on which life actually begins. This is the biggest uncertainty, for if this fraction is zero (with the Earth being a unique exception), then we get nowhere with this entire line of reasoning. Still, the discovery that amino acids can be formed in laboratory simulations of primitive atmospheres, and the discovery of complex molecules in interstellar space, indicate to many astronomers that it is not as difficult as had been thought to form complicated molecules. Amino acids, much less complicated than DNA but also basic to life as we know it, have even been found in meteorites. The discovery from radio observations that molecules are associated with sites of star formation strengthens the idea that these molecules may eventually become part of planets that form. Many astronomers choose to think that the fraction of planetary systems on which life begins may be high.

Analysis of an odd kind of single-celled organism called *archaebacteria* indicates that life may have formed very early in the Earth's history. Some of these living creatures—found in airless places like Yellowstone's hot springs—take in carbon dioxide and hydrogen and give off methane. Their RNA (genetic material) is different from that of other bacteria or of plants or animals. The discovery supports the idea that life evolved before oxygen appeared in the Earth's primeval atmosphere. These and some other living bacteria do not survive in the presence of oxygen. Further, scientists have recently discovered organisms that thrive on sulfur from geothermal sources instead of on solar energy. Some are bacteria that metabolize hydrogen sulfide to reproduce and grow, while others are animals that eat the bacteria. So environments on other planets may not be as hostile to life as we had thought. Even in the most apparently hostile places on Earth, we have now found living things (Fig. 19–7).

Figure 19–7 We see the inside of an Antarctic rock, with a lichen growing safely insulated from the external cold.

If we want to have meaningful conversations with aliens, we must have a situation where not just life but intelligent life has evolved. We cannot converse with algae or paramecia. Furthermore, the life must have developed a technological civilization capable of interstellar communication. These considerations reduce the probabilities somewhat, but it has still been calculated that there may be technologically advanced civilizations within a few hundred light years of the Sun.

Now one comes to the important question of the lifetime of the technological civilization itself. We now have the capability of destroying our civilization either dramatically in a flurry of hydrogen bombs or more slowly by, for example, altering our climate or increasing the level of atmospheric pollution. It is a sobering question to ask whether the lifetime of a technological civilization is measured in decades, or whether all the problems that we have—political, environmental, and otherwise—can be overcome, leaving our civilization to last for millions or billions of years.

19.3b Evaluating the Probability of Life

At this point we can try to estimate (to guess, really) fractions for each of these simpler questions within our chain of reasoning. We can then multiply these fractions together to get an answer to the larger question of the probability of extraterrestrial life. Reasonable assumptions lead to the conclusion that there may be millions or billions of planets in this galaxy on which life may have evolved. The nearest one may be within dozens of light years of the Sun. On this basis, many if not most professional astronomers have come to feel that intelligent life probably exists in many places in the Universe. Carl Sagan's latest estimate is that a million stars in the Milky Way Galaxy may be supporting technological civilizations.

Though this point of view has been gaining increasing favor over the last decade, a reaction has started. Evaluating Drake's equation (Box 19.1) with a more pessimistic set of numbers can lead to the conclusion that we earthlings are alone in our galaxy.

One reason for doubting that the Universe is teeming with life is that extraterrestrials have not established contact with us. Where are they all? To complicate things further, is it necessarily true that if intelligent life evolved, it would choose to explore space or to send out messages?

A

B

Figure 19–8 (*A*) The goldplated copper record bearing two hours' worth of Earth sounds and a coded form of photographs carried by Voyagers 1 and 2. The sounds include a car, a steamboat, a train, a rainstorm, a rocket blastoff, a baby crying, animals in the jungle, and greetings in various languages. Musical selections include Bach, Beethoven, rock, jazz, and folk music. (*B*) The record includes 116 photographs. One of them is this view, which I took when in Australia for an eclipse. It shows Heron Island on the Great Barrier Reef in Australia, in order to illustrate an island, an ocean, waves, beach, and signs of life.

19.4 Interstellar Communication

What are the chances of our visiting or being visited by representatives of these civilizations? Pioneers 10 and 11 and Voyagers 1 and 2 are even now carrying messages out of the solar system in case an alien interstellar traveller should happen to encounter these spacecraft (Fig. 19–8). But it seems unlikely that humans themselves can travel the great distances to the stars, unless someday we develop spaceships to carry whole families and cities on indefinitely long voyages into space.

Yet there is plenty of time in the future for interstellar space travel to develop. The Sun has another 5 billion years to go in its current stable state, while recorded history on Earth has existed for only about 5000 years, one-millionth of the age of the solar system.

We can hope even now to communicate over interstellar distances by means of radio signals. We have known the basic principles of radio for only a hundred years, and powerful radio, television, and radar transmitters have existed for less than 70 years. Even now, though, we are sending out signals into space on the normal broadcast channels. A wave bearing the voice of Caruso is expanding into space, and at present is almost 70 light years from Earth. And once a week a new episode of "The Cosby Show" is carried into the depths of the Universe. Military radars are even stronger, though the waves travel no faster.

From another star, the Sun would appear to be a variable radio source, because the Earth and Sun would be unresolved and the Earth's radio signals would therefore appear to be coming from the Sun. The signal would vary with a period of 24 hours, since it would peak each day as specific concentrations of transmitters rotated to the side of the Earth facing our listener.

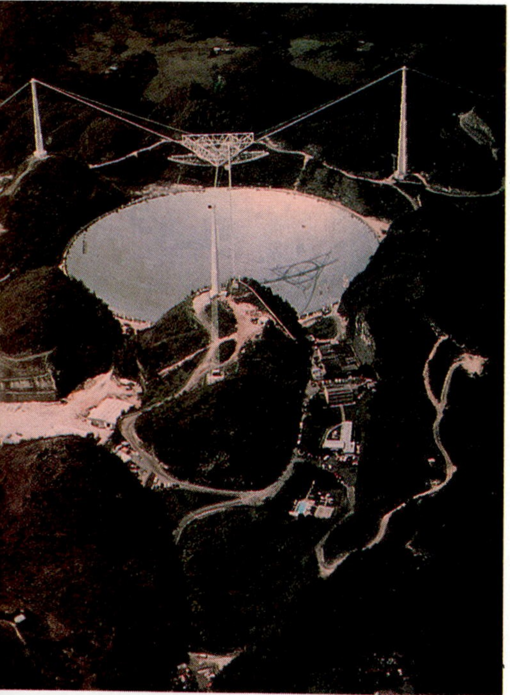

Figure 19–9 The Arecibo telescope in Puerto Rico, used to send a message into space. The telescope is 305 meters across, the largest telescope on Earth.

Focus On

Box 19.1 The Probability of Life in the Universe

Our discussion follows the lines of an equation written out to estimate the number of civilizations in our galaxy that would be able to contact each other, designated by the letter *N*. In 1961, Frank Drake of Cornell wrote

$$N = R_* f_p n_e f_\ell f_i f_c L,$$

where R_* is the rate at which stars form in our galaxy, f_p is the fraction of these stars that have planets, n_e is the number of planets per solar system that are suitable for life to survive (for example, those that have Earth-like atmospheres), f_ℓ is the fraction of these planets on which life actually arises, f_i is the fraction of these life forms that develop intelligence, f_c is the fraction of the intelligent species that choose to communicate with other civilizations and develop adequate technology, and L is the lifetime of such a civilization. Now that we have seen that Jupiter's, Saturn's, and Uranus's moons are worlds with personalities of their own, perhaps we should replace f_p with a new factor f_{mp} (moons and planets) that takes account of the suitable moons in a solar system.

One of the largest uncertainties is f_ℓ, which could be essentially zero or could be close to 1. The result is also very sensitive to the value one chooses for L—is it 1 century or a billion years? Do civilizations destroy themselves? Or perhaps they just go off the airwaves. Depending on the alternatives one chooses, one can predict that there are dozens of communicating civilizations within 100 light years or that the Earth is unique in the galaxy in having one.

In 1974, the giant radio telescope at Arecibo, Puerto Rico (Fig. 19–9), run by Cornell University, was upgraded. At the rededication ceremony, a powerful signal was sent out into space at the fundamental radio frequency of hydrogen radiation, bearing a message from the people on Earth (Fig. 19–10). This is the only major signal that we have purposely sent out as our contribution to the possible interstellar dialogue.

The signal from Arecibo was directed at the globular cluster M13 in the constellation Hercules, on the theory that the presence of 190,000 closely packed stars in that location would increase the chances of our signal being received by a civilization on one of them. But the travel time of the message (at the speed of light) is 24,000 years to M13, so we certainly could not expect to have an answer before twice 24,000, or 48,000 years, have passed. If anybody (or any**thing**) is observing our Sun when the signal arrives, the radio brightness of the Sun will increase by 10 million times for a 3-minute period. A similar signal, if received from a distant star, could be the giveaway that there is intelligent life there.

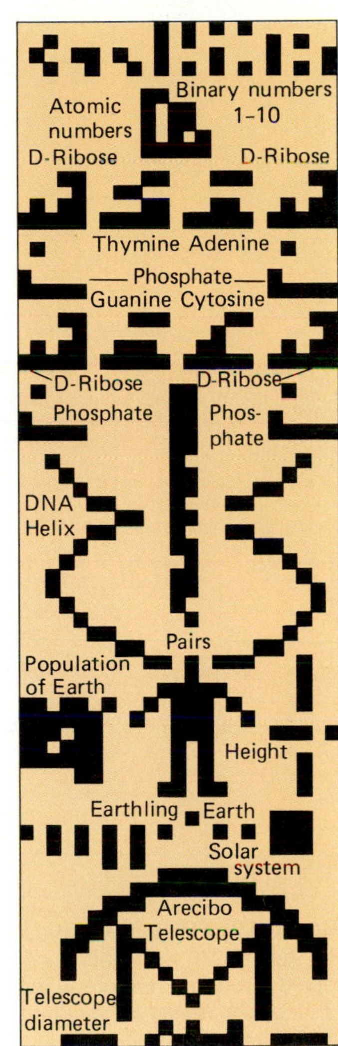

A

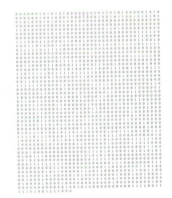

B

Figure 19–10 (*A*) The message sent to M13, plotted out and with a translation into English added. The basic binary-system count at upper right is provided with a position-marking square below each number. (*B*) The message was sent as a string of 1679 consecutive characters, in 73 groups of 23 characters each (two prime numbers). There were two kinds of characters, each represented by a frequency; the two kinds of characters are reproduced here as 0's and 1's.

We mustn't be too quick to close out other possibilities. For example, neutrinos travel at or close to the speed of light. Just because we find neutrinos hard to detect doesn't prove that other civilizations haven't chosen them to carry messages.

19.5 The Search for Life

For planets in our solar system, we can search for life by direct exploration (Fig. 19–11), as we did with Viking on Mars, but electromagnetic waves, which travel at the speed of light, seem a much more sensible way to search for extraterrestrial life outside our solar system. How would you go about trying to find out if there were life on a distant planet? If we could observe the presence of abundant oxygen molecules or discover from the spectrum of a distant star that the molecules there are out of their normal balance, that would tell us that life exists there. The Hubble Space Telescope, for example, should be able to detect the infrared band of radiation from molecular oxygen on a planet orbiting alpha Centauri.

The most promising way to detect the presence of life at great distances appears to many scientists to be a search in the radio part of the spectrum. But it would be too overwhelming a task to listen for signals at all frequencies in all directions at all times. One must make some reasonable guesses on how to proceed.

A few frequencies in the radio spectrum seem especially fundamental, such as the spin-flip line of neutral hydrogen—the simplest element—at 21 cm. This corresponds to 1420 MHz, a frequency over ten times higher than stations at the high end of the normal FM band. We might conclude that creatures on a far-off planet would decide that we would be most likely to listen near this frequency because it is so fundamental.

On the other hand, perhaps the great abundance of radiation from hydrogen itself would clog this frequency, and one or more of the other frequencies that correspond to strong natural radiation should be preferred. The "water hole," the wavelength range between the radio lines of H and OH, has a minimum of radio noise from background celestial sources, the telescope's receiver, and the Earth's atmosphere, and so is another favored possibility. If

Figure 19–11 © Lucasfilm, Ltd. (LFL) 1980. All rights reserved. From the motion picture *The Empire Strikes Back*, courtesy of Lucasfilm, Ltd.

we ever detect radiation at any frequency, we would immediately start to wonder why we had not realized that this frequency was the obvious choice.

In 1960, Frank Drake used a telescope at the National Radio Astronomy Observatory for a few months to listen for signals from two of the nearest stars—tau Ceti and epsilon Eridani. He was searching for any abnormal kind of signal, a sharp burst of energy, for example. He called this investigation Project Ozma after the queen of the land of Oz (Fig. 19–12) in L. Frank Baum's later stories.

Figure 19–12 Dorothy and Ozma climb the magic stairway. From *Glinda of Oz*, by L. Frank Baum, illustrated by Roy Neill, copyright 1920.

The observations were later extended in a more methodical search called Ozma II. Starting in 1973, Ben Zuckerman of the University of Maryland (now at UCLA) and Patrick Palmer of the University of Chicago systematically monitored over 600 of the nearest stars of supposedly suitable spectral type. They used computers to search the data they recorded for any sign of special signals; Ozma II obtained data at 10 million times the rate of the original Ozma. Over the course of several days, they studied each star for a total of about $\frac{1}{2}$ hour. Nothing turned up.

Over a dozen separate searches have been undertaken in the radio spectrum, some concentrating on all-sky coverage and others on coverage of individual stars over time. Paul Horowitz of Harvard tested nearly 12 million combinations of stars and radio frequencies at Arecibo. Radio telescopes at Berkeley's Hat Creek Observatory, Ohio State, and the Algonquin Observatory in Ontario have also been used. Drake and Sagan of Cornell searched five different galaxies with the Arecibo telescope. Their equipment wasn't very sensitive to any individual star's radiation, but had a hundred million stars in its beam at the same time. The telescope would have picked up a civilization if there was even a single one in those galaxies. Even a search for strong ultraviolet radiation from a laser-like source has been made from a telescope in space.

Many astronomers feel that carrying out these brief searches was worthwhile, since we had no idea of what we would find. But now many feel that the early promise has not been justified, and the chance that there is extraterrestrial life seems much lower.

More and more astronomers in both the United States and the Soviet Union are interested in "**c**ommunication with **e**xtra**t**errestrial **i**ntelligence" (CETI) that might follow the detection of a signal in the "**s**earch for **e**xtra**t**errestrial **i**ntelligence" (SETI). The programs include scanning a wide range of frequencies in all directions in the sky and listening for a long time in certain promising directions for suitable strange signals. Paul Horowitz of Harvard, with support from the Planetary Society, has developed a relatively inexpensive way of scanning 8 million narrow-frequency bands simultaneously. He has set up his Project META (Megachannel ExtraTerrestrial Assay) on a 26-m Harvard radio telescope for long-term scanning of certain "magic" frequencies that seem most likely (21 cm, OH, $\frac{1}{2} \times 21$ cm, etc.), covering a wide range of sky. Each channel is ultranarrow in frequency, but the large number of channels allows for some observational uncertainty (perhaps from the Doppler effect of the source's motion) in the "magic" arrival frequency of the signal in the lab. They are doing a megaOzma per minute. After about two years of searching, no clear extraterrestrial signals have been distinguished (Fig. 19–13).

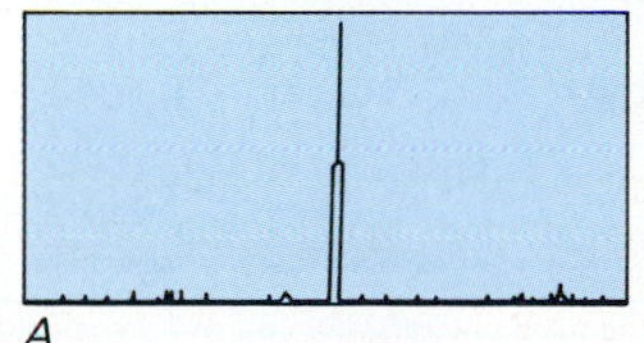

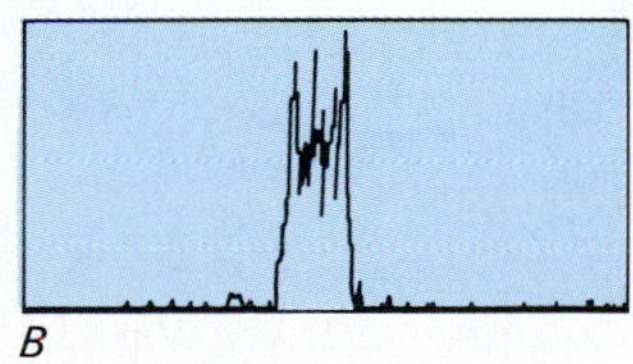

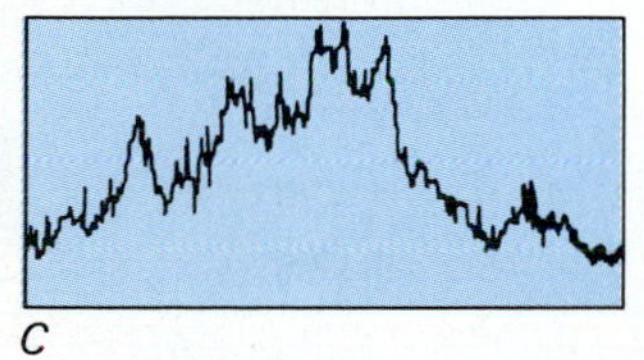

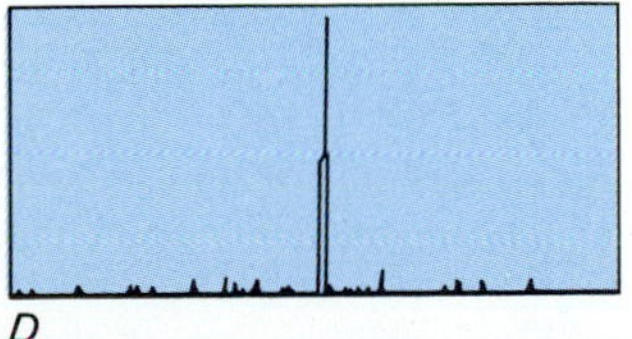

A B C D

Figure 19–13 How the computer screen of Project META would look in various cases, magnifying the 256 channels around the strongest signal out of the 8.4 million channels measured. (*A*) A signal at a fixed, narrow frequency, such as that expected from extraterrestrial signalling. (*B*) Real data recorded in August 1986, spread out like stable terrestrial interference. A signal like this appears about once a week. (*C*) Real data recorded in November 1986, spread out like less stable terrestrial interference. (*D*) Real data received at 17:57 GMT on October 10, 1986. It looks like an extraterrestrial signal would, but it has never been repeated in spite of reobservation of the same position in the sky.

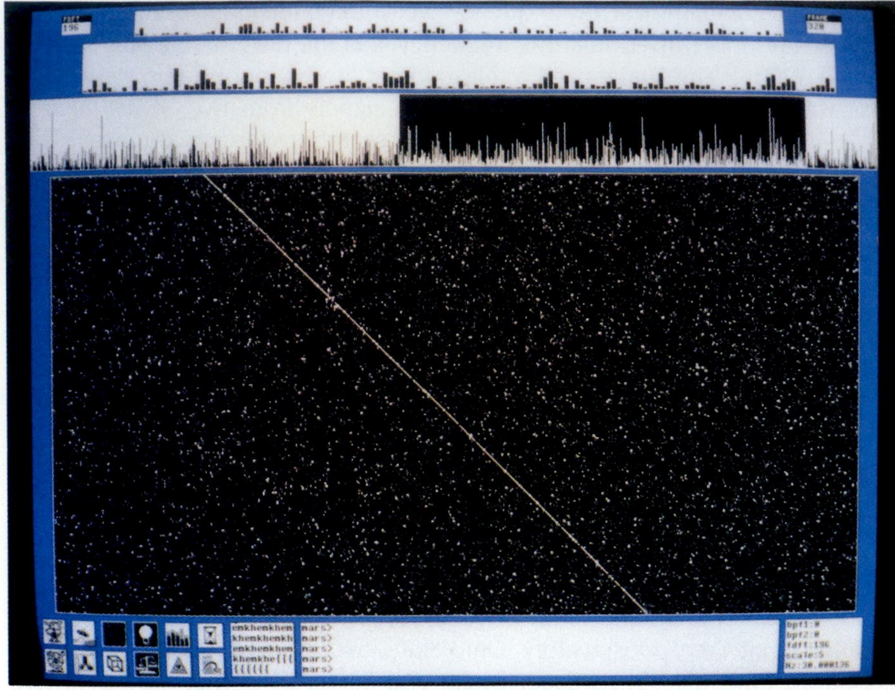

Figure 19–14 A computer screen at NASA's SETI prototype, showing the time history of the signal. The x-axis shows the strength of the radio signal in a subset of the narrowest channels while the y-axis shows time. Thus a narrowband signal that slowly changes frequency shows clearly. Computers would look each second for such nonrandom patterns in the output of the 15-million channels surveyed.

At NASA's Ames Research Center, JPL, and Stanford, Bernard Oliver, Jill Tarter, Ivan Linscott, Allen Peterson, and other scientists are developing a scanner for first 74,000 and later 10 million channels to be used on existing large antennas around the world. They are using sophisticated computer analysis to analyze the signals at 1-sec intervals, systematically covering all frequencies from 1 to 10 GHz, including the water hole. They are starting with 773 nearby solar-type stars within 80 light years and a less sensitive survey of the entire sky. In both the Harvard and the planned NASA searches, very-narrow-frequency signals (Fig. 19–14) could be signs of life, since naturally arising signals cover wider ranges.

Figure 19–15 An artist's conception of a set of radio telescopes on the far side of the Moon, shielded from the Earth's radio signals.

Our improved abilities to observe in the infrared might pick up evidence of advanced civilizations in another way. Freeman Dyson, at the Institute for Advanced Study in Princeton, has suggested that a civilization might redistribute some of its planet's mass around its star to soak up stellar energy that would otherwise escape. IRAS picked up the infrared glow caused by material that blocks even 1 per cent of the sunlight of a star like the Sun at a distance of 190 light years. This distance would include 100,000 stars, so we will see what further data reduction turns up.

The list of searches has grown too long to include completely. The next years, with radio searches under way and the Hubble Space Telescope aloft, will bring us much more thorough studies and precise results than we have had in the quest for planets and life. Someday we may be scanning from radio telescopes on the far side of the Moon (Fig. 19–15), shielded by the Moon's bulk from radio interference from Earth.

19.6 UFO's and the Scientific Method

But why, you may ask, if most astronomers accept the probability that life exists elsewhere in the Universe, do they not accept the idea that unidentified flying objects (UFO's) represent visitation from these other civilizations (Fig. 19–16)? The answer to this question leads us not only to explore the nature of UFO's but also to consider the nature of knowledge and truth. The discussion that follows is a personal view, but one that is shared by many scientists.

The most common UFO is a light that appears in the sky that seems unlike anything manmade or any commonly understood natural phenomenon. UFO's may appear as a point or extended, or may seem to vary. But the observations are usually anecdotal, are not controlled as in a scientific experiment, and are not accessible to sophisticated instrumentation.

19.6a UFO's

First, most of the sightings of UFO's that are reported can be explained in terms of natural phenomena. Astronomers are experts on strange effects that the Earth's atmosphere can display, and many UFO's can be explained by such effects. When Venus shines brightly near the horizon, for example, in my capacity as local astronomer I sometimes get telephone calls from people asking me about the UFO. It is not well known that a planet or star low on the horizon can seem to flash red and green because of atmospheric refraction. Atmospheric effects can affect radar waves as well as visible light.

Sometimes other natural phenomena—flocks of birds, for example—are reported as UFO's. One should not accept explanations that UFO's are flying saucers from other planets before more mundane explanations—including hoaxes, exaggeration, and fraud—are exhausted.

For many of the effects that have been reported, the UFO's would have been defying well-established laws of physics. Where are the sonic booms, for example, from rapidly moving UFO's? Scientists treat challenges to laws of physics very seriously, since our science and technology are based on these laws.

Most professional astronomers feel that UFO's can be so completely explained by natural phenomena that they are not worthy of more of our time. Furthermore, some individuals may ask why we reject the identification of UFO's with flying saucers, when—they may say—that explanation is "just as good an explanation as any other." Let us go on to discover what scientists mean by "truth" and how that applies to the above question.

"Yeeeeeeeeeeeha!"

19.6b Of Truth and Theories

At every instant, we can explain what is happening in a variety of ways. When we flip a light switch, for example, we assume that the switch closes an electric circuit in the wall and allows the electricity to flow. But it is certainly possible, although not very likely, that the switch activates a relay that turns on a radio that broadcasts a message to an alien on Mars. The Martian then might send back a telepathic message to the electricity to flow, making the light go on. The latter explanation sounds so unlikely that we don't seriously consider it. We would even call the former explanation "true," without qualification.

Occam's Razor, sometimes called the Principle of Simplicity, is a razor in the sense that it is a cutting edge that allows a distinction to be made among theories.

We regard as "true" the simplest explanation that satisfies all the data we have about any given thing. This principle is known as *Occam's Razor;* it is named after a 14th-century British philosopher who originally proposed it. Without this rule, we would always be subject to such complicated doubts that we would accept nothing as known.

Science is based on Occam's Razor, though we don't usually bother to think about it. Sometimes, something we call "true" might be more accurately described as a theory. The scientific method is based on hypotheses and theories. A *hypothesis* (plural: hypotheses) is an explanation that is advanced to explain certain facts. When it is shown that the hypothesis actually explains most or all of the facts known, then we may call it a *theory.* We usually test a theory by seeing whether it can predict things that were not previously observed, and then by trying to confirm whether the predictions are valid.

An example of a theory is the Newtonian theory of gravitation, which for many years explained almost all the planetary motions. Only a small discrepancy in the orbit of Mercury, as we describe in Section 23.11, remained unexplained. In 1916, Albert Einstein presented a general theory of relativity as a better explanation of gravitation. The theory explained the discrepancy in Mercury's orbit. It also made certain predictions about the positions of stars during total solar eclipses. When his predictions were verified, his theory was widely accepted.

Is Newton's theory "true"? Yes, in most regions of space. Is Einstein's theory "true"? We say so, although we may also think that one day a new theory will come along that is more general than Einstein's in the same way that Einstein's is more general than Newton's.

How does this view of truth tie in with the previous discussion of UFO's? Scientists have assessed the probability of UFO's being flying saucers from other worlds, and most have decided that the probability is so low that the possibility is not even worth considering. We have better things to do with our time and with our national resources. We have so many other, simpler explanations of the phenomena that are reported as UFO's that when we apply Occam's Razor, we call the identifications of UFO's with extraterrestrial visitation "false." UFO's may be unidentified, but they are probably not flying, nor for the most part are they objects.

Summary and Outline

The origin of life (Section 19.1)

- Earth life based on complex chains of carbon-bearing (organic) molecules; carbon (and silicon) form such chains easily; organic molecules are easily formed under laboratory conditions that simulate primitive atmospheres

Evidence for other solar systems (Section 19.2)

- Search for wobbles in proper motions or radial velocities
- Excess infrared radiation may reveal a planet.

Statistical chances for extraterrestrial life (Section 19.3)

- Problem broken down into stages; major uncertainties include what the chance is that life will form given the component parts, and what the lifetime of a civilization is likely to be

Signals we are sending from Earth (Section 19.4)

Leakage of radio, television, and radar; signal beamed from Arecibo toward globular cluster M13

The search for life (Section 19.5)

Experiments on Viking; Projects Ozma, Ozma II, META

UFO's and the scientific method (Section 19.6)

Definition of "truth"; Occam's Razor

Key Words

exobiology, organic, archaebacteria, Occam's Razor, hypothesis, theory

Questions

1. What is the significance of the Miller-Urey experiments?

2. How have the Miller-Urey experiments been improved upon?

3. Does a star with a large proper motion necessarily have planets? Explain.

4. Discuss the evidence for and against the idea that Barnard's star has planets around it.

5. Describe two ways that the Hubble Space Telescope will aid in the search for planets around distant stars.

6. What is one advantage of the radial-velocity method over the proper-motion method for detecting distant planets?

†7. If 1/10 of all stars are of suitable type for life to develop, 1 per cent of all stars have planets, and 10 per cent of planetary systems have a planet at a suitable distance from the star, what fraction of stars have a planet suitable for life? How many such stars would there be in our galaxy?

†This question requires a numerical answer.

†8. Supply your own numbers in the Drake equation, and calculate the result.

†9. On the message sent to M13, work out the binary value given for the population of Earth and compare it with the actual value.

†10. On the message sent to M13, work out the binary value given for the Arecibo telescope, and compare it with the actual value.

11. Why do we think that intelligent life wouldn't evolve in planets around stars much more massive than the Sun?

12. List three means by which we might detect extraterrestrial intelligence.

13. Why can't we hope to carry on a conversation at a normal rate with extraterrestrials on a distant star?

14. How can Project META distinguish between terrestrial interference and extraterrestrial signals?

15. Describe how Einstein's general theory of relativity serves as an example of the scientific method.

Topics for Discussion

1. Comment on the possibility of scientists in another solar system observing the Sun, suspecting it to be a likely star to have populated planets, and beaming a signal in our direction.

2. Using a grid 37 × 41 (two prime numbers), work out a message you might send. Show the string of 0's and 1's.

III The Stars

When we look up at the sky at night, most of the objects we see are stars. In the daytime, we see the sun, which is itself a star. The moon, the planets, even a comet may give a beautiful show, but, however spectacular, they are only minor actors on the stage of observational astronomy.

From the center of a city, we may not be able to see very many stars, because city light scattered by the earth's atmosphere makes the light level of the sky brighter than most stars. All together, about 6000 stars are bright enough to be seen with the unaided eye under good observing conditions.

The stars we see defining a constellation may be at very different distances from us. One star may appear bright because it is relatively close to us even though it is intrinsically faint; another star may appear bright even though it is far away because it is intrinsically very luminous. When we look at the constellations, we are observing only the directions of the stars and not whether the stars are physically very close together. Nor do the constellations tell us anything about the nature of the stars. Astronomers tend not to be interested in the constellations because the constellations do not give useful information about how the universe works. After all, the constellations merely tell the directions in which stars lie. Most astronomers want to know *why* and *how*. Why is there a star? How does it shine? Such studies of the workings of the universe are called *astrophysics*. Almost all modern-day astronomers are astrophysicists as well, for they not only make observations but also think about their meaning.

Some properties of the stars that can be seen with the naked eye tell us about the natures of the stars themselves. For example, some stars seem blue-white in the sky, while others appear slightly reddish. From information of this nature, astronomers are able to determine the temperatures of the stars. Chapter 20 is devoted to the basic properties of stars and some of the ways we find out such information. Chapter 21 describes some of the ingenious ways we use basic information about stars. We can sort stars into categories and tell how far away they are and how they move in space.

Most stars occur in pairs or in larger groups, and we shall discuss types of groupings and what we learn from their study in Chapter 22. Binary stars are important, for example, in finding the masses of stars. Some stars vary in brightness, a property that astronomers use to tell the scale of the universe itself. The study of star clusters leads to our understanding of the ages of stars and how they evolve.

In Chapter 23 we study an average star in detail. It is the only star we can see close up—the sun. The phenomena we observe on the sun take place on other stars as well, as has recently been verified by telescopes in space.

In recent years, the heavens have been studied in other ways besides observing the ordinary light that is given off by many astronomical objects. Other forms of radiation—radio waves, x-rays, gamma rays, ultraviolet, and infrared—and interstellar particles called cosmic rays are increasingly studied from the earth's surface or from space. Many a contemporary astronomer—even many who consider themselves observers (who mainly carry out

The Rosette Nebula in the constellation Monoceros, with its open cluster of stars NGC 2244.

observations) rather than theoreticians (who do not make observations, but rather construct theories)—has never looked through an optical telescope. Still, astronomy began with optical studies, and our story begins there. This part of the book will tell us how we study stars and what they teach us. Part IV will take up the life stories of stars.

A hot blue star of a type known as Wolf-Rayet, surrounded by a bubble of nebulosity called NGC 2359.

Ordinary Stars

20

Aims: To study the colors and spectra of ordinary stars, and to see how they tell us about the temperatures of the stars' surfaces

What are these stars that we see as twinkling dots of light in the nighttime sky? The stars are luminous balls of gas scattered throughout space. All the stars we see are among the trillion stellar members of a collection of stars and other matter called the Milky Way Galaxy. The sun, which is the star that gives most of the energy to our planet, is so close to us that we can see detail on its surface, but it is just an ordinary star like the rest.

Stars are balls of gas held together by the force of gravity, the same force that keeps us on the ground no matter where we are on the earth. Long ago, gravity compressed large amounts of gas and dust into dense spheres that became stars. The dust vaporized, and balls of gas remained.

Stars generate their own energy and light. It has been only 50 years since the realization that the stars shine by nuclear fusion. Now we are looking to the nearest star, our sun, to be used either as a direct source of energy in the form of its light or as a laboratory to study the physics of hot gases in magnetic fields. The latter study may tell us how to tame the fusion process and re-create it on earth.

We see only the outer layers of the stars; the interiors are hidden from our view. The stars are gaseous through and through; they have no solid parts. The outer layers do not generate energy by themselves, but glow from energy transported outward from the stellar interiors. So when we study light from the stars, we are not observing the processes of energy generation directly. We will study these processes in Chapter 24, where we start to study the life cycles of stars. In this chapter, we shall concentrate on how we study the stars themselves.

20.1 The Colors of Stars

In Section 4.1 we saw that the different colors we perceive result from radiation in certain ranges of wavelength. For example, an object appears blue because most of its visible radiation is at wavelengths between roughly 4300 and 5000 Å. We will now see that how a star's radiation is distributed in wavelength depends on the temperature of the surface of that star. By measuring the color of a star, we can determine the wavelength distribution and thus deduce the star's temperature.

To determine a star's color, we can pass the light through a spectrograph and measure the intensity at each wavelength. More simply, we can measure the light that emerges from filters, each of which allows only a certain group of wavelengths to pass.

When we examine a set of these graphs, we can see that, spectral lines aside, the radiation follows a fairly smooth curve (Fig. 20–1). The radiation "peaks" in intensity at a certain wavelength (that is, has a maximum in its

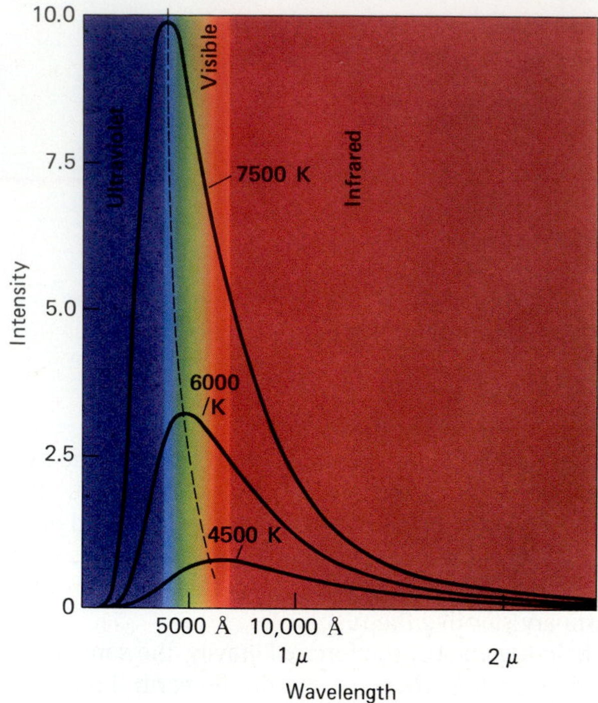

Figure 20–1 The intensity of radiation for different stellar temperatures, according to Planck's law (Section 20.2). The wavelength scale is linear; that is, equal spaces signify equal wavelength intervals. The dotted line shows how the peak of the curve shifts to shorter wavelengths as temperature increases; this is known as Wien's displacement law. The total energy emitted at a given temperature—the area under the curve–is given by the Stefan-Boltzmann law.

intensity at that wavelength), and decreases in amount more slowly on the long wavelength side of the peak than it does on the short wavelength side. Note very carefully that each curve represents a graph of intensity versus wavelength. Don't let the fact that we simultaneously show similar graphs for different temperatures confuse you; each graph shows the intensity of radiation on the vertical axis and the wavelength on the horizontal axis.

The graphs can be best understood by first considering the radiation that represents different temperatures. When you first put an iron poker in the fire, it glows faintly red. Then as it gets hotter it becomes redder. If the poker could be even hotter without melting, it would turn yellow, white, and then blue-white. White hot is hotter than red hot.

A set of physical laws governs the sequence of events when material is heated. As the material gets hotter, the peak of radiation shifts toward shorter wavelengths (the blue). This shift of wavelength of the peak is known as *Wien's displacement law* (Box 20.1 and Table 20–1).

Also, as the material gets hotter, the total energy of the radiation grows quickly. Notice in the figure how the total energy, which is the area under each curve, is much greater for higher temperatures. The energy follows the *Stefan-Boltzmann law* (Box 20.2). This law says that the total energy emitted from each square centimeter of a source in each second grows as the fourth power

Focus On

*Box 20.1 Wien's Displacement Law

$$\lambda_{max} T = \text{constant},$$

where λ_{max} is the **wavelength** (the lowercase Greek letter lambda) at which the energy given off is at a **maximum**, and T is the temperature. The numerical value of the constant is given in Appendix 2.

Table 20–1 Wien's Displacement Law

Temperature	Wavelength of Peak	Spectral Region of Wavelength of Peak
3 K	9,660,000 Å = 0.97 mm	infrared-radio
3,000 K	9660 Å	infrared
6,000 K	4830 Å	blue
12,000 K	2415 Å	ultraviolet
24,000 K	1207 Å	ultraviolet

of the temperature ($T^4 = T \times T \times T \times T$). If we compare the radiation from two same-size regions of a surface at different temperatures, the Stefan-Boltzmann law tells us that the ratio of the energies emitted is the fourth power of the ratio of the temperatures. That is, take the ratio of the temperatures and raise that ratio to the fourth power to find the ratio of the energies emitted. For example, consider the same gas at 10,000 K and at 5000 K. Because 10,000 K is twice 5000 K, the 10,000 K gas gives off $2^4 = 2 \times 2 \times 2 \times 2 = 16$ times more energy than does the same amount of gas at 5000 K. (We must measure from absolute zero, the minimum temperature possible. The Kelvin temperature scale begins there.) Note how even a small increase in temperature makes a star much brighter.

The Stefan-Boltzmann law relates the amount of energy a gas gives off to its temperature if we consider a constant surface area of gas emitting. It is further true that a larger volume of gas at the same temperature gives off more energy. So the actual amount of energy a star emits depends both on how hot its surface is and on how large its surface is.

The overall color that a star appears in the sky depends on the color of the peak of a star's energy distribution. Thus from Wien's displacement law, we see that the colors of stars in the sky tell us the stars' temperatures. The reddish star Betelgeuse, for example, is a comparatively cool star, while the blue-white star Sirius is very hot. By simply measuring the intensity of a star with a set of filters at different wavelengths, we can find out the temperature of the star (Fig. 20–2).

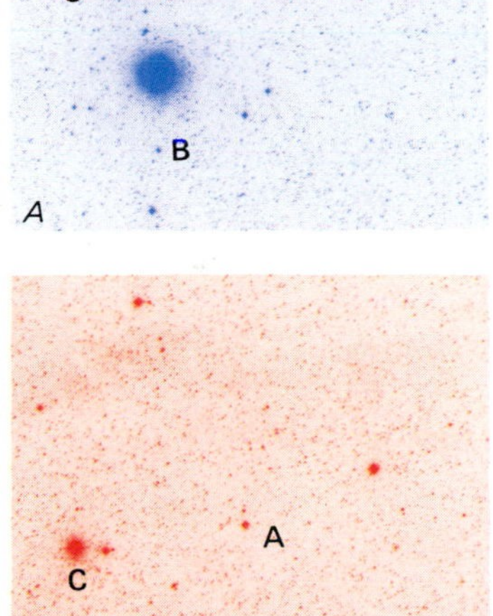

Figure 20–2 The top photograph is a view of several stars in blue light, and the bottom photograph is a view of the same stars in red light. Star A appears about the same brightness in both. Star B, a relatively hot star, appears brighter in the blue photograph. Star C, a relatively cool star, appears brighter in the red photograph.

20.2 Planck's Law and Black Bodies

Wien's displacement law tells you the wavelength of peak intensity and the Stefan-Boltzmann law tells you the area under a graph of intensity versus wavelength, but neither tells you the shape of the curve on the graph. The law that gives the shape is much more general than its application to the stars. In principle, if any gas is heated, its continuous emission follows a certain peaked distribution called *Planck's law,* suggested in 1900 by the German physicist

Focus On

*Box 20.2 Stefan-Boltzmann Law

$$\text{Energy} = \sigma T^4.$$

Energy (E) is the "strength" of the radiation per unit area. The numerical value of σ, the Stefan-Boltzmann constant, is given in Appendix 2.

Figure 20–3 Max Planck (*left*) presented an equation in 1900 that explained the distribution of radiation with wavelength. Five years later, Albert Einstein, in suggesting that light acts as though it is made up of particles, found an important connection between his new theory and the earlier work of Planck. The photograph was taken years later.

Max Planck (Fig. 20–3). Wien's displacement law and the Stefan-Boltzmann law, which had been discovered earlier from study of experimental data, can be derived from Planck's law. Actual gases may not, and usually don't, follow Planck's law exactly. For example, Planck's law governs only the continuous spectrum; any emission or absorption in the form of spectral lines does not follow Planck's law.

Because of these difficulties, scientists consider an idealized case: a fictional object called a *black body,* which is defined as something that absorbs all radiation that falls on it. Note that the "black" in a black body is an idealized black that absorbs 100 per cent of the light that hits it. Ordinary black paint, on the other hand, does not absorb 100 per cent of the light that hits it. Further, something that looks black in visible light might not seem very black at all in the infrared, and so would not be a real black body.

An important physical law that governs the radiation from a black body says that anything that is a good absorber is a good emitter too. The radiation from a black body follows Planck's law. As a black body is heated, the peak of the radiation shifts toward the short wavelength end of the spectrum (Wien's displacement law) and the total amount of radiation grows rapidly (Stefan-Boltzmann law). Note that the curves of energies emitted by black bodies at different temperatures in Figure 20–1 do not cross. In particular, you should realize that if you heat a gas, it emits more radiation at every wavelength. If you trace vertical lines upward on the figure, no matter which wavelength you are investigating, you cross the curve for the cooler gas before you reach upward high enough to reach the curve for the hotter gas.

To a certain extent, the atmospheres of stars appear as black bodies: over a broad region of the spectrum that includes the visible, the continuous radiation from stars follows Planck's law fairly well.

20.3 The Formation of Spectral Lines

Most of what we know about the stars comes from studying their spectral lines. First we shall discuss spectral lines themselves; we will go on to their appearance in stars in a following section.

Spectral lines arise when there is a change in the amount of energy present in any given atom. We can think of an atom (Fig. 20–4) as consisting of particles in its core, called the *nucleus,* surrounded by orbiting particles, called *electrons.* Examples of nuclear particles are the *proton,* which carries one unit of electric charge, and the *neutron,* which has no electric charge. (Protons and neutrons are, in turn, made up of particles called *quarks,* as we discuss further in Chapter 34. The quark structure, though fundamental for understanding the makeup of our world and our universe, does not affect spectra.) An electron, which has one unit of negative electric charge, is only 1/1800 as massive as a proton or a neutron. Normally, an atom has the same number of electrons as it has protons. The negative and positive charges balance, leaving the atom electrically neutral.

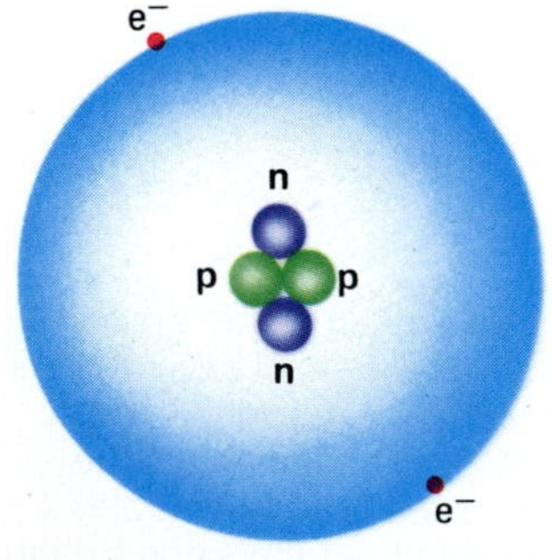

Figure 20–4 A helium atom contains two protons and two neutrons in its nucleus and two orbiting electrons.

The amount of energy that an atom can have is governed by the laws of *quantum mechanics,* a field of physics that was developed in the 1920's and whose applications to spectra were further worked out in the 1930's. According to the laws of quantum mechanics, light (and other electromagnetic radiation) has the properties of waves in some circumstances and the properties of particles under other circumstances. One cannot understand this dual set of properties of light intuitively, and some of the greatest physicists of the time had difficulty adjusting to the idea. Still, quantum mechanics has been worked out to explain experimental observations of spectra, and is now thoroughly accepted. Its development was one of the major scientific achievements of the twentieth century.

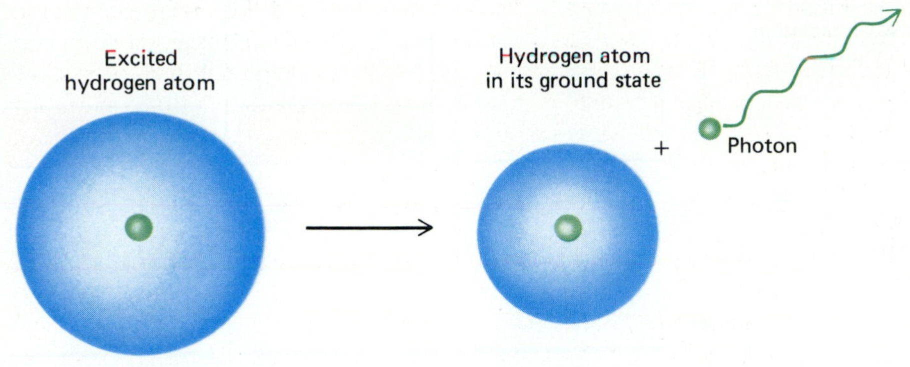

Figure 20–5 Since the hydrogen atom has only a single electron, it is a particularly simple case to study. The lowest possible energy state of an atom is called its *ground state.* All other energy states are called *excited states:* the second energy level, immediately above the ground state, is the "first excited state," the next is the "second excited state," etc. When an atom in an excited state gives off a photon, it drops back to a lower energy state, perhaps even to the ground state. We see the photons for a given transition as an emission line.

According to the laws of quantum mechanics, an atom can exist only in a specific set of energy states, as opposed to a whole continuum (continuous range) of energy states. An atom can have only discrete values of energy; that is, the energy cannot vary continuously. For example, the energy corresponding to one of the states might be 10.2 electron volts (eV). The next energy state allowable by the quantum mechanical rules might be 12.0 eV. We say that these are *allowed states.* The atom **simply cannot** have an energy between 10.2 and 12.0 eV. The energy states are discrete (separate and distinct); we say that they are "quantized," and hence the name "quantum mechanics." The discrete energy states are called *energy levels.* An analogy is that I can stand on one step of a staircase or on another step, but cannot hover in between.

When an atom drops from a higher energy state to a lower energy state without colliding with another atom, the difference in energy is sent off as a bundle of radiation. This bundle of energy, a *quantum,* is also called a *photon,* which may be thought of as a particle of electromagnetic radiation (Fig. 20–5). Photons always travel at the speed of light. Each photon has a specific energy.

A link between the particle version of light and the wave version is seen in the equation that relates the energy of the photon with the wavelength it has: $E = hc/\lambda$. In this equation, E is the energy, h is a constant named after Planck, c is the speed of light (in a vacuum, as always), and λ (the Greek letter lambda) is the wavelength. Since λ is in the denominator on the right-hand side of the equation, a small λ corresponds to a photon of great energy. (We are dividing by a small number, and thus get a large result.) When, on the other hand, λ is large, the photon has less energy. For example, an x-ray photon of wavelength 1 Å has much more energy than does a photon of visible light of wavelength 5000 Å.

20.3a Emission and Absorption Lines

In a cold gas, most of the atoms are on the lowest possible energy level. When a gas is heated, many of its atoms are raised from this lowest energy level to higher energy states. From the higher energy levels, they then spontaneously drop back to their lowest energy levels, emitting photons as they do so. These new photons represent energies at certain wavelengths (Fig. 20–6*A* and *B*). Since photons did not necessarily previously exist at those wavelengths, the new photons appear on the spectrum as wavelengths brighter than neighboring wavelengths. These are the **emission** lines.

As we saw in Section 4.10, when we allow continuous radiation from a body at some relatively high temperature to pass through a cooler gas, the atoms in the gas can take energy out of the continuous radiation at certain discrete wavelengths. We now see that the atoms in the gas are changed into higher energy states that correspond to these wavelengths (Fig. 20–6*C*). Be-

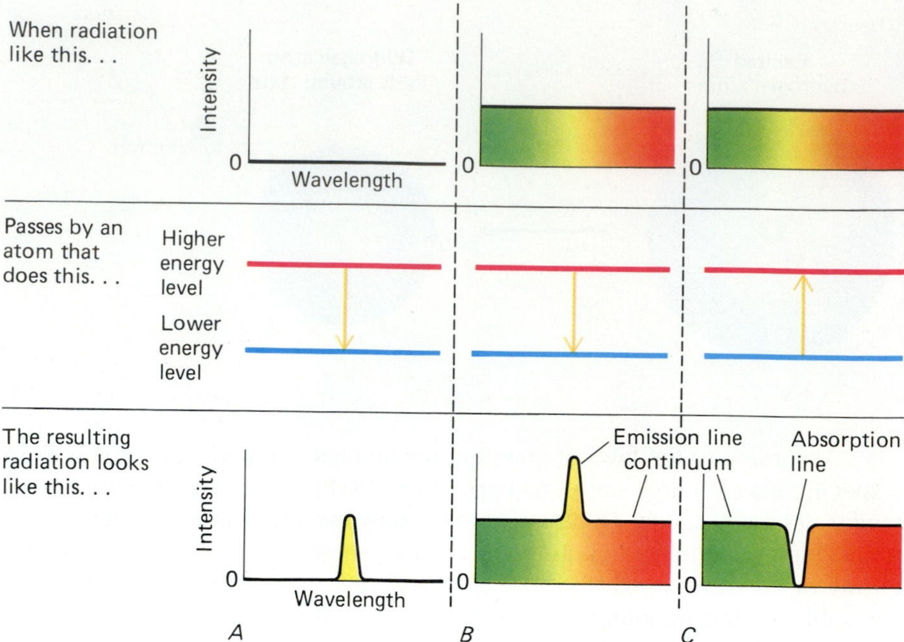

Figure 20–6 When photons are emitted, we see an emission line; a continuum may or may not be present (Cases *A* and *B*). An absorption line must absorb radiation from something. Hence an absorption line necessarily appears in a continuum (Case *C*). The top and bottom horizontal rows are graphs of the spectrum before and after the radiation passes by an atom; a schematic diagram of the atom's energy levels (only two levels are shown to simplify the situation) is shown in the center row.

cause energy is taken up in the gas at these wavelengths, less radiation remains to travel straight ahead at these particular wavelengths. These wavelengths are those of the **absorption** lines. In the visible part of the spectrum, these wavelengths appear dark when we look back through the gas. (The energy absorbed is soon emitted in other directions, which explains why we can see emission lines when we look from the side, as was shown in Figure 4–48.)

Note that emission lines can appear without a continuum or with a continuum, since emission lines are merely energy added to the radiation field at certain wavelengths. But absorption lines must be absorbed **from** something, namely, the continuum, which is just the name for a part of the spectrum where there is some radiation over a continuous range of wavelengths.

A normal stellar spectrum is a continuum with absorption lines. This simple fact leads us to the important conclusion that the temperature in the outer layers of the star is decreasing with distance from the center of the star. The presence of absorption lines tells us that we have been looking through cooler gas at hotter gas. Because the temperature declines as you go outward in a star, stellar spectra rarely show emission lines, which result from the presence of gas hotter than any background continuum. Emission lines can appear, however, when stars are surrounded by hot shells of gas.

20.3b Excited States and Ions

We can think of an atom's energy states as resulting from a change in energy of an atom's electrons. (The production of spectral lines does not involve changes in the nucleus.) If a gas were at a temperature of absolute zero, all the electrons in it would be in the lowest possible energy levels. The lowest energy level that an electron can be in is called the *ground state* of that atom.

As we add energy to the atom, we can "excite" one or more electrons to higher energy levels. The level directly above the ground state is the *first excited state.* If we were to add even more energy, some of the electrons would be given not only sufficient energy to be excited to even higher excited states but also enough to escape entirely from the atom. We then say that the atom is *ionized.* The remnant of the atom with less than its quota of electrons is called an *ion* (Fig. 20–7). Since the remnant has more positive charges in its nucleus

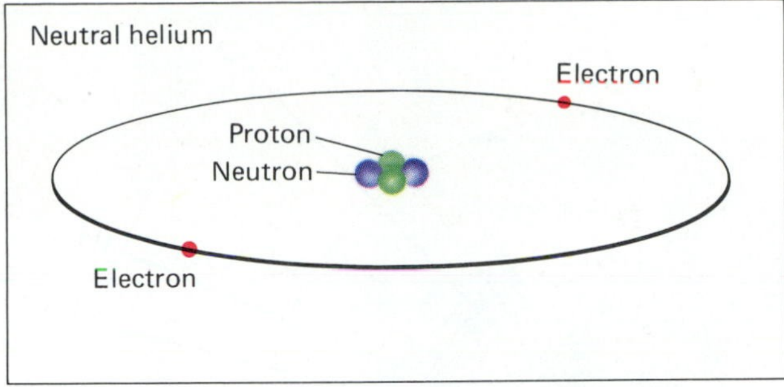

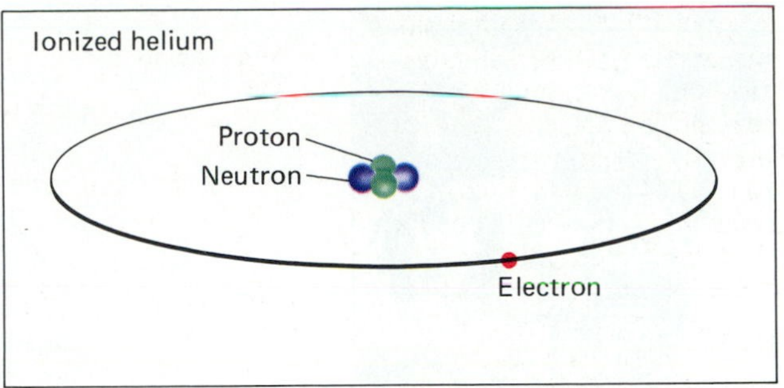

Figure 20–7 An atom missing one or more electrons is *ionized.* Neutral helium is He I and singly ionized helium is He II.

than it has negative charges on its electrons, its net charge is positive, and it is sometimes called a *positive ion.*

Before the atom was ionized, it had the same number of protons and electrons, and so was electrically neutral. Such an atom is called a neutral atom, and is denoted with the Roman numeral I. For example, neutral helium is denoted He I. When the atom has lost one electron (is singly ionized), we use the Roman numeral II. Thus when helium has lost one electron, it is called He II (read "helium two"). If an atom is doubly ionized, that is, has lost two electrons, it is in state III, and so on. The Roman numeral is the number of electrons lost +1.

Atoms can be excited or ionized either by collisions or by radiation. Collisions are the most important when the density is sufficiently high, since then atoms collide frequently with electrons.

20.4 The Hydrogen Spectrum

Hydrogen gas emits, and absorbs, a set of spectral lines that falls across the visible spectrum in a distinctive pattern (Fig. 20–8). This set of hydrogen lines is called the *Balmer series.* The strongest line is in the red, the second strongest is in the blue, and the other lines continue through the blue and ultraviolet, with the spacing between the lines getting smaller and smaller. Interestingly, the lines were first seen (in 1881) as a series in astronomical photographs rather than in a terrestrial laboratory.

It is easy to visualize how the hydrogen spectrum is formed by using the picture that Niels Bohr (Fig. 20–9), the Danish physicist, laid out in 1913. In the *Bohr atom* (Fig. 20–10), electrons can have orbits of different sizes that correspond to different energy levels. Only certain orbits are allowable, which is the same as saying that the energy levels are quantized. This simple picture of the atom has been superseded by the development of quantum mechanics, but the notion that the energy levels are quantized remains.

We use the letter n to label the energy levels. It is called the *principal quantum number.* We call the energy level for $n = 1$ the *ground level* or *ground state,* as it is the lowest possible energy state. The hydrogen atom's series of transitions from or to the ground level is called the Lyman series, after the

Figure 20–8 The Balmer series, representing transitions down to or up from the second energy state of hydrogen. The strongest line in this series, Hα (H-alpha), is in the red. The blue Hβ (H-beta) line is overexposed and therefore looks white on this color photo.

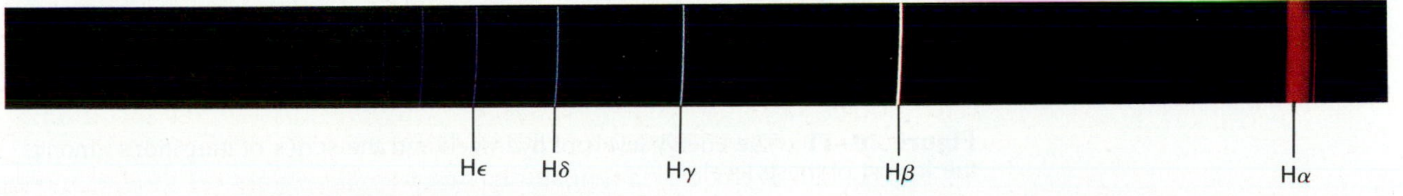

Figure 20–9 Niels Bohr and his wife, Margrethe.

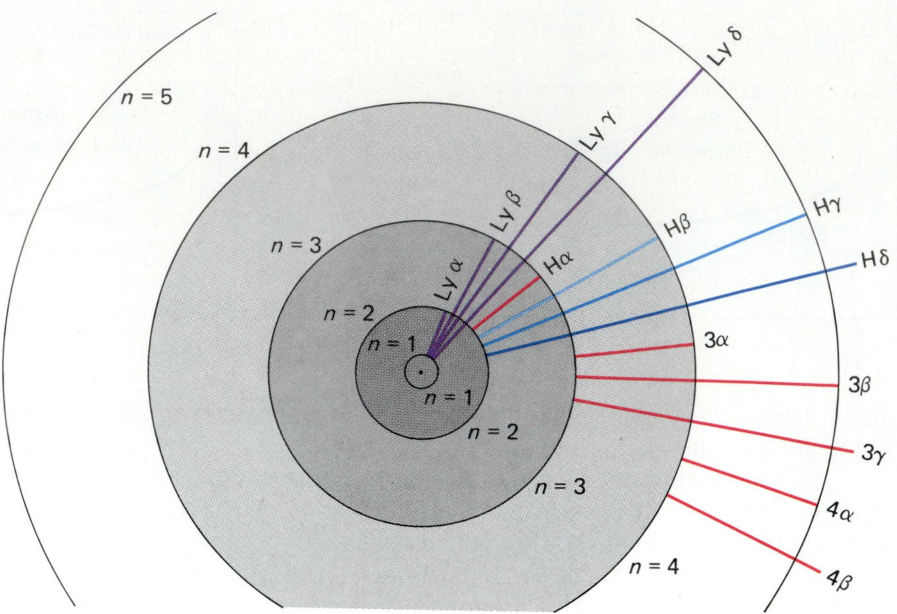

Figure 20–10 The representation of hydrogen energy levels known as the Bohr atom.

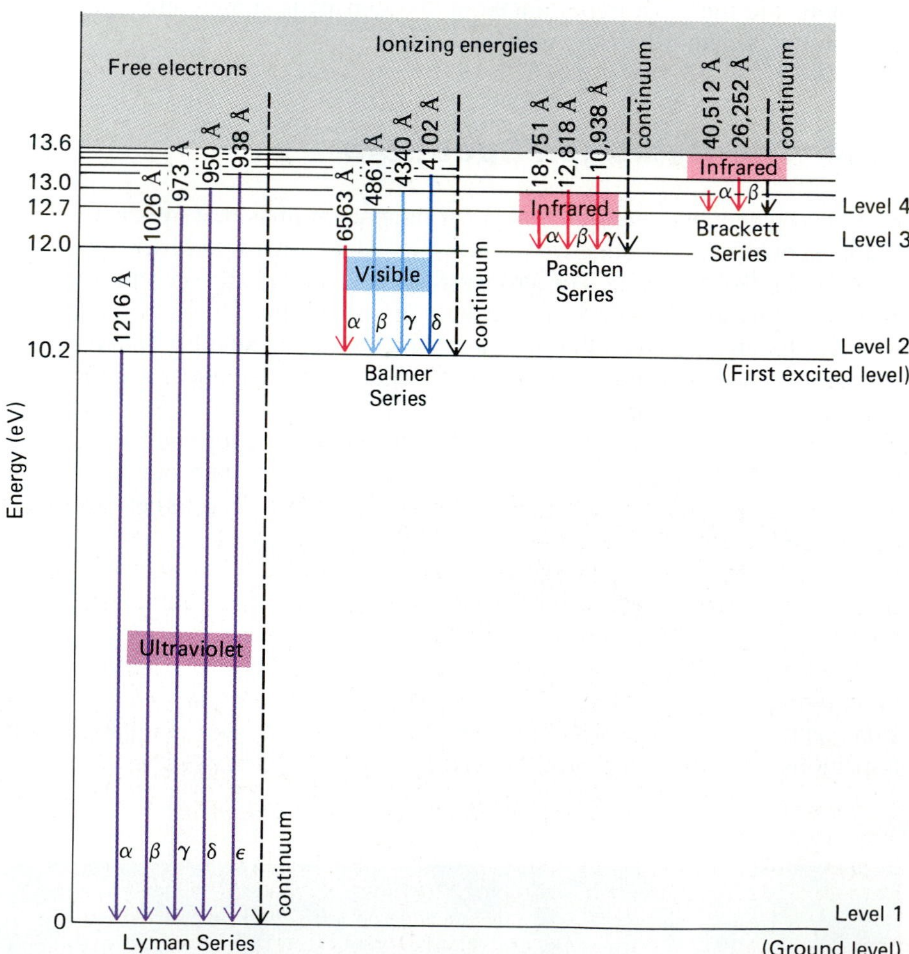

Figure 20–11 The energy levels of hydrogen and the series of transitions among the lowest of these levels.

American physicist Theodore Lyman. The lines fall in the ultraviolet, at wavelengths far too short to pass through the earth's atmosphere. Lyman's observations of what we now call the Lyman lines were made in a laboratory in a vacuum tank. Telescopes aboard earth satellites, such as the International Ultraviolet Explorer (IUE), now enable us to observe Lyman lines from the stars.

The series of transitions with $n = 2$ as the lowest level is the *Balmer series,* which we have already seen is in the visible part of the spectrum. The bright red emission line in the visible part of the hydrogen spectrum, known as the Hα line, arises from the transition from level 3 to level 2. The transition from $n = 4$ to $n = 2$ causes the Hβ line, and so on. Because the series falls in the visible where it is so well observed, we usually call the lines simply H alpha, etc., instead of Balmer alpha, etc. There are many other series of hydrogen lines, each corresponding to transitions to a given lower level (Fig. 20–11).

Lyman went so far as to climb a mountain in 1915, hoping that from its greater altitude there might happen to be a window of transparency that would pass Lyman alpha. Unfortunately, there is no such window, and his photographs of the solar spectrum were blank in the region where ultraviolet radiation would have appeared.

Since the higher energy levels have greater energy, a spectral line caused by a transition from a higher level to a lower level yields an emission line. When, on the other hand, continuous radiation falls on cool hydrogen gas, some of the atoms in the gas can be raised to higher energy levels. Absorption lines result.

Each of the chemical elements has its own sets of energy levels, one set for each stage of ionization. Each thus has its own sets of spectral lines. (The Lyman and Balmer series are only for hydrogen; the lines of most other elements do not form such a regular pattern.)

The hydrogen atom, with its lone electron, is a very simple case. More complicated atoms have more complicated sets of energy levels, and thus their spectral lines have less simple patterns.

20.5 Spectral Types

In the last decades of the 19th century, spectra of thousands of stars were photographed. Ways of classifying the different spectra were developed (Fig. 20–12). The most famous worker at the vital task of classifying spectra was Annie Jump Cannon (Fig. 20–13). Photographic plates with low dispersion (that is, with the wavelengths not spread out very much) were used (Fig. 20–14). Such plates were often taken with "objective prisms," large prisms in front of the telescope's objective.

At first, the stellar spectra were classified only by the strength of certain absorption lines from hydrogen and were lettered alphabetically: A for stars with the strongest hydrogen lines, B for stars with slightly weaker lines, and

Figure 20–12 Part of the computing staff of the Harvard College Observatory in about 1917. Annie Jump Cannon, fifth from the right, classified over 500,000 spectra in the decades following 1896. Her catalogue, called the *Henry Draper catalogue* after the benefactor who made the investigation possible, is still in use today. Many stars are still known by their HD (Henry Draper catalogue) numbers, such as HD 176387. Also in this photograph is Henrietta S. Leavitt, fifth from the left, whose work on variable stars we shall discuss in Sections 22.4b and 31.5.

Figure 20–13 Annie Jump Cannon classifying spectra.

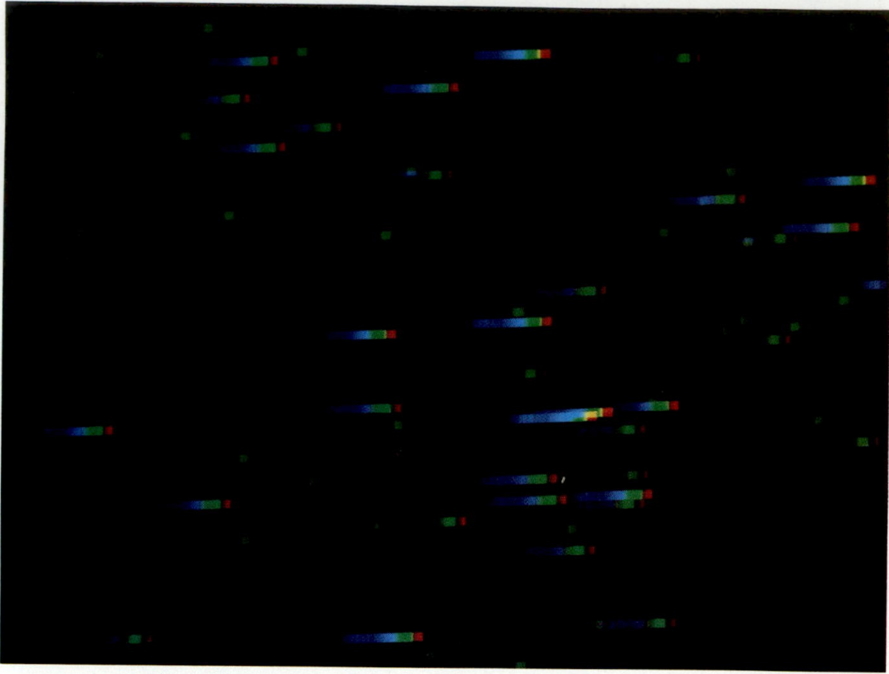

Figure 20–14 An objective-prism spectrogram of the Hyades, a cluster of stars.

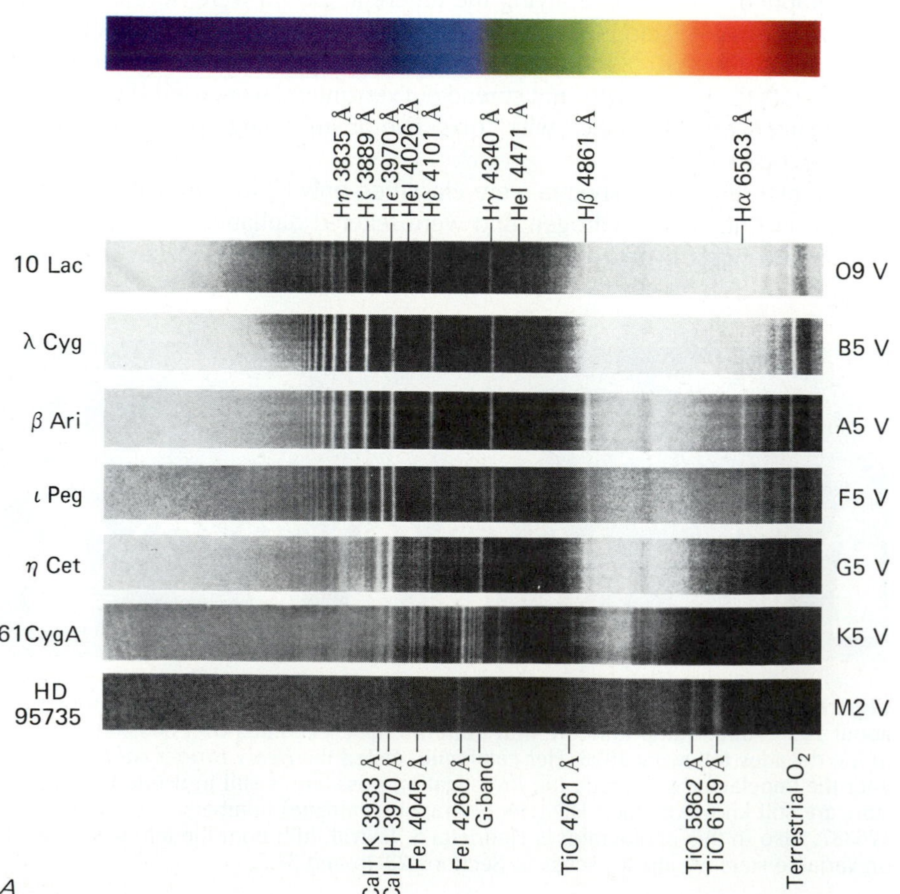

Figure 20–15 (*A*) Spectral types observed with an objective-prism spectrograph, which spreads out the short wavelengths. Note that all the stellar spectra shown have absorption lines. The Roman numeral "V" indicates that the stars are normal ("dwarf") stars, which will be defined in Section 21.4. The numbers following the letters in the right-hand column represent a subdivision of the spectral types into tenths; the step from B8 to B9 is equal to the adjacent step from B9 to A0. The abbreviations of the elements forming the lines and the individual wavelengths (in angstroms) of the most prominent lines appear at the top and bottom of the graph. Notice how the presence or absence of certain spectral lines, like the H and K lines of ionized calcium, can quickly indicate the spectral type. (*B*) A set of spectra drawn on a computer with a linear wavelength scale.

so on. These categories are called *spectral types* (Fig. 20–15). It was later realized that the types of spectra varied primarily because of differing temperatures of the *stellar atmospheres*—the stars' outer layers. The hydrogen lines were strongest in stars in the middle range of stellar temperatures and were weaker at both higher and lower temperatures. In the hottest stars, too much hydrogen is ionized for the lines to be strong. In stars cooler than spectral type A, too few hydrogen atoms have enough energy for electrons to be on levels above the ground state. When we now list the spectral types of stars in order of decreasing temperature, they are no longer in alphabetical order. From hottest to coolest, the spectral types are O B A F G K M.

O stars (that is, stars of type O) are the hottest, and the temperatures of M stars are more than 10 times lower. Generations of American students and teachers have remembered the spectral types by the mnemonic: Oh, Be A Fine Girl, Kiss Me. The lettered types have been subdivided into 10 subcategories each. For example, the hottest B stars are B0, followed by B1, B2, B3, and so on. Spectral type B9 is followed by spectral type A0. It is fairly easy to tell the spectral type of a star by mere inspection of its spectrum.

Historically, the hotter stars are called early types and the cooler stars are called late types. Thus, a B star is "earlier" than an F star, although stars do not change spectral type during the long stable period that takes up most of their lives.

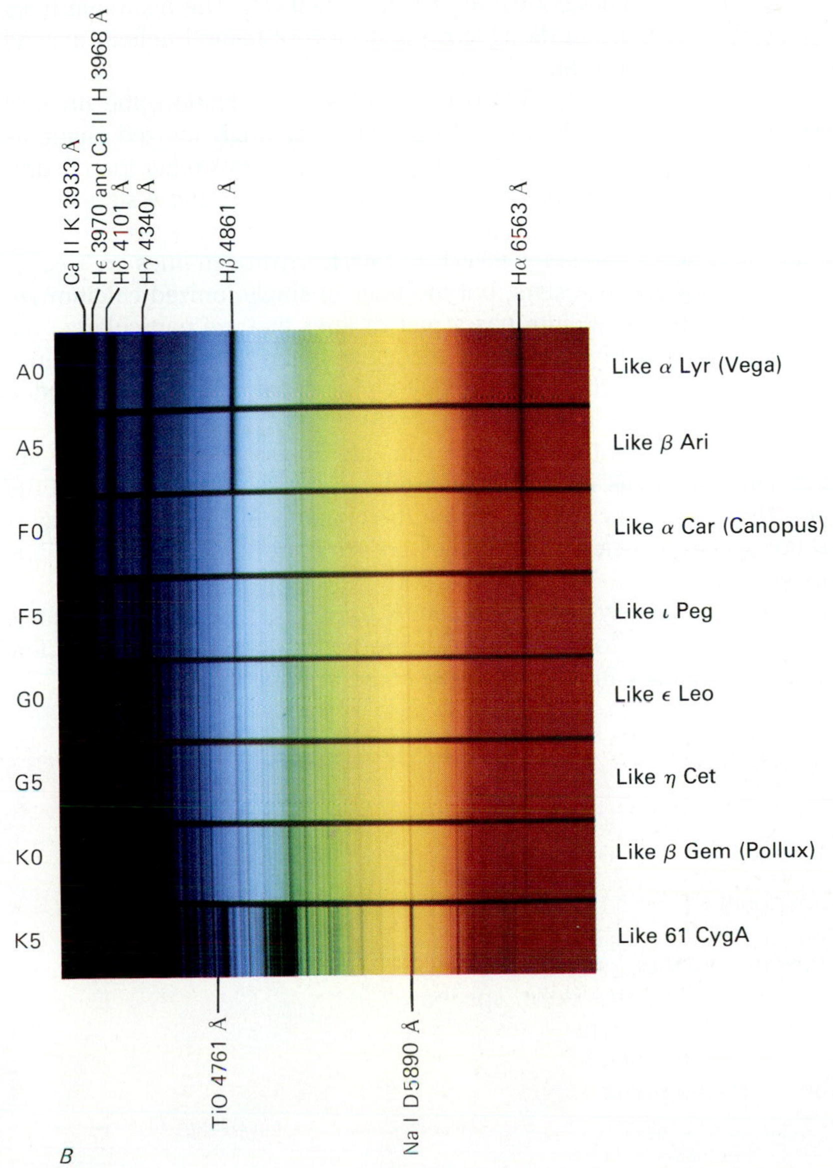

B

The assignment of spectral types follows a particular set of stellar spectra that serve as standards. Stellar spectra are compared with these standards to find the closest match.

20.5a Stellar Spectra

When we observe the light from a star, we are seeing only the radiation from a thin layer of that star's atmosphere. The interior is hidden within.

As we consider stars of different spectral types, we can observe the effect of temperature. For example, the hottest stars, those of type O, have such high temperatures that there is enough energy to remove the outermost electron or electrons from most of the atoms. Since most of the hydrogen is ionized, relatively few neutral hydrogen atoms are left and the hydrogen spectral lines are weak. Helium is not so easily ionized—it takes a large amount of energy to do so—yet even helium appears not in its neutral state but in its singly ionized state in an O star. Temperatures range from 30,000 K to 60,000 K in O stars. O stars are relatively rare, since they have short lifetimes; none of the nearest stars to us are O stars. The system really begins with spectral types O3 and O4, of which only a handful are known at any distance. No O0, O1, or O2 stars apparently exist.

Spectral lines arise from transition of electrons between energy states. Since a hydrogen atom has only one electron to begin with, when it is ionized it then has no electrons at all, and thus has no spectral lines.

B stars are somewhat cooler, 10,000 K to 30,000 K. The hydrogen lines are stronger than they are in the O stars, and lines of neutral helium instead of ionized helium are present.

Rigel and Spica are familiar B stars.

Cooler still are A stars, 7500 K to 10,000 K. By definition, the lines of hydrogen are strongest in this spectral type. Lines of singly ionized elements like magnesium and calcium begin to appear. (All elements other than hydrogen or helium are known as *metals* for this purpose.) O, B, and A stars are all bluish white.

Sirius, Deneb, and Vega are among the brightest A stars in the sky.

F stars have temperatures of 6000 K to 7500 K. Hydrogen lines are weaker in F stars than they are in A stars, but the lines of singly ionized calcium are stronger. Singly ionized calcium has a pair of lines that are particularly conspicuous; they are easy to pick out and recognize in the spectrum. These lines are called H and K (Fig. 20–16), from their alphabetical order in an extended version of Fraunhofer's original list (Fig. 4–46).

Canopus, the second brightest star in the sky (not visible from the latitudes of most of the United States), is a prominent F star, as is Polaris, the north star.

G stars, the spectral type of our sun, are 5000 K to 6000 K. They are yellowish, since the peak in their spectra falls in the yellow/green part of the spectrum. Hydrogen lines are visible, but the H and K lines are the strongest lines in the spectrum. The H and K lines are stronger in G stars than in any other spectral type.

Besides the sun, alpha Centauri (actually the brightest of the three stars that together make up the bright point in the sky that we call alpha Centauri) is also a G star.

K stars are relatively cool, only 3500 K to 5000 K. The spectrum shows many lines, in contrast to the spectra of the hottest stars, which show few spectral lines. The lines in K stars mostly come from neutral metals.

Arcturus and Aldebaran, both visible as reddish points in the sky, are K stars.

Focus On

Box 20.3 K

Do not confuse:
K stars: a spectral class
K: a kelvin, a unit of temperature
K: a strong line in many stellar spectra
K: the element potassium
K: an infrared filter band
k: kilo—a metric prefix

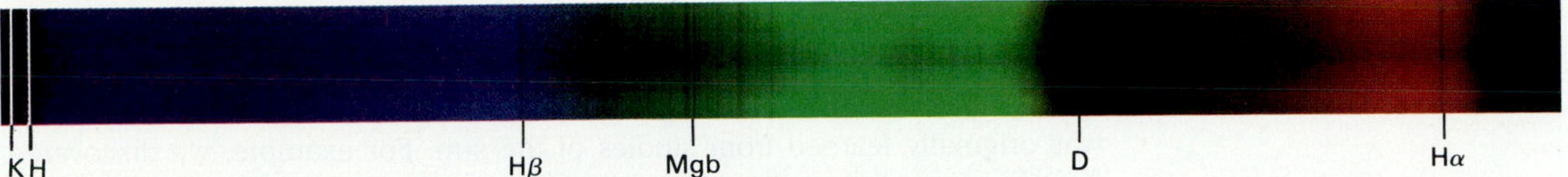

Figure 20–16 The H and K lines of singly ionized calcium are the strongest lines in the visible part of the solar spectrum and are prominent in spectra of stars of spectral types F, G, K, and M. These classes are known as *late spectral types,* while classes O, B, and A are known as *early spectral types,* although no connotations of age are meant.

M stars are cooler yet, with temperatures less than 3500 K, so they look reddish. Their atmospheres are so cool that molecules can exist without being torn apart, and their spectra show many molecular lines. Lines from the molecule titanium oxide are particularly numerous.

Betelgeuse is an example of such a reddish star of spectral class M.

For the very coolest stars, there are alternative spectral types that reflect differences in the relative abundances of various elements. For example, stars that are relatively rich in carbon are called *carbon stars* or *C-type.* (These were originally called R and N stars.) *S-type* stars have, among other things, relatively strong lines of zirconium oxide instead of titanium oxide. Both C- and S-type stars have similar temperatures to M stars.

We have listed the ordinary spectral types of stars, each of which shows an absorption spectrum, that is, a continuum crossed with dark lines. There are many unusual stellar spectra, though. *Wolf-Rayet stars* are O stars that not only show emission lines but also have the emission particularly broad in wavelength. The emission comes from shells of material that the star had ejected into the space surrounding it (see the photograph opening this chapter) and results from strong outflowing "stellar winds."

The amount of *mass loss,* matter flowing out of stars, varies from spectral type to spectral type, but can be considerable (Figs. 20–17 and 20–18). Much of the matter in interstellar space is the result of stellar mass loss.

In one of the first applications of astrophysics to the study of stars, in 1925, Cecilia Payne-Gaposchkin (then Cecilia Payne) graphed the strength of certain absorption lines versus spectral type. She used the graph together with her theoretical calculations to determine the temperature for each spectral type.

Figure 20–17 A hot mass-loss star, with reddish emission coming from twin lobes of gas. We see nebulae NGC 6164-5 around peculiar star HD 148937.

Figure 20–18 A cool mass-loss star, with a less directed stellar wind. The star is surrounded by the nebula it has ejected, IC 2220.

*20.5b Spectra Observed from Space

In the ultraviolet and x-ray regions of the spectrum, we often see radiation from above the star's surface. These higher levels are hotter than the surface, as was originally learned from studies of the sun. For example, we discovered from Einstein Observatory observations that stars of all spectral types give off x-rays. Though it had been expected that the coolest stars, the M stars, would be too cool to be so strong in x-rays, they are actually as much as 10 per cent as strong in the x-ray spectrum as they are in the visible. But this doesn't say that we were wrong about the temperature of these stars' surfaces. The x-rays come from a hot *stellar corona* surrounding the stars at a temperature of over a million kelvins.

Ultraviolet spectra of even cool stars taken with the International Ultraviolet Explorer show spectral lines from gas at temperatures of up to about 100,000 kelvins. These spectral lines are formed in a *stellar chromosphere* between the star's ordinary surface and its corona, or in a *transition zone* between the chromosphere and corona. Only since the launch of these spacecraft could we study chromospheres and coronas in stars other than the sun.

*20.5c The Continuous Spectrum

Where does the continuous spectrum come from? For there to be an absorption line, there must be some continuous radiation to be absorbed. The mechanism that causes continuous emission is different for different spectral types. In the sun and other stars of similar spectral types, the continuous emission is caused by a strange type of ion: the *negative hydrogen ion* (Fig. 20–19). An ordinary ion has one or more electrons missing, and so has a positive charge and could be called a positive ion. A negative ion, on the other hand, has one or more extra electrons. Only a tiny fraction of the hydrogen atoms are in such a state, with one extra electron, but the sun is so overwhelmingly (90 per cent) hydrogen that enough negative hydrogen atoms are present to be important.

When a neutral hydrogen atom takes up an extra electron, the second electron may temporarily become part of the atom, residing in an energy level. But the energy **difference** between the extra electron's energy and its energy as part of the atom is not limited to discrete values, because the electron could have had any amount of energy before it joined the atom (Fig. 20–20). Alternatively, the extra electron starts free of the atom, is affected by the atom, but winds up still free of the atom. In this case, too, the amount of energy involved is not limited to discrete values. Thus a continuous spectrum is formed. The case is thus different from the one in which an electron in an atom jumps between two energy levels, because the electron has a fixed amount of energy in each level, and a fixed number subtracted from another fixed number gives a fixed number. Subtracting one fixed number from another cannot lead to a continuous range of values.

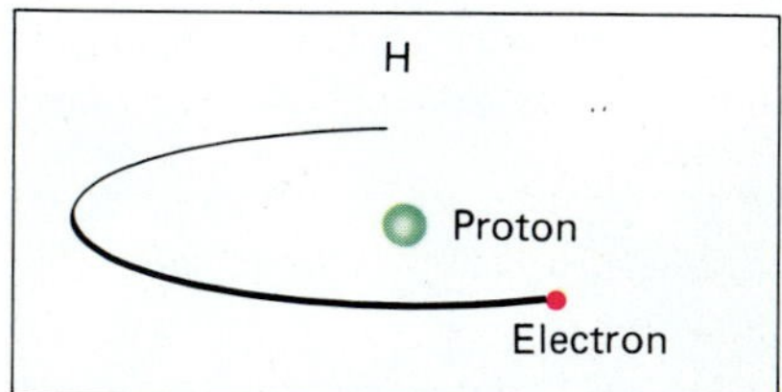

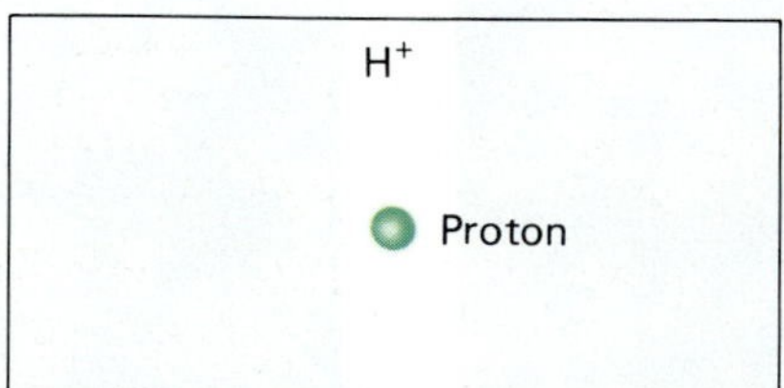

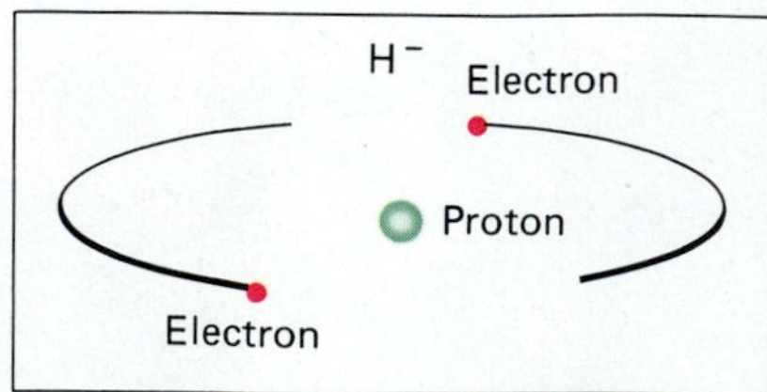

Figure 20–19 Neutral hydrogen, the hydrogen ion (sometimes, though rarely, called the positive hydrogen ion), and the negative hydrogen ion.

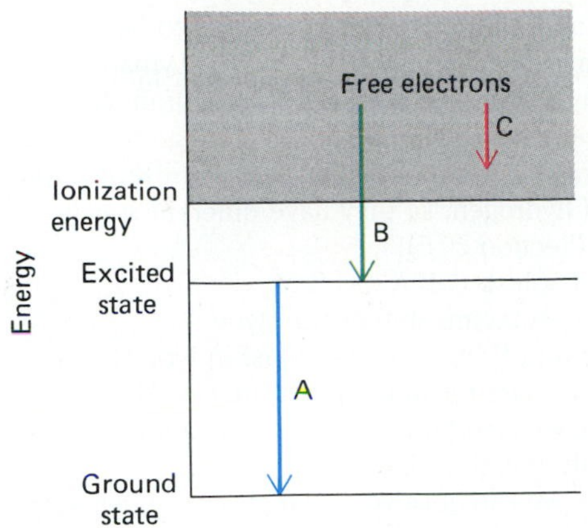

Figure 20–20 A bound-bound transition (*A*) would go from one discrete level to another (that is, from a bound level to another bound level), analogous to jumping on a staircase from one step to another. A free-bound transition (*B*) would be analogous to jumping from some arbitrary height above the staircase (as from the sky with a parachute) to some particular step (that is, from being free to a bound level). In a free-free transition (*C*) the electron never joins the atom. The colors are drawn from bluer to redder as the arrows grow shorter, to indicate the bluer emission from more energetic photons.

Consider the following example. Say that an atom has energy levels of 2, 6, 8, and 9 electron-volts (eV), where electron volts is a unit of energy. And say that 9 eV is the highest energy level of the atom. Then transitions 2–6, 2–8, 2–9, 6–8, 6–9, and 8–9 are each possible, so the atom would have 6 spectral lines. The energy differences of these transitions are 4, 6, 7, 2, 3, and 1 eV, respectively, and these energy transitions would translate into spectral lines by the formula constant/λ = energy difference. But if an electron had more than 9 eV, it wouldn't be part of the atom anymore. It could then have any amount of energy: 9.2, 9.22, 9.222, 9.7, or whatever. Given a lot of atoms, so many of them would be spread through the range above 9 eV that when they were attached themselves to one of the energy levels, the light would be so varied in wavelength that we would see continuous radiation. We might see radiation at 9.22 − 6.00 = 3.22 eV, 9.222 − 6.000 = 3.222 eV, etc. But no matter how many atoms there were, if they had below 9 eV, they would emit only the six spectral lines.

The processes of continuous emission and spectral-line absorption take place together throughout the outer layers of the stars rather than in separate layers. The detailed study of the processes of emission and absorption is called the study of "stellar atmospheres." Nowadays supercomputers are able to handle complicated sets of equations that describe the transfer of radiation from the outer layers of a star into space. Sometimes the most powerful computers, such as a Cray supercomputer at the National Center for Atmospheric Research (NCAR) in Boulder, Colorado, may run for hours to calculate a model of the atmosphere of a single star.

Summary and Outline

Stars are balls of gas held together by gravitation and generating energy by nuclear fusion.

Colors of stars and of other gas radiating a continuum (Sections 20.1 and 20.2)

- As gas gets hotter, it goes from red hot to bluish hot; the peak of the radiation moves to shorter wavelengths (*Wien's displacement law*): $\lambda_{max} T$ = constant.
- As gas gets hotter, the total energy radiated grows rapidly, with the fourth power of the temperature (the *Stefan-Boltzmann law*): $E = \sigma T^4$.

Planck's law gives the amount of radiation at each wavelength, given the temperature. Wien's displacement law and the Stefan-Boltzmann law are included within Planck's law.

> Wien's displacement law: $\lambda_{max} T$ = constant
> Stefan-Boltzmann law: $E = \sigma T^4$
> Planck's law relates λ, E_λ, and T

Matter whose radiation follows Planck's law precisely is called a *black body*.

The structure of an atom (Section 20.3)
- Protons and neutrons with electrons orbiting the nucleus

Spectral lines (Sections 20.3 and 20.4)
- Quantum theory showed that atoms can exist only in discrete states of energy. When an atom drops from a higher energy state to a lower energy state, it emits the difference in energy as a photon of a particular wavelength λ and energy hc/λ. Such photons make an emission line. If, on the other hand, an atom takes up enough energy to raise itself from a lower to a higher state, then such atoms lead to the formation of an absorption line in radiation that has been hitting them.
- Hydrogen's spectral lines fall in a pattern across the visible: the Balmer series.
- The Bohr atom, with each energy level having a different size, is a good way to visualize the hydrogen atom. The Balmer series corresponds to transitions between level 2 and higher levels. The Lyman series, which falls in the ultraviolet, involves transitions to the more basic level, the first level, or ground state.
- Other elements have different sets of energy levels from those of hydrogen, so they have different spectra.

Spectral types (Section 20.5)
- Hottest to coolest: O B A F G K M
- Hydrogen spectra are strongest in type A.
- Calcium H and K lines are strongest in type G.
- Molecular spectra begin to appear in type M.
- All types have absorption spectra. Only a few kinds of stars with emission lines exist.
- From space we can observe stellar chromospheres and coronas.
- Many stars undergo mass loss.
- The sun's continuous spectrum comes from the negative hydrogen ion (H^-).

Key Words

Wien's displacement law, Stefan-Boltzmann law, Planck's law, black body, nucleus, electrons, proton, neutron, quarks, quantum mechanics, allowed states, energy levels, quantum, photon, ground state, first excited state, ionized, ion, positive ion, Balmer series, Bohr atom, principal quantum number, ground level (ground state), spectral types, stellar atmospheres, metals, carbon stars, C-type, S-type, Wolf-Rayet stars, mass loss, stellar corona*, stellar chromosphere*, transition zone*, negative hydrogen ion*

*This term is found in an optional section.

Questions

1. What are two differences between a star and a planet?

†2. (a) By how many times does the energy given off by a bit of gas increase when the gas is heated from 300 K (room temperature) to 600 K? (b) To 6000 K?

†3. The sun's spectrum peaks at 5600 Å. Would the spectrum of a star whose temperature is twice that of the sun peak at a longer or a shorter wavelength? How much more energy than a square centimeter of the sun would a square centimeter of the star give off?

†4. The sun's temperature is about 5800 K and its spectrum peaks at 5600 Å. An O star's temperature may be 40,000 K. At what wavelength does its spectrum peak?

5. For the O star of Question 4, in what part of the spectrum does its spectrum peak? Can the peak be observed with the Palomar telescope? Explain.

†6. One black body peaks at 2000 Å. Another, of the same size, peaks at 10,000 Å. Which gives out more radiation at 2000 Å? Which gives out more radiation at 10,000 Å? What is the ratio of the total radiation given off by the two bodies?

†7. A coolish B star has twice the temperature of the sun in kelvins. (a) How many times more than the sun does each square centimeter of the B star's surface radiate? (b) The B star's diameter is 5 times the sun's. How many times more than the sun does the whole star radiate, given that the surface area of a sphere increases as the radius squared ($A = 4\pi r^2$)?

8. Which contains more information, Wien's displacement law or Planck's law? Explain.

†9. What is the ratio of energy output for a bit of surface of an average O star and a same-sized bit of surface of the sun?

10. Star A appears to have the same brightness through a red and a blue filter. Star B appears brighter in the red than in the blue. Star C appears brighter in the blue than in the red. Rank these stars in order of increasing temperature.

11. (a) From looking at Figure 20–11, draw the Lyman series and the Balmer series on the same wavelength axis, that is, on a horizontal straight line labelled in angstroms every 100 Å from 0 to 10,000 Å. (b) Why is the Balmer series the most observed spectral series of atomic hydrogen?

12. What is the difference between the continuum and an absorption line? The continuum and an emission line? Draw a continuum with absorption lines. Can you draw absorption lines without a continuum? Can you draw emission lines without a continuum? Explain.

†13. Consider a hypothetical atom in which the energy levels are equally spaced from each other, that is, the energy of level n is n. (a) Draw the energy level diagram for the

†This indicates a question requiring a numerical solution.

first five levels. (b) Indicate on the diagram the transition from level 4 to level 2 and from level 4 to level 3. (c) What is the ratio of energies of these two transitions? (d) If the $4 \rightarrow 2$ transition has a wavelength of 4000 Å, what is the wavelength of the $3 \rightarrow 2$ transition? (Hint: do not confuse the energy levels in this hypothetical atom with the distinct pattern of energy levels for the hydrogen atom.)

14. Does the spectrum of the solar surface show emission or absorption lines? Compare and/or distinguish between the sun's spectrum and the spectrum of a B star.

15. (a) Why are singly ionized helium lines detectable only in O stars? (b) Why aren't neutral helium lines prominent in the solar spectrum?

16. Compare Planck curves for stars of spectral types O, G, and M.

17. Why don't we see strong molecular lines in the sun?

18. Make up your own mnemonic for the spectral types. (Send the winning entries of class contests to me, please.)

†19. Using Balmer's formula ($1/\lambda = \text{constant} \times (1/n^2 - 1/m^2)$, where n is the principal quantum number of one level and m is the principal quantum number of the other level) and the fact that the wavelength of Hα is 6563 Å, calculate the wavelength of Hβ. Show your work.

20. What factors determine spectral type? What instrument would you use to determine the spectral type of a star?

21. List spectral types of stars in approximate order of strength, from strongest to weakest, of (a) hydrogen lines, (b) ionized calcium lines, and (c) titanium oxide lines.

22. If we are examining a stellar spectrum, explain how and why comparing the relative intensity of the absorption line at about 3970 Å with the intensity of the calcium K line allows us to judge the star's spectral type.

23. If we take two stones, one twice the diameter of the other, and put them in an oven until they are heated to the same temperature and begin glowing, what will be the relationship between the total energy in the light given off by the stones? (The surface area of a sphere is proportional to the square of the radius.)

24. Why does the negative hydrogen atom have a continuous rather than a line spectrum?

Finding the distances to the stars is not straightforward. Here we see an open cluster of stars in the constellation Crux known as the Jewel Box. Studying clusters of stars has helped us establish methods of finding stellar distances.

Stellar Distances and Motions

21

Aims: To describe the absolute and apparent magnitude scales, the method of trigonometric parallax and how it tells us the distances to the stars, the Hertzsprung-Russell diagram and what it tells us about types of stars and their distances, and how we learn about stellar motions

We have seen in Chapter 20 how we can analyze the spectrum of a star to tell us about the outer layer of that star; for example, we can tell how hot it is. In this chapter, we see how information about the spectra of many stars can be put together to tell us about the overall properties of stars.

We begin by describing the scale in which astronomers give brightness. We also see how we can fairly directly measure the distances to the nearest stars and how these distances, together with classification by spectral type, can give us distances to farther stars. We see how graphing the temperatures and brightnesses of stars gives us an important tool: the Hertzsprung-Russell diagram. Then we study how astronomers tell the speed and direction in which stars are moving.

21.1 Light from the Stars

The most obvious thing we notice about the stars is that they have different brightnesses. Over 2000 years ago, the Greek astronomer Hipparchus (who lived until about 127 B.C.) made a catalogue of stars and divided the stars into classes of brightness, using terms like "bright" and "small." Little else is known of his catalogue. By the early 1st century A.D., a system of six classes of stellar brightness was well known. In the *Almagest* of Ptolemy (written in Alexandria in about A.D. 140), the brightest stars were said to be of the first magnitude, a reasonable idea. Somewhat fainter stars were said to be of the second magnitude, and so on down to the sixth magnitude, which represented the faintest stars that could be seen with the naked eye. Ptolemy often quoted Hipparchus, and it is often not known which ideas belonged to each; most of what we know of Hipparchus comes from Ptolemy's *Almagest.*

21.1a Apparent Magnitude

In the nineteenth century, when astronomers became able to make quantitative measurements of the brightnesses of stars, the magnitude scale was placed on an accurate basis. It was discovered that the brightest stars that could be seen with the naked eye were about 100 times brighter than the faintest stars. This corresponded to a difference of 5 magnitudes between the old first magnitude and sixth magnitude, a number that was used to set up the present magnitude system. Since this type of magnitude tells us how bright a star **appears**, it is called *apparent magnitude.*

The new system is based on the definition that stars that, on the magnitude scale, show a **difference** of five magnitudes are different by a **factor** of exactly

100 times. Thus a star of first magnitude is exactly 100 times brighter than a star of sixth magnitude.

The brightnesses of the stars that had been classified by the Greeks were placed on this new magnitude scale (Fig. 21–1). Many of the stars that had been of the first magnitude in the old system were indeed "first magnitude stars" in the new system. But a few stars were much brighter, and thus corresponded to "zeroth" magnitude, that is, a number less than one. A couple, like Sirius, the brightest star in the night sky, were brighter still. The numerical magnitude scale was easily extended to negative numbers. The new scale, being numerical, admits fractional magnitudes. Sirius is actually magnitude −1.5, for example. Physiologically, this type of measurement makes sense because the human eye happens to perceive equal **factors** of luminosity (such a factor is the energy coming from one object divided by the energy coming from another) as roughly equal **intervals** (additive steps) of brightness, which is just how the magnitude scale is set up.

$\frac{1}{2}$ mag =	1.585	times
1 mag =	2.512	times
2 mag =	6.310	times
3 mag =	15.85	times
4 mag =	39.81	times
5 mag =	100	times
6 mag =	251.2	times
7 mag =	631.0	times
8 mag =	1585	times
9 mag =	3981	times
10 mag =	10^4	times
15 mag =	10^6	times
20 mag =	10^8	times

The magnitude scale was extended not only to stars brighter than first magnitude, but also to stars fainter than 6th magnitude. Since a difference of 5 magnitudes corresponds to a factor of a hundred times in brightness, 11th magnitude is exactly 100 times fainter than 6th magnitude (that is, 1/100 times as bright); 16th magnitude, in turn, is exactly 100 times fainter than 11th magnitude. The faintest stars that can be photographed from the ground are fainter than 24th magnitude.

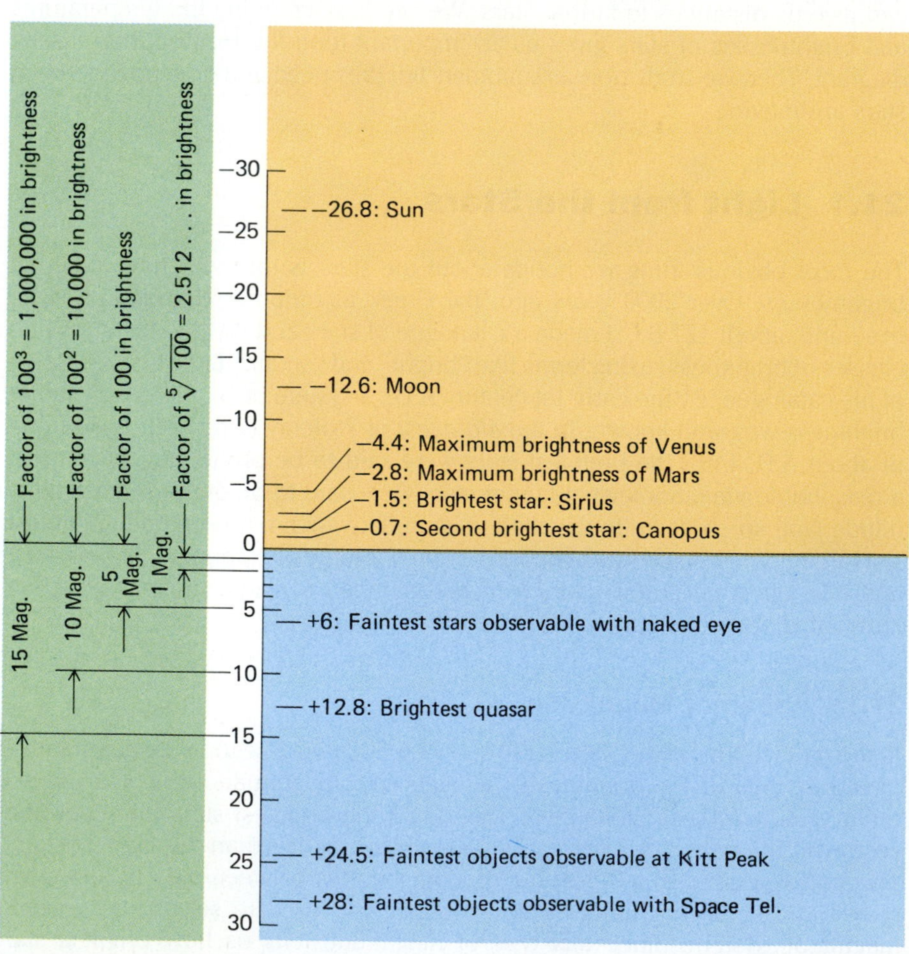

Figure 21–1 The apparent magnitude scale is shown on the vertical axis. At the left, sample intervals of 1, 5, 10, and 15 magnitudes are marked and translated into multiplicative factors.

*21.1b Working with the Magnitude Scale

We have seen how a difference of 5 magnitudes corresponds to a factor of 100 in brightness. What does that indicate about a difference of 1 magnitude? Since an increase of 1 in the magnitude scale corresponds to a decrease in brightness by a certain factor, we need a number that, when 5 of them are multiplied together, will equal 100. This number is just $\sqrt[5]{100}$. Its value is 2.512..., with the dots representing an infinite string of other digits. For many practical purposes, it is sufficient to know that it is approximately 2.5.

Thus a second magnitude star is about 2.5 times fainter than a first magnitude star. A third magnitude star is about 2.5 times fainter than a second magnitude star. Thus, a third magnitude star is approximately $(2.5)^2$ (which is about 6) times fainter than a first magnitude star. (We could easily give more decimal places, writing 6.3 or 6.31, but the additional figures would not be meaningful. We started with only one digit being significant when we wrote down "2nd magnitude," so our data were given to only one *significant figure.* Our result is thus only accurate to one significant figure.)

We must be careful not to fool ourselves that we can gain in accuracy in an arithmetical process; the number of significant figures we put in is the number of significant figures in the result.

In similar fashion, using a factor of 2.5 or 2.512 for the single magnitudes and a factor of 100 for each group of 5 magnitudes, we can easily find the ratio of intensities of stars of any brightness.

Example: By what factor does the brightness of stars of magnitudes -1 and 6 differ (that is, what is the brightness ratio between the brightest and faintest stars, respectively, that are seen with the naked eye)?
Answer: Since the stars differ by 7 magnitudes, the factor is (100)(2.5)(2.5), which is about 600. The star of -1st magnitude is 600 times brighter than the star of 6th magnitude.

One unfortunate thing about the magnitude scale is that it operates in the opposite sense from a direct measure of brightness. Thus the brighter the star, the lower the magnitude (a second magnitude star is brighter than a third magnitude star), which is sometimes a confusing convention. This kind of problem comes up occasionally in an old science like astronomy, for at each stage astronomers have made their new definitions so as to have continuity with the past.

Figure 21–2 "Excuse me for shouting—I thought you were further away." Without clues to indicate distance, we cannot properly estimate size. (Reproduced by special permission of PLAYBOY Magazine; copyright © 1971 by Playboy)

21.2 Stellar Distances

We can tell a lot by looking at a star or by examining its radiation through a spectrograph, but such observations do not tell us directly how far away the star is. Since all the stars are but points of light in the sky even when observed through the biggest telescopes, we have no reference scale to give us their distances (Fig. 21–2).

The best way to find the distance to a star is to use the principle that is also used in rangefinders in cameras. Sight toward the star from different locations, and see how the direction toward the star changes with respect to a more distant background of stars. You can see a similar effect by holding out your thumb at arm's length. Examine it first with one eye closed and then with the other eye closed. Your thumb seems to change in position as projected against a distant background; it appears to move across the background by a certain angle, about 3°. This is because your eyes are a few centimeters apart from each other, so each eye has a different point of view.

In close-up photography, "parallax error" can cause you to mistakenly leave someone's head out of the picture, if you don't take account of the parallax caused by the fact that the lens and your viewfinder are in different places.

Now hold your thumb up closer to your face, just a few centimeters away. Note that the angle your thumb seems to jump across the background as you look through first one eye and then through the other is greater than it was before. Exactly the same effect can be used for finding the distances to nearby stars. The nearer the star is to us, the farther it will appear to move across the background of distant stars, which are so far away that they do not appear to move with respect to each other.

To maximize this effect, we want to observe from two places that are separated from each other as much as possible. For us on earth, that turns out to be the position of the earth at intervals of six months. In that six-month period, the earth moves to the opposite side of the sun. In principle, we observe the position of a star in the sky (photographically), with respect to the background stars, and several months later, when the earth has moved part way around the sun, we repeat the observation. (Six months later, the star would be up in the daytime and could not be seen.)

The straight line joining the points in space from which we observe is called the *baseline.* The distance from the earth to the sun is called an *Astronomical Unit* (A.U.). In the case above, we have observed with a baseline 2 A.U. in length. The angle across the sky that a star seems to move (with respect to the background of other stars) between two observations made from the ends of a baseline of 1 A.U. (half the maximum possible baseline) is called the *parallax* of the star (Fig. 21–3). From the parallax, we can use simple trigonometry to calculate the distance to the star. The process is thus called the method of *trigonometric parallax.*

The basic limitation to the method of trigonometric parallax is that the farther away the star is, the smaller is its parallax. It turns out that this method can be used for only about ten thousand of the stars closest to us; most other stars have parallaxes too small for us to measure. (These other stars, with negligible parallaxes, are the distant stars that provide the unmoving background against which we compare.) Even Proxima Centauri, the star nearest to us beyond the sun, has a parallax of only ¾ arc sec, the angle subtended (taken up across our vision) by a dime at a distance of 5 km! Clearly we must find other methods to measure the distances to most of the stars. But for the cases in which it can be used, the method of trigonometric parallaxes gives the most accurate answers.

Parallaxes are measured by comparing pairs of photographs taken several months apart. Though star images are never smaller than about 1 second of arc in size, their centers can be measured more accurately, to about 1/100 of an arc second. Measurements are repeated over several years to determine the effect of actual motions of stars and to reduce observational errors. Trigonometric parallaxes can now be measured for stars out to about 300 light years from the sun, which is less than 1 per cent of the diameter of our galaxy. Only stars brighter than 20th magnitude can be measured.

A European Space Agency satellite devoted entirely to measuring positions was launched in 1989, but unfortunately did not reach the correct orbit. It is named Hipparcos (**Hi**gh **P**recision **Par**allax **Co**llecting **S**atellite), after the Greek astronomer whose name we often spell Hipparchus. Even with its improper orbit, which will lead to a curtailed lifetime, Hipparcos should lead to improvement in *astrometry,* the study of the positions of stars and their apparent motion across the sky. New astrometric measurements are also necessary for providing accurate positions for faint optical objects so they can be identified with emitters in other parts of the spectrum, such as those identified in radio astronomy or with x-ray satellites.

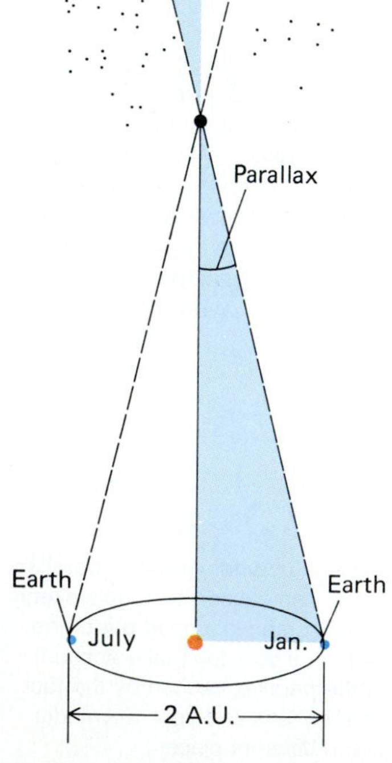

Figure 21–3 The nearer stars seem to be slightly displaced with respect to the farther stars when viewed from different locations in the earth's orbit.

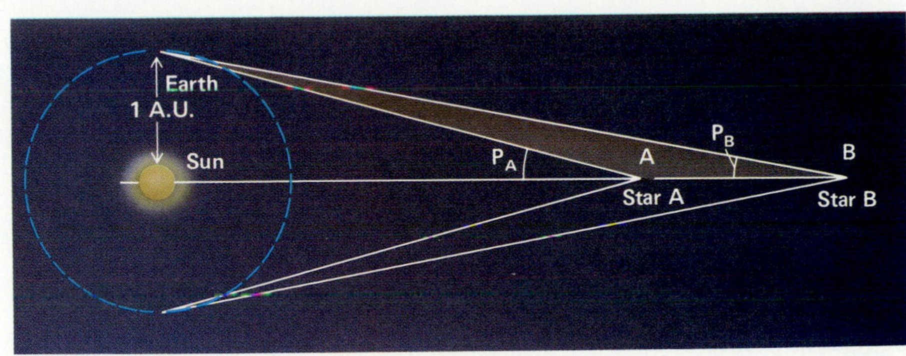

Figure 21–4 Star B is farther from the sun than star A and thus has a smaller parallax. The parallax angles are marked P_A and P_B. To *subtend* is to take up an angle; as seen from point A, 1 A.U. subtends an angle P_A. As seen from point B, the same 1 A.U. subtends an angle P_B.

The Hubble Space Telescope should be able to use its very high resolution to pinpoint the positions of a small number of stars even more accurately than Hipparcos.

21.2a Parsecs

Since parallax measures have such an important place in the history of astronomy, a unit of distance was defined in terms that relate to these measurements. If we were outside the solar system and looked back at the earth and the sun, they would appear to us to be separated in the sky. We could measure the angle by which they are separated. From twice as far away as Pluto, for example, when the earth and the sun were separated by the maximum amount (1 A.U.), they would be approximately 1° apart. As we go farther and farther away from the solar system, 1 A.U. subtends a smaller and smaller angle. When we go about 60 times farther away, 1 A.U. subtends only 1 arc min (60 arc min = 1°). From 60 times still farther, 1 A.U. subtends 1 arc sec (60 arc sec = 1 arc min). Note that the **angle** subtended by the astronomical unit is a measure of the **distance** we have gone (Fig. 21–4), similar to the way that we measure an object's parallax.

The distance at which 1 A.U. subtends only 1 arc sec, we call *1 parsec* (Fig. 21–5). A star that is 1 parsec from the sun has a **par**allax of one arc **sec**. One parsec is a long distance. When talking about stellar distances, most astronomers tend to use parsecs instead of light years, the distance that light travels in a year (= 9.5×10^{12} km). It takes light about 3.26 years to travel 1 parsec; thus 1 parsec = 3.26 light years. Note that parsecs and light years are both **distances**, just like kilometers or miles, even though their names contain references to their definitions in terms of angles or time.

*21.2b Computing with Parsecs

The advantage of using parsecs instead of light years, kilometers, or any other distance unit is that the distance in parsecs is equal to the inverse of the parallax angle in seconds of arc:

$$d(\text{parsecs}) = 1/p \text{ (seconds of arc)}.$$

For example, a star with a parallax angle of 0.5 arc sec is 2 parsecs away (1/0.5 = 2), and would thus be one of the nearest stars to the sun. A star with

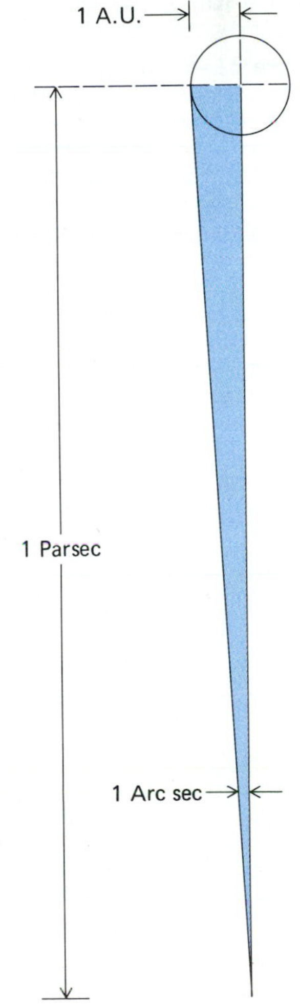

Figure 21–5 A *parsec* is the distance at which 1 A.U. subtends an angle of 1 arc sec. The drawing is not to scale. One arc sec is actually a tiny angle, the width of a dime at a distance of 4 km.

Absolute magnitude	Source
Giant elliptical galaxies	−23
Supernova 1987A	−15.5
Globular clusters	−10 to −6
Brightest supergiant stars	−9
Sun	+4.79
Faintest stars	+20

Figure 21–6 Some stars look bright because they are both intrinsically bright and close to us. A faint star, on the other hand, could be intrinsically bright yet far away. A photograph of the sky shows stars at a wide range of distances and of intrinsic brightnesses. The photograph shows not only stars but also nebulosity between the stars in the constellation Sagittarius.

a parallax of 0.01 is 1/0.01 = 100 parsecs away. Since the parallax angle is the quantity that is measured directly at the telescope, it was convenient to choose a distance unit very closely related to the parallax angle.

21.3 Absolute Magnitudes

We have thus far defined only the apparent magnitudes of stars, how bright the stars **appear** to us. However, stars can appear relatively bright or faint for either of two reasons: they could be intrinsically bright or faint, or they could be relatively close or far away (Fig. 21–6). We can remove the distance effect, giving us a measure of how bright a star actually is, by choosing a standard distance and considering how bright all stars would appear if they were at that standard distance. The standard distance that we choose is 10 parsecs. We define the *absolute magnitude* of a star to be the magnitude that the star would appear to have if it were at a distance of 10 parsecs. Absolute magnitude gives the intrinsic brightness of a star, its *luminosity,* the total amount of energy the star gives off each second.

We normally write absolute magnitude with a capital M, and apparent magnitude with a small m. Sometimes a subscript signifies that we have observed through a special filter: M_V is the absolute magnitude through a special "visual" filter.

If a star happens to be exactly 10 parsecs away from us, its absolute magnitude is exactly the same as its apparent magnitude. If we were to take a star that is farther than 10 parsecs away from us and somehow move it to be 10 parsecs away, then it would appear brighter to us than it does at its actual position (since it is closer). Since the star would be brighter, its absolute magnitude would be a lower number (for example, 2 rather than 6) than its apparent magnitude.

On the other hand, if we were to take a star that is closer to us than 10 parsecs, and move it to 10 parsecs away, it would then be farther away and therefore fainter. Its absolute magnitude would be higher (more positive, for example, 10 instead of 6) than its apparent magnitude.

To assess just how much brighter or fainter that star would appear at 10 parsecs than at its real position, we must realize that the intensity of light from a star follows the *inverse-square law* (Fig. 21–7). That is, the intensity of a star varies inversely with the square of the distance of the star from us. (This law holds for all point sources of radiation, that is, sources that appear as points

Focus On

*Box 21.1 Linking Apparent Magnitude, Absolute Magnitude, and Distance

Astronomers have a formula that allows you to calculate one of the terms apparent magnitude, absolute magnitude, or distance, if you know the other two. But the formula merely does numerically what we have just carried out logically. The formula is $m - M = 5 \log_{10} (r/10)$, where m is the apparent magnitude, M is the absolute magnitude, and r is the distance in parsecs. Note that if $r = 10$, then $r/10 = 1$, $\log 1 = 0$, and $m = M$. If $r = 20$, as in the first example above, $m - M = 5 \log 2 = 5 \times 0.3 = 1.5$ magnitudes. Be careful not to get carried away using the formula without understanding the point of the manipulations.

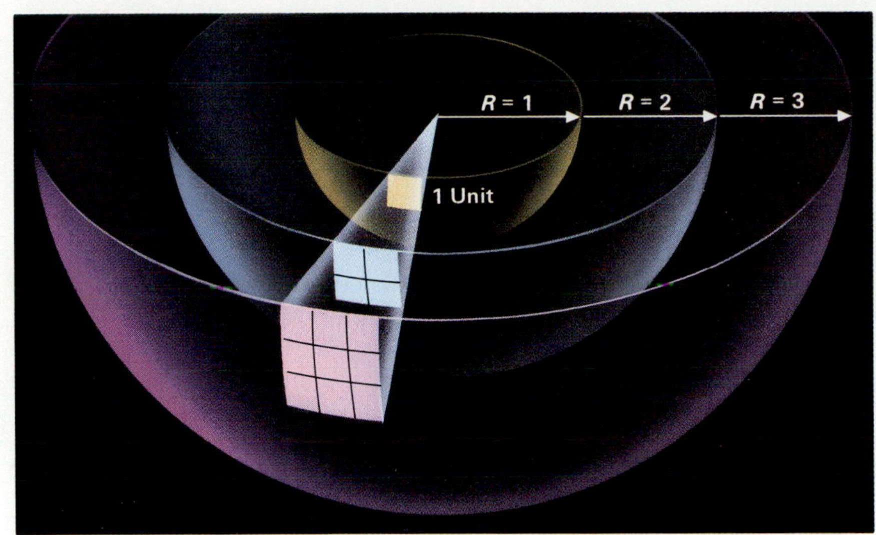

Figure 21–7 The inverse-square law. Radiation passing through a sphere twice as far away as another sphere has spread out so that it covers $2^2 = 4$ times the area; n times farther away it covers n^2 times the area.

without length or breadth.) If we could move a star 2 times as far away, it would grow 4 times fainter. If we could move a star 9 times as far away, it would grow 81 times fainter. Of course, we are not physically moving stars (we would burn our shoulders while pushing), but merely considering how they would appear at different distances. In all this, we are assuming that there is no matter in space between the stars to absorb light. Unfortunately, this is not always a good assumption (as we shall see in Chapter 30), but we make it anyway.

Example: A star is 20 parsecs away from us, and its apparent magnitude is +4. What is its absolute magnitude?
Answer: If the star were moved to the standard distance of 10 parsecs away, it would be twice as close as it was at 20 parsecs and, therefore, by the inverse-square law, would appear four times brighter. Since 2.5 times is one magnitude, and $(2.5)^2 = 6.25$ is two magnitudes, it would be approximately 1½ magnitudes brighter. Since its actual apparent magnitude is 4, its absolute magnitude would be $4 - 1½ = 2½$, equivalent to the apparent magnitude it would have if it were 10 parsecs from us. (We see that we must subtract the magnitude difference, since the star would be closer and therefore brighter at 10 parsecs.)

If you know one type of magnitude and the distance, and want to find the other type of magnitude: (1) Find the ratio of the distances (the distance of the star and the 10-parsec standard distance; make sure the star's distance is also in parsecs); (2) square to find the factor by which the brightness is changed; (3) convert this factor to a difference in magnitudes (using each 1 magnitude = a factor of 2.5; each 5 magnitudes = a factor of 100); (4) add or subtract the magnitude difference to change between absolute and apparent magnitude.

Example: A star has apparent magnitude 10 and absolute magnitude 5. How far away is it?
Answer: The star would grow brighter by 5 magnitudes were it to be moved to the standard distance of 10 parsecs. It would thus brighten by 100 times. By the inverse-square law, it would do so if it were 10 times closer than its real position. Its distance must therefore be 10 times farther away than the

standard distance of 10 parsecs, and 10 × 10 parsecs is 100 parsecs. Note that to find the distance in light years, we must multiply by 3.26 parsecs/light year. The distance here is thus 326 light years.

We have followed the following steps, knowing both types of magnitudes and wanting to find the distance: (1) Find the difference in magnitudes, by subtracting; (2) convert this difference in magnitudes into a factor by which the brightness is changed (using a factor of 2.5 = 1 magnitude; a factor of 100 = 5 magnitudes); (3) find the change in distance by taking the square root of the factor of brightness; (4) multiply or divide the standard distance by this factor of distance, multiplying if the star's apparent magnitude is fainter than its absolute magnitude (and, if necessary, converting from parsecs to light years).

Example: A star has apparent magnitude 10 and absolute magnitude 3. How far away is it? (This is an example in which the numbers don't work out quite as smoothly.)
Answer: The star would grow brighter by 7 magnitudes if we were to move it to the standard distance, making a factor of brightness of $(2.5)^2 \times 100 =$ 600 (approximately). By the inverse-square law, it is thus growing closer by the square root of 600, which is about 25. (After all, $20^2 = 400$, and $30^2 =$ 900, so the answer must be in between 20 and 30.) The star must thus actually be 25 times 10 parsecs, or 250 parsecs away.

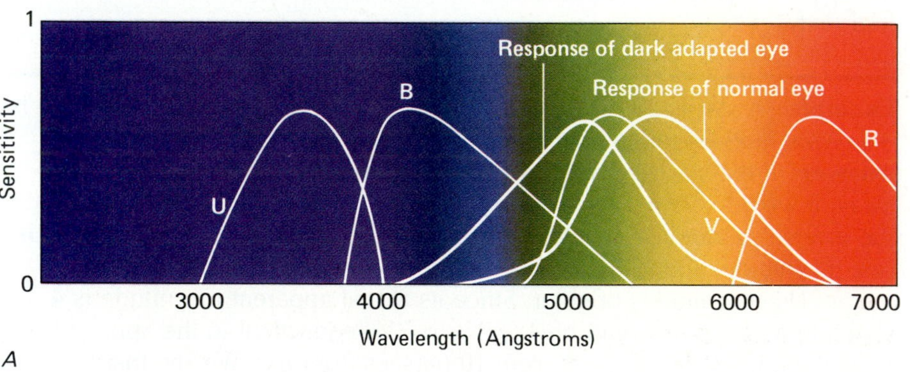

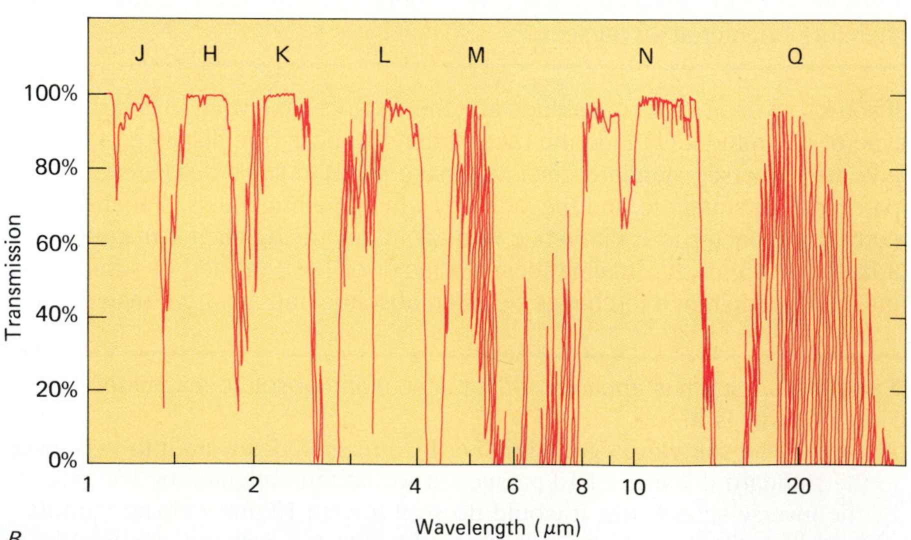

Figure 21–8 (*A*) The U, B, and V curves (ultraviolet, blue, and visual = yellow/green) represent the standard set of filters used by many astronomers. The response of the eye under normal conditions and the response of the dark-adapted eye are also shown. (*B*) The windows of transparency correspond to transmission of 100%; the infrared filters that correspond to the respective windows are marked.

*21.3a Photometry

The actual value of either absolute or apparent magnitude can depend on the wavelength region in which we are observing. Let us consider a blue star and a red star, each of which gives off the same amount of energy. The blue star can seem much brighter than the red star if we observe them both through a blue filter, while the red star can seem much brighter than the blue if we observe them both through a red filter. Astronomers often measure a *color index* (Section 21.3b) by subtracting the magnitude in one spectral range from the magnitude in another. Since the color of a star gives its temperature, color index is a measure of temperature.

A standard set of filters (Fig. 21–8*A*) has been defined for use in photoelectric systems, with one filter in the ultraviolet (called U), one filter in the blue (called B), and one filter in the yellow to more-or-less match the eye (called V for visual). Sets of equivalent filters exist at observatories all over the world. Hundreds of thousands of stars have had their colors measured with this *UBV* set of filters; we call the process *three-color photometry.*

A standard extension of this filter set into the red and infrared has been made (Table 21–1), in almost (but not quite) alphabetical order. The bands correspond to windows of transparency in the earth's atmosphere (Fig. 21–8*B*). The IRAS spacecraft has expanded cataloguing to longer infrared wavelengths.

The quantity of fundamental importance is not the magnitude as observed through any given filter but rather the magnitude that corresponds to the total amount of energy given off by the star over all spectral ranges. This is called the *bolometric magnitude,* M_{bol}.

"Bolometer," the instrument used to measure the total amount of energy arriving in all spectral regions, derives its name from "boli," Greek for "beam of light."

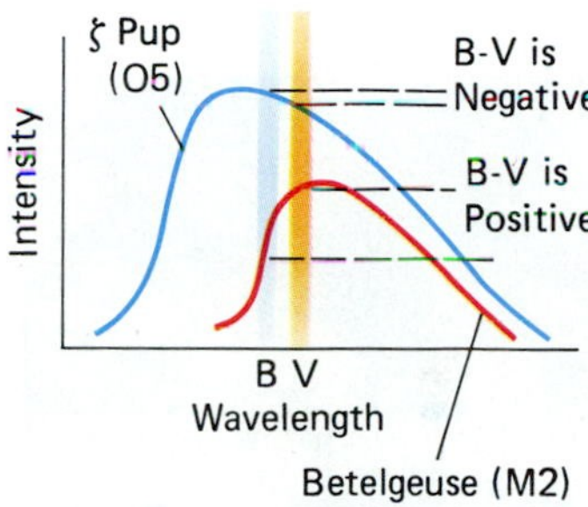

Figure 21–9 Hotter stars have negative color indices while cooler stars have positive color indices. (The star is brighter in the blue than in the visible; that is, B is a lower magnitude than V, so B − V is less than 0.)

21.3b Color Index

A measure of temperature often used in astronomy is the difference between the apparent magnitudes measured in two spectral regions, for example, B − V, blue minus visual. This standard system of filters was defined in the previous subsection.

The difference B − V is called the *color index* (Fig. 21–9). The blue magnitude, B, measures bluer radiation than the visual magnitude, V. A very

Table 21–1 Astronomical Standard Filters

Filter	Central Wavelength (μm)	Band Width (μm)	Corresponds to Peak: *Temperature*	Corresponds to Peak: *Example*
U	0.36 μm	0.07 μm	8500 K	Hot star
B	0.44 μm	0.10 μm	7000 K	
V	0.55 μm	0.08 μm	5500 K	Solar-type star
R	0.70 μm	0.21 μm	4300 K	
I	0.90 μm	0.22 μm	3300 K	Red giant
J	1.25 μm	0.4 μm	2400 K	
H	1.6 μm	0.5 μm	1900 K	Cool star
K	2.2 μm	0.6 μm	1400 K	
L	3.4 μm	0.7 μm	900 K	Circumstellar dust
M	5.0 μm	1.2 μm	600 K	
N	10 μm	5 μm	300 K	Room temperature
(IRAS)	12 μm			
(IRAS)	25 μm		120 K	Cool star shells
(IRAS)	60 μm			
(IRAS)	100 μm		30 K	Dust in galaxies

Figure 21–10 Ejnar Hertzsprung in the 1930's.

hot star is brighter in the blue than in the visible; thus V is fainter, that is, a higher number, than B. Therefore B − V is negative for hot stars.

The B − V color index is zero for an A star of about 10,000 K. It falls in the range form −0.3 for the hottest stars to about +2.0 for the coolest. One can also compute a color index U − B for the U and B (ultraviolet and blue) magnitudes, or indeed a color index for magnitudes measured in any two spectral regions.

21.4 The Hertzsprung-Russell Diagram

In about 1910, Ejnar Hertzsprung (Fig. 21–10) in Denmark and Henry Norris Russell (Fig. 21–11) at Princeton University in the United States independently plotted a new kind of graph (Fig. 21–12). On the horizontal axis, the *x*-axis, each graphed a quantity that measured the temperature of stars. On the vertical axis, the *y*-axis, each graphed a quantity that measured the intrinsic brightness of stars. They found that all the points that they plotted fell in limited regions of the graph rather than being widely distributed over the graph.

Figure 21–11 Henry Norris Russell and his family, circa 1917.

But remember that a star can appear bright either by really being intrinsically bright or, alternatively, by being very close to us. One way to get around this problem is to plot only stars that are at the same distance away from us (Fig. 21–13).

If we somehow knew the absolute magnitudes of the stars, that would also be a good thing to plot. When we know the distances (for relatively close stars, we can get this from trigonometric parallax measurements, which is what Russell originally did), we can calculate the absolute magnitudes, which are actually what is plotted in Figure 21–12.

Such a plot of temperature versus brightness is known as a *Hertzsprung-Russell diagram,* or simply as an *H–R diagram.* Note that since H–R diagrams were sometimes originally plotted by spectral type, from O to M, the hottest stars are on the left side of the graph. Thus temperature increases from right to left. Also, since the brightest stars are on the top, magnitude decreases toward the top. In some sense, thus, both axes are plotted backwards from the way a reasonable person might choose to do it if there were no historical reasons for doing it otherwise. When a color index is used to show temperature, we have a *color-magnitude diagram.*

When plotted on a Hertzsprung-Russell diagram, the stars lie mainly on a diagonal band from upper left to lower right (Fig. 21–14). Thus the hottest stars are normally brighter than the cooler stars. Most stars fall very close to this band, which is called the *main sequence.* Stars on the main sequence are called *dwarf stars,* or *dwarfs.* There is nothing strange about dwarfs; they are the normal kind of stars. The sun is a type G dwarf. Some dwarfs are quite large and bright; the word "dwarf" is used only in the sense that these stars are not a larger, brighter kind of star that we will define below as giant.

Some stars lie above and to the right of the main sequence. That is, for a given spectral type the star is intrinsically brighter than a main-sequence star. These stars are called *giants,* because their luminosities are large compared to dwarfs of their spectral type. Some stars, like Betelgeuse, are even brighter than normal giants, and are called *supergiants.* Since two stars of the same spectral type have the same temperature and, according to the Stefan-Boltzmann law (Section 20.2), the same amount of emission from each area of their surfaces, the brighter star must be bigger than the fainter star of the same spectral type.

Stars in a class of faint hot objects, called *white dwarfs,* are located below and to the left of the main sequence. They are smaller and fainter than main-sequence stars (ordinary dwarfs) of the same spectral type.

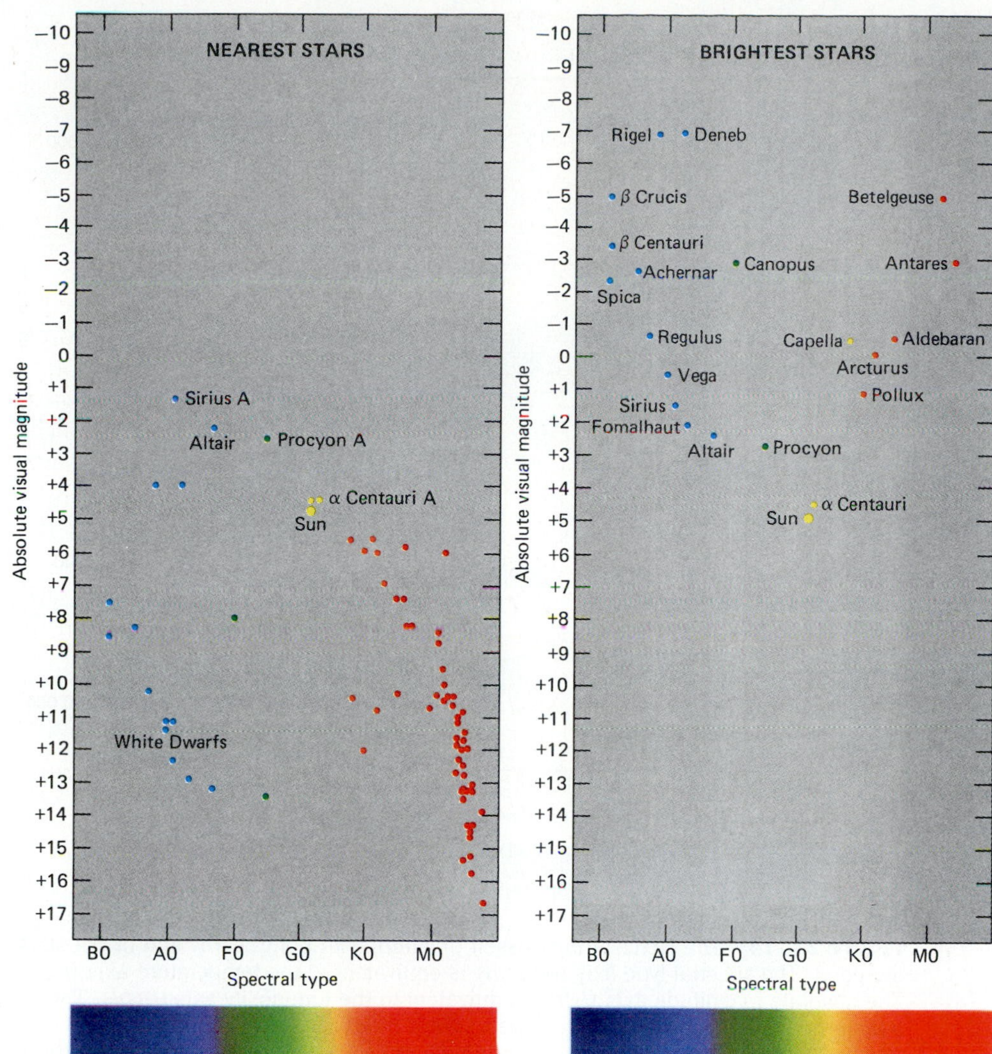

Figure 21–12 Hertzsprung-Russell diagrams (*A*) for the nearest stars in the sky and (*B*) for the brightest stars in the sky. The brightness scale is given in *absolute magnitude*. Because the effect of distance has been removed, the intrinsic properties of the stars can be compared directly on such a diagram.

Note that none of the nearest stars is intrinsically very bright. Also, the brightest stars in the sky are, for the most part, intrinsically very luminous, even though they are not usually the very closest to us. The color bars show the overall color of the star.

Figure 21–13 M13, a globular cluster in Hercules, is the most prominent in the northern sky. It is high in the evening sky from late spring until early autumn. Stars in a given cluster are all essentially the same distance from us.

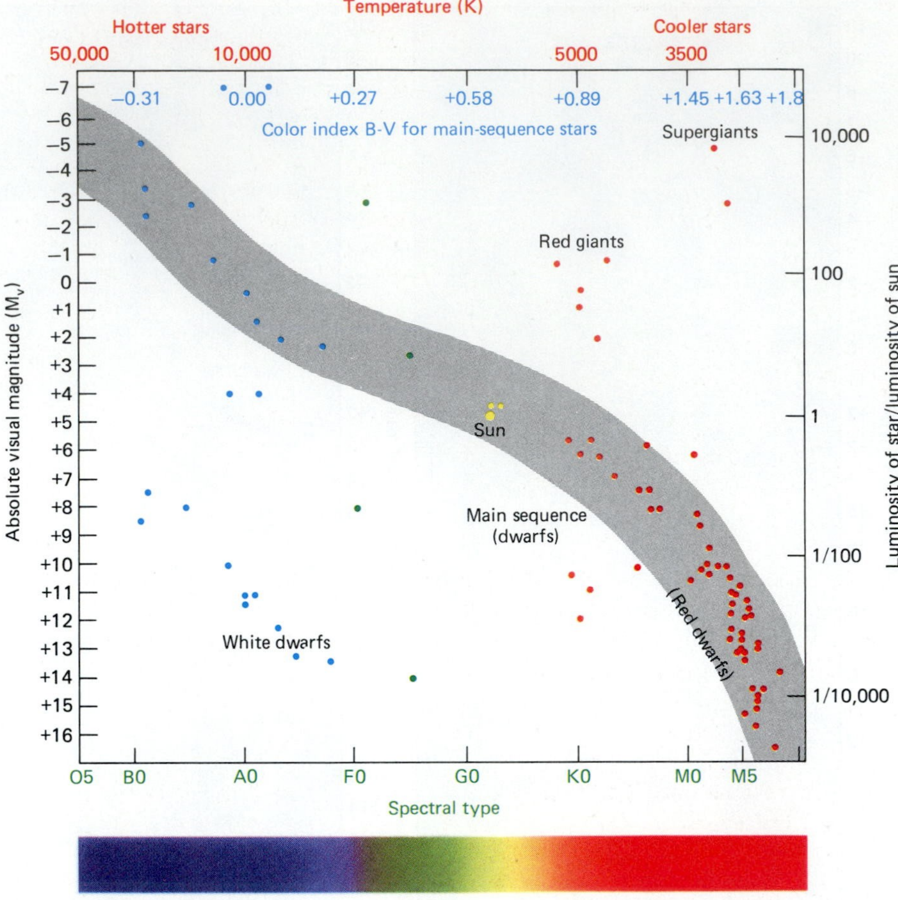

Figure 21–14 The Hertzsprung-Russell diagram, with both nearby and bright stars included. The spectral-type axis (*bottom*) is equivalent to the temperature axis (*top*). The absolute magnitude axis (*left*) is equivalent to the luminosity axis (*right*). The transformation given from spectral type to color index (B-V) is valid for main-sequence stars only; the correspondence is different for giants, supergiants, and other classes of stars.

The use of the Hertzsprung-Russell diagram to link the spectrum of a star with its brightness is a very important tool for stellar astronomers. The H–R diagram provides, for example, another way of measuring the distances to stars, as we shall see below.

21.5 Spectroscopic Parallax

When we take a group of stars whose distances we can measure directly, by some method like that of trigonometric parallax, we can plot a Hertzsprung-Russell diagram. From this standard diagram we can read off the absolute magnitude that corresponds to any star on the main sequence.

We can apply the standard Hertzsprung-Russell diagram to find the distance to any star whose spectrum we can observe, no matter how faint, if we know that the star is on the main sequence. Examining the spectrum in detail reveals to trained eyes whether or not a star is on the main sequence (Fig. 21–15).

For a main-sequence star, we first find the absolute magnitude that corresponds to its spectral type. We must also know the apparent magnitude of

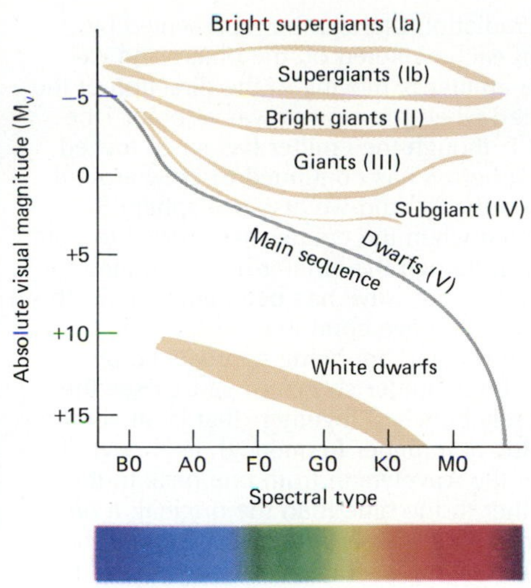

Figure 21–15 One can tell whether a star is a dwarf, giant, or supergiant by looking closely at its spectrum, because slight differences exist for a given spectral type. This procedure is called "luminosity classification." Once we know a star's luminosity class, we can place the star on the H–R diagram and find its spectroscopic parallax. Luminosity classes are specified by numbers (shown in parentheses) or names.

the star, but we can get that easily and directly by simply observing it. Once we know both the apparent magnitude and the absolute magnitude, we have merely to figure out how far the star has to be from the standard distance of 10 parsecs to account for the difference $m - M$. (We get this from the inverse-square law.)

Example: We see a G2 star on the main sequence. Its apparent magnitude is +8. How far away is it?
Answer: From the H–R diagram, we see that $M = +5$. The star appears fainter than it would be if it were at 10 parsecs. It is approximately $(2.5)^3 = 6 \times 2.5 = 15$ times fainter. (More accuracy in carrying out this calculation would not be helpful because the original data were not more accurate.) By the inverse-square law, this means that it is approximately 4 times farther away (exactly 4 times would be a factor of $4^2 = 16$ in brightness). Thus the star is 4 times 10 parsecs, or 40 parsecs, away.

We are measuring a distance, and not actually a parallax, but by analogy with the method of trigonometric parallax for finding distance, the method using the H–R diagram is called finding the *spectroscopic parallax.*

21.6 The Doppler Effect

The Doppler effect is one of the most important tools that astronomers can use to understand the universe. Without having to measure the distance to an object, they can use the Doppler effect to determine its *radial velocity,* its speed toward or away from us (on the radius of an imaginary sphere centered at us). The Doppler effect in sound is familiar to most of us, and its analogue in electromagnetic radiation, including light, is very similar.

21.6a Blueshifts and Redshifts

You may be familiar with how the sound of a train whistle, a jet engine, or a motorcycle motor changes in pitch as the train, plane, or motorcycle first approaches you and then passes you and begins to recede. As the object that is emitting the sound waves approaches, it has moved closer to you by the

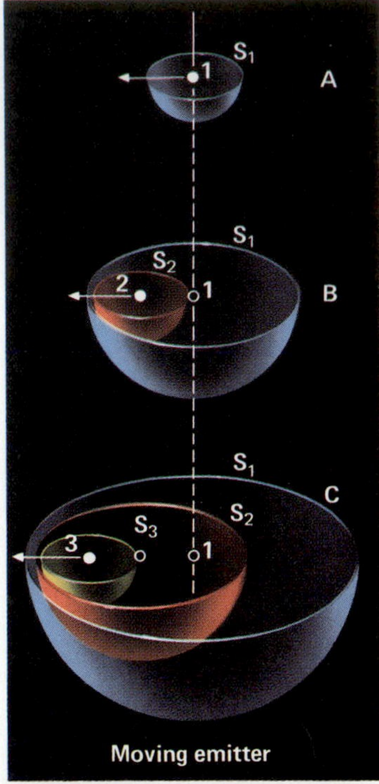

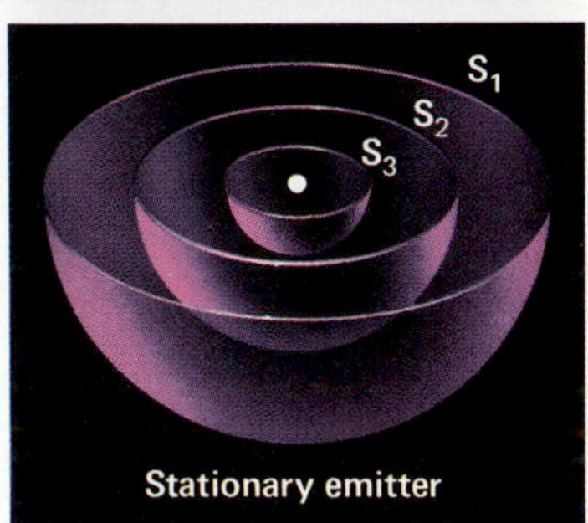

Figure 21–16 An object emits waves of radiation that can be represented by spheres representing the peaks of the wave, each centered on the object and expanding. In the left side of the drawing, the emitter is moving in the direction of the arrow. In part *A*, we see that the peak emitted when the emitter was at point 1 becomes a sphere (labelled S_1) around point 1, though the emitter has since moved toward the left. In part *B*, some time later, sphere S has continued to grow around point 1 (even though the emitter is no longer there), and we also see sphere S_2, which shows the position of the peaks emitted when the emitter had moved to point 2. Sphere S_2 is thus centered on point 2, even though the emitter has continued to move on. In part *C*, still later, yet a third peak of the wave has been emitted, S_3, this time centered on point 3, while spheres S_1 and S_2 have continued to expand.

For the case of the moving emitter, observers who are being approached by the emitting source (those on the left side of the emitter shown in part *C*) see the three peaks drawn coming past them relatively bunched together (that is, at shorter intervals of time, which is the same as saying at a higher frequency), as though the wavelength were shorter, since we measure the wavelength from one peak to the next. This corresponds to a wavelength farther to the blue than the original, a *blueshift.* Observers from whom the emitter is receding (those on the right side of the emitter shown in part *C*) see the three peaks coming past them with decreased frequency (at increased intervals of time), as though the wavelength were longer. This corresponds to a color farther to the red, a *redshift.*

Contrast the case at the bottom, in which the emitter does not move and so all the peaks are centered around the same point. No redshifts or blueshifts arise.

Note that once the light wave is emitted, it travels at a constant speed ("the speed of light," 3×10^{10} cm/sec, equivalent to seven times around the earth in a second), so that a shorter (lower) wavelength corresponds to a higher frequency, and a longer (higher) wavelength corresponds to a lower frequency.

time it emits a second wave than it had been when it emitted the first wave. Thus waves arrive more frequently than they would if the source were not moving. The wavelengths seem compressed, and the pitch is higher. After the emitting source has passed you, the wavelengths are stretched, and the pitch of the sound is lower (Fig. 21–16).

With light waves, the effect is similar. As a body emitting light, or other electromagnetic radiation, approaches you, the wavelengths become slightly shorter than they would be if the body were at rest. Visible radiation is thus shifted slightly in the direction of the blue (it doesn't actually have to become blue, only be shifted in that direction). We say that the radiation is *blueshifted.* Conversely, when the emitting object is receding, the radiation is said to be *redshifted* (Fig. 21–17). We generalize these terms to types of radiation other than light, and say that radiation is blueshifted whenever it changes to shorter wavelengths and redshifted whenever it changes to longer wavelengths.

The point of all this is that we can measure a Doppler shift for any object we can see that has a spectral line or some other feature in its spectrum for which we can measure a wavelength. We can then tell how fast the object is moving along a radius toward or away from us. If we were moving at the same speed and in the same direction as a source of radiation or sound, it would have no net velocity with respect to us. Then no Doppler shift would be observed.

*21.6b Working with Doppler Shifts

The wavelength when the emitter is at rest is called the *rest wavelength.* Let us consider a moving emitter. The fraction of the rest wavelength that the wavelength of light is shifted is the same as the fraction of the speed of light at which the body is travelling toward or away (or, for a sound wave, the fraction of the speed of sound).

Figure 21–17 The Doppler effect in stellar spectra. In each pair of spectra, the position of the spectral line in the laboratory is shown on top and the position observed in the spectrum of the star is shown below it. Lines from approaching stars appear blueshifted, lines from receding stars appear redshifted, and lines from stars that are moving transverse to us are not shifted because the star has no velocity toward or away from us. A short vertical line marks the unshifted position on the spectra showing shifts.

We can write

$$\frac{\text{change in wavelength}}{\text{original wavelength}} = \frac{\text{speed of emitter}}{\text{speed of light}}$$

or

$$\frac{\Delta\lambda}{\lambda_0} = \frac{v}{c},$$

where $\Delta\lambda$ is the change in wavelength (the Greek delta, Δ, usually stands for "the change in"), λ_0 is the original (rest) wavelength, v is the speed of the emitting body along a radius linking us to it, and c is the velocity of light ($= 3 \times 10^5$ km/sec). (We use the symbol v, since velocity is a term that signifies both speed and direction, and here the direction is known to be along a radius.) We define positive speeds as recession (redshifts) and negative speeds as approach (blueshifts). The new wavelength, λ, is equal to the old wavelength plus the change in wavelength, $\lambda_0 + \Delta\lambda$.

Example: A star is approaching at 30 km/sec (that is, $v = -30$ km/sec). At what wavelength do we see a spectral line that was at 6000 Å (which is in the orange part of the spectrum) when the radiation left the star?

Answer:

$$\frac{\Delta\lambda}{\lambda_0} = \frac{v}{c} = \frac{-30 \text{ km/sec} \times 10^5 \text{ cm/km}}{3 \times 10^{10} \text{ cm/sec}} = \frac{-3 \times 10^6 \text{ cm/sec}}{3 \times 10^{10} \text{ cm/sec}} = -10^{-4}.$$

$\Delta\lambda = -10^{-4}\,\lambda_0 = -10^{-4} \times 6000.0$ Å $= -0.6$ Å. (Since the star is approaching, this change of wavelength is a blueshift, and the new wavelength is slightly shorter than the original wavelength.) The new wavelength, λ, is thus $\lambda_0 + \Delta\lambda = 6000.0$ Å $- 0.6$ Å $= 5999.4$ Å. It is still in the orange part of the spectrum. The change would be easily measurable, but would not be apparent to the eye.

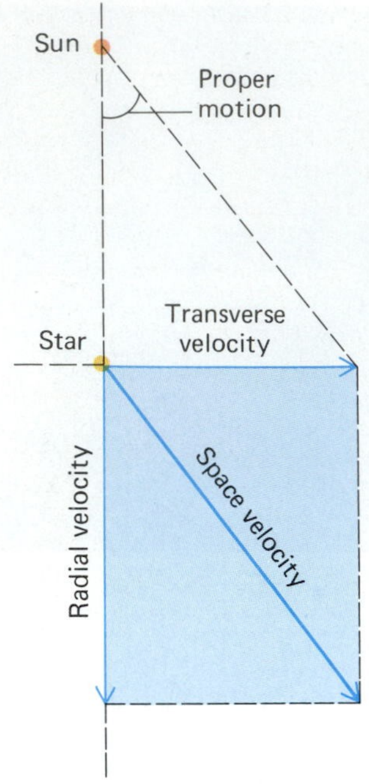

Figure 21–18 If we know a star's proper motion and its distance, we can compute its linear velocity through space in the direction across our field of view. The Doppler effect gives us its linear velocity toward or away from us. These two velocities can be combined to tell us the star's actual velocity through space, its *space velocity*.

Note that since there are proportions on both sides of the equation, if we take care to use v and c in the same units (for example, km/sec, cm/sec, or whatever), the $\Delta\lambda$ will be in the same units that λ is in, no matter whether that is angstroms, centimeters, or whatever.

The 30 km/sec given in the example is typical of the random speeds that stars have with respect to each other. These speeds are small on a universal scale, much too small to change the overall color that the eye perceives, but large compared to terrestrial speeds (30 km/sec = 30 km/sec × 3600 sec/hr = 108,000 km/hr).

21.7 Stellar Motions

On the whole, the network of stars in the sky is fixed. But radial velocities—velocities directly toward or away from us—can be measured for all stars using the Doppler effect, which we discussed in Section 21.6. The Doppler shift allows us to measure only velocities toward or away from us and not side to side. If the object is moving at some angle to the radius of a circle having us at the center, then the Doppler shift measures only the part of the velocity (technically, the component of the velocity) in the radial direction.

Further, some of the stars are seen to move slightly across the sky (that is, apparently from side to side) with respect to the more distant stars. The actual velocity of a star in 3-dimensional space, with respect to the sun, is called its *space velocity* (Fig. 21–18), but when we detect a star's change in position in the sky, we know only through what angle it moved. We cannot tell how far it moved in linear units (like km or light years) unless we also happen to know the distance to the star.

We usually deal separately with the part of the velocity of a star that is toward or away from us (the radial velocity), and the angular velocity of the star across the sky (how fast the object is moving across the sky in units of angle). The angular velocity is called the *proper motion* (Fig. 21–19). Radial velocity (measured from the Doppler shift) and proper motion are perpendicular to each other.

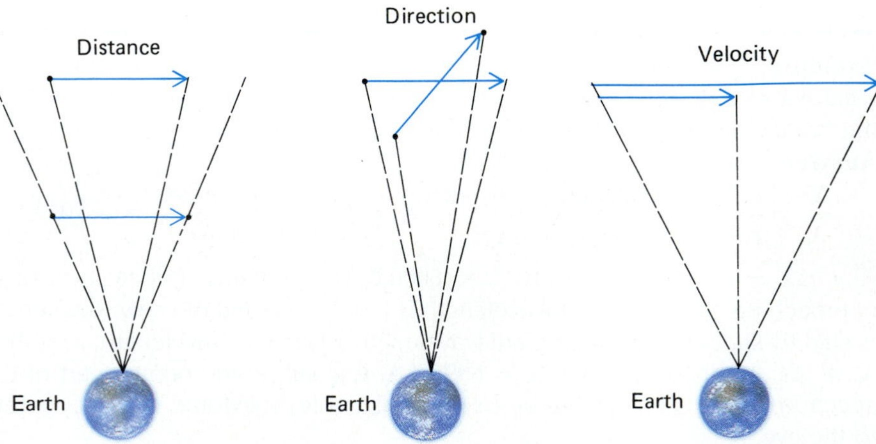

Figure 21–19 The proper motion of a star depends on its distance from us (*left*), the angle at which it travels (*middle*), and its speed in space (*right*).

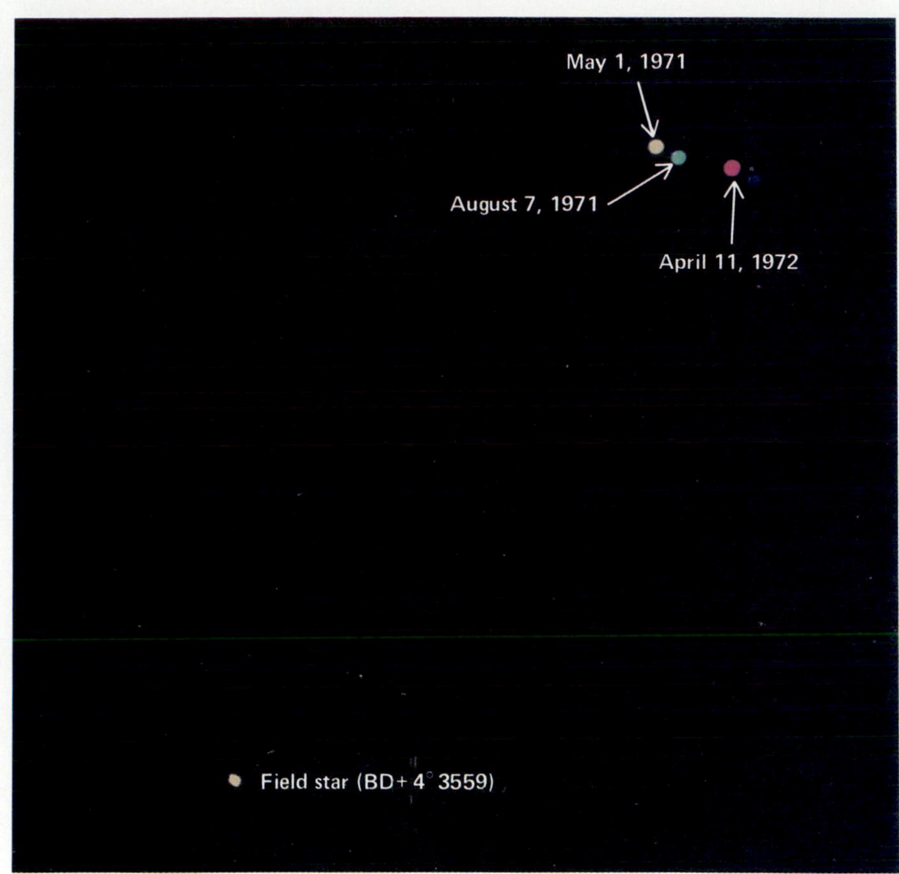

Figure 21–20 A compound photograph of Barnard's star and a reference star taken several months apart with the Thaw Refractor at the Allegheny Observatory. The proper motion of Barnard's star is clearly visible, as is its shift from side to side caused by the changing parallax that results from the earth's orbit around the sun.

The star with the largest proper motion was discovered in 1916 by E. E. Barnard, and is known as Barnard's star (Fig. 21–20). It moves across the sky by 10¼ arc sec per year. Since the moon appears half a degree across, Barnard's star moves across the sky by the equivalent of the diameter of the moon in only 180 years (Fig. 21–21).

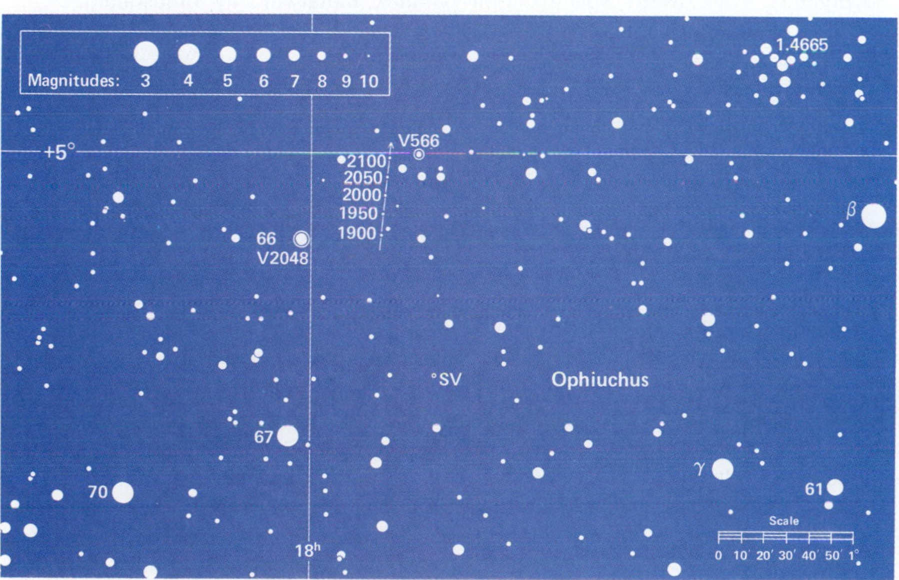

Figure 21–21 The motion of Barnard's star.

Summary and Outline

Magnitudes (Section 21.1)
- Lower (or negative) numbers are the brightest objects.
- Five magnitudes' difference = 100 times in brightness; adding or subtracting magnitudes corresponds to dividing or multiplying brightness.
- Brightest star is magnitude −1.5; faintest stars visible to the naked eye are about sixth magnitude.

Trigonometric parallax (Section 21.2)
- We triangulate, using the earth's orbit as a baseline.
- We find the angle through which a nearby star appears to shift; the more the shift, the nearer the star.
- This method works only for the nearest stars; Space Telescope and Hipparcos will improve our capabilities.
- A *parsec* is the distance of a star from the earth when the radius of the earth's orbit subtends 1 arc sec when viewed from the star.

Absolute magnitude (Section 21.3)
- If we know how bright a star **appears**, and how far away it is, we can calculate how intrinsically bright it is. This can be placed on the same scale as that of apparent magnitudes, and is known as the *absolute magnitude.* The absolute magnitude is defined as the magnitude a star would appear to have if it were at a distance of 10 parsecs.
- The apparent brightness of a point object follows the inverse-square law: brightness decreases with the square of the distance. We can use the inverse-square law to relate apparent magnitude, absolute magnitude, and distance.

The Hertzsprung-Russell diagram (Section 21.4)
- A plot of brightness versus temperature
- Stars fall in only limited regions of the graph.
- The *main sequence* contains most of the stars; these stars are called *dwarfs. Giants* are brighter (and bigger) than dwarfs; *supergiants* are brighter and bigger still. *White dwarfs* are fainter than dwarfs, and so fall below the main sequence.

Method of *spectroscopic parallax* (Section 21.5)
- Observing the spectrum of a star tells us where it falls on an H–R diagram; this tells us its absolute magnitude. Since we can easily observe its apparent magnitude, we can derive its distance.

The Doppler effect (Section 21.6)
- Radiation from objects that are receding is shifted to longer wavelengths: redshifted. Radiation from objects that are approaching is shifted to shorter wavelengths: blueshifted.

Stellar motions (Section 21.7)
- The radial velocity is measured from the Doppler effect.
- The angular velocity from side to side is measured by observing the *proper motion*—the motion of the star across the sky. This can be observed only for the nearest stars.

Key Words

apparent magnitude, significant figure*, baseline, Astronomical Unit, parallax, trigonometric parallax, measuring engines, astrometry, subtend, parsec, absolute magnitude, luminosity, inverse-square law, color index*, UBV*, three-color photometry*, bolometric magnitude*, Hertzsprung-Russell diagram, H–R diagram, color-magnitude diagrams, main sequence, dwarf stars, dwarfs, giants, supergiants, white dwarfs, spectroscopic parallax, radial velocity, blueshifted, redshifted, rest wavelength*, space velocity, proper motion, transverse velocity

*These terms are found in an optional section.

Questions

1. Which star is visible to the naked eye: one of 4th magnitude or one of 8th magnitude?

2. Which is brighter: Mars when it is magnitude +0.5 or Spica, a star whose magnitude is +0.9 (Appendix 5)?

††3. Venus can reach magnitude −4.4, while the brightest star, Sirius, is magnitude −1.4. How many times brighter is Venus at its maximum than is Sirius?

†4. Venus can be brighter than magnitude −4. Antares is a first-magnitude star ($m = +1$). How many times brighter is Venus at magnitude −4 than Antares?

†5. Pluto is about 14th magnitude at most. How many times fainter is it than Venus? ($m_{\text{Venus}} = -4$, approximately)

†6. If a variable star brightens by a factor of 15, by how many magnitudes does it change?

†7. If a variable star starts at 5th magnitude and brightens by a factor of 60, at what magnitude does it appear?

†8. The variable star Mira ranges between magnitudes 9 at minimum and 3 at maximum. How many times brighter is it at maximum than at minimum?

†9. Star A has magnitude +11. Star B appears 10,000

†This indicates a question requiring a numerical solution.
††Answers: 3: $(2.512)^3$ = 15 times; 25: 100 parsecs; 30: 6585 Å.

times brighter. What is the magnitude of star B? Star C appears 10,000 times fainter than star A. What is its magnitude?

†10. Star A has magnitude +10. The magnitude of star B is +5 and of star C is +3. How much brighter does star B appear than star A? How much brighter does C appear than B?

†11. How much brighter is a 0th-magnitude star than a +3rd-magnitude star?

12. You are driving a car, and the speedometer shows that you are going 80 km/hr. The person next to you asks why you are going only 75 km/hr. Explain why you each saw different values on the speedometer.

13. Would the parallax of a nearby star be larger or smaller than the parallax of a more distant star? Explain.

†14. A star has an observed parallax of 0.2 arc sec. Another star has a parallax of 0.02 arc sec. (a) Which star is farther away? (b) How much farther away is it?

†15. (a) What is the distance in parsecs to a star whose parallax is 0.05 arc sec? (b) What is the distance in light years?

†16. Estimate the farthest distance for which you can detect parallax by alternately blinking your eyes and looking at objects at different distances. (You may find it useful to look through a window to outdoor objects, so that you can see them silhouetted against a very distant background.) To what angle does this correspond?

†17. Vega is about 8 parsecs away from us. What is its parallax?

18. What are the two fundamental quantities that are being plotted on the axes of a Hertzsprung-Russell diagram?

19. What is the significance of the existence of the main sequence?

20. What is the observational difference between a dwarf and a white dwarf?

†21. Two stars have the same apparent magnitude and are the same spectral type. One is twice as far away as the other. What is the relative size of the two stars?

†22. Two stars have the same absolute magnitude. One is ten times farther away than the other. What is the difference in apparent magnitudes?

†23. A star has apparent magnitude of +5 and is 100 parsecs away from the sun. If it is a main-sequence star, what is its spectral type? (Hint: refer to Fig. 21–12.)

†24. A star is 30 parsecs from the sun and has apparent magnitude +2. What is its absolute magnitude?

††25. A star has apparent magnitude +9 and absolute magnitude +4. How far away is it?

†26. The nearest star, alpha Centauri, has apparent magnitude 0 and absolute magnitude +4.4. How far away is it in parsecs? In light years?

†27. Betelgeuse is apparent magnitude 0.4 and absolute magnitude −5.6. How far away is it in parsecs? In light years?

†28. The first quasar to be discovered, 3C273 (Chapter 32), was identified with what appeared to be a star of magnitude 13. It turned out to be 1 billion parsecs away. When the discoverer did the calculation, what absolute magnitude did it turn out to be? How many times brighter is it than the sun, whose absolute magnitude is about +5? The result astonished scientists.

†29. A jet plane travels 800 km/hr. Convert this speed to km/sec, and compare it with the speed of light, 3×10^5 km/sec. A red LED on a dial on the plane emits at 6600 Å. At what wavelength would we see it if we could detect it from the ground as the plane recedes from us?

††30. A star is receding from us at 1000 km/sec. The Hα line ordinarily appears at 6563 Å. At what wavelength would it appear in the star's spectrum as seen from earth?

†31. The 5250-Å spectral line from iron appears at 5252 Å in the spectrum of a star. At what velocity is the star moving with respect to us?

32. If a star is moving away from the earth at very high speed, will the star have a continuous spectrum that appears hotter or cooler than it would if the star were at rest? Explain.

33. (a) What does the proper motion of Barnard's star indicate about its distance from us? (b) What would Barnard's star's proper motion be if it were twice as far away from us as it is?

†34. A star has a proper motion of 10 arc sec per century. Can you tell how fast it is moving in space in km/sec? If so, describe how.

†35. Two stars have the same space velocity in the same direction, but star A is 10 parsecs from the earth and star B is 30 parsecs from the earth. (a) Which has the larger radial velocity? (b) Which has the larger proper motion?

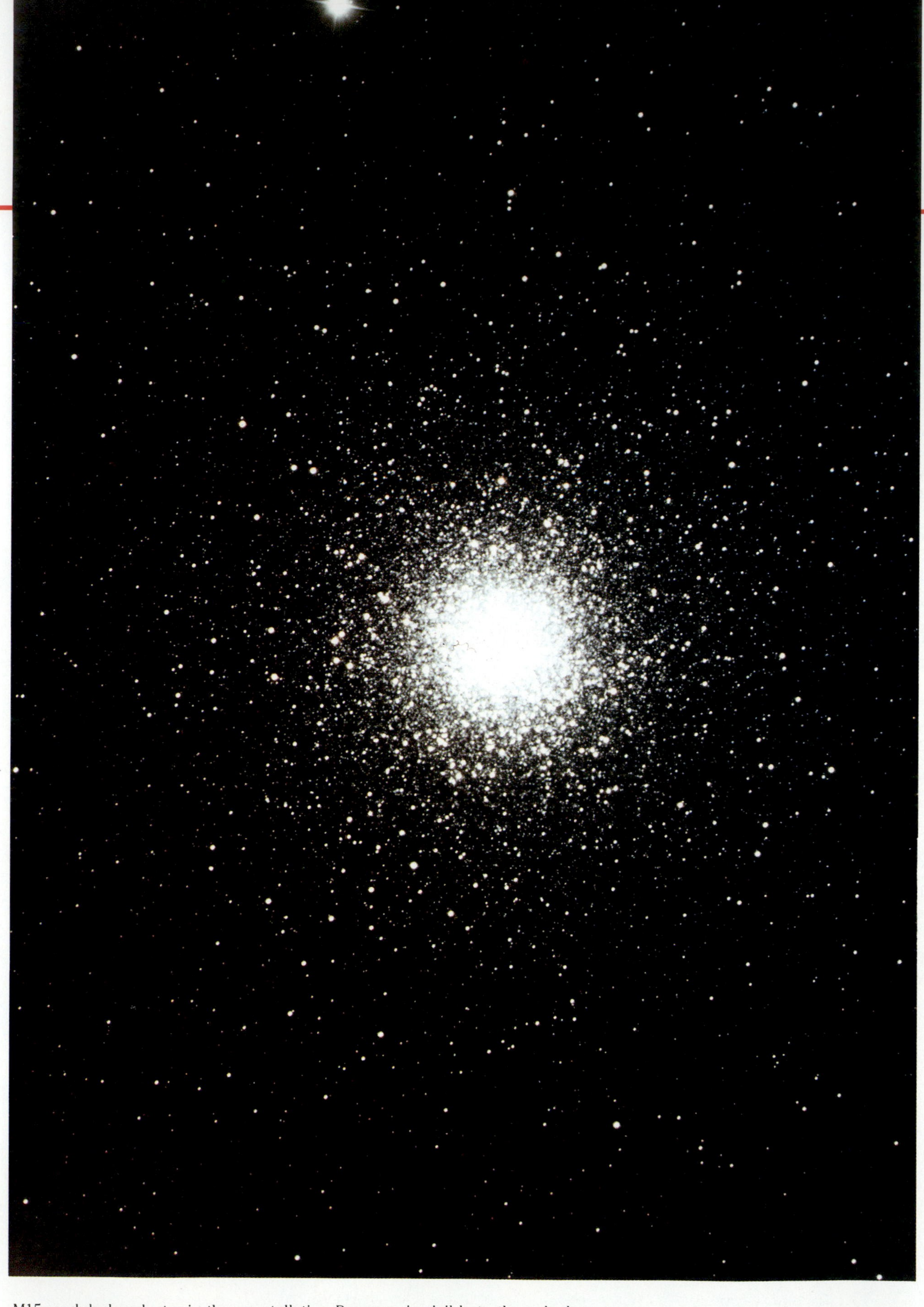

M15, a globular cluster in the constellation Pegasus, is visible to the naked eye as a hazy patch. It contains about 100,000 stars. (CFHT)

Doubles, Variables, and Clusters

22

Aims: To discuss double stars, variable stars, and stellar clusters, and to draw conclusions about the distances to stars, their masses, and about the ages of stars

We often think of stars as individual objects that shine steadily, but many stars vary in brightness and most stars actually have companions close by. Also, many stars appear as members of groupings called clusters. By studying the effects that the stars have on each other, or by studying the nature of all the stars in a star cluster, we can learn much that we could not discover by studying the stars one at a time. This information even leads us to a general understanding of the life history of the stars.

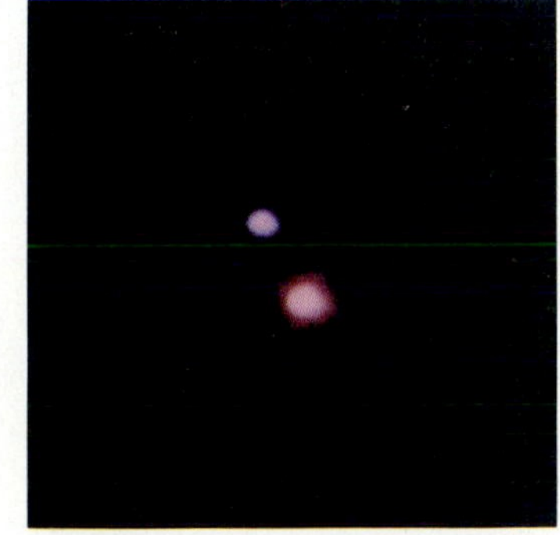

Figure 22–1 Albireo (β Cygni) contains a B star and a K star, which make a particularly beautiful pair because of their different colors. Albireo is high overhead on summer evenings.

22.1 Binary Stars

Most of the objects in the sky that we see as single "stars" really contain two or more component stars. Sometimes a star appears double merely because two stars that are located at different distances from the sun appear in the same line of sight. Such systems are called *optical doubles,* and will not concern us here. We are more interested in stars that are physically associated with each other. We will use the terms *double star* or *binary star* interchangeably to mean two or more stars held together by the gravity they have between each other. (Even systems with three or more stars are usually called "double stars.")

The easiest way to tell that more than one star is present is by looking through a telescope of sufficiently large aperture. Stars that appear double when observed directly are called *visual binaries* (Fig. 22–1). The resolution of small telescopes may not be sufficient to allow the components of a double star to be "separated" from each other; larger telescopes can thus distinguish more double stars. When the components of a double star have different colors, they are particularly beautiful to observe, even with small telescopes. Five to ten per cent of the stars in the sky are visual binaries.

The Hubble Space Telescope needs single stars to guide on while it is taking long exposures. But it will be able to tell that a star is a double when its components are separated by only 0.1 arc sec instead of the nearly 1 arc sec separation detectable from the ground. As a result, it will discover that many stars we now consider single are really double. A major problem has arisen, since a pair of single stars must be in the field of view for each exposure. If the list of "guide stars" available to HST is too short, each of its members may be tossed out for proving to be double, and no exposure could be taken of that field. Scientists at the Space Telescope Science Institute have undertaken major revisions of HST's software to limit the effect of the problem as well as studies of how often duplicity occurs.

Astronomers can detect that stars are double in some cases even when they are not visual binaries—not even the kind that only HST can resolve. Sometimes a star appears as a single object through a telescope, but one can

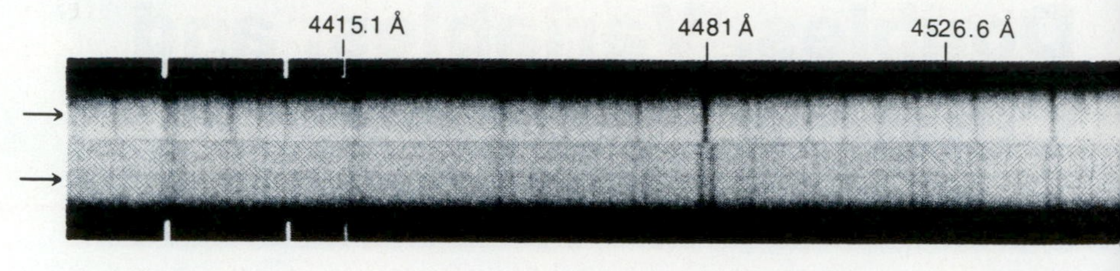

Figure 22–2 Two spectra of Mizar (ζ Ursae Majoris) taken 2 days apart show that it is a spectroscopic binary. The lines of both stars are superimposed in the upper stellar absorption spectrum (*top arrow*) but are separated in the lower spectrum (*bottom arrow*) by 2 Å, which corresponds to a relative velocity of 140 km/sec. Emission lines from a laboratory source are shown at the extreme top and extreme bottom to provide a comparison with a source at rest that has lines at known wavelengths.

see that the spectrum of that "object" actually consists of overlapping spectra of at least two objects. If we can detect the presence of two sets of spectral lines from stars of different spectral types—of, say, one hot star and one cool star—then we say that the object has a *composite spectrum.*

We can tell that a second star is present, even when an image and its spectrum appear single, if the spectral lines we observe change in wavelength with time. We know of no other way such wavelength changes can occur except as the result of variable Doppler shifts, which indicate that the speed of an object along a line linking it with us is changing. We deduce that two or more stars are present and are revolving around each other. Such an object is called a *spectroscopic binary* (Fig. 22–2). As the stars in the system orbit each other, unless we are looking straight down on the orbit from above, each spends half of its orbit approaching us and the other half receding from us, relative to its average space motion. The variations of velocity in spectroscopic binaries are periodic; the spectrum of each component varies separately in wavelength. Note that even if the spectrum of one of the stars is too faint to be seen (making the system a "single-line spectroscopic binary," since only one set of spectral lines is observed), we can still tell that the star is a spectroscopic binary if the radial velocity of the visible component varies (Fig. 22–3).

Careful spectroscopic studies have shown that two-thirds of all solar-type stars have stellar companions. Though the presence of companions of stars of other spectral types has not been studied in such detail, it seems that about 85 per cent of all stars are members of double-star systems. Few stars are single, an idea that is verified by noticing that many of the nearest stars to our sun (Appendix 7), stars we can study in detail, are double. In recent years, we have realized that in some circumstances mass can flow from one member of a binary system to the other, changing the evolution of each. We shall discuss such effects on stellar evolution in Section 25.7.

We can detect visual binaries most easily when the stars are relatively far apart. Then the period of the orbit is relatively long, over 100,000 years in most cases. So we know little about the motions in such systems because of our short human lifetime, even though most double stars discovered are visual

Figure 22–3 Two spectra of Castor B (α Geminorum B) taken at different times show a Doppler shift. Thus the star is a spectroscopic binary, even though lines from only one of the components can be seen. The comparison spectrum of a laboratory source appears at the top and bottom. Note that emission lines from this laboratory source are in the same horizontal positions at extreme top and bottom, while the absorption lines of the stellar spectra (shown with arrows) are shifted laterally (that is, in wavelength) with respect to each other.

binaries. It is harder to detect spectroscopic binaries, so we know fewer of them. But for this group, most of whose periods are between one day and one year, we are able to determine such important details of the orbit as size and period (Fig. 22–4).

Sometimes the components of a double star pass in front of each other, as seen from our viewpoint on the earth. The "double star" then changes in brightness periodically, as one star cuts off the light from the other. Such a pair of stars is called an *eclipsing binary* (Fig. 22–5). The easiest to observe is Algol, β Persei (beta of Perseus), in which the eclipses take place every 69 hours, changing the total brightness of the system within an hour from magnitude 2.3 to 3.5 and back. (A third star is present in the Algol system. It orbits the other two every 1.86 years, and does not participate in the eclipses.)

Note that the way a binary star appears to us depends on the orientation of the two stars not only with respect to each other but also with respect to the earth. If we are looking down at the plane of their mutual orbit (Fig. 22–6), then we might see a composite spectrum; under the most favorable conditions the star might be seen as a visual binary. But in this orientation we would never be able to see the stars eclipse. We would also not be able to see the Doppler shifts typical of spectroscopic binaries, since only radial velocities contribute to the Doppler shift.

It is possible that a star could be a "double," but still not be detectable by any of the above methods. Sometimes, the existence of a double star shows up only as a deviation from a straight line in the proper motion of the "star" across the sky. Such stars are called *astrometric binaries* (Fig. 22–7).

A double star may fall into more than one category. For example, two stars that eclipse each other must be a spectroscopic binary, too.

Many of the celestial sources of x-rays that have been observed in recent years from orbiting telescopes turn out to be binary systems. Matter from one member of the pair falls upon the other member, heats up, and radiates x-rays. We will discuss several such systems in Section 27.8.

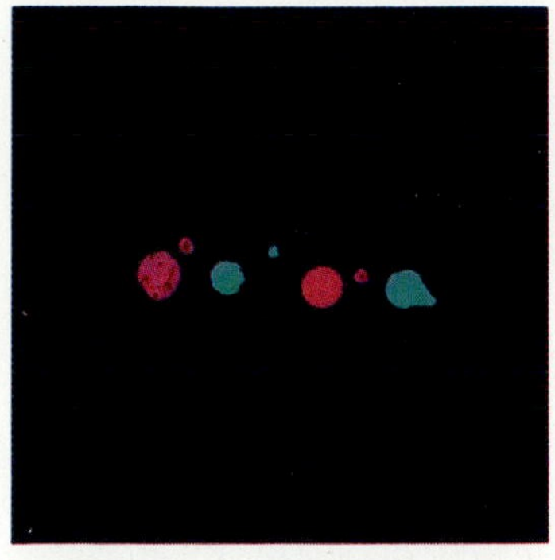

Figure 22–4 The two components of the visual binary 70 Ophiuchi are seen to orbit each other in this composite of photographs covering a 51-year span. The photographs were taken in 1915, 1940, 1955, and 1966, at the Allegheny Observatory. Alternate images are colored differently. The stars range in apparent separation from 1.7 arc seconds to 6.7 arc sec over their 88-year orbital period; they are 17 l.y. from earth.

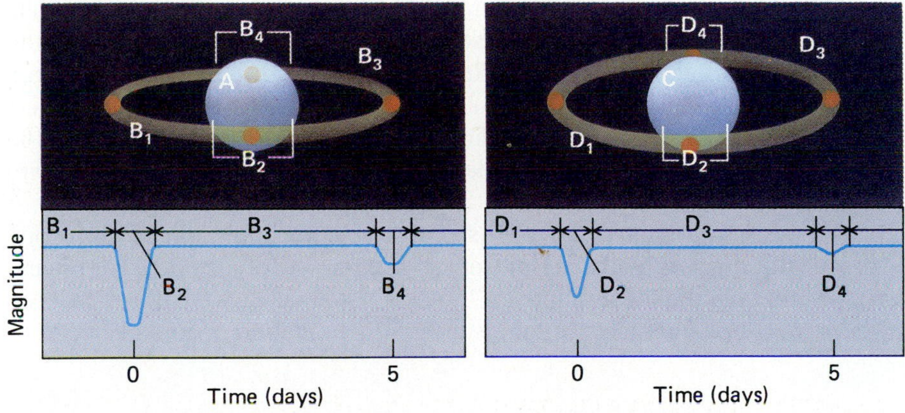

Figure 22–5 The shape of the light curve of an eclipsing binary depends on the sizes of the components and the direction from which we view them. At lower left, we see the light curve that would result for star B orbiting star A, as pictured at upper left. When star B is in the positions shown with subscripts at top, the regions of the light curve marked with the same subscripts result. The eclipse at B_4 is total. At right, we see the appearance of the orbit and the light curve for star D orbiting star C, with the orbit inclined at a greater angle than at left. The eclipse at D_4 is partial. From earth, we observe only the light curves, and use them to determine what the binary system is really like, including the inclination of the orbit and the sizes of the objects.

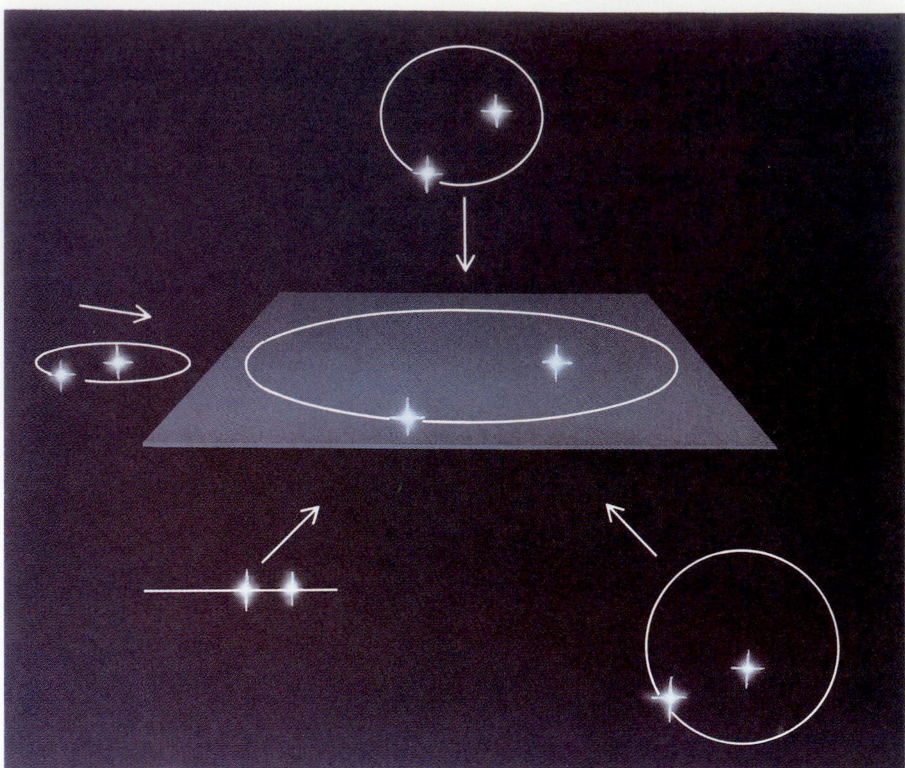

Figure 22–6 The appearance and the Doppler shift of the spectrum of a binary star depend on the angle from which we view the binary. From far above or below the plane of the orbit, we might see a visual binary, as shown on top and at the lower right of the diagram. From close to but not exactly in the plane of the orbit, we might see only a spectroscopic binary, as shown at left. (The stars appear closer together, so might not be visible as a visual binary.) From exactly in the plane of the orbit, we would see an eclipsing binary, as shown at lower left.

The nearest star system to us, α Centauri, is a multiple-star system (Fig. 22–8). It is too far south to be seen from mid-northern latitudes.

22.2 Stellar Masses

The study of binary stars is of fundamental importance in astronomy because it allows us to determine stellar masses. If we can determine the orbits of the stars around each other, we can calculate theoretically the masses of the stars necessary to produce the gravitational effects that lead to those orbits. For a star that is a visual binary with a sufficiently short period (only 20 or even 100 years, for example), we are able to determine the masses of both of the components.

We are more limited if a star is only a spectroscopic binary, even one with spectral lines of both stars present. If the star is not also a visual binary, we can find only the lower limits for the masses—that is, we can say that the masses must be larger than certain values. This limitation occurs because we

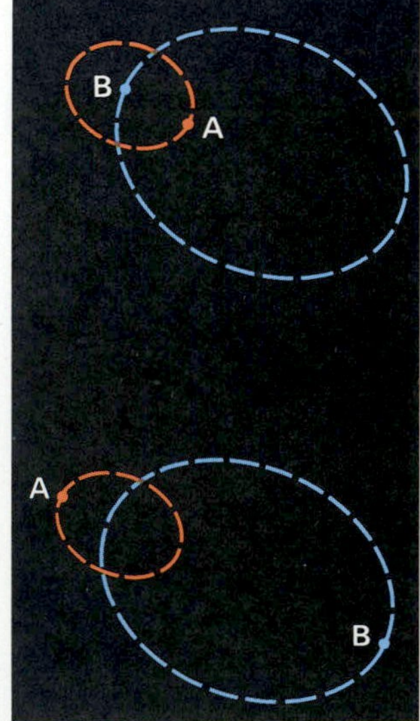

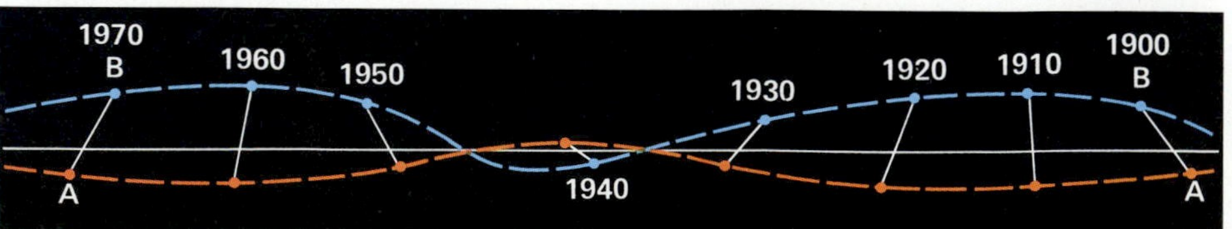

Figure 22–7 Sirius A and B, an astrometric binary. From studying the motion of Sirius A (often called, simply, Sirius) astronomers deduced the presence of Sirius B before it was seen directly. Sirius B's orbit is larger because Sirius A is a more massive star. They contain 2.14 and 1.05 times as much mass as the sun, respectively.

cannot usually tell the angle by which the plane of the orbits of the member stars is inclined. Only if the spectroscopic binary is also an eclipsing binary do we know the angle of inclination, the angle at which the plane of the orbits is inclined to our view, since the eclipses would not take place unless this angle were close to zero. Only if we know this angle can we find the individual masses.

So we do not always know as much as we would like about the masses of stars in binary systems. Of course, we are better off than we are for stars that are not in binary systems, for we cannot directly measure their masses at all! From studies of the several dozen binaries for which we can accurately tell the masses, astronomers have graphed the luminosities (intrinsic brightnesses) of the stars on one axis and the masses on the other axis. Most of the stars turn out to lie on a narrow band, the *mass-luminosity relation* (Fig. 22–9).

The mass-luminosity relation is valid only for stars on the main sequence. The more massive a main-sequence star is, the brighter it is. The mass, in fact, is the prime characteristic that determines where on the main sequence a star will settle down to live its lifetime.

The most massive stars we know are about 50 times more massive than the sun, and the least massive stars are about 7 per cent the mass of the sun.

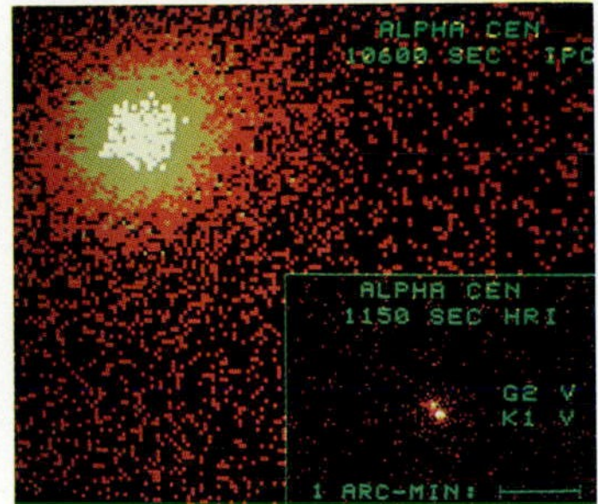

Figure 22–8 An x-ray view of the nearest star system to us, α Centauri, which is only 4.3 light years away. The short exposure (*inset*) shows the two brightest members, main-sequence stars of spectral types G2 and K1. The third member, Proxima Centauri, is an M5 main-sequence star slightly closer to us. In the visible, it is about ten magnitudes fainter than the other two members of the system. X-rays from cool stars like the ones shown here are from their hot outer atmospheres.

*22.3 Stellar Sizes

Stars appear as points to the naked eye and as small, fuzzy disks through large telescopes. Distortion by the earth's atmosphere blurs the stars' images into disks and hides the actual size and structure of the surface of the stars. The sun is the only star whose angular diameter we can measure easily and directly.

Astronomers use an indirect method to find the sizes of most stars. If we know the absolute magnitude of a star (from some type of parallax measurement) and the temperature of the surface of the star (from measuring its spectrum), then we can tell the amount of surface area the star must have. The extent of surface area, of course, depends on the radius.

One type of direct measurement works only for eclipsing binary stars. As the more distant star is hidden behind the nearer star, one can follow the rate at which the intensity of radiation from the farther star declines. From this information together with Doppler measurements of the velocities of the components, one can calculate the sizes of the stars. One can sometimes even tell how the brightness varies across a star's disk.

It is much more difficult to measure the size of a single star directly. One way of measuring the diameter is by *lunar occultation.* As the moon moves with respect to the star background, it occults (hides) stars. By studying the light from a star in the fraction of a second it takes for the moon to completely block it, we can deduce the size of the stellar disk. Unfortunately, the moon passes over only 10 per cent of the sky in the course of a year.

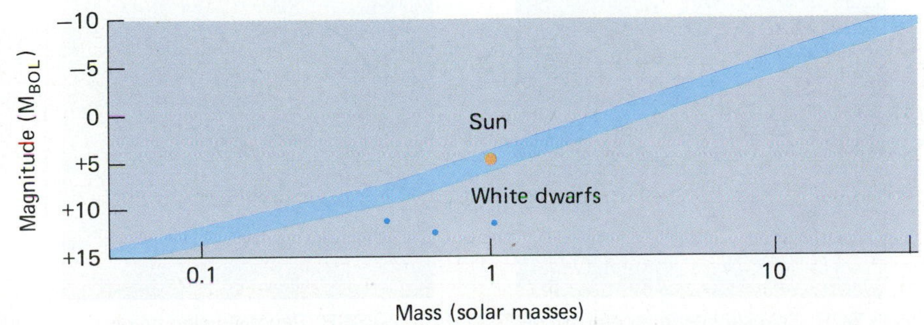

Figure 22–9 The mass-luminosity relation, measured from binary stars. Most of the stars fit within the shaded band. The relation is defined by straight lines of different slopes for stars brighter than and fainter than magnitude +7.5. We do not know if there is a real difference between the brighter and fainter stars that causes this or whether it is an effect introduced when the data are reduced, at which stage we try to take account of the energy that the stars radiate outside the visible part of the spectrum.

Note that white dwarfs, which are relatively faint for their masses, lie below the mass-luminosity relation. Red giants lie above it. Thus we must realize that the mass-luminosity relation holds only for main-sequence stars.

Figure 22–10 Charles Townes with one of the mirrors from his infrared interferometer, set up on Mt. Wilson in 1989. The second mirror is covered at right, and electronics are off the picture to the right. As the system works better and better, the mirrors are moved to greater spacings.

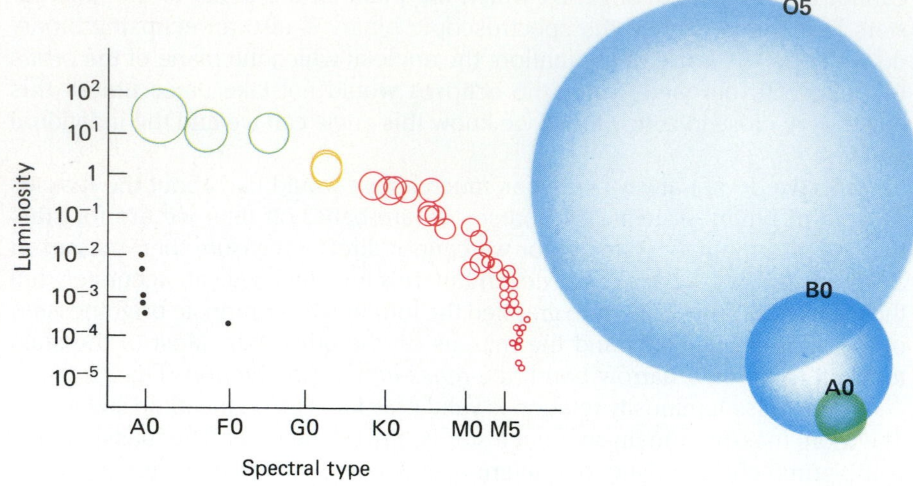

Figure 22–11 The H–R diagram for stars within 17 light years of the sun. The open circles indicate the relative diameters of main-sequence stars; white dwarfs are shown as dots. In addition, relative sizes of O, B, and A stars are shown at right.

Other methods for measuring the diameters of single stars use a principle called *interferometry,* a technique that measures incoming radiation at two different locations and then combines the two signals. This gives the effect, for the purpose of determining the resolution, of a single very large telescope. The method overcomes many of the problems of blurring by the earth's atmosphere.

The first such measurements were carried out at Mt. Wilson 60 years ago; 7 of the stars that subtended the largest angles in the sky had their diameters measured. These were the largest of the closest stars, that is, nearby red giants and supergiants. Modern versions of these techniques are now in use.

The above interferometric methods work best for large, cool stars. R. Hanbury Brown and his associates in Australia have used another type of stellar interferometer that works best for hot, bright stars. They have determined the diameters of three dozen stars whose diameters cannot be measured with other techniques.

New interferometers are being developed, with both optical and infrared projects being tested at the Mt. Wilson Observatory (Fig. 22–10).

Only a few dozen stellar diameters have been measured directly, and they confirm the indirect measurements (Fig. 22–11). Direct and indirect measurements show that the diameters of main-sequence stars decrease as we go from

Figure 22–12 Speckle interferometry takes the set of speckles shown at left and interprets them to show that ψ (psi) Sagittarius is actually the double star shown at right.

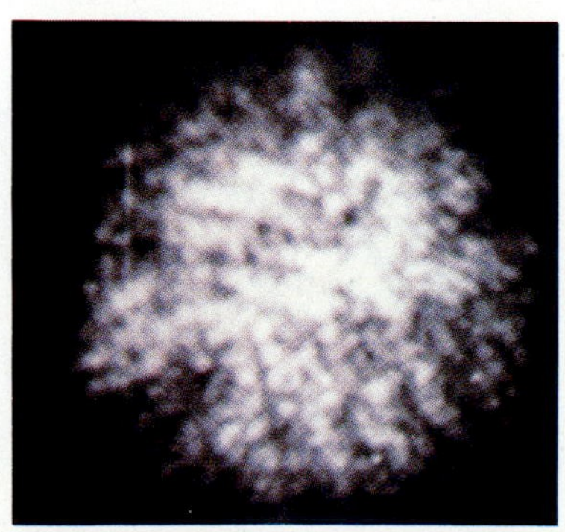

ψ SGR: Speckle interferometry image

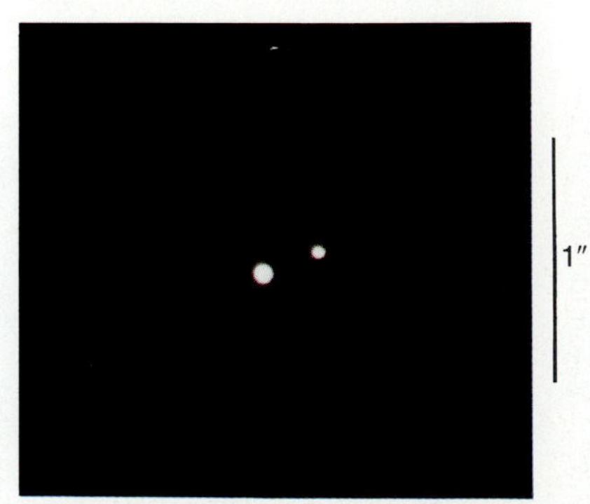

ψ SGR: Reconstructed image

Focus On

*Box 22.1 Speckle Interferometry

A relatively new technique called *speckle interferometry* uses high-speed imaging to overcome the blurring of the earth's atmosphere to a certain extent. At any one instant, the image of a star through a large telescope looks speckled because different parts of the image are affected by different small turbulent areas in the earth's atmosphere. A long exposure blurs these speckles and gives us the fuzzy disk that normally appears on a photograph. Speckle interferometry uses electronic exposures that are so short—perhaps $\frac{1}{50}$ sec—that the atmosphere does not change. Mathematical techniques are used to deduce the properties of the starlight that hit the top of the earth's atmosphere. In this manner, from many short exposures of an image, scientists can measure a star's diameter or find out if two normally unresolved objects are present (Fig. 22–12).

hotter to cooler; that is, O dwarfs are relatively large and F dwarfs are relatively small. Indirect measurements alone show that this trend continues for K and M stars, which are smaller still. As for stars that are not on the main sequence, red giants are indeed giant in size as well as in brightness.

22.4 Variable Stars

Some stars vary in brightness with respect to time. The most basic property of a star's variation is the *period.* The period is the time it takes for a star to go through its entire cycle of variation.

One way for a star to vary in brightness, as we have seen, is for the star to be an eclipsing binary. Sometimes we get the light simultaneously from both members of the binary, and sometimes the light from one of the members is at least partially blocked by the other member. But it is possible for individual stars to vary in brightness all by themselves. Thousands of such stars are known in the sky. The periods of the variations can be seconds for some types of stars or years for others. Plots of the brightness of a star (usually in terms of its magnitude) versus time are called *light curves.*

Besides ordinary variables of various types, some stars occasionally flare up briefly. These stellar flares often make a bigger difference in parts of the spectrum other than the visible. Fig. 22–13 shows an x-ray flare in the nearest star to us, Proxima Centauri.

A

B

C

D

Figure 22–13 An x-ray flare on Proxima Centauri, the nearest star, was observed with the Einstein Observatory by Bernhard M. Haisch of the Lockheed Palo Alto Research Laboratory in this series of 35-minute exposures taken on August 20, 1980. *A* is before the flare, *B* is near the flare's maximum, and *C* and *D* show the flare's decay. The steady x-ray flux probably comes from the star's corona. The flare's x-ray flux was about that of a bright solar flare. A second x-ray source appears at upper right.

Focus On

*Box 22.2 Naming Variable Stars

A naming system has been adopted for variable stars (often simply called "variables") that helps us recognize them on any list of stars. The first variable to be discovered in a constellation is named R, followed by the genitival form ("of...") of the Latin name of the constellation (see Appendix 10), e.g., R Coronae Borealis (R of Corona Borealis). One continues with S, T, U, V, W, X, Y, Z, then RR, RS, etc., up to RZ, then SS (not SR) up to SZ, and so on up to ZZ. Then the system starts over with AA up to AZ, BB (not BA) and so on up to QZ. The letter J was omitted when the method was set up to avoid confusion with I in the German script of that time. This system covers the first 334 variables; after that, one numbers the stars beginning with V for variable (V335...). Variable stars with more commonly known names, such as Polaris and δ Cephei, retain their common names instead of being included in the lettering system.

We will limit our discussion to three types of variables, one of which is especially numerous. The other two have provided important information about the scale of distance in the universe.

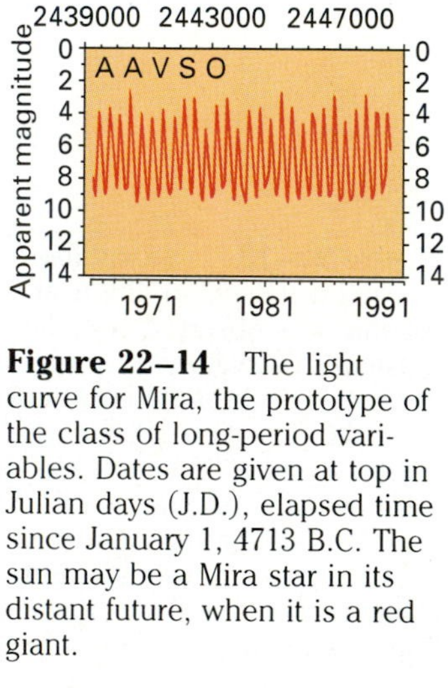

Figure 22–14 The light curve for Mira, the prototype of the class of long-period variables. Dates are given at top in Julian days (J.D.), elapsed time since January 1, 4713 B.C. The sun may be a Mira star in its distant future, when it is a red giant.

22.4a Mira Variables

A type of variable star is often named after its best-known or brightest example. A star named Mira in the constellation Cetus (the star is also known as *o* Ceti, omicron of the Whale) fluctuates in brightness with a long period (Fig. 22–14), about a year. Red stars that share this characteristic are called *Mira variables.* The period of a given Mira-type star can be from three months to about two years. The period of an individual star is not strictly regular; it can vary from the average period.

Mira itself is sometimes of apparent magnitude 9, and is thus invisible to the naked eye. However, it brightens fairly regularly by about six magnitudes, a factor of 250, with a period of about 11 months. At maximum brightness it is quite noticeable in the sky. As it brightens, its spectral type changes from M9 to M5. Thus real changes are taking place at the surface of the star that result in a change of temperature.

Mira stars are giants of spectral type M, and are about 700 times the diameter of the sun. If such a star were in the center of our solar system, it would extend beyond Mars. Mira stars emit most of their radiation in the infrared. They are also the source of strong radio spectral lines from water vapor.

These stars, which are the most numerous type of variable star in the sky, are also known as *long-period variables.*

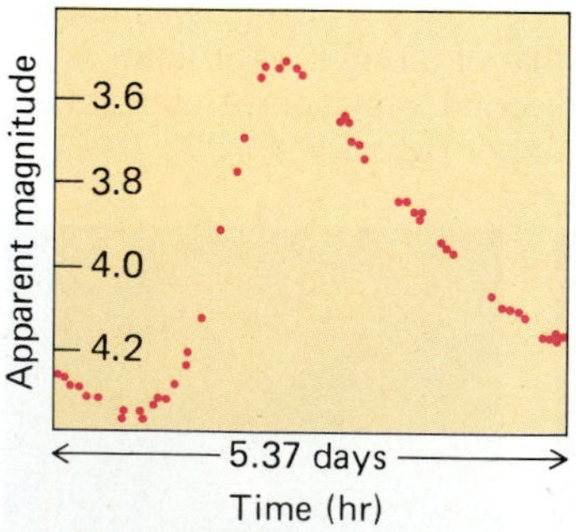

Figure 22–15 The light curve for δ Cephei, the prototype of the class of Cepheid variables.

22.4b Cepheid Variables

The most important variable stars in astronomy are the *Cepheid variables* (cef'e-id). The prototype is δ Cephei (delta of Cepheus). Cepheid variables have very regular periods that, for individual Cepheids (as they are called), can be from 1 to 100 days. δ Cephei itself varies between apparent magnitudes 3.6 and 4.3 with a period of 5.4 days (Fig. 22–15). Polaris, the north star, is a

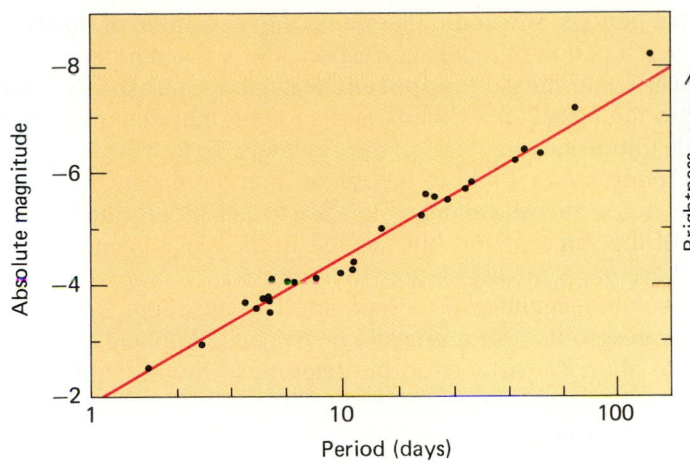

Figure 22–16 The period-luminosity relation for Cepheid variables.

Figure 22–17 Henrietta S. Leavitt, who discovered that Cepheids have a period-luminosity relation, at her desk at the Harvard College Observatory in 1916. Four years earlier, she had published an article reporting her discovery.

Cepheid variable with a period of 4 days; it varies by only 0.1 magnitude. Astronomers can tell Cepheid variables apart from other variables, even when the periods are the same, by the shape and regularity of their light curves.

Cepheids are relatively rare stars in our galaxy; only about 700 are known. Some have been detected in other galaxies.

Cepheids are important because a relation has been found that links the periods of their light changes, which are simple to measure, with their absolute magnitudes (Fig. 22–16). For example, if we measure that the period of a Cepheid is 10 days, we need only look at this period-luminosity relation to see that the star is of absolute magnitude about −4. We can then compare its absolute magnitude to its apparent magnitude, which gives us (by the inverse-square law of brightness) its distance from us.

The study of Cepheids is the key to our current understanding of the distance scale of the universe and has allowed us to determine that the objects in the sky that we now call galaxies (discussed in Chapter 31) are giant systems comparable to that of our own galaxy (discussed in Chapter 29).

The story started out 75 years ago at the Harvard College Observatory with Henrietta Leavitt (Fig. 22–17), who was studying the light curves of variable stars in the southern sky. In particular, she was studying the variables in the Large and Small Magellanic Clouds (Fig. 22–18), two hazy areas in the sky that were discovered by the crew of Magellan's expedition around the world when they sailed far south. Whatever the Magellanic Clouds were—we now know them to be galaxies, but Leavitt did not know this—they looked like concentrated clouds of material. It thus seemed clear that, for each cloud, all its stars were at approximately the same distance from the earth. Thus even though she could plot only the apparent magnitudes, the relation of the absolute magnitudes to each other was exactly the same as the relation of the apparent magnitudes.

By 1912, Leavitt had established the light curves and determined the periods for two dozen stars in the Small Magellanic Cloud. She plotted the magnitude of the stars (actually a median brightness, a value between the maximum and minimum brightness) against the period. She realized that there was a fairly strict relation between the two quantities, and that the Cepheids with longer periods were brighter than the Cepheids with shorter periods. By

Figure 22–18 From the southern hemisphere, the Magellanic Clouds are high in the sky. They are not quite this obvious to the naked eye. Since all stars in a given Magellanic Cloud are essentially the same distance from earth, their absolute magnitudes are in the same relation to each other as their apparent magnitudes.

simply measuring the periods, she could determine the magnitude of one star relative to another; each period uniquely corresponds to a magnitude.

Henrietta Leavitt could measure **apparent** magnitude, but she did not know the distance to the Magellanic Clouds and so could not determine the **absolute** magnitude (intrinsic brightness) of the Cepheids. To find the distance to the Magellanic Clouds, we first had to be able to find the distance to **any** Cepheid—even one not in the Magellanic Clouds—to tell its absolute magnitude. A Cepheid of the same period but located in the Magellanic Clouds would presumably have the same absolute magnitude.

To find the absolute magnitude of a Cepheid, it would seem easiest to start with the one nearest to us. Unfortunately, not a single Cepheid is close enough to the sun to allow its distance to be determined by the method of trigonometric parallax. More complex, statistical methods had to be used to study the relationships between stellar motions and distances. This gave the distance to a nearby Cepheid. Once we have the distance to a Cepheid, we can calculate its absolute magnitude from its apparent magnitude. Then we know the absolute magnitude of all Cepheids of that same period in the Magellanic Clouds, since all Cepheids of the same period have the same absolute magnitude.

From that point on it is easy to tell the absolute magnitude of Cepheids of **any** period in the Magellanic Clouds. After all, if a Cepheid has an **apparent** magnitude that is, say, 2 magnitudes brighter than the apparent magnitude of our Cepheid of known intrinsic brightness, then its **absolute** magnitude is also 2 magnitudes brighter. After this process, we have the period-luminosity relation in a more useful form—period vs. absolute magnitude. We call this process "the calibration" of the period-luminosity relation.

In the 1950's, when overlapping methods of finding distances were applied to some relatively nearby stars and clusters of stars, it was realized that there was a second type of Cepheid whose light curves looked the same but which were about 1.5 magnitudes fainter at each given period than the original type. The distances to many galaxies had to be recalculated because this distinction between ordinary Cepheids and Type II Cepheids had previously not been made. Our estimates of the distances to many distant galaxies doubled.

Thus Cepheids can be employed as indicators of distance: First we identify the star as a Cepheid (by studying its spectrum and the shape of its light curve). Then we measure its period. Third, the period-luminosity relation gives us the absolute magnitude of the Cepheid. And last, we calculate (with the inverse-square law) how far a star of that absolute magnitude would have to be moved from the standard distance of 10 parsecs to appear as a star of the apparent magnitude that we observe.

When the calibration of the period-luminosity relation was worked out quantitatively by the American astronomer Harlow Shapley (the first syllable rhymes with "map," not "cape") in 1917, the distance to the Magellanic Clouds could be calculated. They were very far away, a distance that we now know means that they are not even in our galaxy! Instead, they are galaxies by themselves, two small irregular galaxies that are companions of our own, larger galaxy.

22.4c RR Lyrae Variables

Many stars are known to have short regular periods, less than one day in duration. Certain of these stars, no matter what their periods, have light curves of a specific distinctive shape (Fig. 22–19). All these stars have the same average absolute magnitude.

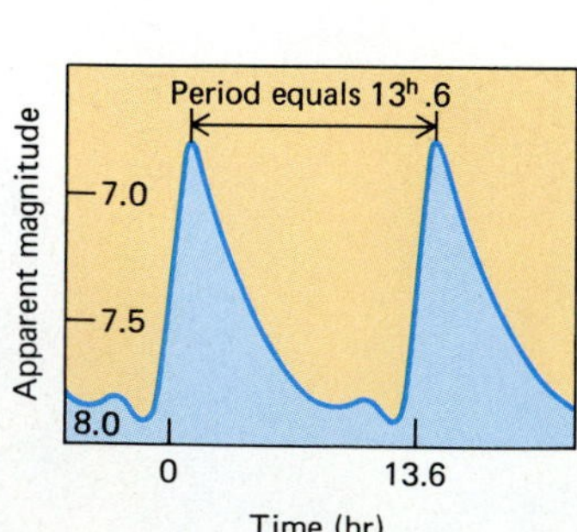

Figure 22–19 The light curve of RR Lyrae.

Such stars are called *RR Lyrae stars* after the prototype of the class. Since many of these stars appear in globular clusters (which will be described in Section 22.5), RR Lyrae stars are also called *cluster variables*.

Once we detect an RR Lyrae star by the shape of its light curve, we immediately know its absolute magnitude, since all the absolute magnitudes are the same (about 0.6). Just as before, we measure the star's apparent magnitude and can thus easily calculate its distance, which is also the distance to the cluster.

Figure 22–20 An open cluster of stars, NGC 3293 in Carina.

Figure 22–21 An open cluster of stars, NGC 2244, surrounded by the Rosette Nebula.

22.5 Clusters and Stellar Populations

Even aside from the hazy band of the Milky Way, the distribution of stars in our sky is not uniform. There are certain areas where the number of stars is very much higher than the number in adjacent areas. Such sections of the sky are called *star clusters.* All the stars in a given cluster are essentially the same distance away from us, and were also formed at about the same time.

One type of star cluster appears only as an increase in the number of stars in that limited area of sky. Such clusters are called *open clusters,* or *galactic clusters* (Figs. 22–20 and 22–21). The most familiar example of a galactic cluster is the Pleiades (Plee′-a-dees), a group of stars visible in the evening sky in the winter (Fig. 22–22). The unaided eye sees at least six stars very close together. With binoculars or the smallest telescopes, dozens more can be seen. A larger telescope reveals hundreds of stars. Another galactic cluster is called the Hyades, which forms the "V" that outlines the face of Taurus, the bull, a constellation best visible in the winter sky. More than a thousand such clusters are known, most of them too faint to be seen except with telescopes.

All the stars in a galactic cluster are packed into a volume not more than 10 parsecs across. Stars in galactic clusters seem to be representative of stars in the spiral arms of our galaxy and of other galaxies. When the spectra of

The Pleiades are often known as the Seven Sisters, after the seven daughters of Atlas who were pursued by Orion and who were given refuge in the sky. That one is missing from the number in the myth—the Lost Pleiad—has long been noticed. Of course, a seventh star is present (and hundreds of others as well), although too faint to be plainly seen with the naked eye. The Pleiades seem to be riding on the back of Taurus.

Figure 22–22 The Infrared Astronomical Observatory clearly shows the dust surrounding the Pleiades, an open cluster. The material is much more extensive than the dust that shows in the visible reflecting the starlight. In this IRAS image, blue shows 12-μ, green shows 60-μ, and red shows 100 μ radiation. The stars themselves are all in the washed-out white region.

We need only know here that ordinary hydrogen contains 1 nuclear particle and so is the least massive element, while ordinary helium contains 4 nuclear particles and is the second least massive element. (See Section 24.3.)

stars in galactic clusters are analyzed to find the relative abundances of the chemical elements in their atmospheres, we find that over 90 per cent of the atoms are hydrogen, most of the rest are helium, and less than 1 per cent are elements heavier than helium. This is similar to the composition of the sun. Such stars are said to belong to stellar *Population I*.

The second major type of star cluster appears in a small telescope as a small, hazy area in the sky. Observing with larger telescopes distinguishes individual stars, and reveals that these clusters are really composed of many thousands of stars packed together in a very limited space. The clusters are spherical, and are known as *globular clusters* (Fig. 22–23).

Globular clusters can contain 10,000 to one million stars, in contrast to the 20 to several hundred stars in a galactic cluster. A globular cluster can fill a volume up to 30 parsecs across. The stars are more closely packed toward the center (Fig. 22–24) than toward the periphery.

Most of the known globular clusters, about three-quarters, are in the galactic *halo*. The remaining quarter are the disk subsystem, and may be similar to or related to the "thick disk." (We do not see many globular clusters in the plane of the galaxy because they are hidden by interstellar dust there.) The abundances of the elements heavier than helium in stars in globular clusters are much lower, by a factor of 10 to 300, than their abundances in the sun. Since the abundances of heavy elements grow over time as the elements are formed and spread through space, these stars are older than Population I stars. Such stars are said to belong to *Population II.*

A typical halo globular cluster passes through the plane of the galaxy every 300 million years, as it orbits the galaxy's center of mass. These passages sweep the globular clusters free of interstellar gas and dust.

Figure 22–23 The globular cluster Omega Centauri, which contains hundreds of thousands of stars. We know of 154 globular clusters in our galaxy.

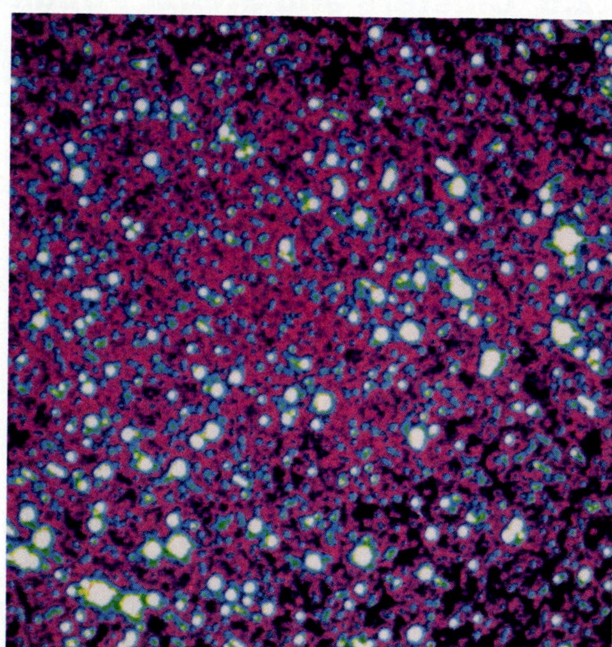

Figure 22–24 A false-color photograph showing a tiny region in the center of Omega Centauri. This whole view would fit within a single star image of the picture at left. The photograph was taken with a CCD on the European Southern Observatory's New Technology Telescope in 1989. The field of view is 47 arc sec square, compared with the 2° square field in the view at left.

22.5a H–R Diagrams and the Ages of Galactic Clusters

The Hertzsprung-Russell diagram for several galactic clusters is shown in Fig. 22–25. For the purposes of this discussion, it is most important to note that the horizontal axis is a measure of temperature and the vertical axis is a measure of brightness. The main part of the figure is actually a set of many individual diagrams like the two on the right, laid on top of each other.

Since all the stars in a cluster were formed at the same time out of the same gas, we can presume that they have similar chemical compositions and differ only in mass. The difference between one cluster and another is principally that the two clusters were formed at different times.

The H–R diagrams for different galactic clusters appear to be similar over the lower part of the main sequence but diverge at the upper part. Comparison

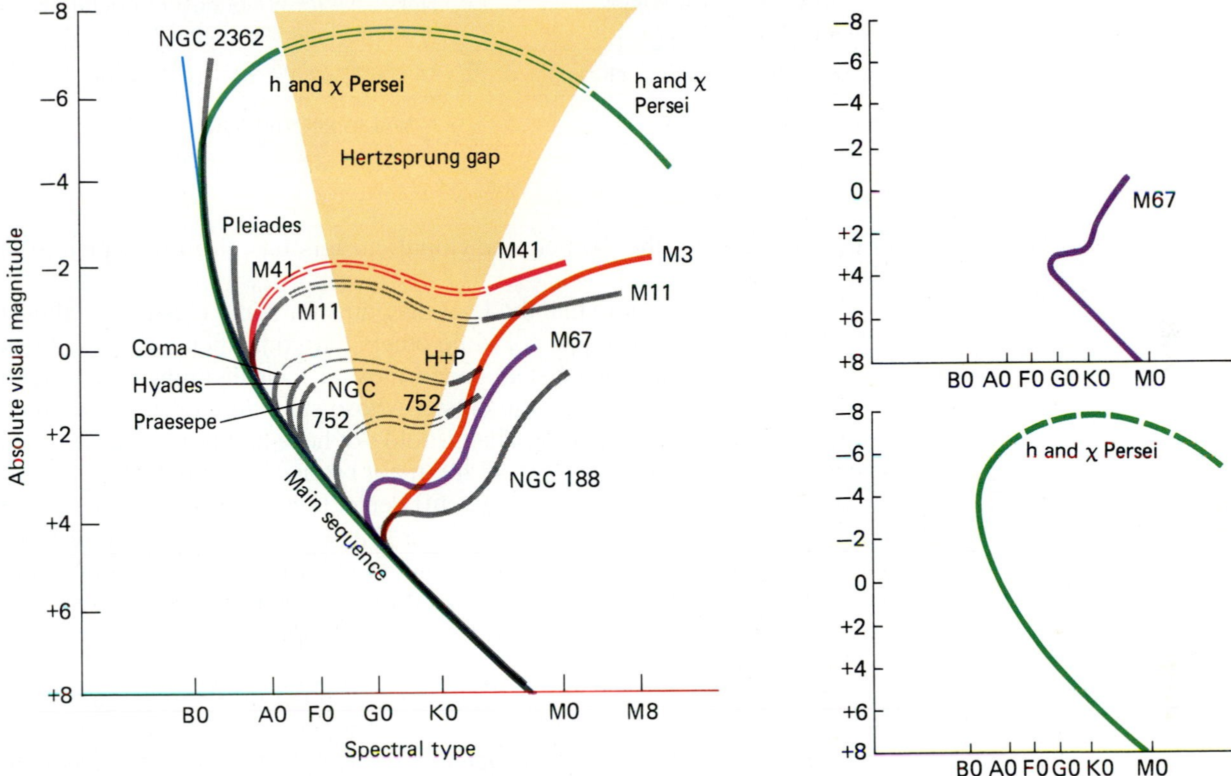

Figure 22–25 (*A*) Hertzsprung-Russell diagrams of several galactic clusters, showing the overlay of several individual diagrams. We see that the fainter stars of galactic clusters are on the main sequence, while the brighter stars are above and to the right of the main sequence. Almost all stars in the younger clusters, like h and χ Persei, follow the main sequence, while more members of older clusters, like M67, have had time to evolve toward the red giant region. By observing the point where the cluster turns off the main sequence, we deduce the length of time since its stars were formed (that is, the age of the cluster). The presence of the *Hertzsprung gap* (*shaded*), a region in which few stars are found, indicates that stars evolve rapidly through this part of the diagram.

Part of the H–R diagram for the globular cluster M3 (*hollow bar*) is graphed for comparison.

An object's M number is its number in the Messier catalogue (Appendix 8), as we shall discuss in the Part V introduction, prior to Chapter 29.

(*B*) Two of the H–R diagrams of individual galactic clusters are shown separately here to illustrate how several of these are put together to make the composite diagram shown in *A*.

Focus On

Box 22.3 Star Clusters

Galactic Clusters	Globular Clusters
No regular shape (also called "open clusters")	Shaped like a ball, stars more closely packed toward center
Many young stars	All old stars
H–R diagrams have long main sequences	H–R diagrams have short main sequences
Where stars leave the main sequence tells the cluster's age	All clusters have the same H–R diagram and thus the same age
Stars have similar composition to sun, Population I	Stars have lower abundances of heavy elements than sun, Population II
Hundreds of stars per cluster	10,000-1,000,000 stars per cluster
Found in galactic plane	Found in galactic halo

of the diagrams for the different individual clusters has given us a picture of how stars evolve.

We see on the figure that the H–R diagrams for some clusters lie almost entirely along the main sequence, while others have only their lowest portions on the main sequence. Since stars spend most of their lifetimes on the main sequence, we can deduce that the cluster whose stars lie mostly on the main sequence must be the youngest. They would not have had time for many stars to have evolved enough to move off the main sequence. Thus NGC 2362 and the pair of clusters known as h and χ (chi) Persei (Fig. 22–26) are the youngest clusters in the composite diagram. (They are known as h Persei and χ Persei—star names rather than ordinary cluster names—because they were once confused with stars.) When a star finishes its main-sequence lifetime, its surface becomes larger and cooler. Thus the star moves upward and to the right on the H–R diagram.

The stars that finish their main-sequence lifetimes most rapidly, and move off the main sequence, are the most massive ones. The most massive stars are the O and B stars on the extreme upper left of the H–R diagram. They are more luminous and use up their nuclear fuel at a faster rate than the cooler, more numerous, ordinary stars like the sun.

In NGC 2362 and h and χ Persei, only the most massive stars have lived long enough to die and move off the main sequence. Theoretical calculations

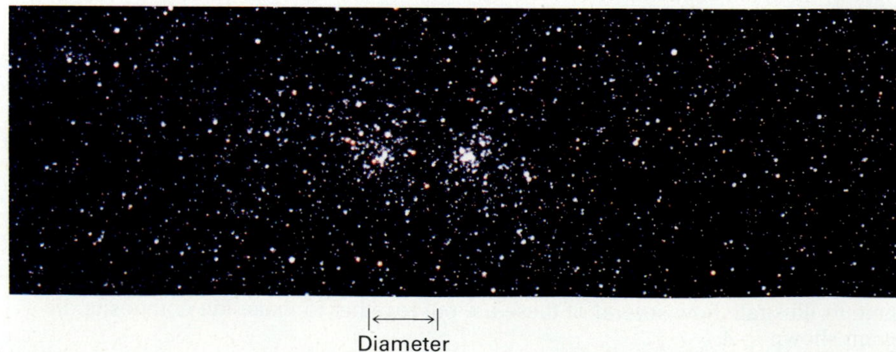

Figure 22–26 The double cluster in Perseus, h and χ Persei, a pair of galactic clusters that are readily visible in a small telescope and close enough together that they appear in the same field of view. Study of the H–R diagram reveals that they are relatively young. Perseus is a northern constellation that is most prominent in the winter sky. In Greek mythology, Perseus slew the Gorgon Medusa and saved Andromeda from a sea monster.

tell us that stars of the spectral type of the point where these two clusters leave the main sequence (about B0) have masses such that their main-sequence lifetimes are 10^7 years. Thus these two clusters must be about 10^7 years old. The stars that live for less time than 10^7 years have died and moved off to the right. The stars that live for longer than 10^7 years are still on the main sequence.

The Pleiades, on the other hand, must be older than 10^7 years, since the stars with lifetimes of 10^7 years have already died. Calculations tell us that stars of the spectral type where the Pleiades' graph turns off the main sequence (about B5) live 10^8 years. Thus the Pleiades must be about 10^8 years old. Similarly, the Hyades must be older than the Pleiades, since still more stars have had time to live their main-sequence lifetimes and die. Calculations tell us that stars at the turnoff point for the Hyades (about A0) have main-sequence lifetimes of 10^9 years, so the Hyades must have been formed 10^9 years ago.

We can thus read a cluster's H–R diagram like a clock, telling how long the cluster has lived by which of its stars (observationally) have turned off the H–R diagram and how old (theoretically) such a star is. Be sure to keep in mind that all the stars in the cluster formed at the same time, so we are telling the age of the cluster and all the stars in it.

Figure 22–27 The globular cluster 47 Tucanae, which is near the Small Magellanic Cloud in the southern sky.

22.5b H–R Diagrams for Globular Clusters

H–R diagrams for all globular clusters (Fig. 22–27) look essentially identical. They all have stubby main sequences. The Hubble Space Telescope is allowing us to see the individual stars in globular clusters more clearly (Fig. 22–28).

From the fact that all globular clusters have very similar H–R diagrams, we can conclude that they are all about the same age. From the fact that they all have H–R diagrams with short main sequences, we can conclude that the globular clusters must be very old. The latest detailed studies assign an age of about 13 billion years, with an uncertainty of about 3 billion years. If we assume that the globular clusters formed when or soon after our galaxy formed, then our galaxy must be about 13 billion years old. Just how different in age individual clusters can be is a matter of debate; a maximum difference of at least two billion years and perhaps even 5 billion years appears likely. The Hubble Space Telescope has found young globular clusters in another galaxy (Fig. 22–29).

The H–R diagrams for globular clusters (Fig. 22–30) have prominent *horizontal branches.* The horizontal branch goes leftward from the stars on the right side of the diagram that have long since turned off the main sequence. It represents very old stars that have evolved past their giant or supergiant phases and are returning leftward. No galactic cluster has stars that old.

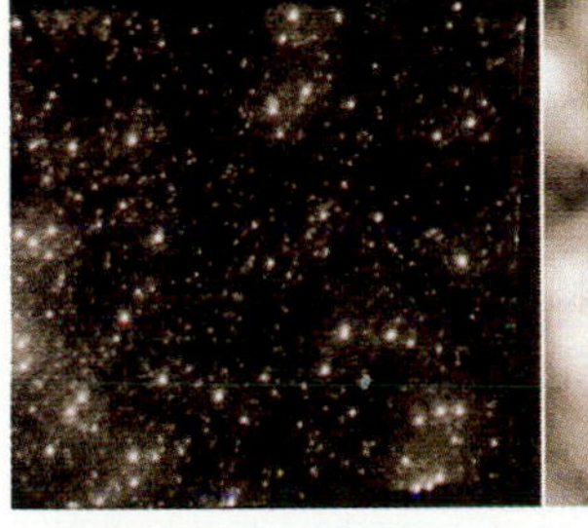

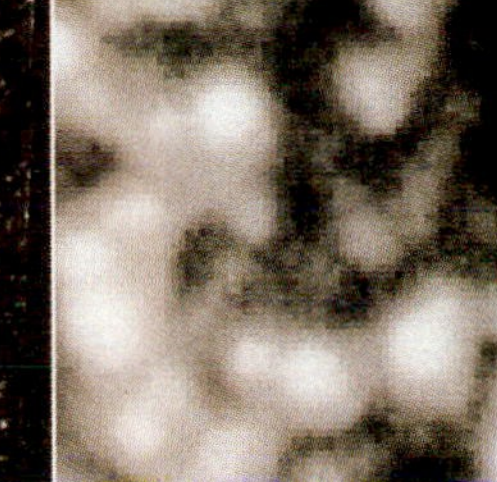

Figure 22–28 The inner region of the globular cluster M14, seen both from the ground (*left*) and from the Hubble Space Telescope (*right*).

Figure 22–29 About 50 massive and compact globular clusters are seen as blue dots in this Hubble Space Telescope image of the giant elliptical galaxy NGC 1275. The white dot is the galaxy's core. Surprisingly, most of the stars in the globular clusters are young, unlike the old stars in globular clusters in our galaxy. The stars in the globular clusters in NGC 1275 are all the same age, though they are all less than a billion years old. These young stars may have formed in a cataclysmic event, perhaps when two galaxies merged.

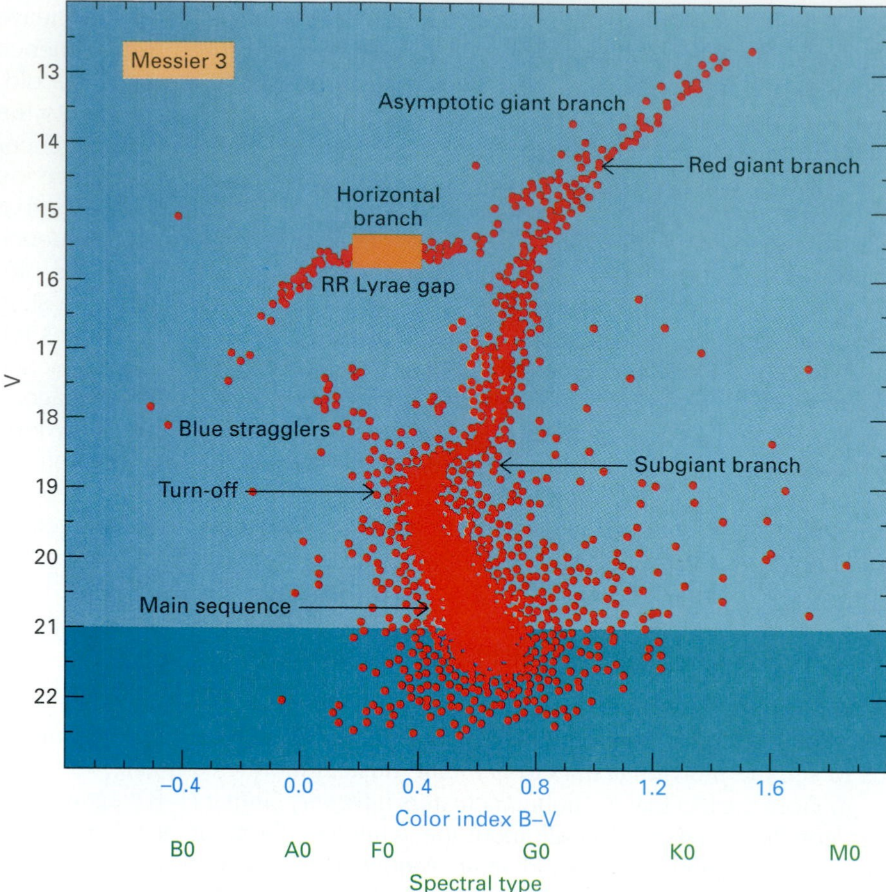

Figure 22–30 The Hertzsprung-Russell diagram for the globular cluster M3, part of which was included in Fig. 22–25*A*. We can observe only the members of the cluster that are brighter than a certain limit. The stars in the more heavily shaded region at the bottom are newly observable because of the use of CCD's.

The positions of stars on the horizontal axis were actually determined from measurements of the colors of the stars made at the telescope by comparing their brightnesses in the blue and in the yellow (which gives the color index B-V). The transformation to spectral type is approximate.

22.5c Galactic Clusters and the Interstellar Medium

Astronomers have long known that the easy-to-find open cluster the Pleiades contains dust that reflects starlight toward us. (See the infrared image Fig. 22–22 and the visible-light image Fig. 29–2*A*.) Only recently, though, have astronomers realized that the dust is present because the stars have made a chance encounter with a cloud of interstellar dust. An image from the IRAS spacecraft shows the wake of the cluster (Fig. 22–31). The hole is about 40 light-years long. Richard White of Smith College and John Bally of the University of Colorado have shown that the Pleiades generates a shock wave that deflects gas and dust. The older idea—that the dust remains from the interstellar cloud out of which the stars were born—is erroneous.

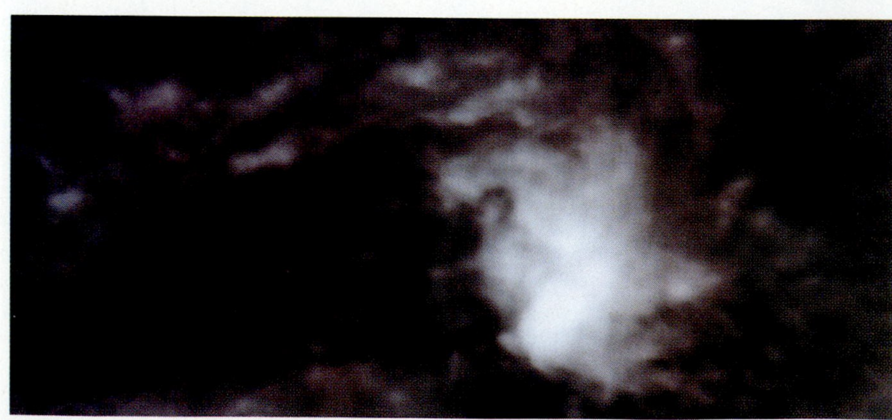

Figure 22–31 The wake of the Pleiades open star cluster through the dust and gas of interstellar space. The stars themselves are in the bright emission to the right of center and give the energy that makes it shine. The wake is the dark region to the left.

Summary and Outline

Binary stars (Section 22.1)
- Optical doubles, visual binaries, binaries with composite spectra, spectroscopic binaries, eclipsing binaries, astrometric binaries

Determination of stellar masses (Section 22.2)
- Mass-luminosity relation

Determination of stellar sizes (Section 22.3)
- Indirect: calculate from absolute magnitude or measure in eclipsing binary system
- Direct: interferometry

Variable stars (Section 22.4)
- Mira variables (Section 22.4a)
- Cepheid variables (Section 22.4b)
  - Period-luminosity relation
  - Uses for determining distances
- RR Lyrae variables (Section 22.4c)
  - All of approximately the same absolute magnitude
  - Uses for determining distances

Clusters and stellar populations (Section 22.5)
- Galactic (open) clusters (Section 22.5a)
  - Population I: relatively high abundance of elements heavier than helium
  - Representative of spiral arms
  - 20 to several hundred members
  - Turnoff on H–R diagram gives age
- Globular clusters (Section 22.5b)
  - Population II: relatively low abundance of metals
  - Representative of galactic halo
  - 10^4 to 10^6 members
  - Old enough to have H–R diagrams with horizontal branches

Key Words

optical doubles, double star, binary star, visual binaries, composite spectrum, spectroscopic binary, eclipsing binary, astrometric binaries, mass-luminosity relation, *lunar occultation, *interferometry, *speckle interferometry, period, light curves, Mira variables, long-period variables, Cepheid variables, RR Lyrae stars, cluster variables, star clusters, open clusters, galactic clusters, Population I, globular clusters, halo, Population II, horizontal branches

*This term is found in an optional section.

Questions

1. Sketch the orbit of a double star that is simultaneously a visual, an eclipsing, and a spectroscopic binary.

2. Define briefly and contrast an astrometric binary and an eclipsing binary.

3. (a) Assume that an eclipsing binary contains two identical stars. Sketch the intensity of light received as a function of time. (b) Sketch to the same scale another curve to show the result if both stars were much larger while the orbit stayed the same.

†4. How much brighter than the sun is a main-sequence star whose mass is 10 times that of the sun?

5. What does the mass-luminosity relationship tell us about the mass of the supergiant star Betelgeuse (spectral type M2)?

†6. A main-sequence star is 3 times the mass of the sun. What is its luminosity relative to that of the sun?

†7. (a) Use the mass-luminosity relation to determine about how many times brighter than the sun are the most massive main-sequence stars we know. (b) How many times fainter are the least massive main-sequence stars?

8. When we look at the Andromeda Galaxy, from what mass range of star does most of the light come? What mass range of star provides most of the mass of the galaxy?

9. When we consider the gravitational effects of a distant galaxy on its neighbors, are we measuring the effects of mostly low- or high-mass stars? Explain.

†This question requires a numerical solution.

†10. A certain Cepheid variable has a period of 30 days. What is its absolute magnitude?

†11. What is the ratio of apparent brightness of two Cepheid variables in the Large Magellanic Cloud, one with a 10-day period and the other with a 30-day period?

†12. An RR Lyrae star has an apparent magnitude of 6. How far is it from the sun?

13. A Cepheid variable with a period of 10 days has an apparent magnitude of 8. How bright would an RR Lyrae star be if it were in a globular cluster near to the Cepheid?

14. An astronomer observes a galaxy and notices that a star in it brightens and dims every 11 days. Sketch the light curve, label the axes, and explain how to find the distance to the galaxy.

†15. What is the absolute magnitude of an RR Lyrae star with an 18-hour period?

16. Briefly distinguish Population I from Population II stars.

17. Cluster X has a higher fraction of main-sequence stars than cluster Y. Which cluster is probably older?

18. What is the advantage of studying the H–R diagram of a cluster, compared to that of the stars in the general field?

19. Which galactic cluster shown in Figure 22–26 is about 8 billion years old?

†20. From the position of the RR Lyrae gap and your knowledge of RR Lyrae stars, how far away is M3, the globular cluster whose H–R diagram is shown in Figure 22–29?

The 1991 total solar eclipse in Baja California.

The Sun: A Star Close Up

23

Aims: To study the sun, which is the nearest star and an example of all the other stars whose surfaces we cannot observe in such detail

We have discussed a range of individual stars of different spectral classes and have discussed groupings of stars in close physical proximity to each other. Studying these distant stars has allowed us to learn a lot about the properties of stars and how they evolve. But not all stars are far away; one is close at hand. By studying the sun, we not only learn about the properties of a particular star but also can study the details of processes that undoubtedly take place in more distant stars as well. We will first discuss the *quiet sun,* the solar phenomena that appear every day. Afterwards, we will discuss the *active sun,* solar phenomena that appear non-uniformly on the sun and vary over time.

23.1 Basic Structure of the Sun

We think of the sun as the bright ball of gas that appears to travel across our sky every day. We are seeing only one layer of the sun, part of its atmosphere; the properties of the solar interior below that layer and of the rest of the solar atmosphere above that layer are very different. The outermost parts of the solar atmosphere even extend through interplanetary space beyond the orbit of the earth.

The layer that we see is called the *photosphere* (Fig. 23–1), which simply means the sphere from which the light comes (from the Greek *photos,* meaning "light"). As is typical of many stars, about 94 per cent of the atoms and nuclei in the outer parts are hydrogen, about 5.9 per cent are helium, and a mixture of all the other elements makes up the remaining one-tenth of one per cent. The overall composition of the interior is not very different.

The sun is an average star, since stars much hotter and much cooler, and stars intrinsically much brighter and much fainter, exist. Radiation from the photosphere peaks (is strongest) in the middle of the visible spectrum; after all, our eyes evolved over time to be sensitive to that region of the spectrum because the greatest amount of the solar radiation occurred there (given that it penetrated our earth's atmosphere). If we lived on a planet orbiting an object that emitted mostly x rays, we, like Superman, might have x-ray vision.

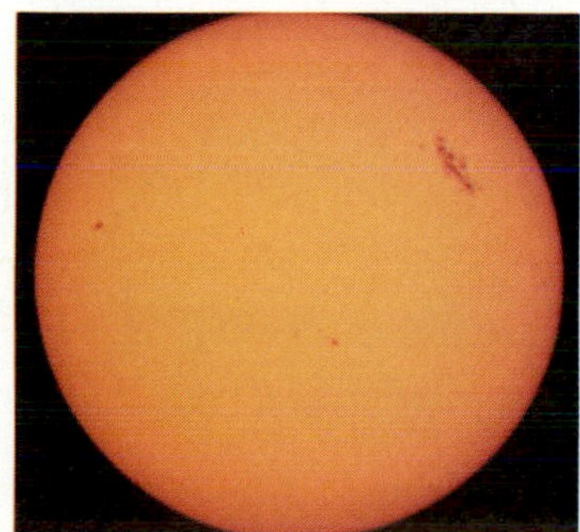

Figure 23–1 The solar photosphere, with sunspots from the 1989-90 solar maximum.

The solar photosphere is about 1.4 million km (1 million miles) across. The disk of the sun (the apparent surface of the photosphere) takes up about one-half a degree across the sky; we say that it *subtends* one-half degree. This angle is large enough for us to see detailed structure on the solar surface, and we shall describe the structure that is always present in Section 23.2. In Section 23.7, we shall discuss sunspots and other structure that changes drastically over time.

Beneath the photosphere is the solar *interior.* All the solar energy is generated there at the solar *core,* which is about 10 per cent of the solar diameter at this stage of the sun's life. The temperature there is about 15 million K. In

Chapter 24 we will discuss how energy is generated in the sun and other stars and what their interiors are like. In the solar core, energy is carried by radiation. Above that region, in the upper part of the solar interior, energy is carried by convection, the way that matter moves when it is heated below and carries energy as it moves upward against gravity. We cannot see the solar interior but are developing ways of learning about it by studying the solar surface.

The photosphere is the lowest level of the *solar atmosphere* (Fig. 23–2). Though the sun is gaseous through and through, with no solid parts, we still use the term "atmosphere" for the upper part of the solar material. The parts of the atmosphere above the photosphere are very tenuous, and contribute only a small fraction to the total mass of the sun. In the visible part of the spectrum, these upper layers are very much fainter than the photosphere, and cannot be seen with the naked eye except during a solar eclipse, when the moon blocks the photospheric radiation from reaching our eyes directly. Now we can also study these upper layers with special instruments on the ground and in orbit around the earth.

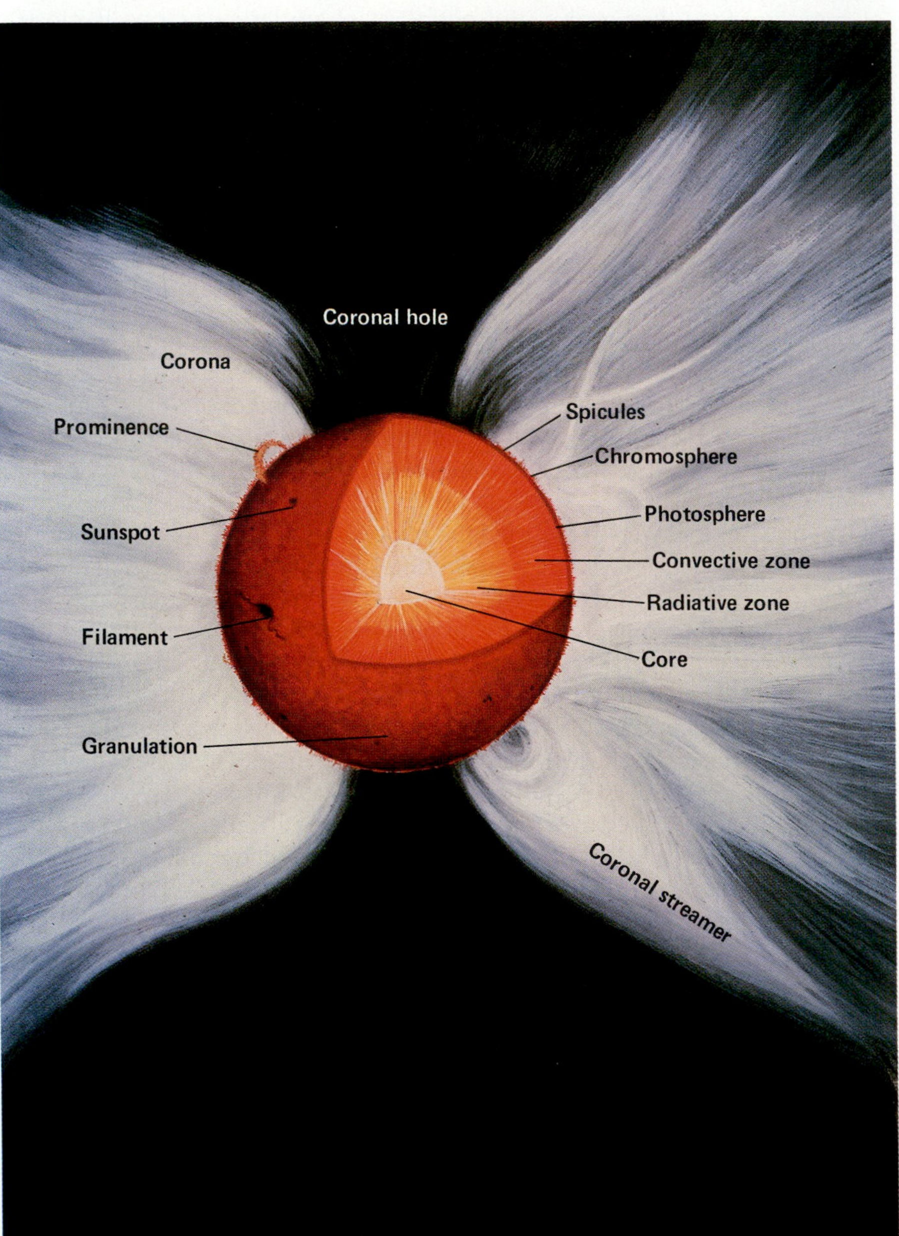

Figure 23–2 The parts of the solar atmosphere and interior. The solar surface is depicted as it appears through a hydrogen filter.

Just above the photosphere is a jagged, spiky layer about 10,000 km thick, only about 1.5 per cent of the solar radius. This layer glows colorfully pinkish when seen at an eclipse, and is thus called the *chromosphere* (from the Greek *chromos*, meaning "color"). Above the chromosphere, a ghostly white halo called the *corona* (from the Latin, meaning "crown") extends tens of millions of kilometers into space. The corona is continually expanding into interplanetary space and in this form is called the *solar wind.* We shall discuss all these phenomena in succeeding sections.

23.2 The Photosphere

The sun is a normal star of spectral type G2, which means that its surface temperature is about 5800 K. But the sun is the only star close enough to allow us to study its surface in detail. In recent years, we have become able to verify directly from the ground and from space that many of its features indeed also appear on other stars.

23.2a High-Resolution Observations of the Photosphere

One major limitation in observing the sun is the turbulence in the earth's atmosphere, which also causes the twinkling of stars. The problem is even more serious for studies of the sun, since the sun is up in the daytime when the atmosphere is heated by the solar radiation and so is more turbulent than it is at night.

Only at the very best observing sites, specially chosen for their steady solar observing characteristics, can one see detail on the sun subtending an angle as small as 1 second of arc. This corresponds to about 700 km on the solar surface, the distance from Boston to Washington, D.C. Occasionally, objects ½ arc second across can be seen.

Sometimes we observe the sun in *white light*—all the visible radiation taken together. When we study the solar surface in white light with 1 arc second resolution, we see a salt-and-pepper texture called *granulation* (Fig. 23–3). The effect is similar to that seen in boiling liquids on earth, which are undergoing *convection.* Convection, the transport of energy as hot matter moves upward and cooler matter falls, is one of the basic ways in which energy can be transported. Convection carries energy to your boiling eggs, for example. Conduction and radiation are the other major methods of energy transport.

Granulation on the sun is an effect of convection. Each granule is only about 1000 km across, and represents a volume of gas that is rising from and falling to a shell of convection, called the *convection zone,* located below the photosphere. The granules are convectively carrying energy from the hot solar interior to the base of the photosphere. But the granules are about the same size as the limit of our resolution, so are difficult to study.

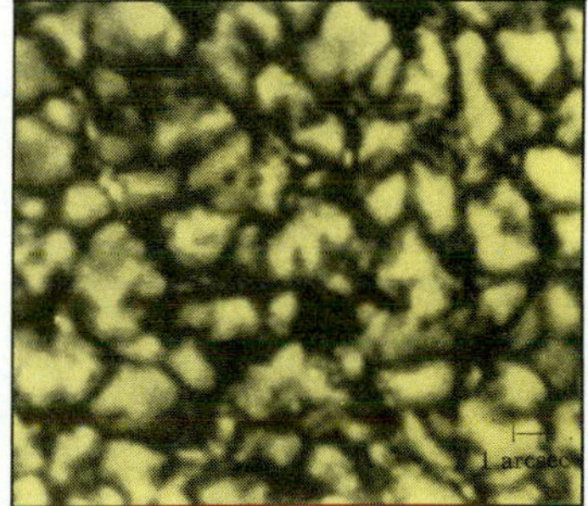

Figure 23–3 Solar granulation, photographed on February 1, 1986, with the new Swedish solar telescope in the Canary Islands. The picture was taken at a wavelength that showed only a narrow band of solar continuous radiation, which gave the same appearance as though a wide white-light band had been viewed. The resolution is extraordinary.

23.2b Solar Seismology

It was discovered in the early 1960's that areas of the upper photosphere are oscillating up and down with a 5-minute period. We have since discovered that the whole sun's surface is oscillating with periods ranging from minutes up to hours. Basically, it is ringing like a bell. However, since the waves that cause the ringing come randomly and continually, the sun is ringing like a bell that is continually hit by particles in a sandstorm rather than a bell hit once. We think that rising and falling motions in the solar convection zone make the waves that cause the ringing.

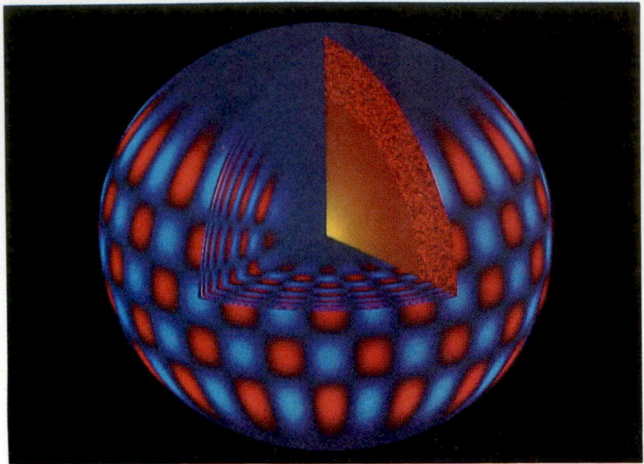

Figure 23–4 One of the thousands of possible modes of the sun's oscillations. Solid lines represent zones of expansion and dotted lines represent zones of contraction.

Figure 23–5 Astronomers work with a solar telescope at the south pole to get a long run of uninterrupted sunlight, in order to study solar oscillations with periods of hours. Their longest run was several days long, during the south pole's summer when the sun never sets.

The 5-minute oscillation we see is really the superposition of perhaps 10 million individual modes of vibration of the sun, each with its own period (Fig. 23–4). The oscillations are studied from their Doppler shifts. Since many of the modes of vibration have nearly but not quite identical periods, astronomers must observe for many days without a break to sort them out. To eliminate the nighttime gaps, the waves have been studied with a telescope at the earth's south pole (Fig. 23–5), where the sun is up for months without setting. Other continuous time-series are being made by coordinating observatories scattered

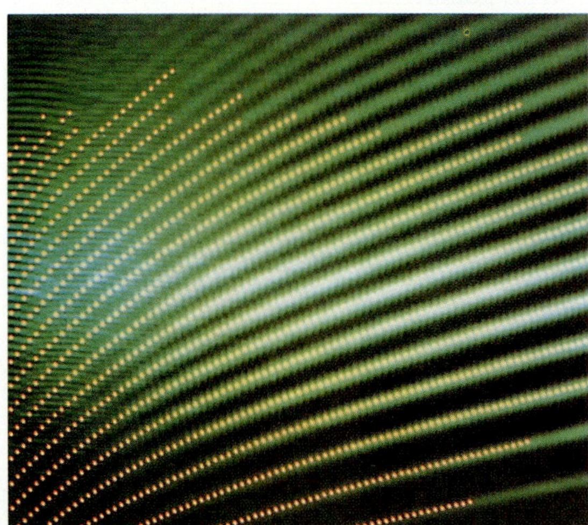

Figure 23–6 Helioseismology data (*points*) agree well with the theoretical predictions (*green lines*). Plotted is the number of waves around the equator (as in Figure 23–4) along the *y*-axis vs the number of waves perpendicular to the equator along the *x*-axis. The agreement is excellent. Since the theoretical predictions depend on the temperature and density inside the sun, we can adjust our model for temperature and density to match the observations as well as possible.

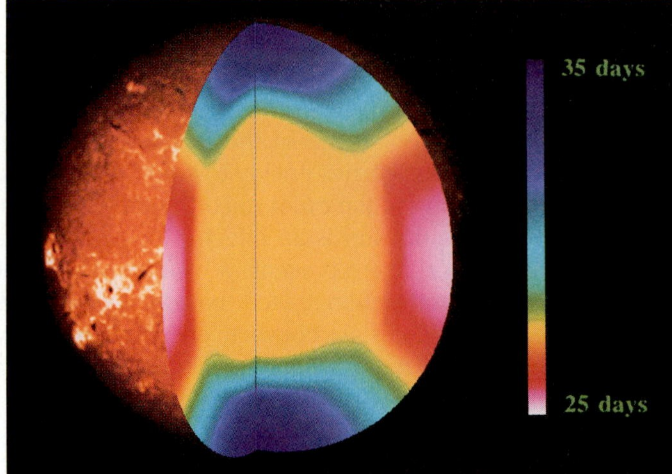

Figure 23–7 The rotation periods of different layers and regions of the sun, based on helioseismology results. Red is the shortest rotation period (26 days), green corresponds to 28 days, and violet corresponds to 36 days. Thus the measurements show that the surface rotation periods, which vary from 25 days at the equator to 36 days at the poles, persist inward through the solar convection zone. Further toward the center, the data show that the sun rotates as a solid body with a 27-day period. The results are still somewhat uncertain, and are more uncertain at deeper depths.

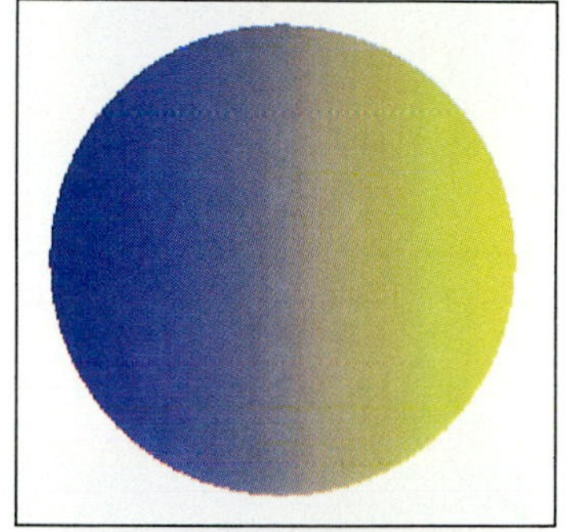
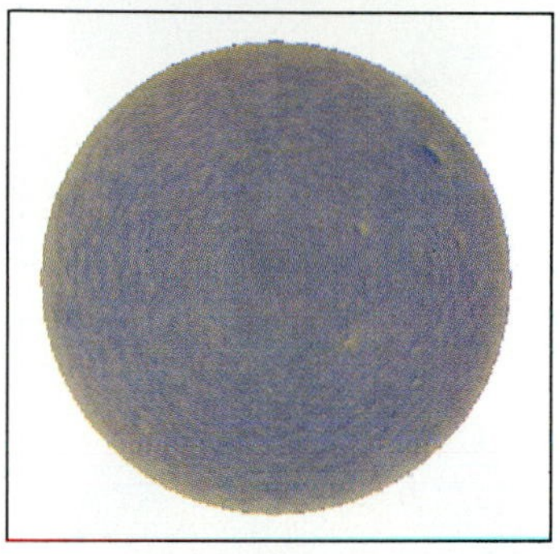

Figure 23–8 Preliminary data from GONG. The top image, with blueshifts shown as blue and redshifts as yellow, minus the middle image (which accounts for the Sun's rotation) equals the bottom image (see Fig. 23–10).

around the world at different longitudes, such as one in Hawaii and one in the Canary Islands.

Interpreting these oscillations in the photosphere is telling us what the inside of the sun is like, similar to the way that geologists find out about the inside of the earth by interpreting seismic waves. The studies are thus called *solar seismology*. The waves travel downward into the sun until the rapid increase of temperature and therefore their speed make them bend back up. They then travel upward until they reflect off the bottom of the photosphere. The waves are trapped between upper and lower levels, and their wavelengths depend on how deep the lower level is. Thus studying different wavelengths gives us information about different levels in the solar interior (Fig. 23–6).

Theoretical astronomers can predict the periods of waves that will exist, but their predictions depend on how temperature, density, pressure, and composition vary under the solar surface. Matching predictions with the observations has already led to the conclusion that the sun's convection zone is deeper than had been thought. It seems to take up the outer 30 per cent of the solar radius. The results also limit how fast the inside of the sun can be rotating relative to the photosphere (Fig. 23–7). The results are leading to new ideas about how sunspots are formed.

To provide continuous coverage of the sun, specially designed telescopes are being installed at a network of sites around the world. Since the sun is ringing like a bell, the project has the acronym GONG—**G**lobal **O**scillation **N**etwork **G**roup (Fig. 23–8). In the years to come, it should provide a series of observations at least three years long, which will lead to good data about the solar interior. The European Space Agency plans to mount helioseismology instruments on its SOHO (**So**lar and **H**eliospheric **O**bservatory) spacecraft to take years of continuous data.

23.2c The Photospheric Spectrum

The spectrum of the solar photosphere, like that of other G stars, is a continuous spectrum crossed by absorption lines (Fig. 23–9). Hundreds of thousands of these absorption lines, which are also called Fraunhofer lines, have been photographed and catalogued. They come from most of the chemical elements, although some of the elements have many lines in their spectra and some have very few. Iron has many lines in the spectrum. The hydrogen Balmer lines are strong but few in number. Helium is not excited at the relatively low temperatures of the photosphere, and so its spectral lines do not appear.

Fraunhofer, in 1814, labelled the strongest of the absorption lines in the solar spectrum with letters from A through H. His C line, in the red, is now known to be the first line in the Balmer series of hydrogen, and is called Hα (H alpha). (We have seen in Section 4.4 that Hα comes from a transition between the 2nd and 3rd energy levels of hydrogen.) We still use Fraunhofer's notation for some of the strong lines: The D lines, a pair of lines close together in the yellow part of the spectrum, are caused by neutral sodium (Na I). The H line and the K line, both in the part of the violet spectrum that is barely visible to the eye, are caused by singly ionized calcium (Ca II).

Fraunhofer's original spectrum is shown in Figure 4.46.

The places where the continuous spectrum (called the *continuum*) and most of the absorption lines are formed are mixed together throughout the photospheric layers. The energy we receive on earth in the form of solar radiation has been transported upward to these layers from the solar interior.

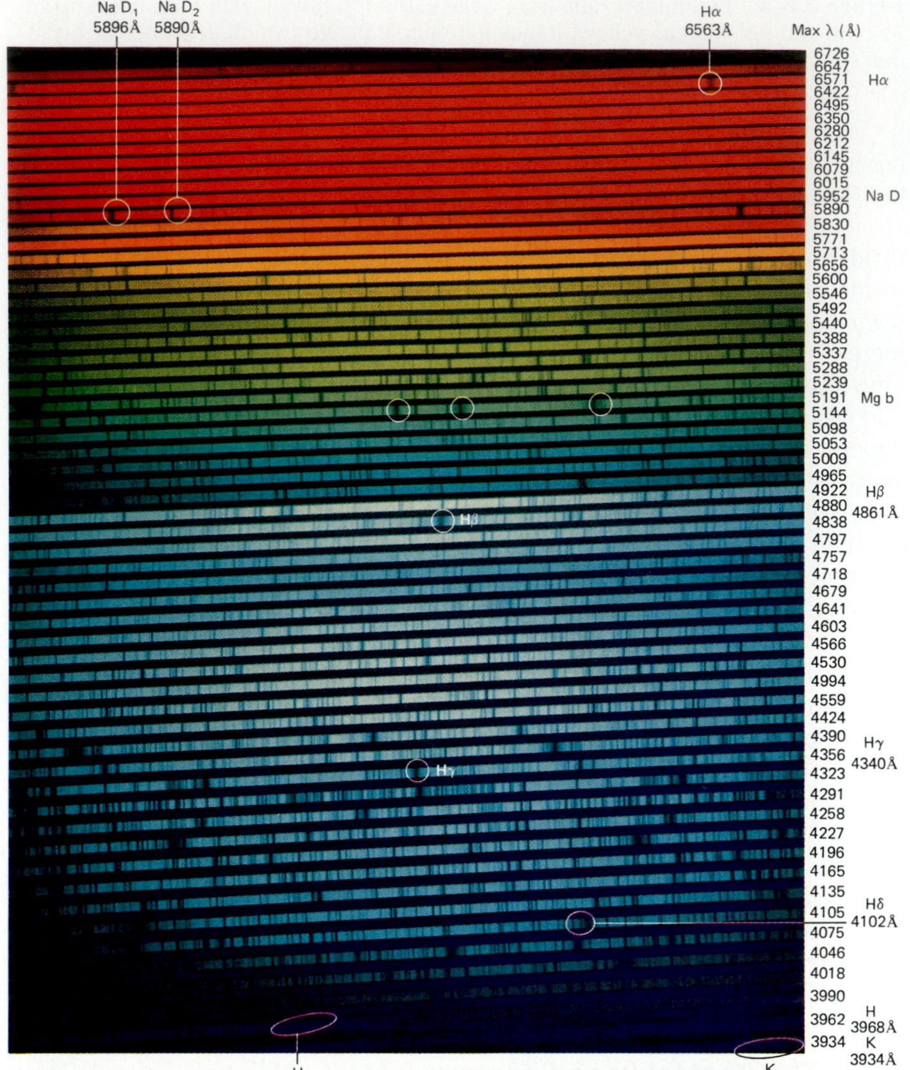

Figure 23–9 The visible part of the solar spectrum, from ultraviolet through red.

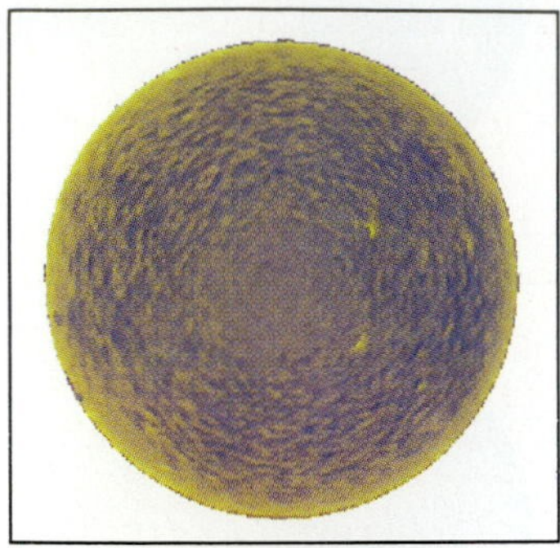

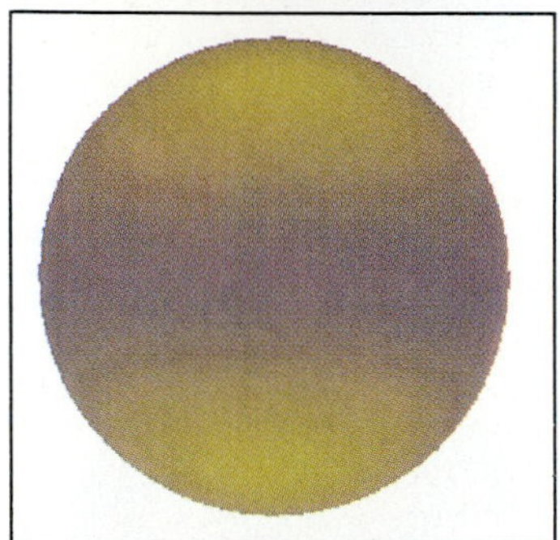

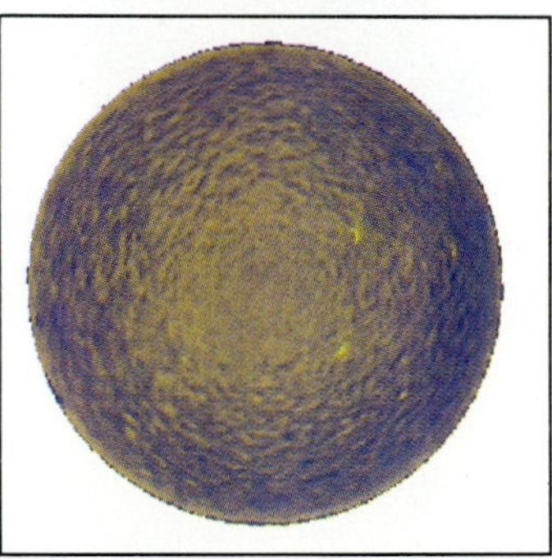

Figure 23–10 Supergranulation is best visible in images of the Sun's velocity field. Dark areas (blue) are receding and light areas (yellow) are approaching us. The supergranules are each less than 1 mm across on this reproduction. Because of the flow from the center to the edge of each supergranule, the Doppler shift makes one side appear dark and other bright on this velocity image (*top*). The image can be further broken down into other components, including a slow flow from the equator to the pole (*middle*) and the supergranule flows (*bottom*), as well as some other effects.

23.3 The Chromosphere

23.3a The Appearance of the Chromosphere

Under high resolution, we see that the chromosphere is not a spherical shell around the sun but rather is composed of small spikes called *spicules.* The spicules rise and fall, and have been compared in appearance to blades of grass or burning prairies.

Spicules are more-or-less cylinders of about 1 arc second in diameter and perhaps ten times that in height, which corresponds to about 700 km across and 7000 km tall. They seem to have lifetimes of about 5 to 15 minutes, and there may be approximately half a million of them on the surface of the sun at any given moment.

Studies of velocities on the solar surface showed the existence of large, organized convection cells, called *supergranulation* (Fig. 23–10). Supergranulation cells look somewhat like polygons of approximately 30,000 km diameter. Supergranulation is an entirely different phenomenon from granulation. Each supergranulation cell may contain hundreds of individual granules.

Matter wells up in the middle of a supergranule and then slowly moves horizontally across the solar surface to the supergranule boundaries. The matter then sinks back down at the boundaries. This slow circulation of matter seems

Figure 23–11 A filter passing only the Hα line of hydrogen shows the chromosphere. Here we see an Hα image of the sun during the 1989 partial eclipse. Dark filaments snake across the sun, and light plages are also visible around active regions. A longer exposure of the prominences above the limb, also in Hα, surrounds the inner image.

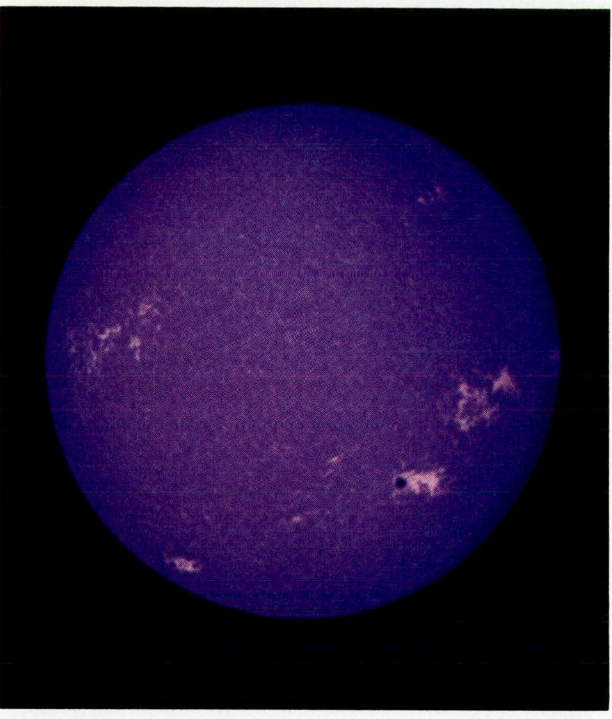

Figure 23–12 A filter passing only the H or K line of ionized calcium also shows the chromosphere. The color is harder to see when looking through an eyepiece at a telescope and the sun is fainter than in Hα, but the supergranulation shows up particularly well.

to be a basic process of the lower part of the solar atmosphere. The network of supergranulation boundaries, called the *chromospheric network,* is visible in the radiation of hydrogen α (Fig. 23–11) or the H and K lines of calcium (Fig. 23–12). In both cases, the solar gas is so absorbing that our line of sight stops at the chromospheric level from seeing farther downward into the sun.

Chromospheric matter appears to be at a temperature of approximately 15,000 K, somewhat higher than the temperature of the photosphere. Ultraviolet spectra of distant stars recorded by the International Ultraviolet Explorer (IUE) have shown unmistakable signs of chromospheres in stars of spectral types like the sun. Thus by studying the solar chromosphere, we are also learning what the chromospheres of other stars are like.

23.3b The Chromospheric Spectrum at the Limb

During the few seconds at an eclipse of the sun that the chromosphere is visible, its spectrum can be taken. This type of observation has been performed ever since the first spectroscopes were taken to eclipses in 1868. Since the chromosphere then appears as hot gas silhouetted against dark sky, the chromospheric spectrum consists of emission lines.

The chromospheric emission lines appear to flash into view at the beginning and at the end of totality, so the visible spectrum of the chromosphere is known as the *flash spectrum* (Fig. 23–13).

Astronomers have been able to study the chromospheric spectrum in the ultraviolet using telescopes on spacecraft; the spectrum in the ultraviolet contains emission lines, even when viewed at the center of the solar disk.

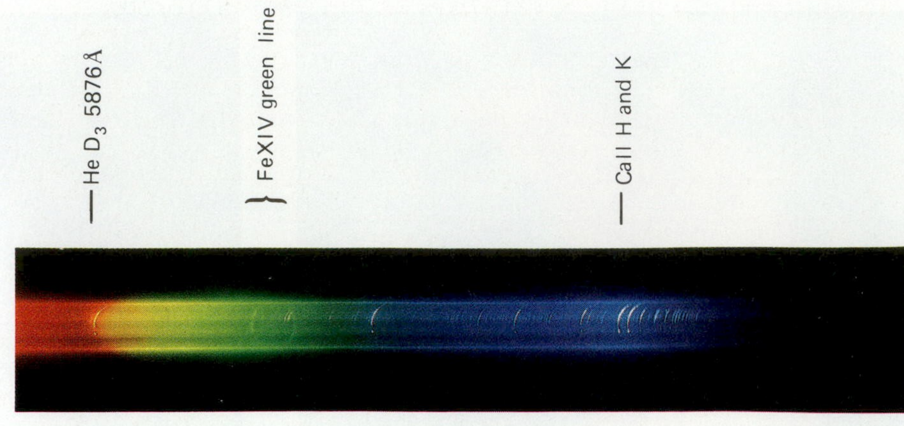

Figure 23–13 The flash spectrum shows the chromospheric spectrum. The curved lines are the shape of the chromosphere peeking out beyond the limb of the moon. The H and K lines of ionized calcium and the D_3 line of helium are prominent in the flash spectrum. (Helium was discovered from this line at an eclipse over 100 years ago.) The coronal green line shows as a complete circle. The spectrum was taken without using a slit in the spectrograph, so the narrow emission shape at the sun is reproduced in the spectrum.

23.4 The Corona

23.4a The Structure of the Corona

During total solar eclipses, when first the photosphere and then the chromosphere are completely hidden from view, a faint white halo around the sun becomes visible. This *corona* (Fig. 23–14) is the outermost part of the solar atmosphere, and extends throughout the solar system. Close to the solar limb, the corona's temperature is about 2 million K.

Even though the temperature of the corona is so high, the actual amount of energy in the solar corona is not large. The temperature quoted is actually a measure of how fast individual particles (electrons, in particular) are moving. There aren't very many coronal particles, even though each particle has a high velocity. The corona has less than one-billionth the density of the earth's atmosphere, and would be considered to be a very good vacuum in a laboratory on earth. For this reason, the corona serves as a unique and valuable celestial laboratory in which we can study gaseous plasmas in a near vacuum.

Figure 23–14 The total solar eclipse of 1988, photographed from the Philippines, with a superimposed picture in the center taken in x-rays from a rocket.

*23.4b Other Ground-Based Observations of the Corona

Figure 23–15 From a few mountain sites, the innermost corona can be photographed without need for an eclipse. The corona shows up best in its green emission line from thirteen-times ionized iron at 5303 Å.

The corona is normally too faint to be seen except at an eclipse of the sun (Section 6.6) because it is fainter than the everyday blue sky. But at certain locations on mountain peaks on the surface of the earth, the sky is especially clear and dust-free, and the innermost part of the corona can be seen (Fig. 23–15). Special telescopes called *coronagraphs* study the corona from such sites—again, only the innermost part is detectable. Coronagraphs are built with special attention to low scattering of light, since their object is not to gather a lot of light but to prevent the strong solar radiation from being scattered about within the telescope.

Several manned and unmanned spacecraft have used coronagraphs to photograph the corona hour by hour in visible light. These satellites studied the corona to much greater distances from the solar surface than can be studied with coronagraphs on earth. Among the major conclusions of the research is that the corona is much more dynamic than we had thought. For example, many blobs of matter were seen to be ejected from the corona into interplanetary space (Fig. 23–16), often in connection with solar flare activity (which will be discussed in Section 23.7b).

In 1984, space shuttle astronauts repaired the Solar Maximum Mission (SMM), a spacecraft bearing a coronagraph (Fig. 23–17) that was launched in 1980 but failed a few months later. The SMM coronagraph hid not only the solar photosphere but also an extra ¾ of a solar radius, so the inner corona was also hidden (Fig. 23–18). So solar eclipses remain the best way to study the inner and lower middle corona. Coordinating eclipse observations with satellite observations is necessary for the most complete picture.

In any case, Solar Maximum Mission fell out of orbit in 1989. A Solar Orbiting Laboratory to provide high-resolution photospheric images, is not yet definitely planned. The European Ulysses mission is, at this writing, scheduled for launch in late 1990. It is to use Jupiter's gravity to throw it out of the ecliptic plane and back over a solar pole.

Figure 23–16 This example of solar ejection was photographed from Skylab. Here we see two pictures superimposed. The eruption at the extreme right, photographed with the coronagraph, is the response of the corona to a lower-level eruption, photographed in the ultraviolet and superimposed on the central dark region occulted by the coronagraph.

Figure 23–17 A coronal mass ejection photographed by the Solar Maximum Mission. The image is six solar radii across.

Figure 23–18 A false-color view of the corona, made from data sent by the Solar Maximum Mission.

23.5 The Heating of the Corona

The gas in the corona is so hot that it emits mainly x-rays, photons of high energy. The photosphere, on the other hand, is too cool to emit x-rays. As a result, when photographs of the sun are taken in the x-ray region of the spectrum, they show the corona and its structure (Fig. 23–19).

X-ray astronomy began with studies of the Sun. Astronauts used an x-ray telescope on Skylab in 1973–1974 to map the corona for months. A solar x-ray telescope is now aloft on the Japanese Yohkoh ("Sunbeam") spacecraft (Fig. 23–19). The telescope, an American/Japanese collaboration, is being used to follow changes in the corona. Other instruments on Yohkoh are dedicated to observing solar flares.

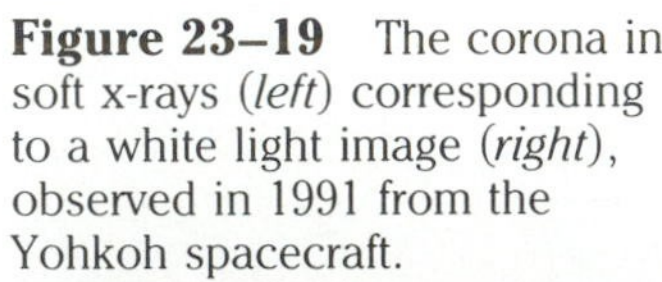

Figure 23–19 The corona in soft x-rays (*left*) corresponding to a white light image (*right*), observed in 1991 from the Yohkoh spacecraft.

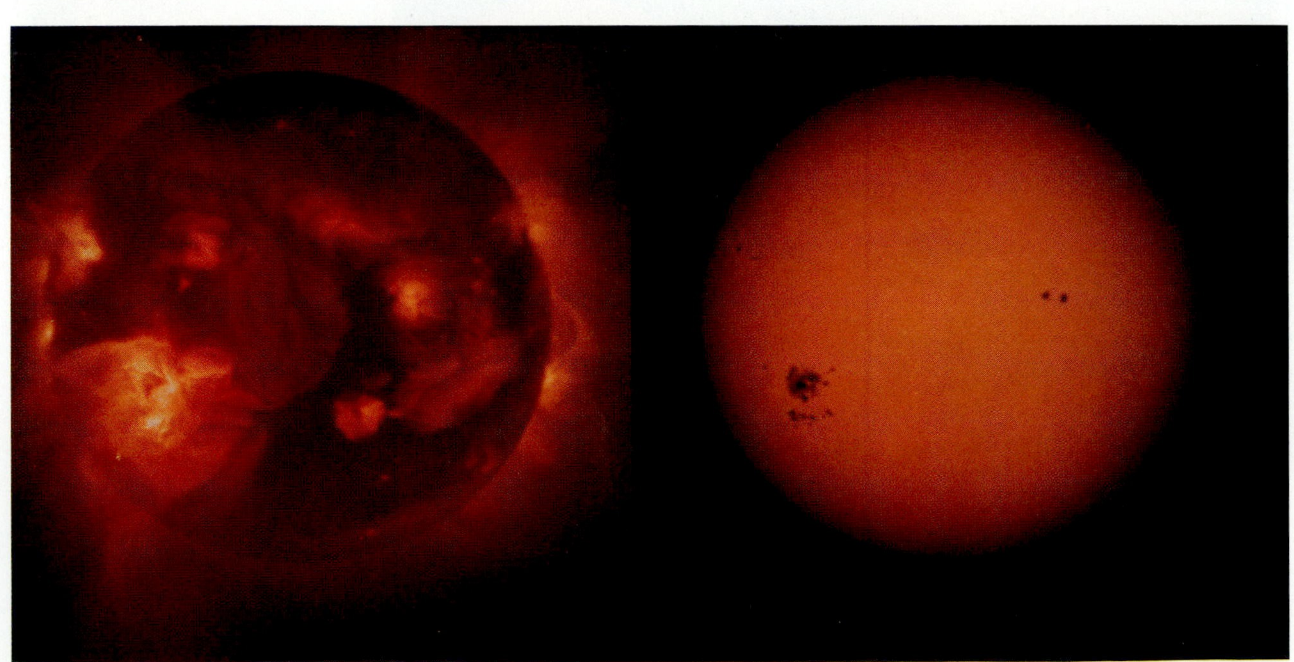

The x-ray pictures show structure that varies with the 11-year cycle of solar activity, which we shall discuss in Section 23.6. The brightest regions visible in x-rays are part of "active regions" that can also be seen in white light. (In white light, these active regions include sunspots.) X-ray images show a higher layer of the solar atmosphere than the white-light images show.

Detailed examination of the x-ray images shows that most, if not all, the radiation of the inner corona is emitted by loops of gas joining points separated from each other on the solar surface (Fig. 23–20). The latest thinking is that the corona is composed entirely of these loops. One important line of thought that follows from this new idea is that we must understand the physics of coronal loops in order to understand how the corona is heated. It is not sufficient to think in terms of a uniform corona, since the corona is obviously so non-uniform.

Figure 23–20 An x-ray photograph of the Sun taken from Skylab on June 1, 1973. The filter passed wavelengths 2–32 Å and 44–54 Å. The dark region across the center is a coronal hole.

The x-ray image also shows a very dark area at the sun's north pole and extending downward across the center of the solar disk. (As the sun rotates from side to side, obviously, the part extending across the center of the disk will not usually be facing us.) These dark locations are *coronal holes*, regions of the corona that are particularly cool and quiet. The density of gas in those areas is lower than the density in adjacent areas.

There is usually a coronal hole at one or both of the solar poles. Less often, we find additional coronal holes at lower solar latitudes. The regions of the coronal holes seem very different from other parts of the sun.

Since the corona is so hot, the peak of its radiation (according to its Planck curve) is in the x-ray region of the spectrum. X-rays don't come through the earth's atmosphere, but x-rays from the sun have been studied from several rockets and spacecraft. The Skylab mission carried a notable x-ray telescope for solar observations (Fig. 23–19) and made the phenomenon of coronal holes known.

Solar x-ray telescopes for a long time used grazing incidence, which was inefficient. A new development is the use of thin films deposited one on top of another that allow x-ray mirrors to be used at the same angles as mirrors for visible light—straight on (Fig. 23–21).

Solar physicists used to think that the corona was heated to millions of degrees by shock waves that carried energy upward from underneath the photosphere. Shock waves are abrupt changes in pressure that occur when motion exceeds the speed of sound at a location. Though the theory of shock-wave heating had seemed very well established and had been worked on continuously since 1946, it was discarded in 1980 on the basis of new evidence. Some of the important new evidence was a set of observations from a spacecraft in the late 1970's. The data showed directly that the shock waves were carrying, as they passed through the chromosphere, 1000 times too little energy to heat the corona.

It was also discovered from the Einstein Observatory x-ray spacecraft that even type O and B stars give off enough x-rays that they must have hot coronas, even though their internal structure is such that they cannot generate shock waves. The x-ray observations show that different mechanisms are necessary to heat the coronas of other stars. Some of these different mechanisms probably heat the solar corona, too. Though the old ideas about coronal heating have been discarded, there is no agreement about what the real mechanism is for heating the corona. Probably the energy is generated in connection with the

Figure 23–21 New techniques of x-ray optics allow imaging the sun with reflections at normal rather than grazing incidence. This rocket photograph from 1989 shows the Sun at 63.5 Å radiation from 15-times-ionized iron. The resolution of ¾ arc sec is 10 times better than that of Skylab's x-ray telescope.

solar magnetic or electric fields. This would explain the x-ray pictures, which show that the corona is hotter above active regions.

23.6 Solar Eclipses

Total eclipses of the sun, which occur somewhere in the world every 18 months or so, are a particularly productive way of studying the outer layers of the sun. In Section 6.3, we discussed them, and surveyed several past and future eclipses. In particular, many people are looking forward to the July 11, 1991, total solar eclipse, and will observe it from Hawaii or Baja California.

My own recent eclipse experiments have been devoted to testing predictions of theories that involve magnetic fields to heat the corona. The results from the 1980 Indian and 1983 Indonesian total solar eclipses verify the predictions of the theory. The eclipse work was discussed in Section 6.3d.

23.7 Sunspots and Other Solar Activity

In this section we discuss the *active* sun.

A host of other time-varying phenomena are superimposed on the basic structure of the sun. Many of them, notably the sunspots, vary with an 11-year cycle, which is called the *solar activity cycle.*

23.7a Sunspots

Sunspots (Fig. 23–22) are the most obvious sign of solar activity. They are areas of the sun that appear relatively dark when seen in white light. Sunspots appear dark because they are giving off less radiation than the photosphere that surrounds them. This implies that they are cooler areas of the solar surface, since cooler gas radiates less than hotter gas. Actually, if we could somehow remove a sunspot from the solar surface and put it off in space, it would appear bright against the dark sky; a large one would give off as much light as the full moon.

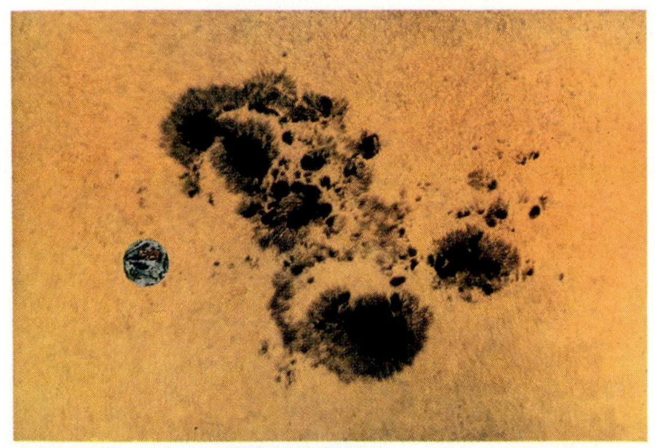

A

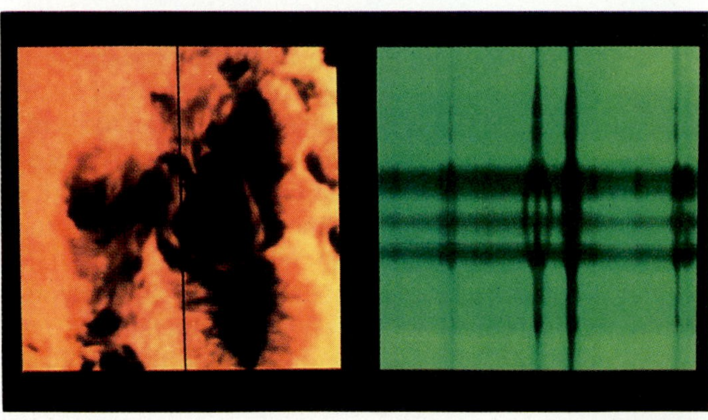

B

Figure 23–22 (*A*) A sunspot, showing the dark *umbra* surrounded by the lighter *penumbra*. Granulation is visible in the surrounding photosphere. A photo of the earth is superimposed to show its relative size. (*B*) The dark vertical line across the sunspot is the slit of the spectrograph onto which the image was projected; the light that we are not seeing in this image is that light that is being analyzed by the spectrograph. At right is the spectrum of the sunspot region. Note that some of the spectral lines are split into several parts at the sunspot umbra. The split results because of the magnetic field, by the *Zeeman effect.* The stronger the magnetic field, the more the lines are split. By measuring the splitting, scientists build up maps of the solar magnetic field, which is highest in sunspot umbrae.

Figure 23–23 Lines of force from a bar magnet are outlined by iron filings. One end of the magnet is called a north pole and the other is called a south pole. Similar poles ("like poles")—a pair of norths or a pair of souths—repel each other, and unlike poles (1 north and 1 south) attract each other. Lines of force go between opposite poles.

A sunspot includes a very dark central region called the *umbra,* from the Latin for "shadow" (plural: *umbrae*). The umbra is surrounded by a *penumbra* (plural: *penumbrae*), which is not as dark (just as during an eclipse the umbra of the shadow is the darkest part and the penumbra is less dark).

To understand sunspots, we must understand magnetic fields. When iron filings are put near a simple bar magnet on earth, the filings show a pattern (Fig. 23–23). The magnet is said to have a north pole and a south pole; the magnetic field linking them is characterized by what we call *magnetic lines of force,* or *magnetic field lines* (after all, the iron filings are spread out in what look like lines). The earth (as well as some other planets) has a magnetic field that has many characteristics in common with that of a bar magnet. The structure seen in the solar corona, including polar plumes and equatorial streamers, results from matter being constrained by the solar magnetic field.

We can measure magnetic fields on the sun by using a spectroscopic method. In the presence of a magnetic field, certain spectral lines are split into a number of components, and the amount of the splitting depends on the strength of the magnetic field.

Measurements of the solar magnetic field were first made by George Ellery Hale. He showed, in 1908, that the sunspots are regions of very high magnetic field strength on the sun, thousands of times more powerful than the earth's magnetic field or than the average solar magnetic field. Sunspots usually occur in pairs, and often these pairs are part of larger groups. In each pair, one sunspot will have a polarity typical of a north magnetic pole and the other will have a polarity typical of a south magnetic pole (Fig. 23–24).

Magnetic fields are able to restrain matter. This restraint of matter is the property we are trying to exploit on earth to contain superheated matter sufficiently long to allow nuclear fusion for energy production to take place. The

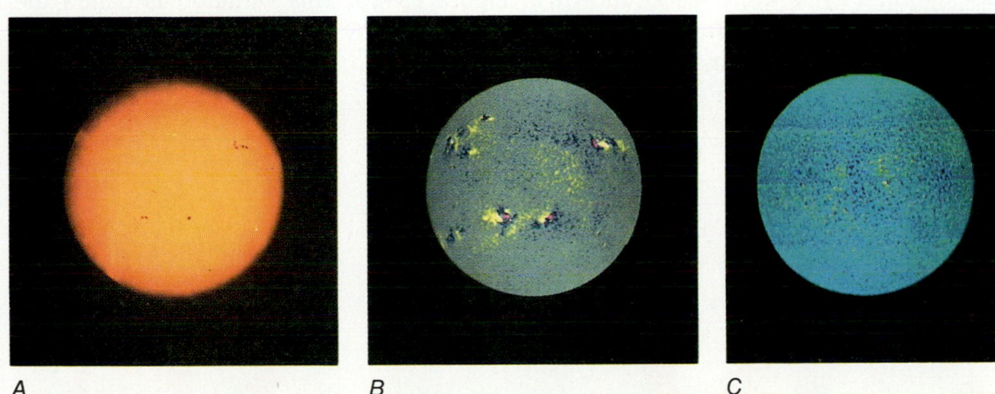

A *B* *C*

Figure 23–24 (*A*) The solar photosphere and (*B*) the photospheric magnetic field at the same time for a day near maximum of the solar-activity cycle. One polarity appears yellow and the other appears purple. Note how the regions of strong fields correspond to the positions of sunspots. Notice also how the two members of a pair of sunspots have polarity opposite to each other, and how which polarity comes first as the sun rotates is different above and below the equator. (*C*) By contrast with the previous photographs, this image was taken at a time nearer minimum of the solar-activity cycle. No large regions of concentrated magnetic field are visible, and, as a result, no sunspots were present on the sun.

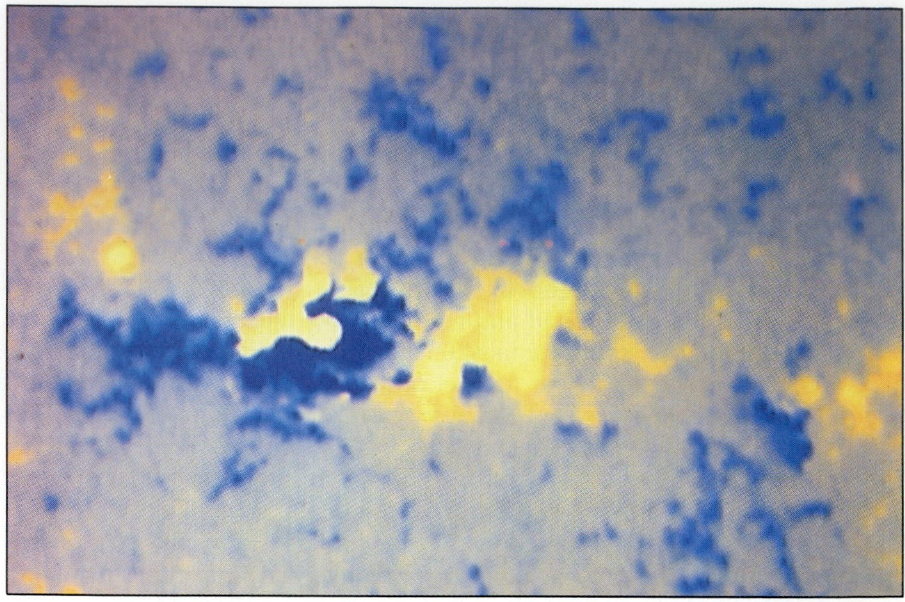

Figure 23–25 North and south magnetic polarities are given yellow and blue coding in this close-up view of the magnetic field of an active region.

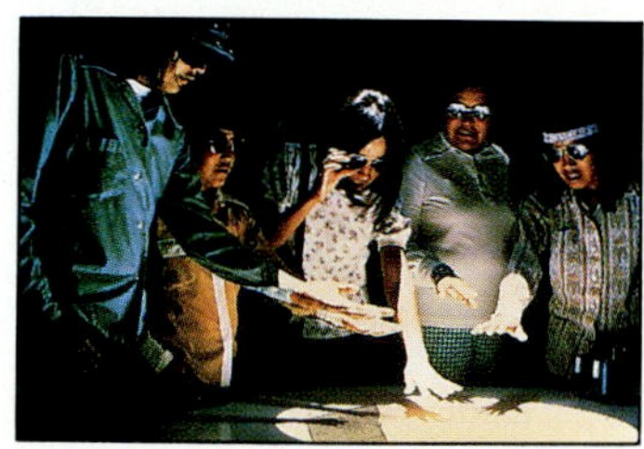

Figure 23–26 The National Solar Observatory's telescopes on Kitt Peak are on the Papago Reservation. Here Navajo students view the giant solar image projected onto an observing table from the largest of the solar telescopes.

strongest magnetic fields in the sun occur in sunspots (Fig. 23–25). The magnetic fields in sunspots restrain the motions of the matter there, and in particular keep convection from carrying energy to photospheric heights from lower, hotter levels. This results in sunspots being cooler and darker, though exactly why they remain so for weeks is not known. Many observatories around the world, including the National Solar Observatory (Fig. 23–26), monitor sunspots and their magnetic field.

The parts of the corona above active regions are hotter and denser than the normal corona. Presumably the energy is guided upward by magnetic fields. These locations are prominent in radio (Fig. 23–27) or x-ray maps of the sun.

Sunspots were discovered in 1610, independently by Galileo in Italy, Fabricius and Christopher Scheiner in Germany, and Thomas Harriot in England. In about 1850, it was realized that the number of sunspots varies with an 11-year-cycle, as is shown in Figure 23–28. This variation is called the *sunspot cycle,* although we now realize that many related signs of solar activity vary with the same period.

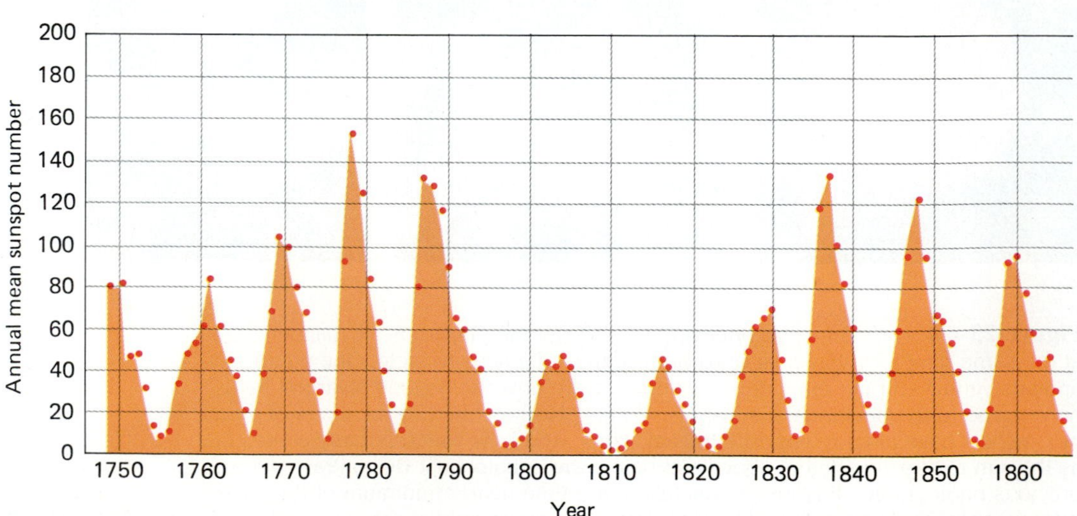

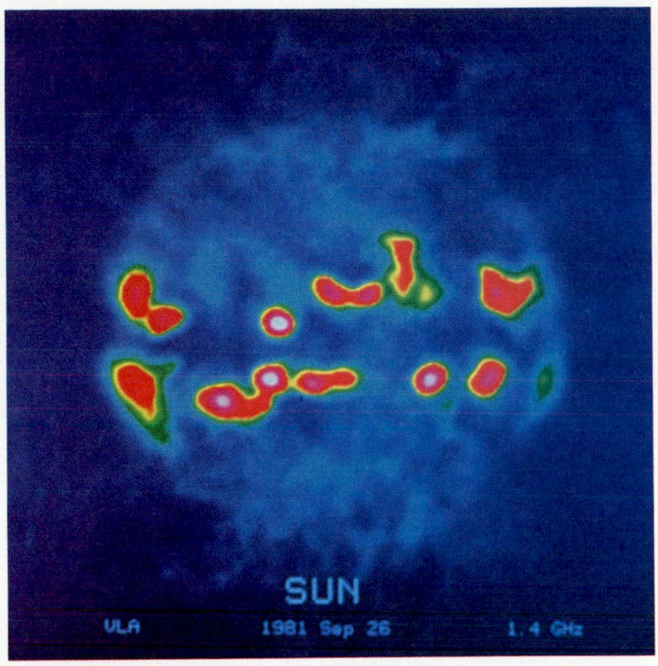

Figure 23–27 Solar active regions viewed by their radio emission from the VLA. The images were made at a wavelength of six cm. The intensity of radiation in the active regions corresponds to a temperature of 2,500,000 K. Regions away from the active regions are transparent to radiation at this wavelength, and we see to lower, cooler regions through these transparent regions of the solar atmosphere.

Besides the specific magnetic fields in sunspots, the sun seems to have a weak overall magnetic field with a north magnetic pole and a south magnetic pole, which may entirely result from the sum of the weak magnetic fields from vanished sunspots. Every 11-year-cycle, the north magnetic pole and south magnetic pole on the sun reverse polarity; what had been a north magnetic pole is then a south magnetic pole and vice versa. This occurs a year or two after the number of sunspots has reached its maximum. For a time during the changeover, the sun may even have two north magnetic poles or two south magnetic poles! But the sun is not a simple bar magnet, so this strange-sounding occurrence is not prohibited. Because of the changeover, it is 22 years before the sun returns to its original configuration, so the real period of the solar activity cycle is 22 years.

At the beginning of a solar cycle, sunspots of that cycle appear at high solar latitudes. (Sunspots of the older cycle may still be visible at lower latitudes.) As a solar cycle advances year by year, sunspots appear closer to the

Figure 23–28 The 11-year sunspot cycle is but one manifestation of the solar activity cycle.

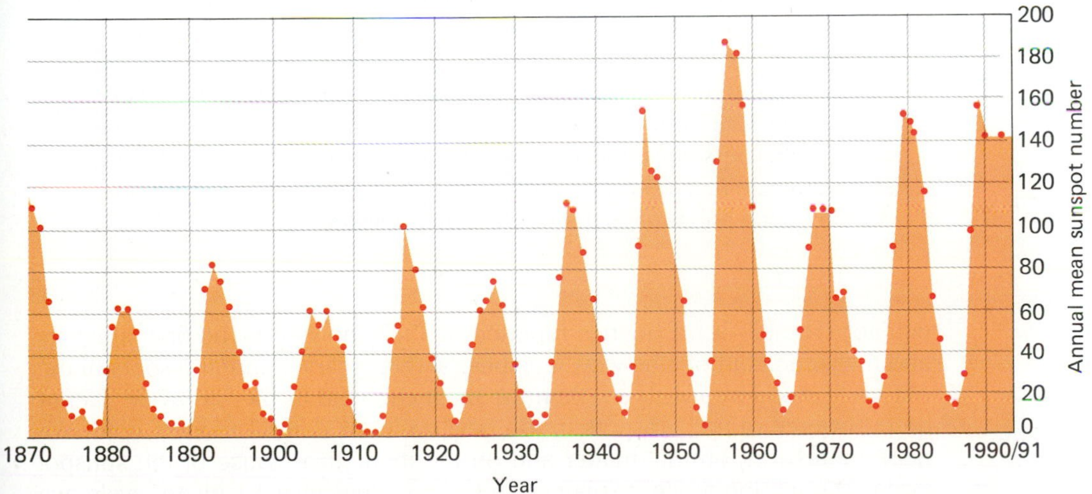

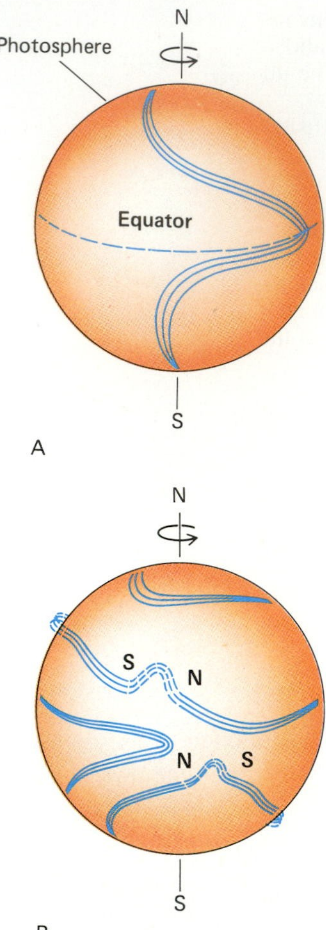

Figure 23–29 A leading model to explain sunspots suggests that the solar differential rotation winds tubes of magnetic flux around and around the sun. When the tubes kink and penetrate the solar surface, we see the sunspots that occur in the areas of strong magnetic field.

Focus On

Box 23.1 The Origin of Sunspots

Although the details of the formation of sunspots are not yet understood, a general picture was suggested in 1961. In this model, just under the solar photosphere the magnetic field lines are bunched in tubes of magnetic field that wind around the sun (Fig. 23–29).

The sun rotates approximately once each earth month. Different latitudes on the sun rotate at different angular speeds (Fig. 23–30)—*differential rotation.* Gas at the equator rotates in 25 days and gas at 40° latitude rotates in about 28 days. The differential rotation shows in the measured Doppler shifts.

A line of force that may have started out north–south on the solar surface is wrapped around the sun by the differential rotation. These lines collect as tubes located not far beneath the solar photosphere. Sometimes buoyant forces carry part of a tube upward until the tube sticks up through the solar photosphere. Where the tube emerges we see a sunspot of one magnetic polarity, and where the tube returns through the surface we see a sunspot of the other polarity.

Because of the differential rotation, the spiral winding of the magnetic lines of force is tighter at higher latitudes on the sun than at lower latitudes. Thus the instability that allows part of a tube to be carried to the surface arises first at higher solar latitudes. As the solar cycle wears on, the differential rotation continues and the tubes rise to the surface at lower and lower latitudes. This explains why spots form at higher latitudes earlier in the sunspot cycle than they do later in the cycle.

Sunspots arise from the interaction of the solar differential rotation with turbulence and motions in the convective zone. Dynamos in factories on earth also depend on the interaction of rotation and magnetic fields; thus these theories for the solar activity are called *dynamo theories,* and one says that the sunspots are generated by the *solar dynamo.*

Unfortunately, mathematical models to explain the motions of sunspots over the solar cycle don't come out right. In particular, they cannot explain why sunspots tend to form closer to the solar equator later in the sunspot cycle, and they do not predict the observed 11-year period. This and other evidence is now indicating to some scientists that the solar-dynamo theories are not sufficient to explain the sunspot cycle.

The dynamo theories limit the formation of sunspots to the solar convection zone. But falling masses of gas may overshoot and descend farther into the solar interior. New theories, invoking the idea that the sunspot cycle results from a global-scale oscillation of the sun controlled by the solar magnetism, are being developed.

equator (Fig. 23–31). Evidence reported in 1987 shows that the first sunspots and magnetic field of a new cycle appear several years before the spots of the old cycle stop forming. If so, then the magnetic field of the old cycle cannot be the source of the sunspots in the new cycle, as current models hold! We need more work on this matter and on the theoretical cause of the sunspot cycle. Though the model explained in the box does give a sunspot cycle, we basically really don't know why the cycle occurs. New models (Fig. 23–32) are being worked on.

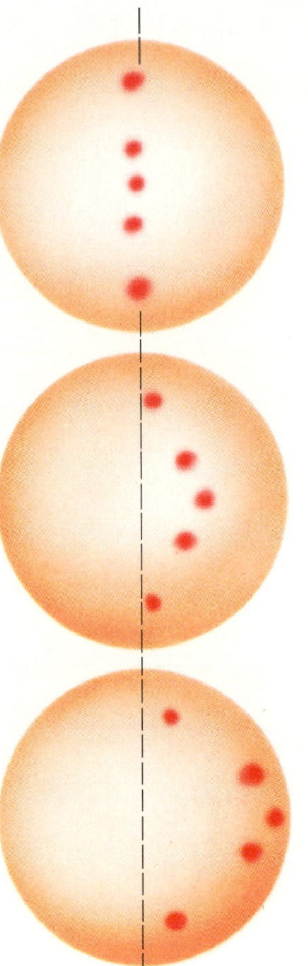

Figure 23–30 The notion that the sun rotates differentially is illustrated by this series, which shows the progress of a schematic line of sunspots month by month. The equator rotates faster than the poles by about 4 days per month.

At the 1979 maximum of the sunspot cycle—the time when there is the greatest number of sunspots—NASA launched the Solar Maximum Mission to study solar activity at that time. SMM soon failed, and a space shuttle mission was devoted to repairing it in space (Fig. 23–33). It then worked well for another half-dozen years, but finally was a casualty of the following solar maximum! At solar maximum, the earth's atmosphere expands a few kilometers, increasing the drag on satellites. SMM could not survive long enough to be reboosted. Worries about this and other spacecraft, such as the Hubble Space Telescope, led to a clear realization of how poor our ability is to predict the time and strength of forthcoming solar maxima.

The solar activity cycle is demonstrated in a variety of ways, including the shape of the corona as seen at eclipses (Fig. 23–34).

23.7b Flares

Violent activity sometimes occurs in the regions around sunspots. Tremendous eruptions called *solar flares* (Fig. 23–35) can eject particles and emit radiation from all parts of the spectrum into space. These solar storms begin in a few seconds and can last up to four hours. A typical flare lifetime is 20 minutes. Temperatures in the flare can reach 5 million kelvins, even hotter than the quiet corona. Flare particles that are ejected reach the earth in a few hours or days and can cause disruptions in radio transmission, cause the aurorae—the *aurora borealis* is the "northern lights" and the *aurora australis* is the "southern

Figure 23–31 The *butterfly diagram* shows the dependence on latitude at which sunspots appear with the phase of the sunspot cycle.

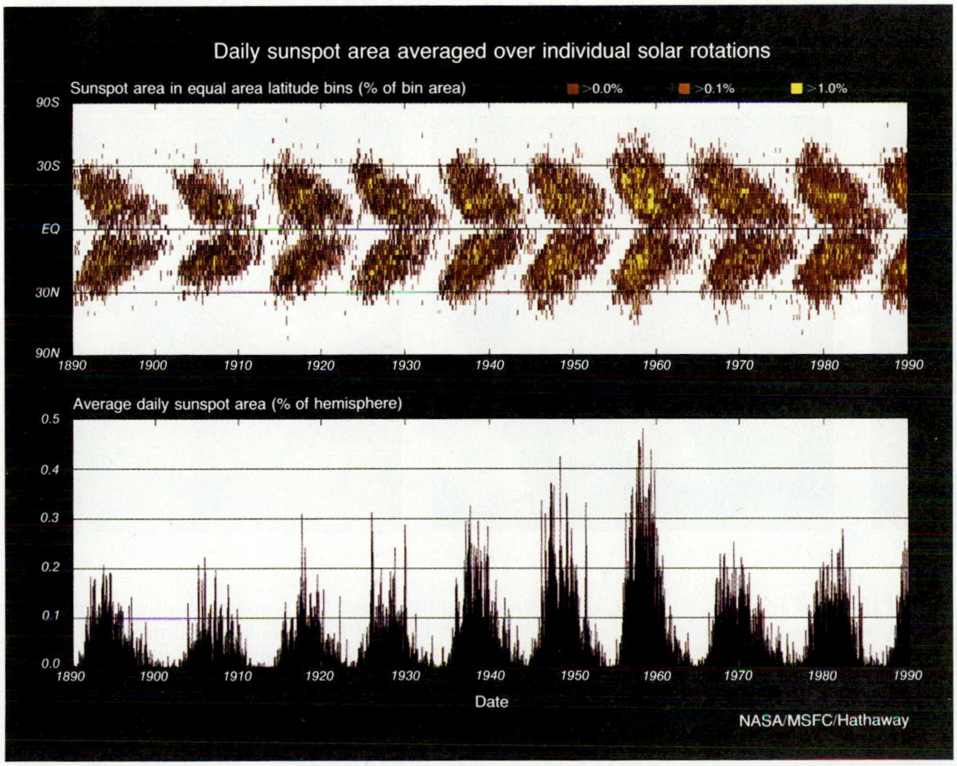

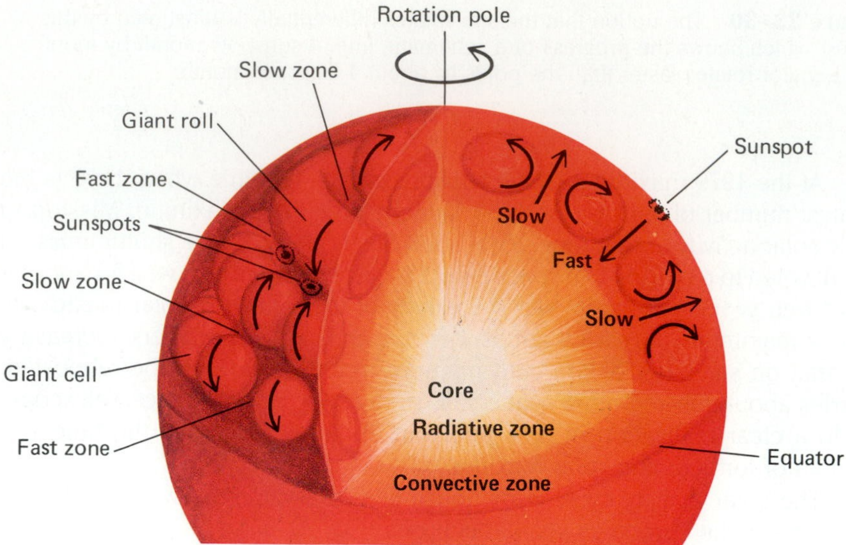

Figure 23–32 Recent observations that show the progression of active regions across the sun in latitude have led to a new model to explain sunspots. The model, shown here, indicates that certain zones around the sun move faster or slower than average at a given latitude. The observations also indicate that these features can be traced for up to 18 or so years, far longer than the 11-year single phase of the sunspot cycle. The model considers giant, doughnut-shaped rolls of gas under the surface of the sun. Adjacent rolls rotate in opposite directions. Sunspots can form where the adjacent rolls meet.

lights"—and even cause surges on power lines. Because of these solar-terrestrial relationships, high priority is placed on understanding solar activity and being able to predict it. The U.S. government even has a solar weather bureau to forecast solar storms, just as it has a terrestrial weather bureau.

Solar flares also emit x-rays, which have been studied from satellites, recently from the Japanese Hinotori spacecraft. Observing flares in the ultraviolet (Fig. 23–36) and x-ray region was one of the Solar Maximum Mission's

A

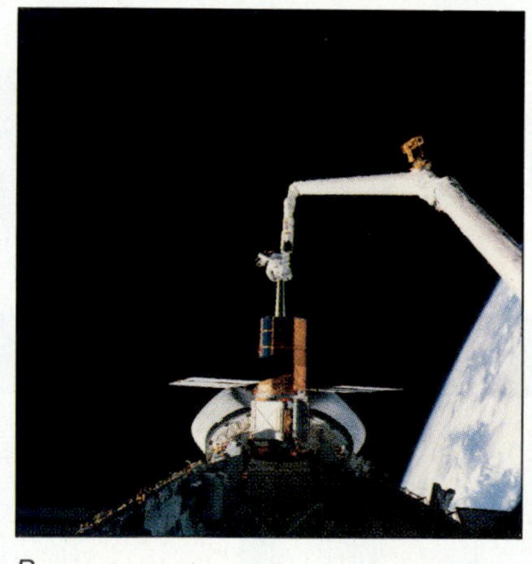

B

C

Figure 23–33 The Solar Maximum Mission repair. (*A*) A space-shuttle astronaut (*lower right*) unsuccessfully trying to attach himself to the Solar Maximum Mission spacecraft. (*B*) SMM was acquired with the Canadian arm on the space shuttle. (*C*) SMM docked in the space shuttle. It was successfully repaired there and relaunched.

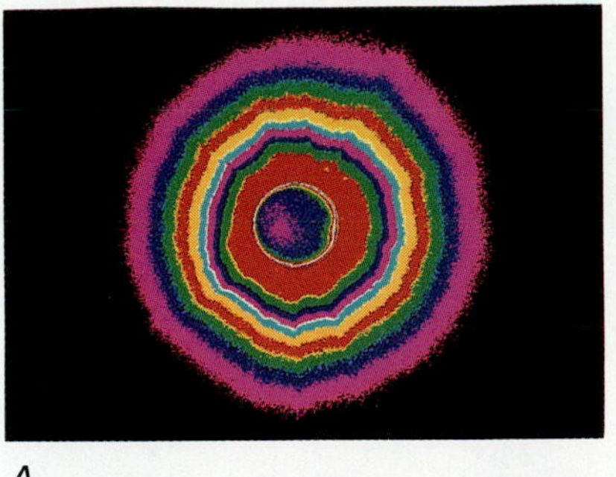
A

B

C

Figure 23–34 The shape of the corona varies with the solar-activity cycle. At solar maximum, there are so many coronal streamers that the coronal shape is very round. Nearer solar minimum, only a few streamers are present, making the coronal shape elongated. (*A*) The eclipse of February 16, 1980, in India. (*B*) The eclipse of June 11, 1983, in Indonesia. (*C*) The eclipse of July 11, 1991, in Baja California, Mexico.

major goals. The radio emission of the sun also increases at the time of a solar flare. X-ray telescopes for studying the sun will be placed, starting in the early 1990's, on the "sail" of U.S. weather satellites that contains the sun-facing solar power cells.

Though the very brightest flares can occur at any time, flares are generally correlated with the solar activity cycle (Fig. 23–37). Flares are much more common at solar maximum.

No specific model is accepted as explaining the eruption of solar flares. But it seems that a tremendous amount of energy is stored in the solar magnetic fields in sunspot regions (Fig. 23–38). Something unknown triggers the release of the energy. Much progress has been made in the last few years in understanding the site of the eruption within a flare; tracing the exact times at which ultraviolet, x-ray, and radio bursts of radiation are detected allows solar astronomers to follow the flare's evolution.

Figure 23–35 A solar flare in Hα.

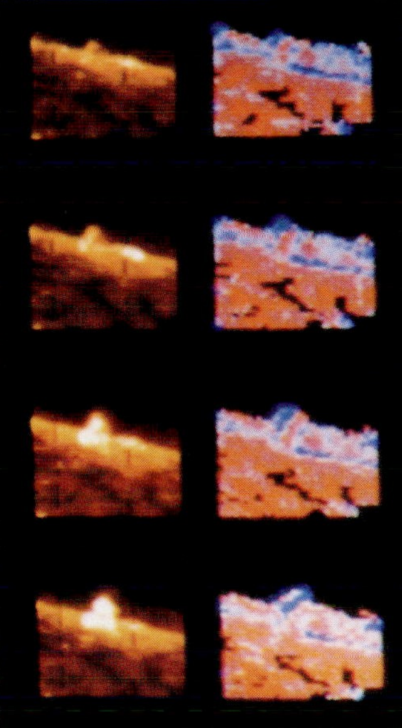

Figure 23–36 A solar flare observed on April 30, 1980, from the Solar Maximum Mission. The sequence on the left, at 2.5-minute intervals, shows the radiation of three-times ionized carbon at 1548 Å. The sequence on the right shows line-of-sight velocities found from the Doppler shift. Red represents material moving away, and blue represents approaching material.

We see a jet of hot gas arising from the base of a magnetic loop and filling the top of the loop. Some of the gas may even go over the top of the loop. The velocity picture supports this last idea, since it shows material coming toward us at the top of the loop. A bright flare was detected in x-rays 3 minutes after the last frame shown.

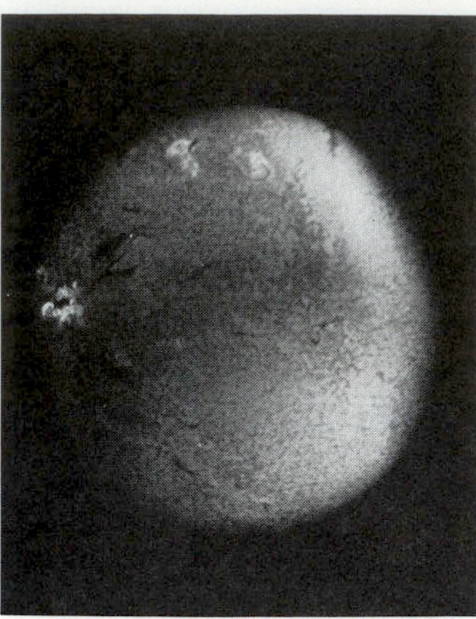

Figure 23–37 Very bright flares with strong x-ray emission marked the current solar maximum when they occurred in March 1989. Here we see an Hα view.

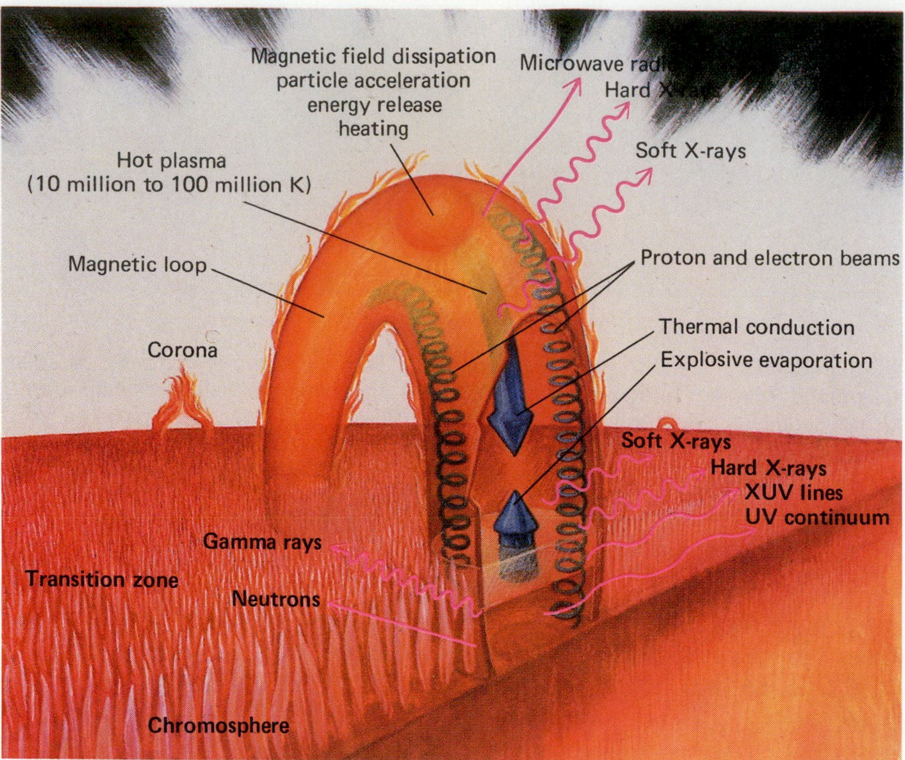

Figure 23–38 A model that explains solar flares, improved from earlier models by use of information from the Solar Max spacecraft. In the model, the temperature in a magnetic loop of gas on the sun is raised quickly to between 10 million and 100 million degrees, perhaps as the result of an interaction with a second loop. The release of energy accelerates electrons and protons, which then flow along the field lines in the loop. At the loop's footpoints, where the loop meets the chromosphere and photosphere, the beams of electrons and protons decelerate abruptly because they encounter denser material. The deceleration changes their kinetic energy into x-rays and gamma rays, which evaporate material that becomes coronal loops.

23.7c Plages, Filaments, and Prominences

Studies of the solar atmosphere in Hα radiation (Fig. 23–39) also reveal other types of solar activity. Bright areas called *plages* (pronounced "plah'jes," an Anglicized pronunciation of the French word for beaches), surround the entire sunspot region. Dark *filaments* are seen threading their way across the sun in the vicinity of sunspots. The longest filaments can extend for 100,000 km. They

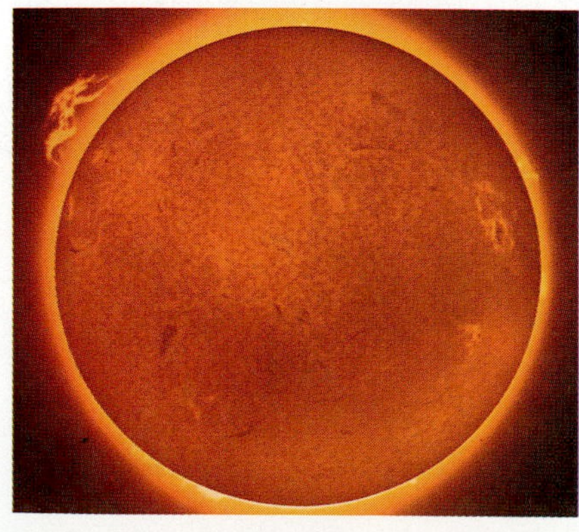

Figure 23–39 The sun, photographed in hydrogen radiation at the last solar maximum, shows bright areas called plages and dark filaments around the many sunspot regions.

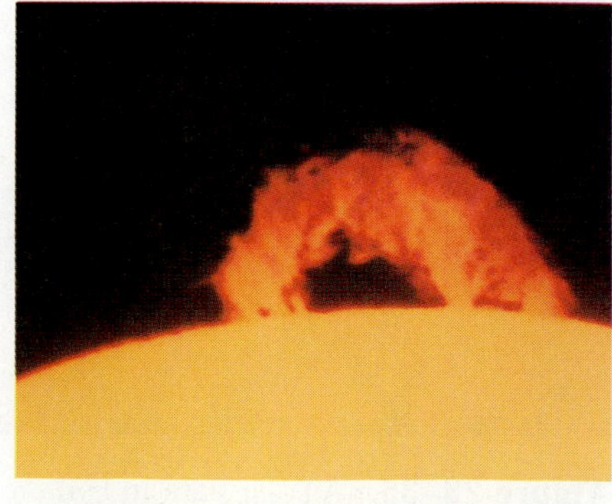

Figure 23–40 A solar prominence.

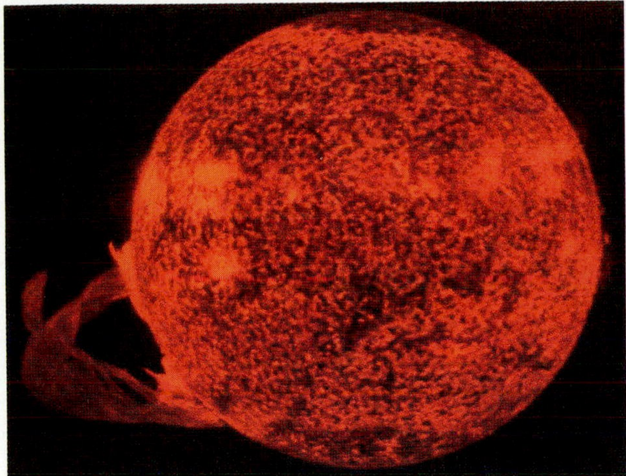

Figure 23–41 An erupting solar prominence observed in a 304 Å ultraviolet spectral line of helium from Skylab.

Figure 23–42 This wider view of a Skylab spectrogram shows how the results are an overlapping series of images of the sun. The helium 304 Å image, with the prominence, is the strongest in the spectral region. Adjacent to it is an image that shows the characteristic hot spots of the corona.

mark the locations of zero magnetic field that separate regions of positive and negative magnetic polarities. When filaments happen to be on the limb of the sun, they can be seen to project into space, often in beautiful shapes. Then they are called *prominences* (Fig. 23–40). Prominences can be seen with the eye at solar eclipses and glow pinkish at that time because of their emission in Hα and a few other spectral lines. They can be observed from the ground even without an eclipse, if an Hα filter is used. They have also been observed in the ultraviolet from Skylab (Figs. 23–41 and 23–42). Skylab data have been used to show the temperature structure (Fig. 23–43). Prominences appear to be composed of matter in a condition of temperature and density similar to matter in the quiet chromosphere, somewhat hotter and less dense than the photosphere.

Sometimes prominences can hover above the sun, supported by magnetic fields, for weeks or months. They are then called *quiescent prominences,* and

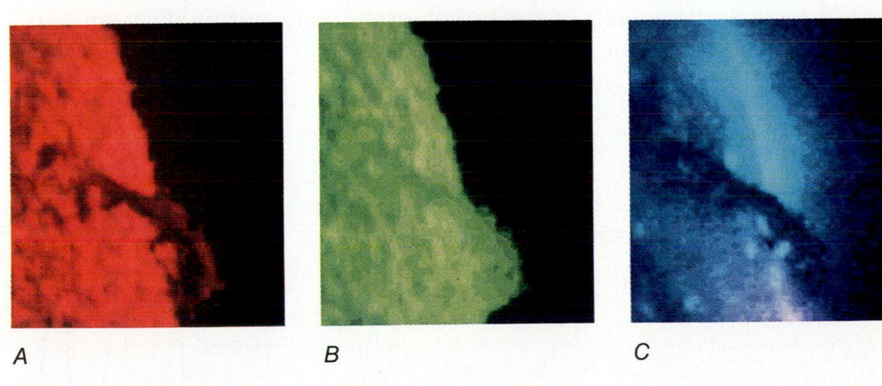

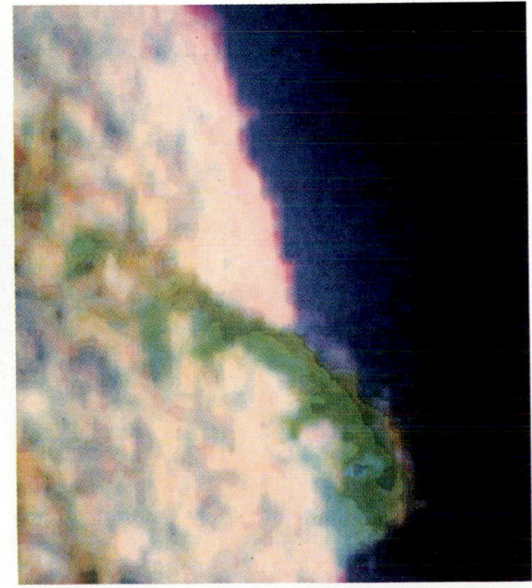

Figure 23–43 This compound shows views of the edge of the sun with a filament and prominence, through filters that pass ultraviolet radiation characteristic of different temperatures. The red image, in Lyman α, shows relatively cool gas. The green image, in radiation of 3-times ionized oxygen at a temperature of 130,000 K, shows the hotter gas of the chromosphere-corona transition region. The blue image shows hot gas; note the void where the prominence is. The relative temperatures show clearly on the composite.

can extend tens of thousands of kilometers above the limb. Other prominences can seem to undergo rapid changes.

In sum, the regions around sunspots show many kinds of solar activity, and so are called *active regions.*

23.8 Solar-Terrestrial Relations

Careful studies of the solar activity cycle are now increasing our understanding of how the sun affects the earth. Although for many years scientists were skeptical of the idea that solar activity could have a direct effect on the earth's weather, scientists currently seem to be accepting more and more the possibility of such a relationship. It was wondered in the early years of this century how the sun might be related to the magnetic disturbances measured on earth. Only with space observations did we discover that the link is through solar coronal holes (Fig. 23–44).

An extreme test of the interaction may be provided by the interesting probability that there were no sunspots at all on the sun from 1645 to 1715! The sunspot cycle may not have been operating for that 70-year period (Fig. 23–45). This period, called the *Maunder minimum,* was known to the British astronomer Walter Maunder and others in the early years of this century but was largely forgotten until its importance was recently noted and stressed by John A. Eddy of the High Altitude Observatory. Although no counts of sunspots exist for most of that period, there is evidence that people were looking for sunspots; it seems reasonable that there were no counts of sunspots because they were not there and not just because nobody was observing. A variety of indirect evidence has also been brought to bear on the question. For example, the solar corona appeared very weak when observed at eclipses during that period.

It may be significant that the anomalous, sunspotless period coincided with a "Little Ice Age" in Europe and with a drought in the southwestern United States. An important conclusion from the existence of this sunspotless period is that the solar activity cycle may be much less regular than we had thought.

The evidence for the Maunder minimum is indirect, and has been challenged. For example, the Little Ice Age could have resulted from volcanic dust in the air or from changing patterns of land use rather than from a lack of sunspots. Also, auroral records may show activity, though this is controversial. Since there was auroral activity, there may have been sunspots after all. It should come as no particular surprise that several mechanisms affect the earth's climate on this time scale, rather than only one.

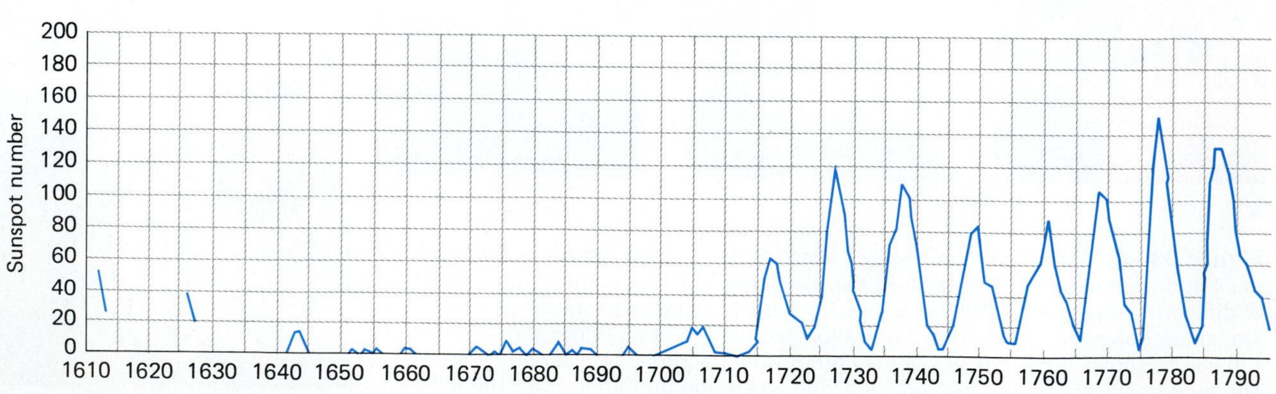

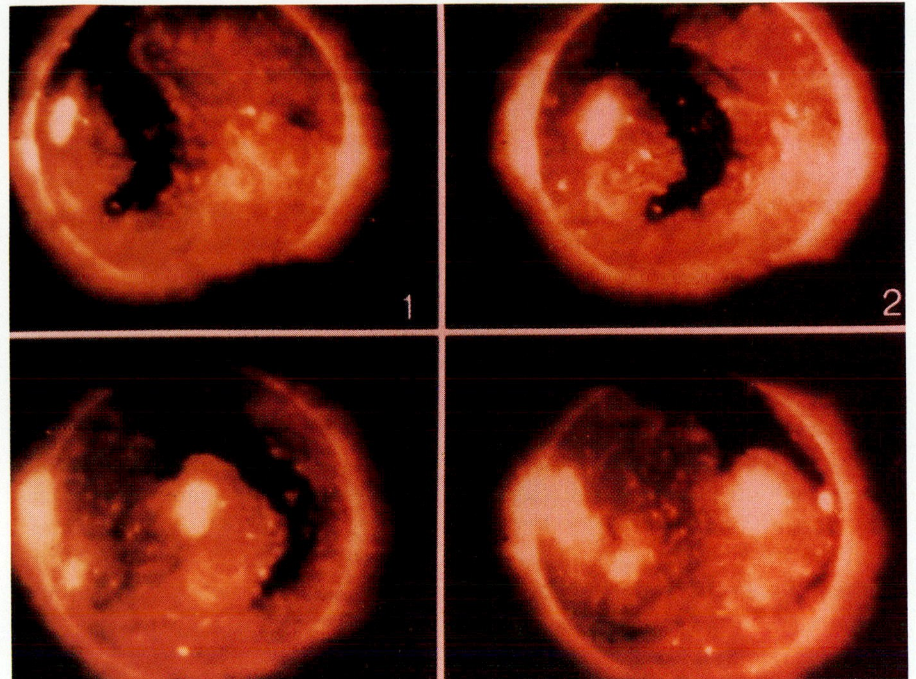

Figure 23–44 A coronal hole rotates across the sun. Gas escaping from coronal holes interacts with the earth. The solar wind flows from coronal holes.

*23.9 The Solar Wind

At about the time of the launch of the first earth satellites, in 1957, it was realized that the corona must be expanding into space. This phenomenon is called the *solar wind.* The expansion causes comet tails always to point away from the sun, an observation that had previously led to suspicions that the solar wind existed. The solar wind apparently emanates from coronal holes. The gas takes about 10 days to reach the earth.

The density of particles in the solar wind, always low, decreases with distance from the sun. There are only about 5 particles in each cubic centimeter at the distance of the earth's orbit. The particles are a mixture of ions and electrons. The solar wind extends into space far beyond the orbit of the earth, possibly even beyond the orbit of Pluto. The Pioneer and Voyager spacecraft are now exploring the region beyond Neptune, and have not yet come to the end of the solar wind.

Figure 23–45 The Maunder minimum (1645–1715), when sunspot activity was negligible for decades, may indicate that the sun does not have as regular a cycle of activity as we had thought. Activity has been extrapolated into the shaded regions with measurements made by indirect means (such as the frequency of auroras).

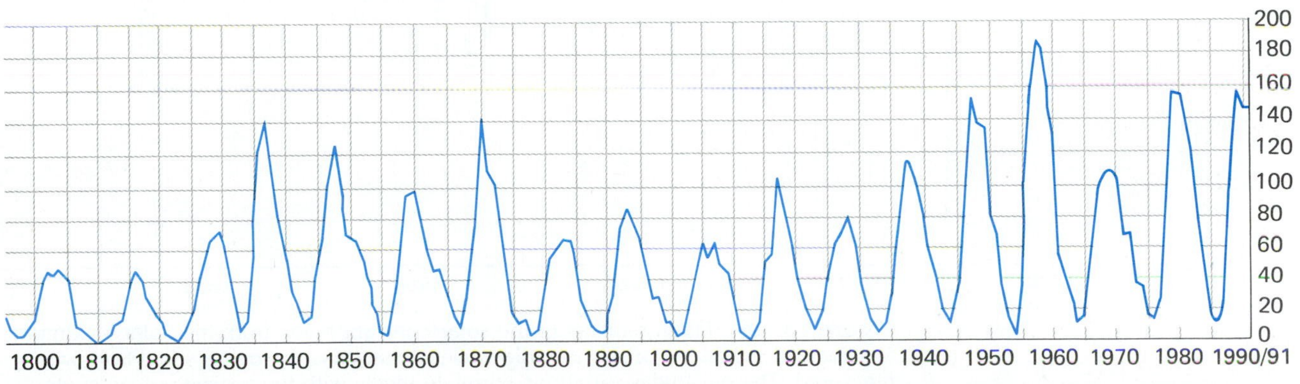

The earth's outer atmosphere is bathed in the solar wind. Thus research on the nature of the solar wind and on the structure of the corona is necessary to understand our environment in the solar system. Ulysses, a European spacecraft, is to fly over one of the solar poles, to give information about the solar wind above the plane in which the planets all lie. Its launch by a space shuttle has been delayed following the Challenger explosion. Earlier, a companion American spacecraft had been cancelled for budgetary reasons. Ulysses is scheduled for launch by a space shuttle in late 1990.

*23.10 The Solar Constant

The solar constant is the amount of energy per second that would hit each square centimeter of the earth at its average distance from the sun if the earth had no atmosphere. 99% of solar energy is in the range from 2760 Å to 49,000 Å (nearly 5 microns), and 99.9% is between 217 Å and 10.94 microns.

Every second a certain amount of solar energy passes through each square centimeter of space at the average distance of the earth from the sun. This quantity is called the *solar constant.* Accurate knowledge of the solar constant is necessary to understand the terrestrial atmosphere, for to interpret our atmosphere completely we must know all the ways in which it can gain and lose energy. Further, knowledge of the solar constant enables us to calculate the amount of energy that the sun itself is giving off, and thus gives us an accurate measurement on which to base our quantitative understanding of the radiation of all the stars.

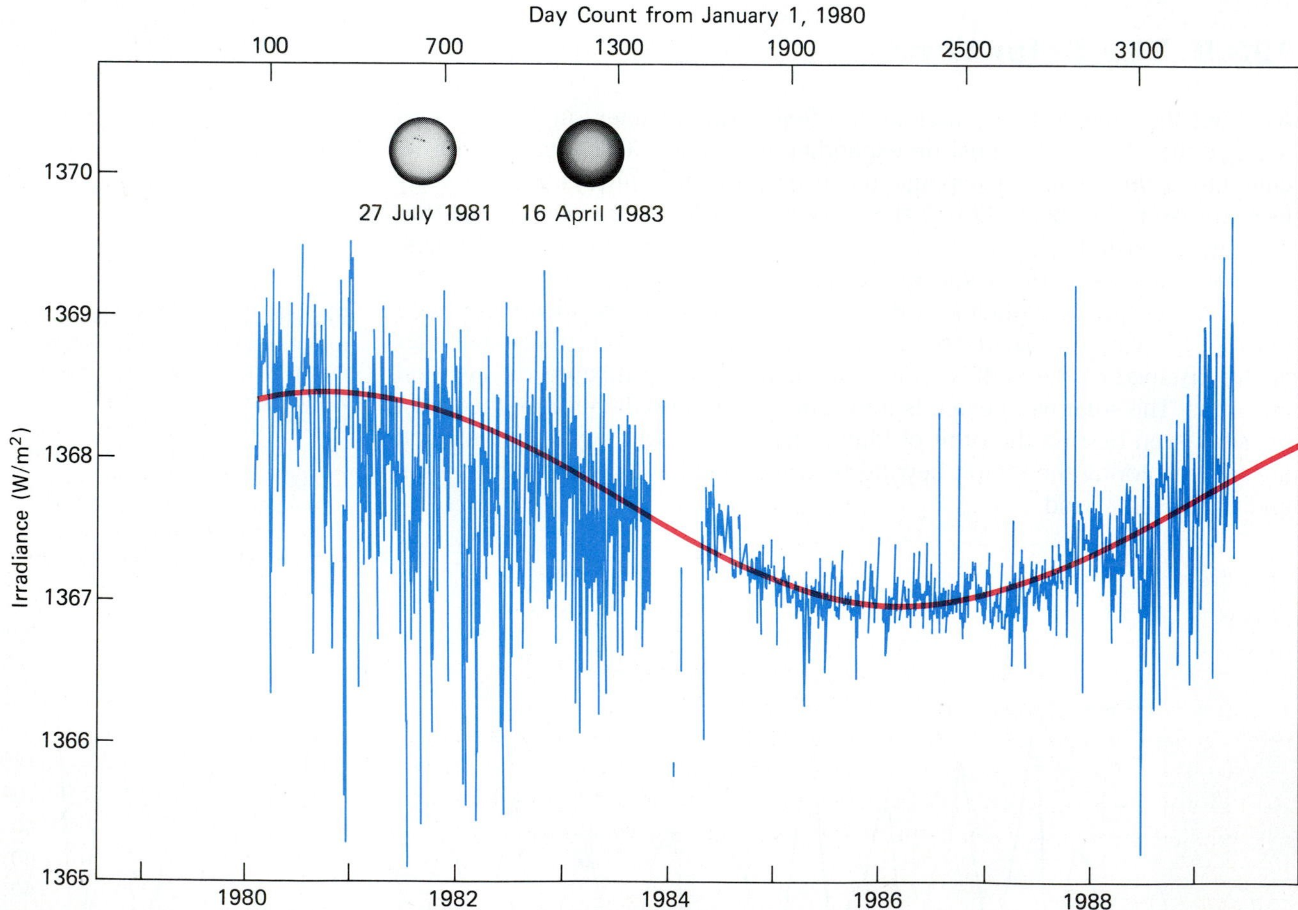

Figure 23–46 Measurements of the solar constant made from the Solar Maximum Mission. All radiation—x-rays to radio waves—that pass through a small hole was measured. The day-to-day variations correlate nicely with the passage of sunspots across the solar disk; the fluctuations are greater near solar maximum. A long-term variation with the solar-activity cycle also shows (colored curve).

The solar constant is about 1368 watts/m^2 (about 2 calories/cm^2/min; watts are the SI unit of power and calories are a non-SI unit of energy). Measurements from the ground gave an accuracy of 1.5 per cent. An instrument on the Solar Maximum Mission gave much higher accuracy—0.1 per cent—between 1980 and 1989. Short-term variations of about 0.2 per cent appeared. These dips in the sun's overall intensity turn out to be correlated with large sunspots crossing the sun (Fig. 23–46). Thus sunspots actually prevent energy from escaping from the solar surface, at least temporarily. A successor instrument is planned for a spacecraft to be launched in 1992.

One interesting aim is to determine if there are long-term changes in the solar constant, perhaps arising from changes in the luminosity of the sun. The Solar Maximum Mission instrument showed a slight but steady decline between 1980 and 1985, but the solar constant then increased as the sun went toward the maximum of the sunspot cycle. Certainly, our climate would be profoundly affected by even small long-term changes. Of course, since the solar constant changes with time, it isn't really a constant at all. Its change over the 11-year sunspot cycle seems to be caused by changes in the appearance of small solar bright regions.

Makers of solar-energy systems for heating and generating electricity start their calculations with the solar constant. But much of the sun's energy is lost passing through the earth's atmosphere and in conversion to a usable form, not to mention the effect of cloudy days. For all these reasons, less solar energy is available to us than the solar constant itself.

23.11 The Sun and the Theory of Relativity

The intuitive notion we have of gravity corresponds to the theory of gravity advanced by Isaac Newton in 1687. We now know, however, that Newton's theory and our intuitive ideas are not sufficient to explain the universe in detail. Theories advanced by Albert Einstein in the first decades of this century now provide us with a more accurate understanding.

The sun, as the nearest star to the earth, has been very important for testing some of the predictions of Albert Einstein's theory of gravitation, which is known as the general theory of relativity. The theory, which Einstein advanced in final form in 1916, made three predictions that depended on the presence of a large mass like the sun for experimental verification. These predictions involved (1) the gravitational deflection of light, (2) the advance of the perihelion of Mercury, and (3) the gravitational redshift. Comparing the predictions with observations provided basic tests of the theory. We shall discuss the gravitational redshift in Section 25.5 and discuss the other two tests here. The theory has recently been further verified by discovery of changes in a binary pulsar (Section 27.6), of giant arcs that appear near two galaxies (Section 31.3), and of multiple images of distant quasars (Section 32.9).

Einstein's theory predicts that the light from a star would act as though it were bent toward the sun by a very small amount (Fig. 23–47). We on earth, looking back, would see the star from which the light was emitted as though it were shifted slightly away from the sun (Fig. 23–48). Only a star whose radiation grazed the edge of the sun would seem to undergo the full deflection; the effect diminishes as one considers stars farther away from the solar limb.

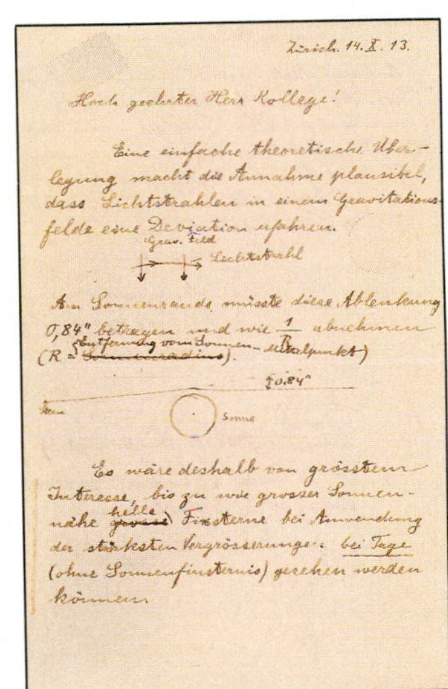
Zürich. 14. X. 13.

Hoch geehrter Herr Kollege!

Eine einfache theoretische Überlegung macht die Annahme plausibel, dass Lichtstrahlen in einem Gravitationsfelde eine Deviation erfahren.

Grav. Feld
Lichtstrahl

Am Sonnenrande müsste diese Ablenkung 0,84" betragen und wie $\frac{1}{R}$ abnehmen (R = ~~Sonnenradius~~ Entfernung vom Sonnen-Mittelpunkt).

0.84"
Sonne

Es wäre deshalb von grösstem Interesse, bis zu wie grosser Sonnennähe ~~grosse~~ helle Fixsterne bei Anwendung der stärksten Vergrösserungen bei Tage (ohne Sonnenfinsternis) gesehen werden können.

A

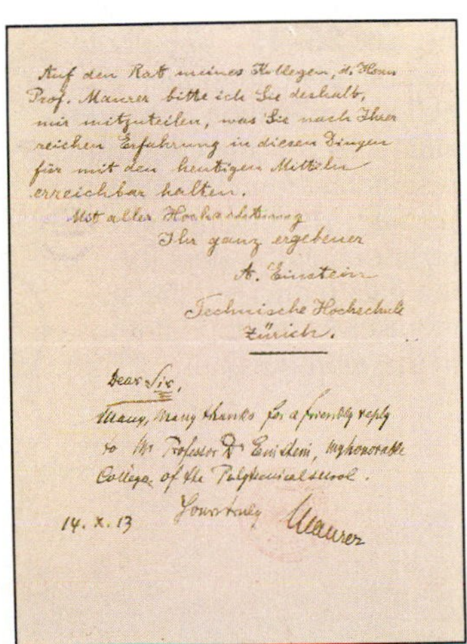
Auf den Rat meines Kollegen, d. Herrn Prof. Maurer bitte ich Sie deshalb, mir mitzuteilen, was Sie nach Ihrer reichen Erfahrung in diesen Dingen für mit den heutigen Mitteln erreichbar halten.

Mit aller Hochachtung
Ihr ganz ergebener
A. Einstein
Technische Hochschule
Zürich.

Dear Sir,
Many, many thanks for a friendly reply to Mr Professor Dr Einstein, my honorable College of the Polytechnical school.
14. X. 13 Yours truly Maurer

B

Figure 23–47 The prediction in Einstein's own handwriting of the deflection of starlight by the sun, taken from a letter from Einstein to Hale at Mount Wilson. "Lichtstrahl" is "light ray." Einstein asked if the effect could be measured without an eclipse, to which Hale replied negatively. The drawing came from a date when Einstein had developed only an early version of the theory, which gave a predicted value half of his later predicted value.

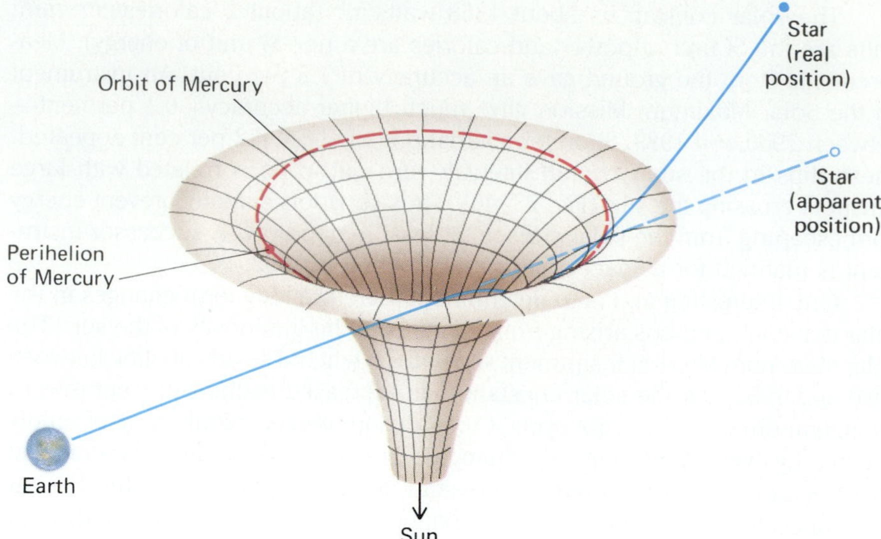

Figure 23–48 Under the general theory of relativity, the presence of a massive body essentially warps the space nearby. This can account for both the bending of light near the sun and the advance of the perihelion point of Mercury by 43 arc sec per century more than would otherwise be expected. The diagram shows how a two-dimensional surface warped into three dimensions can change the direction of a "straight" line that is constrained to its surface; the warping of space is analogous, although with a greater number of dimensions to consider. The effect is similar to a golfer putting on a warped green. Though the ball is hit in a straight line, we see it appear to curve.

To see the effect, one has to look near the sun at a time when the stars are visible, and this could be done only at a total solar eclipse (Fig. 23–49).

The British astronomer Arthur Eddington and other scientists observed the total eclipse of 1919 from sites in Africa and South America. The effect for which they were looking was tiny, and it was not enough merely to observe the stars at the moment of eclipse. One had to know what their positions were when the sun was not present in their midst, so the astronomers had already made photographs of the same field of stars six months earlier when the same stars were in the nighttime sky. The duration of totality was especially long, and the sun was in a rich field of stars, the Hyades, making it a particularly desirable eclipse for this experiment. Even though his observations had been

A

einstein akademie wissenchaften
berlin

Telegraphie des Deutschen Reichs.
Berlin, Haupt-Telegraphenamt

Antw S

Telegramm aus lickobservaory calif 129/12 50 1r 19/16.- westernunion

three pairs australia tahiti eclipse plates measured by cam
trumpler sixty two to eighty four stars each five of six
measurements completely calculated give einstein deflection
between one point fifty nine and one point eighty six seconds
arc mean value one point seventy four seconds = campbell .+

B

Figure 23–49 The photographic plate from the 1922 eclipse expedition to test Einstein's theory; the confirming results first found in 1919 were verified. The circles mark the positions of the stars used in the data reduction; the stars themselves are too faint to see in the reproduction.

Figure 23–50 Albert Einstein visiting California in the 1930's.

limited by clouds, Eddington detected that light was deflected by an amount that agreed with Einstein's revised predictions. Scientists hailed this confirmation of Einstein's theory; from the moment of its official announcement, Einstein was recognized by scientists and the general public alike as the world's greatest scientist (Figs. 23–50 through 23–53).

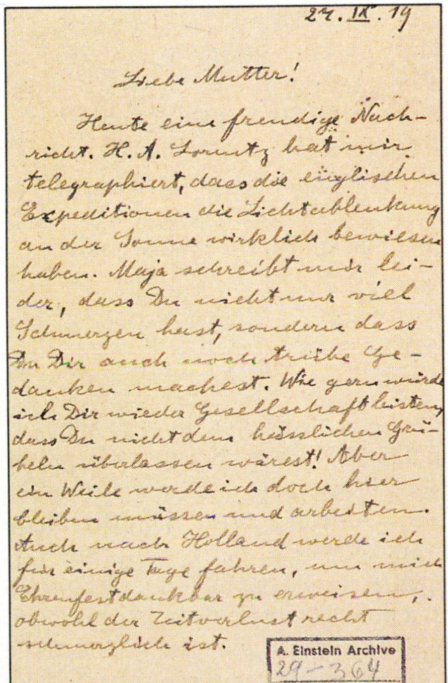

22. IX. 19

Liebe Mutter!

Heute eine freudige Nachricht. H. A. Lorentz hat mir telegraphiert, dass die englischen Expeditionen die Lichtablenkung an der Sonne wirklich bewiesen haben. Maja schreibt mir leider, dass Du nicht nur viel Schmerzen hast, sondern dass Du Dir auch noch trübe Gedanken machest. Wie gern würde ich Dir wieder Gesellschaft leisten, dass Du nicht dem hässlichen Grübeln überlassen wärest! Aber eine Weile werde ich doch hier bleiben müssen und arbeiten. Auch nach Holland werde ich für einige Tage fahren, um mich Ehrenfest dankbar zu erweisen, obwohl der Zeitverlust recht schmerzlich ist.

A. Einstein Archive
29-364

Postkarte

Frau
Pauline Einstein
Sanatorium Rosenau
Luzern (Schweiz)

A. Einstein Archive

Abs. A. Einstein
Haberlandstr. 5
Berlin

Ich wünsche Dir von Herzen gute Tage. Sei innig gegrüsst von Deinem Albert

Grüsse an die Schwester, Frau Dann mit ihrem Sohne, Maja, Pauli und Frl. Doktor. Nächstes Mal ausführlicher!

Figure 23–51 A postal card from Albert Einstein to his mother, reporting to her that the English expedition had found the gravitational deflection that he had predicted.

ECLIPSE SHOWED GRAVITY VARIATION

Diversion of Light Rays Accepted as Affecting Newton's Principles.

HAILED AS EPOCHMAKING

British Scientist Calls the Discovery One of the Greatest of Human Achievements.

Special Cable to THE NEW YORK TIMES.

LONDON, Nov. 8.—What Sir Joseph Thomson, President of the Royal Society, declared was "one of the greatest—perhaps the greatest—of achievements in the history of human thought" was discussed at a joint meeting of the Royal Society and the Royal Astronomical Society in London yesterday, when the results of the British observations of the total solar eclipse of May 29 were made known.

There was a large attendance of astronomers and physicists, and it was generally accepted that the observations were decisive in verifying the prediction of Dr. Einstein, Professor of Physics in the University of Prague, that rays of light from stars, passing close to the sun on their way to the earth, would suffer twice the deflection for which the principles enunciated by Sir Isaac Newton accounted. But there was a difference of opinion as to whether science had to face merely a new and unexplained fact or to reckon with a theory that would completely revolutionize the accepted fundamentals of physics.

The discussion was opened by the Astronomer Royal, Sir Frank Dyson, who described the work of the expeditions sent respectively to Sobral, in Northern Brazil, and the Island of Principe, off the west coast of Africa. At each of these places, if the weather were propitious on the day of the eclipse, it would be possible to take during totality a set of photographs of the obscured sun and a number of bright stars which happened to be in its immediate vicinity.

The desired object was to ascertain whether the light from these stars as it passed by the sun came as directly toward the earth as if the sun were not there, or if there was a deflection due to its presence. And if the deflection did occur the stars would appear on the photographic plates at measurable distances from their theoretical positions. Sir Frank explained in detail the apparatus that had been employed, the corrections that had to be made for various disturbing factors, and the methods by which comparison between the theoretical and observed positions had been made. He convinced the meeting that the results were definite and conclusive, that deflection did take place, and

Figure 23–52 The first report of the events that made Einstein world famous. From *The New York Times* of November 9, 1919.

Focus On

Box 23.2 The Special and General Theories of Relativity

In 1905, Einstein advanced a theory of relative motion that is called the *special theory of relativity.* A basic postulate is that, strangely, the speed of light is the same, no matter how the source of light and the observer are moving. For example, the speed of light is the same even if we are moving toward or away from the object that is emitting the light that we are observing. Ordinary objects, like raindrops, come at us faster if we are moving rapidly toward them.

Einstein's theory shows that the values that we measure for length, mass, and the rate at which time advances depend on how fast we are moving relative to the object we are observing. One consequence of the special theory is that mass and energy are equivalent and that they can be transformed into each other, following a relation $E = mc^2$. The consequences of the special theory of relativity have been tested experimentally in many ways, and the theory has long been well established.

"Special relativity" was limited in that it did not take the effect of gravity into account. Einstein proceeded to work on a more general theory that would explain gravity. Isaac Newton in 1687 had shown the equations that describe motion caused by the force of gravity, but nobody had ever said how gravity actually worked. Special relativity had linked three dimensions of space and one dimension of time to describe a four-dimensional space-time. Einstein's *general theory of relativity* holds that space-time is curved. It explains gravity as the effect we detect when we observe objects moving in curved space. Imagine the two-dimensional analogy to curved space of a golf ball rolling on a warped green (Fig. 23–54). The ball would tend to curve one way or the other. The effect would be the same if the green were flat but there were objects with the equivalent of gravity spaced around it. In four-dimensional space-time, masses warp the space near them and objects or light change their path as a result, an effect that we call gravity. Thus Einstein's theory of general relativity explains gravity as an observational artifact of the curvature of space.

In Einstein's own words, "If you will not take the answer too seriously, and consider it only as a kind of joke, then I can explain it as follows. It was formerly believed that if all material things disappeared out of the universe, time and space would be left. According to the relativity theory, however, time and space disappear together with the things."

The experiment has been repeated, but it is a very difficult one. Attempts at the eclipses in Mexico in 1970 and in India in 1980 failed because of a power failure and because of overexposure of the photographic plates, respectively. A University of Texas expedition to the 1973 African eclipse, hampered by a dust storm, confirmed Einstein's theory, but only to 10 per cent accuracy. A restudy of the original 1919 photographic plates carried out for the 1979 Einstein centennial confirmed the predictions with a precision of only about 6 per cent.

Though the data agree with Einstein's prediction, they are not accurate enough to distinguish between Einstein's theory and newer, more complicated,

Figure 23–53 These reports captured the public's fancy. From *The New York Times* of November 10, 1919.

LIGHTS ALL ASKEW IN THE HEAVENS

Men of Science More or Less Agog Over Results of Eclipse Observations.

EINSTEIN THEORY TRIUMPHS

Stars Not Where They Seemed or Were Calculated to be, but Nobody Need Worry.

A BOOK FOR 12 WISE MEN

No More in All the World Could Comprehend It, Said Einstein When His Daring Publishers Accepted It.

Special Cable to THE NEW YORK TIMES.

LONDON, Nov. 9.—Efforts made to put in words intelligible to the non-scientific public the Einstein theory of light proved by the eclipse expedition so far have not been very successful. The new theory was discussed at a recent meeting of the Royal Society and Royal Astronomical Society, Sir Joseph Thomson, President of the Royal Society, declares it is not possible to put Einstein's theory into really intelligible words, yet at the same time Thomson adds:

"The results of the eclipse expedition demonstrating that the rays of light from the stars are bent or deflected from their normal course by other aerial bodies acting upon them and consequently the inference that light has weight form a most important contribution to the laws of gravity given us since Newton laid down his principles."

Thompson states that the difference between theories of Newton and those of Einstein are infinitesimal in a popular sense, and as they are purely mathematical and can only be expressed in strictly scientific terms it is useless to endeavor to detail them for the man in the street.

rival theories of gravitation. Fortunately, the effect of gravitational deflection is constant throughout the electromagnetic spectrum, and the test can now be performed more accurately by observing how the sun bends radiation from radio sources, especially quasars. The results agree with Einstein's theory to within 1 per cent, enough to make the competing theories very unlikely. The double quasars (Section 32.9) provide strong verification of Einstein's theory.

A related test involves not deflection but a delay in time of signals passing near the sun. This can now be performed with signals from interplanetary spacecraft, most recently with a Viking orbiter near Mars, and these data also agree with Einstein's theory.

Another of the triumphs of Einstein's theory, even in its earliest versions, was that it explained the "advance of the perihelion of Mercury." The orbit of Mercury, like the orbits of all the planets, is elliptical; the point at which the orbit comes closest to the sun is called the *perihelion.* The elliptical orbit is pulled around the sun over the years, mostly by the gravitational attraction on Mercury by the other planets, so that the perihelion point is at a different orientation in space. Each century (!) the perihelion point appears to move around the sun by approximately 5600 seconds of arc (which is less than 2°). Subtracting the effects of precession (the changing orientation of the earth's axis in space, described in Section 3.3) still leaves 574 seconds of arc per century. Most of that shift is accounted for by the gravitational attraction of the planets, as was worked out a hundred years ago. Still, subtracting the gravitational effects of the other planets leaves 43 seconds of arc per century whose origin had not been understood on the basis of Newton's theory of gravitation.

Einstein's general theory predicts that the presence of the sun's mass warps the space near the sun. Since Mercury is sometimes closer to and sometimes farther away from the sun, it is sometimes travelling in space that is warped more than the space it is in at other times. This should have the effect of changing the point of perihelion by 43 seconds of arc per century. The agreement of this prediction with the measured value was an important observational confirmation of Einstein's theory. More recent refinements of solar system observations have shown that the perihelions of Venus and of the earth also advance by even smaller amounts predicted by the theory of relativity. And the verification has since been completely established by studying a pulsar in a binary system (Section 27.6).

Historically, in explaining the perihelion advance, general relativity was explaining an observation that had been known. On the other hand, in the bending of light, it actually predicted a previously unobserved phenomenon.

As a general rule, scientists try to find theories that not only explain the data that are at hand but also make predictions that can be tested. This testing of predictions is an important part of the *scientific method.* Because bending of electromagnetic radiation by a certain amount was a prediction of the general theory of relativity that had not been anticipated, the verification of the prediction is traditionally said to be a more convincing proof of the theory's validity than the theory's ability to explain the perihelion advance.

Stephen Brush, a historian of science at the University of Maryland, showed in 1989 that this traditional view was not the actual case. The light-bending result indeed brought Einstein's work to public attention and brought praise

Figure 23–54 A golf ball rolling on a curving green appears to us to curve, even though it is rolling as straight as possible. Here Arnold Palmer putts.

Figure 23–55 The Einstein statue near the National Academy of Sciences in Washington.

from scientists. However, it was apparently too new a result to be completely convincing; many scientists preferred to see if other theories would prove to explain the result. Indeed, the result stimulated a search for other explanations. It was only ten years or so later that it became clear that no other explanation could be found. The advance of the perihelion of Mercury's orbit, on the other hand, was a well established result that no other theory had succeeded in explaining. Also, the precession result tested a more fundamental part of Einstein's theory than did the light bending. Only with hindsight do scientists really consider the eclipse-bending prediction to be of as much importance as the explanation of the precession of Mercury.

Summary and Outline

- Parts of the sun (Section 23.1)
    - Interior (15 million K), photosphere including granulation (5800 K), chromosphere and spicules (15,000 K), supergranulation, corona (2 million K), coronal holes (perhaps 1,500,000 K)
    - The photosphere (Section 23.2)
        - High-resolution observations: seeing limited, convection causes granulation (Section 23.2a)
        - Solar seismology
            - Sun ringing like a bell; interpretation shows conditions in solar interior; GONG and SOHO planned (Section 23.2b)
- Spectra
    - Photospheric spectrum: continuum with Fraunhofer lines; abundances of the elements (Section 23.2c)
    - Chromospheric spectrum: emission at the limb during eclipses in the visible part of the spectrum (flash spectrum) (Section 23.3b)
    - Coronal spectrum: identification of coronium with highly ionized ions shows the high temperature (Section 23.4c)
- The chromosphere (Section 23.3)
    - Spicules; supergranulation; chromospheric network
- The corona (Section 23.4)
    - Streamers and plumes; eclipses; coronagraphs on the ground and in space
- Heating of the corona (Section 23.5)
    - New theories involving solar magnetism stem from space observations; x-ray observations show link with magnetic fields
- Solar eclipses (Section 23.6)
    - Good way to study chromosphere and corona
- Solar activity cycle (Section 23.7)
    - Sunspots, flares, plages, filaments, and prominences all linked to magnetic field structure
- Solar-terrestrial relations (Section 23.8)
    - Possible effect of solar activity on terrestrial weather
    - May not have been any solar activity during the Maunder minimum
- Solar wind flows from coronal holes (Section 23.9)
- Solar constant and its possible variability (Section 23.10)
- Using the sun to test Einstein's general theory of relativity (Section 23.11)
    - Predicted gravitational bending, then found at eclipses and now also tested in radio region of spectrum
    - Advance of the perihelion of Mercury: phenomenon was known in advance of Einstein's theory
    - Now also verified for the binary pulsar and the multiple quasar

Key Words

quiet sun, active sun, photosphere, subtends, interior, core, solar atmosphere, chromosphere, corona, solar wind, white light, granulation, convection, convection zone, solar seismology, continuum, spicules, supergranulation, chromospheric network, flash spectrum, streamers, plumes, coronagraphs*, coronal holes, solar activity cycle, sunspots, umbra, penumbra, Zeemann effect, magnetic lines of force, magnetic field lines, sunspot cycle, differential rotation, dynamo theories, solar dynamo, solar flares, aurora borealis, aurora australis, plages, filaments, prominences, quiescent prominences, active regions, Maunder minimum, solar wind*, solar constant*, special theory of relativity, general theory of relativity, perihelion, scientific method

*This term is found in an optional section.

Questions

1. Sketch the sun, labelling the interior, the photosphere, the chromosphere, the corona, sunspots, and prominences. Give the approximate temperature of each.
2. What elements make most of the lines in the Fraunhofer spectrum? What elements make the strongest lines? Why?
3. Explain why the photospheric spectrum is an absorption spectrum and the chromospheric spectrum seen at an eclipse is an emission spectrum.
4. Sketch the solar spectrum, including the strongest lines: Hα, Hβ, Hγ, H and K, and the D lines. Your sketch can be either a drawing showing what a photographic spectrum would look like, or a graph of intensity vs. wavelength.
5. Explain why helium lines can be seen in the chromospheric spectrum but not in the photospheric spectrum.
6. Graph the temperature of the interior and atmosphere of the sun as a function of distance from the center.
7. Define and contrast a prominence and a filament.
8. List three phenomena that vary with the solar activity cycle.
9. How do we know that the corona is hot?
10. Describe relative advantages of ground-based eclipse studies and of satellite studies of the corona.
11. Describe the sunspot cycle and a mechanism that explains it.
12. What are the differences between the solar wind and the normal "wind" on earth?
13. (a) What is the solar constant? (b) Why is it difficult to measure precisely?
14. Of what part of the spectrum is it most important to make accurate measurements in order to determine the solar constant?

†15. Referring to Appendix 3 for data about Mars and the earth, what would the solar constant be if we lived on Mars?

16. In what tests of general relativity does the sun play an important role? Describe the current status of these investigations.

†This question requires a numerical solution.

Topics for Discussion

1. Discuss the relative importance of solar observations (a) from the ground, (b) at eclipses, (c) from unmanned satellites, and (d) from manned satellites.
2. Discuss whether you feel the discussion of the sun in this text belongs in the Part of this book on stars, in the Part on the solar system, or in both.

IV Stellar Evolution

We have seen how the Hertzsprung-Russell diagram for a cluster of stars allows us to deduce the age of the cluster and the ages of the stars themselves (Section 22.5). Our human lifetimes are very short compared to the billions of years that a typical star takes to form, live its life, and die. Thus our hope of understanding the life history of an individual star depends on studying large numbers of stars, for presumably we will see them at different stages of their lives.

Though we can't follow an individual star from cradle to grave, studying many different stars shows us enough stages of development to allow us to write out a stellar biography. Chapters 24 through 28 are devoted to such life stories. Similarly, we could study the stages of human life not by watching some individual's aging, but rather by studying people of all ages who are present in a city on a given day.

In Chapter 24, we shall study the birth of stars, and observe some places in the sky where we expect stars to be born very soon, possibly even in our own lifetimes. We shall then consider the properties of stars during the long, stable phase in which they spend most of their histories.

In the chapters that follow, we go on to consider the ways in which stars end their lives. Stars like the sun sometimes eject shells of gas that glow beautifully; the part of such a star that is left behind then contracts until it is as small as the earth (see Chapter 25). Sometimes when a star is newly visible in a location where no star was previously known to exist (a "nova"), we are seeing the interaction of such a dead solar-type star with a companion.

Occasionally, newly visible stars rival whole galaxies in brightness. Then we may be seeing the spectacular death throes of a star (see Chapter 26). What happens to massive stars after they explode is currently a topic of tremendous interest to many astronomers. The light from the explosion of a massive star reached earth in 1987, in what may have been the most exciting thing to happen in astronomy in over 300 years. Some of the massive stars become pulsars (discussed in Chapter 27). Other massive stars, we think, may even wind up as black holes, objects that are invisible and therefore difficult—though not impossible—to detect (see Chapter 28).

When we write a stellar biography from the clues that we get from observation, we are acting like detectives in a novel. As new methods of observation become available, we are able to make better deductions, so the extension of our senses throughout the electromagnetic spectrum has led directly to a better understanding of stellar evolution. Our work in the x-ray part of the spectrum, for example, has told us more about tremendously hot gases like those that result from an exploding star. X-ray astronomy is also intimately connected with our current exciting search for a black hole.

In these chapters, we shall see that an important new tool has also been added: the computer. Calculations that would have taken years or centuries to carry out—and indeed that never would have been carried out because of the time involved or the probability of error—can now be routinely made in minutes. As computers grow faster and cheaper, as they do every year, our capabilities grow.

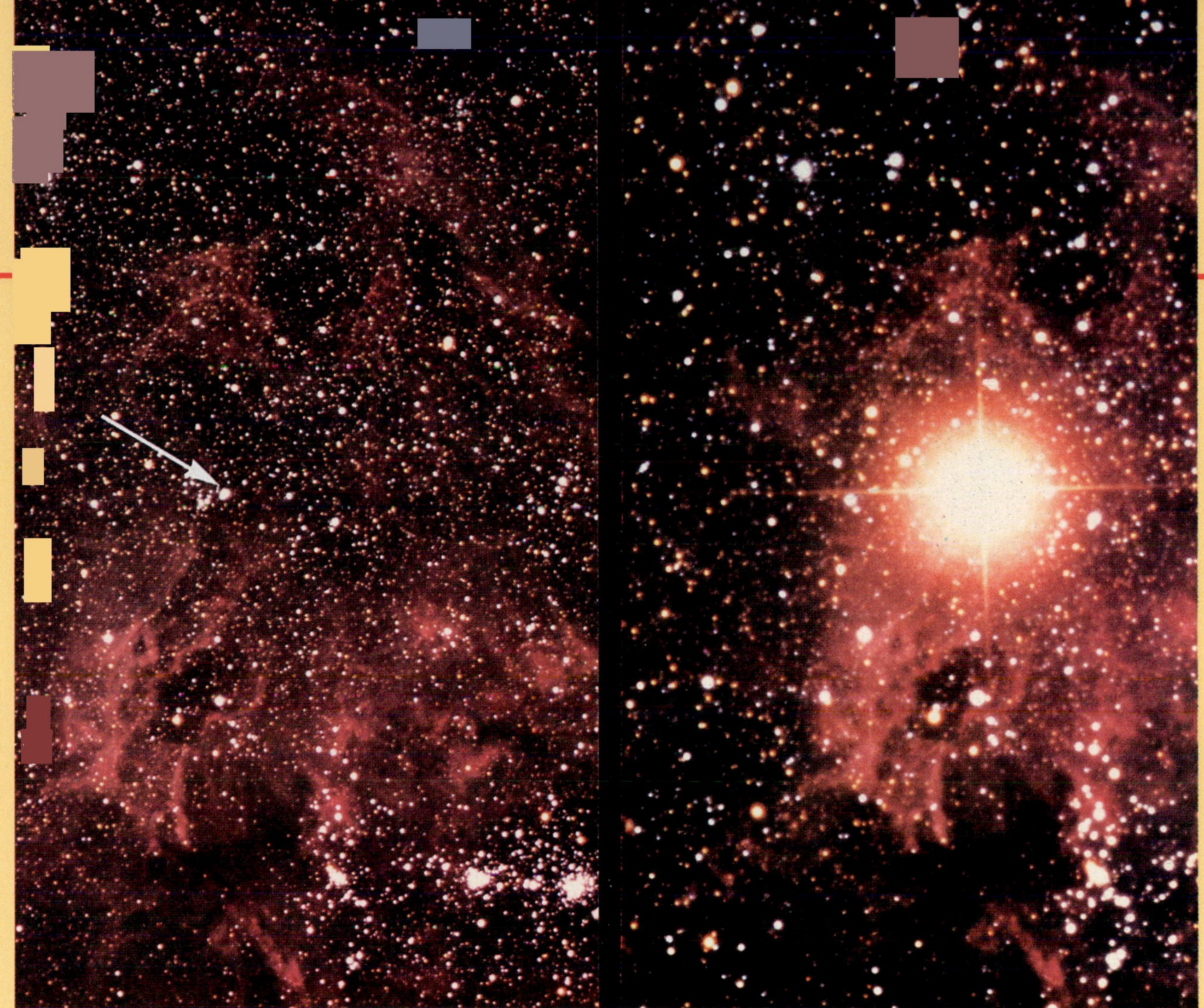

The supernova whose light reached us in 1987 is the closest and brightest supernova we have detected in nearly four hundred years. It is located in the Large Magellanic Cloud. The arrow shows the star before it exploded. We then see the bright, overexposed supernova, SN 1987A, with spikes added to the image by the telescope.

The importance of this Part of the book is based on a theorem that states that all stars of the same mass and of the same chemical composition evolve in the same way, if we neglect the effects of rotation and magnetic field. Since the chemical differences among stars are usually not overwhelming, it is largely the mass that is the chief determinant of stellar evolution. Because of this fact, we need only study a few groups of stars to gain pictures of the evolution of most stars.

We can set up divisions, depending on the mass of stars, using terminology (suggested by Martin Schwarzschild of Princeton) from the sport of boxing. We use the name *featherweight* for collapsing gas that is less than about 7 per cent as massive as the sun; it never reaches stardom, and becomes a brown dwarf. Stars more massive than that but containing less than about 4 solar masses eventually lose some of their mass and become white dwarfs; we call these *lightweight stars. Heavyweight stars,* which contain more than 8 solar masses, explode as one of the types of supernovae. Some wind up as neutron stars, which we may detect as pulsars or in x-ray binaries. Other heavyweight stars, after they become supernovae, wind up withdrawing from the universe in the form of black holes. The fate of *middleweight stars,* between 4 and 8 solar masses, is less clear. Note that in all cases, we are really discussing mass (which is an intrinsic property) rather than weight (which is the force of gravity on a mass).

Let us start in the next chapter with the formation and the main lifetime of stars. The following four chapters will then discuss the death of stars.

A star cluster, NGC 6520 and the dark cloud Barnard 86, seen against the Milky Way.

The Birth, Youth, and Middle Age of Stars

24

Aims: To study the formation of stars, their main-sequence lifetimes, and how they generate their energy

Even though individual stars shine for a relatively long time, they are not eternal. Stars are born out of gas and dust that may exist within a galaxy (Fig. 24–1); they then begin to shine brightly on their own. We now observe the dust and young stars especially well in the infrared (Fig. 24–2). Though we can observe only the outer layers of stars, we can deduce that the temperature at their centers must be millions of kelvins (that is, millions of degrees Centigrade above absolute zero). We can even deduce what it is deep down inside that makes the stars shine.

In this chapter we will discuss the birth of stars, then consider the processes that go on in a stellar interior during a star's life on the main sequence, and then begin the story of the evolution of stars when they finish this stage of their lives. The next three chapters will continue the story of what is called *stellar evolution.*

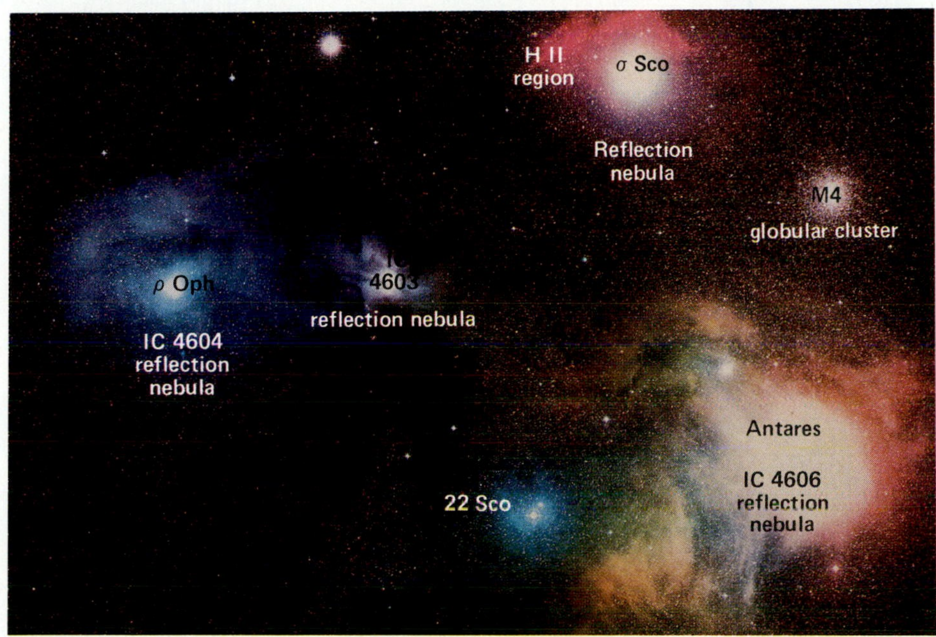

Figure 24–1 The clouds of gas and dust around ρ (rho) Ophiuchi (in the blue reflection nebula). The bright star Antares is surrounded by yellow and red nebulosity. The complex is 500 light years away. M4, the nearest globular cluster to us, is also seen.

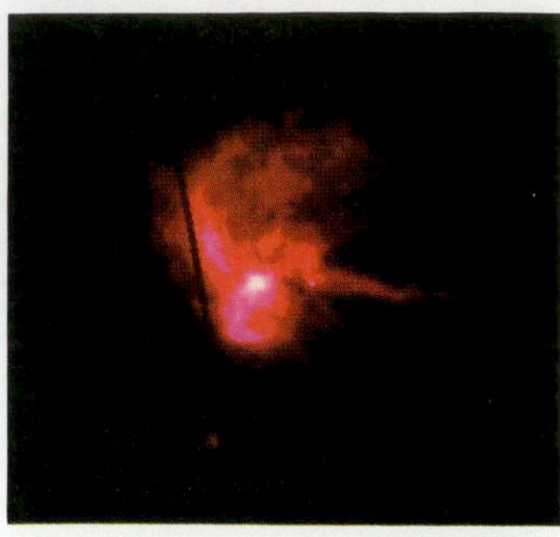

Figure 24–2 This IRAS image of ρ Ophiuchi is a false-color version of data taken at 100 μ, which comes from glowing dust that has been warmed by radiation from the recently formed young stars.

The gas and dust from which stars are forming are best observed in the radio and infrared regions of the spectrum, respectively. We shall have more to say about these regions when we discuss interstellar matter in Chapter 30. In particular, in recent years we have learned of the importance of "giant molecular clouds" in star formation.

24.1 Stars in Formation

The process of star formation starts with a region of gas and dust. The dust—tiny solid particles—may have been given off from the outer atmospheres of giant stars. A region may become slightly denser than its surroundings, perhaps from a random fluctuation in density or because a "density wave" in our galaxy (Section 29.5c) compressed it. Or a star may explode nearby—a "supernova"—sending out a shock wave that compresses gas and dust. In any case, once a minor density enhancement occurs, gravity keeps the gas and dust contracting. As they contract, energy is released; it turns out that half of that energy heats the matter (a result known as the "virial theorem"), causing it to give off an appreciable amount of electromagnetic radiation. Such a not-quite-yet-formed star is called a "protostar" (from the prefix of Greek origin meaning "primitive").

We can plot the position of a star on an H–R diagram for any particular instant of its life. Each of the pairs of luminosity and temperature corresponds to a point on the H–R diagram. Connecting the points representing the entire lifetime of the star gives an *evolutionary track*. A particular star, of course, is only at one point of its track at any given time. And note that an evolutionary track shows how the properties of a star evolve, not how or whether the star is physically moving in space.

Evolutionary tracks of two protostars are shown in Figure 24–3. At first, each protostar brightens, and the track bends upward and toward the left on the diagram. While this is occurring, the central part of the protostar continues

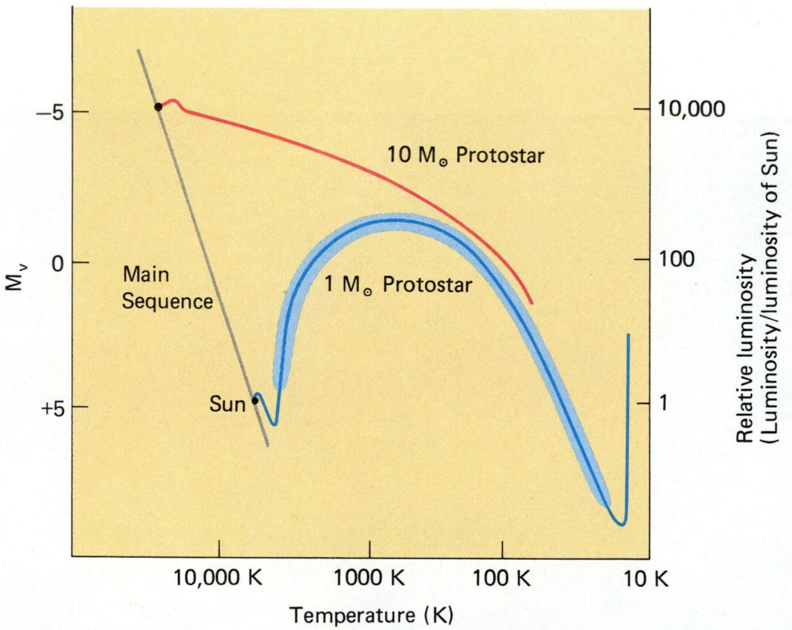

Figure 24–3 Evolutionary tracks for two protostars, one of 1 solar mass and the other of 10 solar masses. ($M_\odot$ is the symbol for the mass of the sun. By saying "a star of 10 solar masses," we mean simply a star that has 10 times the mass of the sun.) The evolution of the shaded portion of the track for the 1-solar-mass star is described in the text. These stages of its life last for about 50 million years. More massive stars whip through their protostar stages more rapidly; the corresponding stages for a 10-solar-mass star may last only 200,000 years. These very massive stars wind up being very luminous.

On the vertical axis, absolute visual magnitude is plotted to the left of the line. An alternative scale showing the star's luminosity relative to that of the sun is plotted to the right of the line.

to contract and the temperature rises. The higher temperature results in a higher pressure, which pushes outward more and more strongly. Eventually a point is reached where this outward force balances the inward force of gravity for the central region.

By this time, the dust has vaporized and the gas has become opaque, so energy emitted from the central region does not escape directly. The outer layers continue to contract. Since the surface area is decreasing, the luminosity decreases and the track of the protostar begins to move downward on the H–R diagram. As the protostar continues to heat up, it also continues to move toward the left on the H–R diagram (that is, it gets hotter).

Note that a given star does not move **along** the main sequence; it stays for a long time at essentially the same place on the H–R diagram. The main sequence shows up only when we plot the properties of a lot of stars.

The time scale of this gravitational contraction depends on the mass of gas in the protostar. Very massive stars contract to approximately the size of our solar system in only 10,000 years or so. These massive objects become O and B stars, and are located at the top of the main sequence. They are sometimes found in groups in space, which are called *O and B associations.*

Less massive stars contract much more leisurely. A star of the same mass as the sun may take tens of millions of years to contract, and a less massive star may take hundreds of millions of years to pass through this stage.

Theoretical analysis shows that the dust surrounding the stellar embryo we call a protostar should absorb much of the radiation that the protostar emits. The radiation from the protostars should heat the dust to temperatures that produce primarily infrared radiation. Infrared astronomers have found many objects that are especially bright in the infrared but that have no known optical counterparts. The Infrared Astronomical Satellite (IRAS), which sent back data in 1983, discovered so many of these that we now think that about one star forms each year in our Milky Way Galaxy. These objects seem to be located in regions where the presence of a lot of dust and gas and other young stars indicates that star formation might be going on.

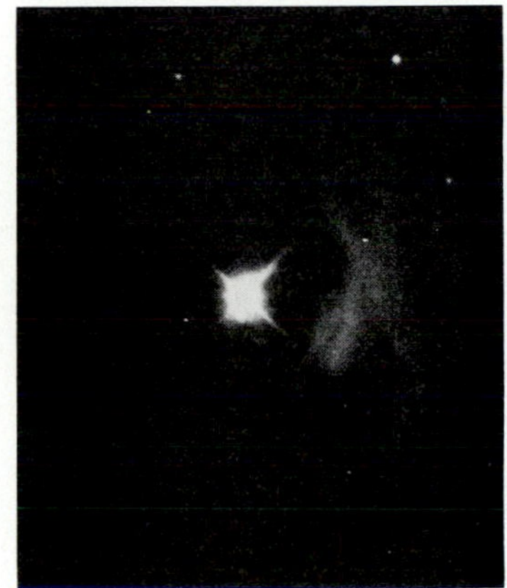

Figure 24–4 T Tauri itself is embedded in a Herbig-Haro object (*center*). The H–H object has changed in brightness considerably in the last century, first fading from view and then brightening considerably. These changes probably result from changes in the angle at which T Tauri illuminated the cloud. The H–H object has now been constant in brightness for decades. A second H–H object is the inner patch of the nebulosity at night; the rest is a reflection nebula. Herbig-Haro objects are named after George Herbig, formerly of the Lick Observatory and now of the University of Hawaii, and Guillermo Haro of the Mexican National Observatory.

The frame shown is about 1 arc min across. It is magnified by 100 times less than the high-resolution radio image (Fig. 24–6), all of which would fit into the white photographic image of T Tauri itself.

In the visible part of the spectrum, several classes of stars that vary erratically in their brightness are found. One of these classes, called *T Tauri stars,* includes stars of spectral types G, K, and M. These spectral types have relatively low masses. T Tauri stars are thought to be less than a few million years old and not to have reached the main sequence. Their visible radiation can vary by as much as several magnitudes. Astronomers work in the infrared to study the dust grains that surround T Tauri stars; observations have included some made with the Kuiper Airborne Observatory farther into the infrared than can be detected from the ground. Presumably, T Tauri stars have not quite settled down to a steady and reliable existence. They fall slightly above the main sequence of the H–R diagram, which is consistent with their being very young. Low-mass stars like these are now known to form more-or-less continuously in dark clouds of gas and dust. High-mass stars, on the other hand, form only in the densest parts of hotter clouds. Further, the high-mass stars form only when shock waves make the clouds collapse.

T Tauri stars are found in close proximity to each other; these groupings are known as *T associations.* The T Tauri stars are embedded in dark dust clouds. Also in these clouds are bright nebulous regions of gas and dust called *Herbig-Haro objects.* T Tauri itself, the prototype of its class of variable stars, is near a typical Herbig-Haro object (Fig. 24–4).

We see Herbig-Haro (H–H) objects (Fig. 24–5) as a result of gas ejected from very young T Tauri stars. The supersonic jet-like gas flows—millions of times more powerful than the solar wind—interact with surrounding material to produce flowing shock waves. Jets of gas are sometimes seen, and probably come from repetitive eruptions out the stars' poles. The heated regions, containing about one earth mass of gas and dust, appear as Herbig-Haro objects. Some H–H objects reflect light from nearby stars. The stars that provide the energy to H–H objects sometimes are detectable only in the infrared.

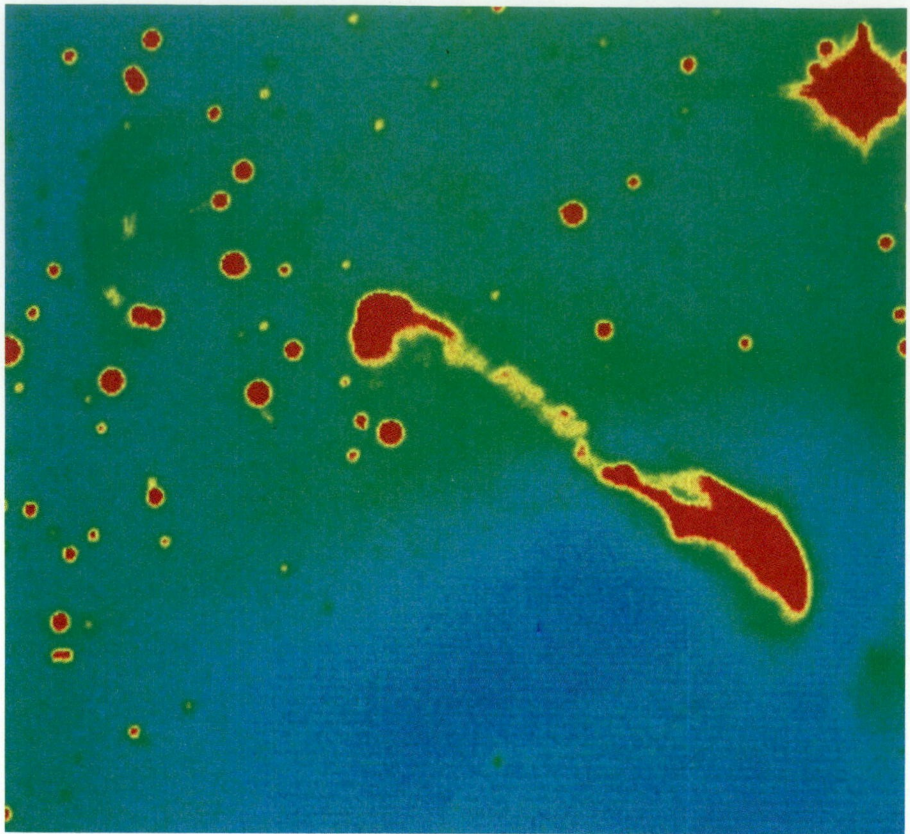

Figure 24–5 The Herbig-Haro objects H–H 46-47.

Radio maps of T Tauri (Fig. 24–6) made with the array of radio telescopes known as the Very Large Array (VLA) show that T Tauri's stellar wind—so strong that we might call it a gale—is not spherical. Observations of many T Tauri stars show that material is sent out in oppositely directed beams. This "bi-polar ejection" (Fig. 24–7) of gas seems common, and may imply that a disk of material orbits these stars, blocking outward flow in the equatorial direction and channelling flow at the poles. The source of energy driving the strong winds is not known, though a leading model uses magnetic-field—related waves to bring energy to the base of the stellar gales. Probably the result of the bipolar ejection, the H–H objects are often aligned as though they were ejected to opposite sides of the T Tauri star (Fig. 24–8).

It was a surprise when T Tauri turned out to have a companion observable only in the infrared. The companion, whose temperature is only 800 K, is the radio source. One reasonable interpretation is that the companion is embedded inside and taking up infalling matter from a disk of gas around T Tauri. The companion might even be too small to be a star, which would shine on its own by the fusion process we discuss in the next section. If so, it would be

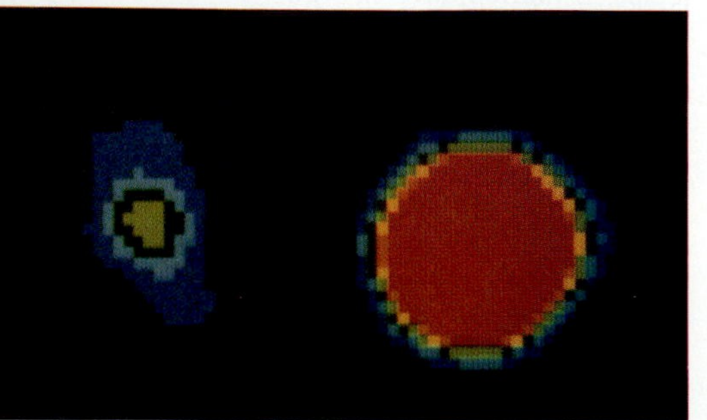

Figure 24–6 A radio map of T Tauri (right), its surrounding matter, and its companion. Note that T Tauri's outer contours are not circular. The companion is hidden by dust, and can be observed and distinguished only in the infrared and radio. With the high resolution of the VLA operating at the radio wavelength of 2 cm, the T Tauri radio source is resolved into these components separated by 0.54 arc sec. The weaker source is T Tauri itself and the brighter source is the infrared companion.

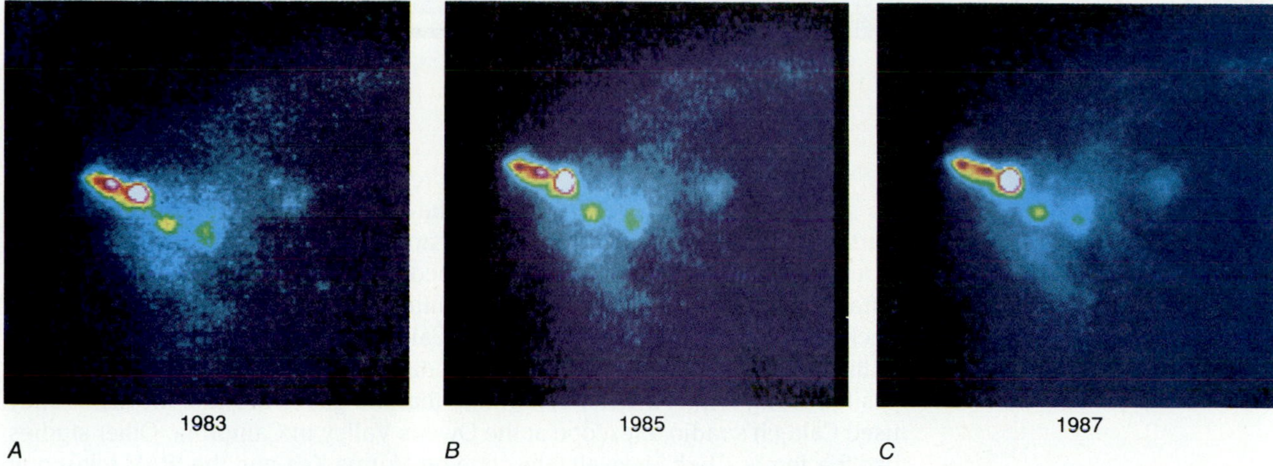

Figure 24–7 A bipolar nebula in the infrared source IRAS 5. It is in L1551, a dark cloud in Taurus 500 light years from us. Herbig-Haro objects show as clumps of gas that form and move downstream. The star itself is invisible at left near the point of the cone.

more like a giant planet in formation. It is, in any case, 30 times farther from T Tauri than Jupiter is from our sun. Another reasonable interpretation is that the companion is heated by outflowing matter. We still have a lot to learn.

Classical (that is, ordinary) T Tauri stars show strong optical emission lines (Fig. 24–9) and show excess radiation in the infrared and ultraviolet. The emission lines and excess radiation come from circumstellar material. Their radiation across the spectrum from x-ray through ultraviolet and visible varies strongly. Frederick Walter of the University of Colorado has, from a survey of x-ray sources, found "naked T Tauri stars," which are also low-mass stars not quite yet on the main sequence but which do not show signs of circumstellar material. They look like normal, cool stars and are distributed throughout regions of star formation. We may come to understand the pre-main-sequence evolution of low-mass stars better from studying the naked T Tauri stars, because we can study the underlying stars directly without the interference of surrounding matter.

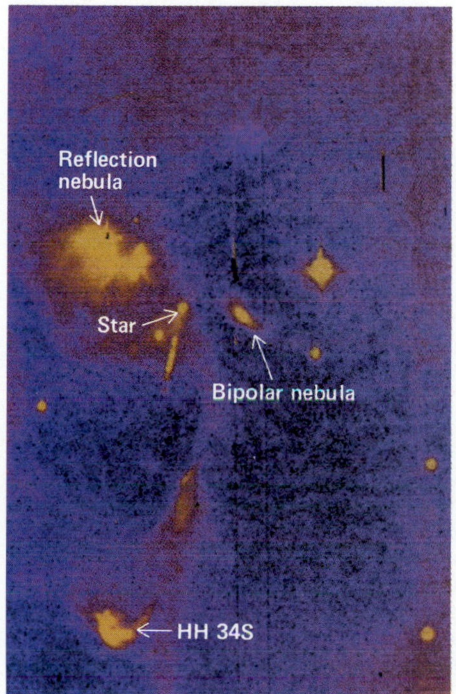

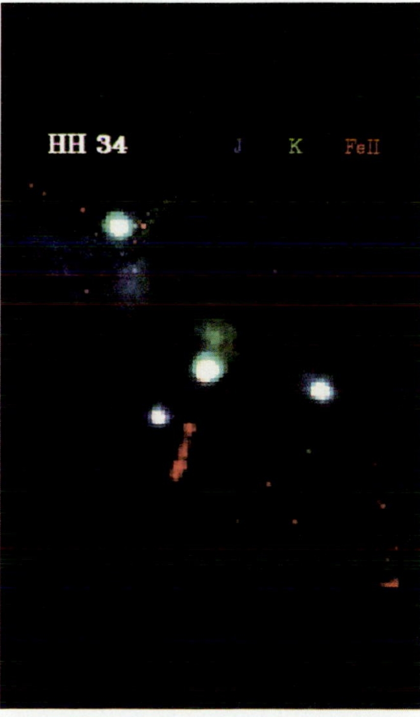

Figure 24–8 (*A*) The jet points from a faint star toward the Herbig-Haro object H–H 34. At its distance from us of 1,600 light years, the angular size of the jet corresponds to 8,000 A.U. The star is a low-mass pre-main-sequence star, about half as intrinsically luminous as the sun. The observations indicate that H–H objects are bow shock waves, resulting when high-velocity beams ram through the interstellar medium. (*B*) This view with the Palomar Prime Focus Infrared Array shows the infrared J band in blue, the infrared K band in green, and the 1.64 μ infrared band that includes a strong ionized iron line in red. The jet shows best in the iron line; at these infrared wavelengths, the parent star shows well. The J band (blue) shows a reflection nebula around another pre-main-sequence star. A deeply embedded reflection nebula or an embedded second source appears in the K band (green).

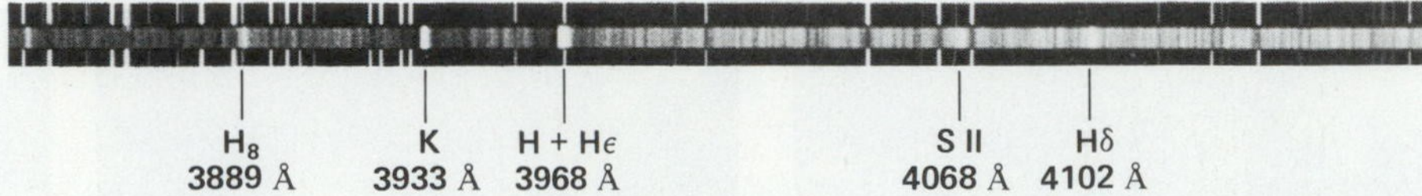

Figure 24–9 T Tauri shows strong, broad, optical emission lines.

Solar-system-sized clouds of dust have been detected around dozens of stars, including HL Tauri (Fig. 24–10) and R Monocerotis. The clouds are elongated, as though they are disk-shaped. The disks are best studied in the infrared, from the IRAS data, and in submillimeter and millimeter waves. Steven Beckwith of Cornell, Anneila Sargent of Caltech, and colleagues have detected millimeter-wavelength emission from the disks. Their measurements show cool gas following Kepler's laws, confirming that the gas is orbiting the stars. They used Caltech's radio telescope at the Owens Valley in California. Other studies use the James Clerk Maxwell telescope on Mauna Kea and the IRAM telescope in Spain. The masses, sizes, and angular momentum of the disks are similar to those we think our solar nebula had when planets started to form. It seems reasonable that planetary systems are also forming around these young stars, some only 100,000 years of age. We discussed more about possible planets formed from dust around other stars in Section 19.2, and in Section 27.6 we will learn about planets formed in a different way, near a pulsar.

24.2 Stellar Energy Generation

All the heat energy in stars that are still contracting toward the main sequence results from the gravitational contraction itself. If this were the only source of energy, though, stars would not shine for very long on an astronomical time scale—only about 30 million years. Yet we know that even rocks on earth are older than that, since rocks over 4 billion years old have been found. We must find some other source of energy to hold the stars up against their own gravitational pull.

The gas in the protostar will continue to heat up until the central portions become hot enough for *nuclear fusion* to take place. Using this process, which we will soon discuss in detail, the star can generate enough energy inside itself to support it during its entire lifetime on the main sequence. The energy makes the particles in the star move around rapidly. For short, we say that the particles have a "high temperature." The particles exert a *thermal pressure* pushing outward, providing a force that balances gravity's inward pull.

The basic fusion processes in most stars fuse four hydrogen nuclei into one helium nucleus, just as hydrogen atoms are combined into helium in a hydrogen bomb here on earth as well as it is controlled in stars. One major experimental approach to fusion on earth uses magnetic fields to hold the

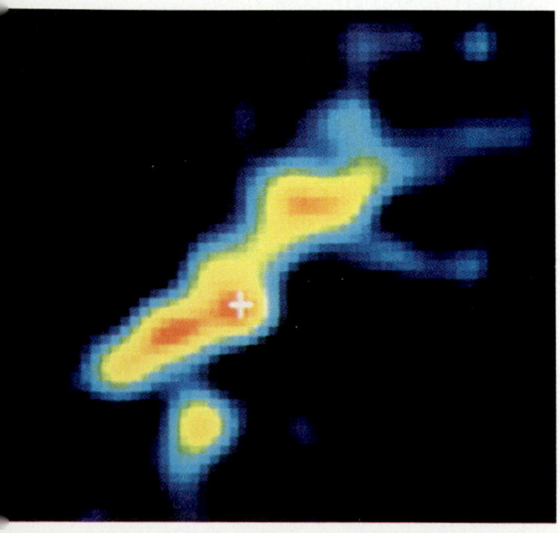

Figure 24–10 The very young star HL Tauri is still embedded in the Taurus dark cloud, the molecular cloud from which it was born. This image shows the region of HL Tau in radiation of a rare isotope of carbon monoxide gas, which is used as a tracer for the hydrogen gas that is much more abundant but harder to see. The image shows a disk 1000 A.U. in radius, perhaps like a protoplanetary disk. This last idea is endorsed by the measurement that the gas revolves following Kepler's laws. The disk contains about one-tenth the mass of the sun, which is about 10 times the mass of the solar system's planets. This mass corresponds with ideas about the formation of solar systems. HL Tau is younger than β Pictoris (Figure 19–5), having been collapsing for only 100,000 to 1 million years. The star itself does not show up in carbon monoxide radiation.

plasma together at a high enough density and for a long enough time for fusion to occur. This "containment" of the plasma is difficult on earth; containment is not a problem for the sun and stars, whose strong gravity keep the plasma dense at their cores. Containment is also not a long-term problem in making a bomb.

A hydrogen nucleus is but a single proton. A helium nucleus is more complex. It consists of two protons and two neutrons (Fig. 24–11). The mass of the helium nucleus that is the final product of the fusion process is slightly less than the sum of the masses of the four hydrogen nuclei that went into it. A small amount of the mass "disappears" in the process: 0.7 per cent of the mass of the four hydrogen nuclei.

The mass does not really simply disappear, but is rather converted into energy according to Albert Einstein's famous formula $E = mc^2$. Now, c, the speed of light, is a large number, and c^2 is even larger. Thus even though m is only a small fraction of the original mass, the amount of energy released is prodigious. The loss of only 0.007 of the mass of the central part of the sun, for example, is enough to allow the sun to radiate at its present rate for a period of at least ten billion (10^{10}) years. This fact, not realized until 1920 and worked out in more detail in the 1930's, solved the long-standing problem of where the sun and the other stars got their energy.

In the interiors of stars, we are dealing with nuclei instead of the atoms we have discussed in stellar atmospheres, because the high temperature strips the electrons off the nuclei. The nuclei have positive electric charge. The electrons are mixed in with the nuclei through the center of the star; there can be no large imbalance of positive and negative charges, or else strong repulsive forces would arise inside the star. Such an ionized gas is known as a "plasma."

All the main-sequence stars are approximately 90 per cent hydrogen (that is, 90 per cent of the atoms are hydrogen), so there is lots of raw material to stoke the nuclear "fires." We speak colloquially of "nuclear burning," although, of course, the processes are quite different from the chemical processes that are involved in the "burning" of logs or of autumn leaves. In order to be able to discuss these processes, we must first discuss the general structure of nuclei and atoms.

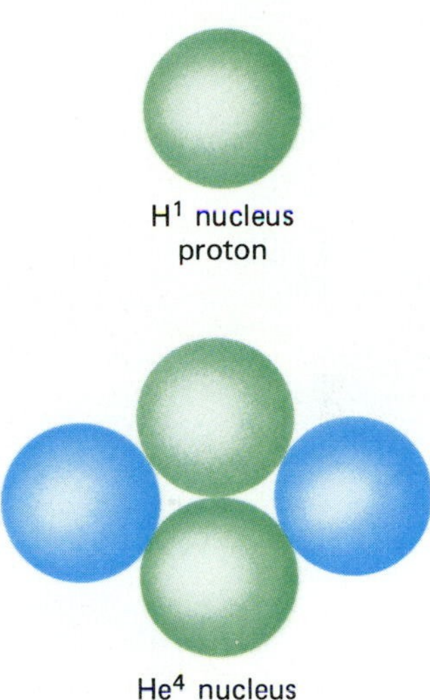

Figure 24–11 The nucleus of hydrogen's most common form is a single proton, while the nucleus of helium's most common form consists of two protons and two neutrons.

24.3 Atoms

An atom consists of a small *nucleus* surrounded by *electrons*. Most of the mass of the atom is in the nucleus, which takes up a very small volume in the center of the atom. The effective size of the atom, the chemical interactions of atoms to form molecules, and the nature of spectra are determined by the electrons.

The nuclear particles with which we need be most familiar are the proton and neutron. Both these particles have nearly the same mass, 1836 times greater than the mass of an electron, though still tiny (Appendix 2). The neutron has no electric charge and the proton has one unit of positive electric charge. The electrons, which surround the nucleus, have one unit each of negative electric charge. When an atom loses an electron, it has a net positive charge of 1 unit for each electron lost. The atom is now a form of *ion* (Fig. 24–12). Astronomers use Roman numerals to show the stage of ionization: I for the electrically

Figure 24–12 Hydrogen and helium ions. The sizes of the nuclei are greatly exaggerated with respect to the sizes of the orbits of the electrons.

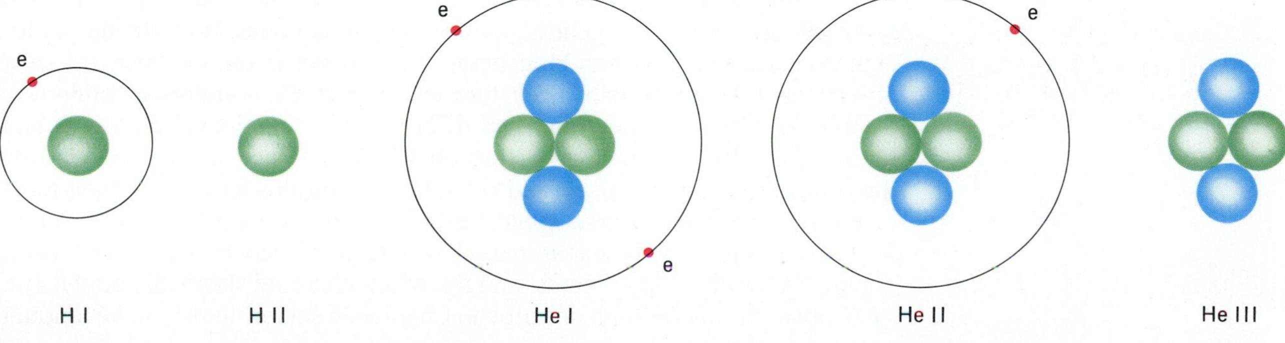

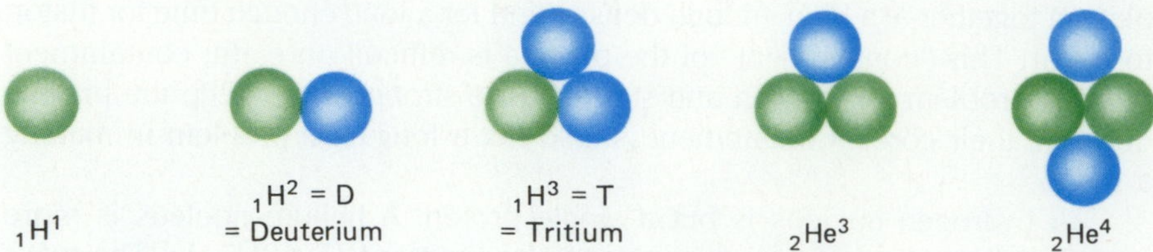

Figure 24–13 Isotopes of hydrogen and helium. $_1H^2$ (deuterium) and $_1H^3$ (tritium) are much rarer than the normal isotope, $_1H^1$. $_2He^3$ is much rarer than $_2He^4$.

neutral state, II for the state in which one electron is lost, etc. Thus Fe I is neutral iron, Fe II is the state of Fe^+ ions, Fe III is the state of Fe^{++} ions, etc. (The Roman numeral is one higher than the number of electrons removed.)

The number of protons in the nucleus determines the quota of electrons that the neutral state of the atom must have, since this number determines the charge of the nucleus. Each *element* (sometimes called "chemical element") is defined by the number of protons in its nucleus. The element with one proton is called hydrogen, that with two protons is called helium, that with three protons is called lithium, and so on.

Though a given element always has the same number of protons in a nucleus, it can have several different numbers of neutrons. (The number is always somewhere between 1 and 2 times the number of protons. Hydrogen, which need have no neutrons, and helium are the only exceptions to this rule.) The possible forms of an element having different numbers of neutrons are called *isotopes.*

For example, the nucleus of ordinary hydrogen contains one proton and no neutrons. An isotope of hydrogen (Fig. 24–13) called deuterium (and sometimes "heavy hydrogen") has one proton and one neutron. Another isotope of hydrogen called tritium has one proton and two neutrons.

Most isotopes do not have specific names, and we keep track of the numbers of protons and neutrons with a system of superscripts and subscripts. The subscript before the symbol denoting the element is the number of protons (called the *atomic number*), and a superscript is the total number of protons and neutrons together (called the *mass number*). For clarity, we write the mass number **after** the symbol, so that you read the numbers and symbol from left to right; the mass number is usually written before the symbol, but spoken after it. For example, $_1H^2$ is deuterium, since deuterium has one proton, which gives the subscript, and a mass number of 2, which gives the superscript. Deuterium has atomic number equal 1 and mass number equal 2. Similarly, $_{92}U^{238}$ is an isotope of uranium with 92 protons (atomic number = 92) and mass number of 238, which is divided into 92 protons and $238 - 92 = 146$ neutrons.

Each element has only certain isotopes. For example, most naturally occurring helium is in the form $_2He^4$, with a lesser amount as $_2He^3$. (We sometimes informally read and write "helium-4" and "helium-3.") Sometimes an isotope is not stable, in that after a time it will spontaneously change—or decay—into another isotope or element; we say that such an isotope is *radioactive.*

During certain types of radioactive decay, a particle called a *neutrino* is given off. A neutrino is a neutral particle (its name comes from the Italian for "little neutral one"). It has long been thought that neutrinos have no "rest mass," the mass they would have if they were at rest. Experiments are underway to find out whether a neutrino may in fact have some small amount of rest mass, about 1/100 million that of a proton. (Einstein's special theory of relativity describes the relation of mass and velocity.) If neutrinos have mass, then most of the mass of the universe could be in the form of neutrinos (see Section 34.2c). But the positive experimental results once reported have not been confirmed by other experiments, and many scientists are skeptical about them. Also, observations we shall describe in Chapter 26 set low limits on the amount of mass a neutrino can have.

24.4 Stellar Energy Cycles

Several sequences of reactions, termed "chains," have been proposed to account for the fusion of four hydrogen atoms into a single helium atom. Hans Bethe (Fig. 24–14), now at Cornell University, suggested some of these processes during the 1930's. The different possible chains that have been proposed are important at different temperatures, so chains that are dominant in the centers of very hot stars may be different chains from the ones that are dominant in the centers of cooler stars.

When the center of a star is at a temperature less than 15×10^6 K, the *proton-proton chain* (Fig. 24–15) dominates. Our sun makes most of its energy in this way. In the proton-proton chain, we put in six hydrogens one at a time, and wind up with one helium plus two hydrogens, a net transformation of four hydrogens into one helium. But the six protons contained more mass than do the final single helium plus two protons. The small fraction of mass that disappears in the process is converted into an amount of energy that we can calculate with the formula $E = mc^2$.

For stellar interiors hotter than that of the sun, the *carbon-nitrogen (CN) cycle* (Fig. 24–16) dominates. The CN cycle begins with the fusion of a hydrogen nucleus with a carbon nucleus. After many steps, and the insertion of four hydrogen nuclei, we are left with one helium nucleus plus a carbon nucleus. Thus as much carbon remains at the end as there was at the beginning, and the carbon can start the cycle again. Again, four hydrogens have been converted into one helium, 0.007 (0.7 per cent) of the mass has been transformed, and an equivalent amount of energy has been released according to $E = mc^2$. It has also been found that alternative cycles involving different oxygen isotopes

Figure 24–14 Hans Bethe.

The Greek letters derive from a former confusion of radiation and particles. We know now that α particles, formerly called α rays, are helium nuclei; β particles, formerly called β rays, are electrons; and γ rays, as we have seen, are electromagnetic radiation of short wavelength.

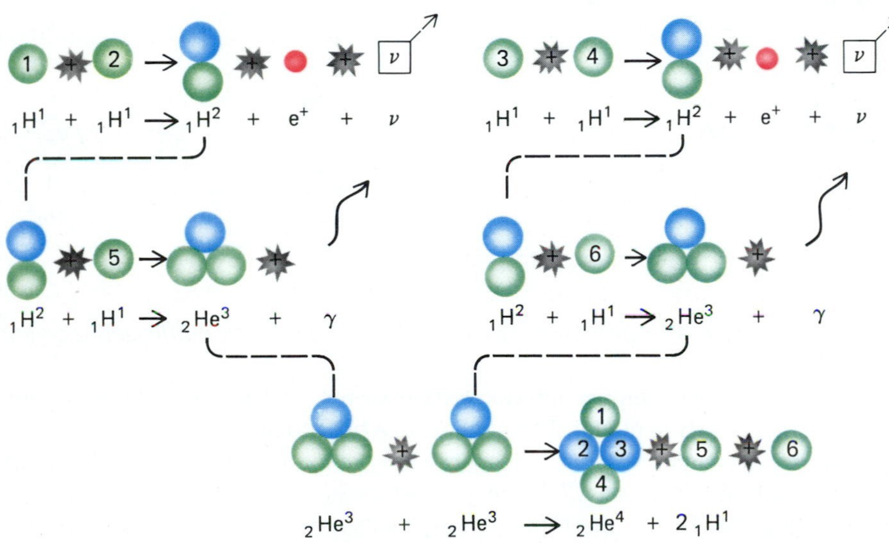

Figure 24–15 The proton-proton chain; e^+ stands for a positron, ν (nu) is a neutrino, and γ (gamma) is electromagnetic radiation at a very short wavelength.

In the first stage, two nuclei of ordinary hydrogen fuse to become a deuterium (heavy hydrogen) nucleus, a positron (the equivalent of an electron, but with a positive charge), and a neutrino. The neutrino immediately escapes from the star, but the positron soon collides with an electron. They annihilate each other, forming gamma rays. (A positron is an anti-electron, an example of antimatter; whenever a particle and its antiparticle meet, they annihilate each other.)

Next, the deuterium nucleus fuses with yet another nucleus of ordinary hydrogen to become an isotope of helium with two protons and one neutron. More gamma rays are released.

Finally, two of these helium isotopes fuse to make one nucleus of ordinary helium plus two nuclei of ordinary hydrogen. The protons are numbered to help you keep track of them.

Figure 24–16 The carbon-nitrogen cycle, also called the carbon cycle. The 4 hydrogen atoms are numbered. Note that the carbon is left over at the end, ready to enter into another cycle.

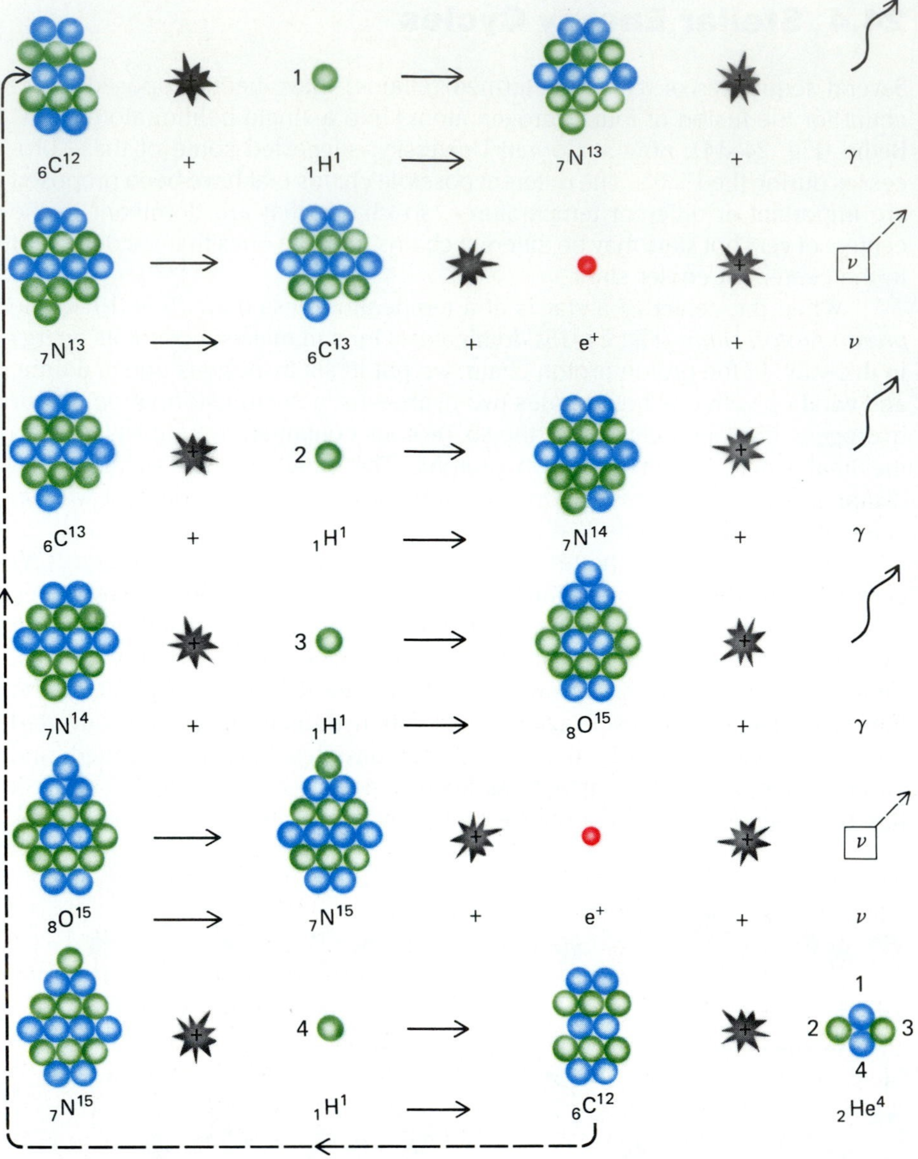

sometimes occur. The processes are often called the *carbon-nitrogen-oxygen (CNO) tri-cycle.*

Stars with even higher interior temperatures, above 10^8 K, fuse helium nuclei to make carbon nuclei. The nucleus of a helium atom is called an "alpha particle" for historical reasons. Since three helium nuclei ($_2He^4$) go into making a single carbon nucleus ($_6C^{12}$), the procedure is known as the *triple-alpha process* (Fig. 24–17). In the triple-alpha process, two alpha particles first interact temporarily to form beryllium-8 ($_4Be^8$). The beryllium-8 is not very stable, and it lasts only long enough for a level of one beryllium-8 atom per billion alpha particles to exist. However, this level is sufficiently high for a third alpha particle to interact with the beryllium-8 nucleus, forming carbon-12. A series of other processes can build still heavier elements (that is, elements with more protons and neutrons in each atom) inside stars.

The processes that build heavier nuclei from lighter ones are called *nucleosynthesis.* The theory of nucleosynthesis can account for the abundances we observe in stars and in the gas between the stars of the elements heavier than helium. Currently, we think that the synthesis of isotopes of hydrogen and helium took place in the first few minutes after the origin of the universe (Section

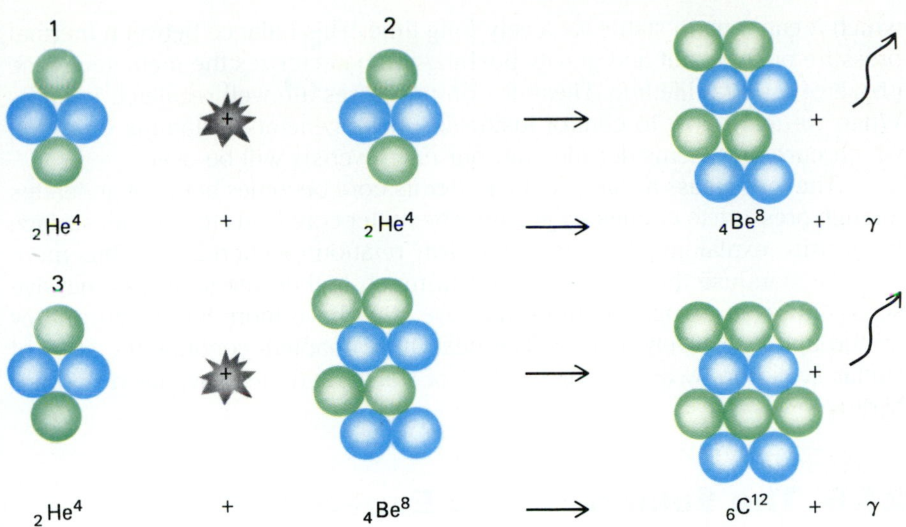

Figure 24–17 The triple-alpha process, which takes place only at temperatures above about 10^8 K. Beryllium, $_4Be^8$, is but an intermediate step.

34.1) and that the heavier elements were formed, along with additional helium, in stars or in supernova explosions (Section 26.2). William A. Fowler of Caltech shared the 1983 Nobel Prize in Physics for his work on nucleosynthesis, including the measurements of the rates of the nuclear reactions that make the stars shine.

24.5 The Stellar Prime of Life

Now that we have discussed basic nuclear processes, let us return to an astronomical situation. We last discussed a protostar in a collapsing phase, with its internal temperature rapidly rising.

One of the most common definitions of temperature describes the temperature as a measure of the velocities of individual atoms or other particles. Since this type of temperature depends on kinetics (motions) of particles, it is called the *kinetic temperature*. A higher kinetic temperature (we shall call it simply "temperature" from now on) corresponds to higher particle velocities.

For a collapsing protostar, the energy from the gravitational collapse goes into giving the individual particles greater velocities; that is, the temperature rises. For nuclear fusion to begin, atomic nuclei must get close enough to each other so that the force that holds nuclei together, the *strong force* (Section 34.3), can play its part. But all nuclei have positive charges, because they are composed of protons (which bear positive charges) and neutrons (which are neutral). The positive charges on any two nuclei cause an electrical repulsion between them, and this force tends to prevent fusion from taking place.

However, at the high temperatures typical of a stellar interior, some nuclei have enough energy to overcome this electrical repulsion and to come sufficiently close to each other for the strong nuclear force to take over. The electrical repulsion that must be overcome is the reason why hydrogen nuclei, which have net positive charges of 1, will fuse at lower temperatures than will helium nuclei, which have net positive charges of 2.

Once nuclear fusion begins, the thermal pressure of the star's gas (the pressure resulting from temperature) provides a force that pushes outward strongly enough to balance gravity's inward pull. In the center of a star, the fusion process is self-regulating. If the nuclear energy production rate increases, then an excess pressure is generated that would tend to make the star grow larger. However, this expansion in turn would cool down the gas and slow down the rate of nuclear fusion. Thus the star finds a temperature and size at which it can remain stable for a very long time. This balance between thermal pressure pushing out and gravity pushing in characterizes the main-sequence phase of a stellar lifetime. These fusion processes are well regulated in stars. When we learn how to control fusion in power-generating stations on earth, which currently seems decades off, our energy crisis will be over.

The more mass a star has, the hotter its core becomes before it generates enough pressure to counteract gravity. The hotter core leads to a higher surface luminosity, explaining the mass-luminosity relation (Section 22.2). Thus more massive stars use their nuclear fuel at a much higher rate than less massive stars, and even though the more massive stars have more fuel to burn, they go through it relatively quickly. The next three chapters continue the story of stellar evolution by discussing the fate of stars when they have used up their hydrogen.

24.6 The Solar Neutrino Experiment

Astronomers can apply the equations that govern matter and energy in a star, and model the star's interior and evolution in a computer. Though the resulting model can look quite nice, nonetheless it would be good to confirm it observationally. Happily, the models are consistent with the conclusions on stellar evolution that one can make from studying different types of Hertzsprung-Russell diagrams. Still, it would be nice to observe a stellar interior directly.

Since the stellar interior lies under opaque layers of gases, we cannot observe directly any electromagnetic radiation it might emit. Only neutrinos escape directly from a stellar interior. Neutrinos interact so weakly with matter that they are hardly affected by the presence of the rest of the solar mass. Once formed, they zip right out into space at or almost at the speed of light. Raymond Davis, Jr., retired from the Brookhaven National Laboratory and working at the University of Pennsylvania, has spent many years studying the neutrinos formed in the solar interior. John Bahcall of the Institute for Advanced Study at Princeton has been a major force behind the theoretical aspect of the program.

Figure 24–18 The neutrino telescope, deep underground in the Homestake Gold Mine in Lead, South Dakota, consists mainly of a tank containing 400,000 liters of perchloroethylene. The tank is now surrounded by water.

Neutrinos formed in the sun's normal proton-proton chain do not have enough energy for Davis's apparatus to detect them. So Davis's experiment can detect only the neutrinos that occur in branches of the chain that are followed less than 1 per cent of the time. Thus theoretical calculations of how often these branches are followed are an important part of the study.

How do we detect the neutrinos? Neutrinos, after all, pass through the earth and sun, barely affected by their mass. At this instant, neutrinos are passing through your body. Davis makes use of the fact that very occasionally a neutrino will interact with the nucleus of an atom of chlorine, and transform it into an isotope of the gas argon. Argon interacts rarely with other atoms (and so is called a "noble gas"); this property makes it relatively easy to separate it from the rest of the matter in the tank so the argon can be studied.

The transformation of chlorine into argon takes place very rarely, so Davis needs a large number of chlorine atoms. He found it best to do this by filling a large tank with liquid cleaning fluid, C_2Cl_4, where the subscripts represent the number of atoms of carbon and chlorine in the molecule, which is called perchloroethylene. One-fourth of the chlorine is the Cl^{37} isotope, which interacts with a neutrino more readily than does the major form of chlorine. Davis now has a large tank containing 400,000 liters (100,000 gallons) of this cleaning fluid (Fig. 24–18). (Davis denies the story that after he bought his cleaning fluid he was besieged by wire-coat-hanger salesmen.) His neutrino telescope is set up 1.5 km underground (in a room in an active gold mine in Lead, South Dakota) to shield it from other particles, none of which pass through that much earth.

Even with this huge tankful of chlorine atoms, calculations show that Davis should expect only about one neutrino-chlorine interaction every day. This experiment is surely one of the most difficult ever attempted. When an interaction occurs, a radioactive argon atom is formed. Any such argon atoms can be removed from the tank by standard chemical techniques involving bubbling helium gas through the fluid (Fig. 24–19). Then the radioactivity of the few resulting argon atoms can be measured, also by standard means.

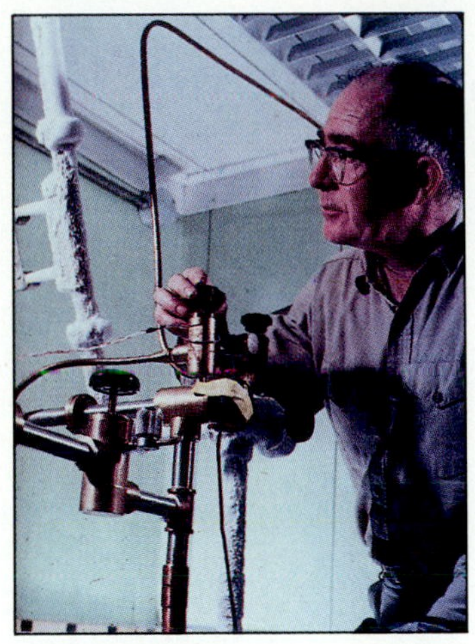

Figure 24–19 Raymond Davis with some of the equipment at his neutrino observatory.

It is fair to say that Davis's results astounded the scientific community, which had confidently expected to hear reports of a number of neutrino interactions in agreement with theoretical predictions. Davis found only about ¼ the number of neutrinos expected. The experiments have continued, with increasing sophistication, for nearly twenty years (Fig. 24–20). The current best observational average is 2.05, while the current best prediction is 7.9 solar neutrino units (SNU's). His measurement is far below the predictions of solar theorists, and is at the very minimum level that any currently reasonable theory of nuclear fusion in stellar interiors can predict, even when pushed.

Several questions immediately come to mind. The first deals with Davis's apparatus, and whether there are some experimental effects that could explain the results in some normal manner. Davis has carried out a series of careful checks of his apparatus, and most astronomers and chemists are convinced that the experimental setup is not the cause of the problem.

Next, perhaps we do not understand the basic physics of nuclear reactions or of neutrinos as well as we thought. Davis's apparatus can detect only the form of neutrinos that the sun emits. Perhaps a neutrino changes in type (there are three "flavors" of neutrinos: electron neutrinos, muon neutrinos, and tau neutrinos) during the 8 minutes it takes to travel at the speed of light from the sun to the earth. Some experimental evidence has been advanced for this change in type taking place in a laboratory but has not been confirmed.

In 1986, two Soviet scientists used a method worked out by an American scientist to point out that electron neutrinos are significantly scattered through the weak interaction as they travel through the very dense regions near the center of the sun. In fact, the theory shows that neutrinos with rest mass can transform themselves from one type to another if they go through a region of sufficiently high density. (Quantum mechanics has shown that neither electron

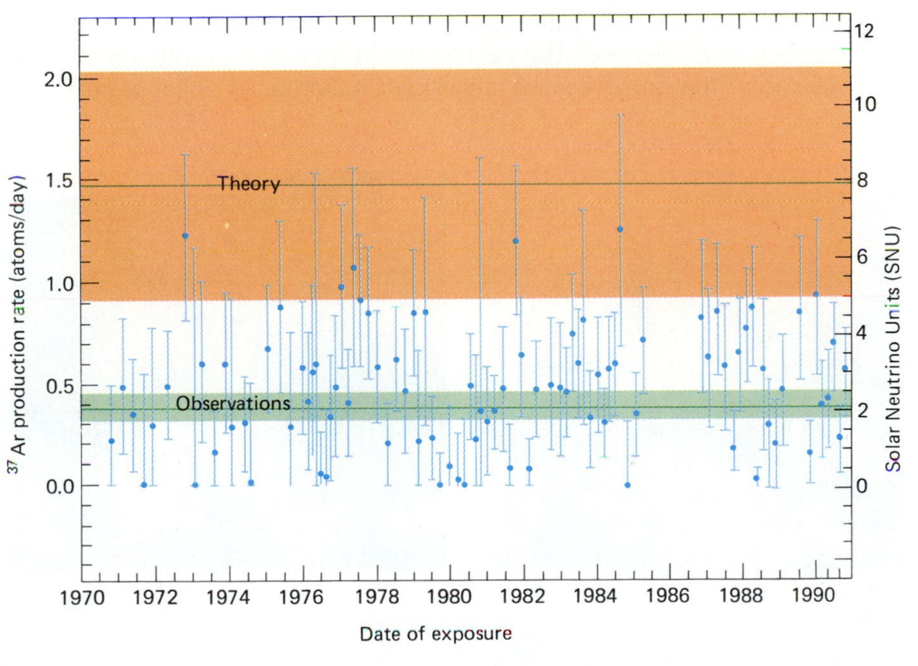

Figure 24–20 The history of solar neutrino observations in chlorine-37, by Raymond Davis, Jr., and collaborators, with vertical error bars showing the uncertainty for each measurement. The green shading shows the uncertainty of the average. For comparison, the best theoretical prediction by John Bahcall of the Institute for Advanced Study, Princeton, is shown at the top, with the orange shaded range indicating its uncertainty.

neutrinos nor muon neutrinos with rest mass travel through space as individual particles; a travelling neutrino is actually a combination of the types of neutrinos.) Since the Davis apparatus can detect only electron neutrinos, and since all neutrinos produced in the sun are electron neutrinos, such a transformation would make the Davis apparatus detect a smaller number of neutrinos. So the neutrino problem may be solved by this method, known after the initials of the scientists as the MSW effect.

Another interesting possibility is that our understanding of stellar interiors is not satisfactory. If, for example, stars are cooler inside than we have predicted, fewer neutrinos would be produced. A hypothetical constituent of matter, a "cosmion"—also known as a "weakly interacting massive particle" (acronym: WIMP)—could explain both the solar-neutrino problem and the missing-mass problem that we shall describe in Chapter 34. If such a particle exists, several could collect near the center of the sun and transport energy out of the central region. They would thus slightly cool the solar core, lowering the neutrino flux. Results from solar seismology (Section 23.2b), though, indicate that the interior is not cool enough for the predictions of neutrino flux to match the observations.

One of the most exciting new developments in the field is that Japanese scientists have used their Kamiokande II detector (to be shown in Figure 26–23) to confirm the deficit in neutrinos from the boron-8 chain found with the chlorine experiment. They indeed measured some neutrinos, though far fewer than the best standard solar model predicts. Though a larger volume of water in the detector was used for detecting neutrinos from Supernova 1987A (Section 26.5), 2100 tons out of the total 3000 tons, only the inner 680 tons is used for the solar-neutrino observations to diminish the contribution of background events. They detect about 0.3 solar neutrino event per day. Also, they can measure the direction of the incoming neutrinos, and have shown that they indeed come from the sun. The name of the detector comes from the Kamioka mine and NDE for **n**ucleon **d**ecay **e**xperiment, with the II representing the upgrade of detectors and electronics that led to the possibility of observing solar neutrinos.

Several searches are now being prepared using the metallic element gallium. Gallium is sensitive to neutrinos of much lower energy than those with which chlorine interacts. It would detect essentially all solar neutrinos, instead of only those with the high energy that the current experiment can detect: gallium-71 interacts with an electron neutrino and the results are germanium-71 plus an electron. The germanium-71 then decays with a half-life of 11.4 days, and the decay can be measured. A European collaborative with

A

B

Figure 24–21 The Gallex gallium solar-neutrino detector in the Italian underground Gran Sasso Laboratory. (*A*) The target tanks, containing the gallium chloride solution, underground in Gran Sasso. (*B*) The counting lab in its protective cage. Only 1 event per day is expected, so the counters must be well shielded to reduce background events.

Figure 24–22 Preparations at CERN (the European research center on the French-Swiss border) for the testing of a prototype liquid argon chamber for the ICARUS solar neutrino detector in the Italian underground Gran Sasso Laboratory.

scientists primarily from Germany, France, and Italy, and with additional participation by scientists from the United States and from Israel, prepared 30 tons of gallium in the form of a water solution of gallium chloride ($GaCl_3$) mixed with hydrochloric acid. The resultant $GeCl_4$ can be extracted from their tank, which is underground in the Gran Sasso Tunnel in Italy (Fig. 24–21). This Gallex consortium expected to see about 1 interaction per day in their target. A Russian experiment with U.S. participation, SAGE (**S**oviet-**A**merican **G**allium **E**xperiment; the word "Soviet" has not yet been changed), is using 60 tons of gallium in metallic form; gallium metal melts at only 30°C, which allows it to be mixed with dilute hydrochloric acid. The experiment is at the Baksan Neutrino Observatory, in a tunnel under a mountain in the Caucasus near Neutrino City, marked on even Russian-language maps with a Greek ν, the symbol for neutrino. Though the initial chemical separation is different between the two experiments, the final counting methods are similar.

The gallium experiment is sensitive to neutrinos from the main part of the proton–proton chain, and a signal of 132 solar neutrino units is expected. The neutrinos should be little transformed from the effect described. By the spring of 1992, the Gallex consortium reported a detection of 83 ± 20 SNU, substantially below the predicted 132 SNU.

As of early 1992, the SAGE experiment had found a level of solar neutrinos that might be as low as zero but that could be one-sixth the expected rate, still a low value. Later, in June 1992, the SAGE group reported a value of 80 ± 20 SNU, similar to that found by Gallex. The low counting rate of the SAGE and Gallex experiments can be meshed with the observed low rate from the chlorine experiments (as calculated by John Bahcall of the Institute of Advanced Study and Edwin Salpeter of Cornell) if neutrinos have some small amount of mass and are transformed by the previously mentioned effect (known as the MSW effect). The values obtained by SAGE and Gallex are apparently high enough to verify that the proton–proton chain fuels the Sun but are low enough to show that some "new physics" beyond the standard model of the neutrino is required.

Solar neutrinos may also be found as a byproduct of other experiments. A liquid argon detector called ICARUS, also for Gran Sasso (Fig. 24–22), will search for the decay of protons predicted by some theories of matter (Section 34.3) with 200 tons of liquid argon but may also detect solar neutrinos. A U.S./Canadian experiment using 1000 tons of "heavy water" (water in which deuterium atoms replace the ordinary hydrogen atoms) in a mine near Sudbury, Ontario, will be sensitive to all flavors of neutrinos when it starts up in 1995. Both of these experiments, though, are sensitive only to high-energy neutrinos, as is the neutrino experiment using chlorine.

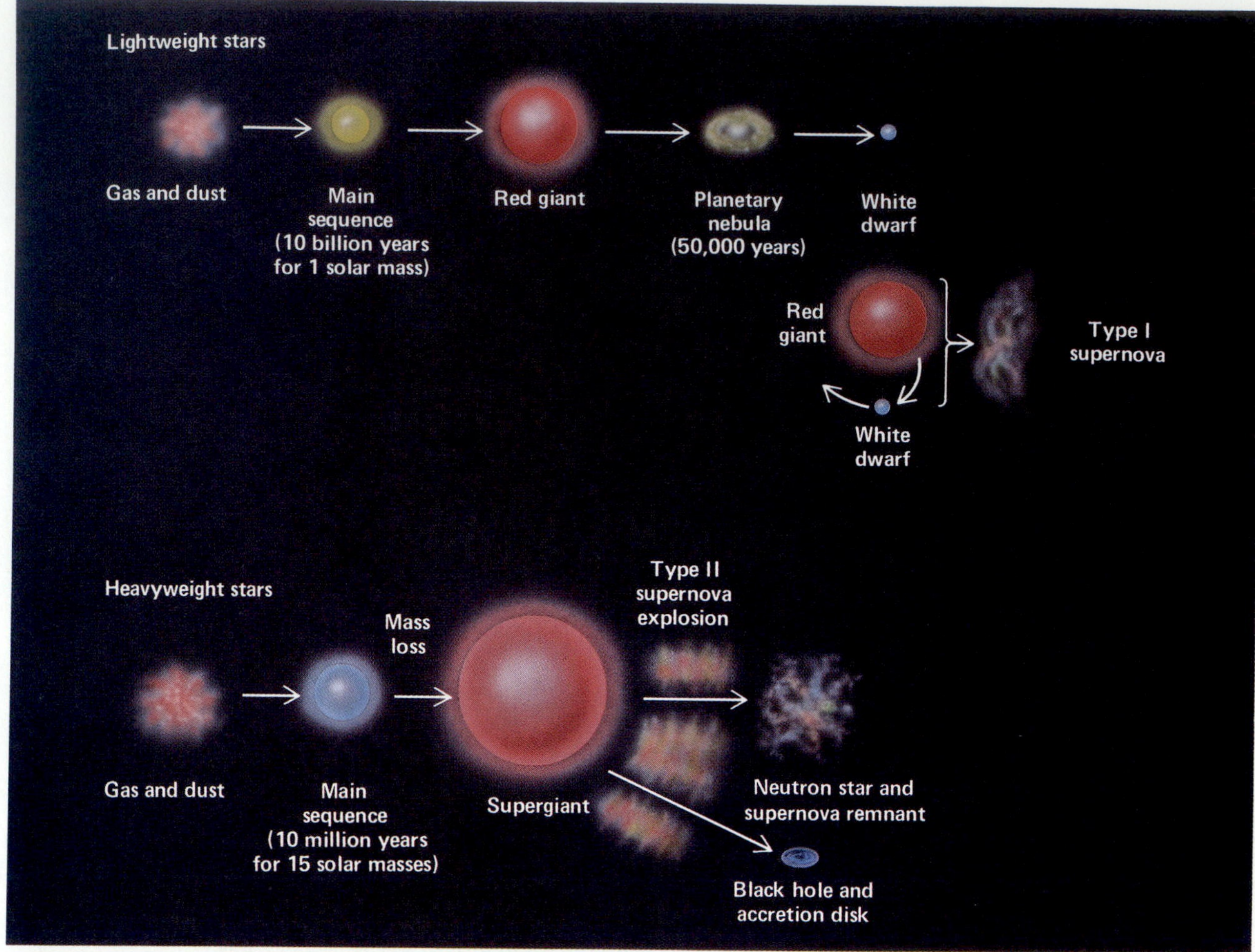

Figure 24–23 A summary of the stages of stellar evolution for stars of different masses; these stages will be described in the following four chapters.

24.7 Dying Stars

We shall devote the next chapters to the various end stages of stellar evolution. The mass of the star determines its fate (Fig. 24–23). First we shall discuss the less massive stars, such as the sun. In Chapter 25, we shall see how such stars swell in size to become giants, possibly become planetary nebulae, and then end their lives as white dwarfs. Some of these white dwarfs have companion stars, and become either novae or supernovae.

More massive stars more regularly come to explosive ends. In Chapter 26, we shall see how these stars are blown to smithereens, and in Chapter 27 we will discuss how strange objects called "pulsars" are sometimes the remnants. Other remnants appear to be part of x-ray-emitting binary systems.

In Chapter 28, we shall encounter the strangest kind of stellar death of all. The most massive stars may become "black holes," and effectively disappear from view.

Summary and Outline

Protostars (Section 24.1)

Their evolutionary tracks on the H–R diagram bring them in from the right; they form when a cloud of gas and dust is compressed; they can be observed in the infrared.

T Tauri stars (Section 24.1)

Irregular variations in magnitude; may not have reached the main sequence; near Herbig-Haro objects

Herbig-Haro objects are the result of jets forming shock waves in the interstellar medium.

Stellar energy generation (Sections 24.2, 24.3, and 24.4)
- Nuclear fusion provides the energy for stars to shine.
- Proton-proton chain dominates in cooler stars, including the sun.
- Carbon-nitrogen cycle is for hotter stars.
- Triple-alpha process takes place at high temperatures.
- Stellar nucleosynthesis accounts for most of the elements.

The main sequence: a balance of pressure and gravity (Section 24.5)

A test of the theory: the neutrino experiment (Section 24.6)
- The theoretical prediction is 3 times the observational result; possible resolutions; uncertainty in the theoretical result must be lowered; further, more definitive experiments planned

Mass determines stellar evolution (Section 24.7)

Key Words

featherweight stars, lightweight stars, heavyweight stars, middleweight stars, stellar evolution, evolutionary track, O and B associations, T Tauri stars, T associations, Herbig-Haro objects, nuclear fusion, thermal pressure, $E = mc^2$, nucleus, electrons, proton, neutron, ion, quarks, element, isotopes, atomic number, mass number, radioactive, neutrino, proton-proton chain, carbon-nitrogen cycle, carbon-nitrogen-oxygen tri-cycle, triple-alpha process, nucleosynthesis, kinetic temperature, strong force

Questions

1. Since individual stars can live for billions of years, how can observations taken at the current time tell us about stellar evolution?

2. What is the source of energy in a protostar? At what point does a protostar become a star?

3. What is the evolutionary track of a star?

4. Describe the evolutionary track of the sun during the 10 billion years it is on the main sequence.

5. Arrange the following in order of development: O and B associations; T Tauri stars; dark clouds; sun; pulsars.

†6. Give the number of protons, the number of neutrons, and the number of electrons in: ordinary hydrogen ($_1H^1$), lithium ($_3Li^6$), iron ($_{26}Fe^{56}$).

7. (a) If you remove one neutron from helium, the remainder is what element? (b) Now remove one proton. What is left? (c) Why is He IV not observed?

8. (a) Explain why nuclear fusion takes place only in the centers of stars rather than on their surfaces as well. (b) What is the major fusion process that takes place in the sun?

9. If you didn't know about nuclear energy, what is one possible energy source you might suggest for stars? What is wrong with these alternative explanations?

10. What forces are in balance for a star to be on the main sequence?

†This question requires a numerical solution.

†11. Use the speed of light and distance between the earth and sun given in Appendix 2 to verify that it takes neutrinos 8 minutes to reach the earth from the sun.

12. What does it mean for the temperature of a gas to be higher?

†13. (a) If all the hydrogen in the sun were converted to helium, what fraction of the solar mass would be lost? (b) How many times the mass of the earth would that be? (Consult Appendix 2.)

14. (a) How does the temperature in a stellar core determine which nuclear reactions will take place? (b) Why do more massive stars have shorter main-sequence lifetimes?

15. In what form is energy carried away in the proton-proton chain?

16. In the proton-proton chain, the products of some reactions serve as input for the next and therefore don't show up in the final result. Identify these intermediate products. What are the **net** input and output of the proton-proton chain?

17. What do you think would happen if nuclear reactions in the sun stopped? How long would it be before we noticed?

18. Why do neutrinos give us different information about the sun than does light?

19. Why are the results of the solar neutrino experiment so important?

20. Why will the gallium experiment for solar neutrinos be superior to the chlorine experiment?

The Helix Nebula, NGC 7293, the nearest planetary nebula to us at a distance of 400 light years. Long exposures show that its diameter in the sky is about the same as the moon's. The shell glows mainly Hα and so is reddish. The violet and green lines of once- and twice-ionized oxygen are superimposed, giving a pinkish cast. The central part glows more with the emission of ions like oxygen and multiply ionized neon; the ions are formed by the central star's ultraviolet radiation and radiate in the ultraviolet and blue. The human eye is not very sensitive to the Hα that makes this photograph appear so red, and the enhanced green sensitivity of the eye's rods brings out the green oxygen radiation, so planetary nebulae appear faintly greenish.

The Death of Stars Like the Sun

25

Aims: To understand what happens to stars of up to about 4 solar masses when they have finished their time on the main sequence of the H–R diagram, including the planetary nebula and white dwarf stages

Let us remember that a star is a continual battleground between gravity pulling inward and pressure pushing outward. In the gas making up main-sequence stars, thermal pressure resulting from energy provided by nuclear fusion balances gravity. Here we will see what happens when fusion stops, and so no longer provides that pressure.

The sun is just an average star, in that it falls in the middle of the main sequence. But the majority of stars have less mass than the sun. So when we discuss the end of the main-sequence lifetimes of low-mass stars, including not only the sun but also all stars containing up to a few times its mass, we are discussing the future of most of the stars in the universe. In this chapter we will discuss the late (post—main-sequence) stages of evolution of stars that, when they are on the main sequence, contain less than about 4 solar masses. We shall call them *lightweight stars.* Let us consider a star like the sun, in particular, remembering that all lightweight stars go through similar stages but at different rates. We shall see their planetary nebula (Fig. 25–1) and white dwarf stages.

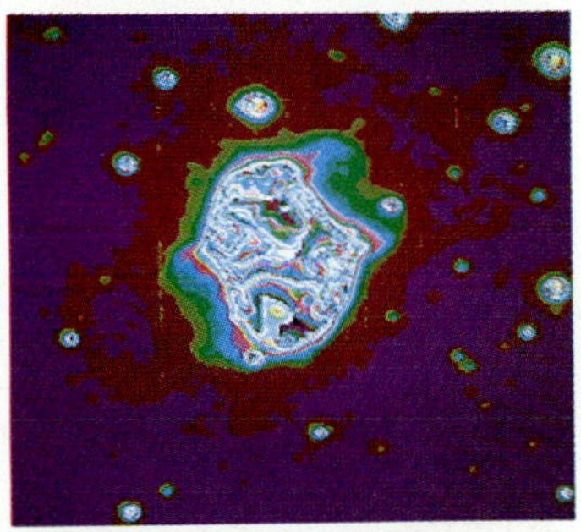

Figure 25–1 A false-color view of internal structure in a planetary nebula.

25.1 Red Giants

During the main-sequence phase (Fig. 25–2) of lightweight stars, as with all stars, hydrogen in the core (the inner 10 per cent or so) gradually fuses into helium. During this time, the position corresponding to the star stays in essentially the same place on the Hertzsprung-Russell diagram, drifting upward only slightly. By about 10 billion (10^{10}) years after a one-solar-mass star first reaches the main sequence, no hydrogen is left in its core, which is thus composed almost entirely of helium. Hydrogen is still undergoing fusion in a shell around the core.

Since no fusion is then taking place in the core, no ongoing nuclear process is replacing the energy that flows out of this hot central region. The core no longer has enough pressure to hold up both itself and the overlying layers against the inward force of gravity. As a result, the core then begins to contract under the force of gravity (we say it contracts *gravitationally*). This gravitational contraction not only replaces the heat lost by the core but also, in fact, heats the core up further. Thus, paradoxically, soon after the hydrogen burning in the core stops, the core becomes hotter than it was before (because of the gravitational contraction).

Half the energy from gravitational contraction always goes into kinetic motion in the interior. This is an important theorem, the virial theorem, whose effects we met before in Section 24.1 when discussing the contraction of protostars. An example of such increased kinetic motion is the rise in temperature that we call heat. Most of the rest of the energy is radiated away.

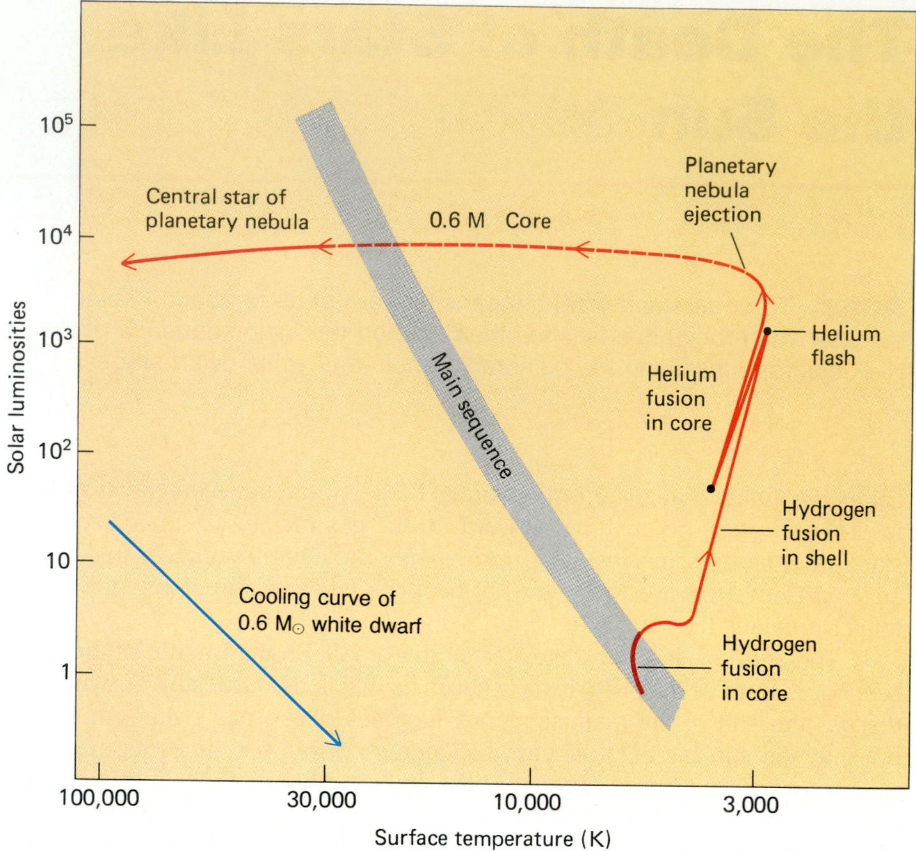

Figure 25–2 A theoretical H–R diagram showing the evolutionary track of a 1-solar-mass star. We will soon describe its evolution after its red-giant stage. The vertical axis is the luminosity of the star, in units of the sun's luminosity; the horizontal axis is the temperature of the star's surface.

As the core becomes hotter, the hydrogen-burning shell around the core becomes hotter too, and the nuclear reactions proceed at a higher rate. A very important phenomenon then occurs: part of the increasing energy production goes into expanding the outer parts of the star. The total amount of energy is spread over the expanded surface area, and the surface temperature of the star decreases.

In sum, detailed calculations show that as soon as a main-sequence star forms a substantial core of helium, the outer layers grow slowly larger and redder. The star winds up giving off more energy (because the core temperatures and nuclear reaction rates are increasing). The star's total luminosity, which is the emission per unit area times the total area, is increased. The luminosity increases even though each bit of the surface is less luminous than before (as a result of its decreasing temperature) because the total surface area that is emitting increases rapidly.

The process proceeds at an ever-accelerating pace, simply because the hotter the core gets, the faster heat flows from it and the more rapidly it contracts and heats up further. The hydrogen shell gives off more and more energy. The layers outside this hydrogen-burning shell continue to expand.

Our star, the sun, is still at the earlier, main-sequence stage, where changes are stately and slow. But in a few billion years, when the hydrogen in its core (about 10 per cent of the hydrogen in the whole sun) is exhausted, the time will come for the sun to brighten and redden faster and faster until it eventually swells and engulfs Mercury and Venus. At this point, the sun will no longer be a dwarf but will rather be what is called a *red giant.* (The region of the H–R diagram where these stars are found is called the "red-giant branch.") The red giant's surface will be so close to earth (Fig. 25–3) that it will sear and char whatever is then here. A very strong stellar wind may form.

Though they may be devastating to the earth because they will come close, the outer layers of red giants are relatively cool for stars. Dust forms in them and shells containing molecules are also sometimes detected.

While the outer layers are expanding, the inner layers continue to contract and heat up, eventually reaching 10^8 K. At this point, the triple-alpha process (Section 24.4) begins for all but the least massive lightweight stars, as groups of three helium nuclei (alpha particles) fuse into single carbon nuclei. For stars about the mass of the sun, the onset of the triple-alpha process happens rapidly. The development of this *helium flash* produces a very large amount of energy in the core, but only for a few years. This input of energy at the center reverses the evolution that took place just prior to the helium flash: now the core expands while the outer portion of the star contracts (because it is no longer receiving as much energy from the core as it was before the helium flash). As the star adjusts to this situation, it becomes smaller and less luminous, moving back down and to the left on the H–R diagram to the "horizontal branch"; we see such a horizontal branch in the H–R diagram of a globular cluster (Section 22.5b). Helium now burns steadily in the core and during the next hundred million years is completely converted to carbon. During this time the star is stabilized, burning hydrogen and helium in separate shells; its core is made of carbon.

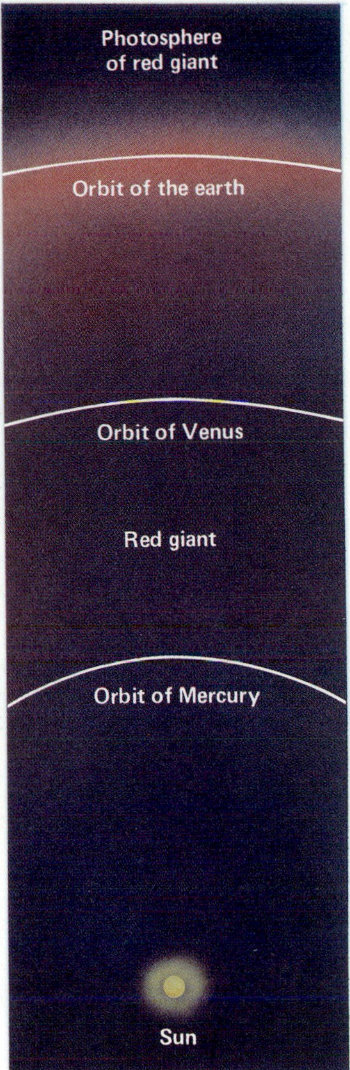

Figure 25–3 A red giant swells so much it can be the diameter of the earth's orbit.

25.2 Planetary Nebulae

When the carbon core forms in a red giant, it contracts and heats up as did the helium core before it. This drives up the rate at which hydrogen is being converted to helium and helium to carbon in shells surrounding the core. The star's parameters then are again at a point at the upper right on the H–R diagram. The star is now said to be on the second red-giant branch, also called the "asymptotic giant branch," asymptotic because in a cluster H–R diagram these stars form a sequence that nearly parallels and converges with the red-giant branch. The hydrogen-burning and helium-burning shells produce energy unstably. These instabilities cause the star to shed mass in a strong stellar wind, and also produce stellar pulsations in radius.

In one model, as the pulsations grow, the star's outer layers also become unstable. They lose a small fraction of their material with each pulse. In a very short time (perhaps 1000 years) the entire envelope has been ejected. On this time scale, the ejected gas has moved several hundred Astronomical Units away from the star, and has spread into a shell thin enough to be transparent. A typical temperature for such a shell is 10,000 K. We see it shining because it is ionized by ultraviolet radiation from the hot exposed core of the original star, now the "central star" of a *planetary nebula.* The light we see is the result of electrons rejoining the ions, forming excited atoms. The excited atoms often give off some of their energy as light.

An alternative model also explains how such bright shells of gas may arise. In this model, the expanding outer layers of the red giant form a stellar wind, an analogy to the solar wind. As the core is bared, the wind speed increases; the inner, faster moving material then piles up against the outer, slower moving material. The dense shell of gas that results from this snowplow effect, ionized by ultraviolet radiation from the central star, is the shell of gas that we observe.

We know of about a thousand such objects in our galaxy that can be explained by the above theoretical models, and there may be 30,000 in all. They were named "planetary nebulae" because two hundred years ago, when they were discovered, they appeared similar to the planet Uranus when viewed in a small telescope. Both the planet and "planetary nebulae" appeared as

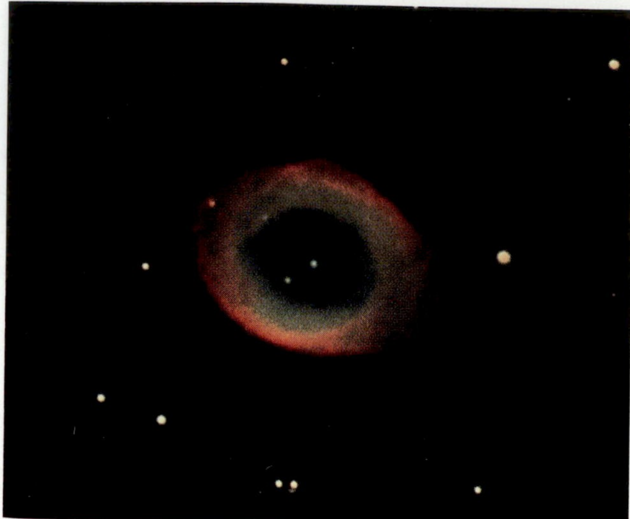

Figure 25–4 The Ring Nebula in Lyra, M57. In a small telescope it looks like a small, hazy smoke ring; long exposures are necessary to bring out the colors. Red hydrogen radiation is visible around its outer edge; green radiation from ionized oxygen shows in the center. Its central star appears distinctly.

Figure 25–5 The Dumbbell Nebula, M27, a planetary nebula in the constellation Vulpecula. Its diameter in the sky is over one-fourth that of the moon. Radiation from the hot blue central star provides the energy for the nebula to shine.

small, greenish disks. The nebulae's color is caused by the presence of certain strong emission lines of multiply (pronounced "mul'ti-plee") ionized oxygen (that is, oxygen that has lost more than one electron) and other elements. We can determine the chemical composition of the nebula by studying these emission lines.

Planetary nebulae are exceedingly beautiful objects (Figs. 25–4 and 25–5). They are actually semitransparent shells of gas. When we look at their edges, we are looking obliquely through the shells of gas, and there is enough gas along our line of sight to be visible. But when we look through the centers of the shells, there is less gas along our line of sight, and the nebulae appear transparent. In the middle of most planetary nebulae, we see the central star from which the nebula was ejected. Central stars are very hot, some as hot as 100,000 K, so they appear high up on the left side of the H–R diagram.

Astronomers are particularly interested in planetary nebulae because they want to study the various means by which stars can eject mass into interstellar space. Cornell scientists, using the large array of radio telescopes known as the VLA (Fig. 25–6), have found Doppler shifts that show the velocities of gas in a planetary nebula. Such velocities are also studied on optical spectra. The material in a planetary nebula contains heavy elements that had been "cooked" inside the original star (Fig. 25–6) and brought to the surface by convection (the transfer of energy from hot gas rising as cooler gas sinks because of gravity) during the red-giant phase. Thus the study of planetaries tells astronomers how a star can reduce its own mass. It also indicates something about the origin of interstellar matter and how it is enriched with heavy elements, since some planetary-nebula shells include material that underwent nuclear processing inside the progenitor star. Each planetary nebula represents the ejection of 10 to 20 per cent of a solar mass, which is only a small fraction of the mass of the star. However, statistically, about one planetary nebula forms in our galaxy per year; hence, within just one century, 10 to 20 solar masses of processed material enrich the interstellar medium.

Cornell scientists, using the large array of radio telescopes known as the VLA (Fig. 25–7), have determined the expansion velocity of a planetary nebula by observing at 5-year intervals. Such expansion velocities are also studied on optical spectra, but not with such high angular resolution.

We can measure the ages of some planetary nebulae by tracing back the shells at their current rate of expansion and calculating when they would have

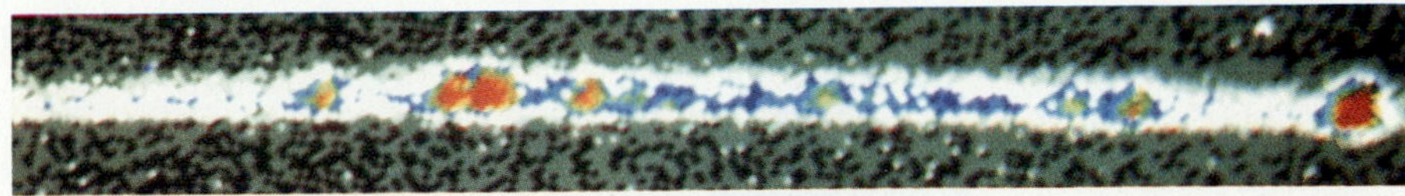

Figure 25–6 An ultraviolet spectrum from the International Ultraviolet Explorer showing emission lines of "metals" (elements heavier than hydrogen and helium) like nitrogen in the spectrum of the planetary nebula NGC 7027.

been ejected from the star. Ages derived in this way have large uncertainties, since many effects could cause the expansion to speed up or slow down. Some planetary nebulae have multiple shells, so may have erupted more than once. Estimates for ages also come from observations of the central stars, interpreted by comparing their measured positions on the H–R diagram with theoretical calculations of how these positions vary with age. Most of the planetary nebulae are less than 50,000 years old. The data fit with our picture: after a longer time, the nebulae will have expanded so much that the gas will be invisible and the central star may have cooled off enough that it is now unable to ionize gas and so cause the nebula to glow. Then only the central, contracted hot star is left to see. We identify it as a white dwarf star, as we will discuss next. The path is particularly interesting since the sun may take it one day, though not for 5 billion years or so.

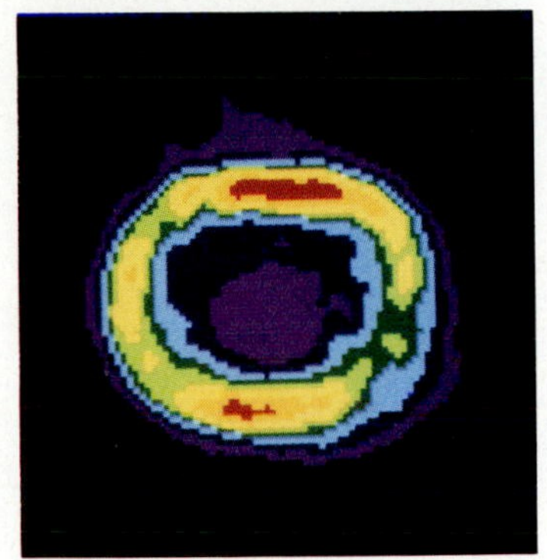

Figure 25–7 A VLA map of the young planetary nebula BD+30°3639, part of an observational project of Yervant Terzian of Cornell. The colors represent intensity, with red as the brightest followed by yellow, green, and blue. The minor axis is only 5 arc sec across.

25.3 White Dwarfs

We have seen a lightweight star transform all the hydrogen in its core to helium and then all the helium in its core into carbon. The shell is ejected as a planetary nebula; now let us consider the core. Because it does not have enough mass, it does not heat up sufficiently to allow the carbon to fuse into still heavier elements. There comes a time when nuclear reactions are no longer generating the energy needed to maintain the internal pressure that balances the force of gravity.

All stars that at this stage in their lives contain less than 1.4 solar masses have the same fate. Many or all stars in this low-mass group come from the red-giant phase, and may pass through the phase of being central stars of planetary nebulae. Other low-mass stars apparently do not go through a planetary-nebula stage, though we do not know why. Those that do not go through a planetary nebula stage lose parts of their mass in some other way, perhaps during the giant stage. In any case, the low-mass stars lose large fractions of their original masses. Stars that contained 4 solar masses when they were on the main sequence probably now contain only 1.4 solar masses.

When their nuclear fires die out for good, the stars with less than 1.4 solar masses remaining shrink in size, and reach a stable condition that we shall describe below. As they shrink, they grow very faint (in the opposite manner to that in which red giants grew brighter as they grew bigger). Whatever their actual color, all these stars are called *white dwarfs.* The white dwarfs occupy a region of the Hertzsprung-Russell diagram that is below and to the left of the main sequence (Fig. 25–8).

Do not confuse the term "white dwarf" with the term " dwarf." The former refers to the dead hulks of stars in the lower left of the H-R diagram, while the latter refers to normal stars on the main sequence.

White dwarfs represent a stable phase in which stars of less than 1.4 solar masses live out their old age. The value of 1.4 solar masses is known as the *Chandrasekhar limit* after the astronomer S. Chandrasekhar (known far and wide as "Chandra"). Chandra derived the concept and the value in 1930 when he was 19 years old, while on a boat from India to England to go to college; he later joined the faculty of the University of Chicago. He shared in the 1983 Nobel Prize in Physics for the discovery.

Chandrasekhar reasoned that something must be holding up the material in the white dwarfs against the force of gravity; nuclear reactions generating thermal pressure no longer take place in their interiors. The property that holds

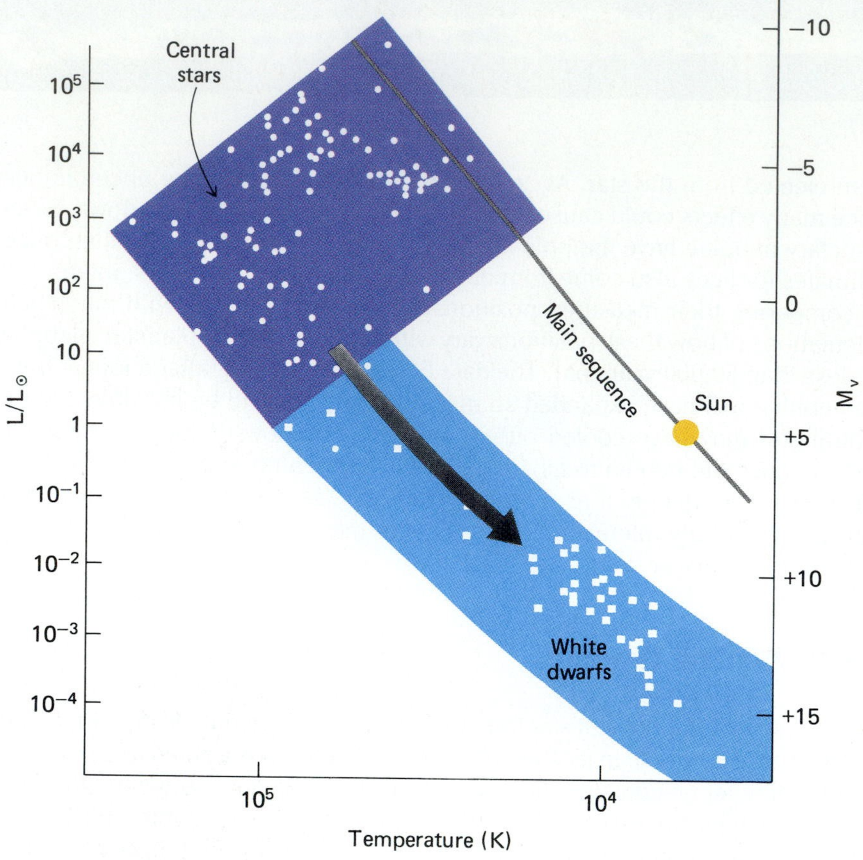

Figure 25–8 When the planetary nebula expands around a red giant, the core of the star becomes visible at the center of the nebula. Since this *central star* has a high temperature, it appears to the left of the main sequence. These stars eventually shrink and cool to become white dwarfs, as shown on this H–R diagram.

up the white dwarfs is a condition called *electron degeneracy;* we thus speak of "degenerate white dwarfs."

Electron degeneracy is a condition that arises in accordance with certain laws of quantum mechanics (and is not something that is intuitively obvious). As the star contracts and the electrons get closer together, there is a continued increase in their resistance to being pushed even closer. This shows up as a pressure. At very great densities, the pressure generated in this way exceeds the normal thermal pressure. When this pressure from the degenerate electrons is sufficiently great, it balances the force of gravity and the star stops contracting.

Thus, the effect of the degenerate electron pressure is to stop the white dwarf from contracting; the gas is then in a very compressed state. In a white dwarf, a mass approximately that of the sun is compressed into a volume only the size of the earth (Fig. 25–9). (This great collapse is possible because atoms are mostly empty space: the nucleus takes up only a very small part of an atom.) A single teaspoonful of a white dwarf weighs 10 tons; it would collapse a table if you somehow tried to put some there. A white dwarf contains matter so dense that it is in a truly incredible state.

What will happen to the white dwarfs with the passage of time? The pressure from degenerate electrons doesn't depend on temperature, so the stars are stable even though no more energy is ever generated within them. Because of their electron degeneracy, they can never contract further. Still, they have some energy stored, and that energy will be radiated away over the next billions of years. Then the star will be a burned-out hulk called a *black dwarf,* though it is probable that no white dwarfs have yet lived long enough to reach that final stage. It will be billions of years before the sun becomes a white dwarf and then many billions more before it reaches the black dwarf stage.

Figure 25–9 The sizes of the white dwarfs are not very different from that of the earth. A white dwarf contains about 300,000 times more mass than does the earth, however.

25.4 Observing White Dwarfs

White dwarfs are very faint and thus are difficult to detect. We find them by looking for bluish (hot) stars with high proper motions or, in binary systems, by noting their gravitational effect on the companion stars.

In the former case, by studying proper motions we find the stars that are close to the sun. Some are fainter than main-sequence stars would be at their distances; they must be white dwarfs. (White dwarfs at greater distances would be too faint for us to see.) To understand the latter case, we must realize that for any system of two masses orbiting each other, we can define an imaginary point, called the *center of mass,* which moves in a straight line across the sky. The individual bodies move around the center of mass, however, so the path in the sky of any of the individual bodies appears wavy. We have already discussed such astrometric binaries in Section 22.1.

At least three of the 40 stars within 5 parsecs of the sun (Appendix 7)—Sirius (Fig. 25–10), 40 Eridani, and Procyon—have white dwarf companions. Another nearby object, known as van Maanen's star, is a white dwarf, although not in a multiple system. Hundreds of white dwarfs are known. So even though we are not able to detect white dwarfs at great distances from the sun, there seems to be a great number of them.

White dwarfs have been observed with temperatures as low as 4000 K and as high as 200,000 K (Fig. 25–11). Since white dwarfs are the cores of stars, revealed as the outer layers were lost, they are made of helium, carbon, and heavier elements. Some have atmospheres of hydrogen, but others have atmospheres composed entirely of helium. Their evolution is not always as straightforward as the path described above. Later in this chapter, we will discuss interchange of matter between members of binary systems. Further, even solitary white dwarfs are affected by mass loss.

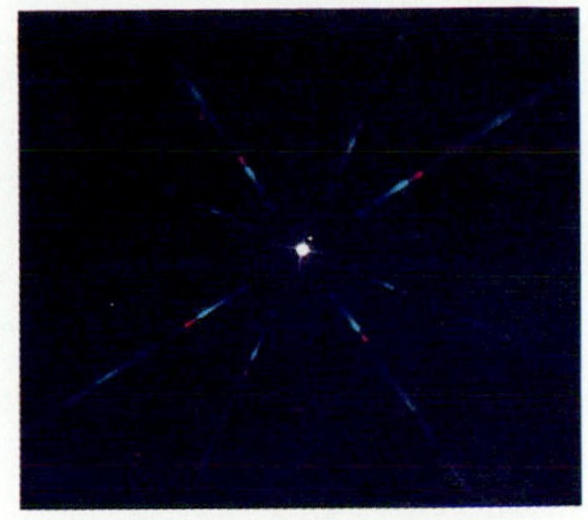

Figure 25–10 Sirius A with its companion white dwarf Sirius B, and diffracted spectra.

The Extreme Ultraviolet Explorer spacecraft, EUVE, was launched by NASA in 1992 to study the spectral region where many white dwarfs put out most of their radiation. Radiation from the hot disks of gas around the white-dwarf member of a binary system should be especially bright.

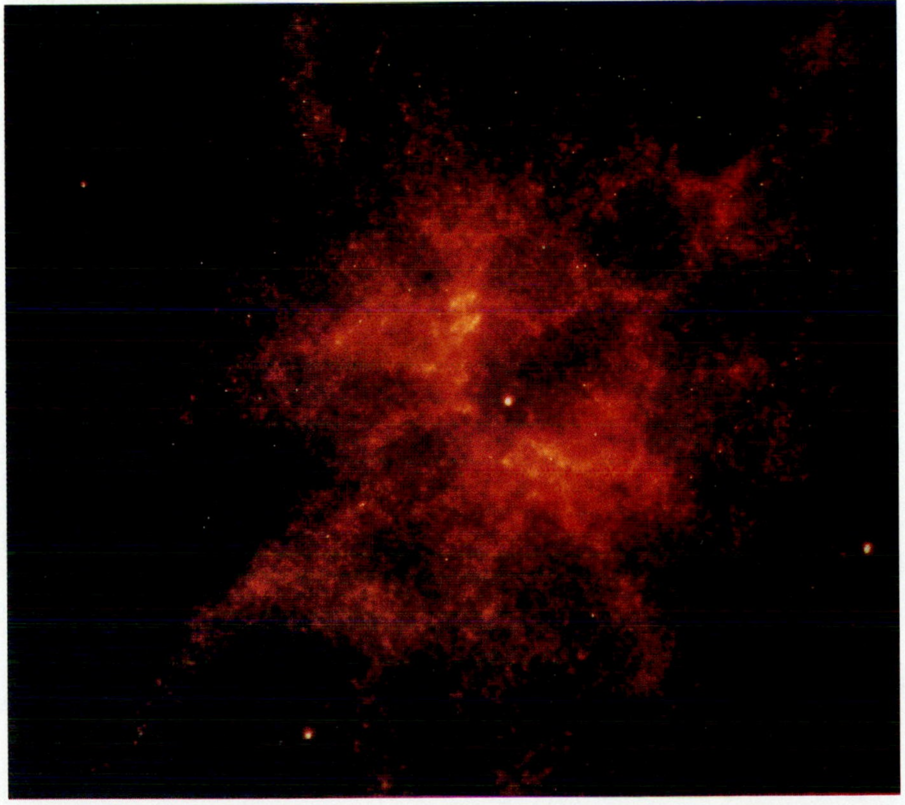

Figure 25–11 The Hubble Space Telescope reveals here one of the hottest known stars, the central star of NGC 2440. This white dwarf, the white dot in the center of the photograph, has a surface temperature of 200,000 K. Previous views of the nebula had blurred the light from the star, preventing the star from being seen.

Section 25.7 describes what it is like to observe with the Hubble Space Telescope.

*25.5 White Dwarfs and the Theory of Relativity

One special reason to study white dwarfs is to make use of them as a laboratory to test extreme physical conditions. It is impossible to create such strong gravity in a laboratory on earth. We have already discussed two tests of general relativity in Section 23.11: the deflection of starlight by the sun, and the advance of the perihelion of the planet Mercury.

A third test, the *gravitational redshift* of light, is best carried out with white dwarfs. Einstein's theory predicts that light leaving a mass will be redshifted by an amount that depends on the amount of mass.

The effect is minuscule on the sun, though it has been detected; the detection has recently been verified at the National Solar Observatory. Furthermore, turbulence in the solar photosphere, such as the rise and fall of granules, distorts and confuses the solar results. Gravity on a white dwarf's surface is much stronger than gravity on the surface of the sun, since the surface of a white dwarf is so much closer to the star's center.

Unfortunately, one must know the mass and size of the white dwarf to perform the test accurately, and the masses and sizes of white dwarfs are not known very well. Still, strong redshifts are found. The results agree with the predictions of the theory of relativity to within the possible error that results from our uncertain knowledge of the masses and sizes of white dwarfs, which we know to about 20 per cent. The test has best been carried out with 40 Eridani B.

25.6 Novae

Although the stars were generally thought to be unchanging on a human time scale, occasionally a "new star," a *nova* (from the Latin for "new"; plural: *novae*), became visible. Such occurrences have been noted for thousands of years; ancient Oriental chronicles report many such events. Only in recent years have we found out that white dwarfs are at the center of the nova phenomenon. (A few of the events historically known as "novae" are now realized to be a grander type of event we call "supernovae"; they will be discussed in the next chapter.)

A nova is a newly visible star rather than actually a new star. A nova is, in fact, a brightening of a star by 5 to 15 magnitudes or more, making it visible to the eye or to the telescope.

A nova (Figs. 25–12 and 25–13*A*) may brighten within a few days or weeks. It ordinarily fades drastically within months, and then continues to fade gradually over the years. Besides these "classical" novae, "dwarf novae" brighten at intervals of months by smaller factors.

Figure 25–12 Nova Herculis 1934 (DQ Her), showing its rapid fading from 3rd magnitude on March 10, 1935, to below 12th magnitude on May 6, 1935.

Novae occur in binary systems in which one member has evolved into a white dwarf while another member is usually on the path toward becoming a red giant. The two are separated by about the distance between the earth and the moon. We have mentioned that the outer layers of a red giant are not held very strongly by the star's gravity. If a white dwarf is nearby, some of the matter originally from the red giant goes into orbit around the white dwarf rather than falling straight onto its surface. The orbiting gas forms a disk around the white dwarf. The disk gives off ultraviolet and x-rays that we detect from satellites, and irregularities in it cause the light from the system to flicker rapidly, on a time scale of seconds.

In a "classical" nova, some of this orbiting material falls onto the white dwarf's surface. After a time, enough builds up and enough energy is deposited to trigger nuclear reactions on the star's surface for a brief time. This runaway

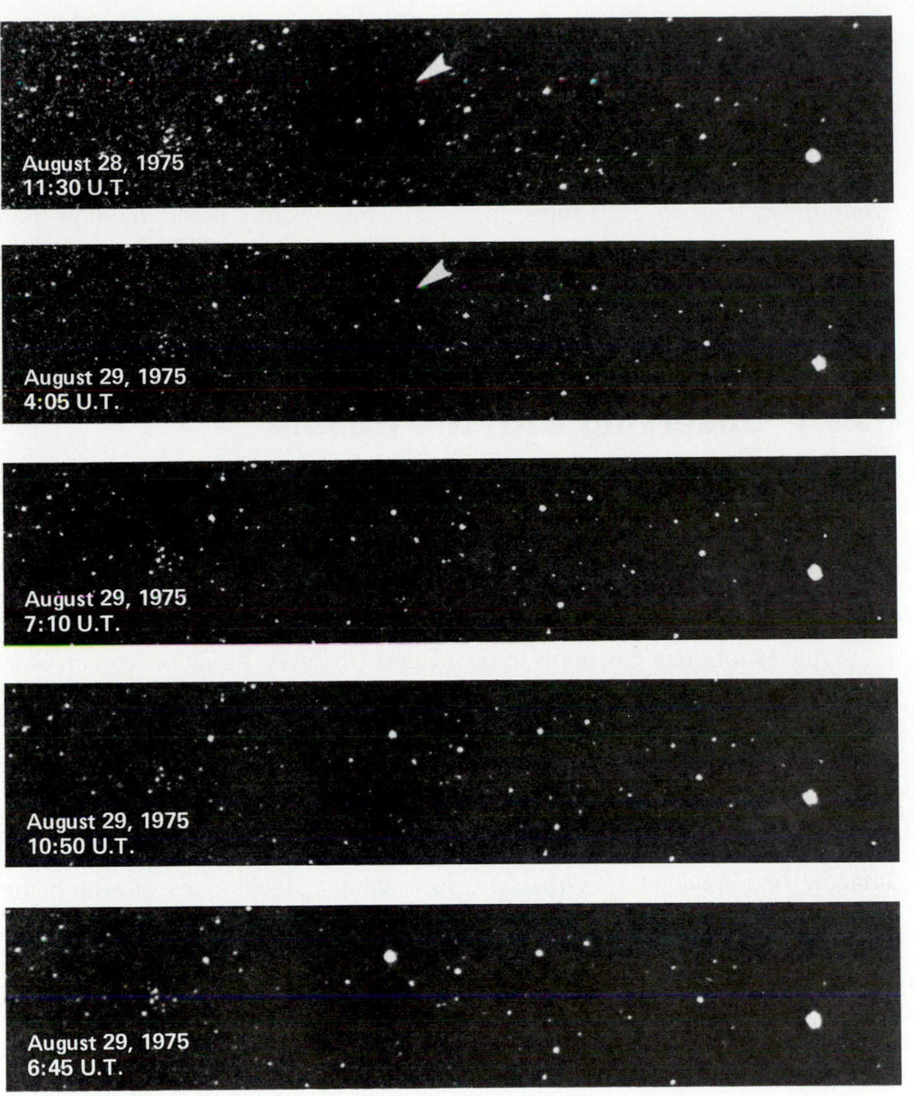

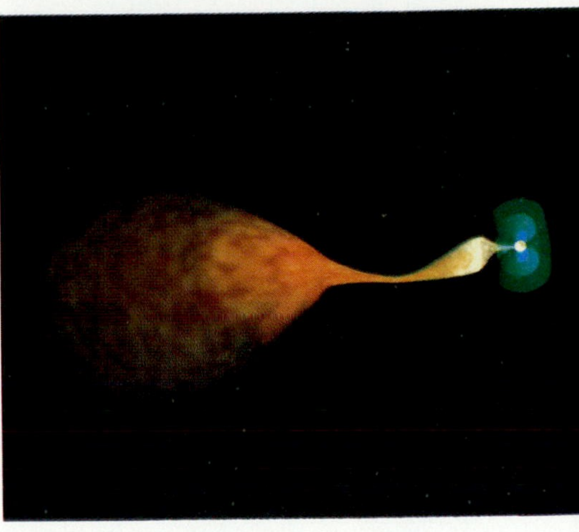

Figure 25–13 (*A*) These photographs are part of a unique series of observations covering the eruption of Nova Cygni 1975 during the period of its brightening. A Los Angeles amateur astronomer, Ben Mayer, was repeatedly photographing this area of the sky at the crucial times to search for meteors. When he heard of the nova, he retrieved his meteor-less film from the wastebasket. Never before had a nova's brightening been so well observed. On August 28 (*top*), no star was visible at the arrow. By August 30 (*bottom*), the nova had reached 2nd magnitude, the brightness of Deneb, which is seen at the right.

Mayer has now organized an international amateur effort, Problicom (**Pro**jection **Bli**nk **Com**parator), in which each participant photographs a region of sky every few weeks and checks over the film using a simple combination of two ordinary slide projectors. Volunteers are welcome.

(*B*) Nova Cygni 1975 was so unusually bright because it is the first nova that we know came from the explosion of a white dwarf with a very high magnetic field. This computer-graphics image shows an accretion stream flowing to the white dwarf from its companion. The white dwarf's magnetic field is shown in green around it.

hydrogen burning causes the brightening we see as a nova; it lasts until all the hydrogen is consumed. Years after the outburst of light, the shell of gas blown off sometimes becomes detectable through optical telescopes.

Nova Cygni 1975, which brightened by 20 magnitudes to its 1st-magnitude maximum, continues to be studied. Spectroscopic studies made when the nebular spectral lines had faded suggest the presence of two objects. One is a main-sequence star of 0.3 solar masses. The other is a white dwarf with mass between 0.9 and 1.4 solar masses.

Only 14 of the hundreds of known binary systems including white dwarfs have high magnetic fields, which probably explains why Nova Cygni 1975 was so unusually bright and why it soon began to oscillate in brightness with a 3-hour 20-minute period. The flow of matter from the companion to a magnetic white dwarf (Fig. 25–13*B*) piles up at the white dwarf's poles. Ignition then occurs at a magnetic pole and matter and radiation shoot off along a magnetic axis. The 3-hour 20-minute rotation period that swept this beam past us explained the periodicity of brightness that was detected on earth soon after the original nova fading began.

Figure 25–14 A cutaway view of the Hubble Space Telescope and its scientific instruments.

*25.7 Observing with the Hubble Space Telescope

Astronomers often thrill to use the best and the latest equipment. Even with its well-publicized faults, the Hubble Space Telescope meets the description (Fig. 25–14). Hundreds of astronomers are using it, and many of the data that result are exciting.

Any astronomer can apply to be a Guest Observer. About once each year, an announcement is sent out from the Space Telescope Science Institute setting an application deadline for the next cycle. The notice gives about four months' warning for the application date. In that time, you must put together a substantial proposal. Of course, the heart of the proposal is the science itself that you propose to do. You must propose to do something that can be done uniquely on the Hubble Space Telescope. So your proposal will involve high-resolution imaging or ultraviolet spectroscopy, for example, techniques at which the Hubble Space Telescope excels.

The proposal must evaluate the details of the observation with the Hubble Space Telescope. You must provide a list of sources that you will observe—a set of quasars, for example—and a calculation of how much time it will take to observe them. To compute the time, you must show how bright the objects are and use measurements of the detailed capabilities of HST for the type of observation intended. You might have to change the detail to which you can obtain a spectrum, for example, if it would take too long in the preferred observing mode. Often, you will consult with some of the scientists on the staff of the Space Telescope Science Institute about the details of the proposal.

At the deadline, about 600 proposals are typically received, asking for six times more observing time than is available. After all, HST is only one telescope, and not a very large one at that by the standards of ground-based observing. Further, because of the main mirror's spherical aberration, even when an observation is possible, it often takes longer to carry out than it would if the light could be better focused. When the applications are received at STScI (the lower case "c" in the acronym emphasizes that it is a science institute, rather than one of the instruments, which may have an acronym entirely in capital letters) they are sent out to several dozen scientists around the world who act as peer "referees." Each may have one or two dozen proposals to evaluate. Back at STScI, a committee considers the referees' evaluations and adds its own ratings of both scientific value and feasibility. Eventually, a final set of acceptances is decided upon and is forwarded to the Director for final approval. These proposals are for the most part divided among the major instruments on board. Other proposals may simply use the star trackers on board or older data that are "archived." All HST data becomes generally available in the archive after a year; the scientist for whom the data were taken is given only that limited window of exclusive availability. Still another type of observation is the "snapshot" mode, which takes advantage of the periods of minutes between scheduled observa-

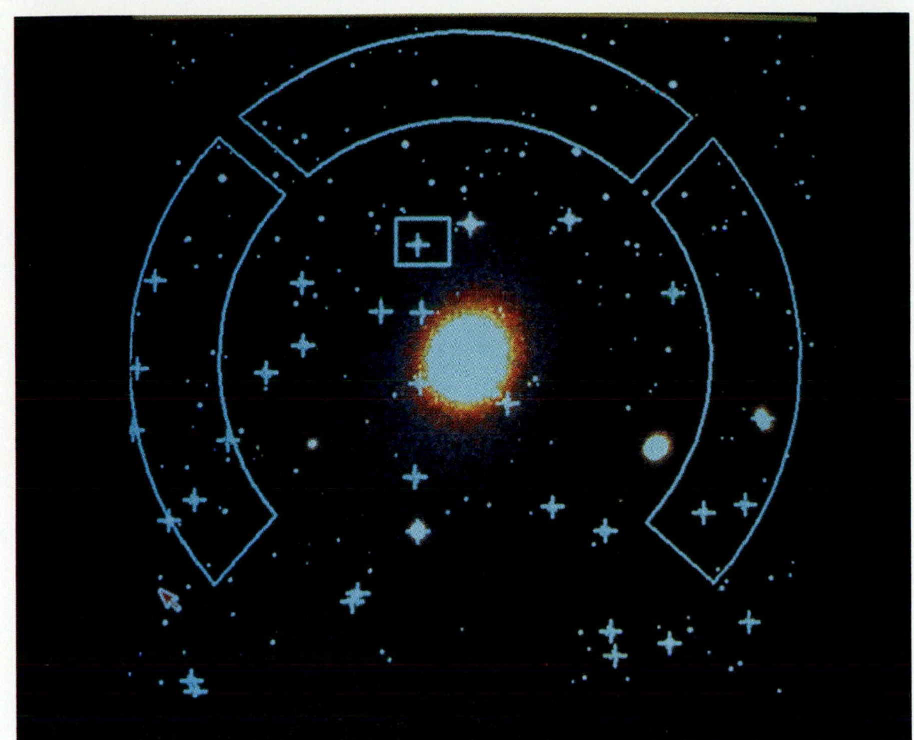

Figure 25–15 The catalogue of guide stars can be displayed on a video screen and compared with an image digitized from an astronomical photograph. Pairs of guide stars must be located in the curved regions.

tions. It might take too long for the telescope to turn to the next object in an observing program; therefore, a set of galaxies and quasars have been tabulated all over the sky so that one of them is always nearby to be looked at briefly.

After acceptance and in the months before you arrive, your plans are considered by various committees and individual scientists at the Space Telescope Science Institute. One important detail is the selection of the guide stars that must be in the field of view for each observation (Fig. 25–15). The telescope locks its view onto them in order to hold itself steady. Major surveys of the sky were undertaken and then digitized to provide the information for choosing the vital guide star.

Eventually your plans are meshed with those of other successful applicants, and the time of the telescope is scheduled second by second. The general ideas that you have submitted must be turned into commands such as "turn telescope 17° west, open shutter, start science tape recorder, shut shutter, dump data to science tape recorder, turn tape recorder off," and so on. The telescope must be programmed not only to find the objects to observe but also, since it is in a low orbit around the Earth, to never look towards the Earth. Looking at the Earth, as well as at other bright objects like the Sun and the Moon, would not only block the view but also damage the observing instruments.

When your observing time actually arrives, you may, like many observers, choose to travel to STScI, which is on the Homewood Campus of the Johns Hopkins University in Baltimore. There, staff scientists and technicians will help you prepare to interpret the data. If you choose, on the other hand, to remain at your own office, you can receive similar aid there via computer link-up.

After your observation, the data are relayed to Earth via intermediate satellites and then taken out of the stream of "housekeeping" and other data being beamed down by HST. Your data should arrive at STScI (Fig. 25–16) within 24 hours. In addition to the copy you are given—often on an 8-mm tape cartridge, the same size cartridge used in many video cameras—copies are archived on a dual set of video disks, one for Baltimore and the other for the European center for Space Telescope.

Figure 25–16 The observing support room at the Space Telescope Science Institute.

STScI's Research Support Branch includes Science Data Analysts who work with Instrument Scientists to draw up a plan to help you reduce and analyze the observations. Most of the data reduction is be on computer workstations, and normally continues at your home institution after you leave Baltimore. Computer methods are available that adjust for the defect in the main mirror and can, in many cases, improve the resolution of your images (Fig. 25–17).

As months pass, you will prepare your data to discuss publicly. In addition to meetings of the American Astronomical Society, workshops and colloquia are held at STScI to disseminate the results. Eventually, as you understand the data, STScI's reduction, and their implications well enough, you can write a

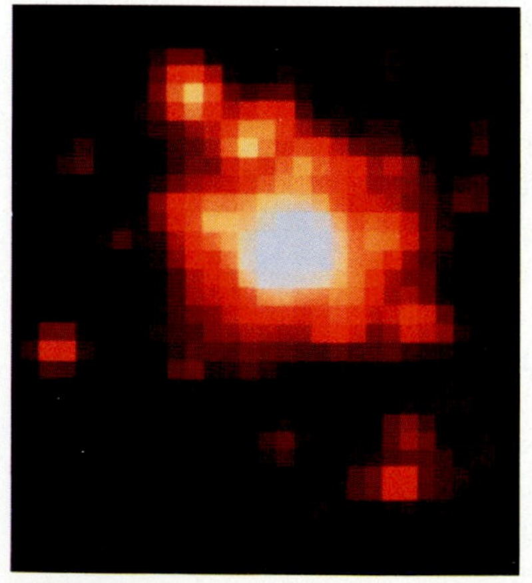

A

B

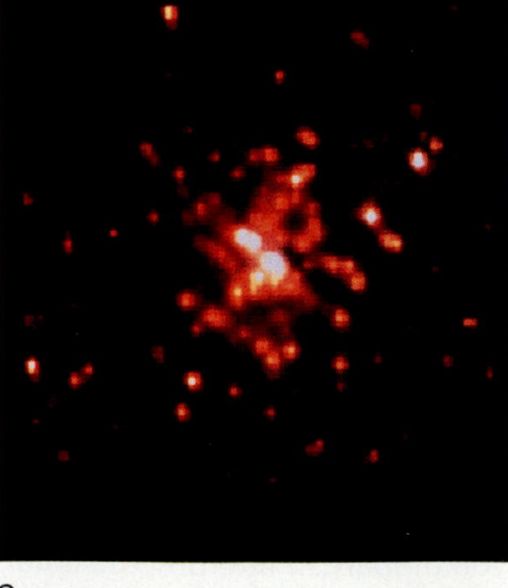

C

Figure 25–17 Stages in the data reduction of R136a, once thought from ground-based observations (*A*) to be possibly the largest, brightest, and most massive star known. The complex is in the Large Magellanic Cloud. Hubble Space Telescope observations (*B*), which have higher resolution than any observations taken with telescopes on the Earth, have shown that R136a is really a cluster of bright stars. The data as originally available are in part *B*. Computer programs have been able to make mathematical solutions that improve the resolution of the images, and so reveal the cluster of stars more clearly (*C*).

formal article for a scientific journal. The *Astrophysical Journal* is the usual choice for American astronomers. With public talks and publications your discoveries find their way to the astronomical community and to the world at large.

Summary and Outline

Red giants (Section 25.1)
- They follow the end of hydrogen burning in the core. The core contracts gravitationally; the core and the hydrogen-burning shell become hotter; the star is red, so each bit of surface has relatively low luminosity but the surface area is very greatly increased, so the total luminosity is greatly increased.
- Helium flash: rapid onset of triple-alpha process

Planetary nebulae (Section 25.2)
- Ages: less than 50,000 years old; the gas is blown off when it absorbs photons; mass loss: 0.1 or 0.2 solar mass

White dwarfs (Sections 25.3, 25.4, and 25.5)
- End result of lightweight stars: less than 1.4 solar masses remaining; supported by electron degeneracy; detected by their proper motion or by their presence in a binary system
- They provide the best test of the gravitational redshift.

Novae (Section 25.6)
- Interaction of a red giant and a white dwarf

Hubble Space Telescope (HST) (Section 25.7)
- This NASA/European spacecraft is sending back exciting data

Key Words

lightweight stars, gravitationally, red giant, helium flash, planetary nebula, white dwarfs, central star, Chandrasekhar limit, electron degeneracy, black dwarf, center of mass, gravitational redshift*, nova (novae), differential effect, tidal force, Roche lobes, contact binaries, detached binaries

*This term is found in an optional section.

Questions

1. What event signals the end of the main-sequence life of a star?

2. When hydrogen burning in the core stops, the core contracts and heats up again. Why doesn't hydrogen burning start again?

3. When the core first starts contracting, what halts the collapse?

4. How can a star reach its red-giant stage twice?

5. Why, physically, is there a Chandrasekhar limit?

†6. If you are outside a spherical mass, the force of gravity varies inversely as the square of the distance from the center. What is the ratio of the force of gravity at the surface of the sun to what it will be when the sun has a radius of one astronomical unit? (Use data in Appendix 2.)

7. Why is helium "flash" an appropriate name?

8. If you compare a photograph of a nearby planetary nebula taken 80 years ago with one taken now, how would you expect them to differ?

9. Why is the surface of a star hotter after the star sheds a planetary nebula?

10. What keeps a white dwarf from collapsing further?

11. What are the differences between the sun and a one-solar-mass white dwarf?

†This question requires a numerical solution.

12. When the sun becomes a white dwarf, approximately how much mass will it have? Where will the rest of the mass have gone?

13. Which has a higher surface temperature, the sun or a white dwarf?

14. Compare the surface temperatures of the hottest white dwarf and an O star. What spectral type of normal dwarf has the same surface temperature as the coolest white dwarf?

15. Sketch an H-R diagram indicating the main sequence. Show the evolution of the sun starting at the time it leaves the main sequence.

16. Compare the desirability of the sun and of 40 Eridani B for testing the gravitational redshift prediction of Einstein's general theory of relativity.

17. When the proton-proton chain starts at the center of a star, it continues for billions of years. When it starts at the surface (as in a nova), it lasts only a few weeks. How can you explain the difference?

18. Compare IUE with the Palomar Observatory, listing two advantages and two disadvantages of IUE.

19. Compare IUE with the Hubble Space Telescope, including two advantages and disadvantages of IUE.

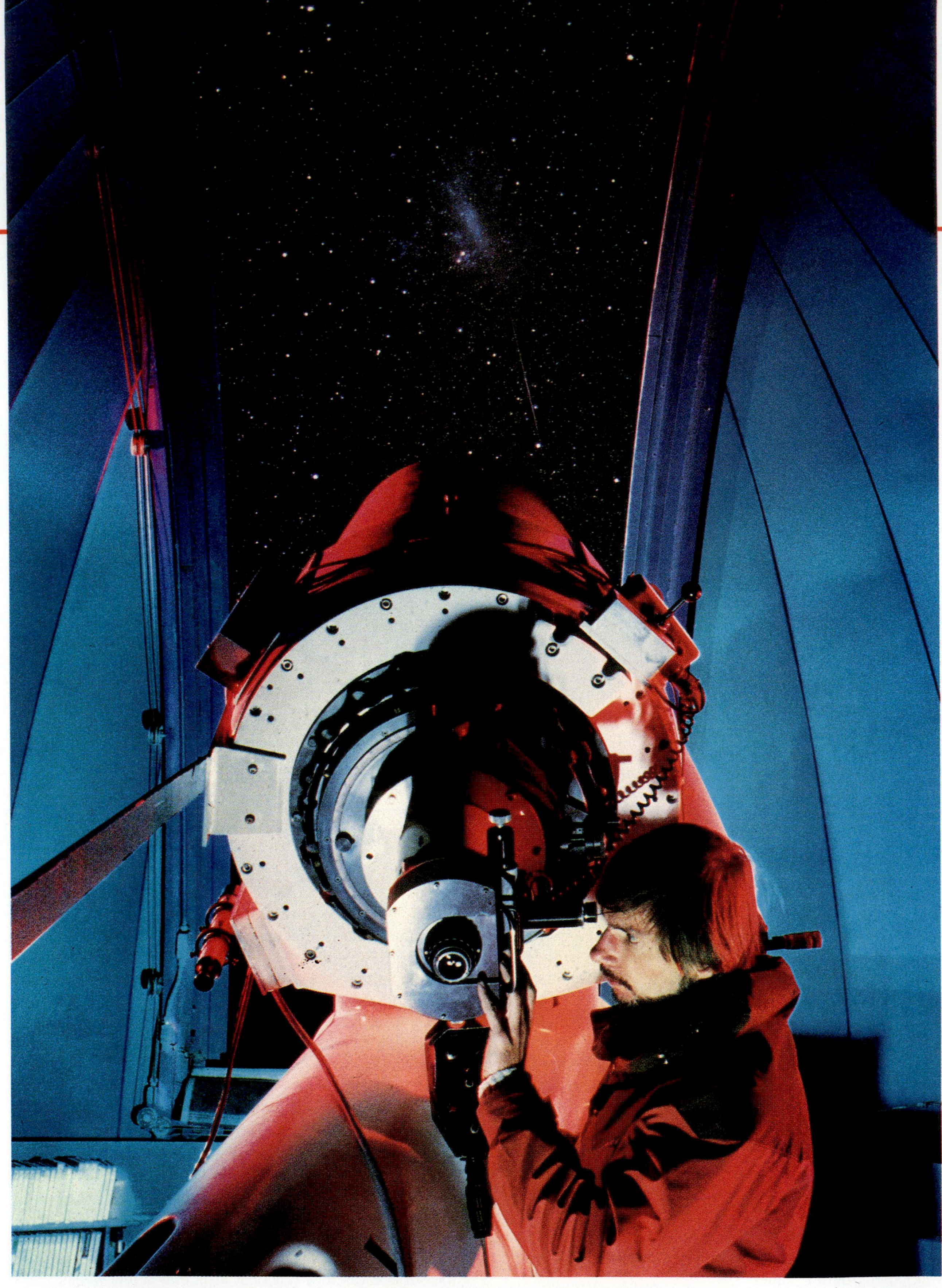

The Large Magellanic Cloud with Supernova 1987A shining brightly just above the Tarantula Nebula. The supernova was the brightest seen on earth since the year 1604. In the foreground is Ian Shelton, its discoverer, at the Las Campanas Observatory in Chile.

Supernovae

26

Aims: To see how stars become supernovae, some as exploding white dwarfs and others as very massive stars whose cores sometimes collapse to become neutron stars, and to learn about the nearby supernova whose light and neutrinos we received in 1987

We have seen how the run-of-the-mill lightweight stars end, not with a bang but a whimper. Still, when the white dwarfs that result are in binary systems, huge explosions called supernovae may yet result.

More massive stars, which contain more than about 8 solar masses when they are on the main sequence, put on a dazzling display more directly. These heavyweight stars cook the heavy elements deep inside. They then blow themselves almost to bits as supernovae, forming still more heavy elements in the process. The matter that is left behind in the stellar core settles down to even stranger states of existence than that of a white dwarf.

In this chapter we shall discuss supernovae in general, and then the appearance, starting in 1987, of the first supernova to be visible to the naked eye in almost 400 years. In the next chapter we shall see that some of the massive stars, after their supernova stage, become neutron stars, which we detect as pulsars and x-ray binaries. In the following chapter, we shall discuss the death of the most massive stars, which become black holes.

26.1 Red Supergiants

Stars that are much more massive than the sun whip through their main-sequence lifetimes at a rapid pace. These prodigal stars use up their store of hydrogen very quickly. A star of 15 solar masses may take only 10 million years from the time it first reaches the main sequence until the time when it has exhausted all the hydrogen in its core. This is a lifetime a thousand times shorter than that of the sun. When the star exhausts the hydrogen in its core, the outer layers expand and the star becomes a red giant.

For these massive stars, the core can then gradually heat up to 100 million degrees, and the triple-alpha process begins to transform helium into carbon. After its ignition, the helium then burns steadily, unlike the helium flash of less massive stars.

By the time helium burning is concluded, the outer layers have expanded even further, and the star has become much brighter than even a red giant. We call it a *red supergiant* (Fig. 26–1); Betelgeuse, the star that marks the shoulder of Orion, is the best-known example (Fig. 26–2). Supergiants are inherently very luminous stars, with absolute magnitudes of up to −10, one million times brighter than the sun. A supergiant's mass is spread out over such a tremendous volume, though, that its average density is less than one-millionth that of the sun.

The carbon core of a supergiant contracts, heats up, and begins fusing into still heavier elements. Eventually, even iron builds up. The iron core is

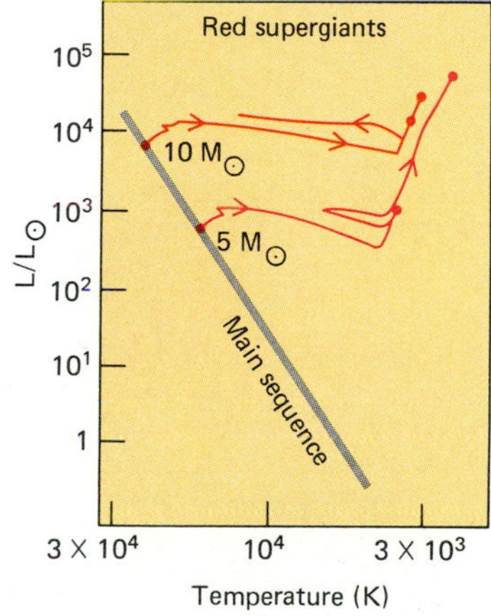

Figure 26–1 An H–R diagram (as in Fig. 25–2 for less massive stars) showing the evolutionary tracks of 5- and 10-solar-mass stars as they evolve from the main sequence to become red supergiants. For each star, the first dot represents the point where hydrogen burning starts, the second dot the point where helium burning starts, and the third dot the point where carbon burning starts.

Figure 26–2 The red supergiant Betelgeuse is the star labelled α in the shoulder of Orion, the Hunter, shown here in Johann Bayer's *Uranometria,* first published in 1603. Betelgeuse, visible in the winter sky, appears reddish to the eye.

Astronomers have recently used speckle interferometry to find two close companion stars to Betelgeuse. One is on an elliptical orbit that brings it within 1 A.U. of Betelgeuse, which may disrupt Betelgeuse enough to account for its variations in brightness.

surrounded by layers of elements of different mass, with the lightest toward the periphery.

26.2 Types of Supernovae

A dying star can explode in a glorious burst (Fig. 26–3) called a *supernova* (plural: *supernovae*). Two quite different situations apparently lead to supernovae (Table 26–1). The best current models indicate that *Type I supernovae* represent the incineration of a white dwarf (which results from a lightweight star). On the other hand, *Type II supernovae* take place in heavyweight stars, and show that stellar evolution can run away with itself and go out of control. (To help remember which is which, recall that the less massive star leads to the lower Roman number—I instead of II.)

The two types can be distinguished by their spectra and by the rate at which their brightnesses change. Rudolph Minkowski pointed out the distinction in the 1940's based on the observation that some supernovae—Type II—

Table 26–1 Types of Supernovae

	Type I	Type II
Source	White dwarf in binary	Massive star
Spectrum	No hydrogen lines	Hydrogen lines
Peak	1.5 magnitudes brighter than Type II	
	Sharper	Broader when graphed vs. time
Light curve	Rapid rise	
	Decay with several-week half-life	
	All same magnitude	Different magnitudes
Location	All type galaxies	Spiral galaxies only
Expansion	10,000 km/sec	5,000 km/sec
Radio radiation	Absent	Present

have prominent hydrogen lines when they are at maximum light, while hydrogen lines are absent in the others—Type I. Type I are more often seen because they are 1.5 magnitudes (4 times) brighter than Type II supernovae and take place in all types of galaxies. Type II supernovae are seen only in spiral galaxies. Indeed, they are usually found near the spiral arms, which endorses the idea that they come from massive stars. (Since the massive stars are short-lived, they are found in the spiral arms near where they were born.) In both cases, Doppler-shifted absorption lines show that gas is expanding rapidly: about 10,000 km/sec for Type I's and half that for Type II's.

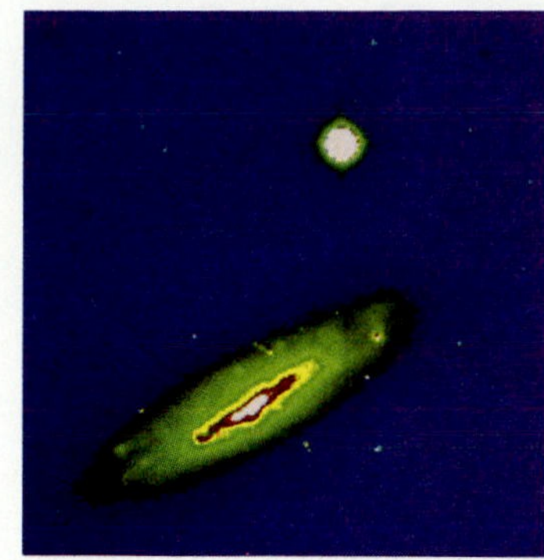

Figure 26–3 A false-color view of Supernova 1987D, which is almost as bright as the galaxy it is in. The faint spiral arms of the galaxy, which include the supernova, are not visible.

In the current leading model for Type II supernovae, a substantial core of heavy elements has formed in a massive star (perhaps 8–12 $M_\odot$), and begins to shrink and heat up. The heavy elements represent the ashes of the previous stages of nuclear burning. The stages of oxygen and magnesium "burning" to form heavier elements take less than 1000 years. Finally, silicon and sulfur "burn" to iron in only a few days. The conditions in the center of the star change so quickly that it becomes very difficult to model them satisfactorily in sets of equations or on computers, but the new supercomputers have enabled astronomers to keep up with the rapidly changing conditions; whether the following model ultimately will prove correct is uncertain. In this model, the temperature becomes high enough for iron to form and then to undergo nuclear reactions. The stage is now set for disaster because iron nuclei have a fundamentally different property from other nuclei when undergoing nuclear reactions. Unlike other nuclei, iron absorbs rather than produces energy in order to undergo either fusion or fission. So processes that release energy no longer can build up heavier nuclei as the core shrinks this time. Iron cores are formed in stars more massive than 12 $M_\odot$ by other processes, and these more massive stars then join the iron-collapse scenario.

$\odot$ is the symbol for the sun. 8 $M_\odot$, read "eight solar masses," means "8 times as much mass as the sun has."

The temperature climbs beyond billions of kelvins, and the iron is broken up into alpha particles (helium nuclei), protons, and neutrons by the high-energy photons of radiation that are generated. Since these processes absorb energy, the pressure at the center of the star diminishes, and the core begins to collapse. The process goes out of control. Within milliseconds—a fantastically short time for a star that has lived for millions and millions of years—the inner core collapses and heats up catastrophically.

As the outer core falls in upon the collapsing inner core, electrons and protons combine to make neutrons and neutrinos (Sections 24.3 and 24.6). The core collapses and (calculations show) bounces outward, since its gas has a natural barrier at nuclear densities against being compressed too far.

The collision of the rebounding inner core with the supersonically collapsing outer core sends off shock waves that cause heavy elements to form and that throw off the outer layers. The shock wave leaves the core and starts the explosion in less than 0.01 second, so the process is called a "prompt explosion." However, if the iron core is sufficiently large, the expanding shock wave loses too much energy to the infalling outer core. The neutrinos have more than enough energy, so they may play a role in blowing off the outer layers or at least in stopping the core from collapsing completely. The effect takes longer, and so is called the "delayed explosion." In any case, perhaps from a combination of prompt and delayed explosions, the star is destroyed. Only the core may be left behind.

The heavy elements formed either near the center of the star or in the supernova explosion are spread out into space, where they enrich the interstellar gas. Though most Type II supernovae come from stars containing between 8 and 12 solar masses, they don't supply as much of the heavy elements as Type II supernovae from still more massive stars do. In any case, when a star forms out of gas enriched by supernovae of both types, the heavy elements are present. Our understanding of element formation in Type II supernovae is

Figure 26–4 The Eta Carinae Nebula, a nebulae around a massive star that may be a supernova in not too long. Eta Carinae is 9000 light years away. A supernova there would be too far away to affect us, but would appear brighter than Venus.

The star η Carinae exploded in 1843; it is in the brightest part of the nebula seen here. The nebulae around it are expanding outwards from the explosion.

The photograph shows both red emission from Hα and blue reflection from hot stars. Some dust regions are seen as dark silhouettes, or by their reddening effect on light from behind it.

circumstantial, since the outburst itself is hidden from our view by the outer layers of the massive star; we can only study the transformed material later on, when it has spread out considerably.

Spectroscopic studies of the star η (eta) Carinae (Fig. 26–4) show that a gas condensation spewed out 150 years ago contains a relatively large abundance of nitrogen. This indicates that the star has processed material in the carbon-nitrogen cycle, brought it to the surface as the star's material mixed up, and ejected it. This is expected from a massive star that will become a supernova soon—though "soon" could be next year or in 10,000 years.

Though both involve white dwarfs, astronomers have no trouble distinguishing between a nova and a Type I supernova. The peak value of the light curve and the rate of fall-off of intensity are very different. The addition of a small amount of mass, which makes a nova, leaves the star basically unchanged, while the addition of enough mass to make a supernova disrupts the entire star.

In contrast to Type II supernovae, a Type I supernova apparently takes place in a binary system containing a white dwarf when mass from the second star (perhaps also a white dwarf) falls onto the white dwarf. Type I supernovae thus take place in the late stages of lightweight stars. The extra mass puts the white dwarf up to the 1.4-solar-mass Chandrasekhar limit, which leads to the star's thermonuclear explosion. After the sudden brightening, the light fades drastically for a month and then begins a slower decline. The energy comes (according to theoretical models) from radioactive decay of material produced in the supernova explosion. In particular, nickel-56, which was formed, decays into cobalt-56 and then iron-56, releasing energy. The optical spectrum of the long, slow decline can be explained as a composite of overlapping lines of ionized iron. NASA's Gamma Ray Observatory should be sensitive enough to test this model by observing Type I supernovae in galaxies some distance away from us, detecting the gamma rays emitted as the nickel, cobalt, and then iron are formed.

Type I supernovae expand so rapidly that they become transparent to gamma rays earlier than Type II's, so their light curves eventually fall below those of Type II's. These gamma rays are formed as a result of the radioactive decay.

The star is apparently entirely disrupted, spreading heavy elements through the galaxy but often leaving no core behind. Theorists think that in some cases, two white dwarfs may merge, leaving an even denser core, a "neutron star," a type of star to be discussed in Section 27.1. The fact that Type I supernovae don't show hydrogen in their spectra fits with these models, since the white dwarf may have shed its entire outer atmosphere before its incineration. Another point in favor of distinguishing models for Type I and Type II supernovae (Fig. 26–5) is that the former take place in elliptical galaxies, which lack young stars but which should have white dwarfs, while the latter occur in spiral galaxies, where we expect massive stars to be forming even now.

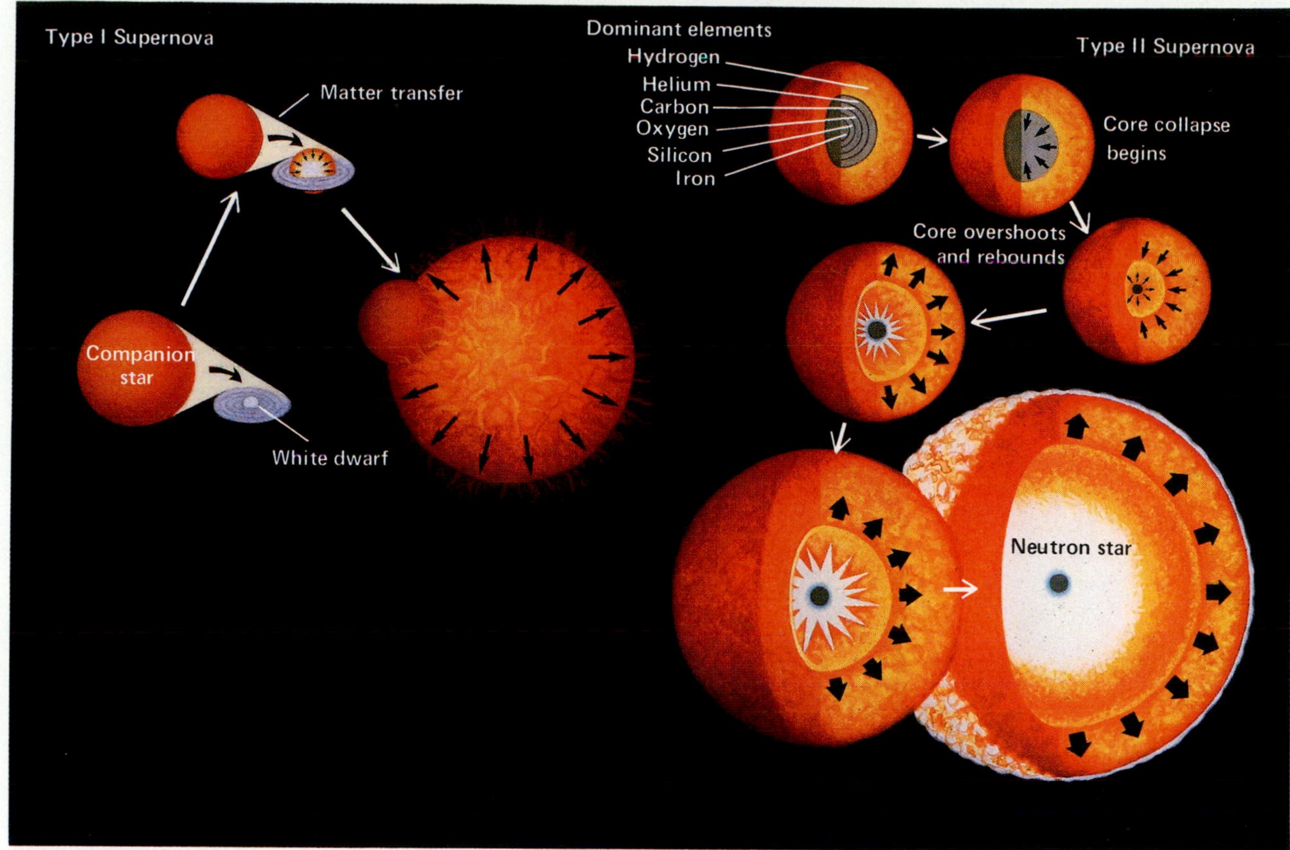

Figure 26–5 Type I supernovae (*left*) come from the incineration of a white dwarf that is accreting matter from a neighboring giant. Type II supernovae (*right*) are the explosions of massive stars, usually from the supergiant phase. When iron forms at the center of the onion-like layers of heavy elements, the star collapses. In this model of the collapse, the core overshoots its final density and rebounds. The shock wave that results blasts off the star's outer layers.

In 1986, Craig Wheeler of the University of Texas realized that not all supernovae without hydrogen lines showing in their spectra—that is, Type I—come from the incineration of white dwarfs, as described above. Some of the Type I supernovae seem to come from supergiant stars after all, though from supergiants that have lost their hydrogen-rich outer layers before their collapse. Aside from the absence of their outer atmospheres, the collapse caused by the buildup of iron atoms proceeds as for Type II supernovae.

Since Type I and Type II supernovae, of which we still see remnants (Fig. 26–6), are the source of the heaviest elements in the universe, they provide many of the heavy elements that are necessary for life to arise. Heavy atoms in each of us have been through such a supernova explosion.

The name "supernova" persists from a time when these events were thought to be merely unusually bright novae. The supernova in the Andromeda Galaxy in 1885 (known by the variable-star name of S Andromedae), since it was assumed to have the same brightness as a nova, gave a value for the distance to that galaxy that misled astronomers for years. They thought the Great Nebula in Andromeda was too close to be an independent galaxy. But in the 1930's, Walter Baade and Fritz Zwicky realized the distinction between novae and supernovae. A supernova explosion represents the death of a star and the scattering of most of its material, while a nova uses up only a small fraction of a stellar mass and can recur. One new similarity does exist: both novae and Type I supernovae are phenomena of binary systems in which one member is a collapsed star.

Figure 26–6 A small part of the Vela supernova remnant, which covers a region 6° square in the southern sky. It came from a supernova that exploded 12,000 years ago. Studies of how the interstellar dust reddens stars, blocking the bluishness on the right side of the image, show that the nebula is about 1500 light years away.

26.3 Detecting Supernovae

Whether Type I or Type II, in the weeks following the explosion, the amount of radiation emitted by the supernova can equal that emitted by the rest of its entire galaxy. It may brighten by over 20 magnitudes, a factor of 10^8 in luminosity. Doppler-shift measurements show high velocities, about 5,000 to 10,000 km/sec, that prove that an explosion has taken place.

In our galaxy, the only supernovae thoroughly reported were in A.D. 1572 and 1604. Tycho's supernova was observed by the great astronomer (Section 2.6) in 1572. Kepler's supernova, observed by that great astronomer in 1604, was slightly farther away from earth than Tycho's, though in another direction. Though these two supernovae were seen only 32 years apart, one actually exploded about 10,000 years after the other. The light curves measured then were accurate enough for us to tell that both were probably Type I. Only faint traces of them remain.

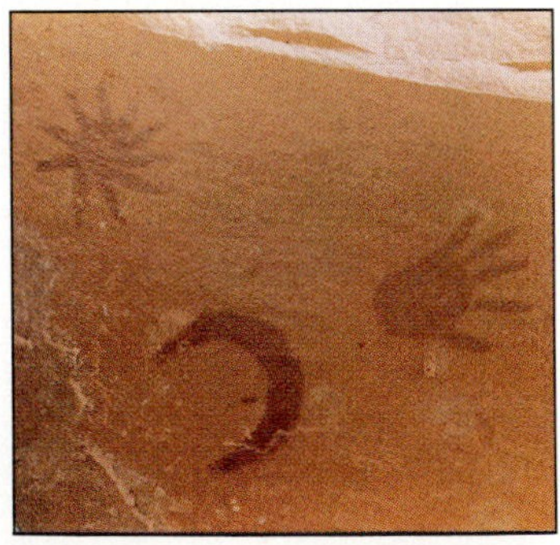

Figure 26–7 An American Indian cave painting discovered in northern Arizona that may depict the supernova explosion 900 years ago that led to the Crab Nebula. It had been wondered why only Oriental astronomers reported the supernova, so searches have been made in other parts of the world for additional observations.

In A.D. 1054, chronicles in China, Japan, and Korea recorded the appearance of a "guest star" in the sky that was sufficiently bright that it could be seen in the daytime. No one is certain why no Western European records of the supernova were made, for the Bayeux tapestry illustrates a comet and thus shows that even in the Middle Ages people were aware of celestial events. A reference to a sighting of the supernova in Constantinople has recently been discovered. Certain cave and rock paintings made by Indians in the American southwest (Fig. 26–7) may show this supernova, though this interpretation is controversial. The light curve was not well enough observed even to determine whether it was a Type I or a Type II supernova. (It may even be an example of some rare additional type.) We shall discuss the remains, the Crab Nebula (Fig. 26–8), further in the next section.

Figure 26–8 The Crab Nebula, in a new color presentation made especially for this book by David Malin of the Anglo-Australian Observatory from three black-and-white plates taken through visible-light filters at the Palomar Observatory.

The sightings of A.D. 185, 393, 1006, and 1181 are now also thought to have probably been supernovae. The 1006 supernova, in the southern constellation Lupus, was apparently the brightest ever, perhaps apparent magnitude −26. It remained visible, even in northern locations, for at least two years.

Not a single supernova has been immediately noticed in our galaxy since the invention of the telescope. Nor do we usually find supernovae in other galaxies before they have passed their peak. We would obviously like to detect a supernova right away, so that we could bring to bear all our modern telescopes in various parts of the spectrum. An attempt a decade ago to set up an automated search program was a bit too far in advance of the technology available. But two new search programs take advantage of the sensitivity of CCD's (Section 4.11) as detectors.

26.4 Supernova Remnants

Optical astronomers have photographed two dozen of the stellar shreds that are left behind, which are known as *supernova remnants* (Fig. 26–9), in our galaxy alone, and others in nearby galaxies. Strong radio radiation and

Figure 26–9 Filaments in the supernova remnant known as the Veil Nebula, NGC 6695, in the Cygnus Loop. It is only 2500 light years from us. The filaments are formed by a shock wave meeting an inhomogeneous interstellar medium (the gas between the stars).

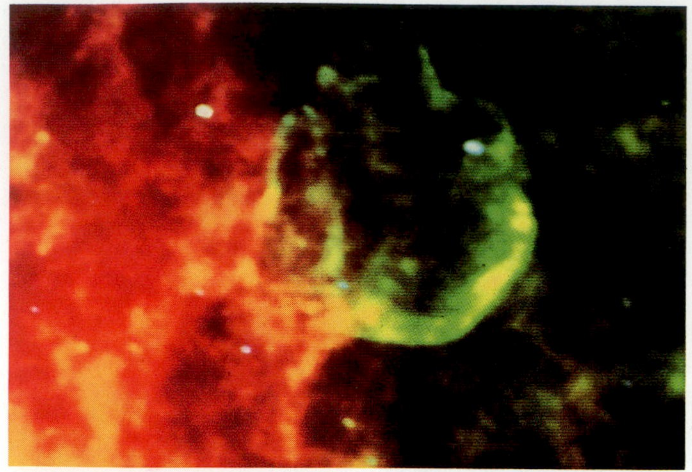
A

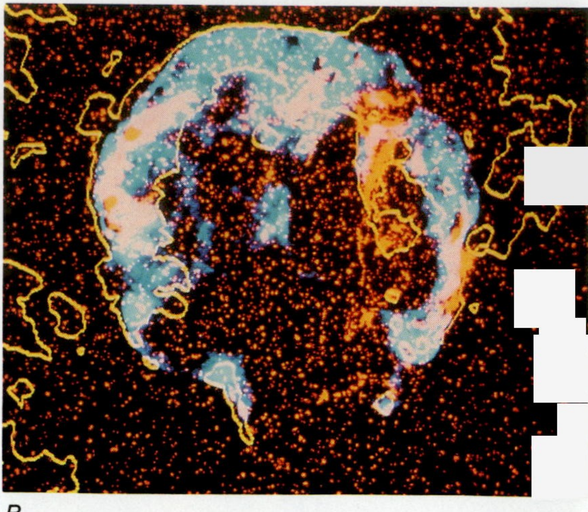
B

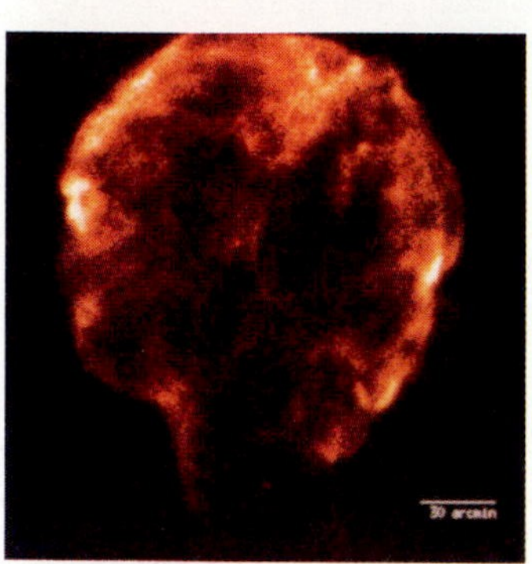

C

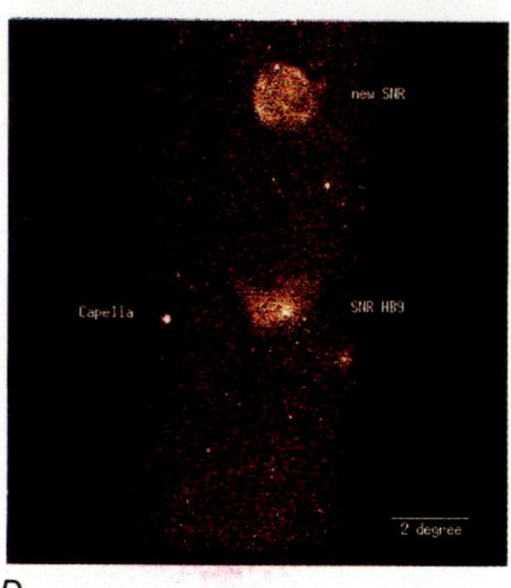

D

Figure 26–10 (*A*) An IRAS view of the Cygnus Loop, of which the previous photograph of the Veil Nebula is a small part. The field of view is about 4.2°, about 8 times the diameter of the moon. The warmest dust appears blue-green and the coolest appears orange-brown. At 25 μ (coded blue), only stars and a uniform background of zodiacal emission from dust can be seen. At 60 μ (green), relatively hot dust at about 50 K appears. The 100 μ emission (red) comes mostly from interstellar dust close to the galactic plane, which runs diagonally across the top left corner of the image. It was a surprise that so much dust survived the supernova blast and that a supernova remnant would be so bright in the infrared. (*B*) The 60-μ emission is shown in yellow, with contours showing widespread emission, and stars in Hα are the orange dots. The x-ray emission measured with the Einstein Observatory is shown in blue. Note the good correlation and that the hot gas seen in x-rays is limited to the boundaries of the dust shell outlined in yellow. In one region, the "carrot" toward the right that appears orange, the infrared, and optical emission are well correlated but no x-ray emission occurs. Presumably this gas is on the face of the remnant nearest us and absorbs the background x-ray emission. (*C*) The Cygnus loop from Rosat. (*D*) A Rosat study of a known supernova remnant in Auriga turned up a new one.

x-radiation can also be detected. Over 100 radio supernova remnants are known. The comparison of optical, infrared, radio, and x-ray observations of supernova remnants (Figs. 26–10 and 26–11) tells us about the supernova itself and about its interaction with interstellar matter. Few supernova remnants are known in the visible; most supernova remnants have been discovered in the radio part of the spectrum (Fig. 26–12).

We find the most famous remnant in the location where Chinese astronomers reported their "guest star" in A.D. 1054. When we look at the reported position in the sky, in the constellation Taurus, we see an object—the Crab Nebula—that clearly looks as though it is a star torn to shreds (Fig. 26–13). The Crab appears to be about ⅕ of the distance from the sun to the center of our galaxy. We find its age by noting the rate at which the filaments in the Crab Nebula are expanding. Tracing the filaments back in time shows that they were at a single point approximately 900 years ago. The Crab's age, location, and current appearance leave us little doubt that it is the remnant of the supernova whose radiation reached earth twelve years before William the Conqueror invaded England. The discovery that the filaments are rich in helium and are expanding at thousands of km/sec as from an explosion further confirms that the Crab is a supernova remnant. And helium is a product of stellar evolution. We will have more to say about the Crab when we discuss pulsars.

From a study of the rate at which supernovae appear in distant galaxies, we estimate that supernovae should appear in our galaxy about once every 30 years. Interstellar matter obscures much of the galaxy from our optical view.

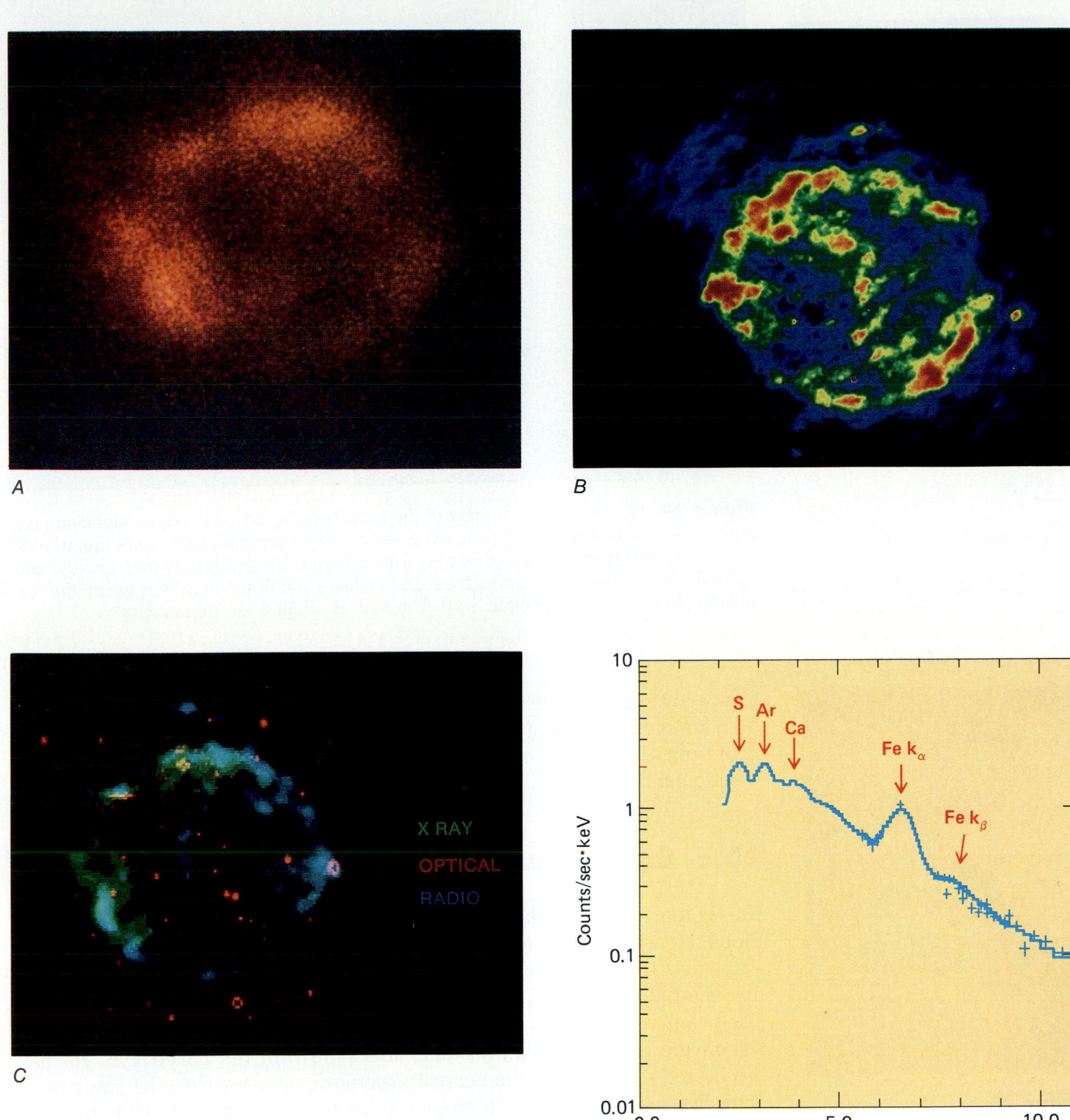

Figure 26–11 The Cassiopeia A supernova remnant. (*A*) An x-ray image made with Rosat. (*B*) A radio image made with the Very Large Array. The source is a relatively recent supernova remnant, but the supernova itself was not observed. We can date it to the late 17th century by studying its expansion. Both images show a bright shell of emission, presumably the supernova matter encountering the interstellar medium. Analysis of the x-ray spectrum shows that the abundances of the elements silicon, sulfur, and argon are higher than the normal solar abundance, demonstrating that heavy elements are indeed formed in supernovae. Other supernova remnants have even greater enhancements of these elements. (*C*) A superposition of x-ray, radio, and optical images. (*D*) The x-ray spectrum of Cas A, measured from the EXOSAT, shows prominent lines of sulfur, argon, calcium, and iron. Studying these spectral lines is leading to information on the abundances and ionization states in the hot gas.

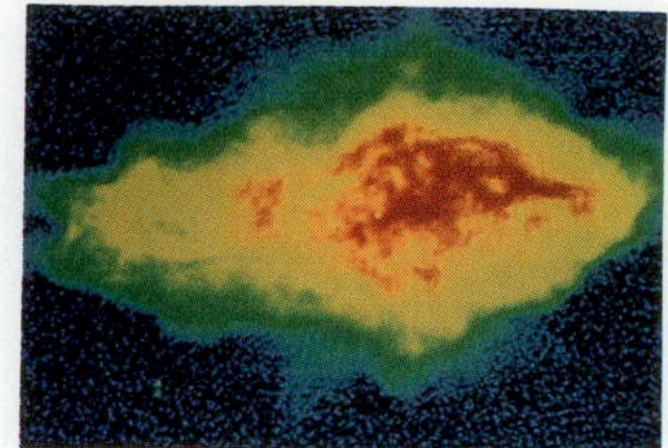

Figure 26–12 The supernova from AD 1181, now called 3C58, imaged from the VLA in the radio spectrum.

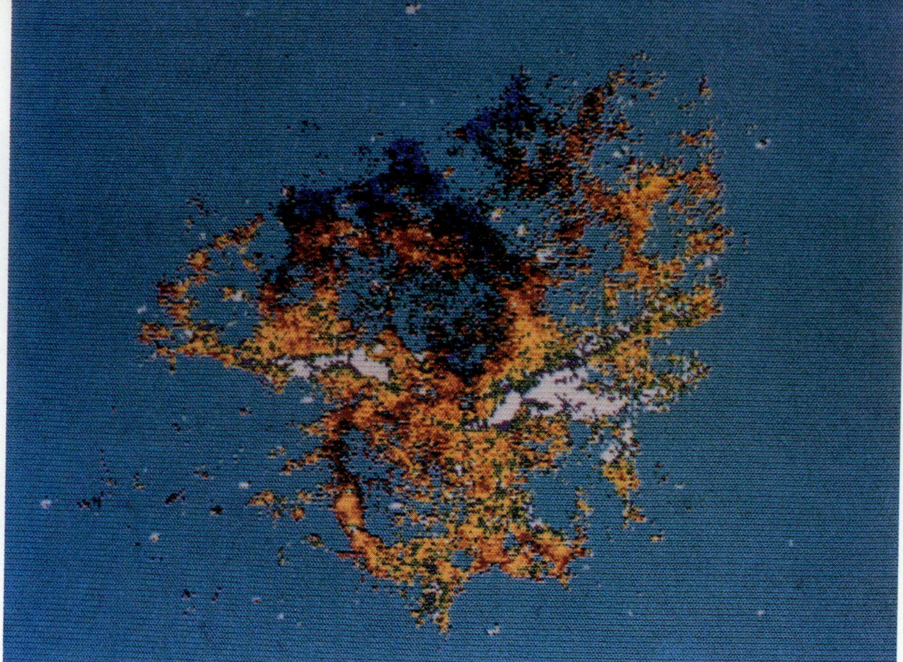

Figure 26–13 A false-color view of the Crab Nebula, with the colors signifying the ratio of helium to hydrogen. From low to high, colors are dark blue, dark green, red/brown, yellow/green, and purple/white. The helium/hydrogen ratio varies greatly, and either is unusually high in some locations or has conditions there that excite the helium spectral lines especially well. The field is about 6 arc min across.

Figure 26–14 Ian Shelton with the plate on which he discovered the supernova.

The "A" in Supernova 1987A means it was the first supernova to be discovered that year. Two years earlier, the IAU had certified the use of capital letters for supernovae in place of the lower-case letters that had been used informally.

Still, we wonder why we haven't seen one in so long. It's about time. Radiation from hundreds of distant extragalactic supernovae is on its way to us now. The light from a supernova in our galaxy may be on its way too. Maybe it will get here tonight.

26.5 Supernova 1987A

On February 24, 1987, Ian Shelton was photographing the Large Magellanic Cloud. He was working for the University of Toronto at their telescope on Las Campanas in Chile. Fortunately, he chose to develop his photographic plate that night. When he looked at it, still in the darkroom, he saw a star where no star belonged (Fig. 26–14). He went outside, looked up, and again saw the star in the Large Magellanic Cloud, this time with his naked eye. He had discovered a supernova, the nearest supernova seen since Kepler saw one in 1604, five years before the telescope was invented. By the next night, the news was all over the world, and all the telescopes that could see the supernova were trained on it (Fig. 26–15). The event was perhaps the most significant to occur in astronomy in centuries. It received its official name from the International Astronomical Union's Central Bureau for Astronomical Telegrams: Supernova 1987A.

Observations made independently by amateur astronomers just before and just after the explosion indicate that the supernova brightened within 3 hours, much faster than the few days of brightening for most other supernovae.

Though the supernova is not in our own galaxy, it is nearly so. The Large Magellanic Cloud is only about 50 thousand parsecs away, which the newspapers often gave with false accuracy as 163,000 light years (multiplying the round number 50,000 by a precise conversion factor with 3 significant digits). Some parts of our own galaxy are farther away. Perhaps it is lucky that the supernova was there instead of closer, for a closer supernova would have been too bright for many telescopes. As it was, detectors are now so sensitive that

Figure 26–15 The supernova (*A*) before and (*B*) after February 24, 1987.

many astronomers had to observe with most of their telescope's mirror masked off or through partially obscuring filters. The only unfortunate thing is that the supernova could not be seen from most of the northern hemisphere; the Large Magellanic Cloud is far enough south in the sky that it can be studied only with telescopes near the equator or in the southern hemisphere. One advantage of the supernova's location is that the Large Magellanic Cloud is circumpolar for southern-hemisphere observatories, so there were no long gaps in coverage.

The first predictions were that the supernova might become as bright as Jupiter. But, strangely, the supernova did not brighten as rapidly as had been expected. Even the subsequent predictions that the supernova might become as bright as Sirius did not come true. The supernova, discovered at 5th magnitude, seemed thereafter to brighten very slowly; over two months later it reached slightly brighter than 3rd magnitude (Fig. 26–16), 20 or more times fainter than would be expected for a supernova at its distance.

Many of the things first said about the supernova did not turn out to be true on further observation. For example, it was first said to be a Type I su-

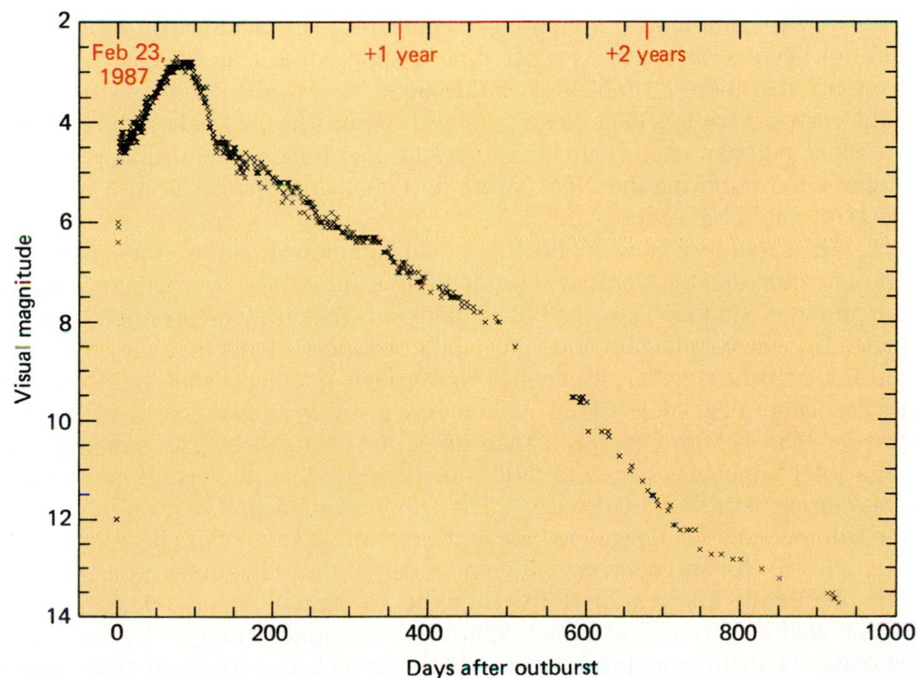

Figure 26–16 The light curve of Supernova 1987A, showing its peak 80 days after its discovery. After day 120, the decay matches the prediction that the supernova light comes from radioactivity. A quick decay of the 0.07 solar masses of radioactive nickel-56 (half-life 6.1 days) to cobalt-56 is then followed by a slower decay (half-life 77.1 days) to iron-56. The light curve is now dropping below the cobalt-56 decay curve, since the supernova matter has thinned out enough for x-rays and gamma rays to escape. Heating and ionization from a buried pulsar may stabilize the supernova for many years at a value predicted to be well within the observable range.

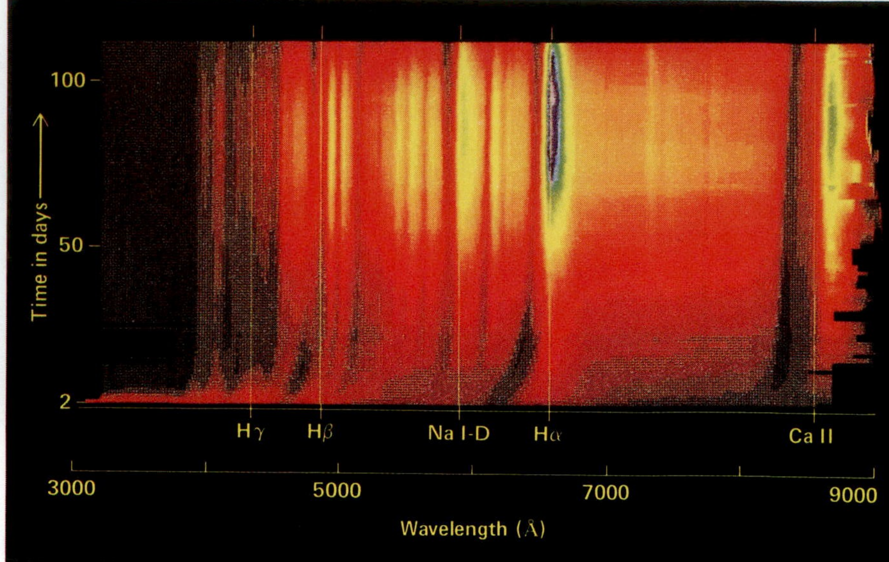

Figure 26–17 The variation of the spectrum of Supernova 1987A, showing especially the hydrogen lines with their strong Doppler shifts.

pernova, but within a few nights it was realized that hydrogen lines were indeed present (Fig. 26–17). That made it a Type II supernova, one that came from a massive star.

We even knew which star had blown up (Fig. 26–18)! It had been star 202 in the −69° declination band of the catalogue of Sanduleak, so is generally known as Sk −69°202. Prior to its eruption, it had seemed ordinary.

Subsequent models put some of the facts together into a coherent picture. B3 supergiants could go supernova after all. Their smaller radius compared with red supergiants made their outer layers denser and more likely to capture the energy produced by the shock wave. Thus a supernova from a B3 supergiant should be less luminous than normal supernovae, as was seen with Supernova 1987A. The energy emitted matched theoretical expectations of the explosion of such a 20-solar-mass star with a 6-solar-mass helium core. The supergiant had already lost perhaps 4 solar masses worth of its outer atmosphere, which is why the supernova did not become as bright as expected. The outer atmosphere had had a high hydrogen content, so its loss explained why the hydrogen spectral lines were not at first detected. The fact that the supernova did not become as bright as most Type II supernovae may indicate why we have not had more examples of similar light curves—the objects with those light curves were too faint, so we missed them. IUE observations found an enriched nitrogen shell, perhaps the result of a hundred thousand years of stellar wind removing the outer layers and changing the star from a red supergiant into a blue supergiant.

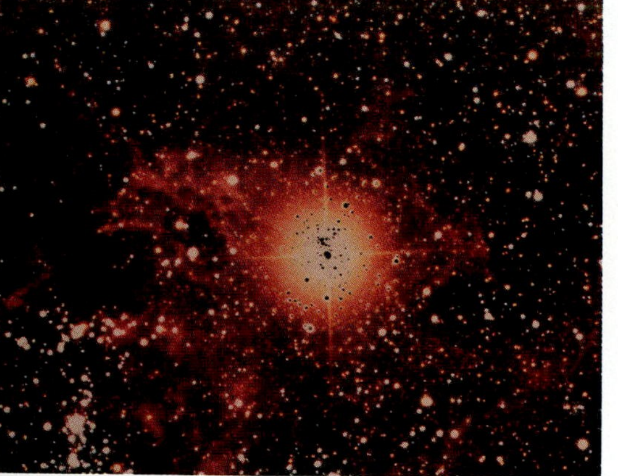

Figure 26–18 A composite photo in which a pre-explosion photograph of the blue supergiant star that exploded, Sk −69°202 (a star from a catalogue of N. Sanduleak, pronounced San-doo'lick), is superimposed on a supernova image. The star had companions 1.5″ and 3″ away, respectively. The closer companion was barely distinguishable but the farther companion shows in this picture as a bump on the upper right of Sk −69°202's image.

The supernova grew brighter in optical radiation for almost three months, an indication that the star that exploded had a substantial hydrogen envelope remaining even though it was a blue supergiant. The envelope presumably was the cause why gamma rays and x-rays did not appear sooner than they did and did not become brighter. It may also have been significant that the object is in the Large Magellanic Cloud, which has a lower abundance of elements heavier than helium ("metals") than does our own galaxy. The supernova's ultraviolet brightness began to fade much sooner than the visible radiation. Thus the supernova turned redder. The supernova faded at one rate through the summer, and then began to fade at the rate of "normal" Type II supernovae (Fig. 26–19). It had gotten over the early loss of the star's outer atmosphere. The rate for the first year or so matched the 77-day half-life of cobalt-56 in its disintegration to iron-56, and the brightness corresponded to the 0.1 solar mass of cobalt-56 that theory predicted would be formed. The cobalt-56 came from nickel-56 formed in the explosion and allows us to calculate that 0.075 solar

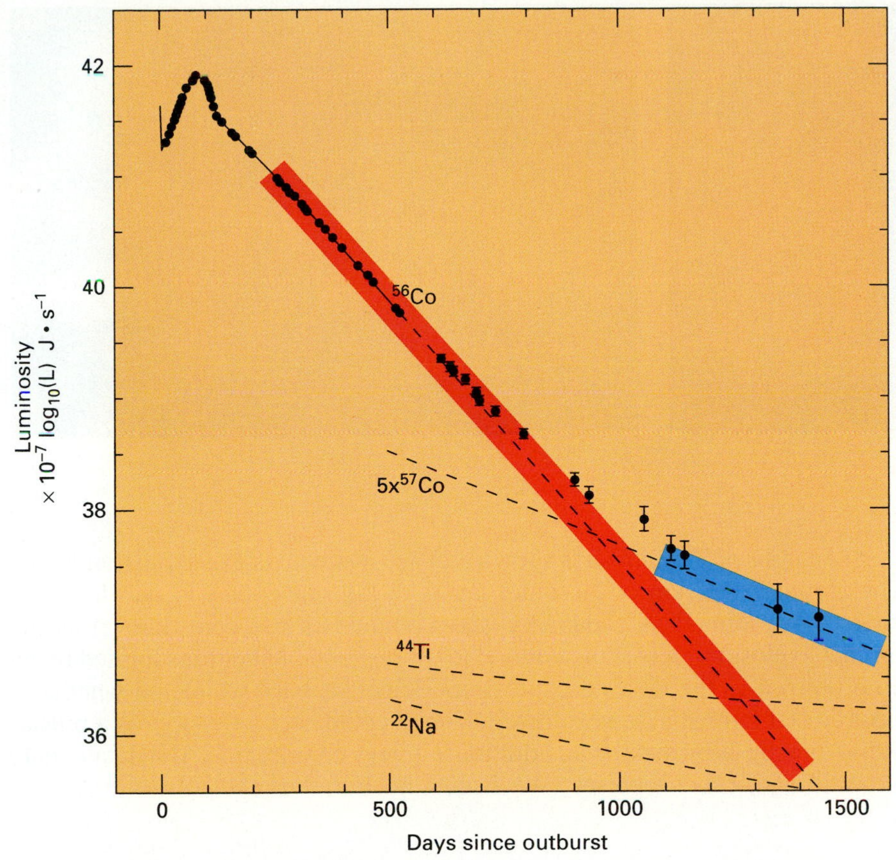

Figure 26–19 Since the blue supergiant that formed Supernova 1987A had less surface area than a red supergiant, the supernova was fainter than normal at first. Eventually, the light curve joined the typical Type II rate of decline. Now it extends far past where other, fainter supernovae have been observed.

The graph shows the total amount of radiation given off by the supernova each second, including contributions to the spectrum outside the visible. About 600 days after the explosion, dust formed and started contributing significantly in the infrared.

From day 40 on, an additional source of energy seemed to be contributing to the light curve, and from day 120 on, it was clear that the source was the radioactive decay of cobalt-56 (*red*), whose half-life is 77 days. After a transition from days 700 to 1100, the source of energy was cobalt-57 (*blue*), whose half-life is 271 days. Eventually, after several years, the radiation should be dominantly from titanium-44, whose half-life is 47 years.

masses of nickel-56 were formed. After about 900 days, the decay rate matched that of cobalt-57, though about 5 times as much cobalt-57 compared with cobalt-56 would have to be present as compared with their ratio in the Sun.

The decay of cobalt-56 produced a positron of high energy and a number of gamma-ray lines. The gamma rays scattered off electrons, becoming x-rays in the process. The x-rays, in turn, were absorbed by heavy elements, and electrons were ejected at high velocity. These fast electrons heated material and led to the ionization of atoms and the emission of optical and ultraviolet spectral lines. Some of the energy provided the continuous radiation (Fig. 26–20).

The theoretical models have now been adjusted to better match the shape of the optical light curve and the time when x-rays and gamma rays were first seen. It seems likely that the onion-layer structure of hydrogen, helium, carbon, oxygen, silicon, and iron is not as discretely separated as had been thought. Some mixing had probably taken place.

The star went to a blue giant stage before exploding from its red giant stage because the abundance of heavy elements in the Large Magellanic Cloud is much lower than that in our own galaxy.

A

B

Figure 26–20 (*A*) The Large Magellanic Cloud with the supernova to the upper left of the Tarantula Nebula. Both appear as reddish spots. Comet Wilson is also in the field of view, near the bottom. (*B*) A more magnified view of Supernova 1987A, the Tarantula Nebula, and Comet Wilson. The image is rotated 90° from the previous view. It was taken on May 3, 1987, over 2 months after the explosion.

Analysis of SN 1987A has shown that two mechanisms of explosion occur. The abrupt expansion of the outer layers of the star comes in part from the shock wave formed as collapsing gases bounce off the core (known as the prompt explosion); this is then added to by the flood of neutrinos formed in the core (known as the delayed explosion). Scientists have wondered whether a neutron star formed in the explosion. If so, it could become visible as a pulsar, either directly or by making an additional energy contribution. The data to date show no sign of such a pulsar. (We discuss neutron stars and pulsars in Chapter 27).

26.5a Ring Around a Supernova

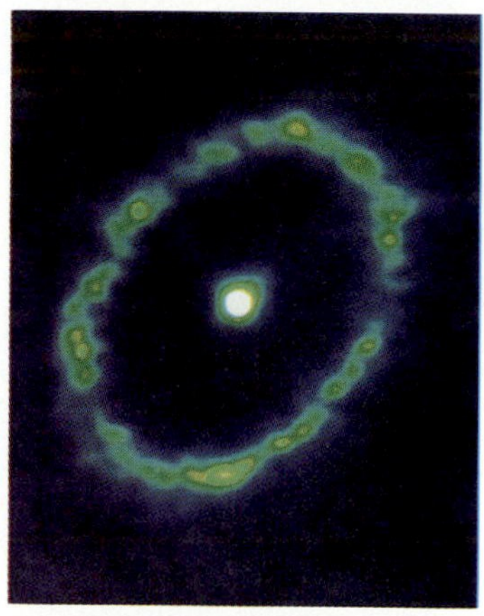

Figure 26–21 Supernova 1987A and its surrounding ring, photographed with the Faint Object Camera of the Hubble Space Telescope on August 24, 1990, 3.5 years after the original explosion. The image was taken through a filter passing only the light of twice-ionized oxygen. North is at upper right and east is at upper left. The ring is only 0.8 arc sec in radius (corresponding to 0.6 light-years), and can be seen clearly only with the Hubble Space Telescope.

The Faint Object Camera on the Hubble Space Telescope was able to image not only the supernova itself but also the gas around it (Fig. 26–21). The ring, imaged 3 years after the explosion, had a radius of only 0.8 arc sec, comparable to the best ground-based seeing; thus it could not have been imaged as clearly from the ground. At the distance to the supernova, the ring is 0.6 light-years across. From the lack of emission across the ring, scientists have concluded that we are seeing an actual ring inclined at 43° rather than the edges of a sphere.

Apparently, gas ejected in the supernova explosion has caught up with the gas that the star had given off as a stellar wind in the hundreds of thousands of years prior to the explosion. Comparing the angular rate of expansion of the ring with its rate in km/s measured from the Doppler effect, scientists have calculated the distance to the supernova (and thus also the Large Magellanic Cloud) to a higher accuracy than had been possible: 169,000 light years, plus or minus only 5 per cent. The ring's velocity of expansion shows that matter had been flowing outward for about 20,000 years. This period is presumably how long it took the star to change from a red supergiant to a blue supergiant.

26.5b Neutrinos from the Supernova

The solar neutrino experiment, with a large tank of cleaning fluid in a mine, was not sensitive enough to the energy range of neutrinos emitted by the supernova. But at least two other experiments were. They had been set up to study the possible decay of the proton (which we discuss in Chapter 34), but were fortunately on line for this event of the century. Both experiments contained large volumes of extremely pure water surrounded by sensitive phototubes to measure any light given off as a result of interactions in the water.

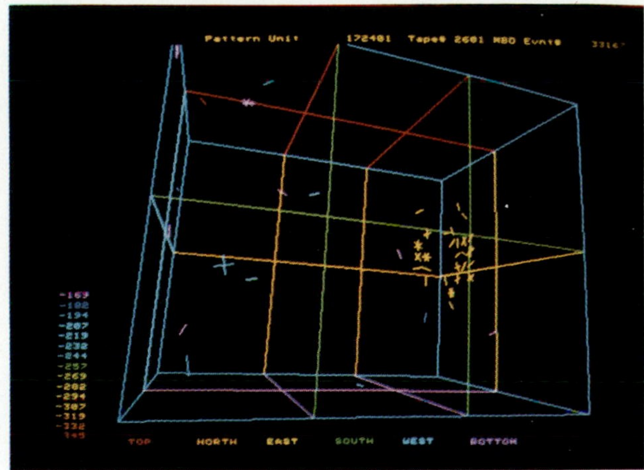

Figure 26–22 A video screen showing one of the neutrinos from the supernova interacting in the tank of water of the U. Cal/Irvine–U. Michigan–Brookhaven neutrino observatory (IMB). More lines on a symbol indicate more interactions. The yellow marks show the intersection of the cone of photons generated by a neutrino interaction in the tank; tracing the path backward shows that the neutrino came from the supernova. Colors show arrival time, with intervals shown in nanoseconds. The neutrino supernova generated photons detected by the photomultipliers within a few nanoseconds. A few random interactions, shown in different colors, occurred at other times.

Figure 26–23 Some of the 1000 photomultipliers to detect the light flashes from neutrinos and other particles at the Kamiokande II detector, which contains 3000 tonnes (1 tonne = 1000 kg) of water. (Photomultipliers are tubes that give off many electrons for individual photons that enter at the other end, multiplying the number of electrons stage by stage.)

The IMB (University of California at Irvine–University of Michigan–Brookhaven National Laboratory) experiment in a salt mine in Ohio reported that 8 neutrinos had arrived and interacted within a 6-second period on February 23 (Fig. 26–22), three hours before the optical burst was seen, whereas otherwise a single neutrino interacted about once every 5 days. This number indicated that 3×10^{16} neutrinos from the supernova had passed through the 5000 tons of water in the detector. The Kamiokande II detector (Fig. 26–23) in a zinc mine in Japan, at the same time, detected a burst of 11 neutrinos. They could even tell that at least some of them were arriving from the direction of the Large Magellanic Cloud. A few other neutrinos were detected by a device in the Soviet Union. The discoveries marked the beginning of a new observational field of astronomy: extrasolar neutrino astronomy.

The neutrino bursts preceded the optical brightening. The sequence matches the theoretical idea that the neutrinos are released as the collapsing core deposits energy, starting nuclear reactions. Also, the amount of energy carried in neutrinos closely matched that of theoretical predictions. Our basic idea of what happens in a supernova appeared verified.

The agreement between the theoretical prediction and the observations matches the scenario of neutron-star formation so well that it seems likely that a neutron star rather than a black hole was formed.

Scientists studied the arrival times of the neutrinos (Fig. 26–24). If neutrinos had mass, they would have spread out in space, since neutrinos could then travel at different speeds. The fact that the neutrinos arrived on earth so close together is consistent with neutrinos being massless and gives a way of figuring out an upper limit on how much mass they may have. The upper limit is very small; in units of energy equivalent to that amount of mass, the supernova observations have shown that the neutrino mass must be less than about 16 electron-volts (much less than, say, the 511-million electron-volts rest mass of an electron). So studies of the supernova have even provided important information for basic particle physics. In Chapter 34, we shall consider the impor-

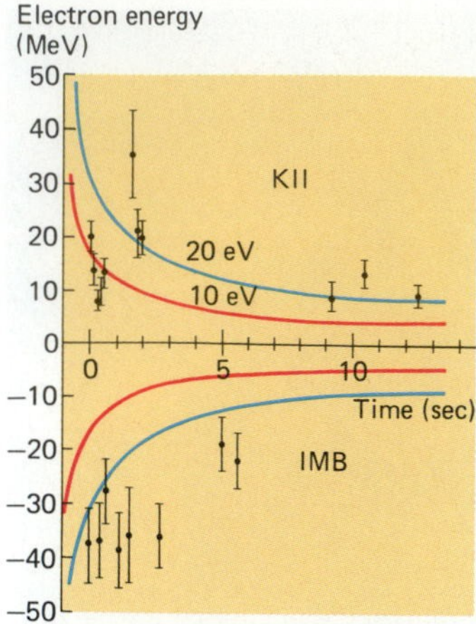

Figure 26–24 The arrival times and energies of the neutrino events at the neutrino observatories U. Cal/Irvine–U. Michigan–Brookhaven and at Kamiokande. The energy threshold for the Kamiokande detector is somewhat lower, 7 MeV, compared with 20 MeV for the IMB detector. (The Kamiokande times were uncertain by about 1 minute, and were adjusted to match the IMB times.) Most of the events came from the absorption of antineutrinos by protons in the water in the detector, resulting in positrons and neutrons. One or more of the events may have come from neutrinos scattering off electrons. There is no evidence that the apparent gap in the arrival times of the Kamiokande neutrinos is other than random. The red lines outline the expected positions if neutrinos have mass 10 eV; the blue curves correspond to 20 eV.

tance of knowing the neutrino mass for understanding the overall composition and future evolution of the universe.

John Bahcall of the Institute for Advanced Study at Princeton has raised the possibility that many supernova explosions produce almost all their energy in neutrinos. These "neutrino bombs" may never become very bright but would show up in continued neutrino observations. If significant numbers are detected, then we would have been underestimating the rate of stellar collapse in our galaxy by a misplaced reliance on optical observations.

26.5c Further Studies of the Supernova

Supernova 1987A will remain visible to large telescopes for the foreseeable future. The supernova shell is now turning transparent, allowing us to view the inner layers where the heavy elements were formed. We first observed x-rays with the Soviet Kvant module on the Mir space station and with the Japanese Ginga spacecraft; the x-rays probably were formed as gamma rays hit the supernova debris. Somewhat later, we observed the gamma rays directly with balloon-borne detectors and with NASA's Solar Maximum Mission. The x-rays and gamma rays were expected only when the supernova gas had thinned out, yet were detected relatively early. The difference in time can be explained if the cobalt-56 were mixed into a hydrogen-rich envelope of gas, an idea that also fits the light curve.

Most of the light we get from the supernova comes to us directly. But some is light that originally went in a slightly different direction and is later reflected toward us (Fig. 26–25). This reflected light tells us about interstellar space in the immediate neighborhood of Sk −69°202.

What will we continue to learn about the formation of the heavy elements in the supernova, as we see to deeper and deeper levels? What about the companion? How will x-ray and radio emission vary? And what even more exciting surprises may be in store? We must wait to see.

26.6 Cosmic Rays

Most of the information we have discussed thus far in this book, aside from the neutrino studies of Supernova 1987A, has been gleaned from the study of electromagnetic radiation, which can be thought of as sometimes having the properties of waves and sometimes having the properties of particles called photons. Certain high-energy particles of matter travel through space in addition to photons and neutrinos. These particles are protons or nuclei of atoms heavier than hydrogen moving at tremendous velocities, and are called *cosmic rays.* Some of the weaker cosmic rays come from the sun, but most cosmic rays come from farther away. Cosmic rays provide about 1/10 of the radiation environment of the earth's surface and of the people on it. (Almost all the radiation we are exposed to comes from cosmic rays, from naturally occurring radioactive elements in the earth or in our bodies, or from medical x-rays. Radon gas in houses has, in recent years, proved to provide about half our exposure.)

The origin of the non-solar "primary cosmic rays," the ones that actually hit the top of the earth's atmosphere as opposed to the "secondary" cosmic

Figure 26–25 A double light echo from Supernova 1987A, reflections in interstellar clouds in the Large Magellanic Cloud of the light from the bright supernova explosion. The light echoes became visible the following year. The clouds making the echoes are 400 and 1000 light years, respectively, from the supernova toward us.

The photo shows two concentric rings around the overexposed image of the supernova itself, which was dimmed by a small obscuring disk that shows at the center.

rays that hit the earth's surface, has long been under debate. Because cosmic rays are charged particles—mostly protons and also some nuclei of atoms heavier than hydrogen—our galaxy's tangled magnetic field bends them. Thus we cannot trace back the paths of cosmic rays we detect to find out their origin. It is possible that most middle-energy cosmic rays were accelerated to their high velocities in the shock waves from supernova explosions. It has been suggested that a significant fraction of the high-energy cosmic rays in our galaxy come apparently from a single source—Cygnus X-3 (Fig. 26–26). This object was a discovery of space research. Its name means that it was the third x-ray source to be detected in the constellation Cygnus. Studies across the spectrum have shown that Cygnus X-3 gives off more energy than all but a couple of other sources in our galaxy, though dust obscures it in visible light. Current evidence indicates that Cygnus X-3 is a neutron star (Chapter 27) in orbit with a larger companion on the edge of our galaxy. The interaction between the two may give the protons very much more energy than we can give protons with even the largest "accelerators" (atom smashers) in laboratories on earth (Fig. 26–27). The idea is being verified, and is not now looking as likely as it did at first mainly because Cygnus X-3's output is not steady. In any case, some rare cosmic rays of hundreds of times higher energy than those that might be coming from Cygnus X-3 may come from sources outside our galaxy.

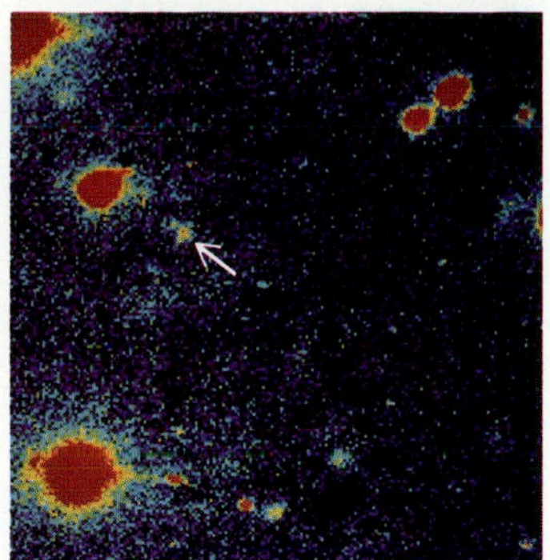

Figure 26–26 The optical counterpart of Cygnus X-3.

Our atmosphere filters out most of the primary cosmic rays. When they interact with atoms in the earth's atmosphere, "secondary cosmic rays" are given off.

To capture cosmic rays from outer space, one must travel above most of the earth's atmosphere in an airplane, balloon, rocket, or satellite. (Only a few of the most energetic cosmic rays reach the ground.) Balloons bear stacks of suitable plastics to altitudes of 50 kilometers, and now the space shuttles carry similar materials higher; the cosmic rays leave marks as they travel through the plastic, and a three-dimensional picture of the track can be built up because of the three-dimensional nature of the stack. The third High-Energy Astronomy Observatory (HEAO-3) carried two cosmic-ray experiments into space in 1979. Cosmic rays were even detected by emulsion chambers mounted for two months in the baggage compartment of a supersonic Concorde as it flew around. And in 1984, a space shuttle carried cosmic-ray detectors aloft to remain in space for what was supposed to be a year on NASA's Long-Duration-Exposure Facility

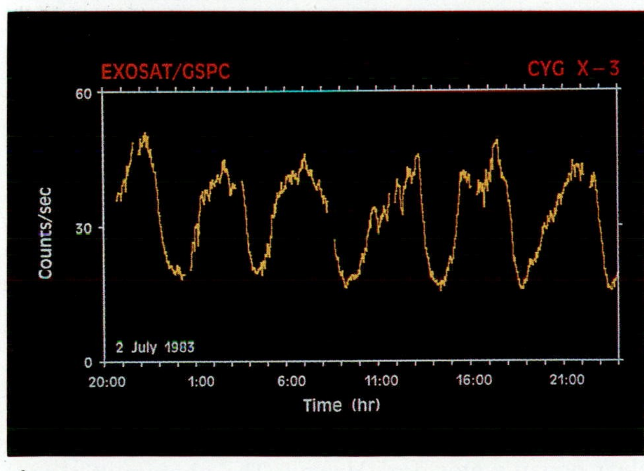

A

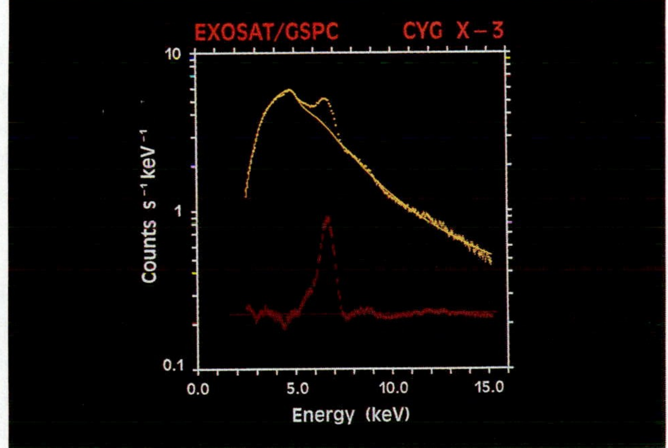

B

Figure 26–27 (*A*) X-rays from Cygnus X-3 have a regular cycle with a period of 4.8 hours, undoubtedly caused by a compact object orbiting an unidentified companion star. The reason for the cycle-to-cycle variations is unknown. (*B*) The spectral line from highly ionized iron (red) is exceptionally strong in Cygnus X-3. The lower curve shows it subtracted from its actual position on a spectral background (yellow). These spectral observations are from EXOSAT.

Figure 26–28 The Long-Duration-Exposure Facility (LDEF) was launched by a space shuttle in 1984, carrying plastic and other materials aloft to be affected by the space environment. LDEF was returned to Earth in 1989; it is being examined to see the effects of the space environment on its surface and contents. Scientific experiments include study of the plastic for cosmic-ray tracks.

Fruit grown from some of the millions of seeds it carried had normal-tasting fruit but often had mutations of leaves and blossoms.

(LDEF) (Fig. 26–28). Unfortunately, because of the Challenger accident, the retrieval of LDEF could not occur until 1990. NASA's Gamma Ray Observatory should be an important mechanism for studying gamma rays which are intimately connected with the charged cosmic rays.

Special ground-based telescopes are now able to see streaks of light in the upper atmosphere formed by giant showers of cosmic rays. Some, at least, of the highest energy cosmic rays seem to come from the direction of the nearest large cluster of galaxies. Scientists suspect that a peculiar galaxy there, M87, may be giving off these particles, perhaps from a giant black hole.

An ingenious proposal to study a variety of cosmic-ray particles and neutrinos is in the preliminary testing stage (Fig. 26–29). Project DUMAND (**D**eep **U**nderwater **M**uon **a**nd **N**eutrino **D**etection) is to set up chains of giant photomultipliers in the clear ocean water near Hawaii. The photomultipliers will detect flashes of light from secondary cosmic-ray particles and from neu-

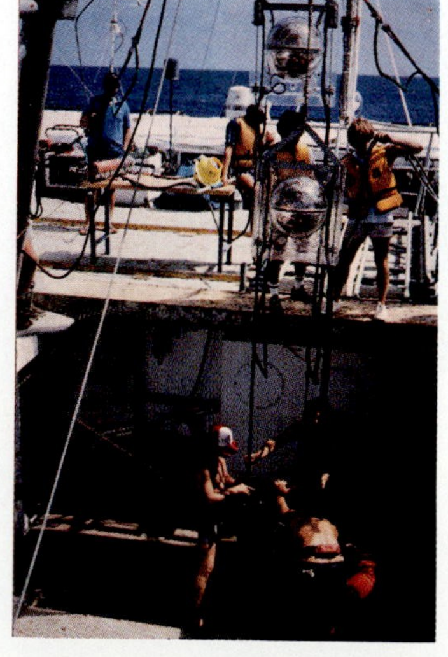

Figure 26–29 The deployment of a test string of 7 photomultipliers for Project DUMAND (Deep Underwater Muon and Neutrino Detection) to a position deep under the Pacific Ocean off Hawaii. The full working DUMAND project, with 9 strings, is hoped for in 1993.

trinos generated by supernovae. The flashes arise as the cosmic-ray particles slow down when they pass through water or the neutrinos interact with protons in the water. A set of 9 strings each carrying 24 detectors 10 m apart is to be lowered 4.7 km into clear water off the island of Hawaii. Collaborators come from the United States, Switzerland, and Japan. They hope that operation will begin in 1993.

A cubic kilometer of the clear ice deep underneath the ice cap at the South Pole may provide a huge detector for neutrinos. Known as the Antarctic Muon Neutrino Detector Array (AMANDA), it will be sensitive to the kind of neutrinos that interacts with muon particles.

Summary and Outline

When stars of more than 8 solar masses exhaust their hydrogen in the core, they become red giants. Later, heavier elements are built up in the core, and the stars become supergiants (Section 26.1).

After cores of heavy elements form, the stars explode as Type II supernovae. Type I supernovae come from less massive stars, and may be the incineration of a white dwarf. Only a few optical supernovae have been seen in our galaxy and a few have been detected in the infrared. Supernova remnants can best be studied with radio and x-ray astronomy (Sections 26.2, 26.3, and 26.4).

The nearest, brightest supernova to be seen since 1604 erupted in 1987. It was in the Large Magellanic Cloud. It reached slightly brighter than 3rd magnitude, and is now declining. Detection of neutrino bursts from the supernova was especially significant, and led not only to verification of our model of supernovae but also to basic information about particle physics (Section 26.5).

Cosmic rays, high-energy particles in space, probably come from supernovae (Section 26.6).

Key Words

red supergiant, supernova, Type I supernovae, Type II supernovae, supernova remnants, cosmic rays (primary, secondary)

Questions

1. Why are red supergiants so bright?
2. What are the basic differences between a nova and a supernova?
3. If we see a massive main-sequence star (a heavyweight star), what can we assume about its age, relative to most stars? Why?
4. What do we know about the core of a star when it leaves the main sequence?
5. What is special about iron in the core of a star?

†6. In a supernova explosion of a 20-solar-mass star, about how much material is blown away?

†7. A supernova can brighten by 20 magnitudes. By what factor of brightness is this? Show your calculations.

8. How do we distinguish observationally between Type I and Type II supernovae?
9. How do the physical models of Type I and Type II supernovae differ?

†This question requires a numerical solution.

†10. A typical galaxy has a luminosity 10^{11} times that of the sun. If a supernova equals this luminosity, and if it began as a B star, by what factor did it brighten? By how many magnitudes?

11. How does η Carinae endorse our model of supergiants?
12. Would you expect the appearance of the Crab Nebula to change in the next 500 years? How?

†13. The light from Supernova 1987A came from the Large Magellanic Cloud, which is 50 kiloparsecs (kpc) away. When did the star actually explode?

14. What type of supernova was Supernova 1987A? Give two reasons why we think so.
15. From what event in Supernova 1987A did the neutrino bursts come?
16. Why was the neutrino burst significant for astronomy? For particle physics?
17. What are cosmic rays?
18. Where are secondary cosmic rays formed?

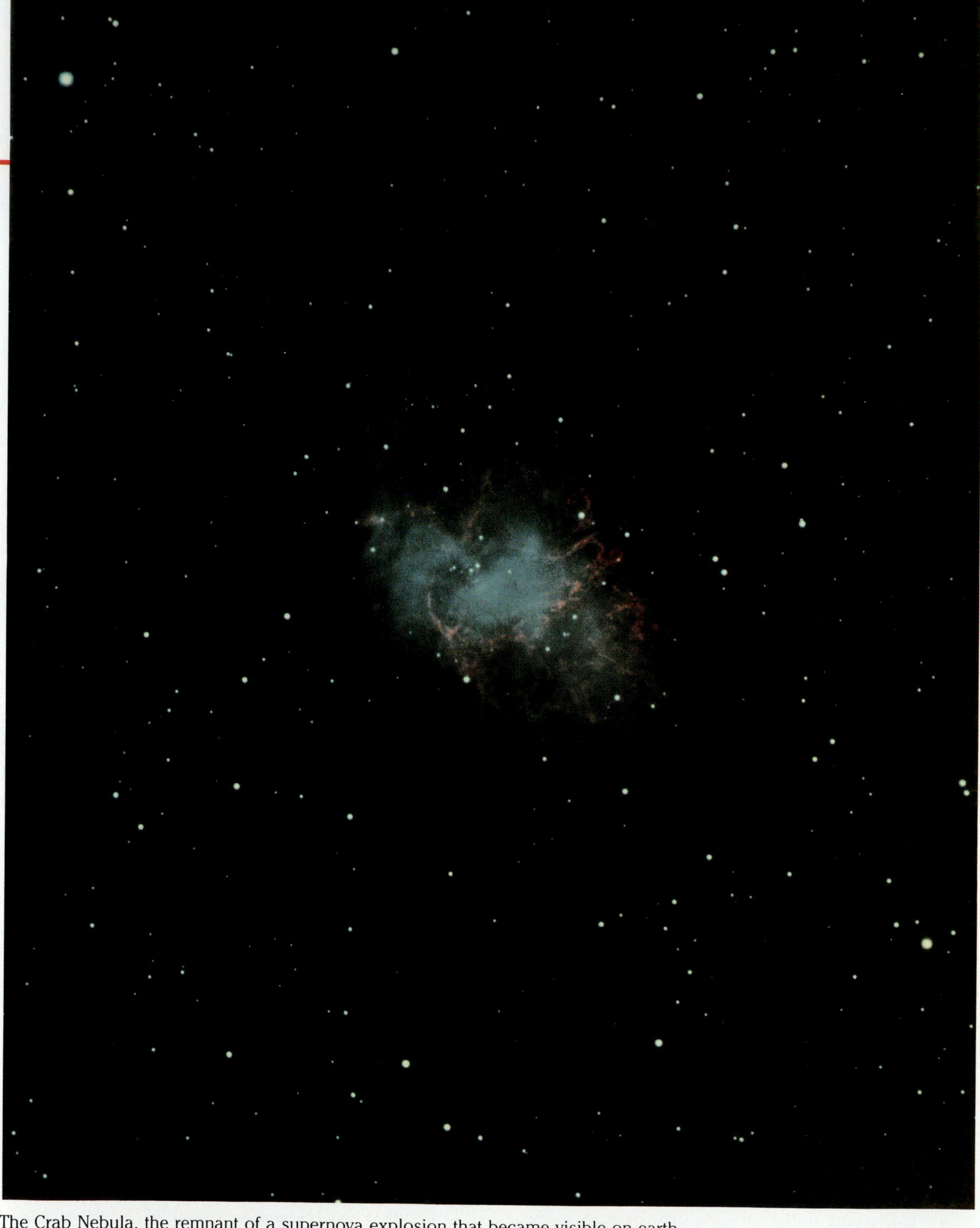

The Crab Nebula, the remnant of a supernova explosion that became visible on earth in A.D. 1054. The pulsar, the first pulsar detected to be blinking on and off in the visible part of the spectrum, is the lower left star of the pair about 1 mm apart at the center of this photo. In the long exposure necessary to take this photograph, the star turns on and off so many times that it appears to be a normal star. Only after its radio pulsation had been detected was it observed to be blinking in optical light. It had long been suspected, however, of being the leftover core of the supernova because of its unusual spectrum, which has only a continuum and no spectral lines, and its unusual blue color.

Pulsars and Other Neutron Stars

27

Aims: To study how the cores of stars more massive than about 8 solar masses sometimes collapse to become neutron stars; to learn how we observe neutron stars as pulsars and as x-ray binaries

In this chapter we study those massive stars that become neutron stars after their supernova stages. We can detect those objects as pulsars and x-ray binaries.

27.1 Neutron Stars

We have discussed the fate of the outer layers of a massive star that explodes as a Type II supernova. Now let us discuss the fate of the star's core.

As iron fills the core of a massive star, the temperatures are so high that the iron nuclei begin to break apart into smaller units like helium nuclei. Energy is carried off, the pressure drops, and the forces holding the star stable are overwhelmed by the force of gravity. The core collapses.

As the density increases, the electrons are squeezed into the nuclei and react with the protons there to produce neutrons and neutrinos. The neutrinos escape, thus stealing still more energy from the core, and may also help eject the outer layers if enough neutrinos interact as they pass through. In any case, a gas composed mainly of neutrons is left behind in the dense core as the outer layers explode as a supernova.

Following the explosion, the core may contain as little as a few tenths of a solar mass or possibly as much as two or three solar masses. This remainder is at an even higher density than that at which electron degeneracy holds up a white dwarf. At this density an analogous condition called *neutron degeneracy,* in which the neutrons cannot be packed any more tightly, appears. The pressure caused by neutron degeneracy balances the gravitational force that tends to collapse the core, and as a result the core reaches equilibrium as a *neutron star.*

If more than about three solar masses remain, then the force of gravity will overwhelm even neutron degeneracy. We shall discuss that case in Chapter 28.

Whereas a white dwarf packs the mass of the sun into a volume the size of the earth, the density of a neutron star is even more extreme. A neutron star may be only 20 kilometers or so across (Fig. 27–1), in which space it may contain the mass of about two suns. In its high density, it is like a single, giant nucleus. A teaspoonful of a neutron star could weigh a billion tons.

Before it collapses, the core (like the sun) has only a weak magnetic field. But as the core collapses, the magnetic field is concentrated. It becomes much stronger than any we can produce on earth.

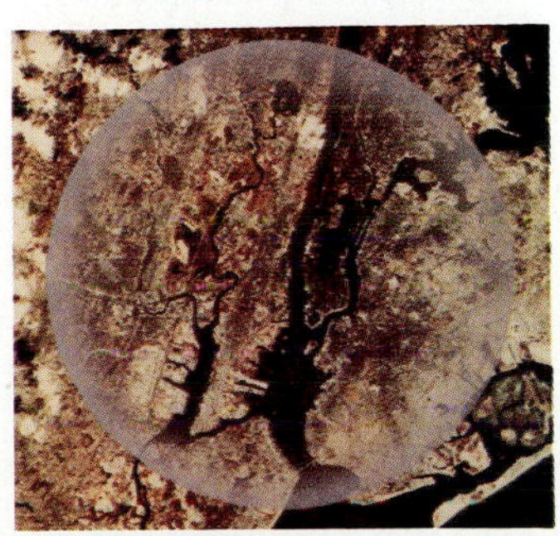

Figure 27–1 A neutron star may be the size of a city, even though it may contain a solar mass or more. A neutron star might have a solid, crystalline crust about a hundred meters thick. Above these outer layers, its atmosphere probably takes up only another few **centimeters.** Since the crust is crystalline, there may be irregular structures like mountains, which would poke up only a few centimeters through the atmosphere.

Figure 27–2 Jocelyn Bell, the discoverer of pulsars. She used a radio telescope—actually a field of aerials—at Cambridge, England. The total collecting area was large, so that it could detect faint sources, and the electronics was set so that rapid variations could be observed.

When neutron stars were discussed in theoretical analyses in the 1930's, there seemed to be no hope of actually observing one. Nobody had a good idea of how to look. Let us jump to consider events of 1967, which will later prove to be related to the search for neutron stars.

27.2 The Discovery of Pulsars

By 1967, radio astronomy had become a flourishing science. But one problem had thus far prevented a major discovery from being made. Like radio receivers in our living rooms, radio telescopes are subject to static—rapid variations in the strength of the signal. It is difficult to measure the average intensity of a signal if the signal strength is jumping up and down many times a second, so radio astronomers usually adjusted their instruments so that they did not record any variation in signal shorter than a second or so. But in 1967 the only rapid variations that astronomers expected besides terrestrial static were those that correspond to the twinkling of stars.

An astronomer at Cambridge University in England, Antony Hewish, wanted to study this "twinkling" of radio sources, which is called *scintillation.* The light from stars twinkles because of effects of our earth's atmosphere. The radio waves from radio sources scintillate not because of terrestrial effects but because radio signals are affected by clouds of electrons in the solar wind. Hewish therefore built a radio telescope uniquely able to detect faint, rapidly varying signals.

One day in 1967, Jocelyn Bell (Fig. 27–2), now Jocelyn Burnell, then a graduate student working on the project, noticed that a set of especially strong variations in the signal appeared in the middle of the night, when scintillations caused by the solar wind are usually weak. In her own words, there was "a bit of scruff" on the tracing of the signal. After a month of observation, it became clear to her that the position of the source of the signals remained fixed with respect to the stars rather than constant in terrestrial time, a sure sign that the source was not terrestrial or solar.

One immediate thought was that the signal represented an interstellar beacon sent out by extraterrestrial life on a planet around another star. For a time the source was called an LGM, for Little Green Men, and the possibility led the Cambridge group to withhold the announcement for a time. The sources were briefly called LGM 1, LGM 2, LGM 3, and LGM 4, but soon it was obvious that the signals had not been sent out by extraterrestrial life. It seemed unlikely that there would be four such beacons at widely spaced locations in our galaxy. Besides, any beings would probably be on a planet orbiting a star, and no effect of a Doppler shift from any orbital motion was detected.

Detailed examination of the signal showed that, surprisingly, it was a rapid set of pulses, with one pulse every 1.3373011 seconds. The pulses were very regularly spaced (Fig. 27–3). Soon, Burnell located three other sources, pulsing with regular periods of 0.253065, 1.187911, and 1.2737635 seconds, respectively.

When the discovery was announced to an astonished astronomical community in 1968, it was immediately apparent that the discovery of these pulsating radio sources, called *pulsars,* was one of the most important astronomical discoveries of the decade. But what were they?

Other observatories turned their radio telescopes to search the heavens for other pulsars. Dozens of new pulsars were found. A typical pulse shape is shown in Fig. 27–4*A*. The time from one pulse to the next pulse is the *period.* The pulse itself lasts only a small fraction of the period. The periods of known pulsars range from thousandths of a second to four seconds. A series of pulses appears in Fig. 27–4*B*.

Figure 27–3 This pulsar, PSR 0329+54, has a period of 0.7145 second. PSR stands for pulsar; 0329+54 stands for 03^h29^m right ascension and 54° declination.

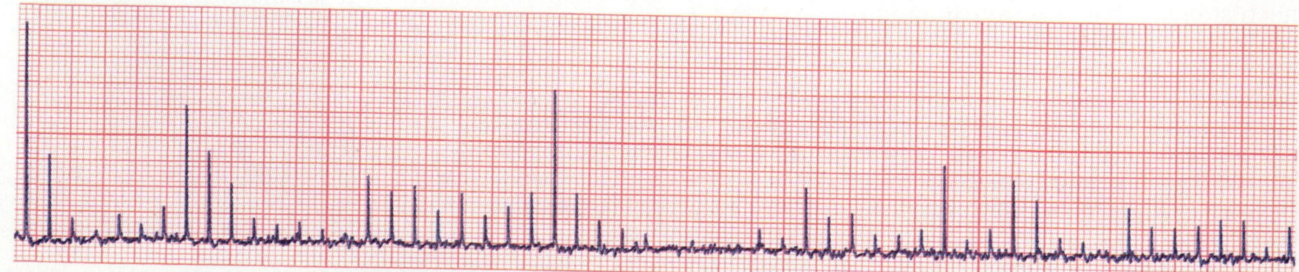

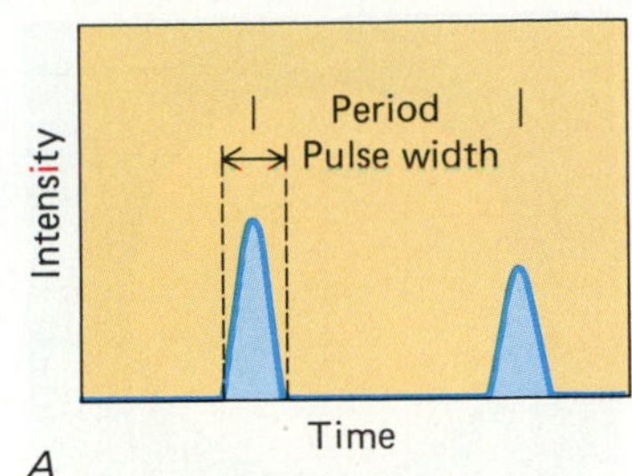

Figure 27–4 (*A*) A schematic series of pulses. (*B*) A series of 400 consecutive pulses from pulsar PSR 0950+08. Each line from bottom to top represents the next 0.253 second. We can see structure within the individual pulses, and an "interpulse" about halfway between some consecutive pulses.

When the positions of all the known pulsars are plotted on a chart of the heavens, it can easily be seen that they are concentrated along the plane of our galaxy (Fig. 27–5). Thus they are clearly objects in our galaxy, for if they were located outside our galaxy, we would expect them to be distributed uniformly (since the Milky Way does not obscure the universe beyond at radio wavelengths). There may be over 100,000 pulsars now active in our galaxy, of which we detect only a tiny fraction.

27.3 What Are Pulsars?

Theoreticians went to work to try to explain the source of the pulsars' signals. The problem had essentially two parts. The first part was to explain what supplies the energy of the signal. The second part was to explain what causes the signal to be regularly timed, that is, what the "clock" mechanism is.

From studies of the second part of the problem, the choices were quickly narrowed down. The signals could not be coming from a pulsation of a normal dwarf star because the rate of pulsation of a star is known to be linked to a star's density. Stars of the density of main-sequence stars pulse too slowly to account for the pulsars. They are too stiff. Also a main-sequence star would be too big. If the sun, for example, were to turn off at one instant, we on earth would not see it go dark at once. The point nearest to us would disappear

Figure 27–5 The distribution of 553 pulsars known in 1992 on a projection that maps the entire sky, with the plane of the Milky Way along the zero-degree horizontal line on the map. The concentration of pulsars near 60° galactic longitude (red labels) on this map merely represents the fact that this section of the sky has been especially carefully searched for pulsars because it is the area of the Milky Way best visible from the Arecibo Observatory. From the concentration of pulsars along the plane of our galaxy we can conclude that pulsars are members of our galaxy; had they been extragalactic we would have expected to see as many near the poles of this map. We can extrapolate that there are at least 100,000 pulsars in our galaxy.

Color indicates whether the pulsars are millisecond pulsars (*red*) with periods less than 100 milliseconds in duration (see Section 27.5), binary pulsars (*blue*; see Section 27.6), or are in known supernova remnants (*green*).

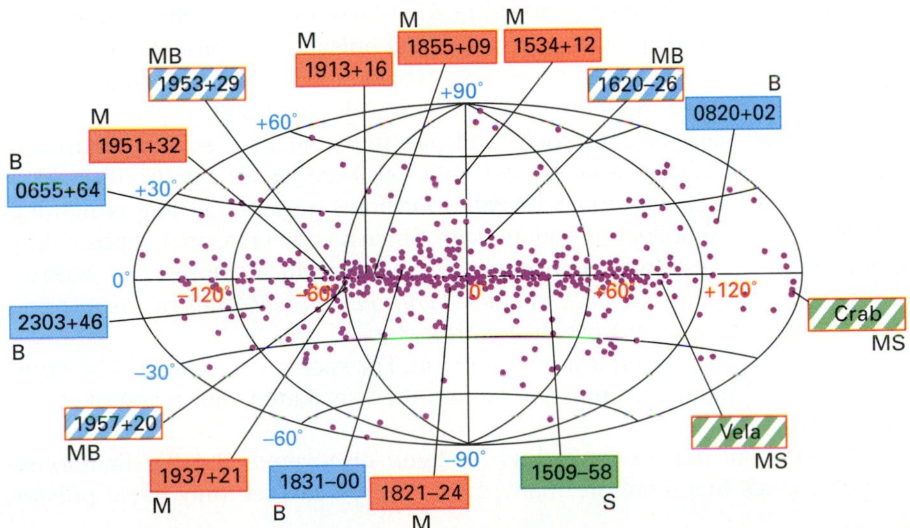

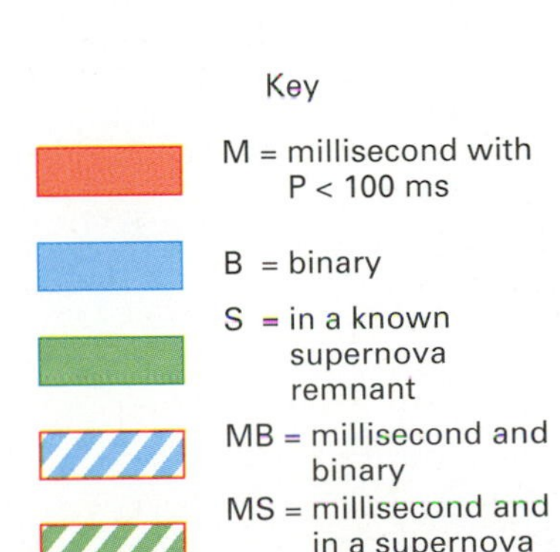

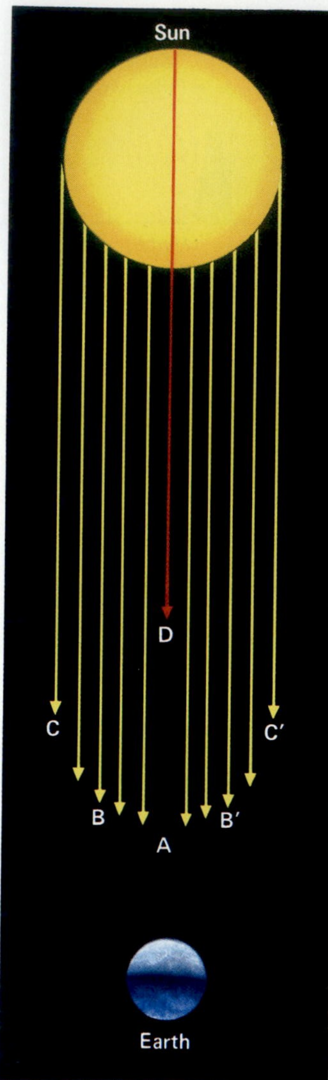

Figure 27–6 Even if a large star were to turn off all at one time, the size of the star is such that it would appear to us to darken over a measurable period of time because radiation travels at the finite speed of light. Normal dwarf stars are too large and not dense enough to account for the rapid pulses from pulsars.

first, and then the darkening would spread farther back around the side of the sun (Fig. 27–6). The sun's radius is 700,000 km, so it would take 5 seconds to go dark. Pulsars pulsed too rapidly to be the size of a normal dwarf star.

That left white dwarfs and neutron stars as candidates. Remember, at that time, neutron stars were merely objects that had been predicted theoretically but had never been detected observationally. Could the pulsars be the more ordinary objects, namely, special kinds of white dwarfs?

Astronomers could conceive of two basic mechanisms as possibilities for the clock. The pulses might be coming from a star that was actually oscillating in size and brightness, or they might be from a star that was rotating. A third alternative was that the emitting mechanism involved two stars orbiting around each other in a binary system. However, it was soon shown that systems of orbiting stars would give off energy in the form of gravitational waves (Section 27.7), making the period change. This was not observed, so orbiting stars were ruled out. This left the rotation or oscillation of white dwarfs or neutron stars.

Theoreticians have found that the denser a star, the more rapidly it oscillates. A white dwarf oscillates in less than a minute. However, it could not oscillate as rapidly as once every second, as would be required if pulsars were oscillating white dwarfs. Neutron stars, however, would indeed oscillate more rapidly, but they would oscillate once every 1/1000 second or so, too rapidly to account for pulsars. Thus oscillating neutron stars were ruled out as well.

What about rotation? Consider a lighthouse whose beacon casts its powerful beam many miles out to sea. As the beacon goes around, the beam sweeps past any ship very quickly, and returns again to illuminate that ship after it has made a complete rotation. Perhaps pulsars do the same thing: they emit a beam of radio waves that sweeps out a path in space. We see a pulse each time the beam passes the earth (Fig. 27–7).

But what type of star could rotate at the speed required to account for the pulsations? We expect a collapsed star to be rotating relatively rapidly, just as ice skaters rotate faster when they pull in their arms to bring all parts of their bodies closer to their axes of rotation (Section 19.10). If the Sun, which rotates once a month, would shrink by a factor that makes it the size of a neutron star, its angular speed would increase greatly. (Angular speed increases with the square of the radius in order to keep an object's angular momentum constant as its distribution of mass changes.) Rotation periods of 1 second are in the range we would expect for collapsed stars.

Could a white dwarf be rotating fast enough? Recall that a white dwarf is approximately the size of the earth. If an object that size were to rotate once every ½ second, the inertial tendency for matter to continue moving straight (which in a rotating body is often loosely called "centrifugal force") would overcome even the immense inward gravitational force of a white dwarf. The outer layers would begin to be torn off. Thus pulsars were probably not white dwarfs; if a pulsar with a period even shorter than ½ second were to be found, then the white dwarf model would be completely ruled out.

Neutron stars, on the other hand, are much smaller, so the centrifugal force would be weaker and the gravitational force would be stronger. As a result, neutron stars can indeed rotate four times a second. There is nothing to rule out their identification with pulsars. Since no other reasonable possibility has been found, astronomers accept the idea that pulsars are in fact neutron stars that are rotating. This is called the *lighthouse model.* Thus by discovering pulsars, we have also discovered neutron stars.

Note that we argued from elimination. For such an argument to be complete and convincing, we must be certain that we started with a complete list (Table 27–1).

About 450 pulsars have thus far been discovered. But the lighthouse model implies that there are many more, for we can see only those pulsars

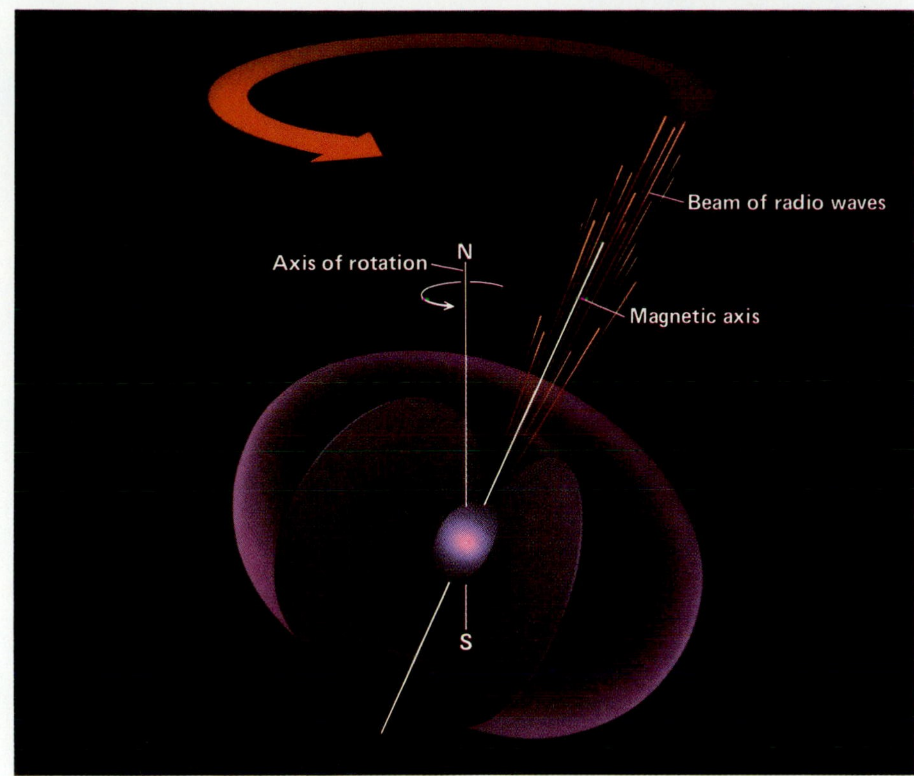

Figure 27–7 In the lighthouse model for pulsars, which is now commonly accepted, a beam of radiation flashes by us once each pulsar period, just as a lighthouse beam appears to flash by a ship at sea. It is believed that the generation of a pulsar beam is related to the neutron star's magnetic axis being aligned in a different direction than the neutron star's axis of rotation. The mechanism by which the beam is generated is not currently understood.

whose beams happen to strike the earth. For pulsars spinning at most orientations, the beams miss us and we do not know that they are there.

We have dealt above with only the second part of the explanation of the emission from pulsars: the clock mechanism. We understand much less well the details of the mechanism by which the radiation is actually emitted in a beam, though we know in general, that we are seeing the effects of electrons moving in spirals around magnetic lines of force. Presumably the beaming has something to do with the extremely powerful magnetic field of the neutron star. The radio waves may be generated by charged particles that have escaped from the neutron star near its magnetic poles. These magnetic poles may not coincide with the neutron star's poles of rotation. After all, the earth's north magnetic pole is not at the north pole, but is rather near Hudson Bay, Canada (Fig. 27–8). The beam seems to have a central cone surrounded by a hollow cone. At higher frequencies, we detect mainly the hollow cone. A different class of models, proposed more recently, considers that the pulsar gets energy from a disk of matter around it, left over from the neutron star's formation. An out-of-favor idea is that the energy may be generated far above the surface of

Table 27–1 Possible Explanations of Pulsars

Hypothesis	Probability
Regular dwarf or giant	Ruled out
System of orbiting white dwarfs	Ruled out
System of orbiting neutron stars	Ruled out
Oscillating white dwarf	Ruled out
Oscillating neutron star	Ruled out
Rotating white dwarf	Ruled out
Rotating neutron star	Most likely

"... when you have excluded the impossible, whatever remains, however improbable, must be the truth."
Spoken by Sherlock Holmes in "The Adventure of the Beryl Coronet"
by Sir Arthur Conan Doyle

Figure 27–8 The magnetic north pole is the point on the earth's surface, about 1400 km from the north geographic pole, to which a magnetic compass would lead you. Interestingly, the magnetic pole is in continuous motion, and its mean position moves about 0.1° per year. Scientists believe that the same fluid motions of the earth's core that create the geomagnetic field are responsible for the movement of the magnetic pole.

the neutron star in the magnetic field, perhaps near the location where the magnetic field lines, which are swept along with the star's rotation, attain velocities that approach that of light.

27.4 The Crab, Pulsars, and Supernovae

Several months after the first pulsars had been discovered, strong bursts of radio energy were discovered coming from the direction of the Crab Nebula. The bursts were sporadic rather than periodic, but astronomers hoped that they were seeing only the strongest pulses and that a periodicity could be found.

Within two weeks of feverish activity at several radio observatories, it was discovered that the pulsar in the Crab Nebula had a period of 0.033 second, the shortest period by far of all the then known pulsars. The fastest pulsar previously known pulsed four times a second, while Crab Nebula's pulsar pulsed at the very rapid rate of thirty times a second. No white dwarf could possibly rotate that fast.

Furthermore, the Crab Nebula is a supernova remnant, and theory predicts that neutron stars should exist at the centers of at least some supernova remnants. Thus the discovery of a pulsar there, exactly where a neutron star would be expected, clinched the identification of pulsars with neutron stars.

Soon after the pulsar was discovered in the Crab Nebula at radio wavelengths, three astronomers at the University of Arizona decided to examine its position to look for an optical pulsar, even though earlier optical searches had failed to detect a pulsar at other sites. They turned their telescope toward a faint star in the midst of the Crab Nebula, and it very soon became apparent that the star was pulsing! They had found an optical pulsar (Fig. 27–9A).

X-ray observations from the Einstein Observatory revealed the structure of hot gas in the Crab Nebula, and clearly showed the pulsar blinking on and off (Fig. 27–9B), since also seen by Rosat (Fig. 27–10). But x-ray observations of other supernova remnants often don't show bright central objects.

Though 20 years have now passed, only three other objects have been found that pulse in the visible spectrum. One is the pulsar in the supernova remnant that covers much of the constellation Vela (which was shown in Figure 26–6). It also gives off x-ray pulses (Fig. 27–11). Several other objects are known that pulse x-rays or gamma rays, but the mechanism by which these pulses are generated is different from the run-of-the-mill radio-pulsing pulsar. When you say "pulsar," you mean that the star pulses radio waves.

Figure 27–9A A display of Hubble Space Telescope photometry of the pulsar in the Crab Nebula. Time goes from left to right; each horizontal line represents the sum of 100 consecutive periods, which takes 3.3 seconds. The main pulse and the interpulse show clearly.

27.5 Slowing Pulsars and the Fast Pulsar

When pulsars were first discovered, their most prominent feature was the extreme regularity of the pulses. It was hoped that they could be used as precise timekeepers, perhaps even more exact than atomic clocks on earth. After they had been observed for some time, however, it was noticed that the pulsars were slowing down very slightly. As a pulsar radiates and accelerates particles to high energy, it gives up some of its rotational energy and slows down.

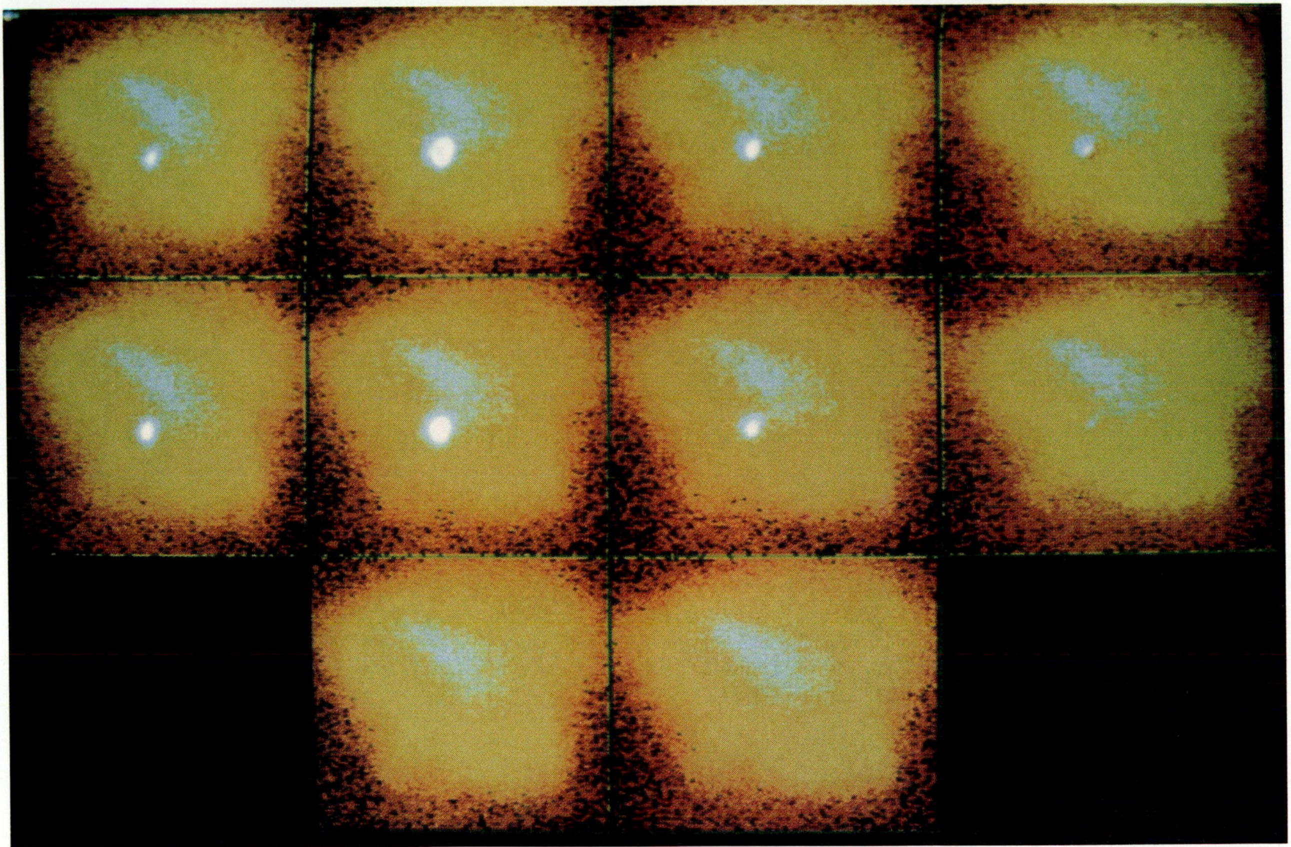

Figure 27–9*B* High-resolution x-ray images of the Crab Nebula and its pulsar, displayed in 10 time intervals covering the 33-millisecond pulse period. A total of 10 hours of observation is included. The main pulse occurs in the second interval, and an in-between pulse called the "interpulse" falls in the sixth. (Einstein Observatory data from F. R. Harnden, Jr., High-Energy Astrophysics Division of the Harvard-Smithsonian Center for Astrophysics)

The filaments in the Crab Nebula still glow brightly, even though over 900 years have passed since the explosion. Furthermore, the Crab Nebula gives off tremendous amounts of energy across the spectrum from x-rays to radio waves. It had long been wondered where the Crab got this energy. Theoretical calculations show that the amount of rotational energy lost by the Crab's neutron star as its rotation slows down is just the right amount to provide the energy radiated by the entire nebula. Thus the discovery of the Crab pulsar solved a long-standing problem in astrophysics: where the energy originates that keeps the Crab Nebula shining.

The Crab pulsar, pulsing 30 times a second, seemed to be spinning incredibly fast. And since it is a young pulsar (only 900 years old) and is slowing down more rapidly than other pulsars, astronomers assumed that the shorter the period, the younger the pulsar and the greater the slow-down rate. But in 1982, a pulsar spinning 20 times faster—642 times per second—was discovered. Even a neutron star rotating at this speed is on the verge of being torn apart. The "fast pulsar," also known as the "millisecond pulsar," since its period is 1.6 milliseconds (0.0016 sec), was found by a University of California team who chose to study in detail a strange radio source whose spectrum resembled that of a pulsar.

This pulsar's period is remaining very constant. Since astronomers think it is the strong magnetic field that slows a pulsar, the magnetic field of a millisecond pulsar must be relatively small for a pulsar. So it must be an old

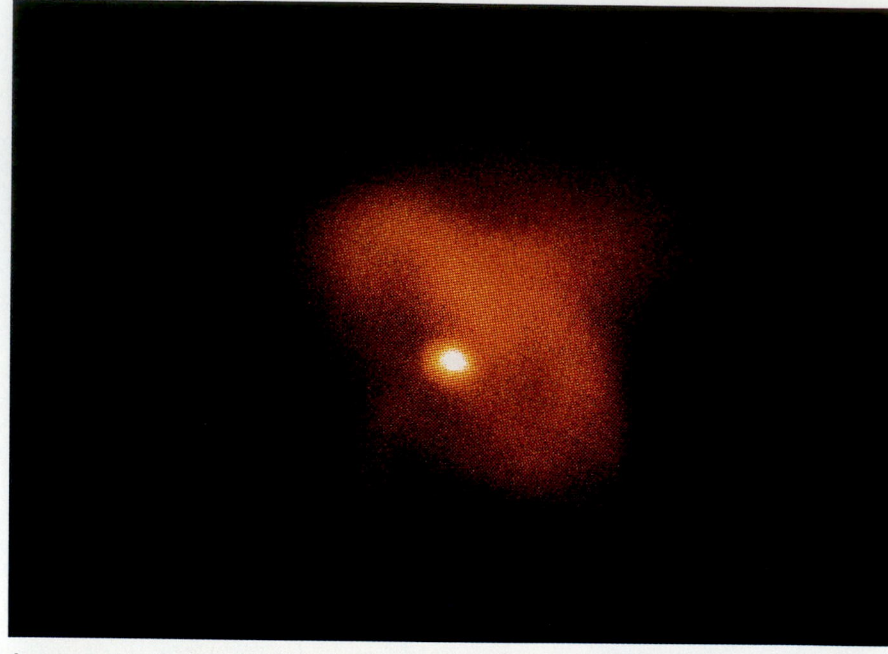

A

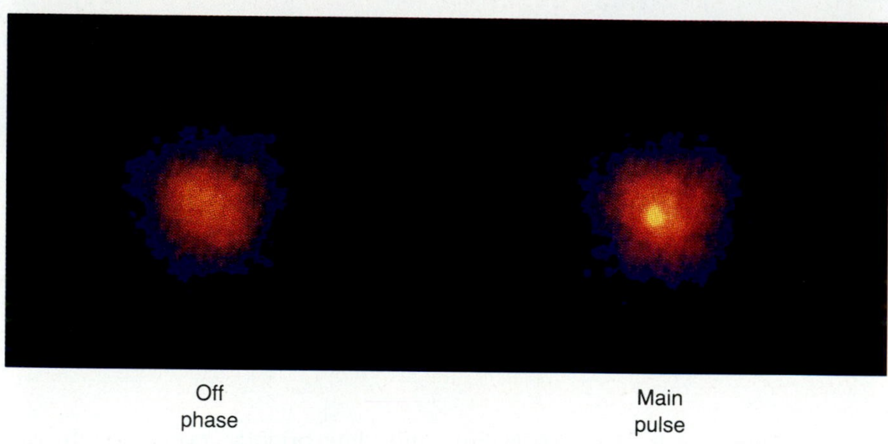

B

Figure 27–10 Rosat views of the pulsar in the Crab Nebula. (*A*) The x-ray view of the Crab Nebula averaged over time. (*B*) An x-ray view of the main pulse (*right*) collected only when the pulsar is on, that is, when the beam of radiation is sweeping by the Earth. The off phase of the pulsar is shown at left.

pulsar, in spite of its short period. About two dozen other "millisecond pulsars" since discovered, ranging in period up to about 300 msec, share its properties (Fig. 27–12); the original millisecond pulsar still has the fastest known spin rate. Roughly half are now in binary systems. Theoretical models can explain the periods of the rest if they were once in binaries. We think they were typically formed over a hundred million years ago (compared with the 5- to 10-million-year age of most pulsars) and their rotation had decayed to very slow rates. Then mass flowing from the companion onto the compact neutron star speeds up the neutron star by supplying additional angular momentum, just as ice skaters speed up by pulling in their arms.

Most of the millisecond pulsars we have detected are in globular clusters; so many stars are packed together in globular clusters that a companion star might have been stripped off in a few of the cases. Supercomputers are used to study the radio observations in searches for periodic signals from millisecond pulsars, and the rate of discovery is increasing. The dense globular cluster

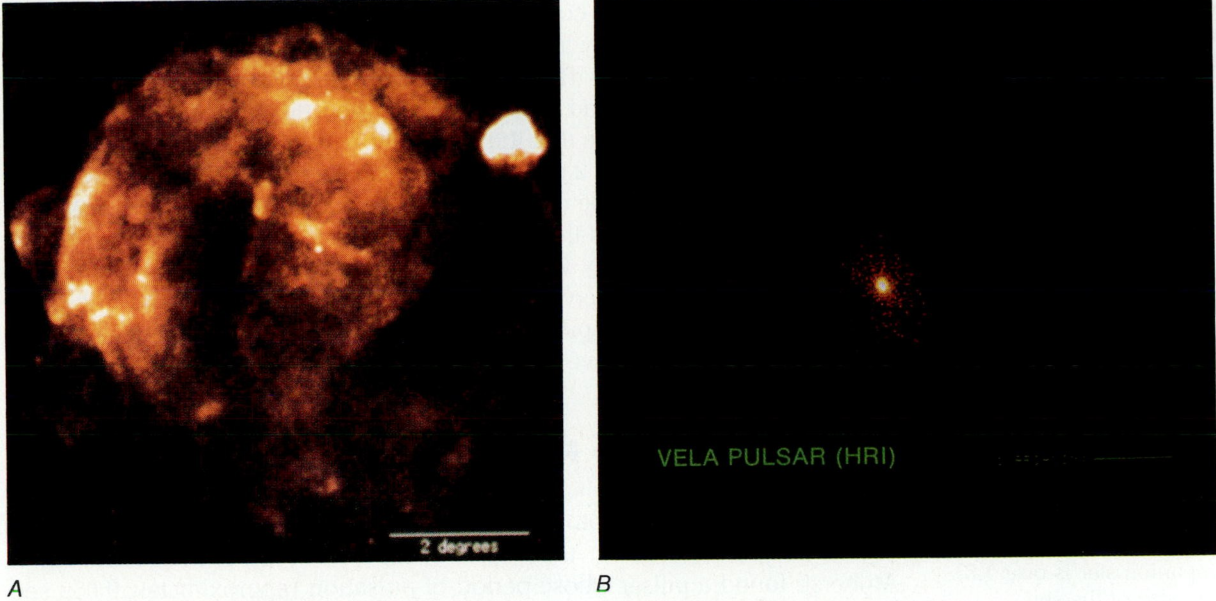

Figure 27–11 (*A*) The Vela and Puppis (smaller) supernova remnants in x-rays from Rosat. Unlike the Crab Nebula, the Vela pulsar is not near the center. (*B*) This x-ray view from the Einstein Observatory shows the Vela pulsar as a point source surrounded by diffuse, weak emission. No x-ray pulses have been detected, though pulses have been detected in the visible and radio regions of the spectrum.

M15 has recently been shown to have three pulsars in its core, with perhaps more waiting to be discovered.

The rotation of millisecond pulsars is so predictable that we may be able to use them to tell time more accurately over long periods than with atomic clocks. They give us time to within a millionth of a second per year. Having a few allows us to compare them, and thus to assess their accuracy as timekeepers. Short period pulsars whose periods are rapidly increasing–appear to be young. Several are associated with the supernova remnants, like the ones in the Crab and in Vela.

Signals from the millisecond pulsar PSR 1957+20 (Fig. 27–13), the second fastest known, disappear abruptly for 50 minutes every 9 hours, indicating that

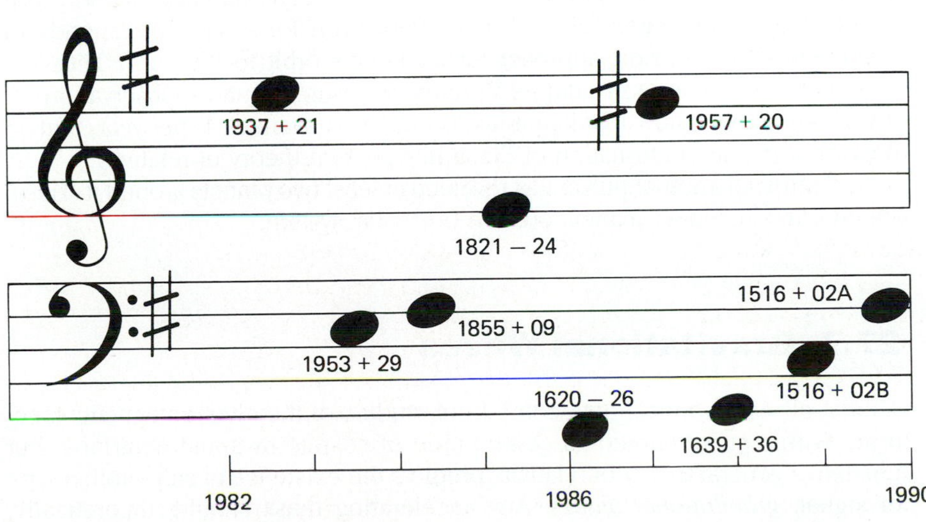

Figure 27–12 Many of the millisecond pulsars have periods fast enough to hear as musical notes when we listen to a signal at the frequency of their pulse rate. The one with the shortest known period, and thus the highest note, was the first to be discovered. Dates of discovery are on the horizontal axis. Dozens more have since been found.

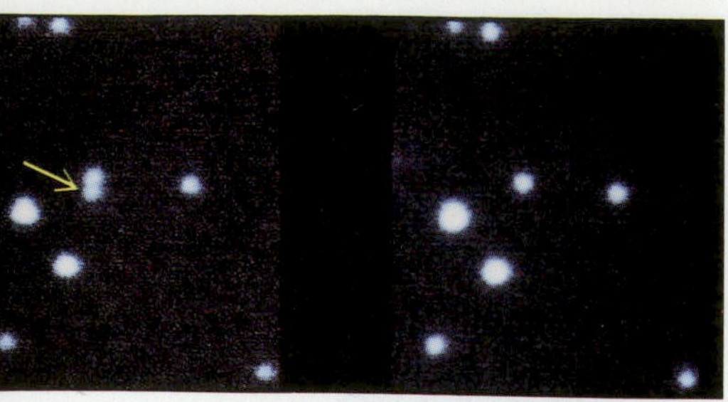

Figure 27–13 A millisecond pulsar, PSR 1957+20, in visible light. The radio and optical radiation disappear abruptly for 50 minutes every 9 hours, indicating that the pulsar is eclipsed by its companion star. The Princeton astronomers who discovered the pulsar have used the eclipse to deduce that the companion star is over 1.5 times the size of the sun, though its mass is only 2% of the sun's; energy from the pulsar has presumably bloated the companion. The companion is being evaporated and may disappear in less than a billion years.

the pulsar is eclipsed by its companion star or from gas coming from it. (Current pulsar names give not only the right ascension but also the sign and number of degrees of declination.) The Princeton University astronomers—Dan Stinebring, Andrew Fruchter, and Joseph Taylor—who discovered the pulsar have used the eclipse to deduce that the companion star is over 1.5 times the size of the sun though it is only 2 per cent of the sun's mass; energy from the pulsar has presumably bloated the companion. The companion is being evaporated, and may disappear in less than a billion years—it is being called the "Black Widow pulsar," in analogy with the black widow spider, which supposedly eats its mate. Optical observations have revealed (Fig. 27–14) that the Black Widow pulsar and its companion are speeding through the galaxy, creating a teardrop-shaped shock wave in the interstellar medium.

27.6 The Binary Pulsar and Pulsar Planets

All pulsars are odd, but some pulsars are odder than others. In 1974, Joseph Taylor and Russell Hulse, then both at the University of Massachusetts at Amherst, found a pulsar whose period of pulsation (approximately 0.059 second, one of the shortest) did not seem very regular. They finally realized that its variation in pulse period could be explained if the pulsar were orbiting another star about every 8 hours. If so, when the pulsar was approaching us, the pulses would be jammed together a bit, and when it was receding, the pulses would be spread apart in time in a type of Doppler effect. The object is a *binary pulsar,* and this particular case is "**the** binary pulsar."

Interpreting the binary pulsar system on the basis of many years of pulse arrival-time data has told us much. The pulses arrive 3 seconds earlier at some times than at others, indicating that the pulsar's orbit is 3 light seconds (1 million km) across, approximately the diameter of the sun. Thus each of the objects has to be smaller than a normal dwarf. The companion object is probably a neutron star too. No pulses have been detected from it, but if it is giving off a radio beacon, it could simply be oriented at an unfavorable angle.

The binary pulsar has allowed us to derive the components' masses, which fall in the range we expected for neutron stars, about 1 to 3 solar masses. But the main interest in the binary pulsar has come about because it provides powerful tests of the general theory of relativity. In Section 23.11, we discussed the excess advance of the perihelion of Mercury, which is only 43 seconds of arc per **century**. The binary pulsar, which is moving in a more elliptical orbit and in a much stronger gravitational field than is Mercury, should have its periastron advance by 4° per **year** (Fig. 27–15). (Periastron for a star corresponds to "perihelion," the closest approach of a planet's orbit to the Sun.) Thus the periastron moves as far in a day as Mercury moves in a century. Observations of the binary pulsar show that its periastron indeed advances 4° per year, and so provide a strong confirmation of Einstein's general theory of relativity.

Timing of another pulsar has revealed at least two planets around it. These are the first accepted planets outside our solar system.

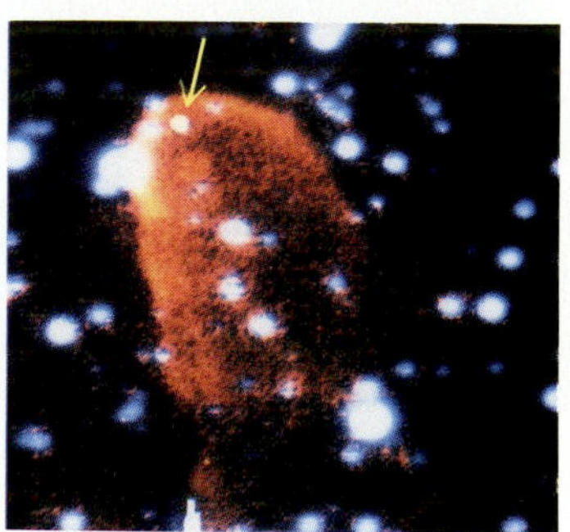

Figure 27–14 Hydrogen emission from the bow shock wave around the eclipsing millisecond pulsar PSR 1957+20, the "Black Widow pulsar." The matter has been blown off the companion, whose side facing the pulsar is heated to 5000 K. The shock waves occur where this stellar wind, which carries the energy from the pulsar's slowing down, encounters the material through which the pulsar is moving. The pulsar, marked with an arrow, is radiating 30 times the energy of the sun. The image is 1 arc minute across.

27.7 Gravitational Waves

Almost all astronomical data come from studies of the electromagnetic spectrum, with a small additional contribution of cosmic rays and neutrinos. But Einstein's general theory of relativity predicts the existence of still another type of signal: *gravitational waves.* Any accelerating mass should, theoretically, cause these waves, which are ripples in the curvature of space-time.

Whereas electromagnetic waves affect only charged particles, gravitational waves should affect all matter. But gravitational waves would be so weak that we could hope to detect them only in situations where large masses were being accelerated rapidly. Two neutron stars revolving in close orbits around each other, as in the binary pulsar, is one possible example. Another would be a supernova or the collapse of a massive star to become a black hole.

Joseph Weber of the University of Maryland was the first to build sensitive apparatus to search for gravitational waves. The detector was a large metal cylinder (Fig. 27–16), delicately suspended from wires and protected from local motion. When a gravitational wave hit the cylinder, it should have begun vibrating; the vibrations could be detected with sensitive apparatus. His apparatus started working in 1969; though it sometimes started vibrating, the vibrations seem to have been from terrestrial causes rather than from gravitational waves. Other scientists have since built more sensitive gravitational-wave apparatus, yet no gravitational waves have been detected.

Current terrestrial gravitational-wave apparatus uses laser interferometers to search for gravitational waves. Both Caltech and MIT are building systems in which laser beams go up and back long arms, so that interference properties of light can be used to measure minute changes in the lengths of the arms. The Caltech instrument is a prototype with two arms each 40 m long; the light is reflected back and forth thousands of times between mirrors in each arm. The hope is to scale it up by a factor of 100, to 4-km arms.

In the meantime, the binary pulsar turned out to be a quicker way to show that gravitational waves exist. The gravitational waves emitted by the system carry away energy. This makes the orbits shrink, which results in a slight decrease of the period of revolution. By 1978 Taylor and colleagues had detected such a slight speedup. The speedup has continued at precisely the rate predicted by the general theory of relativity. The pulsar now reaches periastron more than two seconds earlier than it would if its period had remained

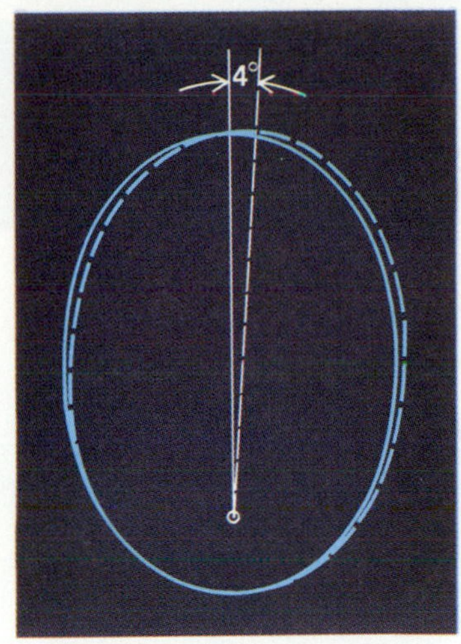

Figure 27–15 The periastron of the binary pulsar PSR 1913+16 advances by 4° per year, providing a strong endorsement of Einstein's general theory of relativity. The angle is shown for the apastron, the farthest point of the pulsar from the other object.

A

B

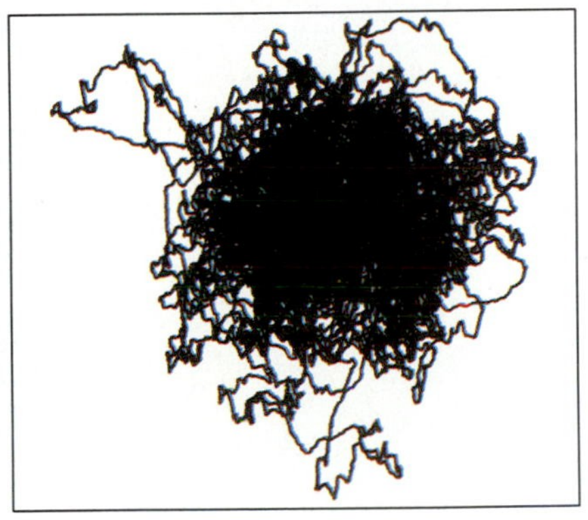

C

Figure 27–16 (*A*) One of Joseph Weber's original gravitational-wave detectors from the 1960's, a large aluminum cylinder weighing 4 tons, delicately suspended so that gravitational waves would set it vibrating 1660 times per second. Weber originally detected many "events" simultaneously affecting bars in both Maryland and Illinois, which could apparently be caused only by a global event like a gravitational wave because they were separated by so great a distance. However, most of the newer, more sensitive experiments have not found events caused by gravitational waves. Events recently affecting widely separated bars in Italy seem to occur too often to be from gravitational waves. In any case, Weber's efforts have stimulated interest in detecting gravitational waves, and even more sensitive detectors are currently in use or under construction at Louisiana State and the Universities of Maryland, Rochester, Rome, and Tokyo. (*B*) A cooled bar, worked on by J. A. Tyson of AT&T Bell Labs. (*C*) The signal from the cooled bar. We see the slowly changing readout over many weeks; a gravitational wave would cause a jump.

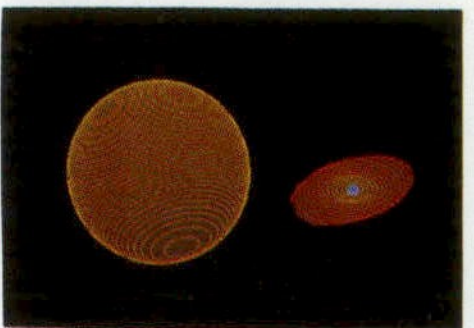
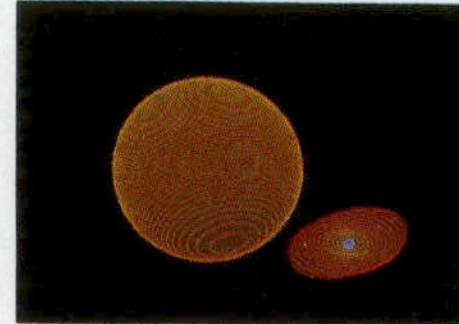
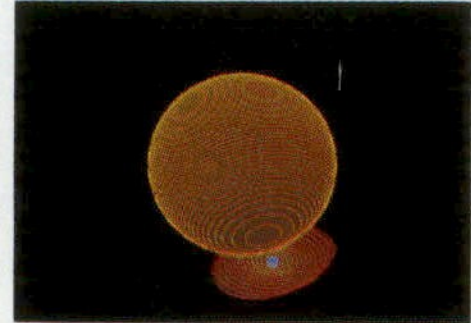
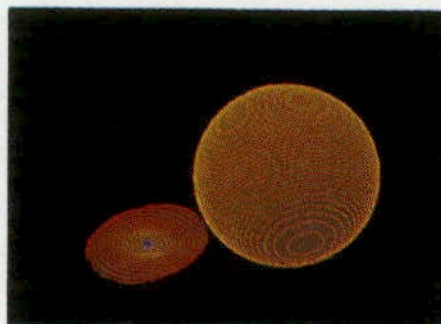
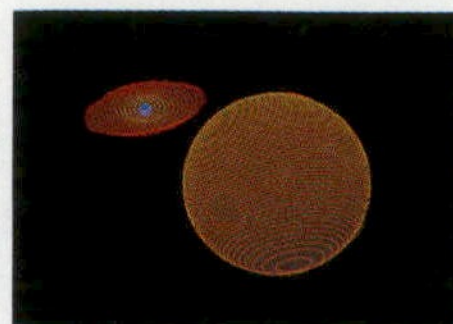

Figure 27–17 A computer simulation of the rotation of a disk of gas accumulated by the gravity of Hercules X-1, that is, the accretion disk.

constant since it was discovered. The agreement of observations with predictions provides strong, though indirect, evidence for the existence of gravitational waves.

*27.8 X-Ray Binaries

Telescopes in orbit that are sensitive to x-rays have detected a number of strong x-ray sources, some of which are pulsating. Most of these objects pulse only in the x-ray region of the spectrum. One of the most interesting is Hercules X-1 (the first x-ray source to be discovered in the constellation Hercules). It pulsates with a 1.24-second period in x-rays and in visible light.

Hercules X-1 (Fig. 27–17) and the other pulsating x-ray sources are apparently examples of the type of evolution of binary systems that we discussed in Section 25.7. Theoreticians think that the x-ray sources are radiating because mass from the companion is being funneled toward the poles of the neutron star by the neutron star's strong magnetic field (Fig. 27–18). Unlike the slowing down of the pulse rate of pulsars (which give off pulses in the radio region of the spectrum), the pulse rate of the binary x-ray sources usually speeds up. The period of Hercules X-1, for example, is growing shorter.

A sudden brightening in 1989 of V404 Cygni, which had last been seen to brighten in 1938, corresponded to the detection of a strong x-ray source by Ginga. The x-ray signals flicker by a factor of 10 within seconds and the visible light flickers on the scale of minutes. The outburst, which was studied across the entire spectrum this time, may have resulted from a glob of material suddenly falling into a neutron star. The last similar outburst detected on earth, from a source in Monoceros, took place in 1975 in an object that is now thought to contain a black hole, so perhaps V404 Cygni also contains a black hole instead of a neutron star; we shall discuss such objects in the following chapter.

Thus we know of two types of systems in which neutron stars are located. First, isolated neutron stars may give off pulses of radio waves, and we detect these stars as pulsars. Second, pulses of x-rays tend to come from neutron stars or other compact objects in binary systems.

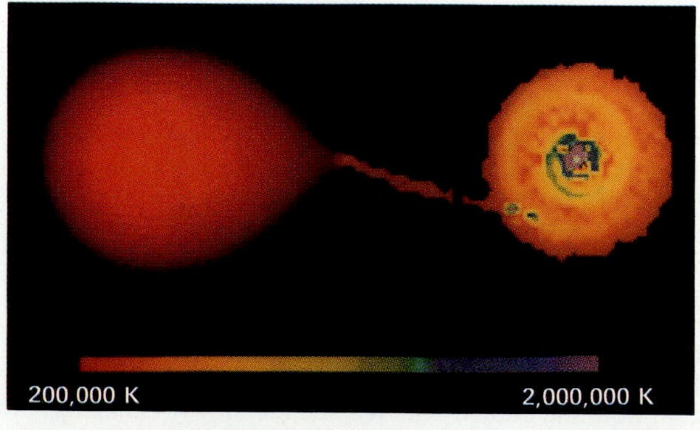

Figure 27–18 A computer calculation showing the flow of gas from a supergiant companion to an accretion disk around a neutron star.

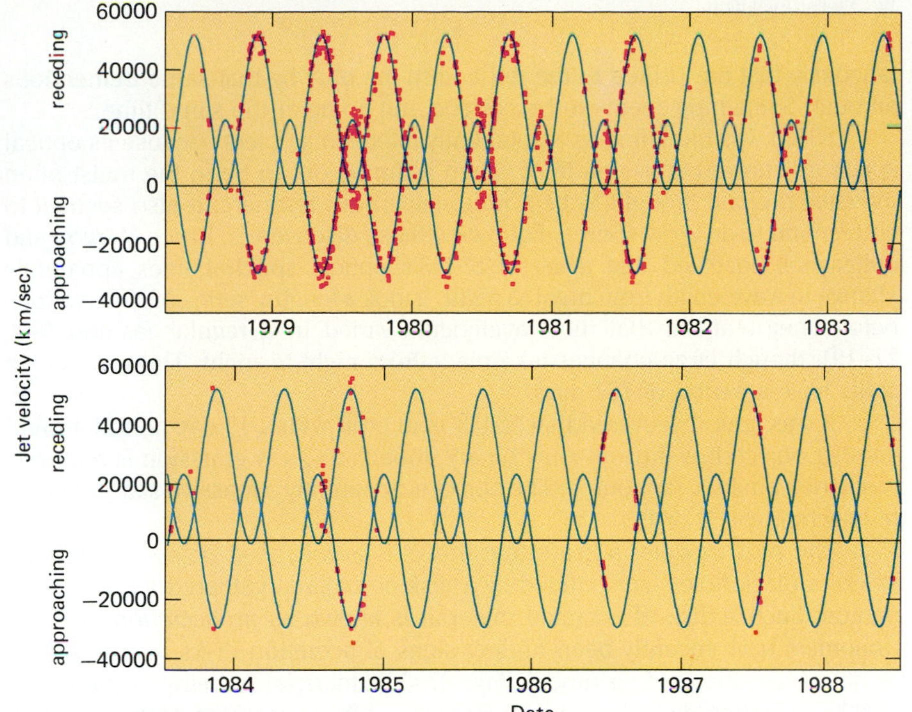

Figure 27–19 The redshifts and blueshifts of spectral lines in SS433 vary in a regular fashion with a 164-day period.

*27.9 SS433: The Star That Is Coming and Going

SS433, one of the most exciting single objects in astronomy, contains spectral lines that change in wavelength by a tremendous amount. Its shifts of hundreds of angstroms correspond to Doppler shifts from velocities up to 50,000 km/sec. Such large shifts, about 15 per cent of the speed of light, are entirely unprecedented for an object in our galaxy. Furthermore, spectral lines appear simul-

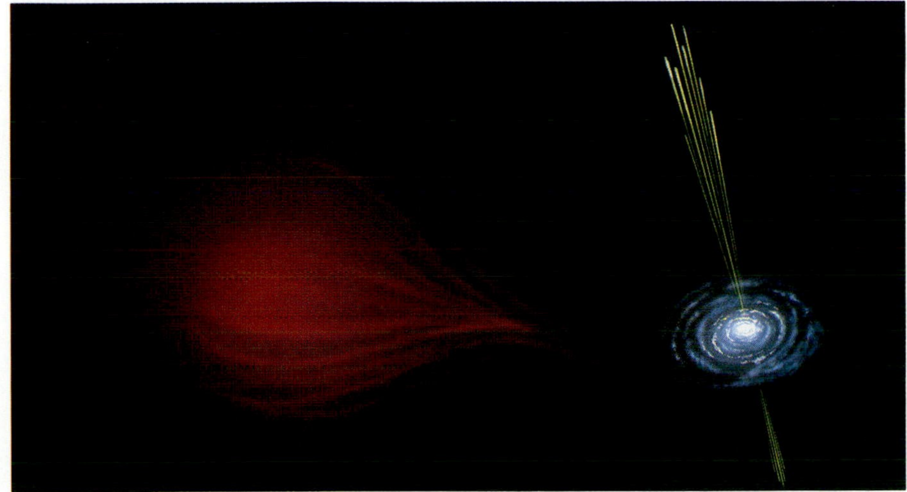

Figure 27–20 A model of SS433 in which the radiation emanates from two narrow beams of matter that are given off by the accretion disk. The model follows ideas of Mordechai Milgrom, Bruce Margon, Jonathan Katz, and George Abell. The velocity of gas emitted is 80,000 km/sec in each direction, though we see a somewhat lower velocity because we do not see the jets head on. In some models, variations in pressure and density of the disk can cause the emission.

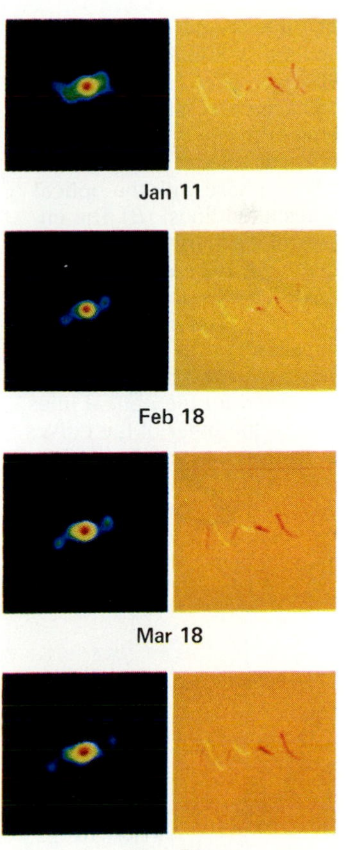

Figure 27–21 SS433 observed at a radio wavelength of 6 cm with the Very Large Array (VLA). SS433 is in the center of a supernova remnant.

The column on the right shows the corkscrew paths followed by individual ejecta blobs over about 3 months of time, with color showing the Doppler shift the blobs had at each position.

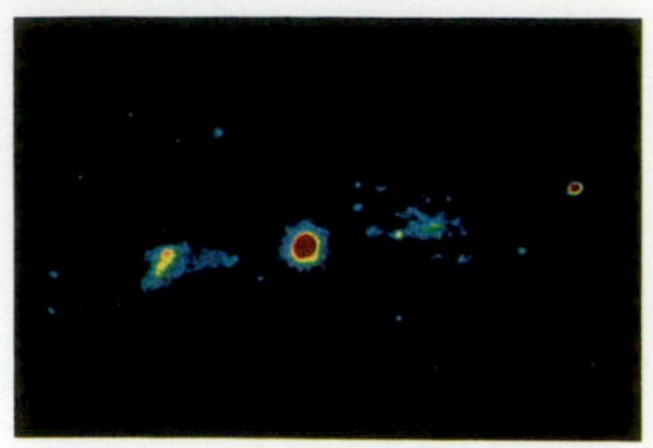

Figure 27–22 An x-ray image of SS433 from the Einstein Observatory. The tunnels carved by the jets extend to the sides of the bright central image. SS433 is not an especially strong source; some of the other binary x-ray sources are 100 times stronger.

taneously that are shifted to the red and to the blue by that same tremendous amount. Something seems to be coming and going at the same time.

SS433 was identified as a potentially interesting object because its optical spectrum showed emission lines. Then it turned out to be in the midst of an interstellar cloud thought to be a radio supernova remnant; it also seemed to correspond to an x-ray source. Following these discoveries, Bruce Margon and colleagues observed that a set of SS433's optical spectral lines apparently change in wavelength from night to night. It took a lengthy series of observations before they realized that the wavelengths varied in a regular fashion (Fig. 27–19), though large changes take place from night to night. The wavelength shifts have a period of 164 days.

It was later discovered that SS433 is in orbit with a 13-day period around another object. It is thus an x-ray binary, though its x-ray emission is relatively weak for a binary x-ray source. The object is apparently a massive star, probably of spectral type O or B.

The best models indicate that SS433 is a collapsed object—a neutron star or a black hole—surrounded by a disk of material gathered from its companion. Such a disk of accreted material is known as an *accretion disk;* astronomers have recently been finding signs of accretion disks in many types of objects. In the leading model (Fig. 27–20), material is being ejected from SS433's accretion disk along a line. Because of the precession of the accretion disk, the line traces out a cone. (Precession is the wobbling motion, as of a child's top, that we have already met for the earth's rotation in Section 5.3.) If we assume that SS433 is oriented with its spin axis more-or-less but not precisely perpendicular to the direction in which we are looking, we see radiation given off by the material ejected upward as redshifted and radiation given off by the material ejected downward as blueshifted. The redshift and blueshift vary as the line traces out the cone. The extended series of observations give the angles, and show that the material is ejected at 26 per cent the speed of light. High-resolution radio observations (Fig. 27–21) and x-ray observations (Fig. 27–22) seem to show the jets, confirming the model. X-ray spectra show that most of the x-ray emission comes from the jets (Fig. 27–23).

Some of the high-resolution radio observations have actually shown knots of emission moving away from the core. Putting these angular measurements

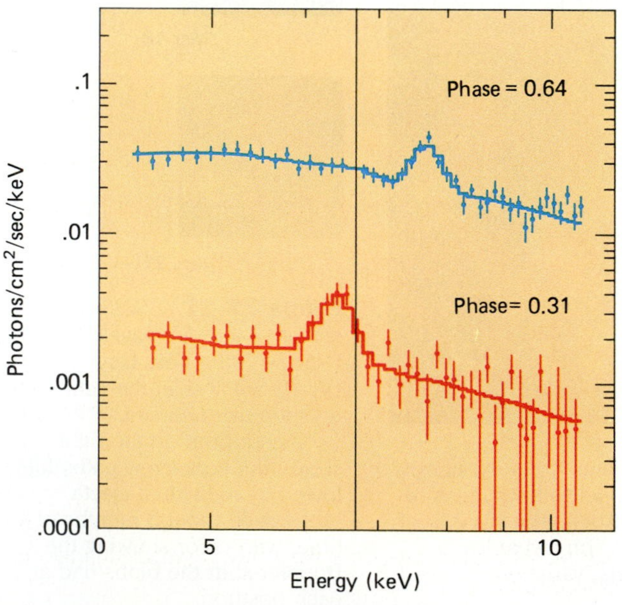

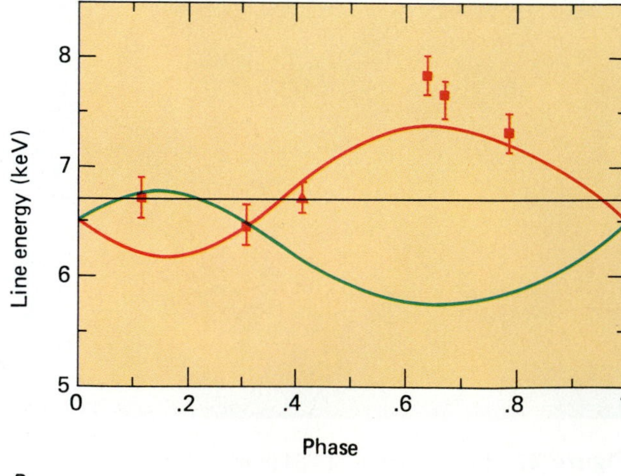

Figure 27–23 (*A*) X-ray spectra of SS433 taken from EXOSAT at different phases in the 162-day precession period. An emission line from iron changes energy from 6.5 to 7.8 keV in phase with the optical blueshifted lines. (*B*) The energy of the iron line over a complete 162-day period from EXOSAT (*squares*) and TENMA (*triangle*). The red and green curves show the predicted variation based on the velocities measured for the optical lines. One or the other of the curves is always obscured by the accretion disk. The observations demonstrate that most of the x-ray emission from SS433 comes from the jets.

together with the velocities in km/sec from the optical spectra, we find that SS433 is about 16,000 light years from us, about half the distance from us to the center of the galaxy. Since other examples of jets occur in distant galaxies, SS433 may be a "nearby" example from which we can generalize to larger, more distant, more powerful cases.

"SS" indicates that it is from a catalogue of stars with hydrogen emission lines compiled in the 1960's by C. Bruce Stephenson and Nicholas Sanduleak of Case Western Reserve University.

Summary and Outline

After a star's fusion stops, neutron degeneracy can support a remaining mass of up to 2 or 3 solar masses. A neutron star may be only 10 km in radius, and so is fantastically dense (Section 27.1).

Pulsars were discovered when a radio telescope was built that did not mask rapid time variations in the signal. Pulsars have very regular signals, and are slowing down very slightly. Pulse periods range from 1.6 milliseconds to 4 sec (Section 27.2).

Scientists have identified pulsars with rotating neutron stars by a process of elimination (Section 27.3), which requires complete, logical reasoning.

A pulsar is observed in the center of the Crab Nebula, a supernova remnant, where we would expect to observe a neutron star.

Pulsars emit mainly in the radio spectrum, although the Crab and one other pulsar also pulse light, x-rays, and γ rays. Other neutron stars pulse x-rays (Section 27.4).

Most pulsars are gradually slowing down as they grow older (Section 27.5). A report of rapid rotation of a pulsar in Supernova 1987A fits with this model. The millisecond pulsars, however, are rotating very rapidly yet slowing only slightly; they have been rejuvenated.

Several binary pulsars have been discovered. The periastron of the Hulse-Taylor binary advances 4°/year, confirming general relativity (Section 27.6).

General relativity predicts the existence of gravitational waves. Attempts to detect them directly have probably been unsuccessful. The orbital period of the binary pulsar is slowing down by just the predicted amount if gravitational waves are indeed being emitted from that system, confirming Einstein's theory (Section 27.7).

Many x-ray sources are caused by the infall of material from its companion onto a neutron-star member of a binary system (Section 27.8).

An object in our galaxy, SS433, has spectral lines that shift periodically to the red and to the blue by about 15 per cent of the speed of light. The Doppler-shifted radiation may come from beams ejected from an accretion disk about the neutron star in an x-ray binary (Section 27.9).

Key Words

neutron degeneracy, neutron star, scintillation, pulsar, period, lighthouse model, binary pulsar*, gravitational waves, accretion disk*

*This term is found in an optional section.

Questions

1. In your own words, replace the last column of Table 27–1 with the reasons why each of the explanations for pulsars involving single stars was ruled out except for the case of rotating neutron stars.

2. In view of current theories about supernovae and pulsars, list the pieces of evidence that indicate that the Crab pulsar is a young one.

†3. A pulsar has a period of 1 second and a pulse lasts 5 per cent of the cycle. What is the pulse's width in seconds? What might that say about the size of the object emitting the pulse on (a) the oscillating model and (b) the rotating model?

4. Why did it seem strange that the fast pulsar is slowing down so slowly? Explain.

5. Explain two important consequences of the discovery of the binary pulsar.

6. Why is the discovery of gravitational waves a test of the general theory of relativity?

7. Sketch a model for an x-ray binary and explain how it leads to x-ray emission.

8. A shift of the Hα line in SS433 of 40,000 km/sec corresponds to a shift of how many Å in wavelength? (Ignore any effects of relativity.)

9. Describe why the precession of an accretion disk in SS433 can cause varying Doppler shifts.

10. Explain why the true velocities of the jets in SS433 may be even greater than the velocities we measure directly from Doppler shifts.

†This question requires a numerical solution.

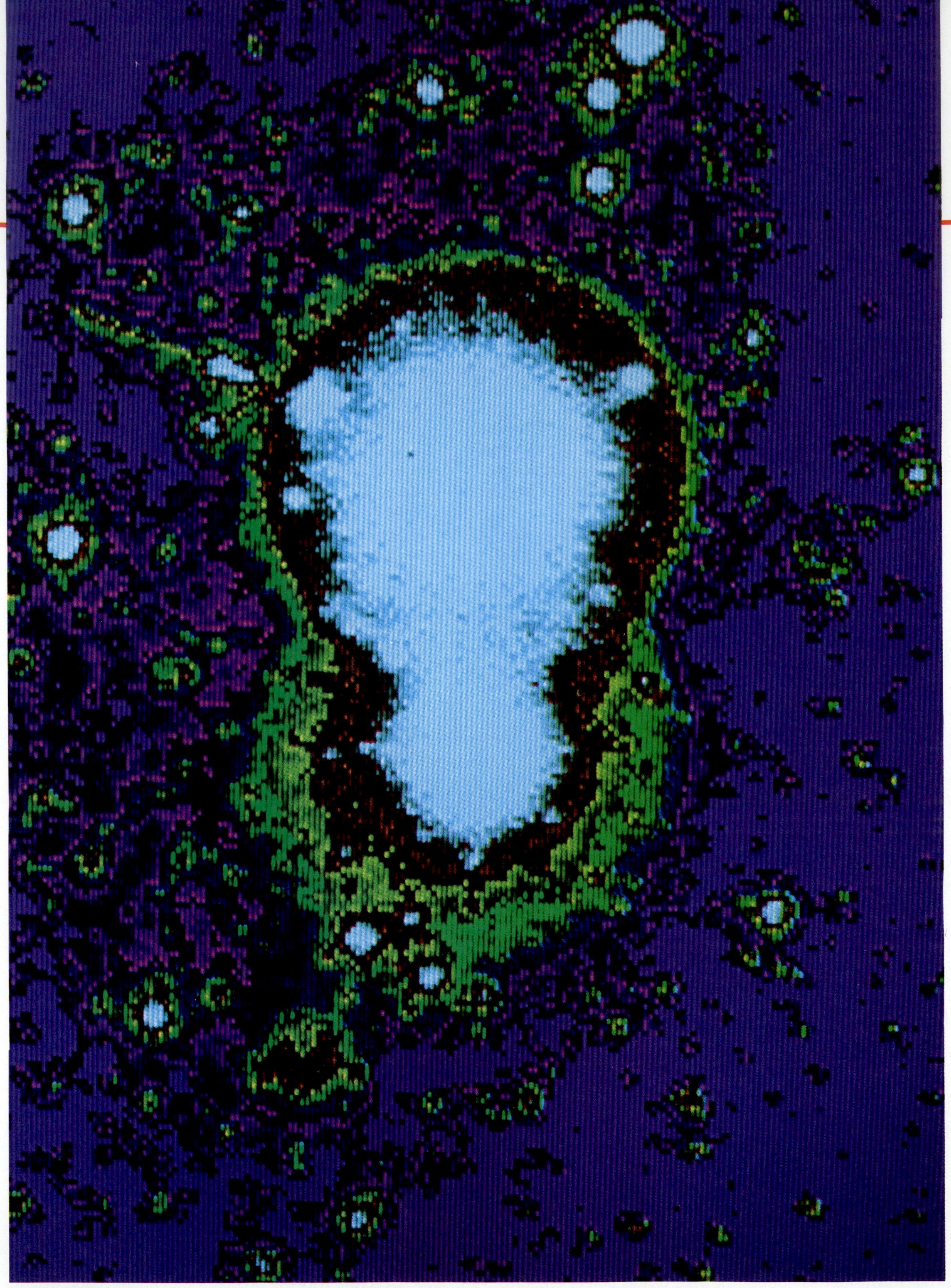

The overexposed object, at the top center of this false-color image, in which brightness corresponds to color, is the supergiant star HDE 226868, which is thought to be the companion of the first black hole to be discovered, Cygnus X-1.

Stellar Black Holes

28

Aims: To understand the black holes that form from stars and how they might be detected, and to comment on black holes much more massive or much less massive than the sun and stars

The strange forces of electron and neutron degeneracy support dying lightweight, middleweight, and some heavyweight stars against gravity. The strangest case of all occurs at the death of the most massive stars, which contained much more than 8 and up to about 60 solar masses when they were on the main sequence. They generally undergo supernova explosions like less massive heavyweight stars, but some of these most massive stars may retain cores of over 2 or 3 solar masses. Nothing in the universe is strong enough to hold up the remaining mass against the force of gravity. The remaining mass collapses, and continues to collapse forever. The result of such a collapse is one type of *black hole,* in which the matter disappears from contact with the rest of the universe. Later, we shall discuss the formation of black holes in processes other than those that result from the collapse of a star.

28.1 The Formation of a Stellar Black Hole

When nuclear fusion ends in the core of a star, gravity causes the star to contract. We have seen in Chapter 25 that a star that has a final mass of less than 1.4 solar masses will end its life as a white dwarf and, when it is in a binary system, as a supernova. In Chapter 26 we saw that a heavier star and some white dwarfs in binary systems will become supernovae, and Chapter 27 showed that if the remaining mass of the core is less than 2 or 3 solar masses, it will wind up as a neutron star. We are now able to observe both white dwarfs and neutron stars and so can study their properties directly.

It now seems reasonable that in some cases more than 2 or 3 solar masses remain after the supernova explosion. The star collapses through the neutron star stage, and we know of no force that can stop the collapse. In some cases, the matter may even have become so dense as the star collapsed that no explosion took place and no supernova resulted.

Astronomers long assumed that the most massive stars would somehow lose enough mass to wind up as white dwarfs. When the discovery of pulsars ended that prejudice, it seemed more reasonable that black holes could exist.

The value for the maximum mass that a neutron star can have is a result of theoretical calculation. We must always be aware of the limits of accuracy of any calculation, particularly of a calculation that deals with matter in a state that is very different from states that we have been able to study experimentally. Still, the best modern calculations show that the limit of a neutron star mass is 2 or 3 solar masses.

We may then ask what happens to a 5- or 10- or 50-solar-mass star as it collapses, if it retains more than 3 solar masses. It must keep collapsing, getting denser and denser. We have seen that Einstein's general theory of relativity predicts that a strong gravitational field will redshift radiation, and that this prediction has been verified both for the sun and for white dwarfs (as discussed in Section 25.5). Also, radiation will be bent by a gravitational field, or at least appear to us on earth as though it were bent (Section 23.11). Further, the general theory of relativity indicates that pressure as well as mass produces gravity. This effect compounds the gravitational contraction of a very massive star.

As the mass contracts and the star's surface gravity increases, radiation is continuously redshifted more and more, and radiation leaving the star other than perpendicularly to the surface is bent more and more. Eventually, when the mass has been compressed to a certain size, radiation from the star can no longer escape into space. The star has withdrawn from our observable universe, in that we can no longer receive radiation from it. We say that the star has become a *black hole.*

Do not confuse a black hole with a black body. A black body is merely a radiating body whose radiation follows Planck's law. A black hole, essentially, does not radiate (though an exception to this rule—really an alternate way of looking at the whole picture—will be discussed in Section 28.6).

Why do we call it a black hole? We think of a black surface as a surface that reflects none of the light that hits it. Similarly, any radiation that hits the surface of a black hole continues into the black hole and is not reflected. In this sense, the object is perfectly black.

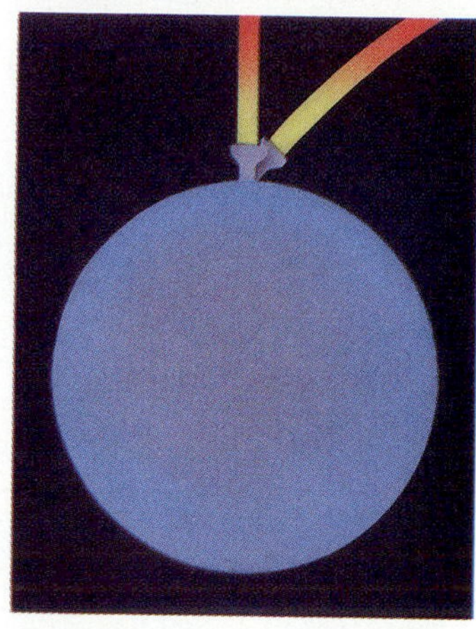

Figure 28–1 As the star contracts, a light beam emitted other than radially outward will be bent.

28.2 The Photon Sphere

Let us consider what happens to radiation emitted by the surface of a star as it contracts. Although what we will discuss affects radiation of all wavelengths, let us simply visualize standing on the surface of the collapsing star while holding a flashlight.

On the surface of a supergiant star, we would note only very small effects of gravity on the light from our flashlight. If we shine the beam at any angle, it seems to go straight out into space.

As the star collapses, two effects begin to occur. Although we on the surface of the star cannot notice them ourselves, a friend on a planet revolving around the star could detect the effects and radio back information to us about them. For one thing, our friend could see that our flashlight beam is redshifted. Second, our flashlight beam would be bent by the gravitational field of the star (Fig. 28–1). If we shined the beam straight up, it would continue to go straight up. But if we shined it away from the vertical, the beam would be bent even farther away from the vertical. When the star reaches a certain size, a horizontal beam of light would not escape (Fig. 28–2).

From this time on, only if the flashlight is pointed within a certain angle of the vertical does the light continue outward. This angle forms a cone, with its apex at the flashlight, and is called the *exit cone* (Fig. 28–3). As the star grows smaller yet, we find that the flashlight has to be pointed more directly upward in order for its light to escape. The exit cone grows smaller as the star shrinks.

Figure 28–2 Light can be bent so that it falls back onto the star.

When we shine our flashlight upward in the exit cone, the light escapes. When we shine our flashlight in a direction outside the exit cone, the light is bent sufficiently that it falls back to the surface of the star. When we shine our flashlight exactly along the side of the exit cone, the light goes into orbit around the star, neither escaping nor falling onto the surface.

The sphere around the star in which the light can orbit is called the *photon sphere.* Its size can be calculated theoretically. Its radius is 13.5 km for a star of 3 solar masses, and varies linearly with the mass; that is, a star of 6 solar masses is 27 km in radius, and so on.

As the star continues to contract, the theory shows that the exit cone gets narrower and narrower. Light emitted within the exit cone still escapes. The photon sphere remains at the same height even though the matter inside it has contracted further, since the total amount of matter within has not changed.

28.3 The Event Horizon

If we were to depend on our intuition, we might think that the exit cone would simply continue to get narrower. But when we apply the general theory of relativity, we find that when the star contracts beyond a certain size the cone vanishes. Light no longer can escape into space, even when it is travelling straight up.

The solutions to Einstein's equations that predict this were worked out by Karl Schwarzschild in 1916, shortly after Einstein advanced his general theory. The radius of the star at the time at which light can no longer escape is called the *Schwarzschild radius* or the *gravitational radius*, and the spherical surface at that radius is called the *event horizon* (Fig. 28–4). The event horizon is not a physical surface. The radius of the photon sphere is exactly $3/2$ times this Schwarzschild radius.

We can visualize the event horizon in another way, by considering a classical picture based on the Newtonian theory of gravitation. The picture is essentially that conceived in 1796 by Laplace, the French astronomer and mathematician. You must have a certain velocity, called the *escape velocity*, to escape from the gravitational pull of another body. For example, we have to launch rockets at 11 km/sec (40,000 km/hr) in order for them to escape from the earth's gravity. For a more massive body of the same size, the escape velocity would be higher. Now imagine that this body contracts. We are drawn closer to the center of the mass. As this happens, the escape velocity rises. When all the mass of the body is within its Schwarzschild radius, the escape velocity becomes equal to the speed of light. Thus even light cannot escape. If we begin to apply the special theory of relativity, we might then reason that since nothing can go faster than the speed of light, nothing can escape. Now let us return to the picture according to the general theory of relativity.

The size of the Schwarzschild radius depends linearly on the amount of mass that is collapsing. A star of 3 solar masses, for example, would have a Schwarzschild radius of 9 km. A star of 6 solar masses would have a Schwarzschild radius of twice 9 km, or 18 km. One can calculate the Schwarzschild radii for less massive stars as well, although the less massive stars would be held up in the white-dwarf or neutron-star stages and not collapse to their Schwarzschild radii (although conceivably some objects could be sufficiently compressed in a supernova). Since one couldn't, theoretically, put a tape measure into and out of a black hole to measure its radius, we would actually measure its circumference and divide by 2π.

Note that anyone or anything on the surface of a star as it passed its event horizon would not be able to survive. An observer would be torn apart by the tremendous difference in gravity between head and foot. (This is called a tidal force, since this kind of difference in gravity also causes the tides on earth.) If the tidal force could be ignored, though, the observer on the surface of the star would not notice anything particularly wrong as the star passed its event horizon. If we chose to stay with the observer at that time, we could still be pointing our flashlights up into space trying to signal our friends on the earth. But once we passed the event horizon, no answer would ever come because our signal would never get out.

Once the star passes inside its event horizon, we lose contact with it. Theories indicate that it continues to contract and that nothing can ever stop its contraction. In fact, the mathematical theory predicts that it will contract to zero radius, a situation that seems impossible to conceive of. The point at which it will have zero radius (it has infinite density there) is called a *singularity*. Strange as it seems, theory predicts that a black hole contains a singularity.

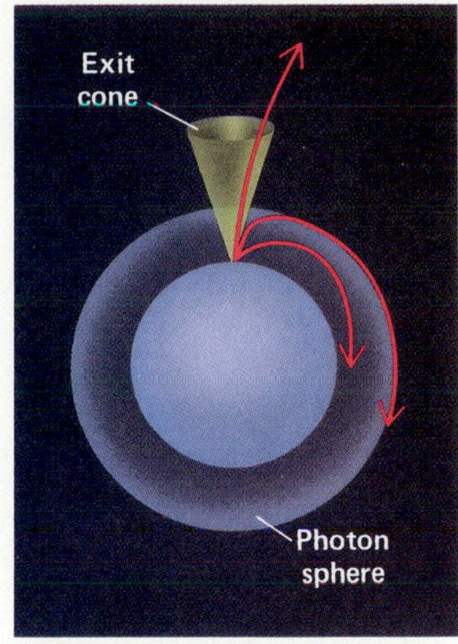

Figure 28–3 When the star has contracted enough (the inner sphere), only light emitted within the exit cone escapes. Light emitted on the exit cone goes into the photon sphere. The further the star contracts within the photon sphere, the narrower the exit cone becomes.

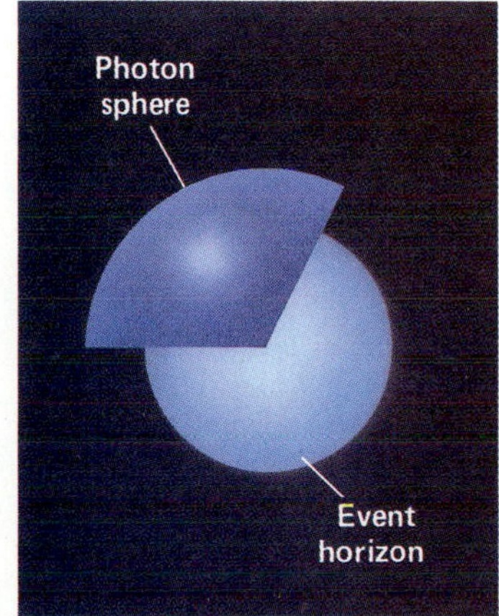

Figure 28–4 When the star becomes smaller than its Schwarzschild radius, we can no longer observe it. We say that it has passed its *event horizon*, by analogy to the statement that we cannot see an object on earth once it has passed our horizon.

Figure 28–5 This drawing by Charles Addams is reprinted with permission of The New Yorker Magazine, Inc., © 1974.

Even though the mass that causes the black hole has contracted further, the event horizon doesn't change. It remains at the same radius forever, as long as the amount of mass inside doesn't change.

*28.4 Rotating Black Holes

Once matter is inside a black hole, it loses its identity in the sense that from outside a black hole, all we can tell are the mass of the black hole, the rate at which it is spinning, and what total electric charge it has. These three quantities are sufficient to completely describe the black hole. Thus, in a sense, black holes are simple objects to describe physically, because we only have to know three numbers to characterize each one. (By contrast, for the earth we have to know shapes, sizes, densities, motions, and other parameters for the interior, the surface, and the atmosphere.) The theorem that describes the simplicity of black holes is often colloquially stated by astronomers active in the field as "a black hole has no hair." All other properties, such as size, can be derived from the three basic properties.

Most of the theoretical calculations about black holes, and the Schwarzschild solutions to general relativity, in particular, are based on the assumption that black holes do not rotate. But this assumption is only a convenience; we think, in fact, that the rotation of a black hole is one of its important properties. It was not until 1963 that Roy P. Kerr solved Einstein's equations for a situation that was later interpreted in terms of the notion that a black hole is rotating. In this more general case, an additional special boundary—the *stationary limit*—appears, with somewhat different properties from the original event horizon. At the stationary limit, space-time is flowing at the speed of light, so particles that would otherwise be moving at the speed of light would remain stationary. Within the stationary limit, no particles can remain at rest even though they are outside the event horizon.

Let us consider a series of non-moving ("static") points at distances close to a black hole and visualize what happens to a sphere representing expanding wavefronts of light emitted from these points. In Figure 28–6*A*, we see the wavefronts a standard time (say, 1 second) after they are emitted from each dot. For a point inside the event horizon, all the emitted waves go inward instead of outward, and no light escapes. In Figure 28–6*B*, we see top views of a series of points near a rotating black hole. The rotation carries the wavefront to the side, and for a point at the stationary limit, some of the wavefront is

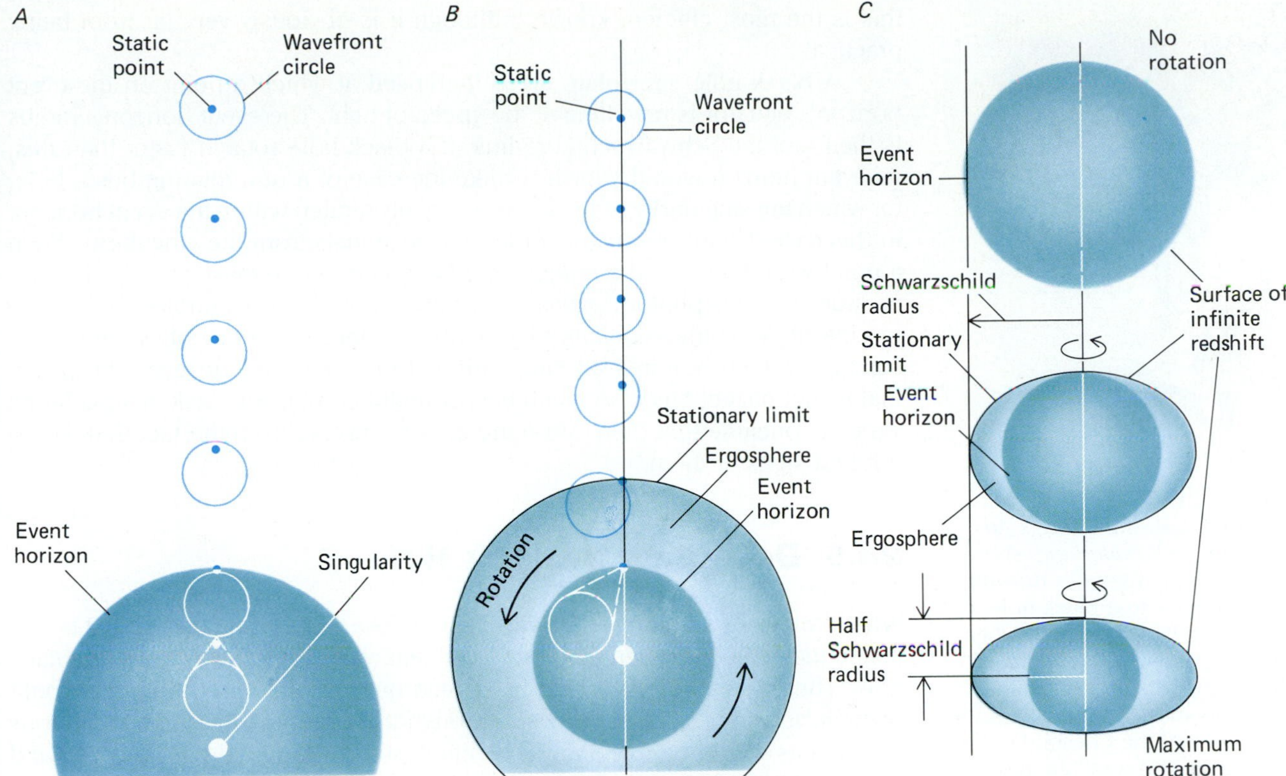

Figure 28–6 (*A*) A static point that emits a pulse of light in all directions, and the circle showing the wavefront a standard time later for locations near a non-rotating black hole. The circle is dragged more toward the black hole the closer the static point is to the black hole. For a static point at the event horizon, the wave circles are entirely inside and no light escapes. (*B*) Near a rotating black hole, the wave circles in this top view are displaced not only inward but also in the direction of rotation. For a static point on the stationary limit, part of the circle is outside the stationary limit, so some light escapes. Some light can escape even from inside the stationary limit. (*C*) Side views of black holes of the same mass and different rotation speeds. The region between the stationary limit and the event horizon of a rotating black hole is called the *ergosphere* (from the Greek word *ergon*, meaning "work") because, in principle, work can be extracted from it.

outside the limit. These light waves can escape. Only for points within the event horizon do all wavefronts move inward.

The equator of a stationary limit of a **rotating** black hole has the same diameter as the event horizon of a **non-rotating** black hole of the same mass. But a rotating black hole's stationary limit is squashed. The event horizon touches the stationary limit at the poles. Since the event horizon remains a sphere, it is smaller than the event horizon of a non-rotating black hole (Fig. 28–6*C*).

The region between the stationary limit and the event horizon is the *ergosphere.* An object can be shot into the ergosphere of a rotating black hole at such an angle that, after it is made to split in two, one part continues into the event horizon while the other part is ejected from the ergosphere with more energy than the incoming particle had. This energy must have come from somewhere; it could only have come from the rotational energy of the black hole. Thus by judicious use of the ergosphere, we could, in principle, tap the energy of the black hole. We are always looking for new energy sources and

Figure 28–7 Lewis Carroll's Cheshire Cat, from *Alice's Adventures in Wonderland,* shown here in John Tenniel's drawing, is analogous to a black hole in that it left its grin behind when it disappeared, while a black hole leaves its gravity behind when its mass disappears. Alice thought that the Cheshire Cat's persisting grin was "the most curious thing I ever saw in my life!" We might say the same about the black hole and its persisting gravity.

this is the most efficient known, although it is obviously very far from being practical.

A black hole can rotate up to the speed at which a point on the event horizon's equator is travelling at the speed of light. The event horizon's radius is then half the Schwarzschild radius. If a black hole rotated faster than this, its event horizon would vanish. Unlike the case of a non-rotating black hole, for which the singularity is always unreachably hidden within the event horizon, in this case distant observers could receive signals from the singularity. Such a point would be called a *naked singularity* and, if one exists, we might have no warning—no photon sphere or orbiting matter, for example—before we ran into it. Most theoreticians assume the existence of a law of "cosmic censorship," which requires all singularities to be "clothed" in event horizons, that is, not naked. Since so much energy might erupt from a naked singularity, we can conclude that there are none in our universe from the fact that we do not find signs of them.

28.5 Detecting a Black Hole

What if we were watching a faraway star collapse? Photons that were just barely inside the exit cone when emitted would leak very slowly away from the black hole. These few photons would be spread out in time, and the image would become very faint. As the star went through its event horizon, the last photons to reach us would be extremely redshifted, so the star would get very red and then blink out. It would do this in a fraction of a second, so the odds are unfavorable that we would actually see a collapsing star as it went through the event horizon.

But all hope is not lost for detecting a black hole, even though we can't hope to see the photons from the moments of collapse. The black hole disappears, but it leaves its gravity behind. It is a bit like the Cheshire Cat from *Alice's Adventures in Wonderland,* which fades away, leaving only its grin behind (Fig. 28–7).

The black hole attracts matter, and the matter accelerates toward it. Some of the matter will be pulled directly into the black hole, never to be seen again.

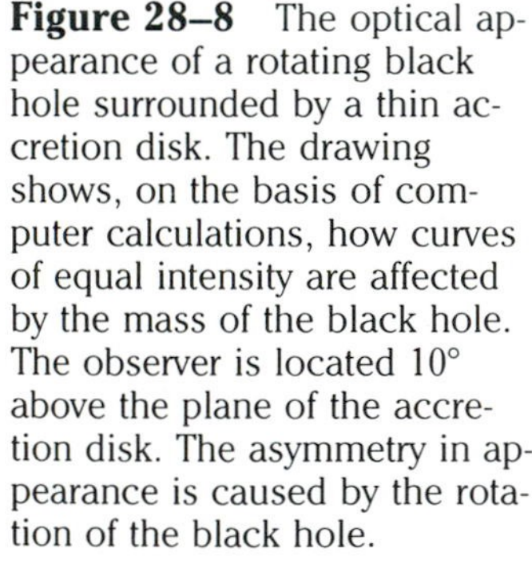

Figure 28–8 The optical appearance of a rotating black hole surrounded by a thin accretion disk. The drawing shows, on the basis of computer calculations, how curves of equal intensity are affected by the mass of the black hole. The observer is located 10° above the plane of the accretion disk. The asymmetry in appearance is caused by the rotation of the black hole.

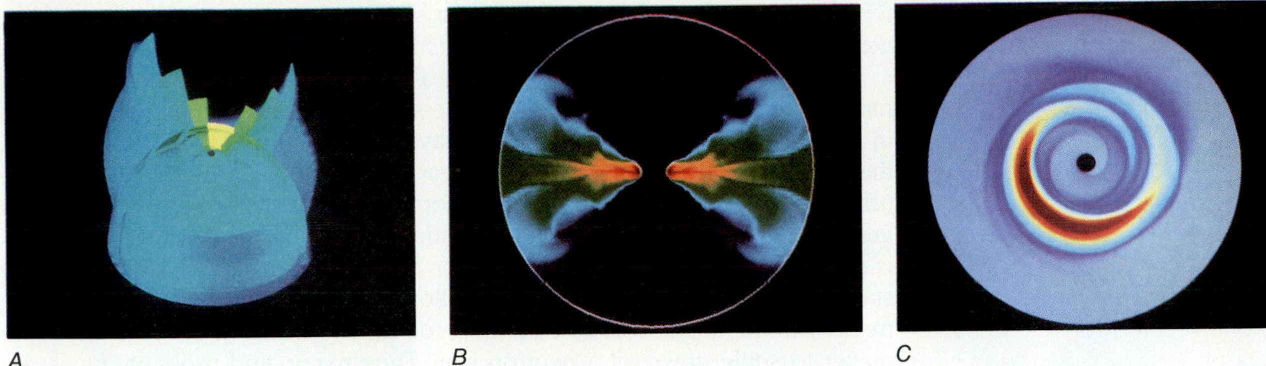

Figure 28–9 Supercomputer simulations showing how a black hole's gravity affects the orbiting gas in the accretion disk that surrounds it. (*A*) A 3-dimensional view showing that when gas flows toward a black hole with high angular momentum, it does not flow into the hole. It is thrown outward, and as it splashes back it opens an empty region that resembles a funnel. (*B*) Effects close to the event horizon show well in this side cross section of density; the calculations showed that the gas sometimes bounces back out. (*C*) This view from above shows gas density, with red being the densest gas. Instabilities cause a ring of gas to be drawn out into an orbiting blob. Narrow disks like this one are too unstable to be important in nature. Thicker disks are less unstable and form spiral waves.

But other matter will go into orbit around the black hole, and will orbit at a high velocity. This accreted matter forms an *accretion disk* (Fig. 28–8). The action of matter in the accretion disk can be modeled with supercomputers (Fig. 28–9).

Calculations show that the gas in orbit will be heated. Theoretical calculations of the friction that will take place between adjacent filaments of gas in orbit around the black hole show that the heating will be so great that the gas will radiate strongly in the x-ray region of the spectrum. The inner 200 km should reach hundreds of millions of kelvins. Thus, though we cannot observe the black hole itself, we can hope to observe x-rays from the gas surrounding it.

In fact, a large number of x-ray sources are known in the sky. They have been surveyed from NASA's Uhuru and HEAO's (High-Energy Astronomy Observatories), the United Kingdom's Ariels, and the European Space Agency's Exosat and Rosat. Russian and Japanese x-ray telescopes are now aloft. Some of the x-ray sources have been identified with galaxies or quasars. Others pulse regularly, like Hercules X-1, and are undoubtedly neutron stars (Section 27.8). Some of the rest, which pulse sporadically, may be related to black holes.

It is not enough to find an x-ray source that gives off sporadic pulses, for one can think of other mechanisms besides matter revolving around a black hole that can lead to such pulses. One would like to show that a collapsed star of greater than 3 solar masses is present. We can determine masses only for certain binary stars. When we search the position of the x-ray sources, we look for a spectroscopic binary (that is, a star whose spectrum shows a Doppler shift that indicates the presence of an invisible companion). Then, if we can show that the companion is too faint to be a normal, main-sequence star, it must be a collapsed star. If, further, the mass of the unobservable companion is greater than 3 solar masses, it must be a black hole.

The most persuasive case is named Cygnus X-1, an x-ray source that varies in intensity on a time scale of milliseconds. In 1971, radio radiation was

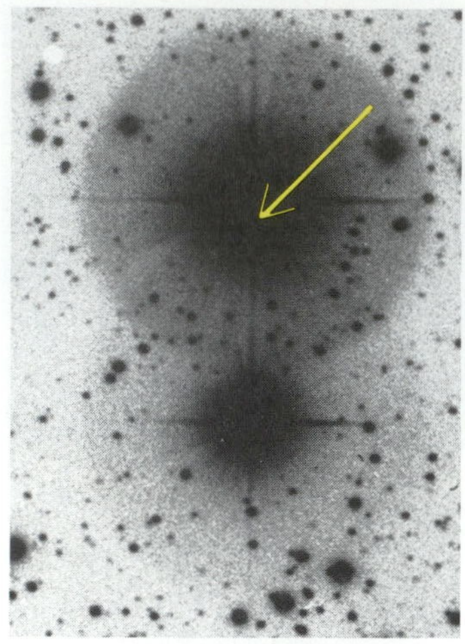

Figure 28–10 A close-up image of the blue supergiant star HDE 226868. The black hole Cygnus X-1 that is thought to be orbiting the supergiant star is not visible. Note the fact that the image of the supergiant appears so large because it is overexposed on the film. The image does not represent the actual angular diameter subtended by the star, which is too small to resolve from the earth.

found to come from the same direction. A 9th magnitude star previously catalogued as HDE 226868 was found at its location (Fig. 28–10).

HDE 226868 has the spectrum of a blue supergiant (type O), and thus has a mass of about 15 times that of the sun. Its spectrum is observed to vary in radial velocity with a period of 5.6 days, indicating that the supergiant and the invisible companion are orbiting each other with that period. From the orbit, it is deduced that the invisible companion must certainly have a mass greater than 4 solar masses; the best estimate is 8 solar masses. Because this is so much greater than the limit of 2 or 3 solar masses above which neutron stars cannot exist, it seems that even allowing for possible errors in measurement or in the theoretical calculations, too much mass is present to allow the matter to settle down as a neutron star. Thus many, and probably most, astronomers believe that a black hole has been found in Cygnus X-1.

Another promising candidate is LMC X-3 (the third x-ray source to be found in the nearby galaxy known as the Large Magellanic Cloud). Its optical counterpart orbits an unseen companion every 1.7 days. Since we know the distance to the Large Magellanic Cloud, our calculations for the invisible object's mass may be more accurate. Again, the invisible object seems to have about 8 solar masses of matter, and so must be a black hole.

A third promising black-hole candidate is A0620-00, with the A standing for the x-ray catalogue of sources studied with the Ariel spacecraft and the digits representing the source's position in the sky, which is in the constellation Monoceros. The source was even brighter in x-rays than Cyg X-1 during an outburst in 1975, but no x-rays have been detected from it since. However, a star that appears like a K dwarf has been detected at its position. Analysis of the star's 7.75-hour period indicates that the invisible companion's mass is 3.20 solar masses, close to but definitely above the apparent neutron-star limit.

The newest, and perhaps the best, candidate for a black hole is in V404 Cygni. This invisible companion to the 404th variable star discovered in Cygnus was reported in 1992. An intense burst of x-rays from the spot was discovered by Ginga in 1989, and the companion star was then monitored. The companion contains 6 solar masses. The distinctive shape of its x-ray spectrum may be a fingerprint that will enable the discovery of other black holes.

Matter in the accretion disk around a black hole (Fig. 28–11) would orbit very quickly. If a hot spot developed somewhere on the disk, it might beam a cone of radiation into space, similar to the lighthouse model of pulsars. If the hot spot lasted for several rotations, we could detect a pulse of x-rays every time the cone swept past the earth. The period of the x-ray pulses would be extremely short, only a few milliseconds. Searches for the short-period pulses

Figure 28–11 An artist's conception of the disk of swirling gas that would develop around a black hole like Cygnus X-1 (*right*) as its gravity pulled matter off the companion supergiant (*left*). The x-radiation would arise in the disk.

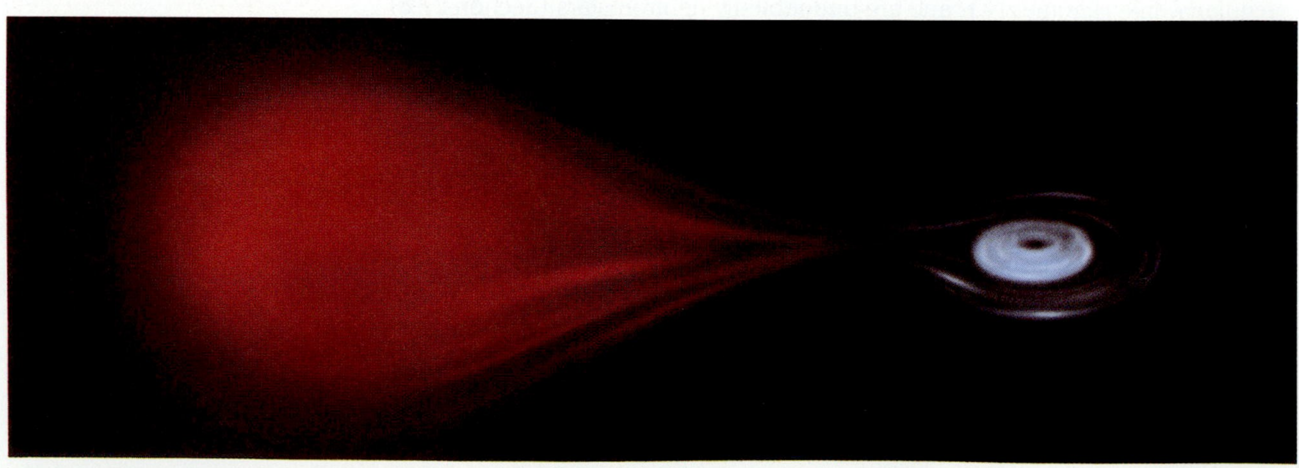

have been made with x-ray satellites (Fig. 28–12). Cygnus X-1 does indeed flicker in x-rays, though current theories do not get the period exactly right. Cyg X-1 shows several additional periods, with a long period of 294 days newly discovered and remaining to be explained. LMC X-3 is so far off and thus so faint that it is hard to see flickering. And no x-rays from A0620-00 are available for current observation, though optical spectra indicate that it has an accretion disk.

Ginga observations of Cygnus X-1 made in 1987 have shown that high-energy x-rays are delayed 2 milliseconds to several seconds from the lower energy x-rays. The length of the delay depends on the repetition rate at which x-rays are received. They are almost but not quite regularly spaced, so are "quasiperiodic" rather than periodic. The delay does not fit with the existing model that the x-rays were formed when lower-energy photons bounced off hot electrons in the accretion disk. A new model of the x-ray formation is needed.

Figure 28–12 An x-ray view of Cygnus X-1, the first picture taken with HEAO-2, the Einstein Observatory. The resolution of the telescope system aboard was about 4 arc sec, whereas the orbiting gas around the black hole—the "accretion disk"—is 100,000 times smaller, according to calculations, so a photograph like this one mainly shows that the telescope optics are focusing properly. The object was allowed to drift into the field of view for this first image, hence the appearance of a "tail." (Courtesy of Riccardo Giacconi, then at the Harvard-Smithsonian Center for Astrophysics)

*28.6 Non-Stellar Black Holes

We have discussed how black holes can form by the collapse of massive stars. But theoretically a black hole should result if a mass of any amount is sufficiently compressed. No object containing less than 2 or 3 solar masses will contract sufficiently under the force of its own gravity in the course of stellar evolution. But the density of matter was so high at the time of the origin of the universe (see Chapter 33) that smaller masses may have been sufficiently compressed to form what are called mini black holes.

Stephen Hawking (Fig. 28–13), an English astrophysicist, has suggested the existence of mini black holes, the size of pinheads (and thus with masses equivalent to those of asteroids). There is no observational evidence for a mini hole, but they are theoretically plausible. Hawking has deduced that small black holes can seem to emit energy in the form of elementary particles (neutrinos and so forth). The mini holes would thus evaporate and disappear. This may seem to be a contradiction to the concept that mass can't escape from a black hole. But when we consider effects of quantum mechanics, the simple picture of a black hole that we have discussed up to this point is not sufficient. Hawking suggests that a black hole so affects space near it that a pair of particles—a nuclear particle and its antiparticle—can form simultaneously. The antiparticle disappears into the black hole, and the remaining particle reaches us. Photons, which are their own antiparticles, appear too.

Emission from a black hole is significant only for the smallest mini black holes, for the amount of radiation increases sharply as we consider less and less massive black holes. Only mini black holes up to the mass of an asteroid—far short of stellar masses—would have had time to disappear since the origin of the universe. Hawking's ideas set a lower limit on the size of black holes now in existence, since we think the mini black holes were formed only in the first second after the origin of the universe by the tremendous pressures that existed then. His derivation that black holes can evaporate was challenged in 1987, and we must wait to see what is correct.

Figure 28–13 Stephen Hawking, who holds Newton's Chair of physics at Cambridge University. Among his many theoretical ideas were black-hole radiation and mini black holes.

On the other extreme of mass, we can consider what a black hole would be like if it contained a very large number, that is, thousands or millions, of solar masses. Thus far, we have considered only black holes the mass of a star or smaller. Such black holes form after a stage of high density. But the more mass involved, the lower the density needed for a black hole to form. For a very massive black hole, one containing hundreds of millions or billions of solar masses, the density would be fairly low when the event horizon formed,

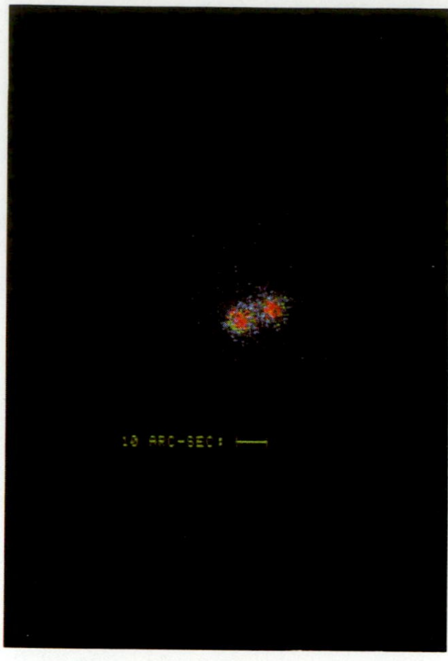

Figure 28–14 The galaxy M87, seen here in an x-ray view from the Einstein Observatory, is thought to have at its center a black hole containing millions of solar masses of matter. The second spot corresponds to part of the jet that is thought to be emitted from the black hole. Jets are found across a wide range of astrophysical objects, from protostars to galaxies and quasars.

Figure 28–15 Directions to a small, deep cove on the shore of the Bay of Fundy. It got its name (probably a century or two ago), because it appears very dark as seen from the sea. Unlike the stellar case, the cove's darkness results from its dark basaltic cliffs and its deep, narrow shape. Nobody lives there.

approaching the density of water. For even higher masses, the density would be lower yet. We think such high masses occur in the centers of active galaxies and quasars (Fig. 28–14).

Thus if we were travelling through the universe in a spaceship, we couldn't count on detecting a black hole by noticing a volume of high density. We could pass through the event horizon of a high-mass black hole without even noticing. We would never be able to get out, but it might be hours on our watches before we would notice that we were being drawn into the center at an accelerating rate.

Where could such a supermassive black hole be located? The center of our galaxy may contain a black hole of a million solar masses (Section 29.3). Though we would not observe radiation from the black hole itself, the gamma

Figure 28–16 Supercomputer calculations of the gravity waves that would result from an oscillating black hole. Two of the three spatial dimensions of the curved space that makes up the universe are shown projected into the kind of flat space that is more familiar to us. The white ring represents the location of the black hole's surface (a sphere in 3D), with its exterior above the ring. Colors show the amplitude of the gravitational wave, with zero amplitude marked in the color bar.

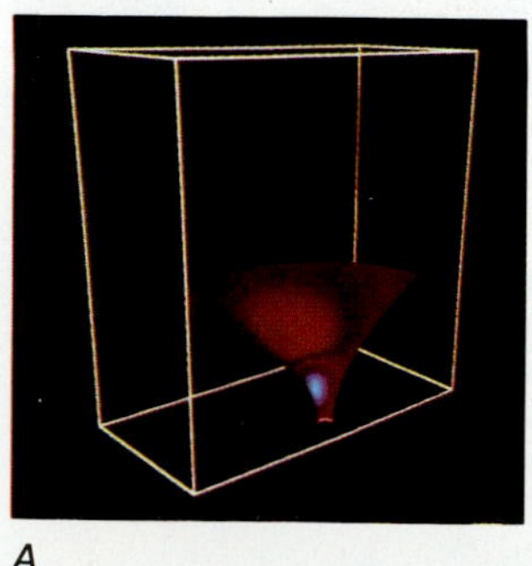

A

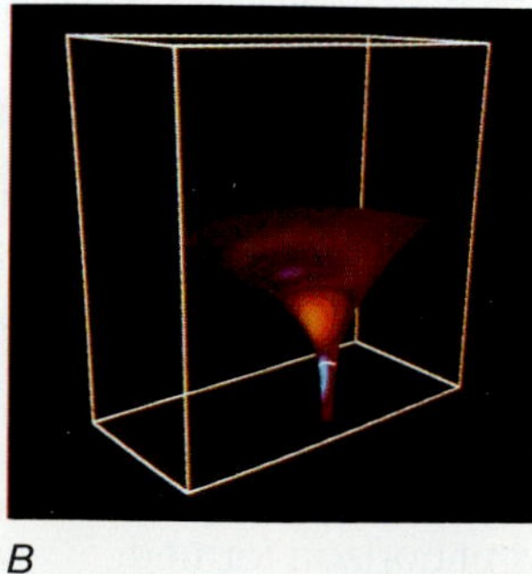

B

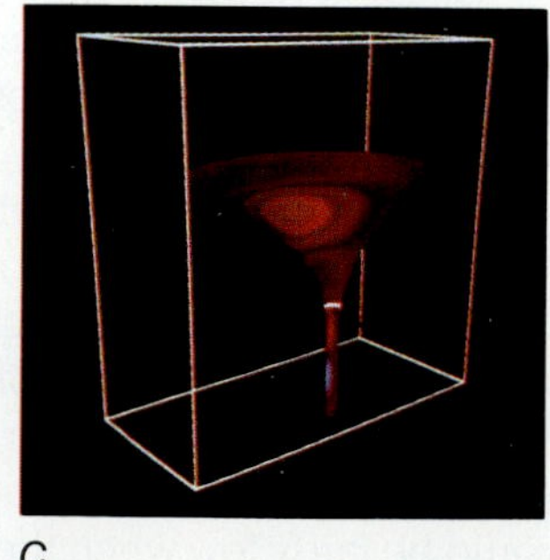

C

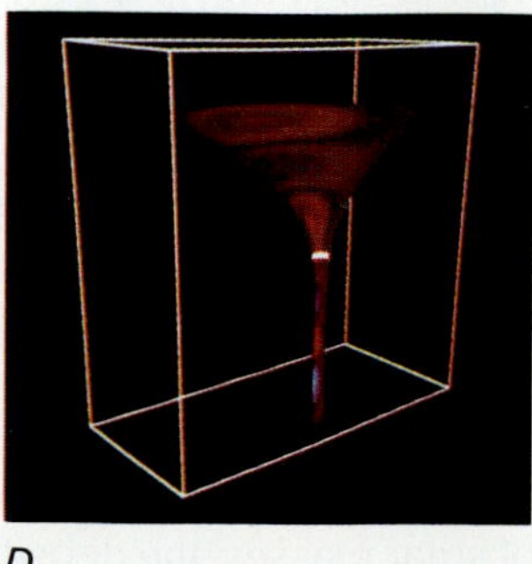

D

rays, x-rays, and infrared radiation we detect would be coming from the gas surrounding the black hole.

We are on the lookout for black holes everywhere (Fig. 28–15). Indeed, the whole universe is in a black hole, if certain ideas about the overall structure of the universe are true.

Though gravitational waves have not yet been detected directly, supercomputer calculations show how they should act (Fig. 28–16). Scientists can calculate how they affect the space they are passing through.

Summary and Outline

Gravitational collapse to a black hole occurs if more than 2 or 3 solar masses remain (Section 28.1).

Critical radii of a black hole
- Photon sphere: exit cones form (Section 28.2).
- Event horizon: exit cones close (Section 28.3).
- Schwarzschild radius (gravitational radius) defines the limit of the black hole, the event horizon.
- Singularity

Rotating black holes (Section 28.4)
- Can get energy out of the ergosphere

Detecting a black hole (Section 28.5)
- Flickering x-radiation expected from accretion disks
- Detection in a spectroscopic binary, as an invisible high-mass companion
- Cygnus X-1, LMC X-3, A0620–00, and V404 Cygni: black holes?

Non-stellar black holes (Section 28.6)
- Mini black holes could have formed in the big bang.
- Very massive black holes do not have high densities.

Key Words

black hole, exit cone, photon sphere, Schwarzschild radius, gravitational radius, event horizon, escape velocity, singularity, stationary limit*, ergosphere*, naked singularity*, accretion disk

*This term is found in an optional section.

Questions

1. Why doesn't electron or neutron degeneracy prevent a star from becoming a black hole?

2. Why is a black hole blacker than a black piece of paper?

3. (a) Is light acting more like a particle or more like a wave when it is bent by gravity? (b) Explain the bending of light as a property of a warping of space (Section 23.11).

4. (a) How does the escape velocity of the moon compare with the escape velocity of the earth? Would a larger rocket engine be necessary to escape from the gravity of the earth or from the gravity of the moon? (b) How will the velocity of escape from the surface of the sun change when the sun becomes a red giant? A white dwarf?

†5. (a) What is the Schwarzschild radius for a 10-solar-mass star? (b) What is your Schwarzschild radius?

†6. What are radii of the photon sphere and the event horizon of a non-rotating black hole of 18 solar masses?

†7. What is the size of the stationary limit and of the event horizon of an 18-solar-mass star that is rotating at the maximum possible rate?

8. What is the relation in size of the photon sphere and the event horizon? If you were an astronaut in space, could you escape from within the photon sphere of a rotating black hole? From within its ergosphere? From within its event horizon?

9. (a) How could the mass of a black hole that results from a collapsed star increase? (b) How could mini black holes, if they exist, lose mass?

10. Would we always notice when we reached a black hole by its high density? Explain.

11. Could we detect a black hole that was not part of a binary system?

12. Under what circumstances does the presence of an x-ray source associated with a spectroscopic binary suggest to astronomers the presence of a black hole? In particular, discuss one deduction from optical observations and one special property of the x-rays.

†This question requires a numerical solution.

V The Milky Way Galaxy

On the clearest nights, when we are far from city lights, we can see a hazy band of light stretched across the sky. This band is the *Milky Way*—the aggregation of dust, gas, and stars that makes up the galaxy in which the sun is located.

Don't be confused by the terminology: the Milky Way itself is the band of light that we can see from the earth, and the Milky Way Galaxy is the whole galaxy in which we live. Like other galaxies, our Milky Way Galaxy is composed of perhaps a trillion stars plus many different types of gas, dust, planets, etc. The Milky Way is that part of the Milky Way Galaxy that we can see with the naked eye in our nighttime sky.

The Milky Way appears very irregular in form when we see it stretched across the sky—there are spurs of luminous material that stick out in one direction or another, and there are dark lanes or patches in which nothing can be seen. This patchiness is simply due to the splotchy distribution of dust, stars, and gas.

Here on earth, we are inside our galaxy together with all of the matter we see as the Milky Way. Because of our position, we see a lot of matter when we look in the plane of our galaxy. On the other hand, when we look "upwards" or "downwards" out of this plane, our view is not obscured by matter, and we can see past the confines of our galaxy. We are able to see distant galaxies only by looking in parts of our sky that are away from the Milky Way, that is, by looking out of the plane of the galaxy.

The gas in our galaxy is more or less transparent to visible light, but the small solid particles that we call "dust" are opaque. So the distance we can see through our galaxy depends mainly on the amount of dust that is present. This is not surprising: we can see great distances through our gaseous air on earth, but if a small amount of particulate material is introduced in the form of smoke or dust thrown up from a road, we find that we can no longer see very far. Similarly, the dust between the stars in our galaxy dims the starlight by absorbing it or by scattering it in different directions.

The abundance of dust in the plane of the Milky Way Galaxy actually prevents us from seeing very far toward its center—with visible light, we can easily observe only 1/10 of the way toward the galactic center itself. The dark lanes across the Milky Way are just areas of dust, obscuring any emitting gas or stars. The net effect is that we can see just about the same distance in any direction we look in the plane of the Milky Way. These direct optical observations fooled scientists at the turn of the century into thinking that the earth was near the center of the universe.

We shall see in the next chapter how the American astronomer Harlow Shapley in the 1920's realized that the Milky Way is part of an isolated galaxy and that the sun is not in its center. This fundamental idea took humanity one step further away from thinking that we were at the center of the universe. Copernicus in 1543, by removing the earth from the center of the solar system, had already taken the first step by removing the earth from the center of the universe.

Some of the dust and gas in our galaxy take a pretty shape and may glow, reflect light, or be visible in silhouette. These shapes are the *nebulae*.

The Origin of the Milky Way by Tintoretto, circa 1578. (The National Gallery, London)

Another class of objects visible in our sky was once known as "spiral nebulae," since they looked like glowing gas with "arms" spiralling away from their centers. (We still speak of the Great Nebula in Andromeda, Figure 31–2.) In the first decades of this century, astronomers debated whether these objects were part of our own galaxy or were independent "island universes." We shall see, in Chapter 31 and in the introduction to Part VI of this book, how we learned that the island-universe idea was correct. Our own Milky Way turned out to be part of a galaxy equal in stature to these other objects, which are now known to be spiral galaxies in their own right.

In recent years astronomers have been able to use wavelengths other than optical ones to study the Milky Way Galaxy. In the 1950's and 1960's especially, radio astronomy gave us a new picture of our galaxy. In the 1980's, we have benefited from infrared observations, most recently from the IRAS spacecraft and from infrared arrays on earth. Infrared and radio radiation can pass through the galaxy's dust and allow us to see our galactic center and beyond.

In this part of the book, we shall first discuss the types of objects that we find in the Milky Way Galaxy. Chapter 29 is devoted to the general structure of the galaxy and its major parts. In Chapter 30, we describe the matter between the stars and what studying this matter has told us about how stars form.

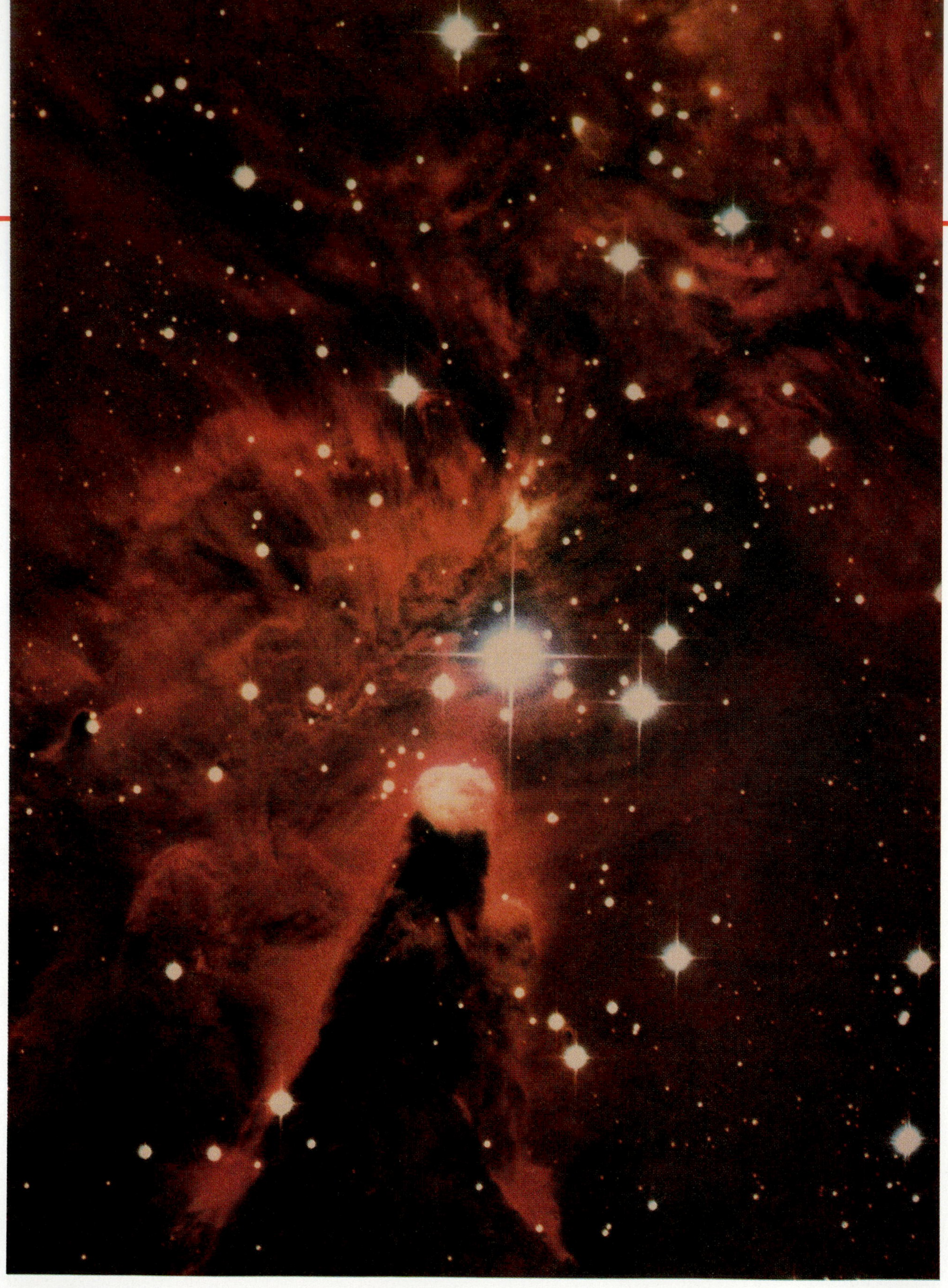

The Cone Nebula, part of a region of star formation in Monoceros. At the tip of the cone, hidden by dust, is a young star that is bright in the infrared. The open cluster NGC 2264 includes this region. Some of the hottest stars in the field are really bluish, though they appear white here because their images are saturated.

The Structure of the Milky Way Galaxy

29

Aims: To describe the basic components of the Milky Way Galaxy, and to understand why spiral structure forms

We have now described the stars, which are important constituents of any galaxy, and how they live and die. And we have seen that many or most stars exist in pairs and clusters. In this chapter, we describe the gas and dust that accompany the stars and that we see in the optical part of the spectrum as nebulae. We also discuss the overall structure of the Milky Way Galaxy and how, from our location inside it, we detect this structure.

Figure 29–1 The Milky Way in and near Sagittarius. The Great Rift across the center may be an overlap of many molecular clouds. The trail of an artificial satellite also appears.

29.1 Nebulae

Not all the gas and dust in our galaxy has coalesced into stars. A *nebula* is a cloud of gas and dust that we see in visible light. When we see the gas actually glowing in the visible part of the spectrum, we call it an *emission nebula.* Sometimes we see a cloud of dust that obscures our vision in some direction in the sky. When we see the dust appear as a dark silhouette against other glowing material, we call the object a *dark nebula* (or, sometimes, an *absorption nebula*). The photo of the Milky Way (Fig. 29–1) shows both emission nebulae (some of the brightest parts) and dark nebulae (some of the dark parts where relatively few stars can be seen). In the heart of the Milky Way, shown in the picture, we also see the great *star clouds* of the galactic center, the bright regions where the stars appear too close together to tell them apart without a powerful telescope.

The clouds of dust in the Milky Way or surrounding some of the stars in the Pleiades (Fig. 29–2) are examples of *reflection nebulae*—they merely reflect

Figure 29–2 (*A*) The Pleiades, M45, is an open cluster in the constellation Taurus, the Bull. Reflection nebulae are visible around many of the brightest stars. The spikes are artifacts caused in the telescope. (*B*) An EXOSAT x-ray view of the Pleiades.

A

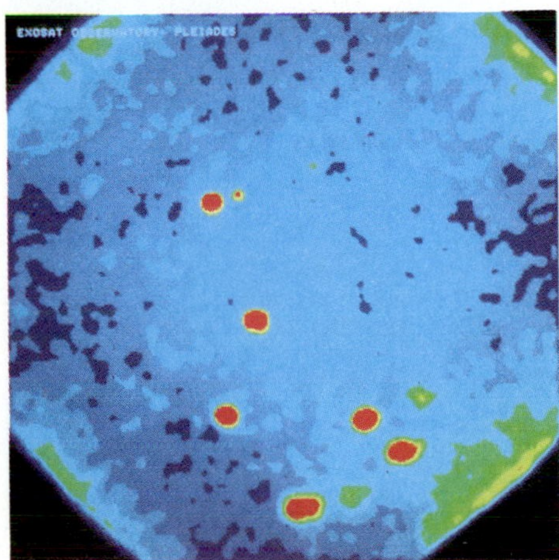

B

Figure 29–3 The Horsehead Nebula, IC 434 in Orion, is dark dust superimposed on glowing gas. The bright star in the nebulosity at top is ζ Orionis, the leftmost star in Orion's belt.

Nebula is Latin for "fog" or "mist." The plural is usually *nebulae* rather than *nebulas.*

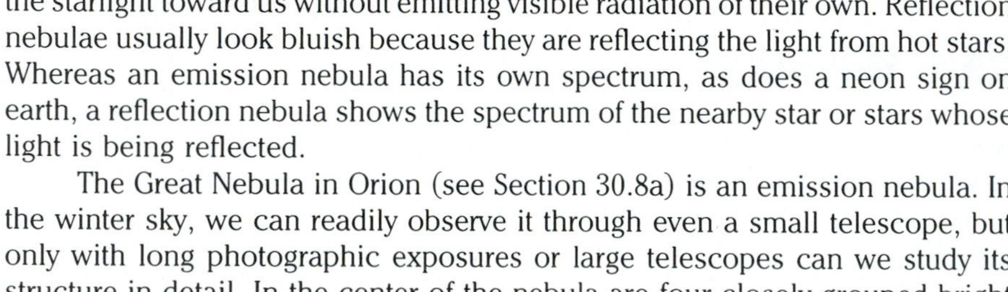

the starlight toward us without emitting visible radiation of their own. Reflection nebulae usually look bluish because they are reflecting the light from hot stars. Whereas an emission nebula has its own spectrum, as does a neon sign on earth, a reflection nebula shows the spectrum of the nearby star or stars whose light is being reflected.

The Great Nebula in Orion (see Section 30.8a) is an emission nebula. In the winter sky, we can readily observe it through even a small telescope, but only with long photographic exposures or large telescopes can we study its structure in detail. In the center of the nebula are four closely grouped bright stars called the Trapezium, which provide the energy to make the nebula glow. In this region of the Orion Nebula we think stars are being born this very minute.

The Horsehead Nebula (Fig. 29–3) is another example of an object that is both an emission and an absorption nebula simultaneously. It also looks red because the hydrogen-alpha line is so strong. A bit of absorbing dust intrudes onto emitting gas, outlining the shape of a horse's head. We can see in the pictures that the horsehead is a continuation of a dark area in which very few stars are visible.

We can show only a few of the beautiful nebulae here (Figs. 29–4 and 29–5). We have already discussed some of the most beautiful nebulae in the sky, composed of gas thrown off in the late stages of stellar evolution. They include planetary nebulae and supernova remnants.

Figure 29–4 The dust lanes in the nebulosity associated with Messier 16 (NGC 6611).

29.2 The New Subdivision of Our Galaxy

It was not until the 1920's that the American astronomer Harlow Shapley realized that we were not in the center of the galaxy. He was studying the distribution of globular clusters and noticed that they were largely in the same general area of the sky as seen from the earth. They mostly appear above or below the galactic plane and thus are not obscured by the dust. When he plotted their distances and directions (using the methods described in Chapter 21), he saw that they formed a spherical halo around a point thousands of light years away from us (Fig. 29–6). Shapley's touch of genius was to realize that this point must be the center of the galaxy.

The picture that we have of our own galaxy has changed in the last few years. In the 1970's, we knew of three components of our galaxy; now we speak of four. We have known of the first three for half a century: (1) the nuclear bulge, (2) the disk, and (3) the halo that contains the globular clusters. Now we also know of an additional component: (4) the galactic corona, though we don't know what it is made of. Let us now discuss these four parts of the galaxy.

1. *The nuclear bulge:* Our galaxy has the general shape of a pancake with a bulge at its center. This *nuclear bulge* is about 5000 parsecs (16,000 light years) in radius, with the galactic *nucleus* at its midst. The nucleus itself is only about 5 parsecs across. The nuclear bulge has the shape of a flattened sphere and does not show spiral structure. From earth, we see it as the broadening of the Milky Way in the direction of the constellation Sagittarius, where the galactic center lies. The nuclear bulge contains densely packed old stars as well as interstellar dust and gas. We shall say more about the nucleus in the next section.

2. *The disk:* The part of the pancake outside the bulge is called the galactic *disk*. It extends out 15,000 parsecs (15 kiloparsecs or 50,000 light years) or so from the center of the galaxy. The sun is located about halfway out, about 8,500 parsecs (8.5 kpc) from the nucleus. The disk contains all the young stars (Population I, as we discussed in Section 22.5) as well as interstellar gas and dust. It is slightly warped at its ends, perhaps by interaction with our satellite galaxies, the Magellanic Clouds. So, in the words of Leo Blitz of the University of Maryland, Michel Fich of the U. of Waterloo/Ontario, and Anthony Stark of Bell Labs, our galaxy looks a bit like a "fedora hat" with a turned-down brim.

It is very difficult for us to tell how the material is arranged in our galaxy's disk, just as it would be difficult to tell how the streets of a city were laid out if we could only stand still on one street corner without moving. Still, other galaxies have similar properties to our own, and their disks are filled with great *spiral arms,* regions of dust, gas, and stars in the shape of a pinwheel. So we

Figure 29–5 Emission and reflection nebulosity in Orion, NGC 1973, 1975, and 1977.

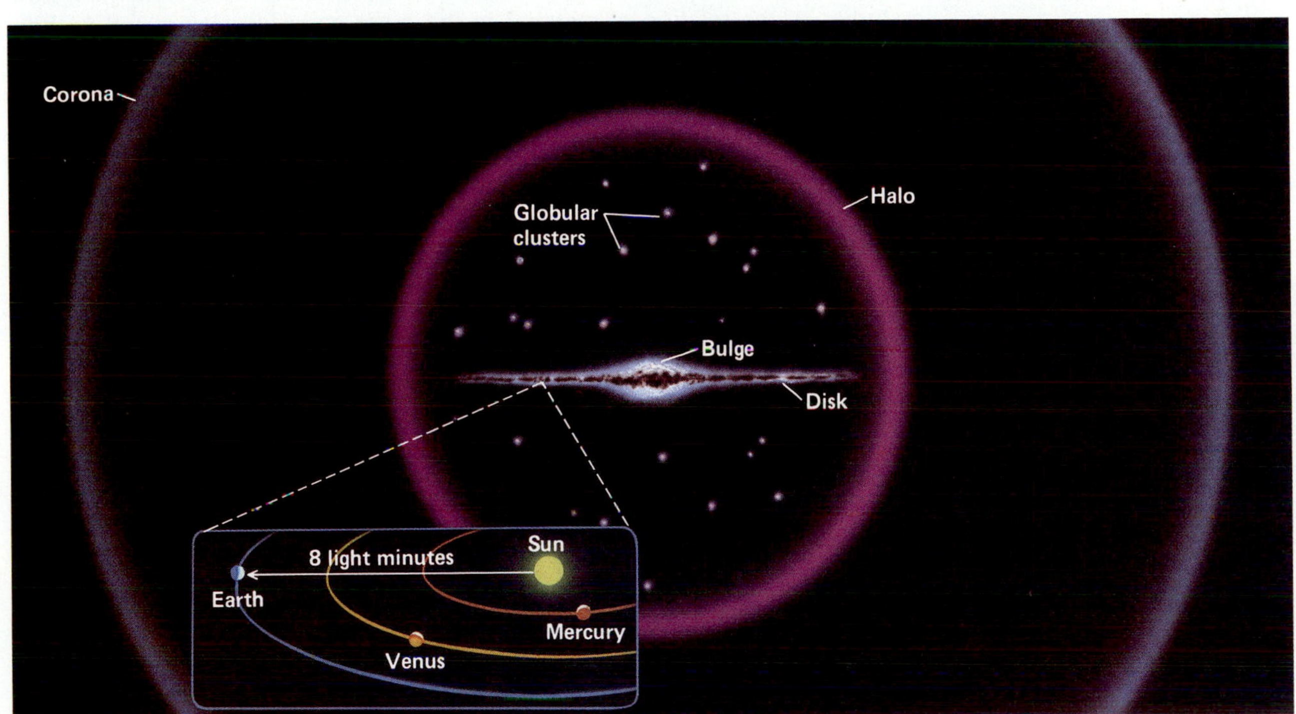

Figure 29–6 The drawing shows the *nuclear bulge* surrounded by the *disk,* which contains the spiral arms. The globular clusters are part of the *halo,* which extends above and below the disk. Extending even farther is the *galactic corona.* From the fact that most of the clusters appear in less than half of our sky, Shapley deduced that the galactic center is in the direction indicated.

Observations from the Pioneer 10 and 11 spacecraft, which are moving out of the solar system, have shown that the sun is about 40 light years above the plane of our galaxy. These spacecraft have moved beyond the region where dust in the ecliptic plane prevents them from seeing far outward along the plane.

assume the disk of our galaxy has spiral arms too, though in Section 29.5 we will see that the direct evidence is ambiguous.

The disk of the galaxy is, classically, very thin, only about 2 per cent of its width, like a phonograph record. The height astronomers assign to it is 325 parsecs. (Technically, it has a scale height of 325 parsecs, which is to say that for every 325 parsecs up from the plane of the galaxy, the number of stars declines by a factor of about 3.)

In recent years, we have learned a lot more about our galaxy, not least by using automatic devices to count the numbers of stars of different brightnesses in different regions of the sky. It now seems that in addition to the classical "thin disk," the galaxy also has a "thick disk" component. The number of stars in this "thick disk" decreases above and below the plane of the galaxy with a height of 1,300 parsecs, that is, four times the height of the "thin disk."

3. *The halo:* Older stars (including the globular clusters) and interstellar matter form a galactic *halo* around the disk. This halo is at least as large across as the disk, perhaps as much as 40 kpc (130,000 light years) in radius, the distance from the center of the most distant globular cluster known. On its inside, the halo gradually merges with the "thick disk." On its outside, the halo extends far above and below the plane of our galaxy in the shape of a flattened spheroid. Spectra from the International Ultraviolet Explorer spacecraft show that a modest amount of the gas in the halo is hot, 100,000 K. IUE discovered the spectra of this hot gas when pointing at more distant objects. The gas in the halo contains only about 2 per cent of the mass of the gas in the disk.

The halo contains relatively old stars of Population II (Section 22.5).

All together, the visible matter in the disk, bulge, and halo shows about 96 per cent stars, 3 per cent interstellar gas, and 1 per cent interstellar dust. That small admixture of dust, however, is enough to severely impede our ability to see distant objects.

4. *The galactic corona:* We shall see in the next chapter (Section 30.6) how studies of the rotation of material in the outer parts of galaxies tell us how much mass is present. These studies have told us of the existence of a lot of mass we had overlooked before because we couldn't see it. This mass extends out 60 or 100 kpc (200,000 or 300,000 light years). Believe it or not, this galactic corona contains 5 or 10 times as much mass as the nucleus, disk, and halo together. And it makes our galaxy about 5 times larger across than we had thought. (The subject is so new that the vocabulary is not settled; "galactic corona" and "outer halo" are among the names for this material.)

The existence of this extra material had been suspected from older gravitational studies, which had led to the "missing-mass problem" (Section 33.2a). But only recently have we accepted the actual presence of so much nonluminous matter (that is, matter that isn't shining) in our own galaxy. If the material in the galactic corona was of an ordinary type, we would have seen it directly—if not in visible light, then in radio waves, x-rays, infrared, etc. But we don't see it at all! We only detect its gravitational properties.

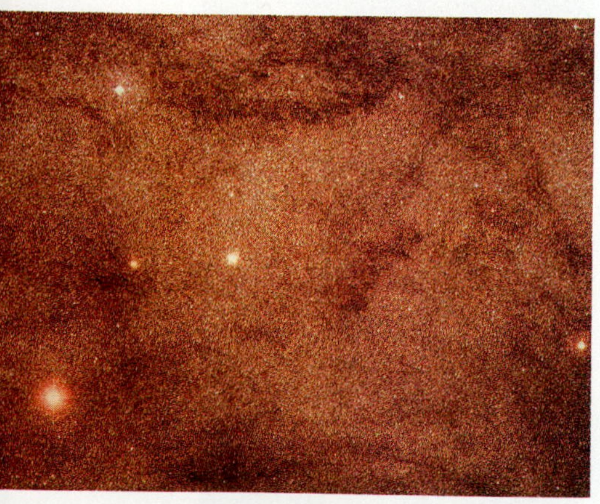

Figure 29–7 Baade's window, stars near the galactic nucleus, with NGC 6522 and γ Sgr.

What is the galactic corona made of? We just don't know. A tremendous number of very faint stars is a possibility, though it seems unlikely, extrapolating from the numbers of the faintest stars that we can study. A large number of small black holes has been suggested. And another possibility is a huge number of neutrinos, the subatomic particles of the type we discussed in Section 24.6 on the solar neutrino problem, if neutrinos have mass. Work is now going on to find out if neutrinos have any mass at all. From the neutrinos received from Supernova 1987A, scientists have been able to deduce very low limits for neutrino mass, making it less likely that neutrinos account for the unseen mass.

If the preceding paragraphs—stating that perhaps 95 per cent of our galaxy's mass is in some unknown form—seem unsatisfactory to you, you may

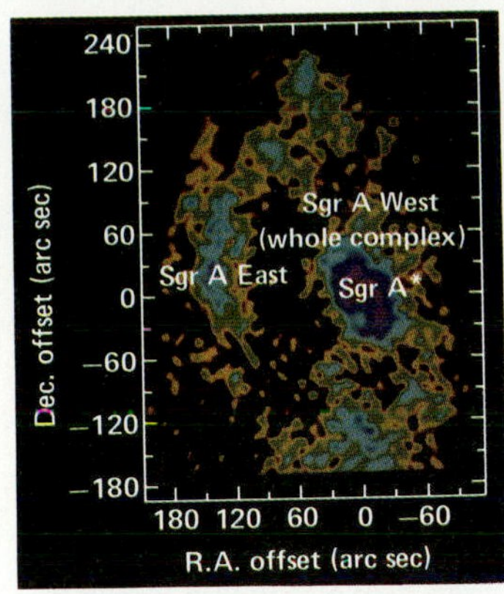

Figure 29–8 A radio map of the galactic center. The whole complex is Sgr A (Sagittarius A). (East is toward the left.) The partial shell of dust emission at left is the eastern part of a dust ring surrounding Sgr A East, which emits synchrotron radiation. The shell coincides with the highest density of molecular hydrogen in a molecular cloud. The shell and dust ring may have resulted from an explosion in a giant molecular cloud.

The purple region at right contains the point source Sgr A*, near or at the galactic nucleus. It is surrounded by free-free emission (emission from free electrons that change direction) from the minispiral that is at the center of the source known as Sgr A West.

The observations show continuum radiation observed with the European IRAM telescope at a wavelength of 1.3 mm.

feel better by knowing that astronomers find the situation unsatisfactory too. But all we can do is go out and do our research, and try to find out more. We just don't know the answers...yet.

29.3 The Center of Our Galaxy and Infrared Studies

We cannot see the center of our galaxy in the visible part of the spectrum because our view is blocked by interstellar dust. We did not even know that the center of our galaxy (Fig. 29–7) lies in the direction of the constellation Sagittarius as seen from the earth until Shapley deduced the fact from his study of the distribution of globular clusters. In recent years, observations in other parts of the electromagnetic spectrum have become increasingly important for the study of the Milky Way Galaxy (Fig. 29–8).

29.3a Infrared and Radio Observations

For many years the sky has been intensively studied at optical wavelengths up to about 8500 Å (0.85 micron) and to a lesser extent up to 1.1 microns, which is in the near infrared. The sky has also been studied for decades at radio wavelengths down to one or two centimeters (10,000 or 20,000 microns). But until the last few years, the sky has been studied very little at the wavelengths in between. We saw in Section 4.13a that the lack of film sensitivity to the infrared, whose photons contain relatively low energy, is a major reason for this limitation. Only recently have especially sensitive electronic detection imaging devices—arrays—been available in the infrared (Fig. 29–9). Atmospheric limitations, caused by the presence of only a few windows of transparency, have been another major factor. Since one can observe better in the infrared from locations where there is little water vapor overhead, the new emphasis on infrared telescopes at high-altitude sites like Mauna Kea in Hawaii should improve the situation considerably.

Astronomers working in the infrared usually use the unit of microns for wavelength. One micron is $1/_{1,000,000}$ m, and is 10,000 Å.

Another difficulty is the fact that the earth's atmosphere radiates conspicuously in the infrared; the radiation coming into a telescope includes infrared radiation from the atmosphere, from the source being observed, and from the telescope itself. To limit the telescopic contribution, equipment is usually bathed in liquid nitrogen, or even in liquid helium, which is colder and more expensive.

A 1969 sky survey in the 2.2-micron window revealed 20,000 infrared sources. IRAS, the NASA/Netherlands/UK **I**nfra**r**ed **A**stronomical **S**atellite, with an 0.6-m telescope, mapped the sky at much longer wavelengths in 1983. With its detectors cooled by liquid helium to only 2 K (2°C above absolute zero) and its telescope cooled to only slightly more than that by keeping the whole

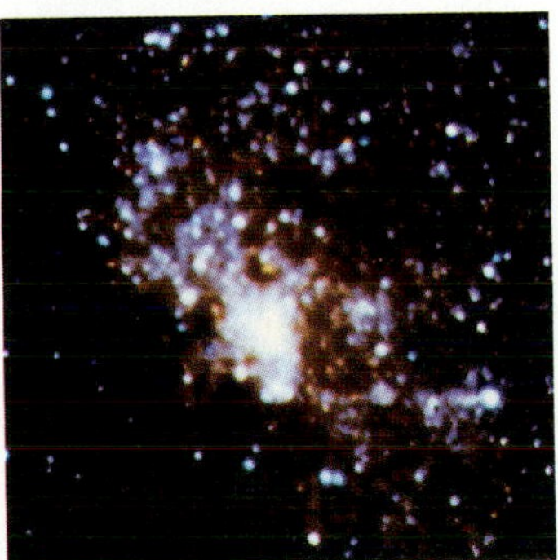

Figure 29–9 A view of the galactic center taken with an infrared array, showing the central 100 light years. One image in the H band at 1.6 μ and one image in the K band at 2.2 μ are assigned different colors. The nucleus of our galaxy shows clearly at the center, as does the plane of our galaxy, which runs from upper left to lower right.

Figure 29–10 An IRAS view of 20° across the sky, showing the Milky Way. The warm dust that IRAS images is close to the plane of the galaxy, making the IRAS image show the galactic plane as narrower than it is in optical images. The bright spots are regions where stars are forming; we see dust heated by the stars. The bright region at the core is about 25 K. Infrared cirrus seems to stream from the plane. Data at 12 μ are in blue, at 60 μ in green, and at 100 μ in red.

One of the instruments on the Cosmic Background Explorer spacecraft (Section 33.4d) is now mapping the sky in many infrared bands.

assembly in a thermos-like bottle, it was 1000 times more sensitive at 10 and 20 microns than past observations, and made the first full survey at 50 and 100 microns. IRAS sent back data for 10 months, until its liquid helium was exhausted. It discovered 245,389 sources—so much data that it will take years to interpret.

Two-thirds of the infrared sources—158,000 objects—are stars within our galaxy. Another 65,000 objects are interstellar objects of the kinds we discuss in the following chapter. The rest, 9 per cent (22,000 objects), are apparently galaxies. Many of these galaxies are much brighter in the infrared than in the visible; long exposures showed their visible images.

IRAS mapped the entire Milky Way Galaxy; its view of our galaxy's disk penetrated to the galactic center (Fig. 29–10). Another of IRAS's discoveries was that the sky is covered with infrared-emitting material, probably outside our solar system but in our galaxy. Since its shape resembles terrestrial cirrus clouds, the material is being called "infrared cirrus" (look back at Figure 4.64). Subsequent ground-based observations have shown that the infrared cirrus corresponds to the distribution of neutral interstellar hydrogen. It also correlates with faint dust features barely detectable on optical photographs. Thus the cirrus is apparently dust grains embedded within the hydrogen gas.

29.3b The Galactic Nucleus

One of the brightest infrared sources in our sky, known before IRAS, is located at the position of the radio source Sagittarius A*, which marks the center of our galaxy; the asterisk in its name distinguishes this point source, pronounced "Sag A star," from the whole complex of sources Sagittarius A. Its position has been pinpointed by simultaneously using a set of radio telescopes scattered around the world. Since the amount of scattering by dust varies with wavelength, we can see further through interstellar space in the infrared than we can in the visible. In particular, in the infrared we can see to the center of our galaxy (Fig. 29–11).

This infrared source subtends 1 arc min, and so is about 4 parsecs (about 13 light years) across. This makes it a very small source for the prodigious amount of energy it emits: as much energy as if there were 80 million suns radiating.

Scientists on the Kuiper Airborne Observatory (KAO) discovered that the infrared source is doughnut shaped, and is centered on two objects close to each other: the compact radio source Sagittarius A* and an infrared source known as IRS-16. Studies from Mauna Kea and other telescopes have shown that the ring of molecules that had been anticipated corresponds to the ring of dust that had been detected by the KAO, and is rotating. Further, spectra

A

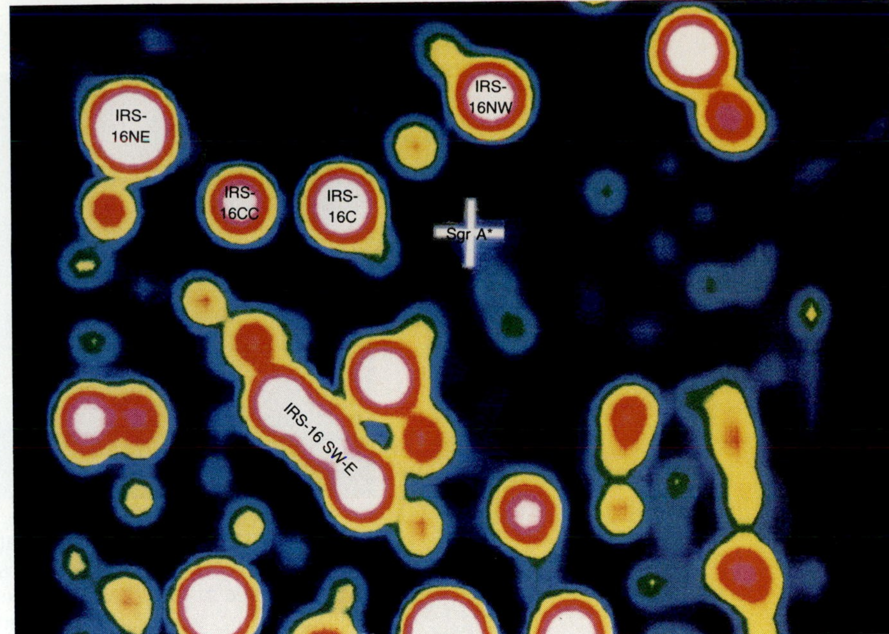

B

Figure 29–11 (*A*) The first VLBA image of the radio source Sagittarius A*. (*B*) The compact, nonthermal radio source Sgr A* is within 1 arcsec of the dynamical and gravitational center of our galaxy. It is marked with a +. These near-infrared observations at 1.6 and 2.2 microns, made with the New Technology Telescope of the European Southern Observatory, reveal the position of Sgr A* with respect to the infrared sources (IRS) that had been previously known and numbered.

show that hydrogen and helium atoms there have high velocities, indicating that a strong wind being given off by the central object or objects is blowing out the hole in the doughnut.

The infrared source is so small for the strength of radiation emitted that the leading models consider it to contain a black hole. Some of the new infrared observations revealed that the infrared source corresponding to Sgr A* is, in fact, not quite at the same location as IRS-16. Thus any black hole at Sgr A* might be much smaller than had been considered—perhaps no more massive than 100 solar masses, though some still say a million.

Some prior evidence had indicated that the central black hole was more massive, but none of it was definitive. Infrared observations of the motion of neon gas had revealed velocities that would cause the gas to escape from the region unless the gravity of 4 million solar masses of material was present. But perhaps the gas is escaping after all. And all the heating necessary in the region can be provided by the hot stars identified with IRS-16.

The infrared observations thus lead to the model that the central parsec of our galaxy contains a sphere of stars about 0.1 parsec in radius, a few small associations or clusters of newly formed B stars within that core, and an accreting black hole of perhaps 100 solar masses. Nonetheless, the subject is unsettled, and some astronomers think a 100-solar-mass black hole is present—the dragon in the center of our galaxy may be but a baby. Others think a much more massive black hole is present. And some think that there is no black hole present at all.

Not only radio and infrared but also gamma rays have been detected from the center of our galaxy. Presumably, matter and antimatter have been annihilating each other there. HEAO-3's gamma-ray spectrometer confirmed the detection of the gamma-ray spectral line that results from electrons and positrons annihilating each other; its strength is variable. The region emitting the line is too small for the gamma rays to arise from cosmic rays interacting with the interstellar medium or from a distribution of many supernovae, novae, or pulsars. Still, the amount of gamma radiation can be accounted for by the presence of a black hole of no more than a few hundred solar masses.

One kind of matter-antimatter annihilation occurs when an electron meets its antiparticle, a *positron.* The two annihilate each other completely, releasing energy in the amount $E = 2mc^2$, where m is the mass of each of the two particles.

The high-resolution radio maps of our galactic center, now made with the Very Large Array, show a small bright spot that could well be the region

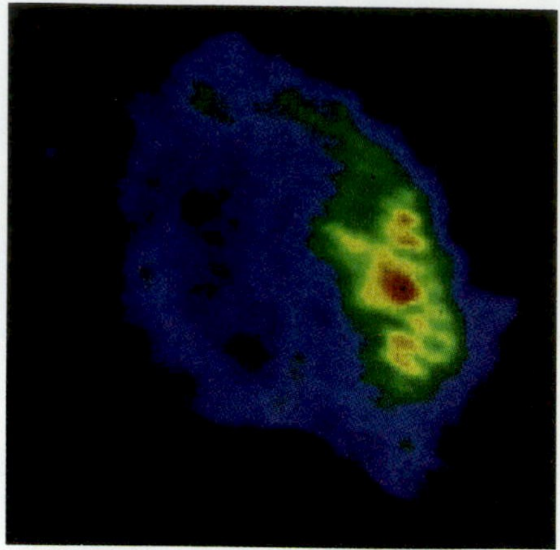

Figure 29–12 The radio source at the galactic center, Sagittarius A West. It was observed at the VLA at a wavelength of 20 cm, a resolution of 8 arc sec, and a field of view 4 arc min across.

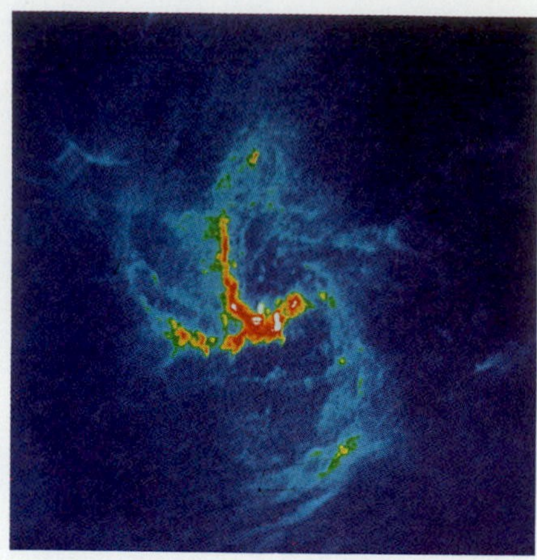

Figure 29–13 Subtracting the compact non-thermal radio source shows a spiral of gas within the central ten light years of our galaxy. The wavelength was 6 cm, the resolution was 1 arc sec, and the field of view was 3 arc min.

of the central giant black hole (Fig. 29–12). But though it appears on the images that a spiral a few parsecs across is present there (Fig. 29–13), this view turns out to be an optical illusion; the "arms" are only apparently superimposed on each other. Doppler shifts seem to show that the streamers that make the apparent "arms" are falling into rather than outward from the nucleus. Fortunately, the streamers are moving sufficiently fast that we have hope that we can actually see their proper motion across the sky in a few years, using high-resolution radio techniques. Then we will have a much better picture of their motion.

In the meantime, it appears that the streamers have come from the doughnut; they would have had to start falling in only 10,000 years ago, an astonishingly recent event. Perhaps some explosion took place in our galaxy's nucleus then, clearing out the hole in the doughnut, and matter is beginning to fall back. If there is a black hole in our galaxy's center maybe the next streamer to hit it would cause another such explosion.

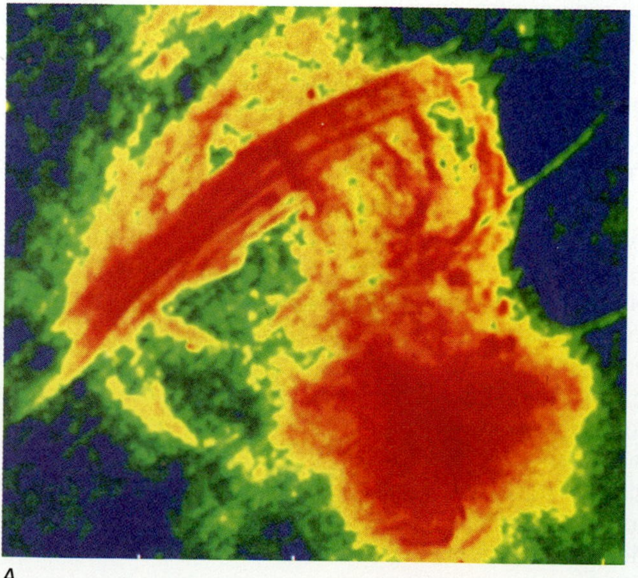

A

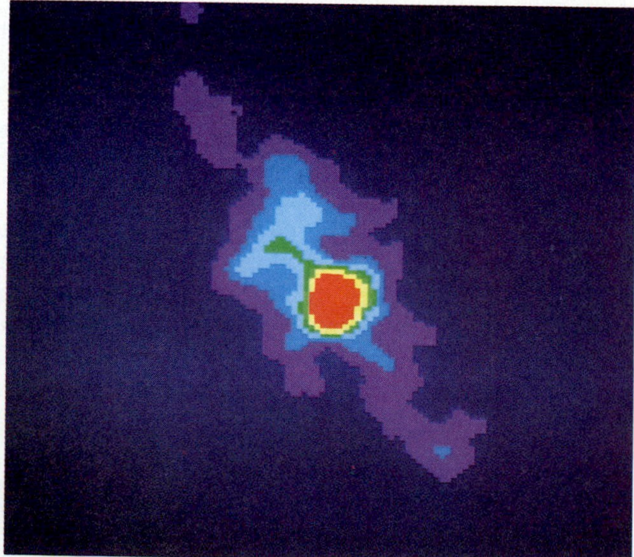

B

Figure 29–14 (*A*) The VLA shows that a vast Arc of parallel filaments stretches over 130 light years perpendicularly to the plane of the galaxy, which runs from upper left to lower right. The field of view is 25 arc min across; the wavelength used was in the continuous emission near 21 cm. (*B*) The Arc and Sgr A at 327 MHz (92 cm), a relatively low frequency mapped by the author and colleagues as a by-product of our study of interstellar deuterium (described in Section 34.2b).

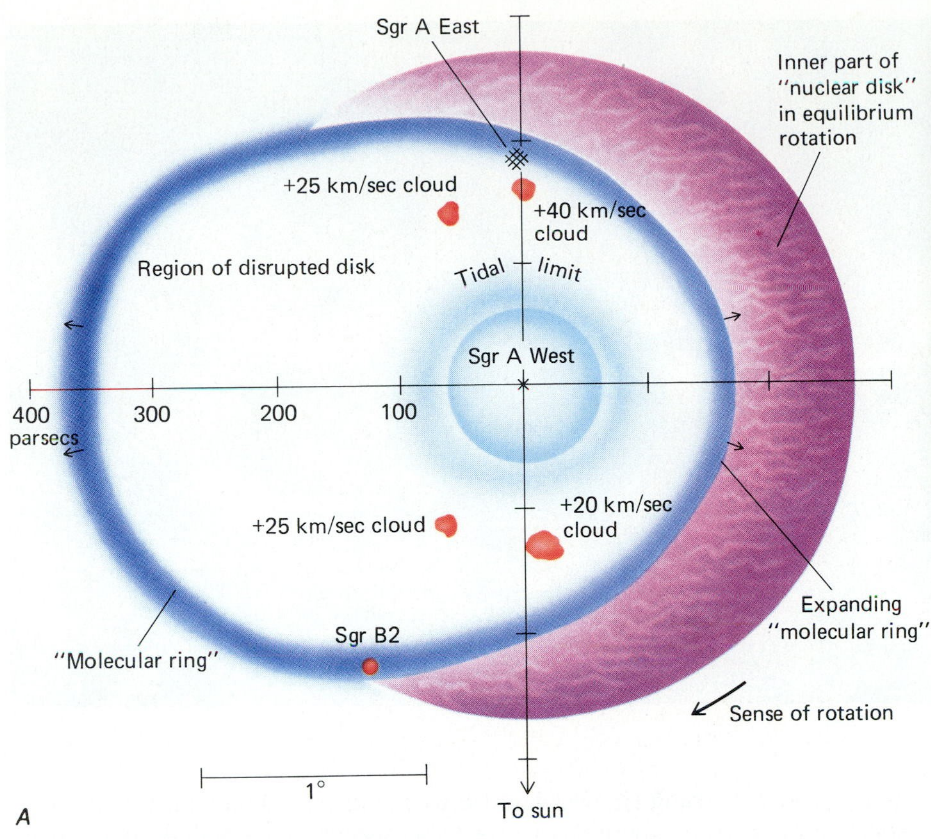

B

Figure 29–15 (*A*) A model of the galactic center. The velocities shown are measured from spectra. (*B*) Surprisingly, a VLA image of the inner 10 light-years of our galaxy in December 1990 showed an extra radio source, visible at lower left. The source is about 4 light-years from our galaxy's nucleus, and lasted for many months.

Continued work with the VLA has revealed parallel filaments stretched perpendicularly to the plane of the galaxy (Fig. 29–14). Because this "Arc" resembles a solar quiescent prominence, the structure seems to indicate that a strong magnetic field is present.

VLA observations have also shown a thin streamer of gas that may link the shell around the central black hole with the nearest gas cloud. The observations may explain how the central black hole continues to be fed.

The center of our galaxy (Fig. 29–15) is a fascinating place under increasing scrutiny.

29.4 All-Sky Maps of Our Galaxy

The study of our galaxy provides us with a wide range of types of sources. Many of these have been known for many years from optical studies (Fig. 29–16). The infrared sky looks quite different (Fig. 29–17). The radio sky provides still a different picture. Technological advances have enabled us to study sources in our galaxy in the x-ray and gamma-ray regions of the spectrum as well.

Non-solar x-ray astronomy began in 1962, when Riccardo Giacconi and colleagues discovered x-rays from a source in the constellation Scorpius (and named it Scorpius X-1). Herbert Friedman and colleagues soon carried out additional x-ray work. The 1960's research was carried out with rockets rather than with satellites. A few dozen sources, including Scorpius X-1, the Crab Nebula, and the Virgo cluster of galaxies, were found.

The first reasonable map of the x-ray sky was made with the American satellite named Uhuru, which observed hundreds of x-ray sources starting in 1970. Most of our current knowledge of x-ray sources came from the U.S. **H**igh-**E**nergy **A**stronomy **O**bservatories. HEAO-1 mapped 1500 x-ray sources (see Fig.

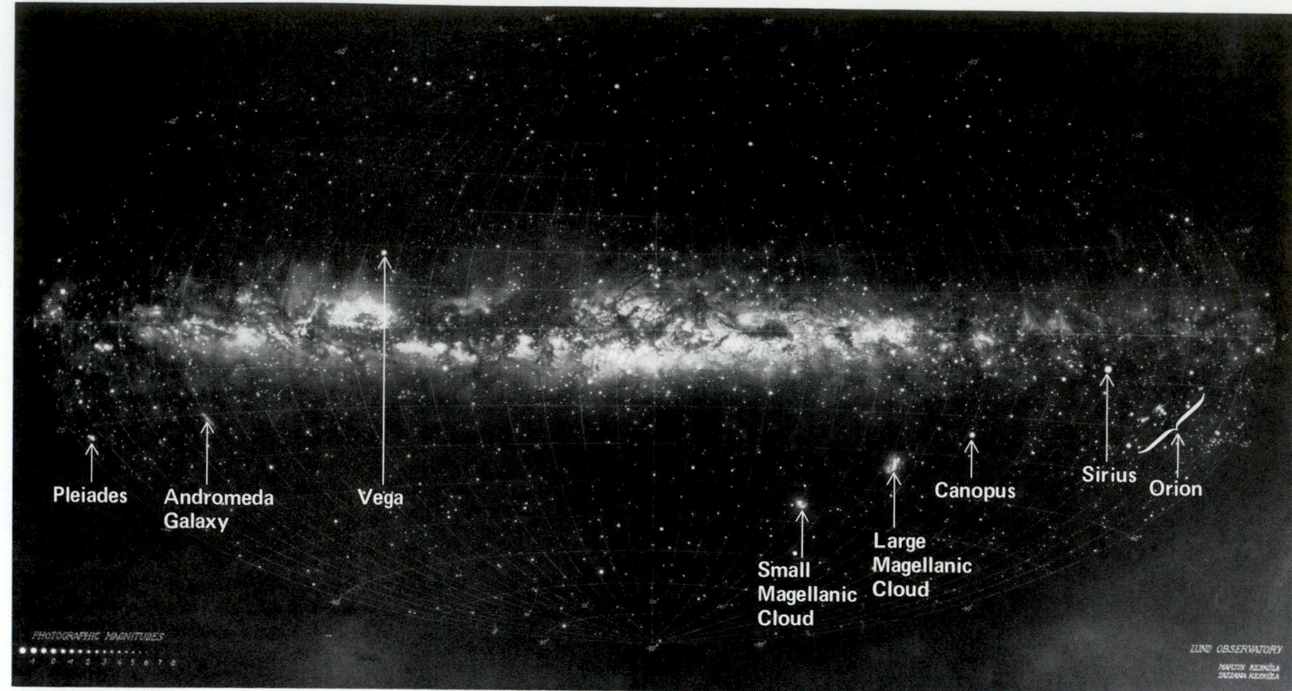

Figure 29–16 A drawing of the Milky Way, made under the supervision of Knut Lundmark at the Lund Observatory in Sweden. Seven thousand stars plus the Milky Way are shown in this panorama, which is in coordinates such that the Milky Way falls along the equator.

4–57). Scientists using HEAO-1 and HEAO-2 (Einstein) studied many of these sources over extended periods of time. Other recent x-ray satellites include the Japanese Ginga ("Silver River" = Milky Way), which re-entered the Earth's atmosphere in 1991, the European Exosat, and the Kvant x-ray telescope on the Mir space station sent up by the former Soviet Union. We discussed x-ray satellites in Section 4.12. Currently, the major x-ray satellite aloft is the German

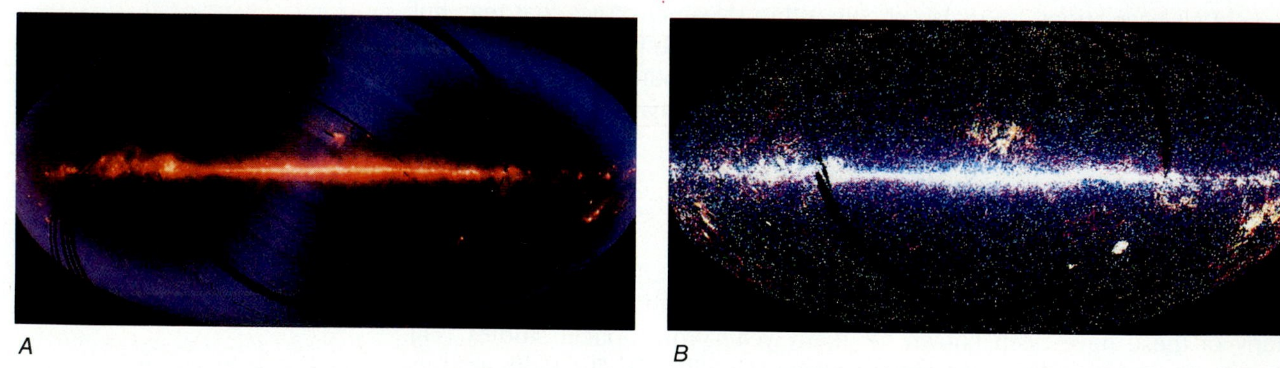

Figure 29–17 The sky observed from the Infrared Astronomical Satellite (IRAS). The 12 μ wavelength band is shown as blue, the 60 μ band is shown as green, and the 100 μ band is shown as red. Black streaks show where data are missing. (*A*) The dust glows all over the sky, rather than being limited to the galactic plane. Much of the emission far from the plane is the infrared cirrus, which is associated with neutral hydrogen gas. The star-forming region in Orion is at lower right. The 12 μ (blue) emission from dust is largely from our own solar system, and so follows the ecliptic plane (the plane of our solar system), which appears in the shape of an S in this projection on to coordinates aligned with the plane of our galaxy. (*B*) An IRAS map of individual sources. Shells of cool material surrounding K and M giant stars appear most prominently at 12 μ (blue). They are concentrated toward the plane of the Milky Way. Galaxies are most prominent in the 60 μ band (green) and are evenly distributed across the sky. Infrared cirrus is cool and so appears most prominently at 100 μ (red). The IRAS Point Source Catalog contains 200,000 infrared sources.

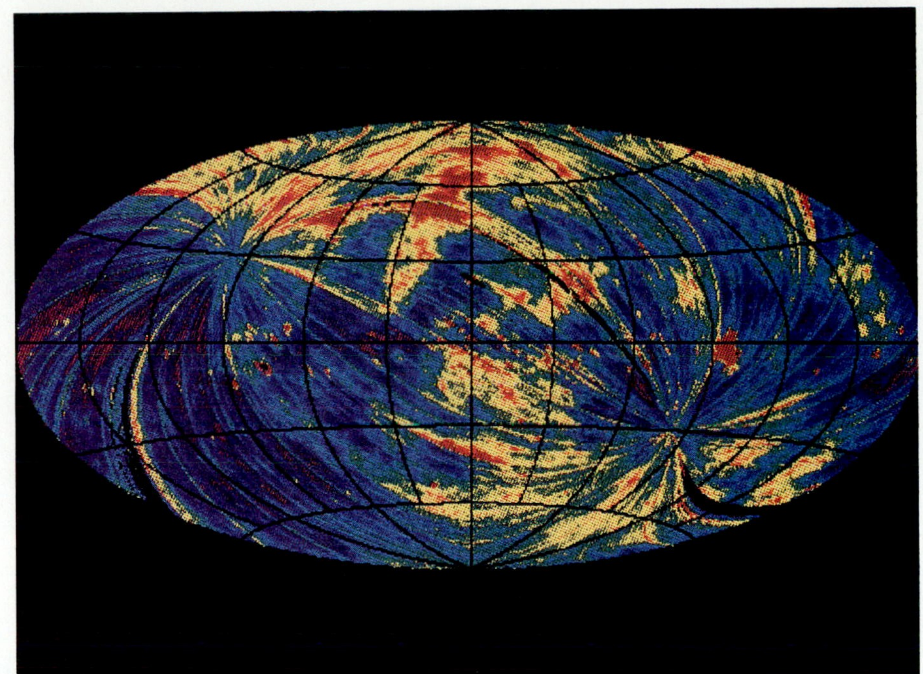

Figure 29–18 An all-sky x-ray map from Rosat, drawn in the same "galactic coordinates" as the preceding IRAS maps.

Rosat (**Ro**entgen **Sat**ellite), which has sent back detailed sky maps (Fig. 29–18).

Until recently, we knew of very few objects in the gamma-ray part of the spectrum. Objects that emitted in every part of the spectrum, like the pulsar in the Crab Nebula, were very rare. The European COS-B spacecraft mapped gamma rays for about 7 years starting in 1975 and found only a few dozen sources (Fig. 29–19). Only the Crab and Vela pulsars, a nearby quasar (3C 273), and a complex of interstellar clouds near ρ (rho) Ophiuchi corresponded to known objects. COS-B also found a diffuse background of gamma rays concentrated along the plane of our galaxy. The gamma rays are presumably caused by the interaction of cosmic rays with interstellar matter.

U.S. satellites in the Vela series, whose prime purpose is to detect nuclear explosions, and HEAO-3 also made gamma-ray observations. Some of the strangest observations were gamma-ray bursts that were detected from time to time. The strongest, a burst observed on March 5, 1979, lasted 1/5000 second and was followed for a few minutes by pulses with an 8-second period. During the short interval of this burst, the object gave off energy at a rate greater than that of the entire Milky Way Galaxy, if we assume that it was in our galaxy.

The gamma-ray situation changed dramatically in 1991 with NASA's launch of the Compton Gamma Ray Observatory, the second of NASA's series of Great Observatories. It was named after Arthur Holly Compton, the American physicist who won the Nobel Prize for his studies of the interaction of photons and matter.

Figure 29–19 This gamma-ray map made by the European COS-B spacecraft shows the plane of the Milky Way and several individual gamma-ray sources. The map includes the region between 25° above and 25° below the plane of the Milky Way.

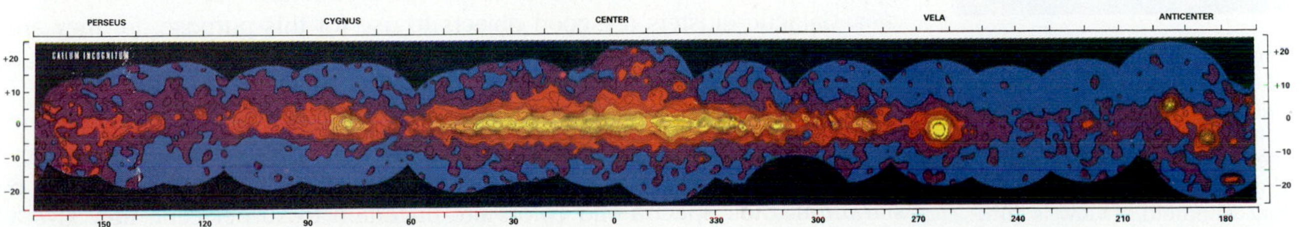

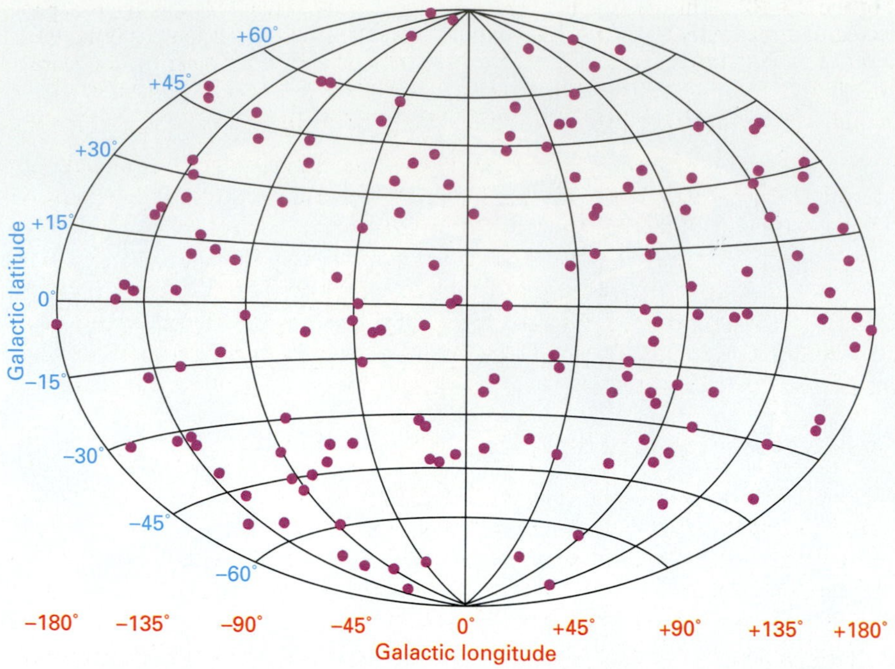

Figure 29–20 Gamma-ray bursts, mapped by the Burst and Transient Source Experiment on the Compton Gamma-Ray Observatory. Surprisingly, the sources seem randomly distributed, so unless they are very close to us, they are not phenomena of our galaxy.

The Italian scientists studying the source chose its name, which is pronounced with a hard second "g" for **Gemini ga**mma-ray source; they found the name suitable, since the source was unidentified for so long and since "geminga" means "it does not exist" in Milanese dialect.

The most studied gamma-ray object is Geminga, a very strong gamma-ray-emitting region in the constellation Gemini. X-ray pulsations were discovered in 1992 using data from Rosat, and these pulsations were confirmed in gamma rays with data from the Compton Gamma-Ray Observatory. Thus, Geminga is a close cousin of the Crab and Vela Nebulae, which have pulsating neutron stars at their cores. Geminga and potentially the other gamma-ray sources in the Milky Way are therefore apparently powered by rotating neutron stars. Geminga is less than 100 light-years from us. Maybe we will find more objects pulsing in gamma rays and yet remaining quiet in the radio spectrum.

29.5 The Spiral Structure of the Galaxy

29.5a Bright Tracers of the Spiral Structure

When we look out past the boundaries of the Milky Way Galaxy, usually by observing in directions above or below the plane of the Milky Way, we can see a number of galaxies with arms that appear to spiral outward from near their centers. In this section we shall discuss some of the evidence that our own Milky Way Galaxy also has spiral structure.

It is always difficult to tell the shape of a system from a position inside it (Fig. 29–22). Think, for example, of being somewhere inside a maze of tall hedges. We might be able to see through some of the foliage and be reasonably certain that layers and layers of hedges surrounded us, but we would find it difficult to trace out the pattern. If we could fly overhead in a helicopter, though, the pattern would become very easy to see.

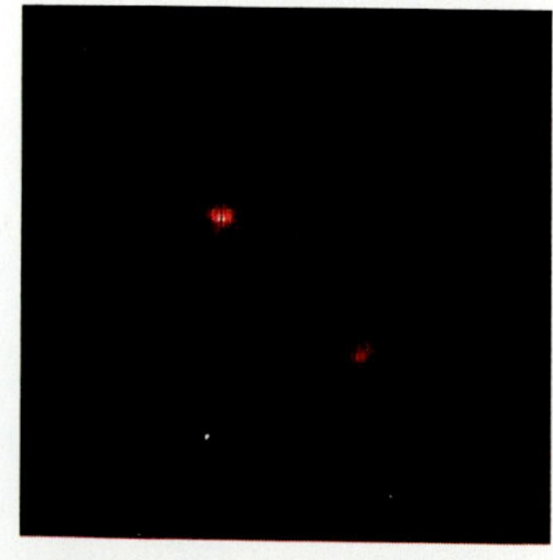

Figure 29–21 The Crab Nebula (*left*) and Geminga (*lower right*), observed with the EGRET (Energetic Gamma-Ray Experiment Telescope) on the Compton Gamma-Ray Observatory. The field of view is 40° square.

Similarly, we have difficulty tracing out the spiral pattern in our own galaxy, even though the pattern would presumably be apparent from outside the galaxy. Still, by noting the distances and directions to objects of various types, we can tell about the Milky Way's spiral structure.

Galactic clusters are good objects to use for this purpose, for they are always located in the spiral arms. We have found the distance to 200 galactic clusters by studying their color-magnitude diagrams.

We think that spiral arms are regions where young stars are found. Some of the young stars are the O and B stars; their lives are so short we know they can't be old. But since our methods of determining the distances to O and B stars from their spectra and colors are uncertain to 10 per cent, they give a fuzzy picture of the distant parts of our galaxy.

Figure 29–22 This view of the "grand design" spiral galaxy M74 has been deprojected in a computer to show what it would look like face on. Also, a continuous level of radiation that decreased radially outward was subtracted to allow the spiral structure to be better seen. Such structure is thought to come from spiral density waves. The photo was made from B (blue) and I (infrared) images.

Other signs of young stars are the presence of regions of ionized hydrogen known as H II regions (pronounced "H two" regions). We know from studies of other galaxies that H II regions are preferentially located in spiral arms. In studying the locations of the H II regions, we are really again studying the locations of the O stars and the hotter B stars, since it is ultraviolet radiation from these hot stars that provides the energy for the H II regions to glow. In Section 29.5c we shall discuss a theory that explains why O and B stars should be found in spiral arms.

*29.5b Differential Rotation

The latest calculations indicate that the sun is approximately 8.5 kiloparsecs from the center of our galaxy. From spectroscopic observations of the Doppler shifts of globular clusters (which do not participate in the galactic rotation) or of distant galaxies, we can tell that the sun is revolving around the center of our galaxy at a speed of approximately 250 kilometers per second. At this velocity, it would take the sun about 250 million years to travel once around the center; this period is called the *galactic year.* But not all stars revolve around the galactic center in the same period of time. The central part of the galaxy rotates like a solid body. Beyond the central part, the stars that are farther out have longer galactic years than stars closer in.

If this system of *differential rotation,* with differing rotation speeds at different distances from the center, has persisted since the origin of the galaxy, we may wonder why there are still only a few spiral arms in our galaxy and in the other galaxies we observe. The sun could have made fifty revolutions during the lifetime of the galaxy, but points closer to the center would have made many more revolutions. Thus the question arises: why haven't the arms wound up very tightly?

*29.5c Why Our Galaxy Has Spiral Arms

The leading current solution to this conundrum is a theory first suggested by the Swedish astronomer B. Lindblad and elaborated mathematically by the American astronomers C. C. Lin at MIT and Frank Shu, now at Berkeley. They say, in effect, that the spiral arms we now see are not the same spiral arms that were previously visible. In their model, the spiral-arm pattern is caused by a spiral *density wave,* a wave of increased density that moves through the stars and gas in the galaxy. This density wave is a wave of compression, not of matter being transported. It rotates more slowly than the actual material, and causes the density of material to build up as it passes. A shock wave—like a sonic boom from an airplane—builds up and compresses the gas. In some galaxies, the compressed gas collapses to form stars. In other galaxies, the increased density leads to an increased rate of collisions of the giant molecular clouds we will discuss in the next chapter. These collisions may lead to the formation of massive stars.

So, in the density-wave model, the spiral arms we see at any given time do not represent the actual motion of individual stars in orbit around the galactic center (Fig. 29–23). We can think of the analogy of a crew of workers painting

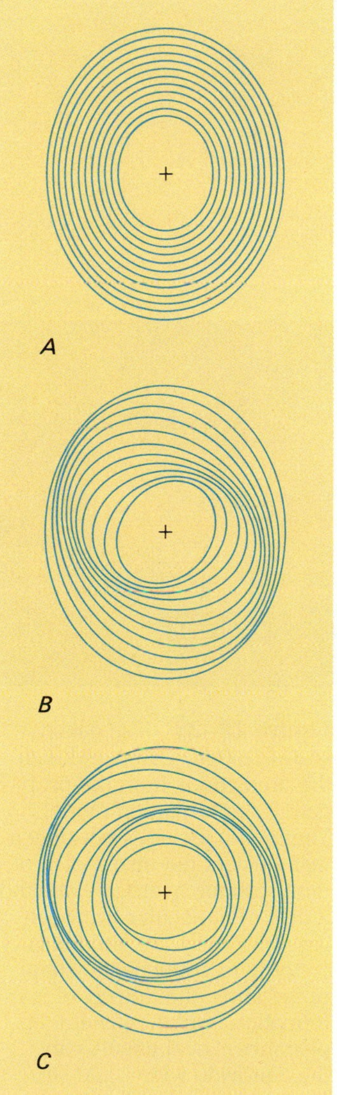

Figure 29–23 Each part of the figure includes the same set of ellipses; the only difference is the relative alignment of their axes. Consider that the axes are rotating slowly and at different rates. The compression of their orbits takes a spiral form, even though no actual spiral exists. The spiral structure of a galaxy may arise from an analogous effect.

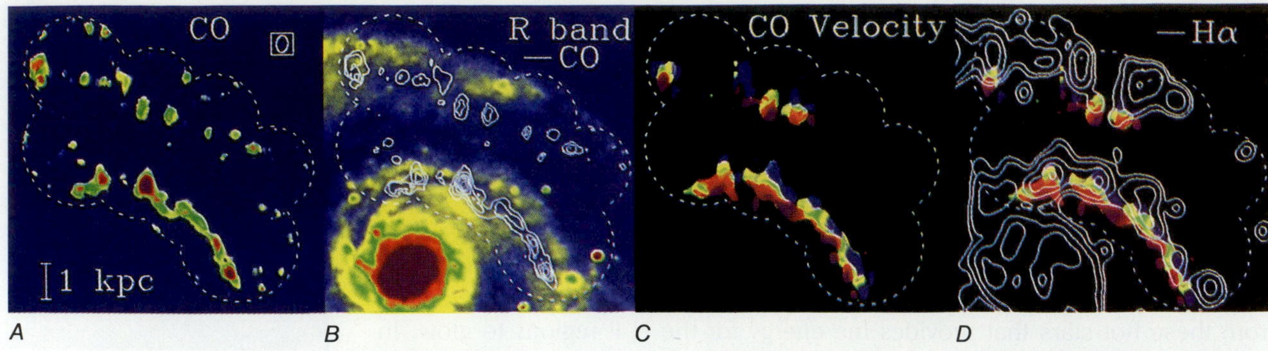

Figure 29–24 Studies of star formation in spiral arms of the grand design galaxy M51. The molecular clouds (Section 30.5) are traced by their emission of carbon monoxide (CO). (*A*) CO emission within the dashed area was mapped with the interferometer of the Owens Valley Radio Observatory. The beam shape and size appear in a box at upper right. (*B*) Background radiation in the R (red) band is compared with the CO emission. A narrow dust lane in the R band, about 300 pc wide, coincides with the CO and therefore with the peak density of the molecular cloud. (*C*) The velocity of the carbon-monoxide gas shifts perpendicularly to the spiral arms. The shifts match predictions of density-wave models. (*D*) The CO as compared with Hα. The CO arm is offset slightly toward the nucleus. If the CO resulted from heating, then it would coincide with the visible arm of OB stars that heats the gas and causes the Hα emission. Since it is displaced, we are seeing the location of maximum molecular density. The result again matches the predictions of density-wave theory.

a white line down the center of a busy highway. A bottleneck occurs at the location of the painters. If we were observers in an airplane, we would see an increase in the number of cars at that place. As the line painters continued slowly down the road, we would seem to see the place of increased density move down the road at that slow speed. We would see the bottleneck move along even if our vision were not clear enough to see the individual cars, which could still speed down the highway, slow down briefly as they cross the region of the bottleneck, and then resume their high speed.

Similarly, we might be viewing only some galactic bottleneck at the spiral arms. Gas can pile up to make spiral arms that are visible without being permanent. New studies with an interferometer operating at millimeter wavelengths have shown that the newly formed stars are slightly downstream from the molecular complexes that the density wave forms from individual molecular clouds. In molecular-rich galaxies, this mechanism provides the luminous stars that define the arms. These stars heat the interstellar gas so that it becomes visible; we see young, hot stars and glowing gas outline the spiral arms (Fig. 29–24).

Figure 29–25 The galaxy NGC 628 (*left*) compared with the predictions from the propagating star-formation model. The model includes not only a disk of stars but also a disk of dilute atomic hydrogen gas and a disk of molecular hydrogen gas. Obviously, the model satisfactorily reproduces the shape of the arms, the density of star formation, and the general appearance. Local effects can, thus, provide wide-scale structure.

Some astronomers do not accept the density-wave theory; one of their major objections is the problem of explaining why the density wave forms. One alternative, very different theory says that stars are produced by a chain reaction and are then spread out into spiral arms by the differential rotation of the galaxy. The chain reaction begins when high-mass stars become supernovae. The expanding shells from the supernovae trigger the formation of stars in nearby regions. Some of these new stars become massive stars, which become supernovae, and so on.

Computer modelling of the *supernova-chain-reaction model* gives values that seem to agree with the observed features of spiral galaxies (Fig. 29–25). Each galaxy has a differential velocity, with regions at different distances from the center rotating at different speeds. At some distance from the center, the speed is a maximum. The degree of winding depends on the value for this maximum velocity.

It may be that the density-wave model and the supernova-chain-reaction model can be important to different degrees in different galaxies.

Summary and Outline

Nebulae (Section 29.1)
- Emission; dark (absorption); reflection; planetary; supernova remnants
- The reddish glow results from Hα emission.

Our galaxy has a nuclear bulge surrounding its nucleus, a disk, a halo, and a corona (Section 29.2).

Infrared observations (Section 29.3)
- Limited by atmospheric transparency and technology
- IRAS mapped the entire sky at relatively long wavelengths.
- Radiation from the galactic nucleus indicates a lot of energy is generated in a small volume there; a very massive black hole may be present.

High-energy astronomy (Section 29.4)
- IRAS maps show cooler material, especially star formation
- Rosat now best x-ray satellite
- Compton Gamma-Ray Observatory aloft

Spiral structure of the galaxy (Section 29.5)
- Bright tracers of the spiral structure (Section 29.5a)
  - Difficult to determine shape of our galaxy because of obscuring gas and dust and because of our vantage point
  - Galactic clusters, O and B stars, and H II regions are used to determine the locations of spiral arms.

Differential rotation (Section 29.5b)
- Sun is approximately 8.5 kpc from center of galaxy.
- Central part of galaxy rotates as a solid body.
- Outer part, including the spiral arms, is in differential rotation.

Theories for spiral structure (Section 29.5c)
- Density wave: we see the effect of a wave of compression; the distribution of mass itself is not spiral in structure
- Chain of supernovae

Key Words

nebula, emission nebula, dark nebula, absorption nebula, star clouds, reflection nebula, nuclear bulge, nucleus, disk, spiral arms, halo, galactic corona, positron, galactic bulge sources*, bursters*, high-energy astrophysics, galactic year*, differential rotation*, density wave*, supernova-chain-reaction model*

*This term is found in an optional section.

Questions

1. Why do we think our galaxy is a spiral?
2. How would the Milky Way appear if the sun were closer to the edge of the galaxy?
3. Sketch (a) a side view and (b) a top view of the Milky Way Galaxy, showing the shapes and relative sizes of the nuclear bulge, the disk, and the halo. Mark the position of the sun. Indicate the location of the galactic corona.
4. Compare (a) absorption (dark) nebulae, (b) reflection nebulae, and (c) emission nebulae.
5. How can something be both an emission and an absorption nebula? Explain and give an example.
6. Sketch the North America Nebula, indicating on the sketch where you see Hα radiation and where you see dust.
7. If you see a red nebula surrounding a blue star, is it an emission or a reflection nebula? Explain.
8. How do we know that the galactic corona isn't made of ordinary stars like the sun?
9. Why may some infrared observations be made from mountain observatories, while all x-ray observations must be made from space?
10. What radio source corresponds to the center of our galaxy?
11. Illustrate, by sketching the appropriate Planck radiation curves, why cooling an infrared telescope from room temperature to 4 K greatly reduces the background noise.
12. Describe infrared and radio results about the center of our galaxy.
13. What are three tracers that we use for the spiral structure of our galaxy? What are two reasons why we expect them to trace spiral structure?
14. Since Einstein's result was that $E = mc^2$, explain how, if m_e is the mass of an electron, $2m_ec^2$ of energy results when an electron and a positron annihilate each other.
15. Why will x-rays expose a photographic plate, while infrared radiation will not?
16. What types of objects give off strong infrared radiation?
17. What types of objects give off x-rays?
18. Discuss how observations from space have added to our knowledge of our galaxy.
19. Comment on our understanding of the gamma-ray bursts.
20. Why does the density-wave theory lead to the formation of stars?

†21. If a spacecraft could travel at 10 per cent of the speed of light, how long would it have to travel to get far enough out to be able to take a photograph showing the spiral structure of our galaxy?

†This question requires a numerical solution.

The center of the Orion Nebula in an enlarged, processed view from the Hubble Space Telescope.

The Interstellar Medium

30

Aims: To discuss the interstellar medium, to describe the techniques (especially radio astronomy) used to study it, and to understand the relation of interstellar clouds to star formation

The gas and the dust between the stars is known as the *interstellar medium*. The nebulae (Fig. 30–1) represent regions of the interstellar medium in which the density of gas and dust is higher than average. The interstellar medium contains the elements in what we call their *cosmic abundances,* that is, the overall abundances they have in the universe (the cosmos).

In this chapter we shall see that the studies of the interstellar medium are vital for understanding the structure of our galaxy and how stars are formed. We shall also discuss the fairly recent realization that units of the interstellar medium called "giant molecular clouds" are basic building blocks of our galaxy.

Figure 30–1 The Lagoon Nebula, M8, an H II region in Sagittarius, glowing red with Hα.

30.1 H I and H II Regions

For many purposes, we may consider interstellar space as being filled with hydrogen at an average density of about 1 atom per cubic centimeter, although individual regions may have densities departing greatly from this average. Regions of higher density in which the atoms of hydrogen are predominantly neutral are called *H I regions* (pronounced "H one regions"; the Roman numeral "I" refers to the first, or basic, state). Where the density of an H I region is high enough, pairs of hydrogen atoms combine to form molecules (H_2). The densest part of the gas associated with the Orion Nebula might have a million or more hydrogen molecules per cubic centimeter. So hydrogen molecules (H_2) are often found in H I regions.

A region of ionized hydrogen, with one electron missing, is known as an *H II region* (from "H two," the second state). Since hydrogen, which makes up the overwhelming proportion of interstellar gas, contains only one proton and one electron, a gas of ionized hydrogen contains individual protons and electrons. Wherever a hot star provides enough energy to ionize hydrogen, an H II region (Fig. 30–2) results. Emission nebulae are such H II regions. They glow because the gas is heated. Emission lines appear.

Studying the optical and radio spectra of H II regions and planetary nebulae tells us the abundances of several of the chemical elements (especially helium, nitrogen, and oxygen). How these abundances vary from place to place in our galaxy and in other galaxies helps us choose between models of element formation and of galaxy formation. The variation of the relative abundances from the center of a galaxy to its outer regions—the gradient of abundances—differs for different galaxies. The abundances of the elements known as "metals" are greatest in the locations where the most star formation took place; this indicates that these elements were indeed produced in stars.

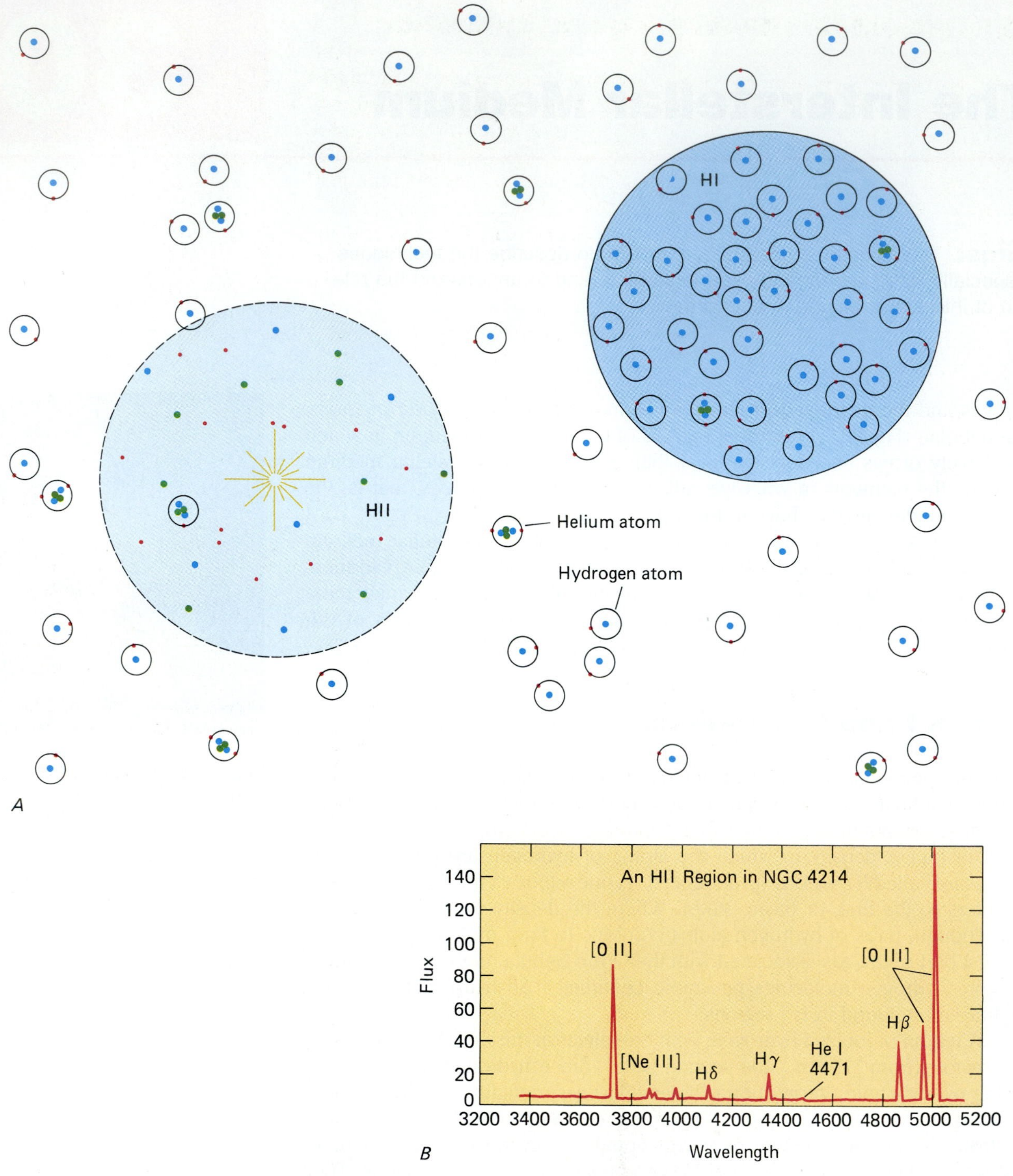

Figure 30–2 (*Top*) *H I regions* are regions of neutral hydrogen of higher density than average, and *H II regions* are regions of ionized hydrogen. The protons and electrons that result from the ionization of hydrogen by ultraviolet radiation from a hot star, and the neutral hydrogen atoms, are shown schematically. The larger dots represent protons or neutrons and the smaller dots represent electrons. The star that provides the energy for the H II region is shown. Though all hydrogen is ionized in an H II region, only some of the helium is; heavier elements have often lost 2 or 3 electrons. (*Bottom*) The emission lines in an H II region, including some of the members of hydrogen's Balmer series, helium, and "forbidden lines" of ionized oxygen and neon. (These "forbidden lines" are enclosed in square brackets to show that they are transitions that happen relatively rarely—they aren't completely forbidden from occurring.)

30.2 Interstellar Reddening and Extinction

Distant stars in the plane of our galaxy are obscured from our vision. In addition, many stars that are still close enough to be visible are partially obscured (Fig. 30–3). The amount of obscuration varies with the wavelength at which we observe. The blue light is *scattered* by dust in space more efficiently than the redder light is; for a given distance through the dust, more of the blue light has been bounced around in every direction. Thus less of the blue light comes through to us than the red light, and the stars look redder—we say that the stars are *reddened.* This reddening is thus a consequence of the scattering properties of the dust. It has nothing to do with "redshifts," since the spectral lines are not shifted in wavelength by reddening.

The amount of scattering together with the amount of actual absorption of visible radiation by dust is known as the *extinction* (Fig. 30–4). In the blue part of the spectrum, the total extinction in the 8.5 kpc between the sun and the center of the galaxy is about 25 magnitudes. Most of this takes place far from the sun, in regions with a high dust content. Even a tremendously bright object located near the center of our galaxy would be dimmed too much—25 magnitudes—to be seen from the earth in the visible part of the spectrum.

What is the dust made of? We know the particles are tiny—smaller than the wavelength of light—or else they would not scatter light this way, with the shorter wavelengths scattered more efficiently. From spectral studies, we know

Figure 30–3 The Trifid Nebula, M20, in Sagittarius, is the red H II region divided into three visible parts by absorbing dust lanes. The red Hα is diluted with blue light scattered from the hot central stars. The blue reflection nebula at the top is unconnected to the Trifid. In the dark regions, the extinction is so high we cannot see stars behind the dust causing the extinction.

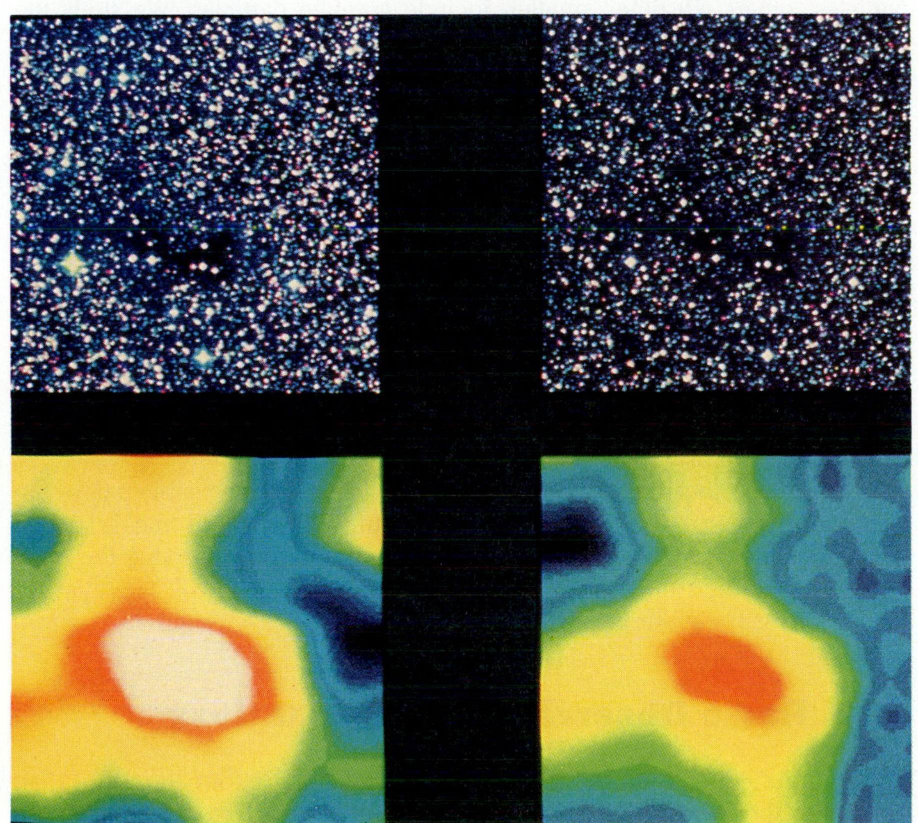

Figure 30–4 A dark globule, C-019. Below each image is a map of the interstellar extinction computed by counting stars in the upper photos. The extinction is higher in the blue (*left*) than in the red (*right*), for reasons similar to interstellar reddening. (Color code: blue = no extinction, yellow = 1 mag. of extinction, red = 2 mag., white = 2.5 mag.)

Focus On

Box 30.1 Why Is the Sky Blue?

Just as interstellar dust scatters blue light more than it scatters red light, scattering by the electrons in air molecules in the terrestrial atmosphere sends blue light preferentially around in all directions. Since we are close to the scattering electrons, our sky appears blue. Further, when the sun is near the horizon, we have to look diagonally through the earth's layer of air. Our line of sight through the air is then longer than a line of sight straight up, and most of the blue light is scattered out before it reaches us. Relatively more red light reaches us, accounting for the reddish color of sunsets.

that at least some of the particles are carbon in the form of graphite. Other particles may be silicates or ices (Fig. 30–5).

Interstellar dust grains form out of chemical elements that result from fusion in stars and supernovae. A typical grain has a core made of magnesium, silicon, or iron that was ejected from a dwarf star of spectral-type M or from a supergiant, and has picked up a surrounding mantle made of compounds of oxygen, carbon, and nitrogen. Some grains may be the 10^{-5}-cm size of smoke particles; others, 10 times smaller, include graphite particles. We study the dust grains by the spectrum of their extinction of starlight, and by laboratory simulation.

Interstellar dust is heated a bit by radiation, thus causing the dust to radiate. Because the energy received is balanced by the dust's radiation at a low temperature, the dust never gets very hot. Thus radiation from interstellar dust peaks in the infrared. The radiation from dust spread out among the stars had been too faint to detect, but the Pioneer 10 and 11 space probes have recently moved beyond so much of the solar system's dust that they have been able to detect, in the visible part of the spectrum, starlight reflected off the spread-out interstellar dust. The radiation coming from clouds of dust surrounding stars has been observed from the ground and from the IRAS spacecraft. IRAS found infrared radiation from so many stars forming in our galaxy that we now think that about one star forms in our galaxy each year, a higher value than previously realized.

Similarly, since the interstellar gas is "invisible" in the visible part of the spectrum (except at the wavelengths of certain weak spectral lines), special techniques are needed to observe the gas in addition to observing the dust. Radio astronomy is the most widely used technique, so we will now discuss its use for mapping our galaxy.

30.3 Radio Observations of Our Galaxy

The rest of the electromagnetic spectrum carries more information in it than do the few thousand angstroms that we call visible light. We will first discuss some basic techniques of radio astronomy (see also Section 4.13b), and then

Figure 30–5 A model of an aggregate of 100 interstellar dust particles, each about 1 μ in size. Each particle has a silicate core, an inner mantle of organic material, and an outer mantle that is mostly water ice with smaller particles embedded. Such particles are found not only in interstellar space but also in comets.

go on to see how radio astronomy joins with other observing methods to investigate interstellar space.

30.3a Continuum Radio Astronomy

All radio-astronomy studies in the early days were of the continuum. In radio astronomy, as in optical astronomy, studying the continuum means that we consider the average intensity of radiation at a given frequency without regard for variations in intensity over small frequency ranges. In short, we ignore any spectral lines. It was immediately apparent that the brightest objects in the radio sky are not identical with the brightest objects in the optical sky. The radio objects were named with letters and with the names of their constellations. Thus Taurus A is the brightest radio object in the constellation Taurus; we now know it to be the Crab Nebula. Sagittarius A is the center of our galaxy; Sagittarius B is another radio source nearby, whose emission is caused by clouds of gas near the galactic center.

*30.3b Synchrotron Radiation

Continuum radio radiation can be generated by several processes. One of the most important is *synchrotron emission* (Fig. 30–6), the process that produces the radiation from Taurus A, a supernova remnant. Lines of magnetic field extend throughout the visible Crab Nebula and beyond. Electrons, which are electrically charged, tend to spiral around magnetic lines of force. Electrons of high energy spiral very rapidly (Fig. 30–7), at speeds close to the speed of light. We say that they move at "relativistic speeds," since the theory of relativity must be used for calculations when the electrons are going that fast. Under these conditions, the electrons radiate very efficiently. (This is the same process that generates the light in electron synchrotrons in some kinds of atom-smashing physics laboratories on earth, hence the name synchrotron radiation.)

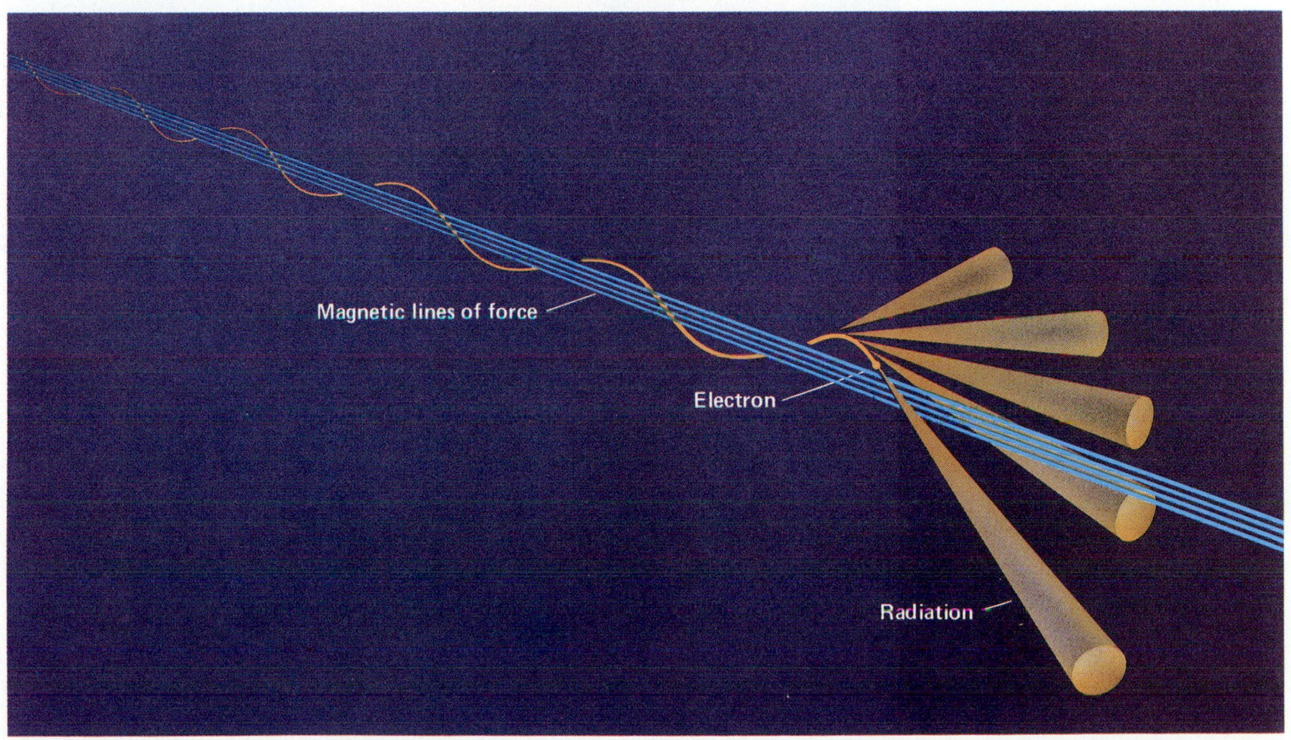

Figure 30–6 Electrons spiralling around magnetic lines of force at velocities near the speed of light (we say "at relativistic velocities") emit radiation in a narrow cone. This radiation, which is continuous and highly polarized, is called synchrotron radiation. Synchrotron radiation has been observed in both optical and radio regions of the spectrum.

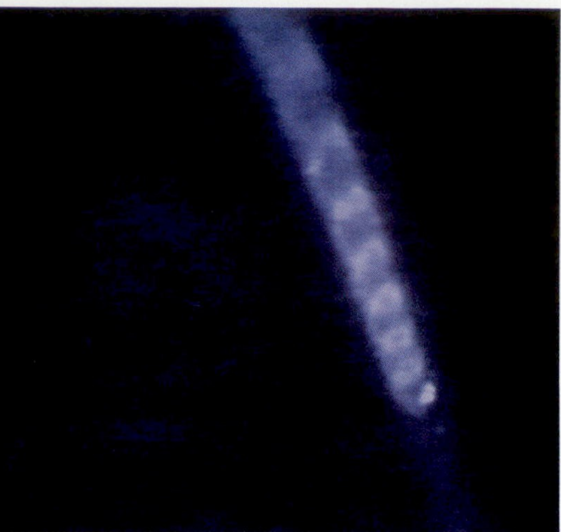

Figure 30–7 Synchrotron radiation viewed in the earth's atmosphere from the Echo-7 rocket. A beam was injected across the earth's magnetic field, and viewed with a TV camera.

A

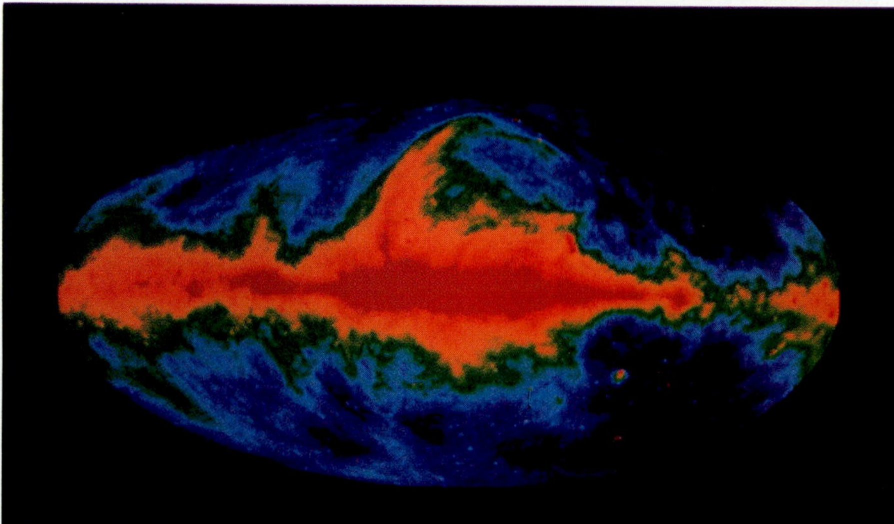

B

Figure 30–8 (*A*) The polarization of the continuous radiation from the Crab Nebula (Figure 26–8). This new color display, made especially for this book by David Malin of the Anglo-Australian Observatory, assigns a color to a photograph taken at each of four angles of polarization: 0° (blue), 45° (black), 90° (red), and 135° (green). The plates were taken with the 5-m Palomar telescope in 1967. We see that the radiation is highly polarized and that different regions are polarized in different directions. The discovery of the polarization verified the prediction that synchrotron radiation was being emitted. (*B*) Synchrotron radiation dominates the view of the radio sky in this image at 408 MHz (73-cm wavelength). On this map in galactic coordinates, the plane of the Milky Way appears horizontally. The Galaxy's magnetic field bends the electrons to cause the synchrotron radiation, which appears both in the plane of the Galaxy and as the "spurs" that extend above the galactic plane. The North Polar Spur, at top center, is about 500 light years away and may have resulted from a supernova 300,000 years ago. Some of the sources on this map, compiled by several observatories together, are background galaxies.

The suggestion that the synchrotron mechanism causes radiation from various astronomical sources (Fig. 30–8) was first made by several Soviet theoreticians about 1950. Synchrotron radiation is highly polarized, and the discovery a few years later that the optical radiation from the Crab Nebula is highly polarized was an important confirmation of this suggestion. The radio radiation from the Crab is also highly polarized.

The intensity of synchrotron radiation is related not to the temperature of the astronomical body that is emitting the radiation, but rather to the strength of its magnetic field and to the number and energy distribution of the electrons caught in that field. Since the temperature of the object cannot be derived from knowledge of the intensity of the radiation, we call this radiation *non-thermal radiation.* Synchrotron radiation is but one example of such non-thermal processes. The synchrotron process can work so efficiently that a relatively cool astronomical body can give off a tremendous amount of such radiation, perhaps so much that it would have to be heated to a few million degrees before it would radiate as much *thermal radiation* at a given frequency.

By *thermal radiation* we mean continuous radiation whose spectrum is directly related to the temperature of the gas.

30.4 The Radio Spectral Line from Interstellar Hydrogen

About 1950, though radio astronomers were very busy with continuum work, there was still a hope that a radio spectral line might be discovered. This discovery would allow Doppler-shift measurements to be made.

What is a radio spectral line? Remember that an optical spectral line corresponds to a wavelength (or frequency) in the optical spectrum that is more (for an emission line) or less (for an absorption line) intense than neighboring wavelengths or frequencies (Fig. 30–9). Similarly, a radio spectral line corresponds to a frequency (or wavelength) at which the radio noise is slightly more, or slightly less, intense. As we saw in Section 20.3a, a spectral line corresponds to an atom's energy change, as any electron changes from one of an atom's energy levels to another.

30.4a The Hydrogen Spin-Flip

The most likely candidate for a radio spectral line that might be discovered was a line from the lowest energy levels of interstellar hydrogen atoms. This line was predicted to be at a wavelength of 21 cm. Since hydrogen is by far

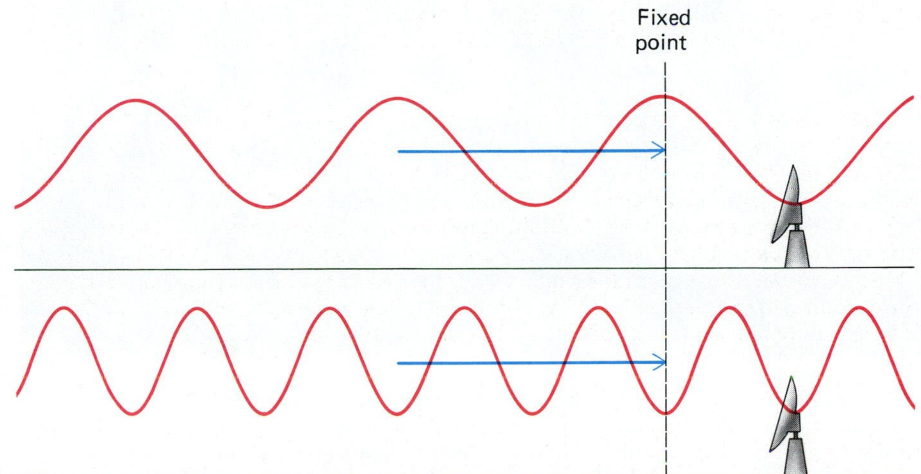

Figure 30–9 Radio astronomers often speak in terms of frequency instead of wavelength. Since all electromagnetic radiation travels at the same speed (the speed of light) in a vacuum, fewer waves of longer wavelength (*top*) pass an observer in a given time interval than do waves of a shorter wavelength (*bottom*). If the wavelength is half as long, twice as many waves pass, that is, the frequency is twice as high. The wavelength λ times the frequency ν is constant, with the constant being the speed of light c: $\lambda\nu = c$.

Frequency is given in hertz (Hz), formerly cycles per second. (Though the name of the unit, "hertz," is not capitalized, the symbol Hz is capitalized to show it is derived from someone's name.) The 21-cm line is at 1420 MHz.

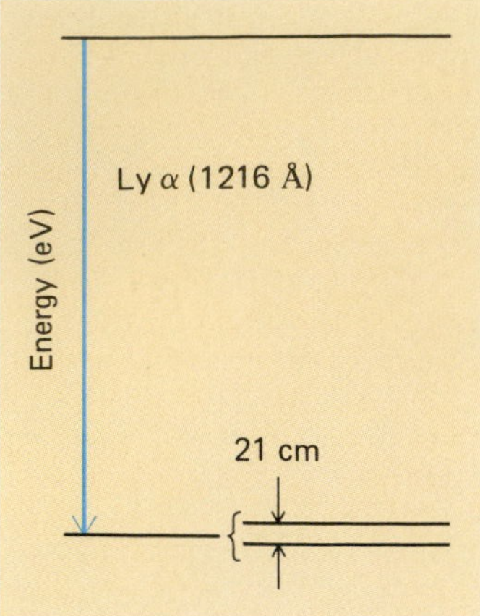

Figure 30–10 21-cm radiation results from an energy difference between two sublevels in the lowest principal energy state of hydrogen. The energy difference is much smaller than the energy difference that leads to Lyman α. So (because $E = h\nu = hc/\lambda$) the wavelength is much longer.

the most abundant element in the universe, it seems reasonable that it should produce a strong spectral line. Furthermore, since most of the interstellar hydrogen has not been heated by stars or otherwise, it is most likely that this hydrogen is in its state of lowest possible energy.

This line at 21 cm comes not from a transition to the ground state from one of the higher states, nor even from level 2 to level 1 as does Lyman alpha, but rather from a transition between the two sublevels into which the ground state of hydrogen is divided (Fig. 30–10).

For this astronomical discussion, it is sufficient to think of a hydrogen atom as an electron orbiting a proton. Both the electron and the proton have the property of spin; each one has angular momentum (Section 6.4) as if it were spinning on its axis.

The spin of the electron can be either in the same direction as the spin of the proton or in the opposite direction. The rules of quantum mechanics prohibit intermediate orientations. If the spins are in opposite directions, the energy state of the atom is very slightly lower than the energy state occurring if the spins are in the same direction. The energy difference between the two states is equal to a photon of 21-cm radiation.

If an atom is sitting alone in space in the upper of these two energy states, with its electron and proton spins aligned in the same direction, there is a certain small probability that the spinning electron will spontaneously flip over to the lower energy state and emit a photon. We thus call this a *spin-flip* transition (Fig. 30–11, *top*). The photon of hydrogen's spin-flip corresponds to radiation at a wavelength of 21 cm—the *21-cm line.*

We have just described how an emission line can arise at 21 cm. But what happens when continuous radiation passes through neutral hydrogen

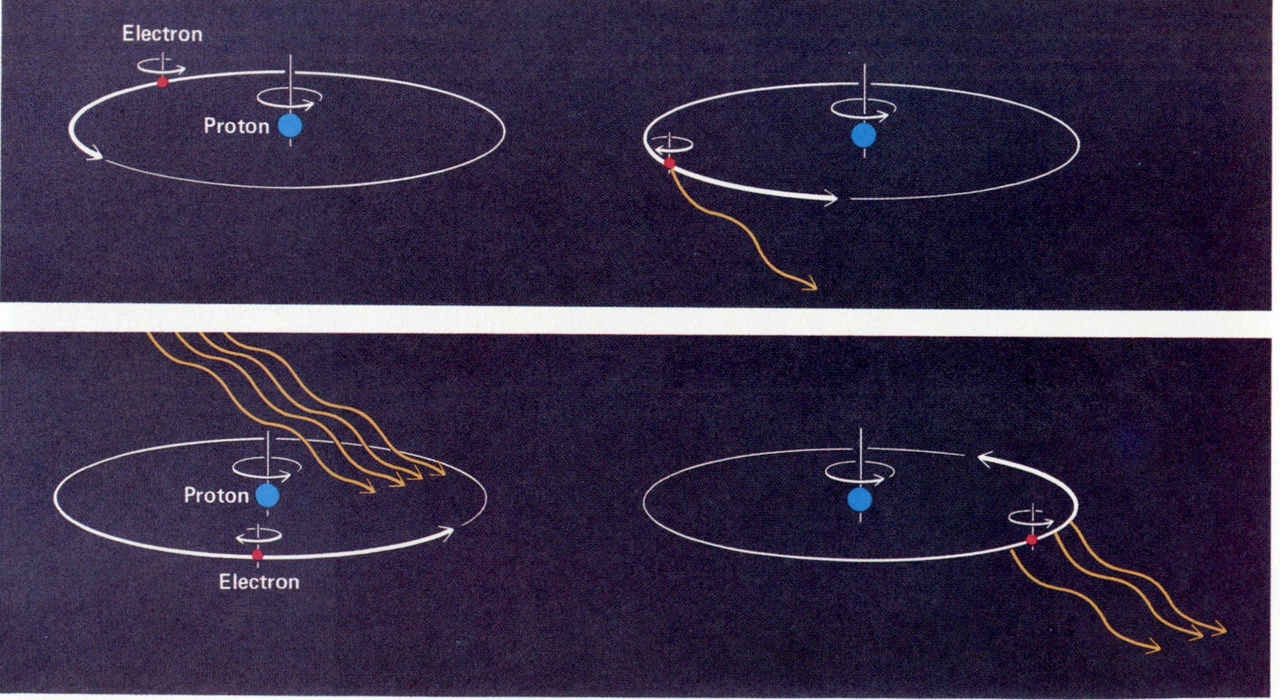

Figure 30–11 When the electron in a hydrogen atom flips over so that it is spinning in the opposite direction from the spin of the proton (*top*), an emission line at a wavelength of 21 cm results. When an electron takes energy from a passing beam of radiation, causing it to flip from spinning in the opposite direction from the proton to spinning in the same direction (*bottom*), then a 21-cm line in absorption results.

gas? In this case, some of the electrons in atoms in the lower state will absorb a 21-cm photon and flip over, putting the atom into the higher state. Then the radiation that emerges from the gas will have a deficiency of such photons and will show the 21-cm line in absorption (Fig. 30–11, *bottom*).

If we were to watch any particular group of hydrogen atoms, we would find that it would take 11 million years before half of the electrons had undergone spin-flips; we say that the *half-life* is 11 million years for this transition. But even though the probability of a transition taking place in a given atom in a given interval of time is very low, there are so many hydrogen atoms in space that enough 21-cm radiation is given off to be detected.

30.4b Mapping Our Galaxy

21-cm hydrogen radiation has proved to be a very important tool for studying our galaxy because this radiation passes unimpeded through the dust that prevents optical observations very far into the plane of the galaxy. Using 21-cm observations, astronomers can study the distribution of gas in the spiral arms. We can detect this radiation from gas located anywhere in our galaxy, even on the far side, whereas light waves penetrate the dust clouds in the galactic plane only about 10 per cent of the way to the galactic center (Fig. 30–12).

But here again we come to the question that bedevils much of astronomy: how do we measure the distances? Given that we detect the 21-cm radiation from a gas cloud (since the cloud contains neutral hydrogen, it is an H I region), how do we know how far away the cloud is from us?

The answer can be found by using a model of rotation for the galaxy, that is, a description of how each part of the galaxy rotates. As we have already learned, the outer regions of galaxies rotate differentially; that is, the gas nearer the center rotates faster than the gas farther away from the center.

Figure 30–13 shows a simplified version of differential rotation. Because of the differential rotation, the distance between us and point A is decreasing. Therefore, from our vantage point at the sun, point A has a net velocity toward us. Thus its 21-cm line is Doppler shifted toward shorter wavelengths. If we were talking about light, this shift would be in the blue direction; even though we are discussing radio waves, we say "blueshifted" anyway. Blueshifted now simply means shifted to a shorter wavelength (higher frequency). If we look from our vantage point toward gas cloud C, we see a redshifted 21-cm line (shifted to a longer wavelength = lower frequency), because its higher speed of rotation is carrying C away from us. But if we look straight toward the center, clouds B_1 and B_2 are both passing across our line of sight in a path parallel to that of our own orbit. They have no net velocity toward or away from us. Thus this method of distance determination does not work when we look in the direction of the center, nor indeed in the opposite direction.

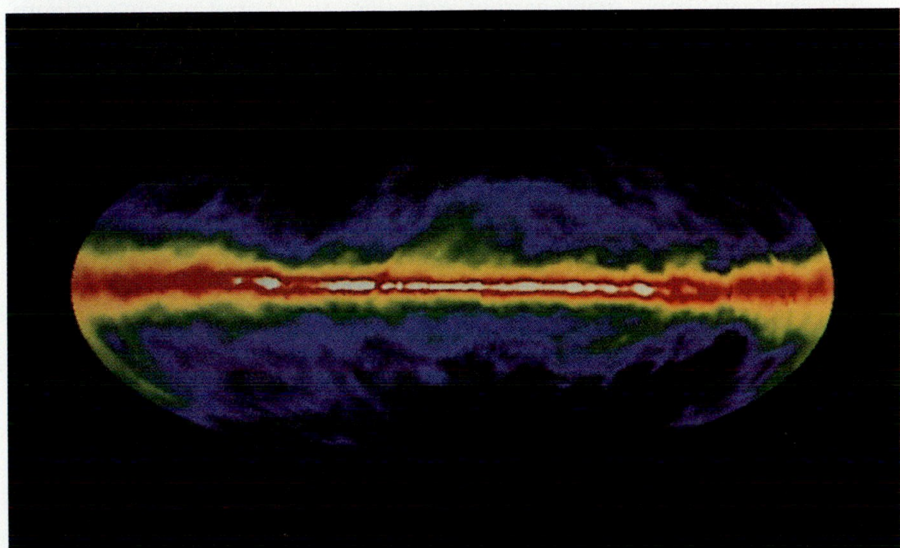

Figure 30–12 The sky at 21 cm, with black and dark blue indicating the smallest and red and white the largest hydrogen densities. The neutral hydrogen is obviously closer to the galactic plane than the sources of synchrotron radiation.

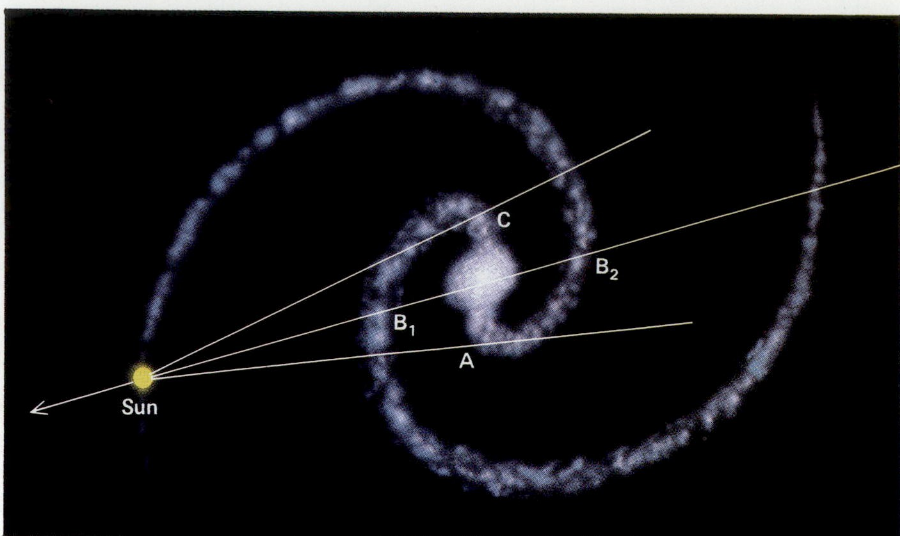

Figure 30–13 Because of the differential rotation, the cloud of gas at point A appears to be approaching the sun, and the cloud of gas at point C appears to be receding. Objects at point B_1 or B_2 have no net velocity with respect to the sun, and therefore show no Doppler shifts in their spectra.

The closer in toward the galactic center a cloud is, the faster it rotates, and so the larger its Doppler shift along any line of sight (other than that toward the center) when observed from the sun. (The difference between the earth's and sun's velocity is known and can be accounted for.)

Clouds farther from the galactic center take longer to revolve than do clouds closer to the center, in a manner similar to Kepler's third law, discussed in Section 3.1c and Box 3.4. Along any line of sight, the cloud with the highest velocity must be closest to the center.

Once we work out the law of differential rotation for the whole galaxy by looking along various lines of sight, then we can apply our knowledge to any particular cloud we observe. From its observed velocity, we can tell how far it is from the center of the galaxy: we figure out where along our line of sight in the direction we are looking we meet the circle of gas having the proper velocity to match the observations. By observing in different directions, we can build up a picture of the spiral arms.

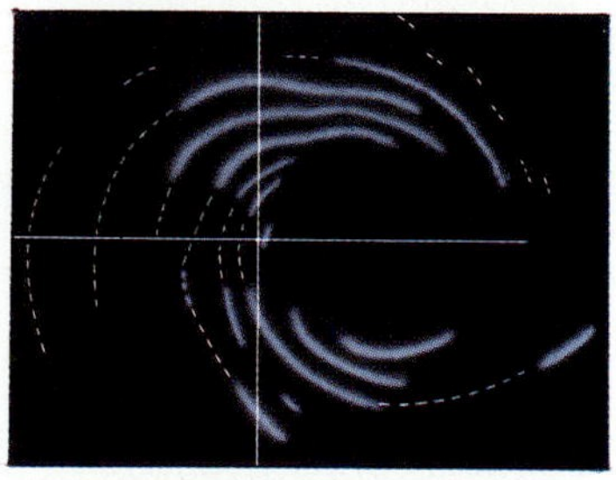

21-cm maps show many narrow arms (Fig. 30–14) but no clear pattern of a few broad spiral arms like those we see in other galaxies. The question emerged: is our galaxy really a spiral at all? Only in recent years, with the additional information from studies of molecules in space that we describe in the next section, have we made further progress. Although questions remain, it is clear that the 21-cm radiation is at the base of our mapping efforts.

Figure 30–14 An artist's impression of the structure of our galaxy based on 21-cm data. Because hydrogen clouds located in the directions either toward or away from the galactic center have no radial velocity with respect to us, we cannot find their distances.

30.5 Radio Spectral Lines from Molecules

For several years after the 1951 discovery of 21-cm radiation, spectral-line radio astronomy continued with just the one spectral line. Astronomers tried to find others. One prime candidate was OH, hydroxyl, a molecule that should be relatively abundant because it is a combination of the most abundant element, hydrogen, with one of the most abundant of the remaining elements, oxygen. OH has four lines close together at about 18 cm in wavelength, and the relative intensities expected for the four lines had been calculated.

It wasn't until 1963 that other radio spectral lines were discovered, and four new lines were indeed found at 18 cm. But the intensity ratios were all wrong to be OH, according to the predicted values, and for a time we spoke of the "mysterium" lines. Mysterium turned out to be OH after all, but with the process of "masering" affecting the excitation of the energy levels of OH and thus amplifying certain of the lines at the expense of others, which are weakened.

The idea of discovering a new element was not unprecedented—after all, unknown lines at the solar eclipse in 1868 had been assigned to an unknown element, "helium," because they occurred only (as far as was known at that time) on the sun. But the periodic table of elements has been filled in during the last 100 years, so we can't expect to find new elements in that way anymore.

Masers and *lasers* are of great practical use on the earth. (Maser is an acronym for **m**icrowave **a**mplification by **s**timulated **e**mission of **r**adiation, and lasers are their analogue using **l**ight instead of **m**icrowaves.) For example, masers are used as sensitive amplifiers. Masers were "invented" on earth not long before they were found in space. For maser action to occur, a large number of electrons are pushed into a higher energy state in which they tend to stay. When "triggered" by a photon of the proper frequency, the electrons jump down together to a lower energy level. (Their emission is "stimulated" by the trigger, leading to the name "maser.") The original radiation is thus amplified, since there are now many photons at that wavelength instead of just one.

Since the interstellar abundance of OH seemed very much lower than that of isolated hydrogen or oxygen atoms (only one OH molecule for every billion H atoms), it seemed quite unlikely that the quantities of any molecules composed of three or more atoms would be great enough to be detected. The chance of three atoms getting together in the same place should be very small.

In 1968, however, Berkeley's Charles Townes, Jack Welch and colleagues observed the radio frequencies that were predicted to be the frequencies of water (H_2O) and ammonia (NH_3). The spectral lines of these molecules proved surprisingly strong, and were easily detected.

Soon afterwards, another group of radio astronomers used a telescope of the National Radio Astronomy Observatory to discover interstellar formaldehyde (H_2CO) at a wavelength of 6 cm. This discovery (by Ben Zuckerman, now of UCLA; Patrick Palmer, now of the University of Chicago; David Buhl, now of NASA; and Lewis Snyder, now of the University of Illinois) was the first molecule that contained two "heavy" atoms, that is, two atoms other than hydrogen.

By this time, it was apparent that the earlier notion that it would be difficult to form molecules in space was wrong. There has been much research on this topic, but the mechanism by which molecules are formed has not yet been satisfactorily determined. For some molecules, including molecular hydrogen, it seems that the presence of dust grains is necessary. In this scenario, one atom hits a dust grain and sticks to it (Fig. 30–15). It may be thousands of years before a second atom hits the same dust grain, and even longer before still more atoms hit. But these atoms may stick to the dust grain rather than bouncing off, which gives them time to join together. Complex reactions may take place on the surface of the dust grain. Then, somehow, the molecule must get off the dust grain, thus being released into space as part of a gas. Perhaps either incident ultraviolet radiation or the energy released in the formation of the molecule allows the molecule to escape from the grain surface.

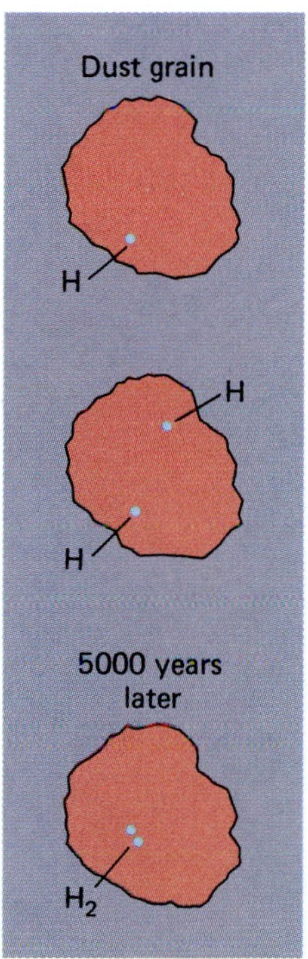

Figure 30–15 Hydrogen molecules are formed in space with the aid of dust grains at an intermediate stage.

Though hydrogen molecules form on dust grains, a strong body of opinion holds that most of the other molecules are formed in the interstellar gas without need for grains. Recent theoretical and laboratory studies indicate that reactions between neutral molecules and ionized molecules may be particularly important. Many of these chains start with molecular hydrogen, formed on grains, being ionized by cosmic rays. There are still many gaps in the theory. It is likely that no single process forms all the interstellar molecules.

The list of molecules discovered expanded gradually from three-atom molecules like ammonia and water, and four-atom molecules like formalde-

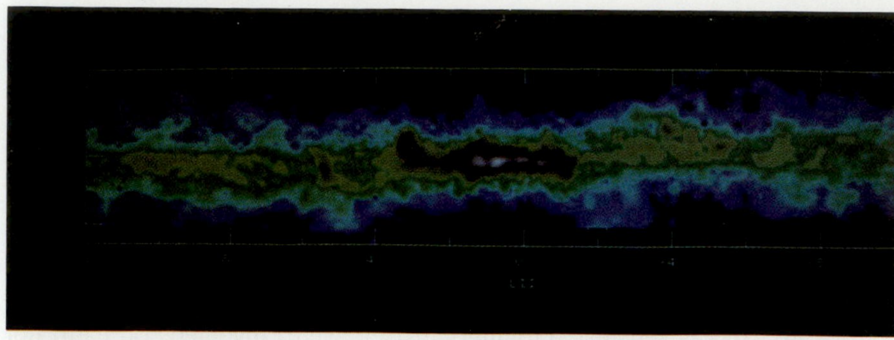

A

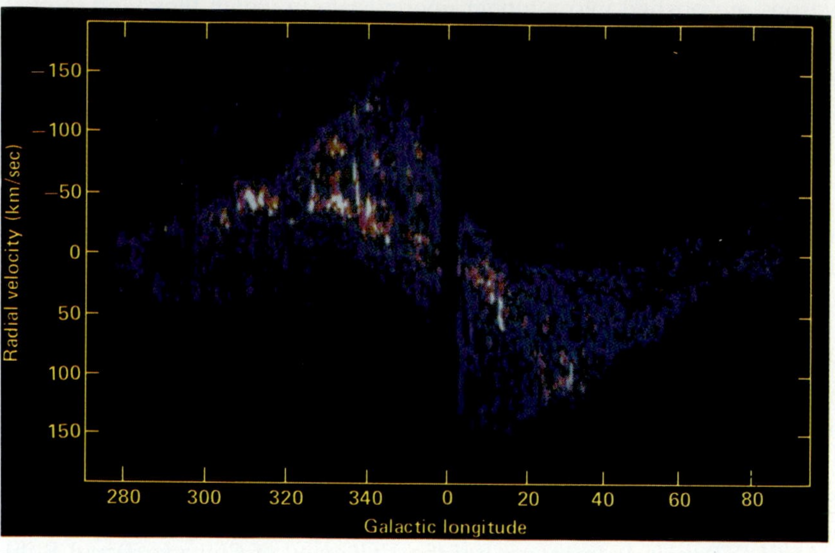

B

Figure 30–16 (*A*) Molecular clouds in the Milky Way all lie close to the galactic plane. This map was made in the 115-GHz radiation from carbon monoxide (2.6-mm wavelength). (*B*) The distances to molecular clouds are calculated from a galactic-rotation model using a longitude-velocity map like this one for the 115-GHz radiation from carbon monoxide (2.6-mm wavelength) in the half of the longitude range that includes the inner galaxy. White is the most intense emission and the darker blue is the weakest. The distribution of carbon monoxide is more clumpy than that of neutral hydrogen, making the large-scale galactic features more readily visible. Data taken in Australia and at the Five College Radio Astronomy Observatory in Massachussetts were combined to make this graph.

hyde, to even more complex molecules. Over 50 molecules have now been discovered in interstellar space—one of the most complex is $HC_{11}N$, which has 12 "heavy" atoms and only 1 hydrogen. Hundreds of spectral lines remain unidentified, some of which are undoubtedly from still other molecules.

Studying the spectral lines provides information about physical conditions—temperature, densities, and motion, for example—in the gas clouds that emit the lines. For example, formaldehyde radiates only when the density is roughly ten times that of the gas when carbon monoxide radiates. The clouds that emit molecular lines are usually so dense that hydrogen atoms have combined into hydrogen molecules and very little 21-cm radiation is emitted.

Studies of molecular spectral lines have been used together with 21-cm observations to improve the maps of the spiral structure of our galaxy. Observations of carbon monoxide (CO) in particular have provided better information about the parts of our galaxy farther out than the distance of the sun from the galaxy's center (Fig. 30–16). The result is a four-armed spiral (Fig. 30–17). But there are still differences to be resolved between the different spirals that have been suggested. In one of the models, the data from the outer part of the galaxy are made to connect with data from interior regions, though the interior regions seem clumpier. This difference between inside and outside may or may not turn out to be real. There are signs, probably real, of short partial arms or spurs, in addition to the four major arms. Our galaxy has been called a "messy spiral."

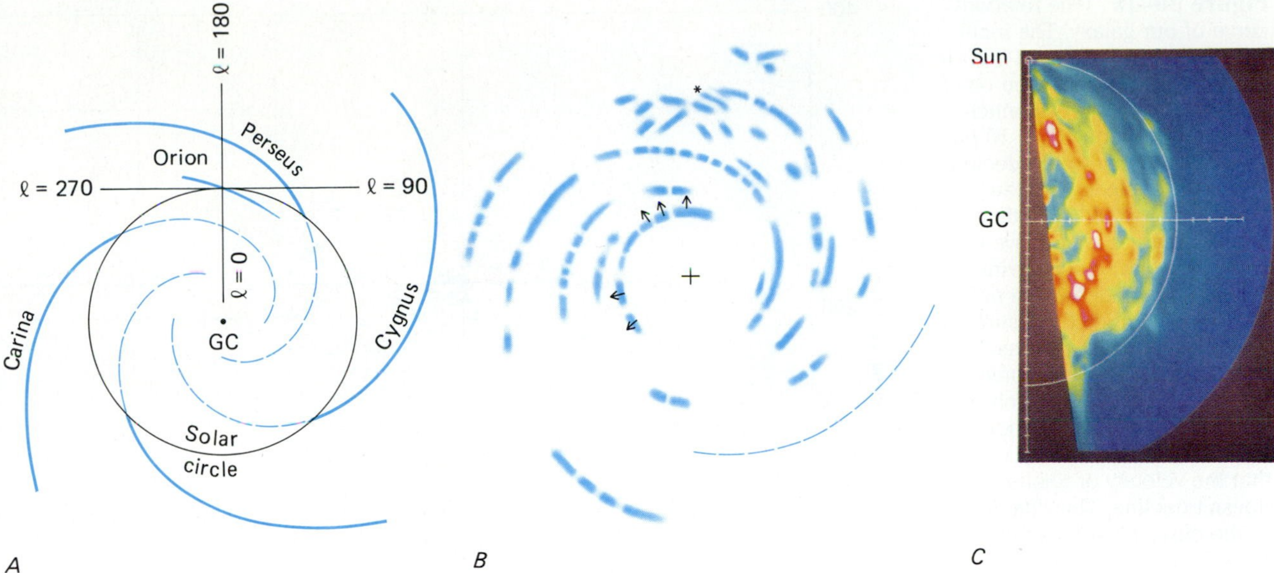

Figure 30–17 (*A*) This four-armed spiral has been fit to 21-cm hydrogen observations and to carbon-monoxide observations to map the spiral structure of our galaxy. The solar circle marks the distance of the sun from the center of our galaxy. The dashed lines show the extrapolation of the solid lines inward to 4 parsecs from the center. Three arms and a spur are based on observations and are given names corresponding to the constellations in which they appear; the fourth arm is drawn assuming a symmetric galaxy. (*B*) The structure of the inner part of our galaxy. The star marks the sun and the + marks the galactic center. The expanding arm at a radius of 3.6 kpc is marked with arrows. The dashed segment is based on 21-cm hydrogen observations, since CO is faint there. Most structure corresponds to four spiral arms; other emission regions appear as spurs and bifurcations of the arms. (Courtesy of B. J. Robinson, J. B. Whiteoak, R. N. Manchester, C.S.I.R.O., Australia, and W. H. McCutcheon, Univ. of British Columbia) (*C*) A map based on higher-resolution CO data does not show clear spiral arms. The regions between 4–6 kpc and near 7 kpc are more ring-like than spiral. Studies of other galaxies are showing that clear spiral arms are common in the outer regions of galaxies but not necessarily in the inner regions, which may be the case with our galaxy as well. (Dan P. Clemens, U. Arizona, and D. Sanders and N. Scoville, Caltech)

30.6 Measuring the Mass of Our Galaxy

We can measure the mass of our galaxy by studying the velocity of rotation of gas clouds. All the mass inside the radius of the cloud's orbit acts about as though it were concentrated at one point; the more mass that is present (and thus the stronger the gravity), the faster a gas cloud has to revolve around the center to keep itself from falling inward. (Objects "revolve" around something, but "rotate" as part of something that is turning around; a gas cloud in a galaxy is an intermediate case, and one should pay attention to when it is suitable to speak of it "revolving" and when "rotating.")

The gravitational effect of the mass outside the cloud's orbit turns out to balance out (assuming spherical symmetry). Thus this exterior mass, however large, does not affect the cloud's velocity. Conversely, the cloud's velocity gives us no information about this mass, so the results described in this section do not give information about the galactic corona (Section 29.2).

From the velocity with which a gas cloud is orbiting, we can thus calculate how much mass is inside the orbit of the cloud. We use Newton's form of Kepler's third law (Box 3.4).

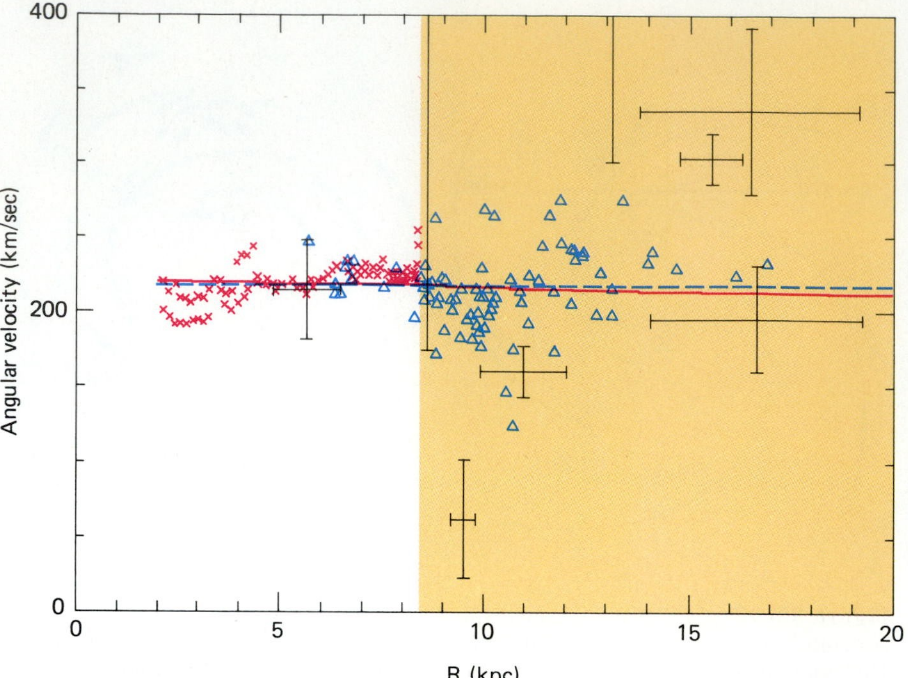

Figure 30–18 The rotation curve of our galaxy. The inner part of the curve (*dotted line*) is based on 21-cm hydrogen observations. It had been anticipated that beyond about 10 parsecs the velocity would decline following Kepler's third law, because there would be little additional mass added as we go farther out. But by observing carbon monoxide we are newly able to make direct measurements of the outer regions (*shaded*). The observations (x's for 21-cm hydrogen observations, triangles for carbon monoxide observations) show that the velocity of rotation doesn't decline. This "flatness" of the curve must mean that there is more mass in the outer regions of the galaxy than we had anticipated.

Black crosses mark the uncertainties for a few carbon monoxide observations. Uncertainties for the 21-cm observations are smaller than the symbols.

It is thus important to measure the velocity of rotation of gas clouds in our galaxy's disk at radii from the center as great as possible. Velocities have been measured inside the sun's orbit from measurements of the 21-cm line, using the method discussed in Section 30.4. Leo Blitz of the University of Maryland has now extended these measurements to much greater radii than was possible with 21-cm radiation by using radiation from carbon monoxide instead. The carbon monoxide, though present in only trace amounts in interstellar clouds, shares in the clouds' motions.

A graph of the velocity of rotation vs. distance from the center is called a *rotation curve* (Fig. 30–18). It had been expected up until about 1980 that our galaxy's rotation curve would stop increasing beyond the sun's distance from the galactic center and begin to decrease, since the galaxy essentially ended. Then little more mass would be included inside as we went to larger radii. But the curve does not stop increasing and begin to decrease, which indicates that the galaxy is larger and contains more mass than we had thought. It now seems that our galaxy is twice as massive as the Andromeda Galaxy, and contains perhaps 10^{12} solar masses; this is at least twice as massive as had previously been calculated. If we assume that an average star has 1 solar mass, then our galaxy contains about 10^{12} stars.

30.7 Molecular Hydrogen

At the low temperature of interstellar space, only the lowest energy levels of molecular hydrogen are excited, and the lines linking these levels fall in the far ultraviolet. Because our atmosphere prevents these lines from reaching us on earth, we had to wait to observe from space in order to observe lines from H_2. H_2 was first observed from a rocket in 1970.

In 1972, a 90-cm telescope was carried into orbit aboard NASA's third Orbiting Astronomical Observatory, named "Copernicus" in honor of that astronomer's 500th birthday (which occurred a year later). The telescope was largely devoted to observing interstellar material. The observers could point

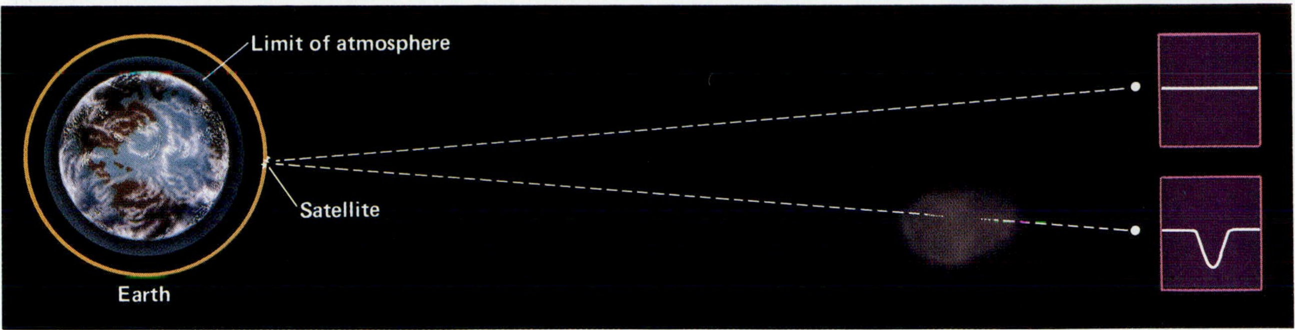

Figure 30–19 From above the atmosphere of the earth, the Copernicus satellite looked toward a star that provides a more or less continuous spectrum. (A rapidly rotating B star was usually chosen, because B stars have few lines and those lines are washed out by Doppler shifts resulting from the velocity of rotation.) When a cloud of gas is in the path, then absorption or emission from molecules or atoms in the cloud results.

the telescope at a star and look for absorption lines caused by gas in interstellar space as the light from the star passed through the gas en route to us. Since it is easiest to pick up interstellar absorption lines if the star itself has no lines of its own, the scientists observed in the direction of B stars, which have few lines and are very bright (Fig. 30–19).

Since the hydrogen molecule is easily torn apart by ultraviolet radiation, the observers did not find much molecular hydrogen in most directions. But whenever they looked in the direction of highly reddened stars, they found a very high fraction of hydrogen in molecular form: more than 50 per cent. Presumably, in these regions some of the dust that causes the reddening shields the hydrogen from being torn apart by ultraviolet radiation. There are also theoretical grounds for believing that the molecular hydrogen is formed on dust grains, so it seems reasonable that the high fraction of H_2 is found in the regions with more dust grains.

30.8 The Formation of Stars

Most radio spectral lines seem to come only from a very limited number of places in the sky—molecular clouds. (Carbon monoxide is the major exception, for it is widely distributed across the sky.) Infrared and radio observations together have provided us with an understanding of how stars are formed from these dense regions of gas and dust.

Giant molecular clouds are 50 to 100 parsecs across. There are a few thousand of them in our galaxy (Fig. 30–20). The largest giant molecular clouds are about 100 parsecs across and contain about 100,000 to 1 million times the mass of the sun. Their internal densities are about 100 times that of the interstellar medium around them. Since giant molecular clouds break up to form stars, they last only 10 million to 100 million years.

Carbon-monoxide observations reveal the giant molecular clouds, but it is molecular hydrogen (H_2) rather than carbon monoxide that is significant in terms of mass. It is difficult to detect the molecular hydrogen directly, though there is over 100 times more molecular hydrogen than dust. So we must satisfy ourselves with observing the tracer, carbon monoxide, whose spectral lines are excited by collisions with hydrogen molecules. The carbon monoxide thus shows us where the hydrogen molecules are.

The Great Rift in the Milky Way, the extensive dark area that can be seen in the midst of the Milky Way, is made up of many overlapping giant molecular clouds.

A giant molecular cloud is fragmented into many denser bits of 1 parsec in size. These bits must become much smaller and denser yet to form a star or, in many cases, several stars.

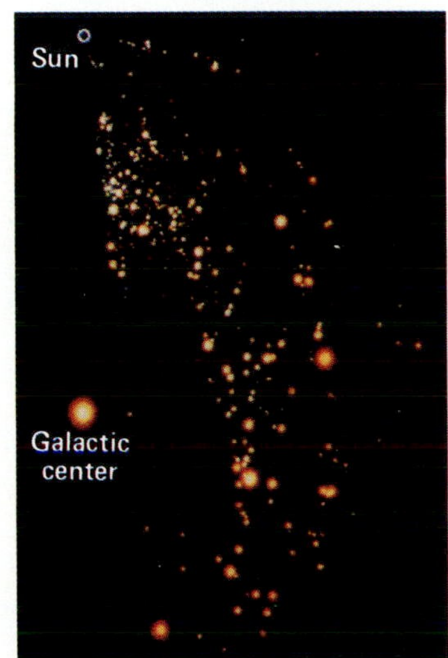

Figure 30–20 A map of the Milky Way as it would appear from above, showing the molecular clouds in one part of our galaxy. Two spiral arms show.

Sometimes we see smaller objects, called globules, or *Bok globules.* (They are named after Bart Bok, who extensively studied these objects.) Some of the smaller globules are visible in silhouette against H II regions. Others are larger, and appear isolated against the stellar background. Two hundred large globules are known to be within 500 parsecs of the sun (5 per cent of the way to the center of the galaxy), and there may be 25,000 in our galaxy. It has been suggested that the globules may be on their way to becoming stars. Molecular-line observations of the larger globules indicate that they may contain sufficient mass for gravitational collapse to be taking place. And though there had been some doubt whether stars were forming in smaller globules, IRAS discovered infrared radiation from at least some that indicates the presence of star formation inside. For example, a few stars, each about the mass of the sun and each only a few hundred thousand years old, were discovered in the globule called Barnard 5.

Observations of interstellar molecules have shown that gas flows rapidly outward from some newly formed stars in two opposing streams. These high-velocity *bipolar flows* typically have carried more than a solar mass of material outward and transport a lot of energy (Fig. 30–21). They are thus clearly very important for the evolution of young stars. The gas flow extends only about 1

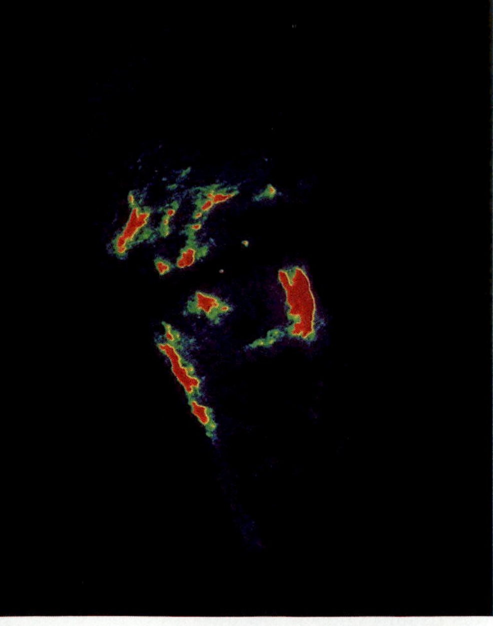

Figure 30–21 The bipolar H II region S106, twin lobes illuminated by a young B0 star. These VLA observations at a wavelength of 6 cm show the star's powerful stellar wind as a point source. The region around the star is free of ionized gas, possibly because a disk or ring of circumstellar material blocks the ultraviolet photons. This region is probably occupied by dense neutral gas. The outer part of the image shows numerous clumps and filaments. Its expansion also indicates that a powerful stellar wind is blowing.

Figure 30–22 The Orion Nebula, an H II region on the side of a molecular cloud.

light year, which can be traversed by gas at the high velocities measured in only about 10,000 years. Herbig-Haro objects (Section 24.1) are sometimes located at the end of one of the jets; the motion of the H–H objects can be projected back to the same location at which the jets originate. In this and in most cases, an infrared source (with often no image in the visible part of the spectrum) is located at the center. We have already met a jet of gas in the model of SS433, and we will learn about other jets in radio galaxies and in quasars. Astrophysicists are finding that signs of violent activity like jets are more common than had been suspected.

Figure 30–23 In the center of the image are the Trapezium stars.

30.8a A Case Study: The Orion Molecular Cloud

Many radio spectral lines have been detected only in a particular cloud of gas located 1600 light years from us in the direction of the constellation Orion, not very far from the main Orion Nebula (Fig. 30–22). This *Orion Molecular Cloud,* which contains about 500 solar masses of material, is itself buried deep in nebulosity. It is relatively accessible to our study because it is only about 500 parsecs from us. Even though less than 1 per cent of the Cloud's mass is dust, that is still a sufficient amount of dust to prevent ultraviolet light from nearby stars from entering and breaking the molecules apart. Thus molecules can accumulate. Most of the cloud is molecular hydrogen, though the cloud is best studied by trace amounts of the molecule carbon dioxide.

We know that young stars are found in this region—the Trapezium (Fig. 30–23), a group of four hot stars readily visible in a small telescope, is the source of ionization and of energy for the Orion Nebula. The Trapezium stars are relatively young, about 100,000 years old. The Orion Nebula, prominent as it is in the visible, is an H II region located as a blister along the near side of the molecular cloud (Fig. 30–24). The nebula contains much less mass than does the molecular cloud. The ultraviolet radiation from the H II region causes a sharp front of ionization at the edge of the molecular cloud. A shock wave forms and concentrates gas and dust in the molecular cloud to make protostars.

Figure 30–24 The structure of the Orion Nebula and the Orion Molecular Cloud, proposed by Ben Zuckerman, now of UCLA.

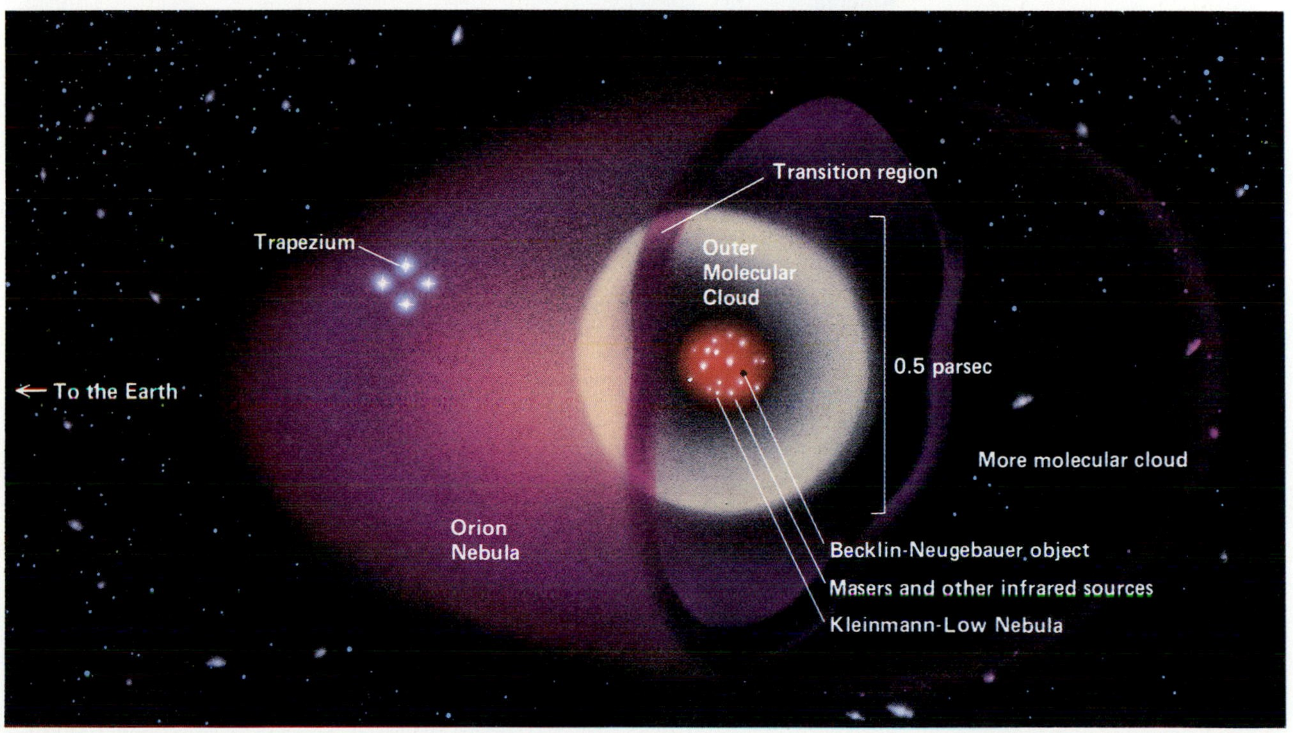

Figure 30–25 The region surrounding the constellation Orion, as seen in this false-color image constructed from data collected by IRAS, includes signs of different temperatures. The circular feature at top can be seen at visible wavelengths but appears different in intensity and size here in the infrared. It corresponds to hot hydrogen gas and dust heated by the star Bellatrix. The brightest regions correspond to regions where stars are forming.

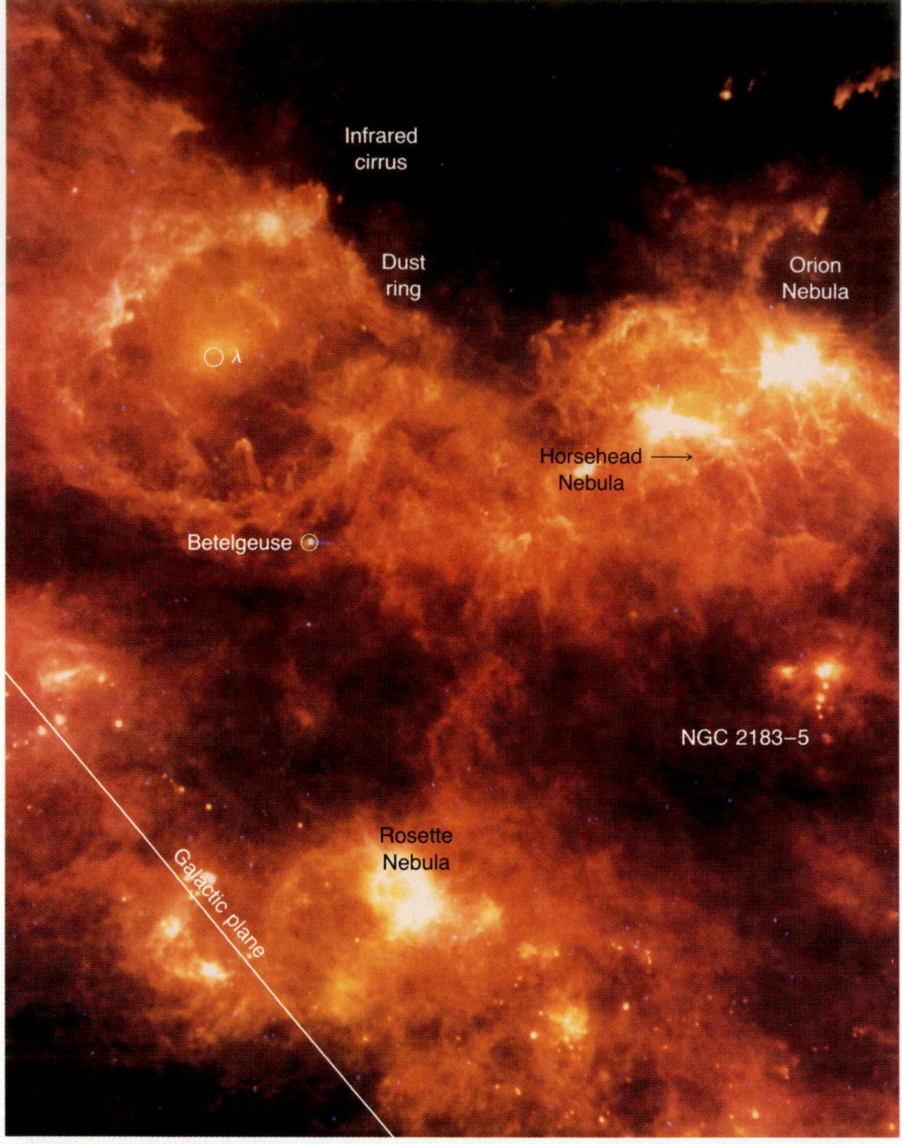

Figure 30–26 Four views of the Orion Nebula taken with the United Kingdom Infrared Telescope on Mauna Kea. (*A*) K-band, at 2.2 μ (*B*) The H II region, from its hydrogen radiation in the infrared Brackett γ line at 2.167 μ. (*C*) Dust radiation, measured at 3.3 μ. (*D*) Molecular hydrogen gas, measured from the emission line at 2.12 μ.

The properties of the molecular cloud can be deduced by comparing the radiation from its various molecules and by studying the radiation from each molecule individually. The main Orion Molecular Cloud is 100,000 times more massive than the sun. Its density, 1000 particles per cubic centimeter in the outer limits at which the cloud is visible to us, increases toward the center. The cloud may actually be as dense as 10^6 particles/cm^3 at its center. This is still billions of times less dense than our earth's atmosphere, and 10^{16} times

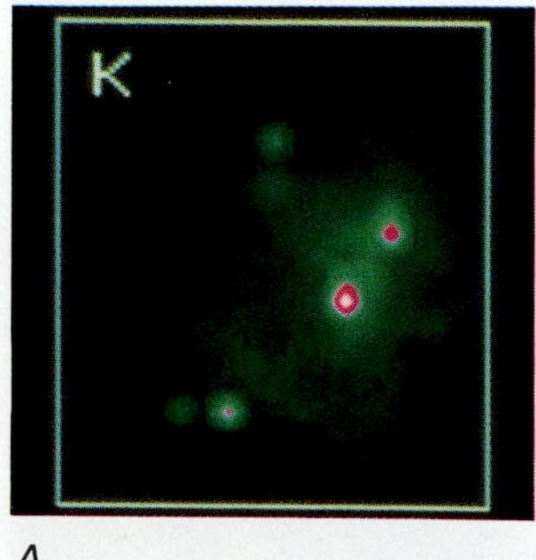

A

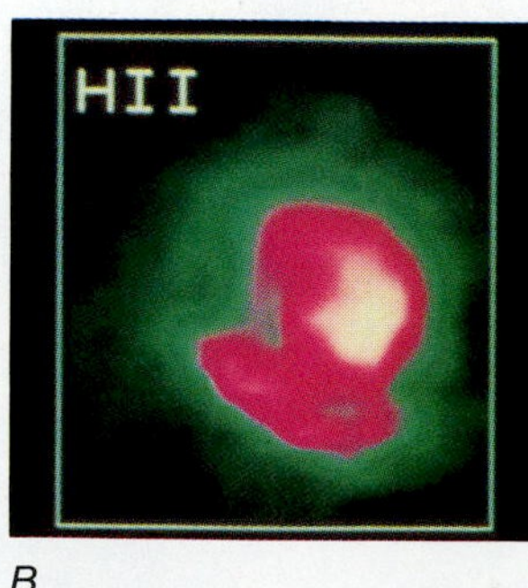

B

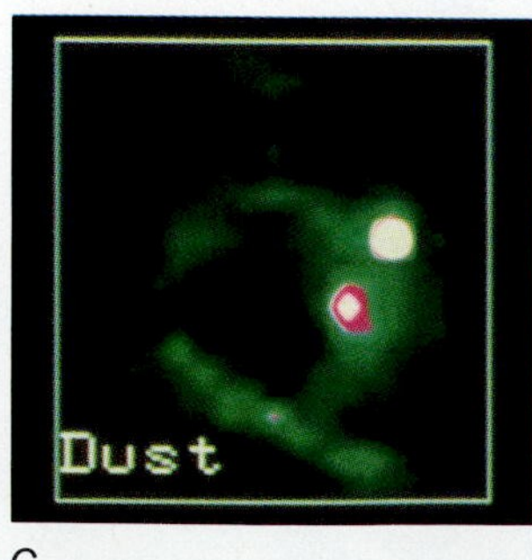

C

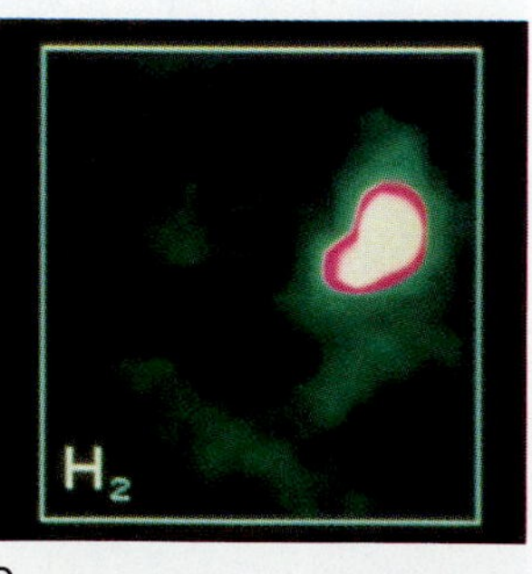

D

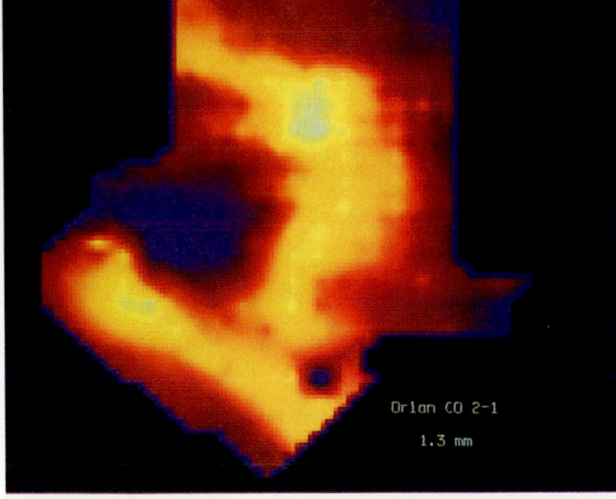

Figure 30–27 The Orion Nebula studied with the James Clerk Maxwell Telescope on Mauna Kea, which operates in the submillimeter and short millimeter wavelength range. This 1.3 μ image in a line from the CO molecule resembles the dust image of the previous figure, showing the association of the molecules with the dust.

less dense than the star that may eventually form, though it is substantially denser than the average interstellar density of about 1 particle per cm^3. This molecular cloud and one in the northern portion of Orion have together caused the Orion OB association—a group of hot stars—to form. The ultraviolet radiation and strong stellar winds from these stars of types O and B, as well as the supernova explosions from the ones that have already died, have made an expanding bubble perhaps 500 light years across in the interstellar medium. One part of this bubble causes a large circular optical feature known as Barnard's Loop. The region can be studied especially well in the infrared (Fig. 30–25), especially with the new infrared arrays (Fig. 30–26). Further, our new capability of observing in the submillimeter region of the spectrum has also been applied to the Orion region (Fig. 30–27).

Scientists have used radio methods to map out the extent of the Orion Molecular Cloud; the maps can be compared with optical views (Fig. 30–28).

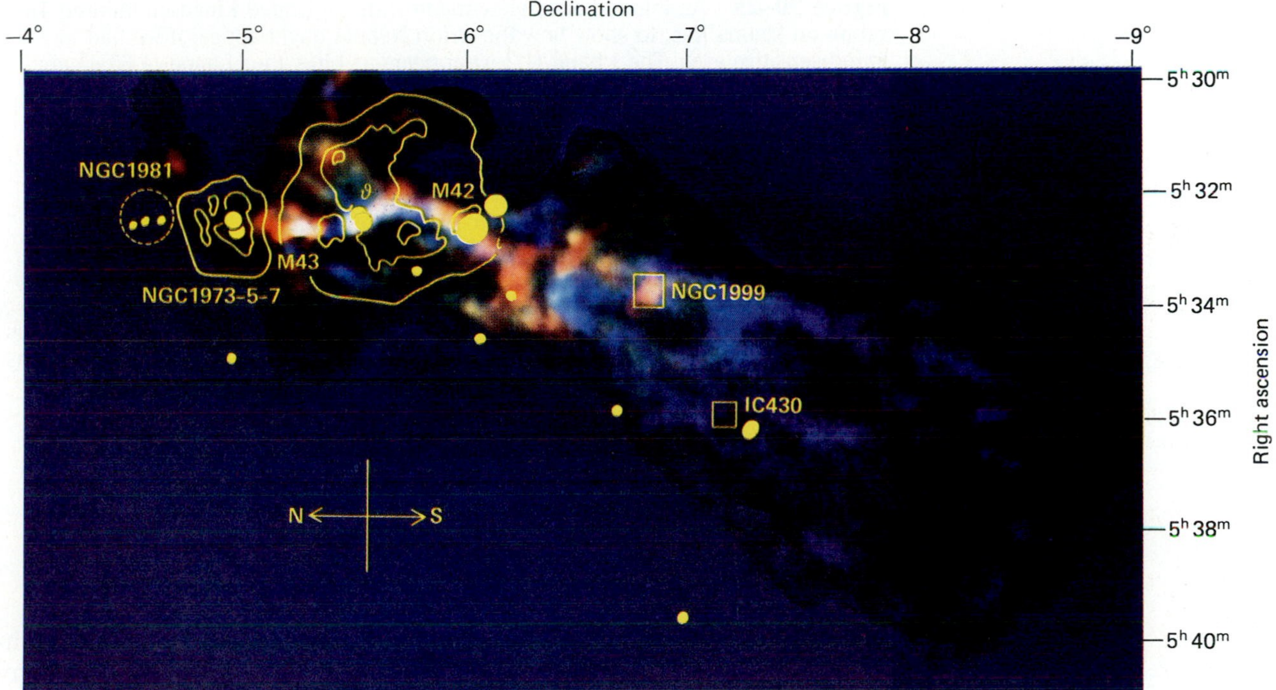

Figure 30–28 The Orion Molecular Cloud. Each color shows an adjacent 1-km/sec-wide velocity band. Intensity is proportional to the density of the carbon-13 isotopic form of carbon monoxide. The position of the Orion Nebula, M42, is outlined, and boxes mark the positions of smaller nebulae.

Figure 30–29 An infrared array view made with the United Kingdom Infrared Telescope on Mauna Kea, to show how the Orion Nebula might appear if we had infrared-sensitive eyes. The J band (1.2 μ) appears as blue, the H band (1.65 μ) appears as green, and the K band (2.2 μ) appears as red. The image is composed of 4000 pixels (individual picture elements), made as a mosaic of 62×58 pixel exposures. Close examination of the star edges show individual pixels.

Hot stars such as the Trapezium appear as blue. Cooler sources and those obscured by dust, like the BN-KL star-forming region at upper right, appear yellow, orange, and red. The upper yellow source is Infrared Compact Source 9 (IRc9) and the lower yellow source is the Becklin-Neugebauer object. The Kleinmann-Low Nebula is a cluster of infrared sources just below the BN object. These sources are on the side nearest us of a molecular cloud only a few light years deep, and may form an H II region like the Orion Nebula within a few thousand years. The source IRc2 is perhaps 10 times more luminous than the BN object but is hidden by even more dust and shows here only as a reddish smudge to the lower left of BN.

The image shows over 500 stars, most of them members of the Trapezium OB association and less than a million years old; most are not visible optically. The image covers 5×5 arc min, which is equivalent to 2.1×2.1 light years at the 1500-light-year distance of the Orion Nebula; thus everything in this image would fit well between the sun and our nearest star.

One of the brightest of all the infrared sources in the sky, an object in the Orion Nebula that was discovered by Eric Becklin, now at UCLA, and Gerry Neugebauer of Caltech, is right in the midst of the Orion Molecular Cloud. This *Becklin-Neugebauer object* (Fig. 30–29), also known for short as the *B–N object*, is about 200 Astronomical Units across. Its temperature is 600 K. The B–N object is the prime candidate for a star about to be born.

The B–N object is behind so much dust as seen from earth that it appears very faint. Comparison of certain of the spectral lines indicated how much the

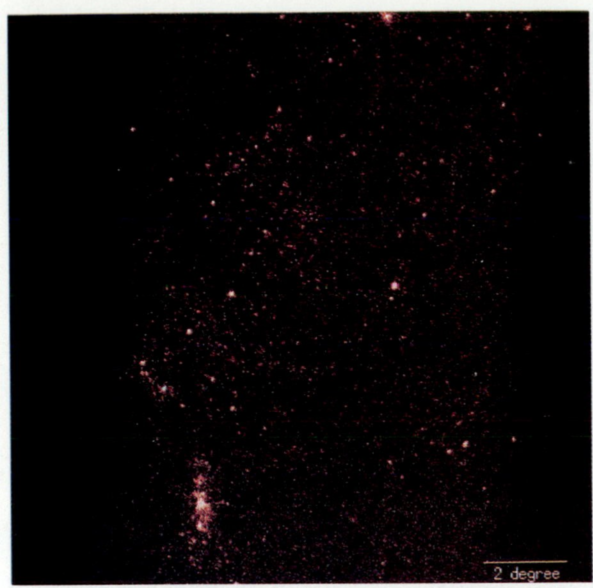

Figure 30–30 X-rays from Orion come from the hot stars. Here we see a Rosat view.

B–N object is reddened. It is apparently a very young star of spectral type B, a massive, hot star. It is already, but just barely, operating on the basis of nuclear fusion.

Other infrared sources are present near the B–N object, and probably also contain young stars or stars in formation. Small intense sources of maser radiation from various molecules also exist nearby. Such interstellar masers can exist only for small sources, about the size of our solar system, and are also signs of stars in formation.

The Orion region, with its active ongoing star formation, is a prime observing direction for all technologies. An x-ray image (Fig. 30–30) shows hot stars that were recently formed.

30.8b IRAS, Infrared Imaging, and Star Formation

The tens of thousands of non-stellar galactic objects discovered by IRAS include H II regions, molecular clouds with pre-main-sequence stars embedded in them, and dense clumps of dust. In molecular clouds studied, dozens of regions of gas containing 5 to 50 solar masses of material were found to have infrared point sources in them. These sources, now only about 100 K in temperature, are probably stars of one-half to one solar mass forming (Fig. 30–31). Some may still be protostars, getting their energy from gravitational contraction. Some may have barely turned on, and are like T Tauri stars.

IRAS may also have detected signs of higher-mass objects forming. There is evidence for loops and rings of star formation surrounding earlier sites of star formation. This evidence endorses the theory of star formation by chain reaction. The formation of high-mass stars is much more disruptive to the interstellar medium than is the formation of low-mass stars.

Another notable IRAS discovery was "starburst galaxies" in which a large amount of star formation went on simultaneously. Thus IRAS has shown us parallels in other galaxies to the interstellar medium of our own. How common such starbursts are is under study. It is already apparent that the starbursts are common in galaxies that are undergoing interactions with other galaxies.

New techniques of ground-based infrared imaging are now clearly showing stars in formation (Fig. 30-32). Astronomers using infrared arrays can observe sources, including those discovered or observed by IRAS, in much more spatial detail than was previously possible.

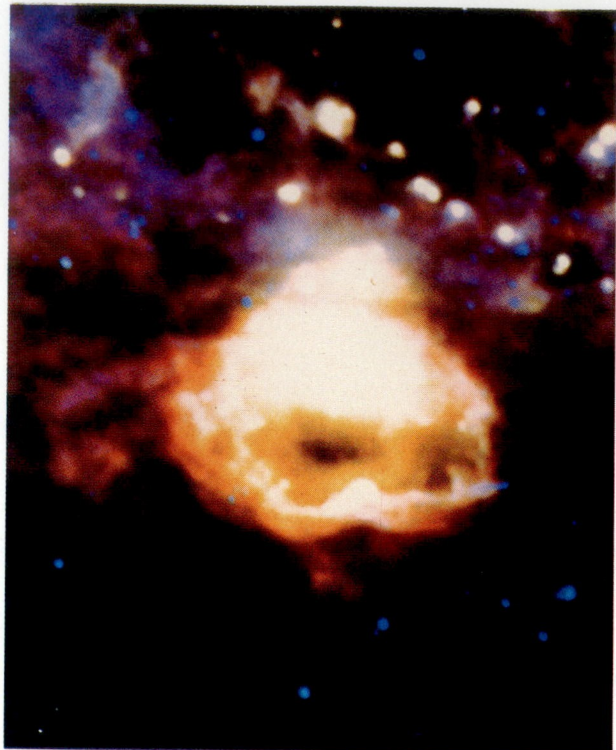

Figure 30–31 The star-forming region Sharpless 171, observed with IRAS.

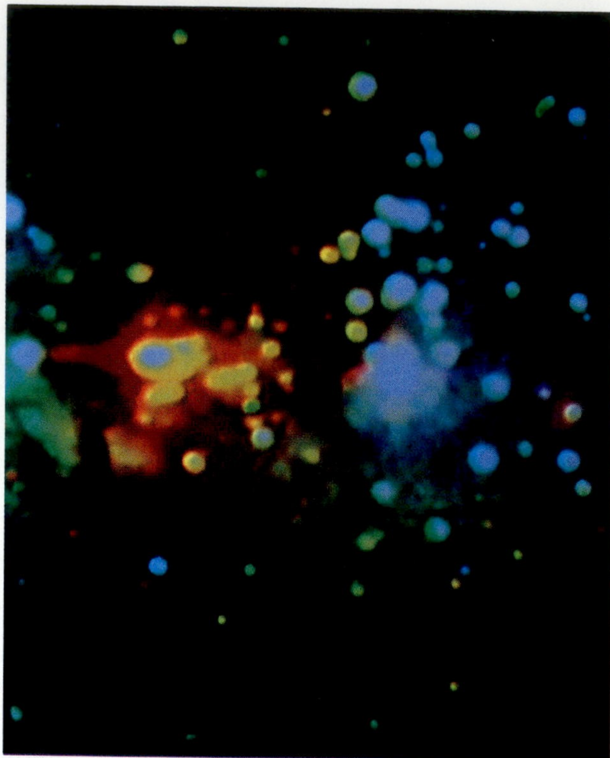

Figure 30–32 The star-forming region NGC 2024, near the Horsehead Nebula in Orion (Fig. 29–3), observed with an infrared array in the J band (1.2 μ, appearing blue), the H band (1.65 μ, appearing green), and the K band (2.2 μ, appearing red). Most of the objects are shielded from our view in the visible.

Summary and Outline

H I and H II regions (Section 30.1)

Interstellar reddening and extinction (Section 30.2)

Radio observations (Section 30.3)

- Continuum radio astronomy
  - Spectra are measured over a broad frequency range.
  - Radiation generated by synchrotron emission (electrons spiralling rapidly in a magnetic field) is found in many non-thermal sources.

Radio spectral line from interstellar hydrogen (Section 30.4)

- 21-cm spectral line from neutral hydrogen was discovered in 1951.
- Line occurs at 21 cm through a spin-flip transition, corresponding to a change between energy subdivisions of the ground level of hydrogen.
- Both emission and absorption have been detected.
- 21-cm radiation used to map the galaxy
- Distances measured using differential rotation and the Doppler effect

Radio spectral lines from molecules (Section 30.5)

- "Mysterium" lines discovered at 18 cm—later identified as OH affected by masering process
- Masers first developed artificially on earth, but later discovered to exist in space
- Dozens of molecules have been discovered in space.
- Analysis of molecular lines tells us physical conditions (for example, temperature, density, and motions).

Measuring mass of our galaxy (Section 30.6)

- Revised measurements have doubled the galaxy's previously accepted size.

Molecular hydrogen (Section 30.7)

- Interstellar ultraviolet absorption lines from H_2, with abundances up to 50 per cent

The formation of stars (Section 30.8)

- Molecular clouds contain gas collapsing to become stars.
- Molecules are associated with dark clouds, such as the Orion Molecular Cloud, where the molecules are shielded by dust from being torn apart by ultraviolet radiation.
- The Becklin-Neugebauer object, a compact infrared source deep inside the Orion Molecular Cloud, has been revealed through infrared spectral observations to be a young star.

Key Words

interstellar medium, cosmic abundances, H I region, H II region, scattered, reddened, extinction, synchrotron emission*, non-thermal radiation*, thermal radiation*, spin-flip, half-life, 21-cm line, masers, lasers, rotation curve, giant molecular cloud, Bok globule, bipolar flows, Orion Molecular Cloud, Becklin-Neugebauer object (B–N object)

*This term is found in an optional section.

Questions

1. List two relative advantages and disadvantages of radio astronomy compared with optical astronomy.
2. What is the ratio of the cosmic abundance of hydrogen to the cosmic abundance of helium?
3. Though cool gas does not give off much continuum radiation, we can detect 21-cm radiation from cool interstellar gas. Explain.
4. Describe the relation of hot stars to H I and H II regions.
5. Briefly define and distinguish between the redshift of a gas cloud and the reddening of that cloud.
6. What determines whether the 21-cm lines will be observed in emission or absorption?
7. Is 21-cm radiation thermal or non-thermal? Explain.
8. If our galaxy rotated like a rigid body, with each point rotating with the same period no matter what the distance from the center, would we be able to use the 21-cm line to determine distances to H I regions? Explain.
9. Describe how a spin-flip transition can lead to a spectral line, using hydrogen as an example. Could deuterium also have a spin-flip line? If so, describe the likely process.
10. Why did it take so long to discover interstellar hydrogen molecules?
11. What was "mysterium"? Why was it thought to be strange?
12. How does a maser work?
13. Why did the abundance of heavy molecules predicted from observations of hydroxyl give too low a value?
14. Why are dust grains important for the formation of interstellar molecules?
15. Compare the roles of molecular hydrogen and of carbon monoxide in giant molecular clouds.
16. Which molecule is found in the most locations in interstellar space?
17. Explain what property of young stars and/or the region of space nearby was observed with the IRAS spacecraft.
18. Describe the relation of the Orion Nebula and the Orion Molecular Cloud.
19. Describe the Becklin-Neugebauer object. Why do scientists find it interesting?
20. How do we detect star formation in another galaxy? Describe an example.
21. Optical astronomers can observe only at night. In what time period can radio astronomers observe? Why?
22. Discuss the choice of false colors for the different wavelength bands in Figure 30–29.
23. Discuss the advantages of infrared arrays over prior abilities to observe in the infrared.
24. List three sciences besides astronomy that have applications to study of interstellar space, and briefly describe one of the connections for each.

Topics for Discussion

1. What does the discovery of fairly complex molecules in space imply to you about the existence of extraterrestrial life?
2. What does the formation of stars like the sun so commonly in our galaxy imply to you about the existence of planets in our galaxy?

VI

Galaxies and Beyond

The individual stars that we see with the naked eye are all part of the Milky Way Galaxy, discussed in the preceding two chapters. But we cannot be so categorical about the conglomerations of gas and stars that can be seen through telescopes. Once they were all called "nebulae," but we now restrict the meaning of this word to gas and dust in our own galaxy. Some of the objects that were originally classified as nebulae turned out to be huge collections of gas, dust, and stars located far from our Milky Way Galaxy and of a scale comparable to that of our galaxy. These objects are galaxies in their own right, and are both fundamental units of the universe and the stepping stones that we use to extend our knowledge to tremendous distances.

In the 1770's, a French astronomer named Charles Messier was interested in discovering comets. To do so, he had to be able to recognize whenever a new fuzzy object appeared in the sky. He thus compiled a list of about 100 diffuse objects that could always be seen. To this day, these objects are commonly known by their *Messier numbers*. Messier's list contains the majority of the most beautiful objects in the sky, including nebulae, star clusters, and galaxies.

Soon after, William Herschel, in England, compiled a list of 1000 nebulae and clusters, which he expanded in subsequent years to include 2500 objects. Herschel's son John continued the work, incorporating observations made in the southern hemisphere. In 1864, he published the *General Catalogue of Nebulae*. In 1888, J. L. E. Dreyer published a still more extensive catalogue, *A New General Catalogue of Nebulae and Clusters of Stars*, the *NGC*, and later published two supplementary *Index Catalogues, IC's*. The 100-odd non-stellar objects that have Messier numbers are known by them, and sometimes also by their numbers in Dreyer's catalogue. Thus the Great Nebula in Andromeda is very often called M31, and is less often called NGC 224. The Crab Nebula = M1 = NGC 1952. Objects without M numbers are known by their NGC or IC numbers, if they have them.

When larger telescopes were turned to the Messier objects, especially by Lord Rosse in Ireland in about 1850, some of the objects showed traces of spiral structure, like pinwheels. They were called "spiral nebulae." But where were they located? Were they close by or relatively far away?

When such telescopes as the 0.9-m reflector at Lick in 1898, and later the 1.5-m and 2.5-m reflectors on Mount Wilson, began to photograph the "spiral nebulae," they revealed many more of them. The shapes and motions of these "nebulae" were carefully studied. Some scientists thought that they were merely in our own galaxy, while others thought that they were very far away, "island universes" in their own right, so far away that the individual stars appeared blurred together. (The name "island universes" had originated with the philosopher Immanuel Kant in 1755.)

The debate raged, and an actual debate on the scale of our galaxy and the nature of the "spiral nebulae" was held on April 16, 1920, as an after-dinner event of the National Academy of Sciences. Harlow Shapley (pronounced to rhyme with "map lee") argued that the Milky Way Galaxy was larger than had been thought, and thus implied that it could contain the spi-

A false-color view of the edge-on spiral galaxy, the Sombrero Galaxy, M104. The color contours show brightness and are taken from visible-light observations.

ral nebulae. Heber Curtis argued for the independence of the "spiral nebulae" from our galaxy. This famous *Shapley-Curtis debate* is an interesting example of the scientific process at work. (It has recently been pointed out that most reports of the Shapley-Curtis debate are based on the published transcript, while the actual words spoken on that evening were less thorough. Shapley, for one, had prepared a low-level introductory talk for this audience.)

Shapley's research on globular clusters had led him to correctly assess our own galaxy's large size. But he also argued that the "spiral nebulae" were close by because proper motion had been detected in some of them by another astronomer. These observations were subsequently shown to be incorrect. He also reasoned that an apparent nova in the Andromeda Galaxy, S Andromedae, had to be close or else it couldn't have been as bright; nobody knew about supernovae then. Curtis's conclusion that the "spiral nebulae" were external to our galaxy was based in large part on an incorrect notion of our galaxy's size. He treated S Andromedae as an anomaly and considered only "normal" novae.

So Curtis's conclusion that the "spiral nebulae" were comparable to our own galaxy was correct, but for the wrong reasons. Shapley, on the other hand, came to the wrong conclusion but followed a proper line of argument that was unfortunately based on incorrect and inadequate data.

The matter was settled in 1924, when observations made at the Mount Wilson Observatory by Edwin Hubble provided distances to some "spiral nebulae" and showed they were so far away they must be galaxies. Thus there were indeed other galaxies in the universe besides our own. In fact, we now think of galaxies and clusters of galaxies as fundamental units in the universe. The galaxies are among the most distant objects we can study. Many quasars are even farther away, and turn out to be certain types of galaxies seen at special times.

Galaxies and quasars can be studied in most parts of the spectrum. Radio astronomy, in particular, has long proved a fruitful method of study. The study across the spectrum of galaxies and quasars provides tests of physical laws at the extremes of their applications and links us to cosmological consideration of the universe on the largest scale. Our notion of the past and future of our universe has changed recently, and we describe this new research.

NGC 2997 in Antlia, a type Sb spiral galaxy whose structure is very similar to that of our own Milky Way Galaxy.

Galaxies

31

Aims: To discuss the different types of galaxies, to see that galaxies are fundamental units of the universe, to study the expansion of the universe, and to consider how interferometry allows radio astronomy to make significant advances in the study of galaxies

The question of the distance to the "spiral nebulae"—the spiral-shaped regions observable in the sky with telescopes—was settled only in 1924 by Edwin Hubble. He used the Mount Wilson telescopes to observe Cepheid variables in two of the "spiral nebulae" and in another object. He concluded (following the line of argument using variable stars that we described in Section 22.4b) that the "spiral nebulae" and the other object were outside our own galaxy; from their observed angular sizes and their distances from us, it followed that they are not overwhelmingly different from the Milky Way Galaxy in size.

Since Hubble's work, there has been no doubt that the spiral forms we observe in the sky are galaxies like our own. For the rest of the book we shall strictly use the term *spiral galaxies;* the currently incorrect, historical term "spiral nebula" often hangs on in certain contexts, chiefly when we discuss the "Great Nebula in Andromeda," which is actually a spiral galaxy.

A

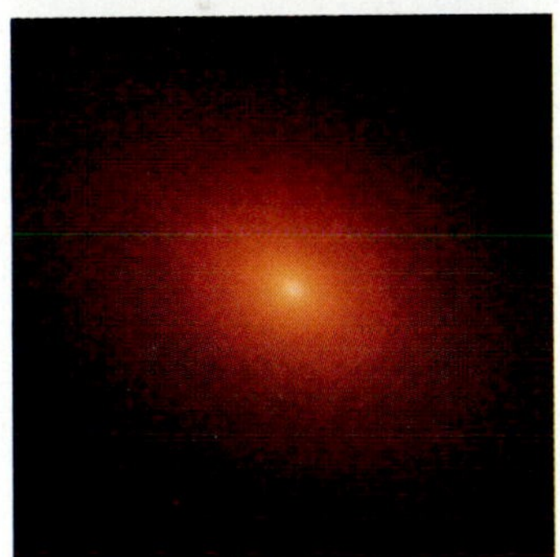

B

Figure 31–1 (*A*) M87 (NGC 4486), a galaxy of Hubble type EO(pec) in the constellation Virgo. The jet that makes M87 a peculiar elliptical galaxy doesn't show in this view (see Fig. 31–50). Globular clusters can be seen in the outer regions. (*B*) The central core of the elliptical galaxy M32 (see also Fig. 31–2), imaged with the Hubble Space Telescope. The high resolution of the image reveals that the brightness is strongly concentrated toward the center of M32, indicating that a massive black hole there is drawing the stars to it. Models are consistent with a 3-million-solar-mass black hole.

31.1 Types of Galaxies

Hubble used the Mount Wilson telescopes to study the different types of galaxies. Actually, spiral galaxies are only one common type of galaxy. Many other galaxies have elliptical shapes, while still others are irregular or abnormal in appearance. In 1925, Hubble set up a system of classification of galaxies that we still use today; we normally describe a galaxy by its *Hubble type,* as we shall discuss below.

31.1a Elliptical Galaxies

About one-third of galaxies are elliptical in shape (Fig. 31–1). The largest of these *elliptical galaxies* contain 10^{13} solar masses and are 10^5 parsecs across (approximately the diameter of our own galaxy); these *giant ellipticals* are rare. Much more common are *dwarf ellipticals,* which contain "only" a few million solar masses and are only 2000 parsecs across.

Elliptical galaxies range from nearly circular in shape, which Hubble called *type E0,* to very elongated, which Hubble called *type E7*. The spiral Andromeda Galaxy, M31 (Fig. 31–2), is accompanied by two elliptical companions of types E2 (for the galaxy closer to M31) and E5, respectively. It is obvious on the photograph that the companions are much smaller than M31 itself.

We assign types based on the optical appearance of a galaxy rather than how elliptical it actually is. After all, we can't change our point of view for such a far-off object. But even a very elliptical galaxy will appear round when

Figure 31–2 The Andromeda Galaxy, also known as M31 and NGC 224, the nearest spiral galaxy to the Milky Way. It is type Sb and is accompanied by two elliptical galaxies, NGC 205, type S0/E5(pec) (*below*), and M32, type E2 (*above*). These galaxies are only 2.2 million light years from earth. The older red and yellow stars give its central regions a yellowish cast, in contrast to the blueness of the spiral arms from the younger stars there.

seen end on. (Just picture looking straight at the end of an egg or a cigar or at the face of a hockey puck or other disk; they look round.) So the ellipticals are, in actuality, at least as elliptical as their Hubble type (which is based only on their appearance) shows; they may actually be more elliptical than they appear.

31.1b Spiral Galaxies

Spiral galaxies, with arms unwinding gracefully from the central regions, are a large fraction of all the bright galaxies in the universe. They form a majority in certain groups of galaxies.

A

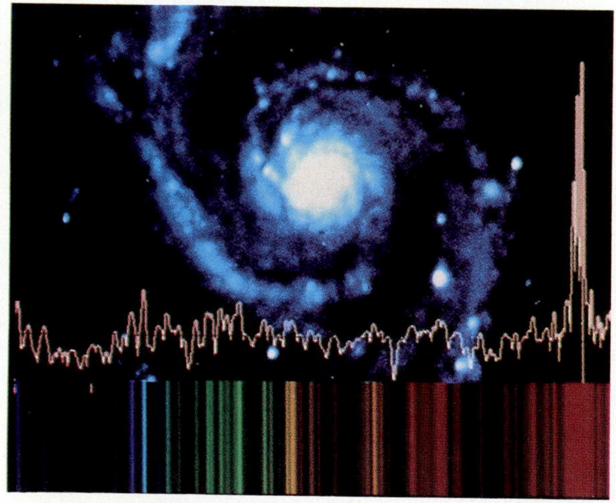

B

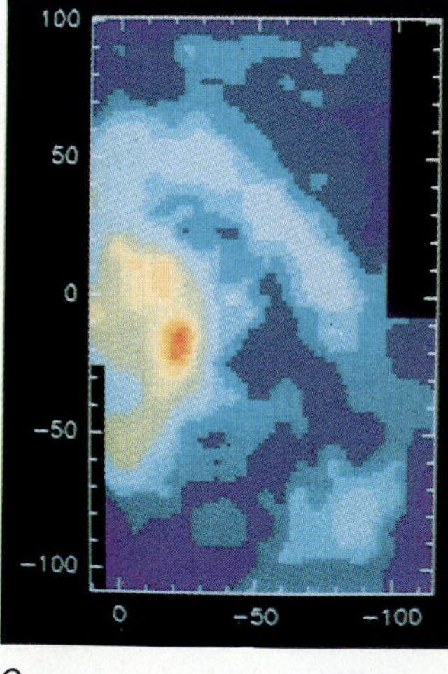

C

Figure 31–3 (*A*) The Whirlpool Galaxy, M51, in Canes Venatici, a type Sc spiral galaxy. At the end of one of its arms, a companion galaxy, NGC 5195, appears. (*B*) The average spectrum of M51, shown both as a visual spectrum and as a trace with intensity on the vertical axis. (*C*) A spiral arm of M51 as traced out from its carbon-monoxide spectral-line radiation, observed with the European IRAM 30-m radio telescope in the millimeter spectral region. The observations that the contrast is high between spiral arm and interarm regions, and that some clouds nevertheless appear between arms, are potentially important for interpreting carbon-monoxide observations of our own galaxy. The observations should lead to improved understanding of the process of star formation. (The axes show seconds of arc.)

Figure 31–4 An edge-on view of NGC 4565, a type Sb spiral galaxy in Coma Berenices. Note its central bulge.

Figure 31–5 NGC 253 in Sculptor, a type Sc galaxy, viewed from a low angle above the plane of its disk. Note that its central bulge is minimal. Its light-absorbing dust lanes show clearly at this angle.

Sometimes the arms are tightly wound around the nucleus; Hubble called this *type Sa,* the S standing for "spiral." Spirals with their arms less and less tightly wound (that is, looser and looser) are called *type Sb* and *type Sc* (Fig. 31–3). The nuclear bulge as seen from edge on (Fig. 31–4) or at a slight angle (Fig. 31–5) is less and less prominent as we go from Sa to Sc. On the other hand, the dust lane—obscuring dust in the disk of the galaxy—becomes more prominent. Some astronomers have a further type, Sd. Spectroscopic measurements from Doppler shifts indicate that galaxies rotate in the sense that the arms trail (Fig. 31–6) (where "sense" means whether the rotation is clockwise or counterclockwise as seen from some location).

Spiral galaxies can be 25,000 to 800,000 parsecs across. They contain 10^9 to over 10^{12} solar masses. Since most stars are of less than 1 solar mass, this means that spirals contain over 10^9 to over 10^{12} stars—we now think our own galaxy has perhaps 10^{12} (1000 billion = 1 trillion). In recent years, we

Figure 31–6 A velocity curve derived from neutral hydrogen spectra of M81, a type Sb spiral galaxy in Ursa Major. Red shows recession. Galaxies rotate in the sense that the arms trail.

Figure 31–7 A comparison of optical, radio continuum, and radio spectral-line hydrogen observations of the spiral galaxy M51, type Sc. The optical image (*reproduced as green*) was taken in blue light to highlight young, hot stars as well as the dust that forms the dust lanes. The continuum radio emission (*reproduced as red*) results partly from emission caused by the high temperature of the H II regions and partly from synchrotron emission. The synchrotron emission comes from electrons moving extremely rapidly (that is, relativistically) in magnetic fields. It is therefore greatest in regions of high compression, namely, the dust lanes. The 21-cm spectral-line observations (*reproduced as blue*) show the distribution of the neutral hydrogen gas (H I).

have learned a lot about galaxies by studying them in different parts of the spectrum (Fig. 31–7).

In about one-third of the spirals, the arms unwind not from the nucleus but rather from a straight *bar* of stars, gas, and dust that extends to both sides of the nucleus (Fig. 31–8). These are similarly classified in the Hubble scheme from *a* to *c* in order of increasing openness of the arms, but with a *B* for "barred" inserted: *SBa, SBb,* and *SBc*.

31.1c Irregular Galaxies

A few per cent of galaxies show no regularity. The Magellanic Clouds, for example, are basically irregular galaxies (Fig. 31–9). Irregular galaxies are classified as *Irr*.

Irregular galaxies like the Magellanic Clouds have relatively little dust, so star formation can be seen even in the visible. Infrared studies of the Magellanic Clouds show their regions of star formation especially well (Fig. 31–10). The ratios of abundances of elements in the interstellar gas in the Magellanic Clouds are somewhat different from the ratios in our own galaxy, and the resulting differences are interesting to contemplate.

From studies of 21-cm radiation, Cornell scientists (including Stephen Schneider, George Helou, Ed Salpeter, and Yervant Terzian) have found a large cloud of gas in the midst of a cluster of distant galaxies. The cloud contains at least a billion and perhaps 10 billion solar masses of material. We don't know yet whether it is a true intergalactic gas cloud or a galaxy in which for some reason stars didn't begin to shine or are fewer in number than we would expect.

Figure 31–8 A barred spiral galaxy, NGC 1365.

31.1d Peculiar Galaxies

In some cases, as in M82 (Fig. 31–11), it appears at first look as though an explosion has taken place in what might have been a regular galaxy. Another possibility, though, is that we are seeing light from the galaxy's nucleus scattered toward us by dust in the filaments. The ring galaxy shown in Figure 31–12 probably resulted from the passage of one galaxy through another. Some other *peculiar galaxies* are ejecting jets of gas.

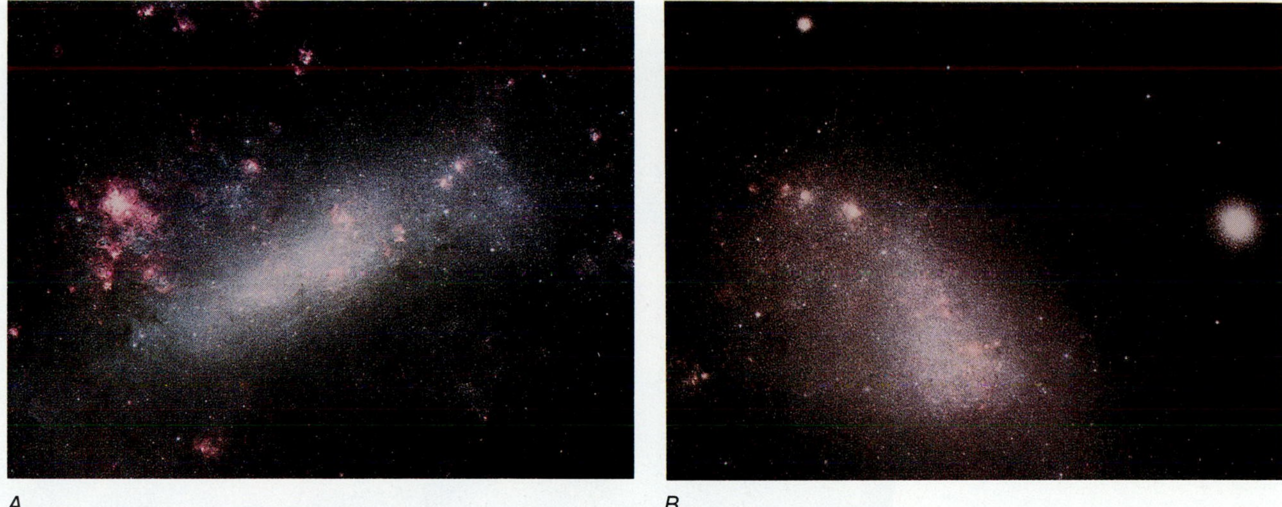

A B

Figure 31–9 The Large Magellanic Cloud (*A*) and the Small Magellanic Cloud (*B*). The LMC is about 50 kpc and the SMC about 65 kpc from us, and are about 20 kpc apart. A few globular clusters, considered part of our Milky Way Galaxy, and some dwarf elliptical galaxies are also that far away. The LMC contains about 1/100 and the SMC about 1/1,000 the mass of our galaxy.

The Clouds were first reported to Europe by 16th-century Portuguese navigators who had travelled to the Cape of Good Hope at the southern tip of Africa. They were named a few decades later in honor of Magellan, who was then circumnavigating the world by that route.

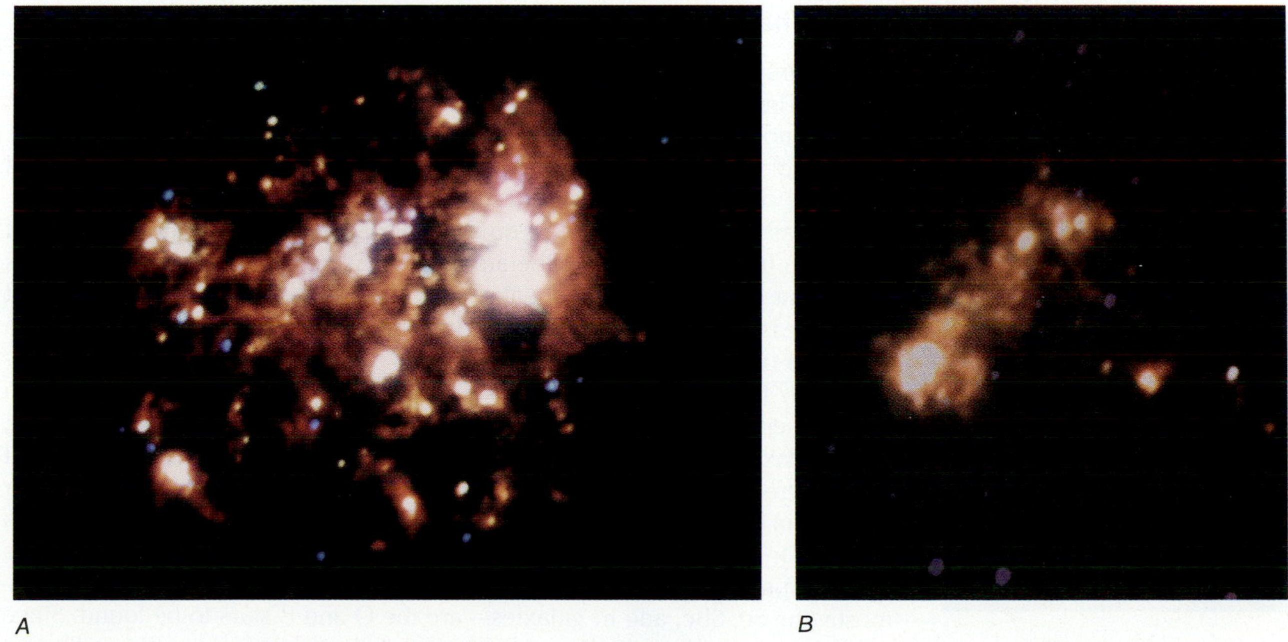

A B

Figure 31–10 (*A*) The Large Magellanic Cloud and (*B*) the Small Magellanic Cloud viewed with IRAS. The warmest sources appear as blue, cooler material appears as green, and the coldest appears as red. The point sources are foreground stars. The regions of active star formation are so bright in each IRAS wavelength band that they appear white. The yellow-green diffuse emission is from cool dust particles heated by the older stars in the galaxies.

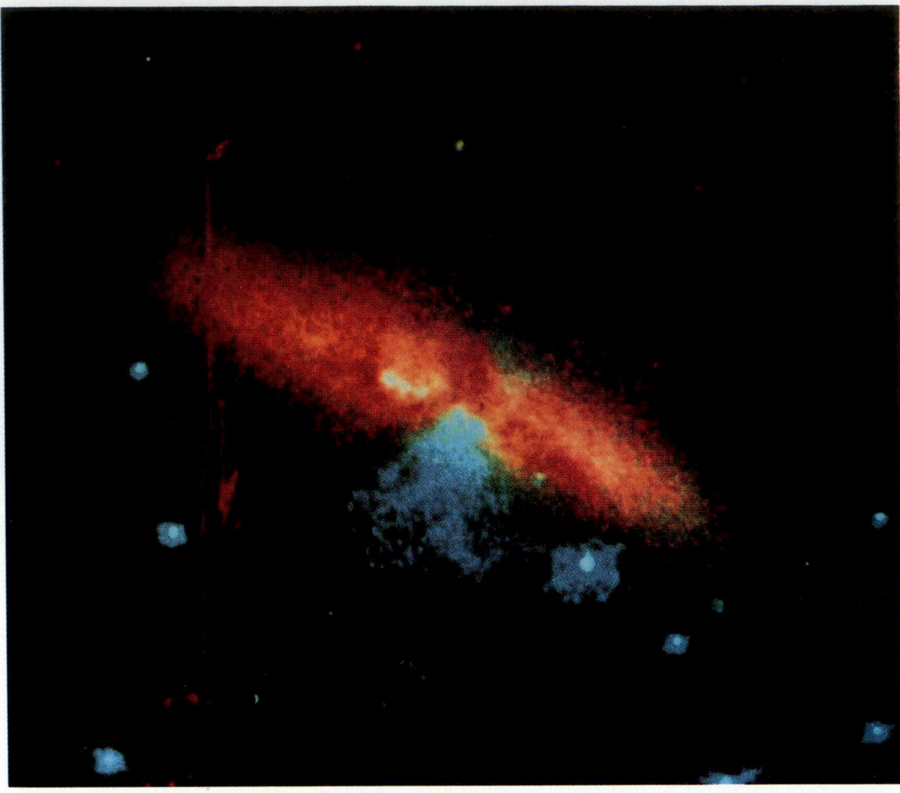

Figure 31–11 M82, a most unusual galaxy that is a powerful source of radio radiation. It was once thought to be exploding, but then it was decided that gentler processes caused its form and non-thermal radiation. Now there is new evidence that it may be exploding after all or the result of a near collision with the galaxy M81.

Peculiar galaxies are classified as the corresponding Hubble type followed by "(pec)": for example, Sa (pec).

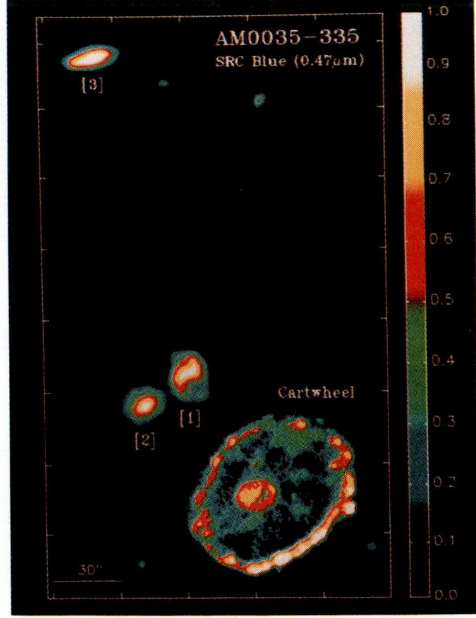

Figure 31–12 A false-color view of the Cartwheel Galaxy (AM 0035-335), a ring galaxy caused by one of its satellite galaxies passing through it. Colors show intensities in the blue.

31.1e The Hubble Classification

Hubble drew out his scheme of classification in a *tuning-fork diagram* (Fig. 31–13). The transition from ellipticals to spirals is represented by *type S0*, which is now thought to be equivalent to E7. S0 galaxies are called *lenticular*. Galaxies of this transition type resemble spirals in having a bright nucleus and the shape of a disk, but do not have spiral arms.

It has since been shown, from optical observations and from studies of the 21-cm hydrogen line, that the amount of gas between the stars in galaxies is different in different types of galaxies. Elliptical galaxies have been thought to have essentially no gas or dust; the new discovery with the IRAM radio telescope of carbon-monoxide spectral-line radiation in several elliptical galaxies may lead to some reevaluation of this point. Recent optical and 21-cm radio observations have also detected gas and dust in ellipticals. In any case, spiral galaxies are known to have a lot of gas and dust. The relative amount of gas increases from types Sa (or SBa) or Sc (or SBc). The gas in the interstellar medium in an irregular galaxy is usually even denser. Though no star formation is found in elliptical or S0 galaxies, the amount of star formation increases toward type Sc. Only in the galaxies with a substantial gas and dust content—mostly the Sc, SBc, and Irr galaxies—are the O and B stars to be found. Since these stars have short lifetimes on a stellar scale, they must have been formed comparatively recently, within the last several million years. Star formation is well observed in the infrared (Fig. 31–14) and radio regions of the spectrum (Fig. 31–15).

New theoretical work has been undertaken to explain the huge, faint shells recently observed on long exposures to surround some elliptical galaxies.

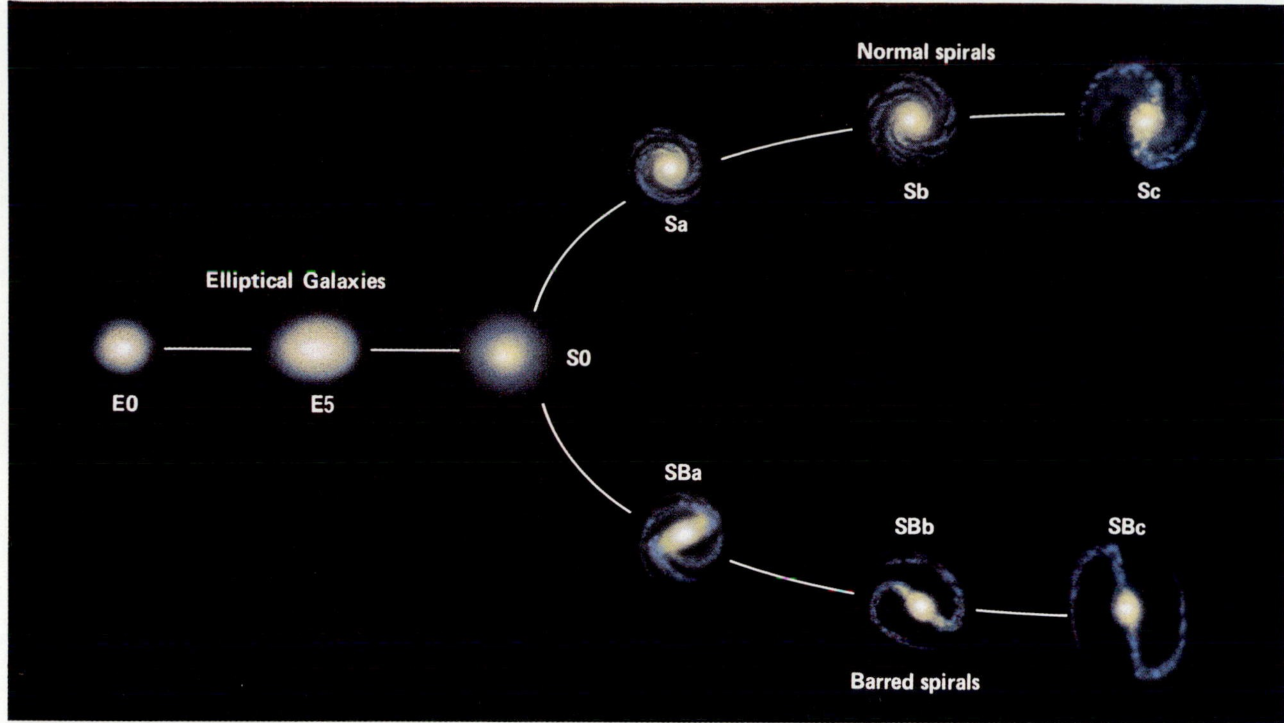

Figure 31–13 The Hubble tuning-fork classification of galaxies. But there are intermediate types between the arms of the tuning fork.

Australian researchers have found that they can explain the shells as stars thrown out by collisions of the ellipticals with spiral galaxies. Thus in this model, some elliptical galaxies are of a later stage of galactic evolution than spirals. The idea that galaxies collide (Fig. 31–16) is of increasing importance.

31.1f The Centers of Galaxies

We have seen the likelihood that a black hole of hundreds or thousands of solar masses may be present in the center of our own galaxy. Studies of nearby galaxies also indicate that giant black holes are present there too. In particular,

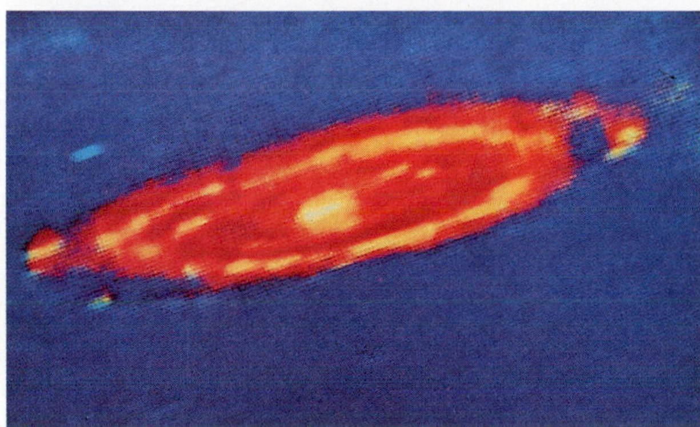

Figure 31–14 An IRAS view of the Andromeda Galaxy, M31. Increasing 60-μ infrared emission is represented by blue to green to yellow to red. The star formation is seen to occur in a ring.

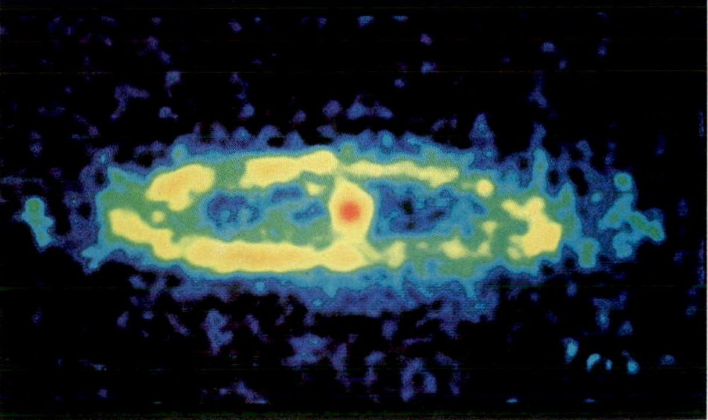

Figure 31–15 A radio map of the Andromeda Galaxy, M31, at 11 cm. As in the infrared view, the strongest emission is in a ring.

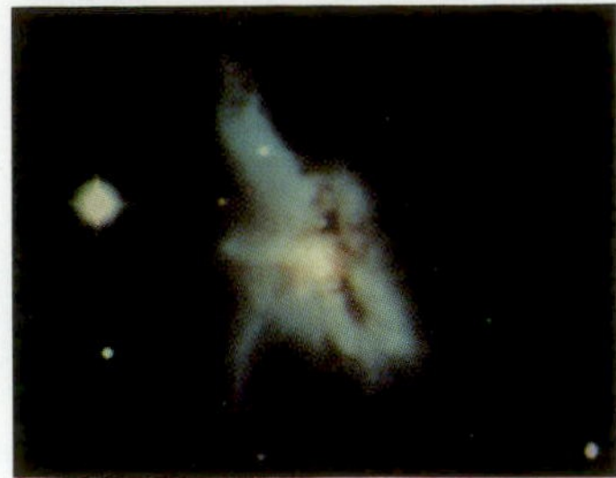

Figure 31–16 NGC 6240, a merger of galaxies in progress. The system is very bright in the infrared.

studies have been carried out of the motions of stars close to the centers of galaxies (Fig. 31–17). The stars' motions in a galaxy reveal the amount of mass that is present between them and the galaxy's center. If enough mass is present in a small enough volume, then it is in the form of a black hole. Black holes containing millions of solar masses of material seem to be present both in the Andromeda Galaxy (M31) and in its elliptical companion M32, and in other nearby galaxies as well.

31.2 The Origin of Galactic Structure

31.2a Interacting Galaxies

In Section 29.5a, we saw how spiral structure can be maintained in a single galaxy by density waves or by a chain of supernovae. In some cases, structure

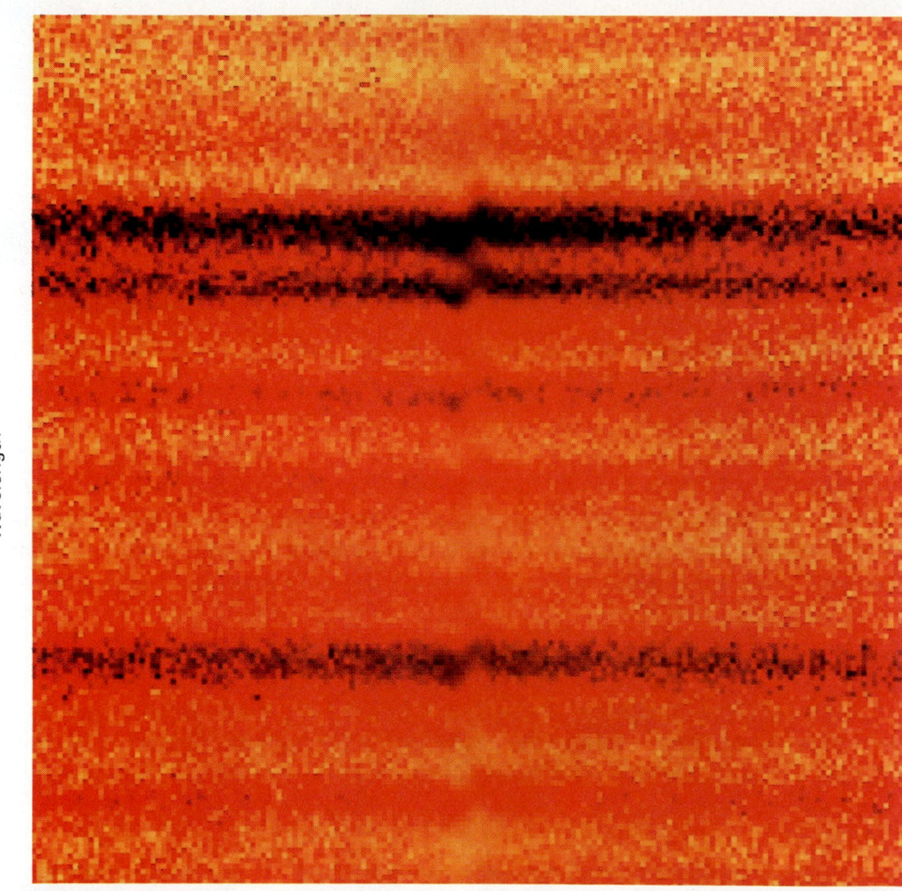

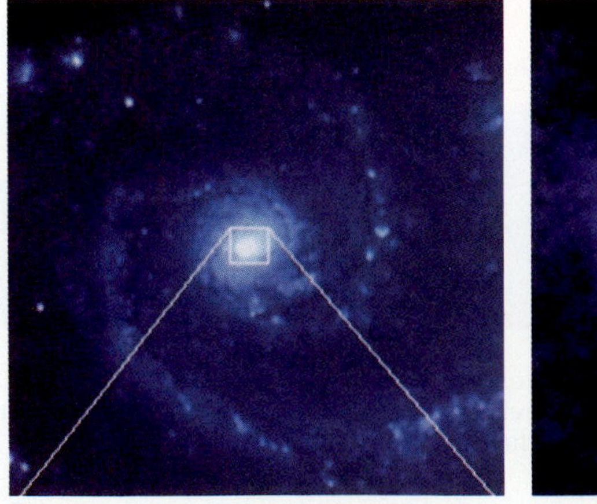

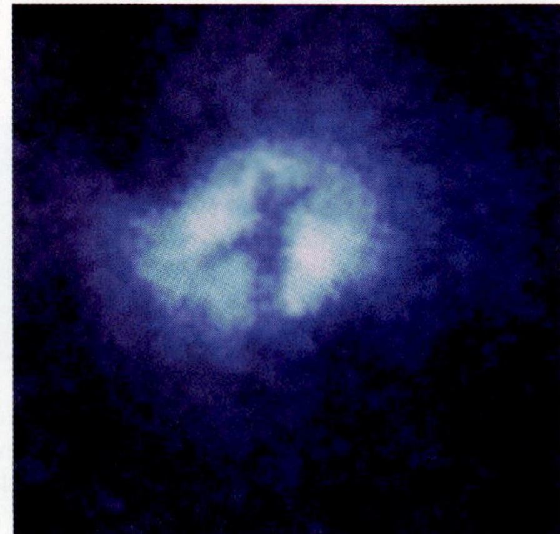

Figure 31–17 (*A*) This spectrum of M31, the Great Galaxy in Andromeda, had the spectrograph's slit laid across the center of the galaxy. Note the abrupt change in wavelength as the slit crossed M31's center, showing an abrupt change in the direction of the velocities there. John Kormendy of the Dominion Astrophysical Observatory, Victoria, British Columbia, has interpreted this spectrum to indicate the presence of a black hole with about 10 million times the mass of the sun. The strongest lines in this spectrum are the *b* lines of neutral magnesium. (*B*) The spiral galaxy M51 (*left*, in a ground-based view) has a central black hole where "X" marks the spot as shown in this image taken with the Hubble Space Telescope (*right*). The thicker band of the "X" may be the side of a doughnut of dust that is hiding the central black hole; there is no accepted theory so far for the other dark band.

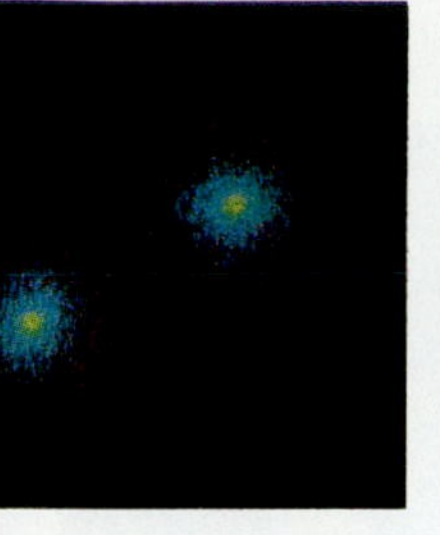
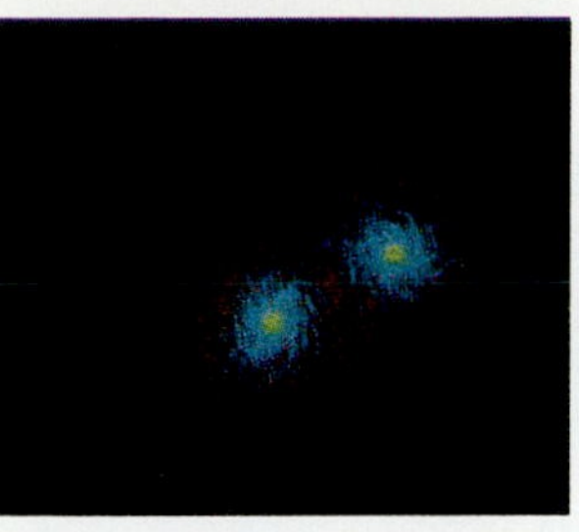
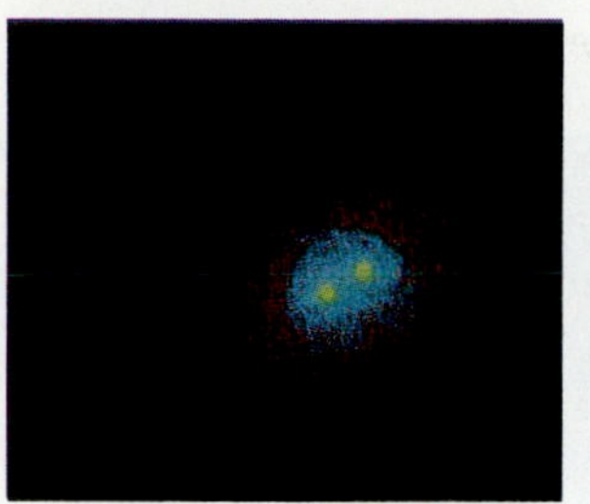
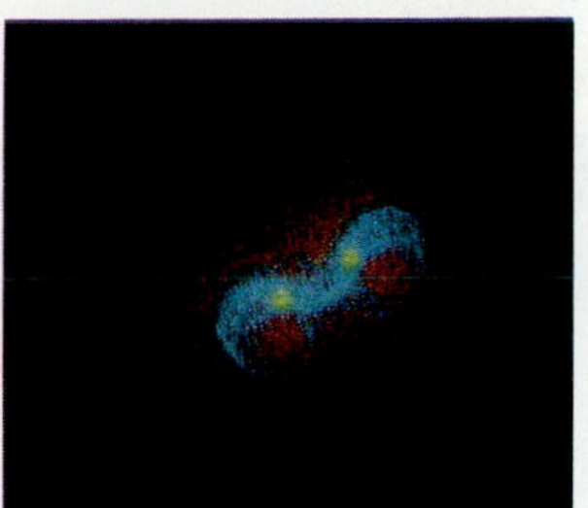
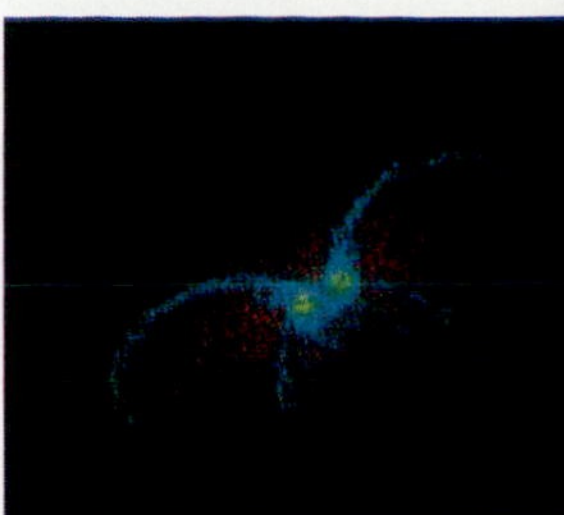

A

Figure 31–18 (*A*) Calculations made with a supercomputer, showing a time sequence of the very close encounter of two identical model galaxies. About ten thousand particles represent each galaxy. The supercomputer enabled Joshua E. Barnes of the Canadian Institute for Theoretical Astrophysics and Lars Hernquist of Princeton University, both then of the Institute for Advanced Study in Princeton, to consider each galaxy to be made of a bulge, a disk, and a massive, dark halo. Further, they considered the interactions of all pairs of particles. The results are in general agreement with and extend the earlier calculations of Alar and Juri Toomre. Time intervals are given in galactic years, which are each 250 million earth years in duration. (*B*) A negative print of the pair of galaxies NGC 4038 and NGC 4039, known as The Antennae. It closely matches the computer drawings of Fig. 31–18, when they are displayed from the proper angle. (*C*) A color close-up of the antennae, showing detail of gas and dust in the two interacting central regions.

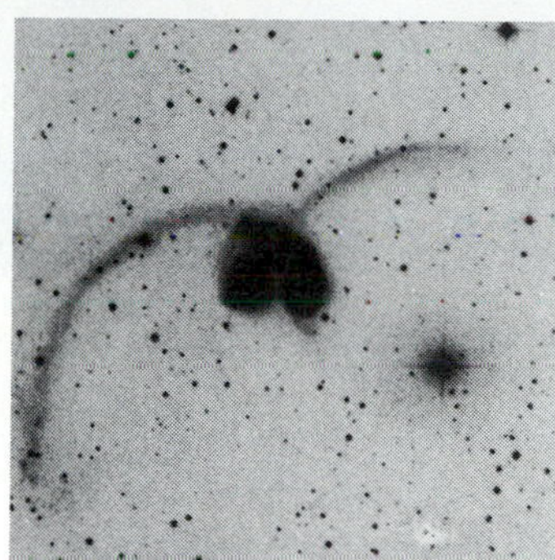

B

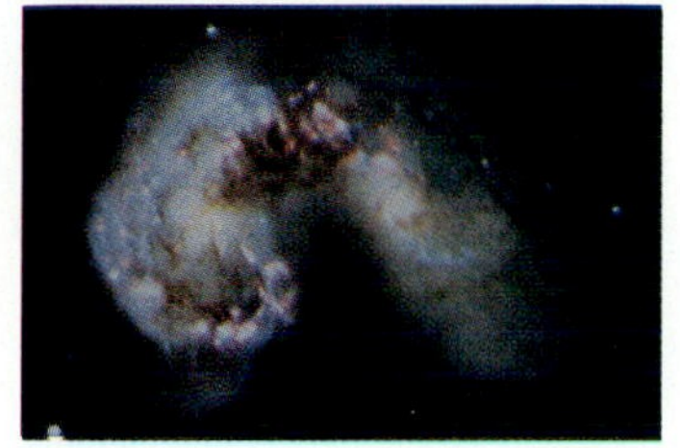

C

can arise from a gravitational interaction of two galaxies. Some scientists have used computers to follow the evolution of a system over time, using the computer to follow the gravitational interaction between many particles, with each particle interacting by gravity with the center of mass of the other galaxy. Long arms, often called "tails," are drawn out by tidal forces (Fig. 31–18*A*). The spin of the original galaxies contributes to the graceful curvature. The shapes can be made to match known galaxies (Fig. 31–18*B* and *C*).

The above example involved galaxies of equal mass. If one galaxy is much more massive than the other, the structure that results from an interaction can resemble that of an ordinary spiral galaxy. Note that the Milky Way Galaxy has close companions—the Magellanic Clouds—and the Andromeda Galaxy has companions as well, so it is possible that there was a gravitational interaction to the spiral structure of our own galaxy and other galaxies. Many examples of interacting galaxies are now known (Fig. 31–19).

The idea is now current that some elliptical galaxies can result from the merger of pairs of spiral galaxies. Computer simulations endorse the possibility. Our galaxy and the Andromeda Galaxy even seem on a line to interact in billions of years, so our sun may wind up in an elliptical galaxy someday.

31.3 Clusters of Galaxies

Careful study of the positions of galaxies and their distances from us has revealed that most galaxies are part of groups or clusters. Groups have just a handful of members, while *clusters of galaxies* may have hundreds or thousands.

31.3a The Local Group

The two dozen or so galaxies nearest us form the *Local Group*. The Local Group contains a typical distribution of types of galaxies and extends over a volume 1 megaparsec in diameter. It contains three spiral galaxies, each at least 31 to 50 kiloparsecs across—the Milky Way, Andromeda, and M33. Several are ellipticals, including four regular ellipticals, two of which are companions to the Andromeda Galaxy. Others are dwarf ellipticals (Fig. 31–20). There are

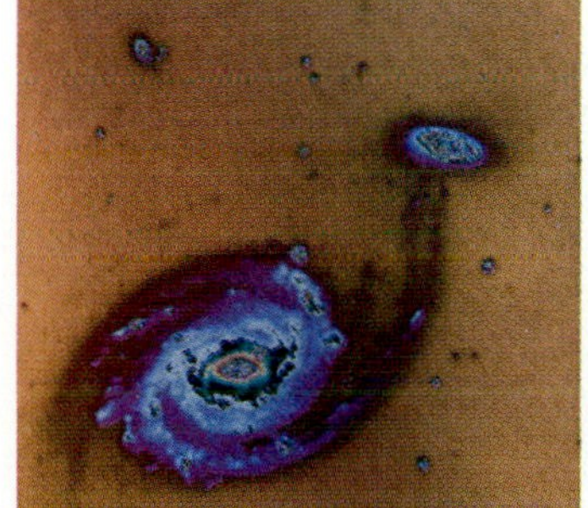

Figure 31–19 A false-color CCD image of the interacting galaxies NGC 7752 and 7753 taken by the prime-focus CCD camera on the Isaac Newton Telescope, Roque de los Muchachos Observatory.

Figure 31–20 The Leo I dwarf spheroidal galaxy, a member of the Local Group.

Figure 31–21 NGC 6822, a galaxy of the irregular Magellanic Cloud type, a member of the Local Group.

The Andromeda Galaxy at 2.2 million light years and perhaps M33 at 2.4 million light years are the farthest objects you can see with your unaided eye; they appear as fuzzy blobs in the sky if you know where to look.

four irregular galaxies, each 3 to 10 kiloparsecs across, including the Large and Small Magellanic Clouds. At least a dozen other dwarf irregulars are known (Fig. 31–21).

Previously unknown candidates for membership in the Local Group are occasionally found. Some of these newly discovered galaxies have been difficult to discover even though they are so close because they lie in the plane of our galaxy and are thus hidden from our view by dust.

31.3b More Distant Clusters of Galaxies

In the vicinity of the Local Group, there are apparently other small groups of galaxies, each containing only a dozen or so members. The nearest cluster of many galaxies (a *rich cluster*, as opposed to a *poor cluster*) can be observed in the constellation Virgo and surrounding regions of the sky; it is called the Virgo Cluster (Fig. 31–22). It covers a region in the sky over 6° in radius, 12

Figure 31–22 M84 (*center*) and M86 (*right*), bright elliptical galaxies, are the most prominent in this view of the center of the Virgo Cluster. NGC 4438 is the distorted galaxy at left, with NGC 4435 above it.

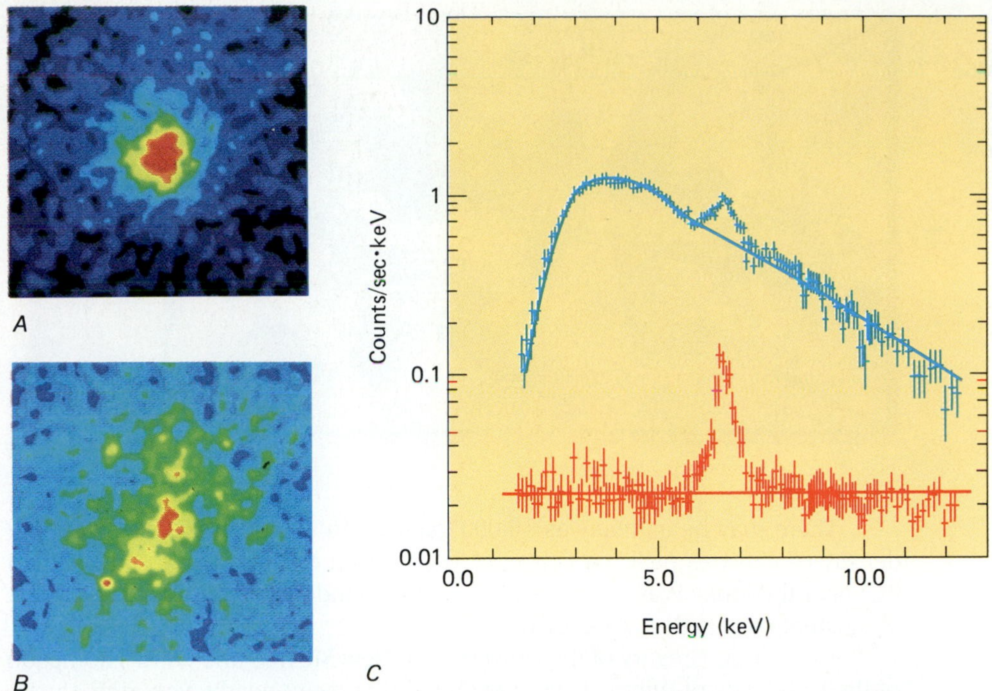

Figure 31–23 (*A*) The Virgo Cluster's x-ray emission is sharply peaked around the central peculiar elliptical galaxy M87. (*B*) X-ray emission from the Coma Cluster is more spread out. (*C*) The x-ray spectrum of the active galaxy NGC 1275, which is embedded in the Perseus Cluster of galaxies, shows an iron emission line at 6.7 keV. The red curve shows the line isolated by subtracting the yellow curves. EXOSAT's small field of view allowed this line to be used to make a temperature map of the cluster. Both images and spectra are from the EXOSAT observatory.

times greater than the angular diameter of the moon. The Virgo Cluster contains hundreds of galaxies of all types. It is about 2 million parsecs across, and most of it is located about 20 million parsecs away from us.

Other rich clusters are known at greater distances, including the Coma Cluster in the constellation Coma Berenices (Berenice's Hair). The Coma Cluster has spherical symmetry; its galaxies are concentrated toward its center, not unlike the distribution of stars in a globular cluster (which is a cluster of **stars**, and is therefore on a much smaller scale). Thus the Coma Cluster is a *regular cluster* as opposed to an *irregular cluster*.

Rich clusters of galaxies are generally x-ray sources. Studies with the Einstein Observatory revealed a hot intergalactic gas containing as much mass as is in the galaxies themselves. The temperature of the gas is 10 to 100 million K. The gas is clumped in some clusters, while in others it is spread out more smoothly with a concentration near the center (Fig. 31–23). This may be an evolutionary effect, with gas being ejected from individual galaxies in younger clusters and spreading out as the clusters age.

X-ray spectra from British and American satellites showed that the intergalactic gas contains iron and other heavy elements with abundances that approximate that found in the sun. This implies that it was ejected from galaxies, in which nucleosynthesis in stars formed the iron. This also indicates that matter apparently flows out of most of the galaxies in a cluster.

The matter is pulled by gravity toward the center of the cluster, although it may be heated and not fall to the center. A giant elliptical galaxy, like M87 at the center of the Virgo Cluster, may be so large because is has gobbled up gas and other galactic debris.

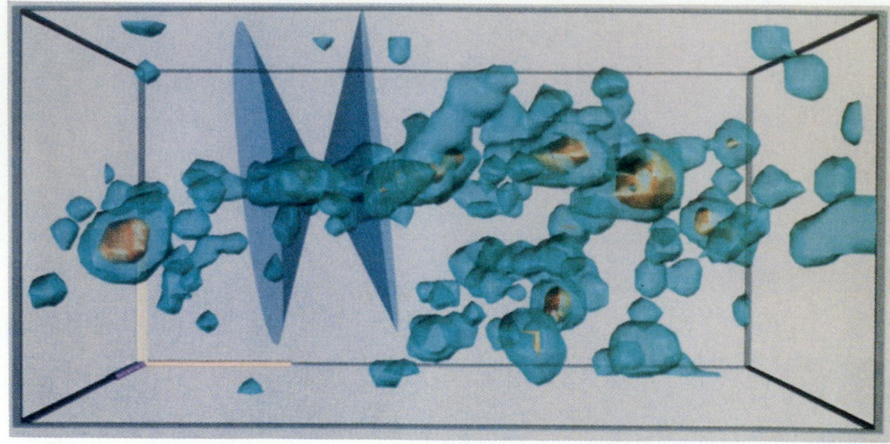

Figure 31–24 The distribution of galaxies in the Local Neighborhood, covering 20 Mpc in the longest dimension. Between the blue cones, which mark latitudes ±20°, is the region of obscuration from absorption in our galaxy. The inner contour shows one galaxy per cubic megaparsec.

There may be as many as 10,000 galaxies in a very rich cluster, and the density of galaxies near the center of such a rich cluster may be higher than that near the Milky Way by a factor of one thousand to one million. Thousands of clusters of galaxies are known (Fig. 31–24).

Stephen A. Gregory of the University of New Mexico and Laird Thompson of the University of Illinois have concluded that every nearby very rich cluster is located in a cluster of clusters, a *supercluster*. A new catalogue of superclusters includes 16 within 2 billion light years of us, each of which contains two or more rich clusters. The Local Group, the several similar groupings nearby, and the Virgo Cluster form the *Local Supercluster*. This cluster of clusters contains 100 member clusters roughly in a pancake shape, on the

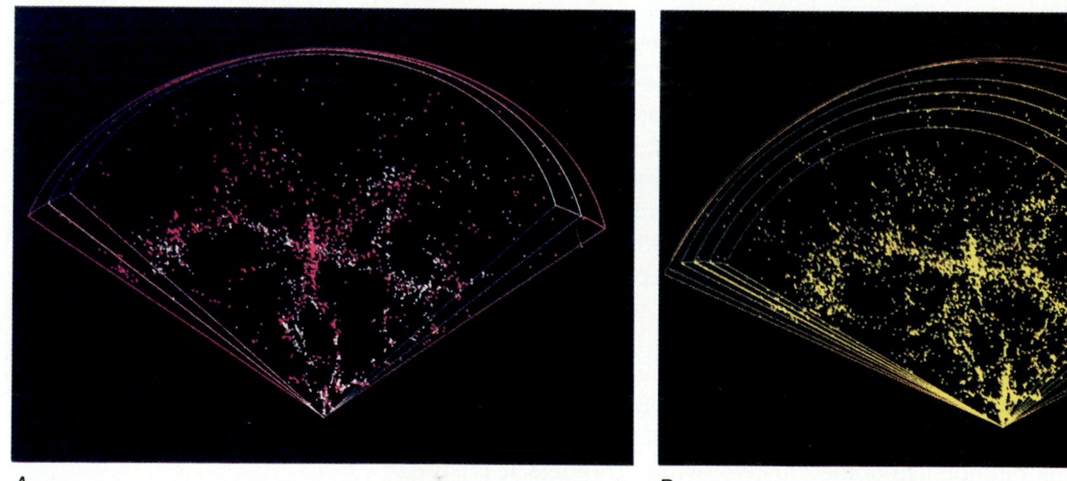

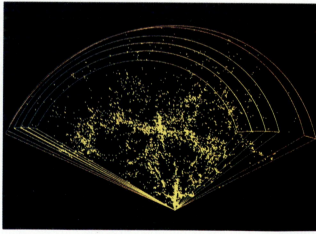

Figure 31–25 Each slice shows distance from earth (going along straight lines outward from the earth, which is at the bottom point) versus position on the sky (in right ascension) for all galaxies in a strip of declination 6° in size. Right ascension corresponds to longitude and declination to latitude in the sky. Each wedge extends out to galaxies with a velocity of recession of 15,000 km/sec, which corresponds to a distance of about 300 Mpc. (The next section describes the correspondence between measured velocity and distance.) (*A*) Velocity versus position in right ascension in pink for all 1065 galaxies in the strip brighter than magnitude 15.5 between declinations (δ) 26.5° and 32.5° and in white for the 702 galaxies in the next 6° strip of declination, between 32.5° and 38.5°. (*B*) Data for all four slices measured so far. Note how the structures apparent in one slice continue to the next, indicating that the galaxies are on the edges of giant bubbles, like the suds in a kitchen sink, or perhaps sponge-like structures in space. The Coma Cluster of galaxies is in the center, apparently at the intersection of several bubbles.

order of 100 million light years across and 10 million light years thick. Superclusters are apparently separated by giant voids.

To get a better overall picture of the universe, we need three-dimensional maps (Fig. 31–25). Margaret Geller, John Huchra, and their collaborators at the Harvard-Smithsonian Center for Astrophysics, using the techniques we describe in the next section to provide the distance dimension (the other two dimensions are those on the sky), are providing such maps of the distribution. They have observed about 6000 galaxies thus far. In the figure, we see the distance dimension plotted against one sky axis for several wedges—"slices"—of the universe. The galaxies appear to be on the surface of bubble-like structures with sharp edges. Now that several slices are available, it has become clear that the structures are elongated with typical sizes of 50 Mpc by 30 Mpc by 5 Mpc (roughly 150 million light years by 100 Mly by 15 Mly). The largest—at least 250 million light years long—has been called the "Great Wall." Calculations based on the recently popular theoretical idea that much of the universe is in the form of invisible particles known as "cold dark matter" (which we will discuss further in Sections 31.8 and 34.2) have been successful in explaining the formation of ordinary clusters of galaxies. But even this theory cannot account for structures of the size and shape of the Great Wall. As for the voids, they are not completely empty, but the density of galaxies in them is only 20 per cent of average. We shall return to these observations when we discuss the formation of galaxies in Section 31.8.

Brent Tully of the University of Hawaii has found about 100 rich clusters of galaxies grouped in a huge flattened structure 1.5 billion light years long and 200 million light years in its shortest dimension (Fig. 31–26). We discuss the relation of these observations to theory in Section 31.8.

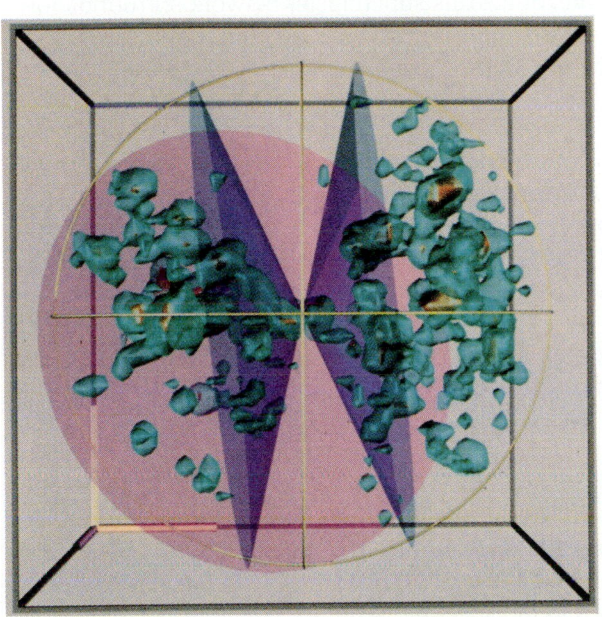

A

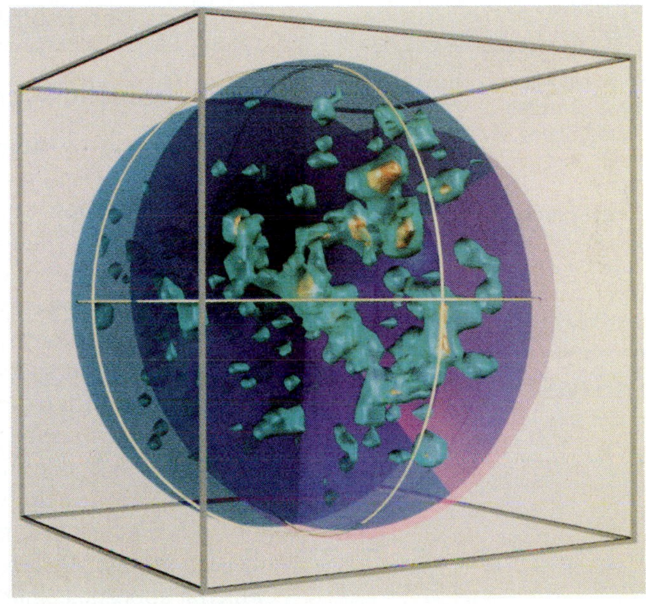

B

Figure 31–26 Two views of the distribution of 382 rich clusters of galaxies. Between the blue cones, which mark latitudes ±20°, is the region of obscuration from absorption in our galaxy. The sphere marks a radius of about 1 billion light years. (*A*) The view is edge on to both the zone of obscuration and the plane of the Local Supercluster. (*B*) The view is inclined 60° from the zone of obscuration.

The Pisces-Cetus complex discovered by the same scientist, Brent Tully of the University of Hawaii, extends for over a billion light years, and includes our Local Supercluster and our neighbors. It is hard to prove that it is real rather than a statistical effect, but there is certainly no sign that the universe is becoming more homogeneous as we go to larger scales.

Figure 31–27 Astonishingly smooth arcs have been discovered in a few galaxy clusters. See also Figure 1–15.

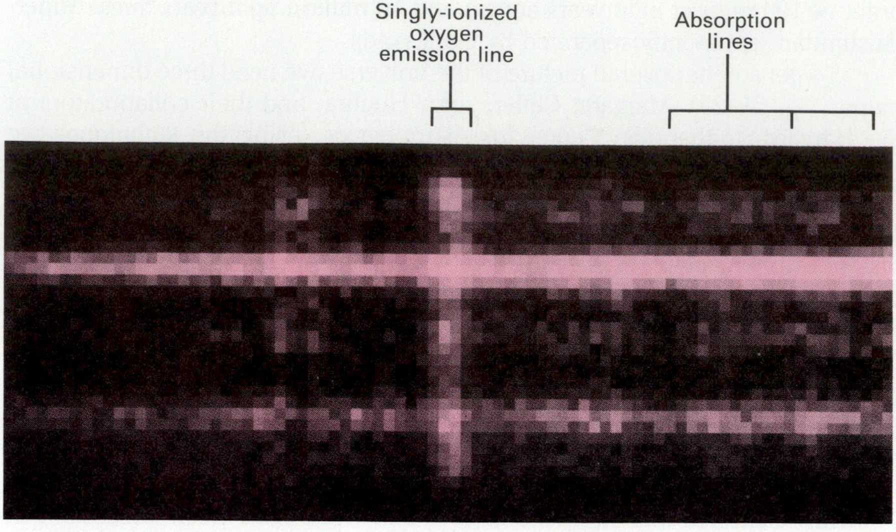

Figure 31–28 The spectrum of one of the arcs shows a bright emission line from singly ionized oxygen, as well as three barely perceptible absorption lines from CN, ionized calcium, and iron. These lines appear at wavelengths 72% longer than their wavelengths in the laboratory; the method we discuss in Section 31.4 shows that the matter emitting the radiation is twice as far away as the galaxy the arc seems to surround. Thus the galaxy is acting as a gravitational lens.

Following the style of the late George Gamow, I can thus write my address as:

Jay M. Pasachoff
Williamstown
Massachusetts
United States of America
North America
Earth
Solar System
Milky Way Galaxy
Local Group
Local Supercluster
Universe

Does the clustering continue in scope? Neta Bahcall of Princeton University has found a network of large-scale superclusters. They extend up to about 500 million light years in scale. Since this supercluster network surrounds low-density regions, Bahcall suggests that the universe is cellular.

31.3c Galactic Arcs

Two examples of strangely symmetrical giant arcs were reported around galaxies in 1987 (Fig. 31–27). At first it was thought that they might be gas in a cluster of galaxies, but now it seems that they are a result of a very strange phenomenon: gravitational lensing. Gravitational lensing is explained with Einstein's general theory of relativity, and results from the bending of light in a strong gravitational field.

Following the discovery of the unusually regular arc, spectra were taken (Fig. 31–28). The spectrum of the brighter of the arcs shows a single strong emission line. Though one can never be certain when identifying just a single spectral line, it seems reasonable to workers in the field that the line is the one at 3727 Å from ionized oxygen, since that line is seen alone in many other sources. Knowing the wavelength at which a line appears gives astronomers the source's distance, as we shall see shortly. The identification of the line as the one from ionized oxygen puts the arc twice as far away as the galaxy around which it appears. Fainter lines in the spectrum seem to have reasonable matches to other known spectral lines. The initial research was carried out by Vahe Petrosian of Stanford University and C. Roger Lynds of the Kitt Peak National Observatory and by a French group.

The gravitational-lensing model can explain why the arc is so regular. An object imaged straight ahead around a lensing object would appear as a ring, but any slight offset turns the ring into arcs. Further, the model places the object so far away that we are seeing the many hot young stars of a distant galaxy, explaining why the arcs appear so blue. Indeed, the arc appears narrower when imaged in the red, which would result because a galaxy's red stars are more concentrated toward the galaxy's nucleus.

We shall meet other examples of gravitational lensing when we discuss the double quasars (Section 32.9).

31.4 The Expansion of the Universe

31.4a Hubble's Law

In the decade before the problem of the location of the "spiral nebulae" was settled, Vesto M. Slipher of the Lowell Observatory took many spectra that indicated that the spirals had large redshifts. This work was to lead to a profound generalization. In 1929 Hubble announced that galaxies in all directions are moving away from us, and that the distance of a galaxy from us is directly proportional to its redshift (that is, when the redshift we observe is greater by a certain factor, the distance is greater by the same factor; Fig. 31–29*A*). The proportionality between redshift and distance is known as *Hubble's law*. Hubble, in collaboration with Milton L. Humason at Mt. Wilson, went on during the 1930's to establish the relation more fully (Fig. 31–29*B*). The redshift (Fig. 31–30) is presumably caused by the Doppler effect. The law is usually stated in terms of the velocity that corresponds (by the Doppler effect) to the measured wavelength, rather than in terms of the redshift itself.

Hubble's law states that the velocity of recession of a galaxy is proportional to its distance. It is written

$$v = H_0 d,$$

where v is the velocity, d is the distance, and H_0 is the present-day value of the constant of proportionality (simply, the constant factor by which you multiply d to get v), which is known as *Hubble's constant* (Fig. 31–31).

Allan Sandage of the Observatories of the Carnegie Institution of Washington and of the Space Telescope Science Institute is Hubble's intellectual heir. He is using the world's largest telescopes to study the distances and redshifts of the farthest galaxies. Working over many years, Sandage and Gustav Tammann of the University of Basel in Switzerland have derived a value of 50 km/sec/Mpc for H_0, as we shall describe in Section 31.5. This is about 10 times lower than the value that Hubble originally announced, but Sandage and Tammann have used new techniques for finding the distance to far-off galaxies, incorporating also earlier corrections to the distance scale. In recent years,

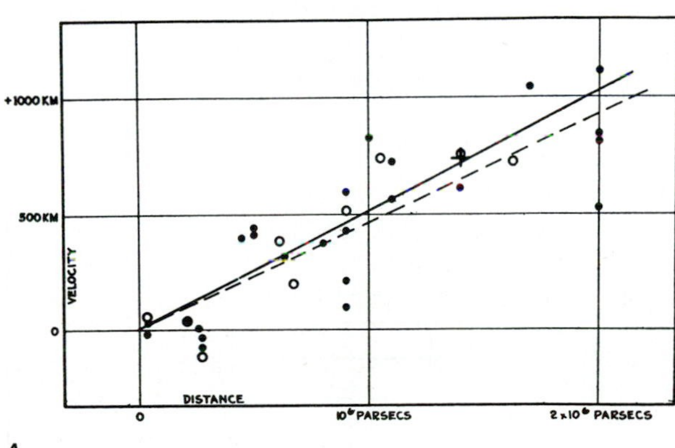

A

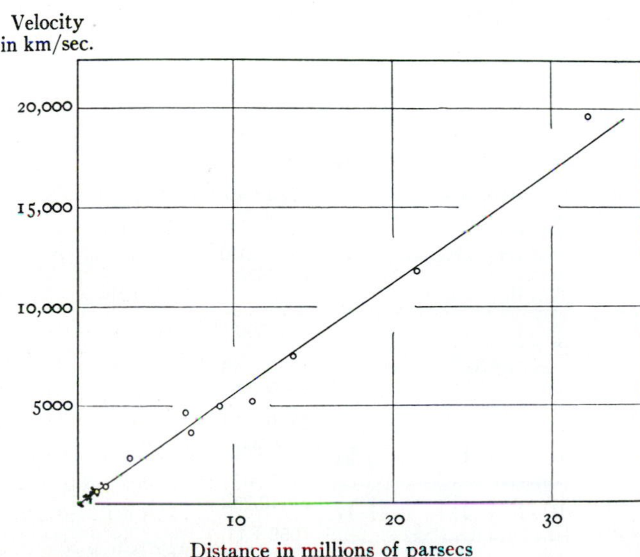

B

Figure 31–29 (*A*) Hubble's original diagram from 1929. Dots are individual galaxies; open circles are from groups of galaxies. The scatter to one side of the line or the other is substantial. (*B*) By 1931, Hubble and Humason had extended the measurements to greater distances, and Hubble's law was well established. All the points shown in the 1929 work appear bunched near the origin of this graph. $v = H_0 d$ represents a straight line of slope H_0. These graphs use older distance measurements than we now use, and so give different values for H_0 than we now derive.

Figure 31–30 (*A*) Spectra are shown at right for the galaxies at left, all reproduced to the same scale. Distances are based on Hubble's constant = 50 km/sec/Mpc. Notice how the farther away a galaxy is, the smaller it looks. The yellow arrow below each horizontal streak of spectrum shows how far the H and K lines of ionized calcium are redshifted.

The spectrum of an emission-line source located inside the telescope building appears as vertical lines above and below each galactic spectrum to provide a comparison with a redshift known to be zero.

The wavelengths of the three brightest comparison lines are 3888 Å (*left*), 4471 Å (*center*), and 5015 Å (*right*).

(*B*) Radio spectra of the 21-cm line appear at right for the optical galaxies, and spectra are at left. The shifted emission line is shown in each case with a blue arrow.

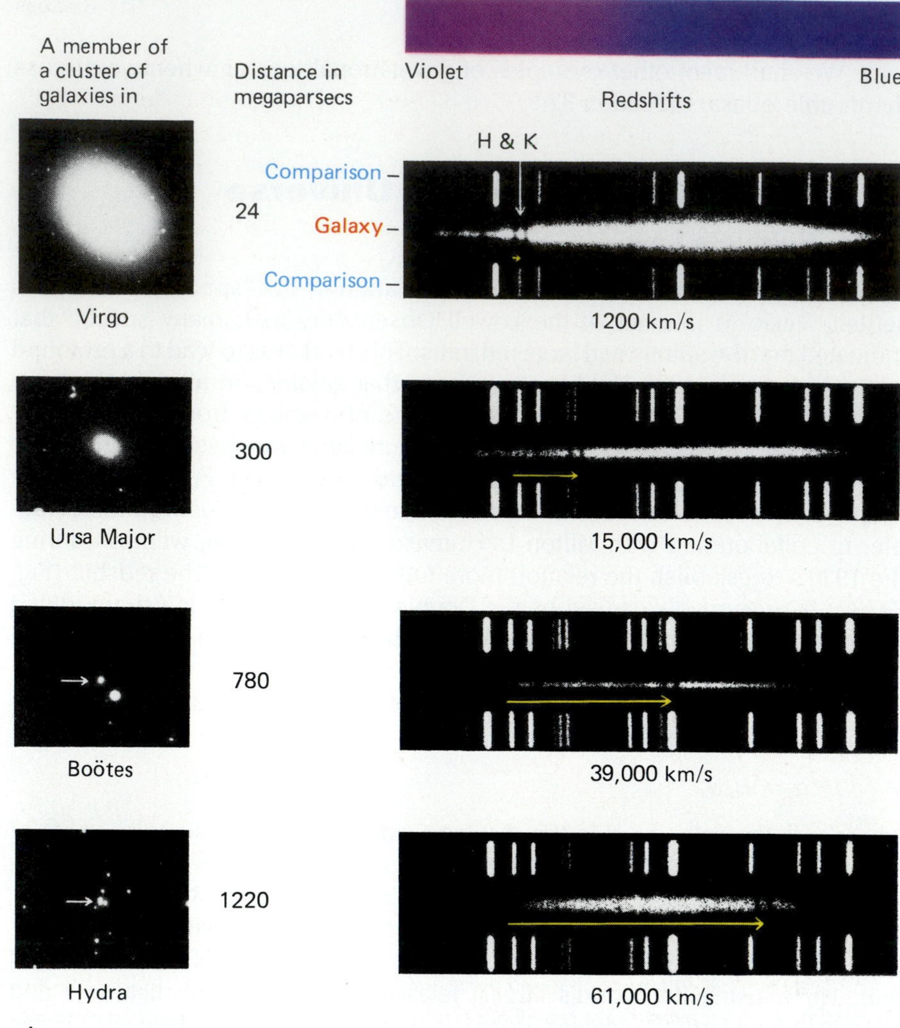

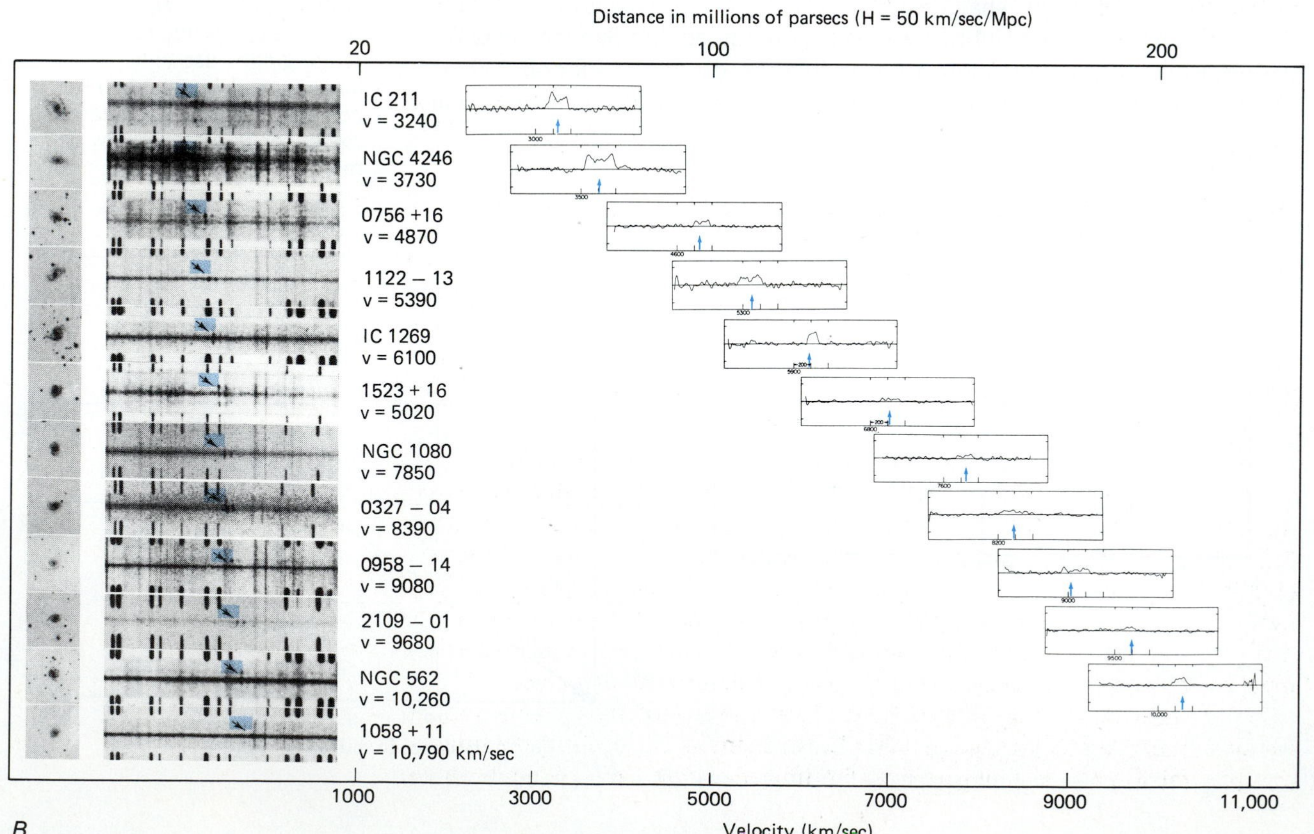

other scientists have made similar sets of observations, and many have derived larger values for H_0. Gerard de Vaucouleurs at Texas, John Huchra of the Harvard-Smithsonian Center for Astrophysics, Marc Aaronson, Jeremy Mould of Caltech, and others have found values closer to 100 km/sec/Mpc. (Tragically, Aaronson was killed in 1987 in an accident at the 4-m telescope at Kitt Peak National Observatory.) Let us use 50 for the rest of this book for convenience, as many astronomers do; there are strong arguments for each value.

Hubble's constant is given in units that may appear strange, but they merely state that for each megaparsec (3.3×10^6 light years) of distance from the sun, the velocity increases by 50 km/sec (thus the units are 50 km/sec per Mpc, spoken "fifty kilometers per second per megaparsec"). From Hubble's law, we see that a galaxy at 10 Mpc would have a redshift corresponding to 500 km/sec; at 20 Mpc the redshift of a galaxy would correspond to 1000 km/sec; and so on. The redshift for a given galaxy is the same no matter in which part of the spectrum we observe.

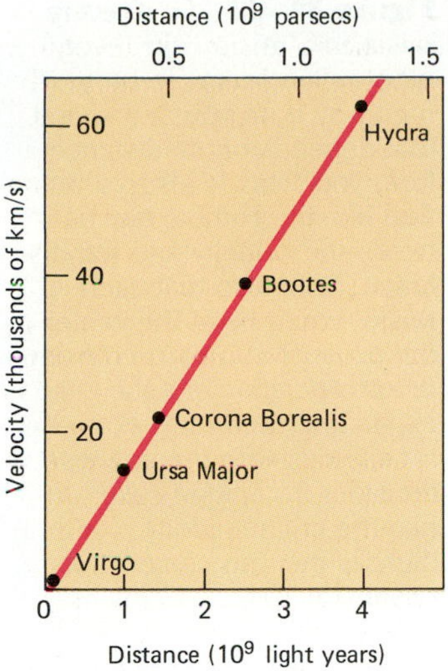

Figure 31–31 The Hubble diagram for the galaxies shown in Figure 31–30.

Example: For $H_0 = 50$ km/sec/Mpc, how far away is a galaxy for which we measure a redshift of 15,000 km/sec?
Answer: Hubble's law can be transformed to

$$d = v/H_0.$$

Thus

$$d = \frac{15{,}000 \text{ km/sec}}{50 \text{ km/sec/Mpc}} = \frac{300}{1/\text{Mpc}} = 300 \text{ Mpc}.$$

Note that if Hubble's constant is 100 instead of 50, then a galaxy whose redshift is measured to be 500 km/sec would be (500 km/sec)/(100 km/sec/Mpc) = 5 Mpc, half the distance that corresponds to the smaller Hubble's constant. So the debate over the size of Hubble's constant has broad effect on the size of the universe. The debate is often heated, and sessions of scientific meetings at which the subject is discussed are well attended. Astronomers hope that new measurements made with the Hubble Space Telescope will settle the controversy.

31.4b The Expanding Universe

The major import of Hubble's law is that since all but the closest galaxies in all directions are moving away from us at a rate proportional to distance, the universe is expanding. Since the time when Copernicus moved the earth out of the center of the universe (and the time when Shapley moved the earth and sun out of even the center of the Milky Way Galaxy), we have not liked to think that we could be at the center of the universe. Fortunately, Hubble's law can be accounted for without our having to be at any such favored location, as we see below.

Imagine a raisin cake (Fig. 31–32) about to go into the oven. The raisins are spaced a certain distance away from each other. Then, as the cake rises, the raisins spread apart from each other. If we were able to sit on any one of those raisins, we would see our neighboring raisins move away from us at a certain speed. It is important to realize that raisins farther from us would be moving away faster, because there is more cake between them and us to expand. No matter in what direction we looked, the raisins would be receding from us, with the velocity of recession proportional to the distance.

The next important point to realize is that it doesn't matter which raisin we sit on; all the other raisins would always seem to be receding. Of course, any real raisin cake is finite in size; the universe may have no limit, so we

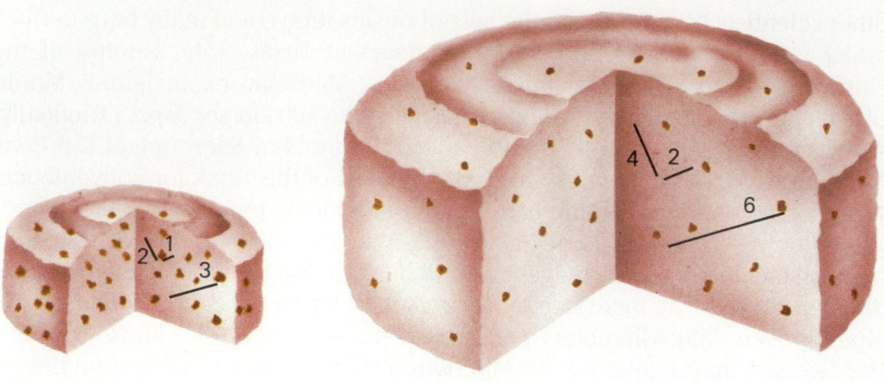

Figure 31–32 From every raisin in a raisin cake, every other raisin seems to be moving away from you at a speed that depends on its distance from you. This leads to a relation like the Hubble law between the velocity and the distance. Note also that each raisin would be at the center of the expansion measured from its own position, yet the cake is expanding uniformly. For a better analogy with the universe, consider an infinite cake; unlike the finite analogy pictured, there is then no center to its expansion.

would never see an edge. The fact that all the galaxies appear to be receding from us does not put us in a unique spot in the universe; there is no center to the universe. Each observer at each location would observe the same effect.

Note that in our analogy the size of the raisins themselves is not changing; only the separations are changing. In the universe, the galaxies themselves and the clusters of galaxies are not expanding; only the distances between the clusters (or, perhaps, the superclusters) are increasing.

Note also that individual stars in our galaxy can appear to have small redshifts or blueshifts, caused either by their peculiar velocities or by the differential galactic rotation. Also, some of the nearer galaxies (such as M31 in Andromeda) have random velocities of sufficient size, or velocities less than our rotational velocity in our galaxy, so that they are approaching us. But all galaxies in distant clusters are receding. Still the velocities of the nearer of those distant galaxies (Fig. 31–33) may be substantially affected by concentrations of mass so as to give misleading answers for measurements of the overall "Hubble flow" that leads to the determination of Hubble's constant.

Figure 31–33 M83, a spiral galaxy in Centaurus, is sufficiently close to earth (3.7 Mpc) that its individual velocity makes it deviate from the "Hubble flow" of expansion. But farther galaxies and clusters of galaxies follow Hubble's law. Some of the H II regions in its arms show types of objects that are our only independent way of measuring distances to farther galaxies. This galaxy has a small bar in its center and is thus intermediate between normal spirals and fully barred spirals.

31.5 The Distance Scale of the Universe

The major problem for setting the Hubble law on the firmest footing is finding the distances to the galaxies for which redshifts are measured. Let us first discuss the distance indicators (Fig. 31–34) used by Sandage and Tammann in deriving their value of 50 km/sec/Mpc for Hubble's constant.

We can't measure trigonometric parallaxes beyond the nearest region of our galaxy. Only for the nearest galaxies can we detect *primary distance indicators:* Cepheid variable stars, for which we derive the distance directly by comparing absolute magnitude (from the period-luminosity relation) with the observed apparent magnitude. RR Lyrae stars are too faint for general use; only in 1985 were any detected in the Andromeda Galaxy, which is relatively nearby. The Hubble Space Telescope should show them farther out.

Beyond those nearest galaxies, we use *secondary distance indicators* such as supergiant stars or H II regions, assuming that the magnitudes of supergiants or sizes of H II regions are more or less the same as they are in our nearby galaxies. (The indicators are secondary because they are calibrated by the primary distance indicators.) We then calculate for a supergiant star its spectroscopic parallax; for an H II region we calculate the distance at which its observed angular size would correspond with the linear dimensions we know for those objects in these nearby galaxies. It has recently been shown that the distribution in brightness of planetary nebulae in a galaxy is an equally useful distance indicator in between 5 million and 50 million light years. Planetary nebulae are found in all types of galaxy, while Cepheids are found only in spiral galaxies.

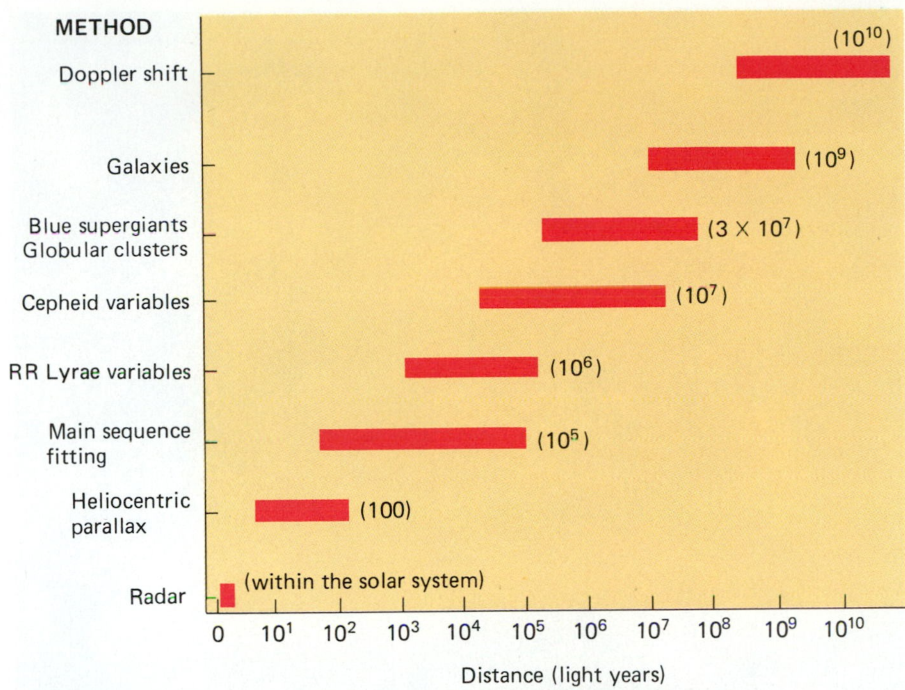

Figure 31–34 Different ways must be used to measure distances at different distances from us. Less desirable ways are used the farther from us we look.

At still greater distances, we must resort to *tertiary distance indicators*. We compare the maximum brightnesses of Type I supernovae, assuming that they all reach the same maximum brightness. Or we assume that the brightest member in a cluster of galaxies has the same absolute magnitude as all other brightest members of other clusters. Sometimes, when one or two members are exceptionally bright, we consider instead the third-brightest member of a cluster, to lessen the possibility of one odd object being the brightest. If this seems a weak way to measure distances, you are right, but it is all we have (Fig. 31–35).

The Tully-Fisher relation, a new and powerful relation discovered by Brent Tully of the University of Hawaii and J. Richard Fisher of the National Radio Astronomy Observatory, is now often used. The relation is based on a link between how rapidly galaxies rotate and how bright they are (absolute magnitude). The speed of rotation is measured from their 21-cm hydrogen emission lines. The best relation is obtained when magnitudes are measured in the infrared.

Some but not all of the difference in measured values for Hubble's constant stems from the fact that the earlier measurements were all made in the northern hemisphere, while we now have many more southern-hemisphere observations available. Our Local Group is moving toward the Virgo Cluster with an appreciable velocity, and this velocity must be taken into account when computing Hubble's constant. The effect of this velocity is different for galaxies observed in the northern and southern sides of the plane of our Milky Way Galaxy.

A surprising recent discovery is that even the Virgo Cluster is moving with respect to the average expansion of the universe. Some otherwise unseen "Great Attractor" is pulling the Local Group, the Virgo Cluster, and even the Hydra-Centaurus supercluster (Fig. 31–36) toward it. Further redshift measurements made by Alan Dressler of the Observatories of the Carnegie Institution of Washington, Sandra Faber of the Lick Observatory, David Burstein of Arizona State U., Gary Wegner of Dartmouth, and colleagues showed the location of the giant mass that must be involved. It includes tens of thousands of galaxies or their equivalent mass. Their measurements even showed galaxies on the far side of

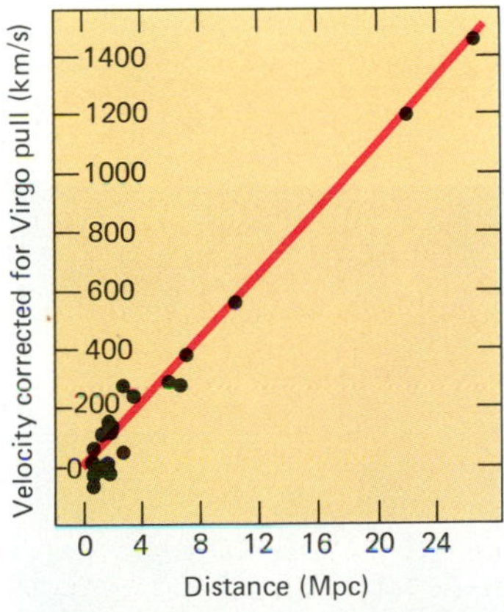

Figure 31–35 A plot by Sandage of recessional velocity, corrected for the gravitational pull of the Virgo Cluster, vs. distance (derived from the period-luminosity relation for Cepheid variables for the nearest points and from Type-I supernovae for the Virgo and Fornax clusters, the two farthest points).

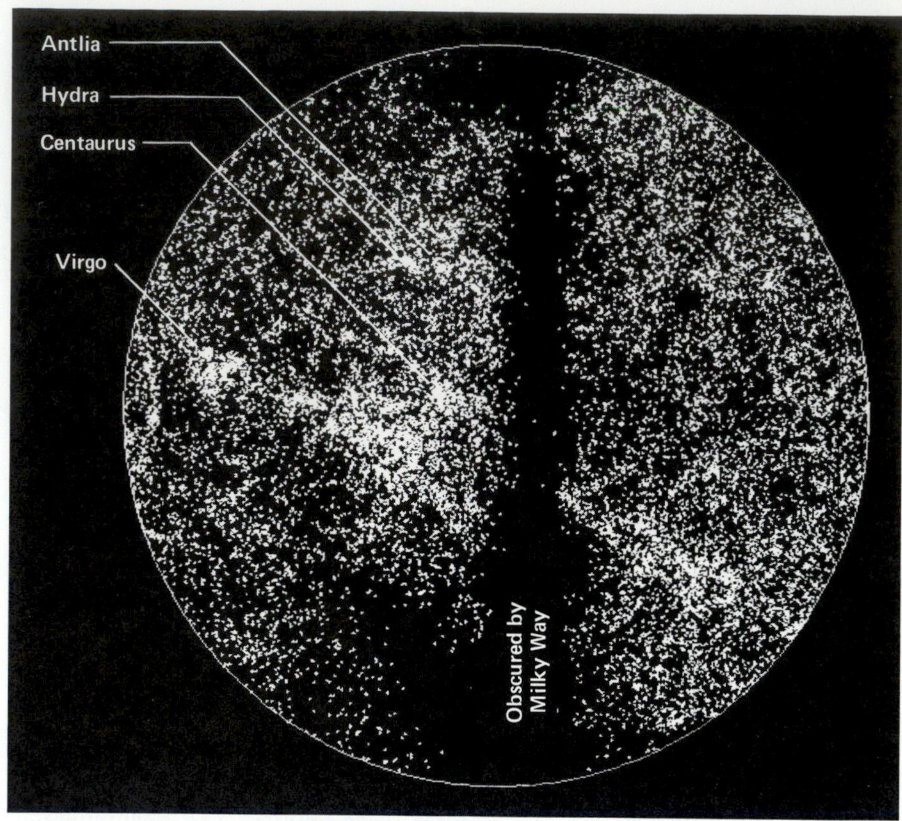

Figure 31–36 The distribution of galaxies over half the sky, centered on the direction of the average motion of nearby elliptical galaxies relative to the cosmic background radiation (Section 33.4). Since redshift measurements show that the Virgo Cluster and the Hydra-Centaurus supercluster share in the motion, there must be a Great Attractor beyond, near the center of this image, exerting tremendous gravitational force. The dark strip across the center shows where the Milky Way prevents distant galaxies from being seen; it unfortunately cuts across the Centaurus band of clusters of galaxies that includes the Great Attractor.

the Great Attractor, which is about 3 times farther from us than the Virgo Cluster, being pulled back by its gravity. The observations on the far side pinpointed the Great Attractor's position. Its gravity makes the surrounding region of the universe expand less rapidly than it otherwise would.

The fact that the Milky Way provides a "zone of avoidance," preventing astronomers from looking out into large regions of the universe, is unfortunate. Radio mapping at the 21-cm hydrogen wavelength is now locating a few galaxies even in this zone, both by examining IRAS sources and by searching blindly.

The methods we have to use grow less precise as we get farther from the sun. In particular, at the very farthest distances we are seeing galaxies that emitted their light very long ago. A galaxy may well have then had a very different brightness from the galaxies nearer to us with which we are comparing it. For example, there has been a belief that the brightest galaxies in clusters may be devouring other galaxies and so growing brighter; this process is called *galactic cannibalism* (Fig. 31–37*A*). One galaxy that is not rotating, a strong radio source, contains a rapidly rotating cloud of ionized hydrogen, something that seems easiest to explain if the cloud were the remnant of another galaxy that was "eaten." The notion that we know little about the evolution of galaxies over long periods of time remains.

Beyond a certain range, we can no longer independently measure distances, and our only method of assessing distance is applying the Hubble law with our current "best value" for Hubble's constant to the observed redshifts. What is the best value for Hubble's constant? The two camps—those deriving 50 and those deriving 100—don't agree. Each seems expert, and the methods will ultimately stand or fall on the basis of the validity of the methods used for each step of the chain. One thing that I can guarantee is that the discussion isn't over (Fig. 31–37*B*).

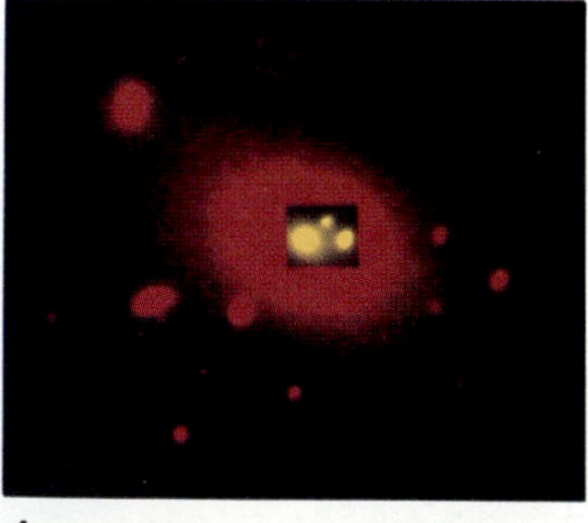

A

Figure 31–37 (*A*) Galactic cannibalism. Computer processing displays the central region of the large galaxy in a yellow box; we see the remains of two victims. The galaxy, NGC 6166, is the central galaxy in the rich cluster A2199.

Robotman

B

Figure 31–37 (*B*) You need accurate measurements of distance to map the universe. (© 1987 United Feature Syndicate, Inc.)

31.5a The Most Distant Galaxies

Several groups of astronomers have been looking for the farthest galaxies, to study how they evolve and what they tell us about the early times of our universe. Scientists at Berkeley, the Space Telescope Science Institute, the Johns Hopkins University, and the University of Hawaii have been especially active in the search. About thirty galaxies have thus far been detected with redshifts larger than 1 (Fig. 31–38). The one shown is perhaps 15 billion light years away. We are seeing the galaxy as it was 15 billion years ago. We say that the *look-back time* for this galaxy is 31 billion years. New observational techniques, like the use of CCD's, have reduced drastically the time to take the spectra necessary for this research. The current record holder was examined spectroscopically because of the steep slope of its radio spectrum, a property that is apparently linked to high luminosity.

Note that redshifts greater than 1 merely mean that the spectral lines are shifted by greater than their original wavelengths. At such great redshifts, the redshifts no longer correspond to v/c; the velocity always remains less than the speed of light. A formula that is part of Einstein's special theory of relativity must be used; we delay discussing it until Section 32.2, since high redshifts are more easily observed in the spectra of quasars than of galaxies.

A major project for the Hubble Space Telescope is to observe distant galaxies, for their study allows us to see what the universe was like as far back in time as possible.

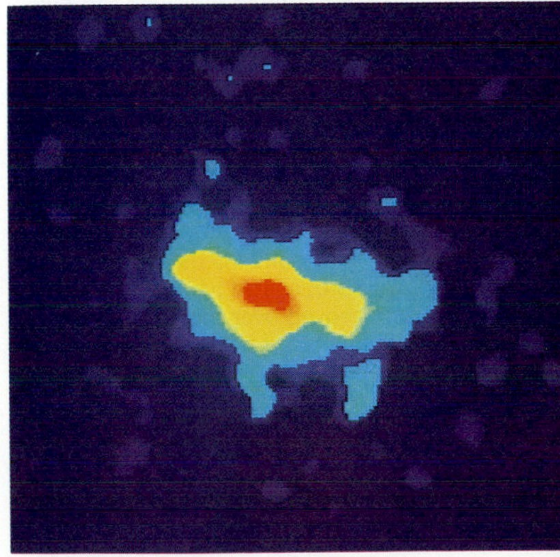

Figure 31–38 This optical counterpart of the radio source 4C41.17 is the most distant galaxy known as of the time of printing of this book. We see a false-color map of its low-contrast Lyman α brightness. Its redshift is 3.8, which means that the spectral lines that are observed (Lyman α and a 5-times-ionized carbon line originally at 1549 Å) are shifted toward the red (actually from the ultraviolet into the yellow and red) by 380 per cent of their original values. The enhanced red sensitivity of CCD's was crucial to this discovery. Radiation has travelled about 95 per cent of the age of the universe to reach us from this object, so we are seeing quite far back in time, to only a billion years after the universe began. Current theories for the formation of galaxies, which hold that the universe contains much cold dark matter that acted as "seeds" for galaxies, can accommodate only a few galaxies this old. If we find many more, new theories will be needed.

The object was found from its unusual radio spectrum, and seems to be different from current galaxies by being brighter and by having its optical emission extend along the direction of its radio structure.

A more nearly normal galaxy, with weak emission lines, has been found at a redshift of 2.4, 84 per cent of the speed of light.

This galaxy was found by George Miley, Ken Chambers, and W. van Breugel of the Space Telescope Science Institute, the Johns Hopkins University, and Lawrence Livermore National Laboratory. Others active in finding and studying the most distant galaxies include Hyron Spinrad at the University of California at Berkeley and Simon Lilley at the University of Hawaii.

31.6 Active Galaxies

Most of the objects that we detect in the radio sky turn out not to be located in our galaxy. The study of these *extragalactic radio sources* is a major subject of this section.

The core of our galaxy, the radio source we call Sagittarius A, is one of the strongest radio sources that we can observe in our galaxy. But if the Milky Way Galaxy were at the distance of other galaxies, its radio emission would be very weak.

Some galaxies emit quite a lot of radio radiation, many orders of magnitude (that is, many powers of ten) more than "normal" galaxies. We shall use the term *radio galaxy* to mean these relatively powerful radio sources. They often appear optically as peculiar giant elliptical galaxies. Radio galaxies, and galaxies that similarly radiate much more strongly in x-rays than normal galaxies, are called *active galaxies*. IRAS detected many of these sources in the infrared, indicating the presence of widespread nuclear dust.

Many of these galaxies have nuclei that are exceptionally bright. These sources are known as *AGN*'s (from *Active Galactic Nuclei*); the whole source and not only the nucleus is often called an AGN. The most plausible and widely accepted source of energy for an AGN is matter falling onto an accretion disk around a massive black hole. In the next chapter, we shall discuss extreme examples of AGN's.

The first distant radio galaxy to be detected, Cygnus A (Fig. 31–39), radiates about a million times more energy in the radio region of the spectrum than does the Milky Way Galaxy. Cygnus A and dozens of other radio galaxies emit radio radiation mostly from two zones, called *lobes,* located far to either side of the optical object. Such *double-lobed structure* is typical of many radio galaxies. New high-resolution radio maps, reported in 1986, indicate that Cygnus A's lobes are only 3 million years old, compared with perhaps 10 billion years old for stars in the central object. No signs of prior ejection of lobes show, so perhaps lobes are one-time, short-lived phenomena.

The optical object that corresponds to Cygnus A—a fuzzy, divided blob or perhaps two fuzzy blobs—has been the subject of much analysis, but its makeup is not yet understood. Perhaps we see a single object partly obscured by dust.

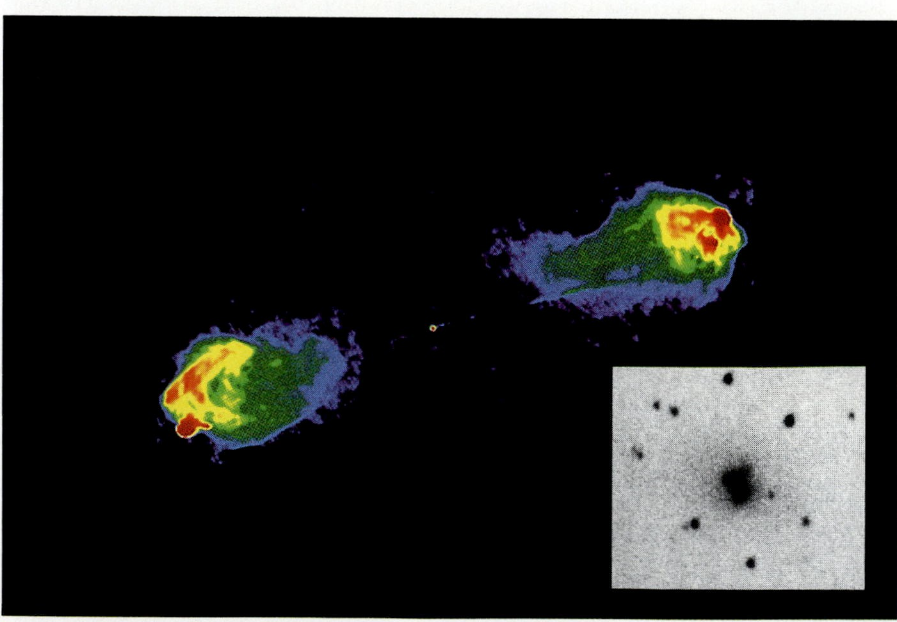

Figure 31–39 A radio map of Cygnus A, with shading and contours indicating the intensity of the radio emission. An electronic image of the faint optical object or objects observable is superimposed at the proper scale. (*Inset*) The especially high resolution of this optical view reveals a third, less prominent feature between the two main structures. This central region contains the central radio source. The resolution was improved over past images by using a computer-controlled monitor of the seeing and telescope guiding to form an image with ⅔-arc-sec resolution on a CCD at Mauna Kea. The overall structure again makes the source look like galaxies in collision.

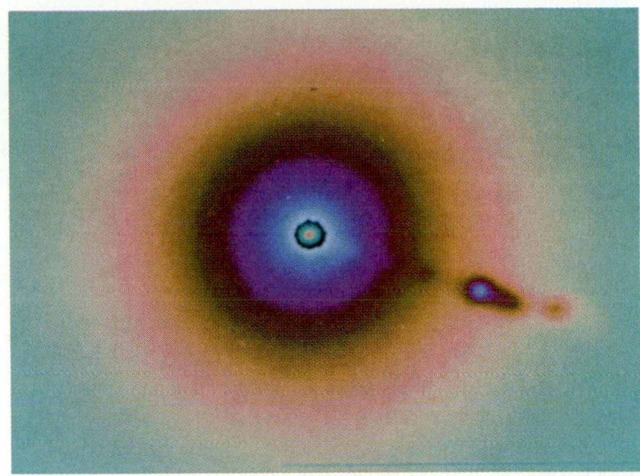

A

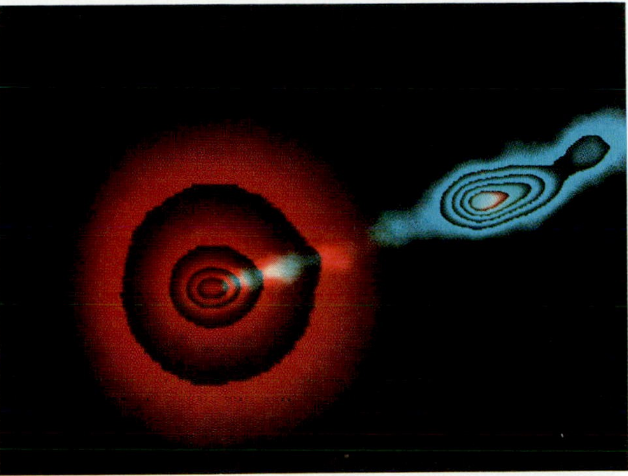

B

Figure 31–40 (*A*) The galaxy M87, which corresponds to the radio source Virgo A. The exposure here is shorter than the one in Figure 31–1 and shows the jet. (*B*) Here the electrons in the jet are shown in blue to contrast with the red display of the reddish stars in the galaxy itself, separated out by their symmetry.

Often the optical images that correspond to radio sources show peculiarities. For example, on short exposures of M87 (Fig. 31–40*A*), which corresponds to the powerful radio galaxy Virgo A, we see (optically) a jet of gas. Light from the jet is polarized, which confirms that the synchrotron process is at work here (Fig. 31–40*B*). High-resolution radio observations have shown that a very small source is present at M87's center. Studies of matter circling this small nucleus show that 5 billion solar masses are present in this small volume, presumably indicating a black hole. (See Box 31.1.)

The radio source Centaurus A has an optical counterpart (Fig. 31–41) that shows unusual wrapping by dust. In recent years, both x-ray (Fig. 31–42) and optical jets, aligned with the radio lobes (Fig. 31–43), have been discovered. Perhaps a radio jet would be discovered as well if a high-resolution, high-sensitivity telescope were available in the southern hemisphere, where it could properly observe this southern source. We thus look forward to the observations with the array to be known as the Australian telescope.

As our observational abilities in radio astronomy have increased, especially with the techniques described in the next section, lobes and jets aligned

Figure 31–41 Centaurus A, NGC 5128, looks like an elliptical or S0 galaxy surrounded by an extensive dust lane. A supernova in its dust lane appears green on this image because the red plate used to make the color composite was taken before it erupted.

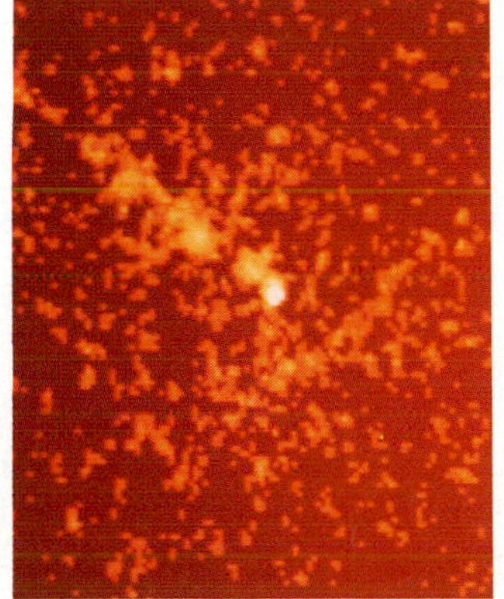

Figure 31–42 An x-ray view of the jet in Centaurus A. The field is 400 arc sec wide.

Figure 31–43 The radio lobes of Centaurus A superimposed on the optical image.

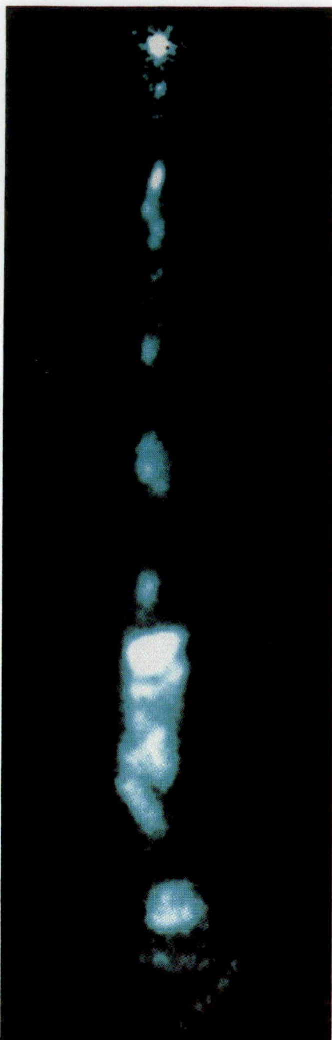

Figure 31–44 The nucleus (bright point at top) and jet of M87, in a computer-processed view from the Faint Object Camera aboard the Hubble Space Telescope. In the jet, which extends 5000 light-years, the image reveals detail as small as 10 light-years across.

There may be a 5-billion-solar-mass black hole at M87's nucleus. Magnetic fields are generated within a spinning accretion disk around the black hole. These magnetic fields spiral around the jet and confine it to a long, narrow tube. High-speed electrons and protons are accelerated near the black hole and race along the tube at almost the speed of light. The bright knot midway along the jet is where the jet becomes more chaotic. Near the bottom of the image, the jet runs into a wall of gas it has brought along ahead of itself.

Focus On

Box 31.1 The Peculiar Galaxy M87

The radio galaxy M87 is a giant elliptical galaxy, one of the brighter members of the Virgo Cluster. The galaxy turns out to have an odd optical appearance on short exposures, for a jet of gas can be seen. The galaxy corresponds to the powerful radio galaxy Virgo A. A radio jet 2 kiloparsecs long has been discovered.

High-resolution radio observations have shown that the nucleus of M87 is only 0.01 arc sec across. Even at its distance of 50 million light years, this makes it a very small source in which to generate so much energy, which is emitted across the spectrum from x-rays to radio waves. Indeed, this galaxy gives off much more energy than do other galaxies of its type; the central region is as bright as 10^8 suns.

Optical studies of the motion of matter circling M87's nucleus have allowed astronomers to estimate the mass of the nucleus. The stars are moving so fast that a huge mass must be present to hold them in. It turns out that 5 billion solar masses of matter must be there. Other optical studies disclosed the presence of an extremely bright point of light in the center of the galaxy. Astronomers continue to gather spectral and spatial information to determine if the source is a giant black hole or is matter under more conventional conditions. Higher-resolution observations from the Hubble Space Telescope should tell us more about this exotic source.

with them have become commonly known. The best current model is that a giant rotating black hole in the center of a radio galaxy is accreting matter. Twin jets carrying matter at a high velocity are given off almost continuously along the poles of rotation. These jets carry energy into the lobes. (We may see only one, depending on the alignment and Doppler shifts.) Calculations show that a not-very-hungry giant black hole would provide the right amount of energy to keep the lobes shining. Optical observations of several galaxies seem to show a sharp concentration of brightness at their centers. A black hole there containing millions of times the mass of the sun would explain this effect, though the black hole itself would still be much too small to detect directly.

*31.7 Radio Interferometry

31.7a Radio Interferometers

The resolution of single radio telescopes is very low, because of the long wavelength of radio radiation. Single radio telescopes may be able to resolve structure only a few minutes of arc or even a degree or so across. The techniques of interferometry are now used in radio astronomy to detect fine detail. Arrays of radio telescopes can now map the sky with resolutions far higher than the 1 arc sec or so that we can get with optical telescopes. Let us first describe how these radio interferometers work, and then discuss some of the high-resolution results.

The resolution of a single-dish radio telescope at a given frequency depends on the diameter of the telescope. (A single reflecting surface of a radio telescope is known as a "dish.") If we could somehow retain only the outer zone of the dish (Fig. 31–45*A*), the resolution would remain the same. (The

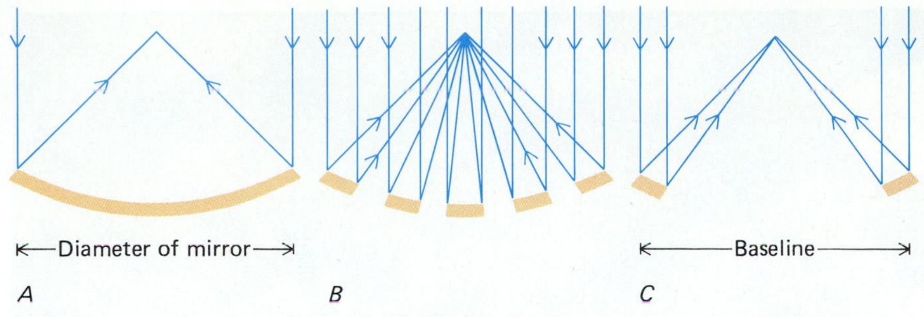

Figure 31–45A A large single mirror (*A*) can be thought of as a set of smaller mirrors (*B*). Since the resolution for radiation of a certain wavelength depends only on the telescope's aperture, retaining only the outermost segments (*C*) matches the resolution of a full-aperture mirror. We can use a property of radiation called *interference* to analyze the incoming radiation. The device is then called an *interferometer.*

collecting area would be decreased, though, so we would have to collect the signal for a longer time to get the same energy.)

Let us picture radiation from a distant source as coming in wavefronts, with the peaks of the waves in step (Fig. 31–45*B*); we say that the waves are "coherent." If we can maintain our knowledge of the relative arrival times of the wavefront at each of two dishes, we can retain the same resolution as though we had one large dish of the spacing of the two small dishes shown. For a single dish, the maximum spacing of the two most distant points from which we can detect radiation is the "diameter." For a two-dish interferometer, we call it the *baseline*.

We study the signals by adding together the signals from the two dishes; we say the signals "interfere," hence the device is an "interferometer." Since the delay in arrival time of a wavefront at the two dishes depends on the angular position of an object in the sky with respect to the baseline, by studying the time delay one can figure out angular information about the object.

If the source were made of two points close together, the wavefronts from the two sources would be at slight angles to each other, and the time interval between the source reaching the two dishes would be slightly different. Thus an interferometer can tell if an object is double, even it is is unresolved by each of the dishes used alone.

31.7b Aperture-Synthesis Techniques

By suitably arranging a set of radio telescopes across a landscape, one can simultaneously make measurements over a variety of baselines, because each pair of telescopes in the set has a different baseline from each other pair (Fig. 31–46) and because various pairs are aligned in different directions. With such an arrangement one can more rapidly map a radio source than one can with two-dish interferometers. Also, using several dishes instead of just two gives that much more collecting area.

Figure 31–45B In the left half of the figure, a given wave peak reaches both dishes simultaneously, so the amplitudes (heights) of the waves add. In the right half, the wave peak reaches one dish while a minimum of the wave reaches the other; the amplitudes subtract and zero total intensity results. Thus "interference" results.

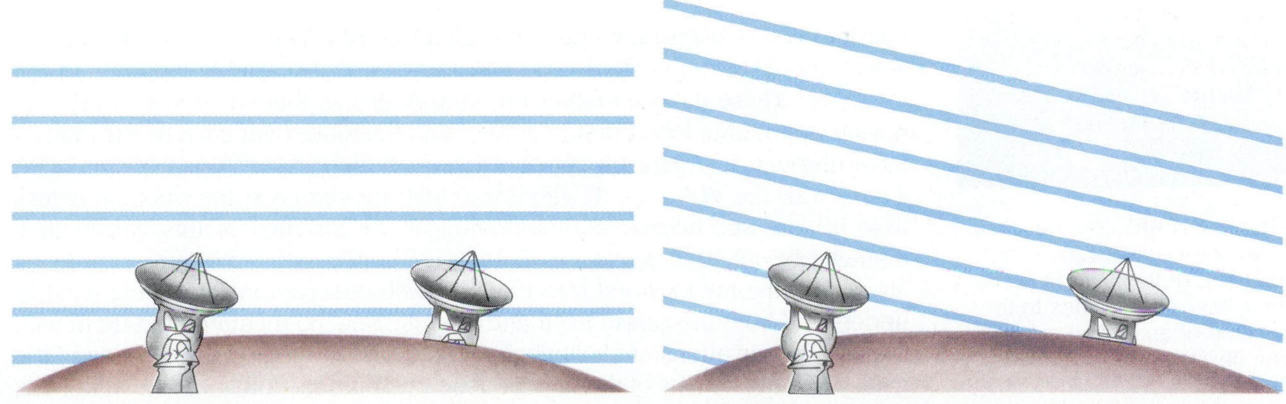

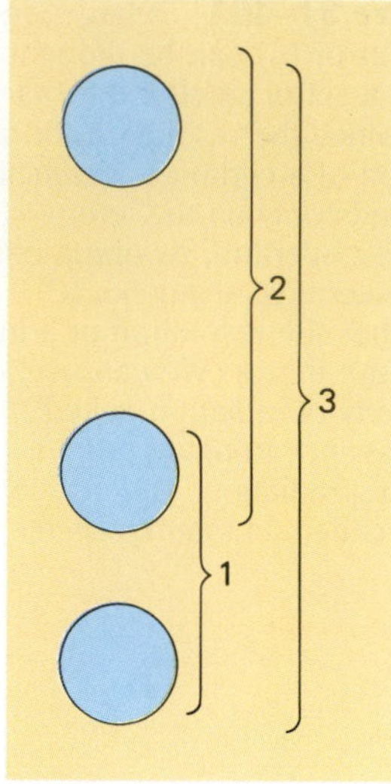

Figure 31–46 The lines joining the several dishes making up an interferometer represent pairs of dishes that give results equivalent to having several baselines simultaneously. With the three dishes shown, one can simultaneously make measurements with three different baselines.

Figure 31–47 The VLA, near Socorro, New Mexico, with its dishes in its most compact configuration.

This interferometric technique is known as *aperture synthesis;* it was used to make the radio image of Cygnus A earlier in this chapter. The technique won a share of the Nobel Prize for Sir Martin Ryle, whose array at Cambridge, England, was the first of the modern configurations. Another array long used to make aperture-synthesis observations is at Westerbork in the Netherlands.

The most fantastic aperture-synthesis radio telescope has been constructed in New Mexico by the National Radio Astronomy Observatory. It is composed of 27 dishes, each 26 m in diameter, arranged in the shape of a "Y" over a flat area that can be as much as 27 km in diameter (Fig. 31–47). Sometimes, lower (though still high) resolution over a larger region of sky is needed; then the Y spreads out only over 1 km. The "Y" is delineated by railroad tracks, on which the telescopes can be transported to 72 possible observing sites; after an incident in which a storm came up on the site while a telescope was in motion, current rules allow moving telescopes only during the daytime and when they can be monitored visually. The control room at the center of the "Y" contains powerful computers to analyze the signals. The system is prosaically called the ***V**ery **L**arge **A**rray (VLA)*. It can operate at several wavelengths between 1.3 cm and 92 cm.

The VLA can make pictures of a field of view a few minutes of arc across, with resolutions comparable to the 1 arc sec of optical observations from large telescopes, in about 10 hours.

31.7c Aperture-Synthesis Observations of Galaxies

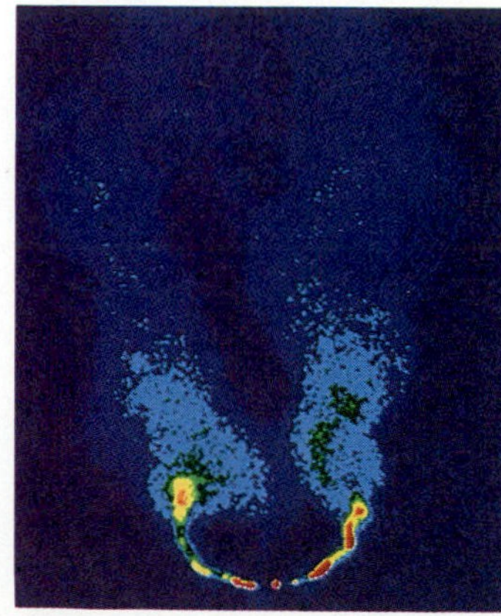

Figure 31–48 The head-tail radio source NGC 1265. The front end of the head of the radio galaxy corresponds to the position of an optical galaxy. The observations were made by the VLA.

Interferometer observations have revealed the existence of a class of galaxies with "tails." They are called *head-tail galaxies,* and resemble tadpoles in appearance. These galaxies expel the clouds of gas that we see as tails. The objects are double-lobed radio sources with the lobes bent back as the objects move through intergalactic space. High-resolution observations of one such galaxy with the VLA (Fig. 31–48) show that the source at the nucleus is less than 0.1 arc sec across, corresponding at the distance of this galaxy to a diameter of only 0.01 parsec. A narrow, continuous stream of emission leads away from the nucleus and into the tail. Such observations are being used to understand both the galaxy itself and the intergalactic medium, and tie in with the x-ray observations of clusters of galaxies. Depending on the velocity of the galaxy and the density of the intergalactic medium, the lobes can be bent back

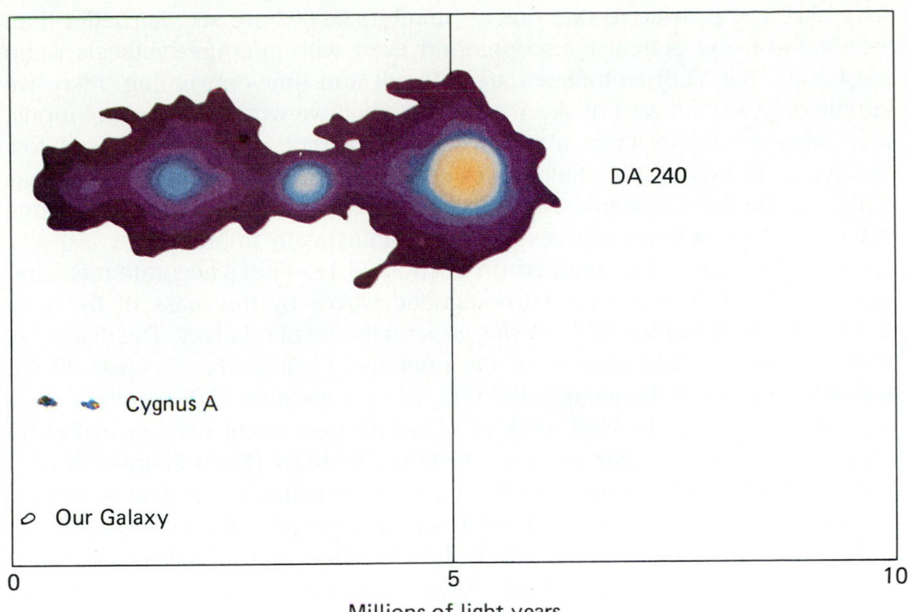

Figure 31–49 A comparison of the sizes of giant double-lobed radio galaxies with our own galaxy. (*A*) DA 240, observed with the Westerbork Synthesis Telescope. (*B*) Cygnus A, observed with the VLA, to the same scale.

by different amounts, so there is a range from lobes opposite each other to lobes slightly bent back to lobes bent back enough to make head-tail galaxies. Thus it makes sense that most head-tail galaxies are found in rich clusters of galaxies.

The discrete blobs that we can see in the tails indicate that the galaxies give off puffs of ionized gas every few million years as they chug through intergalactic space. Perhaps by studying these puffs, we can learn about the main galaxies themselves as they were at earlier stages in their lives. Head-tail galaxies seem to be a common although hitherto unknown type.

Aperture-synthesis observations show many giant double radio sources, much larger than any of the double-lobed sources previously known. Some are hundreds of times larger than our own galaxy (Fig. 31–49). These are the largest single objects currently known in the universe.

31.7d Very-Long-Baseline Interferometry

The first radio interferometers were separated by only hundreds of meters. The signals were sent over wires to a central collecting location, where the signals were combined. Telescopes were freed from the tyranny of wires with the invention of atomic clocks, which drift only three hundred-billionths of a second in a year. A time signal from an atomic clock can be recorded by a tape recorder on one channel of the tape, while the celestial radio signal is recorded on an adjacent tape channel. Since a wide band of frequencies must be recorded, videotape recorders are used, and their development was another necessary event. The radio signal recorded can be compared at any later time with the signal from the other dish, synchronized accurately through comparison of the clock signals.

No longer did dishes have to be near each other to make interference measurements. Now all that is necessary is that the two telescopes observe the same object at the same period of time; the signals can be compared in a computer weeks later. With this ability, astronomers can make up an interferometer of two or more dishes very far apart, even thousands of kilometers. This technique is called ***V**ery-**L**ong-**B**aseline **I**nterferometry (VLBI)*.

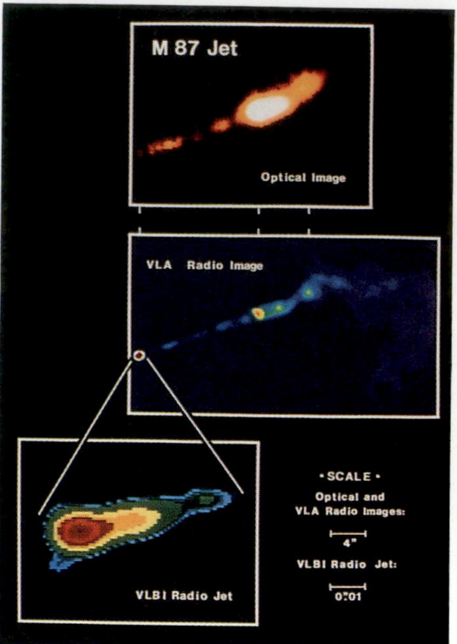

Figure 31–50 VLBI techniques give greatly improved resolution even over a VLA image. Here we see M87, the Virgo A radio source. VLBI has revealed a small jet aligned with the longer jet more familiar in the radio and optical.

VLBI can provide resolutions as small as 0.0002 arc sec, far better than resolutions using optical telescopes and even with aperture-synthesis radio telescopes. But VLBI techniques are difficult and time-consuming. Also, we sample only a small area of sky at any time when we work at high resolutions, so it takes longer to study a region at high resolution than it does at low resolution. Therefore, VLBI techniques can be applied only to very small areas of sky. But for those few areas, chosen for their special interest, our knowledge of the structure of radio sources has been fantastically improved.

We have already mentioned (in Section 23.11) VLBI's accurate measurement for the deflection of electromagnetic waves by the mass of the sun, providing a confirmation of Einstein's general theory of relativity. The discovery of the extremely small sources at the nucleus of our galaxy (Section 29.3b) and at the center of the galaxy M87 (Fig. 31–50) are also VLBI results.

The next stage in VLBI work is to set up permanent networks of radio telescopes devoted to this purpose. Britain's MERLIN (**M**ulti-**E**lement **R**adio-**L**inked **I**nterferometer **N**etwork) with a 64-km—maximum baseline is already in operation. Construction of a larger American network—the **V**ery-**L**ong-**B**aseline **A**rray (VLBA)—has begun, with its headquarters at the Socorro, New Mexico, headquarters of the Very Large Array. The VLBA is to consist of 10 telescopes, each 25 m (about 80 ft) across, spread from Puerto Rico north to Massachusetts and west to Hawaii (Fig. 31–51). Its effective diameter will be 8000 km. Other VLBI plans are under way in the United States, in Canada, in Australia, and elsewhere. At a later date will come the addition of an antenna in space, increasing the baseline 4 times farther. An orbiting satellite has already been used in a test to prove that interference fringes can be obtained with it.

31.7e Observing with the VLA

The Very Large Array of the National Radio Astronomy Observatory is an impressive instrument. Its 27 antennas march across the landscape, and it is hard to say whether it is more impressive when the telescopes are in their most compact status—with all the antennas within 1 km of each other (Fig. 31–52)—or when they are most spread out, with antennas disappearing over the horizon.

Observing time at the VLA is hard to get. Scientists work up detailed proposals, and apply many months in advance. A committee at the VLA assesses all the proposals and gives out as much time as they can for the following quarter. To receive 15 hours of observing would be quite a lot.

My own project deals with the study of deuterium, a form of hydrogen, and what it tells us about the origin and future of the universe. In Chapter 34, I will discuss its significance for the study of cosmology. Here, let me describe what it was like to work at the VLA. I had been waiting years for a more capable telescope at the frequency I needed for the project. Over the last few years, receivers at this 327-MHz frequency have been installed one by one at the VLA. (It is beyond the FM radio band, which covers frequencies 88 to 108 MHz.) By 1987, 16 such receivers had been installed, and I joined Donald Lubowich of Hofstra University and the American Institute of Physics and K. Anantharamaiah of the VLA staff and India's Raman Research Institute to make deuterium observations.

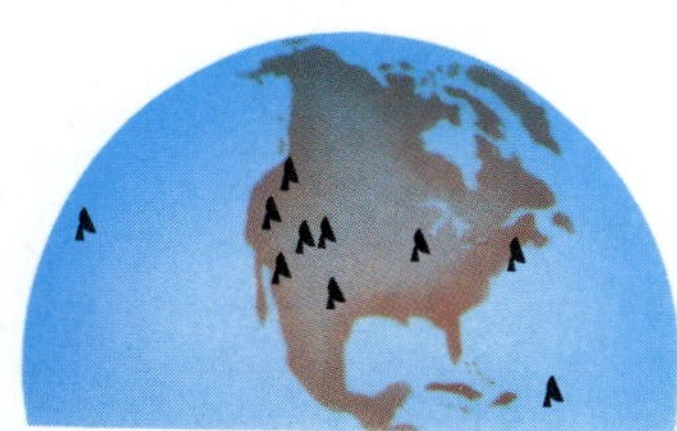

Figure 31–51 The prospective arrangements of telescopes in the VLBA—the Very Long Baseline Array.

We were searching for deuterium in gas that lies between us on earth and the bright source that is at the center of the Milky Way Galaxy. Deuterium, if it were present in measurable quantities, would make an absorption line. The galactic center is marked by the radio source Sagittarius A (Sgr A for short, usually pronounced Saj A). In past observations, we knew only whether or not deuterium appeared in the field of view of our telescope, which was 1½° across, 3 times the angular diameter of the moon in the sky. The VLA would give us

Figure 31–52 The smallest of the four possible configurations in which the VLA is arranged; here the telescopes are all within a 1-km circle. This arrangement provides lower resolution but over a larger field of view than the most spread out configuration.

individual readings with a much smaller scale. We could know about each 2 arc minutes, less than 1⁄10 the angular diameter of the moon.

Finally, the time for our observing came. At the VLA (Fig. 31–53), we checked into our rooms and looked over the schedule. It was mid-afternoon, and we were to observe that night from midnight to 8 a.m. The galactic center was to be in view for 5½ hours out of that time. With Anantha (as Anantharamaiah is known), we sat at a computer terminal. It was clear that we would spend most of our time pointing our telescope at Sgr A, but we had several other tasks to do as well. We had to spend some time looking at sources whose strengths were known and another source that was known to have a smooth spectrum, so that we could assess the details of what our instruments saw. Anantha used his terminal to set up the observing file, which would tell the telescopes where to point and what data to record minute by minute for the long night. We were to start with one of our calibration sources, and then start on the galactic center when it rose. Every 15 minutes we would briefly observe another of our calibration sources. When the galactic center set, we would do more calibrations.

Figure 31–53 The VLA in its most compact configuration.

After dinner, we checked over our observing file. Midnight approached, the time I had been waiting years for. I was eager to have the observations. The telescope operator started on our observing file, and the 27 telescopes swung in unison to our first source. And what did we do next? The three scientists went to bed!

To bed? We had only a little choice. The telescope is so complex, and the types of data that come so involved, that only the most preliminary monitoring can take place while the observing run is going. The rest must be done by computer study, and we had the computer reserved for the morning. If we were to be fresh and awake for the computer work, we had to get some sleep.

I woke up at 5 a.m., and went into the control room. A computer screen blinked updates on the data taking, but there was nothing for me to do there. Everything seemed to be working. I walked out among the telescopes at dawn. They were silhouetted pristinely in the clear sky.

It took a couple of days of computer work (Fig. 31–54) to see what we had. The first stage of our data reduction was on the "pipeline" computer, which was optimized for handling the raw VLA data. Each pair of telescopes had to be treated separately. With two telescopes, you have one pair. With three telescopes you have 3 pairs: 1-2, 1-3, and 2-3. By the time you get to sixteen telescopes you have hundreds of pairs, and the processing takes a long

Figure 31–54 Extensive computer work is necessary at the VLA to put your data in usable order.

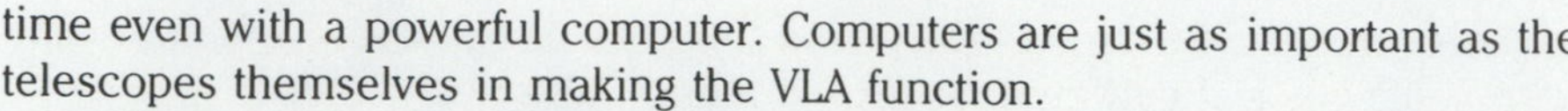

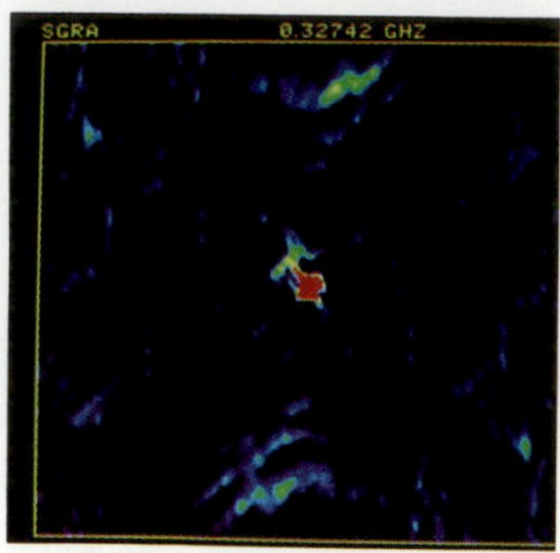

Figure 31–55 The sum of all the maps in our cube showed the arc near the galactic center.

time even with a powerful computer. Computers are just as important as the telescopes themselves in making the VLA function.

After hours of work, we came up with a series of 64 maps, each map showing the strength of the signal in two dimensions. Together, the set is known as a "cube," and we could cut the cube with the computer in any dimension. For example, we could make a display of frequency versus position. Or we could sum point for point (Fig. 31–55).

What did our data mean? We sat and talked about it. We knew that we wouldn't solve the deuterium problem in this one preliminary observing run, but our observing had been a success. We knew that we could do better with more observing time next year. Even this single 5-hour run on the source Sgr A had told us something. We plotted graphs that showed the spectrum. If deuterium formed in small regions, the spectrum would show one or more dips. No dips were seen, so we could place a limit on local formation of deuterium. It is this limit we would like to lower in later years.

If deuterium is not formed locally, then whatever is found was formed cosmically, in the first thousand seconds after the big bang. Our new data fit in with this picture, and we hope to pin it down further. We want to continue to use the VLA in its most compact configuration. It returns to this configuration every 15 months for a period of 3 months or so. That way, the different configurations drift through the seasons of the year (Fig 31–56). We are working on new proposals, delivered a report at a meeting of the American Astronomical Society, and published a scientific paper about our results in the *Astrophysical Journal*. Then, the sunspot number was too high to use the VLA in its compact configuration at the long radio wavelength we need. Now that the sunspot cycle is sufficiently low, we are applying to be back at the VLA.

In the meantime, Lubowich, Peter Schloerb of the University of Massachusetts at Amherst, and I have used the Five College Radio Astronomy Observatory in Massachusetts to study molecules in which a deuterium (D) replaces an ordinary hydrogen (H). We observed DCN and DCO^+ in gas clouds near the center of our galaxy. Lubowich, Stuart Vogel of the University of Maryland, and I are also making observations with the Hat Creek millimeter interferometer. We intend to map the region of the Galactic Center at the DCN and DCO^+ wavelengths.

31.8 The Origin of Galaxies

From the time that it was realized that galaxies were separate units in space, it has generally been believed that galaxies originated through some sort of *gravitational instability*. In this theory, a fluctuation in density either developed or pre-existed in the gas from which the galaxy was to form. This fluctuation grew in mass, collapsed (controlled by the force of gravity), and then cooled until the galaxy was formed.

Figure 31–56 The end of an observing run at the VLA.

Current theories of galaxy formation have to take into account the fact that the universe is expanding, which we discussed in this chapter, and the fact that the universe is filled with a glow called the background radiation, which we will discuss in Chapter 33. This background radiation was much stronger in the era when galaxies were formed than it is now. Also, we understand little about the magnetic fields of galaxies, how they formed, and the role they have played.

To understand the giant voids discovered observationally, theoreticians must assume that galaxies form only in rare, specially high peaks in density. Where galaxies don't quite form we should see relative voids, though the dwarf galaxies we would expect haven't been found. In this case, voids aren't empty of matter; it is just that the matter in them doesn't shine.

In the most widely accepted current theory, the dark matter is composed of exotic types of subatomic particles, none of which has yet been discovered. This dark matter is termed "cold" since it has (by the theory) little or no velocity to and fro with respect to the average expansion of the universe that follows Hubble's law. (Neutrinos, on the other hand, are considered *hot dark matter,* since they would have a large variation of velocity in the apparently unlikely case that they have mass.) Since the *cold dark matter* does not interact well with photons, it could have started collapsing under gravity and forming large structures at early times in the universe when the matter we can see was still being spread out evenly by collisions with photons. Thus cold dark matter has been invoked to explain the formation of galaxies. The theory, though, does not explain the most massive clusters of galaxies, and seems far from explaining huge structures like the Great Wall. Further, the smoothness of the universe early on (which we will discuss in Section 33.4d) as revealed by a new spacecraft has made it even more difficult to understand how the universe could have gone quickly enough from an extremely smooth early state to the large-scale clumpy state we know it assumed within a billion years or so. The cold dark matter theory is being hotly debated.

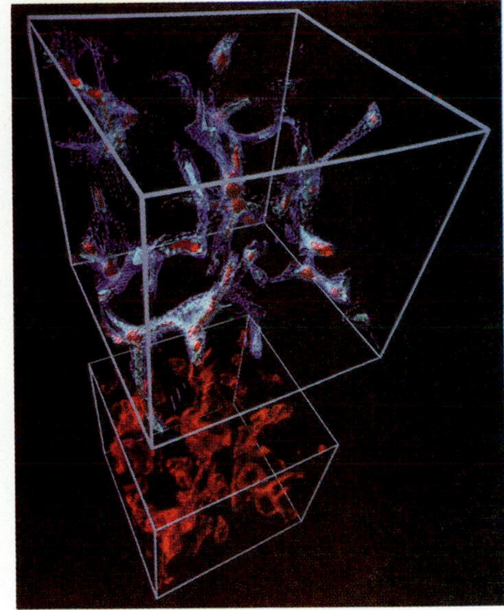

Figure 31–57 These supercomputer calculations of the formation of clusters of galaxies by Joan Centrella show the existence of filaments and voids.

There are still many unanswered questions with the theory that galaxies arose from gravitational instabilities. Theoreticians can't even agree whether individual galaxies formed first and later combined into clusters, or whether clusters of galaxies formed first, with individual galaxies condensing out of the larger bodies. X-ray observations that show subclustering in clusters of galaxies are being interpreted as clusters in formation, and so endorse the former possibility.

The latter possibility, that galaxies condense out of larger conglomerations, follows from ideas of Yakov Zeldovich of the Institute of Applied Mathematics in Moscow. He showed that early in the universe flat sheets of matter could form, each taking approximately the shape of a pancake. In this *pancake model,* the pancake may have fragmented later on into clusters of galaxies and, in turn, into galaxies. In between the pancakes and filaments of galaxies are giant *voids* (Fig. 31–57), making a void the opposite of a supercluster. There is now observational evidence for such giant voids; new techniques are being used to measure redshifts in order to plot three-dimensional distributions of galaxies, as we saw in Section 31.3b. Hot dark matter would explain this. More data and more analysis are needed. Faster supercomputers help in the understanding (Fig. 31–58).

Large flat sheets of matter have been observed and may be the observational discovery of the pancakes.

As an alternative to the gravitational-instability theory, one idea gaining currency is that shock waves are important for starting the matter collapsing. Jeremiah Ostriker of Princeton has suggested that extreme shock waves from massive supernovae started the collapse. He proposes that the core of a galaxy is formed in a rapid chain reaction of very massive supernovae. The chain

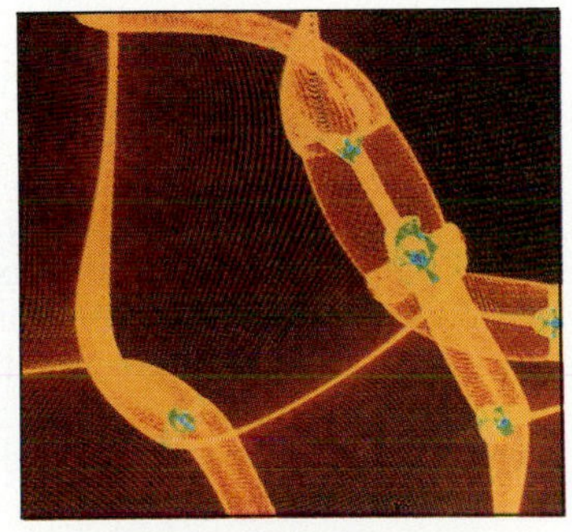

Figure 31–58 The theoretical model of supercluster formation by Adrian L. Melott of the University of Kansas and Sergei Shandaran of the Institute of Physical Problems in Moscow easily generates filaments and so is in reasonable agreement with the observations. The calculation uses a supercomputer to follow trajectories for each of 512×512 particles on a grid. Color coding shows the resulting density of galaxies; darker regions could correspond to filaments of galaxies and the lighter coding could be the voids. The results agree on a large scale with the predictions of Zeldovich's pancake model, but the small-scale structure such as the voids inside the filaments had not been predicted. Large-scale and small-scale features are sometimes aligned, and the distribution never appears uniform even on the largest scale.

reaction would take place in only a million years or so, a very brief time on an astronomical scale. Where several shock waves met, the effect would be enhanced, which could account for the fact that galaxies are found in clusters.

The bubbles found in the Center for Astrophysics redshift survey reported in Section 31.3b appear to be parts of shells rather than the filaments that are predicted by pancake models. The best available model for generating the structure observed in the survey is the explosive galaxy-formation theory advanced by Ostriker and Lennox Cowie, now of the University of Hawaii. In their theory, galaxies form on the surfaces of expanding shock waves. Energetic explosions of billions of supernovae in a newly forming galaxy sweep up gas into shells; the shells then develop instabilities under the force of gravity, forming galaxies. But even with the good points of this theory, it is hard pressed to account for the largest bubbles observed. And can billions of supernovae really explode in such a short time interval?

The formation of galaxies is intimately connected with whatever else was going on in the early years of the universe, and thus is connected with the theories of cosmology we will discuss in Chapters 33 and 34. A new theory that the universe inflated very rapidly in size in one of the first fractions of a second of time implies that any pre-existing structure or turbulence would have been smoothed out. Theoreticians are now working to explain how galaxies might have formed in this "inflationary universe" (Section 33.5) but haven't succeeded yet in predicting fluctuations of the proper size. Still, there is hope. Also in Section 34.6, we will see how "cosmic string" could spur galaxy formation.

Summary and Outline

- Observations and catalogues of non-stellar objects
  - Lord Rosse's early observations of spiral forms
  - Messier's catalogue, General Catalogues by the Herschels, *New General Catalogue* (NGC) and *Index Catalogues* (IC) by Dreyer
  - Galaxies as "island universes"
  - When Hubble observed Cepheids in galaxies, he proved that galaxies were outside our own Milky Way Galaxy
- Hubble classification (Section 31.1)
  - Elliptical galaxies (E0-E7)
  - Spiral galaxies (Sa–Sc) and barred-spiral galaxies (SBa–SBc)
  - Irregular galaxies (Irr); also SBm
  - Peculiar galaxies E(pec), S(pec)
  - Amount of gas, and of star formation, increases toward Sc
  - IRAS infrared observations show star formation
- Origin of galactic structure (Section 31.2)
  - Density waves, chain reactions of supernovae, and gravitational interactions pulling out "tails"
- Clusters of galaxies (Section 31.3)
  - Local Group includes our galaxy, 2 other spirals, and about 2 dozen other galaxies
  - Rich clusters, such as Virgo and Coma, are x-ray sources, containing hot intergalactic gas
  - Local Supercluster exists
- The universe is expanding (Section 31.4)
  - Hubble's law, $v = H_0 d$, expresses how the velocity of expansion increases with increasing distance
  - Best current values for Hubble's constant are 50 and 100 km/sec/Mpc
  - The expansion is universal and has no center
- The cosmic distance scale (Section 31.5)
  - Still controversy over distances
- Active galaxies (Section 31.6)
  - Some objects are powerful sources of radiation in the radio, x-ray, and infrared spectral regions
  - Double-lobed shape is typical of radio galaxies; sometimes a peculiar optical object is present at the center
- Radio interferometry (Section 31.7)
  - Interferometers give resolution higher than optical observations
  - Aperture-synthesis arrays provide quicker maps
  - Current interferometers at Cambridge in England, Westerbork in the Netherlands, and VLA (Very Large Array) in U.S.
  - Giant radio galaxies and head-tail galaxies studied
  - VLBI techniques leading to VLBA
- Origin of galaxies (31.8)
  - Gravitational instability theory: a density fluctuation grew, collapsed, and cooled
  - Opposite possibility: clusters of galaxies formed first, and individual galaxies condensed out of them, as in the pancake model. Possibly endorsed by giant voids.

Key Words

Messier numbers, Shapley-Curtis debate, spiral galaxies, Hubble type, elliptical galaxies, giant ellipticals, dwarf ellipticals, types E0, E7, Sa, Sb, Sc, SBa, SBb, SBc, bar, Irr, peculiar galaxies, tuning-fork diagram, type S0, lenticular galaxies, clusters of galaxies, Local Group, rich cluster, poor cluster, regular cluster, irregular cluster, supercluster, Local Supercluster, Hubble's law, Hubble's constant, primary (secondary, tertiary) distance indicators, galactic cannibalism, look-back time, extragalactic radio sources, radio galaxy, active galaxy, AGN's (Active Galactic Nuclei) lobes, double-lobed structure, baseline*, aperture synthesis*, Very Large Array*, VLA*, head-tail galaxies*, very-long-baseline interferometry*, VLBI*, gravitational instability, pancake model, voids, hot dark matter, cold dark matter, cosmic turbulence

*This term is found in an optional section.

Questions

1. What shape do most galaxies have?

2 Since we see only a two-dimensional outline of an elliptical galaxy's shape, what relation does this outline have to the galaxy's actual three-dimensional shape?

3. Sketch and compare the shapes of types Sb and SBb.

4. Draw side views of types E3, S0, Sa, Sb, and Sc, showing the extent of any nuclear bulge.

5. The sense of rotation of galaxies is determined spectroscopically. How might this be done?

6. Which classes of galaxies are the most likely to have new stars forming? What evidence supports this?

7. Describe 3 pieces of evidence that galaxies collide.

8. How is it possible that galaxies could exist close to our own yet not have been discovered before?

9. To measure the Hubble constant, you must have a means (other than the redshift) to determine the distances to galaxies. What are three methods that are used?

10. Discuss IRAS observations of galaxies.

11. Discuss the distribution of x-ray emission from rich clusters of galaxies, and why different distributions may exist.

12. Does α Centauri, the nearest set of stars to us, show a redshift that follows Hubble's law? Explain.

†This question requires a numerical solution.
††(For H_0 = 50 km/sec/Mpc) 13. 46 km/sec 14. (a) 4 sec; (b) 20 Mpc.

††13. At what velocity is a galaxy 3 million light years from us receding?

††14. A galaxy is receding from us at a velocity of 1000 km/sec. (a) If you could travel at this rate, how long would it take you to travel from New York to California? (b) How far away is the galaxy from us?

†15. (a) At what velocity in km/sec is a galaxy 100,000 parsecs away from us receding? (b) Express this velocity in km/sec, miles/sec, and miles/hr.

†16. At what velocity in km/sec is a galaxy 1 million light years away receding from us?

†17. How far away (in Mpc) is a galaxy with a redshift of 0.2? In km?

18. Compare the radio and x-ray emission of a radio galaxy with that of the Milky Way Galaxy.

19. What comment can generally be made about the optical appearance of active galaxies?

20. Contrast the VLA and the VLBA.

21. Why does interferometry allow you to get finer detail than does a single-dish telescope?

22. Briefly list 4 specific VLBI results, found by looking through the pictures, text, and index of this book.

23. Briefly list 4 specific VLA results, found by looking through the pictures, text, and index of this book.

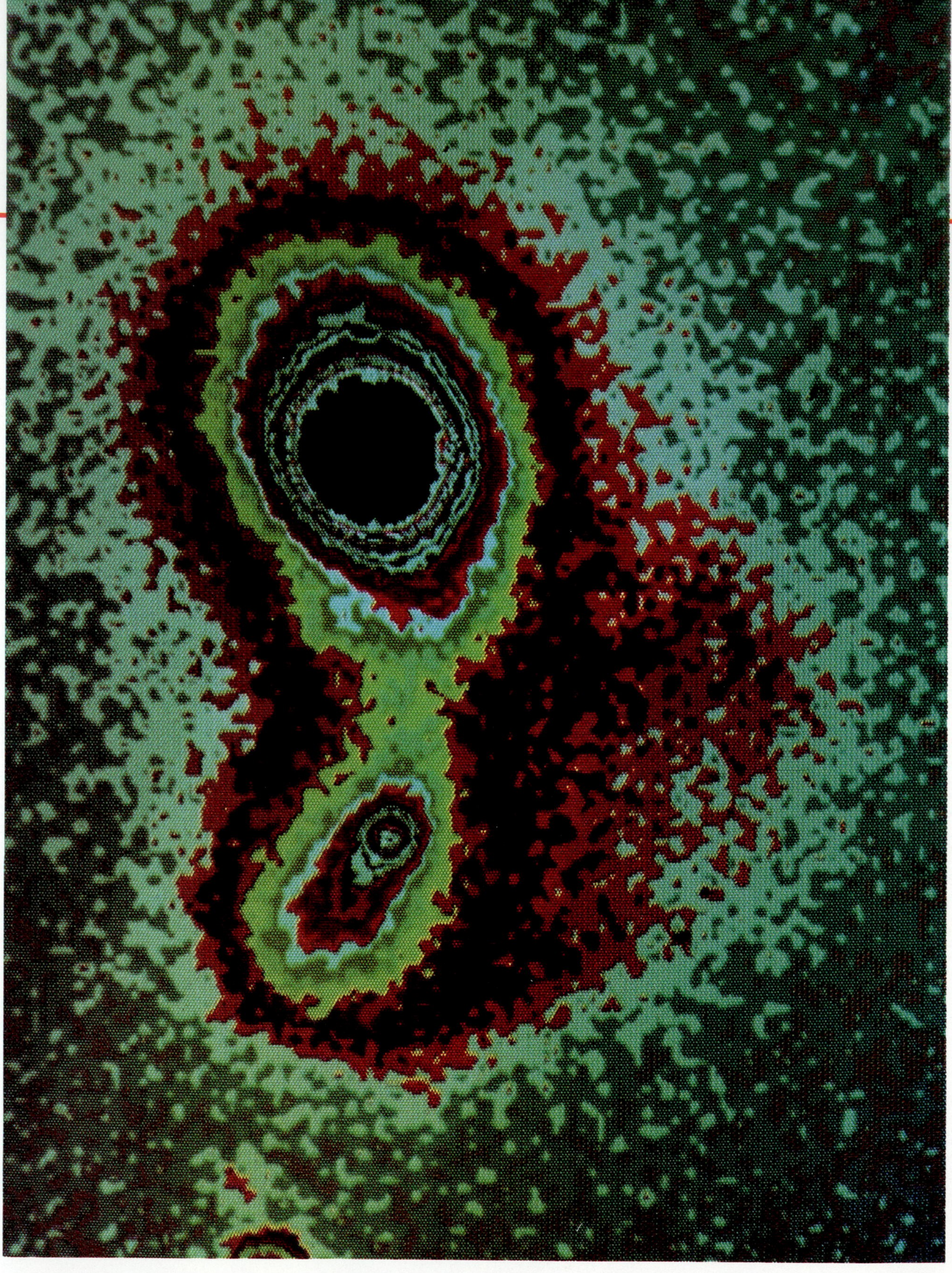

A quasar, 0351+026, interacting with a faint galaxy. The two objects, in the constellation Taurus, have similar redshifts, and so are the same distance from us. The pair have unusual x-ray and radio properties.

32

Quasars

Aims: To describe the discovery of quasars, their significance as probes of the early stages of the universe, their connections with galaxies, and our new knowledge of the giant black holes powering them

Quasars are enigmatic objects that appear almost like stars: points of light in the sky. But unlike the stars that we see, quasars occur in the farthest reaches of the universe and thus may be a key to our understanding the history and structure of space. Although quasars are relatively faint in visible light, they are among the strongest radio sources in our sky, and therefore must be prodigious radiators of energy. Quasars turn out to be small (on a galactic scale), so ordinary methods of generating energy are not sufficient. Similarly, we would be surprised if we got a tremendous explosion out of a tiny firecracker. Consequently, the most exotic and efficient method of generating energy—matter being gobbled by a giant black hole—has been invoked to explain the energy of quasars.

The word *quasar* originated from QSR, a contraction of "quasi-stellar radio source" (*quasi,* meaning "as if," "seemingly"). However, the most important characteristic of quasars is not that they are emitting radio radiation but that they are travelling away from earth at tremendous speeds. After quasars were named, radio-quiet "quasi-stellar objects" (QSO's) were discovered and are also called quasars. So some quasars are radio-quiet, and others are radio-loud. In both cases, astronomers deduce distances by observing the quasar spectra, measuring the Doppler shifts, and applying Hubble's law (Section 32.4). Since some quasars have the largest redshifts known, Hubble's law tells us that they are the most distant objects that we see. We shall discuss the evidence on this point; only a few holdouts don't accept it these days.

After the initial discoveries of hundreds of quasars over two decades ago, advances in understanding quasars came slowly for a while. But in the 1980's, new technologies have enabled us to observe them better, which has led to an improved and revised understanding and a few surprises, as we shall discuss.

32.1 The Discovery of Quasars

Quasars are a discovery resulting from the interaction of optical astronomy and radio astronomy. When maps of the radio sky turned out to be very different in appearance from maps of the optical sky, many astronomers in the 1950's set out to correlate the radio objects with visible ones.

Single-dish radio telescopes did not give sufficiently accurate positions of objects in the sky to allow identifications to be made, so interferometers had to be used. Most of the radio objects catalogued were identified with optical objects, but a few had no clear identifications. At least one of the strong radio sources, 3C 48 (the **48**th source in the **3**rd **C**ambridge catalogue), seemed suspiciously near a faint (16th magnitude) bluish star. At that time, 1960, no stars had been found to emit radio waves, with the sole exception of the sun, whose radio radiation we can detect only because of the sun's proximity to earth.

In Australia, a large radio telescope was used to observe the passage of the moon across the position in the sky of another bright radio source, 3C 273. We know the position of the moon in the sky very accurately. When it occults—hides—a radio source, then we know that at the moment the signal strength decreases, the source must have passed behind the advancing limb of the moon. Later on, the instant the radio source emerges, the position of the lunar limb marks another set of possible positions. The source must be at one of the two points where these two curved lines meet. Three lunar occultations of 3C 273 occurred within a few months, and from the data, a very accurate position for the source and a map of its structure were derived. The optical object (Fig. 32–1), which is 13th magnitude, is not completely star-like in appearance, for a luminous jet appears to be connected to the point nucleus. It is thus "quasi-stellar."

The radio emission from 3C 273 has two components. The discovery that one coincides with the jet (Fig. 32–2), and the other coincides with the bluish stellar object, clinched the identification of the optical object with the radio object. Maarten Schmidt photographed the spectrum of this "quasi-stellar radio source" with the 5-m Hale telescope on Palomar Mountain. The spectra of 3C 273 and 3C 48, both bluish quasi-stellar objects, showed emission lines, but the lines did not agree in wavelength with the spectral lines of any of the elements. The lines had the general appearance of spectral lines emitted by a gas of medium temperature, though.

The breakthrough came in 1963. At that time, Schmidt noted that the spectral lines of 3C 273 (Fig. 32–3) seemed to have the same pattern as lines of hydrogen under normal terrestrial conditions. Schmidt then made a major scientific discovery: he asked himself whether he could simply be observing a hydrogen spectrum that had been greatly shifted in wavelength by the Doppler effect. The Doppler shift required would be huge: each wavelength would have

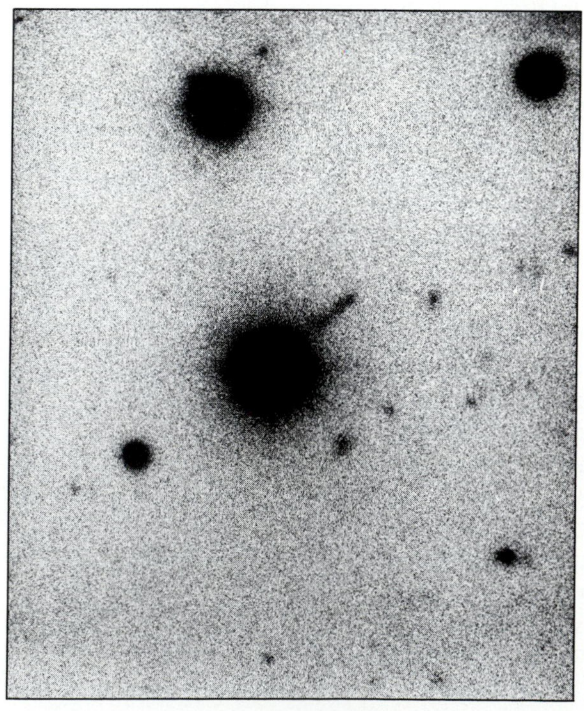

Figure 32–1 An enlarged portion of a negative of the quasi-stellar radio source 3C 273, taken with the 5-m Hale telescope of the Palomar Observatory. The object, the largest of the black disks, looks like any 13th-magnitude star, except for the faint jet that is visible out to about 20 seconds of arc, equivalent to 50,000 parsecs, from the quasi-stellar object.

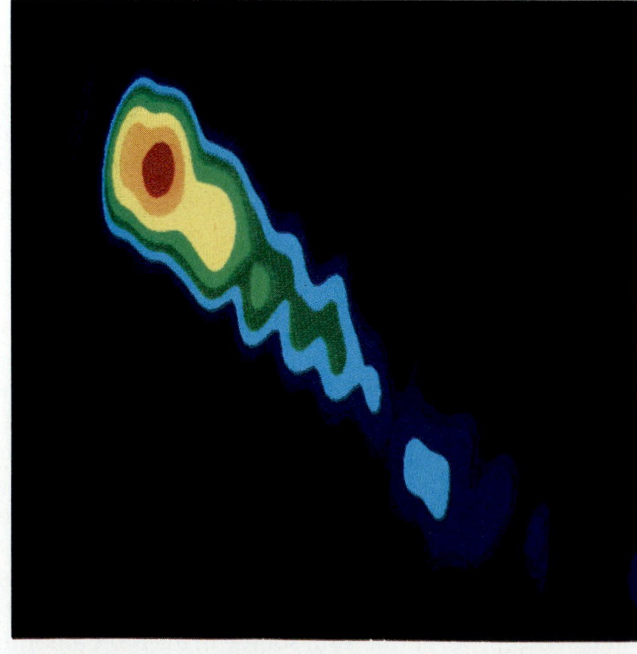

Figure 32–2 The radio jet of the quasar 3C 273.

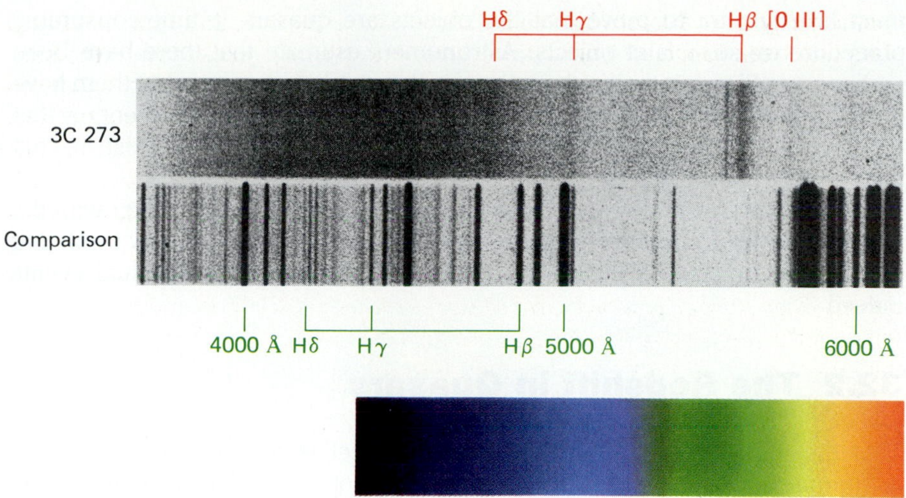

Figure 32–3 Spectrum of the quasar 3C 273. The lower spectrum is a source on earth; this spectrum consists of hydrogen and helium lines. This "comparison spectrum" establishes the scale of wavelength. The upper part is the spectrum of the quasar, an object of 13th magnitude. The Balmer lines Hβ, Hγ, and Hδ in the quasar spectrum are at longer wavelengths (*labels in red*) than in the comparison spectrum (*labels in green*). The redshift of 16 per cent corresponds, according to Hubble's law (H_0 = 50), to a distance of three billion light years. Note that the comparison spectrum represents hydrogen and helium sources on earth (more specifically, located inside the Palomar dome).

to be shifted by 16 per cent toward the red to account for the spectrum of 3C 273. This would mean that 3C 273 is receding from us at approximately 16 per cent of the speed of light. Immediately, Schmidt's colleague Jesse Greenstein recognized that the spectrum of 3C 48 could be similarly explained. All the lines in the spectrum of 3C 48 were shifted by 37 per cent, a still more astounding redshift.

Later, absorption lines were discovered in quasars, in addition to the emission lines already known. Many quasars have several systems of emission and absorption lines of differing redshifts. Some of these lines seem to be formed in clouds of gas of differing velocities surrounding the quasars. Others are formed farther from the quasar as the light travels from the quasar to us. Recently, IUE spectra showed that our galaxy and the Large and Small Magellanic Clouds have extensive halos of gas that cause absorption lines detectable in the ultraviolet. This discovery indicates that the absorption lines of ionized carbon, silicon, magnesium, and other "heavy elements" in quasars might be formed in the halos of otherwise unseen galaxies between the quasars and us. Since each distant galaxy has a different redshift, we can see the same line at different wavelengths.

The basic hydrogen line Lyman alpha in the spectrum of a quasar often appears in absorption at many different redshifts—presumably one for each gas cloud the light is passing through. Thus, through the ultraviolet, we see a "forest" of lines— the *Lyman-alpha forest*. The gas clouds in which these lines are formed seem to be made almost entirely of hydrogen, which indicates that they are located out near the quasars where we are seeing to earlier times in the universe. The nearby quasar 3C 273 shows no such lines, while high-redshift quasars show many, indicating that the neutral hydrogen clouds were more numerous early in the universe's history. The study of quasar absorption lines is one of only 3 "key projects" for the Hubble Space Telescope. Some of the lines will tell us about gas surrounding the quasars themselves, most of the heavy-element lines will tell us about the halos of galaxies very far away, and the hydrogen Lyman-alpha lines will tell us about intergalactic matter or the halos of other galaxies. We do not yet know whether the sources of heavy-element lines and of the Lyman-alpha forest differ in abundance or merely in temperature and density.

Over 1500 quasars have been discovered, with redshifts ranging up to over 489 per cent. Candidate objects that have a high probability of being quasars can be found by looking for star-like objects that seem unusually strong in the blue and ultraviolet. Examining x-ray sources has also proved fruitful. But one

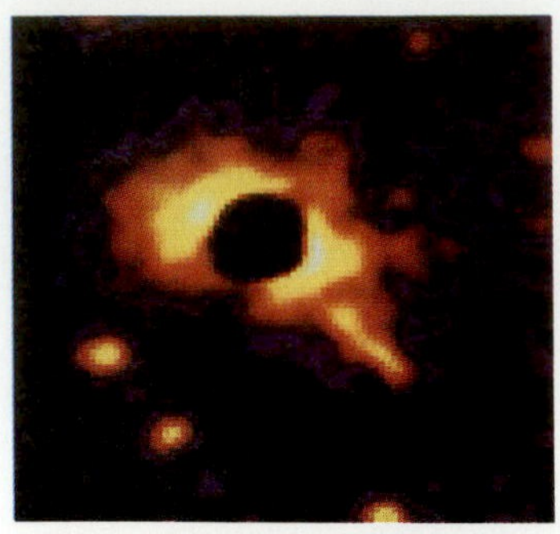

Figure 32–4 Hiding the bright quasar 3C 273 has revealed nebulosity that resembles an elliptical galaxy. Quasars are now thought to be bright events in the centers of galaxies.

must take spectra to prove that the objects are quasars, a time-consuming procedure on such faint objects. Astronomers estimate that there have been perhaps a million quasars in the universe. By now, however, most of them have probably lived out their lifetimes; that is, they have given off so much energy that they are no longer quasars. Perhaps about 35,000 quasars now exist, in that their energy is now reaching us.

New observational capabilities have enabled us to detect matter with the spectra of stars around the bright quasars (Fig. 32–4). Thus we now think that quasars are unusual events in the centers of galaxies—very unusual events indeed.

32.2 The Redshift in Quasars

That the redshift arises from the Doppler effect is accepted by almost all astronomers. Most objects in the universe show some Doppler shift with respect to the earth. For velocities that are small compared to the velocity of light, the amount of shift in the spectrum is written simply, for rest wavelength λ, velocity v, and speed of light c, as

$$\frac{\Delta\lambda}{\lambda} = \frac{v}{c} \text{ (as discussed in Section 21.6).}$$

Astronomers often use the symbol z to stand for $\Delta\lambda/\lambda$, the amount of the redshift.

The Doppler shifts of quasars are much greater than those of most galaxies. Even for 3C 273, the brightest quasar, $z = 0.16$ (read "a redshift of 16 per cent"). Thus Hubble's law implies that 3C 273 is as far away from us as distant galaxies. Only a few galaxies are known to have greater redshifts. To find the velocity of recession for quasars with redshifts larger than about 0.4, we must use a formula based on Einstein's special theory of relativity (Section 32.2b). Some quasars are receding at over 90 per cent of the speed of light. Hubble's law then tells us they are over 15 billion light years away.

*32.2a Working with Non-Relativistic Doppler Shifts

Let us consider a redshift of 0.2, to pick a round number, and calculate its effect on a spectrum (Fig. 32–5). $z = 0.2$ implies that $v/c = 0.2$, and therefore

$$v = 0.2c = 0.2 \times (3 \times 10^5 \text{ km/sec}) = 6 \times 10^4 \text{ km/sec}.$$

If a spectral line were emitted at 4000 Å in the quasar, at what wavelength would we record it on earth on our CCD (or, if we were old-fashioned, on our film)?

$\Delta\lambda/\lambda = \Delta\lambda/4000 = 0.2$. Therefore $\Delta\lambda = 0.2 \times 4000$ Å $= 800$ Å. Note that we have calculated only $\Delta\lambda$, the shift in wavelength. The new wavelength is equal to the old wavelength plus the shift in wavelength,

$$\lambda_{new} = \lambda_{original} + \Delta\lambda = 4000 \text{ Å} + 800 \text{ Å} = 4800 \text{ Å}.$$

Thus the line that was emitted at 4000 Å in the quasar would be recorded at 4800 Å on earth.

Similarly, a line that was emitted at 5000 Å is also shifted by 0.2, and 0.2 of 5000 Å is 1000 Å. Then $\lambda + \Delta\lambda = 5000$ Å $+ 1000$ Å $= 6000$ Å. The spectrum is thus not merely displaced by a constant number of angstroms, but is also stretched more and more toward the higher wavelengths.

*32.2b Working with Relativistic Doppler Shifts

The simple Doppler formula above is valid only for velocities much less than c, the speed of light. For speeds closer to the speed of light, we must use a formula from the special theory of relativity. Considering only radial velocity

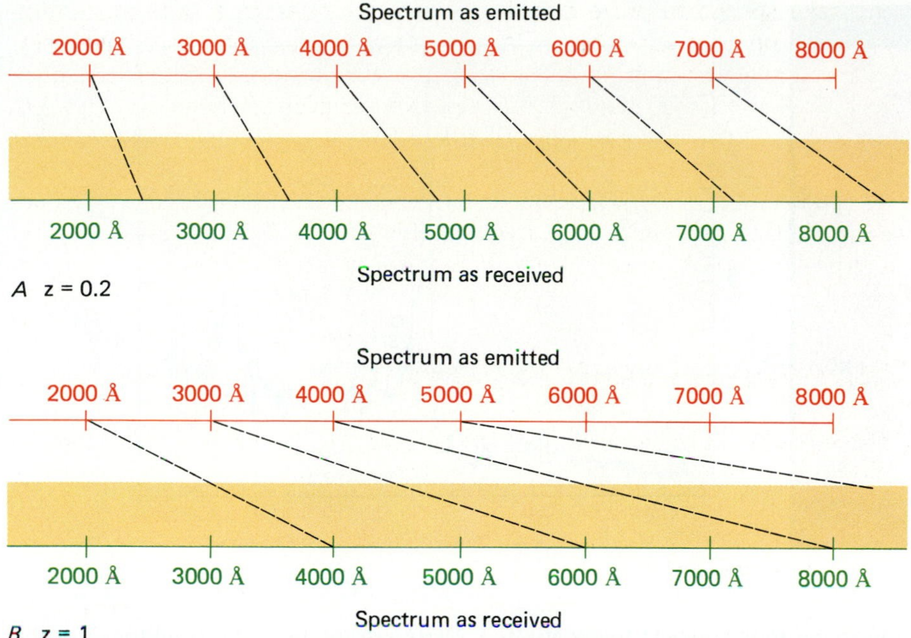

Figure 32–5 (*A*) The wavelengths of the visible part of the spectrum are shown in red at top, and the wavelengths at which spectral lines would appear after undergoing a redshift of 0.2 ($\Delta\lambda/\lambda = 0.2$) are shown in green at bottom. The lines are shifted by 0.2 times their original wavelengths, and wind up at 1.2 times their original wavelengths. The non-relativistic formula $\Delta\lambda/\lambda = v/c$ can be applied to give the approximate velocity; at still smaller redshifts this formula is even more accurate. (*B*) For a redshift of 1 ($\Delta\lambda/\lambda = 1$), as shown here, the lines are shifted (*bottom*) by an amount equal to their original wavelengths (*top*), and wind up at twice their original wavelengths. The velocity is a significant fraction of the speed of light, so the simple non-relativistic formula for the Doppler shift cannot be applied. The relativistic formula allows z to be greater than 1 without the velocity exceeding the speed of light.

(and ignoring any velocity perpendicular to the line of sight, or assuming that the expansion is uniform),

$$\frac{\Delta\lambda}{\lambda} = \sqrt{\frac{1 + v/c}{1 - v/c}} - 1.$$

Positive values of v correspond to receding objects. When v is much less than c, then the relativistic formula approximates the non-relativistic formula. (Note that if you substitute $v = 0$ in the relativistic formula, the redshift derived is indeed 0.) But when v is close to c, $\Delta\lambda/\lambda$ is greater then 1 even though v is still less than c. We still use the letter z to stand for $\Delta\lambda/\lambda$.

For example, if $v = 90$ per cent of c, $v/c = 0.9$.

$$\frac{\Delta\lambda}{\lambda} = \sqrt{\frac{1 + 0.9}{1 - 0.9}} - 1 = \sqrt{\frac{1.9}{0.1}} - 1 = \sqrt{19} - 1 = 4.4 - 1 = 3.4.$$

There is no physical significance for $z = 1$; we merely have the shift, $\Delta\lambda$, equalling the original wavelength, λ. The quasar with the largest known redshift (Fig. 32–6) has $z = 4.89$, which means that its wavelengths are shifted by 489 per cent. This makes the new wavelengths $4.89 + 1 = 5.89$ times the original wavelengths. But its velocity of recession is not much over 90 per cent of the speed of light. The velocities of recession of all quasars are still less than the speed of light, as the special theory of relativity tells us they must always be.

Though quasars had long been found by searching for objects with ultraviolet excesses, automatic scanners for photographic plates are now allow-

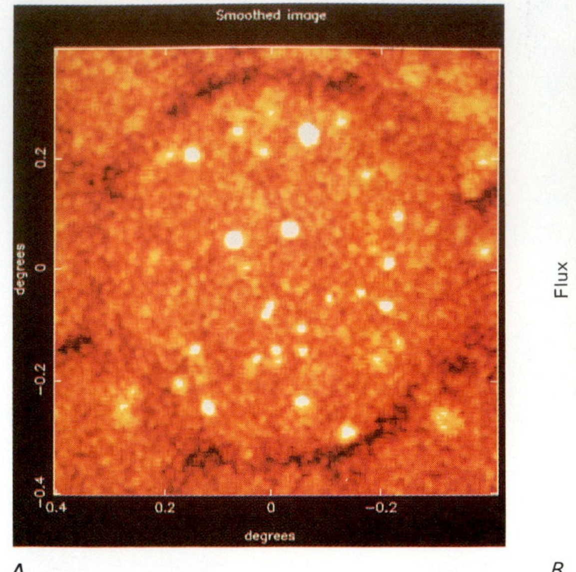

A

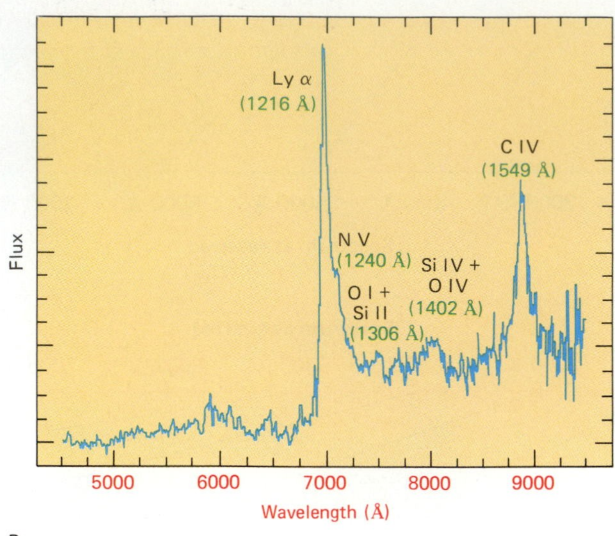

B

Figure 32–6 (*A*) The farthest known quasar, with redshift $z = 4.89$. It is quite faint but still easily bright enough to be photographed. (*B*) The high redshift of the most distant quasars known shifts extreme ultraviolet lines like Lyman alpha, normally 1216 Å, into the yellow and red regions of the spectrum. Original wavelengths appear in green and in parentheses. At this redshift, we are looking back to within 2 billion years of the big bang (Chapter 33), when the scale of size of the universe was only 18% of its current scale.

ing many more objects to be studied. Plots are made of the brightness of the object in one color versus its brightness in another color. Most of the hundreds of thousands of objects scanned fall in the same region of these "color-color diagrams" but a few stand out. Taking spectra of the few that stand out reveals distant quasars. After many years of not finding any quasars with redshifts much greater than 3.5, in 1987 astronomers began finding some with redshifts greater than 4. The technique should even show quasars of redshift 5, looking even farther back in time.

32.3 The Importance of Quasars

If we accept that the redshifts of quasars are caused by the Doppler effect, and we apply Hubble's law, we realize that the quasars with the largest redshifts are the farthest known objects in the universe. Thus we would like to use quasars to test Hubble's law, since deviations from the law would no doubt show up in the farthest objects.

If the quasars were found not to satisfy Hubble's law, on the other hand, then doubt would be cast on all distances derived by Hubble's law. In that case, when we were observing an object, we would never know whether the object satisfied the law or not. We would never be able to trust a distance derived from Hubble's law. At present, the evidence seems overwhelming that quasars do follow Hubble's law, that is, that their redshifts and distances are proportional with the same Hubble constant that we find for other galaxies.

If we accept the quasars as the farthest objects, then they are billions of light years away, and their light has taken billions of years to reach us. Thus we are looking back in time when we observe the quasars, and we hope that they will help us understand the early phases of our universe. For example, a survey of the entire sky to look for quasars has turned up many new ones, and has shown that the number of quasars per volume of space increases as you go outward. Thus there were more quasars in the universe long ago. Among other things, this shows that the universe has evolved.

Schmidt dazzled the astronomical world in 1988 at the General Assembly of the International Astronomical Union in his 20th anniversary lecture about quasars. He concluded (Fig. 32–7) that there was an abrupt epoch of quasar formation, an exciting time in the universe when the quasars were lit up like brilliant candles.

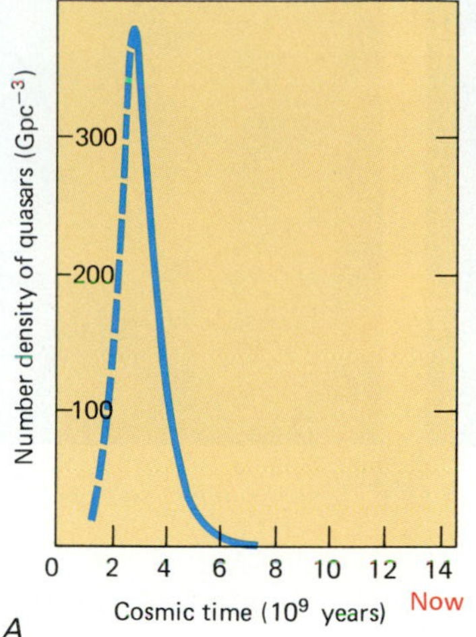

Figure 32–7 (*A*) Maarten Schmidt's conclusion that there was a bright, spectacular era of quasars billions of years ago. (*B*) Maarten Schmidt and the author, at an excursion of the 1988 General Assembly of the International Astronomical Union.

32.4 Feeding the Monster

If the quasars are as far away as the orthodox view holds, then they must be intrinsically very luminous to appear to us at their observed intensities. They are more luminous than entire ordinary galaxies.

But at the same time, the quasars must be very small because of the following argument: The optical brightness of quasars was measured on old collections of photographic plates, and turned out to vary on a time scale of weeks or months. If something is, say, a tenth of a light year across, you would expect that it could not vary in brightness in less than a tenth of a year. After all, one side cannot signal to the other side, so to speak, to join in the variation more quickly than that (Fig. 32–8). Somehow the whole object has to be coordinated, and that ability is limited by the speed of light. So the rapid variations in intensity mean that the quasars are fairly small. Even VLBI techniques have not succeeded in resolving the nuclei of quasars.

So quasars had not only to give off a tremendous amount of energy but also to do so from an exceedingly small volume. The difficulty of giving off so much energy from such a small volume is known as *the energy problem*.

A model of many exploding supernovae explained the observed sporadic variations in the intensity as variations in the number going off from one

Figure 32–8 (*A*) The figure illustrates why a large object can't fluctuate in brightness as rapidly as a smaller object. Say that each object abruptly brightens at one instant. The wave emitted from the top of the object takes somewhat longer to reach us than the wave emitted from the side of the object nearest us, just because of the additional distance it has to travel. We don't see the full effect of the variation in brightness until we have the waves from all parts of the object. This simply takes longer for larger objects than it does for smaller ones. (*B*) Both optical and radio radiation from the quasar 3C 273 vary in step in a matter of weeks or less. The radio radiation is shown here.

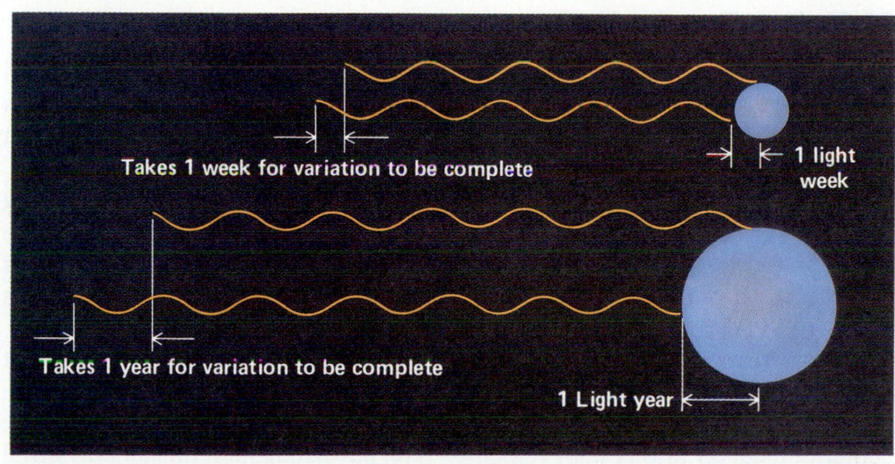

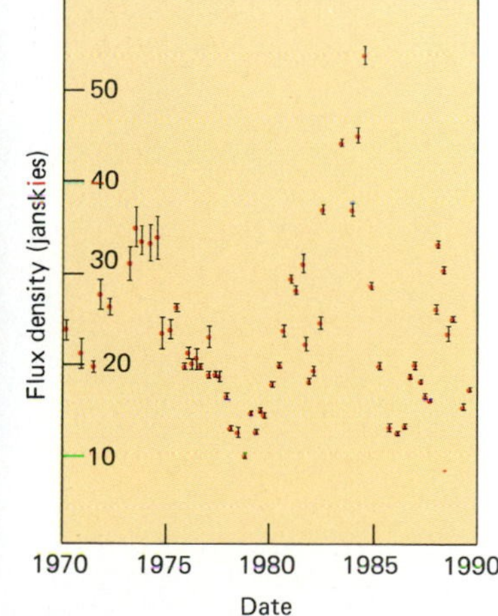

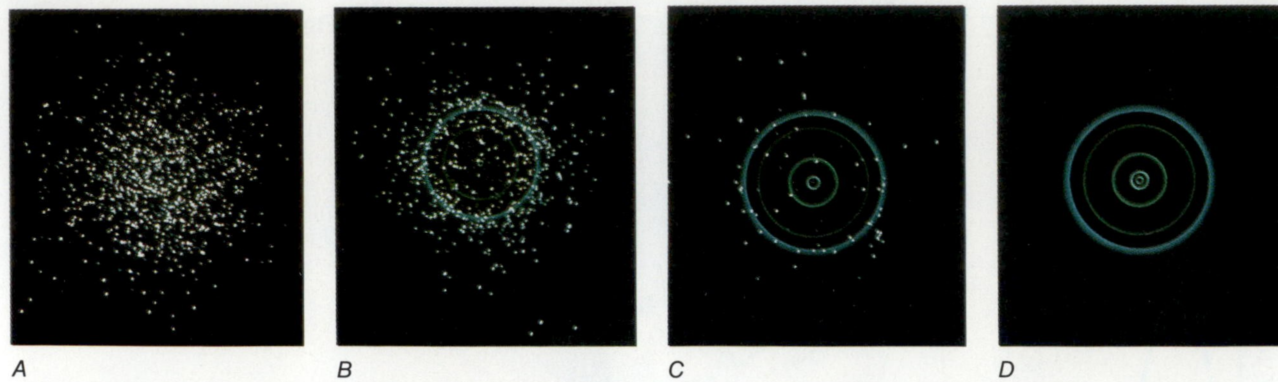

Figure 32–9 The formation of a quasar from a collapsing star cluster, as shown by supercomputer calculations based on Einstein's general theory of relativity. (*A*) The cluster, whose stars are moving at speeds close to the speed of light, is unstable and beginning to collapse. (*B*) During the collapse, a black hole forms at the center of the cluster. (*C*) The concentric circles represent spheres of light that had been sent out from the center of the cluster. (*D*) The light is trapped forever in the supermassive black hole that is at the center of a quasar. More and more of the mass in the central regions of the original cluster is consumed by the black hole.

moment to the next. But this model required too many supernovae, a dozen a day in some cases. Other unconvincing suggestions to explain the source of energy in quasars included the idea that matter and antimatter are annihilating each other there.

At present, the consensus is that quasars have giant black holes in their centers, large-scale versions of the maxi black holes that may well exist in the centers of our own and other galaxies. The black hole would contain millions or billions of times as much mass as the sun (Fig. 32–9). As mass falls into the black hole, energy is given off. (The gravitational potential energy of the mass is converted into kinetic energy and radiation.) Black holes resulting from collapsed stars give off energy in a similar way, but on a much smaller scale. The central black hole in a quasar is known as "the engine" that produces the energy. The gas and dust around the engine are "the fuel." Together, the engine and the fuel are called "the machine."

Theoretical models show how matter falling into such a giant black hole would form an accretion disk. Data across the ultraviolet, visible, and infrared spectrum, including space observations with the International Ultraviolet Explorer (IUE), have supported this picture. The strength of ultraviolet radiation, for example, matches the predictions that the inner part of the accretion disk—where the friction of the gas is especially high—would reach 35,000 K. The matter in the accretion disk would eventually fall into the central black hole, so the process is called "feeding the monster." We will come back to the question of where quasars get their fuel.

32.5 The Origin of the Redshifts

32.5a Non-Doppler Methods

The most obvious explanation of quasar redshifts is the Doppler effect caused by the expansion of the universe, a "cosmological redshift." But the distances implied are so large and the energy problem at first seemed so difficult that other explanations have been investigated very thoroughly.

If the quasars are close, we must think of some other way of accounting for their great redshifts. There are at least three other ways. They are all unattractive to most astronomers because they challenge Hubble's law and thus cast doubt on our knowledge of the whole scale of the universe. If Hubble's law were to fail for the quasars, we would never know when we could trust it.

A first non-cosmological way also relies on velocity and assumes that quasars are close but are going away from us very rapidly. Quasars could be relatively close to us yet show high redshifts if they had been ejected from the center of a galaxy at very high velocities (Fig. 32–10). If such quasars occurred in another galaxy, we would expect to see some of them going away from us and some coming toward us. But all the quasars have redshifts; none of them has a blueshift, so this idea is unlikely.

If quasars exploded from the center of our galaxy or long enough ago from a nearby galaxy, all the quasars would have moved past us and would appear to be receding; no blueshifted quasars would be expected. This matches the observations, but so much energy would be required for such tremendous and perhaps repeated explosions that the energy problem would not be resolved.

A second non-cosmological possibility is the gravitational redshift that has been verified for the sun and for white dwarfs. But even before this method was ruled out by observations, scientists had not been able to work out such a system for quasars that would be free of other, unobserved consequences. And eventually, measurements were made of redshifts from the faint material that surrounds some quasars. The gravitational redshift theory predicts that the redshift would vary with distance from the quasar, but the observations contradict this. Scientists have thus discarded this disproved theory.

A third possibility is that quasars act on principles that we do not yet understand—some new kind of physics. But the basic philosophy of science forbids us from inventing new physical laws as long as we can satisfactorily use the existing ones to explain all our data.

Now that the energy problem has all but disappeared, since giant black holes form a reasonable "engine," there is little need to invoke these non-cosmological methods for redshifts.

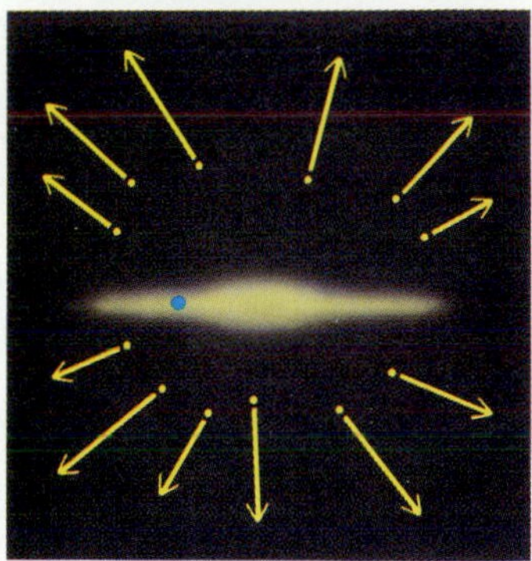

Figure 32–10 The idea that quasars were local, and were ejected from our galaxy, could explain why they all have high redshifts. But a tremendous amount of energy would be necessary to make the ejection. So we might not find ourselves any closer to a solution of the energy problem—where all this energy comes from. The blue dot shows the position of the sun.

32.5b Evidence for the Doppler Effect

Some of the strongest evidence that the quasars are at their *cosmological distances* (their distances according to Hubble's law) instead of being *local* comes from a major survey of the number of quasars carried out over a dozen years by Richard Green, now at the Kitt Peak National Observatory, and Maarten Schmidt of Caltech. The quasars increase in number with distance from us in a manner that is difficult to account for on a local model but occurs naturally if quasars are at their distances according to Hubble's law.

Now there is even stronger evidence that the redshift is a valid estimator of distance for quasars. A strange object named BL Lacertae has been known for years. At first, it was thought to be merely a variable star (as its name shows), but in recent years it was suspected to be stranger than that, partly because of its rapidly varying radio emission. Its spectrum showed only a featureless continuum, with no absorption or emission lines, so little could be learned about it.

Two Caltech astronomers had the idea of blocking out the bright central part of BL Lacertae with a disk held up in the focus of their telescope, thereby obtaining the spectrum of the haze of gas that seems to surround the apparent star. Its spectrum turned out to be a faint continuum with absorption lines typical of the stars in a galaxy. Thus BL Lacertae is apparently a galaxy with a bright central core. Further, the lines are redshifted by 7 per cent. Nobody doubts that a redshift of this amount arises from the expansion of the universe. Yet BL Lacertae, with its point-like appearance and its rapidly varying brightness, appears almost like an extremely nearby quasar! About 100 similar objects (now called Lacertids or BL Lacs) have also been found to have redshifts of this order. Some, at least, appear to be embedded in nebulosity whose spectrum resembles that of an elliptical galaxy.

The Compton Gamma Ray Observatory has detected strong gamma-ray emission from many BL Lacertae objects. These objects can be called "gamma-ray quasars."

The Lacertids appear to be a missing link between galaxies and quasars. The properties of quasars (assuming that they are at their Hubble's law distances)—such as the brightness of the core relative to that of the other gas that may be present—are more extreme than, but a continuation of, the prop-

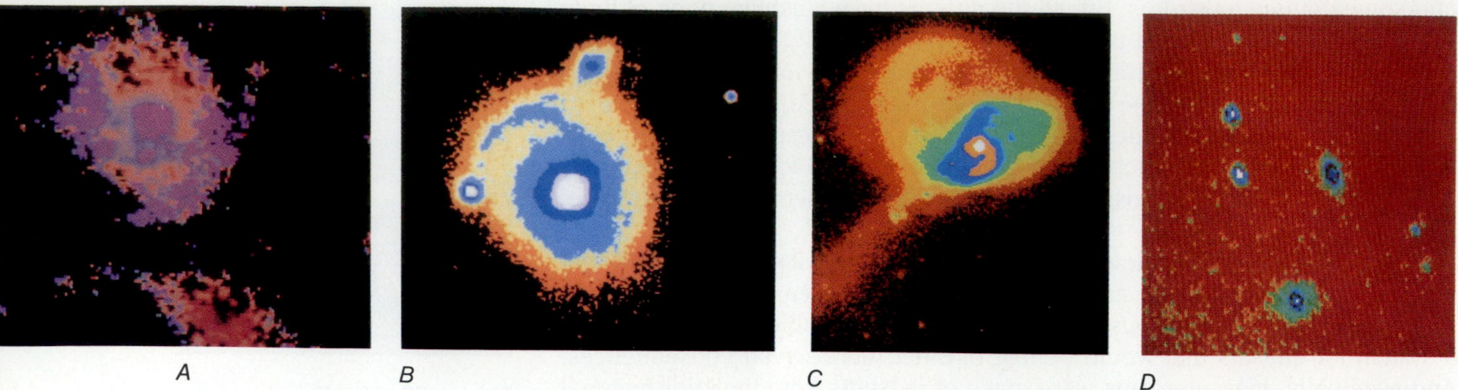

Figure 32–11 (*A*) Quasar 0844+34 in its host galaxy. The system has a redshift of 0.06, which makes the nucleus either a very luminous Seyfert galaxy or a low luminosity quasar. The companion is red and does not show much evidence of interaction while the quasar host is blue, which indicates that star formation is going on. (*B*) The nearby quasar 0050+124 (also known as I Zw 1 from the Zwicky catalogue). The spiral arms of the host galaxy are clearly seen. (*C*) NGC 1614, another merging galaxy with infrared luminosity close to the luminosity of quasars. Hα studies show that the outer curved loops are warped in the sky by tidal effects. (*D*) Note the fuzz around the quasar IE 2344+18, compared with the sharpness of the stellar images.

erties of Lacertids. Since few doubt that the Lacertids are at their Hubble's law distances, the comparison with quasars strengthens the notion that quasars are also at their own Hubble's law distances.

In recent years, new observational methods have succeeded in finding matter around the bright quasar image, and in taking the spectrum of that matter. The studies show that the quasars are embedded in host galaxies (Fig. 32–11), and are active phases in the lives of galaxies. We now know of many intermediate objects between ordinary galaxies and quasars. Thus quasars are now thought to be extreme cases of active galactic nuclei (AGN's).

32.6 Associations of Galaxies with Quasars

32.6a Statistics and Non-Hubble Redshifts

A principal attack on the theory that quasars are at great distances from us has come from Halton Arp of the Mount Wilson and Las Campanas Observatories and now of the Max-Planck Institute for Astrophysics in West Germany. On the basis of his observations, he concludes that quasars might be physically linked to galaxies, both the peculiar and the ordinary types. Arp contends that he has found many examples in which a galaxy and two or more quasars lie on a straight line, with at least one of the quasars on exactly the opposite side of the galaxy from another. He argues that the quasars and the galaxy, which have different redshifts, must therefore be linked. We know the distances to the galaxies from Hubble's law; if the quasars and galaxies are physically linked, then the quasars must be at the same distance from us as the galaxies. These distances are much smaller than the quasars' redshifts indicate.

Any argument that Hubble distances cannot be trusted would show that we cannot rely on Hubble distances for quasars. One such set of observations applies to a group of galaxies rather than to a mixture of galaxies and quasars. The redshift of one of the galaxies in Stephan's Quintet (Fig. 32–12), five apparently linked galaxies, is different from the redshift of the others. Such cases can be explained without rejecting Hubble's law if the object with a discordant redshift only accidentally appears in almost the same line of sight

Figure 32–12 Stephan's Quintet, in a negative print of a photograph taken by Halton Arp with the 5-m telescope at Palomar Observatory. Less exposed photographs show that the dark area to the lower left of the galaxy with the discordant redshift (*arrow*) has two nuclei and so is two interacting galaxies. Thus there are indeed five galaxies in this "quintet."

as do the other objects. Further, associations of a quasar with a distant group of galaxies with redshifts that do agree with each other have been found.

In some cases, it seems that a galaxy and a quasar with different redshifts are actually linked by a bridge of material (Fig. 32–13). But this too can be a projection effect. Even in the "best" examples of galaxy-quasar bridges, different scientists do not agree whether the connection is real. The resolution of the controversy will have to await the Hubble Space Telescope.

Arp's data are subject to several objections, which are mainly statistical. Arp feels that the probability of finding quasars and galaxies so close together in the sky by chance is very small. But statistical methods are valid only before rather than after you know what is actually present. For example, if I flip a coin, show you that it has come up "heads," and then ask you what is the chance that it is "heads," you may answer "50 per cent." But the correct answer is 100 per cent. Once the deed is done, the odds are determined; after all, I showed you that the coin had come up heads. At a racetrack, you can't place your bets after the race is over. Similarly, once we look at a quasar and a galaxy that are apparently linked together, the odds are now 100 per cent that they are apparently linked. We can then no longer apply the simple statistical argument that the odds of finding two objects so close together by accident are so small that they must be physically linked.

The associations of galaxies with a quasar of a different redshift would be harder to explain as a projection effect if we found a cluster of galaxies in which not just one but two discrepant redshifts were found. No such cluster is known. So far, most astronomers do not accept Arp's arguments and feel that quasars are in fact very far away.

Alan Stockton of the University of Hawaii carried out an important statistical test of the association of quasars with galaxies. He selected 27 relatively nearby quasars that were so luminous that the energy problem would be particularly severe. He carefully examined the regions around the quasars, and for 17 of the quasars found one or more galaxies apparently associated with them, as shown by their small angular separations.

For 8 of the quasars, at least one of the galaxies nearby had a redshift identical to that of the quasar (Fig. 32–14). Since Stockton had chosen his sample of quasars before knowing what he would find, he was able to apply

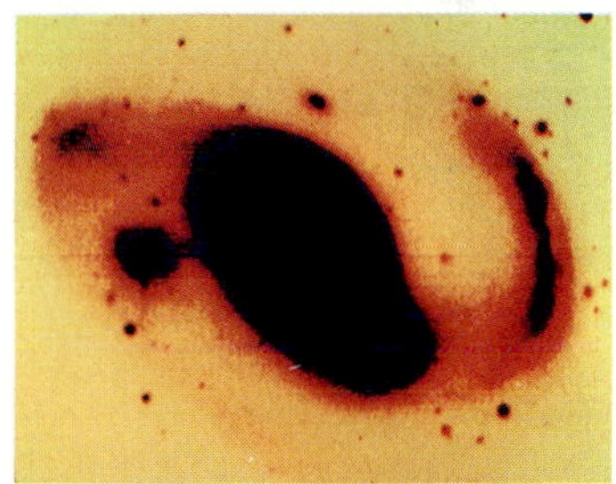

A

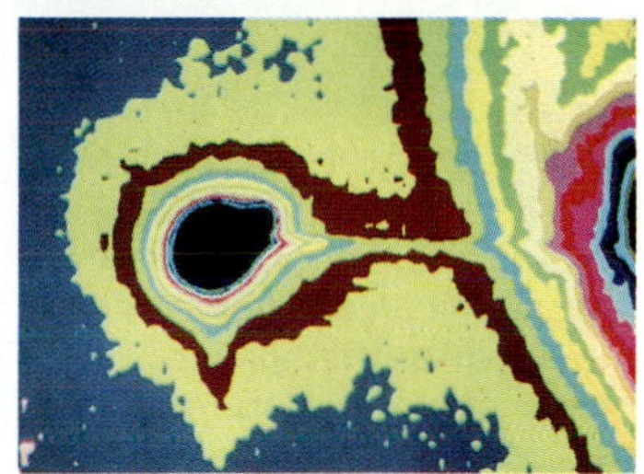

B

Figure 32–13 The quasar seems to be attached to NGC 4319, also known as Markarian 205. The reality of the "bridge" has been carefully assessed, especially in view of the idea that film effects can sometimes cause apparent bridges between bright objects. There is no agreement between the schools of thought on whether the quasar is a chance superposition or not.

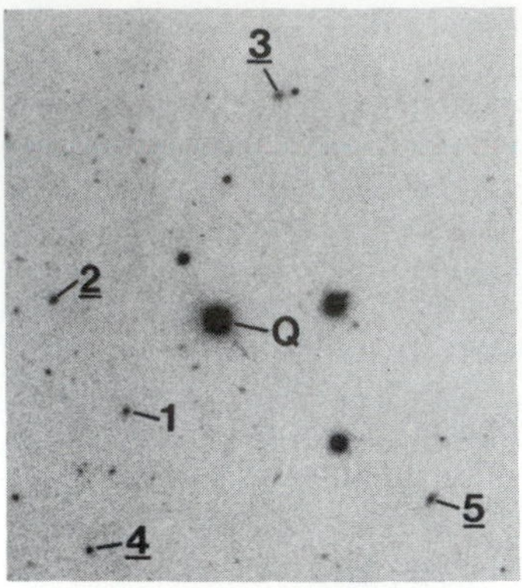

Figure 32–14 In this photograph taken by Alan Stockton on Mauna Kea, the central object is the famous quasar 3C 273, and the numbered objects are galaxies. The redshifts of the galaxies and the quasars agree on his sample of several such images of different regions. Note how the quasars are brighter than galaxies at the same distances.

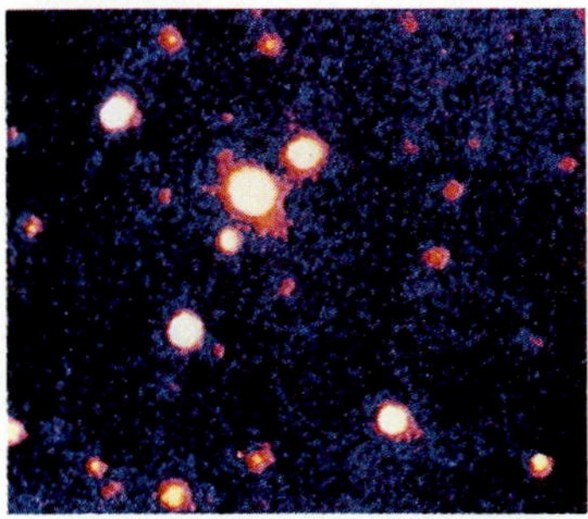

Figure 32–15 The quasar 3C 275.1, the brightest object near the center of this false-color image, is one of the few quasars located at the center of a rich cluster of galaxies. The quasar nucleus is surrounded by a rotating elliptical gas cloud. The quasar and the galaxies have redshift 0.55.

standard statistical tests to assess the probability that he could randomly find galaxies so apparently close in the sky to quasars. The probability of a projection effect making galaxies appear so close to a high fraction of quasars is less than one in a million. This indicates that some of the quasars and their associated galaxies are physically close together in space. Since the quasar and galaxy redshifts agree, the results strongly endorse the idea that the distances to quasars are those we derive from Hubble's law.

In several recent cases, a quasar has been found in a cluster of galaxies (Fig. 32–15). The quasars' redshifts agree with those of the galaxies.

32.6b Quasars and Galaxy Cores

In recent years, evidence has been mounting that quasars are extreme cases of galaxies rather than truly different phenomena. Some spiral galaxies have especially bright nuclei. A *Seyfert galaxy,* a type of active galaxy discovered by Carl Seyfert of the Mount Wilson Observatory in 1943 and named after him, has a very bright nucleus indeed compared with its spiral arms (Fig. 32–16). Another type of galaxy, an *N galaxy,* also has an especially bright core. Can the quasars that we see be only the bright nuclei of galaxies? After all, although we can see both the nuclei and the spiral arms of nearby Seyferts, only the nuclei would be visible if such galaxies were very far away.

Seyfert galaxies not only have relatively bright nuclei but also have broad emission lines from various stages of ionization in their spectra, signs that hot gas is present. The exceptional breadth in wavelength of the lines could be caused by rapidly moving matter in the galaxies' cores, a sign of violent activity there. Two to five per cent of galaxies are Seyferts. Seyferts are quite bright in the infrared, as was confirmed by IRAS.

In the early 1970's, Jerome Kristian of the Hale Observatories studied a sample of about 24 quasars to see whether they could be the bright cores of distant galaxies. He attempted to predict which quasars would be close enough to reveal some structure (nebulosity) around the core, which quasars would be so far away that they appear as points like quasars, and which quasars would be in between. He then found on the best photographs of these objects that almost all objects in which one would expect to see structure or nebulosity did in fact show it, that almost all objects that would not show structure or nebulosity because they were too far away did not show it, and that the middle group was mixed. In the last few years, this work has been carried further by several investigators taking advantage of the exceptionally good seeing available at certain observatories and of new, sensitive equipment like CCD's. The fuzzy structure detectable is technically known as *fuzz.*

In only two cases—3C 273 and one other—is the material near quasars in the form of jets. In at least some other cases, the structure shows the characteristic spectrum of a galaxy. The fuzz around 3C 48 has been found to have spectral lines from hot stars. So star formation must have been going on

Table 32–1 Energies of Galaxies and Quasars

	Relative Luminosity		
	X-ray	*Optical*	*Radio*
Milky Way	1	1	1
Radio galaxy	100–5,000	2	2,000–2,000,000
Seyfert/N galaxy	300–70,000	2	20–2,000,000
Quasar: 3C 273	2,500,00	250	6,000,000

in that source during the past billion years or so, since the hot stars die out on longer time scales. Further, the observation strengthens the case that quasars are objects in spiral rather than normal elliptical galaxies.

The results indicate that quasars are somehow related to the cores of galaxies. Perhaps galaxies go through a quasar stage during which their nuclei are very bright, or quasars are an extreme case somewhat different from but similar to explosive events in galaxy cores. The surrounding material usually seems to resemble material in spiral galaxies, so the quasar phenomenon may usually take place in spirals. Radio-loud quasars may have elliptical galaxies as hosts.

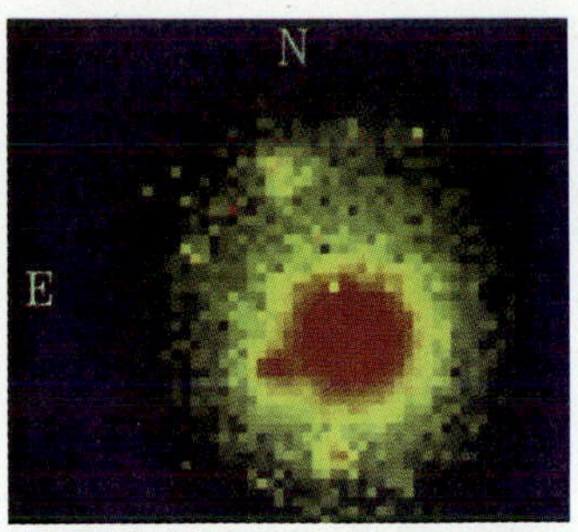

Figure 32–16 The Seyfert galaxy NGC 4151, in false color, with its bright center.

Further, properties of quasars like the types of lines in their spectra are now known to be similar to those of certain types of galaxies. Seyfert nuclei and quasars, for example, have similar ultraviolet and optical emission lines. Infrared observations have shown strong, peaked, central emission in active galaxies that could mean that these active galaxies are a step between normal galaxies and quasars. Indeed, the infrared radiation from quasars may well come from the galaxy surrounding the quasar, given that infrared radiation usually comes from dust heated from some central source. It has recently been found from infrared spectra that at least some quasars and Seyfert galaxies contain clouds of molecular hydrogen. Though the quasars may be extreme cases, they are not as different from other types of known objects as had once been thought.

In sum, the relation between quasars and the cores of active or peculiar galaxies has been consistently strengthened as we have been able to observe finer detail to greater distances. This linkage makes the quasars seem somewhat less strange, but at the same time causes galaxies to seem more exotic. If we had known of BL Lacertae's redshift earlier on, quasars would not have looked as strange to us when they were discovered. And if collapsed objects like neutron stars and black holes had been in our minds when quasars were discovered, we would have had an obvious way to produce a lot of energy in a small space. In that case, we would not have said that "the energy problem" for quasars existed.

32.6c Quasars and Interacting Galaxies

Since almost all quasars are so far away, it seems that whatever provided the fuel to make them so bright must have soon been exhausted. So why can we see a few quasars that are so close? We now newly realize that these closest quasars may be rejuvenated, having received a new supply of fuel.

Some evidence that galaxies interact gravitationally has been known for a while. For example, we discussed (Section 31.2) the tails that have been explained as gravitational interactions. The occurrence of Seyfert galaxies (Fig. 32–17) is more common in galaxy pairs than in individual galaxies.

High-resolution studies of quasars have revealed that some apparent bumps on the quasar images are actually independent objects. In other cases, independent objects were already known from the work of Arp and others. These objects have the same redshifts as the nearby quasars, and statistics now seem to show that the objects and the quasars are actually associated. These could be the objects that have been stripped of their outer layers to provide fuel for the quasar, rejuventating the engine. Close encounters of galaxies with quasars (Fig. 32–18) in a way to provide the fuel for the black holes are rare, which could explain why nearby quasars are few in number. The members of one pair of quasars are so close to each other that they may be interacting. We may be seeing the exchange of gas that turns quasars on. Light has been travelling to us from these quasars for 12 billion years. Perhaps the

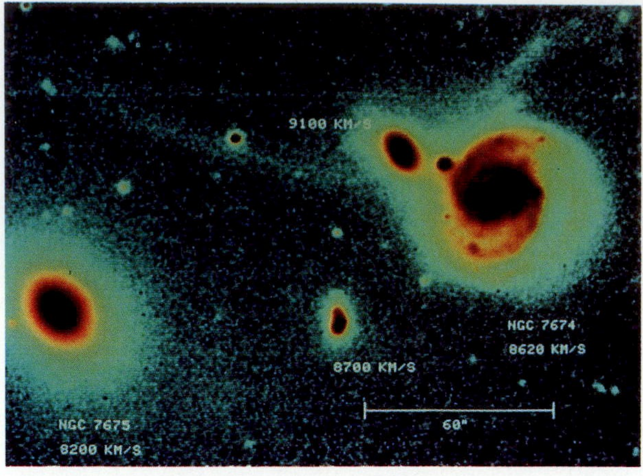

A

B

Figure 32–17 (*A*) An interacting Seyfert galaxy, NGC 7674, in a cluster of galaxies, with similar redshifts. (*B*) An elliptical galaxy interacting with the quasar 1747+684.

collision rate was much higher 12 billion years ago, which would explain why most quasars have such high redshifts.

The quasars that are farthest away, those with the largest redshifts, are using up an original store of fuel. We see these quasars far enough back in time—to when rich clusters were forming. The quasars may have consumed some of the gas present in these clusters. The closer quasars might be having a second youth, with new gas being introduced from encounters with other galaxies. Or, since it takes longer for the small clusters or groups of galaxies (where we see these quasars) to be stripped of their gas, these close quasars could be getting fuel for their black holes for the first time. Observationally, at least 30 per cent of quasars with redshifts up to 0.6—a fair way out into space—appear to be interacting, so the interaction model can be commonly applied.

According to this interaction model, Seyfert galaxies may simply have smaller black holes or less fuel than quasars. Starburst galaxies, Seyfert galaxies, and quasars may be part of an evolutionary sequence. Radio galaxies may be older quasars or objects that never received much fuel. (Some quasars resemble radio galaxies by also being double-lobed radio sources.)

Ground-based observations of a source discovered by IRAS have revealed that it is an extremely luminous colliding-galaxy system. It is emitting more than 95 per cent of its energy at wavelengths detectable only in the far infrared. Much radiation is apparently coming from molecular gas, studied with a ground-based radio telescope from its carbon-monoxide radiation. The molecular gas is fueling a quasar that is surrounded by dust; the hot dust gives off the infrared radiation. Perhaps the presence of a large amount of molecular gas is a prerequisite for quasars to form as galaxies collide.

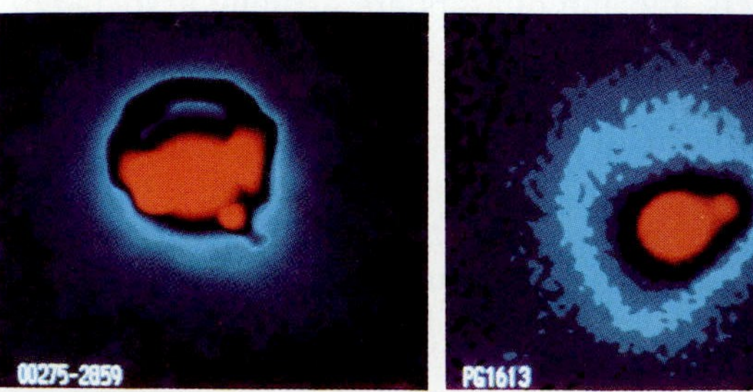

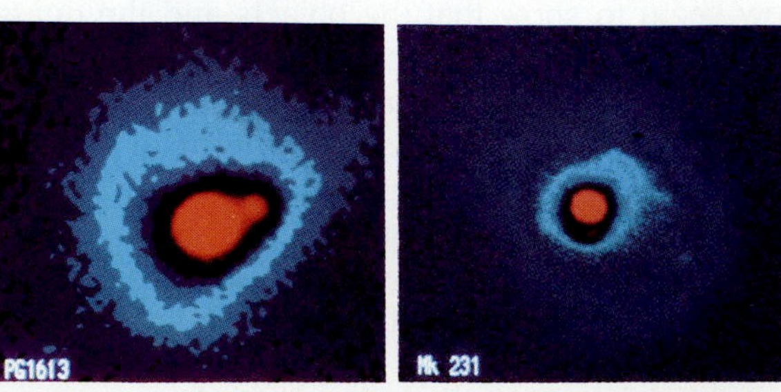

Figure 32–18 Three stages in the ignition of quasars, reproduced at the same scale. In IRAS 00275-2859, a quasar nears collision with a second nucleus, unseen because it is too close to the quasar to distinguish. In PG 1613, the second bright nucleus shows clearly to the upper right of the quasar. In Markarian 231, a tidal tail and nebulous streaks from the gravitational interaction are all that remains of the collision.

Quasars and their surrounding fuzz and galaxies are being carefully studied with the Hubble Space Telescope.

*32.7 Superluminal Velocities

Very-long-baseline interferometry (VLBI) observations with extremely high resolution (0.001 arc sec) have revealed the presence of a few small components in radio images of jets in a few of the quasars. Further, the observations have shown that in some cases these components are separating at angular velocities across the sky that seem to correspond (at the distances of these objects based on Hubble's law) to velocities greater than the speed of light. Since the special theory of relativity tells us that the apparent *superluminal velocities*—velocities greater than the speed of light—could not be real, other theoretical explanations of the data have been sought.

The jets are presumably forced out of the quasar centers above the quasars' poles. The compression that formed the accretion disks would have made such high magnetic fields in the disks that particles couldn't cross them, leaving the poles as the best way out. The material ejected in this way presumably forms the lobes detectable with radio telescopes. The jets may be a larger-scale version of the jets of SS 433 (Section 27.9), a source in our galaxy.

A "Christmas tree" model, in which components are flashing on and off, had seemed possible at first. In this model, no rapid velocities need be implied, since the sources we see this year were not necessarily the same ones we saw last year. But now years of observations (Fig. 32–19) show that the components continue to separate from each other rather than flash on and off. The calculated velocity at which some of the blobs in this source separate is 6*c* (6 times the speed of light), while other sources show velocities ranging up to 45*c*.

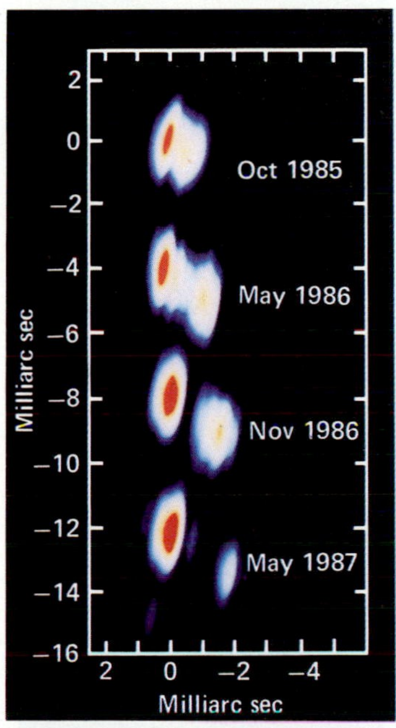

Figure 32–19 A series of views of the quasar 3C 273 with radio interferometry at a wavelength of 1.3 cm (22 GHz). The observations were made in October 1985, May 1986, November 1986, and May 1987. The apparent velocity translates (for H_0 = 100 km/sec/Mpc) to a velocity of 6.2*c*, though the apparent superluminal velocities can be explained in conventional terms.

The current model involves the special theory of relativity, in that it depends on the fact that light travels at a finite speed. Picture a jet of gas that is moving rapidly almost directly toward us. The jet almost (but not quite) keeps up with the light it emits (Fig. 32–20).

Under these circumstances, the apparent separation of the objects is not the actual separation at any one instant of time. For certain angles of view, the apparent rate of separation can look much greater than the real rate of separation. For a correct understanding, we must keep careful account of exactly when light was emitted.

The "superluminal motions" might just give an important key to the understanding of quasars. The model we have just discussed requires that the knot of gas is moving almost right at us (within about 10°), which may seem statistically unlikely. But theory predicts that because of their motion toward us, the radiation from the rapidly moving components may appear to us to be concentrated in beams. We might detect only those quasars that are beaming radiation toward us (which would also explain why we sometimes see only one jet). If quasars concentrate energy in beams this way, then their total energy emission would not be as high as we had calculated on the assumption that they were emitting radiation in all directions at an equal rate. This would make quasar luminosities similar to luminosities of galaxies. But current thinking is that only a few of the quasars are beaming radiation at us, and thus enhancing our measurements of their power.

The apparent superluminal expansion has been detected in several quasars and in at least one galaxy, and so may be a fairly common phenomenon. There presumably would be a second jet of gas oriented away from us, to make the situation symmetric, but this jet would be moving away so fast it would be redshifted too much for us to detect it.

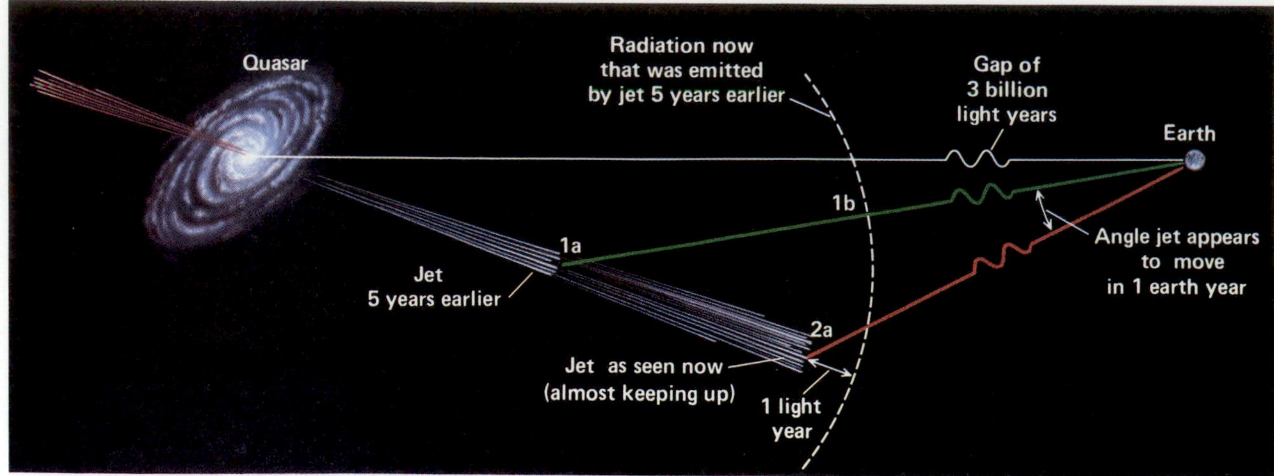

Figure 32–20 The leading model for explaining how a jet of gas emitted from a quasar, shown at left with its accretion disk surrounding a central black hole, can seem to be travelling at greater than the speed of light when seen from earth (shown at right). 1a marks the position of the jet 5 years earlier than the jet we are seeing now. (It then takes another 3 billion years for the radiation from 3C 273 to reach us, an extra duration that we can ignore for the purposes of this example.) After an interval of 5 years, the light emitted from point 1a has reached point 1b, while the jet itself has reached point 2a. The jet has been moving so fast that points 2a and 1b are separated by only 1 light year in this example. The light emitted then from the jet at point 2a is thus only 1 year behind the light emitted 5 years earlier at point 1a. So billions of years later, with a 1-year interval we receive light from two positions that the jet has taken 5 years to go between. The jet actually therefore had five years to move across the sky to make the angular change in position (that is, proper motion) we see over a 1-year interval. We thus think that the jet is moving faster than it actually is.

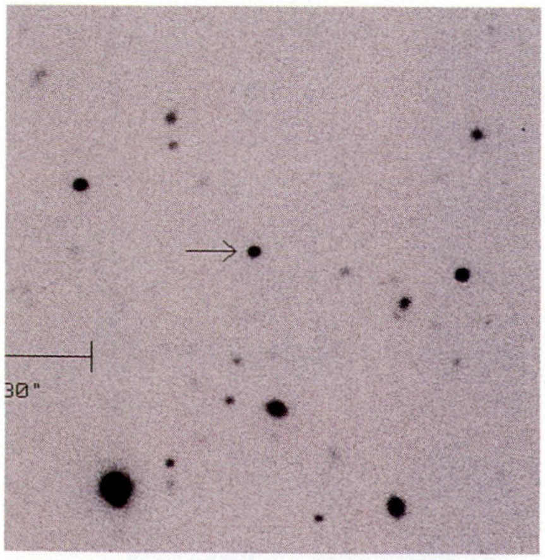

*32.8 Quasars Observed from Space

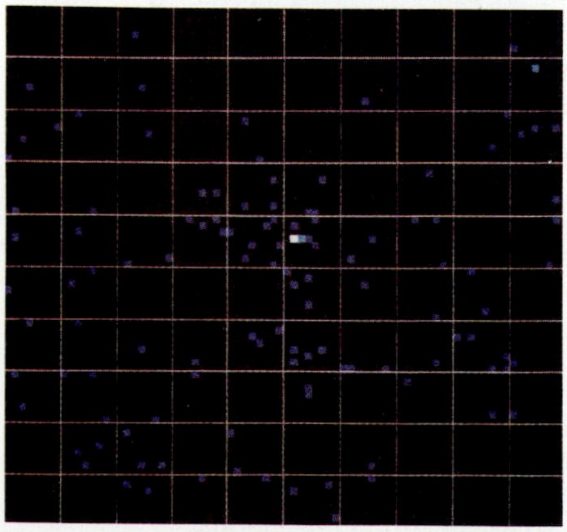

The Earth's atmosphere prevents the ultraviolet and x-ray region of quasar spectra from reaching us, but our ability to launch telescopes in spacecraft has eliminated this handicap. Since many of the spectral lines we observe in the visible spectra of quasars were originally emitted by the quasar in the ultraviolet and have since been redshifted into the visible, it is particularly important to study the ultraviolet spectra of relatively bright nearby quasars.

Studying nearby quasars in the ultraviolet is one of the key projects for the Hubble Space Telescope. Both spectra and images are being obtained. Many of the absorption lines observed in the ultraviolet are the Lyman-alpha line of hydrogen at different wavelengths (the Lyman-alpha forest). Some of the Hubble images are in the "snapshop survey;" these are relatively quick views made whenever there is a gap in other observing programs and thus without taking the time for the fine-guidance sensors to acquire the objects.

The sensitivity of the Einstein Observatory enabled astronomers to observe dozens of x-ray emitting quasars. Now Rosat is finding still more quasars at x-ray and other short wavelengths (Fig. 32–21*A*). The Compton Gamma Ray Observatory has observed quasars at still shorter wavelengths, gamma-ray quasars (Fig. 32–21*B*).

A

B

Figure 32–21 (*A*) Several quasars appear as bright as bright points in this image made in the extreme ultraviolet (between ultraviolet and x-rays) with the Wide Field Camera on Rosat. The dark circle is caused by a support in the Rosat detector. (*B*) The quasar 4C 38.41 in the constellation Hercules appears as the bright spot in this image made with the Energetic Gamma-Ray Experiment Telescope (EGRET) on the Compton Gamma-Ray Observatory.

Figure 32–22 The gravity of a massive object, perhaps a galaxy, in the line of sight can form multiple images of an object. If the alignment of the earth, the intermediate object, and the distant quasar is perfect, we would see a ring. A slight misalignment would make crescents or individual images.

*32.9 Double Quasars

The astronomical world was agog in 1979 at the discovery of a pair of quasars so close to each other that they might be two images of a single object. The story began when inspection of the Palomar Sky Survey charts revealed a close pair of 17th-magnitude objects at the position corresponding to a radio source.

When the spectra of the objects were taken, both objects turned out to be quasars, and their spectra looked identical. Even stranger, their redshifts were essentially identical.

It seemed improbable that two independent quasars should be so similar in both spectral lines and redshifts, so the scientists who took the first spectra suggested that both images showed the same object! They suggested that a gravitational bending of the quasar's radiation was taking place. We have

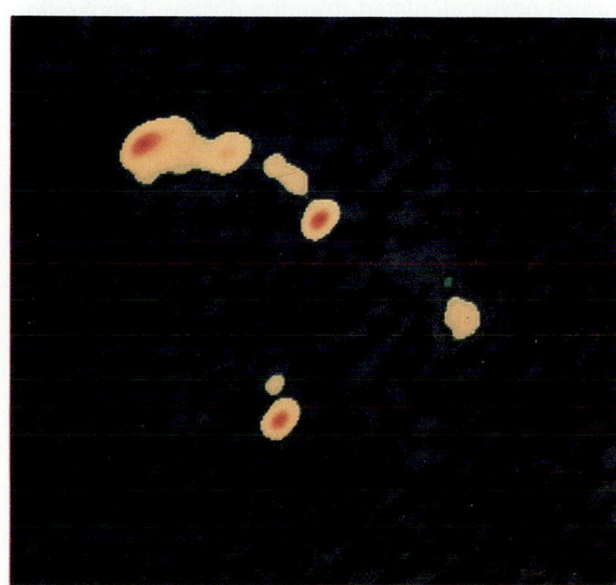

Figure 32–23 A radio map of the double quasar, Q 0957+561 A and B, made with the VLA. The elliptical sources correspond to the optical objects. Additional images are also seen, which could result from an off-center gravitational lens.

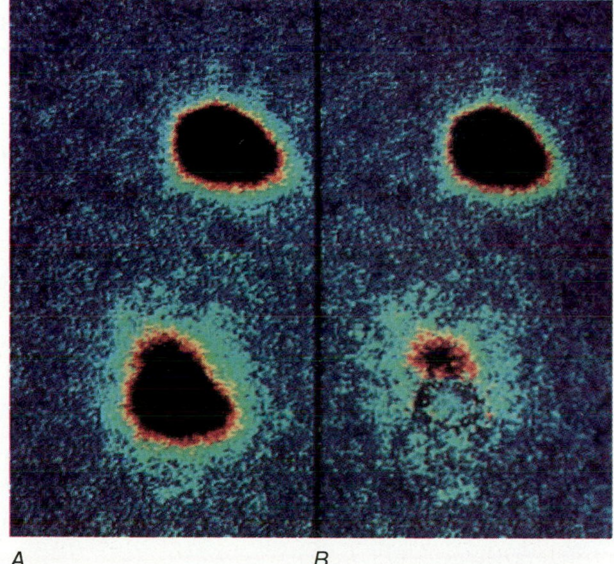

Figure 32–24 This series of optical false-color views of the double quasar, 0957+561A and B, reveals the intervening galaxy. (*A*) The actual images. The 17th-magnitude quasar images are bluish and are separated by only 5.7 seconds of arc. Their redshifts are identical to the third decimal place: $z = 1.4136 \pm 0.0015$. The intervening galaxy is only 1 second of arc from the "B" quasar image. Its redshift is about 0.37. The objects are in the constellation Ursa Major. (*B*) The top image has been subtracted from the bottom image to reveal just the intervening object.

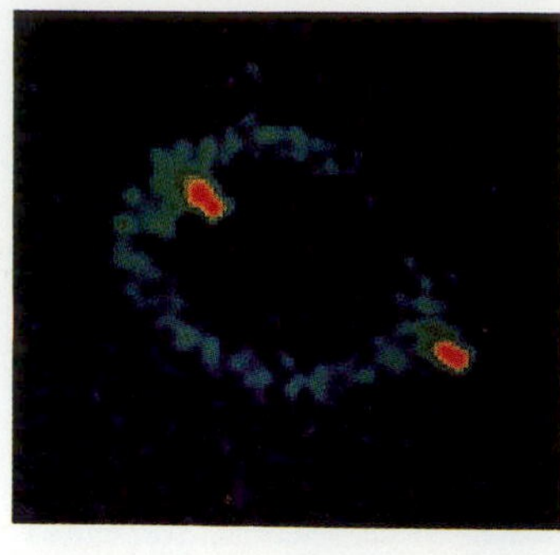

Figure 32–25 A quasar was so exactly aligned with the intervening gravitational lens that its radiation was spread out into an "Einstein ring." A second object is less well aligned; its image appears slightly inside the ring on one side and slightly outside the ring on the other on this VLA image.

discussed such gravitational bending as a consequence of Einstein's general theory of relativity and have seen that it has been verified for the sun (Section 23.11); in 1987, it was also used to explain galactic arcs (Section 31.2b). For the quasars, a massive object between the quasar and us is acting as a gravitational lens, bending the radiation from the quasar one way on one side and the other way on the other side (Fig. 32–22). As a result, we see the image of the quasar in at least two places. Detailed radio maps of the region have been made with several interferometers, most notably with the VLA (Fig. 32–23). The two images detected optically coincide with the two sharpest radio peaks. All the bright regions observable on the radio map cannot be explained—perhaps "dark matter" (Section 34.2a) does the additional lensing.

One test of whether a gravitational lens is present is to see if the brightness fluctuations of the two optical images are similar. Since the light from the two images travels along different paths through space, though, the fluctuations could be displaced in time. There are indications that the delay in fluctuations from one member of the double quasar 0957+561 to the other is 415 days. Further interpretation should give the mass of the lensing galaxy. Comparison of brightness expected for this mass with the observed brightness may eventually give us an independent measure of the Hubble constant.

The gravitational lens model was strongly endorsed by a photograph that seems actually to show the intervening galaxy (Fig. 32–24). The lensing object seems to be a giant elliptical galaxy with a redshift about one-fourth that of the quasar. Theoretical studies have calculated from the observed separation of objects that the lensing galaxy must contain at least 10^{12} solar masses, which is possible for that type of galaxy, though the "dark matter" that helps explain the radio images probably also contributes.

The double quasar was the first known example of a gravitational lens. Ten multiple quasars have since been discovered (Fig. 32–25); at least one also implies the existence of "dark matter." Searches are under way for still more. From the numbers found, we can predict that current techniques would allow us to find a thousand. These objects and the galactic arcs (Section 31.3c) are exciting and valuable verification of a prediction of Einstein's general theory of relativity.

The discovery of a quadruply lensed system (Fig. 32–26) would be so improbable without the lensing explanation that the discovery of gravitational lenses is generally accepted.

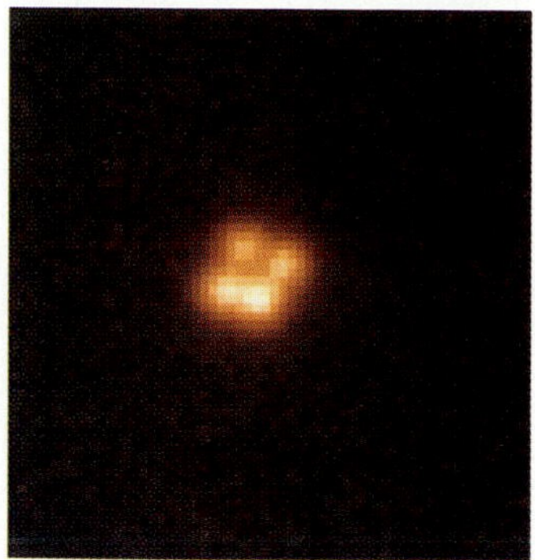

Figure 32–26 The "clover leaf," the quadruply lensed quasar H1413+117. The four images of comparable brightness are only 1 arc second apart. The spectra of two of the images are identical, except for some absorption lines in one that presumably come from different gas clouds than are in the other's line of sight. The redshift is 2.55. The rare configuration and identical spectra show that we are indeed seeing gravitational lensing rather than a cluster of quasars.

Summary and Outline

Discovery of quasars (Section 32.1)
- QSR—quasi-stellar radio source
- Huge Doppler shifts in spectra
- Radio-quiet objects also included as quasars

Doppler shift in quasars (Section 32.2)
- Large redshifts mean large velocities of recession and, by Hubble's law, great distances
- Relativistic formula used for redshifts close to or greater than 1

Importance of quasars (Section 32.3)
- Test of Hubble's law, which is thus a test of the accuracy of the distance scale
- Great distance, which means we view most of them as they were in an early stage of the universe

Energy problem (Section 32.4)
- Too much energy required from a small volume to be accounted for by ordinary processes
- Generally accepted: giant black hole is the engine

Non-Doppler explanations of redshifts (Section 32.5)
- Quasars ejected from cores of galaxies at high velocities; no evidence of blueshifted quasars, however
- Gravitational redshifts; conflict with observations
- Quasars operating under new physical laws?
- All these alternative explanations challenge Hubble's law, but are not now needed since quasars can be understood as being powered by black holes
- BL Lacertae and nearby quasars show that quasars are indeed at their Hubble-law distances

Galaxies and quasars (Section 32.6)
- Question of physical link between galaxies and quasars
- Stephan's Quintet and other sources with discrepant redshifts might be explained as chance alignments
- Other observations and statistical analysis show that quasars are indeed at their cosmological distances
- Seyfert and N galaxies also have bright nuclei; BL Lacertae and other nearby objects known to have distances corresponding to Hubble's law have properties in common with quasars
- Quasars seem to be events in cores of galaxies at some evolutionary stage; many quasars have surrounding fuzz that may be the underlying galaxies; the fuel may come from interactions with other galaxies

Superluminal velocities of quasars (Section 32.7)
- Components detected in several quasars seem to be moving apart at speeds greater than the speed of light
- Current explanation involves special theory of relativity and jets of gas pointing almost at us moving at velocities close to but less than the speed of light

Space observations of quasars (Section 32.8)
- Rocket and IUE studies in ultraviolet
- Einstein Observatory observations of quasars with large redshifts and of numerous faint quasars

Gravitational lenses (Section 32.9)

Key Words

quasar, Lyman-alpha forest, the energy problem, cosmological distances, local, Seyfert galaxy, N galaxy, fuzz, superluminal velocities*

*This term is found in an optional section.

Questions

1. Why is it useful to find the optical objects that correspond in position with radio sources?

†2. (a) You are heading toward a red traffic light so fast that it appears green. How fast are you going? (b) At $1 per mph over the speed limit of 55 mph, what would your fine be in court?

3. A quasar is receding at $\frac{1}{10}$ the speed of light. (a) If its distance is given by the Hubble relation, how far away is it? (b) At what wavelength would the 21-cm line appear?

†4. A quasar has $z = 0.3$. What is its velocity of recession in km/sec, using the non-relativistic formula?

†5. We observe a quasar with a spectral line whose rest wavelength is 3000 Å, but that is observed at 4000 Å. (a) How fast is the quasar receding, using the non-relativistic formula? (b) How far away is it if its distance is given by the Hubble relation? Specify the value you are using for Hubble's constant.

†6. The farthest known quasar has $z = 4.89$. At what wavelength does the Lyman α line, whose wavelength is 1216 Å from a source at rest, appear?

††7. A quasar is receding with a velocity 85 per cent of the speed of light. At what wavelength would a spectral line appear if it appears at 5000 Å when we observe it emitted from a gas in a laboratory on earth? What part of the spectrum is it in when it is emitted, and in what part of the spectrum do we observe it from the quasar?

†8. A quasar is receding with a velocity 95 per cent of the speed of light. At what wavelength would the Lyman-alpha spectral line appear? (It is emitted at 1216 Å.) What part of the spectrum is it in when it is emitted, and in what part of the spectrum do we observe it from the quasar?

†9. A quasar has $z = 2$. At what velocity is it receding from us, and what is its distance from us?

10. What do quasi-stellar radio sources and radio-quiet quasars have in common?

11. Why does the rapid time variation in some quasars make the "energy problem" even more difficult to solve?

12. What are three differences between quasars and pulsars?

13. Briefly list the objections to each of the following "local" explanations of quasars: (a) They are local objects flying around at large velocities. (b) The redshift is gravitational.

14. If quasars were proved to be local objects, would this help solve "the energy problem"? Explain.

15. Explain how parts of a quasar could appear to be moving at greater than the speed of light, without violating the special theory of relativity.

16. Describe the implications of the observations of quasars in the x-ray spectrum. What new ability allowed these observations to be made?

17. What features of some quasars suggest that quasars may be closely related to galaxies?

†This question requires a numerical solution.
††7. 17,500 Å; emitted in visible (blue); observed in infrared.

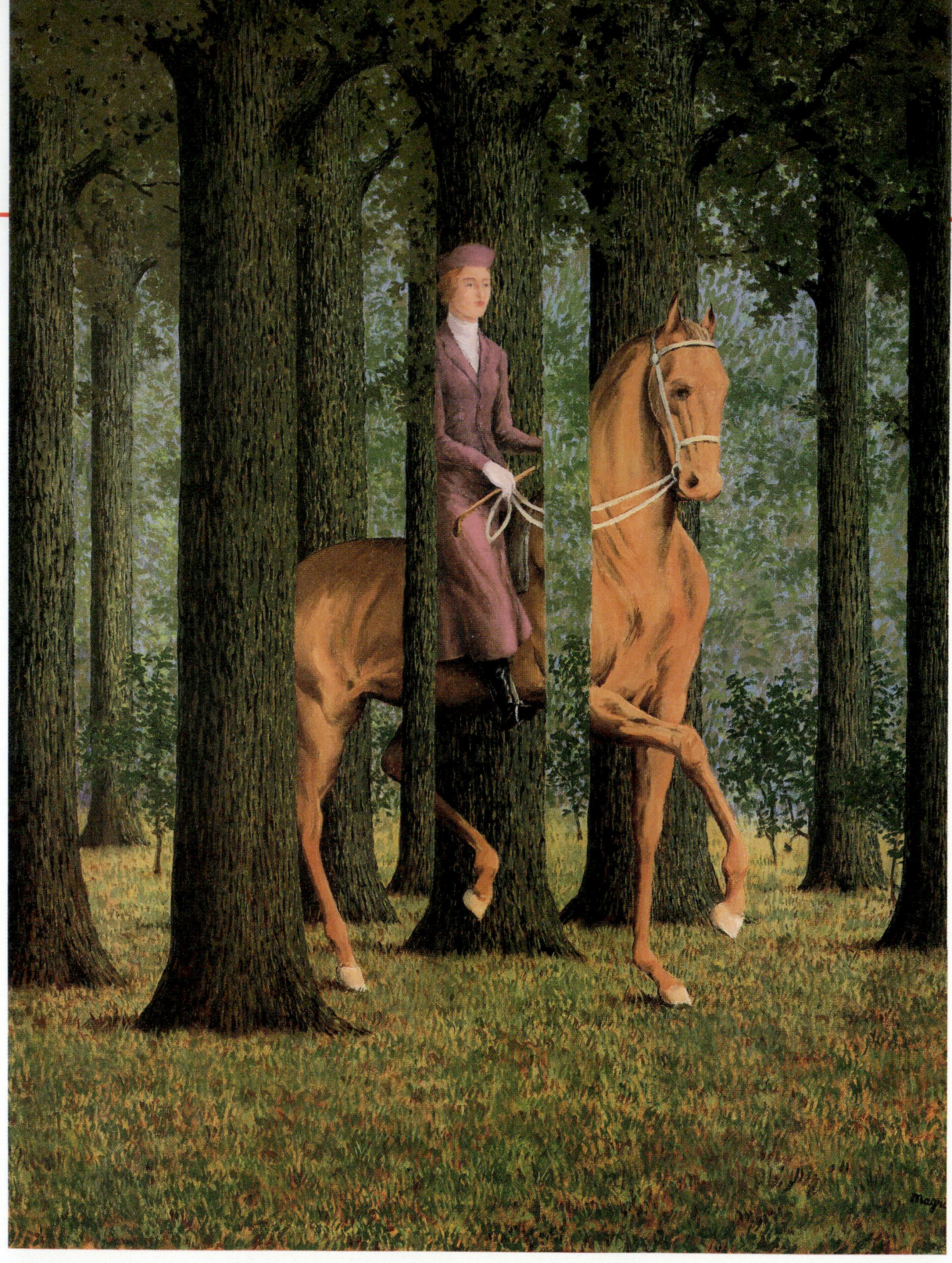

The equivalent of the opposite of Olbers's paradox: Why is the sky dark at night? If we look far enough in any direction, we should see the surface of a star. Why then isn't the universe uniformly bright? This painting by the Belgian surrealist René Magritte shows the opposite, in that our line of sight does not uniformly stop on trees in the forest or on the horse.

Cosmology

33

Aims: To study the origin of the universe and how the background radiation tells us about one of its early stages

In Armagh, Ireland, in the mid-seventeenth century, Bishop Ussher declared that the universe was created the evening preceding Sunday, October 23rd, in the year 4004 B.C. Nowadays we are less certain of the date of our origin (though we do set out a split-second agenda for the first few minutes of time).

We study the origin of the universe as part of the study of the universe as a whole, *cosmology*. Even more than in other parts of astronomy, in order to study cosmological problems we use simultaneously both theoretical calculations and all our abilities to observe a wide variety of celestial objects.

The study of where we have come from and of what the universe is like leads us to consider where we are going. Is the universe now in its infancy, in its prime of life, or in its old age? Will it die? It is difficult for us who, after all, spend a lot of time thinking of topics like "What shall I watch on TV tonight?" or "What's for dinner?" to realize that we can think seriously about the structure of space around us. It is awesome to realize that we can conclude what the future of the universe will be. One must take a little time every day, as Alice was told when she was in Wonderland, to think of "impossible things." By and by we become accustomed to concepts that may seem overwhelming at first. You must sit back and ponder when studying cosmology; only in time will many of the ideas that we shall discuss take shape and form in your mind.

33.1 Olbers's Paradox

Many of the deepest questions of cosmology can be very simply phrased. Why is the sky dark at night? Analysis of this simple observational question leads to profound conclusions about the universe.

We can easily see that the night sky is basically dark, with light from stars and planets scattered about on a dark background. But a bit of analysis shows that if stars are distributed uniformly in space, then the sky shouldn't be dark anywhere. If we look in any direction at all we will eventually see a star (Fig. 33–1), so the sky should appear uniformly bright. We can make the

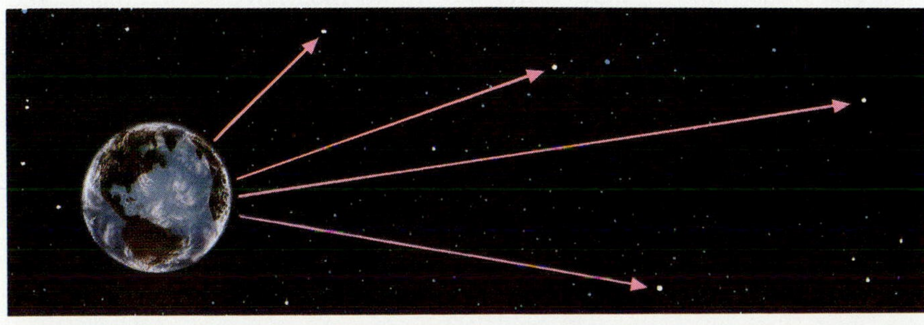

Figure 33–1 If we look far enough in any direction in an infinite universe, our line of sight should hit the surface of a star. This leads to Olbers's paradox.

Figure 33–2 The lower halves of trees are often painted in Mexican parks, and provide an example of a similar phenomenon to seeing a uniform expanse of starlight. Note that the surface brightness of a tree is the same whether the tree is close to us or far away.

analogy to our standing in a forest. There, we would see some trees that are closer to us and some trees that are farther away. But if the forest is big enough, our line of vision will always eventually stop at the surface of a tree. If all the trees were painted white, we would see a white expanse all around us (Fig. 33–2). The white on the trees that are farther away is of the same brightness as the white on the trees that are closer. Similarly, when looking up at the night sky we would expect the sky to have the uniform brightness of the surface of a star.

The fact that this argument implies that the sky is uniformly bright, while observation shows that the sky is dark at night, is called *Olbers's paradox.* (A paradox occurs when you reach two contradictory conclusions, both apparently correct.) Wilhelm Olbers phrased it in 1823, although the question had been discussed at least a hundred years earlier. Solving Olbers's paradox leads us into considering the basic structure of the universe. The solutions could not have been advanced in Olbers's time because they depend on more recent astronomical discoveries.

We have phrased Olbers's paradox in terms of stars, though we know that the stars are actually grouped into galaxies. But we can carry on the same argument with galaxies, and deduce that we must see the average surface brightness of galaxies everywhere. This is patently not what we see.

One might think that the easiest way out of this paradox, as was realized in Olbers's time, is simply to say that interstellar dust is absorbing the light from the distant stars and galaxies. But this doesn't solve the problem: because the sky is generally dark in all directions, the dust would have to be everywhere. This widely distributed dust would soon absorb so much energy that it would heat up and begin glowing. Given a long enough time, all the matter in the universe would begin glowing with the same brightness, and we would have our paradox all over again.

"Olbers" has an "s" on its end, so we write of "Olbers's paradox" (or, alternatively, Olbers' paradox); "Olber's paradox," with the apostrophe before a sole "s," is incorrect, since his name wasn't "Olber."

One solution to Olbers's paradox lies in part in the existence of the redshift, and thus in the expansion of the universe (Section 31.4). This does not mean that the answer to the paradox is simply that visible light from distant galaxies is redshifted out of the visible, for at the same time ultraviolet light is continually being redshifted into the visible. The point is rather that each quantum of light, each photon, undergoes a real diminution of energy as it is redshifted. The energy emitted at the surface of a faraway star or galaxy is diminished by this redshift effect before it reaches us.

But the redshift is not the whole solution of Olbers's paradox. Indeed, new calculations indicate that it would dim the background by only a factor of two. As we look out into space, we are looking back in time, because the light we see has taken a finite amount of time to travel to us. If we could see out far enough, we would possibly see back to a time before the stars were formed. E. R. Harrison of the University of Massachusetts has pointed out that this is the real way out of the dilemma. Harrison calculates that this explanation is a more important contribution to the solution of the paradox than is the existence of the redshift. In most directions, we would not expect to see the surface of a star for 10^{24} light years. We would thus have to be able to see stars 10^{24} years back in time for the sky to appear uniformly bright. However, the stars don't burn that long, and the universe is simply not that old. On the basis of Hubble's law, an age of somewhat over 10^{10} years seems to be the maximum.

The fact that we have to know about the expansion and the age of the universe to answer Olbers's question—why is the sky dark at night?—shows how the most straightforward questions in astronomy can lead to important conclusions. In this case, we find out about the expansion of the universe or about the lifetimes of the universe and the stars in it.

33.2 The Big-Bang Theory

Astronomers looking out into space have assumed that on a sufficiently large scale the universe looks about the same in all directions. That is, ignoring the presence of local effects such as our being in the plane of a particular galaxy and thus seeing a Milky Way across the sky, the universe has no direction that is special. Further, it is generally assumed that there is no change with distance either, except insofar as time and distance are linked.

These notions have been codified as the *cosmological principle:* **the universe is homogeneous and isotropic throughout space.** The assumption of *homogeneity* says that the distribution of matter doesn't vary with position (that is, with distance from the sun), and the assumption of *isotropy* says that the universe looks about the same no matter in which direction we look. Actually, of course, we know that we have to look in certain directions to see out of our galaxy without having our vision ended by the interstellar dust, but remember that we are ignoring inhomogeneities or lack of isotropy on this small scale. The discovery of giant voids in the universe and the possibility of the Great Attractor (Section 31.5) are new evidence that the universe may not, in fact, be homogeneous and isotropic.

The explanations of the universe that almost all astronomers now accept are known as *big-bang theories.* Basically, these cosmologies say that once upon a time there was a great big bang that began the universe. From that instant on, the universe expanded, and as the galaxies formed they shared in the expansion. The big-bang theories satisfy the cosmological principle.

Remember that the galaxies themselves are not expanding; the stars in a galaxy don't tend to move away from each other. It is space that is expanding, enlarging the distances among galaxies.

Many students ask whether the fact that there was a big bang means that there was a center of the universe from which everything expanded. The answer is no; first of all, the big bang may have been the creation of space itself. Furthermore, the matter of this primordial cosmic egg was everywhere at once. There may be an infinite amount of matter in the universe, so it is possible that at the big bang an infinite amount of matter was compressed to an infinite density while taking up all space. We would have to imagine our expanding raisin cake (Section 31.4b) extending infinitely in all directions, with no edge.

Consider a two-dimensional analogy to a universal expansion: the surface of a rubber balloon covered with polka dots (especially a non-expanding kind of polka dot). Though the balloon is three-dimensional, its surface has only two dimensions and we consider only its surface.

Let us consider the view if we are sitting on one dot. As the balloon is blown up, all the other dots seem to recede from us. No matter which dot we are on, all the other dots seem to recede. The **surface** of the balloon is two-dimensional yet has no edge. Even if we go infinitely far in any direction, we never reach a boundary. (Remember, Columbus did not fall off the edge of the earth.) There is no center to the surface from which all dots are actually expanding. It is much more difficult for most of us to visualize a three-dimensional situation like an infinite raisin cake as in Section 31.4b (or, including time as a fourth dimension, a four-dimensional space-time). Yet the above analogy is valid, because our universe can expand uniformly yet have no center to the expansion.

What will happen in the future? One possibility is that the universe will continue to expand forever. This case is called an *open universe.* It corresponds to the case in which the universe is infinite. The other possibility is that at some time in the future the universe will stop expanding and will begin to contract. This case is called a *closed universe* (Fig. 33–3), and corresponds to the case in which the universe is finite (though it may still have no boundary, just as we can continue straight ahead forever on the surface of a balloon). In a closed universe, we would eventually reach a situation that we might call a

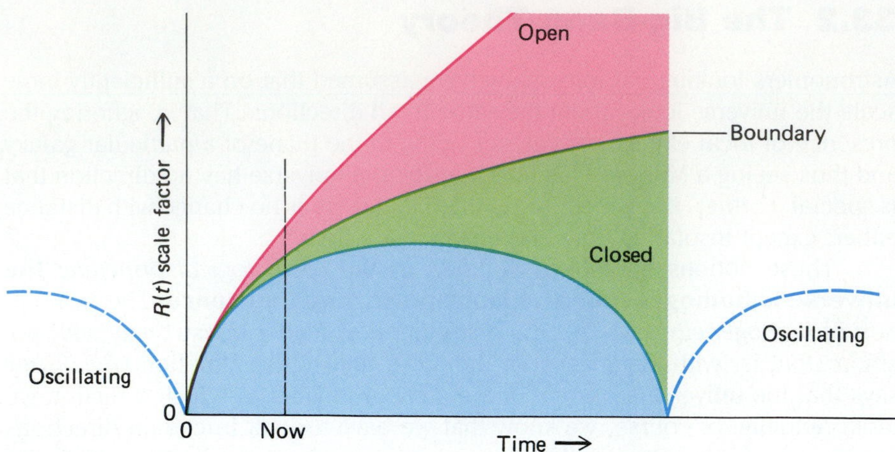

Figure 33–3 To trace back the growth of our universe, we would like to know the rate at which its rate of expansion is changing. Big-bang models of the universe are shown; the vertical axis represents a scale factor, R, that represents some measure of distances and how they change as a function of time, t. The universe could be open and expand forever, or be closed and begin to contract again. If it is closed, we do not know whether it would oscillate or whether we are in the only cycle of expansion or contraction it will ever undergo. We will see later that current evidence favors the open model.

Astronomers use a quantity called the *deceleration parameter,* which is given the symbol q_0, to describe how fast the expansion is slowing down. $q_0 = 1/2$ marks the dividing line between an open universe (q_0 less than $1/2$, including 0) and a closed universe (q_0 greater than $1/2$, including 1). In the flat case—the boundary between the open and closed—the universe would become infinitely large in an infinitely long time, with its rate of expansion approaching zero, and thus never quite collapse.

"big crunch." The situation dividing open and closed is known as a *flat universe.* A flat universe forever expands at a slowing rate that would make it cease expanding only after an infinite time.

The open and closed universe can be considered in an analogy from geometry (Fig. 33–4). Einstein's general theory of relativity predicts that space itself could show properties of being curved, just as the surface of a saddle or the surface of a sphere is fundamentally curved. On none of these surfaces, for example, can we draw straight lines that remain parallel out to infinity, that is, never crossing and always remaining the same distance from each other. Is space in the universe curved positively like a sphere, curved negatively like a saddle, or flat? We do not know. The closed universe would be like the surface of a sphere, which bends inward, while the open universe would be like the surface of a saddle, which bends outward.

In principle, one could decide whether the universe is open or closed by counting the number of radio sources or of other galaxies in shells of constant thickness going outward from our galaxy, since the different geometries imply different changes in the number of sources. But the fact that radio sources and other galaxies may have been brighter or fainter far back in time complicates matters too much. Such studies really tell us mostly about the evolution of radio sources and of other galaxies.

What was present before the big bang? There is no real way to answer this question. For one thing, we can say that time began at the big bang, and that it is meaningless to talk about "before" the big bang because time didn't exist. We do not now think that the universe can remain in a static condition, so it seems unlikely that the universe was always just sitting there in an infinitely compressed state, whatever "always" means.

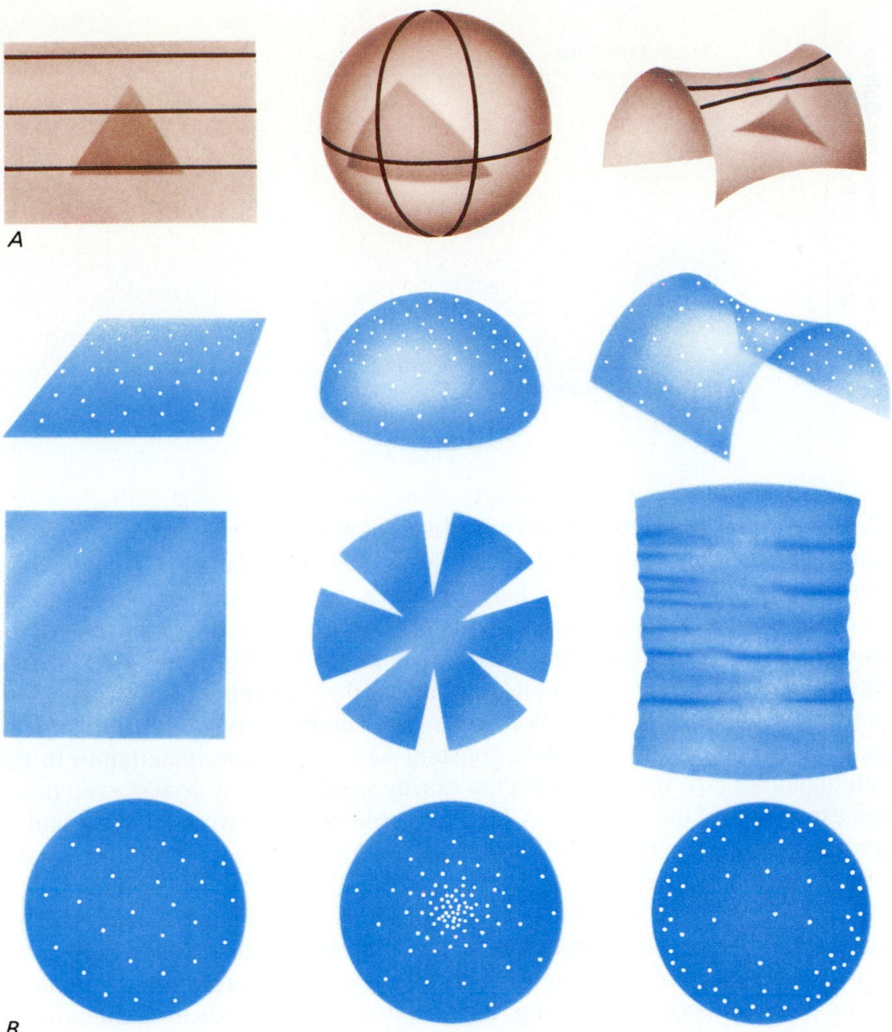

Figure 33–4 Two-dimensional analogues to three-dimensional space. (*A*) A closed universe is like the surface of a sphere in several ways, including the fact that great circles (full diameters) always intersect and cannot be parallel. An open universe is like the surface of a saddle in several ways, including the fact that an infinite number of parallel lines can be drawn through the same point. A flat universe is in between. These concepts of geometry follow from the geometrical calculation of the volume of a shell. (*B*) Consider radio sources uniformly distributed throughout space (*top row*), and count how the number changes with distance from us. If we use the formula for volume from a flat space, we would essentially be flattening out the curvature (*middle row*). The result for the different curvatures is shown at bottom. Though such counts were made for a long time, it now seems that we are seeing so far out into space that such arguments tell us more about how radio sources evolve over time than they do about cosmology.

Of course, these possibilities don't answer the question of why the big bang happened. It is possible that there had been a prior big bang and then a recollapse, and that our current big bang was one in an infinite series. This version of a closed-universe theory is called the *oscillating universe.* But the oscillating-universe theory doesn't really tell us anything about the origin of the universe because in this case there was no origin. The concepts discussed in these paragraphs will not be easy to digest. They may take hours, years, or a lifetime to come to terms with.

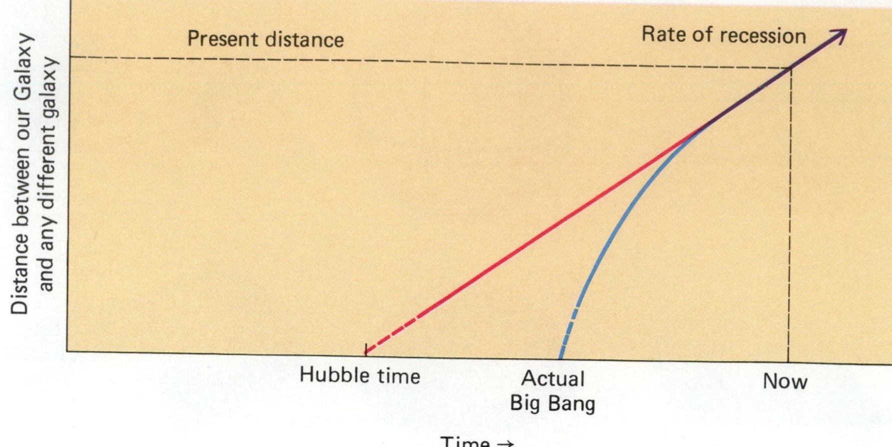

Figure 33–5 If we could ignore the effect of gravity, then we could trace back in time very simply; the Hubble time corresponds to the inverse of the Hubble constant ($1/H_0$). Actually, gravity has been slowing down the expansion. The vertical axis again represents some scale factor.

If we consider Hubble's law with the current value for Hubble's constant—and there is good evidence that Hubble's constant is the same for relatively nearby galaxies as it is for very distant galaxies—we can extrapolate backward in time (Fig. 33–5). We simply calculate when the big bang would have had to take place for the universe to have reached its current state at its current rate of expansion. This calculation indicates that the universe is somewhere between 12 billion and 20 billion years old. The range of values indicates in part the uncertainty in Hubble's constant and in part the uncertainty in the calculation of how much the effect of gravity would have by now slowed down the expansion of the universe. A larger Hubble constant would correspond to a younger age, given the assumption that the universe is expanding at a constant or at a decreasing rate.

Open or closed, the universe will last at least another forty billion years, so we have nothing to worry about for the immediate future. Nevertheless, the study of the future of the universe is an exceptionally interesting investigation. Some of the methods of tackling this question involve measuring the amount of mass in the universe and thus the amount of gravity (as we will discuss in Section 34.2). Other methods of determining our destiny involve looking at the most distant detectable objects to see if any deviation from Hubble's law can be determined. This investigation of distant bodies has been carried out many times, but a deviation from the straight-line relation between velocity and distance known as Hubble's law has not been found conclusively. The deviations have always been within the uncertainty of the measurements. One major uncertainty comes from the fact that the best independent measure of the distances to the farthest galaxies comes from their brightnesses (Fig. 33–6).

We now realize that the farthest galaxies, from which we had hoped to best determine a deviation from Hubble's law, may well have been very different long ago when they emitted the light we are now receiving. This point was made especially by Beatrice Tinsley in the 1970's. Thus we cannot assume that even such basic properties as their size and brightness were similar then to their size and brightness now. Since determining the distance to these distant galaxies depends on understanding their properties, it now seems that such considerations of galaxy evolution are too uncertain to allow us to determine in this way whether the universe is open or closed.

Big-bang theories actually arise as solutions to a set of equations that Einstein advanced as part of his general theory of relativity. Early solutions by Einstein himself (Fig. 33–7) and others were not valid. We now use solutions worked out in the early 1920's by the Soviet mathematician Alexander Friedmann. The Belgian abbé Georges Lemaître discovered similar solutions.

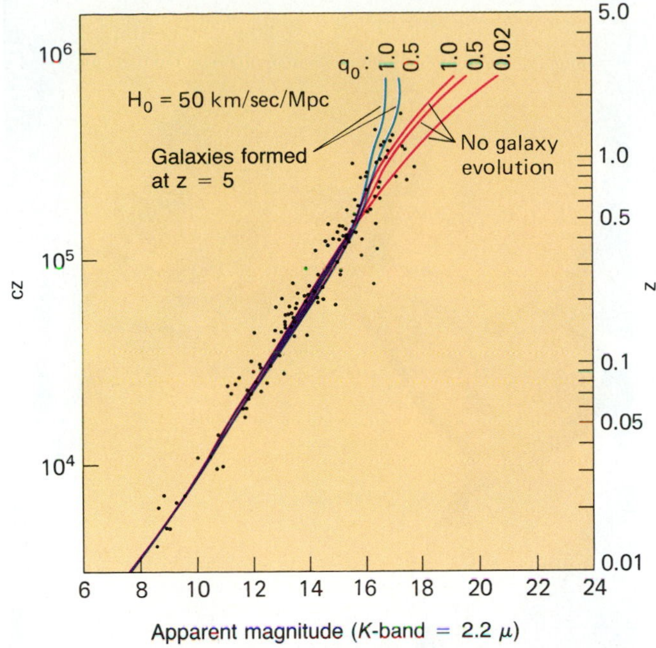

Figure 33–6 This observational Hubble diagram is plotted in terms of redshift and magnitude, since magnitude (a measure of distance) can be directly measured. It plots the magnitude of the brightest galaxy in each cluster of galaxies vs. its redshift. Interstellar absorption is relatively small at the infrared magnitude used; the magnitudes are corrected for the size of the galaxy images in the observing equipment.

In principle, we can determine the future of the universe from the slight deviations at upper right of the curves from the Hubble law, modified from a straight line by instrumental corrections. Curves are labelled with the deceleration parameter q_0. If the points appear to the upper left of the unshown q_0 curves, the expansion is slowing down. If the points fall far enough to the left (to the left of the curve for $q_0 = \frac{1}{2}$), then the universe is closed, is finite, and will eventually begin to contract. If the points are not so far to the left, then the universe is open, is infinite, and will expand forever. A special difficulty that prevents us deciding whether the universe is open or closed is the effect of the galaxies' evolution, which is very uncertain. Blue and red curves show the predictions from different assumptions about galaxy evolution, since evolving galaxies could have had different brightnesses long ago from their current brightnesses. The points do not seem to fit any of the no-evolution curves, so we conclude that evolution took place.

Focus On

*Box 33.1 Einstein's Principle of Equivalence

Our notion of the structure of the universe is based on calculations using the general theory of relativity Albert Einstein advanced in 1916. At the basis of the theory is the idea Einstein formulated showing that gravity and accelerated motion are fundamentally equivalent, a notion known as the *principle of equivalence.*

Einstein explained the idea with a "thought experiment," though we can now have real examples for the situations (Fig. 33–8). Consider people in a closed elevator, "Einstein's elevator," who cannot see out. On the Earth's surface, gravity is a downward force. But even if they were out in space in a location where there is no gravity, there would still be an identical force in one direction as the wall or floor pushes on them if their spacecraft accelerated forward in the opposite direction at the proper rate. Einstein's principle of equivalence holds that there is no way that people in the elevator, without contact with the outside, could tell whether such a pull came from gravity or from an acceleration.

In the vicinity of a massive object, there would be a force toward the object; the force would be in a different direction depending on where we were located with respect to the massive object. We could, following the principle of equivalence, consider ourselves to be in a curved space accelerating down a slope toward the massive object instead of being subject to the object's gravity. Thus we can explain gravity as a curvature of space (Fig. 33–9). The observational tests of general relativity we considered in Section 23.11 are actually tests of only the principle of equivalence.

Figure 33–7 The statue of Albert Einstein at the National Academy of Sciences, Washington, D.C.

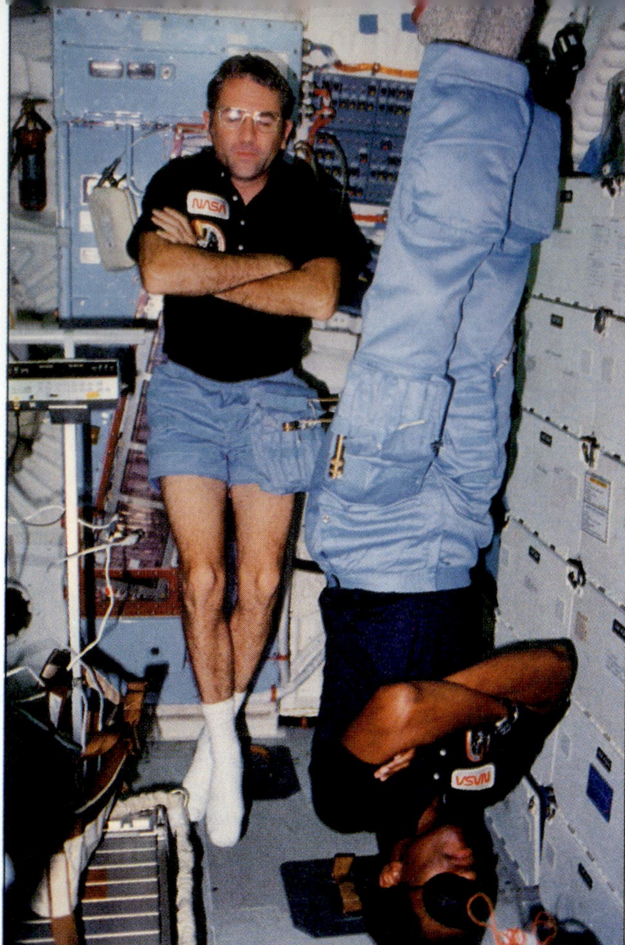

A

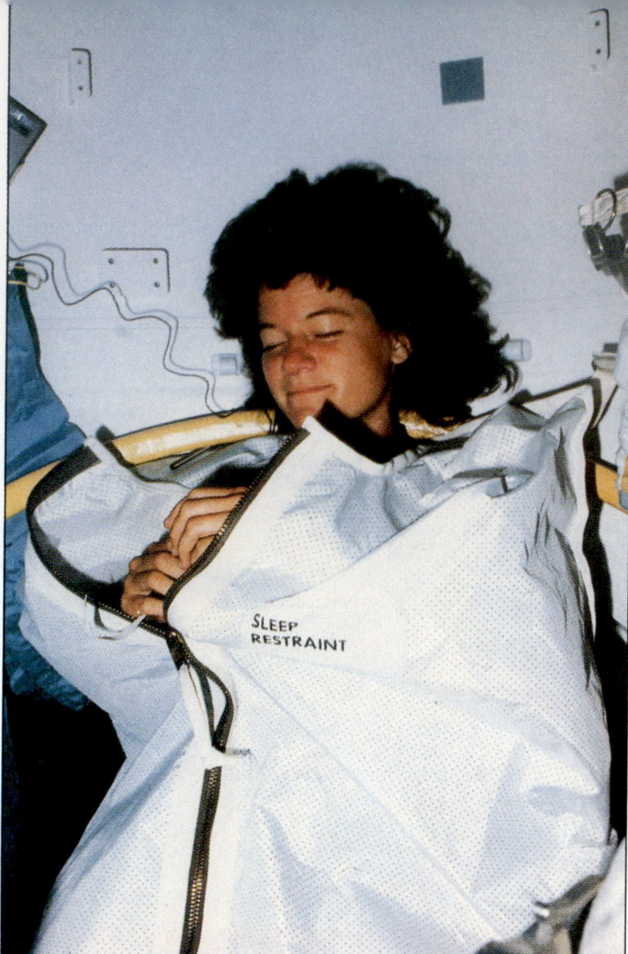

B

Figure 33–8 In "free fall," the spacecraft is accelerating toward earth with the same acceleration that the astronauts have; they thus do not sense any gravity, since they are not pressed against anything. (*A*) Space-shuttle astronauts Richard H. Truly (*left*) and Guy Bluford (*right*) resting. (*B*) Space-shuttle astronaut Sally Ride in a sleep restraint, which keeps her from floating around the cabin.

Recent measurements of the gravity pulling back on galaxies in the huge chain of galaxies known as the Great Wall (Section 31.4) can bear on the question whether the universe is open or closed. Within the Great Wall is the Coma cluster of galaxies, a high-density region 150 million light-years across. Next to it is an empty bubble of the same size, surrounded by galaxies in the bubble wall. Greg Bothun of the University of Michigan joined Margaret Geller and John Huchra of the Harvard-Smithsonian Center for Astrophysics to measure the force pulling the galaxies in the bubble wall toward the Coma cluster. The more force needed, the higher the average mass density of the universe must be. Their measurements, released in 1990, seem to show that the average mass density is relatively low, making the universe open. However, other scientists have measured higher values for the density, so the conclusions are not definitive.

Figure 33–9 A warped golf green is a two-dimensional analogy to curved space. Though the golf ball is rolling "straight," it appears to curve as Arnold Palmer putts here because the surface of the green is curved. Similarly, light travelling through curved space appears to curve even though it is travelling "straight." This test of the principle of equivalence has been verified for light at eclipses and for radio waves by observing quasars apparently passing near the sun.

33.3 The Steady-State Theory

The cosmological principle, that the universe is homogeneous and isotropic, is very general in scope, but starting in the late 1940's, three British scientists began investigating a principle that is even more general. Hermann Bondi, Thomas Gold, and Fred Hoyle considered what they called the *perfect cosmological principle:* the universe is not only homogeneous and isotropic in space but also **unchanging in time.** A criterion for science is that we must accept the simplest theory that agrees with all the observations, but it is a matter of personal preference whether the cosmological principle or the perfect cosmological principle is simpler.

The theory that follows from the perfect cosmological principle is called the *steady-state theory* (Fig. 33–10); it has certain philosophical differences from the big-bang cosmologies. For one thing, according to the steady-state theory, the universe never had a beginning and will never have an end. It always looked just about the way it does now and always will look that way.

The rate at which new matter would have to be created in the steady-state theory works out to be only one hydrogen atom per cubic centimeter of space every 10^{15} years, equivalent to one thousand atoms of hydrogen per year in a volume the size of the Astrodome in Houston. This is far too small for us to be able to measure. The "law of conservation of mass-energy" would thus not be valid at this level, in the steady-state theory.

The steady-state theory must be squared with the fact that the universe is expanding. How can the universe expand continually but not change in its overall appearance? For the density of matter to remain constant, new matter must be created at the same rate that the expansion would decrease the density. Only in this way can the density remain the same.

The matter created in the steady-state theory is not simply matter that is being converted from energy by $E = mc^2$. No, this is **matter that is appearing out of nothing,** and is thus equivalent to energy appearing out of nothing.

For many years a debate raged between proponents of the big-bang theories and proponents of the steady-state theory. The evidence that came in—usually seeming to indicate that distant objects were somehow different from closer ones, which would show that the universe was evolving—seemed to favor the big-bang cosmologies over the steady-state theory. But none of this evidence was conclusive because alternative explanations for the data could be proposed or the steady-state theory itself could be modified (sometimes extensively) to be consistent with the discoveries.

The discovery of quasars provided some of the strongest evidence against the steady-state theory. The quasars are for the most part located far away in space, and so were more numerous at an earlier time. Thus something has been changing in the universe, and change is not acceptable in the steady-state theory. As the evidence that the quasars were indeed at these distances grew, the status of the steady-state theory diminished.

In the next section we shall discuss the still stronger evidence that provided the crushing blow against the steady-state theory.

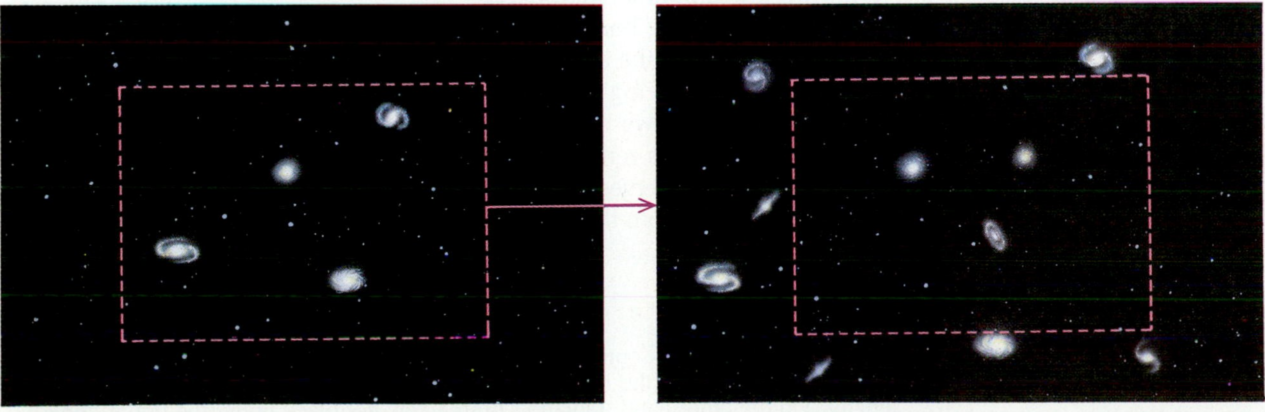

Figure 33–10 In the steady-state theory, as the dotted box at left expands to fill the full box at right, new matter is created to keep the density constant. In the picture, the four galaxies shown at left can all still be seen at right, but new galaxies have been added so that the number of galaxies inside the dotted box is about the same as it was before.

33.4 The Primordial Background Radiation

Figure 33–11 Arno Penzias (*left*) and Robert W. Wilson (*right*) with their horn-shaped antenna in the background. Penzias and Wilson found more radio noise than they expected at the wavelength of 7 cm at which they were observing. After they removed all possible sources of noise (by fixing faulty connections and loose antenna joints, and by removing "sticky white contributions" from nesting pigeons), a certain amount of radiation remained. It was the 3° background radiation. Penzias and Wilson won the 1978 Nobel Prize in Physics for their discovery.

In 1965, a discovery of the greatest importance was made: radiation was detected that is most readily explained as a remnant of the big bang itself.

That much is easy to state, and if we have in fact discovered radiation from the big bang itself, then clearly the steady-state theory is discredited. The discovery was made by Arno A. Penzias and Robert W. Wilson of the Bell Telephone Laboratories in New Jersey (Fig. 33–11). They were testing a radio telescope and receiver system to try to track down all possible sources of static. The discovery, in this way, parallels Jansky's discovery of radio emission from space, which marked the beginning of radio astronomy.

Penzias and Wilson were observing at a wavelength of several centimeters, which is in the radio spectrum. After they had subtracted, from the static they observed, the contributions of all known sources, they were left with a residual signal that they could not explain. The signal was independent of the direction they looked, and did not vary with time of day or season of the year. The remaining signal corresponded to the very small amount of radiation that would be put out at that frequency by a black body at a temperature of only 3 K, 3° above absolute zero.

At the same time, Robert Dicke, P. J. E. Peebles, David Roll, and David Wilkinson at Princeton University had, coincidentally, predicted that radiation from the big bang should be detectable by radio telescopes. First, they concluded that radiation from the big bang permeated the entire universe, so that its present-day remnant should be coming equally from all directions. Second, they concluded that this radiation would have the spectrum of a black body, which just means that the amount of energy coming out at different wavelengths can be described by giving a temperature. Third, they predicted that though the temperature of the radiation was high at some time in the distant past, the radiation would now correspond to a black body at a particular very low temperature, only a few degrees above absolute zero (Fig. 33–12). If radiation could be found that came equally from all directions and had the low-temperature black-body spectrum that matched the prediction, then the theory would be confirmed.

We have discussed black bodies in Section 20.2. Basically, the emission from a black body follows Planck's law of radiation (Fig. 20–1) in that for a given temperature there is an equation that tells us the intensity of radiation at each wavelength. The key fact to remember about Planck's law is that specifying just one number—the temperature—is enough to define the whole Planck curve.

The Princeton group continued the process of building their own receiver to observe at a different radio wavelength. They were soon able to measure the intensity at this wavelength, and they, too, found that it corresponded to that of a black body at 3°. This tended to confirm the idea that the radiation was indeed from a black body, and thus that it resulted from the big bang.

Actually, the Princeton group was not the first to predict that such radiation might be present. Many years earlier, Ralph Alpher and Robert Herman (in 1948) and George Gamow (in 1953) had made similar predictions, but this earlier work was at first overlooked.

Because the big bang took place simultaneously everywhere in the universe, radiation from the big bang filled the whole universe. The radiation thus has the property of being isotropic to a very high degree; that is, it is the same in any direction that we observe. It was generated all through the universe at the same time, so its remnant must seem to come from all around us now. The fact that the observed radiation was highly isotropic was thus strong evidence that it came from the big bang. It also shows that the early universe was very homogeneous.

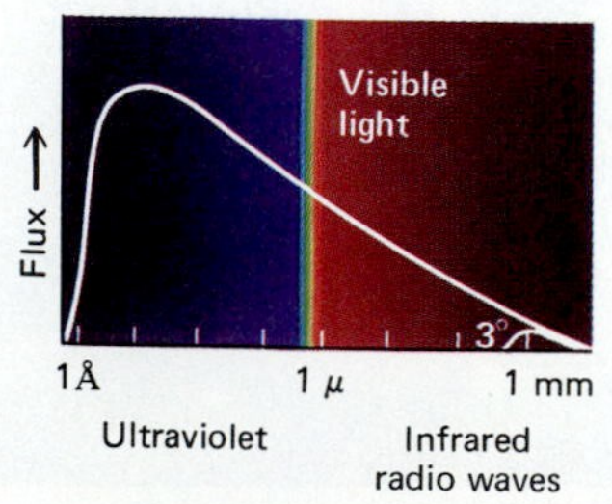

Figure 33–12 Planck curves for black bodies at different temperatures. Radiation from a 3 K black body (the inner curve at lower right) peaks at very long wavelengths. On the other hand, radiation from a very hot black body (the upper curve) peaks at very short wavelengths.

33.4a The Origin of the Background Radiation

The leading models of the big bang consider a hot big bang. The temperature was billions upon billions of degrees in the fractions of a second following the beginning of time.

In the millennia right after the big bang, the universe was opaque. Photons did not travel very far before they were scattered by electrons, the same process that now makes our sky blue. This process results in black-body radiation that corresponds to the temperature of the matter.

Gradually the universe cooled. After about a million years, when the temperature of the universe reached 3000 K, the temperature and density were sufficiently low for the hydrogen ions to combine with electrons to become hydrogen atoms. This *recombination* took place suddenly. Since hydrogen has mainly a spectrum of lines rather than a continuous spectrum and since no electrons remained free, the gas suddenly lost its ability to absorb photons except at a few wavelengths. Thus from this time on, most photons could travel all across the universe without being absorbed by matter. The universe had become transparent.

Since matter rarely interacted with the radiation from that time to the present, it no longer continually recycled the radiation. We were thus left with the radiation at the temperature it had at the instant when the universe became transparent. This radiation is travelling through space forever. Observing it now is like studying a fossil. As the universe continues to expand, the radiation retains the shape of a Planck curve but the curve corresponds to cooler and cooler temperatures.

Since the radiation is present in all directions, no matter what we observe in the foreground, it is known as the *cosmic background radiation.* In reference to its origin, it is also called the *primordial background radiation.*

In the above scenario, the universe has changed from a hot, opaque place to its current cold, transparent state. Such a change is completely inconsistent with the steady-state theory.

33.4b The Temperature of the Background Radiation

Planck curves corresponding to cooler temperatures have lower intensities of radiation and have the peaks of their radiation shifted toward longer wavelengths. We have seen that radiation from the sun, which is 6000 K, peaks in the yellow-green and that radiation from a cool star of 3000 K peaks in the infrared. The universe, much cooler yet, peaks at still longer wavelengths: the peak of the black-body spectrum is at the dividing wavelength between infrared and radio waves. In a moment, we shall see just how cool the universe is.

Now, we recall that Penzias and Wilson measured a particular flux at the wavelength measured by their equipment. This flux corresponds to what a black body at a temperature of only 3 K would emit. Thus we speak of the universe's *3° background radiation**. Following the original measurement at Bell Labs and at Princeton, other groups soon measured values at other radio wavelengths. These other values lay along the slope of one side of the black-body curve, and so showed that the value was indeed about 3°.

Unfortunately, though, for many years we were able to observe only at wavelengths longer than that of the peak, so only the right-hand half of the curve was defined. All the points corresponded to a temperature of approximately 3 K, but still, a black-body curve does have a peak. It would have been more satisfying if some of the points measured lay on the left side of the peak. Points lying on the left side of the peak would prove conclusively that the

*In the Système International, this would be called 3 K radiation, but we shall join the astronomical community in continuing to call it 3° radiation.

radiation followed a black-body curve and was not caused by some other mechanism that could produce a straight line, or some other form that happened to mimic a 3 K black-body curve in the centimeter region of the spectrum.

Though it would have been desirable to measure a few points on the short-wavelength side of the black-body curve, astronomers were faced with a formidable opponent that frustrated their attempts to measure these points: the earth's atmosphere. Our atmosphere absorbs most radiation from the long infrared wavelengths.

In 1975, infrared observations were made from balloons that seem to have proved unequivocally that the radiation follows a black-body curve. More recent measurements appear in Figure 33–13. One point from a rocket experiment seemed to be high, and theoreticians worked out models that would explain this seeming extra infrared flux. For example, an otherwise undetected generation of stars before the ones we know now might have provided the flux. But the high point could have been the result of an imperfect correction of the results for the effects of the earth's atmosphere. Confirmation was needed. In Section 33.4d, we shall discuss the dramatic conclusion of this episode.

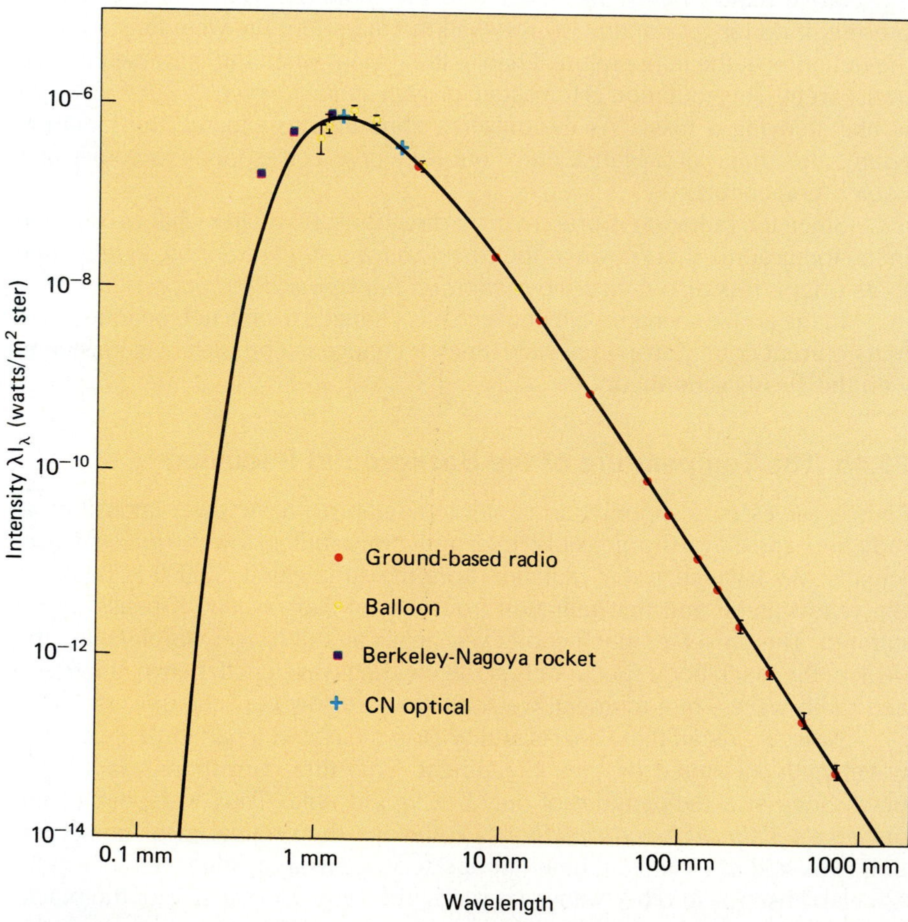

Figure 33–13 Observations of the cosmic background radiation as of the end of 1989. Ground-based radio observations are shown as red filled circles. Balloon observations are shown as green filled circles. The results of a rocket flight made by a collaboration between the University of California at Berkeley and Nagoya University in Japan are shown as orange squares; the apparent deviation from a black-body curve in the submillimeter region of the spectrum was not verified by satellite observations. The results from optical studies of the molecule CN, which can be interpreted to show how radio wavelengths affect the molecules, are shown as blue + signs.

So at present, astronomers consider it settled that radiation has been detected that could only have been produced in a big bang. Accepting this interpretation clearly rules out the steady-state theory. It means that some version of the big-bang theories must hold, though it doesn't settle the question of whether the universe will expand forever or will eventually contract.

33.4c The Background Radiation as a Tool

Now that the spectrum has been measured so precisely on both sides of the peak of intensity, we have moved beyond the point of confirming that the radiation is black-body. We can now study its deviations from the general black-body curve and from isotropy.

The importance of the discovery of the background radiation cannot be overstressed. Dennis Sciama, the British cosmologist, put it succinctly by saying that up to 1965 we carried out all our calculations knowing just one fact: that the universe expanded according to Hubble's law. After 1965, he said, we had a second fact: the existence of the background radiation. That may be a bit oversimplified, but it is essentially true.

A very slight difference in temperature—about one part in one thousand—has been measured from one particular direction in space to the opposite direction. Such a difference is called an *anisotropy*—a deviation from isotropy (Fig. 33–14). To limit the effect of the earth's atmosphere on the measurements, they were made from low-water-vapor regions like the south pole, from high-flying aircraft, and from balloons. The anisotropy measured is what would result from the Doppler effect if our sun was moving away from the red region and toward the blue region at about 600 km/sec with respect to the background radiation. (The blue stripe of variable intensity that runs obliquely around the globe is emission from our own galaxy.) Since we know the velocity with which our sun is moving as it orbits the center of our galaxy, we can remove the effect of our sun's orbital motion. We deduce that our galaxy is moving about 500 km/sec relative to the background about 40° from the direction of the Virgo Cluster. We do not know any particular reason why our galaxy should have such a velocity. A small yearly effect from the earth's motion around the sun

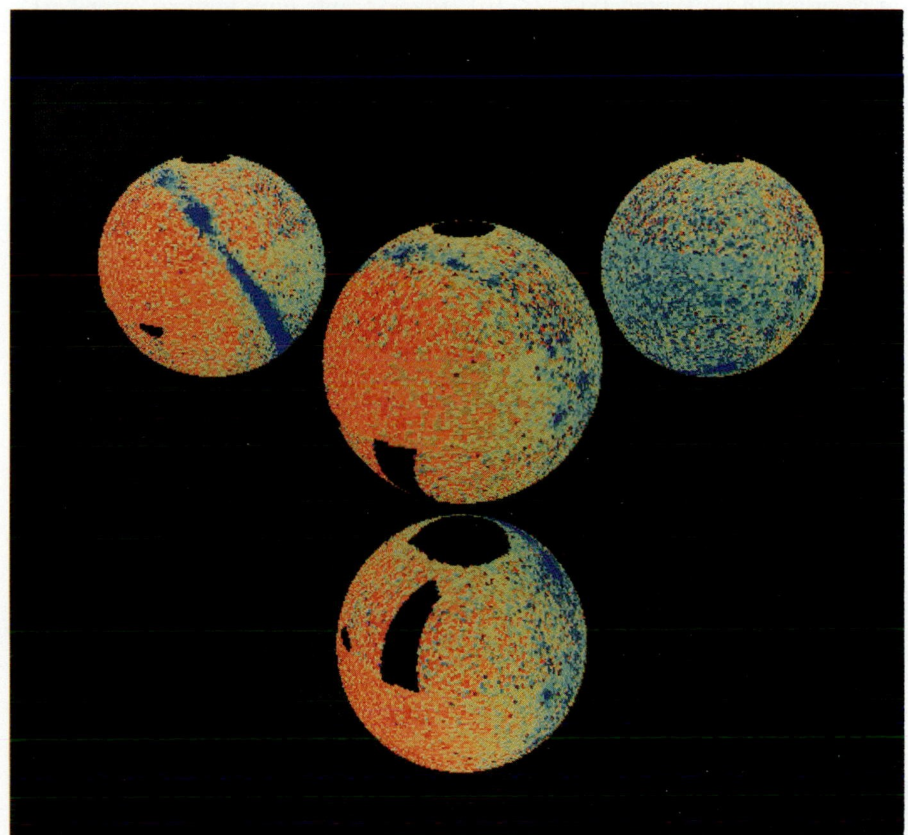

Figure 33–14 The anisotropy of the background radiation is shown by the range of colors in this all-sky map made from four balloon flights. The total range is from +3 millikelvins (*reddish*) at upper left to −3 millikelvins (*bluish*). The center globe is a map of the sky brightness at 1.5 cm (19 GHz), at which the earth's atmosphere contributes relatively little. Three reflecting mirrors show the parts of the globe that are hidden from direct view. No data are available for regions shown in black.

Figure 33–15 The COBE (**Co**smic **B**ackground **E**xplorer) spacecraft, remade smaller to fit on a Delta rocket after the explosion of the space-shuttle Challenger. COBE was launched in 1989. It has greatly improved the accuracy of measurements of the background radiation.

has also been detected. Both effects give the direction we are heading a slightly higher temperature than the direction behind us. The wavelengths are blueshifted, shifting the Planck curve toward the blue; a bluer Planck curve corresponds to a higher temperature. In spite of careful searching, no variations have been found on a smaller scale across the sky.

Observations that relate infrared brightnesses of galaxies with their distances measured from radio studies can also be used to measure overall velocities in different directions. The overall velocity measured in these studies is similar to that deduced from the anisotropy in the background radiation.

Measurements of the anisotropy of the background radiation are now being very effectively made from an observing station at the south pole. The high altitude and exceedingly dry atmosphere allow the background radiation to be observed at shorter wavelengths than are otherwise normally accessible.

33.4d Cosmic Background Explorer

COBE (Fig. 33–15), the Cosmic Background Explorer (pronounced "koh'bee"), is a NASA spacecraft meant to study the background radiation. It was launched in 1989, carrying enough liquid helium to cool its instruments for about a year. It orbits the earth in a polar orbit with the sun always off to the side.

One of its experiments was to study the spectrum over a wide wavelength range to compare with a black body. Indeed, it carried an experimental black body aloft to be placed in the telescope's beam from time to time, for comparison. This set of observations has turned out to be wildly successful. The set of points it measured (Fig. 33–16) (black points with vertical error bars) agrees extremely closely with a black body (colored curve) at a temperature of 2.735 ± 0.06 K, where the uncertainty represents the scientists' estimate of systematic errors that might be cropping up. The error bars will become even

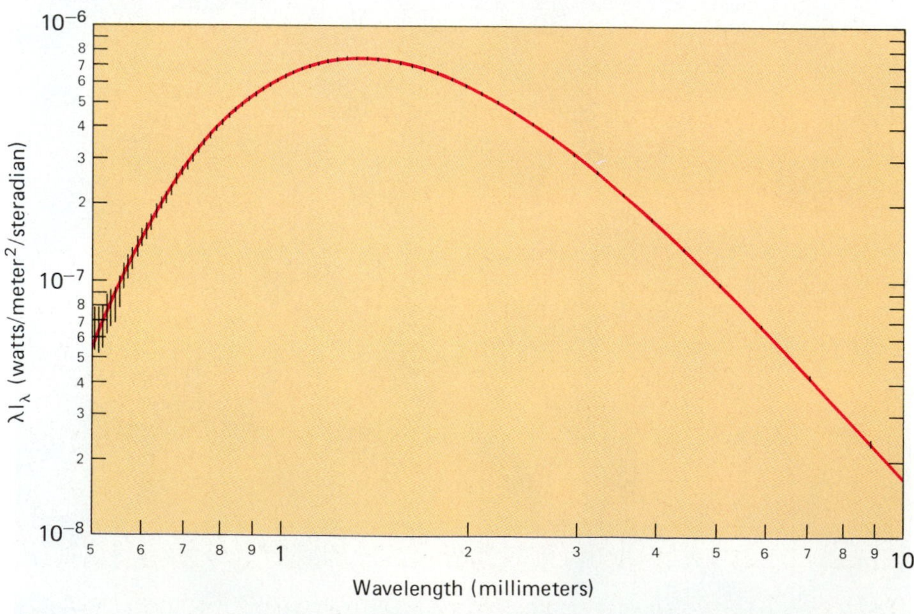

Figure 33–16 The first spectrum based on COBE observations drew an ovation from astronomers when it was first shown at an American Astronomical Society meeting. It represents a proof at astonishingly high accuracy that the cosmic background radiation is a black body. The data points and their error bars (shown in black) are fit precisely by a black body curve (color) for 2.735 ± .06°.

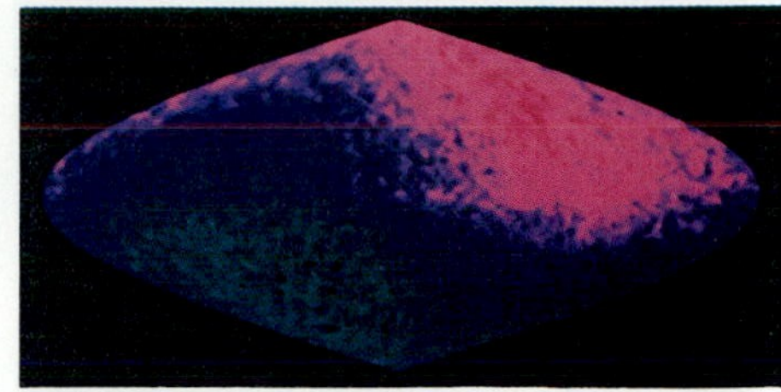
A 3.3 mm

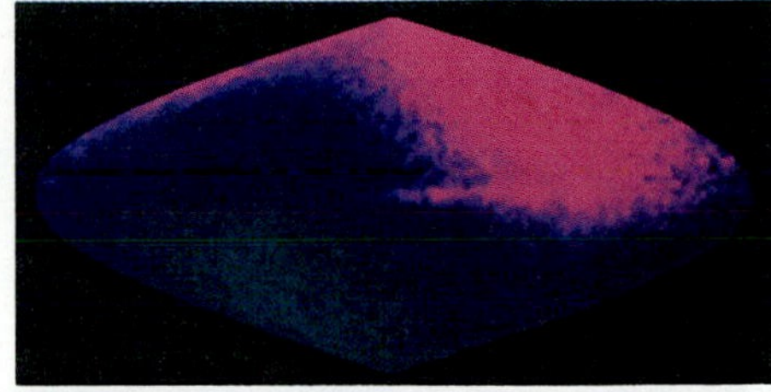
B 5.7 mm

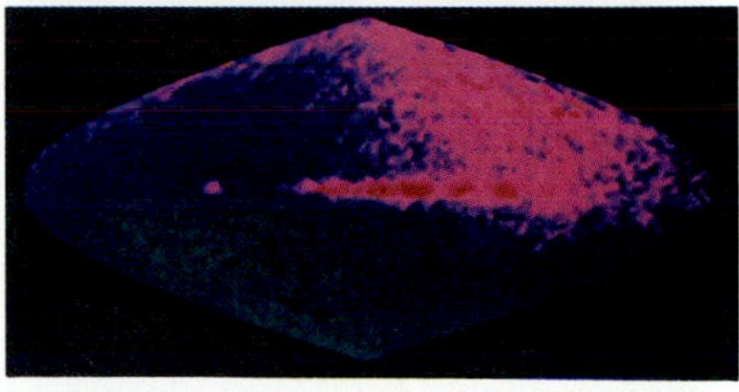
C 9.5 mm

Figure 33–17 Maps of the sky from COBE at radio wavelengths of (*A*) 3.3, (*B*) 5.7, and (*C*) 9.5 mm. At the longest wavelength, we see a galactic source in Cygnus (pink dot at middle left) and a sign of the galactic plane (horizontal dark pink at right). The rest of the signal is from the cosmic background radiation. The asymmetry from bottom left to top right shows the anisotropy from the Sun's motion. The speckling seen here should diminish as more data are obtained. The range shown is ±3 millikelvins.

smaller as more of the data are analyzed. The analysis shows that the existence of infrared excess radiation as suggested by earlier rocket observations was incorrect. COBE's helium ran out on 21 September 1990, 10 months after its launch (November 18, 1989), and this experiment lost all its sensitivity.

COBE is also measuring (Fig. 33–17) how isotropic—uniform from direction to direction—the cosmic background radiation is at three different microwave wavelengths (at the short-wavelength end of the radio spectrum). Scientists are comparing the observations at different wavelengths to determine which part of the radiation comes from our own galaxy. The anisotropy that results from the Sun's motion in space—a Doppler shift—shows clearly as a change in color from upper right to lower left. The range is from +3 millikelvins (+0.003 K = +3 mK), shown in pink, to −3 millikelvins (−3 mK), shown in blue. The direction measured for the Sun's motion agrees with previous measurements to within a few degrees. This type of anisotropy is known as a "dipole anisotropy" because it concerns only two opposite directions. Once this dipole anisotropy is subtracted out of the data, we can then look to see if there are any fluctuations left. These remaining fluctuations would be caused far out in the universe (far back in time), rather than by local effects such as our own motion in space. Various theories of cosmology, including the cold dark matter theory, predict that some small fluctuations should be detectable. In the next subsection, we will discuss the results of the COBE analysis of these data. This type of data did not depend on the instruments' being cooled, so data on anistropy continue to be gathered by COBE.

A third, infrared COBE experiment is searching at various infrared wavelengths for radiation from the first stars and galaxies, the predecessors of our current stars and galaxies. It is showing the Milky Way (Fig. 33–18) and other galactic sources, extending the observations of the IRAS spacecraft. It has already shown that the sky is darker at the long infrared wavelength of 100 microns than IRAS had found. The four short-wavelength channels, between 1 micron and 5 microns, continue to work, but the fact that helium has run out killed the longer wavelength channels and reduced the sensitivity of the working channels.

Figure 33–18 A near-infrared image of the Milky Way Galaxy obtained by COBE. This image combines images obtained at the near-infrared wavelengths of 1.2, 2.2, and 3.4 microns, represented respectively as blue, green, and red colors. The image is presented in galactic coordinates, with the plane of the Milky Way horizontal across the center. The image strikingly shows both the thin disk and central bulge populations of stars closer to the galactic center than our own sun.

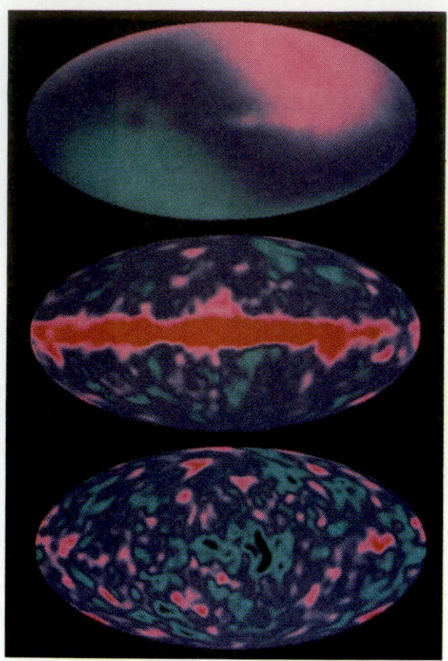

Figure 33–19 Full sky maps of variations in 3 K background radiation as measured by COBE. The top map shows variations of 3 millikelvins (0.12%) before corrections were made. The middle map shows variations of 30 microkelvins (0.01%) when the dipole anisotropy is subtracted. The red band across the center of this map results from the microwave emission of the Milky Way Galaxy. The bottom map, also showing variations of 30 microkelvins (0.01%), results when the galactic emission is then subtracted. Some of these variations—though we don't know which—are fluctuations in the background radiation rather than noise.

33.4e COBE's Great Discovery

COBE's mapping experiment continues to look for anisotropy. On what scale will we find the fluctuations? Probably on all scales, and scientists are using a variety of apparatus to look for such fluctuations. On smaller scales—2°, for example—data are being taken in Antarctica. On the larger scale of 7°, COBE's microwave experiment is the best available. By the scale of 7°, we mean that the resolution is 7°, so that no features smaller than 7° are distinguishable.

By the summer of 1991, it seemed that both ground-based and COBE experiments were detecting signs of anisotropy (aside from, of course, the overall dipole anistropy already detected by COBE). But the results were so important that they were kept secret while they were studied repeatedly to make sure there was no flaw in the analysis. The computer systems of the COBE group were separated from the worldwide network of scientists and some false data were even placed temporarily into draft papers, so that if leaks of information occurred, nobody would really know the truth. Finally, in April 1992, the COBE results were released: fluctuations were detected (Fig. 33–19). The fluctuations were now 30 microkelvins (30 μK), one hundred times smaller than the 3 millikelvins (3 mK) dipole anisotropy we had already seen. Are these the missing link that brings us back to how galaxies formed? We shall return to the question in Section 34.7.

Summary and Outline

Olbers's paradox and its solutions (Section 33.1)
- The universe's expansion diminishes energy in quanta
- We are looking back in time to before the stars formed

Big-bang theory (Section 33.2)
- Cosmological principle: universe is homogeneous and isotropic
- Hubble's constant and the "age" of the universe—12 to 20 billion years

Steady-state theory (Section 33.3)
- Perfect cosmological principle: universe is homogeneous, isotropic, and unchanging in time
- Continuous creation of matter is implied
- Demise of the steady-state theory with discovery of quasars and 3 K black-body radiation

Primordial background radiation (Section 33.4)
- Provides strong support for the big-bang theory
- Direct radio measurements gave the temperature
- Infrared observations confirm peaked curve
- Anisotropy indicates motion of the sun
- COBE, launched in 1989, has been successful
- Data fit black body of 2.735 $\pm$ 0.06 K
- Anisotropy from sun's motion shows; other anisotropies being searched for
- Far infrared background also being mapped

Key Words

cosmology, Olbers's paradox, cosmological principle, homogeneity, isotropy, big-bang theories, open universe, closed universe, deceleration parameter, flat universe, oscillating universe, principle of equivalence*, perfect cosmological principle*, steady-state theory*, recombination, cosmic background radiation, primordial background radiation, 3° background radiation, anisotropy

*This term is found in an optional section.

Questions

1. Hindsight has allowed solutions to be found to Olbers's paradox. For example, knowing that the universe is expanding, we can come up with a solution. However, on the basis of the reasoning in this chapter, do you think it is possible that scientists might have used Olbers's paradox to reach the conclusion that the universe is expanding before it was determined observationally? Explain.

†2. Our universe is about 10^{10} years old. If we were riding on a light beam emitted now, how long would we have to wait on the average to arrive on a star, given that the average line of sight ends 10^{24} years after the big bang? (Assume for the calculation that the stars are still there and that the universe is not expanding.) What percentage of the 10^{24} years is this? The question demonstrates how long 10^{24} years is.

†3. Using the formula for the energy of a photon, how many times less energy does a photon corresponding to a wavelength of 1 mm (the peak of the cosmic background curve) have compared with a photon corresponding to 1 Å in wavelength (which would have been emitted by the background in the early universe)?

4. We say in the chapter that the universe is often assumed to be homogeneous on a large scale. Referring to the discussions in Chapter 31, what is the largest scale on which the universe does not seem to appear homogeneous?

5. List observational evidence in favor of and against each of the following: (a) the big-bang theory, and (b) the steady-state theory.

†This question requires a numerical solution.

6. What is the relation of Einstein's general theory of relativity to the big bang?

7. What is "perfect" about the perfect cosmological principle?

8. If galaxies and radio sources increase in luminosity as they age, then when we look to great distances, we are seeing them when they were less intense than they are now. We thus cannot compare them directly to similar nearby galaxies or radio sources. If this increase in luminosity with time is a valid assumption, does it tend to make the universe seem to decelerate at a greater or lesser rate than the actual rate of deceleration? Explain.

†9. For a Hubble constant of 50 km/sec/Mpc, show how you calculate the Hubble time, the age of the universe ignoring the effect of gravity. (Hint: Take $1/H_0$, and simplify units so that only units of time are left.)

†10. What is the Hubble time if the Hubble constant is 100 km/sec/Mpc? Compare with the answer from Question 9. Comment on the additional effect that gravity would have.

11. In actuality, if current interpretations are correct, the 3 K background radiation is only indirectly the remnant of the big bang, but is directly the remnant of an "event" in the early universe. What event was that?

†12. Apply Wien's displacement law to the solar photospheric temperature and spectral peak in order to show where the spectrum of a 3 K black body peaks.

13. What does the anisotropy of the background radiation tell us?

†14. By how much would the H-alpha line (whose rest wavelength is 6563 Å) be shifted by the sun's velocity of 600 km/sec with respect to the background radiation?

Topics for Discussion

1. Which is more appealing to you: the cosmological principle or the perfect cosmological principle? Discuss why we should adopt one or the other as the basis of our cosmological theory.

2. Who do you think deserved the Nobel Prize: the scientists who first observed the background radiation without having a theory, the scientists who had made rough predictions decades earlier, or the scientists who made theoretical models but hadn't yet carried out their planned observations? Note that the Nobel Prize rules allow the award to go to no more than three individuals.

The central part of the cluster of galaxies that lies far beyond the constellation Fornax. Studies of the motions of the galaxies show that over 10 times more mass is present than is visible.

The Past and Future of the Universe

34

Aims: To study the first second of time in the universe, how the lightest of the elements were formed, and what the future of the universe will be

How did the tremendous explosion we call the big bang result in the universe we now know, with galaxies and stars and planets and people and flowers? Obviously, many complex stages of formation have taken place, and what was torn asunder at the beginning of time has now taken the form of an organized system.

We can trace the expansion backward in time and calculate how long ago all the matter we see would have been so collapsed to a point. The result of this calculation and of other determinations of the universe's age is that the big bang took place some 12 to 20 billion years ago. It is difficult to comprehend that we can meaningfully talk about the first few **seconds** of that time so long ago. But we can indeed set up sets of equations that satisfy the physical laws we have derived, and can make computer simulations and calculations that we think tell us a lot about what happened right after the origin of the universe. In this chapter we will go back in time beyond the point where the background radiation was set free to travel through the universe. This phase of time is known as the *early universe.*

Our knowledge of the structure of the universe and of the nuclear and other particles in it seems to be able to take us back to 10^{-43} second. Before 10^{-43} second, the universe was so compressed that not only the laws of general relativity but also those of quantum mechanics have to be taken into account, and we are not at present able to do so simultaneously. So we can't even say 10^{-43} second after what, since we don't really know that there was a zero of time. But we measure time from the instant at which everything would have been together, extrapolating the measurable part of the expansion backwards. Concepts of time have been changing recently, with Stephen Hawking's book *A Brief History of Time* bringing some of the problems before the general public.

In this chapter, we will first discuss how the chemical elements were formed. Then we will go on to discuss how our studies of the elements enable us to make predictions using the big-bang theories that have been current for the last decades. Next, we will describe our knowledge of the basic physical forces that govern the universe and discuss what we learn about the universe's first second of time. Finally, we will see how our understanding of this early time has given us a new picture of the universe's evolution, and leads to a different understanding of our fate. Many of the ideas described in this chapter about the earliest times are much more speculative than other topics we have discussed.

34.1 The Creation of the Elements

Modern cosmology is merging with *particle physics*—the studies of the particles inside atoms, which are often carried out with giant atom smashers. As we push farther back toward the beginning of time in our understanding of the

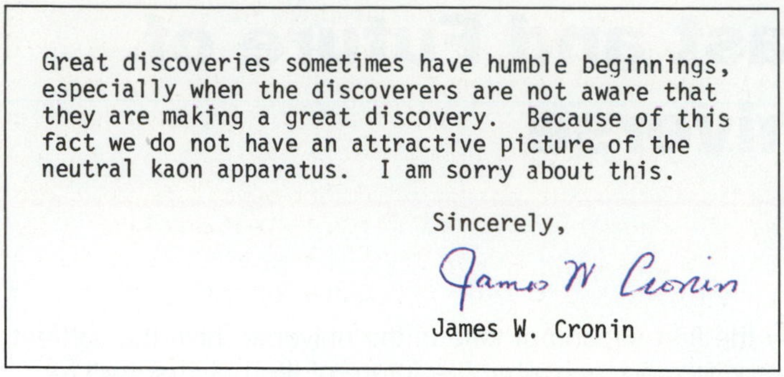

Figure 34–1 James W. Cronin and Val Fitch received the 1980 Nobel Prize in Physics for their experiment that showed an asymmetry in the decay of a certain kind of elementary particle. This result may point the way to the explanation of why there is now more matter than antimatter in the universe.

Great discoveries sometimes have humble beginnings, especially when the discoverers are not aware that they are making a great discovery. Because of this fact we do not have an attractive picture of the neutral kaon apparatus. I am sorry about this.

Sincerely,

James W Cronin

James W. Cronin

universe, it has become necessary to understand what particles were around and how they interacted with each other (Fig. 34–1).

One striking discovery has been that for every subatomic particle, there is a corresponding *antiparticle.* This antiparticle has the same mass as its particle, but is opposite in all other properties. For example, an antiproton has the same mass as a proton, but has negative charge instead of positive charge. Some particles have a property called spin: their antiparticles spin in the opposite direction. Antiparticles together make up *antimatter.* If a particle and its antiparticle meet each other, they annihilate each other; their total mass is 100 per cent transformed into energy in the amount $E = mc^2$.

We know that our universe, up to at least the scale of a cluster of galaxies, is made of matter rather than antimatter. If there were a substantial amount of antimatter anywhere, it would meet and annihilate some matter, and we would see the resulting energy as gamma rays. We do not detect enough gamma rays for this annihilation to be taking place commonly. And there is enough matter between the planets, between the stars, and between the galaxies to provide a continuous chain of matter between us and the Local Group of galaxies, showing that the Local Group is made of matter rather than antimatter. The chain probably extends to the Local Supercluster as well.

Baryons are a kind of subatomic particle that make up nuclei; protons and neutrons are the most familiar examples of baryons. Up to about 10^{-35} second after the big bang, there was a balance between baryons and their antiparticles on the one hand and photons (which can be thought of as particles of light) on the other. Up to this time, the number of particles and antiparticles was the same, and there was a continual annihilation of the two; each time a particle and its antiparticle met, their mass was transformed into energy. Similarly, particle/antiparticle pairs (including protons and antiprotons) kept forming out of the energy that was available.

Since theory indicates that matter and antimatter should have been formed in equal quantities, we must explain why our universe is now made almost entirely out of matter. Only recently have scientists found a reasonable (though speculative) explanation. Though few nuclear reactions produce different amounts of matter and antimatter, enough of these imbalanced reactions could have taken place at a sufficient rate in the early universe to provide a slight imbalance: 100,000,001 particles for each 100,000,000 antiparticles. Then all the antimatter annihilated an equal amount of matter. The relatively small residuum of matter left over is the matter that we find in our universe today! The annihilations created 100,000,000 photons for each baryon, a ratio that—as we can measure from the background radiation—holds true today. The 1980 Nobel Prize in Physics went to the scientists who first detected experimentally that particles can occasionally decay asymmetrically into matter and antimatter.

Following the annihilation of most of the protons and essentially all of the antiprotons, the universe had expanded enough so that less energy was available in any given place. Since $E = mc^2$, no longer was enough energy available to form protons and antiprotons. But until about a second had passed, there was still enough energy to form lighter particles like electrons and positrons (anti-electrons). Then essentially all the electrons and positrons annihilated each other. We were left with a sea of hot radiation, which dominated the universe, and which we now detect as the background radiation. At that time, unlike the present, the photons each had so much energy that most of the universe's energy was in the form of photons. Since neutrinos interact so rarely, they may not have been annihilated and many presumably remain from this era.

The results of computer-aided calculations of what happened after the first few seconds of the universe are less speculative than the reason why we wound up with a universe made of matter. According to these calculations, the universe's fantastically high density and temperature continued to diminish. After about 5 seconds, the universe had cooled to a few billion degrees. Only simple kinds of matter—protons, neutrons, electrons, neutrinos, and photons—were present at this time. The number of protons and neutrons was relatively small, but the number grew so that these particles had a relatively larger share of the universe's energy as the temperature continued to drop.

After about a hundred seconds, the temperature dropped to a billion degrees (which is low enough for a deuterium nucleus to hold together). The protons and neutrons began to combine into heavier assemblages—the nuclei of the heavier isotopes of hydrogen and elements like helium and lithium. The formation of the elements is called *nucleosynthesis.* The standard model of how the lightest elements formed in the big bang (Fig. 34–2) has been improved in recent years by models incorporating results of elementary-particle physics.

The first nuclear amalgam to form was simply a proton and neutron together. We call this a *deuteron;* it is the nucleus of deuterium, an isotope of hydrogen. Then two protons and a neutron could combine to form the nucleus of a helium isotope, and then another neutron could join to form ordinary helium. Within minutes, the temperature dropped to 100 million degrees, too low for most nuclear reactions to continue. Nucleosynthesis stopped, with

Some of the earliest quantitative work on nucleosynthesis in the big bang was described in an article published under the names of Ralph Alpher, Hans Bethe, and George Gamow in 1948. Actually, Alpher and Gamow did the work, and just for fun included Bethe's name in the list of authors so that the names would sound like the first three letters of the Greek alphabet: alpha, beta, gamma. These letters seemed particularly appropriate for an article about the beginning of the universe.

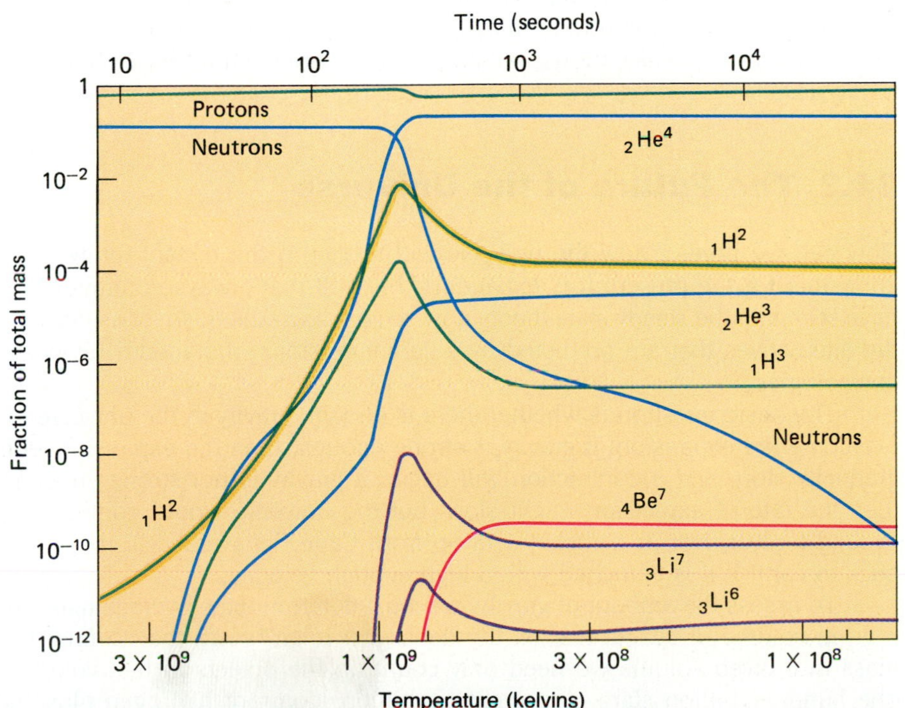

Figure 34–2 The "standard model" is a theoretical model that shows the changing relative abundances in the first minutes after the big bang. Time is shown on the top axis and the corresponding temperature is shown on the bottom axis. The standard model was for years that of William A. Fowler of Caltech, Fred Hoyle, and Robert V. Wagoner of Stanford, updated by Fowler and Wagoner. It has most recently been updated, including the relation of nucleosynthesis to elementary-particle physics, by scientists at Caltech, University of Chicago, Fermilab, and Bartol Research Foundation, including J. Yang, Michael S. Turner, Gary Steigman, David N. Schramm, and K. A. Olive.

Figure 34–3 William A. Fowler, E. Margaret Burbidge, and the author.

about 25 per cent of the mass of the universe in the form of helium. Nearly all the rest was and is hydrogen.

To test this theory, astronomers have studied the distribution of helium around the universe to see if it tends to be approximately this percentage everywhere. But helium is very difficult to observe; it has few convenient spectral lines to study. Furthermore, even when we can observe helium in stars we are observing only the surface layers, which do not necessarily have the same abundances of the elements as the interiors of the stars. "The helium problem"—whether the helium abundance is constant throughout the universe—has been extensively studied. Though the investigation is difficult, the helium is apparently uniformly distributed.

Only the lightest elements were formed in the early stages of the universe. The nuclear state of mass 5 (the lithium isotope with three protons and two neutrons, or else the helium isotope with two protons and three neutrons), which can be reached by adding a proton to a helium-4 nucleus, is not stable; it doesn't hold together long enough to be built upon by the addition of another proton to form still heavier nuclei. Even if this gap is bridged, we get to another extremely unstable nuclear state—mass 8. As a result, heavier nuclei are not formed in the early stages of the universe. Only hydrogen (and deuterium) and helium (and perhaps traces of low-mass isotopes such as those of lithium with nuclear mass of 6 or 7) would be formed in the period soon after the big bang.

The heavier elements must, then, have been formed at times long after the big bang. E. Margaret Burbidge, Geoffrey Burbidge, William A. Fowler, and Fred Hoyle showed in 1957 how both heavier elements and additional amounts of lighter elements can be synthesized in stars (Fig. 34–3). In a stellar interior, processes such as the triple-alpha process (Section 24.4) can get past these mass gaps. Elements are also synthesized in supernovae (Section 26.2), which is now being confirmed by observations of Supernova 1987A.

Astronomers thus believe that element formation took place in two stages. First, light elements were formed soon after the big bang. Later, the heavier elements and additional amounts of most of the lighter elements were formed in stars or in stellar explosions.

The theoretical calculations of the formation of the elements use values measured in the laboratory for the rate at which particles and nuclei interact with each other. For his role in making these measurements and in the theoretical work on nucleosynthesis, William Fowler shared in the 1983 Nobel Prize in Physics.

34.2 The Future of the Universe

How can we predict how the universe will evolve in the distant future? We know that for the present it is expanding, but will that always continue? We have seen that the steady-state theory now seems discredited, so let us discuss the alternatives that are predicted by different versions of standard big-bang cosmologies.

The basic question is whether there is enough gravity in the universe to overcome the expansion. If gravity is strong enough, then the expansion will gradually stop, and a contraction will begin. If gravity is not strong enough, then the rate of expansion might slow, but the universe would continue to expand forever, just as a rocket sent up from Cape Canaveral will never fall back to earth if it is launched with a high enough velocity.

To assess the amount of gravity, we must determine the average mass in a given volume of space, that is, the density. It might seem that to find the mass in a given volume we need only count up the objects in that volume: one hundred billion stars plus umpteen billion atoms of hydrogen plus so much interstellar dust, and so on. But there are severe limitations to this method, because many kinds of mass are invisible to us (if, indeed, mass exists in such forms). How much matter is in black holes or in brown dwarfs or in the form of neutrinos, for example? Until a few years ago, we couldn't measure the molecular hydrogen in space, and until a few decades ago, we couldn't measure the atomic hydrogen in space either.

Still another place mass could be hidden is connected to the fact that an intergalactic medium would be hard to detect if it were hot enough. Above a certain temperature, the matter would be almost entirely ionized, and no 21-cm radiation would be emitted. A hot gas, however, would radiate x-rays, so through x-ray observations or certain others we can set limits on the amount of hot intergalactic gas that can be present. The Cosmic Background Explorer spacecraft's far-infrared observations show no extended matter that could be the emitter of the x-ray background.

We must turn to methods that assess the amount of mass by effects that don't depend on the visibility of the mass. All mass has gravity, for example, so we are led to study the gravitational attraction on a large scale.

34.2a The Missing-Mass Problem

One place to investigate large-scale gravitational attractions is in clusters of galaxies. Clusters of galaxies appear to be large-scale stable configurations that have lasted a long time. But since galaxies have random velocities in various directions with respect to the center of the mass of the cluster, why don't the galaxies escape?

If we assume that the clusters of galaxies are stable in that the individual galaxies don't disperse, we can calculate the amount of gravity that must be present to keep the galaxies bound. Knowing the amount of gravity in turn allows us to calculate the mass that is causing this gravity. When this supposedly simple calculation is carried out for the Virgo Cluster, it turns out that there should be fifty times more mass present than is observed; 98 per cent of the mass expected is not found. This deficiency is called the *missing-mass problem.* Since we cannot see it, the additional mass is *dark matter.*

In what form might the missing mass be present? The idea of a few years ago that neutrinos have mass (Section 34.2c) is not in so much favor now, since experiments have not verified the suggestion. As we saw in Section 31.8, neutrinos with mass would be an example of "hot dark matter." There is much talk now about hypothetical particles, "cosmions." These cosmions would be an example of "cold dark matter." The work of the cosmologists increasingly joins the work of physicists studying elementary particles. At present, the astronomical observations are constraining the properties and numbers of the hypothetical new particles. Other forms of dark matter are also being invoked to explain what may make up 98 per cent or more of the universe.

The *missing-mass problem* is the discrepancy found when the mass derived from consideration of the motions of galaxies in clusters is compared with the mass that we can observe. It is really a "missing-light problem"; the mass **must** be there.

34.2b Deuterium and Cosmology

Perhaps the major method now being used to assess the density of the universe concerns itself with the abundance of the light elements. The light elements include hydrogen, helium, lithium, beryllium, and boron. Since these light elements were formed soon after the big bang (as we saw in Section 34.1), they tell us about conditions at the time of their formation. If we can find what the density was then, we can use our knowledge of the rate that the universe has been expanding to determine what the density is now.

Lots of things are invisible but we don't know how many because we can't see them.

Dennis the Menace

But somehow we must distinguish between the amount of these elements that was formed in the big bang and the amount subsequently formed in stars. This consideration complicates the calculations for helium, for example, be-

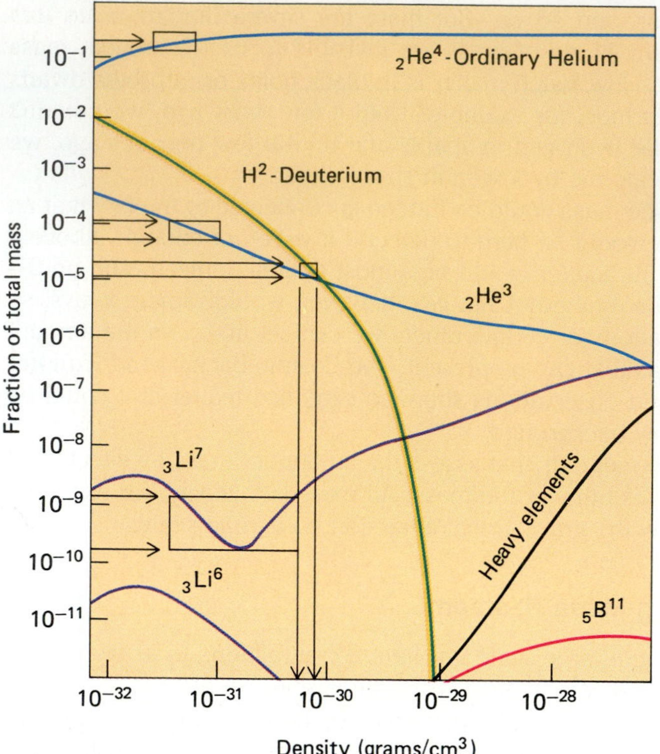

Figure 34–4 The horizontal axis shows the current cosmic density of matter. From our knowledge of the approximate rate of expansion of the universe, we can deduce what the density was long ago. The abundance of deuterium is particularly sensitive to the time when the deuterium was formed. Thus present-day observations of the deuterium abundance tell us what the cosmic density is, by following the arrows on the graph. The measured abundances of helium-3 and of lithium-7 are also useful, though helium-4 is relatively insensitive to nucleosynthesis conditions.

cause helium formed in stellar interiors and then spewed out in supernovae has been added to the primordial helium.

Fortunately, the deuterium isotope of hydrogen is free of this complication. We do not think that any is formed in stars, so all the deuterium now in existence was formed at the time of the big bang. Much of the original deuterium has been destroyed since it was formed, by being used up in stars.

Deuterium has a second property that makes it an important probe of the conditions that existed in the first fifteen minutes after the big bang, when the deuterium was formed. The amount of deuterium that is formed is particularly sensitive to the density of matter at the time of formation. A slight variation in the primordial density makes a larger change in the deuterium abundance than it does in the abundance of other isotopes. We measure the amount of deuterium relative to the amount of hydrogen.

Why is the ratio of deuterium to (ordinary) hydrogen sensitive to density? Deuterium very easily combines with an additional neutron. This combination of one proton and two neutrons is another hydrogen isotope (tritium). The second neutron quickly decays into a proton, leaving us with a combination of two protons and one neutron, which is an isotope of helium. If the universe was very dense in its first few minutes, then it was easy for the deuterium to meet up with neutrons, and almost all the deuterium "cooked" into helium. If, on the other hand, the density of the universe was low, then most of the deuterium that was formed still survives.

Theoretical calculations give the relation of the amount of surviving deuterium and the density of the universe, for a uniform universe (Fig. 34–4). This graph can be used to find the density that would have been present soon after the big bang, if the ratio of deuterium to hydrogen can be determined observationally.

Unfortunately, deuterium is very difficult to observe. It has no spectral lines accessible to optical observation. Deuterium makes up one part in 6600 of the hydrogen in ordinary seawater, but it was not known how this related to the cosmic abundance of deuterium.

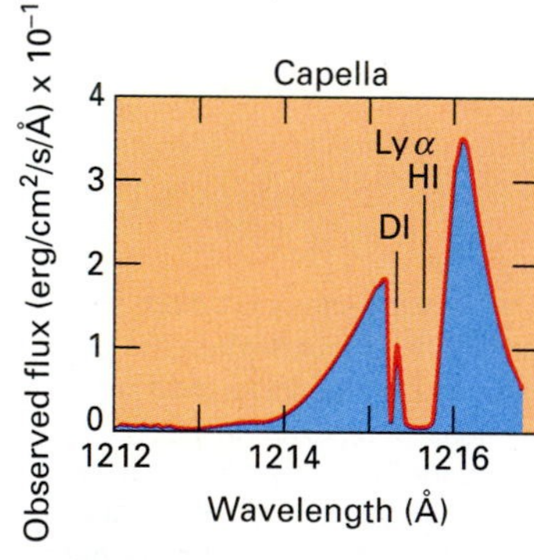

Figure 34–5 Observations from the Hubble Space Telescope of the Lyman α lines of interstellar deuterium and ordinary hydrogen. The broad absorption that takes up the whole graph is from hydrogen in the star Capella. The interstellar deuterium line is the narrow dip marked DI. The interstellar hydrogen line is marked HI. They are narrower than the stellar hydrogen line.

Major uncertainties remain, but the detections of deuterium that have been made so far (Fig. 34–5) agree that the amount of deuterium is such that there is not, and was not, enough density to ever reverse the expansion of the universe. The universe is apparently open. The results indicate, thus, that the universe will **not** fall back on itself in a big crunch. The theoretical calculations until recently assumed that the universe was homogeneous. New calculations allowing for some concentration of matter into dense pockets explain how there could be enough deuterium to detect yet have the universe be flat.

Figure 34–6 Measurements of neutrinos from Supernova 1987A (*lower right*) have given us our best constraint on the mass that neutrinos could have. The limit is so low that it seems unlikely that the missing mass is in the form of neutrinos.

34.2c Neutrinos

Neutrinos have the very interesting property of travelling very rapidly; indeed, we have long thought that they always travel at the speed of light. Now, most matter cannot travel at the speed of light according to Einstein's special theory of relativity, because its mass gets larger and larger and approaches infinity as its speed approaches the speed of light. Relativity theory shows that the mass of an object is equal to a constant quantity called the *rest mass* divided by a quantity that gets smaller and smaller (approaching zero) as the object goes faster and faster in approaching the speed of light. So the mass of an object gets larger and larger as the object goes faster and faster, and at speeds close to the speed of light is so large that it takes a tremendous amount of energy to accelerate it a little more. The mass would become infinite at the speed of light if an object's rest mass is any number other than zero, because the rest mass divided by zero is then infinite.

Since the burst of neutrinos we received on February 23, 1986, from Supernova 1987A (Fig. 34–6) presumably left the star in the Large Magellanic Cloud at the same instant, the spread in arrival times at earth gives us a limit on how much mass neutrinos have. (They would all travel at the same speed—the speed of light—if neutrinos are massless.) The limit is very small, so small that neutrinos probably cannot provide enough mass to close the universe. Experiments in terrestrial laboratories to detect neutrino mass have not succeeded, in spite of occasional reports of success.

34.3 Forces in the Universe

There are four known types of forces in the universe:

1. The *strong force,* also known as the *nuclear force,* is the strongest. It is the force that binds particles together into atomic nuclei. Although it is very strong at close range, it grows weaker rapidly with distance.

Only some particles "feel" the strong force. These particles, which include protons and neutrons, are composed of 6 kinds of particles called *quarks* (Fig. 34–7). (James Joyce used the word "quark" in *Finnegan's Wake,* and scientists

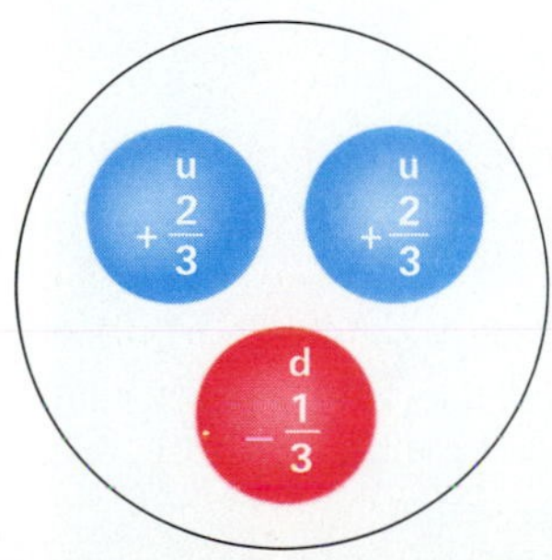

Proton +1 Charge unit

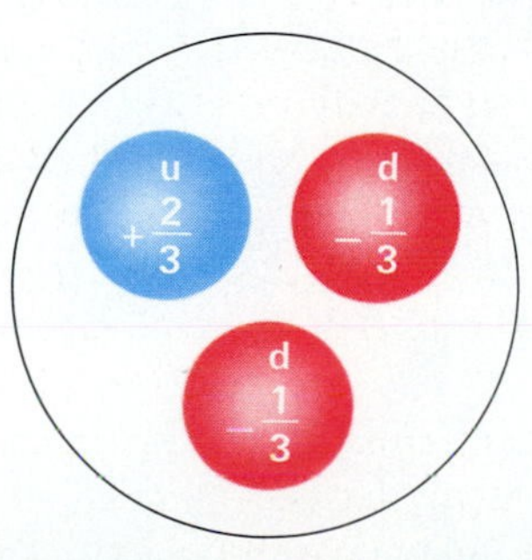

Neutron 0 Charge unit

Figure 34–7 The most common quarks are *up* (*u*) and *down* (*d*); ordinary matter in our world is made of them. The *up* quark has an electric charge of $+2/3$, and the *down* quark has a charge of $-1/3$. This fractional charge is one of the unusual things about quarks; prior to their invention (discovery?), it had been thought that all electric charges came in whole numbers. Note how the charge of the proton and of the neutron is the sum of the charges of their respective quarks.

Figure 34–8 Carlo Rubbia and Simon van der Meer with the apparatus at CERN with which they and their colleagues discovered the W^+, W^-, and Z^0 particles. These discoveries verified predictions of the electroweak theory. Rubbia and van der Meer received the 1984 Nobel Prize in Physics as a result.

have appropriated it.) There are six kinds (*flavors*) of quarks called "up," "down," "strange," "charmed," "truth," and "beauty." ("Truth" and "beauty" are sometimes more prosaically called "top" and "bottom," respectively.) Each kind of quark can have one of three properties called *colors,* often called red, blue, and green. (These names are whimsical and do not have the same meaning that the words have in general speech, though the "strange" quark got its name because it had a property called "strangeness" that made it different from more ordinary particles.) Since each of the six flavors comes in three colors, there are 18 subtypes of quarks; since an antiquark corresponds to each quark, there are really 36. This number is so large that even quarks may not be truly basic.

The strong force is carried between quarks by a particle; since this particle provides the "glue" that holds nuclear particles together, it is known (believe it or not) as a "gluon." Recent studies with atomic accelerators have led to the discovery of evidence for the existence of five of the six quarks and of the gluons. There had been evidence for the sixth quark ("truth" or "top"), but it has not been confirmed.

Theoretical work indicates that the observed cosmic helium abundance could not arise if there were more than four pairs of quark types, which come with four types of neutrinos. Since we currently observe three pairs of quark types (up–down, strange–charmed, truth–beauty) and three types of neutrinos (electron neutrino, muon neutrino, and tau particle), calculations based on the cosmic helium abundance gave hope for discovering one—but only one—more of each.

2. The *electromagnetic force,* 1/137 the strength of the strong force, leads to electromagnetic radiation in the form of photons. So most of the evidence we have discussed in this book, since it was carried by light, x-rays, and so on, was carried by the electromagnetic force. It is also the force involved in chemical reactions.

3. The *weak force,* important only in the decay of certain elementary particles, is currently being carefully studied. It is very weak, only 10^{-13} the strength of the strong force, and also has a very short range.

4. The *gravitational force* is the weakest of all over short distances, only 10^{-39} the strength of the strong force. But the effect from the masses of all the particles is cumulative—it adds up—so that on the scale of the universe gravity dominates the other forces.

Theoretical physicists studying the theory of elementary particles have in the past years made progress in unifying the theory of electromagnetism and the theory of the weak force into a single theory. Thus just as the forces known separately as electricity and magnetism were unified a hundred years ago into the electromagnetic force, the force of electromagnetism and the weak force

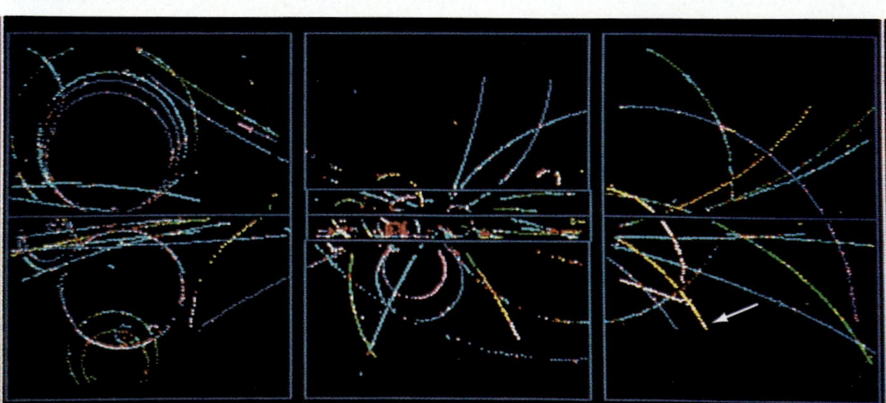

Figure 34–9 By studying the trails left in this electronic detector at CERN by a proton-antiproton collision, scientists concluded that a Z particle had been formed and had disintegrated into an electron and a positron. The electron-positron pair appears in the lower left of the rightmost box.

Figure 34–10 The Stanford Linear Collider, the SLC, which is producing Z particles in California.

have now been unified into the *electroweak force.* (The 1979 Nobel Prize in Physics was awarded for this work.) The forces would be indistinguishable from each other at the extremely high temperatures, above 10^{26} K, that may have existed in the earliest moments of the universe.

The electroweak theory predicted that certain new particles could be discovered if a particle accelerator (informally called an "atom smasher") that was powerful enough could be built. The new particles, known as W and Z, were predicted to be much more massive than the proton, so they could be created out of energy only if a lot of energy were available. The European atom smasher on the Swiss-French border at CERN (formerly the acronym for the European Center for Nuclear Research, but now known as the European Center for Particle Physics, to avoid the "dreaded" word "nuclear") was upgraded for the purpose (Fig. 34–8). It reached such high energies in 1982 and 1983 that the W and the Z particles were indeed created and detected (Fig. 34–9), a great triumph for the electroweak theory. To make Z particles in quantity, an atom smasher at Stanford University in California was modified to make the Stanford Linear Collider (Fig. 34–10), in which two beams collide and make dozens of Z particles each month (Fig. 34–11). In 1989, a new atom smasher at CERN, the Large Electron-Positron accelerator (LEP), started churning out thousands of Z's each month. Studies of the Z's are determining various fundamental nuclear properties, and quickly showed that it is not possible for there to be more than the three already known types of quark pairs.

Progress is even being made in a family of *grand-unified theories* (*GUT's*) that unifies the electroweak and the strong forces. Such a theory would be necessary to provide a thorough explanation of the early universe. (In spite of the overblown name, gravity is still not included in the "grand unification.") The grand-unified theories being considered imply, surprisingly, that protons are not stable—they should decay with a half-life greater than 10^{30} years. So GUT's imply that protons decay. Though this is a much longer time than the lifetime of the universe (10^{10} years), given a sufficiently large quantity of protons,

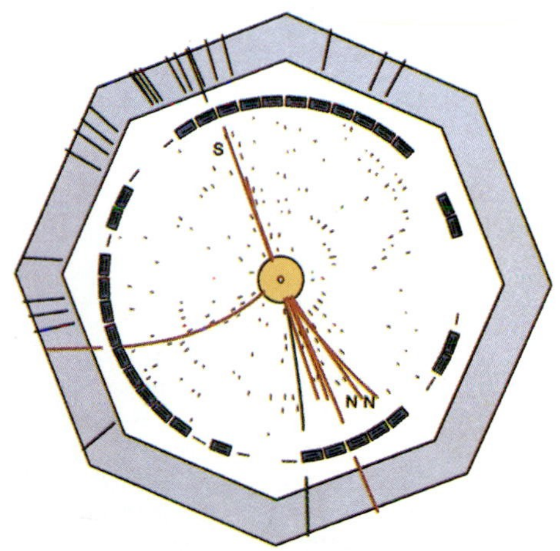

Figure 34–11 In this reconstruction of the first Z particle observed at the SLC, the Z particle was created at the center of the yellow circle and decayed immediately into longer-lived particles that left tracks.

Figure 34–12 In a search for decaying protons, this cavity—deep underground in a salt mine near Cleveland—has been filled with 10,000 tons of water. Water, H_2O, is mostly protons, since an H nucleus is a proton and an oxygen nucleus contains 8 protons. If any of the 2.5×10^{33} neutrons or protons bound in nuclei decay, the resulting particles will give off flashes of light. Twenty-four hundred photomultipliers have been installed to detect these flashes. Though proton decay has not been discovered, a neutrino burst from Supernova 1987A in the Large Magellanic Cloud was.

such a decay should be detectable. After all, 1000 tons of matter contain about 10^{32} protons (Fig. 34–12), so perhaps 10 or 100 could decay each year in that volume. It is just such a change in the total number of baryons in the universe that is necessary to account for the current excess of matter over antimatter. Many scientists had expected to detect signs of proton decay by now, and a few candidate events are being studied, but the detectors have not detected it to a limit of 10^{31} years. This rules out the simplest GUT's, but not more complex ones. Fortunately, however, the same detectors were on line to detect the neutrino burst from Supernova 1987A.

Many theoreticians are working on theories to incorporate gravity with the other forces; *supergravity* is one of the examples. But such theories are not yet satisfactory, and are not as advanced as GUT's.

Basically, scientists believe that all the forces were unified at the extremely high energies that were present early in the universe. As the universe expanded and cooled (Fig. 34–13), basic symmetries among the forces were broken and individual forces became apparent.

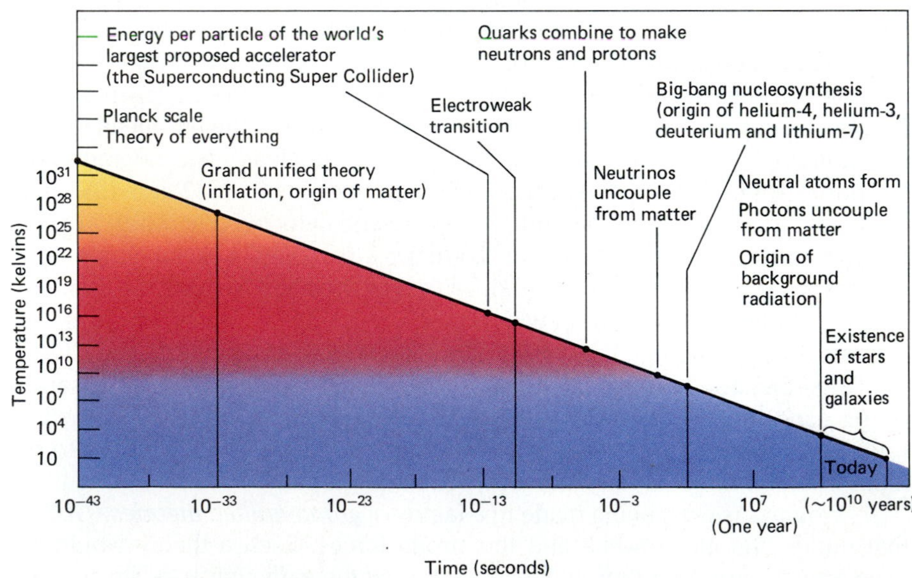

Figure 34–13 As the temperature of the universe cooled, the forces became distinguishable from each other and the universe evolved. Adapted from "Particle Accelerators Test Cosmological Theory" by David N. Schramm and Gary Steigman.

34.4 Superstrings and Shadow Matter

A major line of difficult and speculative theoretical research is now using the concept of *superstring.* Superstring is a one-dimensional analogy to the zero-dimensional points that are often used in theories. Quarks, electrons, photons, and other elementary particles would be made out of one-dimensional elongated objects—"strings"—about 10^{-35} meter long that make loops so small that they seemed to be points. Theories involving superstring thus far seem a promising way to get around theoretical difficulties in the extremely small scale where a quantum theory of gravity is needed. (Such a quantum theory of gravity does not exist.) Thus superstring is considered a prime candidate for a TOE—"theory of everything"—that explains all the forces. We have dealt, in discussing Einstein's general theory of relativity (Section 23.11), with curvature of three spatial dimensions into a fourth dimension. Modern superstring theories are framed in terms of an 11-dimensional space, though all but four of those dimensions have such strong curvature—are curled up so tightly (within 10^{-35} meter)—that we cannot detect them (Fig. 34–14). An analogy might be a garden hose on the other side of your lawn—it has three dimensions but at a distance looks like a one-dimensional line. Such theories are evolving very rapidly at present; they incorporate gravity with the other forces.

Superstring theory implies that "shadow matter," able to interact with normal matter only through gravity, exists. Since shadow matter would not be subject to the electromagnetic force, it would not generate light. It thus could be the dark matter that would solve the missing-mass problem.

At early times in the universe, the exotic particles now being observed by physicists in giant accelerators had been present as well. Even isolated quarks may have been present, though they would have combined to form baryons (protons, neutrons, etc.) by the time 1 microsecond went by. The mini black holes described in Section 28.6 may have been formed in this era.

Figure 34–14 Theorists calculate that superstrings require an 11-dimensional space; we are more familiar with only a four-dimensional space, with three spatial dimensions and one time dimension.

34.5 The Inflationary Universe

The application to cosmology in the last few years of the exciting new theoretical and experimental results of particle physics has led to some big surprises. The results from particle physics have allowed scientists to push our understanding of the universe much farther back in time than previously. In particular, now we can go back beyond the era of element formation.

The new GUT's got scientists thinking about the early universe. A major new theory holds that starting at the first 10^{-35} second of time and lasting another 10^{-32} second, the scale of the existing universe increased rapidly. The rate of this "inflation" of the universe was astounding. During this fraction of a second, the volume of the universe grew perhaps 10^{100} or even more times larger than it would have grown according to the "standard" big-bang theory. The new theory, whose first form was advanced by Alan Guth, now of MIT, is called the *inflationary universe.* A subsequent version, developed by Andrei Linde of the Lebedev Institute in Moscow, Andreas Albrecht, now of the University of Texas, and Paul Steinhardt of the University of Pennsylvania, tackled some major flaws in the first version and is known as "new inflation." Steinhardt is working on a further improved version called "extended inflation." As we shall see below, the inflationary-universe theory solves a number of outstanding problems of cosmology.

Figure 34–15 Paul Steinhardt, who developed the new and extended inflationary-universe models.

*34.5a Explanation of the Inflationary Universe

The crucial assumption of the inflationary-universe model is that the universe underwent a special kind of "phase transition" during its early history. This phase transition was a change in the way elementary particles behave in the universe, and was analogous to the change in the way water molecules behave as they undergo a phase transition from liquid (water) to solid (ice).

In the case of water molecules, what decides whether they form water or ice is the temperature. If one calculates the energy of a group of molecules that have formed a water droplet and compares it to the energy of the same group of molecules that have formed an ice crystal, one finds that the relation between the energies of the water and ice are different at different temperatures. At temperatures above the freezing temperature, the energy of the water droplet is less than that of the ice crystal. Since matter apparently prefers to be in a state with the lowest possible energy, water molecules prefer to form liquid water at such temperatures. At temperatures below the melting temperature, the energy of the ice crystal is lower than that of the water droplet, so water molecules prefer to form ice crystals at such temperatures.

However, as one cools water below the freezing temperature, it may not immediately be transformed into ice, even though the ice has the lower energy. There will generally be an energy barrier that must be crossed before the water can crystallize. In this way, even as one continues to lower the temperature, water molecules can be trapped in the form of liquid water even though water continues to have a greater energy than ice. This phenomenon is known as "supercooling," and is found in nature. (Water can be supercooled several degrees below freezing, a situation that is commonly found in clouds aloft.)

The inflationary-universe model assumes that as the universe expanded and the temperature decreased, the universe underwent such a phase transition in which it supercooled—remained trapped in a state where the forces are symmetric even though that state had a greater energy than the state in which the symmetry among the forces is broken. Because it was trapped in that higher-energy state, the universe had excess energy.

As the universe continued to cool, the excess energy became the dominant contribution to the total energy in the universe. (It was much greater, for example, than the total kinetic and mass energies of the particles contained within the universe.) However, unlike other kinds of energy, the amount of such energy in a unit volume—the energy density—does not change as the universe continues to expand. (This energy is associated with the state in which the universe was and not with the individual energies of the particles contained within it. As the universe continued to expand, the particles got spread out, so the kinetic and mass energy densities decreased, but the state and therefore the state energy density remained the same everywhere in the universe.)

This constant energy density, just as any form of energy, led to a gravitational force, but the gravitational force of a universe with a constant energy density is very peculiar—it causes the universe to expand so rapidly, the so-called "inflation." (This is not at all intuitively obvious, but follows from calculations.) In a period of 10^{-35} second, the universe expanded in volume by a factor of 10^{50} or more than in the hot-big-bang picture (Fig. 34–16).

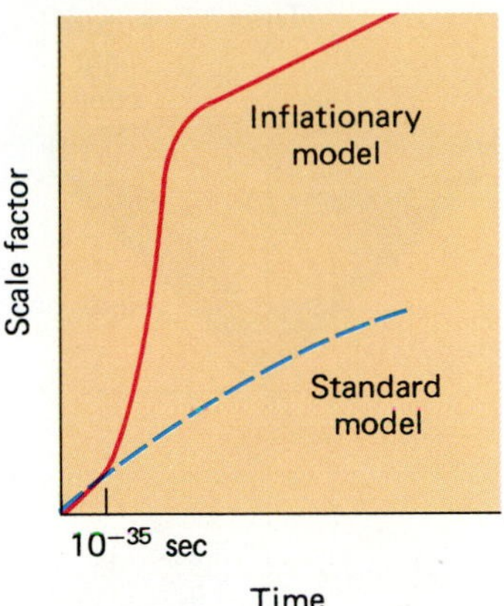

Figure 34–16 The rate at which the universe expands, for both inflationary and non-inflationary models.

The basis of the theory is the relation of the fundamental forces. The temperature of the universe during the first 10^{-35} second was so high that the four forces were not distinguishable from each other. Normally, as the temperature drops, a force can "freeze" out, becoming distinguishable from the other forces. The theory shows that the freezing out of the strong force from the electroweak force could have been delayed, and would have caused the universe to grow rapidly larger.

34.5b The New Inflationary Universe

Of course, to be a useful cosmological picture, the expansion eventually had to stop, because our universe is not inflating today; it is expanding at a much more modest rate. Guth's original model was flawed in that the phase transition would never stop.

In a more recent version of the theory—the "new inflationary universe"—the inflation stops smoothly after a small fraction of a second. At this time, the universe is transformed from the state in which the forces are symmetric into the lower energy state, in which the symmetry among the forces is broken. All the state energy is rapidly converted into ordinary matter and radiation. The universe continues to expand, but at the slower rate we observe today, just as in the standard big-bang theories. (The elements form, the background radiation is set free, and so on, on the same time scale.)

All versions of inflation imply that the universe is very much larger than we had thought, in that the parts with which we could have been in touch have expanded far beyond our sight. The boundary of this volume, our "past light cone," is our "horizon." It is defined by signals travelling at the speed of light (Fig. 34–17), since relativity theory tells us that it is impossible for signals to travel faster. Since we are out of contact with some regions, those outside our past light cone, those regions could even be made of antimatter.

One major question that had been open in cosmology is how the universe became so homogeneous. According to the hot-big-bang model, our universe was supposed to have always been so large that neither light, nor any information about conditions at any given location, had had time since the big bang to travel across it. Thus matter in one part of the observed universe has always been out of contact with matter on the other side, as we see it. It has been a mystery how such widely different regions could have reached identical temperatures and densities. After all, we know that the temperature and density were nearly identical everywhere at the time the cosmic background radiation was emitted, since this background radiation is so isotropic.

In the inflationary-universe picture, our observed universe was much tinier before the phase transition than in the hot-big-bang picture, small enough that signals travelling at the speed of light had time to travel across it. The universe

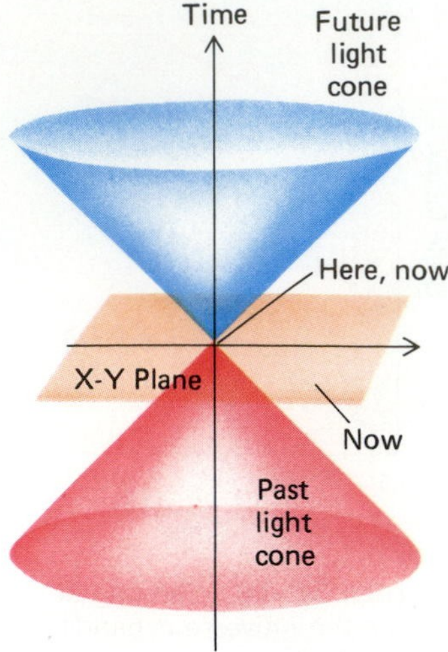

Figure 34–17 If we plot a three-dimensional graph with two spatial dimensions (omitting the third dimension) and a time axis, our location can receive signals only from regions close enough for signals to reach us at the speed of light or less, our "past light cone." We can never be in touch with regions outside our light cone; we say that we are "not causally connected" with those regions. In the inflationary universe, the volume of space with which we are causally connected is much greater than the observable universe. It is easy to explain how a volume that is causally connected can be homogeneous—collisions, for example, can smooth things out.

could thus be smoothed out. The whole part of the universe that we can potentially observe—the visible universe—was, before the inflation, smaller than a single proton. There might have been as many as 10^{15} point-like particles (such as quarks, electrons, etc.) crammed into this tiny region, but essentially all of the approximately 10^{86} particles in the observed universe today were produced from the energy released in the phase transition. The inflation then accounts for how such a tiny region could grow to be the size of our observed universe. It thus explains why the universe is so homogeneous.

Another previously unresolved question is why the density of matter in the universe is so close to the critical amount that marks the difference between the universe's being open or closed—why the universe is so flat. Although we do not yet know the precise value, we do know that the density of matter is within one or two factors of 10 of the critical density. Being even this close to the critical value is rather surprising, since it can be shown that such a situation is highly unstable, like balancing a pencil on its point! Why are we in such an unstable situation? The inflationary-universe theory shows that during the inflation, space-time became very flat (Section 33.2), just as a small bit of the surface of a balloon flattens out as the balloon grows larger.

We may thus have been asking the wrong question all these years when we asked whether the universe is open or closed, wondering what kinds of studies would tell us the answer. The inflationary-universe theory implies that the universe is incredibly close to the borderline case in which it would expand forever at an ever-decreasing rate; the longer it expands, the less its rate of expansion will be. If the theory is right, the universe is so close to this borderline that it would be impossible for any observations to tell us on which side it lies.

The inflationary universe also provides satisfactory answers for another question plaguing some theoreticians: why don't we see magnetic monopoles in the universe? A *magnetic monopole* is a point source of magnetic force; it is analogous to an electron, which is a single particle that has an electric force. But bar magnets always have two opposite poles, and a single-pole magnet—a monopole—has never been discovered. Although scientists have toyed with the notion of monopoles for over a hundred years, it turns out that GUT's predict that there would be monopoles. Further, these monopoles should be very massive and should have been produced in enormous numbers in the early universe. The monopoles never decay and rarely meet up with antimonopoles (in which case they would annihilate each other), so they should be around today. Yet, none are seen. It is crucial to find some explanation of how the universe might have gotten rid of the high density of monopoles.

Since the monopoles are produced before inflation, the inflationary universe can account for the fact that we don't now see any by spreading the magnetic monopoles out. The inflation spreads them out so much that there is at most only a handful in the part of the universe observable by us.

A startling feature of the inflationary-universe theory is that it accounts for the existence of all the ordinary matter and radiation in the universe. In the theory, the energy from the peculiar state of the early universe is transformed,

In Guth's words, the creation of energy in this way was "the ultimate free lunch."

Table 34–1 Advantages of the Inflationary Universe

Explains why the universe is relatively homogeneous
Explains flatness—why we are close to the critical density
Explains why we don't detect magnetic monopoles

after the period of tremendous expansion, into the matter and radiation that we detect today. Almost all the energy we observe in the universe today was produced by this transformation of state energy that occurred at the end of the inflation. The model also shows why there is so much matter in the universe.

There are still some problems with the new inflationary model. For example, though the model can explain the formation of galaxies in principle, the explanation depends on the "fine-tuning," that is, very special choices of the numerical details of the unified-field theory. The more recent "extended inflation" model seeks to eliminate the need for fine-tuning, making the inflationary era and its termination arise naturally. The model is similar to Guth's original flawed version in which the phase transition never stops. The extended model uses the possibility that the gravitational constant G may be forced to vary during the brief period over which the phase transition takes place. The variation of G changes the rate of expansion of the universe in a way that allows the transition to be completed. At completion, G becomes fixed at the value we observe today.

We cannot yet know whether these problems will be overcome through future refinement. But we can predict that the relation between physicists studying subnuclear particles and scientists studying cosmology will become closer and closer. And we trust that our knowledge of the history of the universe (Fig. 34–18) grows more and more accurate.

34.6 Cosmic String and Texture

Another new idea on the interface between physics and astrophysics is the idea that the phase change undergone by the universe was not perfect. If a defect was left, one or more long, thin tubes of space-time may remain in which the old type of space-time remains—a *cosmic string.* Such a tube could be infinitely long or it could form a closed loop. A single such loop could contain 10^{15} times the mass of the sun, equivalent to the mass of 10,000 galaxies. Fortunately, there clearly aren't many of them around now; they do decay as time goes on. The smaller loops have radiated away their energy in the form of gravitational waves, leaving only a few of the longest loops until today.

Theoretical calculation indicates that cosmic strings would be only 10^{-30} centimeters thick. Each centimeter of their length would have a mass of 10^{22} grams. And they might be excellent conductors, carrying a huge current, per-

Focus On

Box 34.1 Keeping Strings Straight

Do not confuse:

A ***string*** of galaxies: galaxies one after the other in space, using the word "string" in its normal sense.

Superstring: the tiniest objects in the universe, only 10^{-35} m long, used in a theory (Section 34.4) that may explain all the known interactions of elementary particles and all the basic forces; in the theory, elementary particles are different vibrational modes of superstrings in an 11-dimensional universe.

Cosmic string: loops or giant strands that would stretch across the whole universe and could result from phase transitions in the early universe.

A

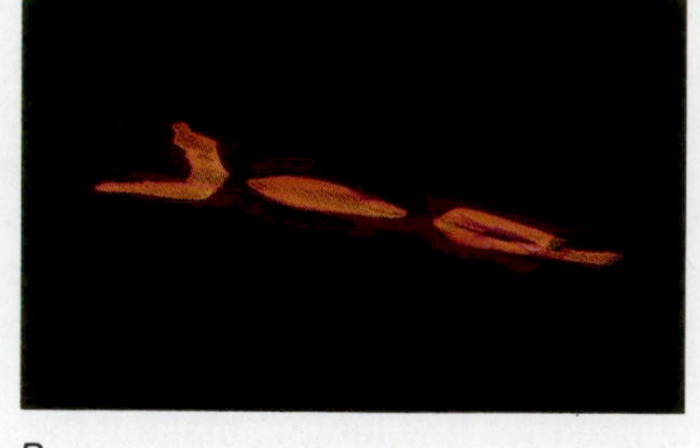

B

C

Figure 34–18 The merger of two cosmic strings, calculated in a supercomputer.

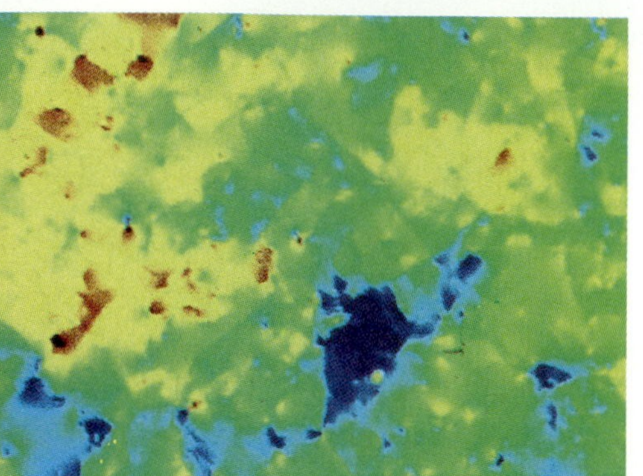

Figure 34–19 If cosmic strings exist, they should generate a certain pattern of temperature contrasts in the cosmic background radiation. They would do so by acting as gravitational lenses. This computer simulation shows such a pattern over a field of view a few degrees across. Blue indicates gas that is hotter than average and red indicates gas that is cooler than average. The study addresses the question of whether cosmic strings acted as seeds to begin galaxy formation.

haps as much as 10^{20} amperes. With these conditions, they might have triggered huge explosions in the early years after the big bang. The wake of the strings might have been enhanced in density, providing the seeds for galaxies to form. These galaxy-forming cosmic strings, though, would give off gravitational radiation. Studies of the steady period of one of Joe Taylor's millisecond pulsars (Sections 27.6 and 27.7) ruled out the simplest form of cosmic string.

Cosmic strings would have enough mass to make a gravitational-lens effect. We may one day see an image of a galaxy cut off by a cosmic string. Such an effect may be seen on images taken for other purposes.

Another kind of defect is cosmic texture. The name *cosmic texture* comes from solid state physics—phase transitions in laboratory materials such as liquid crystals often produce defects similar to those imagined in the early universe. Unlike strings, which are line-like defects, textures are point-like defects. If textures did provide the seeds for galaxy formation, then studies of fluctuations of the microwave background should detect their effects. Improved analysis of the COBE data may soon rule out (or even confirm) this model of galaxy formation.

The universe is a curious place, and the new ideas make it seem possibly more curious still.

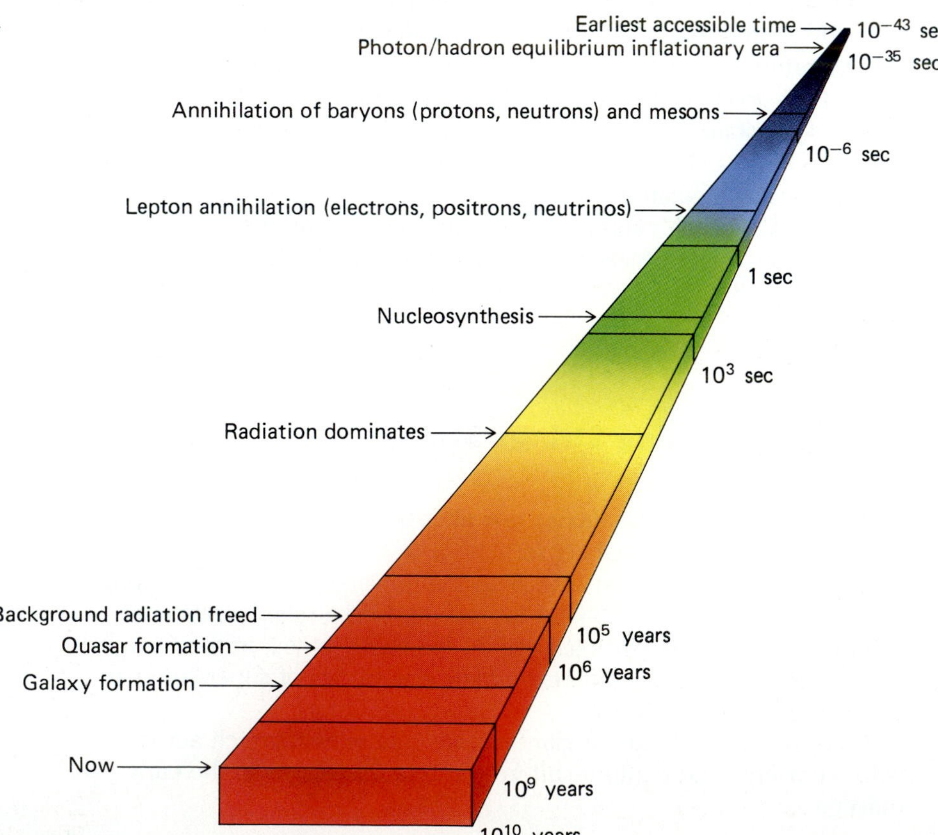

Figure 34–20 Major events in the history of the universe.

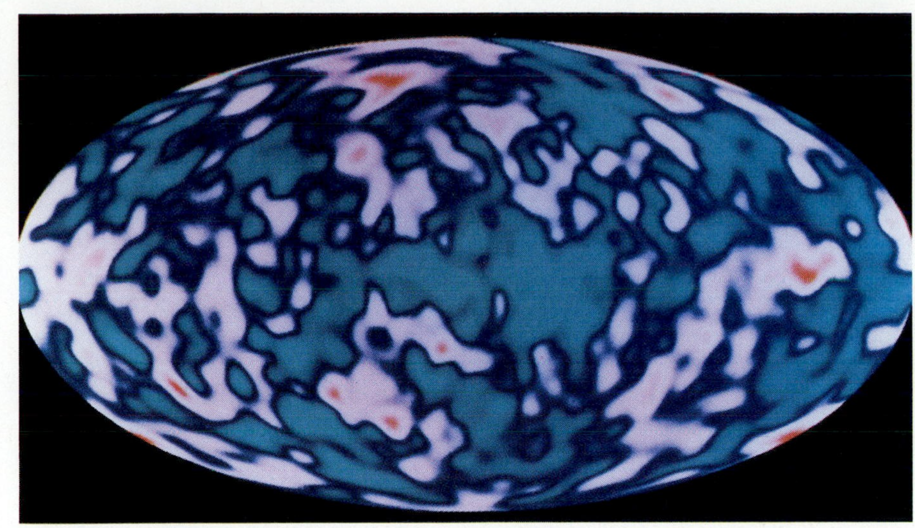

Figure 34–21 Temperature fluctuations on the microkelvin level displayed on an all-sky map, observed from COBE. These fluctuations, about 5 parts per million, show the seeds from which structure in the universe evolved nearly 15 billion years ago. Much of the detail in the image is random noise; one can't point to any one of the features and say it is real. But statistical analysis shows that there are fluctuations present in addition to the noise.

34.7 COBE: Has It Shown the Link to the Big Bang?

When we look out into the universe, we see clusters of galaxies and a structure even as large as the Great Wall and Great Attractor. How did this structure arise? It seemed to scientists analyzing the problem that early on, in the background radiation, there should have been small fluctuations from which the currently observable structure grew. It is these fluctuations, presumably, that COBE has now found. The value of the fluctuations is 30 microkelvins (30 μK).

The fact that there are fluctuations when we look back to 300,000 years after the big bang (Fig. 34–21) seems to endorse our basic ideas about how this structure formed in the universe, the structure that we now see as clusters of galaxies and assemblages of clusters of galaxies. But the detailed value found is too low, about a factor of 10 smaller than was expected for galaxies to form by the standard cold dark matter theory. This observed value and the distribution of sizes of the observed fluctuations are not consistent with the predictions of any of the current theories of formation of the structure in the universe. In particular, theories involving cosmic string and cosmic texture are in trouble.

The level of fluctuations observed by COBE would correspond to an amount of cold dark matter about equal to the amount of visible matter and we can predict how much gravity that cold dark matter would have. That amount of cold dark matter should cause the clusters of galaxies we study to have higher motions than we observe. So there is a disagreement in this important detail. We need a better theory.

The COBE discovery does seem to endorse the inflationary model of cosmology. At the distance to the fluctuations we see, the universe's horizon

Figure 34–22 The evolution of the universe, including the era 300,000 years after the big bang in which the universe turned transparent and the cosmic background radiation was thus set free.

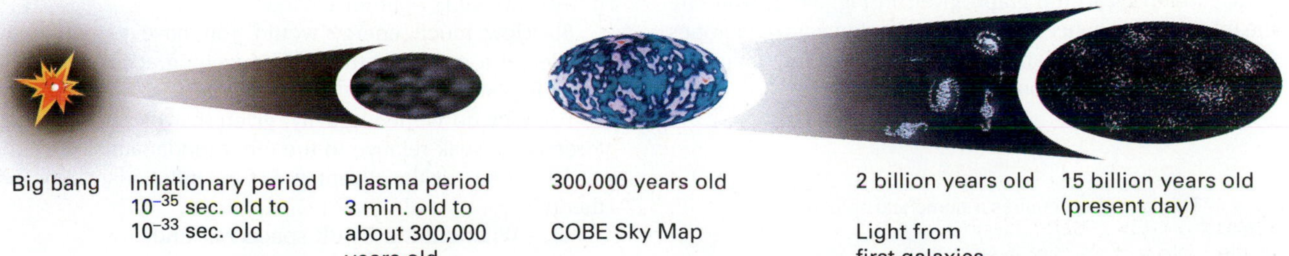

should be only about 2° across. So by looking on a 7° scale, we are seeing units of the universe that are separate from each other. It is the inflationary-universe model that now seems to best understand how such separated parts of the universe can be as alike as they are (Fig. 34–22).

We are making dramatic progress in understanding our universe.

Summary and Outline

Creation of the elements (Section 34.1)
- Pairs of particles and antiparticles created out of energy in the early universe
- Light elements formed in the first minutes after the big bang; heavier elements formed in interiors of stars and in supernovae.

Future of the universe (Section 34.2)
- The missing-mass problem: discrepancy between mass derived by studying motions of galaxies in clusters of galaxies and the visible mass
- Deuterium-to-hydrogen ratio depends on cosmic density: all deuterium was formed in first minutes after the big bang; current evidence from deuterium is that universe is open

Forces in the universe (Section 34.3)
- Four forces of nature, in declining order of strength: strong, electromagnetic, weak, gravitational
- Particles like protons and neutrons are made of quarks
- Theory of electroweak force unifies electromagnetic and weak forces; led to discovery of W and Z particles, which are now being intensively studied
- Grand-unified theories (GUT's) unify electroweak and strong forces; predict proton decay

Shadow matter (Section 34.4)
- Superstring theory implies existence of shadow matter
- Shadow matter would have gravity but no light

The inflationary universe (Section 34.5)
- Based on GUT's
- Universe grew rapidly larger in first 10^{-35} second, and later began expanding as it would from standard big-bang theory
- Explains homogeneity of background radiation, implies that universe is on boundary between being open and closed, and explains why we don't detect magnetic monopoles

Superstring (Section 34.6)
- Defect in space-time

COBE (Section 34.7): Fluctuations detected

Key Words

early universe, particle physics, antiparticle, antimatter, baryons, nucleosynthesis, deuteron, missing-mass problem, dark matter, rest mass, strong force, nuclear force, quarks, flavors, colors, electromagnetic force, weak force, gravitational force, electroweak force, grand-unified theories (GUT's), supergravity, superstring, inflationary universe, magnetic monopole, cosmic string

Questions

1. An electron has a negative charge. What is the charge of a positron, which is an antielectron? What is its mass?

†2. The mass of an electron is given in Appendix 2. What is the energy released in the annihilation of an electron and a positron? (Use the erg for the unit of energy, with 1 erg = 1 g cm²/sec².) Compare this with the energy being given off by the sun each second, also given in Appendix 2.

†3. Referring to the graph given in Fig. 34–2, what are the relative abundances of several light elements or isotopes 1 hour after the big bang? Compare these abundances to the abundance of protons.

4. What is the advantage of studying deuterium over studying helium for assessing the density of the universe?

5. In your own words, explain why the abundance of deuterium is linked with the density of the universe.

6. What are three pieces of evidence that the universe is open, and two that the universe may be closed?

7. Why must we resort to indirect methods to find out if the universe is open or closed?

8. How much energy would you have to put in to accelerate a proton until it was travelling at the speed of light? Explain.

9. Why do we feel gravity, given that the gravitational force is so weak relative to the other fundamental forces?

10. What are the advantages of the inflationary-universe theory?

11. What did the COBE spacecraft find?

†This question requires a numerical solution.

Epilogue

We have had our tour of the universe. We have seen the stars and planets, the matter between the stars, and the distant objects in our universe like quasars and clusters of galaxies. We have learned how our universe is expanding and that it is bathed in a glow of radio waves. We have basked in the glory of a stellar explosion whose light and neutrinos have just reached earth.

Further, we have seen the vitality of contemporary science in general and astronomy in particular. The individual scientists who call themselves astronomers are engaged in fascinating studies, often pushing new technologies to their limits. New telescopes on the ground and in space, new computer capabilities for studying data and carrying out calculations, and new theoretical ideas are linked in research about the universe.

Our knowledge of the universe changes so rapidly that within a few years much of what you read here will be revised. So your study of astronomy shouldn't end here; I hope that, over the years, you will keep up by following astronomical articles and stories in newspapers, magazines, and books (some are listed in the bibliography), and on television. I hope that you will consider the role of scientific research as you vote. And I hope you will remember the methods of science—the mixture of logic and standards of proof by which scientists operate—that you have seen illustrated in this book.

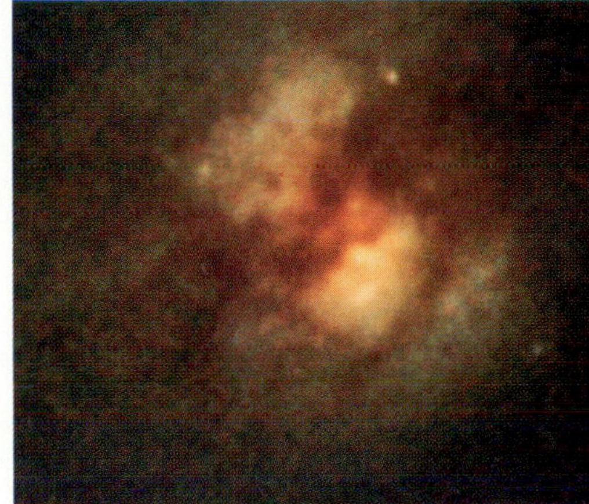

Figure E–1 The innermost part of a starburst galaxy, revealed by the Hubble Space Telescope. This peculiar galaxy, Arp 220, is producing stars at a furious rate from the dust and gas supplied by the collision of two galaxies. This galaxy is especially bright in the infrared. The bright regions near the center are star clusters much brighter than any other clusters known.

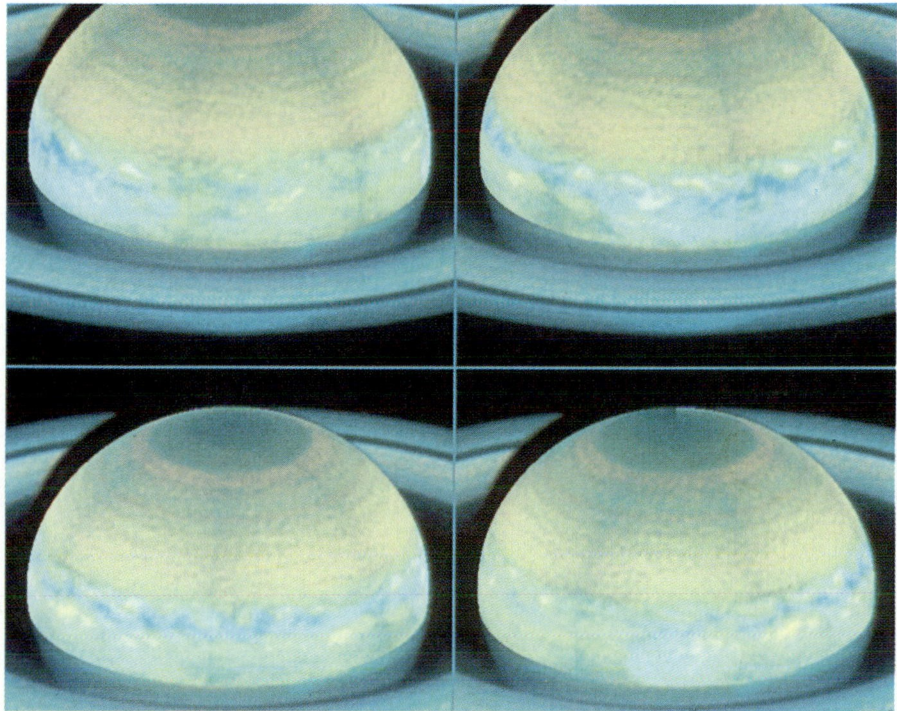

Figure E–2 Saturn, from the Hubble Space Telescope. We hope that the sequence of NASA's Great Observatories, of which the Hubble Space Telescope and the Gamma Ray Observatory are the first two, is to be continued with the Advanced X-Ray Facility and the Space Infrared Telescope Facility.

Appendix 1 Measurement Systems

Système International Units

	SI units	SI Abbrev.	Other Abbrev.
length	meter	m	
volume	liter	L	ℓ
mass	kilogram	kg	kgm
time	second	s	sec
temperature	kelvin	K	°K

Other Metric Units

1 micron (μ) = 1 micrometer (μm) = 10^{-6} meter	μm	μ
1 angstrom (Å or A) = 10^{-10} meter = 10^{-8} cm	0.1 nm	Å

Prefixes for Use With Basic Units of Metric System

Prefix	Symbol	Power		Equivalent
tera	T	10^{12} =	1,000,000,000,000	Trillion
giga	G	10^{9} =	1,000,000,000	Billion
mega	M	10^{6} =	1,000,000	Million
kilo	k	10^{3} =	1,000	Thousand
hecto	h	10^{2} =	100	Hundred
deca	da	10^{1} =	10	Ten
– – –	–	10^{0} =	1	One
deci	d	10^{-1} =	.1	Tenth
centi	c	10^{-2} =	.01	Hundredth
milli	m	10^{-3} =	.001	Thousandth
micro	μ	10^{-6} =	.000001	Millionth
nano	n	10^{-9} =	.000000001	Billionth
pico	p	10^{-12} =	.000000000001	Trillionth
femto	f	10^{-15} =	.000000000000001	
atto	a	10^{-18} =	.000000000000000001	

Examples: 1000 meters = 1 kilometer = 1 km
10^{6} hertz = 1 megahertz = 1 MHz
10^{-3} sec = 1 millisecond = 1 msec

Some Other Units Used in Astronomy

	SI	Other
Energy:	joule (J) = kg · m²/s²	erg, eV
Power:	watt (W) = J/s	joule/sec
Frequency:	hertz (Hz)	cycles per sec

Conversion Factors

1 joule = 10^{7} ergs
1 electron volt (eV) = 1.60207×10^{-19} joules
1 watt = 1 joule per sec
1 cm = 0.3937 in
1 m = 1.0936 yd

1 erg = 10^{-7} joule
1 joule = 6.2419×10^{18} eV
1 in = 25.4 mm = 2.54 cm
1 yd = 0.9144 m

1 km = 0.6214 mi ≃ 5/8 mi	1 mi = 1.6093 km ≃ 8/5 km
1 g = 0.0353 oz	1 oz = 28.3 g
1 kg = 2.2046 lb	1 lb = 0.4536 kg

Appendix 2 Basic Constants

Physical Constants (1986 CODATA Adjustment)

Speed of light*	c	= 299 792 458 m/s (exactly)
Constant of gravitation	G	$= (6.672\ 59 \pm 0.000\ 85) \times 10^{-11}$ m³/kg · sec²
Planck's constant	h	$= (6.626\ 075\ 5 \pm 0.000\ 004\ 0) \times 10^{-34}$ J · s
Boltzmann's constant	k	$= (1.380\ 658 \pm 0.000\ 012) \times 10^{-23}$ J/K
Stefan-Boltzmann constant	σ	$= (5.670\ 51 \pm 0.000\ 19) \times 10^{-8}$ W/m² · K⁴
Wien displacement constant	$\lambda_{max}T$	$= 0.289\ 789$ cm · K $= 28.978\ 9 \times 10^{6}$ Å · K
Mass of hydrogen atom	m_H	$= (1.673\ 534\ 0 \pm 0.000\ 001\ 0) \times 10^{-27}$ kg
Mass of neutron	m_n	$= (1.674\ 928\ 6 \pm 0.000\ 001\ 0) \times 10^{-27}$ kg
Mass of proton	m_p	$= (1.672\ 623\ 1 \pm 0.000\ 001\ 0) \times 10^{-27}$ kg
Mass of electron	m_e	$= (9.109\ 389\ 7 \pm 0.000\ 005\ 4) \times 10^{-31}$ kg
Rydberg's constant	R	$= (1.097\ 373\ 153\ 4 \pm 0.000\ 000\ 001\ 3) \times 10^{7}$/m

Mathematical Constants

π = 3.141 592 653 589 793 238 462 643 383 279 502 884 197 169 399 375 105 820 974 944 592 307 816 406 286 208 998 628 034 825 342 117 067

e = 2.718 281 828 459 045 235 360 287 471 352 662 497 757 247 093 699 959 574 966 967 627 724 076 630 353 547 594 571 382 178 525 166 427

Astronomical Constants

Astronomical Unit*	1 A.U.	$= 1.495\ 978\ 70 \times 10^{11}$ m
Solar parallax*	$\pi_\odot$	= 8.794 148 arc sec
Parsec	1 pc	$= 3.086 \times 10^{16}$ m = 206 264.806 A.U. = 3.261 633 ly
Light year	1 ly	$= (9.460\ 530) \times 10^{15}$ m $= 6.324 \times 10^{4}$ A.U.
Tropical year (1900)*—(equinox to equinox)		= 365.242 198 78 ephemeris days
Julian century*		= 36 525 days
Day*		= 86 400 sec
Sidereal year		= 365.256 366 ephemeris days $= (3.155\ 815) \times 10^{7}$ sec
Mass of sun*	$M_\odot$	$= (1.989\ 1) \times 10^{30}$ kg
Radius of sun*	$R_\odot$	= 696 000 km
Luminosity of sun	$L_\odot$	$= 3.827 \times 10^{26}$ J/sec
Mass of earth*	M_E	$= (5.974\ 2) \times 10^{24}$ kg
Equatorial radius of earth*	R_E	= 6,378.140 km
Center of earth to center of moon (mean)		= 384 403 km
Radius of moon*	R_M	= 1 738 km
Mass of moon*	M_M	$= 7.35 \times 10^{22}$ kg
Solar constant	S	= 1 368 W/m²
Direction of galactic center (2000.0 precession)	α δ	$= 17^h45.6^m$ $= -28°56'$

Precession Formula

Change in δ (dec.) = $\Omega t \sin(23.5°)\cos\alpha$

Change in α (r.a.) = $\Omega t[\cos(23.5°) + \sin(23.5°)\sin\alpha\tan\delta]$,

where Ω is the annual rate of precession. If Ω is used as 50 arc sec/year, declination will be given in arc sec. If Ω is used as 3.3^s/year, right ascension will be given in seconds of r.a. This approximation is not valid for high declination.

*Adopted as "IAU (1976) system of astronomical constants" at the General Assembly of the International Astronomical Union that year. The meter was redefined in 1983 to be the distance travelled by light in a vacuum in 1/299,792,458 second.

Appendix 3 The Planets

Appendix 3a Intrinsic and Rotational Properties

Name	Equatorial Radius *km*	Equatorial Radius ÷ *Earth's*	Mass ÷ Earth's	Mean Density *(g/cm³)*	Oblateness	Surface Gravity *(Earth = 1)*	Sidereal Rotation Period	Inclination of Equator to Orbit	Apparent Magnitude During 1990
Mercury	2,439	0.3824	0.0553	5.43	0	0.378	58.646^d	0.0°	−2.2 to +5.8
Venus	6,052	0.9489	0.8150	5.24	0	0.894	243.01^dR	177.3	−4.6 to −3.9
Earth	6,378.140	1	1	5.515	0.0034	1	$23^h56^m04.1^s$	23.45	—
Mars	3,393.4	0.5326	0.1074	3.94	0.005	0.379	$24^h37^m22.662^s$	25.19	−2.0 to +1.6
Jupiter	71,398	11.194	317.89	1.33	0.064	2.54	9^h50^m to $>9^h55^m$	3.12	−2.7 to −1.8
Saturn	60,000	9.41	95.17	0.70	0.108	1.07	$10^h39.9^m$	26.73	+0.1 to +0.6
Uranus	26,071	4.1	14.56	1.30	0.03	0.8	17^h14^m	97.86	+5.6 to +5.8
Neptune	24,764	3.9	17.15	1.64	0.017	1.2	16^h3^m	29.56	+7.9 to +8.0
Pluto	1,150	0.2	0.002	2.03	?	0.01	$6^d9^h17^m$	120	+13.6 to +13.8

R signifies retrograde rotation.

The masses and radii for Mercury, Venus, Earth, and Mars are the values recommended by the International Astronomical Union in 1976. Surface gravities were calculated from these values. The length of the martian day is from G. de Vaucouleurs (1979). Most densities, oblatenesses, inclinations, and magnitudes are from *The Astronomical Almanac 1990.* Pluto values from David J. Tholen (1990). New Neptune data from *Science,* December 15, 1989.

Appendix 3b Orbital Properties

Name	Semimajor Axis *A.U.*	Semimajor Axis *10^6 km*	Sidereal Period *Years*	Sidereal Period *Days*	Synodic Period *(Days)*	Eccentricity	Inclination to Ecliptic
Mercury	0.3871	57.9	0.24084	87.96	115.9	0.2056	7°00′26″
Venus	0.7233	108.2	0.61515	224.68	584.0	0.0068	3°23′40″
Earth	1	149.6	1.00004	365.25	—	0.0167	0°00′14″
Mars	1.5237	227.9	1.8808	686.95	779.9	0.0934	1°51′09″
Jupiter	5.2028	778.3	11.862	4,337	398.9	0.0483	1°18′29″
Saturn	9.5388	1427.0	29.456	10,760	378.1	0.0560	2°29′17″
Uranus	19.1914	2871.0	84.07	30,700	369.7	0.0461	0°48′26″
Neptune	30.0611	4497.1	164.81	60,200	367.5	0.0100	1°46′27″
Pluto	39.5294	5913.5	248.53	90,780	366.7	0.2484	17°09′03″

Mean elements of planetary orbits for 1980, referred to the mean ecliptic and equinox of 1950 (P. K. Seidelmann, L. E. Doggett, and M. R. De Luccia, *Astronomical Journal* **79,** 57, 1974). Periods are calculated from them.

Appendix 4 **Planetary Satellites**

Satellite	Semimajor Axis of Orbit (km)	Sidereal Revolution Period (d	h	m)	Orbital Eccentricity	Orbital Inclination (°)	Radius (km)	Mass ÷ Mass of Planet	Mean Density (g/cm³)	Discoverer	Visible Magnitude at Mean Opposition Distance
Satellite of the Earth											
The Moon	384,400	27	07	43	0.055	18–29	1738	0.01230002	3.34	—	−12.7
Satellites of Mars											
Phobos	9,378	0	07	39	0.015	1.1	14 × 11 × 9	1.5×10^{-8}	1.95	Hall (1877)	11.8
Deimos	23,459	1	06	18	0.0005	0.9–2.7	8 × 6 × 6	3×10^{-9}	2	Hall (1877)	12.9
Satellites of Jupiter											
XV Adrastea	129,000	0	07	04			13 × 10 × 8	0.1×10^{-10}		Jewett/Voyager 2 (1980)	19.1
XVI Metis	127,000	0	07	06			20	0.5×10^{-10}		Synnott/Voyager (1979)	17.5
V Amalthea	180,000	0	11	57	0.003	0.4	135 × 83 × 75	38×10^{-10}		Barnard (1892)	14.1
XIV Thebe	222,000	0	16	11	0.015	0.8	55 × 45	4×10^{-10}		Synnott/Voyager 1 (1980)	15.6
I Io	422,000	1	18	28	0.004	0.0	1815	4.68×10^{-5}	3.5	Galileo (1610)	5.0
II Europa	671,000	3	13	14	0.009	0.5	1569	2.52×10^{-5}	3.0	Galileo (1610)	5.3
III Ganymede	1,070,000	7	03	43	0.002	0.2	2631	7.80×10^{-5}	1.9	Galileo (1610)	4.6
IV Callisto	1,883,000	16	16	32	0.007	0.5	2400	5.66×10^{-5}	1.8	Galileo (1610)	5.6
XIII Leda	11,094,000	240			0.148	26.1	8	0.03×10^{-10}		Kowal (1974)	20
VI Himalia	11,480,000	251			0.158	27.6	90	50×10^{-10}		Perrine (1904)	14.7
X Lysithea	11,720,000	260			0.107	29.0	20	0.4×10^{-10}		Nicholson (1938)	14.8
VII Elara	11,740,000	260			0.207	24.8	40	4×10^{-10}		Perrine (1905)	14.8
XII Ananke	21,200,000	671R			0.169	147	15	0.2×10^{-10}		Nicholson (1951)	18.9
XI Carme	22,600,000	692R			0.207	164	20	0.5×10^{-10}		Nicholson (1938)	18.0
VIII Pasiphae	23,500,000	735R			0.378	145	20	1×10^{-10}		Melotte (1908)	17.0
IX Sinope	23,700,000	758R			0.275	153	20	0.4×10^{-10}		Nicholson (1914)	18.3
—										Kowal (1975)	20
Satellites of Saturn											
Pan	133,583		13	48			10	8×10^{-12}		Showalter/Voyager 2 (1990)	
Atlas	137,670		14	27	0.000	0.3	20 × 10			Voyager 1	17
Prometheus	139,353		14	43	0.003	0.0	70 × 50 × 40			Voyager 1	16
Pandora	141,700		15	05	0.004	0.0	55 × 45 × 35			Voyager 1	16
Epimetheus	151,400		16	40	0.009	0.3	70 × 60 × 50			Cruikshank/Pioneer 11	14
Janus	151,500		16	40	0.007	0.1	110 × 100 × 80			Smith, Reitsema, Larson, Fountain	15
1 Mimas	185,500		22	37	0.020	1.5	195	8×10^{-8}	1.2	W. Herschel (1789)	12.9
2 Enceladus	238,000	1	08	32	0.005	0.0	250	1.3×10^{-7}	1.1	W. Herschel (1789)	11.7
3 Tethys	294,700	1	22	15	0.000	1.9	530	1.3×10^{-6}	1.0	Cassini (1684)	10.2
Telesto	294,700	1	22	15			17 × 14 × 13			Voyager	19
Calypso	294,700	1	22	15			17 × 11 × 11			Voyager	19
4 Dione	377,400	2	17	36	0.002	0.0	560	1.9×10^{-6}	1.4	Cassini (1684)	10.4
Helene	377,400	2	17	45	0.005	0.0	18 × 16 × 15			Laques and Lecacheux (1980)	19
5 Rhea	527,000	4	12	16	0.001	0.4	765	4.4×10^{-6}	1.3	Cassini (1672)	9.7
6 Titan	1,221,800	15	21	51	0.029	0.3	2575	2.4×10^{-4}	1.9	Huygens (1665)	8.3
7 Hyperion	1,481,000	21	06	45	0.104	0.4	205 × 130 × 110	3×10^{-8}	1.9	Bond (1848)	14.2
8 Iapetus	3,561,300	79	03	43	0.028	14.7	730	3.3×10^{-6}	1.2	Cassini (1671)	11.1
9 Phoebe	12,952,000	549	03	33	0.163	177	110	7×10^{-10}		W. Pickering (1898)	16.5
Satellites of Uranus											
Cordelia	49,771		08	02	<0.001	0.3	13			Voyager 2 (1986)	24.1
Ophelia	53,796		09	02	0.01	<0.5	15			Voyager 2 (1986)	23.8
Bianca	59,173		10	25	<0.001	0.2	21			Voyager 2 (1986)	23.0
Cressida	61,777		11	07	<0.0001	0.2	31			Voyager 2 (1986)	22.2
Desdemona	62,676		11	22	<0.0001	0.2	27			Voyager 2 (1986)	22.5
Juliet	64,352		11	50	0.001	<0.2	42			Voyager 2 (1986)	21.5
Portia	66,085		12	19	<0.005	<0.2	54			Voyager 2 (1986)	21.0
Rosalind	69,942		11	54	<0.0005	0.4	27			Voyager 2 (1986)	22.5
Belinda	75,258		14	57	<0.003	0.1	33			Voyager 2 (1986)	21.1
Puck	86,000		18	17	<0.0003	0.3	77			Voyager 2 (1985)	20.2
5 Miranda	129,783	1	09	56	0.003	3.4	240	0.2×10^{-5}	1.26	Kuiper (1948)	16.3
1 Ariel	191,239	2	12	29	0.003	4.2	579	1.8×10^{-5}	1.65	Lassell (1851)	14.2
2 Umbriel	265,969	4	03	27	0.005	0.4	586	1.2×10^{-5}	1.44	Lassell (1851)	14.8
3 Titania	435,844	8	16	56	0.002	0.1	790	6.8×10^{-5}	1.59	W. Herschel (1787)	13.7
4 Oberon	582,596	13	11	07	0.001	0.1	762	6.9×10^{-5}	1.50	W. Herschel (1787)	14.0

	Satellite	Semimajor Axis of Orbit *(km)*	Sidereal Revolution Period *(d*	*h*	*m)*	Orbital Eccentricity	Orbital Inclination (°)	Radius *(km)*	Mass ÷ Mass of Planet	Mean Density *(g/cm³)*	Discoverer	Visible Magnitude at Mean Opposition Distance
Satellites of Neptune												
N6	Naiad	48,230		7	04	<0.001	4.74	~25			Voyager 2 (1989)	24.7
N5	Thalassa	50,070		7	29	<0.001	0.21	~40			Voyager 2 (1989)	23.8
N3	Despina	52,530		8	02	<0.001	0.07	~75			Voyager 2 (1989)	22.6
N4	Galatea	61,950		10	17	<0.001	0.05	~80			Voyager 2 (1989)	22.3
N2	Larissa	73,550		13	19	0.0014	0.20	104 × 89			Voyager 2 (1989)	22.0
N1	Proteus	117,640	1	02	56	<0.001	0.55	218 × 208 × 201			Voyager 2 (1989)	20.3
	Triton	354,800	5	21	03R	0.03	160.0	1350	1×10^{-3}	2.07	Lassell (1846)	13.5
	Nereid	5,513,400	360	5		0.76	27.4	170	2×10^{-7}	2.03	Kuiper (1949)	18.7
Rings and ring arcs of Neptune												
	Galle	41,900										
	Leverrier	53,200										
	Adams	62,900										
	Liberté	62,900										
	Egalité	62,900										
	Fraternité	62,900										
Satellite of Pluto												
	Charon	19,000	6	9	17	0?	94	590	0.1?		Christy (1978)	17

Based on a table by Joseph Veverka in the *Observer's Handbook 1986 of the Royal Astronomical Society of Canada* and on the *Astronomical Almanac 1991.* Many of Veverka's values are from J. Burns. Charon diameter from David J. Tholen (1990). Density is average of Pluto and Charon. Inclinations greater than 90° are retrograde; R signifies retrograde revolution.

Figure A4–1 A montage of planetary satellites.

Appendix 5 The Brightest Stars

Star	Name	Position (2000.0) R.A.	Position (2000.0) Dec.	Apparent Magnitude (V)	Spectral Type	Absolute Magnitude	Distance D (pc)	Proper Motion R.A.	Proper Motion Dec.	Radial Vel. (km/sec)
1. α CMa A	Sirius	06 45 08.9	−16 42 58	−1.46	A1 V	1.4	2.6	−0.038	−1.21	−8
2. α Car	Canopus	06 23 57.1	−52 41 44	−0.72	F0 Ia	−8.5	360	+0.003	+0.02	+21
3. α Boo	Arcturus	14 15 39.6	+19 10 57	−0.04	K2 IIIp	−0.2	280	−0.077	−2.00	−5
4. α Cen A	Rigil Kentaurus	14 39 36.7	−60 50 02	0.00	G2 V	+4.4	1.3	−0.494	+0.69	−25
5. α Lyr	Vega	18 36 56.2	+38 47 01	0.03	A0 V	+0.5	8.1	+0.017	+0.28	−14
6. α Aur	Capella	05 16 41.3	+45 59 53	0.08	G8 III	+0.3	13	+0.008	−0.42	+30
7. β Ori A	Rigel	05 14 32.2	−08 12 06	0.12	B8 Ia	−7.1	280	0.000	0.00	+21
8. α CMi A	Procyon	07 39 18.1	+05 13 30	0.38	F5 IV	+2.6	3.5	−0.047	−1.03	−3
9. α Ori	Betelgeuse	05 55 10.2	+07 24 26	0.50	M2 Iab	−5.6	95	0.000	0.00	+21
10. α Eri	Achernar	01 37 42.9	−57 14 12	0.46	B5 IV	−1.6	26	+0.013	−0.03	+19
11. β Cen AB	Hadar	14 03 49.4	−60 22 22	0.61	B1 II	−5.1	140	−0.003	−0.02	−12
12. α Aql	Altair	19 50 46.8	+08 52 06	0.77	A7 IV-V	+2.2	5.1	+0.036	+0.39	−26
13. α Tau A	Aldebaran	04 35 55.2	+16 30 33	0.85	K5 III	−0.3	21	+0.005	−0.19	+54
14. α Vir	Spica	13 25 11.5	−11 09 41	0.98	B1 V	−3.5	79	−0.003	−0.03	+1
15. α Sco A	Antares	16 29 24.3	−26 25 55	0.96	M1 Ib	4.7	100	0.000	0.02	−3
16. α PsA	Formalhaut	22 57 38.9	−29 37 20	1.16	A3 V	+2.0	6.7	+0.026	−0.16	+7
17. β Gem	Pollux	07 45 18.9	+28 01 34	1.14	K0 III	+0.2	11	−0.047	−0.05	+3
18. α Cyg	Deneb	20 41 25.8	+45 16 49	1.25	A2 Ia	−7.5	560	0.000	+0.01	−5
19. β Cru	Beta Crucis	12 47 43.2	−59 41 19	1.25	B0 III	−5.0	130	−0.005	−0.02	+20
20. α Leo A	Regulus	10 08 22.2	+11 58 02	1.35	B7 V	−0.6	26	−0.017	0.00	+4
21. α Cru A	Acrux	12 26 35.9	−63 05 56	1.41	B1 IV	−3.9	110	−0.004	−0.02	−11
22. ϵ CMa A	Adhara	06 58 37.5	−28 58 20	1.50	B2 II	−4.4	150	0.000	0.00	+27
23. λ Sco	Shaula	17 33 36.4	−37 06 14	1.63	B2 IV	−3.0	84	0.000	−0.03	0
24. γ Ori	Bellatrix	05 25 07.8	+06 20 59	1.64	B2 III	−3.6	110	−0.001	−0.01	+18
25. β Tau	Elnath	05 26 17.5	+28 36 27	1.65	B7 III	−1.6	40	+0.002	−0.17	+8

Based on a table compiled by Donald A. MacRae in the *Observer's Handbook of the Royal Astronomical Society of Canada,* with 2000.0 positions from *Sky Catalogue 2000.0,* updated with information from W. Gliese (private communications, 1979, 1989).

Appendix 6 Greek Alphabet

Upper Case	Lower Case		Upper Case	Lower Case	
A	α	alpha	N	ν	nu
B	β	beta	Ξ	ξ	xi
Γ	γ	gamma	O	ο	omicron
Δ	δ	delta	Π	π	pi
E	ϵ	epsilon	P	ρ	rho
Z	ζ	zeta	Σ	σ	sigma
H	η	eta	T	τ	tau
Θ	θ	theta	Υ	υ	upsilon
I	ι	iota	Φ	ϕ	phi
K	κ	kappa	X	χ	chi
Λ	λ	lambda	Ψ	ψ	psi
M	μ	mu	Ω	ω	omega

Appendix 7 **The Nearest Stars**

Name	R.A. (2000.0) h	m	Dec. °	′	Parallax π ″	Distance (1/π) pc	Proper Motion μ ″/yr	θ °	Radial Vel km/s	Spectral Type	V	B-V	M_V	Luminosity ($L_\odot$=1)
1 Sun										G2 V	−26.72	0.65	4.85	1.0
2 Proxima Cen	14	30.0	−62	40	0.772	1.30	3.85	283	−16	M5.5 V	11.12	1.90	15.49	0.00006
α Cen A	14	39.6	−60	50	.749	1.34	3.67	281	−22	G2 V	−0.01	0.64	4.37	1.6
α Cen B							3.66	281		K1 V	1.34	0.84	5.71	0.45
3 Barnard's star	17	57.9	+04	41	.545	1.83	10.34	356	−108	M3.8 V	9.56	1.74	13.22	0.00045
4 Wolf 359 (CN Leo)	10	56.7	+07	00	.418	2.39	4.67	235	+13	M5.8 V	13.45	2.00	16.65	0.00002
5 BD +36°2147 = HD95735 (Lalande 21185)	11	03.4	+35	58	.397	2.52	4.78	187	−84	M2.1 V	7.48	1.51	10.50	0.0055
6 L 726−8 = A	1	38.8	−17	57	.381	2.62	3.33	80	+29	M5.6 V	12.56		15.46	0.00006
UV Cet = B									+32		12.96	1.85	15.96	0.00004
7 Sirius A	6	45.1	−16	43	.380	2.63	1.32	204	−8	A1 V	−1.43	0.00	1.42	23.5
Sirius B										DA	8.44	−0.03	11.2	0.003
8 Ross 154 (V1216 Sgr)	18	49.7	−23	49	.341	2.93	0.74	104	−4	M3.6 V	10.45	1.75	13.14	0.00048
9 Ross 248 (HH And)	23	41.9	+44	10	.316	3.16	1.82	176	−81	M4.9 V	12.29	1.91	14.78	0.00011
10 ε Eri	3	32.9	−09	28	.306	3.26	0.98	271	+16	K2 V	3.73	0.88	6.14	0.30
11 Ross 128 (FI Vir)	11	47.6	+00	48	.301	3.32	1.40	151	−13	M4.1 V	11.12	1.75	13.47	0.00036
12 L 789-6	22	38.4	−15	18	.294	3.40	3.27	46	−60	M5.5 V	12.30	1.98	14.49	0.00014
13 BD +43° 44 A (GX And)	00	18.1	+44	00	.290	3.45	2.90	82	+13	M1.3 V	8.08	1.56	10.39	0.0061
+43° 44 B (GQ And) (Groombridge 34 AB)							2.91	83	+20	M3.8 V	11.06	1.79	13.37	0.00039
14 ε Ind	22	03.4	−56	47	.289	3.46	4.69	123	−40	K3 V	4.69	1.06	7.00	0.14
15 61 Cyg A	21	06.9	+38	45	.289	3.46	5.20	52	−64	K3.5 V	5.21	1.18	7.56	0.082
61 Cyg B										K4.7 V	6.03	1.37	8.37	0.039
16 BD +59° 1915 A	18	42.9	+59	37	.286	3.50	2.29	325	0	M3.0 V	8.90	1.52	11.15	0.0030
+59° 1915 B (HD173739/40 = Struve 2398AB = ADS 11632AB)							2.27	323	+10	M3.5 V	9.71	1.59	11.94	0.0015
17 τ Ceti	1	44.1	−15	56	.286	3.50	1.92	297	−16	G8 V	3.49	0.72	5.72	0.45
18 Procyon A	7	39.3	+05	14	.286	3.50	1.25	214	−3	F5 IV-V	0.37	0.42	2.64	7.65
Procyon B										DF	10.7		13.0	0.00055
19 CD −36°15693 = HD217987 (Lacaille 9352)	23	05.9	−35	51	.284	3.52	6.90	79	+10	M1.3 V	7.34	1.49	9.58	0.013
20 G 51-15	8	29.9	+26	46	.276	3.62	1.26	242		M6.6 V	14.90	2.05	17.03	0.00001
21 L 725-32 (YZ Cet)	1	12.4	−16	59	.267	3.75	1.35	62	+28	M4.5 V	12.04	1.83	14.12	0.00020
22 BD + 5°1668	7	27.4	+05	13	.264	3.78	3.76	171	+26	M3.7 V	9.84	1.56	11.94	0.0015
23 CD −39°14192 = HD202560 (Lacaille 8760)	21	17.3	−38	52	.259	3.86	3.47	251	+21	K5.5 V	6.67	1.41	8.74	0.028
24 Kapteyn's star	5	11.2	+45	01	.258	3.88	8.72	131	+245	M0.0 V	8.86	1.55	10.88	0.0039
25 Kruger 60 A	22	28.1	+57	42	.252	3.97	0.87	247	−26	M3.3 V	9.85	1.62	11.87	0.0016
Kruger 60 B (DO Cep)										M5 V	11.3	1.8	13.3	0.0004
26 BD −12°4523	16	30.3	−12	39	.245	4.08	1.17	187	−13	M3.5 V	10.08	1.58	12.07	0.0013
27 Ross 614 A	6	29.3	−02	48	.242	4.13	0.97	131	+24	M4.5 V	11.13	1.71	13.12	0.00049
Ross 614 B (V577 Mon)											14.6		16	0.00004
28 Wolf 424 A (FL Vir A)	12	33.4	+09	01	.232	4.31	1.87	276	−5	M5.3 V	13.07	1.80	14.97	0.00009
Wolf 424 B (FL Vir B)											13.4		15.2	0.00007
29 van Maanen's star	0	49.0	+05	23	.231	4.33	2.98	155	+54	DB	12.38	0.55	14.20	0.00018
30 L 1159-16 (TZ Ari)	2	00.2	+13	03	.224	4.46	2.08	149		M4.5 V	12.28	1.80	14.01	0.00022
31 LHS 288	10	44.5	−61	12	.223	4.48	1.66	348		m	13.87	1.81		
32 CD −37°15492 = HD225213	0	05.1	−37	21	.222	4.50	6.11	112	+23	M2.0 V	8.54	1.46	10.32	0.0065
33 CD −46°11540	17	28.6	−46	54	.220	4.55	1.15	138		M2.7 V	9.37	1.53	11.04	0.0033
34 L 145-141	11	45.4	−64	49	.218	4.59	2.68	97		DC	11.50	0.19	13.07	0.0052
35 LHS 292	10	48.2	−11	20	.217	4.61	1.64	159		M7	15.60	2.10	17.33	
36 CD −49°13515 = HD204961	21	33.5	−49	00	.215	4.65	0.78	184	+8	M1.8 V	8.66	1.50	10.32	0.0065
37 BD + 50°1725 = HD88230A (Groombridge 1618)/ B	10	11.4	+49	27	.213	4.69	1.45	249	−26	K5.0 V	6.59	1.36	8.32	0.041
												8.8		0.013
38 G 158-27	0	06.8	−07	32	.213	4.69	2.06	204		M5.5	13.75	1.98	15.39	0.00006
39 BD +68°946	17	36.5	+68	20	.213	4.69	1.31	196	−22	M3.3	9.18	1.50	10.79	0.0042
40 G208-44 = A	19	53.9	+44	25	.212	4.72	0.75	142		M5.5 V	13.41	1.90	15.03	0.00008
G208-45 = B							0.63	139			14.01	1.98	15.61	0.00005
41 CD −44°11909	17	37.1	−44	19	.212	4.72	1.14	218		M3.9 V	10.95	1.65	12.60	0.00079
42 BD −15°6290	22	53.2	−14	15	.211	4.74	1.12	120	+9	M3.9 V	10.17	1.58	11.77	0.0017
43 o^2 (40) Eri A	4	15.3	−07	39	.207	4.83	4.08	213	−42	K1 V	4.43	0.82	6.01	0.34
40 Eri B							4.07	212	−21	DA	9.52	0.03	11.10	0.0032
40 Eri C (DY Eri)							4.08	213	−45	M4.3 V	11.17	1.67	12.75	0.00069
44 BD +20°2465 (AD Leo)	10	19.6	+19	51	.204	4.90	0.49	265	+11	M3.3 V	9.40	1.54	11.00	0.0035
45 Altair	19	50.8	+08	52	.201	4.98	0.66	54	−26	A7 V	0.77	0.22	2.24	11.1
46 70 Oph A	18	05.5	+02	30	.199	5.03	1.13	167	−7	K0 V	4.21	0.8	5.76	0.43
70 Oph B							1.13	167		K5 V	6.00	1.15	7.54	0.084
47 Gliese 412.0 A	11	05.5	+43	32	.199	5.03	4.53	282	+65	M2Ve	8.74	1.54	10.23	
B: WX UMa	11	05.5	+43	31			4.53	282		M5e	14.40v	2.09v	15.89	
48 BD +43°4305 (EV Lac)	22	46.9	+44	20	.197	5.08	0.84	237	−2	M3.8 V	10.25	1.61	11.7	0.0018
49 AC +79°3888	11	47.5	+78	41	.192	5.21	0.87	57	−119	M3.7 V	10.80	1.60	12.23	0.0011
50 G 9-38 = A	8	58.3	+19	45	.191	5.24	0.89	266		m	14.06	1.84	15.48	0.00006
LP426-40 = B							0.79	263		m	14.92	1.93	16.34	0.000025

Parallaxes and proper motions from the *Yale Trigonometric Parallax Catalogue* (1990), courtesy of John T. Lee and William van Altena and of Wilhelm Gliese (1989). Other information, including V and B-V, courtesy of W. Gliese (1989 and 1990), with spectral types on a consistent scale by Robert F. Wing and Charles A. Dean (1983), and additional comments from Dorrit Hoffleit (1983). Transformed with precession and proper motion corrections to 2000.0 coordinates.

Appendix 8 **Messier Catalogue**

M	NGC	α h	α m (2000.0)	δ °	δ ′	m_v	Description
1	1952	5	34.5	+22	01	8.4	Crab Nebula (Tau)
2	7089	21	33.5	−00	49	6.5	Globular cluster (Aqr)
3	5272	13	42.2	+28	23	6.4	Glob. cluster (CVn)
4	6121	16	23.6	−26	32	5.9	Glob. cluster (Sco)
5	5904	15	18.6	+02	05	5.8	Glob. cluster (Ser)
6	6405	17	40.1	−32	13	4.2	Open cluster (Sco)
7	6475	17	53.9	−34	49	3.3	Open cluster (Sco)
8	6523	18	03.8	−24	23	5.8	Lagoon Nebula (Sgr)
9	6333	17	19.2	−18	31	7.9	Glob. cluster (Oph)
10	6254	16	57.1	−04	06	6.6	Glob. cluster (Oph)
11	6705	18	51.1	−06	16	5.8	Open cluster (Scu)
12	6218	16	47.2	−01	57	6.6	Glob. cluster (Oph)
13	6205	16	41.7	+36	28	5.9	Glob. cluster (Her)
14	6402	17	37.6	−03	15	7.6	Glob. cluster (Oph)
15	7078	21	30.0	+12	10	6.4	Glob. cluster (Peg)
16	6611	18	18.8	−13	47	6.0	Open cl. & nebula (Ser)
17	6618	18	20.8	−16	11	7	Omega nebula (Sgr)
18	6613	18	19.9	−17	08	6.9	Open cluster (Sgr)
19	6273	17	02.6	−26	16	7.2	Glob. cluster (Oph)
20	6514	18	02.6	−23	02	8.5	Trifid Nebula (Sgr)
21	6531	18	04.6	−22	30	5.9	Open cluster (Sgr)
22	6656	18	36.4	−23	54	5.1	Glob cluster (Sgr)
23	6494	17	56.8	−19	01	5.5	Open cluster (Sgr)
24	6603	18	16.9	−18	29	4.5	Open cluster (Sgr)
25	IC4725	18	31.6	−19	15	4.6	Open cluster (Sgr)
26	6694	18	45.2	−09	24	8.0	Open cluster (Scu)
27	6853	19	59.6	+22	43	8.1	Dumbbell N., PN (Vul)
28	6626	18	24.5	−24	52	6.9	Glob. cluster (Sgr)
29	6913	20	23.9	+38	32	6.6	Open cluster (Cyg)
30	7099	21	40.4	−23	11	7.5	Glob. cluster (Cap)
31	224	0	42.7	+41	16	3.4	Andromeda Galaxy (Sb)
32	221	0	42.7	+40	52	8.2	Elliptical galaxy (And)
33	598	1	33.9	+30	39	5.7	Spiral galaxy (Sc) (Tri)
34	1039	2	42.0	+42	47	5.2	Open cluster (Per)
35	2168	6	08.9	+24	20	5.1	Open cluster (Gem)
36	1960	5	36.1	+34	08	6.0	Open cluster (Aur)
37	2099	5	52.4	+32	33	5.6	Open cluster (Aur)
38	1912	5	28.7	+35	50	6.4	Open cluster (Aur)
39	7092	21	32.2	+48	26	4.6	Open cluster (Cyg)
40		12	22.4	+58	05	8	Double star (UMa)
41	2287	6	47.0	−20	44	4.5	Open cluster (CMa)
42	1976	5	35.4	− 5	27	4	Orion Nebula (Ori)
43	1982	5	35.6	− 5	16	9	Orion Nebula; smaller
44	2632	8	40.1	+19	59	3.1	Praesepe; open cl. (Can)
45		3	47.0	+24	07	1.2	Pleiades; open cl. (Tau)
46	2437	7	41.8	−14	49	6.1	Open cluster (Pup)
47	2422	7	36.6	−14	30	4.4	Open cluster (Pup)
48	2548	8	13.8	− 5	48	5.8	Open cluster (Hyd)
49	4472	12	29.8	+ 8	00	8.4	Elliptical galaxy (Vir)
50	2323	7	03.2	− 8	20	5.9	Open cluster (Mon)
51	5194	13	29.9	+47	12	8.1	Whirlpool Galaxy (Sc) (CVn)
52	7654	23	24.2	+61	35	6.9	Open cluster (Cas)
53	5024	13	12.9	+18	10	7.7	Glob. cluster (Com)
54	6715	18	55.1	−30	29	7.7	Glob. cluster (Sgr)
55	6809	19	40.0	−30	58	7.0	Glob. cluster (Sgr)
56	6779	19	16.6	+30	11	8.2	Glob. cluster (Lyr)

M	NGC	α h	α m (2,000.0)	δ °	δ ′	m_v	Description
57	6720	18	53.6	+33	02	9.0	Ring N; planetary (Lyr)
58	4579	12	37.7	+11	49	9.8	Spiral galaxy (SBb) (Vir)
59	4621	12	42.0	+11	39	9.8	Elliptical galaxy (Vir)
60	4649	12	43.7	+11	33	8.8	Elliptical galaxy (Vir)
61	4303	12	21.9	+ 4	28	9.7	Spiral galaxy (Sc) (Vir)
62	6266	17	01.2	−30	07	6.6	Glob. cluster (Sco)
63	5055	13	15.8	+42	02	8.6	Spiral galaxy (Sb) (CVn)
64	4826	12	56.7	+21	41	8.5	Spiral galaxy (Sb) (Com)
65	3623	11	18.9	+13	05	9.3	Spiral galaxy (Sa) (Leo)
66	3627	11	20.2	+12	59	9.0	Spiral galaxy (Sb) (Leo)
67	2682	8	50.4	+11	49	6.9	Open cluster (Can)
68	4590	12	39.5	−26	45	8.2	Glob. cluster (Hyd)
69	6637	18	31.4	−32	21	7.7	Glob. cluster (Sgr)
70	6681	18	43.2	−32	18	8.1	Glob. cluster (Sgr)
71	6838	19	53.8	+18	47	8.3	Glob. cluster (Sgr)
72	6981	20	53.5	−12	32	9.4	Glob. cluster (Aqu)
73	6994	20	58.9	−12	38		Glob. cluster (Aqu)
74	628	1	36.7	+15	47	9.2	Spiral galaxy (Sc) in Pisces
75	6864	20	06.1	−21	55	8.6	Glob. cluster (Sgr)
76	650-1	1	42.4	+51	34	11.5	Planetary nebula (Per)
77	1068	2	42.7	− 0	01	8.8	Spiral galaxy (Sb) (Cet)
78	2068	5	46.7	+ 0	03	8	Small emission nebula (Ori)
79	1904	5	24.5	−24	33	8.0	Glob. cluster (Lep)
80	6093	16	17.0	−22	59	7.2	Glob. cluster (Sco)
81	3031	9	55.6	+69	04	6.8	Spiral galaxy (Sb) (UMa)
82	3034	9	55.8	+69	41	8.4	Irregular galaxy (UMa)
83	5236	13	37.0	−29	52	7.6	Spiral galaxy (Sc) (Hyd)
84	4374	12	25.1	+12	53	9.3	Elliptical galaxy (Vir)
85	4382	12	25.4	+18	11	9.2	S0 galaxy (Com)
86	4406	12	26.2	+12	57	9.2	Elliptical galaxy (Vir)
87	4486	12	30.8	+12	24	8.6	Elliptical galaxy (Ep) (Vir)
88	4501	12	32.0	+14	25	9.5	Spiral galaxy (Sb) (Com)
89	4552	12	35.7	+12	33	9.8	Elliptical galaxy (Vir)
90	4569	12	36.8	+13	10	9.5	Spiral galaxy (SBb) (Vir)
91	4548	12	35.4	+14	30	10.2	M58? (Vir)
92	6341	17	17.1	+43	08	6.5	Glob. cluster (Her)
93	2447	7	44.6	−23	52	6.2	Open cluster in (Pup)
94	4736	12	50.9	+41	07	8.1	Spiral galaxy (Sb) (CVn)
95	3351	10	44.0	+11	42	9.7	Barred spiral g. (SBb) (Leo)
96	3368	10	46.8	+11	49	9.2	Spiral galaxy (Sa) (Leo)
97	3587	11	14.8	+55	01	11.2	Owl Nebula; planetary (UMa)
98	4192	12	13.8	+14	54	10.1	Spiral galaxy (Sb) (Com)
99	4254	12	18.8	+14	25	9.8	Spiral galaxy (Sc) (Com)
100	4321	12	22.9	+15	49	9.4	Spiral galaxy (Sc) (Com)
101	5457	14	03.2	+54	21	7.7	Spiral galaxy (Sc) (UMa)
102							M101; duplication (UMa)
103	581	1	33.2	+60	42	7.4	Open cluster (Cas)
104	4594	12	40.0	−11	37	8.3	Sombrero N.; spiral (Sa) (Vir)
105	3379	10	47.8	+12	35	9.3	Elliptical galaxy (Leo)
106	4258	12	19.0	+47	18	8.3	Spiral galaxy (Sb) (CVn)
107	6171	16	32.5	−13	03	8.1	Glob. cluster (Oph)
108	3556	11	11.5	+55	40	10.0	Spiral galaxy (Sb) (UMa)
109	3992	11	57.6	+53	23	9.8	Barred spiral g. (SBc) (UMa)
110	205	0	40.4	+41	41	8.0	Elliptical galaxy (And)

From Alan Hirshfeld and Roger W. Sinnott, *Sky Catalogue 2000.0*, vol. 2, courtesy of Sky Publishing Co. and Cambridge Univ. Press.

Appendix 9 The Constellations

Latin Name	Genitive	Abbreviation	Translation
Andromeda	Andromedae	And	Andromeda*
Antlia	Antliae	Ant	Pump
Apus	Apodis	Aps	Bird of Paradise
Aquarius	Aquarii	Aqr	Water Bearer
Aquila	Aquilae	Aql	Eagle
Ara	Arae	Ara	Altar
Aries	Arietis	Ari	Ram
Auriga	Aurigae	Aur	Charioteer
Boötes	Boötis	Boo	Herdsman
Caelum	Caeli	Cae	Chisel
Camelopardalis	Camelopardalis	Cam	Giraffe
Cancer	Cancri	Cnc	Crab
Canes Venatici	Canum Venaticorum	CVn	Hunting Dogs
Canis Major	Canis Majoris	CMa	Big Dog
Canis Minor	Canis Minoris	CMi	Little Dog
Capricornus	Capricorni	Cap	Goat
Carina	Carinae	Car	Ship's Keel**
Cassiopeia	Cassiopeiae	Cas	Cassiopeia*
Centaurus	Centauri	Cen	Centaur*
Cepheus	Cephei	Cep	Cepheus*
Cetus	Ceti	Cet	Whale
Chamaeleon	Chamaeleonis	Cha	Chameleon
Circinus	Circini	Cir	Compass
Columba	Columbae	Col	Dove
Coma Berenices	Comae Berencies	Com	Berenice's Hair*
Corona Australis	Coronae Australis	CrA	Southern Crown
Corona Borealis	Coronae Borealis	CrB	Northern Crown
Corvus	Corvi	Crv	Crow
Crater	Crateris	Crt	Cup
Crux	Crucis	Cru	Southern Cross
Cygnus	Cygni	Cyg	Swan
Delphinus	Delphini	Del	Dolphin
Dorado	Doradus	Dor	Swordfish
Draco	Draconis	Dra	Dragon
Equuleus	Equulei	Equ	Little Horse
Eridanus	Eridani	Eri	River Eridanus*
Fornax	Fornacis	For	Furnace
Gemini	Geminorum	Gem	Twins
Grus	Gruis	Gru	Crane
Hercules	Herculis	Her	Hercules*
Horologium	Horologii	Hor	Clock
Hydra	Hydrae	Hya	Hydra* (water monster)
Hydrus	Hydri	Hyi	Sea serpent
Indus	Indi	Ind	Indian
Lacerta	Lacertae	Lac	Lizard
Leo	Leonis	Leo	Lion
Leo Minor	Leonis Minoris	LMi	Little Lion
Lepus	Leporis	Lep	Hare
Libra	Librae	Lib	Scales
Lupus	Lupi	Lup	Wolf
Lynx	Lyncis	Lyn	Lynx
Lyra	Lyrae	Lyr	Harp
Mensa	Mensae	Men	Table (mountain)
Microscopium	Microscopii	Mic	Microscope
Monoceros	Monocerotis	Mon	Unicorn
Musca	Muscae	Mus	Fly
Norma	Normae	Nor	Level (square)
Octans	Octantis	Oct	Octant
Ophiuchus	Ophiuchi	Oph	Ophiuchus* (serpent bearer)
Orion	Orionis	Ori	Orion*
Pavo	Pavonis	Pav	Peacock
Pegasus	Pegasi	Peg	Pegasus* (winged horse)
Perseus	Persei	Per	Perseus*
Phoenix	Phoenicis	Phe	Phoenix
Pictor	Pictoris	Pic	Easel
Pisces	Piscium	Psc	Fish
Piscis Austrinus	Piscis Austrini	PsA	Southern Fish
Puppis	Puppis	Pup	Ship's Stern**
Pyxis	Pyxidis	Pyx	Ship's Compass**
Reticulum	Reticuli	Ret	Net
Sagitta	Sagittae	Sge	Arrow
Sagittarius	Sagittarii	Sgr	Archer
Scorpius	Scorpii	Sco	Scorpion
Sculptor	Sculptoris	Scl	Sculptor
Scutum	Scuti	Sct	Shield
Serpens	Serpentis	Ser	Serpent
Sextans	Sextantis	Sex	Sextant
Taurus	Tauri	Tau	Bull
Telescopium	Telescopii	Tel	Telescope
Triangulum	Trianguli	Tri	Triangle
Triangulum Australe	Trianguli Australis	TrA	Southern Triangle
Tucana	Tucanae	Tuc	Toucan
Ursa Major	Ursae Majoris	UMa	Big Bear
Ursa Minor	Ursae Minoris	UMi	Little Bear
Vela	Velorum	Vel	Ship's Sails**
Virgo	Virginis	Vir	Virgin
Volans	Volantis	Vol	Flying Fish
Vulpecula	Vulpeculae	Vul	Little Fox

*Proper names
**Formerly formed the constellation Argo Navis, the Argonauts' ship.

Appendix 10 Elements and Solar-System Abundances

		Name	Atomic Weight	Solar-System Abundance
1	H	hydrogen	1.01	2.79×10^{10}
2	He	helium	4.00	2.72×10^{9}
3	Li	lithium	6.94	59.1
4	Be	beryllium	9.01	0.73
5	B	boron	10.81	21.2
6	C	carbon	12.01	1.01×10^{7}
7	N	nitrogen	14.01	3.13×10^{6}
8	O	oxygen	16.00	2.38×10^{7}
9	F	fluorine	19.00	843
10	Ne	neon	20.18	3.44×10^{6}
11	Na	sodium	22.99	5.74×10^{4}
12	Mg	magnesium	24.31	1.074×10^{6}
13	Al	aluminum	26.98	8.49×10^{4}
14	Si	silicon	28.09	1.00×10^{6}
15	P	phosphorus	30.97	1.04×10^{4}
16	S	sulfur	32.07	5.15×10^{5}
17	Cl	chlorine	35.45	5240
18	Ar	argon	39.94	1.01×10^{5}
19	K	potassium	39.10	3770
20	Ca	calcium	40.08	6.11×10^{4}
21	Sc	scandium	44.96	34.2
22	Ti	titanium	47.88	2400
23	V	vanadium	50.94	293
24	Cr	chromium	52.00	1.35×10^{4}
25	Mn	manganese	54.94	9550
26	Fe	iron	55.85	9.00×10^{5}
27	Co	cobalt	58.93	2250
28	Ni	nickel	58.69	4.93×10^{4}
29	Cu	copper	63.55	522
30	Zn	zinc	65.39	1260
31	Ga	gallium	69.72	37.8
32	Ge	germanium	72.61	119
33	As	arsenic	74.92	6.56
34	Se	selenium	78.96	62.1
35	Br	bromine	79.90	11.8
36	Kr	krypton	83.80	45
37	Rb	rubidium	85.47	7.09
38	Sr	strontium	87.62	23.5
39	Y	yttrium	88.91	4.64
40	Zr	zirconium	91.22	11.4
41	Nb	niobium	92.91	0.698
42	Mo	molybdenum	95.94	2.55
43	Tc	technetium	(98)	
44	Ru	ruthenium	101.07	1.86
45	Rh	rhodium	102.91	0.344
46	Pd	palladium	106.42	1.39
47	Ag	silver	107.87	0.486
48	Cd	cadmium	112.41	1.61
49	In	indium	114.82	0.184
50	Sn	tin	118.71	3.82
51	Sb	antimony	121.75	0.309
52	Te	tellurium	127.60	4.81
53	I	iodine	126.90	0.90
54	Xe	xenon	131.29	4.7
55	Cs	cesium	132.91	0.372
56	Ba	barium	137.33	4.49
57	La	lanthanum	138.91	0.4460
58	Ce	cerium	140.12	1.136
59	Pr	praseodymium	140.91	0.1669
60	Nd	neodymium	144.24	0.8279
61	Pm	promethium	(145)	
62	Sm	samarium	150.36	0.2582
63	Eu	europium	151.97	0.0973
64	Gd	gadolinium	157.25	0.3300
65	Tb	terbium	158.93	0.0603
66	Dy	dysprosium	162.50	0.3942
67	Ho	holmium	164.93	0.0889
68	Er	erbium	167.26	0.2508
69	Tm	thulium	168.93	0.0378
70	Yb	ytterbium	173.04	0.2479
71	Lu	lutetium	174.97	0.0367
72	Hf	hafnium	178.49	0.154
73	Ta	tantalum	180.95	0.0207
74	W	tungsten	183.85	0.133
75	Re	rhenium	186.21	0.0517
76	Os	osmium	190.2	0.675
77	Ir	iridium	192.22	0.661
78	Pt	platinum	195.08	1.34
79	Au	gold	196.97	0.187
80	Hg	mercury	200.59	0.34
81	Tl	thallium	204.38	0.184
82	Pb	lead	207.2	3.15
83	Bi	bismuth	208.98	0.144
84	Po	polonium	(209)	
85	At	astatine	(210)	
86	Rn	radon	(222)	
87	Fr	francium	(223)	
88	Ra	radium	(226)	
89	Ac	actinium	(227)	
90	Th	thorium	232.04	0.0335
91	Pa	protactinium	231.04	
92	U	uranium	238.03	0.0090
93	Np	neptunium	(237)	
94	Pu	plutonium	(244)	
95	Am	americium	(243)	
96	Cm	curium	(247)	
97	Bk	berkelium	(247)	
98	Cf	californium	(251)	
99	Es	einsteinium	(252)	
100	Fm	fermium	(257)	
101	Md	mendelevium	(258)	
102	No	nobelium	(259)	
103	Lr	lawrencium	(260)	
104	Unq	unnilquadium	(261)	
105	Unp	unnilpentium	(252)	
106	Unh	unnilhexium	(263)	
107	Uns	unnilseptium	(262)	
108	Uno	unniloctium	(265)	
109	Une	unnilennium	(266)	

Atomic mass from 1988 IUPAC report; values in parentheses are the most stable of the important (in availability, etc.) isotopes; values from 106–109 from Peter Armbruster. Abundances based on meteorites from Edward Anders and Nicolas Grevesse, *Geochimica et Cosmochimica Acta*, 53, 197–214 (1989).

Appendix 11

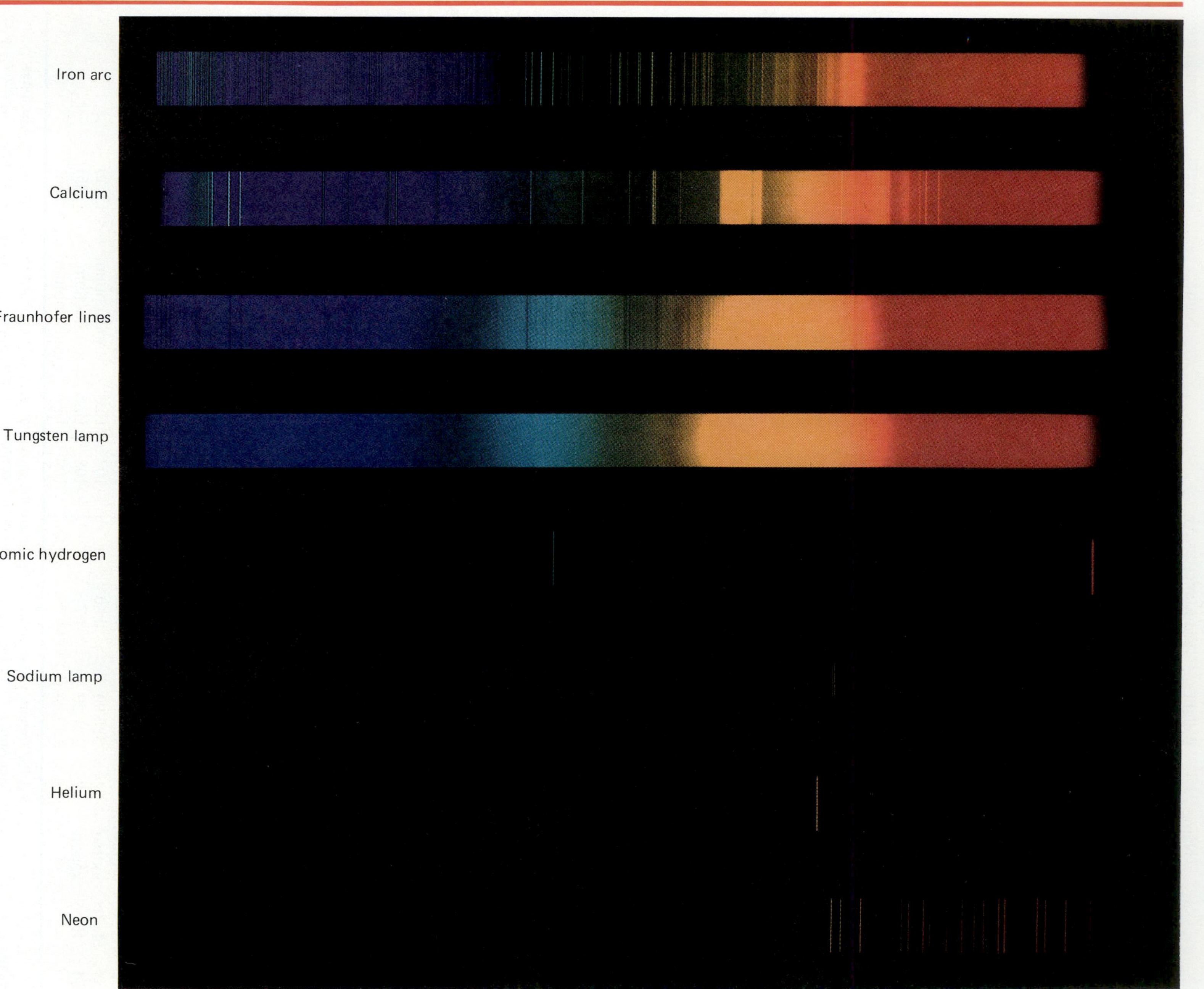
Iron arc
Calcium
Fraunhofer lines
Tungsten lamp
Atomic hydrogen
Sodium lamp
Helium
Neon

Selected Readings

Monthly Non-Technical Magazines on Astronomy

Sky and Telescope, 49 Bay State Road, Cambridge, MA 02138.

Astronomy, 21027 Crossroads Circle, P.O. Box 1612, Waukesha, WI 53187, (800) 446-5489.

Mercury, Astronomical Society of the Pacific, 390 Ashton Ave., San Francisco, CA 94112.

The Griffith Observer, 2800 East Observatory Road, Los Angeles, CA 90027.

Magazines and Annuals Carrying Articles on Astronomy

Science News, 1719 N Street, N.W., Washington, DC 20036. Published weekly.

Scientific American, 415 Madison Avenue, New York, NY 10017. A subject index to astronomy articles from 1979 to 1987 appeared in Mercury, Sept./Oct. 1987, pp. 155–158. A 1960–1981 index appeared in Mercury for March/April 1981.

National Geographic, Washington, DC 20036.

Natural History, Membership Services, Box 4300, Bergenfield, NJ 07621.

Physics Today, American Institute of Physics, 335 East 45 Street, New York, NY 10017.

Science Year (World Book Encyclopedia, Inc., P.O. Box 3073, Evanston, IL 60204). The World Book Science Annual.

Smithsonian, 900 Jefferson Drive, Washington, DC 20560.

Yearbook of Science and the Future (310 S. Michigan Avenue, Chicago, IL 60604: Encyclopaedia Britannica).

Observing Reference Books

Jay M. Pasachoff and Donald H. Menzel, *A Field Guide to the Stars and Planets,* 3rd ed. (Boston: Houghton Mifflin Co., 1992). All kinds of observing information, including monthly maps and the 2000.0 sky atlas by Wil Tirion, and Graphic Timetables to locate planets and special objects like clusters.

Jay M. Pasachoff, *Peterson's First Guide to Astronomy* (Boston: Houghton Mifflin Co., 1988). A brief, beautifully illustrated introduction to observing the sky. Tirion monthly maps.

Jay M. Pasachoff, *Peterson's First Guide to the Solar System* (Boston: Houghton Mifflin Co., 1990). Color illustrations and simple descriptions mark this elementary introduction. Tirion maps of Mars, Jupiter, and Saturn for 1990–2000.

Charles A. Whitney, *Whitney's Star Finder,* 3rd ed. (New York: Alfred A. Knopf, Inc., 1981).

Wil Tirion, *Star Atlas 2000.0* (Cambridge, MA: Sky Publishing Corp., 1981). Accurate and up-to-date. Available in black-on-white, white-on-black, and color versions.

William Liller and Ben Mayer, *The Cambridge Astronomy Guide* (Cambridge Univ. Press, 1985). Alternate chapters by an amateur astronomer and a professional astronomer.

Ben Mayer, *Starwatch* (New York: Perigee/Putnam, 1984). An inspirational introduction to observing.

Ian Ridpath, ed., *Norton's 2000.0 Star Atlas and Reference Handbook* (Longman Scientific & Technical and John Wiley & Sons, 1989). A modern updating of a classic; this is the 18th edition.

The Observer's Handbook (yearly), Royal Astronomical Society of Canada, 136 Dupont Street, Toronto, Ontario M5R 1V2 Canada.

Guy Ottewell, *Astronomical Calendar* (yearly) and *The Astronomical Companion,* Department of Physics, Furman University, Greenville, SC 29613.

The Astronomical Almanac (yearly), U.S. Government Printing Office, Washington, DC 20402.

Hans Vehrenberg, *Atlas of Deep Sky Splendors* (Cambridge, MA: Sky Publishing Corp., 4th ed., 1983). Photographs, descriptions, and finding charts for hundreds of beautiful objects.

Tom Lorenzin, *1000+: The Amateur Astronomer's Field Guide to Deep-Sky Observing* (Davidson, NC, 1987; available from Sky Publishing Corp.).

Alan Hirshfeld and Roger W. Sinnott, *Sky Catalogue 2000.0* (Sky Publishing Corp. and Cambridge Univ. Press, Vol. 1: 1982; Vol. 2: 1985). Vol. 1 is stars and Vol. 2 is full of tables of all other objects.

Roger W. Sinnott, *NGC 2000.0* (Sky Publishing Corp. and Cambridge Univ. Press, 1989). The *NGC* and *IC* updated.

Wil Tirion, Barry Rappaport, and George Lovi, *Uranometria 2000.0* (Richmond, VA: Willmann-Bell, 1986). Vol. 1 covers +90° to −6°. Vol. 2 covers +6° to −90°. Star maps to magnitude 9.5 and an essay on historical atlases.

Kenneth Glyn Jones, *Messier's Nebulae and Star Clusters* (Cambridge Univ. Press, 1991). Messier objects discussed one by one.

Philip S. Harrington, *Touring the Universe Through Binoculars: A Complete Astronomer's Guidebook* (New York: John Wiley & Sons, Inc., 1990).

Laboratory Manuals

Paul Johnson and Ronald Canterna, *Lab Manual for Astronomy* (Philadelphia, Saunders College Publ., 1987).

Leon Palmer, *The Trained Eye: An Introduction to Astronomical Observing* (Philadelphia: Saunders College Publ., 1991). Includes 8 fold-out sky maps.

For Information About Amateur Societies

American Association of Variable Star Observers (AAVSO), 25 Birch St., Cambridge, MA 02138.

American Meteor Society, Dept. of Physics and Astronomy, SUNY, Geneseo, NY 14454.

Association of Lunar and Planetary Observers (ALPO), 8930 Raven Drive, Waco, TX 76710.

The Astronomical League, the umbrella group of amateur societies. For their newsletter, *The Reflector,* write The Astronomical League, Executive Secretary, c/o Science Service Building, 1719 N. St. N.W., Washington, DC 20030.

The Astronomical Society of the Pacific, 390 Ashton Ave., San Francisco, CA 94112.

British Astronomical Association, Burlington House, Piccadilly, London W1V 0NL, England.

Royal Astronomical Society of Canada, 124 Merton St., Toronto, Ontario M4S 2Z2, Canada.

Careers in Astronomy

Mary K. Hemenway, Education Officer, American Astronomical Society, University of Texas, Dept. of Astronomy, Austin, TX 78712. A free booklet, *A Career in Astronomy,* is available on request.

Space for Women, derived from a symposium for women on careers. For free copies, write to: Center for Astrophysics, 60 Garden Street, Cambridge, MA 02138.

Teaching

Jay M. Pasachoff and John R. Percy, *The Teaching of Astronomy* (Cambridge Univ. Press, 1990). The proceedings of International Astronomical Union Colloquium #105.

General Reading

Jean Audouze and Guy Israel, eds., *The Cambridge Atlas of Astronomy,* 2nd ed. (Cambridge Univ. Press, 1988). Coffee-table size with fantastic photos and authoritative text.

David Malin and Paul Murdin, *Colours of the Stars* (Cambridge Univ. Press, 1984). Fantastic color photographs and interesting descriptions of color in celestial objects.

Svend Laustsen, Claus Madsen, and Richard M. West, *Exploring the Southern Sky: A Pictorial Atlas from the European Southern Observatory* (New York: Springer-Verlag, 1987). Beautiful and impressive.

Harlow Shapley, ed., *Source Book in Astronomy 1900–1950* (Cambridge, MA: Harvard University Press, 1960). Reprints.

Otto Struve and Velta Zebergs, *Astronomy of the Twentieth Century* (New York: Macmillan, 1962). A historical view.

J. Trefil, *Space, Time and Infinity* (Washington, DC: Smithsonian Press, 1985).

Stephen P. Maran ed., *The Astronomy and Astrophysics Encyclopedia* (New York: Van Nostrand, 1992). Long entries on all aspects of astronomy.

Jacqueline Mitton, *A Concise Dictionary of Astronomy* (New York: Oxford Univ. Press, 1991). Short entries on all aspects of astronomy.

Some Additional Books

Observatories and Observing

H.T. Kirby-Smith, *U.S. Observatories: A Directory and Travel Guide* (New York: Van Nostrand Reinhold, 1976).

Martin Cohen, *In Quest of Telescopes* (Cambridge, MA: Sky Publishing Corp., 1980). What it is like to be an astronomer.

W. and K. Tucker, *The Cosmic Inquirers* (Cambridge, MA: Harvard University Press, 1986).

Stars

James B. Kaler, *Stars and Their Spectra* (Cambridge Univ. Press, 1989). OBAFGKM.

Lawrence H. Aller, *Atoms, Stars, and Nebulae,* Revised ed. (Cambridge, MA: Harvard University Press, 1971). One of the Harvard Books on Astronomy series of popular works.

Bart J. Bok and Priscilla F. Bok, *The Milky Way,* 5th ed. (Cambridge, MA: Harvard University Press, 1981). A readable and well-illustrated survey from the Harvard Books on Astronomy series; includes good discussions of star clusters and H-R diagrams.

Robert W. Noyes, *The Sun, Our Star* (Cambridge, MA: Harvard University Press, 1982). A part of the Harvard Series.

Harold Zirin, *Astrophysics of the Sun* (Cambridge Univ. Press, 1988). Advanced but fascinating.

Laurence Marschall, *The Supernova Story* (Plenum, 1988). Supernovae in general plus SN 1987A.

Wallace H. Tucker, *The Star Splitters: The High Energy Astronomy Observatories* (NASA SP-466). Readable and well illustrated.

Henry L. Shipman, *Black Holes, Quasars, and the Universe,* 2nd ed. (Boston: Houghton Mifflin Co., 1980). A careful discussion of several topics of great current interest.

David Levy, *Observing Variable Stars* (Cambridge Univ. Press, 1989). A guide for the beginner.

Cecilia Payne-Gaposchkin, *Stars and Clusters* (Cambridge, MA: Harvard University Press, 1979).

Donald A. Cooke, *The Life and Death of Stars* (New York: Crown, 1985). A lavishly illustrated description of stellar evolution.

Ronald Giovanelli, *Secrets of the Sun* (Cambridge Univ. Press, 1984).

Paul and Leslie Murdin, *Supernovae* (Cambridge Univ. Press, 1985).

George Greenstein, *Frozen Star* (New York: New American Library, 1985). Eloquent, prize-winning.

Donat Wentzel, *The Restless Sun* (Washington, DC: Smithsonian Institution Press, 1989). Current views well described.

Fred Espenak, *Fifty Year Canon of Solar Eclipses* (NASA Ref. Pub. 1178, Rev. 1987; order from Sky Publishing Corp.). Maps and tables.

Solar System

Clark R. Chapman, *Planets of Rock and Ice: from Mercury to the Moons of Saturn* (New York: Charles Scribner's Sons, 1982). A non-technical discussion of the planets and their moons.

Fred Espenak, *Fifty Year Canon of Lunar Eclipses: 1868-2035* (NASA Ref. Pub. 1216, 1989; order from Sky Publishing Corp.). Maps and tables.

David Morrison and Tobias Owen, *The Planetary System* (Addison-Wesley, 1988). All about the solar system.

Fred L. Whipple, *Orbiting the Sun: Planets and Satellites of the Solar System,* enlarged edition of *Earth, Moon and Planets* (Cambridge, MA: Harvard University Press, 1981).

J. Kelly Beatty, Brian O'Leary, and Andrew Chaikin, *The New Solar System,* 3rd ed. (Cambridge, MA: Sky Publishing Corp., 1990). Each chapter written by a different expert.

David D. Morrison and Jane Samz, *Voyage to Jupiter* (NASA SP 439, 1980). The story of the Voyager missions to Jupiter and what we learned, masterfully told and profusely illustrated.

David D. Morrison, *Voyages to Saturn* (NASA SP 451, 1982). On to Saturn.

Bevan M. French, *The Moon Book* (New York: Penguin, 1977). A clear, authoritative, thorough, and interesting report on the lunar program and its results.

Donald Goldsmith and Tobias Owen, *The Search for Life in the Universe* (Menlo Park, CA: Benjamin/Cummings, 1980).

James Elliot and Richard Kerr, *Rings* (Cambridge, MA: MIT Press, 1984). Includes first-person and other stories of the discoveries.

Fred L. Whipple, *The Mystery of Comets* (Washington, DC: Smithsonian, 1986). By the master.

Galaxies and the Outer Universe

Richard Berendzen, Richard Hart, and Daniel Seeley, *Man Discovers the Galaxies* (New York: Neale Watson Academic Publications, 1976). A historical review.

Charles A. Whitney, *The Discovery of Our Galaxy* (New York: Alfred A. Knopf, 1971). A historical discussion on a more popular level.

Gerrit L. Verschuur, *The Invisible Universe Revealed* (New York: Springer-Verlag, 1987). A non-technical treatment of radio astronomy.

David A. Allen, *Infrared, The New Astronomy* (New York: John Wiley & Sons, 1975). Includes a personal narrative.

John D. Kraus, *Radio Astronomy,* 2nd ed. (Cygnus-Quasar Books, P.O. Box 85, Powell, OH 43065, 1986).

Halton C. Arp, *Atlas of Pecular Galaxies* (Pasadena, CA: California Institute of Technology, 1966). Worth poring over.

Allan Sandage, *The Hubble Atlas of Galaxies* (Washington, DC: Carnegie Institution of Washington, 1961). Publication No. 618. Beautiful photographs of galaxies and thorough descriptions. Everyone should examine this carefully.

Paul W. Hodge, *Galaxies* (Cambridge, MA: Harvard University Press, 1986). A non-technical study of galaxies, the successor to Harlow Shapley's book. In the Harvard series.

Timothy Ferris, *The Red Limit,* 2nd ed. (New York: Morrow/Quill, 1983). Written for the general reader.

George Gamow, *One, Two, Three . . . Infinity* (New York: Bantam Books, 1971). A reprinting of a wonderful description of the structure of space that has introduced at an early age many a contemporary astronomer to his or her profession.

Steven Weinberg, *The First Three Minutes* (New York: Basic Books, 1977). A readable discussion of the first minutes after the big bang, including a discussion of the background radiation.

Timothy Ferris, *Galaxies* (New York: Stewart, Tabori, and Chang, 1982). Beautifully illustrated; paperback edition.

John D. Barrow and Joseph Silk, *The Left Hand of Creation* (New York: Basic Books, 1983). Cosmology elucidated.

Wallace Tucker and Riccardo Giacconi, *The X-Ray Universe* (Cambridge, MA: Harvard University Press, 1985). In the Harvard non-technical series.

Edward Harrison, *Masks of the Universe* (New York: Macmillan, 1985). Cosmology.

Stephen W. Hawking, *A Brief History of Time* (New York: Bantam Books, 1988). A best-selling discussion of fundamental topics.

Joseph Silk, *The Big Bang* (Freeman, 1989). Cosmology.

Timothy Ferris, *Coming of Age in the Milky Way* (New York: Morrow, 1988). A fascinating historical perspective.

Advanced Textbook

Michael Zeilik, Stephen A. Gregory, and Elske V. P. Smith, *Introductory Astronomy and Astrophysics* 3rd ed. (Philadelphia: Saunders College Publ., 1992). A mathematical treatment of introductory astronomy and astrophysics for science majors.

Computer Planetarium

Voyager (Carina Software, 830 Williams Street, San Leandro, CA 94577, 415 352–7328). For the Mac.

Visible Universe (Parsec Software, 1949 Blair Loop Road, Danville, VA 24541, 804 822–1179). For the IBM-PC and compatibles. The professional version even shows eclipse tracks.

Dance of the Planets (A.R.C. Software, P.O. Box 1974P, Loveland, CO 80539, 303 663–3223. For the IBM PC and compatibles.

Superstar (picoSCIENCE, 41512 Chadbourne Drive, Fremont, CA 94539, 415 498–1095). For the IBM PC and compatibles.

LodeStar (Zephyr Services, 1900 Murray Ave., Pittsburgh, PA 15217, 412 422–6600). For the IBM PC and compatibles.

The_Sky (Software Bisque, 912 12th St., Golden CO 80401, 303 278–4478). For the IBM PC and compatibles.

Glossary

absolute magnitude The magnitude that a star would appear to have if it were at a distance of ten parsecs from us.

absorption line Wavelengths at which the intensity of radiation is less than it is at neighboring wavelengths.

absorption nebula Gas and dust seen in silhouette.

accretion disk Matter that an object has taken up and that has formed a disk around the object.

achondrite A type of stony meteorite without chondrules. (See *chondrite.*)

active galactic nucleus (AGN) A galaxy with an exceptionally bright nucleus in some part of the spectrum; includes radio galaxies, Seyfert galaxies, quasars, and QSO's.

active galaxy A galaxy radiating much more than average in some part of the non-optical spectrum, revealing high-energy processes; radio galaxies and Seyfert galaxies are examples.

active regions Regions on the sun where sunspots, plages, flares, etc., are found.

active sun The group of solar phenomena that vary with time, such as active regions and their phenomena.

AGN Active galactic nucleus

albedo The fraction of light reflected by a body.

allowed states The energy values that atoms can have by the laws of quantum mechanics.

alpha particles A helium nucleus; consists of two protons and two neutrons.

alt-azimuth A two-axis telescope mounting in which motion around one of the axes, which is vertical, provides motion in azimuth, and motion around the perpendicular axis provides up-and-down (altitude) motion.

altitude (a) Height above the surface of a planet, or (b) for a telescope mounting, elevation in angular measure above the horizon.

amino acid A type of molecule containing the group NH_2 (the amino group). Amino acids are fundamental building blocks of life.

Amor asteroids A group of asteroids with semimajor axes greater than earth's and between 1.017 A.U. and 1.3 A.U.; about half cross the earth's orbit.

angstrom A unit of length equal to 10^{-8} cm.

angular momentum An intrinsic property of a system corresponding to the amount of its revolution or spin. The amount of angular momentum of a body orbiting around a point is the mass of the orbiting body times its (linear) velocity of revolution times its distance from the point. The amount of angular momentum of a spinning sphere is the moment of inertia, an intrinsic property of the distribution of mass, times the angular velocity of spin.

angular velocity The rate at which a body rotates or revolves expressed as the angle covered in a given time (for example, in degrees per hour).

anisotropy Deviation from isotropy; changing with direction.

annular eclipse A type of solar eclipse in which a ring (annulus) of solar photosphere remains visible.

anorthosite A type of rock resulting from cooled lava, common in the lunar highlands though rare on earth.

antimatter A type of matter in which each particle (antiproton, antineutron, etc.) is opposite in charge and certain other properties to a corresponding particle (proton, neutron, etc.) of the same mass of the ordinary type of matter from which the solar system is made.

antiparticle The antimatter corresponding to a given particle.

aperture The diameter of the lens or mirror that defines the amount of light focused by an optical system.

aperture synthesis The use of several smaller telescopes together to give some of the properties, such as resolution, of a single larger aperture.

aphelion For an orbit around the sun, the farthest point from the sun.

Apollo asteroids A group of asteroids, with semimajor axes greater than earth's and less than 1.017 A.U., whose orbits overlap the earth's.

apparent magnitude The brightness of a star as seen by an observer, given in a specific system in which a difference of five magnitudes corresponds to a brightness ratio of one hundred times; the scale is fixed by correspondence with a historical background.

apsides (singular: apsis) The points at the ends of the major axis of an elliptical orbit. The *line of apsides* is the line that coincides with the major axis of the orbit.

archaebacteria A primitive type of organism, different from either plants or animals, perhaps surviving from billions of years ago.

association A physical grouping of stars; in particular, we talk of O and B associations or T associations.

asterism A special apparent grouping of stars, part of a constellation.

asteroid A "minor planet," a non-luminous chunk of rock smaller than planet-size but larger than a meteoroid, in orbit around a star.

asteroid belt A region of the solar system, between the orbits of Mars and Jupiter, in which most of the asteroids orbit.

astrometric binary A system of two stars in which the existence of one star can be deduced by study of its gravitational effect on the proper motion of the other star.

astrometry The branch of astronomy that involves the detailed measurement of the positions and motions of stars and other celestial bodies.

Astronomical Unit The average distance from the earth to the sun.

astrophysics The science, now essentially identical with astronomy, applying the laws of physics to the universe.
Aten asteroids A group of asteroids with semimajor axes smaller than 1 A.U.
atom The smallest possible unit of a chemical element. When an atom is subdivided, the parts no longer have properties of any chemical element.
atomic clock A system that uses atomic properties to provide a measure of time.
atomic number The number of protons in an atom.
atomic weight The number of protons and neutrons in an atom, averaged over the abundances of the different isotopes.
aurora Glowing lights visible in the sky, resulting from processes in the earth's upper atmosphere and linked with the earth's magnetic field.
aurora australis The southern aurora.
aurora borealis The northern aurora.
autumnal equinox Of the two locations in the sky where the ecliptic crosses the celestial equator, the one that the sun passes each year when moving from northern to southern declinations.
A.U. Astronomical Unit.
azimuth The angular distance around the horizon from the northern direction, usually expressed in angular measure from 0° for an object in the northern direction, to 180° for an object in the southern direction, around to 360°.

background radiation See *primordial background radiation.*
Baily's beads Beads of light visible around the rim of the moon at the beginning and end of a total solar eclipse. They result from the solar photosphere shining through valleys at the edge of the moon.
Balmer series The set of spectral absorption or emission lines resulting from a transition down to or up from the second energy level (first excited level) of hydrogen.
bar The straight structure across the center of some spiral galaxies, from which the arms unwind.
baryons Nuclear particles (protons, neutrons, etc.) subject to the strong nuclear force; made of quarks.
basalt A type of rock resulting from the cooling of lava.
baseline The distance between points of observation when it determines the accuracy of some measurement.
beam The cone within which a radio telescope is sensitive to radiation.
Becklin-Neugebauer object An object visible only in the infrared in the Orion Molecular Cloud, apparently a very young star.
belts Dark bands around certain planets, notably Jupiter.
beta particle An electron or a positron outside an atom.
big-bang theory A cosmological model, based on Einstein's general theory of relativity, in which the universe was once compressed to infinite density and has been expanding ever since.
binary pulsar A pulsar in a binary system.
binary star Two stars revolving around each other.
bipolar flow A phenomenon in young or forming stars in which streams of matter are ejected from the poles.
black body A hypothetical object that, if it existed, would absorb all radiation that hit it and would emit radiation that exactly followed Planck's law.
black dwarf A non-radiating ball of gas that results either when a white dwarf radiates all its energy or when gas contracts gravitationally but contains too little mass to begin nuclear fusion.
black hole A region of space from which, according to the general theory of relativity, neither radiation nor matter can escape.
blueshifted Wavelengths shifted to the blue; when the shift is caused by motion, from a velocity of approach.
B-N object See *Becklin-Neugebauer object.*
Bohr atom Niels Bohr's model of the hydrogen atom, in which the energy levels are depicted as concentric circles of radii that increase as (level number)^2.
Bok globule A type of round compact absorption nebula.
bolometer A device for measuring the total amount of radiation from an object.
bolometric magnitude The magnitude of a celestial object corrected to take account of the radiation in parts of the spectrum other than the visible.
bound-free transition An atomic transition in which an electron starts bound to the atom and winds up free from it.
breccia A type of rock made up of fragments of several types of rocks. Breccias are common on the moon.
brown dwarf A self-gravitating, self-luminous object, not a satellite and insufficiently massive for nuclear fusion to begin (less than 0.07 solar mass).

calorie A measure of energy in the form of heat, originally corresponding to the amount of heat required to raise the temperature of 1 gram of water by 1°C.
canali Name given years ago to apparent lines on Mars.
capture A model in which a moon was formed elsewhere and then was captured gravitationally by its planet.
carbonaceous Containing a lot of carbon.
carbon cycle A chain of nuclear reactions, involving carbon as a catalyst at some of its intermediate stages, that transforms four hydrogen atoms into one helium atom with a resulting release in energy. The carbon cycle is only important in stars hotter than the sun.
carbon-nitrogen cycle The carbon cycle, acknowledging that nitrogen also plays an intermediary role.
carbon-nitrogen-oxygen tri-cycle A set of variations of the carbon cycle including nitrogen and oxygen isotopes as intermediaries.
carbon stars A spectral type of cool stars whose spectra show a lot of carbon; formerly types R and N.
Cassegrainian (Cassegrain) (telescope) A type of reflecting telescope in which the light focused by the primary mirror is intercepted short of its focal point and refocused and reflected by a secondary mirror through a hole in the center of the primary mirror.
Cassini's division The major division in the rings of Saturn.
catalyst A substance that participates in a reaction but that is left over in its original form at the end.
catastrophe theories Theories of solar-system formation involving a collision with another star.
CCD Charge-coupled device, a solid-state imaging device.
celestial equator The intersection of the celestial sphere with the planet that passes through the earth's equator.
celestial poles The intersection of the celestial sphere with the axis of rotation of the earth.

celestial sphere The hypothetical sphere centered at the center of the earth to which it appears that the stars are affixed.
center of mass The "average" location of mass; the point in a body or system of bodies at which we may consider all the mass to be located for the purpose of calculating the gravitational effect of that mass or its mean motion when a force is applied.
central star The hot object at the center of a planetary nebula, which is the remaining core of the original star.
Cepheid variable A type of supergiant star that oscillates in brightness in a manner similar to the star δ Cephei. The periods of Cepheid variables, which are between 1 and 100 days, are linked to the absolute magnitude of the stars by known relationships; this allows the distances to Cepheids to be found. Type II Cepheids (also known as W Virginis stars) are fainter at each period.
Chandrasekhar limit The limit in mass, about 1.4 solar masses, above which electron degeneracy cannot support a star, and so the limit above which white dwarfs cannot exist.
charm An arbitrary name that corresponds to a property that distinguishes certain elementary particles, including types of quarks, from each other.
chondrite A type of stony meteorite that contains small crystalline spherical particles called chondrules.
chromatic aberration A defect of lens systems in which different colors are focused at different points.
chromosphere The part of the atmosphere of the sun (or another star) between the photosphere and the corona. It is probably entirely composed of spicules and probably roughly corresponds to the region in which mechanical energy is deposited.
chromospheric network An apparent network of lines on the solar surface corresponding to the higher magnetic fields at the boundaries of supergranules; visible especially in the calcium H and K lines.
circle A conic section formed by cutting a cone perpendicularly to its axis.
circumpolar stars For a given observing location, stars that are close enough to the celestial pole that they never set.
classical When discussing atoms, not taking account of quantum mechanical effects.
closed universe A big-bang universe with positive curvature; it has finite volume and will eventually contract.
cluster (a) Of stars, a physical grouping of many stars; (b) of galaxies, a physical grouping of at least a few galaxies.
cluster variable A star that varies in brightness with a period of 0.1 to one day, similar to the star RR Lyrae. They are found in globular clusters. All cluster variables have approximately the same brightness, which allows their distance to be readily found.
CNO tri-cycle See *carbon-nitrogen-oxygen tri-cycle.*
COBE (Cosmic Background Explorer) A spacecraft launched in 1989 to study the cosmic background radiation.
coherent radiation Radiation in which the phases of waves at different locations in a cross-section of radiation have a definite relation to each other; in noncoherent radiation, the phases are random. Only coherent radiation shows interference.
cold dark matter Nonluminous matter with no velocity dispersion compared with the expansion of the universe, such as mini black holes and exotic nuclear particles.
color (a) Of an object, a visual property that depends on wavelength; (b) an arbitrary name assigned to a property that distinguishes three kinds of quarks.
color index The difference, expressed in magnitudes, of the brightness of a star or other celestial object measured at two different wavelengths. The color index is a measure of temperature.
color-magnitude diagram A Hertzsprung-Russell diagram in which the temperature on the horizontal axis is expressed in terms of color index and the vertical axis is in magnitudes.
coma (a) Of a comet, the region surrounding the head; (b) of an optical system, an off-axis aberration in which the images of points appear with comet-like asymmetries.
comet A type of object orbiting the sun, often in a very elongated orbit, that when relatively near to the sun shows a coma and may show a tail.
comparative planetology Studying the properties of solar-system bodies by comparing them.
comparison spectrum A spectrum of known elements on earth usually photographed on the same photographic plate as a stellar spectrum in order to provide a known set of wavelengths for zero Doppler shift.
composite spectrum The spectrum that reveals a star is a binary system, since spectra of more than one object are apparent.
condensation A region of unusually high mass or brightness.
conic sections Geometric shapes obtained by slicing a cone.
conjunction Where two celestial objects reach the same celestial longitude; approximately corresponds to their closest apparent approach in the sky. When only one body is named, it is understood that the second body is the sun.
conservation law A statement that the total amount of some property (angular momentum, energy, etc.) of a body or set of bodies does not change.
constellation One of 88 areas into which the sky has been divided for convenience in referring to the stars or other objects therein.
continental drift The slow motion of the continents across the Earth's surface, explained in the theory of plate tectonics as a set of shifting regions called plates.
continuous spectrum A spectrum with radiation at all wavelengths but with neither absorption nor emission lines.
continuum (pl: **continua**) The continuous spectrum that we would measure from a body if no spectral lines were present.
convection The method of energy transport in which the rising motion of masses from below carries energy upward in a gravitational field. Boiling is an example.
convection zone The subsurface zone in certain types of stars in which convection dominates energy transfer.
convergent point The point in the sky toward which the members of a star cluster appear to be converging or from which they appear to be diverging. Referred to in the moving cluster method of determining distances.
coordinate systems Methods of assigning positions with respect to suitable axes.
core The central region of a star or planet.

corona, galactic The outermost region of our current model of the galaxy, containing most of the mass in some unknown way.
corona, solar or stellar The outermost region of the sun (or of other stars), characterized by temperatures of million of kelvins.
coronagraph A type of telescope with which the corona can be seen in visible light at times other than that of a total solar eclipse.
coronal holes Relatively dark regions of the corona having low density; they result from open field lines.
correcting plate A thin lens of complicated shape at the front of a Schmidt camera that compensates for spherical aberration.
cosmic abundances The overall abundances of elements in the universe.
cosmic background radiation The isotropic glow of 3° radiation.
cosmic rays Nuclear particles or nuclei travelling through space at high velocity.
cosmic turbulence A theory of galaxy formation in which turbulence originally existed.
cosmogony The study of the origin of the universe, usually applied in particular to the origin of the solar system.
cosmological constant A constant arbitrarily added by Einstein to an equation in his general theory of relativity in order to provide a solution in which the universe did not expand. It was only subsequently discovered that the universe did expand after all.
cosmological distances Distances assigned by Hubble's law.
cosmological principle The principle that on the whole the universe looks the same in all directions and in all regions.
cosmology The study of the universe as a whole.
coudé focus A focal point of large telescopes in which the light is reflected by a series of mirrors so that it comes to a point at the end of a polar axis. The image does not move even when the telescope moves, which permits the mounting of heavy equipment.
crust The outermost solid layer of some objects, including neutron stars and some planets.
C-type stars See *carbon stars.*
cytherean Venusian.

dark matter Non-luminous matter. See *hot dark matter* and *cold dark matter.*
dark nebula Dust and gas seen in silhouette.
daughter molecules Relatively simple molecules in comets resulting from the breakup of more complex molecules.
deceleration parameter (q_0) A particular measure of the rate at which the expansion of the universe is slowing down. $q_0 < 1/2$ corresponds to an open universe and $q_0 > 1/2$ corresponds to a closed universe.
declination Celestial latitude, measured in degrees north or south of the celestial equator.
deferent In the Ptolemaic system of the universe, the larger circle, centered at the earth, on which the centers of the epicycles revolve.
degenerate matter Matter whose properties are controlled, and that is prevented from further contraction, by quantum mechanical laws.
density Mass divided by volume.
density wave A circulating region of relatively high density, important, for example, in models of spiral arms.
density-wave theory The explanation of spiral structure of galaxies as the effect of a wave of compression that rotates around the center of the galaxy and causes the formation of stars in the compressed region.
deuterium An isotope of hydrogen that contains 1 proton and 1 neutron.
deuteron A deuterium nucleus, containing 1 proton and 1 neutron.
diamond-ring effect The last Baily's bead glowing brightly at the beginning of the total phase of a solar eclipse, or its counterpart at the end of totality.
differential forces A net force resulting from the difference of two other forces; a tidal force.
differential rotation Rotation of a body in which different parts have different angular velocities (and thus different periods of rotation).
differentiation For a planet, the existence of layers of different structure or composition.
diffraction A phenomenon affecting light as it passes any obstacle, spreading it out in a complicated fashion.
diffraction grating A very closely ruled series of lines that, through diffraction of light, provides a spectrum of radiation that falls on it.
dirty snowball A theory explaining comets as amalgams of ices, dust, and rocks.
discrete Separated; isolated.
disk (a) Of a galaxy, the disk-like flat portion, as opposed to the nucleus or the halo; (b) of a star or planet, the two-dimensional projection of its surface.
dispersion (a) Of light, the effect that different colors are bent by different amounts when passing from one substance to another; (b) of the pulses of a pulsar, the effect that a given pulse, which leaves the pulsar at one instant, arrives at the earth at different times depending on the different wavelength or frequency at which it is observed. Both of these effects arise because light of different wavelengths travels at different speeds except in a vacuum.
D lines A pair of lines from sodium that appear in the yellow part of the spectrum.
DNA Deoxyribonucleic acid, a long chain of molecules that contains the genetic information of life.
Dobsonian An inexpensive type of large-aperture amateur telescope characterized by a thin mirror, composition tube, and Teflon bearings on an alt-azimuth mount.
Doppler effect A change in wavelength that results when a source of waves and the observer are moving relative to each other.
double star A binary star; two or more stars orbiting each other.
double-lobed structure An object in which radio emission comes from a pair of regions on opposite sides.
dust tail The dust left behind a comet, reflecting sunlight.
dwarf ellipticals Small, low-mass elliptical galaxies.
dwarf stars Main-sequence stars.
dwarfs Dwarf stars.
dynamo A device that generates electricity through the effect of motion in the presence of a magnetic field.
dynamo theories Explaining sunspots and the solar activity cycle through an interaction of motion and magnetic fields.

***E* = *mc*²** Einstein's formula (special theory of relativity) for the equivalence of mass and energy.
early universe The universe during its first minutes.
earthshine Sunlight illuminating the moon after having been reflected by the earth.
eccentric Deviating from a circle.
eccentricity A measure of the flatness of an ellipse, defined as half the distance between the foci divided by the semimajor axis.
eclipse The passage of all or part of one astronomical body into the shadow of another.
eclipsing binary A binary star in which one member periodically hides the other.
ecliptic The path followed by the sun across the celestial sphere in the course of a year.
ecliptic plane The plane of the earth's orbit around the sun.
electric field A force field set up by an electric charge.
electromagnetic force One of the four fundamental forces of nature, giving rise to electromagnetic radiation.
electromagnetic radiation Radiation resulting from changing electric and magnetic fields.
electromagnetic spectrum Energy in the form of electromagnetic waves, in order of wavelength.
electromagnetic waves Waves of changing electric and magnetic fields, travelling through space at the speed of light.
electron A particle of one negative charge, 1/1830 the mass of a proton, that is not affected by the strong force. It is a lepton.
electron degeneracy The state in which, following rules of quantum mechanics, the further compression of electrons generates a high pressure that balances gravity, as in white dwarfs.
electron volt (eV) The energy necessary to raise an electron through a potential of one volt.
electroweak force The unified electromagnetic and weak forces, according to a recent theory.
element A kind of atom characterized by a certain number of protons in its nucleus. All atoms of a given element have similar chemical properties.
elementary particle One of the constituents of an atom.
ellipse A curve with the property that the sum of the distances from any point on the curve to two given points, called the foci, is constant.
elliptical galaxy A type of galaxy characterized by elliptical appearance.
emission line Wavelengths (or frequencies) at which the intensity of radiation is greater than it is at neighboring wavelengths (or frequencies).
emission nebula A glowing cloud of interstellar gas.
emulsion A coating whose sensitivity to light allows photographic recording of incident radiation.
energy A fundamental quantity usually defined in terms of the ability of a system to do something that is technically called "work," that is, the ability to move an object by application of force, where the work is the force times the displacement.
energy, law of conservation of Energy is neither created nor destroyed, but may be changed in form.
energy level A state corresponding to an amount of energy that an atom is allowed to have by the laws of quantum mechanics.
energy problem For quasars, how to produce so much energy in such a small emitting volume.
ephemeris A listing of astronomical positions and other data that change with time. From the same root as *ephemeral.*
epicycle In the Ptolemaic theory, a small circle, riding on a larger circle called the deferent, on which a planet moves. The epicycle is used to account for retrograde motion.
equal areas, law of Kepler's second law.
equant In Ptolemaic theory, the point equally distant from the center of the deferent as the earth but on the opposite side, around which the epicycle moves at a uniform angular rate.
equator (a) Of the earth, a great circle on the earth, midway between the poles; (b) celestial, the projection of the earth's equator onto the celestial sphere; (c) galactic, the plane of the disk as projected onto a map.
equatorial mount A type of telescope mounting in which one axis, called the polar axis, points toward the celestial pole and the other axis is perpendicular. Motion around only the polar axis is sufficient to completely counterbalance the effect of the earth's rotation.
equinox An intersection of the ecliptic and the celestial equator. The center of the sun is geometrically above and below the horizon for equal lengths of time on the two days of the year when the sun passes the equinoxes; if the sun were a point and atmospheric refraction were absent, then day and night would be of equal length on those days.
erg A unit of energy in the metric system, corresponding to the work done by a force of one dyne (the force that is required to accelerate one gram by one cm/sec²) producing a displacement of one centimeter.
ergosphere A region surrounding a rotating black hole (or other system satisfying Kerr's solution) from which work can be extracted.
escape velocity The velocity that an object must have to escape the gravitational pull of a mass.
event horizon The sphere around a black hole from within which nothing can escape; the place at which the exit cones close.
evolutionary track The set of points on an H-R diagram showing the changes of a star's temperature and luminosity with time.
excitation The raising of atoms to higher energy states than the lowest possible.
exclusion principle The quantum mechanical rule that certain types of elementary particles cannot exist in completely identical states.
exit cone The cone that, for each point within the photon sphere of a black hole, defines the directions of rays of radiation that escape.
exobiology The study of life located elsewhere than earth.
exponent The "power" representing the number of times a number is multiplied by itself.
exponential notation The writing of numbers as a power of 10 times a number with one digit before the decimal point.
extended objects Objects with detectable angular size.
extinction The dimming of starlight by scattering and absorption as the light traverses interstellar space.
extragalactic Exterior to the Milky Way Galaxy.
extragalactic radio sources Radio sources outside our galaxy.
eyepiece The small combination of lenses at the eye end of a telescope, used to examine the image formed by the objective.

featherweight (stars) A failed "star," contains less than 0.07 solar mass; becomes a brown dwarf.
field of view The angular expanse viewable.
filament A feature of the solar surface seen in Hα as a dark wavy line; a prominence projected on the solar disk.
fireball An exceptionally bright meteor.
fission, nuclear The splitting of an atomic nucleus.
flare An extremely rapid brightening of a small area of the surface of the sun, usually observed in hydrogen-alpha and other strong spectral lines and accompanied by x-ray and radio emission.
flash spectrum The solar chromospheric spectrum seen in the few seconds before or after totality at a solar eclipse.
flavors A way of distinguishing quarks: up, down, strange, charmed, truth (top), beauty (bottom).
fluorescence The transformation of photons of relatively high energy to photons of lower energy through interactions with atoms, and the resulting radiation.
flux The amount of something (such as energy) passing through a surface per unit time.
flux tube A torus through which particles circulate.
focal length The distance from a lens or mirror to the point to which rays from an object at infinity are focused.
focus (pl: **foci**) (a) A point to which radiation is made to converge; (b) of an ellipse, one of the two points the sum of the distances to which remains constant.
force In physics, something that can or does cause change of momentum, measured by the rate of change of momentum with time.
Fraunhofer lines The absorption lines of a solar or other stellar spectrum.
frequency The rate at which waves pass a given point.
full moon The phase of the moon when the side facing the earth is fully illuminated by sunlight.
fusion The amalgamation of nuclei into heavier nuclei.
fuzz The faint light detectable around nearby quasars.

galactic cannibalism The incorporation of one galaxy into another.
galactic cluster An asymmetric type of collection of stars that shared a common origin.
galactic corona The outermost part of our galaxy.
galactic year The length of time the sun takes to complete an orbit of our galactic center.
Galilean satellites The four brightest satellites of Jupiter.
gamma rays Electromagnetic radiation with wavelengths shorter than approximately 1 Å.
gas tail The puffs of ionized gas trailing a comet.
general theory of relativity Einstein's 1916 theory of gravity.
geocentric Earth-centered.
geology The study of the earth, or of other solid bodies.
geothermal energy Energy from under the earth's surface.
giant A star that is larger and brighter than main-sequence stars of the same color.
giant ellipticals Elliptical galaxies that are very large.
giant molecular cloud A basic building block of our galaxy, containing dust, which shields the molecules present.
giant planets Jupiter, Saturn, Uranus, and Neptune.
gibbous moon The phases between half moon and full moon.
globular cluster A spherically symmetric type of collection of stars that shared a common origin.
gluon The particle that carries the color force (and thus the strong nuclear force).
grand unified theories (GUT's) Theories unifying the electroweak force and the strong force.
granulation Convection cells on the sun about 1 arc sec across.
grating A surface ruled with closely spaced lines that, through diffraction, breaks up light into its spectrum.
gravitational force One of the four fundamental forces of nature, the force by which two masses attract each other.
gravitational instability A situation that tends to break up under the force of gravity.
gravitational interlock One body controlling the orbit or rotation of another by gravitational attraction.
gravitational lens In the gravitational lens phenomenon, a massive body changes the path of electromagnetic radiation passing near it so as to make more than one image of an object. The double quasar was the first example to be discovered.
gravitationally Controlled by the force of gravity.
gravitational radius The radius that, according to Schwarzschild's solutions to Einstein's equations of the general theory of relativity, corresponds to the event horizon of a black hole.
gravitational redshift A redshift of light caused by the presence of mass, according to the general theory of relativity.
gravitational waves Waves that many scientists consider to be a consequence, according to the general theory of relativity, of changing distributions of mass.
gravity assist Using the gravity of one celestial body to change a spacecraft's energy.
grazing incidence Striking at a low angle.
great circle The intersection of a plane that passes through the center of a sphere with the surface of that sphere; the largest possible circle that can be drawn on the surface of a sphere.
Great Dark Spot A giant circulating region on Neptune.
Great Red Spot A giant circulating region on Jupiter.
greenhouse effect The effect by which the atmosphere of a planet heats up above its equilibrium temperature because it is transparent to incoming visible radiation but opaque to the infrared radiation that is emitted by the surface of the planet.
Gregorian (telescope) A type of reflecting telescope in which the light focused by the primary mirror passes its prime focus and is then reflected by a secondary mirror through a hole in the center of the primary mirror.
Gregorian calendar The calendar in current use, with normal years that are 365 days long, with leap years every fourth year except for years that are divisible by 100 but not by 400.
ground level An atom's lowest possible energy level.
ground state See *ground level.*
GUT's See *grand unified theories.*

H_0 The Hubble constant.
H I region An interstellar region of neutral hydrogen.
H II region An interstellar region of ionized hydrogen.
H line The spectral line of ionized calcium at 3968 Å.

Hα The first line of the Balmer series of hydrogen, at 6563 Å.
half-life The length of time for half a set of particles to decay through radioactivity or instability.
halo Of a galaxy, the region of the galaxy that extends far above and below the plane of the galaxy, containing globular clusters.
head Of a comet, the nucleus and coma together.
head-tail galaxies Double-lobed radio galaxies whose lobes are so bent that they look like a tadpole.
heat flow The flow of energy from one location to another.
heavyweight stars Stars of more than about 8 solar masses.
heliacal rising The first time in a year that an astronomical body rises sufficiently far ahead of the sun that it can be seen in the morning sky.
heliocentric Sun-centered; using the sun rather than the earth as the point to which we refer. A heliocentric measurement, for example, omits the effect of the Doppler shift caused by the earth's orbital motion.
helium flash The rapid onset of fusion of helium into carbon through the triple-alpha process that takes place in most red-giant stars.
Herbig-Haro objects Blobs of gas ejected in star formation.
hertz The measure of frequency, with units of /sec (per second); formerly called cycles per second.
Hertzsprung gap A region above the main sequence in a Hertzsprung-Russell diagram through which stars evolve rapidly and thus in which few stars are found.
Hertzsprung-Russell diagram A graph of temperature (or equivalent) vs. luminosity (or equivalent) for a group of stars.
high-energy astrophysics The study of x-rays, gamma rays, and cosmic rays, and of the processes that make them.
highlands Regions on the moon or elsewhere that are above the level that may have been smoothed by flowing lava.
homogeneity Uniformity throughout.
horizontal branch A part of the Hertzsprung-Russell diagram of a globular cluster, corresponding to stars that are all at approximately zero absolute magnitude and have evolved past the red-giant stage and are moving leftward on the diagram.
hot dark matter Non-luminous matter with a large velocity dispersion, like neutrinos.
hour angle Of a celestial object as seen from a particular location, the difference between the local sidereal time and the right ascension (H.A. = L.S.T. − R.A.).
hour circle The great circles passing through the celestial poles.
H-R diagram Hertzsprung-Russell diagram.
Hubble constant (H_0) The constant of proportionality in Hubble's law linking the velocity of recession of a distant object and its distance from us.
Hubble type Hubble's galaxy classification scheme: E0, E7, Sa, SBa, etc.
Hubble's law The linear relation between the velocity of recession of a distant object and its distance from us, $V = H_0 d$.
hyperfine level A subdivision of an energy level caused by such relatively minor effects as changes resulting from the interactions among spinning particles in an atom or molecule.
IC *Index Catalogue,* one of the supplements to Dreyer's *New General Catalogue.*
igneous Rock cooled from lava.
image tube An electronic device that receives incident radiation and intensifies it or converts it to a wavelength at which photographic plates are sensitive.
inclination Of an orbit, the angle of the plane of the orbit with respect to the ecliptic plane.
Index Catalogue See *IC.*
inferior conjunction An inferior planet's reaching the same celestial longitude as the sun's.
inferior planet A planet whose orbit around the sun is within the earth's, namely, Mercury and Venus.
inflationary universe A model of the expanding universe involving a brief period of extremely rapid expansion.
infrared Radiation beyond the red, about 7000 Å to 1 mm.
interference The property of radiation, explainable by the wave theory, in which waves in phase can add (constructive interference) and waves out of phase can subtract (destructive interference); for light, this gives alternate light and dark bands.
interferometer A device that uses the property of interference to measure such properties of objects as their positions or structure.
interferometry Observations using an interferometer.
intergalactic medium Material between galaxies in a cluster.
interior The inside of an object.
International Date Line A crooked imaginary line on the earth's surface, roughly corresponding to 180° longitude, at which, when crossed from east to west, the date jumps forward by one day.
interplanetary medium Gas and dust between the planets.
interstellar medium Gas and dust between the stars.
interstellar reddening The relatively greater extinction of blue light by interstellar matter than of red light.
inverse-square law Decreasing with the square of increasing distance.
ion An atom that has lost one or more electrons. See also *negative hydrogen ion.*
ionized Having lost one or more electrons.
ionosphere The highest region of the earth's atmosphere.
ion tail See *gas tail.*
IRAS The Infrared Astronomical Satellite (1983).
iron meteorites Meteorites with a high iron content (about 90%); most of the rest is nickel.
irons Iron meteorites.
irregular cluster A cluster of galaxies showing no symmetry.
irregular galaxy A type of galaxy showing no shape or symmetry.
isotope A form of chemical element with a specific number of neutrons.
isotropy Being the same in all directions.

joule The SI unit of energy, 1 $kg{\cdot}m^2/s^2$.
Jovian planet Same as giant planet.
JPL The Jet Propulsion Laboratory in Pasadena, California, funded by NASA and administered by Caltech; a major space contractor.
Julian calendar The calendar with 365-day years and leap years every fourth year without exception; the predecessor to the Gregorian calendar.

Julian day The number of days since noon on January 1, 713 B.C. Used for keeping track of variable stars or other astronomical events. January 1, 2000, noon, will begin Julian day 2,451,545.

K line The spectral line of ionized calcium at 3933 Å.
Keplerian Following Kepler's law.

laser An acronym for "**l**ight **a**mplification by **s**timulated **e**mission of **r**adiation," a device by which certain energy levels are populated by more electrons than normal, resulting in an especially intense emission of light at a certain frequency when the electrons drop to a lower energy level.
latitude Number of degrees north or south of the equator measured from the center of a coordinate system.
law of equal areas Kepler's second law.
leap year A year in which a 366th day is added.
lens A device that focuses waves by refraction.
lenticular A galaxy of type S0.
libration The effect by which we can see slightly more than half the lunar surface even though the moon basically has one half that always faces us; the back-and-forth motion of an object around one of the stable points in the three-body gravitational problem.
light Electromagnetic radiation between about 3000 and 7000 Å.
light cone The cone on a space-time diagram representing regions that can be in contact, given that nothing can travel faster than the speed of light.
light curve The graph of the magnitude of an object vs. time.
light pollution Excess light in the sky.
light year The distance that light travels in a year.
lighthouse model The explanation of a pulsar as a spinning neutron star whose beam we see as it comes around.
lightweight stars Stars between about 0.07 and 4 solar masses.
limb The edge of a star or planet.
limb darkening The decreasing brightness of the disk of the sun or another star as one looks from the center of the disk closer and closer to the limb.
line profile The graph of the intensity of radiation vs. wavelength for a spectral line.
lithosphere The crust and upper mantle of a planet.
lobes Of a radio source, the regions to the sides of the center from which high-energy particles are radiating.
local In our region of the universe.
Local Group The two dozen or so galaxies, including the Milky Way Galaxy, that form a subcluster.
local standard of rest The system in which the average velocity of nearby stars is zero. We usually refer measurements of velocity of distant objects to the local standard of rest.
Local Supercluster The supercluster of galaxies in which the Virgo Cluster, the Local Group, and other clusters reside.
logarithmic A scale in which equal intervals stand for multiplying by ten or some other base, as opposed to linear, in which increases are additive.
longitude The angular distance around a body measured along the equator from some particular point; for a point not on the equator, it is the angular distance along the equator to a great circle that passes through the poles and through the point.
long-period variables Mira variables.
look-back time The duration over which light from an object has been travelling to reach us.
luminosity The total amount of energy given off by an object per unit time.
luminosity class Different regions of the H-R diagram separating objects of the same spectral type: supergiants (I), bright giants (II), giants (III), subgiants (IV), dwarfs (V).
lunar eclipse The passage of the moon into the earth's shadow.
lunar occultation An occultation by the moon.
lunar soils Dust and other small fragments on the lunar surface.
Lyman alpha The spectral line (1216 Å) that corresponds to a transition between the two lowest major energy levels of a hydrogen atom.
Lyman-alpha forest The many Lyman-alpha lines, each differently Doppler-shifted, visible in the spectra of some quasars.
Lyman lines The spectral lines that correspond to transitions to or from the lowest major energy level of a hydrogen atom.

Magellanic Clouds Two small irregular galaxies, satellites of the Milky Way Galaxy, visible in the southern sky; the SMC may be split.
magnetic field lines Directions mapping out the direction of the force between magnetic poles; the packing of the lines shows the strength of the force.
magnetic lines of force See *magnetic field lines.*
magnetic mirror A situation in which magnetic lines of force meet in a way that they reflect charged particles.
magnetic monopole A single magnetic charge of only one polarity; may or may not exist.
magnetosphere A region of magnetic field around a planet.
magnification An apparent increase in angular size.
magnitude A factor of $\sqrt[5]{100} = 2.511886\ldots$ in brightness. See *absolute magnitude* and *apparent magnitude.* An *order of magnitude* is a power of ten.
main sequence A band on a Hertzsprung-Russell diagram in which stars fall during the main, hydrogen-burning phase of their lifetimes.
major axis The longest diameter of an ellipse; the line from one side of an ellipse to the other that passes through the foci. Also, the length of that line.
mantle The shell of rock separating the core of a differentiated planet from its thin surface crust.
mare (pl: **maria**) One of the smooth areas on the moon or on some of the other planets.
mascon A concentration of mass under the surface of the moon, discovered from its gravitational effect on spacecraft orbiting the moon.
maser An acronym for "**m**icrowave **a**mplification by **s**timulated **e**mission of **r**adiation," a device by which certain energy levels are more populated than normal, resulting in an especially intense emission of radio radiation at a certain frequency when the system drops to a lower energy level.
mass A measure of the inherent amount of matter in a body.
mass-luminosity relation A well-defined relation between the mass and luminosity for main-sequence stars.
mass number The total number of protons and neutrons in a nucleus.

Maunder minimum The period 1645–1715, when there were very few sunspots, and no periodicity, visible.
mean solar day A solar day for the "mean sun," which moves at a constant rate during the year.
meridian The great circle on the celestial sphere that passes through the celestial poles and the observer's zenith.
mesosphere A middle layer of the earth's atmosphere, where the temperature again rises above the stratosphere's decline.
Messier numbers Numbers of non-stellar objects in the 18th-century list of Charles Messier.
metal (a) For stellar abundances, any element higher in atomic number than 2, that is, heavier than helium. (b) In general, neutral matter that is a good conductor of electricity.
meteor A track of light in the sky from rock or dust burning up as it falls through the earth's atmosphere.
meteorite An interplanetary chunk of rock after it impacts on a planet or moon, especially on the earth.
meteoroid An interplanetary chunk of rock smaller than an asteroid.
micrometeorite A tiny meteorite. The micrometeorites that hit the earth's surface are sufficiently slowed down that they can reach the ground without being vaporized.
mid-Atlantic ridge The spreading of the sea floor in the middle of the Atlantic Ocean as upwelling material forces the plates to move apart.
middleweight stars Stars between about 4 and 8 solar masses.
midnight sun The sun seen around the clock from locations sufficiently far north or south at the suitable season.
Milky Way The band of light across the sky from the stars and gas in the plane of the Milky Way Galaxy.
minor axis The shortest diameter of an ellipse; the line from one side of an ellipse to the other that passes midway between the foci and is perpendicular to the major axis. Also, the length of that line.
minor planets Asteroids.
Mira variable A long-period variable star similar to Mira (omicron Ceti).
missing-mass problem The discrepancy between the mass visible and the mass derived from calculating the gravity acting on members of clusters of galaxies.
molecule Bound atoms that make the smallest collection that exhibits a certain set of chemical properties.
momentum A measure of the tendency that a moving body has to keep moving. The momentum in a given direction (the "linear momentum") is equal to the mass of the body times its component of velocity in that direction. See also *angular momentum.*
mountain ranges Sets of mountains on the earth, moon, etc.

naked singularity A singularity that is not surrounded by an event horizon and therefore kept from our view.
neap tides The tides when the gravitational pulls of sun and moon are perpendicular, making them relatively low.
nebula (pl: **nebulae**) Interstellar regions of dust or gas.
nebular hypothesis The particular nebular theory for the formation of the solar system advanced by Laplace.
nebular theories The theories that the sun and the planets formed out of a cloud of gas and dust.
negative hydrogen ion A hydrogen atom with an extra electron.
neutrino A spinning, neutral elementary particle with little or no rest mass, formed in certain radioactive decays.
neutron A massive, neutral elementary particle, one of the fundamental constituents of an atom.
neutron degeneracy A state in which, following rules of quantum mechanics, the further compression of neutrons generates a high pressure that balances gravity.
neutron star A star that has collapsed to the point where it is supported against gravity by neutron degeneracy.
New General Catalogue "A New General Catalogue of Nebulae and Clusters of Stars" by J. L. E. Dreyer, 1888.
new moon The phase when the side of the moon facing the earth is the side that is not illuminated by sunlight.
Newtonian (telescope) A reflecting telescope where the beam from the primary mirror is reflected by a flat secondary mirror to the side.
N galaxy A galaxy (probably elliptical) with a blue nucleus that dominates the galaxy's radiation. The emission lines are generally broader than those from Seyferts.
NGC *New General Catalogue.*
non-thermal radiation Radiation that cannot be characterized by a single number (the temperature). Normally, we derive this number from Planck's law, so that radiation that does not follow Planck's law is called non-thermal.
nova (pl: **novae**) A star that suddenly increases in brightness; an event in a binary system when matter from the giant component falls on the white dwarf component.
nuclear bulge The central region of spiral galaxies.
nuclear burning Nuclear fusion.
nuclear force The strong force, one of the fundamental forces.
nuclear fusion The amalgamation of lighter nuclei into heavier ones.
nucleosynthesis The formation of the elements.
nucleus (a) Of an atom, the core, which has a positive charge, contains most of the mass, and takes up only a small part of the volume; (b) of a comet, the chunks of matter, no more than a few km across, at the center of the head; (c) of a galaxy, the innermost region.

O and B association A group of O and B stars close together.
objective The principal lens or mirror of an optical system.
oblate With equatorial greater than polar diameter.
occultation The hiding of one astronomical body by another.
Olbers's paradox The observation that the sky is dark at night contrasted to a simple argument that shows that the sky should be uniformly bright.
one atmosphere The air pressure at the Earth's surface.
one year The length of time the Earth takes to orbit the Sun.
Oort comet cloud The trillions of incipient comets surrounding the solar system in a 50,000 A.U. sphere.
open cluster A galactic cluster, a type of star cluster.
open universe A big-bang cosmology in which the universe has infinite volume and will expand forever.

opposition An object's having a celestial longitude 180° from that of the sun.
optical In the visible part of the spectrum, 3900–6600 Å, or having to do with reflecting or refracting that radiation.
optical double A pair of stars that appear extremely close together in the sky even though they are at different distances from us and are not physically linked.
organic Containing carbon in its molecular structure.
Orion Molecular Cloud The giant molecular cloud in Orion behind the Orion nebula, containing many young objects.
oscillating universe The version of a closed universe in which our cycle is but one of many.
Ozma One of two projects that searched nearby stars for radio signals from extraterrestrial civilizations.
ozone layer A region in the earth's upper stratosphere and lower mesosphere where O_3 absorbs solar ultraviolet.

pancake model A model of galaxy formation in which large flat structures exist and become clusters of galaxies.
paraboloid A 3-dimensional surface formed by revolving a parabola around its axis.
parallax (a) Trigonometric parallax, half the angle through which a star appears to be displaced when the earth moves from one side of the sun to the other (2 A.U.); it is inversely proportional to the distance. (b) Other ways of measuring distance, as in spectroscopic parallax.
parallel light Light that is neither converging nor diverging.
parent molecules Molecules in a comet that break up into daughter molecules.
parsec The distance from which 1 A.U. subtends one second of arc. (Approximately 3.26 l-y.)
particle physics The study of elementary nuclear particles.
Pauli exclusion principle The quantum-mechanical rule that certain types of elementary particles cannot exist in completely identical states.
peculiar velocity The velocity of a star with respect to the local standard of rest.
penumbra (a) For an eclipse, the part of the shadow from which the sun is only partially occulted; (b) of a sunspot, the outer region, not as dark as the umbra.
perfect cosmological principle The assumption that on a large scale the universe is homogeneous and isotropic in space and unchanging in time.
periastron The near point of the orbit of a body to the star around which it is orbiting.
perihelion The near point to the sun of the orbit of a body orbiting the sun.
period The interval over which something repeats.
phase (a) Of a planet, the varying shape of the lighted part of a planet or moon as seen from some vantage point; (b) the relation of the variations of a set of waves.
photometry The electronic measurement of the amount of light.
photomultiplier An electronic device that through a series of internal stages multiplies a small current that is given off when light is incident on it; a large current results.
photon A packet of energy that can be thought of as a particle travelling at the speed of light.
photon sphere The sphere around a black hole, 3/2 the size of the event horizon, within which exit cones open and in which light can orbit.
photosphere The region of a star from which most of its light is radiated.
plage The part of a solar active region that appears bright when viewed in Hα.
Planck's constant The constant of proportionality between the frequency of an electromagnetic wave and the energy of an equivalent photon. $E = h\nu = hc/\lambda$.
Planck's law The formula that predicts, for gas at a certain temperature, how much radiation there is at every wavelength.
planet A celestial body of substantial size (more than about 1000 km across), basically non-radiating and of insufficient mass for nuclear reactions ever to begin, ordinarily in orbit around a star.
planetary nebulae Shells of matter ejected by low-mass stars after their main-sequence lifetime, ionized by ultraviolet radiation from the star's remaining core.
planetesimal One of the small bodies into which the primeval solar nebula condensed and from which the planets formed.
plasma An electrically neutral gas composed of approximately equal numbers of ions and electrons.
plates Large flat structures making up a planet's crust.
plate tectonics The theory of the earth's crust, explaining it as plates moving because of processes beneath.
plumes Thin structures in the solar corona near the poles.
point objects Objects in which no size is distinguishable.
polar axis The axis of an equatorial telescope mounting that is parallel to the earth's axis of rotation.
pole star A star approximately at a celestial pole; Polaris is now the pole star; there is no south pole star.
poor cluster A cluster of stars or galaxies with few members.
Population I The class of stars set up by Walter Baade to describe the younger stars typical of the spiral arms. These stars have relatively high abundances of metals.
Population II The class of stars set up by Walter Baade to describe the older stars typical of the galactic halo. These stars have very low abundances of metals.
positive ion An atom that has lost one or more electrons.
positron An electron's antiparticle (charge of +1).
precession The slowly changing position of stars in the sky resulting from variations in the orientation of the earth's axis.
precession of the equinoxes The slow variation of the position of the equinoxes (intersections of the ecliptic and celestial equator) resulting from variations in the orientation of the earth's axis.
pressure Force per unit area.
primary cosmic rays The cosmic rays arriving at the top of the earth's atmosphere.
primary distance indicators Ways of measuring distance directly, as in trigonometric parallax.
prime focus The location at which the main lens or mirror of a telescope focuses an image without being reflected or refocused by another mirror or other optical element.

primeval solar nebula The early stage of the solar system in nebular theories.
primordial background radiation Isotropic millimeter and submillimeter radiation following a black-body curve for about 3 K; interpreted as a remnant of the big bang.
primum mobile In Ptolemaic and Aristotelian theory, the outermost sphere around the earth, which gave its natural motion to inner spheres.
principal quantum number The integer n that determines the main energy levels in an atom.
prolate Having the diameter along the axis of rotation longer than the equatorial diameter.
prominence Solar gas protruding over the limb, visible to the naked eye only at eclipses but also observed outside of eclipses by its emission-line spectrum.
proper motion Angular motion across the sky with respect to a framework of galaxies or fixed stars.
proton Elementary particle with positive charge 1, one of the fundamental constituents of an atom.
proton-proton chain A set of nuclear reactions by which four hydrogen nuclei combine one after the other to form one helium nucleus, with a resulting release of energy.
protoplanets The loose collections of particles from which the planets formed.
protosun The sun in formation.
pulsar A celestial object that gives off pulses of radio waves.

q_0 The deceleration parameter, a cosmological parameter that describes the rate at which the expansion of the universe is slowing up.
QSO Quasi-stellar object; a radio-quiet version of a quasar.
quantized Divided into discrete parts.
quantum A bundle of energy.
quantum mechanics The branch of 20th-century physics that describes atoms and radiation.
quark One of the subatomic particles of which modern theoreticians believe such elementary particles as protons and neutrons are composed. The various kinds of quarks have positive or negative charges of ⅓ or ⅔.
quasar One of the very-large-redshift objects that are almost stellar (point-like) in appearance.
quiescent prominence A long-lived and relatively stationary prominence.
quiet sun The collection of solar phenomena that do not vary with the solar activity cycle.

radar The acronym for **ra**dio **d**etection **a**nd **r**anging, an active rather than passive radio technique in which radio signals are transmitted and their reflections received and studied.
radial velocity The velocity of an object along a line (the radius) joining the object and the observer; the component of velocity toward or away from the observer.
radiant The point in the sky from which all the meteors in a meteor shower appear to be coming.
radiation Electromagnetic radiation. Sometimes also particles such as alpha (helium nuclei) or beta (electrons).
radiation belts Belts of charged particles surrounding planets.
radioactive Having the property of spontaneously changing into another isotope or element.
radio galaxy A galaxy that emits radio radiation orders of magnitude stronger than that from normal galaxies.
radio telescope An antenna or set of antennas, often together with a focusing reflecting dish, that is used to detect radio radiation from space.
radio waves Electromagnetic radiation with wavelengths longer than about one millimeter.
red giant A post-main-sequence stage of the lifetime of a star; the star becomes relatively bright and cool.
reddened See *reddening.*
reddening The phenomenon by which the extinction of blue light by interstellar matter is greater than the extinction of red light so that the redder part of the continuous spectrum is relatively enhanced.
redshifted When a spectrum is shifted to longer wavelengths.
red supergiant Extremely bright, cool, and large stars; a post-main-sequence phase of evolution of stars of more than about 4 solar masses.
reflecting telescope A type of telescope that uses a mirror or mirrors to form the primary image.
reflection nebula Interstellar gas and dust that we see because it is reflecting light from a nearby star.
refracting telescope A type of telescope in which the primary image is formed by a lens or lenses.
refraction The bending of electromagnetic radiation as it passes from one medium to another or between parts of a medium that has varying properties.
refractory Having a high melting point.
regolith A planet's or moon's surface rock disintegrating into smaller particles.
regular cluster A cluster of galaxies with spherical symmetry and a central concentration.
relativistic Having a velocity that is such a large fraction of the speed of light that the special theory of relativity must be applied.
resolution The ability of an optical system to distinguish detail.
rest mass The mass an object would have if it were not moving with respect to the observer.
rest wavelength The wavelength radiation would have if its emitter were not moving with respect to the observer.
retrograde motion The apparent motion of the planets when they appear to move backwards (westward) with respect to the stars from the direction that they move ordinarily.
retrograde rotation The rotation of a moon or planet opposite to the dominant direction in which the sun rotates and the planets orbit and rotate.
revolution The orbiting of one body around another.
rich cluster A cluster of many galaxies.
ridges Raised surface features on the moon, apparently volcanic, perhaps having developed on the crust of lava lakes, as volcanic vents, or from faulting; Mercury also has ridges.
right ascension Celestial longitude, measured eastward along the celestial equator in hours of time from the vernal equinox.
rilles Sinuous depressions on the lunar surface, apparently volcanic, perhaps huge collapsed lava tubes or lava channels.
rims The raised edges of craters.
Roche's limit (Roche limit) The sphere for each mass inside of which blobs of gas cannot agglomerate by

gravitational interaction without being torn apart by tidal forces; normally about 2½ times the radius of a planet.

rotation Spin on an axis.

rotation curve A graph of the speed of rotation vs. distance from the center of a rotating object like a galaxy.

RR Lyrae stars A short-period "cluster" variable. All RR Lyrae stars have approximately equal absolute magnitude and so are used to determine distances.

S0 A transition type of galaxy between ellipticals and spirals; has a disk but no arms; all E7's are now known to be S0's.

scarps Lines of cliffs; found on Mercury, earth, the moon, and Mars.

scattered Light absorbed and then reemitted in all directions.

Schmidt camera A telescope that uses a spherical mirror and a thin lens to provide photographs of a wide field.

Schwarzschild radius The radius that, according to Schwarzschild's solutions to Einstein's equations of the general theory of relativity, corresponds to the event horizon of a black hole.

scientific method No easy definition is possible, but it has to do with a way of testing and verifying hypotheses.

scientific notation Exponential notation.

scintillation A flickering of electromagnetic radiation caused by moving volumes of intermediary gas.

secondary cosmic rays High-energy particles generated in the earth's atmosphere by primary cosmic rays.

secondary distance indicators Ways of measuring distances that are calibrated by primary distance indicators.

sector Part of a circle bounded by an arc and two radii.

sedimentary Formed from settling in a liquid or from deposited material, as for a type of rock.

seeing The steadiness of the earth's atmosphere as it affects the resolution that can be obtained in astronomical observations.

seismic waves Waves travelling through a solid planetary body from an earthquake or impact.

seismology The study of waves propagating through a body and the resulting deduction of the internal properties of the body. "Seismo-" comes from the Greek for earthquake.

semimajor axis Half the major axis, that is, for an ellipse, half the longest diameter.

Seyfert galaxy A type of spiral galaxy that has a bright nucleus and whose spectrum shows broad emission lines that cover a wide range of ionization stages.

Shapley-Curtis debate The 1920 debate (and its written version) on the scale of our galaxy and of "spiral nebulae."

shear wave A type of twisting seismic wave.

shock wave A front marked by an abrupt change in pressure caused by an object moving faster than the speed of sound in the medium through which the object is travelling.

shooting stars Meteors.

showers A time of many meteors from a common cause.

sidereal With respect to the stars.

sidereal day A day with respect to the stars.

sidereal rotation period A rotation with respect to the stars.

sidereal time The hour angle of the vernal equinox; equal to the right ascension of objects on your meridian.

sidereal year A circuit of the sun with respect to the stars.

significant figure A digit in a number that is meaningful (within the accuracy of the data).

singularity A point in space where quantities become exactly zero or infinitely large; one is present in a black hole.

slit A long, thin gap through which light is allowed to pass.

solar activity cycle The 11-year cycle with which solar activity like sunspots, flares, and prominences varies.

solar atmosphere The photosphere, chromosphere, and corona.

solar constant The total amount of energy that would hit each square centimeter of the top of the earth's atmosphere at the earth's average distance from the sun.

solar day A full rotation with respect to the sun.

solar dynamo The generation of sunspots by the interaction of convection, turbulence, differential rotation, and magnetic field.

solar flares An explosive release of energy on the sun.

solar rotation period The time for a complete rotation with respect to the sun.

solar time A system of timekeeping with respect to the sun such that the sun is overhead of a given location at noon.

solar wind An outflow of particles from the sun representing the expansion of the corona.

solar year (tropical year) An object's complete circuit of the sun; a tropical year is between vernal equinoxes.

solstice The point on the celestial sphere of northernmost or southernmost declination of the sun in the course of a year; colloquially, the time when the sun reaches that point.

space velocity The velocity of a star with respect to the sun.

spallation The break-up of heavy nuclei that undergo nuclear collisions.

special theory of relativity Einstein's 1905 theory of relative motion.

speckle interferometry A method obtaining higher resolution of an image by analysis of a rapid series of exposures that freeze atmospheric blurring.

spectral classes Spectral types.

spectral lines Wavelengths at which the intensity is abruptly different from intensity at neighboring wavelengths.

spectral type One of the categories O, B, A, F, G, K, M, C, S, into which stars can be classified from study of their spectral lines, or extensions of this system. The sequence of spectral types corresponds to a sequence of temperature.

spectrograph A device to make and photograph a spectrum.

spectrometer A device to make and electronically measure a spectrum.

spectrophotometer A device to measure intensity at given wavelength bands.

spectroscope A device to make and look at a spectrum.

spectroscopic binary A type of binary star that is known to have more than one component because of the changing Doppler shifts of the spectral lines that are observed.

spectroscopic parallax The distance to a star derived by comparing its apparent magnitude with its absolute magnitude deduced from study of its position on an H-R diagram (determined by observing its spectrum—spectral type and luminosity class).
spectroscopy The use of spectrum analysis.
spectrum A display of electromagnetic radiation spread out by wavelength or frequency.
speed of light By Einstein's special theory of relativity, the velocity at which all electromagnetic radiation travels, and the largest possible velocity of an object.
spicule A small jet of gas at the edge of the quiet sun, approximately 1000 km in diameter and 10,000 km high, with a lifetime of about 15 minutes.
spin-flip A change in the relative orientation of the spins of an electron and the nucleus it is orbiting.
spiral arms Bright regions looking like a pinwheel.
spiral galaxy A class of galaxy characterized by arms that appear as though they are unwinding like a pinwheel.
sporadic Not regularly.
sporadic meteor A meteor not associated with a shower.
spring tides The tides at their highest, when the earth, moon, and sun are in a line (from "to spring up").
stable Tending to remain in the same condition.
star A self-luminous ball of gas that shines or has shone because of nuclear reaction in its interior.
star clouds The regions of the Milky Way where the stars are so densely packed that they cannot be seen as separate.
star clusters Groupings of stars of common origin.
stationary limit In a rotating black hole, the location where space-time is flowing at the speed of light, making stationary particles that would be travelling at that speed.
steady-state theory The cosmological theory based on the perfect cosmological principle, in which the universe is unchanging over time.
Stefan-Boltzmann law The radiation law that states that the energy emitted by a black body varies with the fourth power of the temperature.
stellar atmosphere The outer layers of stars not completely hidden from our view.
stellar chromosphere The region above a photosphere that shows an increase in temperature.
stellar corona The outermost region of a star characterized by temperatures of 10^6 K and high ionization.
stellar evolution The changes of a star's properties with time.
stones A stony type of meteorite, including the chondrites.
stratosphere An upper layer of a planet's atmosphere, above the weather, where the temperature begins to increase. The earth's stratosphere is at 20–50 km.
streamers Coronal structures at low solar latitudes.
strong force The nuclear force, the strongest of the four fundamental forces of nature.
strong nuclear force The strong force.
S-type star A red giant showing strong ZrO instead of TiO spectral lines.
subtend The angle that an object appears to take up in your field of view; for example, the full moon subtends 1/2°.
sunspot A region of the solar surface that is dark and relatively cool; it has an extremely high magnetic field.
sunspot cycle The 11-year cycle of variation of the number of sunspots visible on the sun.
superbolt Giant lightning.
supercluster A cluster of clusters of galaxies.
supergiant A post-main-sequence phase of evolution of stars of more than about 4 solar masses. They fall in the upper right of the H-R diagram; luminosity class I.
supergranulation Convection cells on the solar surface about 20,000 km across and vaguely polygonal in shape.
supergravity A theory unifying the four fundamental forces.
superluminal velocity An apparent velocity greater than that of light.
supermassive star A stellar body of more than about 100 $M_{\odot}$.
supernova (pl: **supernovae**) The explosion of a star with the resulting release of tremendous amounts of radiation.
supernova remnants The gaseous remainder of the star destroyed in a supernova.
supernova-chain-reaction model The explanation of spiral structure in terms of a chain of supernova explosions, each one leading to more than one more.
synchronous orbit An orbit of the same period; a satellite in geosynchronous orbit has the same period as the earth's rotation and so appears to hover.
synchronous rotation A rotation of the same period as an orbiting body.
synchrotron radiation Nonthermal radiation emitted by electrons spiralling at relativistic velocities in a magnetic field.
synodic Measured with respect to an alignment of 3 astronomical bodies.
synthetic-aperture radar A radar mounted on a moving object, used with analysis taking the changing perspective into account to give the effect of a larger dish.
syzygy An alignment of three celestial bodies.

T association A grouping of several T Tauri stars, presumably formed out of the same interstellar cloud.
tail Gas and dust left behind as a comet orbits sufficiently close to the sun, illuminated by sunlight.
tektites Small glassy objects found scattered around the southern part of the southern hemisphere of the earth.
terminator The line between night and day on a moon or planet; the edge of the part that is lighted by the sun.
terrestrial planets Mercury, Venus, Earth, and Mars.
tertiary distance indicators Ways to measure distance that are calibrated by secondary distance indicators.
thermal pressure Pressure generated by the motion of particles that can be characterized by a temperature.
thermal radiation Radiation whose distribution of intensity over wavelength can be characterized by a single number (the temperature). Black-body radiation, which follows Planck's law, is thermal radiation.
thermosphere The uppermost layer of the atmosphere of the Earth and some other planets, the ionosphere, where absorption of high-energy radiation heats the gas.
three-color photometry Measurements through U, B, V filters.
3° background radiation The isotropic black-body radiation at 3 K, thought to be a remnant of the big bang.
tidal theory An explanation of solar-system formation in terms of matter being tidally drawn out of the sun by a passing star.
transit The passage of one celestial body in front of another celestial body. When a planet is *in transit,* we understand that it is passing in front of the sun. Also,

transit is the moment when a celestial body crosses an observer's meridian, or the special type of telescope used to study such events.

transition zone The thin region between a chromosphere and a corona.

transparency Clarity of the sky.

transverse velocity Velocity along the plane of the sky.

trigonometric parallax See *parallax.*

triple-alpha process A chain of fusion processes by which three helium nuclei (alpha particles) combine to form a carbon nucleus.

Trojan asteroids A group of asteroids that precede or follow Jupiter in its orbit by 60°.

tropical year The length of time between two successive vernal equinoxes.

troposphere The lowest level of the atmosphere of the Earth and some other planets, in which all weather takes place.

T Tauri star A type of irregularly varying star, like T Tauri, whose spectrum shows broad and very intense emission lines. T Tauri stars have presumably not yet reached the main sequence and are thus very young.

tuning-fork diagram Hubble's arrangement of types of elliptical, spiral, and barred spiral galaxies.

21-cm line The 1420-MHz line from neutral hydrogen's spin-flip.

Type I supernova A supernova whose distribution in all types of galaxies, and the lack of hydrogen in its spectrum, make us think that it is an event in low-mass stars, probably resulting from the collapse and incineration of a white dwarf in a binary system.

Type II supernova A supernova associated with spiral arms, and that has hydrogen in its spectrum, making us think that it is the explosion of a massive star.

UBV system A system of photometry that uses three standard filters to define wavelength regions in the ultraviolet, blue, and green-yellow (visual) regions of the spectrum.

ultraviolet The region of the spectrum 100–4000 Å, also used in the restricted sense of ultraviolet radiation that reaches the ground, namely, 3000–4000 Å.

umbra (pl: **umbrae**) (a) Of a sunspot, the dark central region; (b) of an eclipse shadow, the part from which the sun cannot be seen at all.

uncertainty principle Heisenberg's statement that the product of uncertainties of position and momentum is equal to Planck's constant. Consequently, both position and momentum cannot be known to infinite accuracy.

universal gravitation constant The constant G of Newton's law of gravity: force = Gm_1m_2/r^2.

uvby A system of photometry that uses four standard filters to define wavelength regions in the ultraviolet, violet, blue, and yellow regions of the spectrum.

valleys Depressions in the landscapes of solid objects.

Van Allen belts Regions of high-energy particles trapped by the magnetic field of the earth.

variable star A star whose brightness changes over time.

vernal equinox The equinox crossed by the sun as it moves to northern declinations.

Very Large Array The National Radio Astronomy Observatory's set of radio telescopes in New Mexico, used together for interferometry.

Very-Long Baseline Array The National Radio Astronomy Observatory's set of radio telescopes dedicated to very-long-baseline interferometry and spread over an 8000 km baseline across the United States.

very-long-baseline interferometry The technique using simultaneous measurements made with radio telescopes at widely separated locations to obtain extremely high resolution.

virial theorem In the limited sense here, that half the gravitational energy of contraction goes into heating.

visible light Light to which the eye is sensitive, 3900–6600 Å.

visual binary A binary star that can be seen through a telescope to be double.

VLA See *Very Large Array.*

VLBA See *Very-Long-Baseline Array.*

VLBI See *very-long-baseline interferometry.*

void A giant region of the universe in which no galaxies are found.

volatile Evaporating (changing to a gas) readily.

wave front A plane in which parallel waves are in step.

wavelength The distance over which a wave goes through a complete oscillation.

weak force One of the four fundamental forces of nature, weaker than the strong force and the electromagnetic force. It is important only in the decay of certain elementary particles.

weight The force of the gravitational pull on a mass.

white dwarf The final stage of the evolution of a star of between 0.07 and 1.4 solar masses; a star supported by electron degeneracy. White dwarfs are found to the lower left of the main sequence of the H-R diagram.

white light All the light of the visible spectrum together.

Wien's displacement law The expression of the inverse relationship of the temperature of a black body and the wavelength of the peak of its emission.

winter solstice For northern-hemisphere observers, the southernmost declination of the sun, and its date.

Wolf-Rayet star A type of O star whose spectrum shows very broad emission lines.

W Virginis star A Type II Cepheid, a fainter class of Cepheid variables characteristic of globular clusters.

x-rays Electromagnetic radiation between 1 and 100 Å.

year The period of revolution of a planet around its central star; more particularly, the earth's period of revolution around the sun.

Zeeman effect The splitting of certain spectral lines in the presence of a magnetic field.

zenith The point in the sky directly overhead an observer.

zero-age main sequence The curve on an H-R diagram determined by the locations of stars at the time they begin nuclear fusion.

zodiac The band of constellations through which the sun, moon, and planets move in the course of the year.

zodiacal light A glow in the nighttime sky near the ecliptic from sunlight reflected by interplanetary dust.

zones Bright bands in the clouds of a planet, notably Jupiter's.

Illustration Acknowledgments

Front cover inset: NASA/Goddard Space Flight Center; **Front cover background:** European Southern Observatory, courtesy of Jacques M. Beckers.

Frontmatter—p. ii, iv, x © 1992 Anglo-Australian Telescope Board; **p. xii** NASA/Johnson Space Center, courtesy of Lisa Vazquez and Mike Gentry, Media Services; **p. xiv** prepared for NASA by Stephen P. Meszaros; **p. xv** Torun Museum, courtesy of Marek Demianski; © Royal Greenwich Observatory; **p. xvi** Jay M. Pasachoff; NASA; **p. xvii** Mosaic by Jody Swann, U.S. Geological Survey; NASA/JPL; **p. xviii** NASA/JPL; © 1979 Royal Observatory, Edinburgh/Anglo-Australian Telescope Board, from original U.K. Schmidt plates; **p. xix** Williams College—Hopkins Observatory, photo by Arielle Kagan; © 1987 Anglo-Australian Telescope Board; **p. xx** © 1985 Royal Observatory, Edinburgh/Anglo-Australian Telescope Board, from original U.K. Schmidt plates; © 1980 AATB; **p. xxi** The NASA COBE Science Team, courtesy of Nancy Boggess; **frontispiece** © 1982 Royal Observatory, Edinburgh, and Anglo-Australian Telescope Board, from original U.K. Schmidt plates.

Chapter 1—Opener © 1992 Anglo-Australian Telescope Board; **Fig. 1–1** © 1988 David Scharf; **Fig. 1–2** Jay M. Pasachoff; **Fig. 1–3** Neil Leifer/Sports Illustrated; **Fig. 1–4** Joseph Distefano, City of Boston, and Abrams Aerial Survey Corp.; **Figs. 1–5 and 1–6** NASA; **Figs. 1–7 and 1–9** JPL/NASA; **Fig. 1–12** Akira Fujii; **Fig. 1–13** © 1980 Ben Mayer, Hansen Planetarium; **Fig. 1–14** © 1984 Royal Observatory, Edinburgh, and Anglo-Australian Telescope Board, photograph by David Malin from original U.K. Schmidt plates; **Fig. 1–15** Gary Bernstein (U. Arizona), Tony Tyson (AT&T Bell Labs), and NOAO.

Chapter 2—Opener Jay M. Pasachoff; **Fig. 2–1** British Library; **Fig. 2–2** Geoff Chester, Albert Einstein Planetarium, Smithsonian Institution; **Fig. 2–3** Alinari; **Fig. 2–4** courtesy of The Houghton Library, Harvard University, © 1981 Owen Gingerich; **Fig. 2–5** Biblioteca Nazionale Marciana, Venice; **Fig. 2–8A** Bodleian Library, Oxford, Ms. Marsh 144, p. 326; **Fig. 2–8B** Danish Library; **Fig. 2–8C** Courtesy of the Library of the Jewish Theological Seminary; **Fig. 2–9** Catherine Meisel; **Figs. 2–10, 2–15A, 2–16, and 2–19 to 2–23** Jay M. Pasachoff; **Figs. 2–11 and 2–12** Jay M. Pasachoff and the Chapin Library, Williams College; **Fig. 2–13** Torun Museum, courtesy of Marek Demianski; **Fig. 2–14** Photograph by Charles Eames, reproduced courtesy of The Eames Office and Owen Gingerich, © 1990 The Eames Office; **Fig. 2–15B** Huntington Library, San Marino, CA; **Fig. 2–18** McLaughlin Planetarium, Royal Ontario Museum; **Fig. 2–24** U.S. Forest Service; **Fig. 2–25** Solstice Project; **Fig. 2–26** copyright © Alexander Marshack.

Chapter 3—Opener to Figs. 3–2, 3–13B, 3–17, 3–18, and 3–20 Jay M. Pasachoff; **Fig. 3–8** Jay M. Pasachoff and the Linda Hall Library; **Fig. 3–10** William C. Livingston, National Solar Observatory; **Fig. 3–11** Tintoretto; National Maritime Museum; **Fig. 3–12** Jay M. Pasachoff, after Ewen Whitaker, Lunar and Planetary Laboratory, University of Arizona; **Fig. 3–13A** Alberto Ossola; **Fig. 3–14** University of Michigan Library, Dept. of Rare Books and Special Collections, translation by Stillman Drake, reprinted courtesy of *Scientific American;* **Fig. 3–15A** Akira Fujii; **Fig. 3–15B** drawn from a plot made with *Visible Universe*; **Fig. 3–19** Jay M. Pasachoff, collection of John E. Gaustad; **Fig. 3–21** National Portrait Gallery, London; **Fig. 3–22** Jay M. Pasachoff, collection of the Royal Society.

Chapter 4—Opener © 1991 Roger Ressmeyer—Starlight, all rights reserved; **Fig. 4–5** Chris Jones, Union College; **Figs. 4–6, 4–10, 4–13, 4–17, 4–22, 4–36, 4–42, 4–49, 4–51, 4–54, 4–67, 4–69, 4–71, and 4–75 to 4–78** Jay M. Pasachoff; **Fig. 4–11** Dennis Milon; **Fig. 4–19** © 1986 Columbia Pictures Industries, Inc. All rights reserved; **Fig. 4–21A** © 1977 Anglo-Australian Telescope Board; **Fig. 4–21B** © 1989 Roger Ressmeyer—Starlight; **Fig. 4–23** © 1987 Roger Ressmeyer—Starlight; **Figs. 4–24 and 4–25** Courtesy of I. M. Kopylov, Special Astrophysical Observatory, U.S.S.R.; **Fig. 4–26** © 1972 Gary Ladd; **Fig. 4–27** Dale P. Cruikshank; **Fig. 4–28** From Robert Q. Fugate, Dept. of the Air Force; **Fig. 4–29** © 1979 Gary Ladd; **Fig. 4–30A** California Association for Research in Astronomy/Peter French; **Fig. 4–30B** © 1992 Roger Ressmeyer—Starlight, all rights reserved; **Fig. 4–30C** Itek Optical Systems; **Fig. 4–31** Steward Observatory Mirror Lab, U. Arizona (**A**, **B** photos by Lori Stiles); **Fig. 4–32** © European Southern Observatory; **Fig. 4–34A** © 1982 ROE; **Fig. 4–34B** © 1987 ROE/AATB; **Fig. 4–34C** © 1979 ROE/AATB; **Fig. 4–35A** © James Sugar; **Figs. 4–37, 4–39, and 4–56A** Hughes Danbury Optical Systems; **Fig. 4–38** NASA; **Fig. 4–41** Produced from U.S. Air Force Defense Meteorological Satellite program (DMSP); **Figs. 4–43A and B** Dennis di Cicco; **Fig. 4–43C** Akira Fujii; **Fig. 4–46** Deutsches Museum; **Fig. 4–47** National Optical Astronomy Observatories/Kitt Peak; **Fig. 4–50** Digital Instruments; **Fig. 4–52** Ewe Fink, Alfred B. Schultz, Michael A. DiSanti, and Reiner Fink, University of Arizona; **Fig. 4–56B** Courtesy of Joachim Trümper, Max-Planck-Institut für Physik & Astrophysik; **Fig. 4–57** NASA and TRW Systems Group; **Figs. 4–58 and 4–59** courtesy of Joachim Trümper, Max-Planck-Institut für Physik & Astrophysik; **Fig. 4–60** NASA and TRW Systems Group; **Fig. 4–61** Sten Odenwald; **Fig. 4–62A** NASA/Ames; **Fig. 4–62B** © 1989 Roger Ressmeyer—Starlight; **Figs. 4–63 to 4–65** IPAC at Caltech and JPL; IPAC is funded by NASA as part of the IRAS extended mission program under contract to JPL; **Fig. 4–66 (telescope)** Bruce Elmegreen; **Fig. 4–66 (inset)** courtesy Astronomical Society of the Pacific; **Figs. 4–70A and B** National Radio Astronomy Observatory, courtesy of George A. Seilelstad; **Fig. 4–73** William J. Welch, University of California at Berkeley; **Fig. 4–74** Dennis Downes, IRAM.

Chapter 5—Opener © 1987 Roger Ressmeyer—Starlight; **Fig. 5–2A** Mendillo Collection of Astronomical Prints; **Fig. 5–2B** Von Del Chamberlain, *When Stars Came Down to Earth: Cosmology of the Skidi Pawnee Indians of North America,* Los Altos, CA, and College Park, MD, Ballena Press, and Center for Archaeoastronomy, 1983; **Fig. 5–3** Pekka Parviainen; **Figs. 5–4 and 5–16** Jay M. Pasachoff; **Figs. 5–11 and 5–12** © 1970 United Feature Syndicate, Inc.; **Fig. 5–13** National Optical Astronomy Observatories/NSO; **Fig. 5–14** © Royal Greenwich Observatory; **Fig. 5–18** Emil Schulthess, Black Star; **Figs. 5–20A–G** Dr. Roland Szostak, Institut für Didaktik der Physik, Universität Münster, FRG; **Fig. 5–21** Dennis di Cicco; **Fig. 5–22** Central Intelligence Agency; **Fig. 5–24** Mount Vernon Ladies Association of the Union; **Fig. 5–25** Drawing by Handelsman, © 1978 The New Yorker Magazine, Inc.; **Fig. 5–26** Jay M. Pasachoff and the Chapin Library,

Williams College; **Fig. 5–27** Jay M. Pasachoff and the Chapin Library, Williams College, and FAN Press; **Fig. 5–Puzzle** David Allen, Anglo-Australian Observatory.

Part II—Opener background: Produced from U.S. Air Force Defense Meteorological Satellite program (DMSP).

Fig. 6–1 National Optical Astronomy Observatories/NSO; **Fig. 6–2** Serge Koutchmy, Institut d'Astrophysique, Paris; **Fig. 6–4** Kevin Reardon, Williams College—Hopkins Observatory; **Fig. 6–6** NASA/JPL; **Fig. 6–9** Akira Fujii; **Fig. 6–11** left photos by Jay M. Pasachoff, moon photo by Kevin Reardon, Williams College—Hopkins Observatory; **Figs. 6–13, 6–14A, B, and C, Fig. 6–15, and 6–17A and B** Jay M. Pasachoff; **Fig. 6–16** Fred Espenak, NASA/Goddard Space Flight Center; **Fig. 6–18** Masaharv Suzuki, Goto Optical Mfg. Co.; **Fig. 6–21** Daniel C. Good; **Fig. 6–22** © 1992 Craig Blankenhorn/Black Star; **Fig. 6–24** Harold E. Edgerton, MIT; courtesy of Palm Press; **Fig. 6–25** Alan P. Boss, Carnegie Institution of Washington.

Chapter 7—Opener O. Brown, R. Evans, and M. Carle, University of Miami Rosenstiel School of Marine and Atmospheric Science; **Fig. 7–1** Adapted from Raymond Siever, "The Earth," *Scientific American,* September 1975, and *The Solar System* (New York: W.H. Freeman and Co., 1975), © 1975 by Scientific American, Inc. All rights reserved; **Figs. 7–3 and 7–9A and B** Jay M. Pasachoff; **Figs. 7–4A and B** NASA; **Fig. 7–5** U.S. Geological Survey, photo by R. E. Wallace; **Fig. 7–6** Wilbur Rinehart, National Geophysical and Solar-Terrestrial Data Section, Environmental Data Service, National Oceanic and Atmospheric Administration; **Fig. 7–7A** Courtesy of Stanley N. Williams, Louisiana State U., from *Science,* 13 June 1980, © by the American Association for the Advancement of Science; **Fig. 7–7B** Photograph by James Zollweg; **Fig. 7–7C** Dale P. Cruikshank; **Fig. 7–8** © 1973 National Geographic Society; **Fig. 7–10** photos by NASA and by Kevin Reardon, Williams College—Hopkins Observatory; **Fig. 7–14** False-color composite view of Earth with data from Synchronous Meteorological Satellite 2 and Geostationary Operational Environmental Satellite-East. Geopic™ by Kawana J. Estep, Earth Satellite Corporation, Chevy Chase, MD, courtesy Gary Wade, NOAA, and William A. Lagerroos, Space Science and Engineering Center, University of Wisconsin—Madison; courtesy of National Geographic Society; **Fig. 7–15A** after C. D. Keeling; **Fig. 7–15B** *The New Scientist Magazine* (London), 26 May 1988, p. 58, in an article by Meirion Jones, copyright © New Scientist/IPC Magazines/World Press Network 1990; **Figs. 7–16A and 7–17B** NASA/Goddard Space Flight Center; **Fig. 7–17A** This first appeared in *The New Scientist Magazine* (London), the weekly review of science and technology, 12 November 1987, pp. 52–53, in an article by Joe Farman, British Antarctic Survey; **Figs. 7–19A and B** L. A. Frank, University of Iowa; **Fig. 7–20** Kevin Reardon, Williams College—Hopkins Observatory.

Chapter 8—Opener NASA; **Fig. 8–1** Daniel Good; **Fig. 8–2** Williams College—Hopkins Observatory photo by Kevin Reardon; **Figs. 8–4, 8–5, 8–6A, 8–13, 8–17, 8–18, 8–20, and 8–26** NASA; **Fig. 8–6B** Drawing by Alan Dunn; © 1971 The New Yorker Magazine, Inc.; **Fig. 8–7** Jay M. Pasachoff; **Figs. 8–8 to 8–10 and 8–24** NASA/Johnson Space Center; **Figs. 8–11 and 8–12** R. Wobus, Williams College; **Fig. 8–14** Gerald Wasserburg, Caltech; **Fig. 8–15** Drawings by Donald E. Davis under the guidance of Don E. Wilhelms of the U.S. Geological Survey; **Fig. 8–16** Harold E. Edgerton, MIT; courtesy of Palm Press; **Fig. 8–21** McDonald Observatory, U. Texas; **Fig. 8–22** Stephen Haggerty, U. Mass.; **Fig. 8–23** W. Benz and A. G. W. Cameron, Harvard-Smithsonian Center for Astrophysics, and H. J. Melosh, Lunar and Planetary Lab, U. Arizona; **Fig. 8–25** A. Potter and T. Morgan; **Fig. 8–26** NASA/JPL, courtesy of J. W. Head, Brown U.

Chapter 9—Opener Courtesy of Robert G. Strom, from his *Mercury: The Elusive Planet* (Smithsonian Institution Press), artwork by Karen Denomy; **Figs. 9–4 and 9–6 to 9–8** NASA; **Fig. 9–5** Courtesy of Robert G. Strom, Lunar and Planetary Laboratory, U. Arizona; **Figs. 9–9 and 9–10** Andrew E. Potter, Jr., and Thomas H. Morgan; **Fig. 9–11** Michael J. Ledlow, Jack O. Burns, Galen R. Gisler, Jun-Hui Zhao, Michael Zeilik, and Daniel N. Baker, courtesy of Jack O. Burns, New Mexico State U; **Fig. 9–12** U.S. Geological Survey; **Fig. 9–13** D. O. Muhleman and B. J. Butler, Caltech, and M. A. Slade, J.P.L.

Chapter 10—Opener NASA/Goddard Institute for Space Studies; **Figs. 10–1A and 10–23** NASA/JSC; **Fig. 10–1B** Jay M. Pasachoff; **Fig. 10–3** William Sinton and Klaus Hodapp (Institute for Astronomy, U. Hawaii), David Crisp (Caltech), Boris Ragent (NASA/Ames), and David Allen (Anglo-Australian Obs.); **Fig. 10–8** Courtesy of Donald B. Campbell, Arecibo Observatory; **Figs. 10–9, 10–11 and 10–12, 10–14 to 10–16, and 10–19** NASA/Ames Research Center; **Fig. 10–10** Experiment and data—Massachusetts Institute of Technology; maps—U.S. Geological Survey; NASA/Ames spacecraft; courtesy of Gordon H. Pettingill; **Fig. 10–13** NASA/Ames Research Center; **Fig. 10–17** Courtesy of David J. Diner; **Fig. 10–18** J. T. Schofield and F. W. Taylor, Clarendon Laboratory, Oxford; **Figs. 10–20 and 10–21** Courtesy of Valeriy Barsukov and Yuri Surkov; **Fig. 10–22** Mosaic by Jody Swann, U.S. Geological Survey; **Figs. 10–24 to 10–27** NASA/JPL.

Chapter 11—Opener mosaic by Jody Swann, USGS; **Fig. 11–1** Lowell Observatory; **Fig. 11–2** Jay M. Pasachoff with the 2.5-m telescope at the Mt. Wilson Obs.; **Figs. 11–4A and B** Ch. Buil, P. Laques, J. Lecacheux, and E. Thouvenot at the Pic du Midi Observatory; **Figs. 11–5, 11–6, 11–7A and B, 11–8B, 11–10A, B, C, and D, and 11–12 to 11–22** NASA/JPL; **Figs. 11–8A and 11–11** Image processing by Alfred S. McEwen, USGS, Branch of Astrogeology; **Fig. 11–9** Digital mosaic by Tammy Becker, USGS, Branch of Astrogeology; **Figs. 11–23A and B** Space Research Institute, USSR Academy of Sciences, courtesy of G. Avanesov; **Fig. 11–24** Margarita Naraeva and A. Selivanov, Glavkosmos, USSR; **Fig. 11–25** P. James, U. Toledo. 3

Chapter 12—Opener, Figs. 12–8 to 12–10A, 12–13B, 12–14, 12–17 to 12–19, 12–24 to 12–29, 12–31, and 12–33 to 12–35 NASA/JPL; **Fig. 12–2** Lunar and Planetary Laboratory, U. Arizona; **Fig. 12–4** Imke de Pater, U. Cal./Berkeley, with the VLA of NRAO; **Fig. 12–5** Dennis Milon; **Fig. 12–6** NASA/Goddard SFC; **Fig. 12–7** Jay M. Pasachoff; **Fig. 12–8A and B** NASA/JPL; **Fig. 12–10B** W. Reid Thompson, Space Sciences, Cornell U., using the Cornell National Supercomputer Facility; **Fig. 12–11** Gareth P. Williams and R. John Wilson, Geophysical Fluid Dynamics Laboratory, National Oceanic and Atmospheric Administration, Princeton University; **Fig. 12–12A** Philip S. Marcus, U. Cal./Berkeley; **Fig. 12–12B** J. Sommeria, S. Meyers, H. L. Swinney, U. Texas at Austin, based on a computer model by Philip S. Marcus; **Fig. 12–13A** Palomar Observatory photograph; **Fig. 12–20** USGS; **Fig. 12–21** Ray Batson, USGS; **Fig. 12–22** J. T. Trauger; **Fig 12–24** Reprocessed by Alfred McEwen, USGS; **Fig. 12–30** Mark R. Showalter, NASA/Ames; **Fig. 12–32** NASA/JSC; **Fig. 12–35** STScI.

Chapter 13—Opener, 13–4 to 13–15, and 13–17 to 13–35 NASA/JPL; **Figs. 13–2 and 13–3** STScI; **Fig. 13–16** VLA of NRAO.

Chapter 14—Opener, 14–1, 14–9 to 14–15, and 14–17 to 14–23 NASA/JPL; **Fig. 14–2** W. Liller; **Figs. 14–4, 14–5, and 14–7** James L. Elliot, MIT; **Fig. 14–8** NOAO; **Fig. 14–13** Courtesy of Andre Brahic, Observatoire de Paris; **Fig. 14–16** Information from Norman Ness; style from *Sky & Telescope*; p. 249 Steven Croft, Lunar and Planetary Lab., U. Arizona.

Chapter 15—Opener and 15–6 to 15–22 NASA/JPL; **Fig. 15–1** Master and Fellows of St. Johns College of Cambridge; **Fig. 15–2** Charles Kowal, Space Telescope Science Institute; **Fig. 15–3**

Tobias Owen, Institute for Astronomy, U. Hawaii, after R. Danhy; **Fig. 15–4** Heidi Hammel, JPL; **Fig. 15–5** W. B. Hubbard, A. Brahic, B. Sicardy, E.-R. Elicer, F. Roques, and F. Vilas, *Nature 319,* 636–640, 1986, courtesy of Macmillan Publishing Co.; **Fig. 15–11** Carolyn Porco, U. Arizona; **Fig. 15–12** Reprocessed by USGS; **p. 269 poem** with the permission of John Updike; revised from the version that appeared in LIFE Magazine, 1990.

Chapter 16—Opener Mark V. Sykes and Roc. M. Cutri, Steward Observatory, U. Arizona, Larry A. Lebofsky, Lunar and Planetary Laboratory, U. Arizona, and Richard P. Binzel, Planetary Science Institute, Tucson; **Fig. 16–1** Palomar Observatory photograph; **Fig. 16–2** © 1974 Charles Capen, Hansen Planetarium; **Fig. 16–3** Lowell Observatory photographs; **Fig. 16–4** U.S. Naval Observatory/James W. Christy, U.S. Navy photograph; **Fig. 16–5** Joseph H. Jones, Carol A. Christian and Patrick Waddell, Canada-France-Hawaii Telescope Corp.; **Fig. 16–6** G. Baier and G. Weigelt, Physikalisches Institut, Universität Erlangen-Nürnberg, FRG; **Fig. 16–7** David J. Tholen, Mauna Kea Obs., Institute for Astronomy, U. Hawaii; **Fig. 16–8** Marc W. Buie, David J. Tholen, and Keith Horne; **Figs. 16–9 to 16–11** James L. Elliot, MIT; **Fig. 16–12** STScI; **Fig. 16–13** David J. Eicher.

Chapter 17—Opener, 17–32, and 17–34 Akira Fujii; **Fig. 17–1** Martin Grossmann; **Figs. 17–2, 17–3, 17–11, 17–18, 17–25** Jay M. Pasachoff; **Figs. 17–4A and B** NASA; **Fig. 17–5** Paul Feldman, The Johns Hopkins University; **Fig. 17–6** High Altitude Observatory/NASA; **Fig. 17–7A** Naval Research Laboratory, courtesy of Neil R. Sheeley, Jr.; **Fig. 17–7B** U.S. Air Force photograph; **Fig. 17–8** IPAC, Caltech/JPL; **Fig. 17–9** Hopkins Observatory, Williams College, photo by Marian J. Warren; **Fig. 17–10** National Portrait Gallery, London; **Fig. 17–15** Jay M. Pasachoff and C. Trung Hua at the Canada-France Hawaii Tel.; **Fig. 17–16** Imke de Pater (U. Cal./Berkeley), Patrick Palmer (U. Chicago), and Lewis Snyder (U. Illinois) with the VLA of NRAO; **Figs. 17–17A and B** Kiso Observatory; **Figs. 17–18A and B** James B. Kaler and Karen B. Kwitter; **Fig. 17–19A** NASA; **Fig. 17–19B** NASA/Ames; **Fig. 17–20** © 1986 Royal Observatory Edinburgh, photography by David Malin from U.K. Schmidt plates; **Figs. 17–22A and B** Photography by B. W. Hadley, © Royal Observatory, Edinburgh; **Fig. 17–21** James R. Westlake, Jr.; **Figs. 17–24 and 17–31A and B** Soviet Vega Team; **Fig. 17–26** © 1986 Max Planck Institut für Aeronomie, Lindau/Hartz, FRG; photographed with the Halley Multicolour Camera aboard the European Space Agency's Giotto spacecraft, courtesy of H. U. Keller; **Fig. 17–27** Jay M. Pasachoff and Steven Souza; **Fig. 17–29** European Space Agency; **Fig. 17–30** NASA/Naval Research Laboratory; **Fig. 17–33** Susan Wyckoff, Arizona State U., Tempe; **Fig. 17–35** D. W. E. Green, Harvard-Smithsonian Center for Astrophysics and the *International Comet Quarterly;* **Fig. 17–36** David Jewitt, Inst. for Astron., U. Hawaii; **Fig. 17–37** C. Arpigny, F. Dossin, J. Manfroid, P. Magain, and R. Haefner; **Figs. 17–38 and 17–39** Humberto Campins, U. Florida; **Fig. 17–40** Mark V. Sykes, Steward Obs., U. Arizona.

Chapter 18—Opener Akira Fujii; **Fig. 18–1** © James M. Baker; **Figs. 18–2 and 18–3A** Jay M. Pasachoff; **Figs. 18–3B** John Pazmino; **Fig. 18–4 and 18–5** Allan E. Morton; **Fig. 18–6** Earth Physics Branch, Dept. of Energy, Mines, and Resources, Canada; **Fig. 18–7** Arthur A. Griffin; **Fig. 18–8A, B, and C** C. R. Hammond; **Fig. 18–9** Ian Halliday; **Fig. 18–10** NASA/JSC; **Figs. 18–11 and 18–12** Lawrence Grossman; **Figs. 18–13A, B, and C** Roy S. Lewis and Edward Anders, U. Chicago; **Fig. 18–14** Donald Pearson; **Fig. 18–15A** Dennis Milon; **Fig. 18–15B** Image by Leopoldo Infante during a joint observing run with Christopher Pritchet, both of U. Victoria, at CFHT; asteroid discovered on image by David Balam and processed by Infante and Balam, U. Victoria; **Fig. 18–17** Anna M. Nobili, Università di Pisa, data from the *Ephemerides of Minor Planets, 1987,* from *Asteroids II,* eds. Richard P. Binzel, Tom Gehrels, and Mildred Shapley Matthews (U. Arizona Press, 1989); **Fig. 18–18** Andrew Chaikin; **Fig. 18–19A** Paul D. Maley, International Occultation Timing Association; **Fig. 18–19B** Kevin Moody and Michael Verchota, U. Northern Colorado expedition, courtesy of Richard D. Dietz; **Figs. 18–20A and B** Edward Bowell, Lowell Observatory; **Fig. 18–21** NASA/JPL; **Figs. 18–22A and B** Lucy McFadden; **Fig. 18–23** Walter Alvarez; **Fig. 18–24** Mark Paternostro, courtesy of Encyclopaedia Britannica; **Fig. 18–25** Steven J. Ostro, JPL; **Fig. 18–26** NASA/JPL.

Chapter 19—Opener, 19–11 © Lucasfilm, Ltd. (LFL) 1980. All rights reserved. From the motion picture *The Empire Strikes Back,* courtesy of Lucasfilm, Ltd.; **Fig. 19–1** Drawing by Frank Modell, © 1987 The New Yorker Magazine, Inc.; **Fig. 19–2** NASA/Ames; **Fig. 19–3A** © 1976 National Geographic Society, photo by Bruce Dale; **Fig. 19–3B** Photograph by Andrew Clegg/Cornell University; **Fig. 19–4** Tom Reiland, Allegheny Obs.; **Fig. 19–5** Jonathan Gradie, Hawaii Institute for Geophysics, U. Hawaii; **Fig. 19–6** J. Bally, R. Becker, J. F. Arens, R. Ball/AT&T Bell Laboratories; **Fig. 19–7** © E. Imre Friedmann; **Fig. 19–8A** NASA; **Fig. 19–8B** Jay M. Pasachoff; **Fig. 19–9** Cornell U. photograph; **Fig. 19–12** From "Glinda of Oz," by L. Frank Baum, illustrated by John R. Neill, copyright 1920; **Fig. 19–13** Paul Horowitz, Harvard University, and the Planetary Society; **Figs. 19–14 and 19–15** NASA/Ames; **Fig. 19–16** This FAR SIDE cartoon by Gary Larson is reprinted by permission of Chronicle Features, San Francisco, CA.

Part III Opener © 1983 Royal Observatory Edinburgh/Anglo-Australian Telescope Board, from original U.K. Schmidt plates.

Chapter 20—Opener © 1979 Anglo-Australian Telescope Board; **Figs. 20–2A and B** Palomar Observatory—National Geographic Society Sky Survey. Reproduced by permission from the California Institute of Technology; **Fig. 20–3** Bundesarchiv Koblenz; **Fig. 20–8** Alfred Leitner, Rensselaer Polytechnic Institute; **Fig. 20–9** American Institute of Physics, Niels Bohr Library, Margrethe Bohr Collection; **Figs. 20–12 and 20–13** Harvard College Observatory; **Fig. 20–14** Freeman Miller, University of Michigan; **Fig. 20–15A** Waltraut C. Seitter, *Atlas for Objective Prism Spectra, Bonner Spektral Atlas I,* Ferd. Dummlers Verlag, Bonn, 1970; **Fig. 20–15B** R. A. Bell, University of Maryland; **Fig. 20–16** R. Giovanelli and H. R. Gillett, CSIRO National Measurement Laboratory, Australia; **Fig. 20–17** © 1981 Anglo-Australian Telescope Board; **Fig. 20–18** © 1980 Anglo-Australian Telescope Board.

Chapter 21—Opener © 1981 NOAO/CTIO, photo by Gabriel Martin; **Fig. 21–2** Reproduced by special permission of *Playboy Magazine,* copyright © 1971 by *Playboy;* **Fig. 21–6A** © 1979 Royal Observatory, Edinburgh/Anglo-Australian Telescope Board, from original U.K. Schmidt plates; **Fig. 21–8B** Data plotted by Charles A. Lindsey, Institute for Astronomy, University of Hawaii, from W. Traub and M. T. Stier, *Appl. Optics 15,* 364, 1976; labelled by A. Tokunaga; **Fig. 21–10** Dorrit Hoffleit—Yale U. Observatory; courtesy of American Institute of Physics, Niels Bohr Library; **Fig. 21–11** American Institute of Physics, Niels Bohr Library, Margaret Russell Edmondson Collection; **Fig. 21–13** Daniel Good; **Fig. 21–20** Tom Reiland, Allegheny Observatory; **Fig. 21–21** Wil Tirion.

Chapter 22—Opener Canada-France-Hawaii Telescope Corp., © Regents of the University of Hawaii, photo by Laird Thompson, then Institute for Astronomy, University of Hawaii; **Fig. 22–1** Dan Overcash; **Fig. 22–2** The Observatories of the Carnegie Institution of Washington; **Fig. 22–3** Lick Observatory Photograph; **Fig. 22–4** © 1982, 1989 Tom Reiland, Allegheny Observatory; **Fig. 22–7** Peter van de Kamp; **Fig. 22–8** Harvard-Smithsonian Center for Astrophysics; **Fig. 22–9** Based on data from D. L. Harris, III, K. Aa. Strand, and C. E. Worley in *Basic Astronomical Data,* K. Aa. Strand, ed., U. Chicago Press, © 1963 by the U. Chicago; **Fig. 22–10** Jay M. Pasachoff; **Fig. 22–11** Peter van de Kamp and Sarah Lee Lippincott, from *Vistas in Astronomy 19,* 231 (1975), courtesy Pergamon Press; **Fig. 22–12** G. Baier and G. Weigelt, Physikalisches Institut, Universität Erlangen-Nürnberg, FRG; **Figs. 22–13A, B, C, and D** Bernhard M. Haisch, Lockheed Palo Alto Laboratory; **Fig. 22–14** American Association of Variable Star Observers/Janet Mattei; **Fig. 22–15** Observations obtained for Jay M.

Pasachoff with the Automatic Photoelectric Telescope; **Fig. 22–16** J. D. Fernie and R. McGonegal, reprinted from *Astrophys. J. 275,* 735, with permission of the U. Chicago Press; **Fig. 22–17** Harvard College Observatory; **Fig. 22–18** Akira Fujii; **Figs. 22–19 and 22–25** After B. J. Bok and P. Bok, *The Milky Way,* courtesy Harvard U. Press; **Fig. 22–20** © 1977 Anglo-Australian Telescope Board; **Fig. 22–21** © 1984 Anglo-Australian Telescope Board; **Fig. 22–22** Zaqueline Souras; **Fig. 22–23** Daniel Good; **Figs. 22–24 and 22–27** European Southern Observatory; **Fig. 22–26** Shigeto Hirabayashi; **Figs. 22–27 and 22–29** NASA/ESA/STScI; **Fig. 22–30** R. Buonanno, A. Buzzoni, C. E. Corsi, F. Fusi Pecci, and A. Sandage, in *Globular Cluster Systems in Galaxies,* ed. J. E. Grindlay and A. G. D. Philip (Dordrecht: Reidel, 1987); **Fig. 22–31** Richard E. White, Smith College.

Chapter 23—Opener, Daniel C. Good; **Figs. 22–23, 23–34A, B, and C, 23–54, and 23–55** Jay M. Pasachoff; **Fig. 23–1** Williams College—Hopkins Observatory, photo by Arielle Kagan; **Figs. 23–2, 23–32, and 23–38** Courtesy of Encyclopaedia Britannica, Inc.; illustration by Anne Hoyer Becker; with information for Fig. 23–38 supplied by the High Altitude Observatory/NCAR; from "A New Understanding of Our Sun," by Jay M. Pasachoff, *1989 Britannica Yearbook of Science and the Future;* **Fig. 23–3** Goran Scharmer, Stockholms Observatorium; **Figs. 23–4, 23–5, 23–22B, 23–24A, B, and C to 23–26, 23–35** NOAO/National Solar Observatory; **Figs. 23–6 and 23–7** Big Bear Solar Observatory, Caltech, courtesy of Kenneth Libbrecht; **Fig. 23–8** David Hathaway/Marshall Space Flight Center/NASA; **Fig. 23–9** Optical Science Laboratory, University College London, England; **Fig. 23–10** David Hathaway; **Figs. 23–11, 23–39, and 23–40** R. J. Poole/DayStar Filter Corp; **Fig. 23–12** Del Woods/DayStar Filter Corp.; **Fig. 23–13** W. C. Atkinson, 1970; **Fig. 23–14** National Center for Atmospheric Research; **Fig. 23–15** NOAO/NSO/Sacramento Peak Observatory; **Fig. 23–16** Harvard College Observatory/ Naval Research Laboratory; **Figs. 23–17 and 23–18** Lewis House, Ernest Hildner, William Wagner, and Constance Sawyer/High Altitude Observatory, NCAR, NSF, and NASA; **Fig. 23–19** National Astronomical Observatory of Japan, U. Tokyo, and Lockheed Palo Alto Research Laboratory; **Fig. 23–20** Harvard–Smithsonian Center for Astrophysics; **Fig. 23–21** Leon Golub, Harvard–Smithsonian Center for Astrophysics; **Fig. 23–22A** 7" Starfire refractor, Wolfgang Lille/Baader Planetarium, Germany; **Fig. 23–27** National Radio Astronomy Observatory; **Fig. 23–31** David Hathaway, NASA/ Marshall Space Flight Center; **Figs. 23–33A, B, and C** NASA/ Johnson Space Center; **Fig. 23–34** Jay M. Pasachoff; intensity contours for B made with Optronics. Intl. Colormation, from a Wong Yiu-wan original on film courtesy National Geographic Society and Jay M. Pasachoff; contours for C made by Kevin Reardon, Williams College, from an image by Daniel C. Good; **Fig. 23–36** Bruce E. Woodgate, Einar Tandberg-Hanssen, and colleagues at NASA/Marshall Space Flight Center; **Fig. 23–37** Williams College—Hopkins Observatory, photo by Kevin Reardon; **Figs. 23–41 and 23–42** Naval Research Laboratory and NASA; **Figs. 23–43A, B, C, and D** Harvard College Observatory and NASA; **Fig. 23–44** NASA/Marshall Space Flight Center; **Fig. 23–45** John Eddy, University Corporation for Atmospheric Research; **Fig. 23–46** Richard Willson, Jet Propulsion Laboratory; **Figs. 23–47A and B** Huntington Library, San Marino, CA, and the Albert Einstein Archives, courtesy of The Hebrew University of Jerusalem, Jewish National University and Library; **Fig. 23–49A** Lick Observatory photograph; **Figs. 23–49B and 23–51** Albert Einstein Archives, Courtesy of The Hebrew University of Jerusalem, Jewish National University and Library; **Fig. 23–50** Courtesy of The Archives, California Institute of Technology; **Figs. 23–52 and 23–53** © 1919 by The New York Times Company. Reprinted by permission.

Part IV Opener © 1987 Anglo-Australian Telescope Board, photography by David Malin.

Chapter 24—Opener © 1980 Anglo-Australian Telescope Board, photography by David Malin; **Fig. 24–1** © 1979 Royal Observatory, Edinburgh/Anglo-Australian Telescope Board, photography by David Malin from original U.K. Schmidt plates; **Fig. 24–2** IPAC, Caltech/JPL; **Fig. 24–4** Lick Observatory photograph; **Fig. 24–5** Bo Reipurth, European Southern Observatory; **Fig. 24–6** Philip R. Schwartz (NRL), Theodore Simon (U. Hawaii), Ben Zuckerman (UCLA), and Robert R. Howell (U. Wyoming) at NOAO/VLA and Mauna Kea Observatory; **Figs. 24–7A, B, and C** Thorsten Neckel and Hans J. Staude, Max-Planck-Institut für Astronomie, Heidelberg; **Fig. 24–8A** Bo Reipurth, John Bally, J. A. Graham, A. P. Lane, and W. J. Zealey; **Fig. 24–8B** Charles Beichman and Carl Slopefeldt; **Fig. 24–9** George H. Herbig, then Lick Observatory, now U. Hawaii; **Fig. 24–10** Steven Beckwith, Cornell U., and Anneila Sargent, Caltech; **Fig. 24–14** Courtesy of Hans Bethe; **Figs. 24–18 and 24–19** Brookhaven National Laboratory; **Fig. 24–20** Data supplied by John N. Bahcall, Institute for Advanced Study and updated with data from Kenneth Lande, U. Pennsylvania; **Figs. 24–21A and B** GALLEX Collaboration; **Fig. 24–22** CERN.

Chapter 25—Opener © 1979 Anglo-Australian Telescope Board, photography by David Malin; **Fig. 25–1** Jay M. Pasachoff and C. Trung Hua at the Canada-France-Hawaii Telescope Corp.; **Fig. 25–4** © 1959 California Institute of Technology; **Fig. 25–5** Canada-France-Hawaii Telescope Corp., © Regents of the University of Hawaii, photograph by Laird Thompson, then Institute for Astronomy, U. Hawaii; **Figs. 25–6 and 25–17** NASA/Goddard Space Flight Center; **Fig. 25–7** Yervant Terzian, Cornell U.; **Fig. 25–8** From C. R. O'Dell, in IAU Symposium No. 34, "Planetary Nebulae," eds. D. E. Osterbrock and C. R. O'Dell. Reprinted by permission of the International Astronomical Union; **Fig. 25–10** McDonald Obs., U. Texas; **Figs. 25–11 and 25–13B** G. Dana Berry, Space Telescope Science Institute; **Fig. 25–12** Lick Observatory photos; **Fig. 25–13A** © Ben Mayer, Los Angeles; **Figs. 25–14 to 25–17B and C** NASA/ESA/STScI; **Fig. 25– 17A** G. Meylan, ESO.

Chapter 26—Opener © 1987 Roger Ressmeyer—Starlight; **Fig. 26–1** After Richard L. Sears, *J. Royal Astron. Soc. Canada,* No. 1, Feb. 1974; originally from B. E. Paczyński, *Acta Astron. 20,* 47, 1970; **Fig. 26–2** Mendillo Collection of Astronomical Prints; **Fig. 26–3** Massimo Della Valle, Istituto de Astronomia, Padua; **Fig. 26–4** © 1978 Royal Observatory Edinburgh, from original U.K. Schmidt plates; **Fig. 26–5** Courtesy of Encyclopaedia Britannica, Inc., from the *1989 Britannica Yearbook of Science and the Future;* illustration by Jane Meredith; **Fig. 26–6** © 1979 Royal Observatory Edinburgh/Anglo-Australian Telescope Board, photography by David Malin from original U.K. Schmidt plates; **Fig. 26–7** Michael Zeilik, U. New Mexico; **Fig. 26–8** David F. Malin, Anglo-Australian Telescope Board, and Jay M. Pasachoff from Palomar Observatory plates made available by the California Institute of Technology, courtesy of Robert Brucato; **Fig. 26–9** © 1959 California Institute of Technology; **Figs. 26–10A and B** Eli Dwek, NASA/Goddard Space Flight Center; **Figs. 26–10 and 26–11A** courtesy of Joachim Trümper, Max-Planck Institut für Physik & Astrophysik; **Fig. 26–11B** Philip Angerhofer, Richard A. Perley, Douglas Milne, and Bruce Balick, with the VLA of NRAO; **Fig. 26–11C** Harvard-Smithsonian CfA and NRAO; **Fig. 26–11D** European Space Agency; **Fig. 26–12** Stephen P. Reynolds, N.C. State U./Hugh Aller, U. Michigan, using the VLA of NRAO; **Fig. 26–13** Alan K. Uomoto, The Johns Hopkins U./Gordon M. MacAlpine, U. Michigan; **Fig. 26–14** U. Toronto; **Figs. 26–15A and B** © 1987 Anglo-Australian Telescope Board; **Fig. 26–16** Contributions to IAU Circulars, compiled by Daniel W. E. Green, Smithsonian Astrophysical Obs.; **Fig. 26–17** Ruhr-Universität Bochum Telescope at the European Southern Observatory, Observers: Joachim Dachs, Reinhard W. Hanuschik, and Guido Thimm; February 25—June 14, 1987; **Fig. 26–18** © 1988 Anglo-Australian Telescope Board; **Fig. 26–19** N. B. Suntzeff, M. M. Phillips, J. H. Elias, D. L. De Poy, and A. R. Walker, *Astrophys. J.* **384,** L33-36 (1992); **Fig. 26–20A** © 1987 William Liller; **Fig. 26–20B** Akira Fujii; **Fig. 26–21** Nino Panagia, STScI; **Figs. 26–22 and 26–29** John Learned, U. Hawaii; **Fig. 26–23** Alfred Mann, U. Pennsylvania; **Fig. 26–24** Adam Burrows; **Fig. 26–25** © 1989 Anglo-Australian Telescope Board, photography by David Malin; **Fig. 26–26** John C. Geary, Harvard–Smithsonian Center for Astrophysics; **Figs. 26–27A and B** European Space Agency; **Fig. 26–28** NASA/Johnson Space Flight Center.

Chapter 27—Opener Canada-France-Hawaii Telescope Corp., © Regents of the University of Hawaii, photograph by Laird Thompson, then Institute for Astronomy, U. Hawaii; **Fig. 27–1** background art: NASA; **Fig. 27–2** © Royal Observatory, Edinburgh, Brian Hadley; **Fig. 27–3** Joseph H. Taylor, Jr., Marc Damashek, and Peter Backus, then U. Mass.—Amherst, at the NRAO; **Fig. 27–4B** James Cordes and Arecibo Obs. (operated by Cornell U. under contract with the NSF); **Fig. 27–5** Joseph H. Taylor Jr., at the Princeton Pulsar Physics Laboratory; **Fig. 27–8** Norman W. Peddie, U.S. Geological Survey; **Fig. 27–9A** Jeff Percival and R. C. Bless, U. Wisconsin/NASA; **Fig. 27–9B** F. R. Harnden, Jr., and colleagues, Harvard-Smithsonian Center for Astrophysics; **Figs. 27–10A and B and 27–11A** Courtesy of Joachim Trümper, Max-Planck Institut für Physik & Astrophysik; **Figs. 27–11A and B** Harvard-Smithsonian Center for Astrophysics; **Fig. 27–12** Daniel R. Stinebring, Oberlin College, at the Princeton Pulsar Physics Laboratory; **Fig. 27–13** Paul Callanan, Oxford U.; **Fig. 27–14** Jeff Hester and Shrinivas Kulkarni, Palomar Observatory with the 5-m telescope; **Fig. 27–16A** Jay M. Pasachoff; **Figs. 27–16B and C** Tony Tyson, AT&T Bell Laboratories; **Fig. 27–17** W. Priedhorsky, Los Alamos National Laboratory, and J. Petterson, New Mexico Institute of Mining and Technology; **Fig. 27–18** Hans Ruder, Universität Tübingen; **Fig. 27–19** Bruce Margon, U. Washington; **Fig. 27–21** NRAO, operated by Associated Universities, Inc., under contract with the NSF, courtesy of R. Hjellming; **Fig. 27–22** Courtesy of M. Watson, R. Willingale, Jonathan E. Grindlay, and Frederick D. Seward; *see Astrophys. J. 273,* 688 (1983), courtesy of U. Chicago Press.

Chapter 28—Opener and Fig. 28–10 Palomar Observatory photograph/Jerome Kristian; colorizing for this text by Daniel A. Klinglesmith, NASA Goddard SFC; **Fig. 28–5** Drawing by Charles Addams, © 1974 The New Yorker Magazine, Inc.; **Fig. 28–6** After E. H. Harrison, U. Mass—Amherst; **Fig. 28–7** Chapin Library, Williams College; **Fig. 28–8** Jun Fukue/Osaka Kyoiku U.; **Figs. 28–9A, B, and C** John F. Hawley, U. Virginia, and Larry Smarr, U. Illinois at Urbana-Champaign; **Figs. 28–12 and 28–14** F. R. Harnden, Jr., and colleagues, Harvard-Smithsonian Center for Astrophysics; **Fig. 28–13** © Julian Calder/Woodfin Camp; **Fig. 28–15** Roy Bishop, Acadia U.; **Fig. 28–16** Researchers: David Bernstein, David Hobill, and Larry Smarr on a Cray 2; Visualization: Ray Idaszak and Donna Cox with an Alliant FX/80 and Wavefront Technologies software.

Part V Opener The National Gallery, London.

Chapter 29—Opener © 1981 Anglo-Australian Telescope Board; **Fig. 29–1** Shigeto Hirabayashi; **Fig. 29–2A** © 1985 Royal Observatory, Edinburgh/Anglo-Australian Telescope Board, from original U.K. Schmidt plates; **Fig. 29–2B** EXOSAT Observatory; **Fig. 29–3** © 1979 Royal Observatory, Edinburgh/Anglo-Australian Telescope Board, from original U.K. Schmidt plates; **Fig. 29–4** © 1986 Anglo-Australian Telescope Board; **Fig. 29–5** © 1984 Anglo-Australian Telescope Board; **Fig. 29–7** © 1987 Royal Observatory, Edinburgh/Anglo-Australian Telescope Board, from original U.K. Schmidt plates; **Fig. 29–8** P. G. Mezger, R. Zylka, C. J. Salter, J. E. Wink, R. Chini, E. Kreysa, and R. Tuffs, *Astron. Astrophys. 209,* 337 (1989), at IRAM, courtesy of Dennis Downes; **Fig. 29–9** Ian Gatley, NOAO; **Figs. 29–10 and 29–17** IPAC, Caltech/JPL, Chas. Beichman; **Figs. 29–11A** NOAO/AUI; **Fig. 29–11B** A. Eckart, R. Genzel, A. Krabbe, R. Hofmann, P. P. van der Werf, and S. Drapatz, Max-Planck Institut für extraterrestrische Physik/European Southern Observatory; **Fig. 29–12** VLA of NRAO; **Fig. 29–13** K. Y. Lo, U. Illinois at Urbana-Champaign, with the VLA of NRAO; **Fig. 29–14A** Mark Morris, UCLA, and Farhad Yusef-Zadeh and Don Chance, Columbia U., with the VLA of NRAO; **Fig. 29–14B** Jay M. Pasachoff, Donald A. Lubowich, and K. Anantharamaiah, with the VLA of NRAO; **Fig. 29–15A** After R. Gusten and D. Downes; **Fig. 29–15B** NRAO/AUI, Observers J. H. Zhao, D. A. Roberts, W. M. Goss, D. A. Frail (all NRAO), K. Y. Lo (U. Illinois), and R. D. Ekers, R. Subrahmanyan, and M. Kesteven (all Australian Telescope National Facility); **Fig. 29–16** Lund Observatory, Sweden; **Fig. 29–18** Courtesy of Joachim Trümper, Max-Planck Institut für Physik & Astrophysik; **Fig. 29–19** H. A. Mayer-Hasselwander, K. Bennett, G. F. Bignami, R. Buccheri, N. D'Amico, W. Hermsen, G. Kanbach, F. Lebrun, G. G. Lichti, J. L. Masnou, J. A. Paul, K. Pinkau, L. Scarsi, B. N. Swanenburg, and R. D. Wills; COS-B Observation of the Milky Way in High-Energy Gamma Rays, Ninth Texas Symposium on Relativistic Astrophysics, J. Ehlers, J. J. Perry, and M. Walker, eds., *Annals of the New York Academy of Sciences 336,* 211, 1980; **Fig. 29–20** Gerald J. Fishman, Marshall Space Flight Center; **Fig. 29–21** NASA, Volker Schoenfelder, MPI; **Fig. 29–22** Debra Meloy Elmegreen (also at Vassar College), Bruce G. Elmegreen, and Philip E. Seiden, IBM T. J. Watson Research Center; **Fig. 29–23** After Agris Kalnajs; **Fig. 29–24** Stuart N. Vogel, U. Maryland; **Fig. 29–25** Observations by D. M. Elmegreen (Vassar College) at NOAO/CTIO; paper by Lawrence S. Schulman and Philip E. Seiden, IBM T. J. Watson Research Center.

Chapter 30—Opener NASA/ESA/STScI; **Fig. 30–1** © 1981 Royal Observatory, Edinburgh/Anglo-Australian Telescope Board, from original U.K. Schmidt plates; **Fig. 30–2, inset** David L. Talent, Abilene Christian U.; **Fig. 30–3** © 1977 Anglo-Australian Telescope Board; **Fig. 30–4** J. V. Feitzinger and J. A. Stüwe, Ruhr-Universität, Bochum, FRG; **Fig. 30–5** J. Mayo Greenberg, U. Leiden; **Fig. 30–7** © 1988 Regents of the U. Minnesota; collaboration of U. Minnesota, Air Force Geophysics Laboratory and NASA, John R. Winckler, Perry Malcolm, and colleagues; TV camera experiment by Robert Franz; **Fig. 30–8A** David F. Malin, Anglo-Australian Telescope Board, and Jay M. Pasachoff from Palomar Observatory plates made available by the California Institute of Technology, courtesy of Robert Brucato; **Fig. 30–8B** Jodrell Bank, Max-Planck Institut für Radioastronomie, Bonn, Australian National Radio Astronomy Observatory at Parkes; combined at MPIR and Computing Center of Bonn U.; **Fig. 30–12** Image assembled by C. Jones and W. Forman, Harvard-Smithsonian Center for Astrophysics, from observations from the "Bell Labs H I Survey," A. A. Stark, C. F. Gammie, R. W. Wilson, J. Bally, R. A. Linke, C. Heiles, and M. Hurwitz, *Astrophys. J. Suppl.* (1990); M. N. Cleary, C. Heiles, and C. G. T. Haslam; and F. Kerr, B. Burke, P. Mezger, E. Reifenstein, T. Wilson, R. Vallak, and J. Hindeman; **Fig. 30–14** Gerrit Verschuur; **Figs. 30–16A and B** Tom Dame, Harvard-Smithsonian Center for Astrophysics; **Fig. 30–17A** Leo Blitz, U. Maryland; **Fig. 30–17B** B. J. Robinson, J. B. Whiteoak, R. N. Manchester, C.S.I.R.O., Australia, and W. H. McCutcheon, U. British Columbia; **Fig. 30–17C** Dan P. Clemens, U. Arizona, D. Sanders and N. Scoville, California Institute of Technology; **Fig. 30–18** Leo Blitz, U. Maryland, and Michel Fich, U. Waterloo; **Fig. 30–20** Philip Solomon and Arthur Rivolo based on Stony Brook Catalogue of Giant Molecular Clouds, *Astrophys. J. 319,* 730 (1987); **Fig. 30–21** John Bally (AT&T Bell Labs) and Ronald L. Snell and C. Read Predmore (U. Mass—Amherst); **Fig. 30–22** © 1978 Royal Observatory, Edinburgh/Anglo-Australian Telescope Board, from original U.K. Schmidt plates; **Fig. 30–23** © 1981 Anglo-Australian Telescope Board; **Fig. 30–24** After Ben Zuckerman, UCLA; **Fig. 30–25** IPAC, Caltech/JPL; **Figs. 30–26A, B, C, and D** Ian Gatley, NOAO, at UKIRT, courtesy of Kevin Krisciunas; **Fig. 30–27** Kevin Krisciunas, Joint Astronomy Centre, Hilo, Hawaii; **Fig. 30–28** John Bally, overlay after Wil Tirion; **Fig. 30–29** © 1989 Mark McCaughrean, NASA/Goddard Space Flight Center, based on data from Mark McCaughrean and Colin Aspin, Joint Astronomy Centre, Hilo, Hawaii; **Fig. 30–30** Courtesy of Joachim Trümper, Max-Planck Institut für Physik and Astrophysik; **Fig. 30–31** IPAC/JPL; **Fig. 30–32** R. Probst, NOAO.

Part VI Opener European Southern Observatory, Observer: S. Laustsen.

Chapter 31—Opener © 1980 Anglo-Australian Telescope Board; **Fig. 31–1A** © 1987 Anglo-Australian Telescope Board; **Fig. 31–1B** Tod R. Lauer, NASA; **Fig. 31–2** © 1959 California Institute of Technology; **Fig. 31–3A** Canada-France-Hawaii Telescope Corp., © Regents of the University of Hawaii, photograph by Laird Thompson, then Institute for Astronomy, U. Hawaii; **Fig. 31–3B** Smithsonian Astrophysical Obs./Rudolph Schild and William Wyatt; **Fig. 31–3C** M. Guélin, S. Garcia-Burillo, R. Blundell, J. Cernicharo, D. Despois, and H. Steppe/IRAM, *Highlights of Astronomy 8,* 575–577, © 1989 by the International Astronomical Union; **Fig. 31–4** U.S. Naval Obs.; **Fig. 31–5** © 1980 AATB; **Fig. 31–6** Netherlands Foundation for Radio Astronomy, NRAO, Arnold H. Rots and William W. Shane, Westerbork data, imaged at NRAO; **Fig. 31–7** NRAO/AUI, Observers: A. H. Rots, J. M. van der Hulst, P. E. Seiden, R. C. Kennicutt, P. C. Crane, A. Bosma, L. Athanassoula, and D. M. Elmegreen; **Fig. 31–8** European Southern Observatory; **Figs. 31–9A and B** © 1984 Royal Observatory, Edinburgh/Anglo-Australian Telescope Board; **Figs. 31–10A and B and 31–14** IPAC, Caltech/JPL; **Fig. 31–11** © 1982 Halton C. Arp, based on Palomar Schmidt plates, courtesy of the Astronomical Society of the Pacific; **Fig. 31–12** SRC blue survey image, Marshall Joy/NASA Marshall Space Flight Center, E. V. Tollestrup and P. M. Harvey/U. Texas at Austin, P. McGregor and A. R. Hyland/Mt. Stromlo and Siding Spring Observatories; **Fig. 31–15** R. Wielebinski, R. Beck, E. Berkhuijsen, et al.; **Fig. 31–16** William C. Keel, U. Alabama; **Fig. 31–17A** John Kormendy, Dominion Astrophysical Obs.; **Fig. 31–17B** Holland Ford (Johns Hopkins University/Space Telescope Science Institute), the Faint Object Spectrograph IDT, and NASA; **Fig. 31–18A** Joshua Barnes, Canadian Institute for Theoretical Astrophysics, and Lars Hernquist, U. Cal.—Santa Cruz, then both Institute for Advanced Study; **Fig. 31–18B** Palomar Observatory photograph; **Fig. 31–18C** © 1992 Anglo-Australian Telescope Board; **Fig. 31–19** © Royal Greenwich Observatory; **Fig. 31–20** © 1987 AATB; **Fig. 31–21** © 1981 AATB; **Fig. 31–22** © 1987 Royal Observatory, Edinburgh/Anglo-Australian Telescope Board, from original U.K. Schmidt plates; **Figs. 31–23A, B, and C** EXOSAT Observations courtesy of Michiel van der Klis, Astronomical Institute "Anton Pannekoek"; **Figs. 31–24 and 31–26A and B** Brent Tully, Institute for Astronomy, U. Hawaii; **Figs. 31–25A and B** Margaret Geller and John Huchra, Harvard–Smithsonian Center for Astrophysics; **Figs. 31–27 and 31–28** NOAO/KPNO; **Fig. 31–29A** National Academy of Sciences; **Fig. 31–29B** From E. Hubble and M. L. Humason, *Astrophysical Journal 74,* 77, 1931, courtesy of U. Chicago Press; **Fig. 31–30A** Palomar Observatory photograph; **Fig. 31–30B** Vera C. Rubin, Carnegie Institution of Washington; **Fig. 31–33** © 1986 AATB; **Fig. 31–34** after Donat G. Wentzel and R. Hanisch, Astronomy Program, U. Maryland; **Fig. 31–35** After Allan Sandage; **Fig. 31–36** David Burstein, Sandra Faber, Roger Davies, Alan Dressler, D. Lynden-Bell, Roberto Terlevich, and Gary Wegner; image by Ofer Lahav, Institute of Astronomy, Cambridge U.; **Fig. 31–37A** Rudolph Schild, SAO; **Fig. 31–37B** © 1987 United Feature Syndicate, Inc.; **Fig. 31–38** Ken Chambers, The Johns Hopkins U., George Miley, U. Leiden and Space Telescope Science Institute, and Wil van Breugel, U. Cal.—Berkeley; **Fig. 31–39** radio image from the VLA of NRAO; optical image from Laird Thompson, U. Illinois, then at U. Hawaii, *Astrophysical Journal 279,* L47 (1984), courtesy of U. Chicago Press; **Figs. 31–40A and B** Thomas Stephenson, SAO; **Fig. 31–41** © 1980 AATB; **Fig. 31–42** Harvard–Smithsonian Center for Astrophysics; **Fig. 31–43** VLA of NRAO; **Fig. 31–47** Riccardo Giovanelli and Martha Haynes, Cornell U.; **Fig. 31–44** F. Duccio Macchetto, NASA/ESA; **Fig. 31–48** VLA of NRAO; **Fig. 31–49** Image of DA 240 courtesy of the Max-Planck Institut für Radioastronomie, Bonn, and Richard B. Isaacman, NASA Goddard Space Flight Center/image of Cygnus A from the VLA of NRAO; **Fig. 31–50** NRAO; **Figs. 31–52 and 31–54 to 31–56** Jay M. Pasachoff; **Fig. 31–53** VLA of NRAO; **Fig. 31–57** Joan Centrella, Drexel U.; **Fig. 31–58** Adrian L. Melott, U. Kansas, and Sergei Shandaran, Institute of Physical Problems, Moscow.

Chapter 32—Opener NOAO, also Gregory D. Bothun, CfA; **Figs. 32–1 and 32–3** Maarten Schmidt/Palomar Observatory photograph; **Figs. 32–2 and 32–19** Stephen C. Unwin, Caltech; **Fig. 32–4** Tony Tyson, AT&T Bell Labs, and W. A. Baum and T. Kreidel, Lowell Obs.; **Fig. 32–6** Donald P. Schneider (Institute for Advanced Study), Maarten Schmidt (Caltech), and James E. Gunn (Princeton U.); **Fig. 32–7A** Maarten Schmidt, Caltech; **Fig. 32–7B** Jay M. Pasachoff; **Fig. 32–8B** Thomas Balonek, Colgate U.; **Figs. 32–9A, B, C, and D** Stuart L. Shapiro and Saul Teukolsky, Cornell U.; **Fig. 32–11A** Richard F. Green, NOAO, and Howard K. C. Yee, U. Toronto, on the Steward Obs. 2.3-m tel.; **Figs. 32–11B and C and 32–17B** John B. Hutchings, Dominion Astrophysical Obs.; **Fig. 32–11D** Matthew A. Malkan, UCLA; **Fig. 32–12** Halton C. Arp with the 5-m Hale telescope, courtesy of the Astronomical Society of the Pacific; **Figs. 32–13A and B and 32–18** NOAO; **Figs. 32–14 and 32–24** Alan Stockton, Institute for Astronomy, U. Hawaii; **Fig. 32–15** NOAO, observations by John Stocke and Paul Hintzen; **Fig. 32–16** C. Aspin and M. McCaughrean, CFHT; **Fig. 32–17A** John MacKenty, Space Telescope Science Institute; **Fig. 32–21A** T. Shanks (U. Durham) and G. C. Stewart (U. Leicester), courtesy of M. A. Barstow; **Fig. 32–21B** NASA/GSFC, Carl E. Fichtel, P.I.; **Fig. 32–23** B. F. Burke, P. E. Greenfield, D. H. Roberts, with the VLA of NRAO; **Fig. 32–25** Jacqueline N. Hewitt, Edwin L. Turner, Donald P. Schneider, Bernard F. Burke, Glen I. Langston, and Charles R. Lawrence; **Fig. 32–26** P. Magain, J. Surdej, J.-P. Swings, U. Borgeest, R. Kayser, H. Kurh, S. Refsdal, and M. Remy, observations at ESO.

Chapter 33—Opener From the collection of Mr. and Mrs. Paul Mellon, National Gallery, Washington; **Figs. 33–2, 33–7, and 33–9** Jay M. Pasachoff; **Fig. 33–6** After Allan Sandage, from Y. Yoshii and F. Takahara, from *Astrophys. J. 326,* 1 (1988); **Figs. 33–8A and B** NASA Johnson Space Center, courtesy of Mike Gentry; **Fig. 33–11** Courtesy of AT&T Bell Laboratories; **Fig. 33–13** S. P. Boughn and R. B. Partridge; **Fig. 33–14** Map: E. S. Cheng (NASA Goddard Space Flight Center), D. A. Cottingham (U. Cal.—Berkeley), S. Boughn (Haverford), D. T. Wilkinson (Princeton U.), and D. J. Fixsen (GSFC/USRA); Graphics concept and execution: D. Hon (STX Corp.); supported by a grant from NASA, Astrophysics Division; **Figs. 33–15 to 33–19** The NASA COBE Science Team, courtesy of Nancy Boggess; special graphing for 33–16 courtesy of E. S. Cheng.

Chapter 34—Opener © 1984 ROE/AATB; **Fig. 34–1** James W. Cronin, U. Chicago; **Figs. 34–2 and 34–4** Robert V. Wagoner, Stanford U.; **Fig. 34–3** Scott Sandford, Washington U., St. Louis; **Fig. 34–5** Jeffrey L. Linsky, J.I.L.A., N.I.S.T. and U. Colorado, first appeared in the *Astrophysical Journal* for January 10, 1993; **Fig. 34–6** Photography by B. W. Hadley, © Royal Observatory, Edinburgh; **Fig. 34–8** Philip Crane; **Fig. 34–9** Ralph Crane; **Figs. 34–10 and 34–11** Stanford Linear Accelerator Center and Department of Energy; **Fig. 34–12** Joe Stancampiano and Karl Luttrell/U. Michigan, © National Geographic Society; **Fig. 34–13** Adapted from "Particle Accelerators Test Cosmological Theory" by David N. Schramm and Gary Steigman, copyright © 1988 by Scientific American, Inc. All rights reserved; **Fig. 34–14** © 1989 Sidney Harris—*Physics Today;* **Fig. 34–15** Jay M. Pasachoff; **Fig. 34–19** Richard Matzner; **Fig. 34–20** David Bennett, Francois Bouchet, and Albert Stebbins.

Fig. E-1 E. Shaya, D. Dowling/U. Maryland, the Wide-Field Planetary Camera Team, and NASA/STScI; **Fig. E-2** NASA/STScI.

Appendix 4 Prepared for NASA by Stephen P. Meszaros; **Appendix 11** Photos courtesy of Milton Roy Company.

Index

References to illustrations, either photographs or drawings, are followed by i. References to marginal notes are followed by m. References to tables are followed by t. References to Appendices are prefaced by A.

Significant initial numbers followed by letters are alphabetized under their spellings; for example, 21 cm is alphabetized as *twenty-one*. Other significant numbers appear at the end of the letter group or of index. Less important initial numbers and subscripts are ignored in alphabetizing. For example, 3C 273 appears at the beginning of the C's. M1 appears at the beginning of the M's. Greek letters are alphabetized under their English spellings.

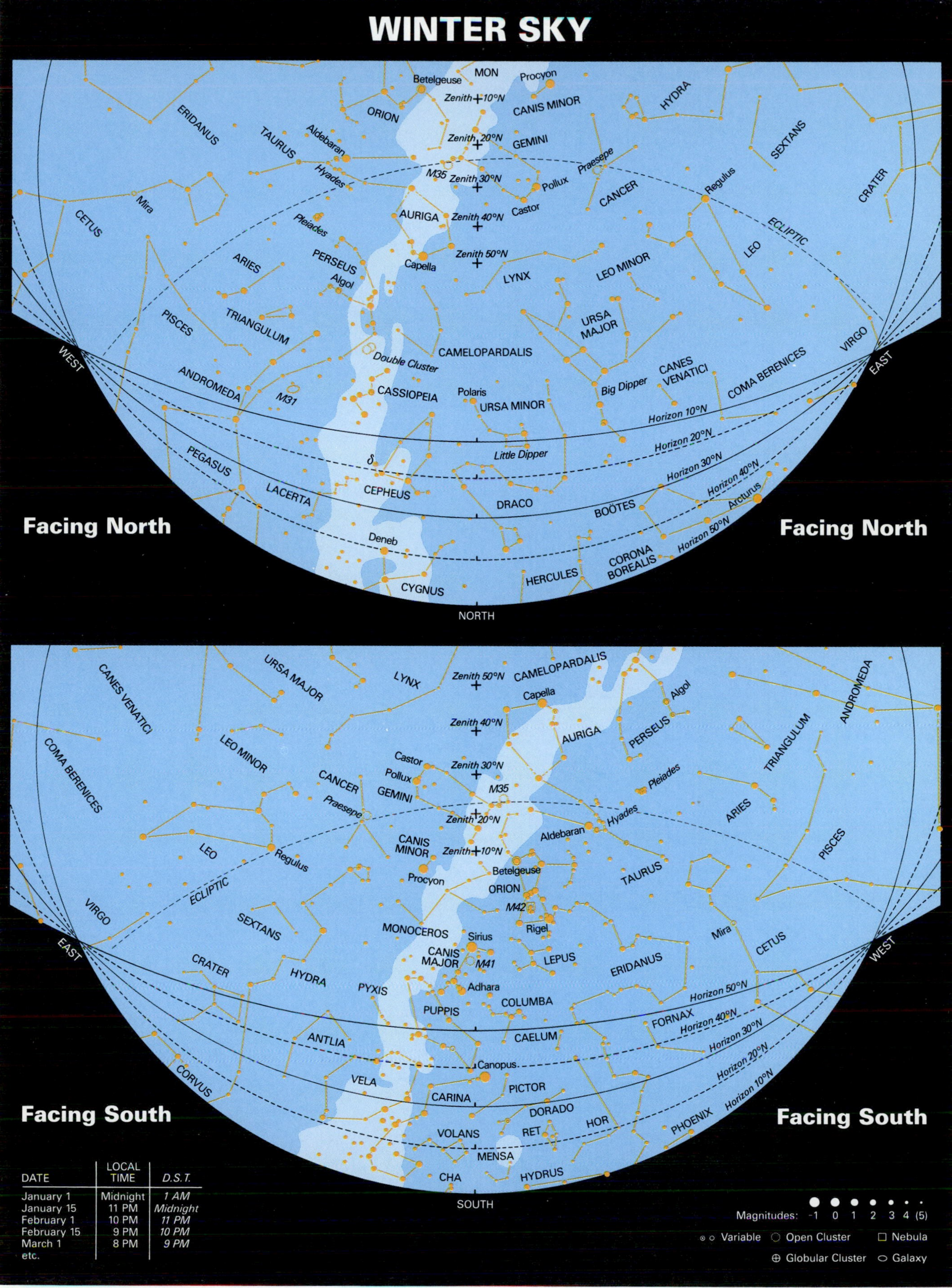

MAP BY WIL TIRION; FOR JAY M. PASACHOFF

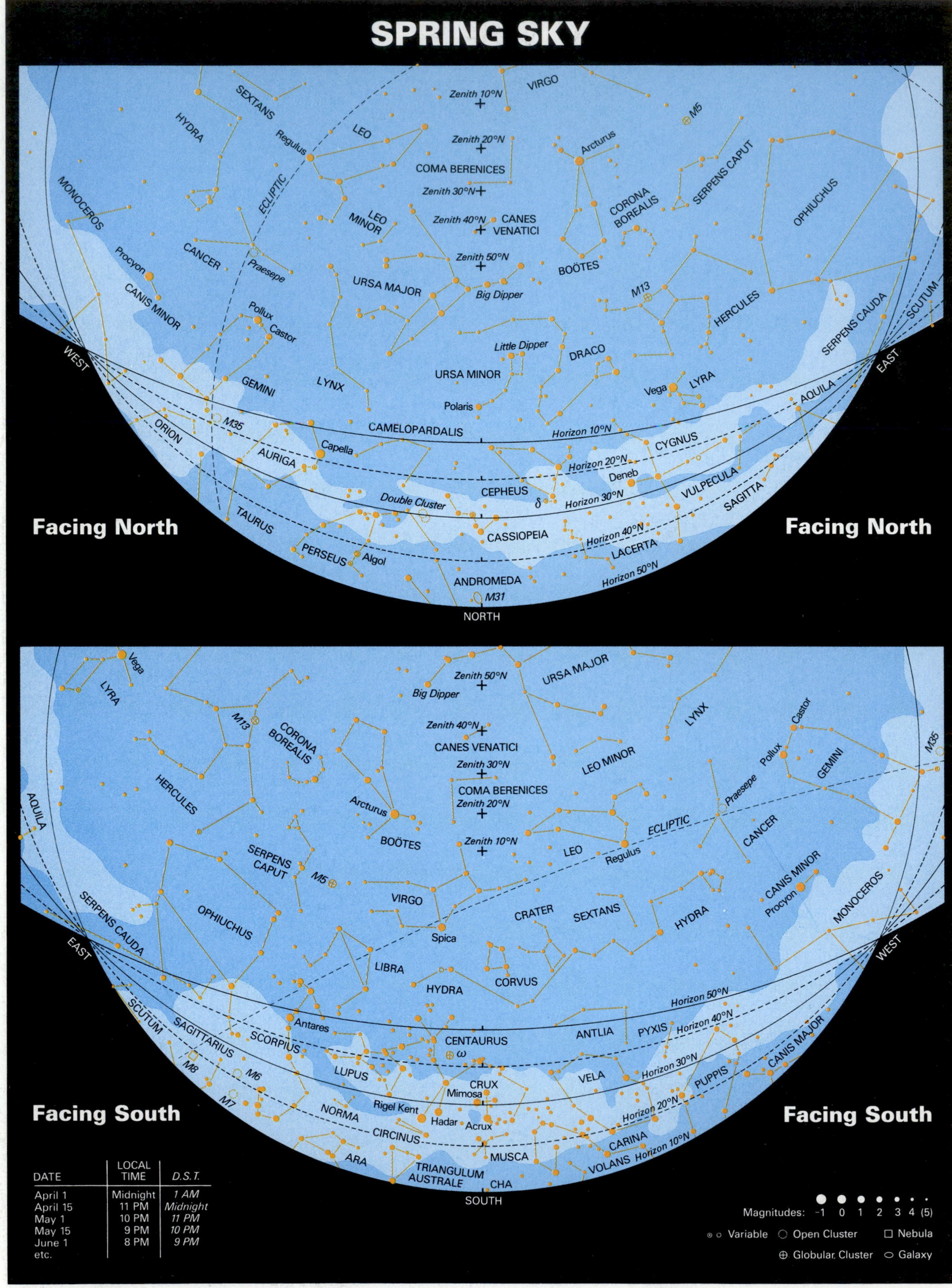

MAP BY WIL TIRION; FOR JAY M. PASACHOFF

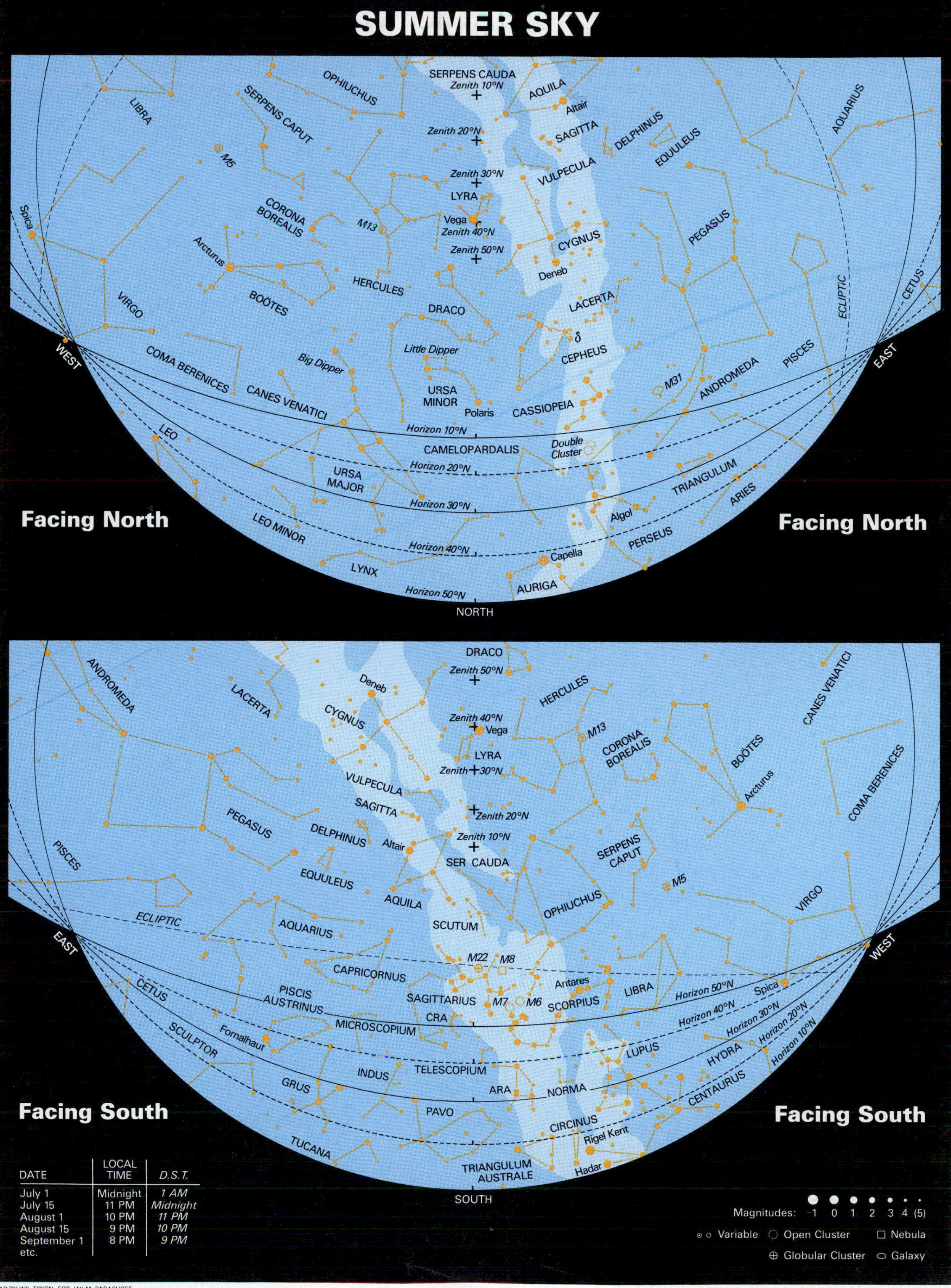

MAP BY WIL TIRION; FOR JAY M. PASACHOFF

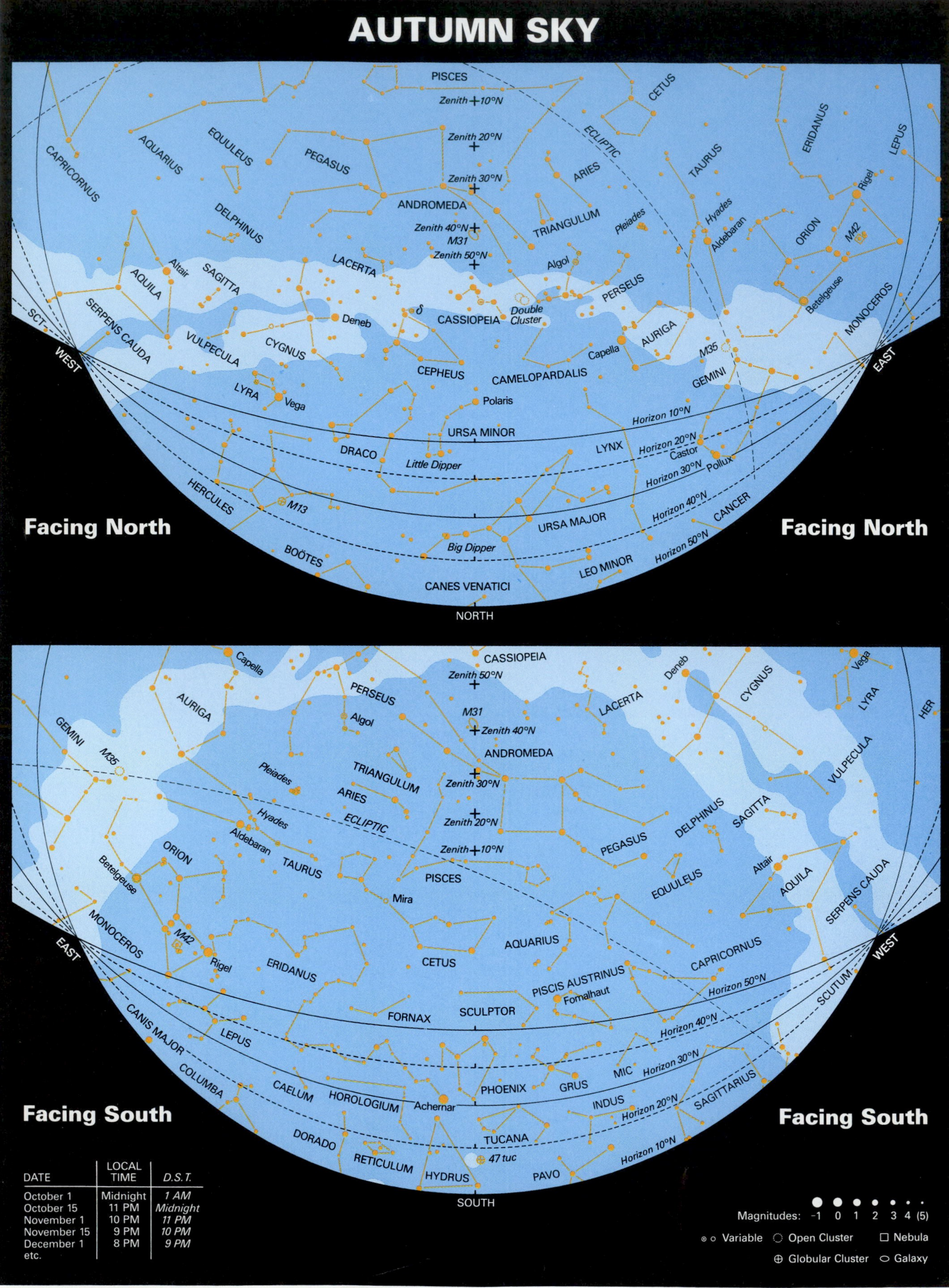

DATE	LOCAL TIME	*D.S.T.*
October 1	Midnight	*1 AM*
October 15	11 PM	*Midnight*
November 1	10 PM	*11 PM*
November 15	9 PM	*10 PM*
December 1	8 PM	*9 PM*
etc.		

MAP BY WIL TIRION; FOR JAY M. PASACHOFF